D0846266

MERRILL

CHEMISTRY

AUTHORS
Robert C. Smoot
Richard G. Smith
Jack Price

Contributing Author
Tom Russo

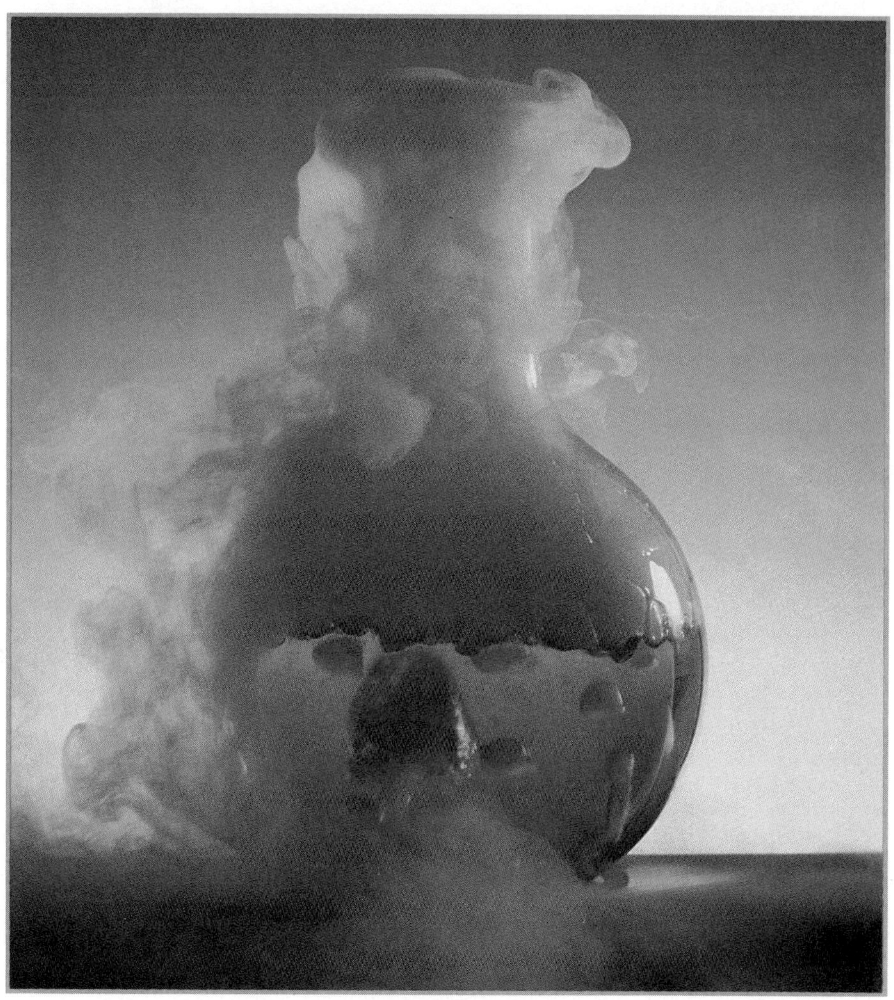

GLENCOE

Macmillan/McGraw-Hill

New York, New York Columbus, Ohio Mission Hills, California Peoria, Illinois

MERRILL

CHEMISTRY A GLENCOE PROGRAM

Student Edition
Teacher Wraparound Edition
Laboratory Manual
Laboratory Manual Teacher Edition
Spanish/English Glossary
Computer Test Bank
Videodisc Correlation
Transparency Package
Problems and Solutions Manual
Solving Problems in Chemistry
Study Guide, Student Edition

Teacher Resource Books:
- **Enrichment**
- **Critical Thinking/Problem Solving**
- **Study Guide**
- **Lesson Plans**
- **Chemistry and Industry**
- **ChemActivities**
- **Vocabulary and Concept Review**
- **Transparency Masters**
- **Reteaching**
- **Applying Scientific Methods in Chemistry**
- **Evaluation**

REVIEWERS

DeWaine Alley
Muncie Southside High School
Muncie, IN 47302

Joanne Bowers
Plainview High School
Plainview, TX 79072

Theodore L. Boydston III
Science Curriculum Coordinator
Dade County Public Schools
Miami, FL 33156

Mae Carr
Corliss High School
Chicago, IL 60628

Victor James DeAlmeida
Alvin High School
Alvin, TX 77511

Julia Shaw Dukes
Mosley High School
Lynn Haven, FL 32444

Sandra Kay Enger
Nettleton High School
Jonesboro, AR 72401

Ronald R. Esman
Abilene High School
Abilene, TX 79603

William S. Frazer
Santa Teresa High School
San Jose, CA 95123

Janet P. Lee
Dr. Phillips High School
Orlando, FL 32819

Patricia Luco
Aldine Contemporary Education Center
Houston, TX 77088

Kathleen Munroe
Osceola High School
Kissimee, FL 34741

Erik Natti
Marblehead High School
Marblehead, MA 01945

Wylie C. Poulos
Cross Keys High School
Atlanta, GA 30319

James Eugene Sims
La Mirada High School
La Mirada, CA 90638

James F. Shannon
Pittsford-Mendon High School
Pittsford, NY 14534

Kelly Wedding
Santa Fe High School
Santa Fe, NM 87505

Earl Wilson
Cass Technical High School
Detroit, MI 49091

Cover Photograph: Filser/The Image Bank

Send all inquiries to:

GLENCOE DIVISION
Macmillan/McGraw-Hill
936 Eastwind Drive
Westerville, OH 43081

ISBN 002-800803-0

Printed in the United States of America.
5 6 7 8 9 0 - VH - 00 99 98 97 96 95 94 93

CONTENTS

AUTHORS

Robert C. Smoot is a chemistry teacher and Rollins Fellow in Science at McDonogh School, McDonogh, Maryland. He has taught chemistry at the high school level for 33 years. He has also taught courses in physics, mathematics, engineering, oceanography, electronics, and astronomy. He earned his B.S. degree in Chemical Engineering from Pennsylvania State University and his M.A. in Teaching from the Johns Hopkins University. He is a Fellow of the American Institute of Chemists and a member of the National Science Teachers Association and the American Chemical Society.

Richard G. Smith is a chemistry teacher and Science Department Chairman at Bexley High School, Bexley, Ohio. He has been teaching chemistry at the high school level for 27 years. He received the outstanding teacher award from the American Chemical Society and has participated in NSF summer institutes in chemistry. Mr. Smith graduated Phi Beta Kappa with a B.S. degree in Education from Ohio University and earned his M.A.T. in Chemistry from Indiana University. He is a member of the American Chemical Society and the National Science Teachers Association.

Jack Price is co-director of the Center for Science and Mathematics Education at California State Polytechnic University, Pomona, California. He taught high school chemistry and mathematics in Detroit for 13 years, and then moved to San Diego to become the Math/Science Coordinator for the county. He earned his B.A. degree at Eastern Michigan University and M.Ed. and Ed.D. degrees at Wayne State University, where he carried out research on organometallic compounds. He has participated in NSF summer institutes at New Mexico State University and the University of Colorado. He is also an author of a high school mathematics textbook.

CONTRIBUTING AUTHOR and MICROCHEMISTRY SPECIALIST

Tom Russo teaches chemistry at the Millburn High School, Millburn, New Jersey. He is also the Science Supervisor for the Millburn Township school system. He received a B.A. in Science Education from Jersey City State College, an M.S. in biology from Seton Hall University, and an M.S. in chemistry from Simmons College in Boston. Mr. Russo has specialized in developing microchemistry methods for classrooms. He is a Woodrow Wilson, Dreyfus Master Teacher in Chemistry, and is the author of the laboratory manual that accompanies this program.

Content Consultants:

Teresa Anne McCowen
Chemistry Department
Butler University
Indianapolis, IN 46208

Mamie W. Moy
Chemistry Department
University of Houston
Houston, TX 77204

William M. Risen, Jr.
Chemistry Department
Brown University
Providence, RI 02912

Safety Consultant:

Joanne Neal Bowers
Plainview High School
Plainview, TX 79072

CONTENTS

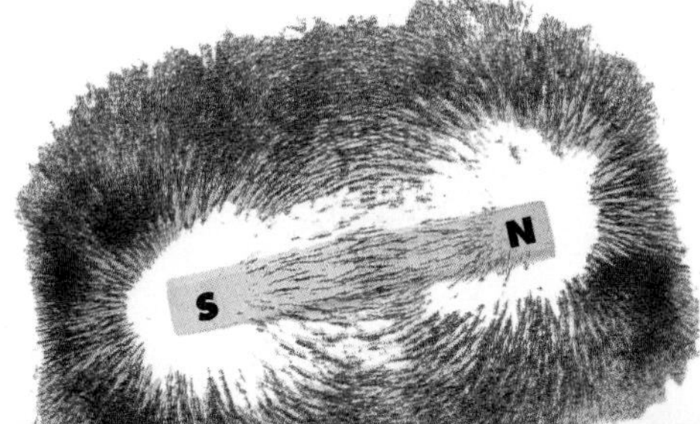

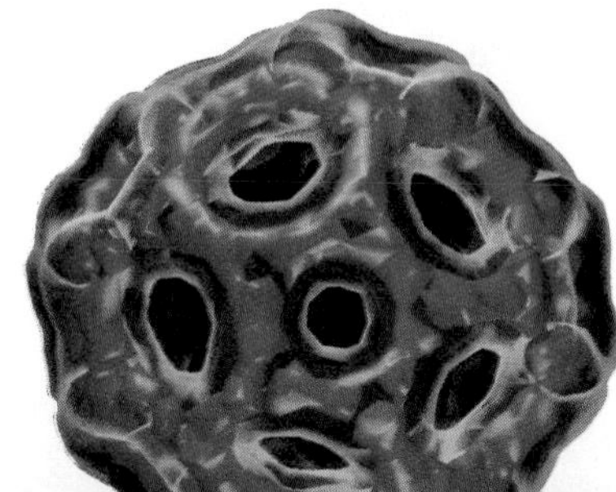

CONTENTS

CONTENTS

CONTENTS

CHEMACTIVITIES

APPENDICES

CHEMISTRY AND SOCIETY

What are some aspects of chemistry that impact society? Develop your own viewpoint on the issues that might be controversial. Learn how to live with chemistry and use it to enrich your life.

CHEMISTRY AND TECHNOLOGY

What new advances have been developed in the last decade from applying the laws of chemistry? Read about these new technologies and discover more advantages of a chemical awareness.

CAREERS IN CHEMISTRY

How can you use chemistry to develop a career? Find an idea from the career features presented in your textbook. Then plan for an exciting future of discovery.

FRONTIERS

Discover strange uses for chemicals and their reactions along with the scientists of your own time. Find out how chemistry is a living, exciting world of discovery.

Bridges to OTHER SCIENCES

How is chemistry connected to the other sciences? Much of what we know is based on an understanding of chemistry. Chemistry provides fundamental information for many of the other sciences.

CHEMISTS AND THEIR WORK

Many men and women have influenced your life by their dedication to discovery and explanation of the world of chemistry. Let them be your models for a possible future in chemistry.

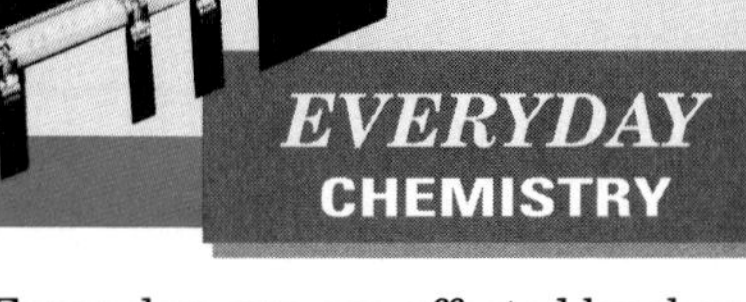

EVERYDAY CHEMISTRY

Every day, you are affected by chemistry. Read about familiar topics, but now learn how to explain them to your friends and family.

MINUTE LABS

Want to see it for yourself? Then try it! Discover for yourself how something works. Just as a picture is worth a thousand words, a minute lab is worth an hour of lecture!

INTERCURRICULAR CONNECTIONS

It may not have occurred to you that chemistry is an integral part of all your courses. Learn in these features how chemistry is connected to your subjects and hobbies.

A quick tour of

MERRILL

CHEMISTRY

Have you ever wondered how a color TV produces color or how a cellular phone works? Have you ever watched bread dough rise or eaten a pickle? Have you ever looked at the deposits that clog up a discarded spark plug or revved a car engine to warm it up on a chilly morning?

If you've done any of these things, you've already got some experience in chemistry. Communications, cooking, and cars depend on chemistry for their effectiveness. In this course, you'll find out why.

The next few pages describe the many features of this book that will help you understand the exciting world of chemistry.

The chapters of ***Merrill CHEMISTRY*** are organized to keep you centered on the topic at hand.

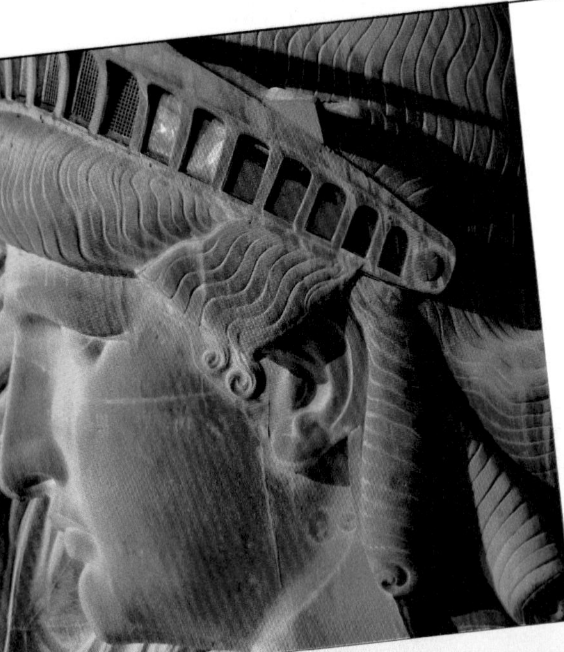

CHAPTER 25

Oxidation-Reduction

The Statue of Liberty has become a symbol of freedom. Manufactured from copper sheets, it was erected in 1886 in New York harbor, where the light in the torch was considered an aid to navigation. More importantly, the statue became one of the first sights seen by immigrants seeking a better life in the United States and came to symbolize freedom and friendship.

Over the decades, the copper sheets of the Statue of Liberty, as well as the steel skeleton over which they were stretched, gradually corroded. Inspectors found evidence of decay, especially where rivets penetrated the copper sheets. In the mid-1980s, the statue was restored. The process of corrosion involves a kind of chemical reaction called oxidation-reduction. In this chapter, you will learn about these reactions and practice balancing the equations that represent them.

Preview

CONTENTS

.1 Oxidation and Reduction Processes
Types of Reactions
Oxidation and Reduction
Oxidizing and Reducing Agents
Oxidation Numbers
Assigning Oxidation Numbers

EXTENSIONS AND CONNECTIONS

History Connection
Politics and Chemistry—Do They Mix? page

actions in print, page 642

The photos and text on the first two pages of each chapter focus you on the chapter topic and often relate chemistry to the world around you. A detailed table of contents at the bottom of these pages gives you a quick preview of what's in each chapter.

Grasp the important ideas of chemistry

The text of ***Merrill CHEMISTRY*** develops ideas with descriptions, analogies, and examples. Ideas build on one another throughout the course to give you a cohesive knowledge of chemistry. Some sections, labeled "Chemistry in Depth," explore certain chemistry topics in much greater detail.

Photographs, illustrations, and **diagrams** are vital to the presentation of each topic. They will show you examples of chemistry concepts and help you visualize important processes.

Table 10.5

Electron Affinities (kilojoules per mole)

H 72.766							He (−21)
Li 59.8	Be (−241)	B 23	C 122	N 0	O 141	F 328	Ne (−29)
Na 52.9	Mg (−230)	Al 44	Si 120	P 74	S 200.42	Cl 348.7	Ar (−34)
K 46.36	Ca (−156)	Ga 36	Ge 116	As 77	Se 194.91	Br 324.5	Kr (−39)
Rb 46.88	Sr (−167)	In 34	Sn 121	Sb 101	Te 190.16	I 295.3	Xe (−40)
Cs 45.5	Ba (−52)	Tl 48	Pb 101	Bi 101	Po (170)	At (270)	Rn (−41)

Parentheses indicate a calculated rather than an experimental value.

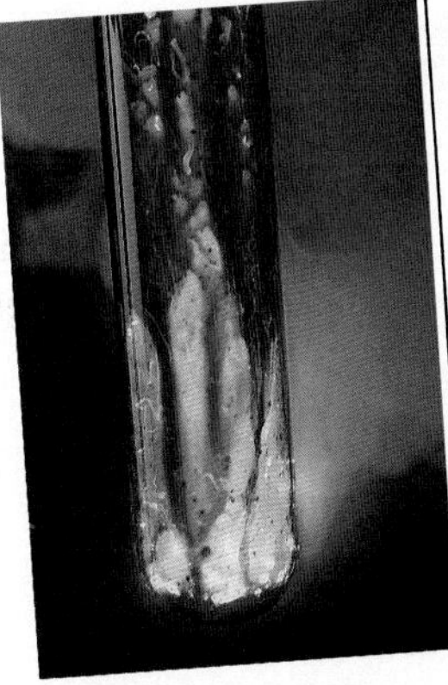

Figure 10.11 The reaction between aluminum, an element with a low electron affinity, and bromine, which has a high electron affinity, is shown here.

Electron Affinities

Now consider an atom's attraction for additional electrons. The attraction of an atom for an electron is called **electron affinity.** The same factors that affect ionization energy also affect electron affinity. In general, as electron affinity increases, an increase in ionization energy can be expected. *Metals have low electron affinities. Nonmetals have high electron affinities,* as shown in Table 10.5. Although not as regular as ionization energies, electron affinities still show periodic trends. Look at the column headed by hydrogen. The general trend as we go down the column is a decreasing tendency to gain electrons. We should expect this trend since the atoms farther down the column are larger. As a consequence, the nucleus is farther from the surface and attracts the outer electrons less strongly.

Look at the period beginning with lithium. The general trend as we go across is a greater attraction for electrons. The increased nuclear charge of each successive nucleus accounts for the trend. These elements with high electron affinities will tend to gain electrons and form negative ions.

How do we account for the exceptions of beryllium, nitrogen, and neon in the lithium period? The more stable an atom is, the less tendency i... value for beryl... sublevel. If be... would be less ... has a negative ... electrons mor... 2*p* sublevel. N... level. Becaus...

Rules of Thumb highlight key generalizations that you will use often when studying or solving problems.

28.1 CONCEPT REVIEW

9. What are the ways in which scientists investigate the structure and properties of the nucleus? Describe the action of a linear particle accelerator as it is used in ...nvestigation.

... three rules allow us to estimate rel- ...nuclide stability? Would ${}^{212}_{82}Pb$ be ...cted to be stable or unstable?

11. Explain the functions of the mo... and the containment vessel in a... reactor.

12. Complete the following nuclear ... ${}^{231}_{91}Pa \rightarrow {}^{227}_{89}Ac + ?$

13. Apply A sample of ${}^{129}_{55}Cs$ contai... 10^{18} atoms. How many of these ... remain after 96.3 hours?

Nuclear Chemistry

Concept Review questions help you check your progress as you complete each section.

Solve problems successfully

Chemistry is not something you learn just by reading. In ***Merrill CHEMISTRY*** you'll solve mathematical problems and apply the solutions to real-world situations. You'll discover and work with the mathematical relationships that are fundamental to chemical reactions and the structure of matter.

SAMPLE PROBLEM

Enthalpy Change

Calculate the enthalpy change in the following reaction.

carbon monoxide + oxygen → carbon dioxide

Solving Process:

First, write a balanced equation. Include all the reactants and products.

$$2CO(g) + O_2(g) \rightarrow 2CO_2(g)$$

Each formula unit represents one mole. Remember that free elements have zero enthalpy by definition. Using the table of enthalpies of formation, Appendix Table A-6, the total enthalpy of the reactants is

$$\Sigma\Delta H_{f\,(reactants)}^{\circ} = \frac{2\ \cancel{mol\ CO}}{} \left| \frac{-110.5\ kJ}{\cancel{mol\ CO}} \right. + 0\ kJ = -221.0\ kJ$$

The total enthalpy of the product ($2CO_2$) is

$$\Sigma\Delta H_{f\,(products)}^{\circ} = \frac{2\ \cancel{mol\ CO_2}}{} \left| \frac{-393.5\ kJ}{\cancel{mol\ CO_2}} \right. = -787.0\ kJ$$

The difference between the enthalpy of the reactants and the enthalpy of the product is

$$\Delta H_r^{\circ} = \Sigma\Delta H_{f\,(products)}^{\circ} - \Sigma\Delta H_{f\,(reactants)}^{\circ}$$
$$\Delta H_r^{\circ} = -787.0\ kJ - (-221.0\ kJ) = -566.0\ kJ$$

This difference between the enthalpy of the products and the reactants (−566.0 kJ) is released as the enthalpy of reaction.

$$2CO(g) + O_2(g) \rightarrow 2CO_2(g) + \textit{enthalpy of reaction}$$

PRACTICE PROBLEMS

5. Compute ΔH_r° for the following reaction. $2NO(g) + O_2(g) \rightarrow 2NO_2(g)$

STEP BY STEP PROBLEM SOLVING

As you encounter each new problem-solving situation, you are given a carefully worked **Sample Problem** that describes the procedure and often gives you problem-solving hints. Then, you are given immediate practice in solving that type of problem and some of its typical variations. This format lets you continue to refer to the sample problem as you begin to solve problems on your own.

PROBLEM SOLVING HINT

Be careful with signs. Most enthalpies of formation are negative. Likewise, enthalpies of reaction for most spontaneous processes are negative.

You'll find solutions to the Practice Problems worked out for you in Appendix C!

Chapter 24 REVIEW

Summary

24.1 Water Equilibria

1. The solubility product constant, K_{sp}, for an ionic substance is equal to the product of the concentrations of the dissociated ions found in solution.
2. The ion product constant for water, K_w, is 1.00×10^{-14} at 25°C and is equal to the product of $[H_3O^+]$ and $[OH^-]$.
3. The pH of a solution is equal to −log $[H_3O^+]$. The pH of pure water and neutral solutions is 7. Acid solutions have a pH ... solutions have a pH

Rev

31. Calcium c... ble in wat...
a. Write ... disso...
b. Writ... the ... tion...
32. The K_s... the co... rated...
33. The ... satu... 3.4... su...

PRACTICE and MORE PRACTICE

Each chapter ends with a **Chapter Review,** where you'll get even more problem-solving practice. In the Chapter Review, you'll answer many different types of questions, working with both words and numbers to reinforce your newly-gained knowledge of chemistry.

Is it possible for you to do some of the same experiments that led to major breakthroughs in scientific knowledge? Can you learn some of the same techniques that are used every day in industry and research labs?

Merrill CHEMISTRY supplies a number of opportunities for you to explore and experiment, to *do* chemistry. This book gives you lots of hands-on learning experience—using **Minute Labs** and **ChemActivities**—to strengthen your understanding of chemistry.

MINUTE  LAB

Air Pressure vs. Gravity

Put on goggles and an apron. Pour water into a small fruit-juice glass until it overflows. Place a 3″ × 5″ index card over the mouth of the glass. Hold the card across the top of the glass while you invert the glass over a sink. Remove your hand from the card. What did you observe? Explain. Which force appears to be stronger, gravity or air pressure? What is the pressure of one standard atmosphere?

ChemActivities

ChemActivities for each chapter are placed together at the back of the book. These lab activities will get you involved with the methods and concepts of chemistry. Many of them employ micro-scale techniques, which conserve chemicals and time and are safer to do.

MINUTE LABS

Minute Labs appear throughout the book. They are quick, simple to set up, and often require no special equipment. Minute Labs are placed in the text margins so you can easily relate the activity to the concept it illustrates.

Micro

Chapter 25
ChemActivity

Oxidation/Reduction of Vanadium

Vanadium, element 23 on the periodic table, is one of the transition elements of period 4. One characteristic of transition elements is that they have variable oxidation states. This means that elements 21 through 29 and the elements directly below them on the periodic table have several stable oxidation states. For example, the vanadium atom in an ionic form can have oxidation states of 5+, 4+, 3+ and 2+. The purpose of this activity is to prepare ions of an element that are in different oxidation states.

Objectives

- **Prepare** ions of the ...

3. Make a microscoop by cutting ... of a micropipet, as shown in F...
4. Place a half microscoop of zinc ... well A1.
5. Using a microtip pipet, place ... NH_4VO_3 in well A2. Ammonium ... contains vanadium in the 5+ ... A2 will serve as a control wel... diagram of your microplate in F... a guide to the steps that follow...

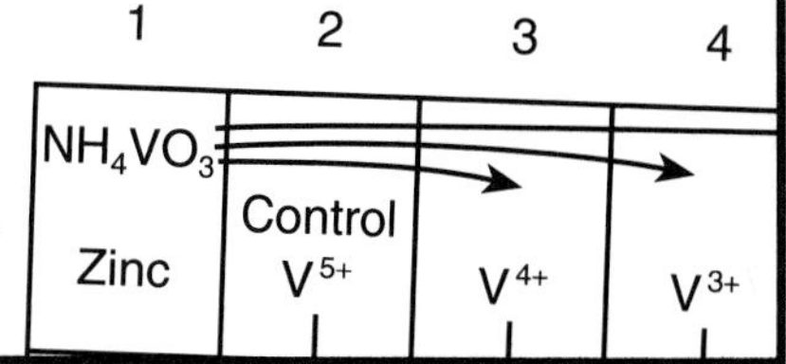

Discover the

Chemistry Connections

How does chemistry relate to the world you live in and to other subjects you're studying? Who are the people involved? What areas that chemists are researching now will provide breakthroughs in the future? How does the technology made possible by chemistry discoveries affect you? You'll find the answers to these and other questions in the numerous articles and special features scattered throughout the text.

CHEMISTRY AND TECHNOLOGY

Bridge to PHYSICS

These features report on basic research and current applications of chemistry and show how chemistry relates to the other sciences.

These three features focus on the connections between chemistry and the world you live in—from daily life experiences, to art and literature, to issues of concern to society at large.

CHEMISTRY AND SOCIETY

EVERYDAY CHEMISTRY

MUSIC CONNECTION

CAREERS IN CHEMISTRY

CHEMISTS AND THEIR WORK

Chemists and their careers are the focus of these brief margin features.

Read on for a message from the author of *Merrill CHEMISTRY*.

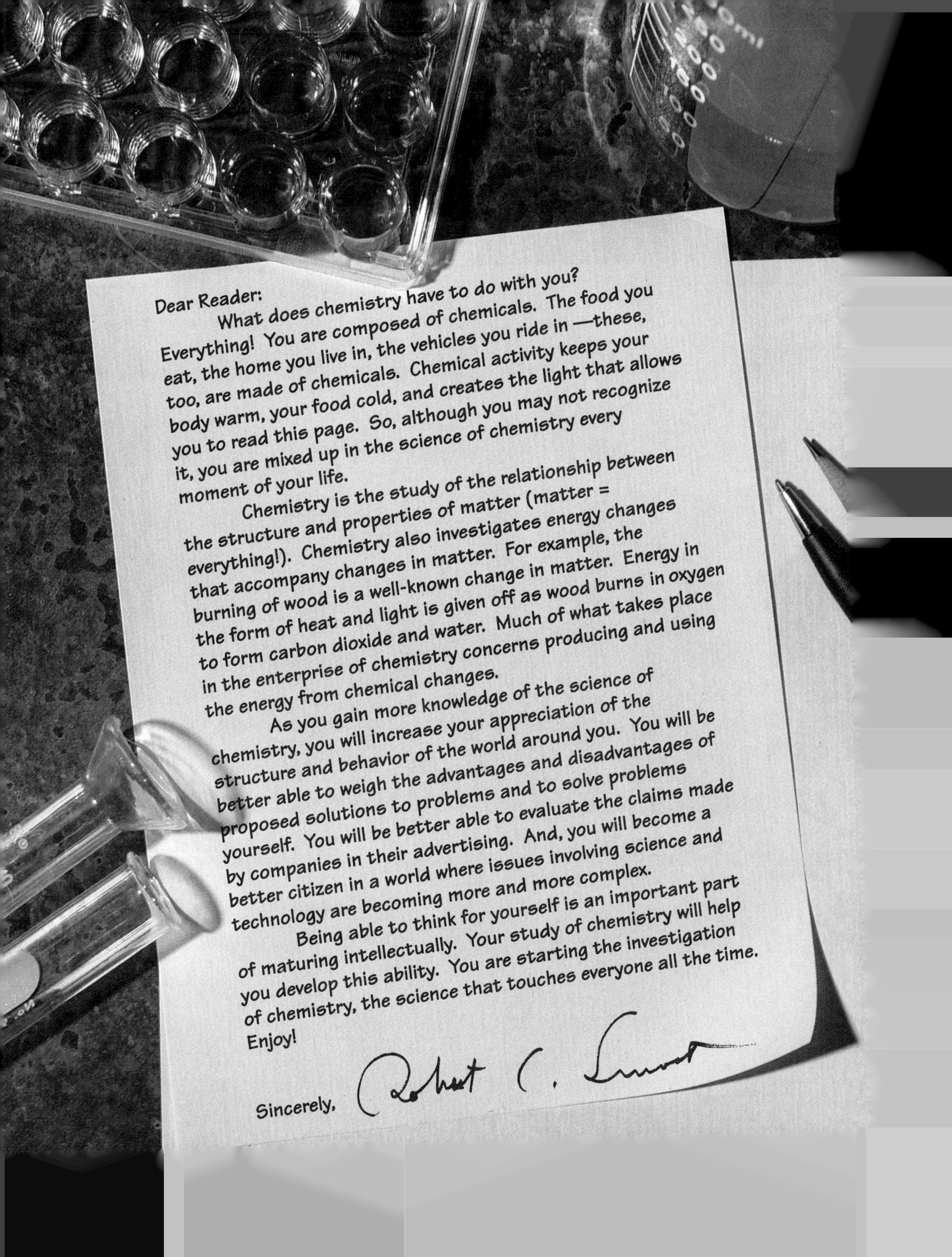

Dear Reader:

What does chemistry have to do with you? Everything! You are composed of chemicals. The food you eat, the home you live in, the vehicles you ride in —these, too, are made of chemicals. Chemical activity keeps your body warm, your food cold, and creates the light that allows you to read this page. So, although you may not recognize it, you are mixed up in the science of chemistry every moment of your life.

Chemistry is the study of the relationship between the structure and properties of matter (matter = everything!). Chemistry also investigates energy changes that accompany changes in matter. For example, the burning of wood is a well-known change in matter. Energy in the form of heat and light is given off as wood burns in oxygen to form carbon dioxide and water. Much of what takes place in the enterprise of chemistry concerns producing and using the energy from chemical changes.

As you gain more knowledge of the science of chemistry, you will increase your appreciation of the structure and behavior of the world around you. You will be better able to weigh the advantages and disadvantages of proposed solutions to problems and to solve problems yourself. You will be better able to evaluate the claims made by companies in their advertising. And, you will become a better citizen in a world where issues involving science and technology are becoming more and more complex.

Being able to think for yourself is an important part of maturing intellectually. Your study of chemistry will help you develop this ability. You are starting the investigation of chemistry, the science that touches everyone all the time.

Enjoy!

Sincerely, Robert C. Smoot

CHAPTER Preview

CONTENTS

ChemActivities

The Enterprise of Chemistry

CHAPTER 1

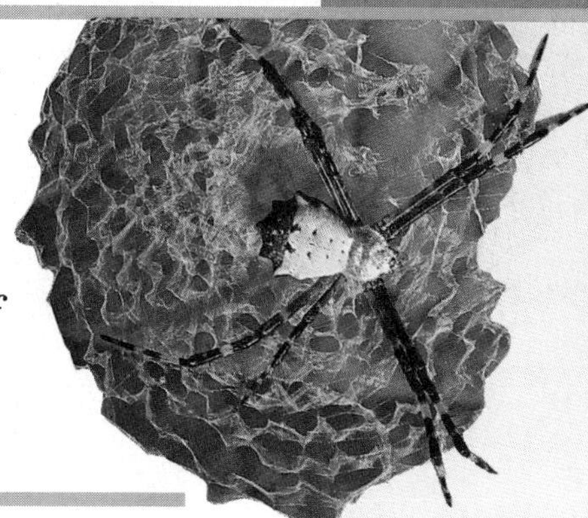

Silkworms and spiders spin natural fibers that are stronger than nylon, stronger even than steel. Chemists have learned what there is in the chemical structure of spider silk that makes it so strong. Engineers are now able to develop a method of manufacturing synthetic spider silk.

Did you ride to school today in a vehicle? If so, it had synthetic rubber tires, a product of the chemical industry. You are probably wearing some clothing made of a synthetic fiber such as nylon, Orlon, or Dacron. At lunch you may eat something that came packaged in a plastic film also produced by the chemical industry. You may be taking an antibiotic to control an infection. Everything in your life is related to chemistry. In fact, your life processes are chemical processes. You are about to begin an exciting study of the principles on which the enterprise of chemistry is based. Some of the materials used and some of the products made by the enterprise of chemistry are shown on the facing page.

EXTENSIONS AND CONNECTIONS

OBJECTIVES

- describe the role of chemists and some of the procedures that chemists use in their studies of matter and energy.
- define matter, energy, and the forms of energy.
- explain the law of conservation of mass-energy and its importance to chemistry.

" 'What does chemistry mean to me?' said Mr. Averageman, as he looked at this page printed with ink made by a chemical process, and laced his shoes, made of leather tanned by a chemical process. He glanced through a pane of glass, made by a chemical process, and saw a baker's cart full of bread leavened by a chemical process, and a draper's wagon delivering a parcel of silk, made by a chemical process.

"He put on his hat, dyed by a chemical process, and stepped out upon the pavement of asphalt, compounded by a chemical process, bought a daily paper with a penny refined by a chemical process.

" 'No,' he added, 'of course not, chemistry has nothing to do with me.' "

(Herbert Newton Casson, 1869–1964)

The People

What do chemists do? Chemists solve problems that have to do with all the "stuff" of which our world is made. This "stuff" is called *matter*. Chemists have found that there is a very close relationship between the properties of matter (how it behaves) and the structure of matter (how it is put together). For instance, a diamond is very hard. It is hard because of the way its atoms are held together.

Nylon is one of the most common synthetic materials we encounter: nylon stockings, nylon fishing line, nylon bearings in the motor of a hairdryer. Nylon was developed by a chemist, Wallace H. Carothers. The pattern of Carothers' discovery is typical of what chemists do. He studied the structure and properties of silk. Once he understood how the properties of silk depended upon its structure, he could predict that a similar structure would

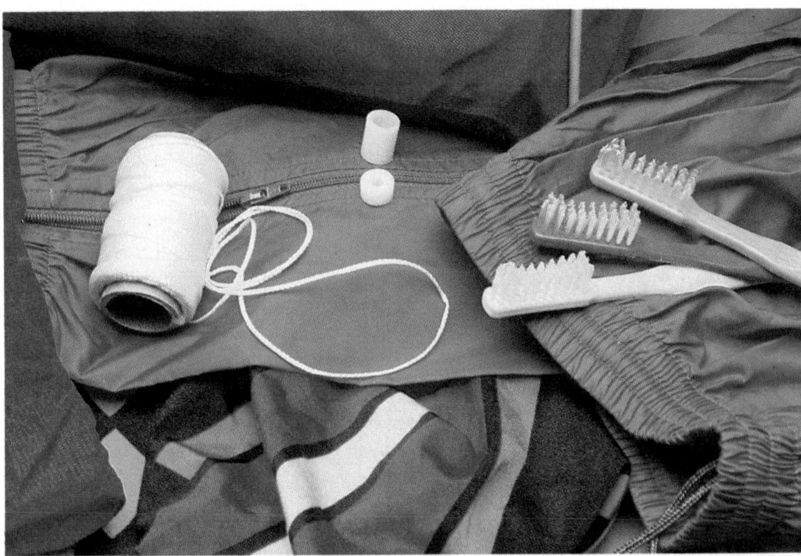

Figure 1.1 Nylon is mainly used in fibers and fabrics because of its strength, its low shrinkage, and its silky look. Carpets, swimsuits, parachutes, bristles in brushes, fishing line, bearings, gears, and even automobile body panels are made of nylon.

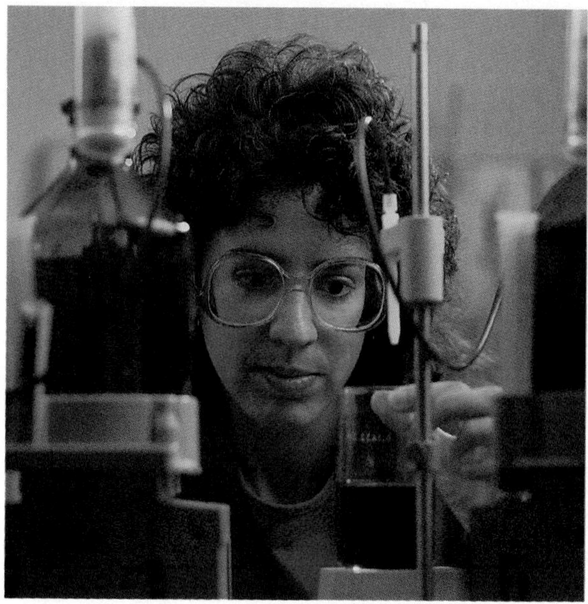

Figure 1.2 Chemists (left) develop new materials, and chemical engineers (right) design the plants and processes to produce these materials.

lead to similar properties. By reacting two chemicals, Carothers produced a material similar to silk in structure and behavior—nylon.

Once chemists like Carothers have developed a new material, they work with chemical engineers to produce the material. Together they develop a good production process. The chemical engineers then design, build, and operate the chemical plant. Chemists and chemical engineers are the principal operators of the chemical enterprise.

The things a chemist studies may be as different as the structure of the materials that transmit nerve impulses in the human brain and the bonding of rubber molecules in a car tire. In these studies, a chemist uses all of the sciences, especially mathematics and physics. The relationships between properties and structure can be organized into several basic principles, facts, and theories. A theory is simply an explanation of a phenomenon. These basic principles, facts, and theories are the foundations of chemistry. Therefore, it is not necessary for you to study the properties of *all* known materials in order to gain a knowledge of chemistry.

The Ingredients

You know that some materials are chemicals—for example, salt, copper, tin, chlorine, and petroleum. What is a chemical? Everything! All matter. Even you are a mixture of chemicals! **Matter** is anything that has the property of inertia. **Inertia** is a property of matter that shows itself as a resistance to any change in motion. This change can be in either the direction or the rate of motion, or in both. For example, suppose you are riding in a moving car. When the car is stopped suddenly, your body tends to continue moving forward. If the car makes a sharp turn, your body tends to continue to move in its original direction. Then, your body moves toward the side of the car opposite from the direction of the turn. In both cases, your body is showing the property of inertia. All matter has the property of inertia.

Word Origins

en: (GK) in
ergon: (GK) work or action
energy—the capacity for doing work

Matter also has energy. In their work with the structure and properties of materials, chemists are also interested in the energy changes that take place. **Energy** is a property possessed by all matter, and, in the proper circumstances, can be made to do work. All objects possess energy. A hockey stick, an automobile, an atom, and an electron have energy. An object has two general forms of energy: potential and kinetic. **Potential energy** depends upon the position of the object with respect to some reference point. A book on a table has a greater potential energy than the same book on the floor because the table is farther from Earth. The gravitational attraction between Earth and the book gives the book greater potential energy on the table. The book could fall farther from the table to the surface of Earth than from the floor to the surface of Earth. The book could, therefore, do more work by falling from the top of the table. For the same reason, an electron close to its nucleus has less potential energy than when it is farther away. Here, however, the attraction between the electron and nucleus produces the potential energy.

Figure 1.3 The potential energy of the book on the table is greater than that of the book on the floor with respect to their distances from Earth.

Kinetic energy is the energy possessed by an object because of its motion. An airplane traveling 700 kilometers per hour has a greater kinetic energy than when it is traveling 500 kilometers per hour.

Energy can be transferred between objects in two ways: through direct contact and through electromagnetic waves. An example of direct energy transfer is the collision of two billiard balls. Kinetic energy is transferred directly from one ball to another. An example of energy transfer by electromagnetic waves

Figure 1.4 The chemical energy in gasoline is converted to mechanical energy in the engine of a car.

is the transfer of energy from the sun to Earth. Energy being transferred by electromagnetic waves is often called **radiant energy**.

Many other terms we use for energy are special cases or combinations of potential, kinetic, and radiant energy. Energy can be transformed from one kind to another. For instance, think about the battery-alternator system of a car. As the starter switch is turned on, the chemical energy in the battery is converted to electric energy. The electric energy is converted by the starter to mechanical energy used to start the car. When the car starts, chemical energy in the gasoline is converted into mechanical energy of the moving car. As the crankshaft gains speed, some of this mechanical energy is transferred by belt and pulley to the alternator. In the alternator, the mechanical energy is converted to electric energy. This electric energy is transferred to the battery, where it is converted to chemical energy as it recharges the battery. During this time, other energy transformations produce sound and change the temperature of engine parts.

MINUTE LAB

Risk vs. Benefit of Inertia

Place a small stuffed animal in a small wagon or cart. Place a board across a table to act as a barrier. Make a prediction of what will happen to the stuffed animal if you push the cart toward the board. Then, test your idea. What did you observe? Next, use a large rubber band as a seat belt for your stuffed animal. Again, make a prediction and then test it. Did the restraint make a difference? How would you define inertia?

The Rules

If you toss a burning match into a pile of crumpled paper, the paper will ignite. Paper is combustible. Chemists investigate the changes that matter undergoes, such as the burning of paper. You will begin studying these changes in Chapter 3. Whenever matter undergoes a change, there is an energy transfer. These changes are summarized by rules, or, as chemists refer to them, laws. Two of these laws are the *law of conservation of mass* and the *law of conservation of energy*. As you will study in Chapter 2, mass is a measure of the amount of matter in an object.

The **law of conservation of mass** states that matter is always conserved. This statement means that the total amount of matter in the universe remains constant. Matter is neither created nor destroyed. It is changed only in form. The **law of conservation of energy** states that energy is always conserved. This statement means that the total amount of energy in the universe remains the same. Energy is neither created nor destroyed. It, too, is changed only in form.

For almost two hundred years, scientists believed these two laws to be true under all circumstances. However, in the early 1900s, Albert Einstein showed that matter can be changed to energy and that energy can be changed to matter. Einstein expressed this relationship in his famous equation

$$E = mc^2$$

In this equation, E is energy, m is mass, and c is the speed of light in a vacuum (a constant).

According to Einstein's equation, mass and energy are equivalent. Thus, we see that the two conservation laws can be combined in just one law. This law is known as the **law of conservation of mass-energy**. It states that the sum of mass and

Figure 1.5 This mushroom-shaped cloud is the result of a test nuclear explosion. The large amount of energy comes from the conversion of a small amount of mass.

Figure 1.6 Much of the plastic used to make consumer products, such as this telephone, is made from petroleum like that in the watch glass. The plastic pellets are an intermediate in the manufacturing process.

energy is conserved. Mass and energy can be changed from one to the other, but their sum remains constant; it cannot be increased or decreased. Changes of energy to mass and mass to energy are observable only in nuclear reactions. Because chemical reactions do not involve nuclear changes, we may assume the original laws are correct.

> **MINUTE LAB**
>
> **Potential and Kinetic Energy**
>
> Make a ramp by raising one end of a long, wide board with a brick. Place a coin on top of a toy car. Roll the car down the hill. Allow the car to strike a barrier a short distance from the bottom and observe what happens to the coin. Increase the elevation of the ramp by using two and then three bricks. Make predictions before rolling the car with the coin down steeper hills. How do the bricks relate to potential energy? What is the observed relationship between potential and kinetic energy? Did the inertia of the coin change when it was thrown a greater distance?

The Industry

Many products of the chemical industry are for sale in drugstores (aspirin), hardware stores (vinyl siding), and other kinds of stores. These products are called consumer products. When chemists and chemical engineers begin developing a process for producing a new consumer product, they must decide where to start the process. For example, nylon is made from adipic acid and hexamethylenediamine. These two materials are called intermediates. **Intermediates** are not consumer products, but neither are they raw materials. Raw materials, such as coal, salt, air, petroleum, and metal ores, are found in nature. Intermedi-

Figure 1.7 The sulfur depot shown here is part of a sulfuric acid plant that produces large quantities of this acid. Sulfuric acid, H_2SO_4, is used in the preparation of many other chemicals.

ates such as adipic acid and hexamethylenediamine are made from raw materials. If you wanted to build a plant to produce nylon (or most other consumer products), you would have to make a decision about whether to start with raw materials or to buy intermediates from another company. When you watch commercials on television, you should realize that there is a huge chemical industry that provides you with many of the comforts of modern life. Yet, much of the chemical industry consists of large companies that the general public never hears about. These companies produce intermediates for other companies.

The substance sulfuric acid is one of the major industrial chemicals of commerce. It is used as an intermediate in hundreds of other industries. In an industrial economy, sulfuric acid is so important that its annual production can be used to estimate a country's extent of industrialization. The fifty most important chemicals produced by the United States chemical industry, ranked by tonnage produced, are listed in Table 1.1.

Table 1.1

The Top 50 Chemicals	
1. Sulfuric acid	26. Xylene
2. Nitrogen	27. Ethylene oxide
3. Oxygen	28. *p*-Xylene
4. Ethylene	29. Ethylene glycol
5. Lime	30. Ammonium sulfate
6. Ammonia	31. Hydrochloric acid
7. Phosphoric acid	32. Cumene
8. Sodium hydroxide	33. Acetic acid
9. Propylene	34. Potash
10. Chlorine	35. Phenol
11. Sodium carbonate	36. Propylene oxide
12. Urea	37. Butadiene
13. Nitric acid	38. Acrylonitrile
14. Ammonium nitrate	39. Carbon black
15. Ethylene dichloride	40. Vinyl acetate
16. Benzene	41. Cyclohexane
17. Carbon dioxide	42. Aluminum sulfate
18. Vinyl chloride	43. Acetone
19. Ethylbenzene	44. Titanium dioxide
20. Styrene	45. Sodium silicate
21. Methanol	46. Adipic acid
22. Terephthalic acid	47. Sodium sulfate
23. Formaldehyde	48. Calcium chloride
24. Methyl *tert*-butyl ether	49. Isopropyl alcohol
25. Toluene	50. Caprolactam

CAREERS IN CHEMISTRY

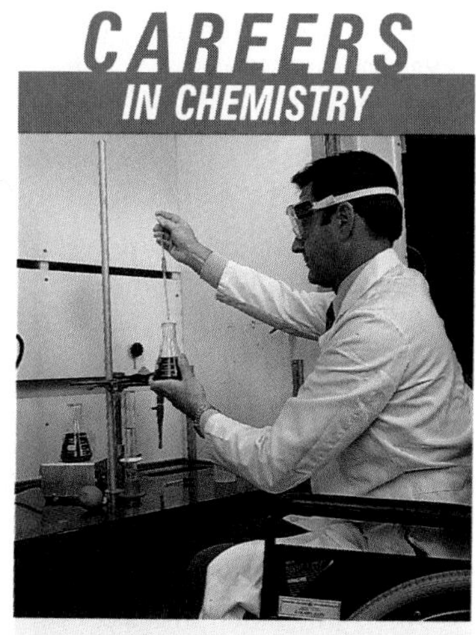

Chemist

The "nitro" fuel used in race cars, the bubble-forming agents in bubble gum and soap, the artificial sweetener in your diet cola, the aspirin you take for a headache, the fibers in your nylon jacket—all of these materials were developed by chemists. Chemists work in an enormous variety of jobs. They work in industry, government, universities, and private consulting firms.

Chemists usually specialize in a particular branch of chemistry, such as analytical, organic, or nuclear chemistry. Within a particular branch of chemistry and a particular institution, a chemist may supervise production of a product, provide marketing expertise, or carry out basic research. Throughout his or her career, a chemist may work in several different settings performing a diverse set of functions. There's one thing a chemist need never be—bored!

CHEMISTRY AND

SOCIETY

Essential Materials

The United States is dependent upon a number of countries as sources for materials used in advanced technology. The table on the right lists metals for which the United States must import more than 50% of its needs. Most of the metals in the list are probably already familiar to you. The names of others, such as germanium, indium, and tantalum, you may be seeing for the first time.

Germanium is recovered from the wastes of refining zinc ores. It also occurs to a slight extent in some coal and can be recovered from the exhaust gases in the combustion of the coal. By far, the majority of germanium is used to make semiconductor devices such as transistors, computer chips, and other solid state devices. Germanium is found in small quantities in Germany and Bolivia. In the United States, Missouri, Kansas, and Oklahoma have minute quantities of germanium available.

Indium is a very soft metal found in the ores of other metals, particularly zinc. The amount of indium in these ores is usually less than 0.001%. Very little indium is found in the United States. The major producers are Canada, Peru, Japan, and Europe, including Russia. The major use for indium is in coating bearings to give them tarnish resistance. The metal is also used in producing semiconductor devices, solder, and nuclear reactor control rods.

Tantalum is not found in useable amounts in the United States. It is imported from Canada, Thailand, Malaysia, and Brazil. Tantalum is extremely resistant to corrosion. As a consequence, it is used to make chemical-handling equipment, electronic components, dental and surgical instruments, tools, and body implants.

Many of the metals shown in the table come from Russia and the southern half of Africa. When you consider the political problems in these parts of the world, you realize that our supply could be drastically reduced without any notice. Consequently, there are industrial and government leaders who advocate stockpiling these necessary materials. Stockpiling, however, would require the investment of considerable sums of money by industry, government, or both, without any immediate likelihood of return on the investment. There are also tax laws that make it uneconomical to stockpile materials. The solution to this problem is one our elected representatives must address in the near future.

Analyzing the Issue

1. Do you think the United States should stockpile the materials it needs? Investigate the pros and cons of stockpiling critical materials and prepare a report for the class explaining your view.
2. Can you think of another solution to the problem of possible shortages of essential materials?
3. Use reference materials from the library to find out about any elements in the table with which you are not familiar. Write a short report about these elements.

Imported Metals	
Aluminum	Nickel
Beryllium	Platinum
Chromium	Silver
Cobalt	Tantalum
Germanium	Tin
Gold	Titanium
Indium	Tungsten
Manganese	Zinc
Mercury	Zirconium

Figure 1.8 Penicillin inhibits the growth of staphylococcus bacteria, a fact discovered by Alexander Fleming in 1928. The white paper disk in the center of this petri dish has been soaked in penicillin. No bacteria colonies were able to grow in the circular region around the disk.

The Process

In 1928, Alexander Fleming returned to his laboratory at the University of London after being on vacation. He found a peculiar pattern in a dish in which he had been growing staphylococcus bacteria. The dish had become contaminated with a blue mold. He observed that the staphylococcus had failed to grow all around the mold. He then hypothesized that the mold was secreting a material deadly to staphylococcus. Fleming designed an experiment to test his hypothesis and found that he had been correct. The secreted material is what we call penicillin. Observation, hypothesis, and experiment are problem-solving processes used by all types of scientists. There are many other scientific procedures, such as classification, identification, and verification. You will come across these techniques and others as you proceed in your study of chemistry.

Sometimes accident plays a part in scientific discovery. The growth of mold in Fleming's petri dish is an example. However, there is no substitute for hard work. Even accidental discoveries must be recognized, analyzed, and expanded upon. As Sherwin B. Nuland says in his book *Doctors: The Biography of Medicine,* ". . . science is awash with serendipity; science is hard work when done properly, but in the hard work there is joy and in the discovery there is abundant reward. . . ."

Scientists often develop **models** in their minds to help deal with abstract ideas and objects. When you play a video game, in your mind you imagine you are in a real situation. You build a mental model of the real situation. Chemists often build mental models. In the early part of this century, the British scientist J.J. Thomson developed an idea of the atom. He pictured the atom as a big ball of positive charge with little electrons embedded in it. J.J. Thomson's model is often called the "plum pudding" model. Plum pudding is a ball of sweet bread with pieces of fruit embedded in it. Thomson compared the ball of positive charge to pudding and the electrons to fruits. In the United States, Thomson's model could be called the "chocolate chip ice cream" model.

MINUTE LAB

Observe and Hypothesize

Add whole milk to a petri dish or flat-bottomed bowl to a height of 0.5 cm. Place one drop each of four different food colorings in the milk in four different locations. Don't put a drop of coloring in the center. Dip the end of a toothpick in liquid dishwashing detergent. Touch the tip of the toothpick to the milk at the center of the dish. Then, hold the toothpick just off the bottom of the dish, but still in the milk, and observe what happens. Based on your observations and knowledge that milk contains fats, what do you hypothesize about the effect the detergent had on the fats? Can you think of an experiment to test your hypothesis?

Figure 1.9 A variety of chemical pesticides are used to control plant damage caused by insect populations (right). Pesticides can be applied to large areas by cropdusting planes (left).

CHEMISTS AND THEIR WORK

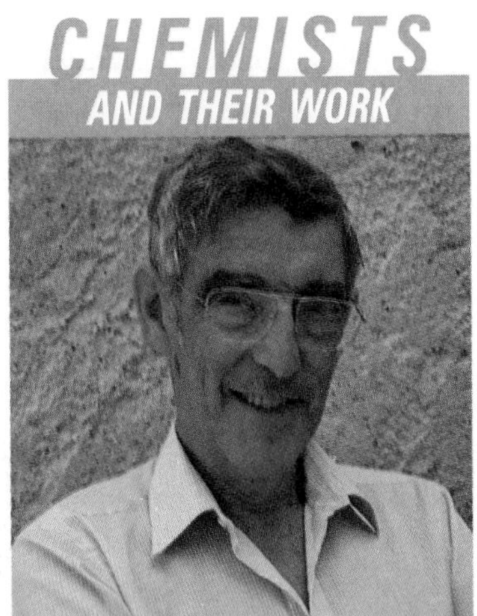

Bruce N. Ames (1928-)
Some pesticides have been classified by government agencies as possible carcinogens; that is, they may cause cancer. There are agricultural and industrial groups, however, that dispute the findings of the government agencies. The disagreement comes about in the attempt to predict what a substance will do to humans on the basis of what that substance has done to laboratory test animals. Dr. Bruce N. Ames, of the University of California at Berkeley, has devoted much of his professional career to developing reliable tests that predict the carcinogenic effects of chemicals on humans. He has spent much of his own time trying to bring a voice of reason to the public's fear of chemicals and cancer-causing materials.

The Practice

What did you have for lunch yesterday? Was it enough? Many people, in all countries of the world, did not get enough to eat yesterday. They are in varying stages of starvation. The chief competitor of humans for food is the insect. Insects reduce worldwide food production by 25-35%. Insects are also carriers of a large number of diseases that make us miserable: malaria, typhus, yellow fever, bubonic plague, and African sleeping sickness. The human race has been battling insects for thousands of years.

Many chemicals have been used to kill insects. Such chemicals are called pesticides. The beginning of modern pesticide chemistry was a discovery, in 1939, by the Swiss chemist Paul Müller. He produced a chemical with a very long name that came to be shortened to the initials DDT. The chemical contained many different kinds of atoms, among which were atoms of the element chlorine. The development of DDT was followed within a few years by the discovery, development, and production by chemists and the chemical industry of three other chlorine-containing pesticides: lindane, dieldrin, and chlordane.

At first, these materials appeared to be doing a great job of eliminating unwanted insects. DDT usage was the most widespread, especially against mosquitoes that spread malaria. The disease was virtually eliminated in some areas of the world. Crop production soared in areas of the world formerly plagued by large insect infestations.

But then, the insect populations started to increase again in those same areas where they had been controlled. Even greater application of pesticides did not seem to slow down the growth of the insect population very much. What had happened?

The Reality

In every population of a species, there will be a few individuals that are resistant to a pesticide. Once the susceptible population is killed off, the resistant organisms will multiply rapidly. These resistant organisms have plenty of food and ample opportunity to reproduce. The usefulness of the pesticide is now limited.

Why not just switch to another pesticide? In some cases, that change is just what was attempted. The cycle then repeated itself, and it was necessary to try a third pesticide. Eventually, there would be no more pesticides to try. Resistant strains of insects were developing faster than chemists could devise new pesticides. Actually, other events interrupted the repetitive process just described.

All of the above chlorine-containing pesticides are persistent. That is, they remain in the environment without being destroyed by rain, sunlight, or any other natural occurrence. Chemists discovered that these persistent chemicals were being washed by rainwater into streams and creeks. These bodies of water gradually carried the pesticides to major rivers, bays, and eventually to the open sea. In the water, microscopic plants and animals called plankton absorbed these chemicals. Many small fish feed on plankton. Big fish eat small fish. Birds eat fish of all sizes. The pesticides are soluble in the fatty tissues of animals. Consequently, many fish and birds built up considerable amounts of pesticides in their bodies.

Figure 1.10 Many insects have become resistant to pesticides. This housefly is being treated with DDT but is unaffected.

It appears that many affected bird species laid eggs with thinned shells. As a result, many chicks died before hatching. The levels of pesticides in certain fish were high enough for authorities to question whether people should eat them. It is still not known what the final consequences of this situation will be.

Another harmful effect of pesticides was observed. They killed insects indiscriminately. That is, they killed good insects as well as bad. Is there such a thing as a good insect? There certainly is. Insects pollinate most of our flowering plants. Bees produce honey and beeswax. Some silk is an insect product. Insects help keep the soil aerated. There are many insects that feed on only the insects we call pests. These predators should be encouraged, not killed. The introduction of predators to fields where their natural prey are pests is one method now in use to control insects.

As you can see, there are two sides to the use of pesticides. On the one side is the necessity of increasing crop yields to keep pace with the growth in the world's population. On the other side is the upsetting of the balance of nature.

The Future

What to do? Science cannot answer that question, but it can provide some facts to be used in coming to a decision. The decision must weigh the benefits against the risks. Even that consideration is not always easy. For instance, the ability of analytical

Figure 1.11 Ladybugs are used as a nonchemical means of pest control. Aphids are the ladybug's favorite food, but ladybugs also eat boll weevils, mealybugs, scale insects, corn earworms, and other insects and worms.

chemists to detect the presence of extremely small amounts of chemicals (parts per trillion) exceeds our ability to predict what effect these small quantities will have. There are some scientists who believe any amount of a harmful chemical, no matter how small, is bad. There are other scientists who believe that below a certain amount, called a threshold, a chemical can do no harm. Note that the expression is "believe." Neither viewpoint has been established by scientific methods.

There are a number of other ways to control pests. Some insects undergo one or more changes in body form during their lifetime. This process is called metamorphosis. In these insects, a chemical called a juvenile hormone prevents the insect from maturing until conditions are right. In a process that took ten

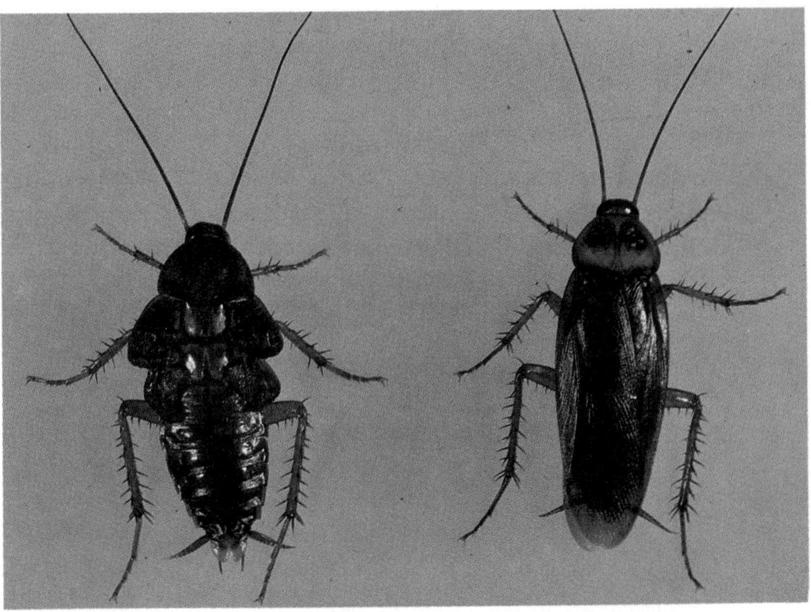

Figure 1.12 Scientists are using synthetic juvenile hormone to prevent insects from maturing and reproducing. Eventually, the population size decreases, resulting in less damage to crops. The photo shows a malformed adult American cockroach on the left and a normal adult on the right. The malformed adult, treated with juvenile hormone, cannot reproduce.

years, chemists isolated and found the structure of an insect juvenile hormone. They then learned to make, or synthesize, the material. By manufacturing and using juvenile hormone, scientists can prevent insects from maturing and reproducing. These hormones are not persistent. Some decompose so quickly that they must be encapsulated to be released slowly if they are to be effective. They are fairly specific to particular insects so that they do not harm useful species. In some cases, by using a molecule very similar to the natural hormone, but differing slightly, the result can still be achieved without the disadvantages.

Insects do not have an internal skeleton as we do. Their skeletons are outside their bodies. The only way for insects to grow is for them to shed this skeleton when it becomes too small. This molting process is also triggered by a hormone. Chemists took 15 years to isolate and synthesize the first molting hormone. Its use can disrupt the natural molting cycle of an insect and help control insect populations.

Figure 1.13 This bag has been treated with the same sex attractant that female Japanese beetles use to attract mates. The males are lured to the trap, where they are unable to land without sliding into the bag below.

Scientists have observed that certain plants do not seem to be affected by particular insects. They infer that these plants secrete a chemical that repels the insects. Investigation by chemists has confirmed the inference. Botanists are always trying to develop insect-resistant crops through selective breeding. Genetic engineering also offers promise of developing insect-resistant plants.

Other chemicals that have been isolated and studied structurally are sex-attractants. Female insects secrete a chemical that can be detected by males at a great distance. The males then follow the scent to mate with the females. By producing the sex-attractant and placing it in a trap, scientists have captured males and prevented them from mating with the females.

Several insect pests have been controlled by raising millions of males in captivity and then sterilizing them with nuclear radiation. The males are then released at just the right time to mate with females of the species. Being sterile, the males can produce no offspring and the population rapidly declines.

Other organisms can be used to attack insects. Fungi, viruses, bacteria, and nematodes (roundworms) can all be used to infest insect populations and produce diseases that devastate the insect population. Fortunately, most organisms used in this manner are harmless to humans.

Finally, a more recent attack on insects is to use integrated pest management, IPM. This technique makes use of several of the control measures mentioned above. Each application must be evaluated separately because of different crops, different pests, and different environments. However, pesticides are still the most effective agent in the battle against insects.

The story of pesticides is typical of major problems that chemists try to solve. Each time it appeared that the problem was solved, a negative side effect was noted. It is common in science for an investigation or experiment to raise more questions than it answers. It is often difficult to extrapolate results from the laboratory to the world.

EVERYDAY CHEMISTRY

Bread Making

From the earliest times, bread making has been an essential art of civilization. Good bread owes its existence to chemistry and chemical reactions. The major ingredients in bread are yeast, flour, water, and salt. Each is included in the recipe for a reason.

Flour contains starch and protein. Flour and water are mixed with yeast to produce a dough. As the dough is mixed, water and protein in the dough form tangled molecular chains called gluten. When the dough is kneaded, the chains align and the dough becomes smooth. The starch forms a jelly-like material with the water and gives structure to the dough.

Yeast are single-celled organisms that are related to molds. When activated by water, yeast digest starch in the flour and release carbon dioxide and alcohol. In bread making, the carbon dioxide bubbles are trapped in the dough by the gluten. As the yeast produce carbon dioxide, the dough "rises," or leavens.

Salt adds flavor and prevents the gluten from breaking down and leavening the dough too rapidly.

When the bread is baked, the trapped gas bubbles expand and cause the dough to rise even further. During baking, yeast cells are killed and the alcohol evaporates, giving off the tempting aroma of baking bread.

Exploring Further

1. Baking soda and baking powder are common household chemicals also used to leaven baked products. Find these on your shelf at home or at the grocery. List the ingredients from the containers.
2. Put a small amount of baking soda in a flat saucer. Pour a few drops of vinegar on the baking soda. What happens? This chemical reaction is similar to the one that takes place when baking cakes.

The Course

The first part of your work in chemistry will be to learn to communicate with your teacher and with each other by the use of very careful measurement and by understanding how chemists classify matter. You need to master the skills of classification and measurement to proceed with your studies in chemistry.

Second, you will begin your study of structure by examining the atom, the basic construction unit with which chemists work. Third, you will find out about how chemists represent matter and its changes with symbols. Then you will learn about putting atoms together to make molecules and other particles larger than

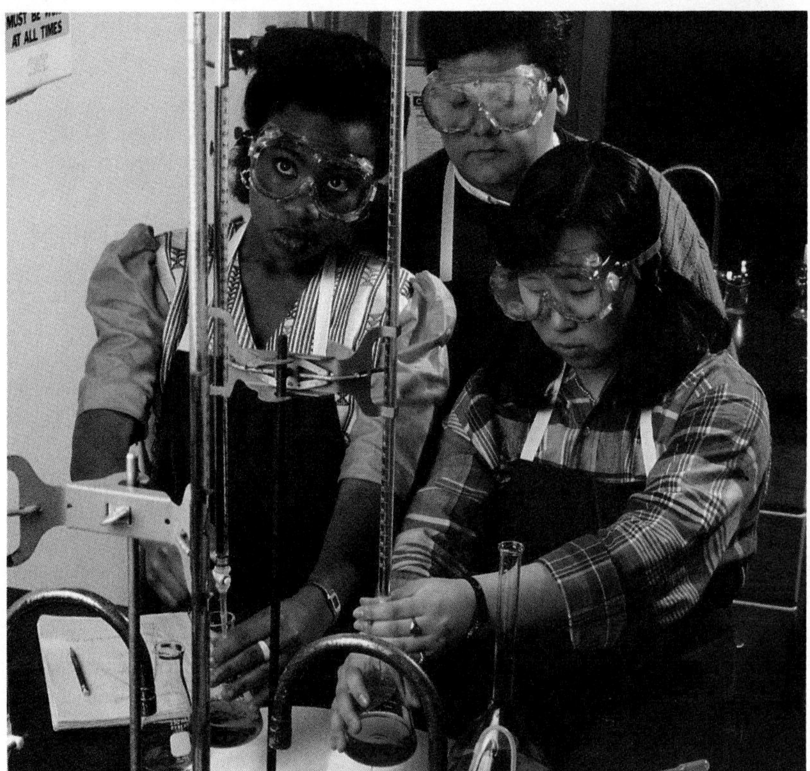

Figure 1.14 In chemistry, a student uses scientific problem-solving skills to learn about the structure and behavior of matter.

atoms. You will also learn about how groups of similar particles behave. Finally, you will study mixtures of different particles, both mixtures that do react and those that do not.

As you study chemistry, you will be finding out about the structure and the properties of matter. Once you have learned about the connection between a particular structure and its associated properties, you can make predictions about matter. For example, if you found a material whose atoms were held together in much the same way as those in a diamond, you could predict that material would be hard.

Chemists use this predictive ability (a scientific procedure) to make new materials with certain desirable properties. Knowing that these properties are related to certain structures, the chemist then develops a process for making new materials that have these structures. What is chemistry? **Chemistry** is the study of the structure and properties of matter. Chemistry is what chemists do! They discover the relationships between structure and properties of matter and use these to produce new materials.

CONCEPT REVIEW

1. How is inertia related to matter?
2. Distinguish between potential and kinetic energy.
3. Is the light produced by fluorescent lights potential, kinetic, or radiant energy?
4. Why do chemists need to keep developing new pesticides?
5. **Apply** Why might a chemist need to understand the law of conservation of mass-energy?

Chapter 1 REVIEW

Summary

1. Chemists investigate the relationship between structure and properties in matter. Chemical engineers design, build, and operate chemical plants after a production process has been developed.
2. Matter is anything with the property of inertia. Inertia is resistance to changes in motion.
3. Energy is the capacity to do work under the proper circumstances.
4. Potential energy is the energy an object has because of its position. Kinetic energy is the energy an object has because of its motion. Radiant energy is energy being transferred as electromagnetic waves.
5. The law of conservation of mass states that matter is always conserved. That is, it cannot be created or destroyed. The law of conservation of energy states that energy is always conserved. That is, energy cannot be created or destroyed. Both matter and energy can change form.
6. Since Einstein showed that mass can be changed to energy and vice-versa, scientists now use the law of conservation of mass-energy, which states that the sum of mass and energy cannot be increased or decreased.
7. The chemical industry converts raw materials into intermediates and consumer products.
8. Scientists tend to follow certain patterns in their work. These patterns are called scientific processes or procedures.
9. Scientists often use the results of scientific processes to build mental models to help deal with abstract ideas and objects such as atoms.
10. Chemistry is the study of how the properties of matter are a result of its structure and the use of this knowledge to make new materials.

Key Terms

matter
inertia
energy
potential energy
kinetic energy
radiant energy
law of conservation of mass
law of conservation of energy
law of conservation of mass-energy
intermediate
model
chemistry

Review and Practice

6. What is inertia?
7. Define potential, kinetic, and radiant energy.
8. What are the conservation laws of mass, energy, and mass-energy?
9. What is chemistry?
10. What is the difference between a chemist and a chemical engineer?

Concept Mastery

11. Differentiate between matter and energy.
12. How do intermediates differ from consumer products and raw materials?
13. List five scientific processes or procedures, and describe each of them.
14. What are models? Why are models important to scientists?
15. Make a list of at least five different forms of energy. Use reference materials in your school library, particularly physics texts, to help you.

Application

16. Using a dictionary, find out what aspects of nature are investigated by each of the following scientists: agronomist, astronomer, biochemist, biologist, botanist, ecologist, entomologist, geochemist, geolo-

gist, geophysicist, horticulturalist, limnologist, metallurgist, meteorologist, physicist, and zoologist.

17. Find out what kinds of careers require a knowledge of chemistry. Make use of the career education materials that your guidance counselor may have, including college catalogs, the *Dictionary of Occupational Titles*, and the *Occupational Outlook Handbook*.

18. Investigate one product from the list of top 50 chemicals, Table 1.1. Find the raw materials and intermediates from which it is made, the manufacturing process, the properties and structure of the product, and its uses.

19. In March 1989, the oil tanker *Exxon Valdez* struck a reef after leaving the port of Valdez, Alaska. A very large oil spill resulted. Describe any chemical processes that were used in the cleanup.

Critical Thinking

20. You are given a sealed box about the size of a shoe box containing an unknown object. Describe some things you might do to decide what is inside the box. Think about tests that might eliminate some classes of objects.

21. If 50 cm^3 of water are added to 50 cm^3 of ethanol, 95 cm^3 of solution result. Can you think of a model that explains this property?

Readings

Eberhart, Jonathan. "The Saying of Science." *Science News* 133, no. 5 (January 30, 1989): 72-73.

Heitz, James R. "Photoactivated Pesticides." *CHEMTECH* 18, no. 8 (August 1988): 484-488.

Lamb, James C. "Regulating Carcinogenic Pesticides." *CHEMTECH* 21, no. 1 (January 1991): 42-49.

Leng, Marguerite. "Reregistration: Its Consequences." *CHEMTECH* 21, no. 7 (July 1991): 408-413.

Levi, Primo. "The Mark of the Chemist." *Discover* 10, no. 2 (February 1989): 70-75.

Marco, Gino J., *et al.* "Silent Spring Revisited." *CHEMTECH* 18, no. 6 (June 1988): 350-353.

McClintock, J. Thomas, *et. al.* "Are Genetically Engineered Pesticides Different?" *CHEMTECH* 21, no. 8 (August 1991): 490-494.

Rhodes, Richard. *The Making of the Atomic Bomb: Part I.* New York: Simon and Schuster, 1986.

Rochow, Eugene G. "Choices." *Journal of Chemical Education* 63, no. 5 (May 1986): 400-405.

Sime, Ruth L. "The Discovery of Protactinium." *Journal of Chemical Education* 63, no. 8 (August 1986): 653-657.

Treptow, Richard S. "Conservation of Mass: Fact or Fiction?" *Journal of Chemical Education* 63, no. 2 (February 1986): 103-105.

Woods, Michael. "Nature Makes its Own Toxins." *CHEMECOLOGY* 20, no. 5 (July/August 1991): 12-13.

CHAPTER **Preview**

CONTENTS

Measuring and Calculating

CHAPTER 2

The farmer who planted this cornfield had to solve problems to know how much seed corn and fertilizer to order and how little expensive weed killer and insecticide to buy. To make good decisions, farmers and chemists alike must use measurements and solve problems.

Why do people continue to use hazardous pesticides? The food exporting nations want to increase production in order to have more food to sell. Many developing countries must increase production in order to reduce the amount of food they import. Consequently, most nations use pesticides in agriculture. How to use pesticides efficiently and safely is a problem that must be faced by almost all countries. In this chapter, you will learn how to approach problems in a general way and how to approach chemistry problems in a more specific way.

EXTENSIONS AND CONNECTIONS

OBJECTIVES

- compare and apply strategies for solving problems in chemistry.
- realize that success in solving chemistry problems lies in knowledge and practice.

2.1 Decision Making

Solving problems is more than just following set procedures established by someone else. There will be times in chemistry when a set procedure is useful. However, there is no substitute for thinking through a problem on your own. What is a problem? One dictionary definition includes the words ". . .an unsettled matter demanding solution or decision. . . ." Problem solving is something many of us would rather not do. It places a burden on us for making decisions and being accountable for outcomes. Whether the decision is who to invite to your next party or how much of a substance to use in an experiment in the chemistry laboratory, you must learn to solve the problem at hand and have confidence in your solution. The development of that ability and confidence will serve you well now and in the future.

General Problem Solving

You have certainly come across the word *creativity*. As generally used, *creativity* refers to the ability in a *creator* (musician, writer, artist) to generate something original and interesting. However, problem solving is also a creative activity. You must learn to think for yourself in order to arrive at solutions.

In order to solve problems, you will often have to substitute one thing for another, combine two or more things, or adapt a solution method you have used before in another connection. Good problem solvers learn to modify their thought processes for each new situation they encounter. They learn how to sort out useless information, eliminate it, organize the remaining material so that it can be put to good use, and reverse procedures they have used before. How do you know when to do these things?

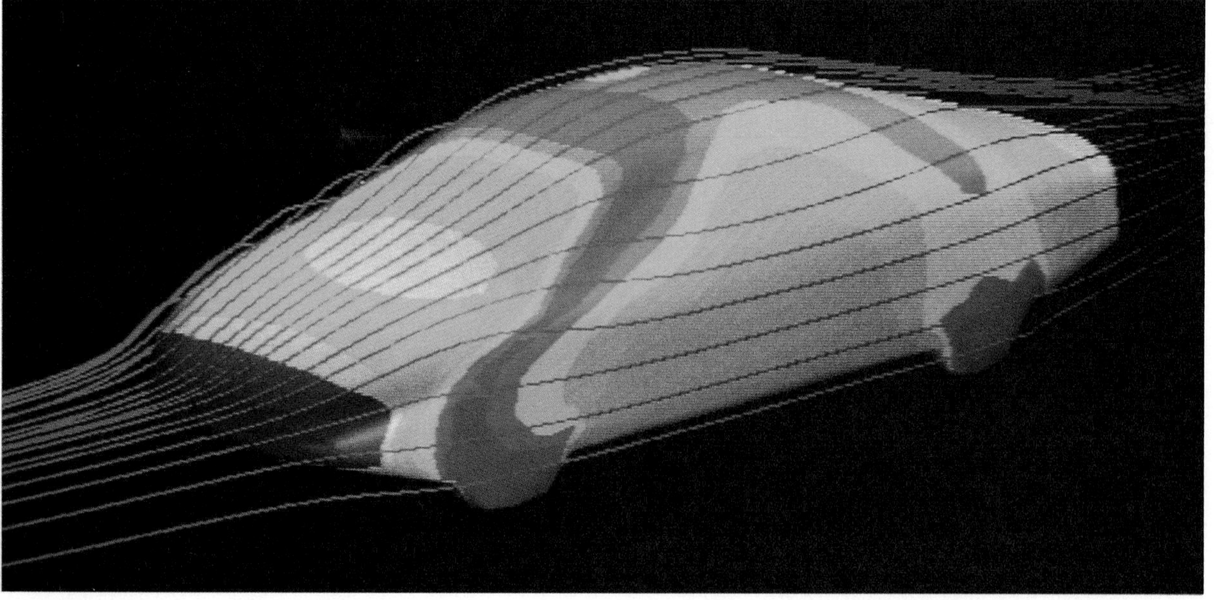

Figure 2.1 Automobile designers must solve problems in order to produce cars of an aerodynamic shape for best fuel efficiency.

Figure 2.2 The problem this student must solve is to decide which solution and in what concentration she should add to zinc in order to produce hydrogen gas. Making this decision requires not only problem-solving skills, but also a knowledge of chemistry.

There is no set answer to that question. However, there are strategies and techniques that you can learn to help you apply these ideas to the solving of problems.

Experience is the best teacher. The more you practice your problem-solving skills, the better they will become. Practice is valuable only if it is done correctly. If the basketball coach teaches your team a new play on Monday, and the team practices it correctly all week, chances are the play will be successful on game night.

As you practice problem solving, it is important to think logically. Because problem solving is not always a fun activity, it is important to practice self-discipline and not quit before a correct solution is reached. Learn to ask questions of others, both fellow students and teachers. From them, you may learn of alternate approaches to solving a problem. That new approach may prove valuable in the future for the solution of other problems that are similar to the one that puzzled you.

MATH CONNECTION

Problem Solver

George Polya (1887–1985), a Hungarian-born American mathematician, devised a four-step plan for problem solving. His steps were (1) understand the problem, (2) devise a plan, (3) carry out the plan, and (4) look back. He wanted students to examine a problem carefully to determine what information was given and what needed to be found, to devise and carry out strategies for solving the problem, and then to check the results in terms of the problem. Polya's method is similar to the one presented in this section.

Problem Solving Techniques

Read carefully! Few pieces of advice on problem solving are more important. Read rapidly through the statement of each problem when you first meet it in order to get the sense of the problem. Make sure you understand all the words used in the problem. If unfamiliar words appear in the problem, look up their meaning in the glossary or in a dictionary. Then, read the problem again, but carefully this time, to determine the facts presented. Then, read it a third time! This third time you are reading in order to determine just exactly what sort of answer is required. Once you are confident that you have found the facts presented and the determination to be made, you may find it useful to write out the relevant information under these two headings. At this point, you may even find it worthwhile to restate the problem in your own words. It is frequently useful to organize the data given by writing it in a table or by graphing it.

In solving the problem you will need to use all relevant factors. How do you know what is relevant? Does a piece of information have a bearing on the solution you are seeking? If not, it is irrelevant information and may be discarded. For instance, suppose you are asked to find the cost of 7 candy bars that weigh 3 ounces and cost 35¢ each. For the purposes of this problem, the weight of the candy bars is irrelevant and can be disregarded. In order to decide on the relevancy of information in a problem, you must have in mind a possible path of solution. At first, you may have several possible solutions in mind. What you must do is consider the consequences of each solution path and choose the one that produces the answer you are looking for. Any information that does not contribute to the chosen path is extra and can be ignored. In a similar manner, it is important to decide if vital information is missing. Perhaps an important piece of information must be looked up in the textbook or a reference book.

What happens if the proposed solution doesn't work? Go back and consider all the factors again to devise a new solution path. This procedure of going back and starting over again requires self-discipline and patience. In analyzing the various possible solutions, it is usually useful to predict the outcome of the different possibilities. Only paths that promise a reasonable outcome need be considered. Finally, once a solution is attained, that solution should be evaluated for reasonableness.

Figure 2.3 When you finish a problem, always check to see that your answer makes sense. If it doesn't, you will have a valuable clue that will help you find your error.

Problem Solving in Chemistry

All the techniques just discussed for general problem solving apply to solving problems in chemistry. In addition, there are techniques that apply specifically to chemistry. A general rule followed by scientists is to proceed from the simple to the complex. Many problems in chemistry can be broken down into a series of simpler steps. Solving each simpler step leads readily to the solution to the overall, complex problem.

Many times, you will want to try a method called "guess and check," but better known as "trial and error." You think you know how to solve the problem but are not sure. Try it! If the answer you get seems reasonable, then your solution is probably correct. In evaluating numerical solutions, you should practice estimating an answer. In the candy bar problem on page 24, if your answer came to \$735.00, you would quickly recognize that sum as a ridiculous cost for 7 candy bars. You obviously made a mistake in your solution, and must devise a better one.

Using set procedures is especially valuable in solving numerical problems in science. As you learn about different kinds of numerical problems, learn to recognize the types of problems. One of the first things you can do with a new problem is to try to classify it into one of the categories of problems you have already learned to solve. When you learn to solve a new kind of problem, look for the pattern of steps used to solve that particular kind of problem. The pattern is useable in any other problem of that kind. Such a pattern is often called an algorithm. You must be careful not to become a slave to algorithmic problem solving. There is no substitute for thinking for yourself.

Another useful technique is to solve a problem by working backwards through it. In other words, by knowing what the answer should be like, think about what the last step would have to be. Then, once you have decided on the last step, proceed to think about the next-to-last step, and so on.

Strategies for Problem Solving in Chemistry

Skill in solving problems can be developed by practice and by using the strategies described here. The strategies discussed for solving chemistry problems can be summarized as follows.

1. Identify the known facts. (Where am I?)
2. Define the answer required. (Where do I want to be?)
3. Develop possible solutions. (What paths might take me where I want to be?)
4. Analyze these solutions and choose the correct one. (Which path is most likely to be the correct one?)
5. Develop the individual steps to arrive at the answer. (Plan the trip.)
6. Solve the problem. (Travel along the selected path.)
7. Evaluate the results. (Did I reach the place I expected?)

CHEMISTS AND THEIR WORK

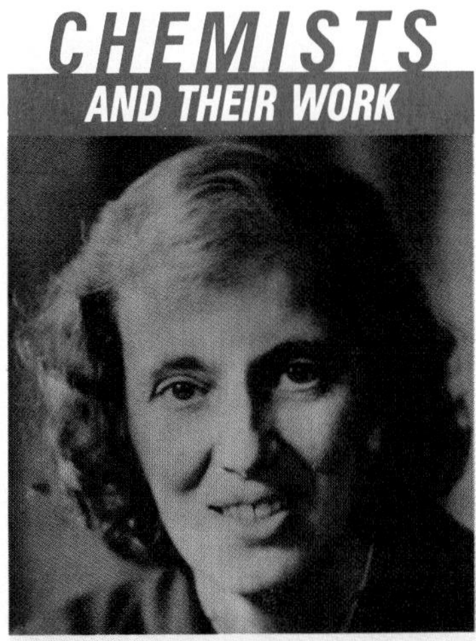

Dorothy Crowfoot Hodgkin (1910–)

Dorothy Crowfoot Hodgkin, an English chemist, is one of the few women to receive a Nobel prize for chemistry. Her work was principally in the determination of molecular structures through the use of X-ray diffraction measurements. She studied at both Oxford and Cambridge Universities and later became a faculty member at Oxford. In her research, she determined the molecular structures of pepsin, penicillin, and vitamin B_{12}. Vitamin B_{12} is a very complex compound, and it was for this work that she received the Nobel prize in 1964.

Figure 2.4 Practice enabled this marching band to turn in a precision show. Chemistry students tune up their problem-solving abilities by practice. When you practice by doing chemistry problems, your performance will impress everyone—including yourself.

You generate the correct solution mentioned in step 4 on page 25 with your general knowledge and the knowledge you will learn in your study of chemistry. Perhaps your chemical knowledge will not lead you to an immediate method of solving the problem. However, the knowledge you already have will be a guide to using the textbook, including the Table of Contents, the Appendices, and the Index. The Sample Problems and descriptions will help you to solve problems. Remember to look for patterns in solving different types of problems. Then, apply the pattern to solving the problem at hand.

Solving chemistry problems is not an inherited talent. Everyone can learn to do it. Practice, be patient, concentrate, and you will develop the skill. Remember: practice is the key. You may understand a worked example in the text, and you may understand perfectly as your instructor works through a problem of the same type on the chalkboard. However, *you* must work through a number of practice problems yourself if you are to master problems of that type.

Also, always remember to check to make sure your answer makes sense. For example, if you were calculating the mass of a grain of sand and the answer you obtained was 3000 kg, you would know you had made a mistake! Make an estimate of the magnitude of the answer when you begin solving a problem.

2.1 CONCEPT REVIEW

1. State an advantage and a disadvantage to the use of algorithms in solving problems.
2. List five steps that must be accomplished before beginning calculation in a chemistry problem.
3. **Apply** Explain why an understanding of the concepts of chemistry is essential in order to be successful in solving chemistry problems.

CHEMISTRY AND
SOCIETY

Chemical Research

Because the field of chemistry touches almost all aspects of our lives, the variety of research projects carried out by chemists is mind-boggling. Many current projects are in biochemistry—the chemistry of living organisms. Biochemical researchers continue an age-old quest to understand exactly how the human body operates on a molecular level. In other words, they want to know what chemical reactions are taking place in human cells. The better we understand how a properly functioning human body operates, the better able we will be to design chemical procedures for the diagnosis and treatment of illness.

Drug research occupies many chemists. Pharmaceutical companies are constantly trying to find substances that will treat illnesses presently considered incurable, such as AIDS. There is also the pursuit of improved treatments for ailments that are already controllable, but for which treatment is expensive, painful, or accompanied by high risk, such as some forms of cancer.

Chemists are also closely involved in the search for ways to improve the quality of the environment. Continued research on the chemistry of the atmosphere will lead to cleaner air and to the use of products that will not harm the protective ozone layer in the stratosphere. Investigation of the chemistry of seawater will enable us to clean up many waterways that have become badly contaminated through years of misuse and neglect. Chemists work very closely with U.S. Environmental Protection Agency technicians to find methods of cleaning up dumps and landfills which were improperly or illegally maintained.

In industry, chemists look for ways to produce better products at lower prices. For instance, use of improved plastics can greatly reduce the weight of an automobile. Less weight means greater gas mileage. Greater gas mileage means lower operating cost as well as the lowered consumption of petroleum, an important natural resource. Other new plastics are used in the manufacture of fiber for high-tech clothing that can keep us warm and dry in rain. Materials developed by chemists are used in strong, lightweight bikes and other sports equipment.

The areas of research detailed above barely scratch the surface. The opportunities available to those who study chemistry are tremendously varied and are sure to increase.

Analyzing the Issue

1. Review the annual reports of several manufacturing corporations and report on their research aims. Concentrate on one industry.
2. People often complain that the cost of prescription drugs is too high. Pharmaceutical companies answer that research and development is very expensive and they must try to recover their investment. Organize a debate on this issue. Find out what factors other than research costs affect the price of prescription drugs.

OBJECTIVES

- list and use the SI base units for mass, length, time, and temperature.
- express and convert quantities using the common SI prefixes.
- use significant digits to express the exactness of measurements.
- perform calculations using density measurements.

2.2 Numerical Problem Solving

When people first began to describe the properties of materials, they talked about scruples, gills, hundredweights, minims, and drams. The same measurement did not always mean the same quantity in different countries. The king, queen, or parliament of a country ruled what a measurement should be. King Henry VIII of England decreed a *hand* to be four inches. The hand is used to this day as a unit to measure the height of horses. In Germany, a unit called the *rute* was defined as the total length of the left feet of 16 men. Sometimes measurements were changed to make a country's exports more attractive. By making a bushel a little smaller, one country could offer more bushels of wheat than another country for the same price.

The International System (SI)

Chemistry involves measuring and calculating. It is a quantitative science. When you describe a property without measurements you are characterizing the object **qualitatively**. For example, you might say the weather is hot and humid. When the property is measured and described by a number of standard units, you have characterized the object **quantitatively**. An oven temperature of 425°F is a quantitative property.

When you refer to properties as you describe materials, it is helpful to measure the property and state the result quantitatively. In order to make a measurement, you must meet three requirements.

1. Know exactly what property you are trying to measure.
2. Have some standard with which to compare whatever you are measuring.
3. Have some method of making this comparison.

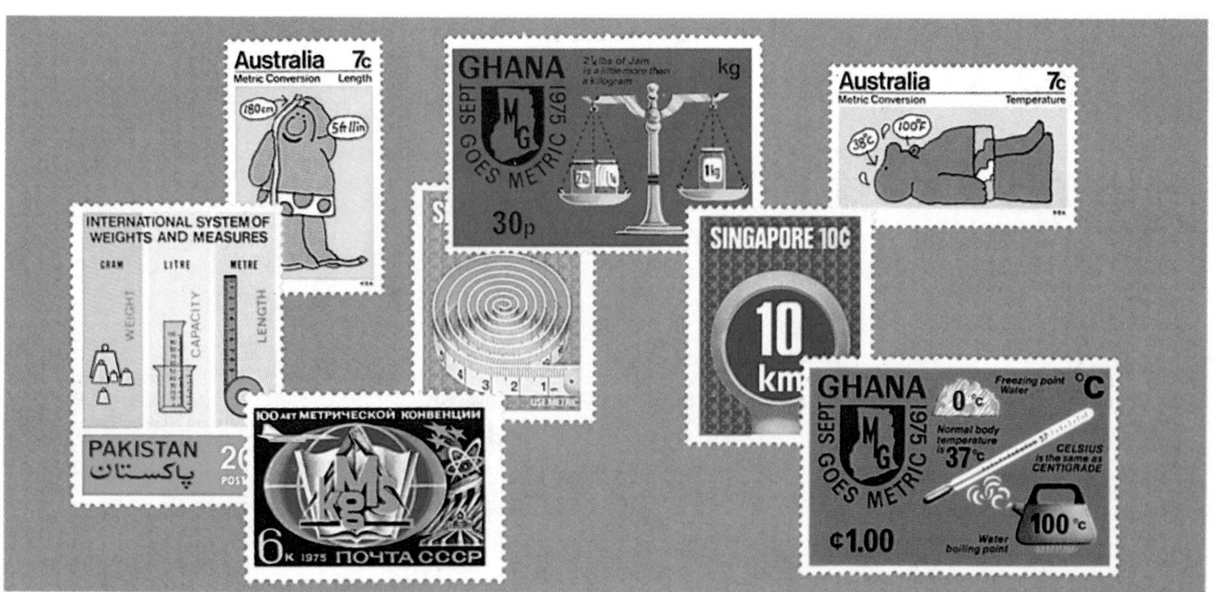

Figure 2.5 SI units are used by almost all countries in the world. Some of these countries have issued stamps to help the general public become familiar with SI units of measurement.

The standard units of measurement in science are part of a measuring system called the International System (SI). The letters are reversed in the symbol because they are taken from the French name Le Système International d'Unités. SI is used by all scientists throughout the world. It is a modern version of the metric system. The people in most countries use SI in everyday life or are in the process of converting to SI. This measurement system will be used in this text.

One important feature of SI is its simplicity. Seven base units are the foundation of the International System. These units are shown in Table 2.1. Detailed definitions of these units are found in Appendix Table A-1.

Table 2.1

SI Base Units		
Quantity	**Unit**	**Unit Symbol**
Length	meter	m
Mass	kilogram	kg
Time	second	s
Electric current	ampere	A
Thermodynamic temperature	kelvin	K
Amount of substance	mole	mol
Luminous intensity	candela	cd

In SI, prefixes are added to the base units to obtain different units of a convenient size for measuring larger or smaller quantities. A *kilo*meter is one thousand meters while a *milli*meter is one thousandth of a meter. A kilometer is a convenient unit for measuring the distance between cities. Millimeters are useful for measuring the thickness of a board. The commonly used SI prefixes and their equivalents are listed in Table 2.2. These prefixes are used throughout this book and should be memorized. A complete list of SI prefixes is in Appendix Table A-2.

Table 2.2

SI Prefixes				
Prefix	**Symbol**	**Meaning**	**Multiplier** (Numerical)	**Multiplier** (Exponential)
		Greater than 1		
mega	M	million	*1 000 000	1×10^{6}
kilo	k	thousand	1 000	1×10^{3}
		Less than 1		
deci	d	tenth	0.1	1×10^{-1}
centi	c	hundredth	0.01	1×10^{-2}
milli	m	thousandth	0.001	1×10^{-3}
micro	μ	millionth	0.000 001	1×10^{-6}
nano	n	billionth	0.000 000 001	1×10^{-9}
pico	p	trillionth	0.000 000 000 001	1×10^{-12}

*Spaces are used to group digits in long numbers. In some countries, a comma indicates a decimal point. Therefore, commas will not be used.

PRACTICE PROBLEMS

4. Which of the units in each of the following pairs represents the larger quantity?
   a. millimeter, centimeter
   b. picometer, micrometer
   c. kilogram, centigram
   d. deciliter, milliliter
5. A decigram is 0.1 g. Give the name of each of the following quantities.
   a. 0.001 m
   b. 0.000 001 s
   c. 0.01 g
   d. 0.000 000 000 001 s

Mass and Weight

In chemistry, we will often need to know the amount of matter in an object. For instance, we may wish to measure the amount of wood in a small block. One way of measuring the block is to weigh it. Suppose we weigh such a block on a spring scale and find that its weight is one newton (N). The newton is the measurement standard for weight or force. Now suppose we take the scale and the block to the top of a high mountain. There we weigh the block again. The weight now will be slightly less than one newton. The weight has changed because the weight of an object depends on its distance from the center of Earth. **Weight** is a measure of the force of gravity between two objects. In this example, these two objects are the block and Earth. The force of gravity changes when the distance between the object and the center of Earth changes.

The weight of an object can vary from place to place. In scientific work, we need a measurement that does not change from place to place. This measurement is called mass. **Mass** is a measure of the quantity of matter. Therefore, as we saw in Chapter 1, mass is a measure of the inertia of an object.

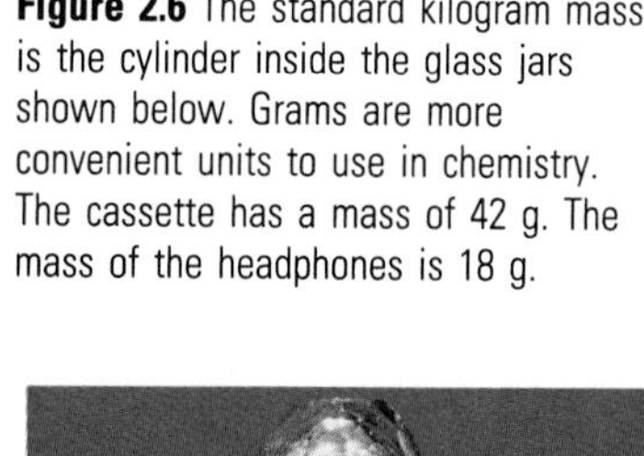

Figure 2.6 The standard kilogram mass is the cylinder inside the glass jars shown below. Grams are more convenient units to use in chemistry. The cassette has a mass of 42 g. The mass of the headphones is 18 g.

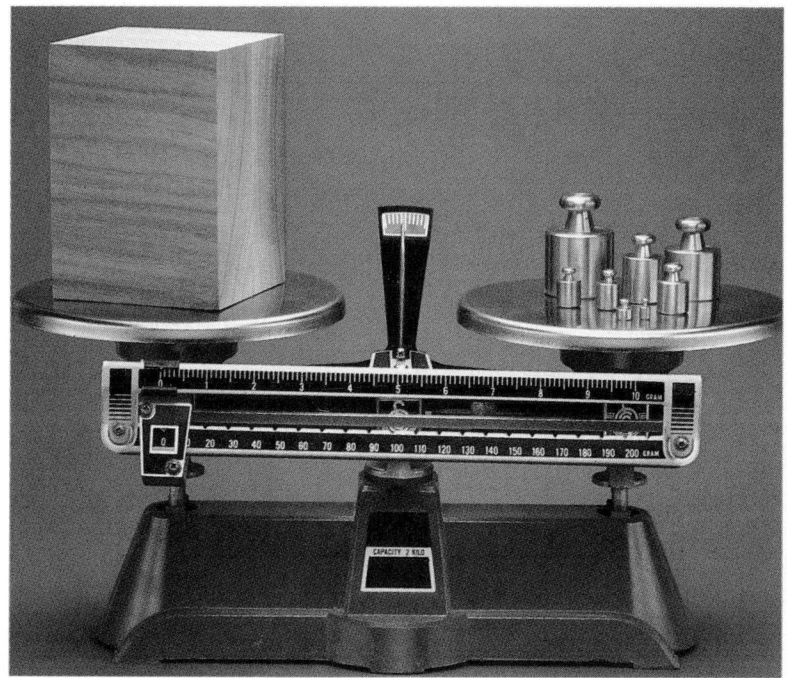

Figure 2.7 A double-pan balance (left) measures mass by comparing the mass of an object placed on the pan to standard masses. The electronic balance (right) works automatically using the same principle, but provides a digital readout.

The standard for mass is a piece of metal kept at the International Bureau of Weights and Measures in Sèvres, France. This object is called the International Prototype Kilogram. Its mass is defined as one kilogram. The SI standard of mass is the **kilogram (kg).** However, the kilogram is too large a unit for convenient use in the chemical laboratory. For this reason, the gram (g), one-thousandth of a kilogram, is commonly used. The mass of a large paperclip is approximately 1.5 grams. An ordinary nickel has a mass of about 5 grams.

So far, you know what property you are going to measure—mass—and what your standard of comparison will be—the gram. Now you must compare the object with a standard by using a balance. A **balance** is an instrument used to determine the mass of an object by comparing the object's mass to known mass. To compare these masses, first place the object with unknown mass on the balance pan. Then add known masses to the beams until the masses are equal. The known masses and the unknown mass (the object) are the same distance from the center of Earth. They are, therefore, subject to the same attraction by Earth. At the top of a mountain, the unknown mass and known masses will still be equally distant from the center of Earth. Thus, Earth's attraction for each mass will still be equal. Thus, the balance will indicate no change in mass.

The process of measuring mass by comparing masses on a balance is called massing. Unfortunately, many terms that refer to weighing are often incorrectly applied to the measurement of mass. For example, standard masses used on the balance are often incorrectly called "weights." Also, using a balance to compare masses is often incorrectly called "weighing."

EVERYDAY CHEMISTRY

Metric Measurement

The metric system of measurement was adopted in the United States in 1875, but we still use the English system of measurement in much of our daily lives. Recently, several government agencies have specified that they will purchase only metric goods. As a result, manufacturers must convert their operations to metric standards to continue doing business with the government. Goods to be exported to other countries must meet metric specifications, particularly if the items are to be repaired, adjusted, or used in manufacturing. The U.S. automobile industry now manufactures some engines using metric measurements. Packaged food products must have metric labeling if they are to be understood abroad. Tire pressure gauges are now calibrated in kilopascals, and the tires themselves have metric specifications. In pharmacy and medicine, the metric system has been used for many years. Dosages are given in milligrams and liquids are measured in cubic centimeters.

Complete conversion to metric measurement isn't welcomed by everyone. "Thinking metric" is hard for those who grew up with the older system. Also, conversion to metric is expensive for manufacturers, who must change all their dies and industrial tools to metric standards and retrain workers.

Exploring Further

1. Look up the factors for converting between milliliters and fluid ounces. How many milliliters are in a standard cup (8 oz.)? How many fluid ounces are in a liter?
2. The claim is sometimes made that the English system is a more "natural" system, and is therefore easier to learn. Present arguments for and against this idea.

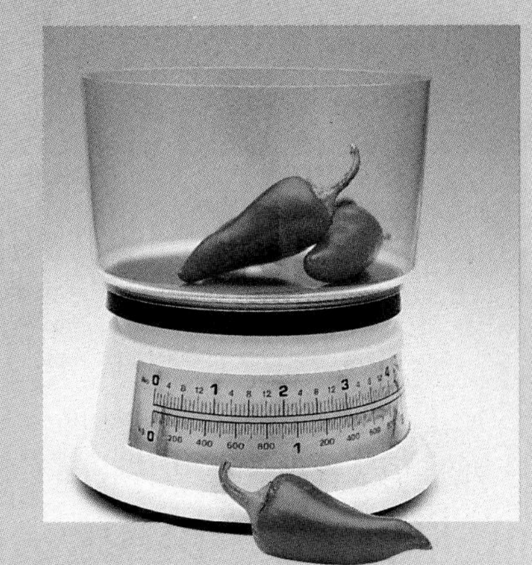

Length

A second important measurement is that of length. **Length** is the distance covered by a straight line segment connecting two points. The standard for measuring length is defined in terms of the distance traveled by light in a unit of time (Appendix Table A-1). The standard unit of length is the **meter (m).** Length is usually measured with a ruler or similar device. A nickel has a diameter of about two centimeters. Your spiral-bound notebook is about 2.5 decimeters, or 25 centimeters, long.

Time

A third basic measurement is time. **Time** is the interval between two occurrences. Our present standard of time is defined in terms of an electron transition in an atom. In practical terms, the unit of time, the **second (s),** is 1/86 400 of an average day. The most common device for measuring time is a watch or clock. More accurate timepieces include the chronometer, the atomic clock, and the solid-state digital timer.

Temperature

Matter is composed of small particles called atoms, ions, and molecules. These particles are in constant motion and, therefore, possess kinetic energy. The average kinetic energy of a group of particles determines the group's temperature. The **temperature** of a sample of matter is a measure of the average kinetic energy of the particles that make up the sample. If we add energy in the form of heat to an object, the kinetic energy of its particles increases. The greater the average kinetic energy of the particles, the higher the temperature of the material.

A thermometer is used to measure temperature. When the bulb is heated, the liquid inside expands and rises in the tube. When the bulb is cooled, the liquid contracts and the height of the liquid column decreases. The height of the liquid column can thus be used to measure temperature. The temperature can be read directly from the scale on the tube.

The SI unit of temperature is the **kelvin (K).** We will define this unit later in the textbook. The kelvin has a direct connection with a more familiar unit, the Celsius degree (C°).

The Celsius temperature scale is based on the fact that the freezing and boiling temperatures of pure water under normal atmospheric pressure are constant. The difference between the boiling and freezing points of water is divided into 100 equal intervals. Each interval is called a Celsius degree. The point at which water freezes is labeled zero degrees Celsius (0°C). The point at which water boils is labeled 100°C. The size of a Celsius degree is exactly equal to that of a kelvin. Thus, there are also 100 kelvins between the freezing point and the boiling point of water. Note that Celsius temperatures are expressed in °C, whereas temperature intervals and changes are expressed in C°.

Some data tables indicate that the values were measured at 25°C (Appendix Table A-10). Room temperatures are usually between 20°C and 25°C. Your average body temperature is 37°C.

HISTORY CONNECTION

Nautical Knots

The speed of ships and aircraft is generally reported in knots, (nautical miles per hour). The International Nautical Mile is a little more than 6000 feet. The term *knot* comes from the early sailing days. A line with a weight on it was knotted every 47 feet 3 inches and rolled onto a reel on board ship. The weight was allowed to drag through the water for 28 seconds unreeling the line. The number of knots unreeled in that time was the speed of the ship. Like other units the knot is measured today in a much more precise way.

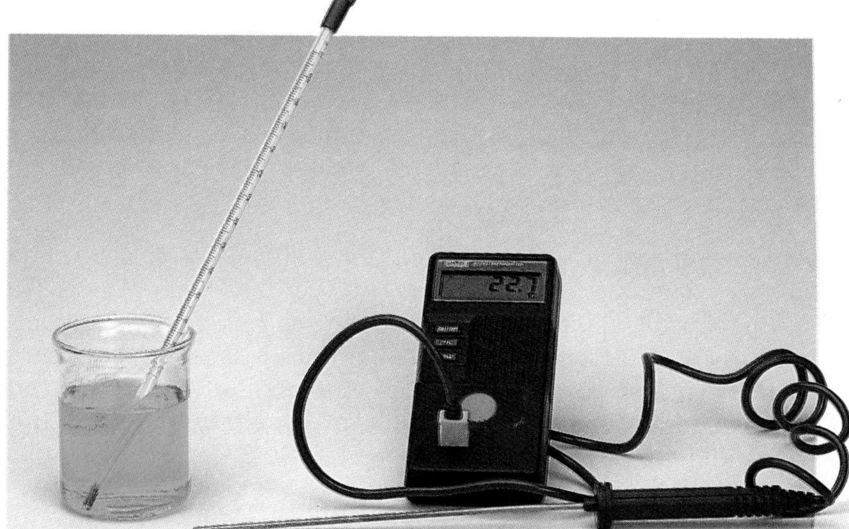

Figure 2.8 As the bulb in the mercury thermometer is warmed, the mercury expands into the capillary tube. The length of the mercury column indicates the temperature. Digital thermometers provide accurate temperature measurement with less threat of glass breakage.

CAREERS IN CHEMISTRY

Surveyor

Accurate measurement is absolutely necessary in surveying. The surveyor sets boundaries of property, finds elevations and depressions of land to assist in mapmaking, determines where pipes can be laid and tunnels dug, and helps in the placement of homes, commercial buildings, bridges, roads, and waterways. Very little building of any kind can be done without a surveyor. A surveyor uses precise measurements, solves problems, and carries out calculations involving geometry and trigonometry. A surveyor does not require a college degree, although he or she may take specialized courses in a community college. Often, surveyors move into civil engineering, which requires a college degree. As long as buildings are built or earth moved, surveyors will be needed.

Accuracy and Precision

Every measurement is a comparison of the physical quantity being measured with a fixed standard of measurement, such as the second or the centimeter. In describing the *reliability* of a measurement, the terms *accuracy* and *precision* are often used. **Accuracy** refers to how close a measurement is to the true or correct value for the quantity. **Precision** refers to how close a set of measurements for a quantity are to one another, regardless of whether the measurements are correct.

Suppose you check your car's odometer against mileposts on a highway. If the odometer has increased by 1.0 mile at the second milepost, you can say the odometer is accurate. If it gives this reading on every trial, you can say that the odometer is both accurate and precise.

Usually measurements that have precision are also accurate. However, it is possible to have a source of error that repeats throughout a set of measurements. Suppose, for example, that you are to find the mass of a piece of copper wire. You happen to be using a laboratory balance that is improperly adjusted and always reads 0.50 gram too low. You make five measurements of the mass of the wire and obtain the values 5.52 g, 5.54 g, 5.51 g, 5.52 g, and 5.51 g. Your values differ by at most ± 0.02 g from the average value, 5.52 g. The values are, therefore, precise. They are not accurate, however, because of the 0.50-g error in the balance. The actual mass of the piece of copper wire is closer to 6.00 g.

PRACTICE PROBLEMS

6. Which student's data are precise? Explain your answer.
   Student 1: 72.75 g, 73.34 g, 73.02 g, 73.25 g
   Student 2: 72.01 g, 71.99 g, 72.00 g, 71.98 g
7. Using a balance that always reads 0.50 g too low, a student obtained the mass of a beaker to be 50.62 g. The student then added some sugar to the beaker and, using the same balance, obtained a total mass of 69.88 g. The student recorded the mass of the sugar as 19.26 g. Is the mass of the sugar inaccurate by 0.50 gram? Why or why not?

Percent Error

At times in your laboratory work, you will use data to determine an experimental value for a quantity that is already known. To evaluate your results, you may need to use percent error to compare your result to a value listed in an appropriate literature source. To find percent error, the absolute difference between your value and the literature value is divided by the literature value and the quotient is multiplied by 100%.

$$\text{Percent error} = \frac{|\text{your value} - \text{literature value}|}{\text{literature value}} \times 100\%$$

SAMPLE PROBLEM

Percent Error

A student determines the atomic mass of aluminum to be 28.9. What is her percent error?

Solving Process:

Consult the table of atomic masses on the inside back cover of your book. The atomic mass of aluminum is 27.0.

$$\text{Percent error} = \frac{|28.9 - 27.0|}{27.0} \times 100\%$$
$$= 7.0\%$$

PROBLEM SOLVING HINT

Because absolute value is used, percent error is always positive.

PRACTICE PROBLEMS

8. An experiment performed to determine the density of lead yields a value of 10.95 g/cm^3. The literature value for the density of lead is 11.342 g/cm^3. Find the percent error.

9. Find the percent error in a measurement of the boiling point of bromine if the laboratory figure is 40.6°C and the literature value is 59.35°C.

Significant Digits

Suppose we want to measure the length of a strip of metal. We have two rulers. One ruler is graduated only in whole centimeters. The other is graduated in tenths of centimeters. Which do you suppose can give the more accurate measurement? Look at Figure 2.9. The length of the strip when measured with the ruler graduated in tenths of centimeters is the more significant measurement because it is closer to the actual length of the strip. We say that this latter measurement has more significant digits than the measurement in whole centimeters.

The number of significant digits in a measurement depends on the instrument used in making the measurement. Look again at Figure 2.9. The measurement on the ruler marked in whole

Figure 2.9 The length of the copper strip can be read as 8.63 cm on the top ruler but only as 8.6 cm on the bottom ruler. You can see that the number of significant digits in the measurement depends on the accuracy of the ruler used.

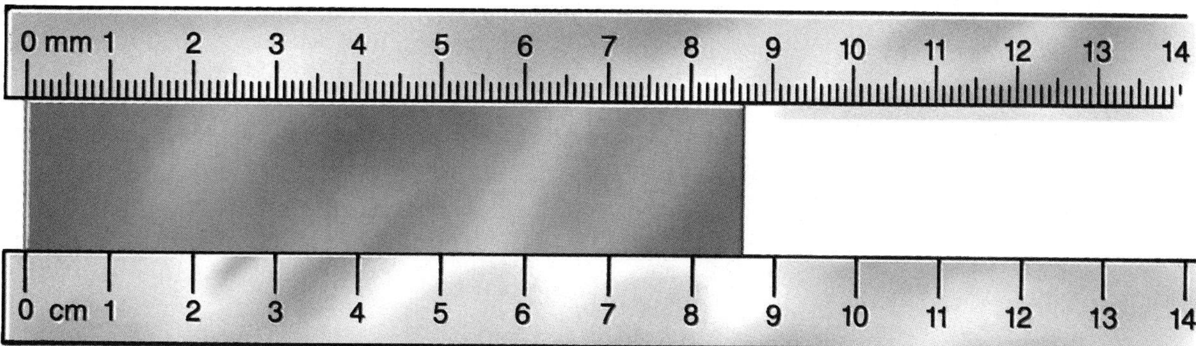

centimeters lies approximately 6/10 of the way from the 8-cm mark to the 9-cm mark. This length is recorded as 8.6 cm. On the ruler marked in tenths of a centimeter, that is, millimeters, the length lies approximately 3/10 of the way from the 8.6-cm mark to the 8.7-cm mark. This length is recorded as 8.63 cm. The measurement 8.6 cm has two significant digits. The measurement 8.63 cm has three significant digits. The last digit of each measurement is an estimate. All digits that occupy places for which actual measurement was made are referred to as **significant digits**. The places actually measured include the one uncertain, or estimated, digit.

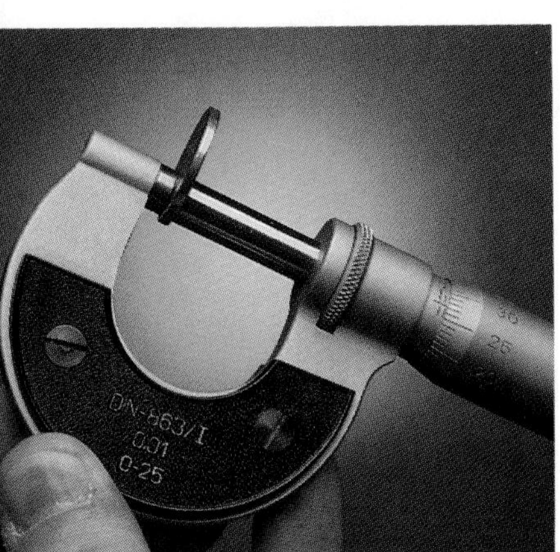

Figure 2.10 This micrometer caliper will measure lengths up to 2.5 cm to the nearest 0.001 cm. The thickness of a nickel is about 2 mm. The nickel's diameter is about 2 cm.

The exactness of measurements is an important part of experimentation. The observer and anyone reading the results of an experiment want to know how exact a measurement is. The exactness of a measurement is indicated by the number of significant digits in that measurement. The following rules are used to determine the number of significant digits in a measurement.

1. *Digits other than zero are always significant.*

96 g	2 significant digits
61.4 g	3 significant digits
0.52 g	2 significant digits

2. *One or more final zeros used after the decimal point are always significant.*

4.72 km	3 significant digits
4.7200 km	5 significant digits
82.0 m	3 significant digits

3. *Zeros between two other significant digits are always significant.*

5.029 m	4 significant digits
306 km	3 significant digits

4. *Zeros used solely for spacing the decimal point are not significant. The zeros are placeholders only.*

7000 g	1 significant digit
0.007 83 kg	3 significant digits

If the quantity 7000 grams has been measured on a balance that is accurate to the nearest gram, all four digits are significant. We must show this in a way that follows the rules for representing significant digits. To do this, we use scientific notation. Scientific notation is a type of exponential notation. It is discussed in detail in Chapter 8, on pages 194–196. For now, note that in scientific notation, you would represent 7000 g measured to the nearest gram as 7.000×10^3 g. This method shows that the mass was measured to four significant digits.

Not all numbers represent measurements. For instance, suppose there are 23 students in a chemistry class. How many significant digits are in 23? The 23 is not a measurement. We do not *measure* the number of students in a class. We *count* them. Students come in natural numbers, or **counting numbers**. We cannot have 23.4 or 22.8 students. Since counted objects occur in *exact* numbers, we consider that these numbers contain an infinite number of significant digits. We can, therefore, ignore them when determining the number of significant digits in the answer to a problem. The same is true of a number such as the 2 in the equation for the area of a triangle, $A = bh/2$. It is also true in the case of a definition; for example, 1000 millimeters = 1 meter. Both the 1000 and the 1 are exact.

PRACTICE PROBLEMS

10. How many significant digits are there in each of the following quantities?

a. 20 kg	**d.** 0.010 s	**g.** 0.089 kg	**j.** 20 cars
b. 0.0051 g	**e.** 90.4°C	**h.** 0.009 00 L	
c. 11 m	**f.** 0.004 cm	**i.** 100.0°C	

Derived Units

By combining the SI base units, we obtain measurement units used to express other quantities. Distance divided by time equals speed. If we multiply length by length, we get area. Area multiplied by length produces volume. The SI unit of volume is the cubic meter (m^3). However, this quantity is too large to be practical for the laboratory. Chemists often use cubic decimeters (dm^3) as the unit of volume. One cubic decimeter is given another name, the liter (L). The liter is a unit of volume. One liter equals 1000 milliliters (mL) and 1000 cubic centimeters (cm^3). From these facts, you can see that

$$1\ dm^3 = 1\ L = 1000\ mL = 1000\ cm^3$$

Much of your laboratory equipment is marked in milliliters.

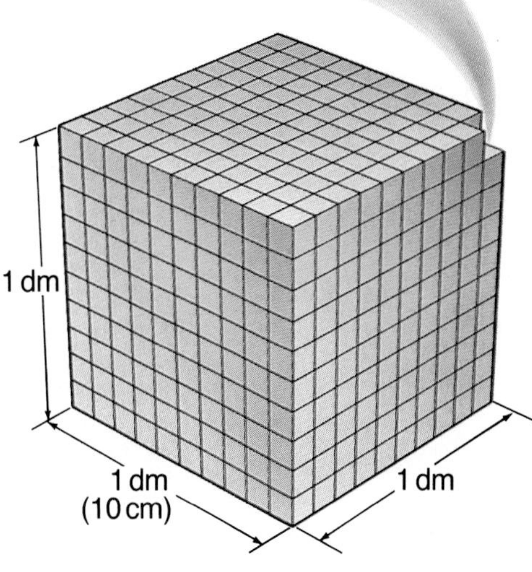

Figure 2.11 Volume is an example of a quantity that is measured in derived units. The volumes shown here all equal 1 dm^3. A cubic decimeter is the volume of a cube that is 1 dm (10 cm) on an edge. It contains $(10\ cm)^3$ or 1000 cm^3, which equals 1 L.

The units used to express measurements of speed, area, and volume are called derived units. You saw that area and volume measurements are expressed using length units. The units used to express speed, such as kilometers per hour or meters per second, combine fundamental units of length and time.

Density

Another common scientific measurement is density. **Density** is mass per unit of volume. To measure the density of a material, we must be able to measure both its mass (m) and its volume (V). Mass can be measured on a balance. The volume of a solid can be measured in different ways. For instance, the volume of a cube is the length of one edge cubed. The volume of a rectangular solid is the length times the width times the height. Width and height are simply different names for length. The volume of a liquid can be measured in a clear container graduated to indicate units of volume. You will use graduated cylinders in the laboratory to measure liquid volumes. In measurements of the densities of liquids and solids, volume is usually measured in cubic centimeters (cm^3). Density, then, is expressed in grams per cubic centimeter (g/cm^3). The units to express density are derived units. In equation form, density can be expressed as

$$\text{Density} = \frac{\text{mass}}{\text{volume}}; \quad D = \frac{m}{V}$$

People sometimes say that lead is heavier than feathers. However, a truckload of feathers is heavier than a single piece of lead buckshot. To be exact, we should say that the density of lead is greater than the density of feathers. Densities of some common materials are listed in Table 2.3. Such a table offers a convenient and accurate means of comparing the masses of equal volumes of different materials. Note that the values given for densities of gases are quite small. It is more practical to express the density of gases in g/dm^3.

Changes in temperature change the volume of an object. Therefore, the density of a substance changes when temperature changes. Densities of most materials are given at 25°C.

MINUTE LAB

Diet Density

Density differences are used to separate ripe tomatoes from green tomatoes and coal from shale. To see how this works, fill an aquarium with water. Place a can of regular soda and a can of diet soda in the water. Note which can floats and which can sinks. Remove the cans, dry them, and mass them on a balance. Calculate the density of each type of soda using the volume printed on the can. How do your calculations explain your observations? What do you think accounts for the difference in density?

Table 2.3

Densities of Some Common Materials at 25°C

Material	Density
natural gas	0.000 656 g/cm^3
air	0.001 18 g/cm^3
ethanol	0.789 48 g/cm^3
sucrose (table sugar)	1.587 g/cm^3
sodium chloride (table salt)	2.164 g/cm^3
stainless steel	8.037 g/cm^3
copper	8.94 g/cm^3
mercury	13.545 g/cm^3

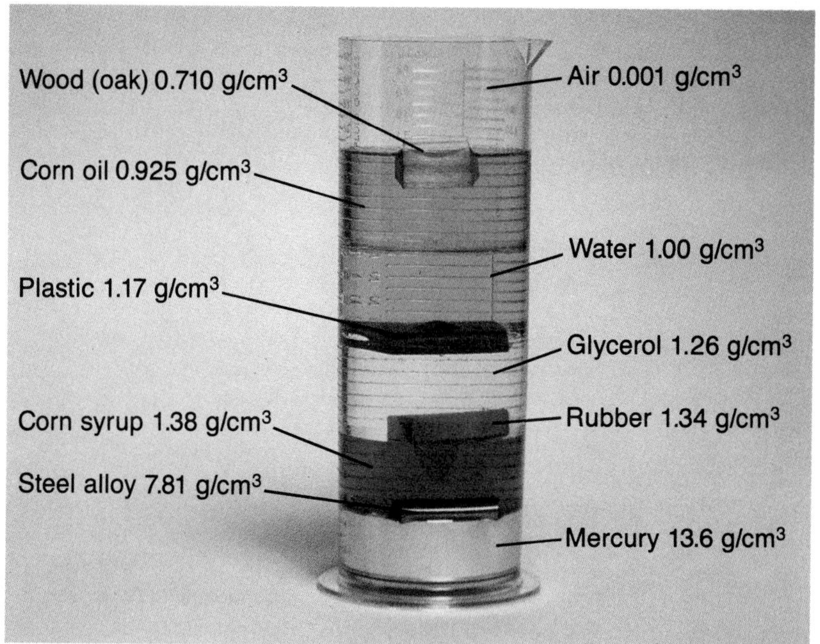

Figure 2.12 Materials of lower density will float on other materials of greater density. Because of differences in density, the materials shown here float at different levels in the container.

SAMPLE PROBLEM

Calculating Density

A piece of beeswax with a volume of 8.50 cm^3 is found to have a mass of 8.06 g. What is the density of the beeswax?

Solving Process:

Density is mass per unit volume, or

$$D = \frac{m}{V}$$

Substituting the known information, we obtain

$$D = \frac{8.06 \text{ g}}{8.50 \text{ cm}^3} = 0.948 \frac{\text{g}}{\text{cm}^3}$$

The fact that the answer is expressed in grams per cubic centimeter is a check on the work.

PROBLEM SOLVING HINT

Decide whether beeswax would be more or less dense than water (1 g/cm^3).

PRACTICE PROBLEMS

11. What is the density of a piece of concrete that has a mass of 8.76 g and a volume of 3.07 cm^3?

12. Illegal ivory is sometimes detected on the basis of density. What is the density of a sample of ivory whose volume is 14.5 cm^3 and whose mass is 26.8 g?

13. An archeologist finds that a piece of ancient pottery has a mass of 0.61 g and a volume of 0.26 cm^3. What is the density of the pottery?

SAMPLE PROBLEMS

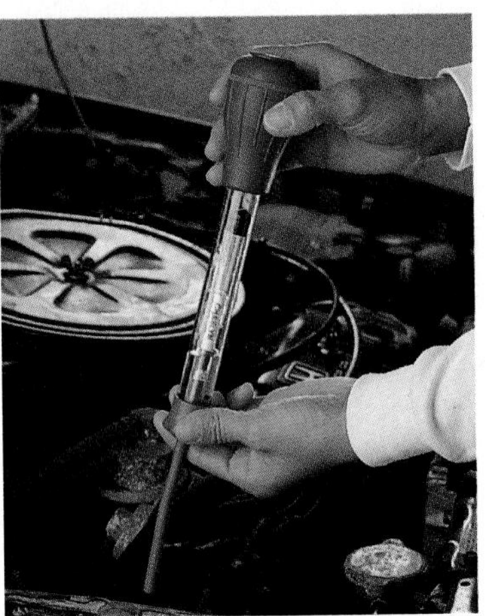

Figure 2.13 You can determine the concentration of acid in a car battery by using a hydrometer. The higher the float rises in the solution, the greater the solution's density. Greater density indicates a higher concentration of battery acid.

1. Using Density to Find Volume

Cobalt is a hard magnetic metal that resembles iron in appearance. It has a density of 8.90 g/cm^3. What volume would 17.8 g of cobalt have?

Solving Process:

You know that density can be expressed as

$$D = \frac{m}{V}$$

Solving this equation for volume and substituting the known information in the resulting equation, you obtain

$$V = \frac{m}{D} = \frac{17.8\ \cancel{g}}{8.90\ \cancel{g}/\text{cm}^3} = 2.00\ \text{cm}^3$$

The answer is in cm^3, a unit of volume, which is another check on the accuracy of the problem solving approach.

2. Using Density to Find Mass

What is the mass of 19.9 cm^3 of coal that has a density of 1.50 g/cm^3?

Solving Process:

The mathematical relationship relating density, volume, and mass is

$$D = \frac{m}{V}$$

Solving the equation for m gives

$$m = DV$$

Substituting known values produces

$$m = \left(1.50\ \frac{\text{g}}{\cancel{\text{cm}^3}}\right) \times (19.9\ \cancel{\text{cm}^3}) = 29.9\ \text{g}$$

PROBLEM SOLVING HINT

Be sure your setup will result in units of mass.

PRACTICE PROBLEMS

14. Limestone has a density of 2.72 g/cm^3. What is the mass of 24.9 cm^3 of limestone?
15. Calcium chloride is used as a deicer on roads in winter. It has a density of 2.50 g/cm^3. What is the volume of 7.91 g of this substance?
16. Ammonium magnesium chromate has a density of 1.84 g/cm^3. What is the mass of 7.62 cm^3 of this substance?
17. What is the minimum volume of a tank that can hold 795 kg of methanol whose density is 0.788 g/cm^3?

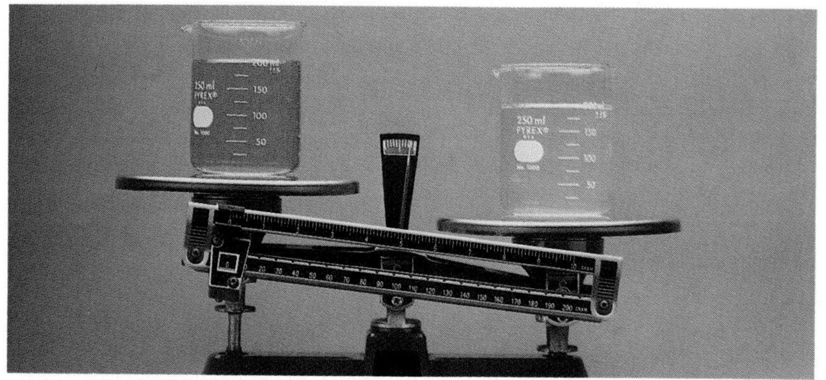

Figure 2.14 The beaker on the left contains corn oil. The beaker on the right contains water. Even though the volumes of both liquids are equal, the balance shows that their masses are not the same. Water has more mass per cubic centimeter than corn oil. That is, water is more dense than corn oil.

One type of density problem lends itself readily to the use of the memory function found on most calculators. In cases where density must be found from a mass and the dimensions of an object, the memory can be used to hold the calculated volume.

SAMPLE PROBLEM

Solving Density Problems with a Calculator

What is the density of a rectangular block of granite of mass 40.4 g and dimensions 2.00 cm by 1.09 cm by 7.04 cm?

Solving Process:

Calculate the volume of the block by multiplying length by width by height. Press keys [2] [.] [0] [0] [×] [1] [.] [0] [9] [×] [7] [.] [0] [4] [=] The result will appear as 15.3472. Don't round this result. When doing chemistry problems with a calculator, you should round off values only at the end. Press the [Min] key to store the calculated volume.

The equation for density is $D = \frac{m}{V}$. Therefore, you can complete the calculation by pressing keys [4] [0] [.] [4] [÷] [MR] [=]. The result will appear as 2.632 402 002. The data used had only three significant digits. Therefore, the answer also should be rounded to three significant digits.

$$D_{\text{granite}} = 2.63\ \frac{\text{g}}{\text{cm}^3}$$

Figure 2.15 Although your calculator will give the result 2.632402002, the number must be rounded to three significant digits because the data you used to make your calculation had only three significant digits. You should report the density as 2.63 g/cm^3.

PRACTICE PROBLEMS

18. Use a calculator to determine the density of a pine board whose dimensions are 4.05 cm by 8.85 cm by 164 cm and whose mass is 2580 g.

19. Use a calculator to find the density of a 51.6 g cylindrical steel rod of diameter 0.622 cm and length 22.1 cm.

2.2 CONCEPT REVIEW

20. Two students measure the mass of the same beaker on two different balances. The first student reports the beaker's mass to be 47.0 g. The second student reports the mass to be 47 g. Assuming the measurements were correctly reported, how can you explain the difference?

21. Express each of the following quantities in the unit named.
   - **a.** 5 grams in milligrams
   - **b.** 15 centimeters in meters
   - **c.** 200 milliseconds in seconds
   - **d.** 0.5 decimeters in centimeters
   - **e.** 60 grams in kilograms
   - **f.** 0.2 centimeters in micrometers

22. **Apply** Add the appropriate SI base units to the measurements in the following passage.

   On a day when the thermometer read 21, we wrapped and mailed a package that required almost 3 of tape to seal. When the postal clerk put it on a balance, the dial read 1.4. We were in and out of the post office in less than 90.

Bridge to METALLURGY

Froth Flotation Separation

Mining engineers use a process involving density differences to concentrate minerals. Some ores of copper, lead, and zinc are concentrated in this way. When minerals are removed from Earth's crust, they are mixed with dirt, rock, and other contaminants. Copper sulfide ores, for example, usually contain less than 10% copper. In the froth-flotation process, the output of the mine is first ground to a powder. Then it is mixed by a motor-driven agitator in a tank with water to which oil has been added. The oil is carefully selected to work with the mineral being separated. A froth forms when air at high pressure is blasted through the mixture. The minerals, such as the copper sulfides, have little or no attraction for the water, but they are attracted to the oil and coated by it. The oily particles stick to the air bubbles in the froth. The froth is less dense than the water because it contains a lot of air. The froth floats to the top of the tank. There, the froth can be floated off, the oil removed, and the concentrated mineral recovered. The denser dirt and rock, meanwhile, fall to the bottom of the tank. The flotation tank is flushed from time to time to remove the refuse.

Exploring Further

1. Find out the names and formulas of two ores concentrated by the froth-flotation process.
2. Why do you think the oil sticks to the ore particles, but the water does not?

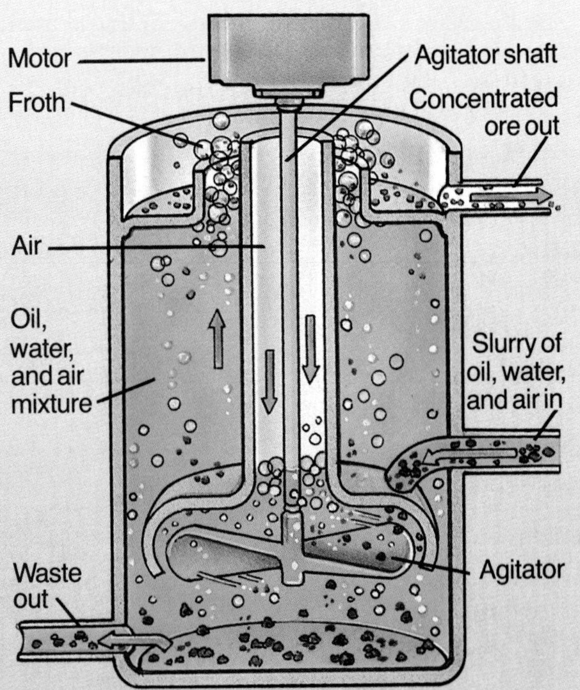

Chapter 2

REVIEW

Summary

2.1 Decision Making

1. Problem solving can be learned and uses certain techniques that are adaptable to different situations.
2. In solving problems, it is important to know exactly what information is given and what information is desired.
3. A result should always be checked against the original problem to make sure the answer makes sense.
4. Algorithms are frequently useful in solving numerical chemical problems.

2.2 Numerical Problem Solving

5. Scientific work requires a quantitative approach. In other words, to investigate a phenomenon, certain characteristics must be measured.
6. Scientists use SI measurements. This system of measurement consists of seven base units. In SI, convenient units for any measurement are obtained by using prefixes.
7. Mass is a measure of the quantity of matter. Weight is a measure of the force of gravity between two objects.
8. The SI standard unit of mass is the kilogram. For length, the unit is the meter; for time, the second; and for temperature, the kelvin.
9. Accuracy of a measurement refers to the closeness of the measurement to the true value. Precision refers to reproducibility within a group of measurements.
10. The exactness of a measurement is indicated by the number of significant digits in that measurement.
11. SI base units may be combined to form derived units such as the cubic meter, the SI standard unit of volume.
12. Density is the mass of a material in a unit volume of that material.

Key Terms

qualitative	second
quantitative	temperature
weight	kelvin
mass	accuracy
kilogram	precision
balance	significant digit
length	counting number
meter	density
time	

Review and Practice

23. What is an algorithm?
24. How is the guess-and-check method of problem solving different from the algorithm method?
25. What is the density of a piece of cork that has a mass of 0.650 g and a volume of 2.71 cm^3?
26. Barium perchlorate has a density of 2.74 g/cm^3. What is the mass of 27.2 cm^3 of this substance?
27. Bismuth phosphate has a density of 6.32 g/cm^3. What is the mass of 25.9 cm^3 of this substance?
28. Cerium sulfate has a density of 3.17 g/cm^3. Calculate the volume of 599 g of this substance.
29. Chromium silicide has a density of 5.50 g/cm^3. Calculate the volume of 35.9 g of this substance.
30. What are the seven base units in SI?
31. How is the Celsius temperature scale defined?
32. What is the SI fundamental unit of mass?
33. List the number of significant digits for each of the following.
    a. 1 km **c.** 2.15 000 cm^2
    b. 1.5 mL **d.** 5.380 000 0 s
34. What is the difference between counting and measured numbers? Which are expressed in significant digits?

35. What is a "derived" SI unit? Give two examples.

36. What is the difference between accuracy and precision?

37. Find the density in g/cm^3 of the following.
   a. concrete, if a rectangular piece 2.00 cm × 2.00 cm × 9.00 cm has a mass of 108 g
   b. granite, if a rectangular piece 5.00 cm × 10.0 cm × 23.0 cm has a mass of 3.22 kg
   c. gasoline, if 9.00 dm^3 have a mass of 6120 g

38. An automobile is traveling at the rate of 30.0 kilometers per hour. What is its rate in centimeters per second (cm/s)?

39. The density of nitrogen gas is 1.25 g/dm^3. What is the mass of 1.00 m^3 of N_2?

40. Bismuth has a density of 9.80 g/cm^3. What is the mass of 3.74 cm^3 of Bi?

41. Iron has a density of 7.87 g/cm^3. What volume would 26.3 g of Fe occupy?

42. If 1.00 km^3 of air has a mass of 1.2 billion kg, what is the density of air in g/cm^3?

43. Magnesium has a density of 1.74 g/cm^3. What is the volume of 56.6 g of Mg?

44. Tin has a density of 7.28 g/cm^3. What is the volume of 2.32 kg of Sn?

Concept Mastery

45. Why is it a good idea to estimate the answer to a problem before beginning your calculation?

46. What is the last thing you should do after calculating an answer to a chemistry problem?

47. Why is mass used instead of weight for scientific work?

48. Why is it important to use the correct number of significant digits when expressing measurements?

49. Would an astronaut's weight change on the moon? Would an astronaut's mass change on the moon? Why or why not?

50. Does a pound of feathers "weigh" more than a pound of iron? What truly is different between the two?

51. A wood block has dimensions of 2.0 cm × 1.5 cm × 5.6 mm. What is its volume in cubic centimeters? In liters? In cubic decimeters?

52. Vanadium hardens steel. Vanadium steel is used in armor plate on army tanks, for example. If its density is 5.96 g/cm^3, a cubic meter would have a mass of how many kg?

Application

53. A chemist has been given a piece of an unusual meteorite for analysis. The meteorite fell into a field 220 km from the chemist's laboratory. The meteorite had a mass of 14.9 kg. The piece sent to the chemist is irregularly shaped and is 3 cm long. A scale shows that the piece weighs 1.6 newtons. A balance shows its mass to be 163 g. The chemist decides to determine the volume of the fragment by seeing how much water it will displace when submerged. The chemist adds water to a graduated cylinder that can hold 50.0 cm^3. The level of the water in the cylinder shows that it contains 25.8 cm^3. After the meteorite piece is submerged, the water level rises to 45.9 cm^3. The density of water is 1.00 g/cm^3. What density will the chemist calculate for the meteorite sample? Organize the information just given into two lists—relevant and irrelevant data. Then, carry out the calculation.

54. A person who was on a diet measured his mass one week and found it to be 82 kilograms. Two weeks later he used the same scale again and found that his mass was 79 kilograms. (Assume that his scale has an uncertainty of 1 kilogram.)

a. What is the person's maximum possible loss of mass?
b. What is the minimum possible loss of mass?
c. Explain the mass loss using the ± notation.

55. Name the SI unit that would be the best measure of the size of each of the following items.
a. a glass of milk
b. a tablespoon of sugar
c. the mass of an elephant
d. the surface area of a dollar bill
e. the volume of an automobile gasoline tank
f. the volume of air in a balloon

56. Aluminum, because of its light density and strength, has been used for making aircraft. It has a density of 2.70 g/cm^3. Magnesium has similar properties, and it has a density of 1.74 g/cm^3. What is the difference in mass if 1000 cm^3 of magnesium are used rather than 1000 cm^3 of aluminum?

57. The use of hydrogen for filling blimps and dirigibles led to disasters because of its flammability. Helium does not burn and is now used instead of hydrogen. Hydrogen has a density of 0.089 9 g/dm^3 and helium has a density of 0.178 g/dm^3, about twice that of hydrogen. What is the mass of helium in a 10 000-m^3 blimp?

58. The diameter of the sun is estimated at about 865 000 miles and its average density is 1.4 g/cm^3. What is its mass in kilograms? (Use $V = 4\pi r^3/3$.)

Critical Thinking

59. It is sometimes said that the calculation itself is the least important step in solving a problem while studying chemistry. Do you think that statement is valid? Explain your answer.

60. The volume of one block of wood is greater than that of a second block of wood, but the mass of the second is greater. Which has the greater density and why?

61. The definition of the liter was originally based on the volume of a certain mass of water. What is the difficulty with such a definition?

62. The mineral quartz has a density of 2.65 g/cm^3 and the mineral zircon has a density of 4.50 g/cm^3. Suppose a rock composed of quartz and zircon has a density of 3.00 g/cm^3. Find the percentage by mass and the percentage by volume of quartz in the rock.

Readings

Deavor, James P. "Chemistry: The Ultimate Liberal Art." *Journal of Chemical Education* 67, no. 10 (October 1990): 881-882.

Jakuba, Stan R. "Go Metric? Now?" *CHEMTECH* 18, no. 7 (July 1988): 424-425.

Kotz, John C., and Keith F. Purcell. *Chemistry & Chemical Reactivity*. New York: Saunders College Publishing, 1987.

Smith, Richard G., and Gary K. Himes. *Solving Problems in Chemistry*. Columbus, OH: Merrill Publishing Company, 1993.

Cumulative Review

1. List five scientific procedures.
2. What are the two general forms of energy an object can possess?
3. What is the principal aim of chemists in their work?
4. State the law of conservation of mass-energy.
5. What is the difference between a law and a theory?

CHAPTER **P**review

CONTENTS

CHAPTER

3

Matter

The blacksmith shapes a piece of glowing iron by hammering it on an anvil. After shaping the iron, the blacksmith may plunge it into a tub of cold water. The blacksmith's work is possible because of the properties of both iron and water.

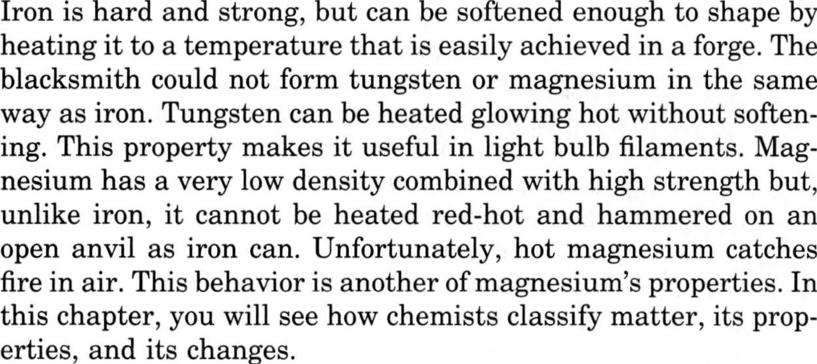

Iron is hard and strong, but can be softened enough to shape by heating it to a temperature that is easily achieved in a forge. The blacksmith could not form tungsten or magnesium in the same way as iron. Tungsten can be heated glowing hot without softening. This property makes it useful in light bulb filaments. Magnesium has a very low density combined with high strength but, unlike iron, it cannot be heated red-hot and hammered on an open anvil as iron can. Unfortunately, hot magnesium catches fire in air. This behavior is another of magnesium's properties. In this chapter, you will see how chemists classify matter, its properties, and its changes.

EXTENSIONS AND CONNECTIONS

OBJECTIVES

- describe and distinguish heterogeneous and homogeneous materials, substances, mixtures, and solutions.
- describe and give examples of elements and compounds.
- classify examples of matter.

3.1 Classification of Matter

The world around us is filled with objects of many kinds. There are people, chairs, books, trees, lumps of sugar, ice cubes, drinking glasses, doorknobs—an endless number of familiar objects. Each of these objects may be characterized by its size, shape, use, color, and texture. Many unlike objects have certain things in common. For example, a tree and some chairs are made of wood. A car body, a soda bottle, and costume jewelry may all be made of plastic. Wood and plastic are examples of materials.

Heterogeneous Materials

The word **material** is used when referring to a specific kind of matter, such as wood. Familiar materials include wood, steel, air, copper, sugar, salt, nickel, marble, concrete, and milk. Most of the things you see around you are mixtures. A **mixture** is matter that contains two or more different materials. Wood, granite, concrete, air, and milk are examples. Sometimes it is necessary to use a microscope to distinguish between the different materials in a mixture. If you look closely at granite, you can see at least three minerals. If a piece of granite is crushed into sand-sized particles, it is possible to pick out the minerals quartz, biotite, and feldspar. Nonuniform materials such as granite are called heterogeneous (het uh roh JEE nee us) materials. Milk appears to be uniform. Under a microscope, however, you can see particles suspended in water. One type of material can be separated from the other material in milk. Fat globules can be removed using a cream separator. Milk is not uniform.

Any physically separate part of a material is called a phase. A **phase** is any region with a uniform set of properties. Each of the minerals in granite is a separate phase. In milk, for example, the watery part is one phase while the fat is a second phase. You can also distinguish between different phases of the same material.

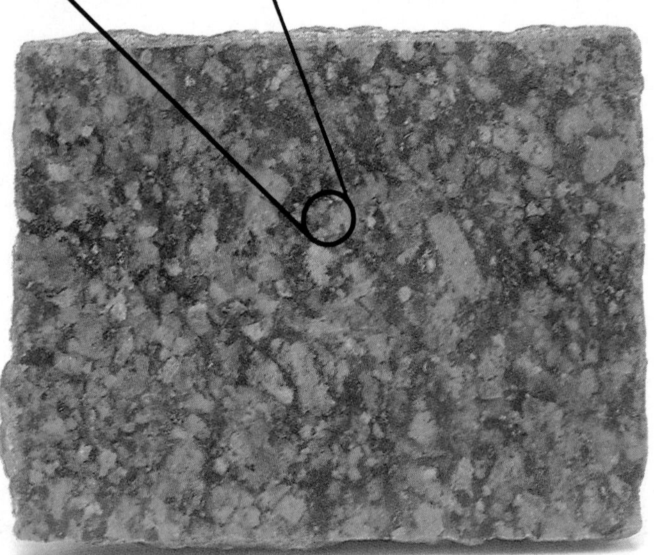

Figure 3.1 You can see that granite (left) is heterogeneous because it is actually composed of several distinct minerals. Although milk appears uniform, it will separate into two visible phases (right). You can see that milk, too, is heterogeneous.

Figure 3.2 Ice and seawater form a definite boundary between phases. This boundary is called an interface.

For example, ice and water are different phases of the same material. All the material in the water region has the same set of properties. Likewise, all the material in the ice region has the same set of properties. Ice and water also represent different states, solid and liquid. In milk, on the other hand, both the water phase and the fat phase are in the liquid state.

A **heterogeneous mixture** is a mixture that is composed of more than one phase. The different phases in a heterogeneous mixture are separated from each other by definite boundaries called **interfaces.** In the two-phase system of ice and water, the surfaces of the ice and water are the interfaces.

Word Origins

heteros: (GK) other, different
genea: (GK) origin, source
heterogeneous—of different sources, nonuniform

Homogeneous Materials

Materials that consist of only one phase are called **homogeneous** (hoh moh JEE nee us) materials. If you break a piece of homogeneous matter into smaller pieces, each piece will have the same properties as every other small piece. If you look at one of the pieces under a microscope, it is impossible to distinguish one part as being a different material from any other part. Examples of homogeneous materials are sugar, salt, seawater, quartz, window glass, and air.

Heterogeneous matter is always composed of more than one phase and, as we have seen, is always a mixture. One type of homogeneous matter can be classified as a mixture. This type of homogeneous matter is composed of more than one material and is called a **solution.**

A solution consists of a **solute** (dissolved material) in a **solvent** (dissolving material). In the case of two liquids in solution, the solvent is the component that is the larger proportion of the whole solution. The solute is scattered in the solvent as very small particles. Thus, the solution appears uniform, even under the most powerful optical microscope. Because the scattering of particles appears to be uniform, there are no separate phases. Thus, solutions are classified as homogeneous materials.

Room Enough to Spare

Put on an apron and goggles. Fill a small clear juice glass with water until it is ready to overflow. Predict the number of metal paper clips that can be added before the water in the glass overflows. Add one paper clip at a time to the glass. How did your prediction compare with the actual number of paper clips required to make the water overflow the glass? What phases are present at the conclusion of the lab? Give two examples of interfaces that are present in the glass.

Word Origins

omios: (GK) like, same
genea: (GK) origin, source

homogeneous—of the same source, uniform

The composition of homogeneous mixtures (solutions) can vary. If you put a small amount of pure salt into pure water, stir it, and let it stand, you have a solution, or homogeneous mixture. If you add a larger amount of pure salt to the same amount of pure water, you again have a solution. The composition of the second sample would differ from the first. The second sample contains more salt in an equal volume of water. In each case, the resulting material is homogeneous. A solution may also be defined as a single phase that can vary in composition. Thus, solutions such as antifreeze, seawater, and window glass vary in composition from sample to sample.

Solutions are not necessarily liquid. To a chemist, air is a homogeneous material composed of nitrogen, oxygen, and smaller quantities of other gases. Its composition varies from place to place. However, the ordinary air we breathe usually contains heterogeneous particles such as soot, spores, and pollen. Different types of window glass have different compositions, yet each type is homogeneous. Both air and glass are solutions. So is automobile radiator antifreeze. In antifreeze, the solvent is an organic liquid called 1,2-ethanediol (ethylene glycol). The solutes are various dyes, rust inhibitors, and rubber hose conditioners.

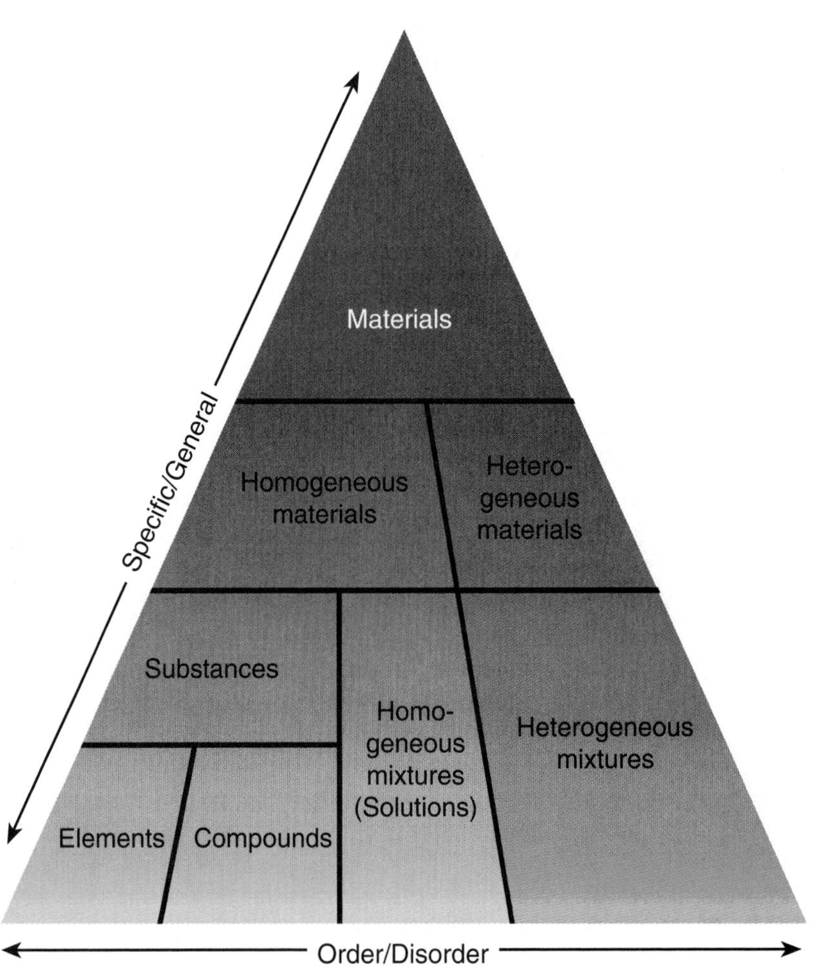

Figure 3.3 All matter can be classified from general to specific types according to composition and properties.

In your laboratory work, you will be using solutions labeled with a number followed by the letter "*M*." The symbol represents the term *molarity*. The exact meaning of molarity will be studied in Chapter 8. In the meantime, you should keep in mind that molarity is used to indicate the amount of solute in a specific amount of solution. A 6*M* (six molar) solution contains 6 times as much solute as a 1*M* (one molar) solution of the same volume. Concentrated solutions have a higher ratio of solute to solvent than dilute solutions.

Percent by volume, percent by mass, molality, and normality are other methods used to express solution concentration. Normality is used by some analytical chemists and technicians.

HISTORY CONNECTION

The Village Blacksmith

At one time, every community needed a blacksmith to make hinges, hasps, weather vanes, and other pieces of necessary hardware. Even though there are now machines that manufacture these items, blacksmiths are still in great demand to shoe horses. The workers who shoe horses are called *farriers,* a term that was derived from the Latin word for iron, *ferrum.*

Substances

Pure salt, pure sugar, and pure sulfur are homogeneous materials and always have the same composition. Such materials are called **substances.** A large part of chemistry is the study of the processes by which substances change into other substances.

According to the atomic theory, matter is made of very tiny particles called atoms. Substances are divided into two classes based on the atoms they contain. Substances composed of only one kind of atom are called **elements.** Examples are sulfur, oxygen, hydrogen, nitrogen, copper, gold, and chlorine. Substances composed of more than one kind of atom are called **compounds.** The atoms in the particles of compounds are always in definite ratios. For example, water contains hydrogen atoms and oxygen atoms combined in a ratio of 2 to 1.

To summarize, all matter can be classified as heterogeneous mixture, solution, compound, or element. The development of this system of classification played a significant role in the early history of chemistry. Early chemists spent much time and energy sorting the pure substances from the mixtures.

Figure 3.4 Substances are either elements or compounds. Shown here are the elements sulfur (left), copper (top center), iodine (crystals in beaker), and bromine (red liquid in flask). Also shown are the compounds sodium dichromate (orange crystals), sodium chromate (yellow crystals), quartz (silicon dioxide, clear crystal), and galena (lead(II) sulfide, gray chunk).

You have probably seen a periodic table of the elements. There may be one visible in your classroom right now. The periodic table is an indispensable tool of the chemist because of the amount of information it contains. Later in the course, you will learn more about this table and the reasons for its arrangement. Table 3.1 on pages 54–55 is a special version of the periodic table which pictures samples of nearly all the naturally-occurring elements. A complete list of the elements is found in Table 4.2 on page 84. Note that the second column of Table 4.2 also gives a symbol that is used to represent each element.

Chemists know of 88 naturally-occurring elements. The natural element with the most complex atoms is uranium. Two elements, technetium and promethium, which have simpler atoms than uranium, are not found in nature. Astatine and francium have been detected in nature. However, they are present in such small amounts that they cannot be easily separated from their ores. These four elements are not normally counted among the natural elements. These synthetic elements can be produced by scientists through nuclear reactions. Synthetic elements and nuclear reactions will be discussed later.

FRONTIERS

Crystal Coolers

This shimmering diamond was not mined from the earth. Instead, it was produced in a laboratory and is almost 99.9% pure. Because of its purity, it has properties that differ significantly from those of natural diamonds. One such property is its thermal conductivity.

A natural diamond conducts heat about six times better than copper wire at room temperature. However, synthetic diamond is almost 50% more conductive than natural diamond, making it the world's best conductor of heat.

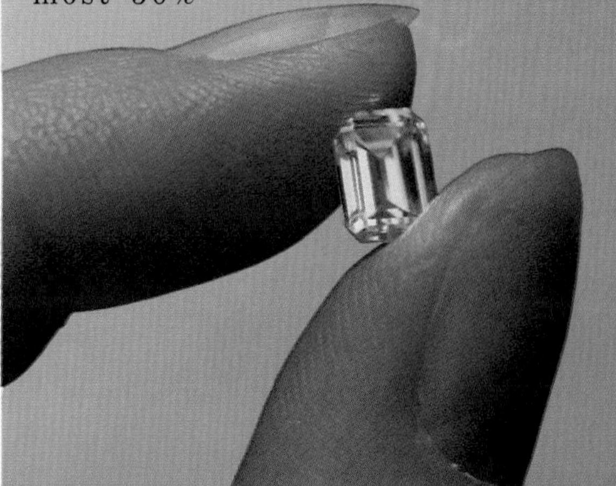

In microelectronic circuits, even minute quantities of heat may damage delicate components. To prevent damage, heat sinks are built into the circuits. Heat sinks are devices that carry away unwanted heat. Because of the synthetic diamond's high thermal conductivity, engineers will be able to replace miniature heat sinks made of natural diamond with even smaller slabs of synthetic diamond. Thus it will be possible to miniaturize circuits even further.

The high thermal conductivity and transparency of synthetic diamond may also find use in the field of laser optics. Because synthetic diamond can remove heat at a high rate, crystals of synthetic diamond can transmit intense laser light without damage. Engineers are trying to fashion mirrors and windows for lasers out of synthetic diamond that will allow them to construct more powerful and complex lasers. Perhaps in the future, these gems will make possible complex computers operated only by light.

Figure 3.5 All organic substances contain the element carbon. Organic materials shown here include rubber, plastic, wood, and paper. Typical inorganic substances do not contain carbon. Inorganic materials shown here are glass, ceramic, metal, and water.

Chemists are interested in the reactions of elements and compounds, the analysis of compounds into their component elements, and the synthesis of compounds from elements or other compounds. These properties and processes depend upon the structure of the elements and compounds. The development of new pharmaceutical products is pursued largely on the basis of known reactions of the human body to molecules with particular structures.

There are more than 10 000 000 substances known to chemists. Obviously, no one scientist knows the characteristics of all of these substances. Thus, chemists classify them into different groups with common features. Chemists divide all substances into two large classes called *organic* and *inorganic*. **Organic substances** are compounds that contain the element carbon. **Inorganic substances** are the elements and the compounds of all elements other than carbon. These categories are defined only for convenience. Some carbon compounds are considered inorganic, for reasons we will consider later. Chemists generally concentrate their efforts in one of the two fields, *organic chemistry* or *inorganic chemistry*.

3.1 CONCEPT REVIEW

1. Classify the following materials as heterogeneous mixtures, solutions, compounds, or elements. Use a dictionary to identify any unfamiliar materials.
   a. air
   b. india ink
   c. paper
   d. table salt
   e. wood alcohol
   f. apple
   g. milk
   h. plutonium
   i. water
2. How would you determine if a piece of cloth advertised as 50% wool and 50% synthetic fiber was a heterogeneous mixture or a homogeneous mixture?
3. Distinguish between elements and compounds by contrasting examples of each.
4. **Apply** How many phases are present in a glass of soda on ice?

Table 3.1

Periodic Table of the Elements

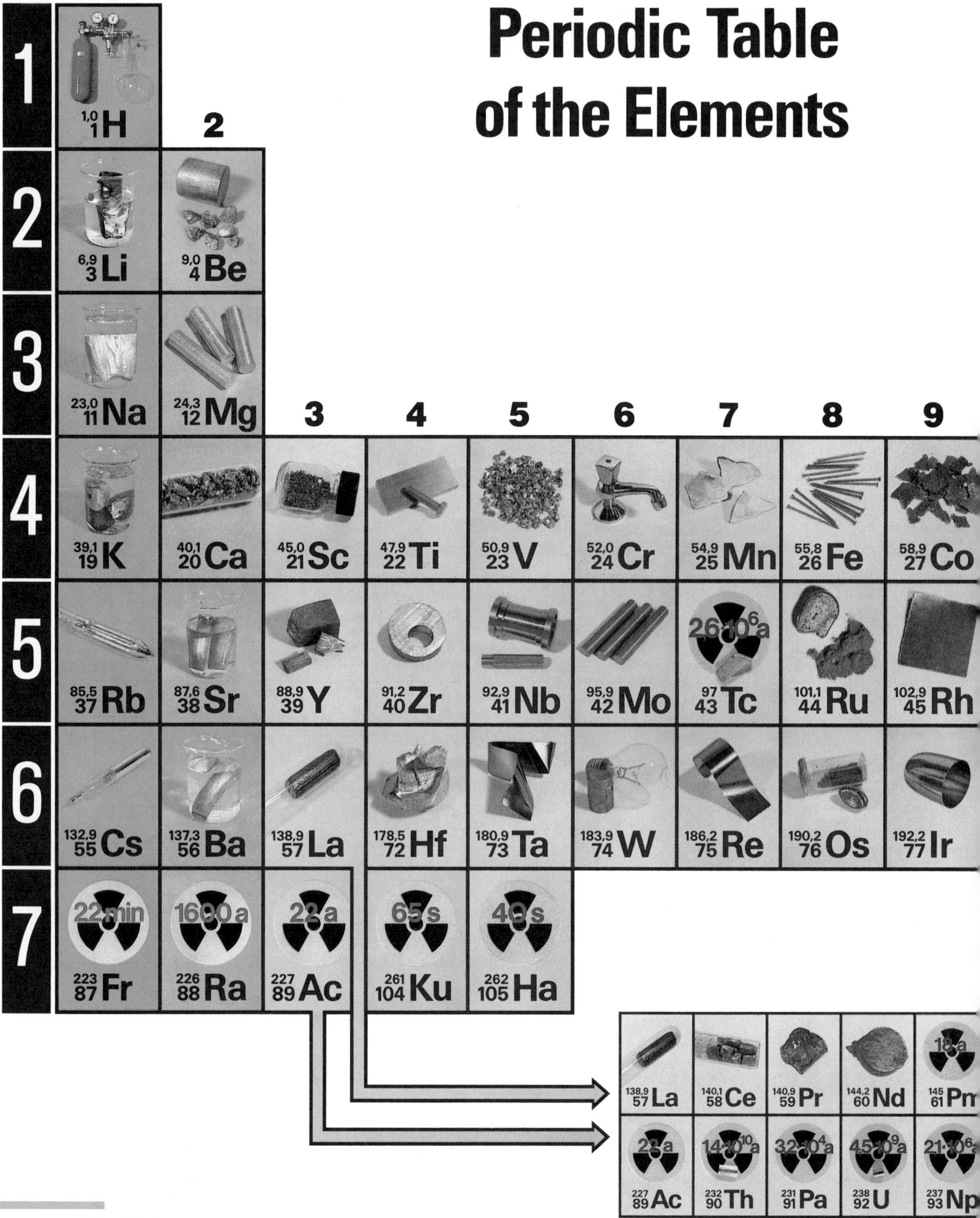

Period	1	2	3	4	5	6	7	8	9
1	1,0 1 H								
2	6,9 3 Li	9,0 4 Be							
3	23,0 11 Na	24,3 12 Mg							
4	39,1 19 K	40,1 20 Ca	45,0 21 Sc	47,9 22 Ti	50,9 23 V	52,0 24 Cr	54,9 25 Mn	55,8 26 Fe	58,9 27 Co
5	85,5 37 Rb	87,6 38 Sr	88,9 39 Y	91,2 40 Zr	92,9 41 Nb	95,9 42 Mo	97 43 Tc ($2{,}6 \cdot 10^6$ a)	101,1 44 Ru	102,9 45 Rh
6	132,9 55 Cs	137,3 56 Ba	138,9 57 La	178,5 72 Hf	180,9 73 Ta	183,9 74 W	186,2 75 Re	190,2 76 Os	192,2 77 Ir
7	223 87 Fr (22 min)	226 88 Ra (1600 a)	227 89 Ac (22 a)	261 104 Ku (65 s)	262 105 Ha (40 s)				

(from La)	138,9 57 La	140,1 58 Ce	140,9 59 Pr	144,2 60 Nd	145 61 Pm (18 a)
(from Ac)	227 89 Ac (22 a)	232 90 Th ($1{,}4 \cdot 10^{10}$ a)	231 91 Pa ($3{,}2 \cdot 10^4$ a)	238 92 U ($4{,}5 \cdot 10^9$ a)	237 93 Np ($2{,}1 \cdot 10^6$ a)

10	11	12	13	14	15	16	17	18
								4,0 2 He
			10,8 5 B	12,0 6 C	14,0 7 N	16,0 8 O	19,0 9 F	20,2 10 Ne
			27,0 13 Al	28,1 14 Si	31,0 15 P	32,1 16 S	35,5 17 Cl	39,9 18 Ar
58,7 28 Ni	63,5 29 Cu	65,4 30 Zn	69,7 31 Ga	72,6 32 Ge	74,9 33 As	79,0 34 Se	79,9 35 Br	83,8 36 Kr
106,4 46 Pd	107,9 47 Ag	112,4 48 Cd	114,8 49 In	118,7 50 Sn	121,8 51 Sb	127,6 52 Te	126,9 53 I	131,3 54 Xe
195,1 78 Pt	197,0 79 Au	200,6 80 Hg	204,4 81 Tl	207,2 82 Pb	209,0 83 Bi	103 a 209 84 Po	8,3 h 210 85 At	3,8 d 222 86 Rn

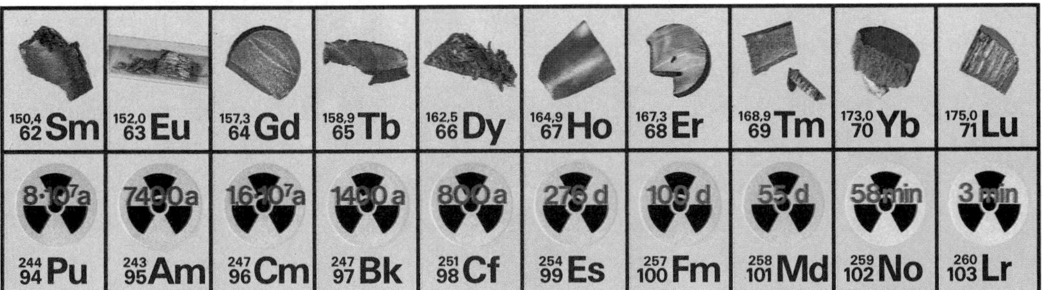

OBJECTIVES

- classify changes in matter as physical or chemical.
- obtain information from a graph.
- distinguish among extensive, intensive, physical, and chemical properties.

3.2 Changes in Properties

The iron that the blacksmith is shaping in the chapter opening photograph has some characteristics by which we can identify it. It is solid and gray. It also reacts with hydrochloric acid. The characteristics of solidity and color are called physical properties, while the ability to react with hydrochloric acid is called a chemical property. In this part of the chapter, you will learn about the different properties of matter and about the different kinds of changes matter can undergo.

Physical and Chemical Changes

Substances undergo changes when their conditions are changed. A change in condition could be an increase in temperature, a mechanical force, exposure to another substance, or any of a number of other alterations. If the same substance remains after the change, a **physical change** has taken place. If a new substance has appeared, a **chemical change** has occurred.

Physical Changes

Changes like pounding, pulling, or cutting do not usually change the chemical character of a substance. Pounded copper is still copper. Only its shape is changed. Cutting a piece of wood into smaller pieces, tearing paper, dissolving sugar in water, and pouring a liquid from one container to another are other examples of physical changes.

Physical changes occur when a substance melts or boils. At its melting point, a substance changes from the solid phase to the liquid phase. At its boiling point, a liquid changes to a gas. Such physical changes are called *changes of state,* because the substance is not altered except for its physical state.

Figure 3.6 Hammering copper changes only its shape. This is an example of a physical change because the substance remains copper.

A knowledge of physical changes and the conditions under which they occur can be used to separate mixtures. Separating the substances in mixtures by distillation is a change-of-state operation. It is used to separate substances with different boiling points. For instance, you can separate a solution of salt in water by heating the solution to its boiling temperature. The water will then be turned to steam and escape from the container. The steam can be cooled to turn it into pure water. The salt, whose boiling point is very high compared with that of water, will remain behind. Distillation is a method that is used to produce drinking water from seawater in a few locations in the world. However, distillation is a very expensive way to produce pure water because it requires a great deal of energy, so the process is used only when necessary.

Bridge to *MINERALOGY*

Salts from the Dead Sea

The Dead Sea is located between the countries of Israel and Jordan. It is called the Dead Sea because almost nothing can live in its waters, which contain large quantities of dissolved salts. Far from being worthless, the Dead Sea's salts are valuable to industry and agriculture. Both Jordan and Israel are extracting these substances from the waters of the Dead Sea as a commercial enterprise. The Israeli plant was begun in the 1930s, and the Jordanian operation began in the early 1980s.

The substances extracted are sodium chloride, potassium chloride, magnesium chloride, and magnesium bromide. Of these chemicals, potassium chloride provides the most income. Two factors enable the substances in solution to be separated by fractional crystallization. One factor is the difference in the solubilities of the substances in solution. The second factor is the rate at which the solubility changes as the temperature changes. Thus, by a carefully controlled combination of evaporation and cooling, the various salts can be separated and purified.

Exploring Further

1. The water of the Dead Sea contains many dissolved materials. Would you classify it as heterogeneous or homogeneous? As a material or a substance?
2. Look up the solubilities of the four substances sodium chloride, potassium chloride, magnesium chloride, and magnesium bromide at 20°C and at 100°C. Suggest a way to separate these substances.

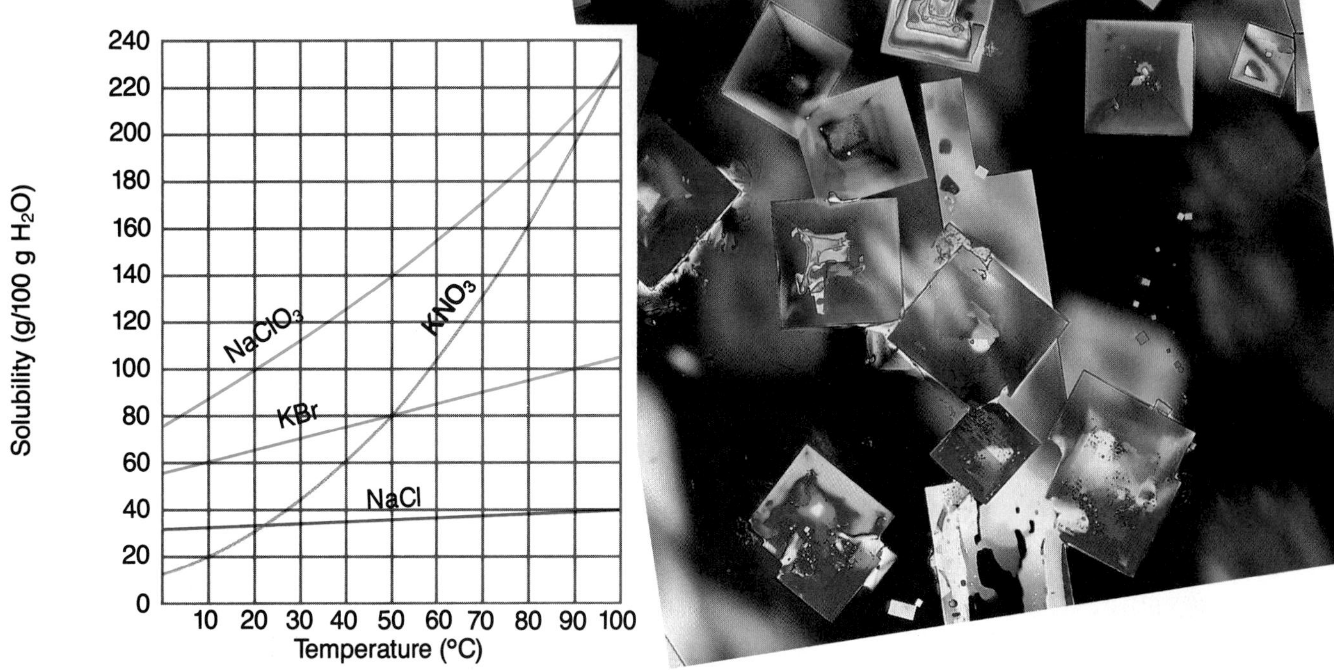

Figure 3.7 This solubility graph shows the amount of each substance that will dissolve in water at any temperature between 0°C and 100°C. When temperature is reduced or water is evaporated, crystals like those shown (above right) will form.

Another separation based on phase difference depends on the solubility of one substance in another. Most substances have a specific solubility (amount of solute that dissolves) in water at a given temperature. Therefore, it is usually possible to separate two substances that are dissolved in the same solution by a process called fractional crystallization. The less soluble substance drops out of solution first when the amount of water (the solvent) is reduced. A solid substance that forms from a solution is called a **precipitate**. This process is often called crystallization because most substances form crystals when they precipitate.

Notice in Figure 3.7 that, at 70°C, potassium bromide (KBr) is less soluble than potassium nitrate (KNO_3). If a water solution containing equal amounts of both substances is allowed to evaporate at 70°C, the potassium bromide will crystallize first. This process is used quite often in industry in the purification of many crystalline items, such as sugar, salt, and medicines.

PRACTICE PROBLEMS

5. Obtain information from the graph in Figure 3.7 to determine the solubilities of the following.

a. NaCl at 70°C
b. $NaClO_3$ at 100°C
c. KNO_3 at 30°C
d. KBr at 80°C

Chemical Changes

Suppose you are given two test tubes containing colorless liquids. One test tube contains water. The other contains nitric acid. If you place a pinch of sugar in the water, the sugar will disappear and the liquid will remain colorless. The sugar has dissolved and you now have a sugar-water solution. A physical change has occurred. Neither the sugar nor the water has changed to a different substance. If you put a small piece of copper into the other test tube, the copper will also disappear. However, the liquid will turn blue and give off a brown gas. You now have a solution of copper(II) nitrate and some nitrogen dioxide gas, both of which are new substances. A chemical change has occurred.

Let us take another example. Sodium is a silvery, soft metal that reacts vigorously with water. Chlorine is a yellow-green gas that is highly corrosive and poisonous. However, if these two elements are brought together, they combine to produce a different substance, a white crystalline solid. This new substance is common table salt, sodium chloride, which neither reacts with water nor is poisonous. The behavior of the product is quite different from that of either reactant.

Whenever a substance undergoes a change so that one or more new substances with different characteristics are formed, a chemical change (reaction) has taken place. Burning, digestion, and fermenting are examples of chemical changes.

There are many instances in chemistry when a general statement can be made to cover most, but not all, cases of a particular phenomenon. Such a statement is called a rule of thumb. You come across rules of thumb every day. For example, one rule of thumb for driving a car is that you leave one car-length between you and the car ahead for each 10 miles per hour of speed. That statement is a rule of thumb because there are exceptions. Likewise, in chemistry, rules of thumb are generally true, but keep in mind that there may be exceptions.

Figure 3.8 When copper reacts with nitric acid, a change produces a blue solution and a brown gas (below left). Because new substances with different properties are formed, a chemical change has occurred. Sodium chloride (table salt) is the new substance resulting from the chemical change that takes place when sodium metal and chlorine gas are combined (below).

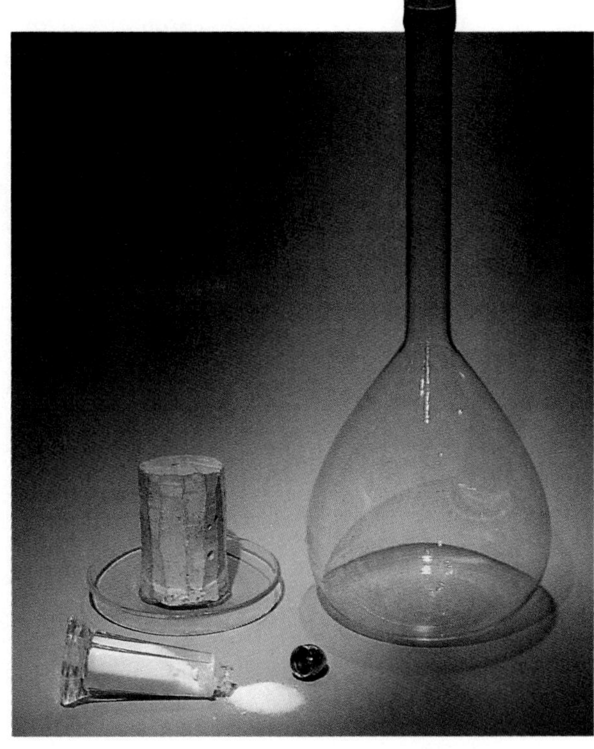

RULE OF THUMB

In this textbook, useful rules of thumb will be indicated by a logo in the margin. The following is a rule of thumb for deciding whether or not a chemical reaction has taken place. *If a precipitate, gas, color change, or energy change occurs, a chemical change has taken place.* Why is this only a rule of thumb? It is because there are other circumstances that can cause these changes. For instance, cooling a solution may cause some of the solute to precipitate. This would not be a chemical reaction.

Decomposing chemical compounds into their component elements always requires a chemical change. On the other hand, mixtures can be separated by physical means. However, it is sometimes more convenient to separate mixtures by a chemical change. For example, brass is a mixture of copper and zinc. It should be possible to separate the two elements, copper and zinc, by grinding the brass into fine particles. Then, using a microscope and a very fine pair of tweezers, we could pick out the zinc particles from among the copper particles. How much easier it is to react the brass with hydrochloric acid. The copper is unaffected, but the zinc is converted to zinc chloride, which dissolves in the acid. The copper is then recovered. The zinc chloride solution can be treated with aluminum to regenerate the zinc metal.

Physical and Chemical Properties

A description of the behavior of a substance undergoing a physical change is one type of **physical property,** while a **chemical property** describes the behavior of a substance undergoing a chemical change. For example, chlorine has the chemical property that it reacts with sodium, as described earlier. The yellow-green color of chlorine gas is a physical property. Length, color, and temperature are typical descriptive terms that are also physical properties of a substance.

Figure 3.9 A physical property of lead is that it is very malleable. It can be hammered thinner without breaking. Glass becomes pliable like taffy when it is very hot. Pliability is a physical property of glass at elevated temperatures.

Figure 3.10 A mixture of sugar, sand, iron filings, and gold dust (a) is clearly heterogeneous when viewed under a microscope (b). The mixture can be separated using physical properties, such as the magnetism of iron (c). The pure, separated components of the mixture are shown in (d).

Physical properties may be divided into extensive properties and intensive properties. **Extensive properties** depend on the amount of matter present. Some of these properties are mass, length, and volume. **Intensive properties** do not depend on the amount of matter present. Intensive properties are useful for identifying substances. For example, density is an intensive property. Each sample of a substance, regardless of its size, has the same density. Other intensive properties include malleability, ductility, and conductivity. For example, copper can be hammered into thin sheets. It is more malleable than iron. Copper can also be drawn out into fine wire; it is quite ductile. Both copper and silver conduct heat and electric current well. That is, they offer little resistance to the flow of energy or electricity. A silver spoon will become hot if left in a hot pan of soup because the silver is a good conductor of heat. Most house wiring is made of copper because copper has a high electrical conductivity.

In addition to density, the intensive properties most important to the chemist are color, crystalline shape, melting point, boiling point, and refractive index (ability of material to bend light).

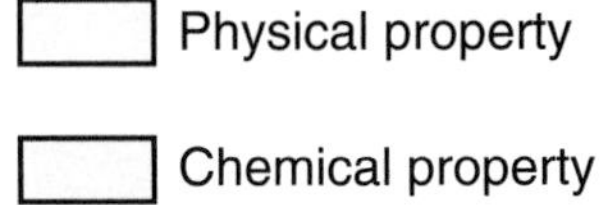

Uranium. U: atomic mass 238.029; atomic number 92; oxidation states 6+, 5+, 4+, 3+. Occurrence in the earth's crust 2×10^{-5}%; melting point 1132.3°C; boiling point 3818°C; density 19.05 g/cm³. Silver-white radioactive metal, softer than glass. Uranium is malleable, ductile, and can be polished. Half-life of the U-238 isotope is 4.51×10^{9} years. Specific heat is 0.117 J/g•C°; heat of fusion 12.1 kJ/mol; heat of vaporization 460 kJ/mol. Burns in air at 150–175°C to form U_3O_8. When finely powdered, it slowly decomposes in cold water, more quickly in hot water. Burns in fluorine to form a green, volatile tetrafluoride; also burns in chlorine, bromine, and iodine. Reacts with acids with the liberation of hydrogen and the formation of salts with the 4+ oxidation state. Not attacked by alkalies.

Figure 3.11 Information concerning the physical and chemical properties of a substance can be found in a chemical handbook. The entry shown here describes both the physical and chemical (highlighted) properties of uranium.

Chemical properties describe the reaction of a substance with other materials such as air, water, acids, or a reaction within the substance itself. For example, iron reacts with air and water to form rust. Compounds can react in several ways. Thus, the organic compound paraquat has two chemical properties that make it useful as a herbicide. It is toxic to plant tissue, so it will kill certain plants. On the other hand, it is decomposed by water in soil particles, particularly clays. Thus, it does not poison the soil permanently.

For a chemist, it is just as important to find out if a particular substance *does not* react as it is to discover if it *does* react. Thus, the fact that baking soda is not flammable makes it a valuable substance to use in an emergency to extinguish a kitchen fire.

3.2 CONCEPT REVIEW

6. Classify the following properties as extensive or intensive.
- **a.** mass
- **b.** color
- **c.** ductility
- **d.** length
- **e.** melting point

7. Classify each of the following changes as chemical or physical.
- **a.** fading of dye in cloth
- **b.** growth of a plant
- **c.** melting of ice
- **d.** digestion of food
- **e.** formation of clouds in the air
- **f.** healing of a wound
- **g.** making of rock candy by evaporating water from a sugar solution
- **h.** production of light by an electric lamp

8. Classify the following properties as chemical or physical. Use a dictionary to identify any unfamiliar properties.
- **a.** color
- **b.** reactivity
- **c.** flammability
- **d.** odor
- **e.** porosity
- **f.** stability
- **g.** ductility
- **h.** solubility
- **i.** thermal expansion
- **j.** melting point
- **k.** catches fire in air

9. Apply Helium reacts with no known substance and forms no chemical compounds. Would it be accurate to say that helium has no chemical properties? Explain.

3.3 Energy

OBJECTIVES

- describe conditions under which heat is transferred.
- convert between units used to measure energy.
- describe endothermic and exothermic processes and state the function of activation energy.
- perform calculations involving specific heat.

Physical and chemical changes are always accompanied by energy changes. The energy changes occur between a system and its surroundings. Exactly what is a system? A **system** is that "piece" of the universe under consideration. Everything else is the surroundings. For example, a system may be as simple as a single molecule, or even a single atom, if that is the object under consideration. On the other hand, a system may be a laboratory setup involving several pieces of apparatus and two or three chemicals, or it could be an entire chemical refinery.

Energy Transfer

The most common form of energy change involves heat. What is heat? Consider a system of two objects, A and B. Assume that object A is at a temperature of 25°C and object B is at a temperature of 20°C. What happens when we put A into contact with B? Energy transfers from the matter at higher temperature, object A, to the matter at lower temperature, object B. The energy transferred as a result of a temperature difference is called **heat** and is represented by the letter, q. If the system remains undisturbed, energy will continue to transfer until the temperatures of objects A and B are equal.

There is another way that energy can be transferred between a system and its surroundings. The surroundings may do work on the system, or the system may do work on the surroundings. For example, if a strip of copper is the system under consideration, by hammering the strip of copper we would do work on it. On the other hand, if we consider the gases produced in an automobile engine cylinder as a system, the system does work on the engine as it expands and pushes the piston down. Quantitative measurements of energy changes are expressed in joules. The **joule** (J) is a derived SI unit rather than a base unit.

There are still many references to an older energy unit called a *calorie*. One calorie is equivalent to 4.184 joules. Undoubtedly, you have heard or read of calorie values for various foods. You may have noticed that these caloric values are given in Calories with an uppercase "C." These dietary Calories are a thousand times as large as a calorie and so they are, in fact, kilocalories. Caloric values are the amount of energy the human body can obtain by chemically breaking down food. Average caloric values for the basic food groups are listed in Table 3.2.

Table 3.2

Caloric Values

Food	joules/gram	calories/gram	Calories/gram
protein	17 000	4000	4
fat	38 000	9000	9
carbohydrate	17 000	4000	4

CHEMISTS AND THEIR WORK

Gerty Theresa Cori (1896-1957)

A chemical change that takes place in your cells is the conversion of glucose to glycogen. Whenever you need glucose for energy, another change converts glycogen back to glucose. The research of Gerty Cori and her husband, Carl, revealed the details of these important chemical processes. Understanding how the body stores and releases energy reserves paved the way for discoveries on the causes and treatment of metabolic disorders. Cori was born in Prague, Czechoslovakia, but came to the United States with her husband in 1922 to conduct research. The Coris were awarded a Nobel Prize in 1947 for their work.

SAMPLE PROBLEM

Energy Units

An average-sized baked potato (200 g) has an energy value of 686 000 joules. What is that value expressed in Calories?

Solving Process:

We know that 4.184 joules = 1 calorie. Therefore,

$$\frac{686\ 000\ \not{J}}{4.184\ \not{J}/\text{cal}} = 164\ 000\ \text{cal}$$

We also know that 1000 cal = 1 Cal. Therefore,

$$\frac{164\ 000\ \not{\text{cal}}}{1000\ \not{\text{cal}}/\text{Cal}} = 164\ \text{Calories}$$

PRACTICE PROBLEMS

10. Make the following conversions.
   a. 1980 joules to calories
   b. 1.11 Calories to joules
   c. 800 calories to Calories
   d. 3.40 joules to Calories
   e. 47.0 calories to joules

Word Origins

exo: (GK) out, outside of
endon: (GK) in, within
therme: (GK) heat
exothermic—giving out heat
endothermic—taking in heat

Energy and Chemical Change

Chemical changes are always accompanied by a change in energy. If energy is absorbed in a reaction, the reaction is **endothermic.** Since energy is taken in by the system, the products of the reaction are at a higher energy level than the substances that reacted. On the other hand, if energy is given off by a reaction, the reaction is **exothermic.** Since the system has lost energy, the products of the reaction have less energy than the substances that reacted. Look at Figure 3.12. It is convenient to think of an endothermic reaction as a system running uphill, while an exothermic reaction is a system running downhill. It is a rule of thumb for reacting systems that *nature tends to run downhill. Exothermic reactions tend to take place spontaneously, that is, without outside help.* The opposite is generally true of endothermic reactions. That is, they must have some external source of energy to take place.

RULE OF THUMB

Calcium hydroxide and phosphoric acid react with each other readily. According to our rule of thumb, we would predict that the reaction is exothermic. If it is exothermic, what would we observe about the system? Exothermic reactions give off energy, usually in the form of heat. As calcium hydroxide and phosphoric acid react, the container in which they react should get hotter if our prediction is true. Measurement of the temperature in the reaction vessel would confirm that the prediction is correct.

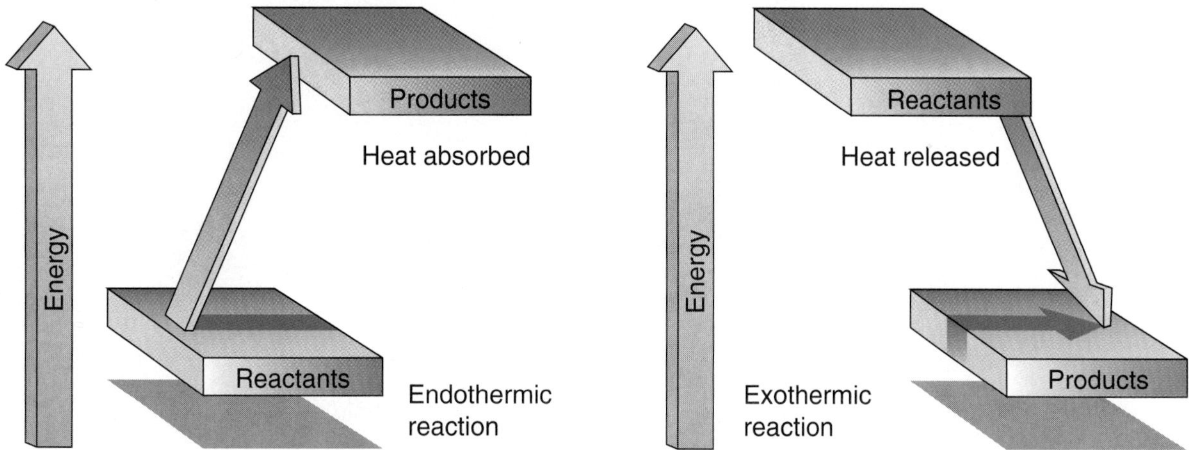

Figure 3.12 An endothermic reaction (left) absorbs energy because the products are at a higher level of energy than the reactants. An exothermic reaction (right) gives off energy because the products are lower in energy than the reactants.

Both endothermic and exothermic reactions require a certain minimum amount of energy to get started. This minimum amount of energy is called the **activation energy.** Without it, the reactant atoms or molecules will not unite to form the product and the reaction does not occur.

Chemical reactions have many possible sources for their activation energy. When you strike a match, friction produces enough heat to activate the reactants on the match head. As a result, the match ignites. Even though a small amount of energy is put in to activate the reaction, the reaction is exothermic because it produces more energy than was put in. In photography, light is the source of activation energy.

Measuring Energy Changes

A **calorimeter** is a device used to measure the energy given off or absorbed during chemical or physical changes. See Figure 3.13. There are several kinds of calorimeters. A calorimeter containing water is often used to measure the heat absorbed or released by a chemical reaction. Heat from a chemical reaction in the cup in the calorimeter causes a change in temperature of the water in the calorimeter. The temperature change of the water is used to measure the heat absorbed or released by the reaction.

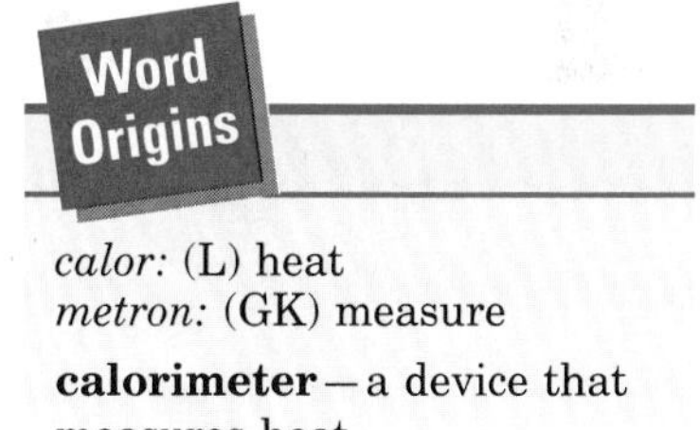

calor: (L) heat
metron: (GK) measure

calorimeter—a device that measures heat

To change the temperature of a substance such as water, heat must be added or removed. Some substances require little heat to cause a change in their temperature. Other substances require a great deal of heat to cause the same temperature change. For example, one gram of liquid water requires 4.184 joules of heat to cause a temperature change of one Celsius degree. It takes only 0.902 joule to raise the temperature of one gram of aluminum one Celsius degree. The heat needed to raise the temperature of one gram of a substance by one Celsius degree is called the **specific heat** (C_p) of the substance. Every substance has its own specific heat. The heat required to raise the temperature of one gram of water one Celsius degree is 4.184 joules. The specific heat of water is 4.184 J/g·C°.

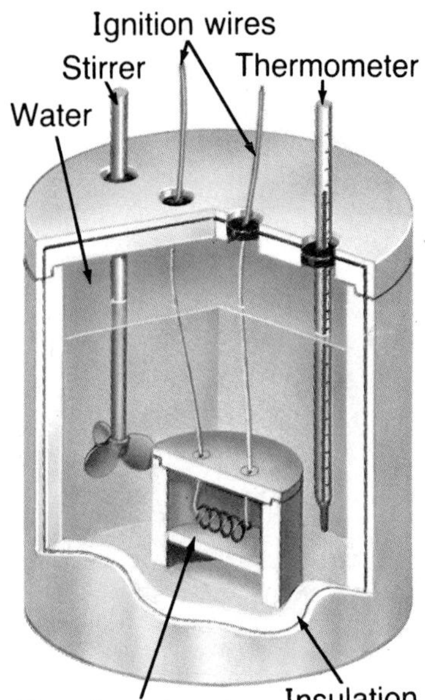

Figure 3.13 A bomb calorimeter (left) is used to measure the heat released or absorbed in the reaction taking place inside the reaction chamber. The heat released or absorbed changes the temperature of the water. The temperature change is measured with a thermometer. Energy released by chemical reactions is used to put the space shuttle (right) in orbit.

Specific heats are given in joules per gram-Celsius degree (J/g·C°). Appendix Tables A-3 and A-5 list the specific heats of some substances. Specific heats can be used in calculations involving the change in temperature of a specific mass of substance.

The law of conservation of energy means that in an insulated system, any heat lost by one quantity of matter must be gained by another. The transfer of energy takes place between two quantities of matter that are at different temperatures until the two reach the same temperature. Further, the amount of energy transferred can be calculated from the relationship:

$$\begin{pmatrix} \textit{heat gained or} \\ \textit{lost by water} \end{pmatrix} = \begin{pmatrix} \textit{mass} \\ \textit{in grams} \end{pmatrix} \begin{pmatrix} \textit{change in} \\ \textit{temperature} \end{pmatrix} \begin{pmatrix} \textit{specific} \\ \textit{heat} \end{pmatrix}$$

$$q = (m)(\Delta T)(C_p)$$

The term ΔT refers to the change in temperature. When heat is gained by water, $\Delta T = T_f - T_i$ where T_f is the final temperature and T_i is the initial temperature. When heat is lost by water, $\Delta T = T_i - T_f$.

In the calorimeter, the product of the specific heat of the water, the temperature change of the water, and the mass of the water gives the heat transferred. The heat released or absorbed by water is calculated using the following equation.

$$q = (m)(\Delta T)(4.184 \text{ J/g} \cdot \text{C}°)$$

The same method can be used to calculate the transfer of heat when two dilute solutions react. The solutions are placed in a cup in the calorimeter. As the chemical reaction occurs, the temperature change is measured. Because the solutions are dilute, we can assume that the mixture has the same specific heat as water, 4.184 J/g·C°. By multiplying the specific heat by the temperature change and the mass of the solutions, we can calculate the heat produced by the reaction.

EVERYDAY CHEMISTRY

Smog

Smog was originally defined as a mixture of *sm*oke and f*og*. Today, the term *smog* is applied most specifically to photochemical smog. This smog is created by the breaking down of pollutants in the air by sunlight. Photochemical smog owes many of its properties to reactions involving atmospheric oxygen, oxides of nitrogen, and hydrocarbons (compounds of hydrogen and carbon such as gasoline vapors).

Automobiles are the primary contributors of the raw materials that produce smog. Nitrogen oxides, particularly nitrogen dioxide, and hydrocarbon fumes mix with air and other pollutants. With sunlight, this mixture produces ozone, other oxides of nitrogen, and sulfur oxides.

When these products are trapped in a city surrounded by hills or mountains, the smog is particularly intense. This intensity is often increased by a temperature inversion layer in the atmosphere. The inversion results from a layer of cold air overlying the area, preventing the hot air and gases from rising and dispersing. Because blue light is scattered more by smog particles, the air often appears murky blue or brown. The color is accentuated by the brown nitrogen dioxide in the air. In addition to unhealthful air, smog also causes eye irritation and tearing probably brought on by the organic compounds formaldehyde, acrolein, and peroxyacetyl nitrate (PAN). It is also damaging to plants, affecting agriculture and forests.

Over the years, there have been many attempts to control the major source of smog, auto exhaust. Among these are engine modifications, catalytic converters, and afterburners that complete the burning of waste gases. Even if there were no photochemical reactions, auto exhaust would still produce carbon monoxide and some particles. Besides, all other forms of combustion, including smoking, also contribute to smog. Industrial operations such as refineries and service stations emit hydrocarbon vapors. Even backyard cookouts with charcoal and charcoal starter contribute particles and hydrocarbons to the atmosphere.

Obviously it is not possible to eliminate all air pollutants; but legislation and self-regulation have done much to reduce smog levels. Much, however, remains to be done.

Exploring Further

1. Find out how catalytic converters work. Why are they called "catalytic" and what do they convert?
2. Electric automobiles have been promoted as a possible solution to air pollution from internal-combustion engines. Can you think of ways in which electric cars might cause air pollution?

Notice that throughout our discussion of calorimetry, we have assumed that the calorimeter itself does not absorb any heat. We also have assumed that no heat escapes from the calorimeter. Neither of these assumptions is completely true. However, we will continue to make these assumptions in order to simplify our calculations. The error from actual losses of heat should be considered in any laboratory exercise using calorimeters.

SAMPLE PROBLEMS

1. Calculating Transfer of Heat

How much heat is lost when a solid aluminum ingot with a mass of 4110 g cools from 660.0°C to 25°C?

Solving Process:

From Appendix Table A-3, we find that the specific heat of aluminum is 0.903 J/g · C°. We also know that

$$q = (m)(\Delta T)(C_p) \quad \text{and} \quad \Delta T = 660.0°\text{C} - 25°\text{C} = 635\ \text{C}°;$$

$$\text{so } q = (4110\ \cancel{\text{g}})(635\ \cancel{\text{C}°})\left(0.903\ \frac{\text{J}}{\cancel{\text{g}}\cdot\cancel{\text{C}°}}\right) = 2\ 360\ 000\ \text{J}$$

Note that the result is rounded to three significant digits because the data had only three significant digits.

2. Calculating Temperature

Suppose a piece of iron with a mass of 21.5 grams at a temperature of 100.0°C is dropped into an insulated container of water. The mass of the water is 132 grams and its temperature before adding the iron is 20.0°C. What will be the final temperature of the system?

Solving Process:

We know that the heat lost must equal the heat gained. Since the iron is at a higher temperature than the water, the iron will lose energy. The water will gain an equivalent amount of energy.

(a) The heat lost by the iron is

$$q = (m)(\Delta T)(C_p) = (21.5\ \cancel{\text{g}})(100.0°\text{C} - T_f)\left(0.449\frac{\text{J}}{\cancel{\text{g}}\cdot\text{C}°}\right)$$

(b) The heat gained by the water is

$$q = (m)(\Delta T)(C_p) = (132\ \cancel{\text{g}})(T_f - 20.0°\text{C})\left(4.184\ \frac{\text{J}}{\cancel{\text{g}}\cdot\text{C}°}\right)$$

(c) The heat gained must equal the heat lost

$$(132\ \cancel{\text{g}})(T_f - 20.0°\text{C})\left(4.184\frac{\cancel{\text{J}}}{\cancel{\text{g}}\cdot\text{C}°}\right) =$$

$$(21.5\ \cancel{\text{g}})(100.0°\text{C} - T_f)\left(0.449\frac{\cancel{\text{J}}}{\cancel{\text{g}}\cdot\text{C}°}\right)$$

$$T_f = 21.4°\text{C}$$

PROBLEM SOLVING HINT

Note that the iron is at 100°C and the water is at 20°C. The final temperature will fall somewhere between these two values.

The same type of calculation may be used to determine the specific heat of an unknown metal. When a warm piece of metal is dropped into a container of cool water, the energy gained by the water will be the same as that lost by the metal. We can measure the mass and the initial and final temperatures of the water, and we know its specific heat. From these data, the heat gained by the water can be calculated. For the metal, we can also measure mass and initial and final temperatures. We can then compute the specific heat, C_p, using the equation, $q = (m)(\Delta T)(C_p)$. If you have an unknown metal that must be identified, measuring its specific heat is a simple task, and the result is an important physical property that aids in identifying the metal.

The values for the specific heats of metals change little over a wide range of temperatures. In fact, the specific heats of all solids and liquids are fairly constant. In contrast, the specific heats of gases vary widely with temperature.

PRACTICE PROBLEMS

11. How much heat is required to raise the temperature of 854 g H_2O from 23.5°C to 85.0°C?

12. Phosphorus trichloride, PCl_3, is a compound used in the manufacture of pesticides and gasoline additives. How much heat is required to raise the temperature of 96.7 g PCl_3 from 31.7°C to 69.2°C? The specific heat of PCl_3 is 0.874 J/g·C°.

13. Carbon tetrachloride, CCl_4, was a very popular organic solvent until it was found to be toxic. How much heat is required to raise the temperature of 10.35 g CCl_4 from 32.1°C to 56.4°C? (See Appendix Table A-5.)

14. If a piece of aluminum with mass 3.90 g and a temperature of 99.3°C is dropped into 10.0 cm^3 of water at 22.6°C, what will be the final temperature of the system? (Recall the density of water is 1.00 g/cm^3.)

15. The color of many ceramic glazes comes from cadmium compounds. If a piece of cadmium with mass 65.6 g and a temperature of 100.0°C is dropped into 25.0 cm^3 of water at 23.0°C, what will be the final temperature of the system?

16. A piece of an unknown metal with mass 23.8 g is heated to 100.0°C and dropped into 50.0 cm^3 of water at 24.0°C. The final temperature of the system is 32.5°C. What is the specific heat of the metal?

3.3 CONCEPT REVIEW

17. How do joules, calories, and Calories differ?

18. Under what conditions does heat move from one object to another?

19. Suppose you combine two solutions in a test tube and they react. How would you determine whether the reaction was endothermic or exothermic?

20. Apply A classmate argues that the burning of charcoal must be an endothermic process because getting the charcoal started requires the input of a large amount of heat. Would you agree? Explain your answer.

CHEMISTRY AND

SOCIETY

Energy Problems

The United States imports over three billion barrels of crude oil a year. Eventually, the world will run out of oil. The resources of natural gas and coal will also be exhausted someday. What will we do for energy sources then? Scientists are doing research in a number of areas. One area you have undoubtedly heard about is solar energy. Others are geothermal energy, wind energy, and tidal energy. All of these sources can provide some energy, but with our present technology none will be able to supply anticipated world needs.

Another area of research is the nuclear fusion reaction you will study in Chapter 28. The fusion reaction is the same process that is taking place in the sun and is the origin of most of the energy we use today. As you will learn later, there are distinct advantages of the fusion reactor over the nuclear fission reactor presently in use throughout the world. Unfortunately, scientists have not yet learned to operate a reactor powered by fusion reactions.

What part do chemists play in the solution of the energy problem? Actually, chemists are involved in almost all aspects of the problem. Let's look at some special areas where chemists can be of immediate and significant help.

Solar energy can be harnessed in several ways, but one of the most important is by the use of a device called a photovoltaic cell. This cell can take in the energy radiated by the sun and create an electric current. In order to be a useful technology, photovoltaic cells must meet two criteria. First, the cell must be capable of producing in its lifetime more energy than is consumed in producing the cell. Second, the cell must be economical and practical to manufacture. Chemists are hard at work trying to meet both of these requirements. The production of these cells involves the use of a highly purified form of the element silicon. Chemists are investigating economical, energy-efficient ways of producing the silicon.

Another attack on our energy problems is to make more efficient use of present energy resources. Here, too, chemists are making a major contribution. For instance, the process of combustion is very complex, and not fully understood at present. Chemists are active in studying combustion at the molecular and atomic level in order to find out what is really taking place. Only then can they recommend ways to make the process more efficient.

Chemists are also researching ways to produce fuels from non-traditional sources. These processes include the production of methane (natural gas) from garbage and animal waste, synthetic petroleum from water and charcoal, and the extraction of hydrocarbons from certain types of rock formations found in the western section of the United States and Canada.

Analyzing the Issue

1. In most processes, energy is neither created nor destroyed. Why, then, do we need to develop new energy sources?
2. Nearly all fuels are organic substances. What element do they contain? What substances are released into the air as a result of their combustion?
3. Research the advantages and limitations of geothermal, wind, and tidal energy sources. Debate the utility of these sources as alternatives to combustion.

Summary

3.1 Classification of Matter

1. A mixture is a combination of two or more substances that retain their individual properties.
2. A phase consists of a region of uniform matter. Phases are separated by boundaries called interfaces.
3. Heterogeneous matter is made of more than one phase.
4. Homogeneous matter consists of only one phase.
5. A solution is a homogeneous mixture consisting of a solute dissolved in a solvent. The component parts need not be present in specific ratios.
6. Elements are substances made of one kind of atom.
7. Compounds are substances made of more than one kind of atom. The component atoms are present in definite ratios.
8. Organic substances are compounds that contain the element carbon. Inorganic substances are the elements and the compounds that contain no carbon.

3.2 Changes in Properties

9. A substance has undergone a physical change if the same substance remains after the change. A change of state is a physical change from one state—solid, liquid, or gas—to another.
10. In a chemical change, new substances with different properties are formed. Chemical changes must be used to separate the elements of a compound.
11. Physical properties are classified as either extensive or intensive. Extensive properties, such as mass and length, depend on the amount of the substance present. Intensive properties, such as ductility and melting point, depend on the nature of the substance itself.
12. The chemical properties of a substance describe the reaction of the substance with other substances. Lack of reactivity is also a chemical property.

3.3 Energy

13. Physical and chemical changes always involve energy transfer, either in the form of work or heat, between a system and its surroundings.
14. Energy is measured in joules.
15. Energy is absorbed in an endothermic reaction and given off in an exothermic reaction.
16. The specific heat of a substance is the heat required to raise the temperature of 1 g of the substance 1 C°. The specific heat of water is 4.184 J/g·C°.
17. The energy transferred when matter changes temperature is given by $q = (m)(\Delta T)(C_p)$.

Key Terms

material	physical change
mixture	chemical change
phase	precipitate
heterogeneous mixture	physical property
interface	chemical property
homogeneous	extensive property
solution	intensive property
solute	system
solvent	heat
substance	joule
element	endothermic
compound	exothermic
organic substance	activation energy
inorganic substance	calorimeter
	specific heat

Review and Practice

21. Estimate the number of phases and interfaces present in an ice-cream soda complete with whipped cream and candied cherry.

22. Ethanol and water will form a solution. How is it determined which one is the solute and which one is the solvent?

23. Table sugar is dissolved in a hot cup of coffee. What is the solvent of the resulting solution? What is the solute?

24. Explain the terms "concentrated" and "dilute" in terms of solvent and solute.

25. An emerald is formed when chromium atoms replace aluminum atoms in certain aluminum compounds. Differentiate between an element and a compound. How is it possible for an element to occur within a compound?

26. How many naturally-occurring elements exist? At this time, how many synthetic elements have been produced?

27. Human blood is composed of many materials, such as proteins, that are organic and other substances, such as iron and water, that are inorganic. What is the basic difference between organic substances and inorganic substances?

28. Classify the following changes as chemical or physical.
   a. burning of coal
   b. tearing of a piece of paper
   c. kicking of a football
   d. excavating of soil
   e. exploding of TNT

29. Classify the following properties as chemical or physical.
   a. density
   b. melting point
   c. length
   d. flammability

30. What is the difference between endothermic reactions and exothermic reactions?

31. What is the relationship between joules and calories?

32. Heated bricks or blocks of iron were used long ago to warm beds. A 1.49-kg block of iron heated to 155°C would release how many joules of heat as it cooled to 22°C?

33. How many joules are required to heat 692 g of nickel from 22°C to 318°C?

34. How many joules are required to heat 18.2 g of tin from 14.7°C to 47.7°C?

35. Dysprosium was discovered in 1886. Its freezing point is 1400°C and its boiling point is 2600°C. If its specific heat is 0.1733 J/g·C°, how many joules are required to heat 10.0 g from its freezing point to its boiling point?

36. Copper has a specific heat of 0.384 52 J/g·C°. A 105-g sample is exposed to 15.2 kJ in an insulated container. How many degrees will the temperature of the copper sample increase?

37. Glass, which is mostly SiO_2, is not a good insulator. How much energy does a 1400-g pane of glass lose as it cools from a room temperature of 25°C to an outside temperature of 5.0°C? Use data from Appendix Table A-5.

38. Use reference materials to classify the following materials as heterogeneous mixture, solution, compound, or element.
   a. paint
   b. orthoclase
   c. granite
   d. leather
   e. corn syrup
   f. gold

39. Many alloys are heterogeneous mixtures. An alloy of zinc and nickel, which is a heterogeneous mixture, is placed in a beaker containing a solution of calcium chloride in water. What phases and interfaces are present?

40. A swimming pool, 10.0 m by 4.00 m, is filled to a depth of 2.50 m with water at a temperature of 20.5°C. How much energy is required to raise the temperature of the water to a more comfortable 30.0°C?

41. Fatty tissue is 15% water and 85% fat. When fat is completely broken down to carbon dioxide and water, each gram releases 9.0 kilocalories of energy. How many kilocalories are released by the loss of 2.2 kilograms of fatty tissue? (Assume that the fat is completely broken down.)

Concept Mastery

42. How does a phase differ from a state?

43. Explain how a mixture may be either heterogeneous or homogeneous. Use examples in your explanation.

44. Why is a solution always a mixture, but not every mixture a solution?

45. When would fractional crystallization be used to separate the components of a solution?

46. How is distillation used to separate a solution of two liquids? What is the main physical property that must be considered when performing a distillation?

47. In an experiment, two clear liquids are combined. A white precipitate forms and the temperature in the beaker rises.
   a. Is this reaction endothermic or exothermic?
   b. Is heat released or absorbed?
   c. Are the products higher or lower in energy than the reactants?

48. Imagine you are working in a lab and your boss gives you a sample of an unknown metal and a calorimeter. Your boss instructs you to use the calorimeter to gather data on the sample.
   a. Explain what a calorimeter measures.
   b. Describe how to use the calorimeter.
   c. What two units could be used to report your data?

49. Methane is often used as home heating fuel. When it burns, carbon dioxide and water are formed. What elements does methane contain if the oxygen in these two products comes from the air?

50. Which type of substance needs more energy to undergo a ten-degree rise in temperature, one with a high specific heat or one with a low specific heat? What factor, other than the type of substance, must you know before you can be sure of your answer to this question?

Application

51. Using Figure 3.7, determine the solubility of each of the following.
   a. NaCl at 10°C c. KBr at 60°C
   b. KNO_3 at 40°C d. $NaClO_3$ at 80°C

52. A solution contains equal amounts of KNO_3 and NaCl.
   a. If this solution is allowed to evaporate at 60°C, which compound will crystallize first?
   b. Describe the results if the evaporation had occurred at 10°C.

53. Which of the following involve a change in state of a substance?
   a. grinding beef into hamburger
   b. soldering a computer circuit board
   c. pouring milk into a glass
   d. allowing soup to cool in a bowl

54. The steps in the combustion process for a four-stroke engine are shown. Is the change that takes place in the gasoline-air mixture in the cylinder in each of steps 2 and 3 chemical or physical?

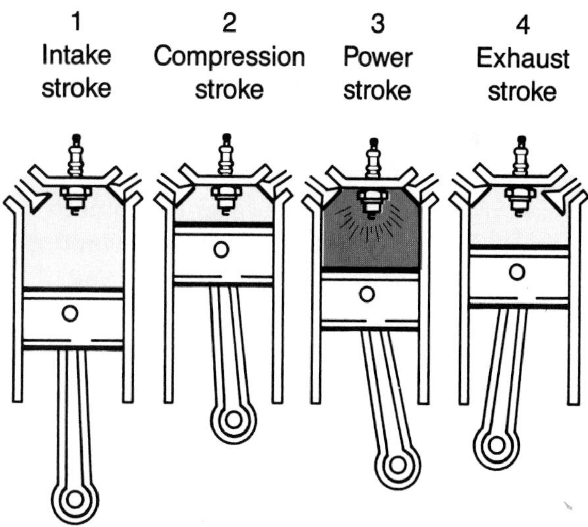

55. People often call wood, paper, milk, and paint "substances." How would you explain that paint is not a substance? How could you demonstrate the difference between paint and a true substance?

56. What are the properties of glass and plastic that make them interchangeable for some uses? How are these materials different?

57. Aluminum, magnesium, and titanium have similar properties that make them useful for aircraft production. What are these properties?

58. Could distillation be used to separate air into its component parts? Explain.

59. The burning of wood is a chemical change that can be used to heat a home. The reaction proceeds by itself as long as an adequate amount of oxygen reaches the wood. However, wood does not just catch fire spontaneously. Instead, matches, paper, kindling, or other materials must be used to start the fire. Explain why these materials are needed.

60. Think of the different methods used to cook food. In what sort of cooking might you want to use cooking utensils with low specific heat? Where might utensils with high specific heat be useful?

61. A blacksmith heated an iron bar to 1445°C. The blacksmith then tempered the metal by dropping it into 42 800 cm^3 of water that had a temperature of 22°C. The final temperature of the system was 45°C. What was the mass of the bar?

62. A 752-cm^3 sample of water was placed in a 1.00-kg aluminum pan. The initial temperature of the pan was 26°C, and the final temperature of the system was 39°C. What was the initial temperature of the water?

63. Assume that samples of equal mass of the metals listed in the following table are heated in boiling water. Each sample is then dropped into a different beaker of water. All the beakers contain equal volumes of water at the same temperature. Arrange the beakers in order from lowest to highest final water temperature. Identify each beaker by the metal it contains.

Metal	Specific Heat (J/g · C°)
copper	0.384 52
gold	0.129 05
iron	0.449 4
silver	0.235 02

64. The fuel value of peanuts is 25 kJ/g. If an average adult needs 2800 kilocalories of energy a day, what mass of peanuts would meet an average adult's energy needs for the day? Assume all of the fuel value of the peanuts can be converted to useful energy.

Critical Thinking

65. The joule is an SI derived unit. Express joules in terms of SI base units.

66. A solution contains equal amounts of table salt and sugar. How might the salt be separated from the sugar?

67. Originally, organic compounds were those compounds found in or given off by living organisms, hence the name, organic. It was believed that these compounds possessed a special property called the "vital principle" that prevented humans from ever synthesizing them in the laboratory. Use reference books to find out what event changed this notion. Even today, materials produced by living things are often called organic materials. What does the use of the term *organic* tell you about the elemental composition of living organisms?

68. Physical changes are often used to separate two or more substances in a mixture. What physical change and what properties could be used to separate each of the following pairs?
   a. salt and iron filings
   b. iron filings and aluminum filings
   c. sand and water
   d. rubies and emeralds

69. The energy that our bodies need to continue life functions comes primarily from the carbohydrates that we eat. The decomposition of carbohydrates is a multistep process that includes the oxidation of glucose to form carbon dioxide and water. Predict whether this reaction is endothermic or exothermic and explain your reasoning.

70. Many athletic trainers use cold packs for injuries that occur during practice or at sporting events. In one type of cold pack, a thin lining between two chemicals breaks when the cold pack is bent in half. When the chemicals mix, the resulting solution freezes. Predict whether this reaction is endothermic or exothermic, and explain your reasoning.

71. Why are some foods, beverages, and other materials stored in brown bottles? In your answer, consider what you learned about energy changes in reactions.

72. Why are air temperatures less variable in locations near bodies of water?

73. A geologist at a mining company is trying to identify a metal sample obtained from a core sample. The unknown metal with mass 5.05 g is heated to 100.00°C and dropped into 10.0 cm^3 of water at 22.00°C. The final temperature of the system is 23.83°C. What is the specific heat of the metal? Using the data in Appendix Table A-3, determine what this metal might be.

74. Carbon has the highest melting point (3620°C) of any element and a specific heat of 0.7099 J/g·C°. Tungsten has a specific heat of 0.1320 J/g·C°. An 11.2-g sample of carbon is heated to its melting point and allowed to radiate heat to a 165.3-g sample of tungsten. The initial temperature of the tungsten is 31.0°C. How many Celsius degrees will the temperature of the tungsten sample increase?

75. A woman runs the New York Marathon in 3 hours, 42 minutes, and 18 seconds. Marathon runners need about 600 Calories per hour to supply their energy needs. How much pasta should the woman have eaten to supply her energy needs? Assume the food value of pasta to be 16 700 joules/gram.

Readings

Gordon, J. E. *The Science and Structures of Materials.* New York: Scientific American Library, 1988.

Laudise, Robert A. "Hydrothermal Synthesis of Crystals." *Chemical and Engineering News* 65, no. 39 (September 28, 1987): 30-43.

Layman, Patricia L. "Preservation of 16th Century Ship Stretches Conservation Chemistry." *Chemical and Engineering News* 65, no. 22 (June 1, 1987): 19-21.

Scientific American 255, no. 4 (October 1986). The whole issue is devoted to materials.

Cumulative Review

1. Compute the density of each of the following materials.
- **a.** clay, if 42.0 g occupy 19.1 cm^3
- **b.** cork, if 8.17 g occupy 34.0 cm^3
- **c.** linoleum, if 6120 g occupy 5100 cm^3
- **d.** ebony wood, if 201 g occupy 165 cm^3

2. What is the property possessed by all matter?

3. What is weight? How is weight measured?

4. How many significant digits are in each of the following quantities?
- **a.** 26.66 m
- **b.** 0.00402 kg
- **c.** 900 cm^3
- **d.** 100.00 cm

CHAPTER **Preview**

CONTENTS

CHAPTER 4

Atomic Structure

Many of our "current" conveniences depend on electricity. Electricity will travel through many different materials, but some of these are too expensive for common use. Other materials may be too impractical or inefficient. Most of our electricity travels from place to place along wires made of the element copper.

Let's look more closely at copper. If we cut a piece of copper wire into very small pieces, would these pieces still be copper? What do we have when we have cut the copper into the smallest possible pieces? We found earlier that the smallest piece of matter that would still be copper is an atom—the smallest particle of an element that has the properties of the element. An atom, however, is made of smaller particles. What evidence do we have that these small particles exist? Our modern concept of the atom is a result of generations of work. Even today, our knowledge of the atom is not complete.

EXTENSIONS AND CONNECTIONS

OBJECTIVES

- discuss early developments in atomic theory.
- explain the laws of multiple proportions and definite proportions and give examples.
- determine the atomic number (Z) and mass number (A) of given isotopes of elements.

4.1 Early Atomic Models

The first three chapters have been devoted to learning some of the vocabulary of chemistry. In this chapter, you will study atoms and their parts. In learning about the structure of the atom you will gain a better understanding of the properties of atoms. You will follow the historical development of ideas about atoms and see how concepts about atomic structure have changed as scientists have accumulated more experimental evidence.

Early Atomic Theory

Early thoughts concerning atoms were proposed by a Greek philosopher, Democritus, about 400 B.C. He suggested that the world was made of two things—empty space and tiny particles he called "atoms." He thought of atoms as the smallest possible particles of matter. He also thought there were different types of atoms for each material in the world. His belief was very general and was not supported by experimental evidence.

Soon after Democritus proposed his belief in the existence of atoms, Aristotle proposed that matter was continuous and was not made up of smaller particles. He called this continuous substance *hyle.* Aristotle's teachings were accepted until the 17th century. Then, many people began to express doubts and objections to his ideas.

Two of these individuals were Isaac Newton and Robert Boyle. They published articles stating their belief in the atomic nature of elements. However, their works offered no proof. They were attempted explanations of the known, with no predictions of the unknown. It was an English chemist, John Dalton, who offered a logical hypothesis about the existence of atoms.

During the early 1800s, Dalton studied experimental observations made by others concerning chemical reactions. Antoine Lavoisier, a French chemist, had made one of these discoveries. He found that when a chemical change occurred in a closed system, the mass after a chemical change equaled the mass before the change. A closed system is a system that cannot exchange matter with its surroundings. In all tests of chemical changes in a closed system, he found that mass remained constant. He proposed that, *in ordinary chemical reactions, matter can be changed in many ways, but it cannot be created or destroyed.* This proposal is the law of conservation of mass.

The work of another French chemist, Joseph Proust, also came to the attention of Dalton. Proust had observed that *specific substances always contain elements in the same ratio by mass.* Consider table salt as an example. It is made of sodium and chlorine. The ratio of the mass of sodium to the mass of chlorine in any sample of pure salt is always the same. No matter where the sample is obtained, how it is obtained, or how large it is, the ratio of the mass of sodium to the mass of chlorine never changes. This principle is known as the **law of definite proportions.**

PHILOSOPHY CONNECTION

Ancient Greeks

The Greeks had highly developed ideas about nature. However, they were just that—ideas. The ancient Greeks did not believe it was necessary to investigate nature directly. Consequently, they never performed experiments to test their ideas.

Word Origins

a-: (GK) negates what comes after it
temnein: (GK) to cut

atom—not cuttable, indivisible

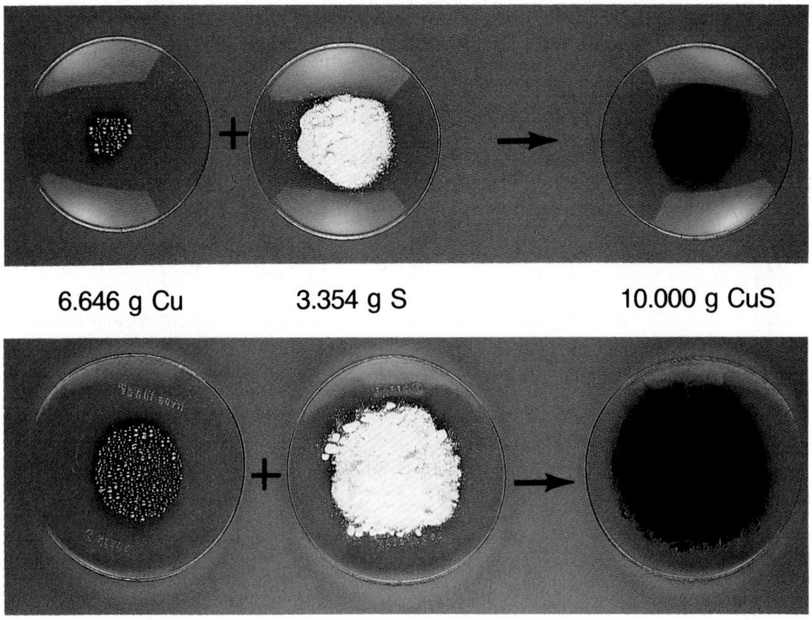

6.646 g Cu 3.354 g S 10.000 g CuS

66.46 g Cu 33.54 g S 100.00 g CuS

Figure 4.1 Copper and sulfur combined in copper(II) sulfide are in a copper-to-sulfur mass ratio of almost 2 to 1 (1.982 g to 1.000 g).

Dalton's Hypothesis

Dalton was trying to explain the findings of Lavoisier and Proust when he formed the basis of our present atomic theory. He stated that all matter is composed of very small particles called atoms, and that these atoms cannot be broken into smaller particles. Dalton's ideas were like those of Democritus. However, Dalton believed that atoms were simpler than particles of air or rock. He also believed that all atoms of an element were exactly alike and that atoms of different elements were quite unlike. He stated that atoms can combine in simple ratios to form compounds.

Dalton's ideas explain the laws of conservation of mass and definite proportions. If atoms cannot be destroyed, then they must simply be rearranged in a chemical change. The total number and kind of atoms must remain the same. Therefore, the mass before a reaction must equal the mass after a reaction. If the atoms of an element are always alike, then all atoms of a particular element must have the same mass.

How does Dalton's theory apply to sodium chloride? According to Dalton, all sodium atoms have the same mass and all chlorine atoms have the same mass. When a sodium atom combines with a chlorine atom, table salt is formed. The same is true of any pair of these atoms. Therefore, the ratio of the mass of sodium to the mass of chlorine must be the same for any sample of table salt. Dalton believed this reasoning would hold true for any given material. Experiments have shown that Dalton's ideas are not entirely correct. As you will see later, not all atoms of the same element have exactly the same mass. However, by changing the word "mass" to "average mass," we can use Dalton's ideas today.

MINUTE LAB

Think Small, Very Small

Using a dropper, place 15 to 20 drops of vanilla flavoring into a balloon. Inflate the balloon and tie it closed. What do you notice when you feel the outside of the balloon where the vanilla is located on the inside? Place the balloon in a small enclosed area such as a closet or school locker for an hour. What did you notice? How do you explain your observations? Do you think helium will leak more rapidly from a fully inflated balloon or from a half inflated balloon? Explain your answer.

CHEMISTS AND THEIR WORK

John Dalton (1766-1844)
John Dalton's major contribution to science was his statement of the first atomic theory based on experimental evidence. He then used this theory to predict the existence of the law of multiple proportions. When chemists demonstrated that his prediction was correct, the acceptance of the theory was accelerated. The rapid acceptance of the atomic theory started chemists on an explosive investigation of matter. That investigation continues today.

Dalton stated a second law based on his own atomic theory but not based on experimental data. It concerns elements that form more than one compound with each other. For example, oxygen can combine with tin in different mass ratios forming more than one compound, as shown in Table 4.1. According to Dalton's second law, *the ratio of masses of one element that combine with a constant mass of another element can be expressed in small whole numbers.* This statement is known as the **law of multiple proportions.** Look at Table 4.1. Note the whole number ratio of oxygen in the two compounds. This ratio indicates that atoms react as whole units, and they are not divided into smaller parts in chemical reactions.

Table 4.1

Proportions of Tin to Oxygen

Compound	Mass of Sn in a sample	Mass of O in a sample	Ratio of O masses combined with constant mass (119 g) of Sn
tin(II) oxide, SnO	119 g	16 g	1
tin(IV) oxide, SnO_2	119 g	32 g	2

At about the same time that Dalton formed his atomic theory, J. L. Gay-Lussac, a French chemist, made an interesting observation. He was studying gas reactions at constant temperature and pressure. He noted that, under constant conditions, the volumes of reacting gases and gaseous products are in the ratio of small whole numbers.

This observation was explained a few years later by Amadeo Avogadro, an Italian physicist, using Dalton's theory. Avogadro's hypothesis also concerned gases at the same temperature and pressure. He stated that *equal volumes of gases, under the same conditions, have the same number of molecules.* These observations sped the acceptance of Dalton's theory.

The atomic theory and the law of multiple proportions, as stated by Dalton, have been tested and are accepted as correct. There are, however, major exceptions to some of Dalton's statements. These differences will be considered as we form our modern model of the atom in the sections that follow.

Early Research on Atomic Particles

Experiments by several scientists in the middle of the 19th century led to the conclusion that the atom is made up of smaller particles. With the use of a tube such as the one in Figure 4.2, it was possible to learn more about atoms. In each end of the tube, there is a metal piece called an electrode. When connected to a source of high voltage electricity, the electrodes become charged. The positive electrode is called the **anode.** The negative electrode

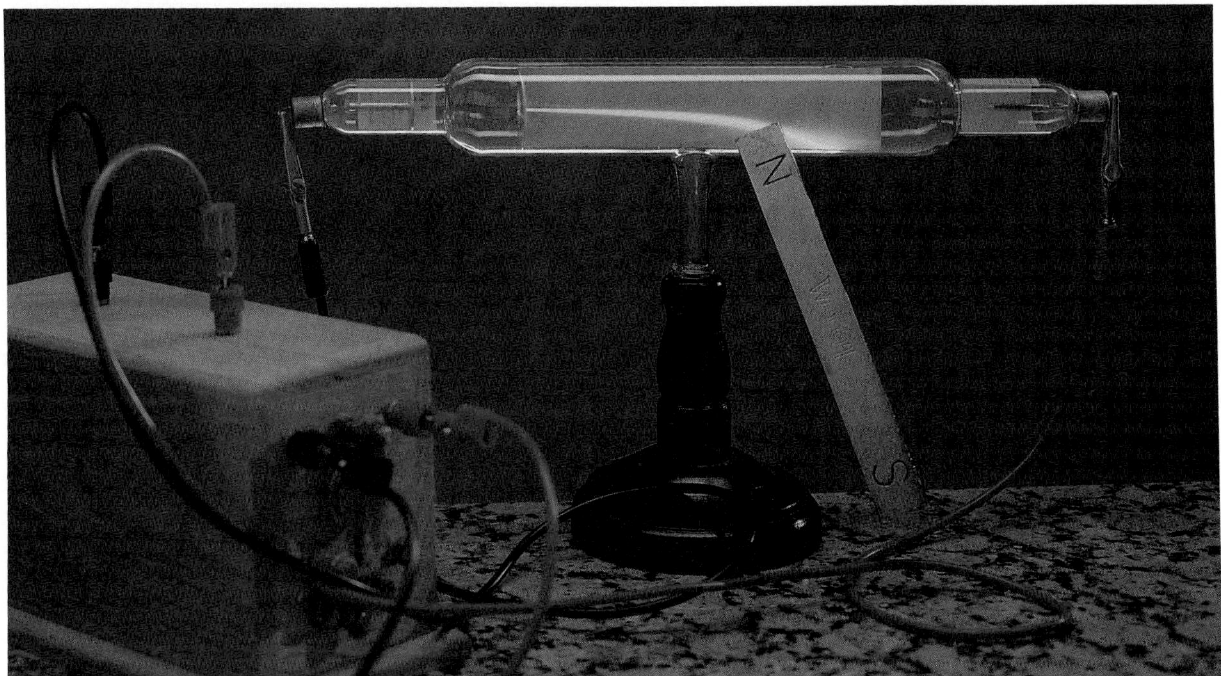

is called the **cathode.** Careful observation revealed rays in the tube. Because these rays appeared to begin at the cathode and travel toward the anode, they were called **cathode rays.**

In 1897, J. J. Thomson, an English scientist, did some skillful research on cathode rays. As a result, Thomson is generally credited with the discovery that cathode rays consist of electrons. Thomson built a cathode-ray tube to subject the rays to both a magnetic field and an electric field. He measured the bending of the path of the cathode rays and was able to determine the ratio of an electron's charge to its mass.

Robert Millikan, an American scientist, obtained the first accurate measurement of an electron's charge. He used a device like the one in Figure 4.3. As oil was sprayed from a brass atomizer, electrons were transferred from the atomizer to the oil droplets. These negatively charged droplets, under the influence of

Figure 4.2 J.J. Thomson used a cathode-ray tube similar to this one to study the properties of electrons. In this tube, the left electrode is negatively charged and the right electrode is positively charged. Note how the path of the ray is affected by the magnet.

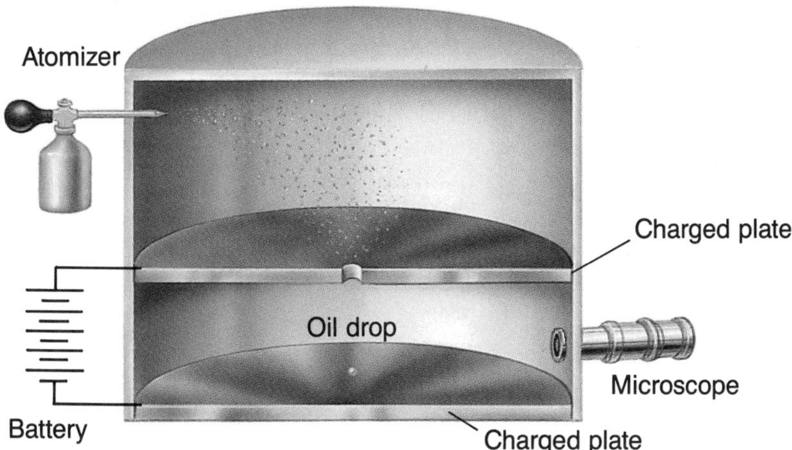

Figure 4.3 This illustration shows a cross-sectional view of Millikan's apparatus for determining the charge on an electron.

Figure 4.4 In 1885, E. Goldstein, using a modified cathode-ray tube, discovered a beam consisting of positive particles.

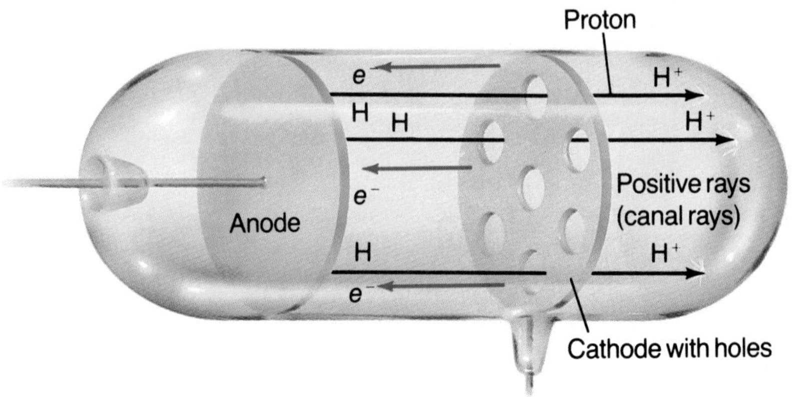

CHEMISTS AND THEIR WORK

Robert Andrews Millikan (1868-1953)

Educated at Oberlin College and Columbia University, Robert Millikan began his scientific career at the University of Chicago. Later, he moved to the California Institute of Technology. His most famous contribution to science was the accurate measuring of the charge on the electron. However, he also made accurate measurement of Planck's constant and, with that value, verified Einstein's equations explaining the photoelectric effect. From 1925 until his retirement, he devoted much of his research time to investigating cosmic rays, which he named.

gravity, fell through a vacuum chamber. The charge on the plates was adjusted to just offset the gravitational force on the droplet. Millikan calculated the charge on the droplet. He found that the charges on the droplets varied. However, each charge was a multiple of one small charge. He correctly concluded that this small charge must be the charge on a single electron. This charge is now the standard unit of negative charge (1−). The electron and its charge may be represented by the symbol e^-.

Using the data of Thomson and Millikan, it was possible to calculate the actual mass of the electron. Its mass is only $1/1837$ the mass of the lightest atom known, the hydrogen atom.

Protons were also discovered in an experiment involving hydrogen in a modified cathode-ray tube such as the one in Figure 4.4. Rays were discovered traveling in the direction opposite to that traveled by the cathode rays. Later it was shown that these rays possessed a positive charge. Thomson showed they consisted of particles. The particles he found have the same amount of electric charge as an electron has. However, the charge is opposite in sign to that on the electron. These particles are now called protons. The atoms of hydrogen gas he used in the tube consist of a single proton and a single electron. Thomson calculated that the mass of the proton was just about 1836 times that of the electron. The proton is now the standard unit of positive charge (1+).

A third particle in an atom remained unobserved for a long time. However, its existence had been predicted by Lord Rutherford, an English physicist, in 1920. The first evidence of the particle was obtained by Walter Bothe in 1930. Another English scientist, James Chadwick, repeated Bothe's work in 1932. He found high energy particles with no charge and with essentially the same mass as the proton. These particles are now known as neutrons. Neutrons, as well as protons, are now known to be made up of still smaller particles. These smaller particles are discussed in a later section of this chapter.

Dalton had assumed that atoms could not be broken into smaller particles. The discovery of subatomic particles led to a major revision of Dalton's atomic theory to include this new information.

Isotopes and Atomic Number

While working with neon, J. J. Thomson observed what seemed to be two kinds of neon atoms. They were exactly alike chemically, but different in mass. Atoms of the same element that differ in mass are called **isotopes.** Isotopes have the same number of protons but a different number of neutrons. For example, the isotopes of neon are neon-20, which has 10 neutrons; neon-21, which has 11 neutrons; and neon-22, which has 12 neutrons.

In 1913, another English scientist, Henry Moseley, studied X rays produced in X-ray tubes with anodes of different metals. He found that the wavelength of the X rays is characteristic of the metal used as the anode. The wavelength depends on the number of protons in the nucleus of the atom and is always the same for a given element. This number of protons is known as the **atomic number** of the element and is represented by the symbol Z. Because an atom is electrically neutral, the number of electrons must equal the number of protons. The mass difference of isotopes must then be due to the different numbers of neutrons in the nucleus. Thus, *the number of protons determines the identity of the element* and *the number of neutrons determines the particular isotope of the element.* The atomic numbers of the elements are given in Table 4.2.

Dalton's atomic theory was again changed. It now states that all atoms of an element contain the same number of protons but they can contain different numbers of neutrons. For example, the element hydrogen has three isotopes. All hydrogen isotopes have one proton but different numbers of neutrons, as shown in Table

isos-: (GK) equal, same
topos: (GK) place

isotope—found in the same place (in the periodic table, Chapter 6)

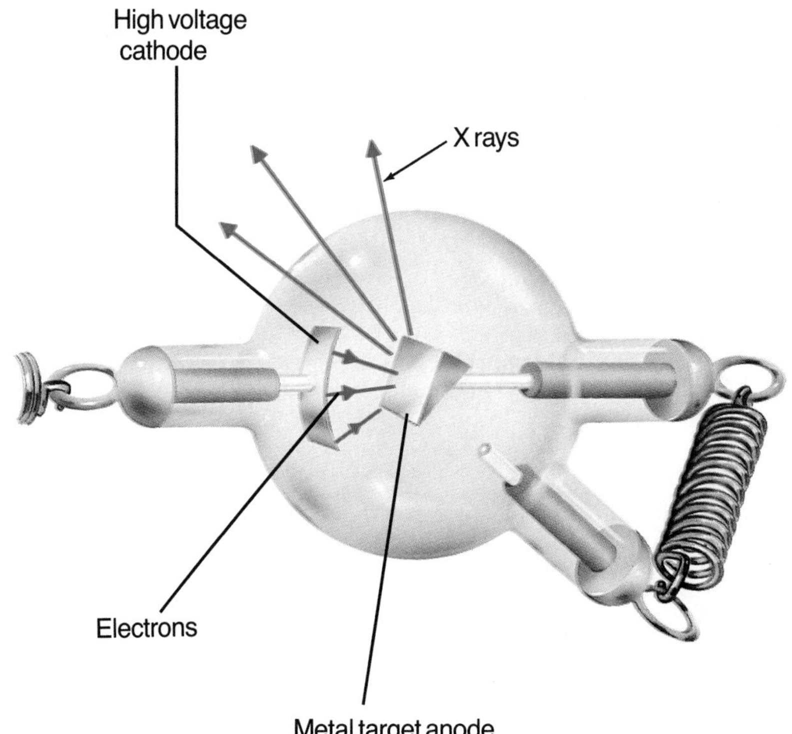

Figure 4.5 The metal target in the X-ray tube can be changed to produce X rays of different wavelengths.

Table 4.2

International Atomic Masses

Element	Symbol	Atomic number	Atomic mass	Element	Symbol	Atomic number	Atomic mass
Actinium	Ac	89	227.027 8*	Neon	Ne	10	20.179 7
Aluminum	Al	13	26.981 539	Neptunium	Np	93	237.048 2
Americium	Am	95	243.061 4*	Nickel	Ni	28	58.6934
Antimony	Sb	51	121.757	Niobium	Nb	41	92.906 38
Argon	Ar	18	39.948	Nitrogen	N	7	14.006 74
Arsenic	As	33	74.921 59	Nobelium	No	102	259.100 9*
Astatine	At	85	209.987 1*	Osmium	Os	76	190.2
Barium	Ba	56	137.327	Oxygen	O	8	15.999 4
Berkelium	Bk	97	247.070 3*	Palladium	Pd	46	106.42
Beryllium	Be	4	9.012 182	Phosphorus	P	15	30.973 762
Bismuth	Bi	83	208.980 37	Platinum	Pt	78	195.08
Boron	B	5	10.811	Plutonium	Pu	94	244.064 2*
Bromine	Br	35	79.904	Polonium	Po	84	208.982 4*
Cadmium	Cd	48	112.411	Potassium	K	19	39.098 3
Calcium	Ca	20	40.078	Praseodymium	Pr	59	140.907 65
Californium	Cf	98	251.079 6*	Promethium	Pm	61	144.912 8*
Carbon	C	6	12.011	Protactinium	Pa	91	231.035 88
Cerium	Ce	58	140.115	Radium	Ra	88	226.025 4
Cesium	Cs	55	132.905 43	Radon	Rn	86	222.017 6*
Chlorine	Cl	17	35.452 7	Rhenium	Re	75	186.207
Chromium	Cr	24	51.996 1	Rhodium	Rh	45	102.905 50
Cobalt	Co	27	58.933 20	Rubidium	Rb	37	85.467 8
Copper	Cu	29	63.546	Ruthenium	Ru	44	101.07
Curium	Cm	96	247.070 3*	Samarium	Sm	62	150.36
Dysprosium	Dy	66	162.50	Scandium	Sc	21	44.955 910
Einsteinium	Es	99	252.082 8*	Selenium	Se	34	78.96
Erbium	Er	68	167.26	Silicon	Si	14	28.085 5
Europium	Eu	63	151.965	Silver	Ag	47	107.868 2
Fermium	Fm	100	257.095 1*	Sodium	Na	11	22.989 768
Fluorine	F	9	18.998 403 2	Strontium	Sr	38	87.62
Francium	Fr	87	223.019 7*	Sulfur	S	16	32.066
Gadolinium	Gd	64	157.25	Tantalum	Ta	73	180.947 9
Gallium	Ga	31	69.723	Technetium	Tc	43	97.907 2*
Germanium	Ge	32	72.61	Tellurium	Te	52	127.60
Gold	Au	79	196.966 54	Terbium	Tb	65	158.925 34
Hafnium	Hf	72	178.49	Thallium	Tl	81	204.383 3
Helium	He	2	4.002 602	Thorium	Th	90	232.038 1
Holmium	Ho	67	164.930 32	Thulium	Tm	69	168.934 21
Hydrogen	H	1	1.007 94	Tin	Sn	50	118.710
Indium	In	49	114.82	Titanium	Ti	22	47.88
Iodine	I	53	126.904 47	Tungsten	W	74	183.85
Iridium	Ir	77	192.22	Unnilennium†	Une	109	266*
Iron	Fe	26	55.847	Unnilhexium†	Unh	106	263*
Krypton	Kr	36	83.80	Unniloctium†	Uno	108	265*
Lanthanum	La	57	138.905 5	Unnilpentium†	Unp	105	262*
Lawrencium	Lr	103	260.105 4*	Unnilquadium†	Unq	104	261*
Lead	Pb	82	207.2	Unnilseptium†	Uns	107	262*
Lithium	Li	3	6.941	Uranium	U	92	238.028 9
Lutetium	Lu	71	174.967	Vanadium	V	23	50.941 5
Magnesium	Mg	12	24.305 0	Xenon	Xe	54	131.29
Manganese	Mn	25	54.938 05	Ytterbium	Yb	70	173.04
Mendelevium	Md	101	258.098 6*	Yttrium	Y	39	88.905 85
Mercury	Hg	80	200.59	Zinc	Zn	30	65.39
Molybdenum	Mo	42	95.94	Zirconium	Zr	40	91.224
Neodymium	Nd	60	144.24				

*The mass of the isotope with the longest known half-life.
†Names for elements 104-109 have been approved for temporary use by the IUPAC. The USSR has proposed Kurchatovium (Ku) for element 104, and Bohrium (Bh) for element 105. The United States has proposed Rutherfordium (Rf) for element 104, and Hahnium (Ha) for element 105.

Table 4.3

Isotopes of Hydrogen			
Nuclide	**Protons**	**Neutrons**	**Mass Number**
protium	1	0	1
deuterium	1	1	2
tritium	1	2	3

4.3. A particular kind of atom containing a definite number of protons and neutrons is called a **nuclide.** For example, protium, hydrogen-1, is a nuclide of hydrogen.

The particles that make up the atomic nucleus are called **nucleons.** These particles are protons and neutrons. The total number of nucleons in an atom is called the **mass number** of that atom. The symbol for the mass number is A. The number of neutrons for any nuclide may be found by subtracting the atomic number from the mass number.

$$\textit{number of neutrons} = A - Z$$

In this way, you can calculate the number of neutrons in any nuclide if you know its mass number and atomic number.

PROBLEM SOLVING HINT

To check your work, add the numbers of protons and neutrons. Your total should equal the mass number.

SAMPLE PROBLEM

Number of Neutrons

How many electrons, neutrons, and protons are found in a copper atom of mass number 65?

Solving Process:

Using Table 4.2, you can find the atomic number of copper to be 29. Since the atomic number of an element is the number of protons in the nucleus, there are 29 protons in the nucleus of this copper atom. Atoms are electrically neutral, so an atom having 29 protons in the nucleus must also have 29 electrons. The mass number is the number of nucleons in the atom. This copper atom has 65 nucleons, of which 29 are protons. There must, therefore, be $65 - 29 = 36$ neutrons in the nucleus.

PRACTICE PROBLEMS

1. An atom of vanadium contains 23 electrons. How many protons does it contain?

2. An atom of silver contains 47 protons. What is its atomic number?

3. An atom of sodium contains 11 electrons. What is the atomic number of this atom?

Use Table 4.2 to complete the following questions.

4. An atom contains 37 protons. What element is it?

5. An atom contains 56 electrons. What element is it?

6. How many electrons, protons, and neutrons are contained in each of the following nuclides?
   a. actinium of mass number 221
   b. magnesium of mass number 25
   c. rhodium of mass number 105
   d. lanthanum of mass number 133

FRONTIERS

Arsenic and Stardust

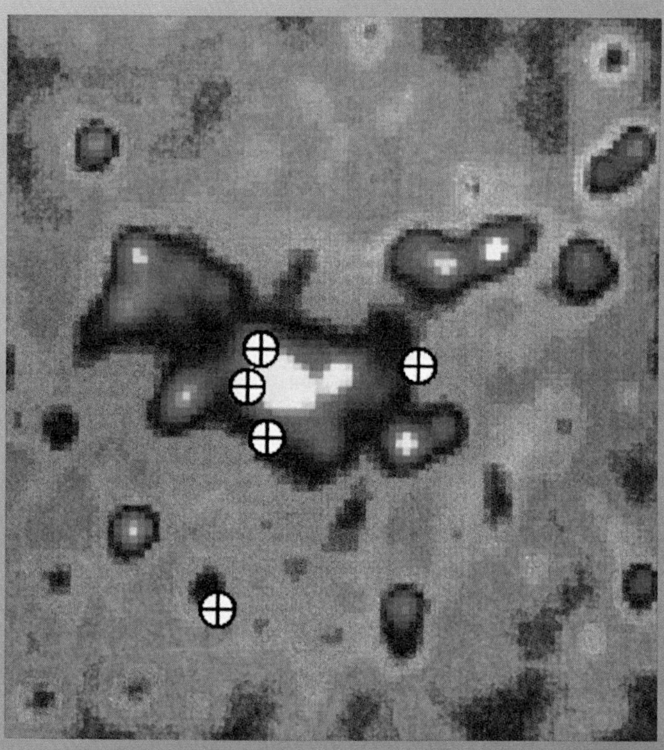

X-ray telescopes can detect a neutron star, shown here as a cross in this image of the Milky Way, by bursts of energy produced at the star's surface. Astrophysicists have hypothesized a model that would account for these X-ray bursts. They predicted that a yet undiscovered isotope of arsenic, arsenic-65, would play a crucial role in this model. Researchers have now found this isotope.

Unlike Earth's sun, a star composed mostly of the nuclei of the element hydrogen, neutron stars are composed of neutrons. According to the model, a neutron star would attract protons and nuclei of elements of lower atomic masses from a nearby hydrogen-rich star. Then a series of proton-captures would take place. In a proton capture, a nucleus of an element of low atomic number captures a proton and produces a nucleus of an element of greater atomic number. After a series of proton captures, elements of greater atomic number and mass, such as arsenic-65, would evolve. According to the model, the energy produced from proton-captures would be great enough to generate bursts of X rays.

Researchers have now produced arsenic-65 in a laboratory. Arsenic-65, and five other new isotopes, have been detected among the shower of atomic nuclei formed when a beam of high-energy ions of krypton-78 crashes into a nickel target. The nuclei that are produced pass through a detector that measures their masses and charges. Although the isotopes exist for only about 150 nanoseconds within the detector, scientists have identified the isotopes as gallium-61, arsenic-65, bromine-69, strontium-75, and two isotopes of germanium, germanium-62 and germanium-63.

The discovery of arsenic-65 adds credibility to the model. Researchers are designing experiments that will isolate and measure the properties of arsenic-65. One such property is half-life, an indication of how long the nuclei of isotopes can exist. Establishing the half-life of arsenic-65 will allow astrophysicists to determine if the isotope is stable enough to be one step in the model.

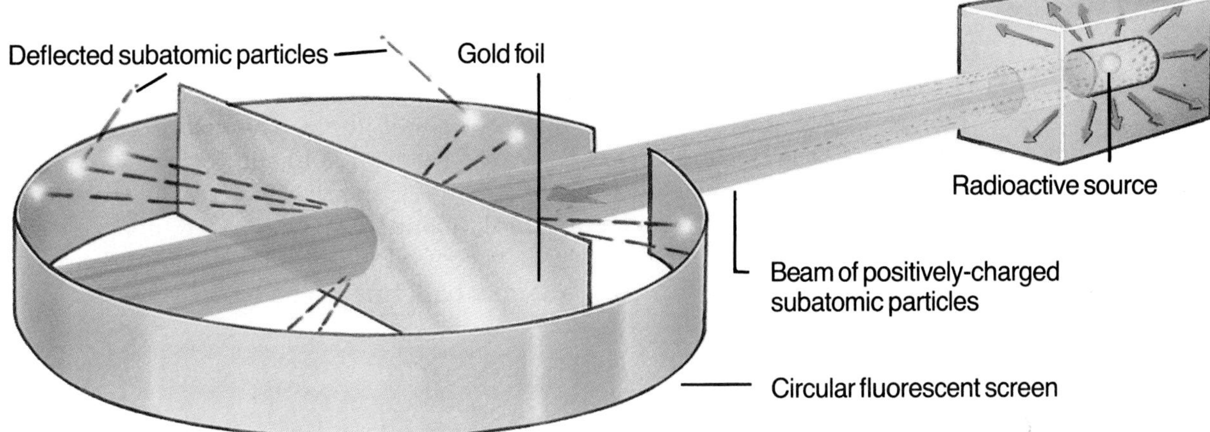

Figure 4.6 Most of the positively-charged subatomic particles pass through the gold foil undeflected. The deflections of particles indicate that a collision has occurred between a subatomic particle and a gold nucleus.

The Nuclear Atom

During the period 1912-1913, Lord Rutherford brought together a brilliant team of physicists. Included in this group were Niels Bohr, Hans Geiger, and Ernest Marsden. The beginnings of our modern concept of atomic structure were developed by this group through experiments and hypotheses under Rutherford's direction. Geiger and Marsden subjected a very thin sheet of gold foil to a stream of positively-charged subatomic particles. They found that most of the particles passed right through the sheet. From this observation, Rutherford concluded that the atom is mostly empty space. They also found that a few particles were deflected at large angles. Some of these (about 1 in 8000) bounced back in almost the opposite direction from which they started. Rutherford explained this observation as meaning that there was a very small "core" to the atom. The core contained all the positive charge and almost all of the mass of the atom. This core is now called the nucleus.

Figure 4.7 The results of the gold foil experiment indicate that the gold atom consists of a positively-charged nucleus surrounded by electrons.

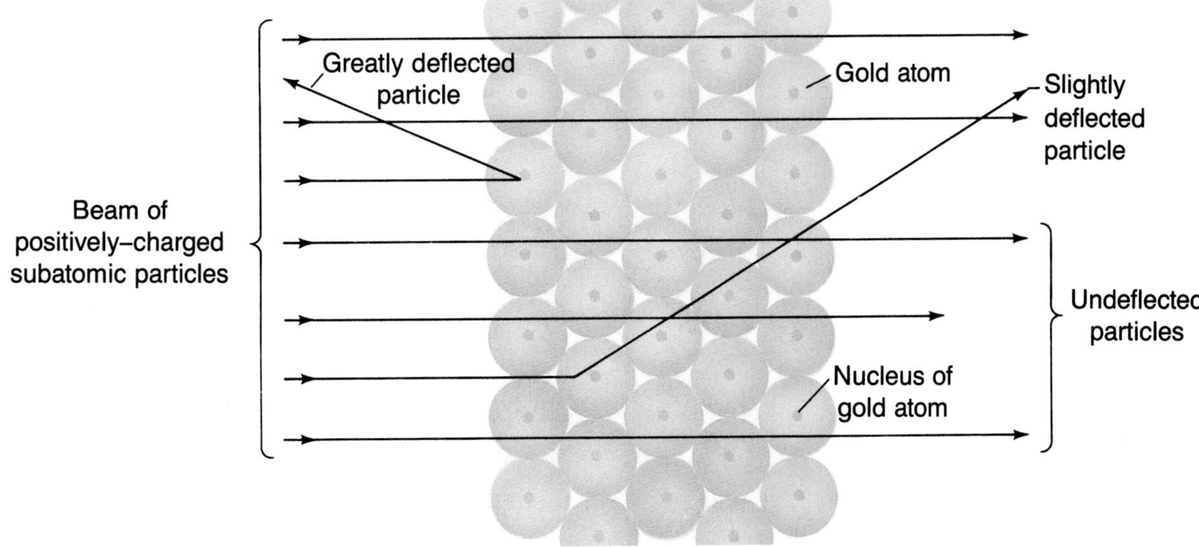

PHOTOGRAPHY CONNECTION

Unexpected Exposure

Becquerel initially thought that exposure to sunlight was necessary for uranium ore to emit rays. One day when the sun did not shine he threw the uranium ore samples he was using into a drawer with a photographic plate encased in two layers of heavy black paper. A day later he developed the plate on a hunch and found the plate exposed to the same degree as when the ore was exposed to light.

Geiger and Marsden found that most of the atom is empty space. Most atoms have a diameter between 100 and 500 pm (1 pm = 1×10^{-12} m). However, the radii of the nuclei of atoms vary between 1.2×10^{-3} and 7.5×10^{-3} pm. The radius of the electron is about 2.82×10^{-3} pm. In small atoms, the distance between the nucleus and the nearest electron is about 50 pm. Thus, the nucleus occupies only about one trillionth (10^{-12}) of the volume of an atom. To help you think about this relationship, imagine the hydrogen nucleus as the size of a Ping-Pong ball. The electron is roughly the size of a tennis ball, and about 1.35 km away.

Radioactivity

In 1896, Henri Becquerel, a French physicist, found that matter containing uranium exposes sealed photographic film. This fact led another scientist, Marie Curie, and her husband, Pierre, to an important discovery. They found that rays are given off by the elements uranium and radium.

Uranium and radium can be found in nature in an ore called pitchblende. The rays from this ore have a noticeable effect on a charged electroscope. An electroscope, as shown in Figure 4.8, contains two thin, free-hanging metal leaves attached to a metal rod. When a charge is applied to the rod, the leaves become charged and repel each other. If some of this ore is placed near, but not touching, a charged electroscope, the leaves become discharged. Substances that have this effect are called radioactive substances. **Radioactivity** is the phenomenon of rays being produced spontaneously by unstable atomic nuclei.

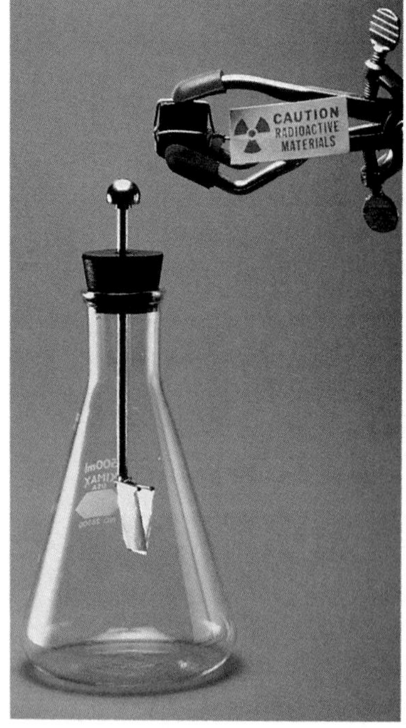

Figure 4.8 The charged electroscope on the left will become discharged if a radioactive substance is brought near the electroscope.

Figure 4.9 Uranium ore is mined in both underground and surface mines. This photograph shows an aerial view of part of a surface mine.

The rays produced by radioactive materials can be particles or energy or a mixture of both. The particles and energy are given off by the nuclei of radioactive atoms during spontaneous nuclear decay, a process covered in Chapter 28. We say the decay is spontaneous because it occurs without external influence. The amount of energy released in a nuclear change is very large. It is so large that it cannot be a result of an ordinary chemical change.

Albert Einstein, in the early 1900s, explained the origin of the energy released during nuclear changes. Einstein theorized that mass and energy are equivalent. This statement can be expressed in the equation

$$E = mc^2$$

where E is the energy released (in joules), m is the mass (in kilograms) of matter involved, and c is a constant, the speed of light (in meters per second).

The large amount of energy released in splitting uranium nuclei is measurable. If 1.00 g of the uranium nuclide with mass number 235 undergoes nuclear fission (splitting of the nucleus), 8.09×10^7 kJ are produced. In contrast, the energy released in normal chemical reactions is considerably less. When 1.00 g of the same uranium nuclide reacts with hydrochloric acid in a typical chemical change, only 3.25 kJ are produced.

4.1 *CONCEPT REVIEW*

7. Explain the difference between the law of definite proportions and the law of multiple proportions.

8. What is the difference between atomic number and mass number?

9. Find the number of electrons, neutrons, and protons in an atom of the nuclide strontium-88.

10. What contribution to atomic theory was made by Geiger and Marsden in their gold foil experiment?

11. Apply Uranium-238 undergoes a nuclear change and becomes thorium-234 and helium-4. How do you know that this change is nuclear and not chemical or physical?

OBJECTIVES

- differentiate among the major subatomic particles.
- discuss the development of modern atomic theory.
- calculate the average atomic mass of a mixture of isotopes of an element.

4.2 Parts of the Atom

Continued research has shown that the proton and neutron are not truly "elementary" particles. That is, they appear to be composed of yet simpler particles. All evidence about the electron, however, indicates that it is truly elementary. As you study the section on nuclear structure, refer frequently to Table 4.4 and Figure 4.10, which list subatomic particles.

CHEMISTRY IN DEPTH

Nuclear Structure

The uranium nuclide with mass number 238 contains 92 protons and 146 neutrons. These particles are bound tightly in the nucleus. Electrostatic attraction alone would not be enough to prevent the uncharged neutrons from floating away. Furthermore, because all protons have the same positive charge, electrostatic forces should cause them to fly apart. Gravitational force, which keeps us at the surface of Earth, is not strong enough to hold the nucleus together. The force that holds the protons and neutrons together in the nucleus is called the **nuclear force**. It is effective for very short distances only (about 10^{-3} pm). This distance is about the same as the diameter of the nucleus. Ideas explaining nuclear structure differ, but scientists agree on certain facts.

(1) Nucleons (protons and neutrons) have a property that corresponds to spinning on an axis.

(2) Electrons do not exist in the nucleus, yet they can be emitted from the nucleus.

The particles composing atoms are called **subatomic particles.** Nuclear scientists divide subatomic particles into two broad classes, leptons and hadrons. Current theory holds that **leptons** ("light" particles) are truly elementary particles. The electron is the best known lepton. **Hadrons,** on the other hand, appear to be made of even smaller particles. Neutrons and protons are the best known hadrons.

For every particle, a mirror-image particle called an **antiparticle** exists, or is believed to exist. Thus, there is an antielectron, called a positron, which is like an electron in every way except that it has a positive charge. Positrons are not common. They exist only until they collide with an electron. Such a collision is very likely in our world. When the collision occurs, both particles are destroyed and energy is produced. It is interesting to note that there is at least one particle that is its own antiparticle. This particle is the neutral pion.

There are several other leptons. In order to account for a certain kind of nuclear decay, a neutral particle called a **neutrino** was postulated. The neutrino has been identified and found to be essentially massless. The muon and the tau, both much more massive than the electron, make up the rest of the lepton family. A neutrino has been discovered for the muon and for the tau. There are antiparticles for all of these particles.

Table 4.4

Hadrons
Baryons
Protons
Neutrons
Other, short-lived particles
Mesons
Pions
Kaons
Other, short-lived particles

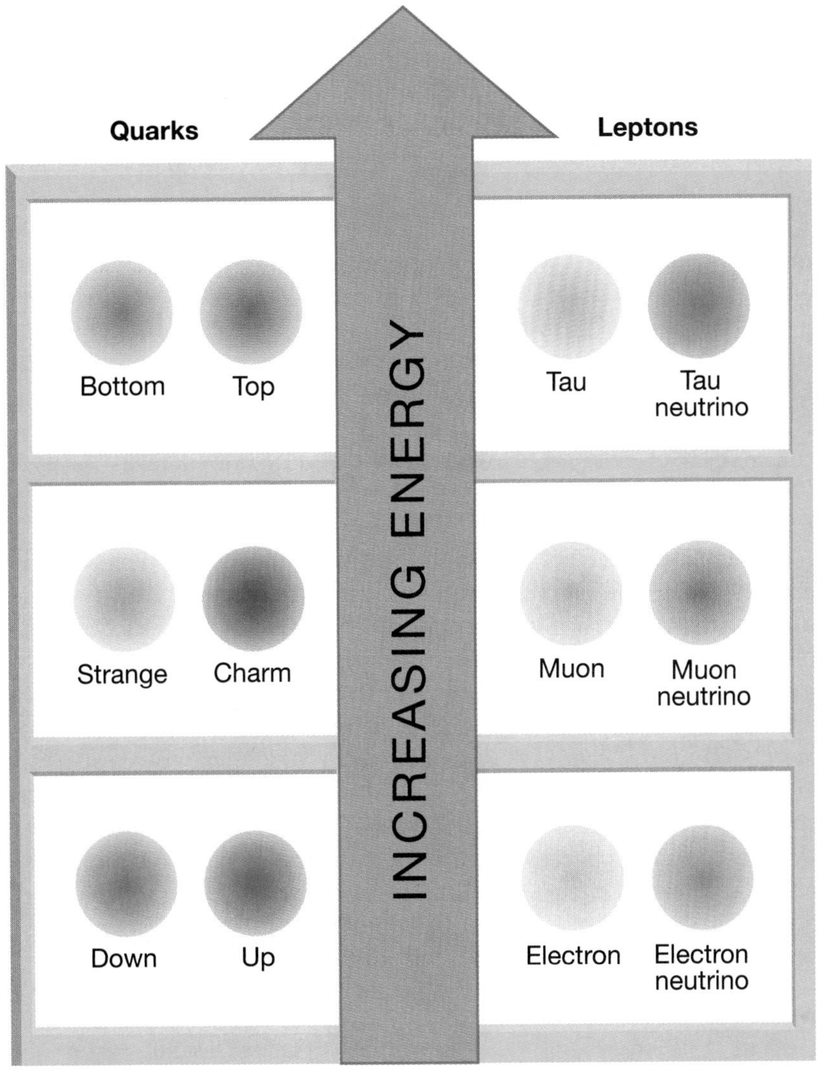

Figure 4.10 The known quarks and leptons are divided into three families. The everyday world is made from particles in the family at the bottom of the chart. Particles in the middle group are found in cosmic rays and are routinely produced in particle accelerators. Particles in the upper two families are created in high-energy collisions. This chart provides a list of known quarks and leptons. The colors, sizes, and shapes shown are not meant to represent actual particles.

The hadrons are subdivided into two groups, the mesons and the baryons. Protons and neutrons are baryons, as are a number of short-lived particles. There are several kinds of mesons.

Mesons and baryons are made of smaller particles called **quarks.** There are six kinds of quarks: "up," "down," "charm," "strange," "top," and "bottom." The names convey nothing about the properties of the quarks. They are just identification labels. Each quark comes in three "colors," red, blue, and green. Again, the color label refers only to a distinguishing property, not an appearance. Each kind of quark has an antimatter counterpart, an antiquark. **Baryons** are composed of three quarks, each of a different color. **Mesons** are composed of a quark and an antiquark of complementary color. It is believed that quarks hold together by exchanging "particles" called **gluons.** There are believed to be eight gluons, each of which is characterized by one color and one anti-color. In a similar manner, the nucleons are held together in the nucleus by the exchange of pions.

▲

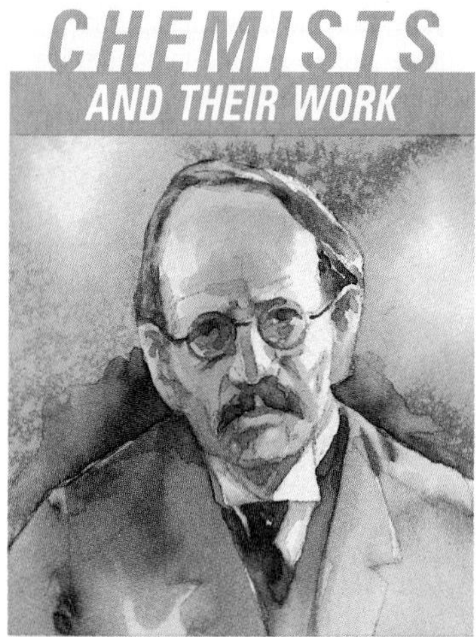

J. J. Thomson (1856-1940)
Sir Joseph John Thomson was educated at Cambridge University in England and spent his entire professional life there as professor and director of the physics laboratories. In his own research, he is credited with the discovery of the electron and the proton, and for producing the first experimental evidence for the existence of isotopes. He was awarded the Nobel Prize in Physics in 1906. One of his greatest scientific legacies was the training of a new generation of scientists. Seven of his research assistants eventually won Nobel Prizes.

Radiation

If the structure of a nucleus is not stable, the nucleus will eject either a particle or energy until it reaches a more stable arrangement. Some nuclei are unstable as found in nature. Other nuclei can be made artificially unstable (radioactive) by bombardment in a particle accelerator.

Three forms of radiation can come from naturally radioactive nuclei. Two forms consist of particles. The third consists of energy. The particles are alpha (α) and beta (β) particles. The energy consists of gamma (γ) rays. An **alpha particle** is a helium nucleus and consists of two protons and two neutrons. A **beta particle** is a high-speed electron emitted from radioactive nuclei. Electrons do not exist as such in the nucleus but are generated at the instant of decay. **Gamma rays** are very high-energy X rays. Unstable nuclei that emit rays or particles to become stable nuclei are said to decay.

Scientists have created many radioactive nuclides that do not exist in nature on Earth. These substances are generated by bombarding stable nuclei with accelerated particles or exposing stable nuclei to neutrons in a nuclear reactor (see Chapter 28). These artificially radioactive nuclides can decay by emitting α, β, and γ rays as well as by several other methods such as positron emission. Artificially radioactive nuclides can also decay by capturing one of the electrons outside the nucleus. The electrons closest to the nucleus are sometimes called the *K* electrons. For this reason, this radioactive process is called *K*-capture. It is also sometimes called electron capture. Most antiparticles have been formed and observed during the bombardment of normal nuclei in particle accelerators. They may also be formed in other ways.

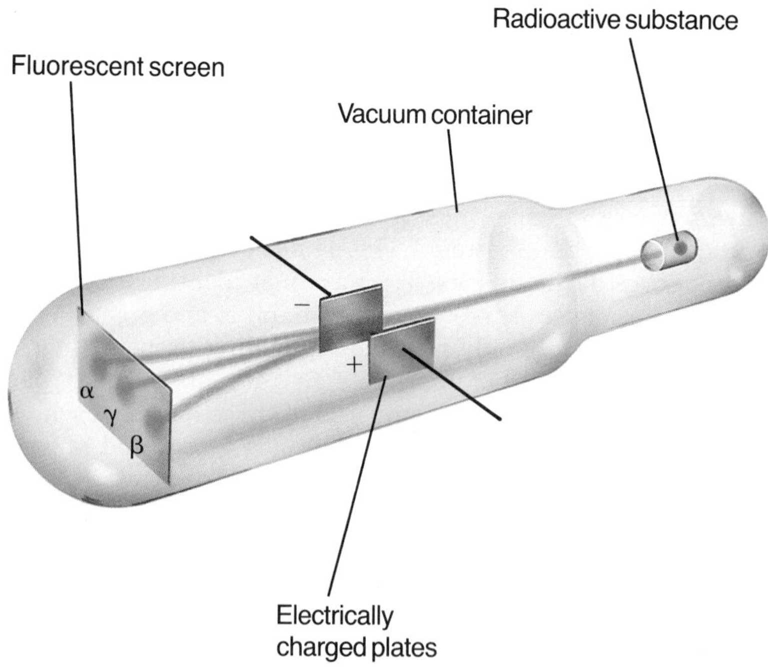

Figure 4.11 Alpha, beta, and gamma emissions behave differently in an electric field. Alpha and beta particles are deflected because of their charge.

Bridge to *HEALTH*

Ultraviolet Radiation

When ultraviolet light of wavelength less than 193 nm strikes an oxygen molecule, the molecule will form two oxygen atoms.

$$O_2 + h\nu \rightarrow 2O$$

These oxygen atoms, in turn, will react with other oxygen molecules to form ozone.

$$O + O_2 \rightarrow O_3$$

This process takes place in the band of the atmosphere between 15 and 30 km above the surface of Earth.

The maximum ozone concentration is reached at an altitude of approximately 25 km. This layer of ozone in the atmosphere is of vital importance to the health of living organisms on Earth because it absorbs much of the ultraviolet light in the wavelength range 280-320 nm. This part of the ultraviolet spectrum is the energy that can produce a sunburn or skin cancer in humans.

This ultraviolet energy can also damage eyesight, and, if strong enough, can kill many microorganisms at the low end of the food chain. These organisms include many sea-living creatures called phytoplankton that produce much of the oxygen we breathe. The ultraviolet energy has a devastating effect on the DNA (deoxyribonucleic acid) molecules that control the function of living cells.

Exploring Further

1. Chlorine atoms from certain refrigerants and aerosol propellants destroy ozone in the atmosphere. What steps can be taken to prevent this from happening?
2. Explain why all atmospheric oxygen does not change to ozone.

The Rutherford-Bohr Atom

Electrons are negatively charged and attracted to the positive nucleus. What prevents the electrons from being pulled into the nucleus? The discussion of this question, led by Rutherford and Bohr, resulted in a new idea. They thought of electrons as being in "orbit" around the nucleus in much the same manner as Earth is in orbit around the sun. They suggested that the relationship between the electrons and the nucleus is similar to that between the planets and the sun. The Rutherford-Bohr model of the atom is sometimes called the planetary atomic model. Thus, according to the planetary model, the hydrogen atom should be similar to a solar system consisting of a sun and one planet.

In order to improve his description of atomic structure, Bohr used the experimental evidence of atoms exposed to radiant energy. When a substance is exposed to a certain intensity of light or some other form of energy, the atoms absorb some of the energy. Such atoms are said to be excited. When atoms and molecules are in an excited state, unique energy changes occur that can be used to identify the atom or molecule. Radiant energy of several different types can be emitted (given off) or absorbed (taken up) by excited atoms and molecules. The methods of studying substances that are exposed to some sort of exciting energy are called **spectroscopy.** A pattern of radiant energy studied in spectroscopy is called a **spectrum.**

Visible light is one form of radiant or **electromagnetic energy.** Other forms of electromagnetic energy are radio, infrared, ultraviolet, and X ray. This energy consists of variation in electric and magnetic fields taking place in a regular, repeating fashion. If we plot the strength of the variation against time, our graph shows "waves" of energy. The number of wave peaks that occur in a unit of time is called the **frequency** of the wave. Frequency is represented by the Greek letter *nu* (ν) and is measured in units of **hertz** (Hz). A hertz is one peak, or cycle, per second. All electromagnetic energy travels at the speed of light. The speed of light is 3.00×10^8 m/s in a vacuum and is represented by the symbol c. Another important characteristic of waves is the distance between peaks. This distance between peaks is called the **wavelength** and is represented by the Greek letter *lambda* (λ). These characteristics of waves are related by the statement

$$c = \lambda\nu$$

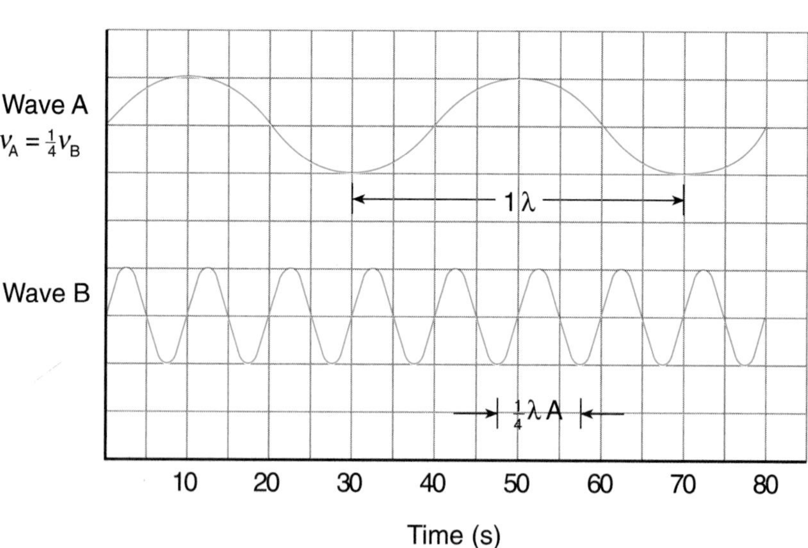

Figure 4.12 Wave A has the longer wavelength; however, wave B has the larger frequency. Short electromagnetic waves have higher frequencies than do long electromagnetic waves.

Another wave property that is of importance is the amplitude of a wave, or its maximum displacement from zero. In Figure 4.13, two waves are plotted on the same axes. Note that the amplitude of wave A is twice that of wave B, even though they have the same wavelength.

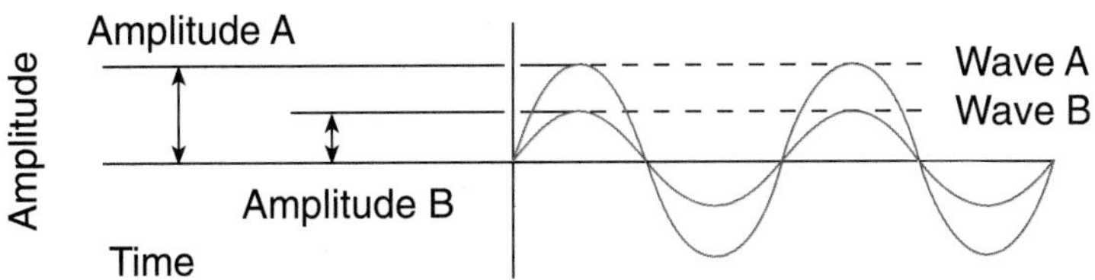

Figure 4.13 Waves A and B have the same wavelength, even though their amplitudes differ.

Excited atoms soon lose the energy they have gained. The energy emitted by gaseous atoms occurs at specific points (or lines) in a spectrum. The lines in an emission spectrum as seen in Figure 4.16 are characteristic of the element being excited. If these same atoms are exposed to light of all wavelengths, they will absorb energy. If this light is examined after it passes through the gaseous form of an element, some of the incident light will be missing. This missing light is absorbed by the gaseous atoms. The spectrum of the light leaving the gaseous atoms has lines missing. Such a spectrum is seen in Figure 4.14 and is called an absorption spectrum. These lines will be at the same wavelengths as the bright lines in the emission spectrum. The collection of lines, absorption or emission, for any element is the spectrum of that element and is unique. Spectroscopy can, therefore, be used as a means of identifying elements. Absorption and emission spectra are the fingerprints of the elements.

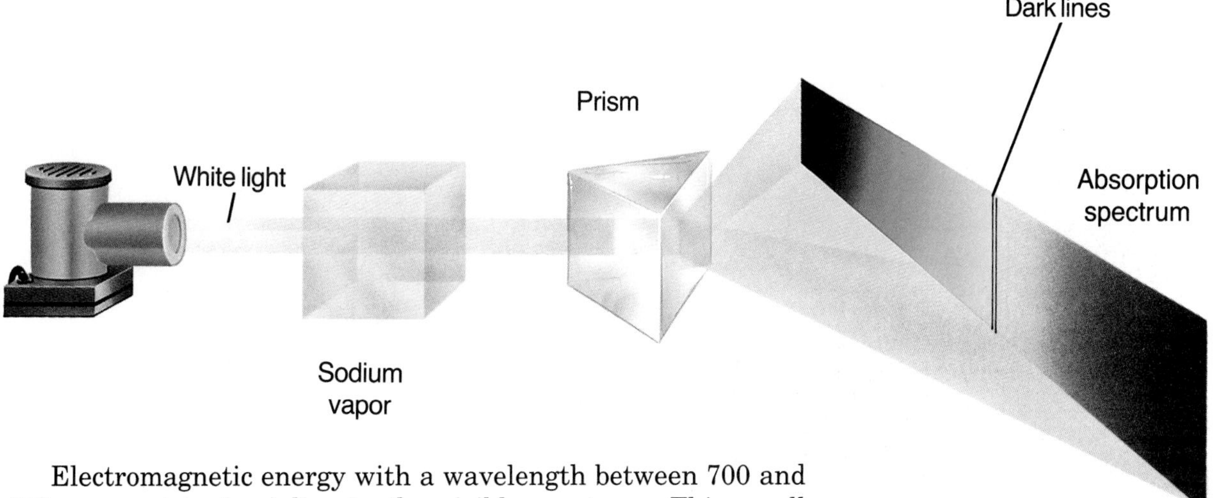

Figure 4.14 This apparatus is used to produce the absorption spectrum of sodium.

Electromagnetic energy with a wavelength between 700 and 400 nanometers (nm) lies in the visible spectrum. This small band of visible radiation has given chemists and physicists much information about the elements. Some elements (rubidium, cesium, helium, and hafnium) were discovered only after their spectra were observed. The visible spectrum may also be used for finding the concentration of substances, and for analyzing mixtures. Almost any change involving color can be measured using visible spectroscopy.

As is shown in Figure 4.15, the electromagnetic spectrum includes wavelengths much longer and wavelengths much shorter than those of visible light. Radio waves have the longest wavelengths, gamma rays the shortest. Ultraviolet radiation (UV) has wavelengths between 400 and 200 nm. Like visible light, ultraviolet radiation can be used to study atomic and molecular structure. Both ultraviolet spectra and visible spectra are produced by changes in the energy states of electrons. The ultraviolet spectrum of an element or compound consists of bands rather than lines. Ultraviolet radiation has such high energy, it violently excites the electrons. The transition of the electrons from the normal state to such highly excited states causes changes in the molecule being studied. Bonds between atoms may even be broken. Visible radiations are not as destructive because they have less energy than ultraviolet radiation has. Ultraviolet and visible spectroscopy are used for the same types of analyses. In order to describe completely the electronic structure of a substance, both types of analysis must be used. Infrared spectroscopy is also useful to the chemist when studying whole molecules. This type of spectroscopy will be discussed in Chapter 12.

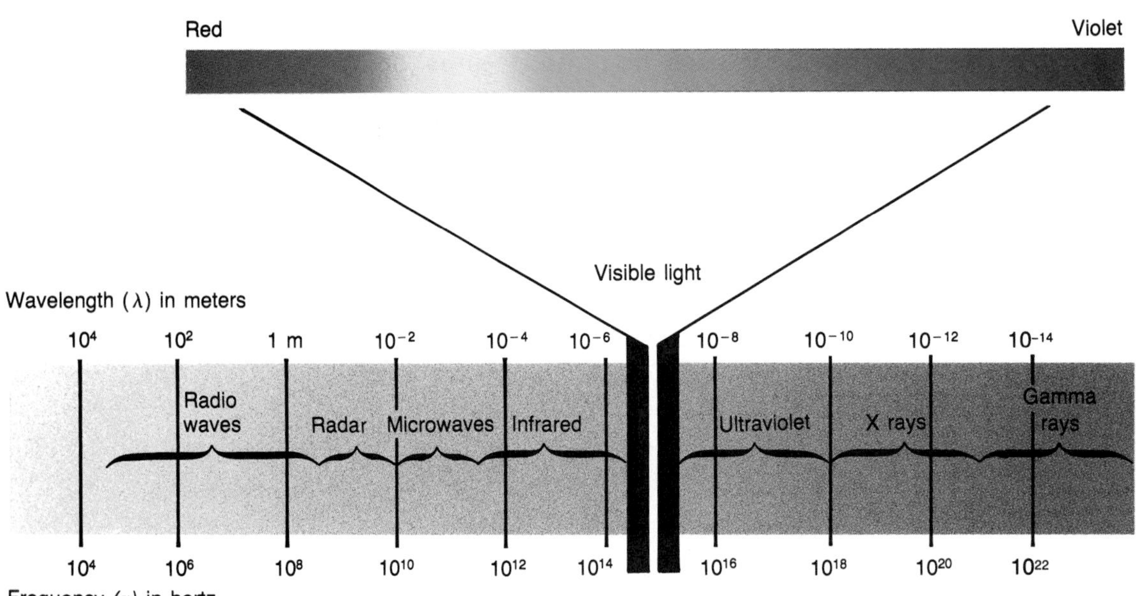

Figure 4.15 Visible light is only a small part of the electromagnetic spectrum. Note that the waves with high frequency have short wavelengths.

Planck's Hypothesis

In his attempt to explain the hydrogen spectrum, Bohr developed his planetary model of the atom. Bohr used the **quantum theory,** a theory of energy emission that had been stated by a German physicist, Max Planck. Planck assumed that energy, instead of being given off continuously, is given off in little packets, or **quanta.** Quanta of radiant energy are often called **photons.** He further stated that the amount of energy given off is directly related to the frequency of the light emitted.

Planck's idea was that one quantum of energy (light) was related to the frequency by the equation $E = h\nu$, where h is a constant. The constant is known as Planck's constant. Its value is $6.626\ 075\ 5 \times 10^{-34}$ joules per hertz.

The Hydrogen Atom and Quantum Theory

Planck's hypothesis stated that energy is given off in quanta instead of continuously. Bohr pointed out that the absorption of light by hydrogen at definite wavelengths corresponds to definite changes in the energy of the electron. He reasoned that the orbits of the electrons surrounding a nucleus must have a definite diameter. Furthermore, electrons could occupy only certain orbits. The only orbits allowed were those whose differences in energy equaled the energy absorbed when the atom was excited. Bohr thought that the electrons in an atom could absorb or emit energy only in whole numbers of photons. In other words, an electron could emit energy in one quantum or two quanta, but not in 1¼ or 3½ quanta.

Bohr pictured the hydrogen atom as an electron circling a nucleus at a distance of about 53 pm. He also imagined that this electron could absorb a quantum of energy and move to a larger orbit. Because a quantum represents a certain amount of energy, the next orbit must be some definite distance away from the first. If still more energy is added to the electron, it moves into a still larger orbit, and it continues to move to larger orbits as more energy is added. When an electron drops from a larger orbit to a smaller one, energy is emitted. Because these orbits represent definite energy levels, a definite amount of energy is radiated.

The size of the smallest orbit an electron can occupy, the one closest to the nucleus, can be calculated. This smallest orbit is called the **ground state** of the electron. Bohr calculated the

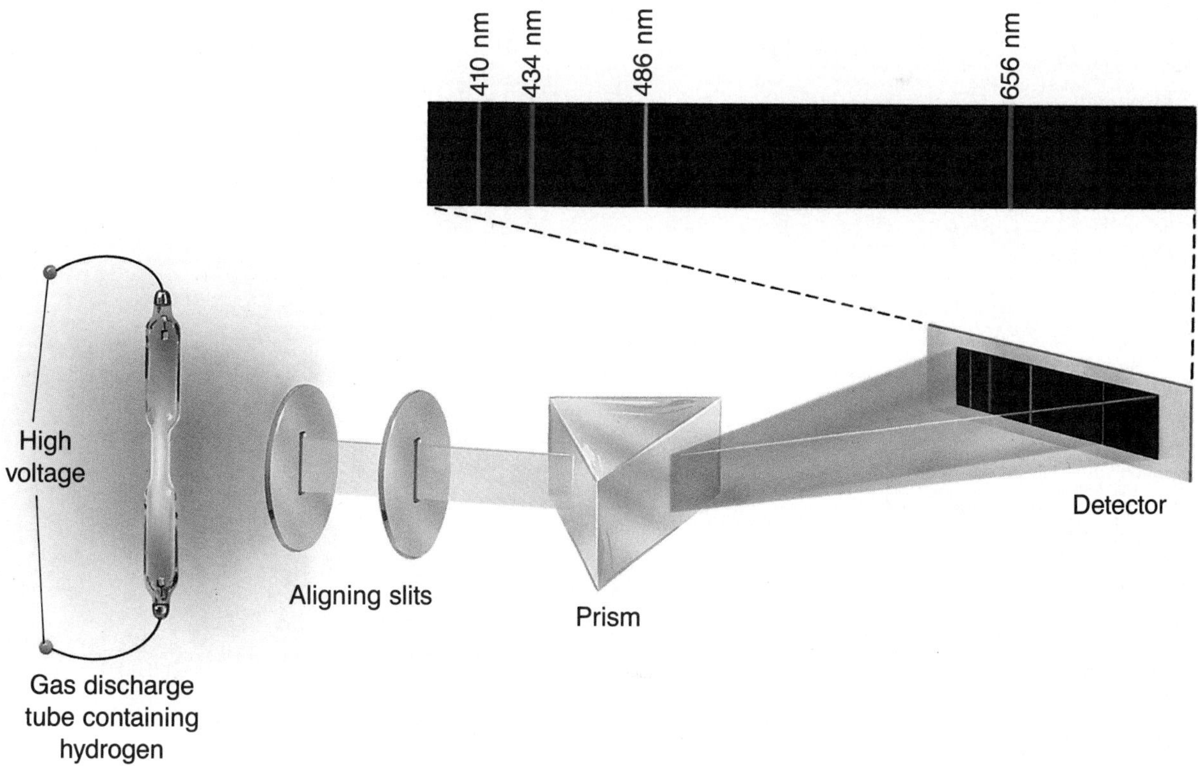

Figure 4.16 A prism spectroscope can be used to observe emission spectra. The visible region of the emission spectrum of hydrogen is a series of discrete lines.

ground state of the hydrogen electron. Using the quantum theory, he calculated the frequencies for the lines that should appear in the hydrogen spectrum. His results agreed almost perfectly with the actual hydrogen spectrum. Although today we use a model of the atom that differs from the Bohr model, many aspects of his theory are still retained. The major difference is that electrons do not move around the nucleus as the planets orbit the sun. We will explore this difference in the next chapter. However, the idea of energy levels is still the basis of atomic theory. The energy level values calculated by Bohr for the hydrogen atom are still basically correct.

We can summarize the relationship between electromagnetic energy and an electron as follows: An electromagnetic wave of a certain frequency has only one possible wavelength, given by $\lambda = c/\nu$. It has only one possible amount of energy, given by $E = h\nu$. Since both c and h are constants, if frequency, wavelength, or energy is known, we can calculate the other two.

The visible and ultraviolet spectra produced by a compound can be used to determine the elements in the compound. Each line in a spectrum represents one frequency of light. Because the velocity of light is always constant, each frequency is associated with a certain energy. This energy is determined by the movement of electrons between energy levels that are specific for each element. The same set of energy levels will always produce the same spectrum.

Figure 4.17 The lasers used in this colorful light show emit colors of light that are characteristic of the emission spectra of the elements used. The colors of neon signs also result from emission spectra.

CHEMISTRY AND

TECHNOLOGY

Photoelectric Effect

Many stores have doors that open automatically. Some of these doors are actuated by a device whose operation depends on light. In front of the door on one side is a source of light. Opposite the source of light is a light detector. The beam of light falling on the detector causes a substance in the detector to give off electrons and an electric current to flow in a circuit. This emission of electrons is called the photoelectric effect. When you break the beam of light by stepping between the light source and the detector, the detector stops emitting electrons. Thus, the flow of electric current stops. This interruption in the flow of electric current actuates a mechanism that opens the door.

Einstein received a Nobel Prize in 1921 for his explanation of the photoelectric effect. It had been known for some time that light falling on the surface of certain substances caused electrons to be emitted. There is, however, a puzzling fact about this change. When the intensity of light (the number of photons per unit time) is reduced, the electrons emitted still have the same energy. However, fewer electrons are emitted. Einstein pointed out that Planck's hypothesis explains this observation.

A certain amount of energy is needed to remove an electron from the surface of a substance. If a photon of greater energy strikes the electron, the electron will also move away from the surface. Since it is in motion, the electron has some kinetic energy. Some of the energy of the photon is used to free the electron from the surface. The remainder of the energy becomes the kinetic energy of the electron. If light of one frequency is used, then the electrons escaping from the surface of the substance will all have the same energy.

If the light intensity is increased, but the frequency remains the same, the number of electrons being emitted will increase. If the frequency of the light is increased, the energy of the photon is increased. The amount of energy that must be used to free the electron from the atom is constant for a given substance. Thus, the electrons now leave the surface with a higher kinetic energy than they did with the lower frequency. Planck's hypothesis, together with Einstein's explanation, confirmed the particle nature of light.

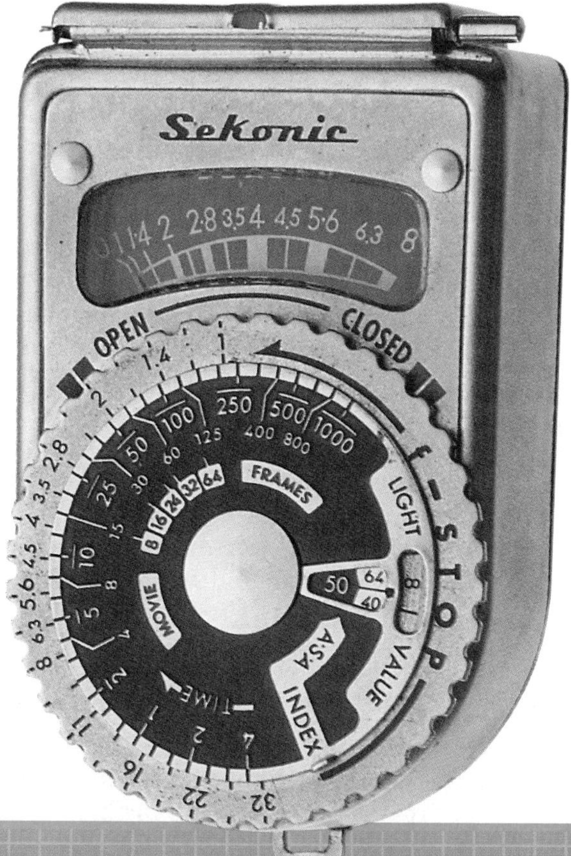

Thinking Critically

1. Explain how the photoelectric effect can be used in construction of a security system.
2. How is the photoelectric effect used in a photographic exposure meter?

Figure 4.18 This rhinoceros and bird have about the same mass ratio as that of a proton and an electron.

Atomic Mass

The proton and neutron are essentially equal in mass. The mass of the electron is extremely small, so almost all of the mass of an atom is located in the nucleus. Even the simplest atom, which contains only one proton and one electron, has $^{1836}\!/_{1837}$ of its mass in the nucleus. In other atoms that have neutrons in the nucleus, an even higher fraction of the total mass of the atom is in the nucleus.

It is possible to discuss the mass of a single atom. However, chemists use the masses of large groups of atoms. They do so because of the very small size of the particles in the atom. Chemists measure the mass of one atom in atomic mass units (u). You know that the unit gram was defined as $^{1}\!/_{1000}$ the mass of the International Prototype Kilogram. What is an atomic mass unit? To measure atomic masses, an atom of one element was chosen as a standard, and the other elements were compared with it. Scientists used a carbon nuclide with mass number 12 as the standard for the atomic mass scale. This carbon nuclide, carbon-12, has 6 protons and 6 neutrons in the nucleus. *One carbon-12 atom is defined as having a mass of 12 atomic mass units.* An atomic mass unit is defined to be $^{1}\!/_{12}$ the mass of the carbon-12 nuclide.

The most important subatomic particles you have studied thus far have the following masses.

$$\begin{aligned} \text{electron} &= 9.109\,53 \times 10^{-28}\text{ g} = 0.000\,549\text{ u} \\ \text{proton} &= 1.672\,65 \times 10^{-24}\text{ g} = 1.0073\text{ u} \\ \text{neutron} &= 1.674\,95 \times 10^{-24}\text{ g} = 1.0087\text{ u} \end{aligned}$$

Look carefully at Table 4.2. Notice that many of the elements have a mass in atomic mass units that is close to the total number of protons and neutrons in their nuclei. However, some do not.

What causes the mass of chlorine or copper, for example, to be about halfway between whole numbers? The numbers in the table are based on the "average atom" of an element.

Most elements have many isotopic forms that occur naturally. It is difficult and costly to collect a large amount of a single nuclide of an element. Thus, for most calculations, the average atomic mass of the element is used.

Average Atomic Mass

Using a standard nuclide, there are two ways of determining masses for atoms of other elements. One method is by reacting the standard element with the element to be determined. Using accurate masses of the two elements and a known ratio, the atomic mass of the second element may be calculated.

Higher accuracy can be obtained by using a physical method of measurement in a device called a mass spectrometer. The mass spectrometer has many uses. Geologists, biologists, petroleum chemists, and many other research workers use the mass spectrometer as an analytical tool. Its development was based on the design of the early tubes J. J. Thomson used to find the charge/mass ratio of the electron.

Using a mass spectrometer, we can determine the relative amounts and masses of the nuclides for all isotopes of an element. The element sample, which is in gaseous form, enters a chamber where it is charged by a beam of electrons. These charged particles are then propelled by electric and magnetic fields. As in the Thomson tube, the fields bend the path of the charged particles as shown in Figure 4.19. The paths of the heavier particles are bent slightly as they pass through the fields. The paths of the lighter particles are curved more. Thus, the paths of the particles are separated by relative mass. The particles are both caught and

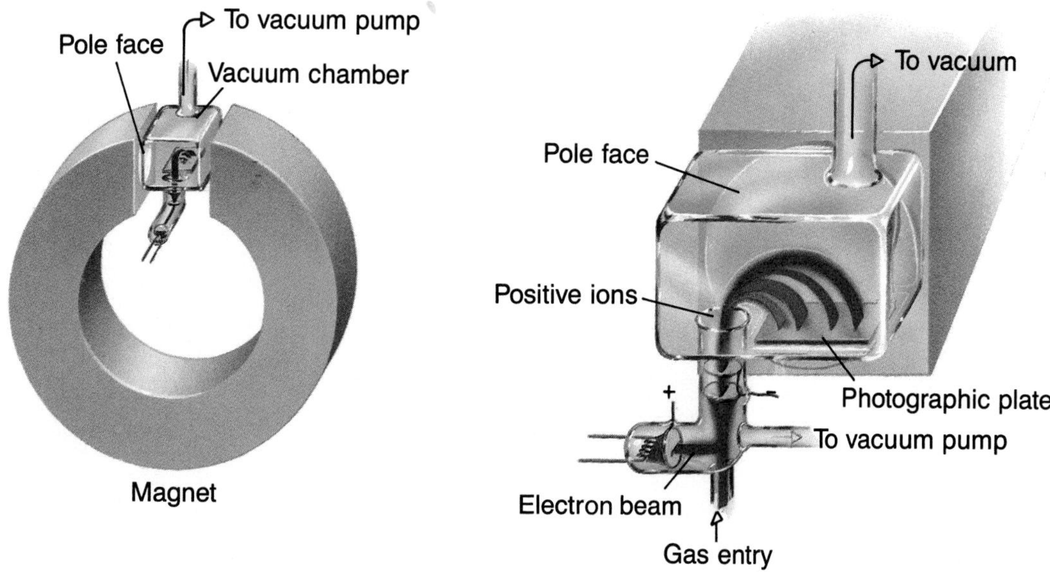

Figure 4.19 The mass spectrometer is used extensively as an analytical tool. Inside the spectrometer, a magnet (left) causes the positive ions to be deflected according to their mass. In the vacuum chamber (right), the process is recorded on a photographic plate or a solid-state detector.

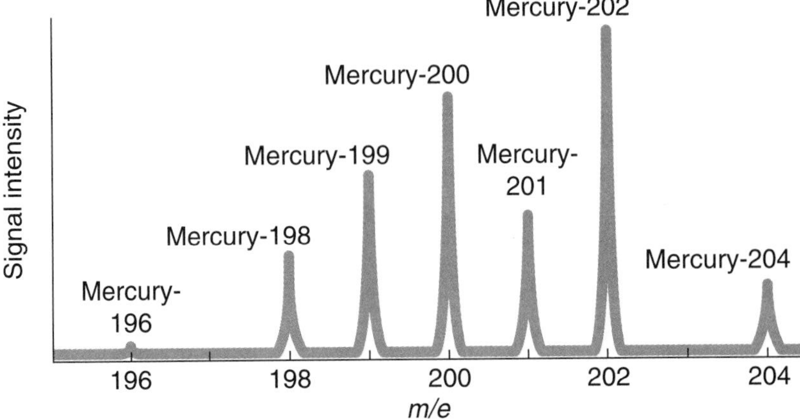

Figure 4.20 This readout from a mass spectrometer shows the different isotopes of mercury. Mercury-202 is the most abundant isotope.

recorded electronically. One drawback of the instrument is that the charging chamber, field tube, and detection device must all be in a vacuum. The vacuum containing these devices must be equal to about one hundred-millionth (1/100 000 000) of normal atmospheric pressure. An example of a readout of an analysis by mass spectroscopy can be seen in Figure 4.20.

Because the strength of the magnetic and electric fields as well as the speeds and paths of the particles are known, the mass of the particles can be calculated. Once the masses of the isotopes and their relative amounts have been found, the average atomic mass can be calculated. This average mass is called the **atomic mass** of an element.

Not all isotopes of an element are present in the same amount. Thus, in finding average atomic masses we use what is called a weighted average. A similar process is used by a teacher in finding the class average from a set of test grades. For example, in a class of 24 students, suppose that five students earn a grade of 100% on a quiz while the other 19 students receive a grade of 80%. Is the class average 90%? No, the average is closer to 80% because there were many more students with a grade of 80% than with a grade of 100%. To get the weighted average, the teacher would add five grades of 100% and 19 grades of 80%, and

Figure 4.21 Materials containing magnesium are used in many items that need to be lightweight and strong. Magnesium contains 78.99% $^{24}_{12}Mg$ atoms, 10.00% $^{25}_{12}Mg$ atoms, and 11.01% $^{26}_{12}Mg$ atoms, resulting in a weighted average atomic mass of 24.305 u.

then would divide that sum by 24, the total number of students in the class.

$$\text{Weighted grade average} = \frac{(5 \text{ \sout{students}} \times 100\%) + (19 \text{ \sout{students}} \times 80\%)}{24 \text{ \sout{students}}}$$
$$= 84\%$$

In finding average atomic masses, a similar calculation can be done. First, multiply the mass of each isotope by its abundance. The resulting products are then added and the sum is divided by the total abundance to get the weighted average.

SAMPLE PROBLEM

Average Atomic Mass

Neon has two isotopes. Neon-20 has a mass of 19.992 u and neon-22 has a mass of 21.991 u. In an average sample of 100 neon atoms, 90 will be neon-20 and 10 will be neon-22. Calculate the average atomic mass of neon.

Solving Process:

Each of the isotopic masses is multiplied by its abundance. The products are added, and the sum is divided by the total abundance.

$$\text{Average atomic mass of neon} = \frac{(90 \times 19.992 \text{ u}) + (10 \times 21.991 \text{ u})}{100}$$
$$= 20.192 \text{ u}$$

PROBLEM SOLVING HINT

Remember that 90 out of 100 gives the same relationship as 90% out of 100%.

PRACTICE PROBLEMS

12. What is the atomic mass of hafnium if, out of every 100 atoms, 5 have mass 176, 19 have mass 177, 27 have mass 178, 14 have mass 179, and 35 have mass 180.0?
13. What is the atomic mass of silicon if 92.21% of its atoms have mass 27.977 u, 4.70% have mass 28.976 u, and 3.09% have mass 29.974 u?

4.2 CONCEPT REVIEW

14. How do hadrons and leptons differ? baryons and mesons? quarks and gluons?
15. Explain how absorption and emission spectra can be used to identify elements.
16. How do alpha, beta, and gamma radiation differ?
17. A quantum of light has a wavelength of 5.00×10^2 nm. Another quantum has a wavelength of 7.00×10^2 nm. Which quantum has the greater frequency? Justify your answer.
18. **Apply** An element occurs naturally in two isotopes with atomic masses of 187 and 188. The nuclide of mass 187 is radioactive, decaying by emitting a beta particle. What happens to the average atomic mass of the element over time?

Chapter 4
REVIEW

Summary

4.1 Early Atomic Models

1. Democritus proposed the earliest recorded atomic theory. He believed that matter is composed of tiny particles, or atoms.
2. Before Dalton proposed the atomic theory, Lavoisier stated the law of conservation of mass—matter is not created or destroyed in ordinary chemical reactions. Proust's law of definite proportions stated that substances always contain elements in the same ratio by mass.
3. John Dalton used the law of conservation of mass and the law of definite proportions to state that all matter is formed of indivisible particles called atoms; all atoms of one element are the same; atoms of different elements are unlike; and atoms can unite with one another in simple whole-number ratios.
4. Modern atomic theory differs from Dalton's atomic theory due to the discovery of subatomic particles and isotopes.
5. An electron is a negatively charged particle with a very small mass. A proton is a positively charged particle with a mass 1836 times that of an electron. A neutron is an uncharged particle with a mass about the same as that of a proton.
6. The atomic number, Z, of an atom is the number of protons in its nucleus.
7. All atoms of an element contain the same number of protons in their nuclei. Atoms of the same element having different numbers of neutrons are isotopes.
8. The mass number, A, of an atom is the number of particles in its nucleus.
9. From experiments, especially those of Geiger and Marsden, Rutherford concluded that atoms consist mostly of empty space and have a small nucleus.
10. Radioactivity is particle or energy emission due to nuclear disintegration.
11. Radioactive decay is spontaneous; that is, it cannot be controlled.
12. Einstein proposed that matter and energy are equivalent.

4.2 Parts of the Atom

13. Subatomic particles are either elementary particles called leptons or complex particles called hadrons.
14. Hadrons are believed to be made of particles called quarks. Quarks and hadrons are thought to be held together by exchanging gluons and pions, respectively.
15. Naturally radioactive nuclides emit three kinds of radiation: alpha (helium nuclei), beta (electrons), and gamma (energy).
16. Many subatomic particles in addition to electrons and nucleons have been discovered. Some of these are antimatter.
17. Rutherford and Bohr pictured the atom as consisting of a central nucleus surrounded by electrons in orbits.
18. Atoms excited by energy both absorb and emit definite wavelengths of electromagnetic radiation. The unique collection of lines, absorption or emission, for any element is the spectrum of that element. Visible and ultraviolet spectroscopy are used to study structure.
19. Planck stated that energy is radiated in discrete units called quanta. A photon is a quantum of light energy. The energy of a quantum of radiation varies directly as the frequency of the radiation ($E = h\nu$).
20. Bohr theorized that the energies in the hydrogen spectrum corresponded to certain quanta emitted or absorbed when the electron moved from one orbit to another. Thus, he was able to calculate the orbits for the hydrogen atom.
21. The atomic mass of an element is the weighted average mass of all its natural isotopes compared with $\frac{1}{12}$ the mass of the carbon-12 atom. Atomic mass can be measured with the mass spectrometer.

Key Terms

law of definite proportions
law of multiple proportions
anode
cathode
cathode ray
isotope
atomic number
nuclide
nucleon
mass number
radioactivity
nuclear force
subatomic particle
lepton
hadron
antiparticle
neutrino
quark
baryon
meson
gluon
alpha particle
beta particle
gamma ray
spectroscopy
spectrum
electromagnetic energy
frequency
hertz
wavelength
quantum theory
quanta
photon
ground state
atomic mass

Review and Practice

19. What did each of the following scientists contribute in forming the modern model of the atom?

a. Dalton
b. Thomson
c. Rutherford
d. Chadwick
e. Proust
f. Democritus
g. Millikan
h. Einstein
i. Moseley
j. Bohr
k. Planck
l. Avogadro
m. Lavoisier
n. Gay-Lussac
o. Becquerel
p. Geiger and Marsden

20. What are the major points in Dalton's atomic theory?

21. State Avogadro's hypothesis.

22. What are cathode rays? Why are they called cathode rays?

23. What are the differences in charge and mass among protons, neutrons, and electrons?

24. What nuclide is used as the reference standard in defining the atomic mass unit?

25. A particular atom of potassium contains 19 protons, 19 electrons, and 20 neutrons. What is the atomic number of this atom? What is its mass number? Write the symbol for this potassium nucleus.

26. How many electrons, neutrons, and protons are in atoms of chlorine with mass number 35? How many of each are in the atoms of thorium with mass number 232?

27. Yttrium was discovered in 1794. It is one of the elements used in superconductors. How many electrons, protons, and neutrons are in an atom of yttrium-88?

28. Compare the amount of energy involved in chemical changes to the amount of energy resulting from nuclear changes.

29. What are the two broad classes of subatomic particles? What are the differences between these two classes? How do protons, neutrons, and electrons fall into these classes?

30. What are the differences among the three types of natural radiation?

31. How many neutrons and protons are in each of the following nuclides?

a. carbon-14
b. phosphorus-32
c. nickel-63
d. iridium-192
e. iron-54
f. neptunium-235

32. How does a mass spectrometer separate different types of atoms?

33. Find the average atomic mass of silver if 51.83% of the silver atoms occurring in nature have mass 106.905 u and 48.17% of the atoms have mass 108.905 u.

34. Find the average atomic mass of krypton if the relative amounts are as follows.

Isotopic mass	Percentage
77.920 u	0.350
79.916 u	2.27
81.913 u	11.56
82.914 u	11.55
83.912 u	56.90
85.911 u	17.37

Chapter 4

REVIEW

Concept Mastery

35. How did discovery of subatomic particles and isotopes affect Dalton's theories?

36. What was Rutherford's interpretation of the Geiger-Marsden experiment?

37. Describe the force that holds the particles of an atom's nucleus together. Does electrostatic force help or oppose the effect of this force?

38. Describe the way that excited atoms produce (a) an emission spectrum and (b) an absorption spectrum.

39. How did Bohr use spectroscopic data to formulate his model of the atom?

40. According to Bohr, what happened when an electron absorbed a photon?

41. How are the law of conservation of mass and the law of definite proportions explained by Dalton's atomic theory?

42. Why is it necessary to ionize an element sample before it can be separated into its isotopic components with a mass spectrometer?

43. Studies with gas discharge tubes had been conducted for more than 70 years before J. J. Thomson discovered the true nature of the cathode rays produced in these tubes. It was Thomson who ultimately proved that these rays consist of elementary negatively charged particles that are identical regardless of the element from which they are produced. He called these negative particles electrons. With the information that the cathode rays are actually a stream of electrons that are negatively charged, answer the following questions.

a. To which terminal of the electric current source should the cathode of the gas discharge tube be connected?

b. Why do electrons move toward the anode?

c. Why do you think the cathode rays produce a colored glow in the tube?

Application

44. What is the energy of a quantum of light of frequency 4.31×10^{14} Hz?

45. A certain violet light has a wavelength of 413 nm. What is the frequency of the light? The velocity of light is equal to 3.00×10^8 m/s.

46. A certain green light has a frequency of 6.26×10^{14} Hz. What is its wavelength?

47. What is the energy content of one quantum of the light in Problem 45?

48. What is the energy content of one quantum of the light in Problem 46?

49. Hydrazine, N_2H_4, is a fuming, corrosive liquid used in rocket and jet fuels. Ammonia, NH_3, is a gas that dissolves in water to form a solution that can be used as a cleaning agent. How do hydrazine and ammonia illustrate the law of multiple proportions?

50. Explain why people need protection from ultraviolet radiation, particularly short-wavelength ultraviolet radiation.

51. What is the energy of light with wavelength 662 nm? First find the frequency in hertz of this wavelength of light.

52. Photoelectric devices rely on the ability of light to remove electrons from the surface of some substances. The energy required to release an electron from atoms on the surface of a certain substance is 3.60×10^{-19} J. What wavelength of light would be necessary to cause electrons to leave the surface of this substance?

53. How is an element's emission spectrum related to its absorption spectrum?

54. How could you demonstrate and explain the law of conservation of mass with a burning candle?

55. Design an experiment to demonstrate the law of definite proportions.

Critical Thinking

56. What effect did the discovery of radioactivity and Einstein's explanation have on the law of conservation of mass?

57. Suppose calcium-40 had been used as the standard for atomic mass and had been defined as having an atomic mass of 1.00 × 10^2 u. How would this standard change the periodic table as we know it?

58. According to the standard stated in Problem 57, what would be the atomic mass of the atom we know as carbon-12?

59. If Geiger and Marsden had used lithium instead of gold, would they have come to the same conclusion? Explain.

60. The element boron has two naturally occurring isotopes. One has atomic mass 10.0129 u, while the other has atomic mass 11.0093 u. The average atomic mass of boron is 10.811 u. Determine the percentage occurrence of each isotope.

61. The picture tube in a television set is a cathode-ray tube. A heated cathode in the neck of the tube produces a beam of electrons. (Some color tubes emit three beams.) Attached to the tube are coils of wire that form electromagnets. The magnetic field produced by these magnets can be varied by varying the current flowing through the coils. Explain the function of these electromagnets. Use reference books to find out how an electron beam forms a television picture. In some electronic instruments, such as oscilloscopes, electrically charged plates inside the tube are used instead of electromagnets. Explain why these can produce results similar to those produced by magnets.

Readings

Crawford, Mark. "Racing after the Z Particle." *Science* 241, no. 4869 (August 26, 1988): 1031-1032.

Fisher, A. "Hunting Neutrinos." *Popular Science* 232, no. 5 (May 1988): 72-74, 115.

Rayner-Canham, Geoffrey W., and Marelene F. Rayner-Canham. "The Shell Model of the Nucleus." *The Science Teacher* 54, no. 1 (January 1987): 19-21.

Thomsen, Dietrick E. "Seeking Neutrinos Under the Ocean." *Science News* 133, no. 16 (April 16, 1988): 246.

Weisburd, Stefi. "All Charged Up for the Positron Microscope." *Science News* 133, no. 8 (February 20, 1988): 124-125.

Cumulative Review

1. Classify each object or material as a heterogeneous mixture, solution, compound, or element.

a. plastic garbage bag
b. automobile
c. seawater
d. helium gas
e. maple syrup
f. newsprint
g. milk
h. air
i. diamond
j. soda pop
k. paper
l. blood

2. Calculate the density of a material that has a mass of 7.13 grams and occupies a volume of 7.77 cm^3.

3. What is the number of significant digits in each of the following measurements?

a. 0.558 g
b. 7.3 m
c. 410 cm
d. 0.0094 mg
e. 19.000 g
f. 75.0 s

4. What is the difference between a physical change and a chemical change?

5. What are the seven SI base units?

6. A 50.0-g piece of iron is heated to 150°C. It is placed in 100.0 cm^3 of water at 0°C. What will be the final temperature of the water?

CHAPTER **Preview**

CONTENTS

Electron Clouds and Probability

CHAPTER

5

Often, we consider chance to be the same as luck or fate, and it seems as though we have no control over chances in our lives. Consider a game spinner, such as the one pictured. If you spin the spinner would the results be entirely random, or can you determine what chance there is of a desired outcome?

"Chance" is a limited possibility of achieving a desired outcome. For instance, if a game spinner has 10 numbers on it, a person who has chosen a particular number has one chance in 10 of winning. The probability of winning is 1/10. Chemists and physicists usually have to deal with the location of electrons in terms of the chances, or probability, of finding the electron at a particular location. Current ideas about the electronic structure of the atom are best incorporated in a mental model that will lay the groundwork for understanding chemical changes that are likely to take place.

EXTENSIONS AND CONNECTIONS

Careers in Chemistry

Electron Microscopist, page 111

Psychology Connection

Uncertainty and Measurement, page 112

Mathematics Connection

Probability and Statistics, page 114

Everyday Chemistry

Fireworks: A look at how colored lights in fireworks relate to the energy of electrons, **page 118**

Chemists and Their Work

Wolfgang Pauli, page 122

Bridge to Astronomy

Spectra of Stars: How scientists determine composition and evolution of stars by examining their spectra, **page 125**

Chemistry and Technology

Thin-Layer Magnetism: Making smaller, stronger magnets, **page 126**

OBJECTIVES

- describe the wave-mechanical view of the hydrogen atom.
- characterize the position and velocity of an electron in an atom.
- describe an electron cloud.

5.1 Modern Atomic Structure

In the attempts to refine our model of the atom, you will see that the division between matter and energy has become less clear. Radiant energy is found to have many properties of particles. Small particles of matter are found to display the characteristics of wave motion. The purpose of this section is to look more closely at this wave-particle problem.

The de Broglie Hypothesis

You saw in Chapter 4 that the frequencies predicted by Bohr for the hydrogen spectrum are "essentially" correct. Note that we did not use the word "exactly." Improved instrumentation has shown that the hydrogen spectral lines predicted by Bohr are not single lines. With a better spectroscope, you see instead that there are several lines closely spaced. To account for this "fine structure" of the hydrogen spectrum, scientists again had to refine their model of the atom.

In 1923, a French physicist, Louis de Broglie, proposed a hypothesis that led the way to the present theory of atomic structure. De Broglie knew of Planck's ideas concerning radiation being made of discrete amounts of energy called quanta. This theory seemed to give waves the properties of particles. De Broglie thought if Planck were correct, then it might be possible for particles to have some of the properties of waves.

Using Einstein's relationship between matter and energy

$$E = mc^2$$

and Planck's quantum theory

$$E = h\nu$$

de Broglie equated the two expressions for energy as follows.

$$mc^2 = h\nu$$

He then substituted $\boldsymbol{v}$*, a general velocity, for c, the velocity of light

$$m\boldsymbol{v}^2 = h\nu$$

and $\boldsymbol{v}/\lambda$ for ν because the frequency of a wave is equal to its velocity divided by its wavelength.

$$m\boldsymbol{v}^2 = \frac{h\boldsymbol{v}}{\lambda}$$

$$\lambda = \frac{h\boldsymbol{v}}{m\boldsymbol{v}^2} = \frac{h}{m\boldsymbol{v}}$$

This final expression enabled de Broglie to predict the wavelength of a particle of mass m and velocity $\boldsymbol{v}$. Within two years, de Broglie's hypothesis was proved correct. Scientists found by

*Symbols in boldface are vector quantities. Vectors have both magnitude and direction.

Figure 5.1 Diffraction is a wave property. In the electron microscope the diffraction of electrons is used to obtain high resolving power as shown in this electron micrograph of fatty acid molecules on a graphite base.

experiment that, in some ways, an electron stream acted in the same way as a ray of light. They further showed that the wavelength of the electrons was exactly that predicted by de Broglie.

The Apparent Contradiction

Waves can act as particles, and particles can act as waves. You saw how Bohr's atom model explained light in terms of particle properties. There are also properties of light that can be explained by wave behavior.

Like light, electrons also have properties of both waves and particles. However, one cannot observe both the particle and wave properties of an electron by the same experiment. If an experiment is done to show an electron's wave properties, the electron exhibits the behavior of a wave. Another experiment, carried out to show the electron as a particle, will show that the electron exhibits the behavior of a particle. The whole idea of the two-sided nature of waves and particles is referred to as the **wave-particle duality of nature.** The duality applies to all waves and all particles. Scientists are not always interested in duality. For example, when scientists study the motion of a space shuttle, wave characteristics do not enter into their study. They are only interested in the shuttle as a particle. However, with very small particles they cannot ignore wave properties. For an electron, a study of its wave characteristics can tell as much about its behavior as a study of its particle characteristics.

CAREERS IN CHEMISTRY

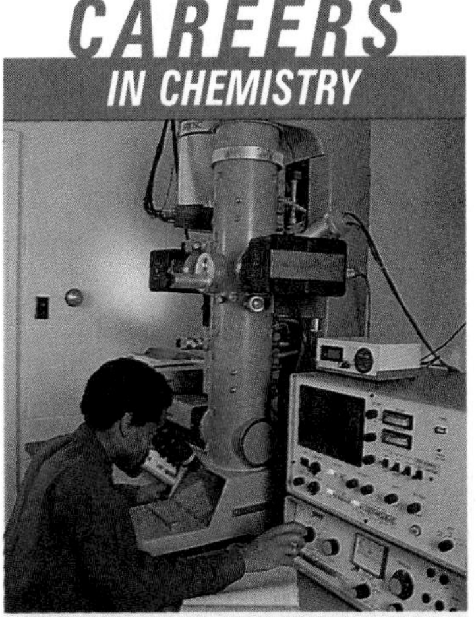

Electron Microscopist

Electron microscopists use the wave properties of electrons in electron microscopes to produce images of specimens too small to be viewed by ordinary microscopes. Beams of electrons are focused by electromagnetic fields to reproduce a well-defined image of a specimen that may be too small to be investigated by ordinary microscopy. One use of electron microscopy in chemistry is in the investigation of crystal structure. Because electron microscopy is most commonly used in biology, chemistry, and engineering, microscopists must have strong backgrounds in those areas, as well as physics, mathematics, and computer science.

CHEMISTRY IN DEPTH

Momentum

The product of the mass and velocity of an object is the **momentum** of the object. In equation form, $mv = p$, where m is the mass, v is the velocity, and p is the symbol for momentum. Note that velocity includes not only the speed but also, because it is a vector quantity, the direction of motion. Substituting momentum (p) for mv in the de Broglie equation, we can then write

$$\lambda = \frac{h}{p}$$

Note that the wavelength is inversely proportional to the momentum. The wave properties of all objects in motion are not always of interest. There is a basic difference between Newtonian mechanics and quantum mechanics. **Newtonian mechanics,** or classical mechanics, describes the behavior of visible objects at ordinary velocities. **Quantum mechanics** describes the behavior of extremely small particles at velocities near that of light.

To the chemist, the behavior of the electrons in an atom is of great interest. To be able to give a full description of an electron, you must know where it is and where it is going. In other words, you must know the electron's present position and its momentum. From the velocity and position of an electron at one time, we can calculate where the electron will be some time later.

Measuring Position and Momentum

Werner Heisenberg, a German scientist, further refined the ideas about atomic structure. He pointed out that *it is impossible to know both the exact position and the exact momentum of an object at the same time.* Let us take a closer look at Heisenberg's ideas. To locate the exact position of an electron, we must be able to "look" at it. When we look at an object large enough to see with our eyes, we actually see the light waves the object has reflected. When radar detects an object, the radar receiver is actually "seeing" the radar waves reflected by the object. In other words, for us to see an object, it must be hit by a photon of radiant energy. A photon hitting an airplane has negligible effect on the airplane. However, a collision between a photon and an electron results in a large change in the energy of the electron. Let us assume we have "seen" an electron, using some sort of radiant energy as "illumination." We have found the exact position of the electron. However, we would have little idea of the electron's velocity. The collision between it and the photons used to see it has caused its velocity to change. Thus, we would know the position of the electron, but not its velocity. On the other hand, if we measure an electron's velocity, we will change the electron's position. We would know the velocity fairly well, but not the position. Heisenberg stated that there is always some uncertainty about the position and momentum of an electron. This statement is known as the **Heisenberg uncertainty principle.**

PSYCHOLOGY CONNECTION

Uncertainty and Measurement

The effect of the observer on the observed as indicated in the Heisenberg uncertainty principle is also a problem in evaluation and measurement. This has led to the use of "unobtrusive" measures of assessment. An example of an unobtrusive measure is determining the wear patterns on the floor covering in front of certain exhibits at a museum. If the floor covering has to be replaced more often at one exhibit than at another, the first is a more popular exhibit—more feet have walked by it.

The uncertainty of the position and the uncertainty of the momentum of an electron are related by Planck's constant. If $\Delta \boldsymbol{p}$* is the uncertainty in the momentum and Δx is the uncertainty in the location, then

$$\Delta \boldsymbol{p} \Delta x \geq h$$

Because h is constant, $\Delta \boldsymbol{p}$ and Δx are inversely proportional to each other. Therefore, the more certain we are of the position of the electron, the less certain we are of its momentum. Conversely, the more certain we are of its momentum, the less certain we are of its position.

*Recall the Greek letter delta (Δ) means *change in*.

Schrödinger's Work

Chemists and physicists found themselves unable to adequately describe the exact structure of the atom. Heisenberg had, in effect, stated that the exact motion of an electron was unknown and could never be determined. Notice, however, that Heisenberg's principle of uncertainty treats the electron as a particle. What happens if a moving electron is treated as a wave? The wave nature of the electron was investigated by the Austrian physicist, Erwin Schrödinger.

Schrödinger treated the electron as a wave and developed a mathematical equation to describe its wave-like behavior. Schrödinger's equation related the amplitude of the electron-wave, ψ (psi), to any point in space around the nucleus. He pointed out that there is no physical meaning to the values of ψ. Terms for the total energy and for the potential energy of the electron are also part of this equation. In computing the total energy and the potential energy, certain numbers must be used. For example, the term for the total energy is

$$2\pi^2 me^4/h^2n^2$$

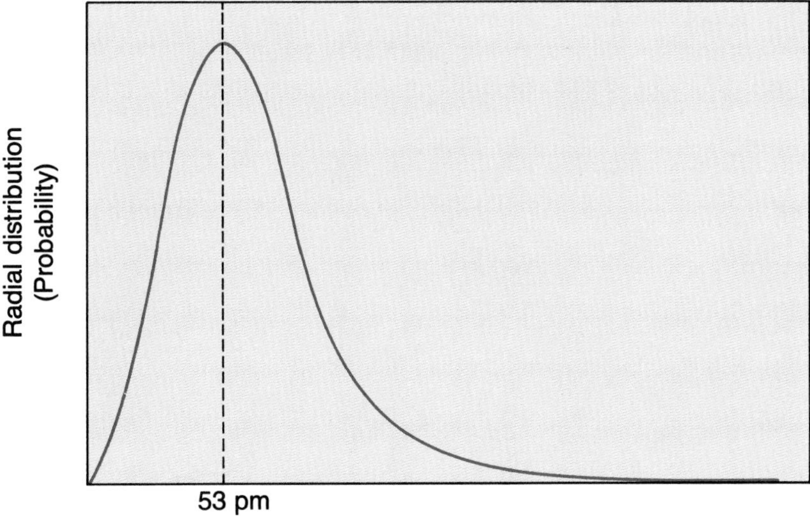

Figure 5.2 This graph shows the probability of finding the electron at a specific distance from a hydrogen nucleus. Bohr predicted the hydrogen electron to be approximately 53 pm from the nucleus. Born showed that 53 pm is merely the point of highest probability.

MATHEMATICS CONNECTION

Probability and Statistics

The study of probability and statistics is part of the branch of mathematics called discrete, or finite, mathematics. Statistics are used in many fields besides science. For example, marketing executives study buying and population statistics carefully in deciding on new products or where to market certain items.

In this expression, m is the mass of the electron, e is the charge on the electron, h is Planck's constant, and n can take positive whole number values. The symbol n represents the first of four **quantum numbers**, which are used in describing electron behavior. They will be studied in detail in later sections. The actual wave equation involves mathematics with which you are probably not familiar, so it will not be given.

The physical significance of all this mathematics was pointed out by Max Born. He worked with the square of the absolute value of the amplitude, $|\psi|^2$. He showed that $|\psi|^2$ gave the probability of finding the electron at the point in space for which the equation was solved. This **probability** is also the ratio between the number of times the electron is in that certain position and the total number of times it is at all possible positions. The higher the probability, the more likely the electron will be found in a given position.

Wave-Mechanical View of the Hydrogen Atom

Schrödinger's wave equation is used to determine the probability of finding the hydrogen electron in any given place. The probabilities can be computed for finding the electron at different points along a given line away from the nucleus. One point will have a higher probability than any other, as shown in Figure 5.2. Note that the point of highest probability, 53 pm, is the same as that calculated by Neils Bohr. In Bohr's model, however, the electron is assumed to have a circular path and always to be found at this distance from the nucleus. To carry the process even further, computers can be used to calculate the probabilities of the location of an electron for thousands of points in space. There will be many points of equal probability. If all the points of highest probability are connected, some three-dimensional shape is formed. These shapes will be shown later in the chapter. The most probable place to find the electron will be some place on the surface of this calculated shape. Remember that this shape is only a mental model and does not actually exist. It is something we use in our minds to help locate the most probable position of the electron.

Figure 5.3 The fan blades on the left appear to occupy the total volume through which they turn. An electron effectively occupies a three-dimensional space to form a cloud of negative charge (right).

Figure 5.4 A spray of water from a lawn sprinkler can also be compared to an electron cloud.

There is another way of looking at this probability. The electron moves about the nucleus, passing through the points of high probability more often than through any other points. The electron is traveling at high speed. If the electron were visible to the eye, its rapid motion would cause it to appear much like a cloud. Think of an electric fan as shown in Figure 5.3. When the fan is turning, it appears to fill the complete circle through which it turns. If you try to place an object between the blades while the fan is turning, the result will show that the fan is effectively filling the entire circle. So it is with the electron. It effectively fills all the space. At any given time, it is likely to be somewhere on the surface of the shape described by the points of highest probability. The probability of finding the fan blade outside its volume is zero. However, it is possible to find the electron outside of its high probability surface. Therefore, since the volume occupied by an electron is somewhat vague, it is better to refer to it as an **electron cloud.** Let's look more closely at this electron cloud to learn more about its size and shape.

5.1 *CONCEPT REVIEW*

1. What caused de Broglie to start thinking about particles and waves?
2. Explain how measuring the position of an electron changes its velocity.
3. Why can you sometimes consider an electron as a cloud of negative charge?
4. If two particles are traveling at the same speed, does the more massive particle have a longer or shorter wavelength than the less massive particle?
5. **Apply** Why is the wave-particle duality ignored in everyday life?

OBJECTIVES

- characterize the four quantum numbers.
- use the Pauli exclusion principle and quantum numbers to describe an electron in an atom.

5.2 Quantum Theory

Think for a moment about the chemical behavior of neon gas, atomic number 10, and sodium metal, atomic number 11. Neon gas does not react with water or with oxygen in air, yet sodium metal reacts vigorously with water and also with oxygen in the air. The chemically changed sodium atom produced in these reactions behaves more like neon gas than it does like sodium metal. The key to understanding this behavior lies in the fact that the chemical behavior of an atom is determined by the number and arrangement of electrons around the nucleus. The chemically changed sodium atom now has a number and arrangement of electrons identical to that of neon. Therefore, it is important to continue to refine our understanding of the electronic structure of the atom.

Figure 5.5 The exact location of these moving cars cannot be pinpointed. Likewise, only the most probable location of an electron can be determined.

Schrödinger's Equation

Prior to the use of computers, any solutions to the Schrödinger equation proved difficult for even the best mathematicians. It has been solved exactly only for the hydrogen atom and other one-electron systems. The key feature of the solution of the wave equation is the use of quantum numbers. These numbers represent different energy states of the electron. In Schrödinger's atomic model, changes between energy states must take place by emission or absorption of whole numbers of photons. For the hydrogen atom, solution of the wave equation gives accurate energy states. The differences between these energy states correspond to the lines observed in the hydrogen spectrum. Thus, when an electron in a hydrogen atom moves from a higher to a lower energy state, the energy change is emitted as a quantum of light. This light is one line in the emission spectrum of hydrogen.

With more complex atoms, however, more than one electron is present, and interactions between electrons make exact solution of the equation impossible. Recall that electrons all have the same charge. Thus, they tend to repel each other. In spite of this difficulty, it is possible to approximate the electronic structure of multielectron atoms. This approximation is made by first calculating the various energy states in the hydrogen atom. A quantum number is assigned different values to arrive at the energy for each state. Then, *it is assumed that the various electrons in a multielectron atom occupy these same energy states without affecting each other.* Experimental evidence has shown that these values are very close approximations to actual values.

To completely describe an electron in an atom, four quantum numbers are needed and are identified by the letters n, l, m, and s. Each electron within an atom can be described by a unique set of these four quantum numbers. We will discuss each quantum number starting with n.

Principal Quantum Number

An electron can occupy only specific energy levels. These energy levels are numbered, starting with 1 and proceeding to the higher integers. The number of the energy level, referred to as n, is called the **principal quantum number.** The principal quantum number corresponds to the energy levels 1, 2, 3, . . . n calculated for the hydrogen atom.

Electrons may be found in each energy level of an atom. The greatest number of electrons possible in any one level is $2n^2$. Figure 5.6 shows the relative energies of the various levels. It also indicates the maximum number of electrons that may be contained in each level.

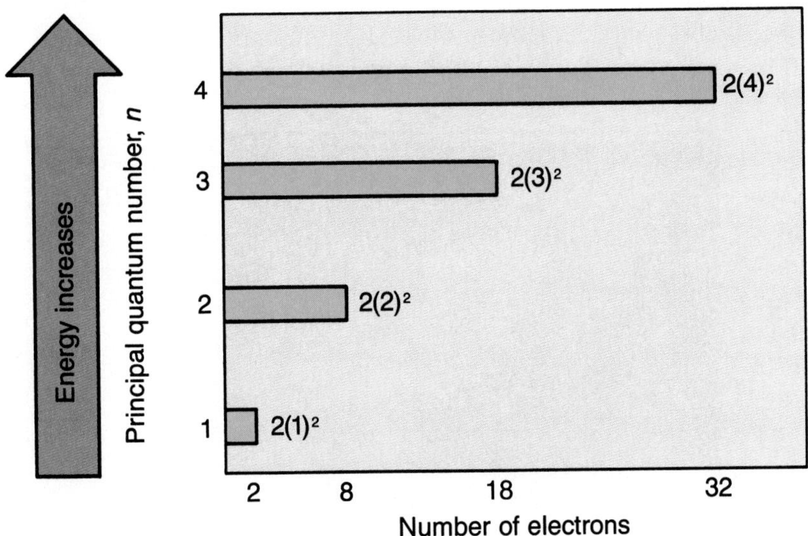

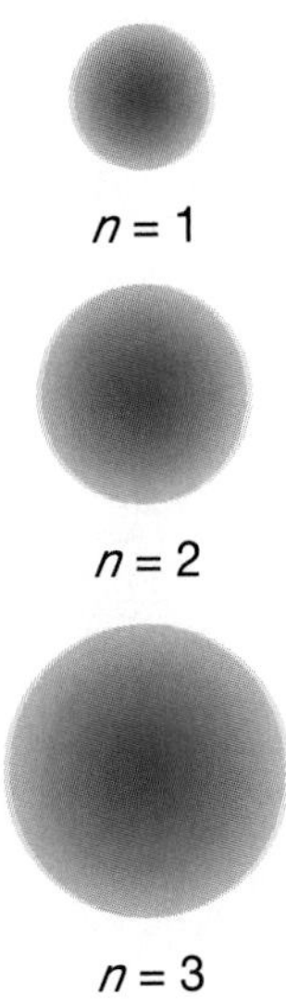

Figure 5.6 The relationship among energy, principal quantum number, and number of electrons is shown on the left. The relationship between size of electron cloud and principal quantum number is shown on the right.

SAMPLE PROBLEM

Capacity of Energy Levels

What is the maximum number of electrons that can occupy the first energy level in an atom? The fourth energy level?

PROBLEM SOLVING HINT

Be sure to find the value of n^2 before multiplying by 2.

Solving Process:

The maximum electron capacity of an energy level is given by $2n^2$ where n is the number of the energy level (principal quantum number).

For the first level, $2n^2 = 2(1)^2 = 2$.

For the fourth level, $2n^2 = 2(4)^2 = 32$.

PRACTICE PROBLEMS

6. Calculate the maximum number of electrons that can occupy the energy levels in an atom when the principal quantum number has the following values.

a. $n = 2$ **b.** $n = 3$ **c.** $n = 5$ **d.** $n = 7$

EVERYDAY CHEMISTRY

Fireworks

Much of what you see and enjoy at fireworks displays is a result of the emission spectrum of certain metal atoms. During the explosion of fireworks, a great deal of energy is released. When this energy is absorbed by metal atoms, the electrons in these atoms are raised to higher energy levels. This higher energy arrangement is not stable and the electrons quickly return to lower energy levels. The difference in energy between high and low energy arrangements is given off as brilliant light. Red lights are produced by strontium-containing compounds such as strontium nitrate. Green lights are produced by barium-containing compounds such as barium chlorate. Sodium oxalate is often included for its bright yellow light. The presence of copper sulfate will produce a bright blue-green light.

Exploring Further

1. Are the lights of fireworks more like absorption spectra or emission spectra? Explain your answer.
2. How can one Roman candle have two different explosions of two different colors?

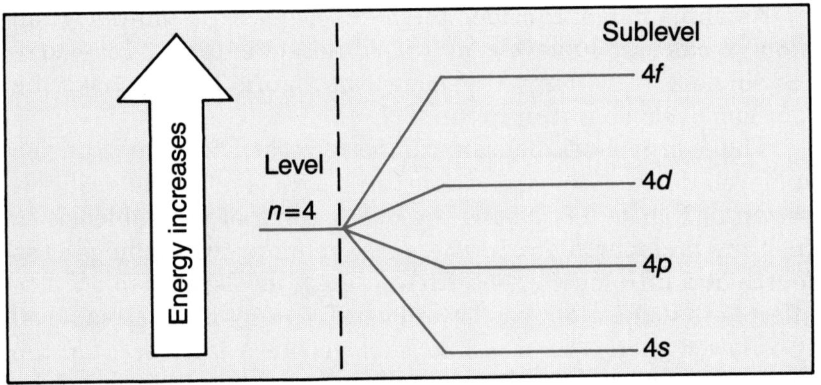

Figure 5.7 The relationship between energy and the sublevels in $n = 4$ is shown by this model.

Energy Sublevels and Orbitals

The second quantum number is l. In a hydrogen atom, there is only one energy state per level. This statement is not true for any other atom. Spectral studies have shown that in multielectron systems what may have initially appeared to be a single line in a spectrum is actually several lines closely grouped together. Therefore, an energy level is actually made of many energy states that are also closely grouped together. We refer to these states as **sublevels.** The l quantum number describes the sublevels within an energy level.

The number of sublevels for each energy level equals the value of the principal quantum number of that level. Thus, there is one sublevel in the first level, two sublevels in the second level, and three sublevels in the third level. The lowest sublevel in each level has been named *s,* and, for this sublevel, l has been assigned a value of 0. The second sublevel is named *p,* in which l has a value of 1. The third sublevel is *d,* in which l is 2. In the fourth sublevel, *f,* l has a value of 3. Thus, the first level has only an *s* sublevel. The second energy level has *s* and *p* sublevels. The third energy level has *s, p,* and *d* sublevels.

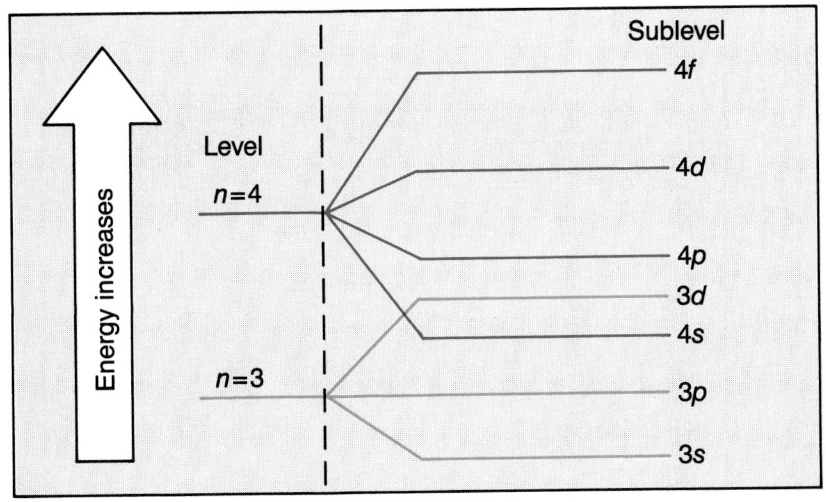

Figure 5.8 This energy level diagram shows the overlapping of sublevels that occurs between $n = 3$ and $n = 4$.

As an example, consider the $n = 4$ level. In a multielectron atom, it has four sublevels, all with different energies. Instead of the single $n = 4$ energy line shown in Figure 5.6, you have the four sublevels shown in Figure 5.7.

The energy level diagram can be reworked to include all the sublevels for the third and fourth energy levels. The result is shown in Figure 5.8. Notice the overlapping of the sublevels in the third and fourth levels. There is even more overlapping in the fourth and fifth levels, the fifth and sixth levels, and so on. The effect is that the atom has lower overall energy if the 4*s* sublevel is of lower energy than the 3*d* sublevel. In a later section, the reason for the overlap will be considered and we will explain why the overlap is so important in understanding atomic structure.

Calculation has shown that any *s* sublevel may contain one pair of electrons; any *p* sublevel, three pairs; any *d* sublevel, five pairs; and any *f* sublevel, seven pairs. Each pair in a given sublevel has a different place in space. This space occupied by one pair of electrons is called an **orbital.** Orbitals are designated by the third quantum number, *m*. Each orbital can hold a pair of electrons. For example, a *p* sublevel has three orbitals. The quantum number *m* is used to distinguish between orbitals that are within the same sublevel.

Always an Integer

An electron can occupy only specific energy levels. These levels are numbered 1 through 7. A toy called a Slinky can be used to illustrate this concept. Go to the top of a set of stairs. **CAUTION:** *Be careful when working on stairs.* Place a Slinky on the top step near the front edge of the stair tread. Take hold of the top edge of the Slinky and lift it up and then over, placing it on the next stair tread. The Slinky will continue to loop over and "walk" down the stairs. Can the Slinky stop between stair treads number 2 and 3 and be in a position 2.5? Can an electron be found in between two energy levels of an atom? Which letter is used to represent the principal quantum number?

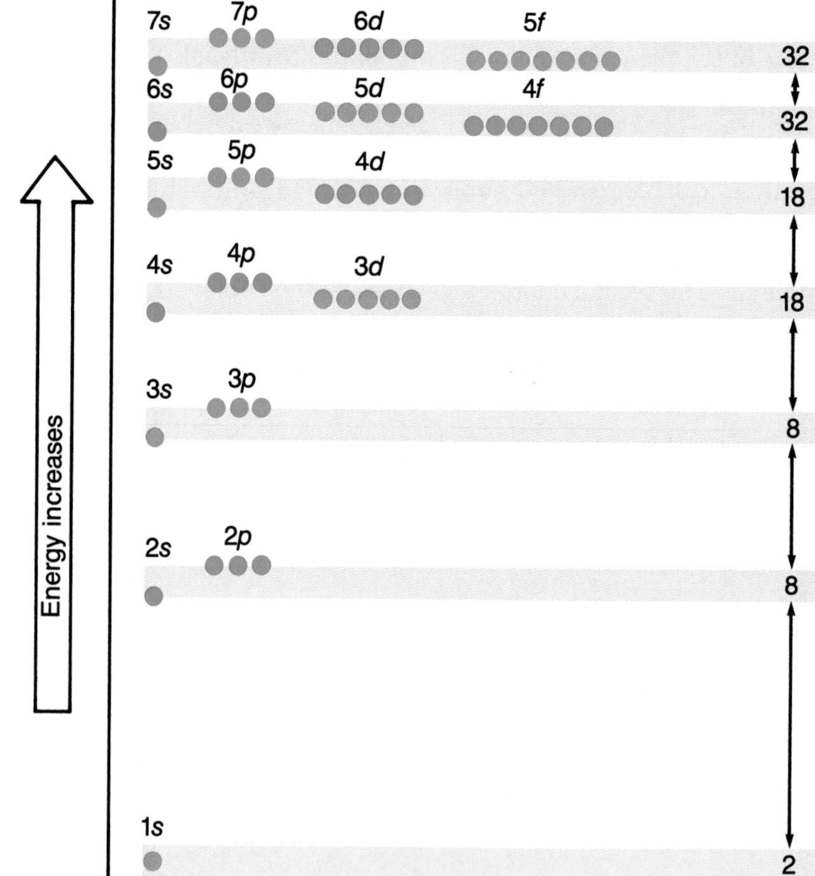

Figure 5.9 This diagram shows the energy relationships of the orbitals in the various sublevels.

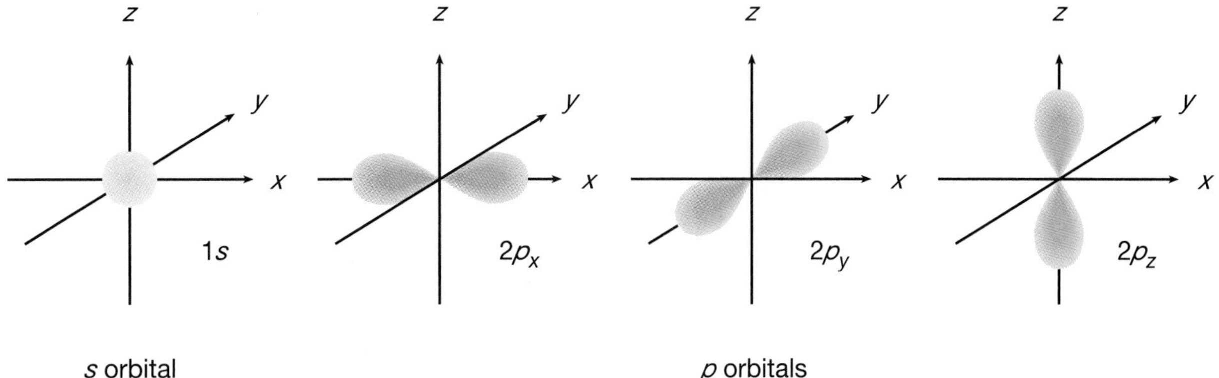

Figure 5.10 Orbitals have characteristic probability shapes, given by quantum number *l*. The orientation in space, for example, a *p* orbital along the *x* axis, is given by quantum number *m*.

Shape of the Charge Cloud

We now want to see how these various energy levels and sublevels affect the electron charge cloud. In general, the size of the charge cloud is related to n, the principal quantum number. The larger the value of n, the larger the cloud; see Figure 5.6. However, there are also other factors that govern the size of the cloud. Electrons are repelled by each other; they are also attracted by the positively charged nucleus. At the same time, other electrons serve to screen the effect of the nucleus. For example, an electron A between the nucleus and electron B reduces the attraction of the nucleus for electron B. Thus, the size of the charge cloud is not controlled by any single factor.

The sum of all electron clouds in any sublevel (or energy level) is a spherical cloud. Orbitals have characteristic probability shapes of their own, given by quantum number l. The values for l are integers 0 through $n - 1$. The three p orbitals can be directed along the three perpendicular x, y, and z axes, Figure 5.10. They are sometimes labeled as p_x, p_y, and p_z.

As shown in Figure 5.11, the five d orbitals also have characteristic probability shapes. When filled and combined, both the d and the p orbitals form a spherical charge cloud. A large, simple shape (such as 4s) represents a lower energy state in an atom than a small, complex shape (such as 3d).

Figure 5.11 The five *d* orbitals have characteristic probability shapes and spatial directions.

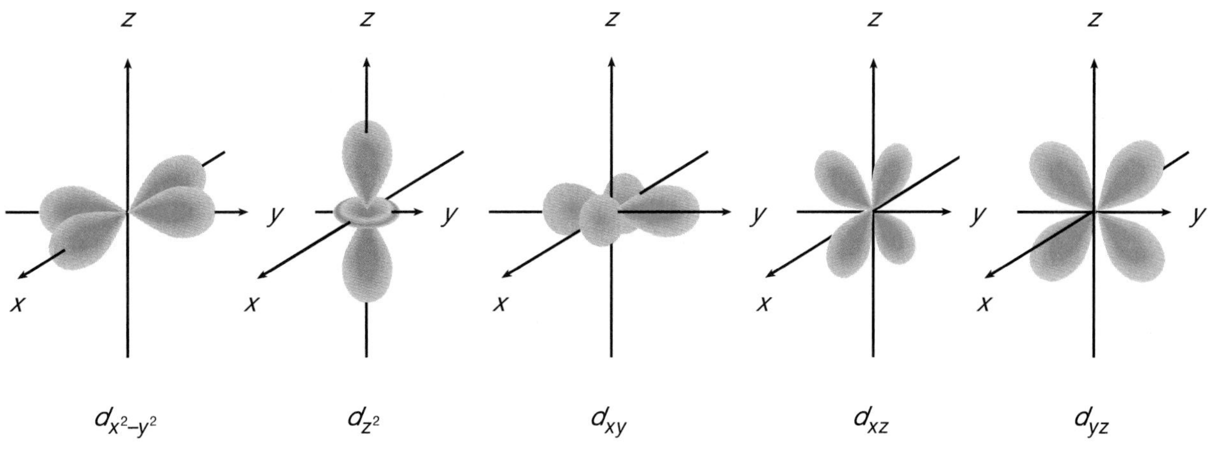

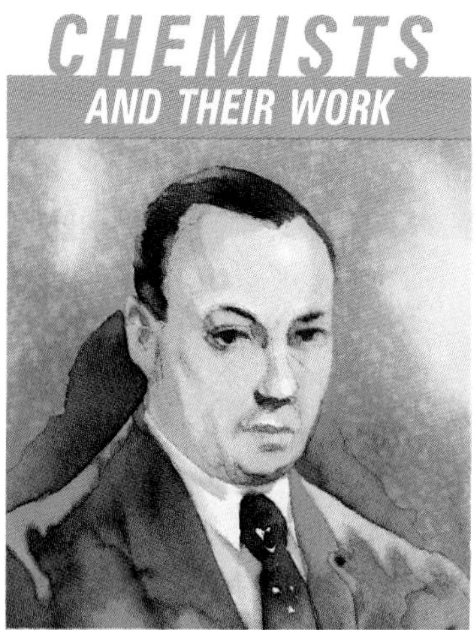

Wolfgang Pauli
(1900-1958)
The Pauli exclusion principle is named for its originator, Wolfgang Pauli. As a young man, Pauli studied under some of the giants of science, including Born and Bohr. Pauli introduced the concept of a fourth quantum number to describe electron spin, and proposed the existence of neutrinos. A native of Austria, he spent his early career teaching in Germany and Switzerland. He was awarded the Nobel Prize in physics in 1945. Just before World War II, he emigrated to the United States and became a citizen in 1946.

The third quantum number, m, defines each orbital more precisely by indicating its direction in space. The values for m equal integers ranging from $-l$ through $+l$. For a p sublevel there are three possible values for m. The numbers designate orbitals aligned along the x, y, and z axes. There are five values for m in a d sublevel and seven values for m in an f sublevel. Orbitals that are alike in size and shape and differ only in direction have the same energy. Orbitals of the same energy are said to be **degenerate.** Therefore, the three $2p$ orbitals, $2p_x$, $2p_y$, and $2p_z$, are degenerate, as are the five $3d$ orbitals.

Chemical reactions involve electrons. In some cases, electrons are removed from atoms. In other cases, electrons are added to atoms or shared between atoms. These electrons are in atomic orbitals whose energies depend on the structure of the atom. The orbital energies can change when one atom interacts with another. Thus, when atoms react, degenerate orbitals may change energy by different amounts. If this energy change occurs, the orbitals will no longer be degenerate.

PRACTICE PROBLEMS

7. How many sublevels are in the energy levels with the following principal quantum numbers?
a. $n = 1$ **b.** $n = 3$ **c.** $n = 4$

8. How many orbitals are in each of the following?
a. s sublevel **c.** d sublevel
b. p sublevel **d.** f sublevel

Distribution of Electrons

How are electrons of a particular atom arranged among the energy levels? Recall that the atom is electrically neutral. For each proton in the nucleus, there is one electron in the charge cloud. Thus, as the atomic number increases, the number of electrons increases.

The energy levels in an atom can be thought of as a rooming house in which the better-quality double rooms are on the ground level. Electrons, as tenants, will tend to fill the better (lower) rooms first, one at a time. Then, they double up in these better rooms before going to the next level. In other words, electrons occupy the energy level and sublevel that produces the arrangement with the lowest energy. Let's consider as the first "tenant" the electron from the hydrogen atom. It will occupy the position of least energy, the 1s orbital.

What about a different set of room tenants, the two electrons of the helium atom? Before we can answer this question, we must consider a principle that helps to explain the arrangement of electrons. It has been found that *no two electrons in an atom have the same set of four quantum numbers.* This behavior was first observed and stated by Wolfgang Pauli and is called the **Pauli exclusion principle.** The quantum numbers *n*, *l*, and *m* describe relative cloud size (*n*), shape of the cloud (*l*), and orientation in space of the cloud (*m*). The fourth quantum number, *s*, is used to distinguish between electrons in the same orbital and describes the spin of the electron, clockwise or counterclockwise. We know that an electron can have two spin directions because of the observation that atomic spectra change when a magnetic field is present. Spinning electrons act like tiny bar magnets. An electron that spins in one direction generates a magnetic field that is attracted to the north pole of a magnet. An electron spinning in the opposite direction is attracted to the south pole of a magnet. If two electrons occupy the same orbital, they must have opposite spins. If they did not, these two electrons would have identical quantum numbers; see Figure 5.12.

Table 5.1 summarizes the quantum states for the hydrogen atom.

1. The principal quantum number, *n*, is the energy level number. It can have values of 1, 2, 3, and so on. The principal quantum number gives information about relative electron cloud size.

2. The second quantum number, *l*, is the energy sublevel number and can range in value from 0 to $n - 1$. It gives information about the shape of the electron cloud. A sublevel with $l = 0$ is an *s* sublevel, $l = 1$ is a *p* sublevel, $l = 2$ is a *d* sublevel, and $l = 3$ is an *f* sublevel.

3. The third quantum number, *m*, is the orbital quantum number and can have integral values from $-l$ through $+l$. It gives information about the orientation in space of an orbital.

4. The fourth quantum number, *s*, is the spin quantum number. Spin is either clockwise or counterclockwise and is designated $+1/2$ or $-1/2$.

Now we can consider the helium atom. Both electrons of a helium atom take positions in the 1*s* orbital, and thus have opposing spins. We call this arrangement a $1s^2$ (read as *one-s-two*) electron configuration. We can represent the electron configuration

Figure 5.12 On the left, a spinning electron generates a magnetic field. On the right, an external magnetic field is applied, and the electron assumes one of two spins. Two electrons in the same orbital have opposite spins.

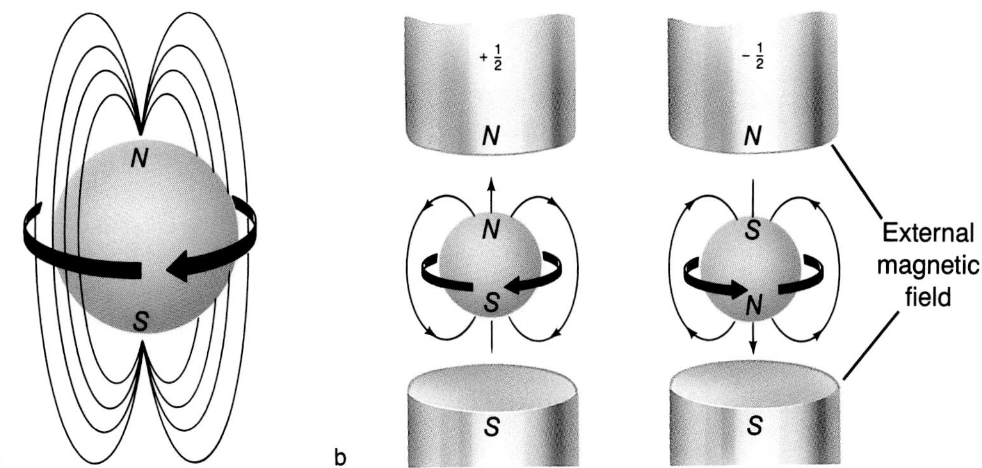

Table 5.1

Quantum States for the Hydrogen Atom

n	l	m	s	Sublevel	Sublevel Electron Capacity	Level Electron Total
1	0	0	±½	1s	2	2
2	0	0	±½	2s	2	8
2	1	−1	±½	2p	6	
		0	±½			
		+1	±½			
3	0	0	±½	3s	2	18
3	1	−1	±½	3p	6	
		0	±½			
		+1	±½			
3	2	−2	±½	3d	10	
		−1	±½			
		0	±½			
		+1	±½			
		+2	±½			
4	0	0	±½	4s	2	32
4	1	−1	±½	4p	6	
		0	±½			
		+1	±½			
4	2	−2	±½	4d	10	
		−1	±½			
		0	±½			
		+1	±½			
		+2	±½			
4	3	−3	±½	4f	14	
		−2	±½			
		−1	±½			
		0	±½			
		+1	±½			
		+2	±½			
		+3	±½			

by a diagram like that shown below. In the diagram, the circle stands for an orbital. Arrows are used to indicate the direction of the spin of the electrons in the orbital.

1s

Helium (⇅) $1s^2$

The electron configuration of lithium ($Z = 3$) shows both positions in the 1s orbital and one position in the 2s orbital filled. We designate this arrangement as $1s^22s^1$.

1s 2s

Lithium (⇅) (↑) $1s^22s^1$

Now, let's see how the seven electrons of a nitrogen atom are distributed. We would expect two electrons in the 1s orbital, two in the 2s, and three in the 2p. This arrangement can be indicated as follows, where each arrow represents an electron.

1s 2s 2p

Nitrogen (⇅) (⇅) (↑)(↑)(↑) $1s^22s^22p^3$

The opposing arrows indicate opposite spins. Notice that each electron takes an empty orbital within a sublevel, if possible, rather than pair with another. This arrangement is reasonable because the negative electrons repel each other. It requires no more energy because all orbitals in a sublevel of an atom are degenerate. Since each orbital has a different orientation in space, electrons in different orbitals are farther apart than electrons in the same orbital. The electrons in the oxygen atom are arranged as shown below. The eighth electron enters a partially filled $2p$ orbital.

	$1s$	$2s$	$2p$	
Oxygen	⇅	⇅	⇅ ↑ ↑	$1s^2 2s^2 2p^4$

5.2 CONCEPT REVIEW

9. Define orbital. How many orbitals are in an *f* sublevel?
10. State Pauli's exclusion principle.
11. List and explain each of the four quantum numbers.
12. If two orbitals are degenerate, what do they have in common?
13. **Apply** How many sublevels will be found in level x?

Bridge to ASTRONOMY

Spectra of Stars

One of the principal studies of astronomers is the investigation of the birth, normal life, and death of stars. Obviously, astronomers cannot experiment on stars in the laboratory. Also, a typical star may live 10 000 000 000 years! How, then, can scientists determine the life cycle of stars? As stars are born and age, the proportions of the different elements composing them change. You will find out more about this process in Chapter 28. By carefully examining the spectra of the light produced by stars, astronomers can tell what elements are present in stars and in what proportions these elements occur. The astronomers construct a mental model for the evolution of stars and then see if the spectra that would be produced by that model agree with the observed spectra of stars of various ages. If they do, then the model is a reasonable model. If not, then a new model is constructed and again compared with observations. As methods of stellar observation are constantly being improved, astronomers are constantly refining their model of stellar evolution. There are many other details of star structure and behavior that astronomers can deduce from stellar spectra. More and more astronomical instruments are placed in Earth orbit where interference from outside sources is minimized. Thus, the part of the electromagnetic spectrum available to astronomers is increased and more information on stellar composition is gained.

Exploring Further

1. How would a spectrum of a star change with time in regard to the star's composition?
2. Astronomers can deduce the relative density of a star from its spectrum. How is this analysis done?

CHEMISTRY AND
TECHNOLOGY

Thin-Layer Magnetism

Magnetism in metals results from relationships among the electrons that surround the nuclei in the atoms of the metal. In most metals, electrons are paired, and the magnetic field that is set up by one spinning electron is canceled by an opposite magnetic field set up by the oppositely-spinning electron paired with it. In some metals, however, there are unpaired electrons. These electrons lead to the magnetic properties of the metals. The degree of magnetic behavior comes from the relative numbers of unpaired electrons that are spinning in the same direction in the atoms.

The greater the number of electrons spinning in one direction, the greater the magnetic behavior of the metal. Technology has long searched for ways to make stronger magnets in smaller packages. Recent research has shown that a thin layer of iron, one atom thick, can produce a stronger magnetic field than a thicker layer of iron. In a crystal, every iron atom is surrounded by eight neighboring atoms and each of these atoms has twenty-six electrons surrounding it. The interference among electrons is great, and the ability of electrons to be induced to spin in the same direction is reduced in a bar or thick magnet. In a single layer, however, the atoms are not surrounded in three dimensions. As a result, the electrons have more freedom to move and are more easily induced to spin in the same direction.

The major problem with one-atom thin films, or monolayers, is that they cannot survive in the real world. Therefore, computer simulations were used to determine whether monolayers could be coated on another metal. Scientists had guessed that if it were possible to lay a monolayer on another metal, then that metal would become magnetic. Neither silver nor gold has unpaired electrons. However, the simulation showed that both would accept an overlay of the iron monolayer and both would then exhibit the same magnetism as the iron itself. Later actual experiments, carried out in a high vacuum to avoid contamination, showed that the computer simulation was correct.

One interesting aspect of this monolayer deposit is that its magnetic field was oriented vertically rather than horizontally. This fact has led to some interesting possibilities for magnetic tape storage of information. The usual orientation of "magnets" along a magnetic tape is horizontal. The number of magnets that can be placed along a tape limits the amount of information that can be stored. Imagine the amount that could be stored if the magnets stood on end, a one-atom end, rather than horizontally. In experiments, information density has been increased by at least 40 times.

The discovery and use of one-atom thin layers is just one more example of scientists using a prepared mind and contemporary technology to produce a technological advance from scientific curiosity.

Thinking Critically

1. What other uses might there be for thin layer magnets coated over other metals?
2. Why is the use of the computer significant in this experiment?

5.3 Distributing Electrons

OBJECTIVES

- determine the electron configurations of the elements.
- write electron dot diagrams for the elements.

By studying the arrangement of electrons in atoms, you will better understand chemical reactivity. Particular arrangements of electrons lead to specific reaction behaviors. It is, therefore, important to know how to find the electron arrangements of an atom in order to predict its reaction behavior. In this section, you will see how to predict the ground state electron arrangements for any atom for which you know the atomic number. In most of your subsequent work you will assume that each atom has the predicted electron arrangement. You should be aware, however, that there are several important exceptions to these predictions. A number of exceptions will be pointed out in Chapter 6.

Order of Filling Sublevels

In predicting electron arrangements, everything works well until you finish the electron configuration of argon: $1s^22s^22p^63s^23p^6$. Where will the electrons of the next element, potassium ($Z = 19$), go? You may remember that in the energy level diagram in Figure 5.8, the 4*s* level was shown below the 3*d* level. Thus, in this case, the 4*s* level fills first because that order produces an atom with lower energy. The electron configuration for potassium, therefore, is $1s^22s^22p^63s^23p^64s^1$. Calcium atoms ($Z = 20$) have the electron configuration $1s^22s^22p^63s^23p^64s^2$. Scandium, however, begins filling the 3*d* orbitals, and has a configuration of $1s^22s^22p^63s^23p^64s^23d^1$.

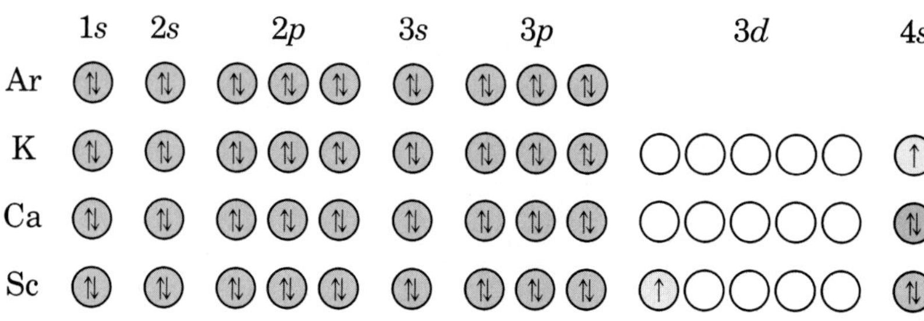

In many atoms with higher atomic numbers, the sublevels are not regularly filled. It is of little value to memorize each configuration. However, there is a rule of thumb that gives a correct configuration for most atoms in the ground state. It is a rule of thumb because it makes an assumption that is not always true. The order of increasing energy sublevels is figured for the one-electron hydrogen atom. In a multielectron atom, each electron affects the energy of the others. Consequently, there are a number of exceptions to the rule of thumb. We will explore these exceptions more fully in the next chapter. This rule of thumb is summarized in Figure 5.13. *If you follow the arrows listing the orbitals passed, you can find the electron configuration of most atoms.*

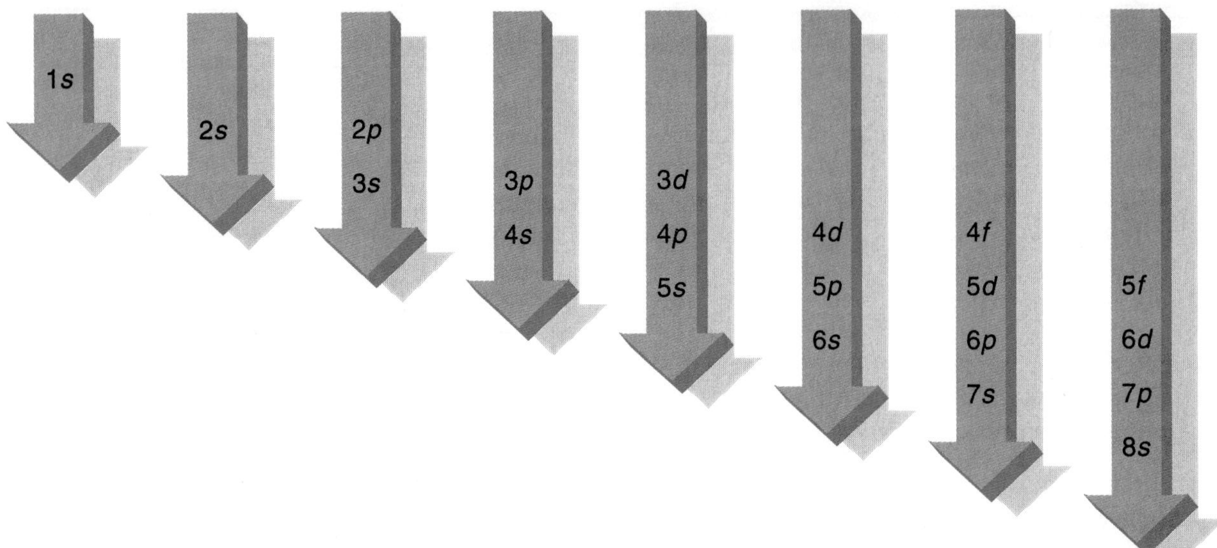

Figure 5.13 This arrow diagram provides a convenient method for writing electron configurations. Begin at top left, follow the arrow, then move right and down, always moving from the head of one arrow to the tail of the next.

PROBLEM SOLVING HINT

Add the superscripts of the electron configuration as a problem check. The total should equal Z.

SAMPLE PROBLEM

Writing Electron Configurations

Use the arrow diagram in Figure 5.13 to find the electron configuration of zirconium ($Z = 40$).

Solving Process:

Begin with the 1*s* level, move over to the 2*s*, and then over to the 2*p*. Follow the arrow down to 3*s* and over to 3*p*. Follow the arrow down to 4*s*. Move back up to the tail of the next arrow at 3*d* and follow it down through 4*p* and 5*s*. Move back to the tail of the next arrow, 4*d*, and place the remaining two electrons there. Thus, the electron configuration is

$$1s^22s^22p^63s^23p^64s^23d^{10}4p^65s^24d^2$$

A quick addition of the superscripts gives a total of 40, which is the atomic number.

PRACTICE PROBLEMS

14. Write the electron configurations for elements with atomic numbers $Z = 1$ through $Z = 20$.

Electron Dot Diagrams

In studying electrons in atoms in the following chapters, of primary concern will be the electrons in the outer energy level. These are the electrons most often involved in chemical reactions. It is often useful to draw these outer electrons around the symbol of an element. This notation is referred to as a **Lewis electron dot diagram,** see Table 5.2.

Table 5.2

Electron Dot Diagrams for Some Elements					
Element	**Configuration Ending**	**Dot Diagram**	**Element**	**Configuration Ending**	**Dot Diagram**
carbon	$2s^22p^2$	·Ċ:	bromine	$4s^23d^{10}4p^5$	:B̈r:
sodium	$3s^1$	Na·	xenon	$5s^24d^{10}5p^6$	:Ẍe:
magnesium	$3s^2$	Mg:	cerium	$6s^24f^2$	Ce:
aluminum	$3s^23p^1$	Ȧl:	tungsten	$6s^24f^{14}5d^4$	W:
phosphorus	$3s^23p^3$	·Ṗ̣:	osmium	$6s^24f^{14}5d^6$	Os:
zinc	$4s^23d^{10}$	Zn:	uranium	$7s^25f^36d^1$	U:

The procedure for drawing electron dot diagrams follows.

Step 1. Let the symbol of the element represent the nucleus and all electrons except those in the outer level.

Step 2. Write the electron configuration of the element. From the configuration, select the electrons that are in the outer energy level. Those electrons in the outer level are the ones with the largest principal quantum number. Remember that n represents electron cloud size.

Step 3. Each "side" (top, bottom, left, right) of the symbol represents an orbital. Draw dots on the appropriate sides to represent the electrons in that orbital. It is important to remember which electrons are paired and which are not. *It is not important which side represents which orbital.*

Note the following special considerations. First, when choosing the outer electrons, ignore the highest energy electrons in the atom if they are not in the outermost energy level. Their principal quantum number is less than that of the electrons in the outermost energy level. Second, note that the s electrons are always shown as paired, or on the same side of the symbol. Also, remember from page 124 that electrons repel each other because of their negative charge. Therefore, electrons will occupy an empty degenerate orbital before pairing with other electrons. For example, four p electrons are distributed as one pair of electrons and two single electrons. No pairing takes place until the fourth electron enters the p sublevel.

SAMPLE PROBLEM

Electron Dot Diagrams

Write the electron dot diagrams for hydrogen, helium, oxygen, calcium, and cadmium.

Solving Process:

(a) Begin with the symbols for the elements required.

H, He, O, Ca, and Cd

(b) Write the electron configurations for each atom and determine the number of outer level electrons in the configuration.

H = $1s^1$ (outer electron = $1s^1$)
He = $1s^2$ (outer electrons = $1s^2$)
O = $1s^22s^22p^4$ (outer electrons = $2s^22p^4$)
Ca = $1s^22s^22p^63s^23p^64s^2$ (outer electrons = $4s^2$)
Cd = $1s^22s^22p^63s^23p^64s^23d^{10}4p^65s^24d^{10}$ (outer electrons = $5s^2$)

(c) Use the outer configuration to determine the number of dots required for each element. Remember that each symbol represents the nucleus and inner level electrons for that atom. Then correctly place a dot for each outer electron in an orbital around the symbol.

H·, He:, :Ö·, Ca:, and Cd:

PROBLEM SOLVING HINT

To avoid misplacement of dots, think of each symbol being surrounded by a square. An orbital is represented by each side of the square.

PRACTICE PROBLEMS

15. Predict electron configurations using the arrow diagram, and draw electron dot diagrams for the following elements.

a. $Z = 28$
b. $Z = 18$
c. $Z = 16$
d. $Z = 47$
e. $Z = 19$
f. $Z = 32$

Electron Summary

We have treated the electron as a particle, a wave, and a cloud of negative charge. Which approach is correct? They all are. As pointed out earlier in the chapter, electrons can behave as particles or as waves. There are times when it is useful to consider the electron as a particle. At that time, the electron exhibits its particle characteristics. At other times, the wave properties of the electron are of the greatest importance. At still other times, it is best to consider the electrons in an atom as a cloud of negative

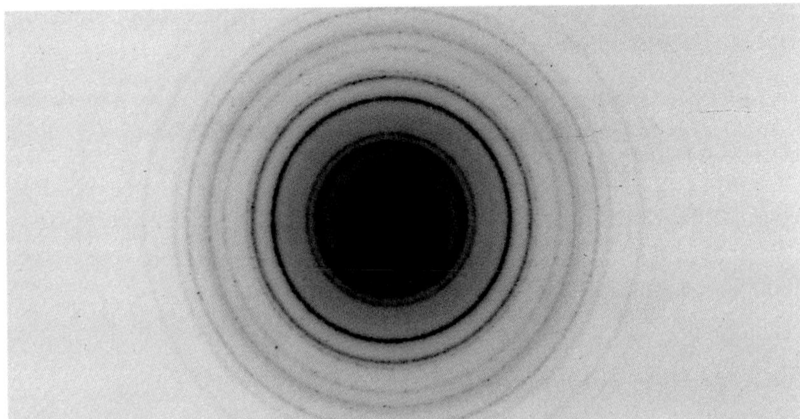

Figure 5.14 Electron diffraction patterns of aluminum demonstrate the wave characteristics of particles.

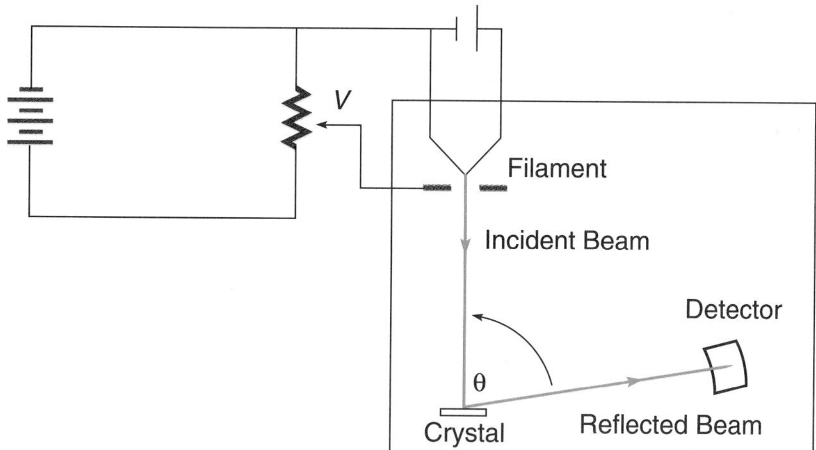

Figure 5.15 In this apparatus a beam of electrons from a hot cathode is directed at a crystal, and angles of the scattered electrons are detected. In this experiment, the electrons act as particles.

charge. Scientists do not have a single, completely satisfactory description of the structure of atoms. Consequently, they make use of more than one explanation for the properties they observe. The particular explanation that best fits each situation is the one applied to it.

You are now able to describe the electron configurations of the atoms of the elements. Your next study will be of a system of arranging elements based on their electronic structure—the periodic table.

Figure 5.16 A free-electron laser shows the wave properties of an electron. An electron beam is generated, then contained by mirrors in the optical cavity. An alternating magnetic field uses this electron beam and generates a laser beam.

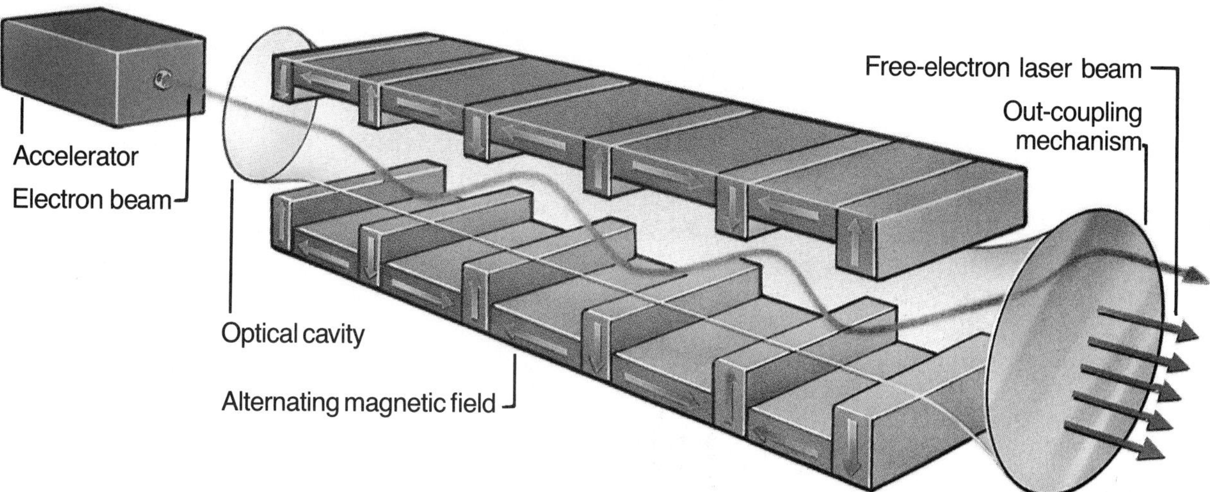

5.3 *CONCEPT REVIEW*

16. Write electron configurations for aluminum, hydrogen, and manganese.

17. Write the part of the electron configuration of sulfur that would be represented by dots in an electron dot diagram.

18. Draw electron dot diagrams for helium, rubidium, and silicon.

19. Apply Explain why inner-level electrons are not drawn as dots placed in orbitals when drawing a Lewis electron dot diagram of an atom.

Chapter 5 REVIEW

Summary

5.1 Modern Atomic Structure

1. De Broglie proposed that electrons and other particles of matter have both particle and wave properties. Two years later his hypothesis was proved correct.
2. All particles exhibit wave properties and all waves exhibit particle properties. This principle is known as the wave-particle duality of nature.
3. Chemists are interested in knowing the position and velocity of electrons in an atom.
4. Heisenberg's uncertainty principle states that it is impossible to know accurately both the position and the momentum of an electron at the same time. Measuring one quantity changes the other.
5. Schrödinger developed a mathematical equation that describes the behavior of the electron as a wave. The solution of the wave equation can be used to calculate the probability of finding an electron at a particular point around the nucleus.
6. Because of the electron's high speed, it effectively occupies all the volume defined by the path through which it moves. This volume is called the electron cloud.

5.2 Quantum Theory

7. Each electron in an atom can be described by a unique set of four quantum numbers, n, l, m, and s.
8. The principal quantum number ($n = 1, 2, 3, \ldots$) is the number of the energy level and describes the relative electron cloud size.
9. Each energy level has as many sublevels as the principal quantum number. The second quantum number ($l = 0, 1, 2, \ldots n - 1$) describes the shape of the cloud.
10. The third quantum number, m, describes the orientation in space of each orbital.
11. The fourth quantum number ($s = +\frac{1}{2}$ or $-\frac{1}{2}$) describes the spin direction of the electron.
12. Pauli's exclusion principle states that no two electrons in an atom can have the same set of quantum numbers. Each orbital may contain a maximum of one pair of electrons. Electrons in the same orbital have opposite spins.
13. Electrons normally occupy the set of orbitals that give the atom the lowest overall energy.

5.3 Distributing Electrons

14. The arrow diagram can be used to provide the correct electron configuration for most atoms.
15. The chemist is primarily concerned with the electrons in the outer energy level. Electron dot diagrams are useful in representing these outer level electrons.

Key Terms

wave-particle duality of nature
momentum
Newtonian mechanics
quantum mechanics
Heisenberg uncertainty principle
quantum number
probability
electron cloud
principal quantum number
sublevel
orbital
degenerate
Pauli exclusion principle
Lewis electron dot diagram

Review and Practice

20. How did de Broglie show the relationship between waves and particles?
21. What observation did scientists make that confirmed de Broglie's hypothesis?
22. What is meant by the wave-particle duality of nature?
23. What is the relationship between momentum and wavelength?

24. Explain the following statement: The more certain the position of an electron is, the less certain is its momentum.

25. What was the main assumption that Schrödinger made that enabled him to develop his equation?

26. Explain how the movement of an electron is similar to that of electric fan blades.

27. Explain what it is that quantum numbers represent.

28. Why is it difficult to apply the Schrödinger equation to multielectron atoms?

29. What does the principal quantum number, n, designate?

30. How many electrons can exist in the third energy level? Show the expression by which this value is calculated from the principal quantum number.

31. What does the second quantum number, l, designate? How many values can l have in a given energy level?

32. Copy the following table and complete it for energy levels 1-4.

Level	Number of sublevels	Sublevel letter(s)
1		
2		
3		
4		

33. What is an orbital? How many orbitals are possible at each sublevel?

34. Explain the factors that determine the size of an electron charge cloud.

35. How are degenerate orbitals alike? How do they differ from one another?

36. What are the possible values for the fourth quantum number (s)? What do these numbers signify?

37. State the Pauli exclusion principle. What does this principle tell us about two electrons occupying the same orbital?

38. The electron configuration of carbon is $1s^22s^22p^2$. Explain how this configuration is built up, electron by electron, using the Pauli exclusion principle at each step.

39. Write the electron configuration for uranium. How many electrons are in each level? Which levels are not filled?

40. What elements are composed of atoms having the following electron configurations?

a. $1s^22s^22p^63s^23p^4$
b. $1s^22s^22p^63s^23p^64s^23d^5$
c. $1s^22s^22p^63s^23p^64s^23d^{10}4p^3$
d. $1s^22s^22p^63s^23p^64s^23d^{10}4p^65s^24d^4$

41. Write the electron configurations for titanium and gallium.

42. Write the electron configurations for bismuth and ruthenium.

43. Write orbital-filling diagrams for atoms of boron, fluorine, sulfur, germanium, and krypton.

44. Why does the 4s sublevel fill with electrons before electrons are found in the 3d sublevel?

45. Draw electron dot diagrams for the elements with Z equal to 7, 15, 33, 51, and 83.

46. Selenium ($Z = 34$) was discovered by Berzelius in 1817. Write its electron configuration and its electron dot diagram.

47. Terbium (Tb), erbium (Er), ytterbium (Yb), and yttrium (Y) were all found near the town of Ytterby, Sweden, and named for it. What are the electron dot diagrams for all four of these elements?

48. Neon was first found in 1898. It occurs naturally in the air in trace amounts. Show how the four quantum numbers are used to describe the electron structure of neon.

49. An atom's electron configuration ends in $5s^24d^{10}5p^4$. Identify the element and write its electron dot diagram.

50. Explain how quantum numbers relate to charge clouds and energy.

51. In the fifth energy level ($n = 5$) unnilquadium (Unq, $Z = 104$) has 32 electrons. How are they arranged in sublevels? How many more electrons would the fifth energy level accommodate theoretically?

Concept Mastery

52. Why is it impossible to demonstrate the wave nature and the particle nature of electrons in the same experiment?

53. Explain why Heisenberg's uncertainty principle has a great effect on our ability to observe electrons and other very small particles.

54. What is the significance of the Schrödinger wave equation as interpreted by Max Born? How does this idea lead to the concept of electron clouds?

55. When an electron moves from the $n = 2$ to the $n = 4$ state, does it gain or lose energy? Is a photon emitted or absorbed?

56. In any energy level, the charge cloud is spherical. How, then, can orbitals have nonspherical shapes?

57. In the electron dot diagram for nitrogen, electrons from which orbitals are shown? Which electrons are not shown?

58. Niobium (Nb) and tantalum (Ta) are usually found together in nature. Both are used to make abrasives. How many paired electrons are in atoms of each?

Application

59. Determine whether each of the following is best described by Newtonian mechanics or by quantum mechanics.
   a. a car traveling 90 km/h
   b. gamma rays given off during nuclear decay
   c. an electron in an atom
   d. Earth orbiting the sun

60. Why is Heisenberg's uncertainty principle considered when a chemist examines electrons in an atom but is negligible when a traffic engineer in a helicopter observes traffic patterns at a highway intersection?

Critical Thinking

61. Classical physics would predict that only the first three quantum numbers should be necessary to describe the motion of the electron in three-dimensional space. What necessitates a fourth?

62. Draw electron dot diagrams for sodium ($Z = 11$) and chlorine ($Z = 17$). Can you explain how they might combine?

63. Shown here are orbital-filling diagrams for the elements named. Each diagram is incorrect in some way. Explain the error in each and write a correct diagram.

a. carbon
1s (↑↓) 2s (↑↓) 2p (↑↓)( )( )

b. calcium
1s (↑↓) 2s (↑↓) 2p (↑↓)(↑↓)(↑↓) 3s (↑↓)
3p (↑↓)(↑↓)(↑↓) 3d (↑)(↑)( )( )( )

c. iron
1s (↑↓) 2s (↑↓) 2p (↑↓)(↑↓)(↑↓) 3 s (↑↓)
3p (↑↓)(↑↓)(↑↓) 3d (↑)(↑)(↑)(↑)(↑)
4s (↑↓)

d. bromine
1s (↑↓) 2s (↑↓) 2p (↑↓)(↑↓)(↑↓) 3s (↑↓)
3p (↑↓)(↑↓)(↑↓) 3d (↑↓)(↑↓)(↑↓)(↑↓)(↑↓)
4s (↑↓) 4p (↑↓)(↑↓)(↑↓)

64. Use what you have learned in this chapter to account for the complex spectra of multielectron atoms.

Readings

Freeman, Robert D. " 'New' Schemes for Applying the Aufbau Principle." *Journal of Chemical Education* 67, no. 7 (July 1990): 576.

Peterson, Ivars. "Electron Excitement in Three Dimensions." *Science News* 134, no. 16 (October 15, 1988): 252.

Peterson, Ivars. "Imitating Iron's Magnetism." *Science News* 131, no. 16 (April 18, 1987): 252-253.

Salem, Lionel. *Marvels of the Molecule*. New York: VCH Publishers, 1987.

Thomsen, Dietrick E. "A Midrash Upon Quantum Mechanics." *Science News* 132, no. 2 (July 11, 1987): 26-27.

Thomsen, Dietrick E. "Violating a Not-So-Exclusive Exclusion Principle." *Science News* 132, no. 9 (February 27, 1988): 132.

Tykodi, R. J. "The Ground State Electronic Structure for Atoms and Monatomic Ions." *Journal of Chemical Education* 64, no. 11 (November 1987): 943.

Cumulative Review

1. What is the average atomic mass of the element molybdenum if it has this isotopic composition?

Isotope	Mass	Abundance
92	91.906 808 u	15.84%
94	93.905 090 u	9.04%
95	94.905 837 u	15.72%
96	95.904 674 u	16.53%
97	96.906 023 u	9.46%
98	97.905 409 u	23.78%
100	99.907 478 u	9.63%

2. What is a nuclide? What are isotopes?

3. What is the difference between atomic number and atomic mass?

4. How much energy is needed to heat 10.0 g of tin from 25°C to 225°C? Use Appendix Table A-3.

5. What data led Bohr to formulate the planetary model of the atom?

6. What is Planck's contribution to the development of atomic theory?

7. What observations led John Dalton to formulate his atomic theory?

8. How many electrons, neutrons, and protons are in mendelevium ($A = 256$)?

9. Acetone, a common solvent, has a density of 0.788 g/cm^3. What is the volume of 2.50×10^2 grams of acetone?

10. The density of ethyl acetate is 0.898 g/cm^3. Calculate the mass of 0.588 dm^3 of ethyl acetate.

11. Distinguish between homogeneous and heterogeneous mixtures. In which category do solutions fall?

12. If you wanted to cook in a metal container that heats rapidly, would you choose an iron pan or a copper pan? Explain your answer.

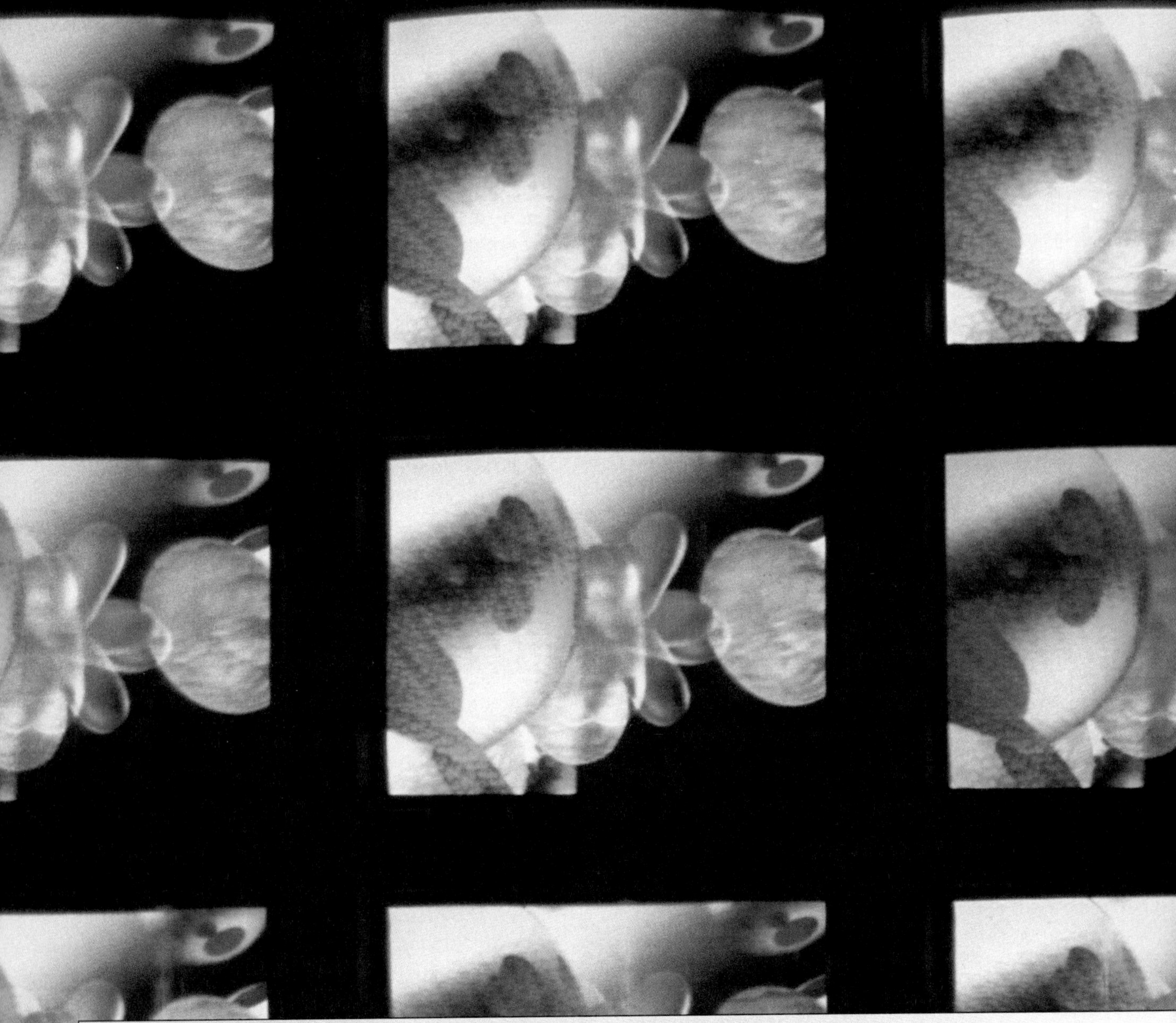

CHAPTER **Preview**

CONTENTS

CHAPTER

6

Periodic Table

What is the Great American Pastime? It used to be baseball, but today it may be watching the game on television. The amount of time many people spend watching television surpasses what is good for their health. However, if programs are chosen selectively, television viewing can be beneficial. To make wise selections, people usually consult a television schedule.

If the schedule is in the form of a table, you can tell at a glance what you need to know. A table is an efficient way to display large amounts of related information. Chemists (and chemistry students) need a way to organize information about the 109 elements in a manner that is as efficient for study and use as a TV schedule. In this chapter you will learn about a table that organizes information about the elements. It is called the periodic table.

EXTENSIONS AND CONNECTIONS

Music Connection

Octaves, page 139

Chemists and Their Work

Dmitri Ivanovich Mendeleev, page 140
Henry Gwyn-Jeffreys Moseley, page 141
Percy Lavon Julian, page 146
Gertrude Belle Elion, page 155

Frontiers

Easy as Un-, Bi-, Tri-: A look at a new system for naming new elements, **page 149**

Chemistry and Society

The Strontium-90 Hazard: An investigation of the dangers of strontium-90 in radioactive fallout, page 150

Everyday Chemistry

Beverage Cans: Steel versus aluminum for the can of the future, **page 157**

OBJECTIVES

- describe the early attempts at classifying elements.
- use the periodic table to predict the electron configurations of elements.
- explain the basis for the arrangement of the modern periodic table.

6.1 Developing the Periodic Table

Scientists have studied many properties of the elements and have found that there are groups of elements that have similar chemical and physical properties. For example, the element sodium reacts vigorously with water. Potassium reacts even more vigorously with water. A chemist can predict that the elements rubidium and cesium will react in a similar manner. In addition, potassium and sodium are silvery white metals and are very soft. A similar prediction can be made about rubidium and cesium.

What information enables scientists to make these predictions? They know that all of these elements have the same number of electrons in their outer energy levels. An understanding of the electron configuration of each of the elements enables scientists to use the periodic table of the elements in a systematic way. At a glance, the arrangement of the elements in this table gives a great deal of information about each of them. Let's look more closely at this periodic table and find out how it is constructed and how it can be used.

Early Attempts at Classification: Dobereiner and Newlands

Early in the nineteenth century, scientists began to seek ways to classify the elements. One attempt at classification was by Johann Dobereiner, a German chemist, in 1817. Dobereiner found that the properties of the metals calcium, barium, and strontium were very similar. He also noted that the atomic mass of strontium was about midway between those of calcium and barium. He grouped these three elements into what he termed a **triad.** Later, Dobereiner found several other groups of three elements with similar properties. Table 6.1 shows three of Dobereiner's triads. In each case, the second element has an atomic mass about halfway between those of the first and third elements in the triad.

In 1863, John Newlands, an English chemist, suggested another classification. He arranged the elements in order of their increasing atomic masses. He noted that there appeared to be a repetition of similar properties every eighth element. Therefore,

Table 6.1

Examples of Dobereiner's Triads						
	Triad 1		Triad 2		Triad 3	
	Name	Atomic Mass	Name	Atomic Mass	Name	Atomic Mass
First element	Calcium	40.1	Chlorine	35.5	Sulfur	32.1
Third element	Barium	137.3	Iodine	126.9	Tellurium	127.6
	Average 88.7		**Average** 81.2		**Average** 79.9	
Second element	Strontium	87.6	Bromine	79.9	Selenium	79.0

he placed seven elements in each group. He then arranged the 49 elements known at that time into seven groups of seven each. Newlands referred to his arrangement as the **law of octaves** because the same properties repeated every eight elements.

Table 6.2

Newlands' Law of Octaves

1	2	3	4	5	6	7
Li	Be	B	C	N	O	F
Na	Mg	Al	Si	P	S	Cl
K						

MUSIC CONNECTION

Octaves

The octave is the standard interval of eight notes on the diatonic scale. If the frequency of note X is, say, 400 Hz, then the frequency of the eighth note after X, called X′, is 800 Hz.

Mendeleev's Periodic Table

Just six years after Newlands' proposal, Dmitri Mendeleev, a Russian chemist, proposed a similar idea. He suggested, as had Newlands, that the properties of the elements were a function of their atomic masses. However, Mendeleev believed that similar properties occurred after periods (horizontal rows) that could vary in length. Although he placed seven elements each in his first two periods, he placed seventeen elements each in the next two.

In the 1860s, Mendeleev and the German chemist Lothar Meyer, each working alone, made an eight-column table of the elements. However, Mendeleev had to leave some blank spots in order to group all the elements with similar properties in the same column. To explain these blank spots, Mendeleev suggested there must be other elements that had not yet been discovered. On the basis of his arrangement, Mendeleev predicted the properties and atomic masses of several elements that were unknown at the time. One of the blank spaces in Mendeleev's table was below silicon. Mendeleev assumed such an element existed but had not yet been discovered. Table 6.3 shows Mendeleev's predictions for the properties of ekasilicon (later named germanium). Today, the missing elements have been discovered, and Mendeleev's predictions have been found to be very nearly correct.

Table 6.3

Mendeleev's Predictions

Ekasilicon (Es)	Germanium (Ge)
Predicted properties	*Actual properties*
1. Atomic mass = 72	1. Atomic mass = 72.61
2. High melting point	2. Melting point = 945°C
3. Density = 5.5 g/cm^3	3. Density = 5.323 g/cm^3
4. Dark gray metal	4. Gray metal
5. Will obtain from K_2EsF_6	5. Obtain from K_2GeF_6
6. Slightly dissolved by HCl	6. Not dissolved by HCl
7. Will form EsO_2	7. Forms oxide (GeO_2)
8. Density of EsO_2 = 4.7 g/cm^3	8. Density of GeO_2 = 4.70 g/cm^3

Figure 6.1 Mendeleev's first arrangement of the elements was published in 1869. Chemists became particularly interested in his periodic law when Mendeleev's predictions were confirmed by the discovery of gallium in 1875 and the discovery of scandium in 1879.

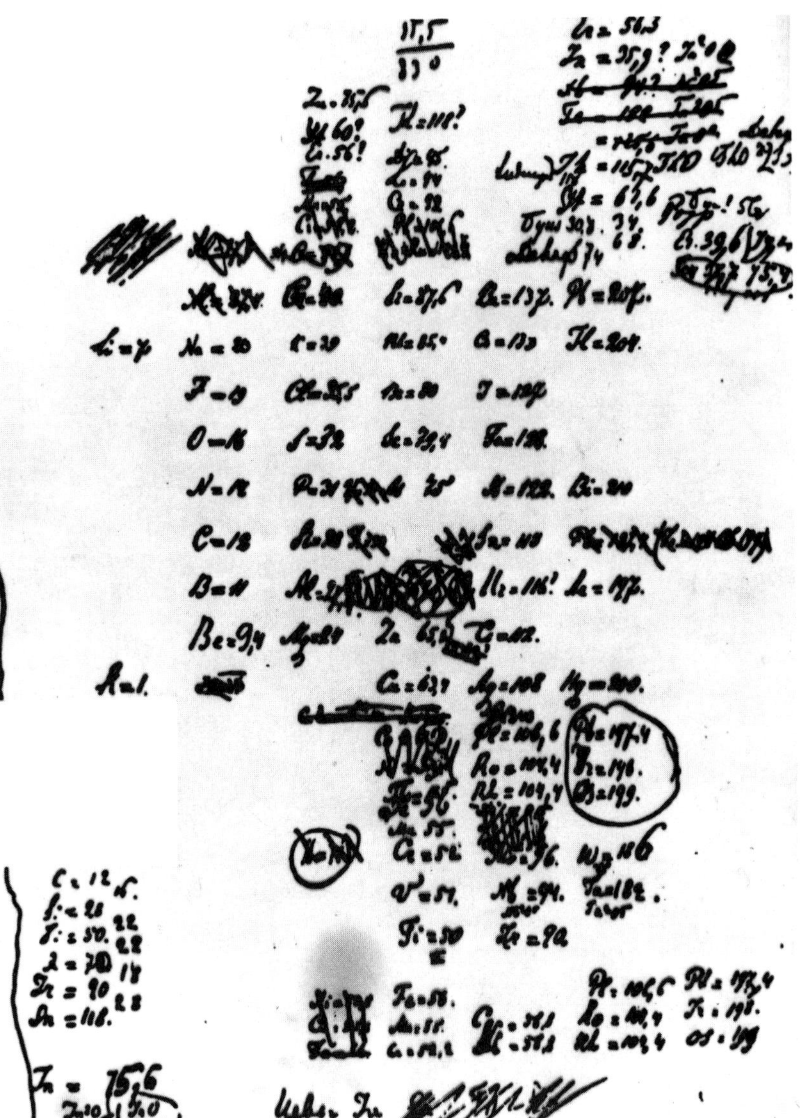

CHEMISTS AND THEIR WORK

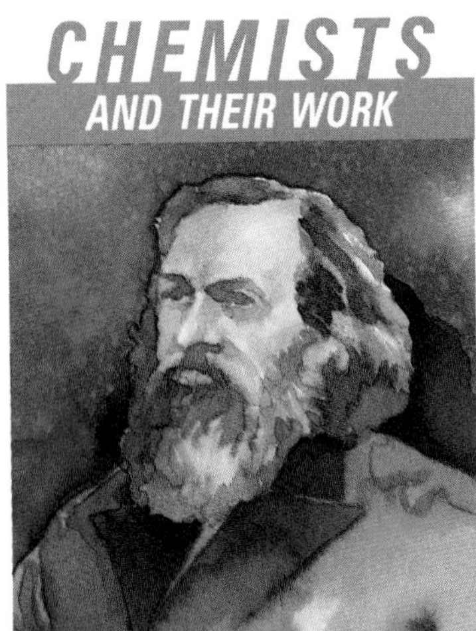

Dmitri Ivanovich Mendeleev (1834-1907)

The periodic table is Mendeleev's lasting contribution to chemistry, but his investigations were much broader in scope throughout his scientific career. Mendeleev was born in Siberia and educated there and at the University of St. Petersburg. He also studied in France and Germany. He then returned to the University of St. Petersburg and worked as a professor until his resignation over political matters in 1890. Mendeleev was a Russian patriot who worked hard for his country. He wrote and continually updated a superb textbook for chemistry students. He received many honors and awards from foreign governments and societies, and element 101 is named in his honor.

In Mendeleev's table, the elements were arranged in order of their increasing atomic masses. The table showed that the properties of the elements are repeated in an orderly way. Such a repeating pattern is periodic. Mendeleev stated that the properties of the elements are a periodic function of their atomic masses. This statement was called the periodic law.

Modern Periodic Law

There was a problem with Mendeleev's table of elements. If the elements were arranged according to increasing atomic masses, tellurium and iodine seemed to be in the wrong columns. Their properties were different from those of other elements in the same column but were similar to those of elements in adjacent columns. Switching their positions put them in the columns where they belonged, according to their properties. If the switch were made, Mendeleev's assumption that the properties of the elements were a periodic function of their atomic masses would be wrong. Mendeleev assumed that the atomic masses of these two elements had been poorly measured, and he placed these two elements according to their properties. He thought that new mass measurements would prove his hypothesis to be correct. However, new measurements simply confirmed the original masses.

Soon, new elements were discovered, and two other pairs showed the same kind of reversal. Cobalt and nickel were known by Mendeleev, but their atomic masses had not been accurately measured. When such a determination was made, it was found that their positions in the table were also reversed. When argon was discovered, the atomic mass was found to be greater than that of potassium. If argon and potassium were put in the table on the basis of atomic masses, their positions would have been reversed.

Henry Moseley found the reason for these apparent exceptions to Mendeleev's periodic law. Moseley's X-ray experiments, in 1913, showed that the nucleus of each element has an integral positive charge, the atomic number. Iodine, nickel, and potassium have greater atomic numbers than do tellurium, cobalt, and argon, respectively. As a result of Moseley's work, the periodic law was revised. It now has as its basis the atomic numbers of the elements instead of the atomic masses. The modern statement of the **periodic law** is *the properties of the elements are a periodic function of their atomic numbers.*

CHEMISTS AND THEIR WORK

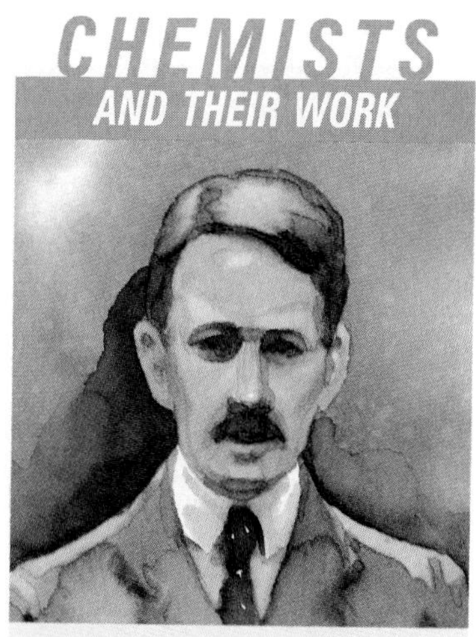

Henry Gwyn-Jeffreys Moseley (1887-1915)

Henry Moseley was a British scientist whose work led to the use of atomic number rather than atomic mass as the basis for the periodic table. His announcement led directly to support for Mendeleev's periodic table. The concept of atomic number also enabled chemists to confirm Mendeleev's claim about missing elements. It could then be demonstrated that there were not any more missing elements.

Modern Periodic Table

The atomic number of an element indicates the number of protons in the nucleus of each atom of the element. Because an atom is electrically neutral, the atomic number also indicates the number of electrons surrounding the nucleus.

Certain electron arrangements are repeated periodically as atoms increase in atomic number. Elements with similar electron configurations can be placed in the same column. The elements in the column can also be listed in order of their increasing principal quantum numbers. Thus, a table of the elements like that on pages 144 and 145 can be formed. This table is the modern **periodic table** of the elements.

The periodic table is constructed in the following manner. Use the arrow diagram on page 128 to determine the order of filling the sublevels. Each *s* sublevel can contain two electrons, as shown in Table 6.4. Each *p* sublevel can contain six electrons arranged in three pairs, or orbitals. Each *d* sublevel can contain ten electrons in five orbitals. Each *f* sublevel can contain fourteen electrons in seven orbitals. Then align the elements with similar outer electron configurations. As you read about the arrangement of the periodic table, refer to Tables 6.5 through 6.9 that show electron configurations. Also refer to the diagrams that indicate the position of each element in the periodic table.

Table 6.4

Sublevel	e^- capacity
s	2
p	6
d	10
f	14

Table 6.5

Elements 1-10				
		1	2	
Z	Element	s	s	p
1	H	1		
2	He	2		
3	Li	2	1	
4	Be	2	2	
5	B	2	2	1
6	C	2	2	2
7	N	2	2	3
8	O	2	2	4
9	F	2	2	5
10	Ne	2	2	6

The first configuration in Table 6.5, hydrogen ($Z = 1$), consists of one electron in the 1*s* sublevel. The second configuration, helium ($Z = 2$), consists of two electrons in the 1*s* sublevel. These two electrons completely fill the 1*s* sublevel. Because the first energy level has only one sublevel (*s*), the first level is now full. The electron configurations of hydrogen and helium are not similar, so each element is in a separate column of the periodic table. Thus, the first row of the periodic table has two elements, just as the first energy level has only two electrons. The third element, lithium ($Z = 3$), has two electrons in the 1*s* sublevel and one electron in the 2*s* sublevel. Lithium is similar to hydrogen in that it has only one electron in its outermost level. Therefore, it is placed in the same column as hydrogen. The next element, beryllium ($Z = 4$), has two electrons in the 1*s* sublevel and two electrons in the 2*s* sublevel. It might seem to belong in the column with helium. However, the two electrons in helium's outermost level fill that level. Recall from Chapter 5 that the $n = 2$ level has a *p* sublevel as well as an *s* sublevel. The two electrons in the 2*s* sublevel of beryllium do not fill the second level. Therefore, beryllium starts a new column next to lithium. Boron ($Z = 5$) has a configuration composed of two 1*s* electrons, two 2*s* electrons, and one 2*p* electron, so boron heads a new column. Carbon ($Z = 6$), nitrogen ($Z = 7$), oxygen ($Z = 8$), and fluorine ($Z = 9$) atoms come next. These atoms have structures containing two, three, four, and five electrons, respectively, in the 2*p* sublevel. Each of these elements heads a new column. The atoms of neon ($Z = 10$), the tenth element, contain six 2*p* electrons. The second level ($n = 2$) is now full. Thus, neon is placed in the same column as helium.

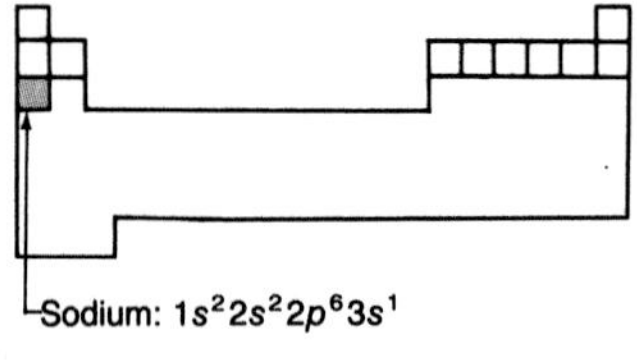

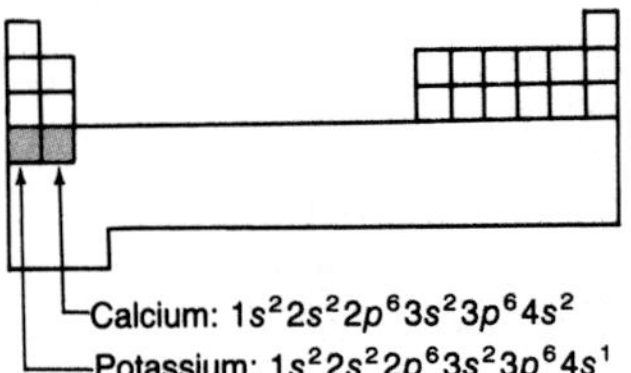

The Group 18 (VIIIA) elements, the noble gases, have full outer energy levels. The symbol of a noble gas, in brackets, can be used to simplify writing electron configurations. For example, the electron configuration of Na could be written $[Ne]3s^1$. This notation is used in Tables 6.6 through 6.9.

Table 6.6

Elements 11-20									
		1	2		3			4
Z	Element	s	s	p	s	p	d	s
		2	2	6				
11	Na	[Ne]			1			
12	Mg				2			
13	Al				2	1		
14	Si				2	2		
15	P				2	3		
16	S				2	4		
17	Cl				2	5		
18	Ar				2	6		
19	K				2	6		1
20	Ca				2	6		2

Sodium atoms ($Z = 11$) have the same outer level configuration as lithium atoms, one s electron ($3s^1$). Thus, sodium is placed under lithium in the same column of the periodic table. The elements magnesium ($Z = 12$) through argon ($Z = 18$) have the same outer structures as the elements beryllium through neon have. They are also placed in the appropriate columns under atoms with similar configurations. Atoms of potassium ($Z = 19$) and calcium ($Z = 20$) have outer structures that are similar to the atoms of sodium and magnesium. Look closely at Tables 6.5 and 6.6 and the positions of these elements in the periodic table, pages 144–145. Note that each time a new energy level is started, a new row in the periodic table begins.

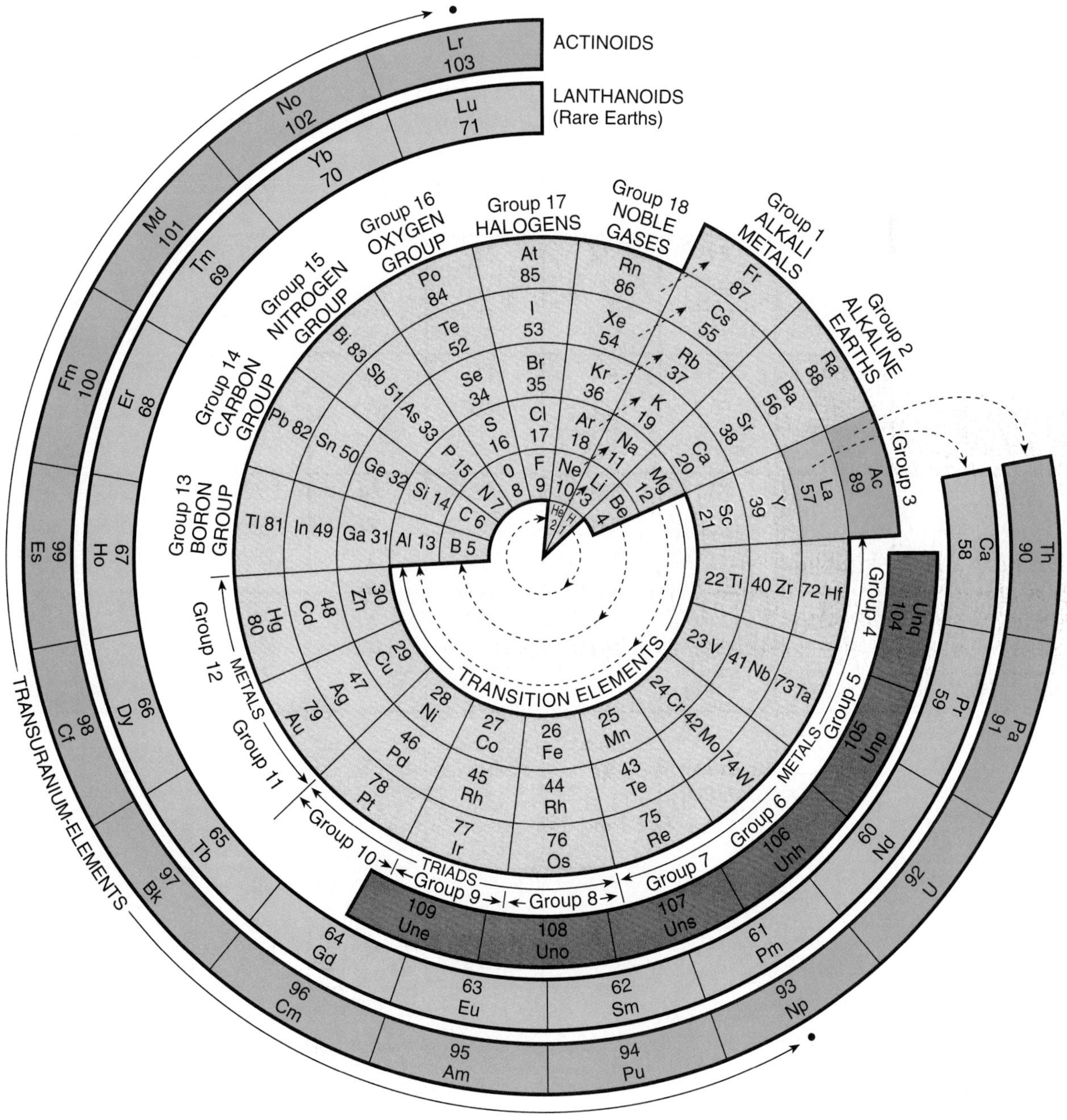

Figure 6.2 This representation of the periodic table also places elements with similar electron structures in the same column.

Periodic Table

(Based on Carbon 12 = 12.000)

Metals

Period	1* IA*	2 IIA	3 IIIB	4 IVB	5 VB	6 VIB	7 VIIB	8	9 VIIIB
1	1 **H** Hydrogen 1.007 94								
2	3 **Li** Lithium 6.941	4 **Be** Beryllium 9.012 182							
3	11 **Na** Sodium 22.989 768	12 **Mg** Magnesium 24.305 0							
4	19 **K** Potassium 39.098 3	20 **Ca** Calcium 40.078	21 **Sc** Scandium 44.955 910	22 **Ti** Titanium 47.88	23 **V** Vanadium 50.941 5	24 **Cr** Chromium 51.996 1	25 **Mn** Manganese 54.938 05	26 **Fe** Iron 55.847	27 **Co** Cobalt 58.933 20
5	37 **Rb** Rubidium 85.467 8	38 **Sr** Strontium 87.62	39 **Y** Yttrium 88.905 85	40 **Zr** Zirconium 91.224	41 **Nb** Niobium 92.906 38	42 **Mo** Molybdenum 95.94	43 **Tc** Technetium 97.907 2	44 **Ru** Ruthenium 101.07	45 **Rh** Rhodium 102.905 50
6	55 **Cs** Cesium 132.905 43	56 **Ba** Barium 137.327	71 **Lu** Lutetium 174.967	72 **Hf** Hafnium 178.49	73 **Ta** Tantalum 180.947 9	74 **W** Tungsten 183.85	75 **Re** Rhenium 186.207	76 **Os** Osmium 190.2	77 **Ir** Iridium 192.22
7	87 **Fr** Francium 223.019 7	88 **Ra** Radium 226.025 4	103 **Lr** Lawrencium 260.105 4	104 **Unq** — 261	105 **Unp** — 262	106 **Unh** — 263	107 **Uns** — 262	108 **Uno** — 265	109 **Une** — 266

Series						
LANTHANOID SERIES	57 **La** Lanthanum 138.905 5	58 **Ce** Cerium 140.115	59 **Pr** Praseodymium 140.907 65	60 **Nd** Neodymium 144.24	61 **Pm** Promethium 144.912 8	62 **Sm** Samarium 150.36
ACTINOID SERIES	89 **Ac** Actinium 227.027 8	90 **Th** Thorium 232.038 1	91 **Pa** Protactinium 231.035 88	92 **U** Uranium 238.028 9	93 **Np** Neptunium 237.048 2	94 **Pu** Plutonium 244.064 2

Gases—green, Liquids—blue, Solids—yellow, Synthetics—orange
(State is at room temperature and standard atmospheric pressure.)

Nonmetals

10	11 IB	12 IIB	13 IIIA	14 IVA	15 VA	16 VIA	17 VIIA	18 VIIIA
								2 **He** Helium 4.002 602
			5 **B** Boron 10.811	6 **C** Carbon 12.011	7 **N** Nitrogen 14.006 74	8 **O** Oxygen 15.999 4	9 **F** Fluorine 18.998 403 2	10 **Ne** Neon 20.179 7
			13 **Al** Aluminum 26.981 539	14 **Si** Silicon 28.085 5	15 **P** Phosphorus 30.973 762	16 **S** Sulfur 32.066	17 **Cl** Chlorine 35.452 7	18 **Ar** Argon 39.948
28 **Ni** Nickel 58.6934	29 **Cu** Copper 63.546	30 **Zn** Zinc 65.39	31 **Ga** Gallium 69.723	32 **Ge** Germanium 72.61	33 **As** Arsenic 74.921 59	34 **Se** Selenium 78.96	35 **Br** Bromine 79.904	36 **Kr** Krypton 83.80
46 **Pd** Palladium 106.42	47 **Ag** Silver 107.868 2	48 **Cd** Cadmium 112.411	49 **In** Indium 114.82	50 **Sn** Tin 118.710	51 **Sb** Antimony 121.757	52 **Te** Tellurium 127.60	53 **I** Iodine 126.904 47	54 **Xe** Xenon 131.290
78 **Pt** Platinum 195.08	79 **Au** Gold 196.966 54	80 **Hg** Mercury 200.59	81 **Tl** Thallium 204.383 3	82 **Pb** Lead 207.2	83 **Bi** Bismuth 208.980 37	84 **Po** Polonium 208.982 4	85 **At** Astatine 209.987 1	86 **Rn** Radon 222.017 6

†Metalloids lie along this heavy stairstep line.

63 **Eu** Europium 151.965	64 **Gd** Gadolinium 157.25	65 **Tb** Terbium 158.925 34	66 **Dy** Dysprosium 162.50	67 **Ho** Holmium 164.930 32	68 **Er** Erbium 167.26	69 **Tm** Thulium 168.934 21	70 **Yb** Ytterbium 173.04
95 **Am** Americium 243.061 4	96 **Cm** Curium 247.070 3	97 **Bk** Berkelium 247.070 3	98 **Cf** Californium 251.079 6	99 **Es** Einsteinium 252.082 8	100 **Fm** Fermium 257.095 1	101 **Md** Mendelevium 258.098 6	102 **No** Nobelium 259.100 9

*Currently there are two systems of labeling groups on the periodic table. A traditional system uses Roman numerals I through VIII with letters A and B. A more current system uses Arabic numerals 1 through 18, with no A and B designations. Throughout this text the current system will be used with the traditional heading following in parenthesis, for example, Group 1(IA).

Table 6.7

Transition Elements																			
		1	2		3			4				5				6			7
Z	Element	s	s	p	s	p	d	s	p	d	f	s	p	d	f	s	p	d	s
		2	2	6	2	6													
21	Sc						1	2											
22	Ti						2	2											
23	V						3	2											
24	Cr						5	1											
25	Mn						5	2											
26	Fe			[Ar]			6	2											
27	Co						7	2											
28	Ni						8	2											
29	Cu						10	1											
30	Zn						10	2											
		2	2	6	2	6	10	2	6										
39	Y									1		2							
40	Zr									2		2							
41	Nb									4		1							
42	Mo									5		1							
43	Tc									5		2							
44	Ru									7		1							
45	Rh									8		1							
46	Pd									10									
47	Ag									10		1							
48	Cd									10		2							
71	Lu									10	14	2	6	1		2			
72	Hf					[Kr]				10	14	2	6	2		2			
73	Ta									10	14	2	6	3		2			
74	W									10	14	2	6	4		2			
75	Re									10	14	2	6	5		2			
76	Os									10	14	2	6	6		2			
77	Ir									10	14	2	6	7		2			
78	Pt									10	14	2	6	9		1			
79	Au									10	14	2	6	10		1			
80	Hg									10	14	2	6	10		2			
103	Lr									10	14	2	6	10	14	2	6	1	2
104	Unq									10	14	2	6	10	14	2	6	2	2?
105	Unp									10	14	2	6	10	14	2	6	3	2?

CHEMISTS AND THEIR WORK

Percy Lavon Julian
(1899-1975)

U.S. chemist Percy Julian had to overcome stiff prejudice against African Americans in his search for both education and employment. Eventually, he was accepted at DePauw University from which he graduated Phi Beta Kappa. Julian received masters and doctorate degrees. During his varied career, Julian synthesized several medically important compounds found naturally in soybeans and yams and made them available at much lower cost. One of the best-known compounds Julian synthesized was cortisone, used in treating arthritis.

Transition Elements

The scandium (Z = 21) configuration introduces a new factor into the arrangement. It has two electrons in the outer level ($4s^2$) and is similar to the calcium configuration. However, in addition to a filled $4s$ sublevel, the scandium atom has one electron in the $3d$ sublevel. It is therefore placed in a new column, which is labeled 3 (IIIB). For titanium (Z = 22) through nickel (Z = 28), elements

Figure 6.3 Shown here are samples of the transition elements from $Z = 21$ through $Z = 30$. From left to right in the first row are Sc, Ti, V, Cr, and Mn. From left to right in the back row are Fe, Co, Ni, Cu, and Zn.

have additional electrons in the 3*d* sublevel. The electron configuration of each of these elements is shown in Table 6.7. For these elements, the outer level is the 4th level, so they are placed in the fourth row. Each of these elements heads a new column. These are columns 3 through 10 (IIIB-VIIIB). Note that the atoms of copper and zinc have filled inner levels. All structures in column 11 (IB) have filled inner levels and one electron in the outer level. All structures in column 12 (IIB) have filled inner levels and two electrons in the outer level. The elements in columns 3 through 12 (IIIB-IIB) are called the **transition elements.**

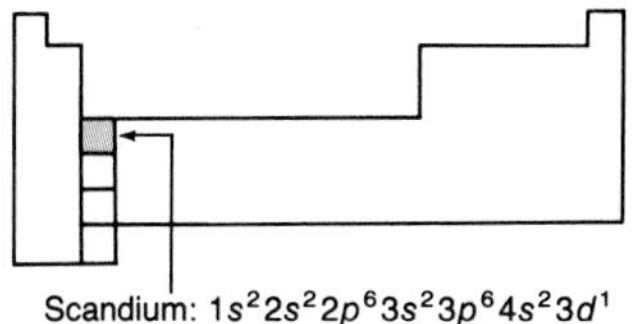

In columns 13 through 18 (IIIA-VIIIA), each element has one more electron in the *p* sublevel. The elements in these columns are not shown in Table 6.7 because they follow the same pattern as elements 13 through 18, which you have already seen. Except for helium, elements in column 18 (VIIIA) have a total of eight electrons in the outer level. The next electron is in the next *s* sublevel and begins a new energy level, whether the inner level is filled or not. This process is continued until all of the elements are placed in the main part of the table.

The Lanthanoids and Actinoids

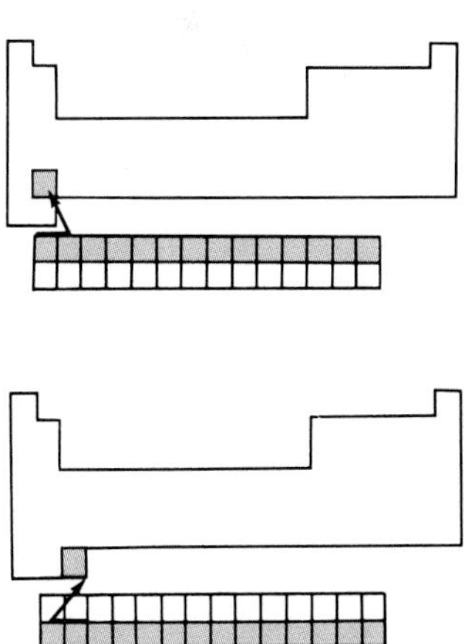

The **lanthanoid series** contains the elements lanthanum ($Z = 57$) through ytterbium ($Z = 70$). These elements have a predicted structure with two electrons in the outer level. In general, in this series, electrons are being added to the 4*f* sublevel instead of to the sixth or outer level, as shown in Table 6.8.

The **actinoid series** contains actinium ($Z = 89$) through nobelium ($Z = 102$). In general, elements in this series have an increasing number of electrons in the 5*f* sublevel, as shown in Table 6.9.

The electron configurations of some elements in these periods do not fit the pattern of the arrow diagram. However, the differences involve only one or two electrons. For example, note the configuration of gadolinium ($Z = 64$). The electron that you would expect to find in the 4*f* sublevel is in the 5*d* sublevel. However, for purposes of constructing the periodic table, we assume that all elements have the predicted configurations.

The two rows of fourteen blocks representing the lanthanoids and the actinoids could be placed between elements 56 and 71 and 88 and 103, respectively. However, to do so in the main part of the periodic table would make the table very wide and awkward to use. For this reason, the two rows representing the actinoids and the lanthanoids are placed below the main body of the table.

Table 6.8

Lanthanoid Series															
Z	**Element**	**1**	**2**		**3**			**4**				**5**			**6**
		s	*s*	*p*	*s*	*p*	*d*	*s*	*p*	*d*	*f*	*s*	*p*	*d*	*s*
		2	2	6	2	6	10	2	6						
57	La									10		2	6	1	2
58	Ce									10	1	2	6	1	2
59	Pr									10	3	2	6		2
60	Nd									10	4	2	6		2
61	Pm									10	5	2	6		2
62	Sm									10	6	2	6		2
63	Eu									10	7	2	6		2
64	Gd	[Kr]								10	7	2	6	1	2
65	Tb									10	9	2	6		2
66	Dy									10	10	2	6		2
67	Ho									10	11	2	6		2
68	Er									10	12	2	6		2
69	Tm									10	13	2	6		2
70	Yb									10	14	2	6		2

Table 6.9

Actinoid Series																			
Z	**Element**	**1**	**2**		**3**			**4**				**5**				**6**			**7**
		s	*s*	*p*	*s*	*p*	*d*	*s*	*p*	*d*	*f*	*s*	*p*	*d*	*f*	*s*	*p*	*d*	*s*
		2	2	6	2	6	10	2	6										
89	Ac									10	14	2	6	10		2	6	1	2
90	Th									10	14	2	6	10		2	6	2	2
91	Pa									10	14	2	6	10	2	2	6	1	2
92	U									10	14	2	6	10	3	2	6	1	2
93	Np									10	14	2	6	10	4	2	6	1	2
94	Pu									10	14	2	6	10	6	2	6		2
95	Am	[Kr]								10	14	2	6	10	7	2	6		2
96	Cm									10	14	2	6	10	7	2	6	1	2
97	Bk									10	14	2	6	10	8	2	6	1	2
98	Cf									10	14	2	6	10	10	2	6		2
99	Es									10	14	2	6	10	11	2	6		2
100	Fm									10	14	2	6	10	12	2	6		2
101	Md									10	14	2	6	10	13	2	6		2
102	No									10	14	2	6	10	14	2	6		2

In the periodic table, all elements in a horizontal row are referred to as a **period.** All elements in the same vertical column are referred to as a **group.** Note that there are 18 groups in the periodic table and that the group number is at the top of each column in the table.

6.1 CONCEPT REVIEW

1. For the lanthanoid and actinoid elements, what is the most common deviation of electron configurations from the predicted arrow diagram configurations?
2. State the modern periodic law.
3. Predict the ending of the electron configuration of each of the following elements using only the periodic table as a guide: Ne, Pr, Sc, Tl, Zr.
4. Trace the development of periodic organization of the elements from Dobereiner through Mendeleev.
5. **Apply** Which two elements from the following list would have similar chemical properties?
   a. magnesium
   b. sodium
   c. nickel
   d. barium
   e. phosphorus
   f. chlorine

FRONTIERS

Easy as Un-, Bi-, Tri-

Scientists from both the United States and the former Soviet Union lay claim to the discovery of element 104. American scientists named it rutherfordium, but Russian scientists named it kurchatovium. The element is a synthetic element, created in linear accelerators such as the one shown. In the accelerators, ions traveling at high speeds crash into targets of massive elements, producing new elements. Because similar research was simultaneously going on worldwide, evaluating the discovery claims is difficult.

One of the responsibilities of the International Union of Pure and Applied Chemistry, IUPAC, is to standardize the naming of chemical substances.

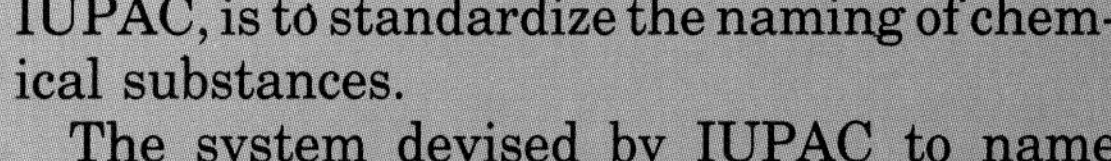

The system devised by IUPAC to name elements of atomic number 104 and greater is simple. The name of the element is derived directly from its atomic number using the Latin and Greek roots shown below. The roots are arranged in the order of the digits of the atomic number with the suffix *-ium* added to complete the name. Thus, element 104 becomes *unnilquadium* (oon nihl KWAH dee um), literally, *one-zero-four*. The chemical symbol of an element in this system is composed of the initial letters of the numerical roots which make up its name.

0	*nil*	2	*bi*	4	*quad*	6	*hex*	8	*oct*
1	*un*	3	*tri*	5	*pent*	7	*sept*	9	*enn*

CHEMISTRY AND SOCIETY

The Strontium-90 Hazard

On April 26, 1986, there was a meltdown at the nuclear power plant in Chernobyl, Ukraine. As a result of the accident, about eight metric tons of material were sprayed into the atmosphere. The waste products of a nuclear reactor contain a mixture of substances, most of them radioactive. A common waste material is cesium-137. Another substance found in nuclear waste is an isotope of strontium, strontium-90. Stron-Strontium-90 is radioactive, but its chemistry is what makes it particularly hazardous to people.

If you locate strontium on the periodic table, you will see that it is in Group 2 (IIA), just below calcium. As you know, members of the same group in the periodic table react in similar ways. Thus, strontium can take the place of calcium in chemical reactions. In the human body, calcium is found in the bones and teeth. The major constituent of the mineral part of bone is calcium phosphate, $Ca_3(PO_4)_2$. By mass, calcium phosphate is 38.76% calcium. If strontium-90 replaced some of the calcium in bones, the bones would be exposed to radiation from the strontium-90.

How does this replacement happen? Think about how you get calcium into your body. Calcium compounds in the soil are absorbed by grass, which is then eaten by a cow. You drink the cow's milk or eat milk products such as cheese. Strontium released into the atmosphere will fall to the ground. There, like calcium, it will move into the soil and be absorbed by plants. Because strontium can take the place of calcium in reactions, the strontium-90 follows the same biochemical pathways that calcium takes.

After the Chernobyl accident, areas of northern Europe received fallout from the atmosphere. In these places, the strontium-90 level in milk was carefully monitored. In some places, the milk had to be discarded because it contained too much of the radioactive isotope. Because scientists understand how one element can take the place of another, they were able to help protect the people against this potential hazard. Many other foods were also discarded because of the presence of radioactive isotopes that posed health hazards. Many people believe that these and other risks associated with the use of nuclear power plants outweigh any benefits achieved from this form of energy.

Supporters of the use of nuclear power point out that the use of nuclear power plants conserves fossil fuels, provides large amounts of energy, and contributes little to air pollution. Even some wastes have positive uses. Strontium-90 has been used to treat some kinds of bone diseases and power experiments on spacecraft.

Analyzing the Issue

1. Research the present safeguards used in nuclear power plants in the United States. Do you think these safeguards are adequate? Justify your answer.
2. How would you feel about a nuclear power plant being built close to your home? Support your opinion.

6.2 Using the Periodic Table

> **OBJECTIVES**
> - identify metals, nonmetals, and metalloids on the periodic table.
> - give examples of the relationship between an element's electron configuration and its placement on the periodic table.
> - predict the chemical stability of atoms using the octet rule.

Now that you have studied the construction of the periodic table, you are in a position to use the table. Because the table was constructed on the basis of the electron configurations of the elements, anything that depends on those configurations may be related to the position of an element in the table. As you learn about the variation of properties of the elements, always remember that the differences in properties are due to the differing electron configurations, not to the position of the element in the table.

Surveying the Table: Electron Configurations

The periodic table was originally constructed by placing elements with similar properties in a column. We now know that an atom's chemical properties are determined by its electron configuration. Therefore, the modern periodic table has been constructed on the basis of electron configurations of the atoms of each element. By knowing the basis on which the table is constructed, you can use the table to "read" the electron configuration of an element. Remember that elements in columns 1 (IA), 2 (IIA), and 13 through 18 (IIIA through VIIIA) have their highest energy electron in an outer *s* or *p* sublevel. Those elements in columns 3 through 12 (IIIB through IIB) have their highest energy electron in a *d* sublevel, which is one level below the outer level.

Thus, the written configuration of any element in Group 1 (IA) will end in s^1. This configuration means that the outer level of each atom of Group 1 (IA) elements contains one electron. The coefficient of s^1 is easily found from the table because the number of the period indicates the outer energy level. For example, potassium is in the fourth period of Group 1 (IA). Thus, the written electron configuration for the outer level of potassium is $4s^1$. The expression s^1 indicates the group, and the coefficient, 4, indicates the period number. Find lithium in the periodic table. How does

Figure 6.4 Silver, gold, and copper are known as the coinage metals. Their properties are similar because their electron structures are similar.

s Comes Before *d*

Put on an apron and goggles. Place about 1 g of iron filings into a test tube. Add 4 cm^3 of 6*M* hydrochloric acid to the test tube. **CAUTION:** *Acid is corrosive.* Place the test tube in a test-tube rack. Place approximately 1 g of iron(III) chloride, $FeCl_3$, into a second test tube. **CAUTION:** *$FeCl_3$ is a skin irritant.* Add 4 cm^3 of water to the test tube. Stopper the test tube and shake until some of the solid has dissolved. Allow the contents of both test tubes to settle, then record your observations. What is the electron configuration for iron metal? Iron loses two electrons to form the pale green substance in the first test tube. Predict what two electrons are lost. Iron loses three electrons to form the yellow iron(III) chloride in the second test tube. Which additional electron do you think was lost? Other transition elements can lose up to eight electrons. Why can iron lose only two or three electrons?

its written electron configuration end? Find Group 2 (IIA) in the periodic table. How does the written electron configuration for all elements in this group end? The same procedure can be used for Groups 13 through 18 (IIIA through VIIIA). There the endings, instead of s^1 or s^2, are p^1 through p^6 preceded by a coefficient that is the same as the number of the period.

For Groups 3 through 12 (IIIB through IIB), the endings are d^1 through d^{10}, preceded by a coefficient that is one less than the period number (for example, ... 4*s*3*d*). For these transition elements, remember that the *d* sublevel is always preceded by an *s* sublevel that is one quantum number higher. For example, iron is in the fourth period. You should expect iron to have two electrons in the 4*s* sublevel, and it does. Iron is in the sixth group, or column, of the transition elements. You should expect iron to have six electrons in the 3*d* sublevel, and it does. For the lanthanoids and actinoids, the endings are f^1 through f^{14} preceded by a coefficient that is two less than the period number (for example, ... 6*s*4*f*). The coefficient is 4 for the lanthanoids and 5 for the actinoids.

Some electron configurations are not what you might predict. To understand some of the exceptions to the arrow diagram, it is necessary to know that there appears to be a special chemical stability associated with certain electron configurations in an atom.

Relative Stability of Electron Configurations

Look at the order for filling the energy sublevels predicted by the arrow diagram on page 128:

1*s* 2*s* 2*p* 3*s* 3*p* 4*s* 3*d* **4*p* 5*s*** 4*d* **5*p* 6*s*** 4*f* 5*d* **6*p* 7*s*** 5*f* 6*d*...

Note that the outer energy level can have electrons in only the *s* and *p* sublevels. For example, if an atom has full 3*s* and 3*p* sublevels, the 4*s* sublevel will fill before electrons can enter the 3*d* sublevel. As soon as an electron enters the fourth energy level, the third energy level is no longer the outer level. In a similar

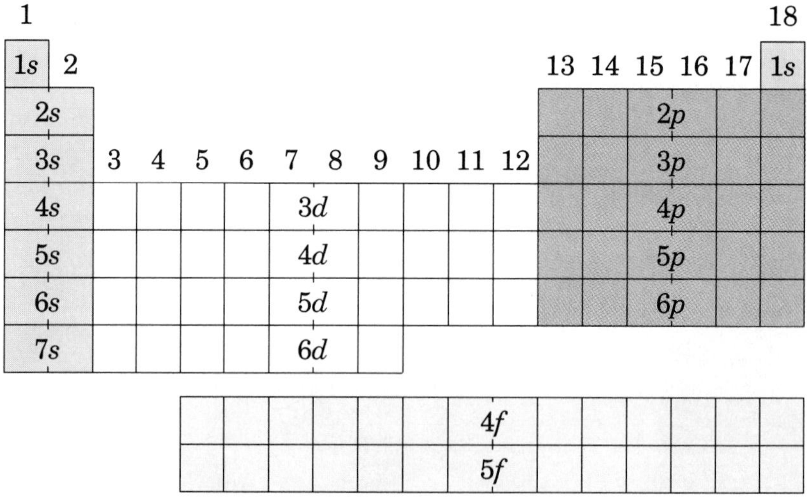

1	2	3	4	5	6	7	8	9	10	11	12	13	14	15	16	17	18
1*s*																	1*s*
2*s*														2*p*			
3*s*														3*p*			
4*s*						3*d*								4*p*			
5*s*						4*d*								5*p*			
6*s*						5*d*								6*p*			
7*s*						6*d*											

						4*f*							
						5*f*							

Figure 6.5 The position of an element in the periodic table can be used to determine the element's electron configuration. The outermost electron in an element is assigned to the indicated orbital.

Figure 6.6 The noble gas xenon reacts with fluorine to form xenon tetrafluoride, seen here in crystalline form.

way, 4*d* is not begun until 5*s* is filled, 4*f* and 5*d* are not begun until 6*s* is filled, and 5*f* and 6*d* are not begun until 7*s* is filled. Although any energy level greater than 2 has more than the *s* and *p* sublevels, an outer energy level is considered filled when the *s* and *p* sublevels are filled. The energy level then is considered to have the maximum number of electrons that can be contained in a normal outer level.

One of the primary rules in chemistry is that atoms with full outer levels are particularly stable (less reactive). For all such elements except helium, the outer level contains eight electrons, two in the outer *s* sublevel and six in the outer *p* sublevel. These eight outer electrons are called an octet. The fact that *eight electrons in the outer level render an atom unreactive* is called the **octet rule.** Although the helium atom has only two electrons in its outer level, it, too, is one of these particularly stable elements. Helium's outer level is the first level, and it can hold only two electrons. Thus, it has a full outer level; the octet rule includes helium. Under some circumstances, it is possible to force the outer level of an atom in the third or higher period to hold more than eight electrons.

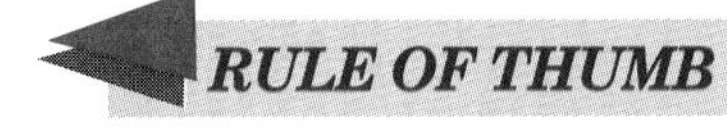

In addition to the outer octet, there are other electron configurations of high relative stability. *An atom having a filled or half-filled sublevel is slightly more stable (less reactive) than an atom without a filled or half-filled sublevel.* Thus, chromium is predicted to have two electrons in its 4*s* sublevel and four electrons in its 3*d* sublevel. Actually, as is shown in the diagram on page 154, chromium has one electron in its 4*s* sublevel and five electrons in its 3*d* sublevel. Note that one electron is shifted between two very closely spaced sublevels. The atom thus has two half-full sublevels instead of one full sublevel and one with no special arrangement. Copper has a similar variation from the prediction made by using the arrow diagram. Copper is predicted to have two 4*s* electrons and nine 3*d* electrons. Actually, it has one electron in its 4*s* sublevel and ten electrons in its 3*d* sublevel. One full and one half-full sublevel make an atom more stable than do one full

Figure 6.7 The relative stability of an atom can be predicted by its electron configuration.

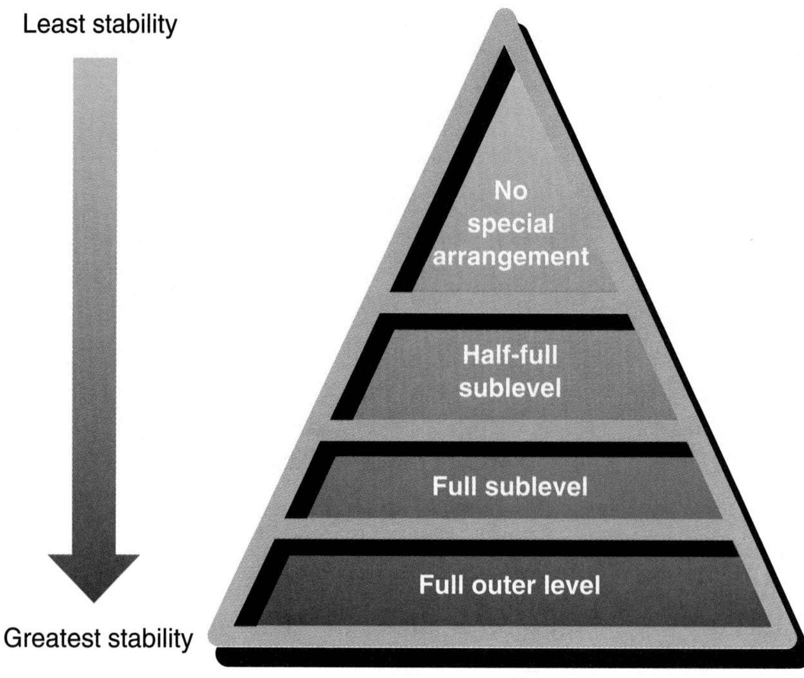

sublevel and one sublevel with no special arrangement as predicted. Most of the exceptions from predicted configurations can be explained in this way.

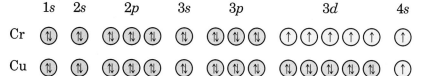

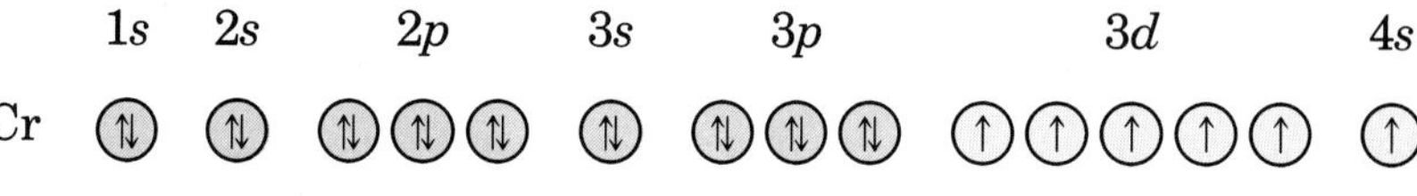

When atoms react with each other, they do so because the reacting system is more stable after the reaction than it was before the reaction. There are a number of factors that determine the stability of reacting systems, and you will learn more about some of them later in this course. However, one of the factors contributing to stability is the electron configurations of the reacting atoms and of their reaction products. In general, atoms will react to produce a system in which all the atoms end up with full outer electron energy levels. There are several ways in which full outer electron energy levels may be achieved. One obvious way is to add electrons to a partially filled outer level. Another way is to share electrons with another atom. A third method is for an atom to lose all the electrons in its outer level, thereby exposing an underlying level that is already full. You will examine each of these processes in greater detail later.

As you look at the electron structure of an atom for the purpose of predicting how that atom will react, remember that achievement of a full outer level is only one of many factors determining reactivity. Consider, for example, the chlorine atom. Chlorine has seven electrons in its outer level. If electron structure were the only consideration, you would expect that all chlorine

Bubbles That Burn

Put on an apron and goggles. Half-fill a small beaker with water. Add a few drops of liquid dishwashing detergent and stir. Place a small piece of calcium carbide, CaC_2, in the water. Use tongs to hold a burning splint. **CAUTION:** *Fire risk.* Use the burning splint to pop the bubbles that form. Record your observations. Ethyne gas (acetylene), C_2H_2, is produced in the reaction. Draw electron dot diagrams for the two carbon atoms and the two hydrogen atoms in acetylene. Predict how the atoms can share electrons to demonstrate the octet rule. How do you know this gas is reactive? If enough oxygen were available for complete combustion, what would the products be?

atoms would instantly gain one electron to achieve an outer level of eight electrons. Such is not the case. Although chlorine does gain one electron readily, it also undergoes other kinds of reactions that result in stable products. When chlorine does gain one electron, that electron must come from some other atom. Thus, not only chlorine's stability must be considered, but also the stability of the atom that loses the electron.

Metals and Nonmetals

Groups 1 (IA) and 2 (IIA) of the periodic table contain the most active metals. Because the elements of a group have similar characteristics, a group is sometimes called a **family.** Many groups have family names. Group 1 (IA), except hydrogen, is called the alkali metal family. Group 2 (IIA) is called the alkaline earth metal family.

The nonmetals are on the other side of the table in Groups 16 (VIA), 17 (VIIA), and 18 (VIIIA). Group 16 (VIA) is called the chalcogen (KAL kuh juhn) family. Group 17 (VIIA) is known as the halogen family. The elements in Group 18 (VIIIA) are called the noble gases.

You're probably familiar with typical metallic properties. Generally, **metals** are hard and shiny. They conduct heat and electricity well. **Nonmetals** are generally gases or brittle solids at room temperature. If solid, their surfaces are dull, and they are insulators. Elements are classified as metals or nonmetals on the basis of their electronic structure. One characteristic of a metal atom is that it has only a few electrons in the outer level. In forming compounds, metals have a tendency to lose these outer electrons. Nonmetals have more electrons in the outer level. When forming compounds, nonmetals frequently gain electrons. They can also share their outer electrons with other atoms.

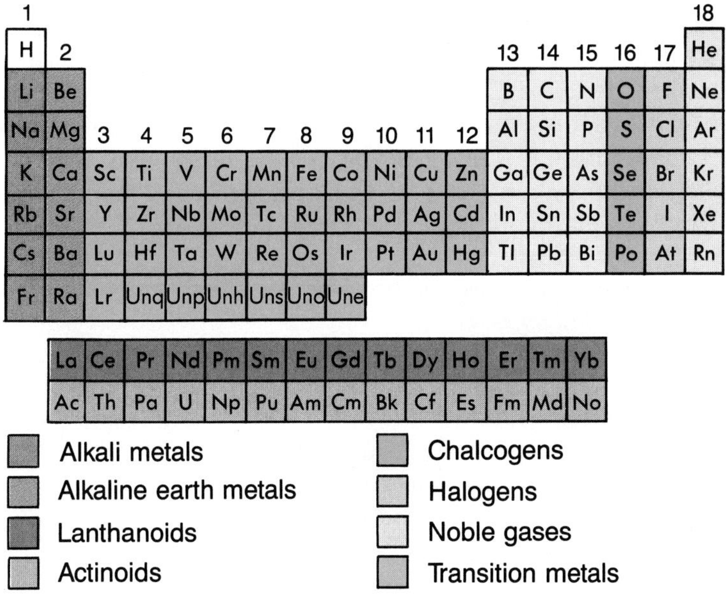

1	2	3	4	5	6	7	8	9	10	11	12	13	14	15	16	17	18
H																	He
Li	Be											B	C	N	O	F	Ne
Na	Mg											Al	Si	P	S	Cl	Ar
K	Ca	Sc	Ti	V	Cr	Mn	Fe	Co	Ni	Cu	Zn	Ga	Ge	As	Se	Br	Kr
Rb	Sr	Y	Zr	Nb	Mo	Tc	Ru	Rh	Pd	Ag	Cd	In	Sn	Sb	Te	I	Xe
Cs	Ba	Lu	Hf	Ta	W	Re	Os	Ir	Pt	Au	Hg	Tl	Pb	Bi	Po	At	Rn
Fr	Ra	Lr	Unq	Unp	Unh	Uns	Uno	Une									

La	Ce	Pr	Nd	Pm	Sm	Eu	Gd	Tb	Dy	Ho	Er	Tm	Yb
Ac	Th	Pa	U	Np	Pu	Am	Cm	Bk	Cf	Es	Fm	Md	No

- Alkali metals
- Alkaline earth metals
- Lanthanoids
- Actinoids
- Chalcogens
- Halogens
- Noble gases
- Transition metals

Figure 6.8 Groups of elements are often called families.

CHEMISTS AND THEIR WORK

Gertrude Belle Elion
(1918-)

Educated at Hunter College and New York University, American chemist Gertrude Elion faced considerable bias in her attempts to get a job as a professional chemist. She taught in high school and took a nonpaying laboratory job to gain experience in her field. Finally, when World War II caused a shortage of male chemists, Elion got a job in pharmaceutical research. She synthesized several new drugs that are still in use and eventually became director of experimental therapy in her firm. She has received honorary degrees from universities and honors from professional societies. In 1988 she was one of the three recipients of the Nobel Prize for Physiology and Medicine.

Figure 6.9 The metal copper, the metalloid arsenic, and the nonmetal sulfur are pictured left to right. Each element has properties characteristic of its classification.

RULE OF THUMB 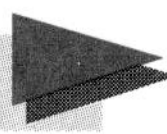

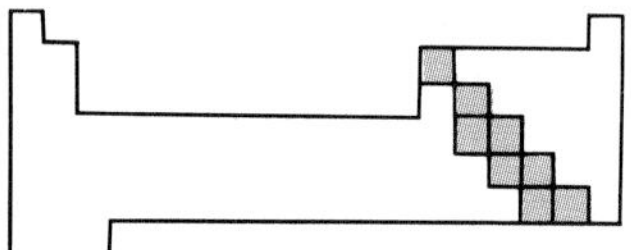

As a general rule, elements with three or fewer electrons in the outer level are considered to be metals. Elements with five or more electrons in the outer level are considered to be nonmetals. There are exceptions. Some elements have properties of both metals and nonmetals. These elements are called the **metalloids.** Silicon, an element used in the manufacture of microcomputer chips, is a metalloid. On the periodic table, you will note a heavy, stairstep line toward the right side. This line is a rough dividing line between metals and nonmetals. As you might expect, the elements that lie along this line are generally metalloids.

The elements of Groups 3 through 12 (IB through VIIIB), the transition metals, all have one or two electrons in the outer level. They all show metallic properties. The lanthanoids and actinoids also have two electrons in the outer level as predicted and are therefore classified as metals.

The elements of Groups 13 (IIIA) through 15 (VA) include both metals and nonmetals. At the top of the table, each of these groups contains nonmetallic elements. The metallic character of the elements increases toward the bottom of the table, and the last member of each family is distinctly metallic.

We can now look at the periodic table as a whole. Metals are located on the left and nonmetals on the right. Note again that most of the elements are metallic; that is, their atoms contain one, two, or three electrons in the outer energy level. The most unreactive atoms are those of the noble gases. Their chemical stability is explained by the octet rule.

PRACTICE PROBLEMS

6. Classify the following elements as metals, metalloids, or nonmetals.

a. oxygen	**e.** europium
b. scandium	**f.** cerium
c. silicon	**g.** mercury
d. lithium	

6.2 CONCEPT REVIEW

7. Classify the following elements as metals, metalloids, or nonmetals: calcium, phosphorus, tellurium, tungsten, yttrium.
8. Find two transition elements whose actual electron configurations differ from those predicted by the arrow diagram. Explain each of these deviations.
9. State the octet rule in terms of electron pairs.
10. **Apply** The elements mercury and bromine are both liquids at room temperature. Why is mercury a metal and bromine a nonmetal?

EVERYDAY CHEMISTRY

Beverage Cans

In the past few years more than 90% of the beverage cans in the United States have been made of aluminum. The aluminum can manufacturers achieved this high percentage of the beverage can market because aluminum cans are manufactured more cheaply from single sheets of aluminum, aluminum is fairly easy to recycle, and the cans themselves have a low density, thus reducing shipping costs.

Thirty years ago most cans were made of tin-plated steel, the old "tin can." They were heavy, had a tendency to leak, and gave a metallic taste to the contents. The one advantage was that iron—from which steel is made—rusts.

Steel companies have now improved production facilities and believe that they can challenge the use of aluminum cans. They are now able to produce a much thinner steel can. Because the price of aluminum has increased, the steel can is actually less expensive to make. A major problem, however, is that steel cans still need aluminum lids. A steel "flip-top" has not yet been developed.

Although a switch to steel cans might save millions of dollars annually in materials costs, assembly lines would have to be retooled and a recycling program would need to be set up. The public expects to recycle cans. More than 40 billion aluminum cans, more than half of those produced, are recycled each year. Most steel manufacturers say that it is more costly to recycle a steel can than to produce a new one. Until the steel manufacturers can solve the "flip-top" problem and cope with recycling, aluminum cans will probably continue to have an advantage.

Exploring Further

1. What properties of aluminum make it easier for cans to be made of aluminum than steel?
2. Find the mass of an aluminum can. Using its density, estimate the number of cm^3 used in the manufacture of the can. Assuming the steel manufacturers could use the same volume of steel to make a steel can, how much mass would that can have? Assume steel has approximately the same density as iron. Refer to Appendix Table A-3 for densities.

Summary

6.1 Developing the Periodic Table

1. Many attempts have been made to classify the elements in a systematic manner. These attempts include Dobereiner's triads, Newlands' law of octaves, and Mendeleev's and Meyer's tables.
2. Mendeleev arranged his periodic table on the basis of the similar properties of elements. He concluded that the properties of elements were a function of atomic mass.
3. The modern periodic law states that the properties of the elements are a periodic function of their atomic numbers.
4. Today's periodic table reflects electron configurations of atoms, indicating that similarity in properties of different elements is related to similarity in their electron configurations.
5. In the transition elements, electrons are being added to a *d* sublevel. The placement of the transition elements on the periodic table reflects the overlap of energy levels.
6. In lanthanoids, electrons are being added to the 4*f* sublevel. In actinoids, electrons are filling the 5*f* sublevel.
7. All elements in a horizontal line of the table are called a period. All elements in a vertical line are called a group.

6.2 Using the Periodic Table

8. With a few exceptions, correct electron structures for atoms can be derived from examining the element's position on the periodic table.
9. According to the arrow diagram, only *s* and *p* electrons can be in the outer level of atoms, making a possible total of eight electrons in the outer level. An atom with eight electrons in the outer level is chemically stable or unreactive.
10. The most chemically stable atoms, the noble gases, have eight electrons in the outer level. Helium atoms are stable with two electrons in the outer level.
11. Filled and half-filled sublevels represent atoms in states of special chemical stability. Electrons in certain atoms are shifted between closely spaced sublevels to produce one of these more stable configurations.
12. Elements with one, two, or three electrons in the outer level usually have metallic properties. Elements with five, six, or seven outer-level electrons usually have nonmetallic properties. Elements that have both metallic and nonmetallic properties are called metalloids.

Key Terms

triad
law of octaves
periodic law
periodic table
transition element
lanthanoid series
actinoid series
period
group
octet rule
family
metal
nonmetal
metalloid

Review and Practice

11. What observations led Mendeleev to conclude that there were several undiscovered elements?
12. Why did Mendeleev and other chemists of his time arrange elements in the periodic table in order of atomic masses? What events changed this method? In what order are the elements arranged in the modern table?
13. Why does the first period of the periodic table contain only two elements while all other periods have eight or more elements in them?
14. To which sublevel are electrons being added across the first row of transition

elements? Why do transition elements begin in the fourth period of the table rather than in the third period?

15. Elements whose highest-energy electrons are in *s* orbitals are found in what groups?

16. List the groups of elements that have family names. List the elements in each of these families.

17. Each of the following represents the end of an element's electron configuration. For each, tell which period and group the element is in, identify the element, and state whether the element is a metal, a nonmetal, or a metalloid.

a. $3s^23p^2$
b. $5s^24d^2$
c. $4s^23d^7$
d. $3s^23p^6$
e. $5s^24d^{10}5p^5$
f. $6s^14f^{14}5d^{10}$

18. Classify the following elements as metals or nonmetals.

a. manganese
b. fluorine
c. silver
d. mercury
e. cobalt
f. praseodymium
g. nitrogen
h. niobium
i. hydrogen
j. lithium
k. radium
l. carbon

19. In general, where are the metals found on the periodic table? Where are the nonmetals found?

20. Why are all of the transition elements metallic?

21. Iodine is used in many commercial chemicals and dyes. To what family does iodine belong? What are the other members of this family? How many electrons are in the outer energy level of each of these elements?

22. The existence of scandium was predicted by Mendeleev in 1871. It is the first of the transition elements. What are the other members of the group containing scandium? How are the predicted electron configurations of these elements similar?

23. List the elements generally considered to be metalloids.

24. Why are there many more metallic elements than nonmetallic elements?

Concept Mastery

25. What feature of electron configuration is unique to actinoids and lanthanoids?

26. Known elements may contain as many as 32 electrons in an energy level. If this is so, why are electrons in *s* and *p* sublevels considered so important? Why are these the only electrons considered in the octet rule?

27. What factors determine an element's chemical properties?

28. Why does the element molybdenum have an electron configuration ending $5s^14d^5$ rather than $5s^24d^4$? Explain the principle involved.

29. What are the major differences in properties among metals, nonmetals, and metalloids?

30. Why do you think the lanthanoid elements have close similarity in chemical properties?

31. If a lithium atom becomes more stable by losing an electron, does it have a stable octet? Can you say it follows the octet rule? Explain your answer.

32. Copper, silver, and gold do not follow the arrow diagram for electron configurations. Explain why they do not.

Application

33. Element X has the electron configuration $1s^22s^22p^63s^23p^64s^23d^{10}4p^4$. Use this information to answer each of the following questions.

a. Use the periodic table to identify element X.
b. To what group and to what period does this element belong?
c. Classify the element as a metal, nonmetal, or metalloid.

d. List the properties associated with the classification you chose.
e. Draw the electron dot diagram for an atom of element X.
f. List two other elements that are likely to be similar in properties to element X.
g. If this element gained electrons to achieve a stable octet, how many electrons would it gain? To which element would its outer electron arrangement then be similar?

34. The symbols of some elements are written on the periodic table shown. For each element whose symbol is shown, answer the following items.
   a. To what group and what period does this element belong?
   b. Classify the element as a metal, nonmetal, or metalloid.
   c. List the properties this element should exhibit, based on the classification you chose.
   d. Write the electron configuration of the element.
   e. Draw the electron dot diagram for an atom of this element.

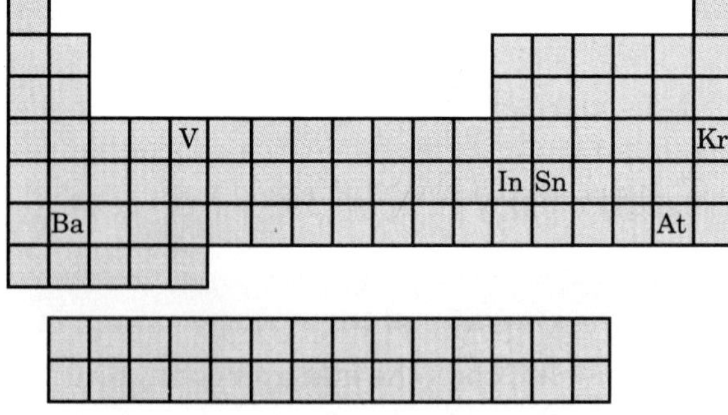

35. What conclusions do you think Dobereiner would have made if he had tried to compare F, Cl, and Br as a triad instead of Cl, Br, and I? Using the modern table, explain why F, Cl, and Br would not have fit his scheme. Would he have found P, As, and Sb to be a triad? What about As, Sb, and Bi? Explain your answers.

36. Without using a periodic table, and by working out the electron configurations, determine which elements from the following lists belong to the same group.
   a. Z = 2, 8, 16, 24
   b. Z = 3, 11, 27, 37
   c. Z = 4, 10, 18, 26
   d. Z = 24, 29, 42, 47, 55

37. Using the periodic table on pages 144 and 145, find two examples of triads other than those mentioned on page 138. Explain why each set of elements you chose is considered to be a triad.

Critical Thinking

38. The noble gases were among the last naturally-occurring elements to be discovered, even though some of them, particularly argon and neon, occur in moderate abundance on Earth. Why were the noble gases so difficult to detect and isolate?

39. In the system of numbering groups of elements with Roman numerals and letters, what is the significance of the Roman numeral in the A groups?

40. The octet rule applies to atoms within compounds as well as individual atoms. Elements unite to form a more stable configuration. How is this stability shown in the case of sodium and chlorine combining to form sodium chloride?

41. Predict the atomic mass, melting point, and boiling point for element 118. What would you expect its chemical properties to be? Use Appendix Table A-3 as needed.

42. Lithium and potassium are both elements in Group 1 (IA) and have the same electron configuration in the outer energy level. Both elements react vigorously with water, but potassium reacts more violently than does lithium. Why does potassium react more violently than lithium reacts?

Chapter 6 REVIEW

Readings

Ciparek, Joseph D. "Element X." *ChemMatters* 6, no. 4 (December 1987): 8-9.

Fernelius, W. Conrad. "Some Reflections on the Periodic Table and Its Use." *Journal of Chemical Education* 63, no. 3 (March 1986): 263-266.

Guenther, William B. "An Upward View of the Periodic Table." *Journal of Chemical Education* 64, no. 1 (January 1987): 9-10.

Osorio, Hanán von Marttens. "A Numerical Periodic Table and the *f* Series Chemical Elements." *Journal of Chemical Education* 67, no. 7 (July 1990): 563-565.

Peterson, Ivars. "A Superconducting Banquet From the Periodic Table." *Science News* 133, no. 10 (March 5, 1988): 148.

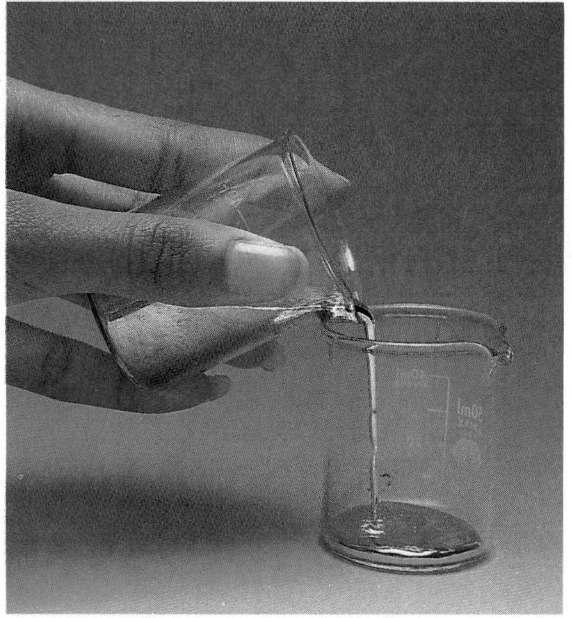

Cumulative Review

1. How does Schrödinger's analysis of the atom differ from that of Bohr?

2. How much energy is needed to raise the temperature of 5.24 g of ruthenium from 25°C to 202°C? See Appendix Table A-3.

3. Write the electron configurations and draw electron dot diagrams for the following elements.
   a. Al (Z = 13) **f.** Mn (Z = 25)
   b. S (Z = 16) **g.** Co (Z = 27)
   c. Ca (Z = 20) **h.** Ge (Z = 32)
   d. Ti (Z = 22) **i.** Br (Z = 35)
   e. V (Z = 23) **j.** Sr (Z = 38)

4. Explain the meaning of each symbol in the de Broglie equation.

5. Compute the average atomic mass of tungsten if its isotopes occur as follows.

Isotopic mass	Percentage
179.946 7 u	0.140%
181.948 25 u	26.41%
182.950 27 u	14.40%
183.950 97 u	30.64%
185.954 40 u	28.41%

6. How many significant digits are in the measurement 26°C?

7. What volume is occupied by 42.5 g of a substance of density 3.15 g/cm^3?

8. What are the four quantum numbers associated with each electron in an atom? What information do the first three quantum numbers give us about the electron cloud? What does the fourth tell us about the behavior of the electron?

9. An iron bar of mass 2.08 kg is heated to 150°C and is then quenched by plunging it into 50.5 kg of water at 12.0°C. What will be the final temperature of the water and iron?

10. After several years exposure to the weather, an aluminum screen door that was originally bright has become dark gray, pitted, and has a thin coating that can be scraped off to a powder. Is this change chemical or physical? Distinguish between chemical and physical changes in your answer.

CHAPTER **Preview**

CONTENTS

CHAPTER 7

Chemical Formulas

Look at the reproduction of the music on this page. Can you make sense of the information presented? If you can read music, the strange symbols will mean something. For many other people, the cryptic entries carry no information. Musicians have evolved a system of symbols that can be used to convey a great deal of information in a small space.

Like the musician on the opposite page, chemists use a system of symbols to convey information. Early chemists developed a kind of shorthand to represent substances and chemical changes. Modern chemists continue to add to this symbolic language as they make new discoveries. In this chapter, you will learn about the symbolic representation of substances. Much of the information in this book, as well as in reference materials such as your laboratory manual, will use this same system of shorthand.

EXTENSIONS AND CONNECTIONS

OBJECTIVES

- demonstrate proficiency in writing chemical formulas.
- define *oxidation number* and state oxidation numbers for common monatomic ions and charges for common polyatomic ions.

7.1 Symbols and Formulas

Although there are only 109 elements, there are about ten million compounds known to chemists. It is convenient to represent elements by the use of symbols. Compounds are represented by combinations of symbols called formulas. This shorthand system improves communications among all parts of the scientific and technological community. These representations of substances are another way to help us classify these substances quickly. The ability to classify simplifies the study of the vast number of substances with which chemists work.

Names of Elements

Is there an element named for a headache? Yes, sodium. Sodium was named for one of its compounds, soda, or sodium carbonate, which was once used as a headache remedy. Soda comes from the Arabic word for headache, *suda*. The names of the elements are generally bestowed by their discoverers. However, some elements, for example gold and tin, have been known since prehistoric times, and we do not know how their names were derived.

The most common source for an element name is a property of the element. An example is nitrogen. Its name comes from the Greek words *nitron* (niter) and *genes* (to be born), meaning "niter forming." Niter was the name for naturally occurring substances that contain nitrogen. Protactinium is a radioactive element that decays to actinium. The name comes from the Greek *protos,* which means first. Protactinium comes before actinium.

After properties, the next most popular source for an element name is the place of discovery. Hafnium, for example, was discovered in Copenhagen, Denmark. The Latin name for Copenhagen was *Hafnia*.

Another source of names is the mineral in which the element was found. Lithium was found in a mineral and is named from the Greek word for stone, *lithos*. Tungsten comes from the Swedish *tung sten,* heavy stone. Boron is named for a property and a mineral. Its name is a combination of borax and carbon. Boron is

Figure 7.1 The name of the element boron (right) is taken from the mineral ***bor***ax (left) and the physical resemblance of boron to carb***on***.

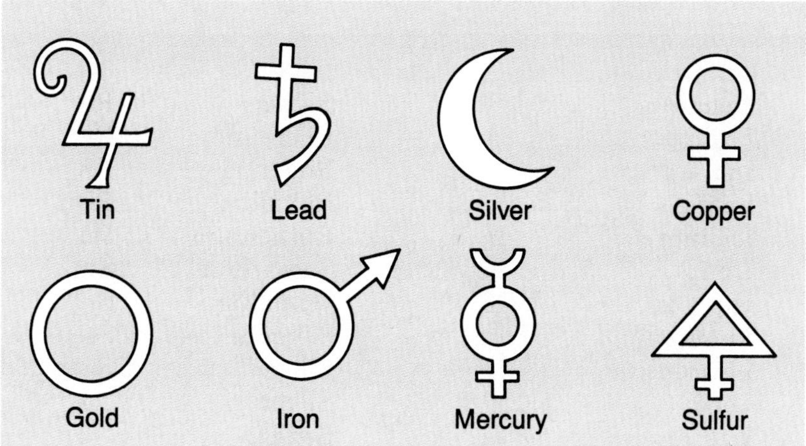

Figure 7.2 The alchemists used symbols like these to represent elements.

found in the mineral borax, and is black like carbon. The name borax is derived from the Arabic word for *glisten* because the mineral is shiny.

Finally, there are several elements named to honor a place or a person. Curium, for instance, is a radioactive element named to honor Marie and Pierre Curie, the discoverers of radium.

Symbols for Elements

The symbols for elements are derived from their names. Scientists throughout the world have agreed to represent one atom of aluminum by the symbol Al. Oxygen is given the symbol O, hydrogen H. The **chemical symbols** of the elements are shorthand representations of the elements. They take the place of the complete names of the elements.

Ancient symbols for some elements are shown in Figure 7.2. J. J. Berzelius, a Swedish chemist, is generally given credit for creating the modern symbols for elements. Berzelius proposed that all elements be given a symbol corresponding to the first letter of their names. In the case of two elements that began with the same letter, the second letter or an important letter in the name was added. In some cases, the Latin name of the element was used. Thus, the symbol for sulfur is S; selenium, Se; strontium, Sr; and sodium, Na (Latin *natrium*).

The symbols that have been agreed upon for the elements are listed in Table 7.1. Notice that they contain capital and lowercase letters. Names for elements 104 through 109 and their three-letter symbols, shown in the last column, are the result of a system adopted by the International Union of Pure and Applied Chemistry (IUPAC). In this system, Latin and Greek stems representing the atomic numbers of the elements are used for both the name and the symbol. Scientists in both the United States and Russia claim discovery of elements 104 and 105. Because these scientists cannot agree on names for these two elements, the IUPAC systematic naming was adopted for them and for all elements discovered after them.

CHEMISTS AND THEIR WORK

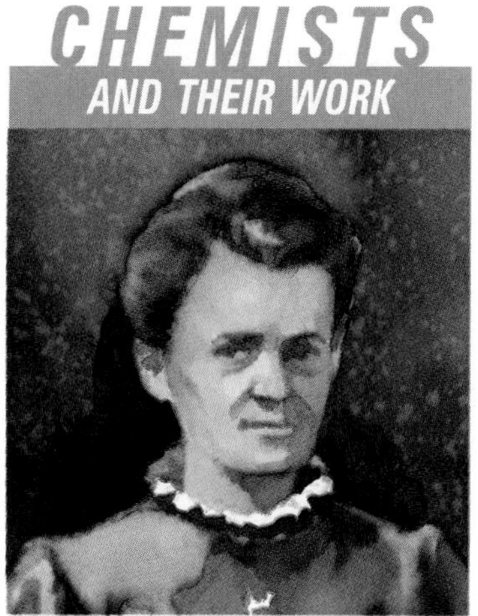

Marie Curie
(1867-1934)

A Nobel Prize usually caps the career of a scientist. Marie Curie won two. In 1903, she was awarded the prize in physics for her joint research with her husband Pierre on the radiation phenomena discovered by Becquerel. In 1911, she won the chemistry prize for the discovery of radium and polonium and for the study of radium compounds. Madame Curie was devoted not only to scientific research but also to the applications of the research. The use of the radioactivity of radium in the treatment of cancer is but one example of this concern.

Table 7.1

Elements and Their Symbols					
Actinium	Ac	Holmium	Ho	Radon	Rn
*Aluminum	Al	*Hydrogen	H	Rhenium	Re
Americium	Am	Indium	In	Rhodium	Rh
Antimony	Sb	*Iodine	I	Rubidium	Rb
*Argon	Ar	Iridium	Ir	Ruthenium	Ru
Arsenic	As	*Iron	Fe	Samarium	Sm
Astatine	At	Krypton	Kr	Scandium	Sc
*Barium	Ba	Lanthanum	La	Selenium	Se
Berkelium	Bk	Lawrencium	Lr	*Silicon	Si
*Beryllium	Be	*Lead	Pb	*Silver	Ag
Bismuth	Bi	*Lithium	Li	*Sodium	Na
*Boron	B	Lutetium	Lu	*Strontium	Sr
*Bromine	Br	*Magnesium	Mg	*Sulfur	S
*Cadmium	Cd	*Manganese	Mn	Tantalum	Ta
*Calcium	Ca	Mendelevium	Md	Technetium	Tc
Californium	Cf	*Mercury	Hg	Tellurium	Te
*Carbon	C	Molybdenum	Mo	Terbium	Tb
Cerium	Ce	Neodymium	Nd	Thallium	Tl
Cesium	Cs	*Neon	Ne	Thorium	Th
*Chlorine	Cl	Neptunium	Np	Thulium	Tm
*Chromium	Cr	*Nickel	Ni	*Tin	Sn
*Cobalt	Co	Niobium	Nb	Titanium	Ti
*Copper	Cu	*Nitrogen	N	Tungsten	W
Curium	Cm	Nobelium	No	Unnilennium	Une
Dysprosium	Dy	Osmium	Os	Unnilhexium	Unh
Einsteinium	Es	*Oxygen	O	Unniloctium	Uno
Erbium	Er	Palladium	Pd	Unnilpentium	Unp
Europium	Eu	*Phosphorus	P	Unnilquadium	Unq
Fermium	Fm	Platinum	Pt	Unnilseptium	Uns
*Fluorine	F	Plutonium	Pu	Uranium	U
Francium	Fr	Polonium	Po	Vanadium	V
Gadolinium	Gd	*Potassium	K	Xenon	Xe
Gallium	Ga	Praseodymium	Pr	Ytterbium	Yb
Germanium	Ge	Promethium	Pm	Yttrium	Y
Gold	Au	Protactinium	Pa	*Zinc	Zn
Hafnium	Hf	Radium	Ra	Zirconium	Zr
*Helium	He				

*You should memorize those symbols marked with an asterisk as they will be used often throughout the text.

How Symbols Are Used

You saw in the last chapter that atoms can react by gaining or losing electrons. Atoms, by definition, are electrically neutral. They have equal numbers of electrons and protons. Electrons, on the other hand, are negatively charged. If an atom gains or loses electrons, it must, then, become electrically charged. An atom that has become charged is called an **ion.** Because electrons are negative, an atom that gains electrons becomes a negatively charged ion. An atom that loses electrons becomes a positively charged ion.

Scientists use a shorthand method of representing information about an atom or ion. Each "corner" of a symbol for an element is used to show some property of that atom or ion. Let's look at the example in Figure 7.3. The upper right-hand corner is used to show the electric charge on an ion. The lower right-hand corner is used to show the number of atoms in a formula. The upper left-hand corner is used for the mass number, or number of nucleons, in an atom. In the lower left-hand corner, the charge on the nucleus of an atom is shown. Thus, a typical sulfur nucleus or atom is represented as ${}^{32}_{16}S$. Fluorine would be ${}^{19}_{9}F$. The same system is used with subatomic particles. An electron, which has a negligible mass and a 1− charge, is represented as ${}^{0}_{-1}e$. An alpha particle is ${}^{4}_{2}He$.

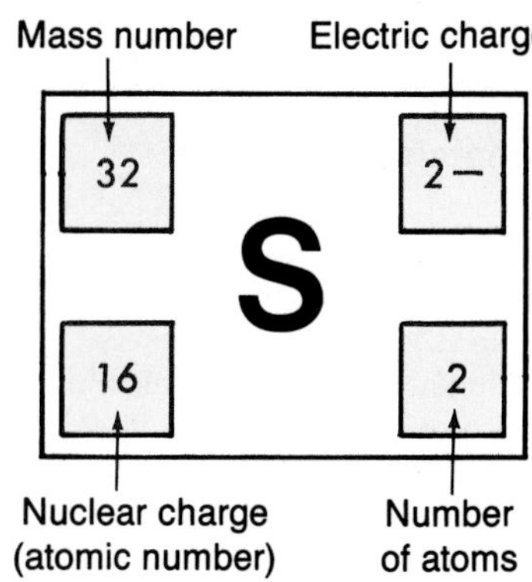

Figure 7.3 The diagram shows how the shorthand form for representing sulfur as ${}^{32}_{16}S$ is derived.

Chemical Formulas

Chemists combine symbols in chemical formulas to represent compounds. A **chemical formula** is a combination of symbols that represents the composition of a compound. A formula shows two things. It indicates the elements present in the compound and the relative number of atoms of each element in the compound. Formulas often contain numerals to indicate the ratio of elements in a compound. For example, chemists have learned from experiments that water is composed of the elements hydrogen and oxygen. It is also known that two atoms of hydrogen will react (combine chemically) with one atom of oxygen to form one molecule of water. Thus, the formula for water is written as H_2O. The small subscript, $_2$, after the H indicates that there are two atoms of hydrogen in one molecule of water. Note that there is no subscript after the oxygen. If a symbol has no subscript, it is understood that only one atom of that element is present.

Figure 7.4 A water molecule is composed of two atoms of hydrogen and one atom of oxygen.

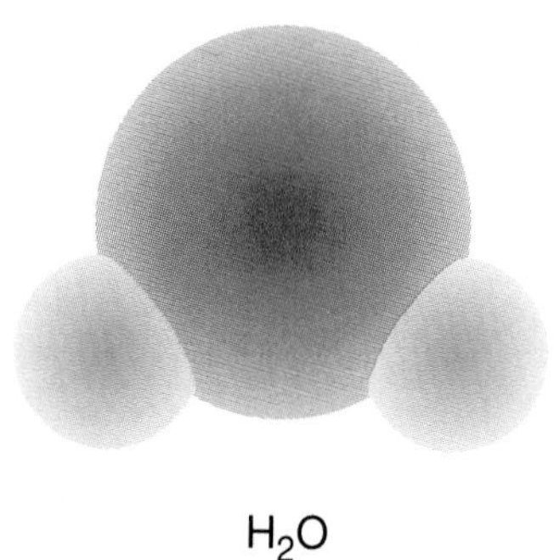

MINUTE LAB

Chemical Synthesis

Put on an apron and goggles. Separately mass out 1.00 g of powdered sulfur and 1.75 g of iron filings. Examine each element separately, recording its physical properties of texture, color, odor, and magnetism. Then, mix the two elements on a piece of paper and pour them into a test tube. In a fume hood, use a burner to heat the test tube until the contents glow red. Immediately plunge the hot tube into a beaker of cold water. **CAUTION:** *The test tube will break.* Remove the product from the water with forceps, making sure no broken glass adheres to the product. Record its physical properties. Does the product retain the color of the sulfur? Is the product affected by a magnet? When iron reacts with sulfur, it has a 2+ oxidation number. Write the formula for the product. What is its chemical name?

Formulas for organic compounds are written according to a different set of rules. For example, the formula for acetic acid, the acid in vinegar, is written as CH_3COOH*. The actual structure of the acetic acid molecule is represented by

```
    H   O
    |   ||
H—C—C—O—H
    |
    H
```

It shows how the atoms are joined, as well as the kind and number of atoms present. From this structural formula, you can see how the shorthand formula is derived. Other common compounds and their formulas are shown in Table 7.2.

Table 7.2

Some Common Compounds and Their Formulas

Compound	Formula	Elements
ammonia	NH_3	nitrogen, hydrogen
rust	Fe_2O_3	iron, oxygen
sucrose	$C_{12}H_{22}O_{11}$	carbon, hydrogen, oxygen
table salt	NaCl	sodium, chlorine
water	H_2O	hydrogen, oxygen

Oxidation Number

Through experiments, chemists have determined the ratios in which most elements combine. They have also learned that these ratios depend on the structure of the atoms of the elements. When an atom reacts to form an ion, the stability of an octet of electrons enables you to predict the number of electrons to be gained or lost. You can then predict the charge on the ion formed. When a single atom takes on a charge, it is called a monatomic ion. The charge on a monatomic ion is known as the **oxidation number** of the atom. Most of these charges can be verified by applying the octet rule. An ion made of more than one atom, for example, OH^-, is called a **polyatomic ion.**

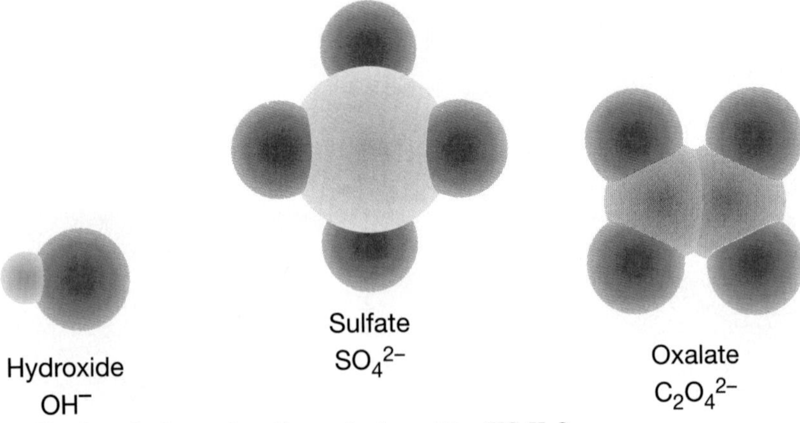

Figure 7.5 Polyatomic ions, such as hydroxide (left), sulfate (center), and oxalate (right) are composed of more than one atom.

*The formula for acetic acid can also be written $HC_2H_3O_2$.

Table 7.3 lists the oxidation numbers of some elements. Table 7.4 lists the charges of several polyatomic ions. You should memorize these charges. They will be important throughout your study of chemistry. You will use this information to write correct chemical formulas. Atoms and ions combine chemically in definite ratios. Oxidation numbers of elements and the charges on polyatomic ions tell us these combining ratios. The way to determine the ratio of elements in a compound is to add the charges algebraically. If the charges add up to zero, the formula for the compound has been written correctly. Not all the formulas you can write represent compounds that actually exist.

Table 7.3

Oxidation Numbers of Some Monatomic Ions*

1+	2+	
hydrogen, H^+	barium, Ba^{2+}	magnesium, Mg^{2+}
lithium, Li^+	cadmium, Cd^{2+}	manganese(II), Mn^{2+}
potassium, K^+	calcium, Ca^{2+}	mercury(II), Hg^{2+}
silver, Ag^+	cobalt(II), Co^{2+}	nickel(II), Ni^{2+}
sodium, Na^+	copper(II), Cu^{2+}	strontium, Sr^{2+}
	iron(II), Fe^{2+}	tin(II), Sn^{2+}
	lead(II), Pb^{2+}	zinc, Zn^{2+}

3+	4+	1−	2−
aluminum, Al^{3+}	lead(IV), Pb^{4+}	bromide, Br^-	oxide, O^{2-}
chromium(III), Cr^{3+}		chloride, Cl^-	sulfide, S^{2-}
iron(III), Fe^{3+}		fluoride, F^-	
		hydride, H^-	
		iodide, I^-	

*Appendix Table A-3 lists additional monatomic ions and their charges.

Table 7.4

Charges of Common Polyatomic Ions*

1+	2+
ammonium, NH_4^+	mercury(I), Hg_2^{2+}

1−	2−	3−
acetate, CH_3COO^-	carbonate, CO_3^{2-}	phosphate, PO_4^{3-}
chlorate, ClO_3^-	chromate, CrO_4^{2-}	
chlorite, ClO_2^-	dichromate, $Cr_2O_7^{2-}$	
cyanide, CN^-	oxalate, $C_2O_4^{2-}$	
hydroxide, OH^-	peroxide, O_2^{2-}	
hypochlorite, ClO^-	silicate, SiO_3^{2-}	
iodate, IO_3^-	sulfate, SO_4^{2-}	
nitrate, NO_3^-	sulfite, SO_3^{2-}	
nitrite, NO_2^-	tartrate, $C_4H_4O_6^{2-}$	
perchlorate, ClO_4^-	tetraborate, $B_4O_7^{2-}$	
permanganate, MnO_4^-	thiosulfate, $S_2O_3^{2-}$	

*Appendix Table A-4 lists additional polyatomic ions and their charges.

Common table salt, NaCl, is made from sodium, Na, and chlorine, Cl. Table 7.3 shows a 1+ charge for sodium ions and a 1− charge for chloride ions.

$$Na^{+}Cl^{-}$$

Adding these charges, you see that 1 + (1−) = 0. Therefore, the formula for salt is NaCl. This formula indicates that a one-to-one ratio exists between sodium ions and chloride ions in a crystal of salt. Sodium chloride is an **ionic compound;** that is, it is composed of ions. Note that in the formula for NaCl the positive ion is written first. This is true of all ionic compounds.

Elements also combine in another way. Neutral atoms combine to form neutral particles called **molecules.** Compounds formed from molecules are called molecular compounds. In Chapter 12, you will study ionic and molecular substances.

Table 7.5

Diatomic Elements

Name	Formula
bromine	Br_2
chlorine	Cl_2
fluorine	F_2
hydrogen	H_2
iodine	I_2
nitrogen	N_2
oxygen	O_2

The element chlorine is a gas composed of molecules that are diatomic. A diatomic molecule is made of two atoms of the same element. One chlorine molecule contains two chlorine atoms. Chlorine gas is represented by the formula Cl_2. Six other common elements also occur as diatomic molecules. Hydrogen (H_2), nitrogen (N_2), oxygen (O_2), and fluorine (F_2) are diatomic gases under normal conditions. Bromine (Br_2) is a gas above 58.8°C. Iodine (I_2) is a gas above 184°C. Table 7.5 lists the diatomic elements.

The formula of a substance represents a specific amount of the substance. If a substance is composed of molecules, the formula represents one molecule. If a substance is ionic, the formula represents the ions in their lowest combining ratio.

Figure 7.6 Iron pyrite (fool's gold) is composed of the elements iron and sulfur. The formula for this compound is FeS_2.

SAMPLE PROBLEMS

1. Writing a Simple Formula

Write a formula for a compound of calcium and bromine.

Solving Process:

Using Table 7.3, you see the oxidation states of calcium and bromide are as follows:

$$Ca^{2+} \qquad Br^{-}$$

Because the sum of the charges must equal zero, two Br^- are needed to balance the one Ca^{2+}. The correct formula is $CaBr_2$.

PROBLEM SOLVING HINT
Becoming proficient in formula writing will help you throughout your study of chemistry.

2. Writing a More Complex Formula

Write the formula for a compound made from the aluminum ion and the sulfate ion.

Solving Process:

Using Tables 7.3 and 7.4, you see that the oxidation states of aluminum and the sulfate ion are as follows:

$$Al^{3+} \qquad SO_4^{2-}$$

To make the sum of the charges equal zero, you must find the least common multiple of 3 and 2. The least common multiple is 6. Because 6/3 = 2 and 6/2 = 3, it is necessary to have two Al^{3+} and three SO_4^{2-} in the compound to maintain neutrality. Writing two aluminum ions in the formula is simple.

$$Al_2$$

For the sulfate, the entire polyatomic ion must be placed in parentheses to indicate that three sulfate ions are required.

$$(SO_4)_3$$

Thus, aluminum sulfate has the formula $Al_2(SO_4)_3$. Parentheses are used in a formula only when you are expressing multiples of a polyatomic ion. If only one sulfate ion were needed in writing a formula, parentheses would not be used. For example, the formula for the compound made from calcium, Ca^{2+}, and the sulfate ion, SO_4^{2-}, is $CaSO_4$.

PRACTICE PROBLEMS

1. Write the formula for each of the following compounds.
   - **a.** calcium chloride
   - **b.** sodium cyanide
   - **c.** magnesium oxide
   - **d.** barium oxide
   - **e.** sodium fluoride
   - **f.** aluminum nitrate
   - **g.** zinc iodide
   - **h.** cobalt(II) carbonate
2. Write the formula for the compound made from each of the following pairs.
   - **a.** silver and fluorine
   - **b.** nickel(II) and sulfur
   - **c.** chromium(III) and bromine
   - **d.** lead(II) and phosphate ion
   - **e.** ammonium and oxalate ions
   - **f.** strontium and iodine
   - **g.** lithium and oxygen
3. Write the formula for each of the following compounds.
   - **a.** potassium hydride
   - **b.** mercury(II) cyanide
   - **c.** zinc tartrate
   - **d.** cadmium silicate
   - **e.** ammonium dichromate
   - **f.** lead(II) nitrate
   - **g.** copper(II) perchlorate
   - **h.** sodium tetraborate

Figure 7.7 The elements chlorine (yellow-green), bromine (red-brown), and iodine (purple) exist as diatomic molecules.

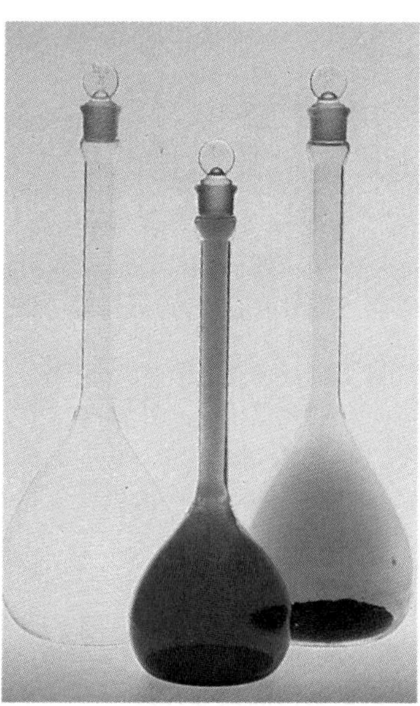

EVERYDAY CHEMISTRY

Hazardous Materials at Home

Some of the most hazardous materials, as well as those that are most polluting to the environment, are found right in your home. Hazardous materials are those that are poisonous, corrosive, or flammable. Typical poisonous substances are insecticides, some medicines, antifreeze, and rubbing alcohol. Corrosive compounds destroy tissue, metals, and other materials. Some corrosives are toilet and shower cleaners, bleach, battery acid, and oven cleaners. Flammable compounds, those that burn easily, include gasoline, lighter fluids, and some aerosols.

These and other materials are hazardous to the health and safety of people and pets in the home. For example, the common plant killer 2,4-D has been shown to cause lymphoma in dogs. The dogs pick up the chemical while in the yard and then lick it off. A single quart of oil poured into the ground can contaminate a water supply. Some propellants from aerosol cans escape into the atmosphere and contribute to the destruction of the ozone layer.

It is possible to use alternatives to some of these materials. For example, in place of the window-cleaning products that contain phosphates or ammonia, water containing vinegar can be used. Vinegar is simply a mild solution of acetic acid and is nontoxic. Products that are packaged as pump sprays or are sold as lotions can replace those that come only in harmful aerosol cans.

Many communities are starting to schedule regular collections of toxic household materials to make sure they are disposed of properly. The complete removal of hazardous materials from the home will probably not be accomplished, but where changes can be made, they should be.

Exploring Further

1. Check the shelves in your home to find out what cleaning products are there. Make a list of the active ingredients in each.

2. Aside from the alternatives given above, what other ideas do you have for reducing the use of hazardous materials?

7.1 *CONCEPT REVIEW*

4. Write formulas for the following.
   a. barium sulfate
   b. calcium sulfide
   c. magnesium phosphate
   d. strontium bromide
   e. chromium(III) acetate

5. How many protons are there in an atom represented by the symbol $^{207}_{82}Pb$? How many electrons? How many neutrons?

6. What is the charge on an atom of copper that has lost two electrons? How is this ion represented?

7. Apply An insecticide lists sodium selenate as one of its ingredients. What is the formula for sodium selenate?

7.2 Nomenclature

OBJECTIVES

- demonstrate proficiency in naming chemical compounds.
- distinguish between molecular and empirical formulas.
- demonstrate the use of coefficients to represent the number of formula units of a substance.

There are many times when chemists prefer to describe a compound by a name rather than a formula. The rules for naming compounds are called the *nomenclature* of chemistry. Nomenclature can also refer to the names themselves. The rules have been adopted internationally under the sponsorship of IUPAC. There are several volumes of such rules, but fortunately, just a few rules are needed for the compounds discussed in this book.

Naming Compounds

Unfortunately, some compounds have both a common name and a chemical name. For example, $NaHCO_3$ is commonly known as baking soda. However, its chemical name is sodium hydrogen carbonate. We will use a systematic method of naming practically all the compounds in this book. The names of only a few compounds, particularly acids, will not be included in this system.

Naming Binary Inorganic Compounds

Compounds containing only two elements are called **binary compounds.** To name a binary compound, first write the name of the element having a positive charge. Then add the name of the negative element. The name of the negative element must be modified to end in *-ide*. For example, the compound formed by aluminum (Al^{3+}) and nitrogen (N^{3-}), with the formula AlN, is named aluminum nitride.

Table 7.6

Formulas and Names of Some Binary Compounds

Formula	Name
Al_2S_3	aluminum sulfide
$CaBr_2$	calcium bromide
H_2O	hydrogen oxide (water)
H_2Se	hydrogen selenide

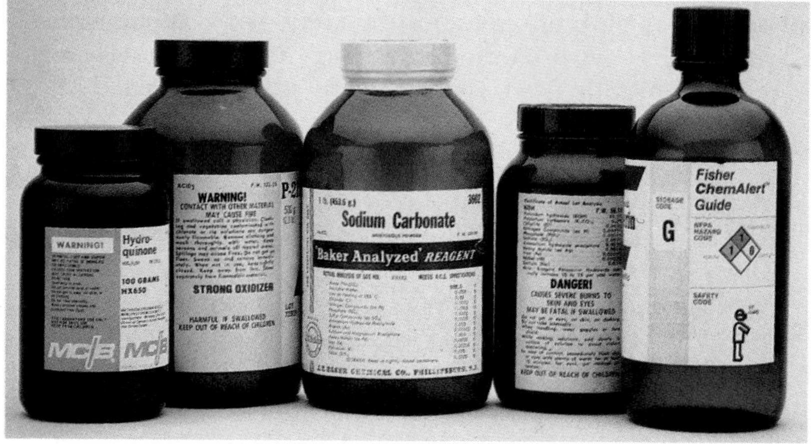

Figure 7.8 It is important to know the names and formulas of chemical compounds in order to distinguish the many different substances used in the lab and the hazards associated with them.

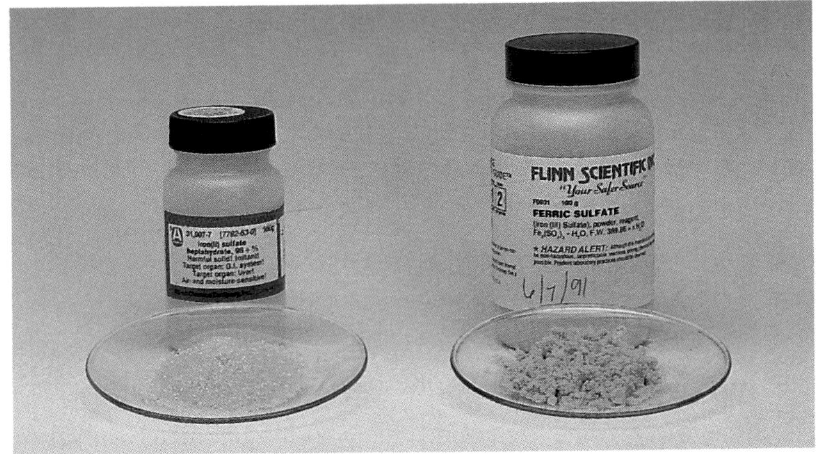

Figure 7.9 Iron exists in the 2+ and 3+ oxidation states and forms two compounds with the sulfate ion. Iron(II) sulfate (left) is blue-green and iron(III) sulfate (right) is yellow-orange. Their properties are different.

In looking at Table 7.3, you see that some elements, such as iron, have more than one possible charge. Therefore, they may form more than one compound with an element. The elements nitrogen and oxygen together form five binary compounds.

Chemists must have a way of distinguishing the names of these compounds. They tell the difference by writing the oxidation number of the positively charged element after the name of that element. The oxidation number is written as a Roman numeral in parentheses. Examples of some of these compounds are listed in Table 7.7.

Table 7.7

Formulas and Names of Some Binary Molecular Compounds Having Variable Oxidation States

Formula	Name	Formula	Name
N_2O	nitrogen(I) oxide	Cu_2S	copper(I) sulfide
NO	nitrogen(II) oxide	CuS	copper(II) sulfide
N_2O_3	nitrogen(III) oxide	SO_2	sulfur(IV) oxide
NO_2	nitrogen(IV) oxide	SO_3	sulfur(VI) oxide
N_2O_5	nitrogen(V) oxide		

There are many compounds that have been named by an older system in which prefixes indicate the number of atoms present. These names have been used for so long that these more common names are usually used. Examples are listed in Table 7.8.

Table 7.8

Formulas and Common Names of Some Binary Molecular Compounds

Formula	Common Name	Formula	Common Name
CS_2	carbon disulfide	SF_6	sulfur hexafluoride
CO	carbon monoxide	SO_2	sulfur dioxide
CO_2	carbon dioxide	SO_3	sulfur trioxide
CCl_4	carbon tetrachloride		

CAREERS IN CHEMISTRY

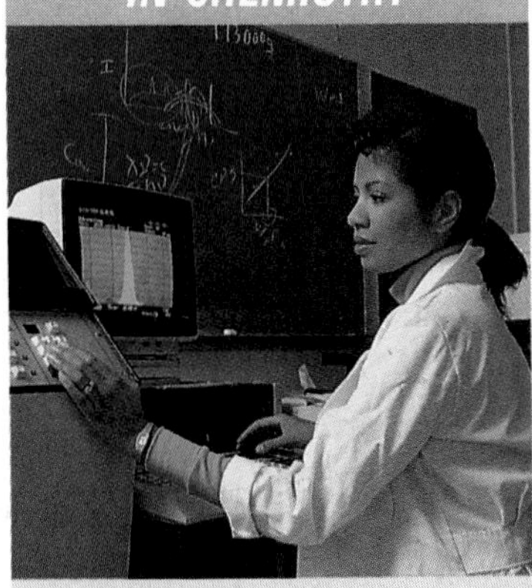

Analytical Chemist

Analytical chemists are concerned chiefly with identifying the type and quantity of substances present in a material. Quality control in such fields as foods, pharmaceuticals, and safe water is one of the jobs performed by some analytical chemists. Others develop new techniques for analyzing materials. Analytical chemists may also investigate the chemical and physical properties of substances.

Not all compounds ending in *-ide* are binary. Notice in Table 7.4 that the names of some polyatomic ions end in *-ide:* OH^- (hydroxide) and CN^- (cyanide).

Table 7.9

Numerical Prefixes	
Prefix	**Number of atoms**
mono-	1
di-	2
tri-	3
tetra-	4
penta-	5
hexa-	6
hepta-	7
octa-	8

SAMPLE PROBLEM

Naming Binary Compounds

Name the compound BeI_2.

Solving Process:

Be is the symbol for beryllium. I is the symbol for iodine. The positive part of the compound retains its name unchanged, beryllium. The ending of the negative part of the compound is changed to *-ide,* iodide. The name of the compound is beryllium iodide.

PRACTICE PROBLEMS

8. Name the following binary compounds.
a. BaS **c.** Mg_3N_2 **e.** ZnF_2
b. BiI_3 **d.** $PbBr_2$

9. Name the following binary compounds.
a. CaH_2 **c.** CaS **e.** $CoBr_2$
b. Na_3P **d.** TlI

Naming Other Inorganic Compounds

For naming compounds containing more than two elements, several rules apply. The simplest of these compounds is formed from one element and a polyatomic ion or from two polyatomic ions. These compounds are named in the same way as binary compounds. However, the ending of the name of the polyatomic ion is not changed. An example is $AlPO_4$, which is named aluminum phosphate. Other examples are listed in Table 7.10.

Table 7.10

Formulas and Names for Some Compounds Containing Polyatomic Ions

Formula	Name	Formula	Name
$AlAsO_4$	aluminum arsenate	$CuSO_4$	copper(II) sulfate
$(NH_4)_2SO_4$	ammonium sulfate	$Ni(OH)_2$	nickel(II) hydroxide
$Cr_2(C_2O_4)_3$	chromium(III) oxalate	$ZnCO_3$	zinc carbonate

Other rules for naming compounds and writing formulas will be discussed when the need arises. The names of the common acids, for example, do not normally follow these rules. Table 7.11 lists names and formulas for acids that you should memorize.

Table 7.11

Acids			
Formula	**Name**	**Formula**	**Name**
CH_3COOH	acetic	$(COOH)_2$	oxalic
H_2CO_3	carbonic	$HClO_4$	perchloric
HCl	hydrochloric	H_3PO_4	phosphoric
HNO_3	nitric	H_2SO_4	sulfuric

Naming Organic Compounds

Organic compounds are named using a different set of rules. The simplest group of organic compounds is the hydrocarbons. As their name suggests, these compounds are composed solely of the elements hydrogen and carbon. Carbon atoms can link to each other in chains and in rings. The first step in naming these compounds is to count the number of carbon atoms in a chain or ring. The stem of the compound name is then chosen from Table 7.12.

Table 7.12

Hydrocarbon Stems			
Number of Carbon Atoms	**Stem**	**Number of Carbon Atoms**	**Stem**
1	meth-	6	hex-
2	eth-	7	hept-
3	prop-	8	oct-
4	but-	9	non-
5	pent-	10	dec-

A suffix is then added to indicate how the carbon atoms are linked to each other. For the time being, we will use only the suffix *-ane*. Thus, a compound named propane would contain three carbon atoms. One named heptane would contain seven carbon atoms. If these atoms are linked in a ring rather than in a chain, the name is prefixed with *cyclo-*. Cyclopentane has five carbon atoms linked in a ring.

Figure 7.10 Propane is used as a fuel for some gas grills.

Bridge to
MEDICINE

Cyclopentamine for Allergies

If you suffer from allergies, your physician may prescribe for you a drug with the common name *cyclopentamine*. Common names are often called generic names in the pharmaceutical industry to distinguish them from brand names. Cyclopentamine has the chemical name 1-cyclopentyl-2-(methylamino)propane. If you look at the structure of this compound, you can see the source of both the common and the chemical names.

Note that there are five carbon atoms linked in a ring. That group is the source of the "cyclopent-" in both the common and the chemical names. The nitrogen atom is the source of the amine in both names. In the chemical name, you may recognize two stems, *meth-* and *prop-*. In the structural diagram, see if you can find a one-carbon group and a three-carbon group that lead to these parts of the chemical name.

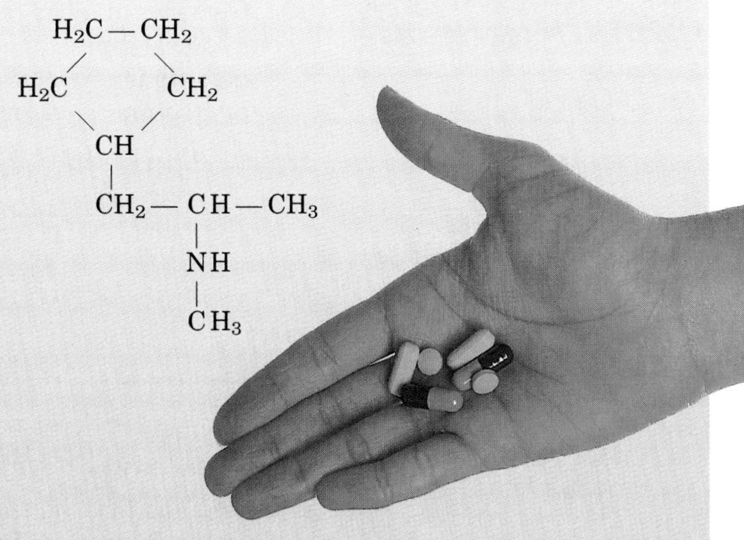

Exploring Further

1. Why don't manufacturers use strict chemical names for their products?
2. Suggest a name for the following molecule based on the rules of nomenclature in this chapter.

```
CH2 — CH — NH2
 |     |
CH2 — CH2
```

SAMPLE PROBLEM

Naming Simple Hydrocarbons

Write the name of each of the following hydrocarbons: C_5H_{12} and $CH_3CH_2CH_2CH_2CH_2CH_2CH_3$. Use Table 7.12 and the suffix *-ane*.

Solving Process:

The hydrocarbon C_5H_{12} contains five carbon atoms. The stem for five carbons is listed in Table 7.12 as *pent-*. Combining this stem with the suffix *-ane* gives the correct name for C_5H_{12}, pentane.

The second hydrocarbon contains seven carbon atoms in a chain. The stem for seven listed in Table 7.12 is *hept-*. Thus, the name of this compound is heptane.

CHEMISTS AND THEIR WORK

Todd Blumenkopf
(1956-)
Todd Blumenkopf is a senior research chemist at Burroughs Wellcome Co., a research-based pharmaceutical company. Blumenkopf's job involves designing and synthesizing molecules intended for therapeutic purposes. It is challenging work—more challenging when one realizes that Blumenkopf is a wheelchair user. Through adaptation of the laboratory equipment and fixtures, he has been able to carry out important work in chemistry. Blumenkopf is head of the American Chemical Society's Committee on Chemists with Disabilities and is preparing a book profiling disabled scientists.

PRACTICE PROBLEMS

10. Write the name of each of these compounds.
a. $BaCl_2$ **c.** AlF_3 **e.** KCl **g.** $Zn(NO_3)_2$
b. $NaBr$ **d.** Li_2CO_3 **f.** HgI_2 **h.** $Ba(OH)_2$

11. Use inorganic numerical prefixes to name each of the following compounds.
a. P_2O_5 **b.** PCl_5 **c.** SF_6 **d.** PCl_3

12. Name each of the following compounds.
a. NH_4NO_3 **c.** Na_3PO_4 **e.** HNO_3 **g.** K_2O
b. CH_3COOH **d.** HCl **f.** $Cu(CH_3COO)_2$ **h.** H_2SO_4

13. Give the number of carbon atoms in each of the following hydrocarbons.
a. butane **c.** cyclopropane **e.** octane
b. heptane **d.** nonane **f.** cyclopentane

14. Name these organic compounds.

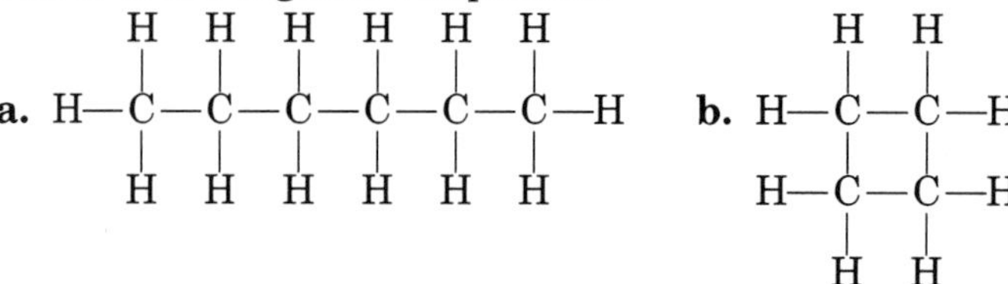

Molecular and Empirical Formulas

The formulas for compounds that exist as molecules are called **molecular formulas.** For instance, one compound of hydrogen and oxygen is hydrogen peroxide (H_2O_2). The formula H_2O_2 is a molecular formula because one molecule of hydrogen peroxide contains two atoms of hydrogen and two atoms of oxygen. However, chemists also use another kind of formula. The atomic ratio of hydrogen to oxygen in hydrogen peroxide is one to one. Therefore, the simplest formula that would indicate the ratio between hydrogen and oxygen is HO. This simplest formula is called an **empirical formula.** As another example, both benzene (C_6H_6) and ethyne (C_2H_2) have the same empirical formula, CH. Chem-

Ethyne C_2H_2

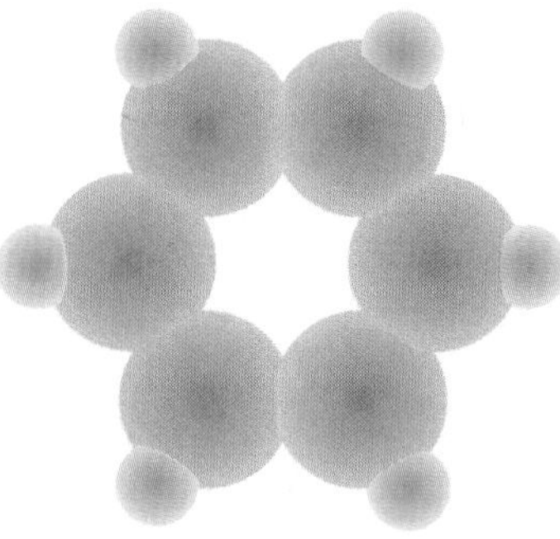

Benzene C_6H_6

Figure 7.11 Benzene (C_6H_6) on the right and ethyne (C_2H_2) on the left have the same empirical formula (CH).

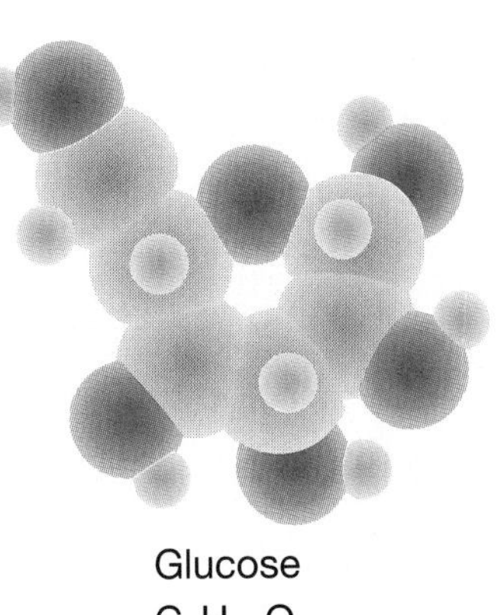

Glucose
$C_6H_{12}O_6$

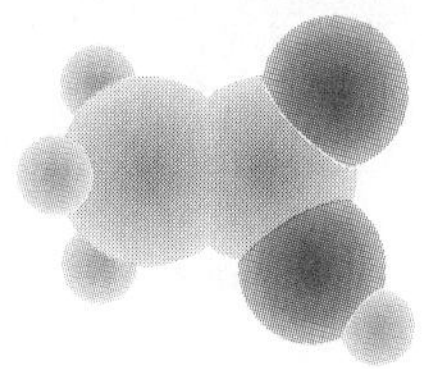

Acetic Acid
CH_3COOH

Figure 7.12 Glucose (left) and acetic acid (right) have the same empirical formula, CH_2O, but they look very different and have different molecular formulas.

ists can determine the empirical formula of an unknown substance through analysis. The empirical formula may then be used to help identify the molecular formula of the substance.

For many substances, the empirical formula is the only formula possible. The formulas for ionic compounds are almost all empirical formulas. The molecular formula of a compound is always some whole-number multiple of the empirical formula. In the above example, if we multiply CH by two, we have the formula for ethyne, C_2H_2. If we multiply by six, we have the formula for benzene, C_6H_6. The calculation of empirical formulas will be presented in Chapter 8.

Table 7.13

Molecular and Empirical Formulas for Two Common Substances

	Glucose	Acetic Acid
Empirical Formula	CH_2O	CH_2O
Molecular Formula	$C_6H_{12}O_6$	CH_3COOH
Use	Food supplement	Making plastics
Physical Properties:		
State at 25°C	Colorless solid	Colorless liquid
Molecular mass	180.16 u	60.05 u
Melting point	146°C	16.7°C
Boiling point	Decomposes	118°C

Coefficients

The formula of a compound represents a definite amount of that compound. This amount may be called a **formula unit.** One molecule of water is represented by H_2O. How do we represent two molecules of water? You use the same system as you would use in mathematics, that is, coefficients. When you wish to represent two x, you write $2x$. When you wish to represent two molecules of water, you write $2H_2O$. Each formula unit of table salt consists of a sodium ion and a chloride ion, NaCl. Three formula units of sodium chloride would be written as 3NaCl.

PROBLEM SOLVING HINT

If the subscripts in a formula have no common divisor, then the formula and the empirical formula are the same.

SAMPLE PROBLEMS

1. Empirical Formula

What is the empirical formula for a compound with the formula $Cs_2C_4H_4O_6$?

Solving Process:

All of the subscripts are divisible by two. Therefore, the empirical formula is $CsC_2H_2O_3$.

2. Coefficients

What information is conveyed by the formula $3(NH_4)_2SiF_6$?

Solving Process:

The coefficient of 3 means that the quantity of the substance represented is three formula units. The name of the substance is ammonium, NH_4^+ (Table 7.4), hexafluorosilicate, SiF_6^{2-} (Appendix Table A-4). The entire representation, then, is three formula units of ammonium hexafluorosilicate.

PRACTICE PROBLEMS

15. Write the empirical formula for each of the following compounds.

a. N_2O_4 **c.** C_2H_6 **e.** Hg_2I_2
b. $C_6H_8O_6$ **d.** CH_4 **f.** C_8H_{18}

16. State the number of formula units that is represented by each of the following.

a. Ag_2CO_3 **e.** $6Ba_3(PO_4)_2$
b. 3HBr **f.** $5SnBr_4$
c. $2Fe(NO_3)_2$ **g.** $3H_3PO_4$
d. $4AlBr_3$ **h.** $3CH_3COOH$

7.2 CONCEPT REVIEW

17. Name the following compounds.

a. $BaC_4H_4O_6$ **d.** $NaIO_3$
b. $CaSiO_3$ **e.** $CoSO_4$
c. MgS

18. Name the following compounds.

a. $Hg(NO_3)_2$ **d.** $CaC_4H_4O_6$
b. $Al(ClO_3)_3$ **e.** $MgSiO_3$
c. CoS

19. Determine the empirical formula for ammonium oxalate, $(NH_4)_2C_2O_4$.

20. What does the expression $4Ni(IO_3)_2$ mean?

21. Apply A throw-away cigarette lighter lists only one substance in its contents, $CH_3CH_2CH_2CH_3$. What is the name of the substance?

Making the Bone More Binding

The implanting of artificial joints has been done for more than a decade. Still, the binding of bone to implants continues to create problems for doctors and technicians, as well as for the patients. Manufacturers have attempted many different mechanical methods for easing the problem of wobbling artificial joints. Now they believe there may be a chemical solution to the problem.

Prosthetic replacement for ball of femur

If the manufacturers are correct, they may have found a way for bone cells to grow right up to the implant, making it nearly as strong as the original bone. Bioceramic coatings whose structures mimic those of actual bone may provide a solution at last. Hydroxyapatite (HA) is a crystalline form of calcium phosphate that has properties very much like a component of bone. The use of HA is not new; it has been used for ten years in dentistry to fill tooth sockets and build up jaw bones. Calcium, phosphorus, and oxygen atoms also combine in other proportions to form tricalcium phosphate (TCP), a substance that encourages the growth of bone.

In research studies, bioceramic coatings have acted as frameworks for new bone growth in animals. However, an animal's bones grow much faster than a human's, so the two may not be completely analogous with regard to bone growth. Further, the coatings are often unstable and may not remain on the artificial joint. Disintegration of the coating could result in abrasions to the bone and the implant, and cause inflammation of the joint. It is difficult to determine the real corrosion and wear rates in the artificial joints.

Manufacturers of implants most often apply the bioceramic coatings directly to the implants. Powdered HA is ionized by plasma temperatures, 10 000-15 000°C, and sprayed on the joints. When the implant is used, some calcium and phosphorus is released and used by the body to produce a strong interface between the bone and the implant coating. However, some researchers believe that it is the HA itself that may eventually cause the coating to fail. These coatings have been in use only since 1986, so there is no long-term research on their ability to hold up over time.

Scientists and technologists are continuing to find out more about synthetic bone, real bone, the implant, the coatings, and the interactions among all of these. They hope that this knowledge will lead to the development of better implants.

Thinking Critically

1. Find the date of the first artificial hip implant and compare it with the time of the first artificial heart. Give a possible reason for this difference.
2. What responsibility do manufacturers have to make certain that the coatings used are not injurious to the health of the patient?
3. Find out more about the structure of bone tissue.

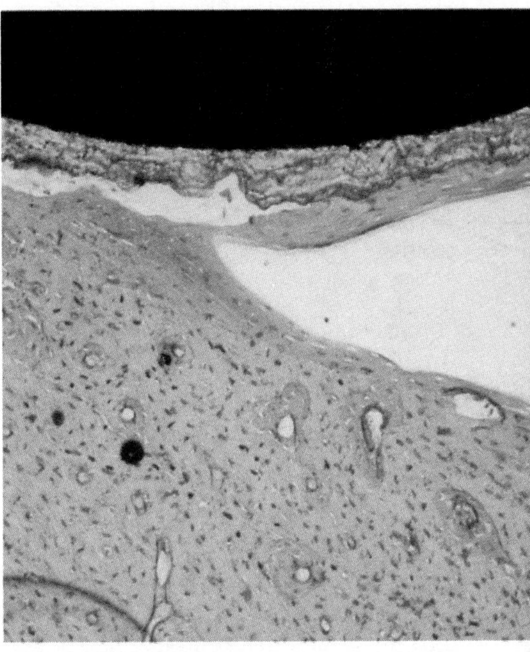

Chapter 7 REVIEW

Summary

7.1 Symbols and Formulas

1. A chemical symbol for an element represents one atom of that element when it appears in a formula.
2. A chemical formula is a statement in chemical symbols of the composition of one formula unit of a compound. A subscript in a formula represents the relative number of atoms of an element in the compound.
3. An ion is a charged particle that is formed when an atom gains or loses electrons.
4. A polyatomic ion is a stable, charged group of atoms.
5. The combining capacity of an atom or polyatomic ion is indicated by its oxidation number or charge.
6. In chemical compounds, atoms combine in definite ratios. Their combined charges add to zero.
7. A binary compound is composed of two elements. Its name is the name of the positive element followed by the name of the negative element modified to end in *-ide*.
8. Some elements have more than one possible oxidation number. A compound containing such an element is named by showing the oxidation number as Roman numerals in parentheses after the element.

7.2 Nomenclature

9. A compound containing a polyatomic ion is named in the same way as a binary compound. However, the ending of the name of the polyatomic ion is not changed.
10. Hydrocarbons are named using a stem that indicates the number of carbon atoms and a suffix indicating how the carbon atoms are linked.
11. A molecular formula shows the actual number of each kind of atom in a single molecule of a compound. The molecular formula is always a whole-number multiple of the empirical formula.
12. An empirical formula represents the simplest whole-number ratio between atoms in a compound.
13. A formula unit represents one molecule or the smallest number of particles giving the ratio of the elements in the compound.
14. A coefficient written before a formula indicates the number of formula units of a substance.

Key Terms

chemical symbol
ion
chemical formula
oxidation number
polyatomic ion
ionic compound
molecule
binary compound
molecular formula
empirical formula
formula unit

Review and Practice

Use tables from this chapter and Appendix Tables A-3 and A-4 when answering the following questions.

22. Why are some chemical symbols so different from the names of the elements that they represent?
23. Why do some symbols have one letter, some have two letters, and some have three letters?
24. What elements exist as diatomic molecules?

25. Write the formula for each of the following compounds.
    a. zinc tartrate
    b. silver nitrate
    c. sodium oxalate
    d. barium nitrate

26. Write the formula for each of the following compounds.
    a. magnesium hydroxide
    b. cadmium hydroxide
    c. silver acetate
    d. magnesium nitrate
    e. aluminum sulfate
    f. potassium cyanide

27. Write the name for each of the following compounds.
    a. $AgBr$
    b. $Ca_3(PO_4)_2$
    c. LiH
    d. $RaCl_2$

28. Using Appendix Table A-3, find all possible formulas for binary compounds of chromium and oxygen.

29. Classify each of the following ions as either monatomic or polyatomic.
    a. Al^{3+}
    b. Hg_2^{2+}
    c. $C_2O_4^{2-}$
    d. H^-
    e. O_2^{2-}
    f. MnO_4^-

30. Which of the following is not a binary compound?
    a. potassium chloride
    b. magnesium hydroxide
    c. calcium bromide
    d. carbon dioxide

31. Why is it necessary to use Roman numerals in writing the name for the compound $Fe(OH)_3$?

32. Copper forms two different compounds with the chloride ion, $CuCl$ and $CuCl_2$. In writing their names, how do we distinguish between them?

33. Write the formula for each of the following compounds.
    a. iron(II) sulfate
    b. manganese(II) nitrate
    c. chromium(III) nitrate
    d. silver perchlorate

34. Write the name for each of the following compounds.
    a. $NaNO_3$
    b. K_2SO_4
    c. $Fe(NO_3)_3$
    d. NH_4CH_3COO

35. Write the formula for each of the following compounds.
    a. tin(IV) chloride
    b. magnesium iodate
    c. ammonium sulfide
    d. cobalt(II) sulfate

36. Write the name for each of the following compounds.
    a. $Pb(CN)_2$
    b. MgH_2
    c. MnO
    d. HCl

37. Write the formula for each of the following compounds.
    a. cadmium nitrate
    b. cobalt(II) nitrate
    c. nickel(II) nitrate
    d. magnesium perchlorate
    e. nitric acid

38. Write the name for each of the following compounds.
    a. $Sr(NO_3)_2$
    b. CaC_2O_4
    c. $H_2C_2O_4$
    d. HgF_2

39. Write the formula for each of the following compounds.
    a. cobalt(II) hydroxide
    b. magnesium carbonate
    c. calcium oxide
    d. lithium acetate

40. Write the name for each of the following compounds.
    a. SrI_2
    b. NH_4F
    c. Ag_2S
    d. H_2SO_4

41. Write the formula for each of the following compounds.
    a. nickel(II) phosphate
    b. calcium peroxide
    c. cobalt(II) perchlorate
    d. barium perchlorate

42. Write the name for each of the following compounds.
    a. $Cr_2(SO_4)_3$
    b. $Hg(IO_3)_2$
    c. $CdBr_2$
    d. CH_3COOH

43. Write the formula for each compound.
- **a.** copper(I) nitride
- **b.** chromium(III) chloride
- **c.** tin(II) oxide
- **d.** cobalt(II) sulfite

44. Write the name for each compound.
- **a.** $PbCr_2O_7$
- **b.** $Mg(CN)_2$
- **c.** $(NH_4)_3AsO_4$
- **d.** CaSe

45. Write the formula for each compound.
- **a.** zinc iodate
- **b.** magnesium selenate
- **c.** cerium(III) oxide
- **d.** strontium bromate

46. Write the formula for each of the following compounds.
- **a.** lithium peroxide
- **b.** lead(II) chromate
- **c.** sodium hypochlorite
- **d.** thallium(I) fluoride

47. Hydrocarbons are compounds of hydrogen and carbon. They are named by the number of carbons in the chain. If C_5H_{12} is pentane, what is C_6H_{14}? C_8H_{18}?

48. Name the following organic compounds according to the number of carbon atoms in the chain or ring.

a. CH_4

b.
```
        H  H  H  H
         \ |  | /
    H     C — C     H
     \   /     \   /
       C         C
     /   \     /   \
    H     C — C     H
         / |  | \
        H  H  H  H
```

c.
```
     H   H   H   H   H   H   H   H   H
     |   |   |   |   |   |   |   |   |
  H— C — C — C — C — C — C — C — C — C —H
     |   |   |   |   |   |   |   |   |
     H   H   H   H   H   H   H   H   H
```

d.
```
      H   H   H
      |   |   |
  H — C — C — C — H
      |   |   |
      H   H   H
```

49. Write the empirical formula for each of the following.
- **a.** Na_2O_2
- **b.** Li_2SO_4
- **c.** Hg_2F_2
- **d.** $Cu_2C_2O_4$
- **e.** $CaC_4H_4O_6$
- **f.** C_6H_6

50. CH is the empirical formula for ethyne (C_2H_2—also called acetylene) and benzene (C_6H_6). What is the empirical formula for $C_{10}H_{22}$? C_6H_{10}? C_6H_{12}?

51. Define formula unit. How is a formula unit different from a molecule?

52. When iron changes oxidation number from iron(II) to iron(III), what happens to the charge on the iron ion?

53. What is the most important factor you must consider when writing a correct formula for an ionic compound?

54. Write the name for each of the following compounds.
- **a.** NH_4HCOO
- **b.** $Hg_2(CH_3COO)_2$
- **c.** Li_2MoO_4
- **d.** $RbBrO_3$

55. In each of the following examples, assume that an element, X, forms an ionic compound with the formula shown. What is the charge on the X ion in each compound?
- **a.** XBr
- **b.** XSO_3
- **c.** Mg_3X_2
- **d.** X_2CO_3
- **e.** $(NH_4)_2X$
- **f.** $X_3(PO_4)_2$

56. Write a formula for each of the following compounds.
   a. calcium azide
   b. aluminum oxalate
   c. cobalt(III) hydroxide
   d. sodium periodate

57. Write the name for each of the following compounds.
   a. $Bi(IO_3)_3$
   b. $CaSiO_3$
   c. $Pb_3[Fe(CN)_6]_2$
   d. KNH_2

58. For simple hydrocarbons, if there are n carbon atoms in a chain, there will be $2n + 2$ hydrogen atoms also in the compound. For example, methane has one carbon atom ($n = 1$) in a molecule, so $2(1) + 2 = 4$ hydrogen atoms will also be present. The formula for methane is CH_4. Write the formula for each of the following hydrocarbons.
   a. pentane
   b. hexane
   c. propane
   d. decane

59. If a simple hydrocarbon is cyclic, and there are n carbon atoms in the ring, there will also be $2n$ hydrogen atoms in a molecule of the compound. For example, cyclodecane would have the formula $C_{10}H_{20}$. Write the formula for each of the following compounds.
   a. cyclobutane
   b. cycloheptane
   c. cyclooctane
   d. cyclononane

Concept Mastery

60. Why is it necessary to use parentheses in writing the formula for zinc phosphate?

61. Element X has oxidation numbers 2+ and 3+. Element Y has oxidation numbers 1− and 2−. List the possible formulas for compounds of X and Y.

62. How do the terms "charge" and "oxidation number" differ?

63. What is the difference between Co_3 and CO_3 when written in a formula? Between ClO_3 and $(ClO)_3$?

64. Why would it be difficult to write formulas if the symbols for all the elements were written in uppercase letters?

65. The formula for baking soda is $NaHCO_3$. What information about the makeup of baking soda is conveyed by this chemical formula?

66. Why is the formula for calcium bromate written $Ca(BrO_3)_2$ rather than $CaBr_2O_6$? After all, both formulas show the same ratios of atoms and represent the composition of the compound.

67. NaCl is an ionic compound, while HCl is a molecular compound. How do these compounds differ?

68. To indicate four molecules of carbon dioxide, we write $4CO_2$. What would be wrong with writing $(CO_2)_4$ or C_4O_8?

69. Three formula units of aluminum sulfate contain how many atoms of oxygen?

70. How many atoms of chlorine are in $7Al(ClO_3)_3$? How many oxygen atoms?

Application

71. When lead storage batteries, such as those found in automobiles, are being charged, lead sulfate is changed, in the presence of sulfuric acid, to lead or lead(IV) oxide. We can represent this reaction by using chemical symbols and chemical formulas.
   a. What is the difference between a chemical symbol and a chemical formula?
   b. Determine whether each of the following is a chemical symbol or a chemical formula.

   $PbSO_4$ H_2SO_4 Pb PbO_2

72. The earliest cosmetics were made from natural compounds. Green eye shadow was made from malachite, $Cu_2CO_3(OH)_2$, a copper ore. How many hydrogen atoms are there in $17Cu_2CO_3(OH)_2$? How many oxygen atoms?

73. When stored, sodium is usually kept in kerosene because it reacts violently with water.

$$2Na + 2H_2O \rightarrow 2Na^+ + 2OH^- + H_2$$

Classify each of the following as an atom, a molecule, or an ion.

a. Na
b. H_2O
c. Na^+
d. H_2
e. OH^-

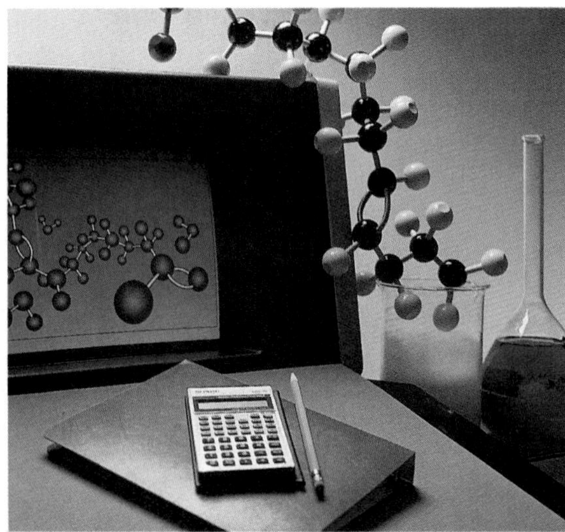

74. Chromium is probably best known as a rust-inhibiting plating on car parts and tools. Chromium can be produced from one of its compounds by the following reaction.

$$Cr_2O_3 + 2Al \rightarrow Al_2O_3 + 2Cr$$

a. List all of the substances that are written using coefficients.
b. List all of the substances that are written using subscripts.
c. List the number of atoms of each element shown on the left side of the equation. List the number of atoms of each element shown on the right side of the equation. What can you conclude from examining these two sets of numbers?

75. How many atoms of nitrogen are in one formula unit of $Pb_3[Fe(CN)_6]_2$?

76. Why is sulfuric acid considered to be such an important industrial chemical? Use references to find four specific industrial processes that use sulfuric acid.

77. Potassium permanganate may be used to get rid of many types of organic odors. A solution may be sprayed over a feedlot, or gases from breweries, paper mills, or meatpacking plants may be bubbled through a $KMnO_4$ solution.

a. How many oxygen atoms are in $11KMnO_4$?
b. In a chemistry book, you read that $2KMnO_4$ react with $5H_2SO_4$. What does this statement mean?

Critical Thinking

78. Hydrofluoric acid is such a reactive acid that it cannot even be stored in a glass container. It is used to etch glass for decorative purposes. From the formulas you know for other acids, what element must be present for a compound to be an acid?

79. Crude oil is a complex mixture mostly of hydrocarbons. What elements would you expect to be most abundant in oil? Describe the probable structure of most of the molecules present.

80. Why are there no compounds named cyclomethane or cycloethane?

81. Ammonia, whose solution is used for cleaning, has the formula NH_3. What is the oxidation number of nitrogen in this compound?

82. The mercury(I) ion is unusual. Look up the charge of this ion in the appropriate table. Suggest a possible explanation for its charge.

83. A compound made up of elements X and Y has the formula X_3Y_2. What does this tell you about the possible oxidation states of elements X and Y?

84. An elaborate system for naming organic compounds has been devised in recent years. Even so, this system must be revised from time to time. Why do you think a complex naming system exists and why must it undergo revision?

85. Cycloalkanes are sometimes found in petroleum and petroleum products.

a. Write the molecular formulas for cyclopentane, cyclohexane, and cyclopropane.

b. What is the empirical formula for each?

c. What conclusion can you reach about cycloalkanes and their empirical formulas?

d. What similar conclusion can you reach about the empirical formulas for butane, heptane, and octane?

86. A compound has the empirical formula CH_2O. Two molecules of this compound combine to form a substance with the molecular formula, $C_{12}H_{22}O_{11}$, and a molecule of water, H_2O. What is the molecular formula of the first compound?

Readings

Block, B. Peter, *et al. Inorganic Chemical Nomenclature.* Washington, D.C.: American Chemical Society, 1990.

Fletcher, John H., *et. al. Nomenclature of Organic Compounds.* Washington, D.C.: American Chemical Society, 1974.

Leigh, G. J. *Nomenclature of Inorganic Chemistry.* Oxford: Blackwell Scientific Publications (IUPAC), 1990.

Cumulative Review

1. Use the arrow diagram to predict electron configurations for niobium, radon, and tin.

2. Find the atomic mass of magnesium if the element has the following isotopic composition in nature.

Isotopic mass	Abundance
23.985 044	78.70%
24.985 839	10.13%
25.982 593	11.17%

3. Identify the group(s) or class(es) of elements that are most clearly identified with the following characteristics.

a. have partially filled *d* sublevel

b. have filled *s* and *p* sublevels

c. have loosely held single *s* electron

d. have half-filled *p* sublevel

e. have partially filled *f* sublevel

f. gain one electron to achieve a noble gas configuration

g. have electron configurations ending in np^4, where n is the principal quantum number

h. have electron configurations ending in $(n - 1)d^7$, where n is the principal quantum number

i. have an outer energy level with a principal quantum number of 4

j. have 1, 2, or 3 electrons in the outer energy level

k. have 5, 6, or 7 electrons in the outer energy level

l. generally lose electrons to satisfy the octet rule

m. is a group of elements that was unknown to Mendeleev

n. have properties of both metallic and nonmetallic elements

o. generally gain electrons to satisfy the octet rule

p. have a Lewis electron dot structure with seven dots

q. react to form positively charged ions

r. have an incomplete 5*f* sublevel

CHAPTER Preview

CONTENTS

CHAPTER

8

The Mole

How many candies in a jar? You could dump out the candies and count them, but what a tedious job! An alternative approach would be to mass a jar empty and then again with its contents. The difference in masses divided by the mass of one candy, would provide you with the number of candies in the jar.

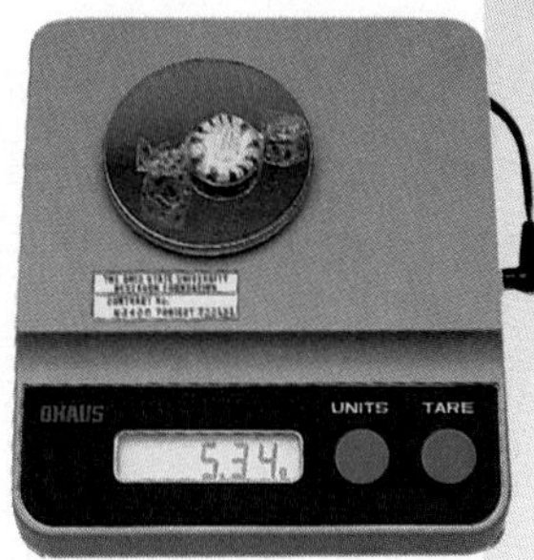

Just as a manufacturer needs to know how many candies will fill each jar, chemists often need to know the number of atoms in a sample of a substance. Because atoms are so small, it would be impossible to count them. How do chemists know how many atoms are in a sample of a substance? In this chapter you will explore a concept that allows you to answer that question.

EXTENSIONS AND CONNECTIONS

Chemistry and Society

Economic Trade-offs: Examines factors used to make difficult decisions in our society, **page 197**

Chemists and Their Work

Amadeo Avogadro, page 200

Bridge to Medicine

Making Medically Effective Products: A look at how chemists begin the task of synthesizing naturally occurring materials for medical use, **page 209**

Everyday Chemistry

Trees, Natural Regulators: Using trees to reduce atmospheric CO_2, **page 211**

OBJECTIVES

- use the factor-label method in calculations.
- use scientific notation to express and evaluate large and small measurements.

8.1 Factor-Label Method

To use mathematical relationships in solving problems in chemistry, you must first learn how to generate the relationships you need. You also must learn to handle very large and very small numbers, which are frequently used in chemistry. You will often relate quantities by using conversion factors, relationships that convert one quantity to another.

Conversion Factors

As you recall from Chapter 2, the relationships between various units that measure the same quantity can be determined from the prefixes listed in Table 2.2. For example, the relationship between the centimeter and the meter is 1 m = 100 cm. The kilogram and the gram are related by the equation 1 kg = 1000 g. Using these and similar relationships, we can form conversion factors to convert a unit to any other related unit. A conversion factor is a ratio equivalent to one.

PROBLEM SOLVING HINT

Visualize the relative sizes of the units you are converting. Since you are converting from a smaller unit to a larger unit, the numerical value of the quantity will decrease.

SAMPLE PROBLEMS

1. Determining a Conversion Factor

What conversion factor can be used to convert centimeters to meters?

Solving Process:

(a) The given quantity is in cm. The desired quantity must be in meters. We know that

$$100 \text{ cm} = 1 \text{ m}$$

(b) By dividing both sides of this equation by 100 cm, the equation becomes

$$\frac{100 \text{ cm}}{100 \text{ cm}} = \frac{1 \text{ m}}{100 \text{ cm}}; \quad 1 = \frac{1 \text{ m}}{100 \text{ cm}}$$

The fraction that is equal to one is called a conversion factor, because we multiply by it to convert one unit into another. Always make certain that the unit you wish to eliminate is properly placed in the conversion factor. If the unit to be eliminated is in the numerator of the given information, then that unit should appear in the denominator of the conversion factor. If the unit to be eliminated is in the denominator of the given information, then that unit should appear in the numerator of the conversion factor. Only then will the units divide out properly.

2. Conversion

How many cubic centimeters are there in 5 cubic decimeters of gold?

Solving Process:
The given quantity is 5 dm^3. The required quantity is expressed in cm^3. The relationship we must find is that which will take us from cm^3 to dm^3.

(a) We know that 1 dm is equal to 10 cm. The relationship between dm and cm is

$$1 \text{ dm} = 10 \text{ cm}$$

(b) We need to find a relationship between dm^3 and cm^3. If we cube both sides of the equation we get

$$1 \text{ dm}^3 = 1000 \text{ cm}^3$$

(c) Both sides of the equation can be divided by the quantity 1 dm^3. Now the equation appears as

$$1 = \frac{1000 \text{ cm}^3}{1 \text{ dm}^3}$$

The expression on the right side of the equation is the conversion factor for changing dm^3 to cm^3.

(d) Using the known quantity of 5 dm^3 that is given in the problem, we can write the equation

$$5 \text{ dm}^3 = 5 \text{ dm}^3$$

(e) We can now multiply the expression on the right side of the equation in (d) by the fraction

$$\frac{1000 \text{ cm}^3}{1 \text{ dm}^3}$$

Since this fraction equals *one,* the value of the expression on the right side of the equation is not changed. (Recall that the value of any quantity multiplied by 1 is unchanged.) The equation then becomes

$$5 \text{ dm}^3 = 5 \text{ dm}^3\left(\frac{1000 \text{ cm}^3}{1 \text{ dm}^3}\right) = \frac{(5 \text{ dm}^3)(1000 \text{ cm}^3)}{(1 \text{ dm}^3)}$$

The equation then reduces to

$$5 \text{ dm}^3 = (5)(1000 \text{ cm}^3) = 5000 \text{ cm}^3$$

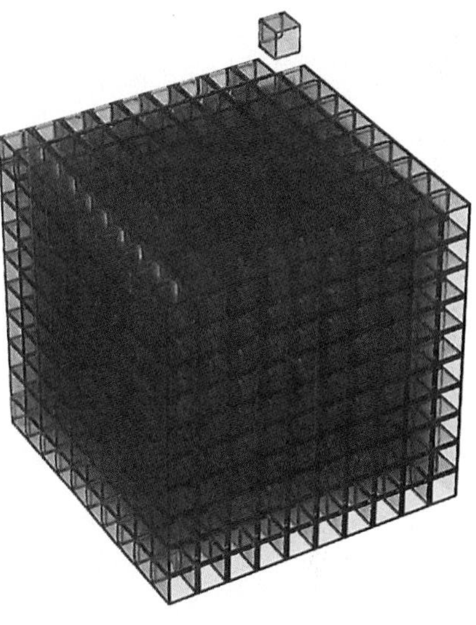

Figure 8.1 One cubic decimeter of any material contains $(10 \text{ cm})^3$ or 1000 cm^3.

PRACTICE PROBLEMS

Convert each of the following quantities.

1. 0.143 hour to seconds
2. 0.84 meter to centimeters
3. 31.5 centigrams to milligrams
4. 65.22 mg to g
5. 531 cm^3 to dm^3
6. 718 nm to cm

The principles in the procedure just described can be used to solve many kinds of problems. In order to simplify the working of chemistry problems, we will use a different notation. In the last Sample Problem, we had the equation

$$5\ \text{dm}^3 = 5\ \text{dm}^3\left(\frac{1000\ \text{cm}^3}{1\ \text{dm}^3}\right)$$

That equation is equivalent to the equation

$$5\ \text{dm}^3 = \left(\frac{5\ \text{dm}^3}{1}\right)\left(\frac{1000\ \text{cm}^3}{1\ \text{dm}^3}\right)$$

Instead of enclosing every factor in parentheses, we will simply set off each factor of the equation by a vertical line. Our equation is now written as

$$5\ \text{dm}^3 = \frac{5\ \text{dm}^3 \mid 1000\ \text{cm}^3}{\phantom{5\ \text{dm}^3} \mid 1\ \text{dm}^3} = 5000\ \text{cm}^3$$

Factor-Label Method

The problem-solving method discussed in the previous section is called the **factor-label method.** In effect, unit labels are treated as factors. As common factors, these labels may be divided out. The correct solution to a problem will have the correct unit label. Thus, this method aids in solving a problem and provides a check on mathematical operations. The individual conversion factors (ratios whose value is equivalent to one) may usually be written by inspection. Let us see how the method can be applied to more complex measurements.

Many measurements are simply combinations of the elementary measurements as discussed in Chapter 2. You are probably familiar with the measurement of speed. Speed is the distance covered during a single unit of time. The speed of an automobile is measured in kilometers (length) per hour (time). Most scientific measurements of speed are made in centimeters per second or meters per second.

Figure 8.2 Speed is measured in derived units composed of units of length and units of time. The speedometer reading of an automobile (left) indicates the speed, that is, the rate at which the automobile is moving at that moment (right).

SAMPLE PROBLEM

Conversion of Units

Express 60.0 kilometers per hour as centimeters per second.

Solving Process:

(a) Write 60.0 kilometers per hour as a ratio. $\frac{60.0\text{ km}}{1\text{ h}}$

(b) We wish to convert kilometers to meters. Since 1000 m = 1 km, we can use the conversion factor

$$\frac{1000\text{ m}}{1\text{ km}}$$

because its value equals 1.

(c) In the same manner, we will use

$$100\text{ cm} = 1\text{ m, or } \frac{100\text{ cm}}{1\text{ m}} \quad \text{(to convert m to cm)}$$

$$1\text{ h} = 60\text{ min, or } \frac{1\text{ h}}{60\text{ min}} \quad \text{(to convert h to min)}$$

$$1\text{ min} = 60\text{ s, or } \frac{1\text{ min}}{60\text{ s}} \quad \text{(to convert min to s)}$$

(d) Because any number may be multiplied by 1 or its equivalent without changing its value, we can now write

$$\frac{60.0\text{ km}}{1\text{ h}} = \frac{60.0\text{ }\cancel{\text{km}}}{1\text{ }\cancel{\text{h}}} \Bigg| \overset{\text{km to m}}{\frac{1000\text{ }\cancel{\text{m}}}{1\text{ }\cancel{\text{km}}}} \Bigg| \overset{\text{m to cm}}{\frac{100\text{ cm}}{1\text{ }\cancel{\text{m}}}} \Bigg| \overset{\text{h to min}}{\frac{1\text{ }\cancel{\text{h}}}{60\text{ }\cancel{\text{min}}}} \Bigg| \overset{\text{min to s}}{\frac{1\text{ }\cancel{\text{min}}}{60\text{ s}}}$$

$$= 1670\text{ cm/s}$$

Notice that ratios are arranged so that units can be divided out as factors. This procedure leaves the correct units in the answer and provides a check on the method used.

The dividing out of units in this Sample Problem will occur in almost all sample problems. Follow the same procedure in your own problem solutions.

PRACTICE PROBLEMS

Use the factor-label method to convert the following quantities.

7. 0.032 g to mg

8. 0.436 m^3 to cm^3

9. 302.1 mL to cm^3

10. 0.693 dm^3 to cm^3

11. 9.06 km/h to m/min

12. 0.307 mg/cm^3 to g/cm^3

13. 822 dm^3/s to L/min

14. 0.78 L/min to cm^3/s

15. 0.848 kg/L to mg/cm^3

16. 81.42 nm/s to cm/min

17. 7.56 mm^3/s to dm^3/min

18. 0.03 cm/s to km/h

19. 0.0775 cg/cm^3 to g/m^3

20. 0.95 kg/cm^3 to mg/mm^3

Figure 8.3 For extremely small measurements, such as the diameter of a red blood cell (left), and for very large measurements, such as the distance to the Andromeda galaxy (right), it is convenient to use scientific notation. Showing all the digits in measurements like these makes calculations difficult.

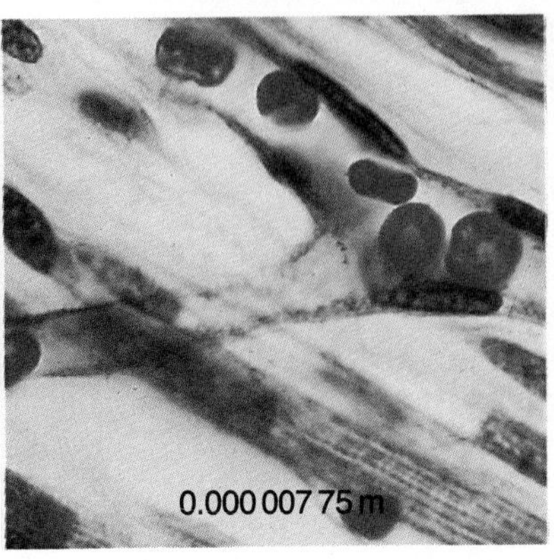

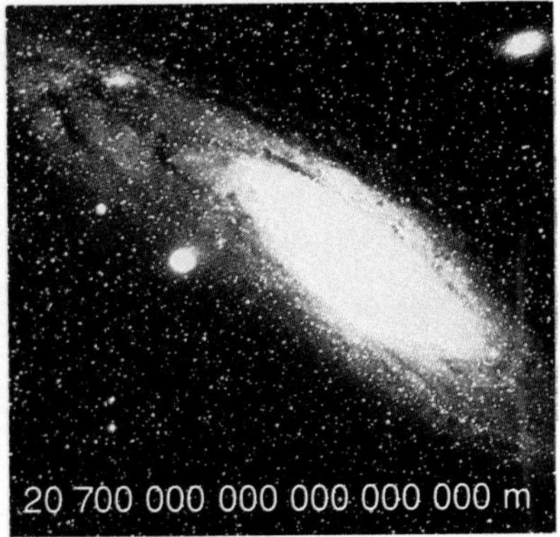

Handling Numbers in Science

In this course we will sometimes use very large numbers. For example, later in this chapter you will learn about the Avogadro constant, which is 602 214 000 000 000 000 000 000. We will also use very small numbers as in describing the distance between the particles in a sodium chloride crystal as 0.000 000 002 814 cm. In working with such numbers it is easy to make a mistake by overlooking a zero or misplacing the decimal point.

Scientific notation makes it easier to work with very large or small numbers. In **scientific notation,** all numbers are expressed as the product of a number between 1 and 10 and a whole-number power of 10.

$$M \times 10^n$$

In this expression, $1 \leq M < 10$, and n is an integer. This number is read as "M times ten to the nth". For example, 5.2×10^5 is read "five point two times ten to the fifth."

One advantage of scientific notation is that it removes any doubt about the number of significant digits in a measurement. Suppose the volume of a gas is expressed as 8000 cm^3. As written, this number contains only one significant digit. Suppose the measurement was actually made to the nearest cm^3. Then the volume 8000 cm^3 must be expressed to four significant digits. In scientific notation, we can indicate the additional significant digits by placing zeros to the right of the decimal point. Thus, 8×10^3 cm^3 has only one significant digit, while 8.000×10^3 cm^3 has four significant digits.

Once we have recorded measurements to the correct number of significant digits and expressed them in scientific notation, we can use them in calculations. To determine the number of digits that should appear in the answer to a calculation, we will use two rules.

1. *In addition and subtraction, the answer may contain only as many decimal places as the measurement having the least number of decimal places.* For example, if 345 g is added to 27.6 g, the answer must be given to the nearest whole number. In adding a column of figures such as

$$\begin{array}{r l} 677.1 & \text{cm} \\ 39.24 & \text{cm} \\ 6.232 & \text{cm} \\ \hline 722.572 & \text{cm} \end{array}$$

the answer should be rounded off to the nearest tenth. The answer to the problem above is 722.6 cm.

2. *In multiplication and division, the answer may contain only as many significant digits as the measurement with the least number of significant digits.* For example, in the following problem the answer has three significant digits.

$$\underset{(3)}{(1.13\text{ m})}\ \underset{(7)}{(5.126\ 122\text{ m})} = 5.792\ 517\ 86\text{ m}^2 = \underset{(3)}{5.79\text{ m}^2}$$

The answer to the next problem has four significant digits.

$$\frac{\overset{(6)}{49.600\ 0\text{ g}}}{\underset{(4)}{47.40\text{ cm}^3}} = 1.046\ 413\ 5\ \frac{\text{g}}{\text{cm}^3} = \overset{(4)}{1.046}\ \frac{\text{g}}{\text{cm}^3}$$

If you are using a calculator to obtain your numerical answer, you must be very careful. Remember, you are calculating with measured quantities, not just numbers. For example, in the problem 49.600 0 g/47.40 cm^3, the calculator may display the numerical answer 1.046413502. This answer must be rounded to the proper number of significant digits. In this case, the number of significant digits is four. Thus, the answer is 1.046 g/cm^3. Always double-check your answer against the data given. Report the answer only to the number of significant digits justified by the data. Remember, calculated accuracy cannot exceed measured accuracy.

Figure 8.4 Not all the digits that appear on a calculator are significant. In performing the calculation 49.600 g ÷ 47.40 cm^3 the calculator displays 1.046413502. The answer should be reported as 1.046 g/cm^3 because the factor 47.40 cm^3 has only four significant digits.

SAMPLE PROBLEM

Multiplying in Scientific Notation

Find the product of

$$(4.0 \times 10^{-2}\text{ cm})(3.0 \times 10^{-4}\text{ cm})(2.0 \times 10^{1}\text{ cm}).$$

Solving Process:

To multiply numbers expressed in scientific notation ($M \times 10^n$), multiply the values of M and add the exponents. The exponents do not need to be alike. When using a scientific calculator, use the EXP key, which stands for the "× 10" part of the number.

Multiply the values of M. $4.0 \times 3.0 \times 2.0 = 24$

Multiply the units. $\text{cm} \times \text{cm} \times \text{cm} = \text{cm}^3$

Add the exponents. $-2 + (-4) + 1 = -5$

Thus, $(4.0 \times 10^{-2})(3.0 \times 10^{-4})(2.0 \times 10^{1}) =$

$$(4.0 \times 3.0 \times 2.0) \times 10^{-2+(-4)+1}\text{ cm}^3 = 24 \times 10^{-5}\text{ cm}^3$$
$$= 2.4 \times 10^{-4}\text{ cm}^3$$

SAMPLE PROBLEM

Dividing in Scientific Notation

Divide. $\dfrac{(4 \times 10^3 \text{ m})(6 \times 10^{-1} \text{ m})}{(8 \times 10^2 \text{ m})}$

Solving Process:

In division the exponents are subtracted. To divide numbers in scientific notation ($M \times 10^n$), divide the values of M and subtract the sum of the exponents of the denominator from the sum of the exponents of the numerator.

$$\frac{(4 \times 6) \times 10^{3+(-1)} \text{ m}^2}{8 \times 10^2 \text{ m}} =$$

$$\frac{24}{8} \times \frac{10^2}{10^2} \times \frac{\text{m}^2}{\text{m}} = 3 \times 10^0 \text{ m}$$

$$= 3 \text{ m}$$

PRACTICE PROBLEMS

Perform the following operations. Express your answers to the correct number of significant digits. Units should be handled in the same way that numbers are.

21. (10.18 m)(0.007 40 m)

22. (302 000 mm)(3.5 mm)

23. $\dfrac{3120.14 \text{ g}}{4.2 \text{ dm}^3}$

24. 701 g + 4.24 g + 57.397 g

25. 9.66 kg − 0.256 92 kg

26. $(2.63 \times 10^{-9} \text{ cm})(5.06 \times 10^3 \text{ cm})$

27. $\dfrac{(6.12 \times 10^2 \text{ cm})(4.92 \times 10^3 \text{ cm})}{(8.38 \times 10^{-7} \text{ cm})}$

8.1 CONCEPT REVIEW

28. Make the following conversions.
- **a.** 895 picoseconds to seconds
- **b.** 1.098 km to m
- **c.** 0.924 kilogram to milligrams
- **d.** 0.705 milligram to grams
- **e.** 491 dm^3 to liters
- **f.** 136 dm^3 to m^3
- **g.** 0.852 m^3 to cm^3

29. Perform the following operations. Express each answer in scientific notation.
- **a.** 1100 m + 596 m
- **b.** 32 144°C − 800°C
- **c.** 32 235 kg/1991 m^3
- **d.** 191 $\text{cm}^2 \times 0.0024 \text{ cm}^2$
- **e.** (32 395 kg/m^3)(32 161 m^3)/32 143 m^2

30. If the speedometer reading of a car is 55 miles per hour, what is the car's speed in km/h? in m/s?

31. Apply Explain which quantity, 1.0×10^3 g or 1.00×10^3 g, more nearly expresses the mass of exactly one kilogram.

Economic Trade-Offs

Did you know that silver is the best electrical conductor of any element? However, you won't find it in any wiring in your house. Why not? You'll know the answer if you've ever looked at the price of silver jewelry. The answer, of course, is expense. The electrical wire found in most houses is made of copper. Copper is about 10% less efficient in conducting electrical current than silver. However, it is far less expensive than silver.

The use of copper wire rather than silver wire in electrical installations is an example of what is often called an *economic trade-off*. In this circumstance, consumers have traded one set of costs for another. Even though the use of copper wire might add a small cost to an electric bill every month because it is less conductive than silver, the cost of wiring a house using silver would have been far greater.

In a technological society, company managers, political organizations, citizen groups, and individuals are constantly having to make decisions that involve economic tradeoffs.

A number of communities across the United States are now facing a decision concerning the disposition of hazardous wastes. Some of these communities are presently suffering from low employment levels. The added industry would provide much-needed jobs. However, there are in these communities large numbers of people who do not wish to have hazardous waste disposal facilities near them because they are afraid of contamination of water supplies and the local atmosphere.

Another trade-off decision made every day by millions of citizens is what type of gasoline to buy for their automobiles. Premium gasolines cost significantly more than regular gasoline. However, they also deliver better mileage.

Whenever you are considering a question involving the choice of materials or processes, it will always be wise to consider the economic consequences of your decision. In each instance, you will have to consider what the trade-off involves: what are the benefits, what are the costs, and what are the risks?

Analyzing the Issue

1. Find out what precautions are taken to prevent hazardous waste landfills from contaminating water supplies.
2. Find the price of regular and super unleaded gasolines in your area. If the regular provides 19 miles per gallon and the super provides 26 miles per gallon, which gasoline would you buy?

OBJECTIVES

- use the Avogadro constant to define the mole and to calculate molecular and molar mass.
- use the molar mass to calculate the molarity of solutions, percentage composition, and empirical formulas.
- determine the formulas of hydrates.

8.2 Formula-Based Problems

Chemical symbols and formulas (such as H and H_2O) are shorthand signs for chemical elements and compounds. The symbol of an element may represent one atom of the element. The formula of a compound may represent one molecule or one formula unit of the compound. Symbols and formulas may also represent a group of atoms or formula units. Since atoms are so very small, chemists usually deal with large groups of atoms. This section is about a group, called a mole, containing a specific number of units. Counting items in groups is a familiar thing to do. We count eggs in dozens (12), and merchants often count items in a group called a gross (144). A mole is just a particular number of atoms, ions, molecules, or formula units. It is a large number because these particles are so small, and we want these particles in a group that is big enough to obtain the mass conveniently.

Molecular and Formula Mass

A hydrogen atom has a mass of 1.67×10^{-24} gram. The mass of an oxygen atom is 16 times that of hydrogen, or 2.66×10^{-23} gram, and the mass of a carbon atom is 12 times that of hydrogen. Because these are very small numbers, it is convenient to use a mass scale defined on the atomic level as described in Chapter 4. A list of atomic masses of the elements is found inside the back cover of this book. The units associated with the numbers in the atomic mass column are atomic mass units/atom.

The atomic mass of hydrogen in atomic mass units is 1 u, and the atomic mass of oxygen is 16 u. Therefore, the total mass of a water molecule, H_2O, is 1 + 1 + 16, or 18 u. If the atomic masses of all the atoms in a molecule are added, the sum is the mass of that molecule. Such a mass is called a **molecular mass.** This name is incorrect when applied to an ionic substance. Sodium chloride, NaCl, is an ionic substance and does not exist in molecular form. A better name for the mass of ionic substances is

Figure 8.5 It is often convenient to count items by arranging them in a group, such as a dozen (12) or a gross (12×12).

formula mass. The sum of the atomic masses of all atoms in the formula unit of an ionic compound is called the **formula mass.** To calculate a formula mass, add the masses of all the atoms in the formula.

In this text, atomic masses are used that will ensure three significant digits in the result. Thus, to obtain the formula mass of NaCl, the values are rounded to tenths: 23.0 + 35.5 = 58.5. For H_2, the values are rounded to hundredths: 1.01 + 1.01 = 2.02.

SAMPLE PROBLEMS

1. Molecular Mass

Find the molecular mass of 2,3-butanedione, $CH_3COCOCH_3$.

Solving Process:

Add the atomic masses of all the atoms in the $C_4H_6O_2$ formula unit.

4 C atoms	4 × 12.0 = 48.0 u
6 H atoms	6 × 1.01 = 6.06 u
2 O atoms	2 × 16.0 = 32.0 u
molecular mass of $C_4H_6O_2$	86.1 u

2. Formula Mass

Find the formula mass of calcium nitrate, $Ca(NO_3)_2$, a compound used in making explosives, fertilizers, and matches.

Solving Process:

Add the atomic masses of all the atoms in the $Ca(NO_3)_2$ formula unit. Remember that the subscript applies to the entire polyatomic ion.

1 Ca atom	1 × 40.1 = 40.1 u
2 N atoms	2 × 14.0 = 28.0 u
6 O atoms	6 × 16.0 = 96.0 u
formula mass of $Ca(NO_3)_2$	164.1 u

PRACTICE PROBLEMS

32. Calculate the molecular mass of each of the following organic compounds.

a. methanol, CH_3OH
b. ethane, C_2H_6
c. sucrose, $C_{12}H_{22}O_{11}$
d. 1-propanol, $CH_3CH_2CH_2OH$

33. Calculate the molecular or formula mass of each of the following compounds.

a. tantalum carbide, TaC, used to make tools and dies
b. aluminum nitride, AlN
c. tetraphosphorus trisulfide, P_4S_3, used in the tips of strike-anywhere matches
d. calcium phosphate, $Ca_3(PO_4)_2$
e. barium hypochlorite, $Ba(ClO)_2$

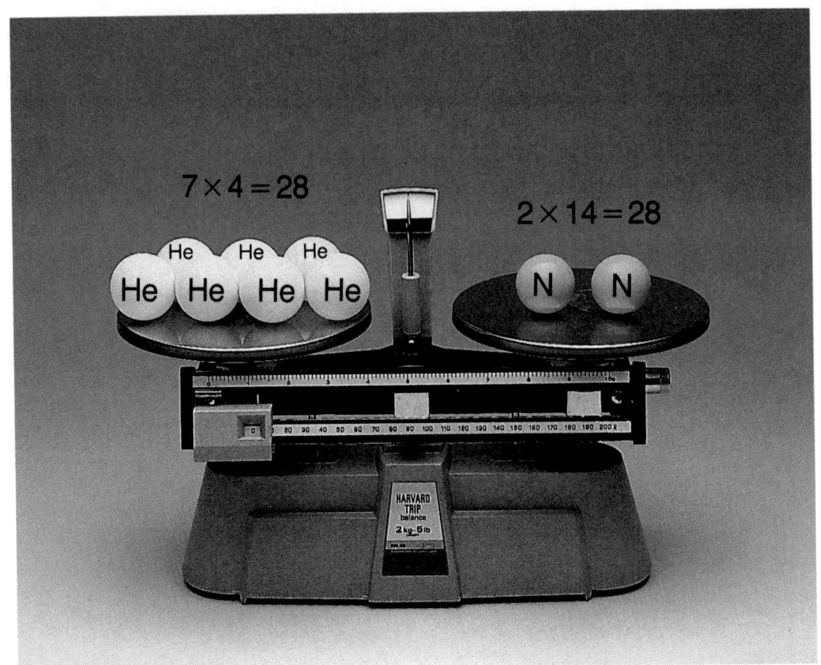

Figure 8.6 If it were possible to mass single atoms, the mass of two nitrogen atoms would equal the mass of seven helium atoms. Thus, the mass ratio of N to He is 7 to 2.

CHEMISTS AND THEIR WORK

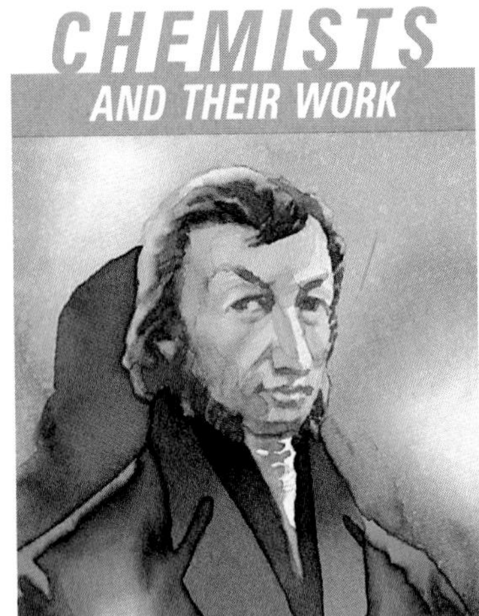

Amadeo Avogadro
(1776-1856)

Avogadro initially studied and practiced law in his native Italy. After three years of law practice, he turned to science and spent the remainder of his life as a professor of science at the University of Turin. Avogadro is best known for the hypothesis that equal volumes of gases contain equal number of particles, which he proposed in 1811. He was the first person to distinguish between atoms and molecules. In fact, he coined the word "molecule." The ideas first proposed by Avogadro enabled chemists to determine a table of realistic atomic masses—something they could not do until they understood the difference between atoms and molecules.

The Avogadro Constant

Chemists do not deal with amounts of substances by counting out atoms or molecules. Rather, they obtain the masses of quantities of substances. Thus, it is important to obtain a relationship between mass and number of particles.

There is a problem in using the molecular masses of substances. Molecular masses are in atomic mass units. An atomic mass unit is only 1.66×10^{-24} gram. The mass of a single molecule is so small that it is impossible to measure by ordinary means in the laboratory. For everyday use in chemistry, a larger unit, such as a gram, is needed.

One helium atom has a mass of 4 u, and one nitrogen atom has a mass of 14 u. The ratio of the mass of one helium atom to one nitrogen atom is 4 to 14, or 2 to 7. Let us compare the mass of two helium atoms to that of two nitrogen atoms. The ratio would be 2×4 to 2×14, or 2 to 7. If we compare the mass of 10 atoms of each element, we will still get a 2 to 7 ratio. No matter what number of atoms we compare, equal numbers of helium and nitrogen atoms will have a mass ratio of 2 to 7. In other words, the numbers in the atomic mass table give us the relative masses of the atoms of the elements.

The laboratory unit of mass you will use is the gram. We would like to choose a number of atoms that would have a mass in grams equivalent to the mass of one atom in atomic mass units. The same number would fit all elements, because equal numbers of different atoms always have the same mass ratio. Chemists have found that 6.02×10^{23} atoms of an element have a mass in grams equivalent to the mass of one atom in atomic mass units. For example, a single atom of hydrogen has a mass of 1.0079 u;

6.02×10^{23} atoms of hydrogen have a mass of 1.0079 g. This number, 6.02×10^{23}, is called the **Avogadro constant** in honor of a 19th century Italian scientist.

As you recall, the atomic mass unit is defined as 1/12 the mass of a carbon-12 atom. Thus, the atomic mass of a carbon-12 atom is exactly 12 u. Suppose we were to count the number of atoms in exactly 12 grams of carbon-12. The number of atoms would have the value of the Avogadro constant.

The Mole

The Avogadro constant is an accepted SI standard. We may add it to the others we have studied: the kilogram, the meter, the second, and the kelvin. The symbol used to represent the Avogadro constant is N_A. This quantity can be expressed as $6.022\,1367 \times 10^{23}$ to be more precise. This number of things is called one **mole (mol)** of the things. Chemists have chosen one mole as a standard unit for large numbers of atoms, ions, or molecules. Thus, the mole is the SI base unit representing the chemical quantity of a substance. One mole of particles (atoms, ions, molecules) has a mass in grams equivalent to that of one particle in atomic mass units. Thus, if a mole of any type of particle has a mass of 4.02 grams, then a single particle has a mass of 4.02 u. In the same manner, if a single particle has a mass of 54.03 u, then a mole of particles will have a mass of 54.03 grams. Chemists chose the value of the Avogadro constant so that the mass of N_A atoms in grams is equivalent to the mass of one atom in atomic mass units. Because of this relationship, we may associate a second set of units with the numbers in the atomic mass column: grams/mole of atoms.

Figure 8.7 Shown here are molar quantities of several substances. Clockwise from left are 32.1 g of sulfur, 342 g of sucrose (table sugar), 249 g of copper (II) sulfate pentahydrate, 58.5 g of sodium chloride (table salt), 63.5 g of copper, and 216 g of mercury(II) oxide.

N_A is a number that has been experimentally determined. In 1905, two English scientists, William Ramsay and Frederick Soddy, devised an experiment to determine the Avogadro constant using radium. Radioactive radium decays by emitting an alpha particle, which then gains two electrons and forms a helium atom. If the volume of the helium produced by a sample of radium is collected after a period of time and measured, it is possible to compute the value for N_A. Using modern instruments, the value of N_A has been found to be 6.02×10^{23} atoms. If a mole of iodine molecules is crystallized and then inspected by X-ray diffraction, the number of I_2 molecules in the crystal can be determined. Again, the number is 6.02×10^{23} molecules. The Avogadro constant has also been determined by the scattering of light and by Millikan's oil drop experiment, which was described in Chapter 4.

It is important to note that *one mole of atoms contains 6.02×10^{23} atoms. One mole of molecules contains 6.02×10^{23} molecules. One mole of formula units contains 6.02×10^{23} formula units. One mole of ions contains 6.02×10^{23} ions.* N_A, therefore, can have any of these units.

$$\frac{\text{atoms}}{\text{mole}} \quad \text{or} \quad \frac{\text{molecules}}{\text{mole}} \quad \text{or} \quad \frac{\text{formula units}}{\text{mole}} \quad \text{or} \quad \frac{\text{ions}}{\text{mole}}$$

The mass of one mole of molecules, atoms, ions, or formula units is called the **molar mass** of that species.

SAMPLE PROBLEMS

1. Conversion of Grams to Moles

Use the factor-label method to determine how many moles are represented by 11.5 g of ethanol, C_2H_5OH.

Solving Process:

We have grams, and we wish to convert to moles. Using the atomic mass table inside the back cover, we can determine that one mole of C_2H_5OH has a mass of 46.1 g. Therefore,

$$1 \text{ mol} = 46.1 \text{ g} \quad \text{or} \quad \frac{1 \text{ mol}}{46.1 \text{ g}} = 1$$

$$\frac{11.5 \cancel{\text{g } C_2H_5OH}}{} \left| \frac{1 \text{ mol } C_2H_5OH}{46.1 \cancel{\text{g } C_2H_5OH}} \right. = 0.249 \text{ mol } C_2H_5OH$$

2. Conversion of Molecules to Moles and Grams

How many moles is 1.20×10^{25} molecules of ammonia, NH_3? What mass is this number of molecules?

Solving Process:

(a) We have molecules and wish to convert to moles. One mole equals 6.02×10^{23} molecules. Use the ratio

$$\frac{1 \text{ mol}}{6.02 \times 10^{23} \text{ molecules}}$$

$$\frac{1.20 \times 10^{25} \text{ molecules } NH_3}{} \left| \frac{1 \text{ mol}}{6.02 \times 10^{23} \text{ molecules}} \right.$$

$$= 1.99 \times 10^1 \text{ mol } NH_3$$
$$= 19.9 \text{ mol } NH_3$$

(b) Using the table inside the back cover, the molar mass of NH_3 is calculated to be 17.0 g. Use

$$\frac{17.0 \text{ g } NH_3}{1 \text{ mol } NH_3}$$

$$\frac{19.9 \text{ mol } NH_3}{} \left| \frac{17.0 \text{ g } NH_3}{1 \text{ mol } NH_3} \right. = 338 \text{ g of } NH_3$$

PRACTICE PROBLEMS

Make each of the following conversions.

34. 0.638 mole $Ba(CN)_2$ to grams

35. 50.4 grams $CaBr_2$ to moles

36. 1.26 moles NbI_5 to grams

37. 86.2 grams C_2H_4 to moles

SAMPLE PROBLEM

Conversion of Grams to Atoms

Determine the number of atoms that are in a 10.0-g sample of calcium metal.

Solving Process:

First, find the number of moles in 10.0 g Ca, and then convert to atoms. 1 molar mass of calcium is 40.1 g. Therefore, use the ratios

$$\frac{1 \text{ mol Ca}}{40.1 \text{ g Ca}} \quad \text{and} \quad \frac{6.02 \times 10^{23} \text{ atoms}}{1 \text{ mol}}$$

$$\frac{10.0 \text{ g Ca}}{} \left| \frac{1 \text{ mol Ca}}{40.1 \text{ g Ca}} \right| \frac{6.02 \times 10^{23} \text{ atoms}}{1 \text{ mol}}$$

$$= 1.50 \times 10^{23} \text{ atoms Ca}$$

PROBLEM SOLVING HINT

Use the concept of the mole to relate molar mass and the mass of the sample. Use the concept of a mole again to relate the number of atoms to the molar mass.

PRACTICE PROBLEMS

Make each of the following conversions.

38. 0.943 mole H_2O to molecules

39. 7.74×10^{26} formula units Al_2O_3 to moles

40. 91.9 grams NH_4IO_3 to formula units

41. 6.63×10^{23} molecules $C_6H_{12}O_6$ to moles

Figure 8.8 A $0.1000M$ solution of $CuSO_4 \cdot 5H_2O$ is made by dissolving 24.97 g of solid $CuSO_4 \cdot 5H_2O$ in enough water to make exactly 1000 cm^3 of solution.

Moles in Solution

Most chemical reactions take place in solution. There are several methods of expressing the relationship between the dissolved substance and the solution. The method most often used by chemists is molarity (M). **Molarity** is the ratio between the moles of dissolved substance and the volume of solution expressed in cubic decimeters.

$$\text{Molarity} = \frac{\text{moles of solute}}{\text{volume of solution in dm}^3}$$

A one-molar ($1M$) solution of nitric acid contains one mole of nitric acid molecules in one cubic decimeter of solution. A $0.372M$ solution of $Ba(NO_3)_2$ contains 0.372 mole of $Ba(NO_3)_2$ in 1 dm^3 of solution.

Assume that you wish to try a reaction using 0.1 mole of glucose ($C_6H_{12}O_6$). If the solution of glucose in your laboratory is $1M$, then 0.1 mole of glucose would be contained in 0.1 dm^3, or 100 cm^3 of the glucose solution. Expressing the composition of solutions in units of molarity is a convenient way of measuring a number of particles.

Note that when the concentration of a solution is expressed in molarity, the expression uses cubic decimeters of *solution* and not cubic decimeters of solvent. For example, to make a $1.0M$ glucose solution, you do not add one cubic decimeter of water to one mole (180 g) of glucose. The final volume might not be one cubic decimeter. Rather, dissolve the glucose in water, and then add water until you reach one cubic decimeter of solution.

It is important to be able to compute the molarity of solutions if you are given their composition. It is also important to be able to compute the amount of substance you need to produce a specific solution. You will be using these calculations in your laboratory work and later in solving problems in this book.

SAMPLE PROBLEM

Molarity

What is the molarity of 2.50×10^2 cm^3 of solution containing 9.46 g CsBr, cesium bromide, a compound that is used to make optical devices such as prisms and spectrophotometer cells?

Solving Process:

The units of molarity are moles of substance per (divided by) dm^3 of solution. Thus, to obtain molarity we must divide the substance, expressed in moles, by the solution, expressed in dm^3. The table inside the back cover can be used to determine the molar mass of cesium bromide. Recall from Chapter 2 that 1 dm^3 = 1000 cm^3.

$$\frac{9.46\ \cancel{\text{g CsBr}}}{} \left| \frac{1\ \text{mol CsBr}}{213\ \cancel{\text{g CsBr}}} \right. = 0.0444\ \text{mol CsBr}$$

$$\frac{2.50 \times 10^2 \text{ cm}^3}{} \left| \frac{1 \text{ dm}^3}{1000 \text{ cm}^3} \right. = 0.250 \text{ dm}^3$$

$$\frac{0.0444 \text{ mol CsBr}}{0.250 \text{ dm}^3} = 0.178M \text{ CsBr}$$

To save time and to reduce the chance for error, these calculations can be combined in one continuous chain. Note that the factors are arranged to yield an answer in mol/dm^3, which is molarity.

$$\frac{9.46 \text{ g CsBr}}{2.50 \times 10^2 \text{ cm}^3} \left| \frac{1 \text{ mol CsBr}}{213 \text{ g CsBr}} \right| \frac{1000 \text{ cm}^3}{1 \text{ dm}^3} = 0.178M \text{ CsBr}$$

PRACTICE PROBLEMS

Compute the molarity of each of the following solutions.

42. 5.23 g $Fe(NO_3)_2$ in 100.0 cm^3 of solution

43. 8.55 g NH_4I in 50.0 cm^3 of solution

44. 9.94 g $CoSO_4$ in 2.50×10^2 cm^3 of solution

45. 44.3 g $Pb(ClO_4)_2$ in 250.0 cm^3 of solution

SAMPLE PROBLEM

Making a Solution

How would you make 5.00×10^2 cm^3 of a 0.133M solution of $MnSeO_4$, manganese(II) selenate, and water?

Solving Process:

First, use the given quantity of solution, 5.00×10^2 cm^3, to find the part of a dm^3 desired. Then use the molarity to convert from solution to moles of substance. Finally, convert the moles of substance to grams.

$$\frac{5.00 \times 10^2 \text{ cm}^3}{} \left| \frac{1 \text{ dm}^3}{1000 \text{ cm}^3} \right| \frac{0.133 \text{ mol MnSeO}_4}{1 \text{ dm}^3} \left| \frac{198 \text{ g MnSeO}_4}{1 \text{ mol MnSeO}_4} \right. = 13.2 \text{ g MnSeO}_4$$

To make the solution, then, you would dissolve 13.2 g $MnSeO_4$ in sufficient water to make 5.00×10^2 cm^3 of solution.

PRACTICE PROBLEMS

Describe the preparation of each of the following solutions.

46. 1.00 dm^3 of 3.00M $NiCl_2$

47. 2.50×10^2 cm^3 of 4.00M $CoCl_2$

48. 0.500 dm^3 of 1.50M AgF

49. 2.50×10^2 cm^3 of 0.002 00M $Cd(IO_3)_2$

Percentage Composition

The **percentage composition** of a compound is a statement of the relative mass each element contributes to the mass of the compound as a whole. A chemist often compares the percentage composition of an unknown compound with the percentage composition calculated from an assumed formula. If the percentages agree, it will help to confirm the identity of the unknown.

Consider the percentage composition of the following substances: copper, sodium chloride, and ethanol. For copper, the percentage composition is 100 percent Cu because it is composed of a single element. Salt is composed of two elements, sodium and chlorine, in a 1 to 1 atomic ratio. We know they are always present in the same ratio by mass. The ratio in which they are present is the ratio of their atomic masses. Therefore, the percentage of sodium in any sample of sodium chloride would be the atomic mass of the element divided by the formula mass and multiplied by 100%.

$$\frac{\text{mass Na}}{\text{mass NaCl}} \times 100\% = \frac{23.0 \text{ u}}{(23.0 + 35.5) \text{ u}} \times 100\% = 39.3\% \text{ Na}$$

The percentage of chlorine is (35.5 u/58.5 u) × 100, or 60.7%. The two percentages should total 100%.

It is just as easy to calculate the percentage composition of a compound such as ethanol where more than one atom of an element appears. The formula for ethanol is C_2H_5OH and its molecular mass is 46.1 u. It can be seen that one ethanol molecule contains two carbon atoms with a combined atomic mass of 24.0 u. Therefore, the percentage of carbon in the compound is

$$\frac{\text{mass 2C}}{\text{mass } C_2H_5OH} \times 100\% = \frac{24.0 \text{ u}}{46.1 \text{ u}} \times 100\% = 52.1\% \text{ C}$$

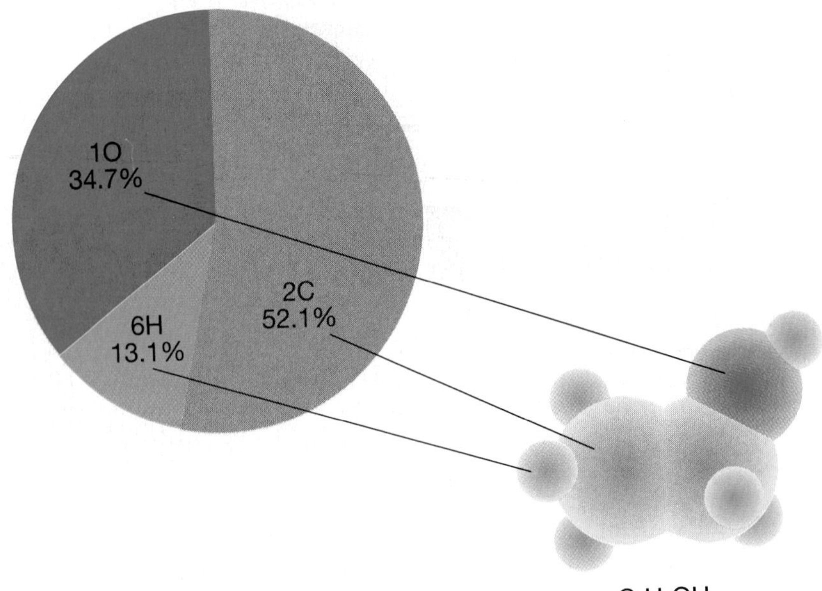

Figure 8.9 The percentage composition of a compound indicates the relative mass that each element contributes to the mass of the compound in the same way that a pie graph shows the relative contribution of each of its segments.

Figure 8.10 Double salts of aluminum sulfate, called alums, are used in pickling, to treat sewage and purify water, to waterproof cement, to make paper, and to act as foaming agents in fire foams.

The percentage of hydrogen is (6.06 u/46.1 u) × 100%, and that of oxygen is (16.0 u/46.1 u) × 100%. The three percentages should add to 100%. Often, because we round off answers, we may find that the percentages total one- or two-tenths more or less than 100%. That is, the total will often be (100 ± 0.2)%.

SAMPLE PROBLEM

Percentage Composition

Find the percentage composition of aluminum sulfate. Aluminum sulfate is a substance used in water treatment plants to help purify water.

Solving Process:

The formula for aluminum sulfate is $Al_2(SO_4)_3$. The formula mass is

2 Al atoms	2 × 27.0 =	54.0 u
3 S atoms	3 × 32.1 =	96.3 u
12 O atoms	12 × 16.0 =	192 u
		342 u

$$\text{Percentage of Al} = \frac{\text{mass 2Al}}{\text{mass } Al_2(SO_4)_3} \times 100\% = \frac{54.0\ \text{u}}{342\ \text{u}} \times 100\% = 15.8\%$$

$$\text{Percentage of S} = \frac{\text{mass 3S}}{\text{mass } Al_2(SO_4)_3} \times 100\% = \frac{96.3\ \text{u}}{342\ \text{u}} \times 100\% = 28.2\%$$

$$\text{Percentage of O} = \frac{\text{mass 12O}}{\text{mass } Al_2(SO_4)_3} \times 100\% = \frac{192\ \text{u}}{342\ \text{u}} \times 100\% = 56.1\%$$

Thus, a 100-g sample of aluminum sulfate would contain 15.8 g of aluminum, 28.2 g of sulfur, and 56.1 g of oxygen.

PROBLEM SOLVING HINT

Check that the sum of the percentages is 100%.

PRACTICE PROBLEMS

50. Two important nitrogen fertilizers are ammonia, NH_3, and urea, $CO(NH_2)_2$. Calculate the percentage of nitrogen in each of these compounds.

Determine the percentage composition of each of these compounds.

51. aluminum sulfide, Al_2S_3
52. nickel(II) iodide, NiI_2
53. calcium cyanide, $Ca(CN)_2$
54. copper(II) perchlorate, $Cu(ClO_4)_2$
55. ammonium oxalate, $(NH_4)_2C_2O_4$

Empirical Formulas

As you recall, elements are made of atoms, and compounds are made of elements. Half an atom does not exist. Therefore, we can state that the elements in a compound combine in simple whole-number ratios such as 1 to 1, 1 to 2, 2 to 3, and so on. If the atoms of the elements are present in simple ratios, then the moles of atoms for each element in the substance will also be in small whole-number ratios.

Consider the following example. We find that a 2.50-gram sample of a compound contains 0.900 gram of calcium and 1.60 grams of chlorine. The compound is composed of only two elements. We can calculate the number of moles of calcium and the number of moles of chlorine in the compound. Then, we can find the ratio of the number of moles of calcium atoms to the number of moles of chlorine atoms. From this ratio, we can find the empirical formula of calcium chloride. You may recall that the empirical formula of a compound is the simplest ratio of the elements in that compound. We can also calculate empirical formulas from percentage composition data.

Figure 8.11 Liquid ammonia is used as a fertilizer and is applied directly to the soil before crops are planted (left). Ammonia is liquefied under pressure and delivered to the fields in a pressurized tank (right).

Bridge to
MEDICINE

Making Medically Effective Products

A major area of pharmaceutical research concerns the search for medically effective natural products. Flowering plants, fungi, and bacteria have all been found to produce medically useful products. For example, the potent painkiller morphine is found in a species of poppy and the heart stimulant digitalis in foxglove. Because substances like these are usually present only in minute amounts, the cost of extracting them from their natural sources is often prohibitive. Consequently, chemists are constantly challenged to discover synthetic pathways for the production of these useful products. The first step for the chemist is the determination of the formula for the compound, using percentage composition data obtained in an analysis. Once the formula is known, the chemist can investigate structure and find a likely method of synthesis.

Exploring Further

1. Use the library to research plants that were used by Native Americans for medicinal purposes.
2. What roles do you think natural drugs such as morphine and digitalis play in the lives of the plants that produce them?

SAMPLE PROBLEMS

1. Empirical Formula

What is the empirical formula for a compound if a 2.50-g sample contains 0.900 g of calcium and 1.60 g of chlorine?

Solving Process:

(a) Determine the number of moles of each element in the compound. Ca has a mass of 0.900 g, and atomic mass 40.1 g. Cl has a mass of 1.60 g, and atomic mass 35.5 g.

$$\frac{0.900\ \cancel{g\ Ca}}{} \left| \frac{1\ mol\ Ca}{40.1\ \cancel{g\ Ca}} \right. = 0.0224\ mol\ Ca$$

$$\frac{1.60\ \cancel{g\ Cl}}{} \left| \frac{1\ mol\ Cl}{35.5\ \cancel{g\ Cl}} \right. = 0.0451\ mol\ Cl$$

(b) *To obtain the simplest ratio, divide both numbers of moles by the smaller number of moles (0.0224 mol).*

$$Ca\ \ \frac{0.0224\ \cancel{mol}}{0.0224\ \cancel{mol}} = 1.00 \qquad Cl\ \ \frac{0.0451\ \cancel{mol}}{0.0224\ \cancel{mol}} = 2.01$$

The ratio must be in whole numbers, so round off to 1 and 2. For each mole of calcium, there are 2 moles of chlorine. The empirical formula is $CaCl_2$.

MINUTE LAB

Bite the Bubble

Work together in groups of five. Use a balance to determine the mass of a clean paper cup. Unwrap five pieces of bubble gum containing sugar and place them in the cup. Determine the mass of the cup and the gum. Each person in the group should chew a piece of gum to dissolve and remove the sugar from the gum. After about five minutes, collect the chewed gum in the cup that you used previously. Wash your hands after handling the chewed gum. Determine the mass of the cup and the chewed gum. From your data, determine the mass of the sugar that was dissolved from the gum. Calculate the percentage of sugar in the gum by dividing the mass of the dissolved sugar by the mass of the unchewed gum and multiplying this value by 100. What is the percentage of sugar? Do you think a dentist would recommend chewing this gum? Why? What assumption are you making about the mass of the sugar and the difference in the masses of the gum before and after chewing? How would using more pieces of gum affect the results of this Minute Lab?

2. Empirical Formula from Percentage Composition

A compound has a percentage composition of 40.0% C, 6.71% H, and 53.3% O. What is the empirical formula?

Solving Process:

To calculate the ratio of moles of these elements, we assume a convenient amount of compound, usually 100 g. Then the percentages of the elements have the same numerical value in grams. In the present example, a 100-g sample would have 40.0 g of C, 6.71 g of H, and 53.3 g of O. We then change the quantities to moles.

$$\frac{40.0 \text{ g C}}{} \left| \frac{1 \text{ mol C}}{12.0 \text{ g C}} \right. = 3.33 \text{ mol C}$$

$$\frac{6.71 \text{ g H}}{} \left| \frac{1 \text{ mol H}}{1.01 \text{ g H}} \right. = 6.64 \text{ mol H}$$

$$\frac{53.3 \text{ g O}}{} \left| \frac{1 \text{ mol O}}{16.0 \text{ g O}} \right. = 3.33 \text{ mol O}$$

Dividing each result by 3.33 mol, we get 1 to 1.99 to 1. We round off to 1 to 2 to 1. The empirical formula is CH_2O.

PRACTICE PROBLEMS

56. Calculate the empirical formula of a compound that contains 1.67 g of cerium, Ce, and 4.54 g of iodine, I.

57. 2-Methylpropene is a compound used to make synthetic rubber. A sample of 2-methylpropene contains 0.556 g C and 0.0933 g H. Determine its empirical formula.

Sometimes dividing by the smallest number of moles does not yield a ratio close to a whole number. Suppose we have an empirical formula problem that produces a ratio of 2.33, or $2\frac{1}{3}$, to 1. What is the whole number ratio? If we multiply each member of the ratio by 3, the ratio becomes 7 to 3. Other common ratios are 1.33 to 1, or 4 to 3, and 1.67 to 1, or 5 to 3. When these ratios occur, another step is needed in the calculation of an empirical formula.

SAMPLE PROBLEM

Empirical Formula from Percentage Composition—Another Example

What is the empirical formula of a compound that is 66.0% Ca and 34.0% P?

Solving Process:

Assume a 100-g sample so that we have 66.0 g Ca and 34.0 g P. Convert these quantities to moles of atoms.

$$\frac{66.0\ \cancel{g\ Ca}}{} \Big| \frac{1\ mol\ Ca}{40.1\ \cancel{g\ Ca}} = 1.65\ mol\ Ca$$

$$\frac{34.0\ \cancel{g\ P}}{} \Big| \frac{1\ mol\ P}{31.0\ \cancel{g\ P}} = 1.10\ mol\ P$$

Dividing both results by 1.10 mol, we obtain 1.50 to 1. The result is not close to a whole number. When we substitute the fractional form of 1.50, we have

$$\frac{1.5}{1} = \frac{3/2}{1}$$

We then multiply by 2/2:

$$\frac{3/2 \times 2}{1 \times 2} = \frac{3}{2}$$

Thus, the empirical formula is Ca_3P_2.

EVERYDAY CHEMISTRY

Trees, Natural Regulators

One problem facing scientists today is the buildup of carbon dioxide in the atmosphere. Nearly 20% of all carbon dioxide emissions come from electrical energy production for homes. About 0.35 kg of carbon products, including carbon dioxide, are released into the atmosphere for each kilowatt-hour (kWh) of electrical energy that is produced.

Trees absorb carbon dioxide and store it as carbon compounds in their biomass—trunks, branches, leaves, and roots. One tree can absorb about 0.6 kg of carbon dioxide for every cubic decimeter of wood in the tree. Further, through energy conservation, the tree can prevent 35 times that amount of carbon dioxide from entering the atmosphere. Planting trees around homes can save from 10-50% for cooling and 4-22% for heating.

Planting trees improves our environment in many ways. Trees provide a habitat for living creatures, beautify urban landscapes, and give shade. In the future, their role as regulators of carbon dioxide in the atmosphere will only increase our appreciation for trees.

Exploring Further

1. Estimate the volume of a tree near your home or school. How much carbon dioxide could this tree absorb?
2. What is happening to the Amazon rain forest? What effect do you think this will have on the atmosphere?
3. Find out how trees use carbon dioxide. Discuss the chemistry involved.

PRACTICE PROBLEMS

58. Benzoic acid is a compound used as a food preservative. The compound contains 68.8% C, 4.95% H, and 26.2% O by mass. What is its empirical formula?

59. Freons are gaseous compounds used in refrigeration. A particular Freon contains 9.93% carbon, 58.6% chlorine, and 31.4% fluorine by mass. What is the empirical formula of this Freon?

CHEMISTRY IN DEPTH

Molecular Formulas

We have, thus far, calculated empirical formulas from experimental data. In order to calculate a molecular formula, we need one additional piece of data, the molecular or molar mass. In one of the examples in the previous section, the empirical formula was calculated to be CH_2O. If we know that the molecular mass of the compound is 180, how can we find the molecular formula? The molecular formula shows the actual number of atoms of each element in a molecule, as well as the ratio of atoms. Knowing that the elements will always be present in the ratio 1:2:1, we can calculate the mass of the empirical formula. Then we can find the number of these empirical units present in one molecular formula. In the substance CH_2O, the empirical unit has a formula mass of

$$12.0 + 2(1.01) + 16.0 = 30.0$$

It will, therefore, take six of these units to equal 180 or one molecular formula. Thus, the molecular formula is $C_6H_{12}O_6$.

Formula mass: 30.0

Empirical unit: CH_2O

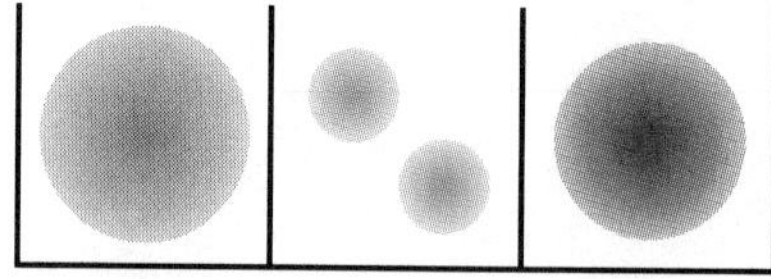

Molecular mass: 180.0

Molecular formula: $C_6H_{12}O_6$

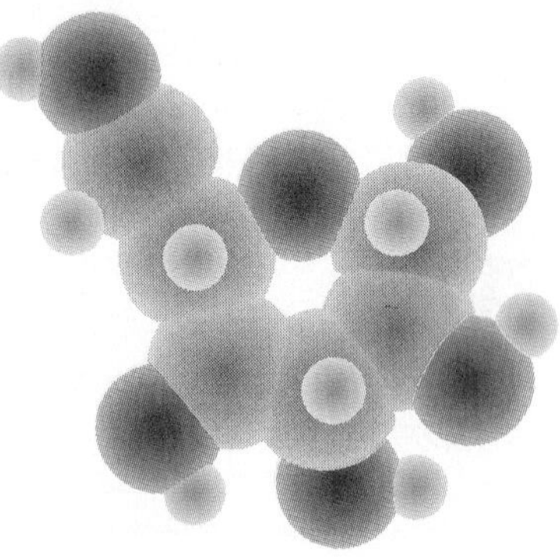

Figure 8.12 The empirical unit of a compound indicates the ratio of the atoms of each element in a compound. The molecular formula shows the actual number of atoms in a molecule of the compound.

SAMPLE PROBLEM

Molecular Formula

What is the molecular formula of a substance that has an empirical formula of $AgCO_2$ and a formula mass of 304?

Solving Process:

The formula mass of the empirical unit, $AgCO_2$, is 152. If we divide 304 by 152, we get 2. Therefore, the molecular formula must be 2 times the empirical formula, or $Ag_2C_2O_4$.

PRACTICE PROBLEMS

60. The molecular mass of benzene, an important industrial solvent, is 78.0 u and its empirical formula is CH. What is the molecular formula for benzene? Benzene is a proven carcinogen.

61. What is the molecular formula of dichloroacetic acid, if the empirical formula is CHOCl and the molecular mass of the acid is 129 u? Dichloroacetic acid is corrosive to the skin and has been used to remove skin blemishes.

62. What is the molecular formula of cyanuric chloride, a compound used in the production of some dyes, if the empirical formula is CClN and the molecular mass is 184.5 u?

63. Ascorbic acid, also known as vitamin C, has a percentage composition of 40.9% carbon, 4.58% hydrogen, and 54.5% oxygen. Its molecular mass is 176.1 u. What is its molecular formula?

64. Aspirin contains 60.0% carbon, 4.48% hydrogen, and 35.5% oxygen. It has a molecular mass of 180 u. What are its empirical and molecular formulas?

Hydrates

There are many compounds that crystallize from a water solution with water molecules adhering to the ions or molecules and becoming part of the crystal. These **hydrates**, as they are called, usually contain a specific ratio of water to compound. Formulas for hydrated compounds place the water of hydration following a dot after the regular formula. For example, $CuSO_4 \cdot 5H_2O$ is the formula for a hydrate of copper(II) sulfate that contains 5 moles of water for each mole of copper(II) sulfate. The name of the compound is copper(II) sulfate pentahydrate. Such compounds are named just as regular compounds except that the number of water molecules is included. The regular name is followed by the word hydrate to which a prefix has been added to indicate the relative molar proportions of water and compound. The prefixes are listed in Table 8.1.

hydor: (GK) water

hydrate—a crystalline compound in which the ions are attached to one or more water molecules

Table 8.1

Prefixes Used in Naming Hydrates			
Prefix	**Moles of Water**	**Name**	**Formula**
mono-	1	monohydrate	$XY \cdot H_2O$
di-	2	dihydrate	$XY \cdot 2H_2O$
tri-	3	trihydrate	$XY \cdot 3H_2O$
tetra-	4	tetrahydrate	$XY \cdot 4H_2O$
penta-	5	pentahydrate	$XY \cdot 5H_2O$
hexa-	6	hexahydrate	$XY \cdot 6H_2O$
hepta-	7	heptahydrate	$XY \cdot 7H_2O$
octa-	8	octahydrate	$XY \cdot 8H_2O$
nona-	9	nonahydrate	$XY \cdot 9H_2O$
deca-	10	decahydrate	$XY \cdot 10H_2O$

PRACTICE PROBLEMS

65. Name each of the following hydrates.
  a. $Na_2S_2O_3 \cdot 5H_2O$
  b. $CaSO_4 \cdot 2H_2O$

66. Write the formula for sodium tetraborate decahydrate.

An example of a hydrate is $NiSO_3 \cdot 6H_2O$. The dot shows that 6 molecules of water adhere to 1 formula unit of $NiSO_3$. To calculate the formula mass, we add the formula mass of the compound and water. For $NiSO_3$ we obtain 139 u. We multiply the 18.0 u for water by 6 and add to the 139 u. The formula mass of $NiSO_3 \cdot 6H_2O$ is then 139 u + 6(18.0 u), or 247 u.

To analyze hydrates, we can dry the compounds by heating them to drive off the water. We can then measure how much smaller the mass is without the water.

Figure 8.13 When a hydrate, such as $CuSO_4 \cdot 5H_2O$ (left), is heated (center), the water is driven off, leaving the anhydrous salt (right).

SAMPLE PROBLEM

Hydrate Calculation

We have a 10.407-g sample of hydrated barium iodide. The sample is heated to drive off the water. The dry sample has a mass of 9.520 g. What is the mole ratio between barium iodide, BaI_2, and water, H_2O? What is the formula of the hydrate?

Solving Process:

The difference between the initial mass and that of the dry sample is the mass of water that was driven off.

mass of hydrate	10.407 g
mass of dry sample	− 9.520 g
mass of water	0.887 g

The mass of water and mass of dry BaI_2 are converted to moles.

$$\frac{9.520\ \cancel{\text{g BaI}_2}}{} \left| \frac{1\ \text{mol BaI}_2}{391\ \cancel{\text{g BaI}_2}} \right. = 0.0243\ \text{mol BaI}_2$$

$$\frac{0.887\ \cancel{\text{g H}_2\text{O}}}{} \left| \frac{1\ \text{mol H}_2\text{O}}{18.0\ \cancel{\text{g H}_2\text{O}}} \right. = 0.0493\ \text{mol H}_2\text{O}$$

Dividing both results by 0.0243 mol, we obtain a ratio of 1 to 2.03, or 1 to 2. Thus, for every 1 mole of BaI_2, there are 2 moles of H_2O. The formula for the hydrate is written as $BaI_2 \cdot 2H_2O$.

PRACTICE PROBLEMS

Find the formulas for the following hydrates.

67. 0.391 g Li_2SiF_6, 0.0903 g H_2O

68. 0.737 g $MgSO_3$, 0.763 g H_2O

69. 76.9% $CaSO_3$, 23.1% H_2O

70. 89.2% $BaBr_2$, 10.8% H_2O

8.2 CONCEPT REVIEW

71. What is the molar mass of Na_2SiF_6?

72. What mass of CsOH is needed to make 500.0 cm^3 of a 1.00*M* solution?

73. Calculate the percentage composition of $AlAsO_4$.

74. Find the empirical formula of a substance with the composition 87.5% Zn and 12.5% N.

75. Apply While in a grocery store, you happen to pick up two identical, unopened boxes of washing soda labeled "1 lb (0.454 kg)." You notice that their masses differ. After purchasing the two boxes, you determine that the mass of one box is less than that indicated on the label. Later, you find that the formula for washing soda is $Na_2CO_3 \cdot 10H_2O$. How can you use this information to account for the difference in the masses of the two boxes?

Summary

8.1 Factor-Label Method

1. Conversion factors always have a value equivalent to 1 and are used to convert from one unit to a related unit.
2. In the factor-label method of problem solving, unit labels are treated as factors.
3. Any number can be expressed in scientific notation as $M \times 10^n$, where M is some number ≥ 1 and < 10 and n is an integer.

8.2 Formula-Based Problems

4. The symbol for an element represents one atom or one mole of the element.
5. The formula for a compound can represent one molecule, one formula unit, or one mole of the compound.
6. The molecular mass of a molecule is found by adding the atomic masses of all the atoms in one molecule.
7. For ionic compounds, adding the atomic masses of all the atoms in a formula unit of the compound gives the formula mass.
8. Avogadro constant represents 6.02×10^{23} things and is called one mole.
9. The number of moles in a given mass of substance can be found by dividing the total mass by the molar mass expressed in units of g/mol.
10. Molarity is the ratio of moles of solute to cubic decimeters of solution.
11. The percentage composition of a compound is a statement of the relative mass each element contributes to the mass of the compound as a whole.
12. The empirical formula of a compound indicates the simplest whole-number ratio of atoms in the compound.
13. The molecular formula of a compound is some whole-number multiple of the empirical formula.
14. Hydrates are crystalline compounds that contain water molecules attached to the particles of the compound.

Key Terms

factor-label method	mole
scientific notation	molar mass
molecular mass	molarity
formula mass	percentage composition
Avogadro constant	hydrate

Review and Practice

76. Perform the following operations. Express your answers to the correct number of significant digits. Be sure to include units in your calculation and express the resulting unit in your answer.
    a. (0.086 85 m)(28 000 m)
    b. (0.003 600 0 cm³)(10.74 g/cm³)
    c. $\dfrac{0.6717 \text{ g}}{3.99 \text{ cm}^3}$
    d. $\dfrac{(7.77 \times 10^3 \text{ kg})(9.76 \times 10^9 \text{ m})}{(8.67 \times 10^9 \text{ s})(3.43 \times 10^3 \text{ s})}$
77. Perform the following operations.
    a. $\dfrac{(5.77 \times 10^4 \text{ N/m}^3)(5.38 \times 10^5 \text{ m}^3)}{(2.67 \times 10^9 \text{ m})(4.83 \times 10^2 \text{ m})}$
    b. $\dfrac{(3.43 \times 10^5 \text{ g})(9.60 \times 10^2 \text{ cm})}{7.90 \times 10^8 \text{ s}^2}$
    c. $\dfrac{(5.32 \times 10^3 \text{ mm}^2)(1.83 \times 10^8 \text{ mm})}{7.15 \times 10^8 \text{ s}}$
    d. $\dfrac{(1.7 \times 10^6 \text{ mg})(9.468 \times 10^{-7} \text{ N})}{(3.84 \times 10^{-10} \text{ N})(4 \times 10^{-1} \text{ dm}^3)}$
78. Convert each of the following.
    a. 1.36 cm^3 to dm^3
    b. 0.761 kilometer to centimeters
    c. 0.725 m^2 to cm^2
    d. 948 cm^3 to dm^3
    e. 0.605 dm^3 to cm^3

f. 146 dm^3 to cm^3
g. 574 milligrams to grams
h. 8.91 mL to dm^3
i. 462 mL to L
j. 1.40 L to mL

79. Make the following conversions.
a. 950 dm^3 to m^3
b. 55.3 nm to cm
c. 0.0400 h to s
d. 0.0674 L to mL

80. Make the following conversions.
a. 3.5 m/s to cm/s
b. 18.8 g/cm^3 to kg/dm^3
c. 0.0572 mg/L to ng/dm^3
d. 1.352 km/h to mm/s

81. Express in scientific notation.
a. 77.9 cL
b. 0.0853 kg
c. 0.000 496 9 m^3
d. 255 130 000 000 nm

82. Express as a whole number or decimal.
a. 1.72×10^{-3} h **c.** 5.96×10^{4} K
b. 5.20×10^{-6} m **d.** 3.19×10^{6} μm

83. Perform the following operations.
a. 7.75×10^4 m + 4.79×10^3 m
b. 6.32×10^{-6} kg − 7.99×10^{-7} kg
c. $(9.6 \times 10^1$ mL$)(2.2 \times 10^{-5}$ g/mL$)$
d. $(39.9 \times 10^3$ cm$)(0.978 \times 10^{-7}$ cm$) \div (3.7 \times 10^7$ cm$)$

84. Give the number of significant digits that should appear in the answer to each of the following problems.
a. (0.30 kg/L)(0.0022 L)
b. (0.8 kg)(1.99×10^2 dollars/kg)
c. 0.90 m divided by 0.027 s
d. 0.5 kg divided by 4.13×10^9 cm^2

85. Express the number 56 000 in scientific notation to show (a) two significant digits; (b) four significant digits.

86. Solve the following problems. Round your answers to the proper number of decimal places.

a.
  0.089 90 dL
+ 52.dL

b.
  0.9 s
− 0.000 05 s

c.
  63 km
+ 93 km

d.
  4 dm^3
+ 6 dm^3

87. What is the density of sugar in g/cm^3 if its density is 1590 kg/m^3?

88. Several skin divers, wearing thermometers, recorded ocean temperatures while searching for a certain kind of fish. The values recorded were 10.5°C, 11.5°C, 11°C, 12.5°C, and 12°C. In how many significant digits should the average temperature be expressed?

89. To what kinds of substances does the term formula mass refer?

90. Calculate the formula mass of each of the following compounds.
a. K_2SO_4 **c.** $Ba_3(PO_4)_2$
b. CuO **d.** CoF_2

91. Calculate the molecular or formula mass of each of the following.
a. $CuCl_2$ **c.** NiF_2
b. $Hg(CH_3COO)_2$ **d.** $CdSiO_3$

92. Calculate the molecular or formula mass of each of the following compounds.
a. ethyl ether, $C_4H_{10}O$, once used as an anesthetic
b. titanium(IV) oxide, TiO_2, a covering medium used in house paint
c. silver nitrate, $AgNO_3$, used to kill bacteria
d. acetone, CH_3COCH_3, used in nail polish remover

93. Make the following conversions.
a. 64.3 g of $PbBr_2$ to moles
b. 3.33×10^{25} molecules of I_2 to grams
c. 0.846 mole of HF to molecules
d. 39.8 moles of CO_2 to molecules

94. Make the following conversions.
a. 8.50 grams $Hg(NO_3)_2$ to moles
b. 0.35 mole benzene, C_6H_6, to grams
c. 4.85×10^{25} formula units Cu_2S to grams
d. 308 formula units of MnC_2O_4 to moles

95. Compute the molarities for the indicated volumes of solution.
 a. 5.70 g $Cr_2(SO_4)_3$ in 5.00×10^2 cm^3
 b. 7.90 g NH_4Br in 2.50×10^2 cm^3
 c. 6.44 g CuF_2 in 1.00×10^3 cm^3
 d. 13.1 g $CdCl_2$ in 1.00×10^3 cm^3

96. Describe the preparation of each of the following solutions.
 a. 1.00 dm^3 of 4.00*M* $MnBr_2$
 b. 1.00 dm^3 of 6.00*M* LiBr
 c. 5.00×10^2 cm^3 of 2.00*M* $MgBr_2$
 d. 1.00×10^3 cm^3 of 1.50*M* $MnSO_4$

97. How would you prepare 100.0 cm^3 of 0.10*M* potassium sulfate, K_2SO_4?

98. What is the molarity of a solution of silver nitrate that contains 42.5 g $AgNO_3$ in 100.0 mL of solution?

99. What is the molarity of a solution of sodium iodide that contains 3.75 g NaI in 0.500 L of solution?

100. How would you prepare 50.0 cm^3 of 0.1*M* potassium thiocyanate, KSCN?

101. Find the percentage composition of each element in the following compounds.
 a. NaBr c. HgF_2
 b. CH_3COOH d. ZnS

102. Give the empirical formula of each.
 a. Na_2O_2 c. KBr
 b. C_4H_{10} d. $Al_2(SO_4)_3$

103. Find the empirical formulas of compounds with the following composition.
 a. 63.0 g Rb, 5.90 g O
 b. 0.159 g U, 0.119 g Cl
 c. 7.22 g Ni, 2.53 g P, 5.25 g O
 d. 0.295 g Ca, 0.236 g S, 0.469 g O

104. Find the empirical formulas of compounds with the following analyses.
 a. 32.8% Cr, 67.2% Cl
 b. 42.7% Co, 57.3% Se
 c. 56.6% La, 43.4% Cl
 d. 81.9% Ta, 18.1% O

105. The percentage composition of a compound is 92.3% C and 7.7% H. If the molecular mass is 78 u, what is the molecular formula?

106. Find the molecular formula of a compound with percentage composition 26.7% P, 12.1% N, and 61.2% Cl and molecular mass 695 u.

107. Find the formulas for the hydrates with the following analyses.
 a. 5.262 g $Tl(NO_3)_3$ and 0.728 g H_2O
 b. 2.94 g $Sn(NO_3)_2$ and 4.37 g H_2O
 c. 8.351 g $UO_2(HCOO)_2$ and 0.421 g H_2O
 d. 17.02 g UO_2HPO_4 and 3.42 g H_2O

108. A compound is found to have a percentage composition of 22.1% aluminum, 25.4% phosphorus, and 52.5% oxygen. Which of the following compounds could it be?
 a. $AlPO_4$ c. $Al_4(P_2O_7)_3$
 b. $Al(PO_3)_3$ d. $AlPO_3$

109. A compound is found to have a percentage composition of 64.3% barium, 13.2% silicon, and 22.5% oxygen. Which of the following compounds could it be?
 a. Ba_2SiO_4 b. $Ba_3Si_2O_7$ c. $BaSiO_3$

110. A compound is found to have a percentage composition of 80.4% bismuth, 18.5% oxygen, and 1.16% hydrogen. Which of the following compounds could it be?
 a. $Bi(OH)_5$ b. $Bi(OH)_3$ c. H_3BiO_4

111. A compound is found to contain 33.3% calcium, 40.0% oxygen, and 26.7% sulfur. Its formula mass is 120 u. What is the formula for the compound?

112. What is the meaning of the dot in the formula $CuSO_4 \cdot 5H_2O$?

113. Write formulas for the following hydrates.
 a. magnesium nitrate hexahydrate
 b. iron(II) sulfate heptahydrate
 c. copper(II) nitrate trihydrate
 d. tin(II) chloride dihydrate

114. Name each of the following hydrates.
 a. $MgSO_4 \cdot H_2O$
 b. $Mn(NO_3)_2 \cdot 4H_2O$
 c. $CaCrO_4 \cdot 2H_2O$
 d. $Na_2SO_4 \cdot 10H_2O$

115. What is the formula for a hydrate that is 90.7% SrC_2O_4 and 9.30% H_2O?

116. What is the formula for a hydrate that is 76.9% $La_2(CO_3)_3$ and 23.9% H_2O?

117. What is the formula for a hydrate that is 86.7% Mo_2S_5 and 13.3% H_2O?

Concept Mastery

118. Why do chemists use 6.02×10^{23} as the number of things in a mole?

119. Why is it not practical to measure the mass of substances in atomic mass units?

120. Explain the difference between the terms mole and molarity.

121. A chemist finds that a certain compound has an empirical formula CH_3O. What does the chemist know about the compound on the basis of this formula?

122. Explain the difference between empirical formula and molecular formula.

123. Why do we not make the Avogadro constant some convenient round number such as 1×10^{25}?

124. Why is it incorrect to apply the term *molecular mass* to ionic substances?

125. The empirical formulas of acetylene and benzene are the same. How could we tell which was which?

Application

126. In a routine hospital lab test, a student is found to have a cholesterol level of 2.3 g/dm^3. Cholesterol levels are usually expressed in milligrams per deciliter of blood. Express the student's cholesterol level in mg/dL.

127. The speed of light is 2.9979×10^8 m/s. How far does light travel in a nanosecond?

128. A Tollycraft 61 Motor Yacht travels at 23.7 knots at full speed. A knot is 1.852 kilometers per hour. What is the full speed of the yacht in meters per second?

129. The magnitude (brightness) of the star Nova Cygni varies over a period of 0.140 day. Express this period in seconds.

130. How many significant digits are in the quantity, 5.67×10^3 m? What does this expression tell you about how the measurement was made? Write the quantity without using scientific notation. Why is scientific notation necessary to express this measurement appropriately?

131. A spaceship from another planet travels at a speed of 4.27 googs/mulm. There are 256 googs in a meter and 8000 mulm in one hour.What is the spaceship speed in meters per second?

132. During the late 19th century cobalt blue glassware was popular. Cobalt blue is considered to be the most durable of all blue pigments. It is a mixture of compounds including the compound $Co(AlO_2)_2$. What is the percentage composition of this compound?

133. The compound mannitol, $C_6H_8(OH)_6$, is used as a sweetener in some dietetic foods. What is the percentage composition of this compound?

134. The compound 2,3,7,8-tetrachlorodioxin, also known as TCDD or, simply, dioxin, has been implicated as a cause of a number of physical ailments. Its formula is $C_{12}H_4Cl_4O_2$. What is the percentage of chlorine in the compound?

135. Caffeine, $C_8H_{10}N_4O_2$, and theophylline, $C_7H_8N_4O_2$, are both found in tea leaves, and both are used as medicines affecting hormone action in the human body. Compare the percentage of nitrogen found in the two compounds.

136. Indoleacetic acid, $C_{10}H_9NO_2$, and gibberellic acid, $C_{19}H_{22}O_6$, are compounds that occur naturally in plants and affect plant development. Indoleacetic acid increases growth rates of plants, while

gibberellic acid causes plants to flower earlier than normal. What is the percentage of carbon in each compound? Which substance has the greater percentage of oxygen?

137. A laboratory technician ran an analysis of a germicide stocked by a janitorial service. The technician determined that the effective ingredient in the germicide was a compound that was 33% sodium, 36% arsenic, and 31% oxygen. What is the formula for this compound?

138. One problem reported by the nickel-plating industry is that some workers develop "nickel itch," a form of dermatitis that occurs when a certain compound comes in contact with skin. What is the empirical formula for this compound if a sample contains 5.13 g Ni, 2.45 g N, and 8.39 g O?

139. Satellite measurements have shown a large section of the stratosphere over Antarctica where the ozone has been depleted. A compound used to test for the presence of ozone contains 96.2% thallium and 3.77% oxygen. What is the empirical formula for this compound?

140. One student in an art class was interested in the many different effects that ceramic glazes create. One glaze that gave an iridescent effect contained 48.8% cadmium, 20.8% carbon, 2.62% hydrogen, and 27.8% oxygen. What is the empirical formula for this compound?

141. In areas where temperatures get extremely cold, people must take special precautions to make sure machinery runs properly. One compound containing 83% rubidium, 16% oxygen, and 1% hydrogen is used in low-temperature storage batteries. What is the empirical formula for this compound?

142. How could you find out which of two solutions of a salt had the greater molarity? *Never taste unknown substances.*

143. The percentage of cadmium in $CdSiF_6$ is 44.2%. How many grams of Cd are in 22.8 g $CdSiF_6$?

144. The percentage of oxygen in $CaSeO_4$ is 35.0%. How many grams of oxygen are in 34.2 g $CaSeO_4$?

145. Magnesium phosphate forms two different hydrates as white crystalline powders. One has a ratio of 1 mole compound to 5 moles water, and the other has a ratio of 1 mole compound to 8 moles water. How could you determine which form you had if you were given a sample of one of these two hydrates?

146. Both urea $CO(NH_2)_2$ and calcium cyanamide $CaCN_2$ can be used as fertilizers. Which produces the greater amount of nitrogen from 20 kg of substance?

147. A solution of vinegar is usually 5% acetic acid (CH_3COOH). What is the mass of acetic acid in 100 g of vinegar?

148. Iron has a density of 7.874 g/cm^3. How many atoms are in a cube of iron 0.10 cm on a side?

Critical Thinking

149. Why would it be difficult to construct a balance that measured the amounts of substances in moles?

150. Without calculating, which of the following compounds has the greater percentage of calcium: CaS or $CaSO_4$? How do you know?

151. What is the molarity of water?

152. A mixture of salt and pepper analyzes at 50% chlorine by mass. If pepper contains no chlorine, what percent of the mixture is pepper?

Readings

Poskozim, Paul S. "Analogies for Avogadro's Number." *Journal of Chemical Education* 63, no. 2 (February 1986): 125-126.

Chapter 8

REVIEW

Cumulative Review

1. Write the names of the following compounds.
   a. $Sr(CH_3COO)_2$ c. CdC_2O_4
   b. $Mn(OH)_2$ d. Li_3AsO_4

2. Write formulas for the following compounds.
   a. sodium nitride
   b. cerium(III) sulfide
   c. barium iodate
   d. hydrogen telluride

3. What is the uncertainty in the momentum of an electron if its position has been determined to an accuracy of ± 10 picometers?

4. How much heat is required to raise the temperature of 91.0 grams of PbI_2 by 26.8°C? See Appendix Table A-5.

5. Name each of the following compounds.
   a. Be_3N_2 c. $Hg_3(AsO_4)_2$
   b. $CeCl_3$ d. Ca_3P_2

6. Write the formula for each of the following compounds.
   a. silicon(IV) fluoride
   b. thallium(I) perchlorate
   c. nickel(II) cyanide
   d. calcium diphosphate

7. Write the names for the following compounds.
   a. $ZnC_4H_4O_6$ c. $Cr(CH_3COO)_3$
   b. PbSe d. Mn_3P_2

8. What is the wavelength of an automobile having a mass of 1.50×10^3 kilograms traveling at 1.00×10^2 kilometers an hour?

9. Express 81.2 g as mg.

10. How many significant digits are in the measurement 41.02 m?

11. Classify the following properties as physical or chemical.
    a. flammability
    b. electrical conductivity
    c. ability to displace hydrogen from water
    d. ability to react with acids

12. Classify the following changes as physical or chemical.
    a. distillation c. crystallization
    b. fermentation d. dissolving

13. Element D has oxidation numbers 1+ and 2+. Element E has oxidation numbers 1− and 2−. List the possible formulas of compounds of these elements.

14. Density and molarity are expressed in derived SI units. Give these units and tell why they are called derived units.

15. The specific heat of aluminum is 0.902 J/g·C° and that of silver is 0.235 J/g·C°. You have two spoons of equal masses. One is made of aluminum; the other is made of silver. Which spoon would increase in temperature faster in a pot of hot soup? Why?

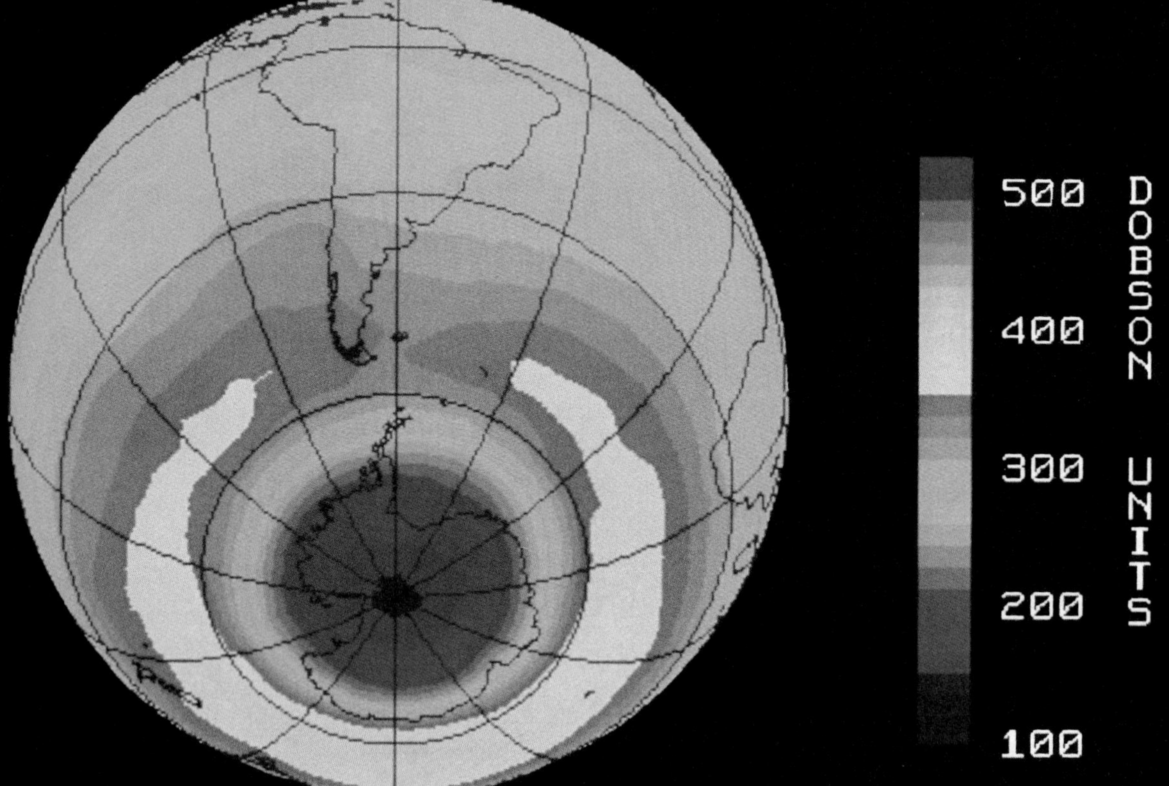

CHAPTER Preview

CONTENTS

CHAPTER 9

Chemical Reactions

What is destroying the ozone layer surrounding our planet? Can the thinning of the ozone layer, shown on the opposite page, be fixed? Both industrial and environmental scientists are beginning to work together to answer these important questions and to find a solution to these problems.

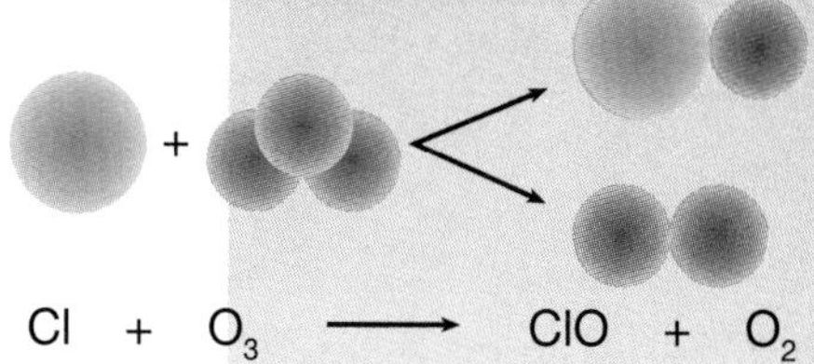

The depletion of the ozone content of the stratosphere has become a major concern to citizens throughout the world during the last few years. (See page 470 for more information about this problem.) The photograph opposite is the result of an assay of this ozone from the Nimbus 7 satellite. You can see that the area in the center has less ozone than other areas. In order to deal with this problem, chemists have had to investigate the many chemical reactions taking place in the stratosphere. Here you see the symbolic representation of one of these reactions. In this chapter, you will study how to represent reactions by equations and how to calculate the quantities of substances involved in those reactions.

EXTENSIONS AND CONNECTIONS

Chemistry and Technology

Developing Self-Fertilizing Plants: New developments in the chemical reaction that converts atmospheric nitrogen into a form that plants can use, page 232

Careers in Chemistry

Chemical Engineer, page 234

Chemists and Their Work

St. Elmo Brady, page 236

Frontiers

Gas Stations: A look at a new type of automotive fuel, page 238

Bridge to Biology

Soaking Up the Rays: How chlorophyll in plants converts sunlight to chemical energy, page 242

OBJECTIVES

- write chemical equations to represent reactions.
- use coefficients to balance chemical equations.
- differentiate among five general types of chemical reactions.

9.1 Chemical Equations

You already know how to use symbols to represent elements and formulas to represent compounds. Now you can use that shorthand to describe the chemical changes substances undergo. That is, you can describe chemical reactions. Consider the following statement: "Two molecules of acetylene gas will react with five molecules of oxygen gas to produce four molecules of carbon dioxide gas and two molecules of water vapor." It is much easier to write this as

$$2C_2H_2(g) + 5O_2(g) \rightarrow 4CO_2(g) + 2H_2O(g)$$

Representing Chemical Changes

The formulas of compounds are used in certain combinations to represent the chemical changes that occur in a chemical reaction. A **chemical reaction** is the process by which one or more substances are changed into one or more different substances. A chemical reaction can be represented by an equation. A correct chemical equation shows what changes take place. It also shows the relative amounts of the various elements and compounds that take part in these changes. The starting substances in a chemical reaction are the **reactants**. The substances that are formed by the chemical reaction are the **products**.

$$\underbrace{2C_2H_2(g) + 5O_2(g)}_{\text{reactants}} \underbrace{\rightarrow}_{\text{yield}} \underbrace{4CO_2(g) + 2H_2O(g)}_{\text{products}}$$

The letters in parentheses indicate the physical state of each substance involved. The symbol (g) after a formula means that the substance is a gas. Liquids are indicated by the symbol (l), and solids by the symbol (cr). The symbol (cr) for solid indicates that the solid is crystalline. We can see that in the reaction described above, C_2H_2, O_2, CO_2, and H_2O are all gases.

Figure 9.1 Acetylene gas is burned in the presence of oxygen in a welder's torch.

Since many chemical reactions take place in water solution, a substance dissolved in water is shown by the symbol (aq). This symbol comes from the word *aqueous*. For example, if a water solution of sulfurous acid is warmed, it decomposes. The products of this reaction are water and sulfur dioxide, a gas.

$$H_2SO_3(aq) \rightarrow H_2O(l) + SO_2(g)$$

Word Origins

aqua: (L) water

aqueous—made from, with, or by water

Writing Balanced Equations

A chemical reaction can be represented by a chemical equation. To write an equation that accurately represents the reaction, you must correctly perform three steps.

Step 1. *Determine the reactants and the products.* For example, when propane gas burns in air, the reactants are propane (C_3H_8) and oxygen (O_2). The products formed are carbon dioxide (CO_2) and water (H_2O).

Step 2. *Assemble the parts of the chemical equation.* Write the formulas for the reactants on one side of the equation, usually on the left, and connect them with plus signs. Write the formulas for the products on the right side of the equation. Connect the two sides using an arrow to show the direction of the reaction. Thus,

C_3H_8 + O_2	$\longrightarrow$	CO_2 + H_2O
propane + oxygen	yield	carbon dioxide + water
reactants	*yield*	*products*

The symbols and formulas must be correct. If not, Step 3 will be useless. Use Appendix Tables A-3 and A-4 and Tables 7.3 and 7.4 in Chapter 7 when trying to write correct formulas. We will omit the symbols that indicate physical state while we go through the procedure for writing balanced equations.

Step 3. *Balance the equation.* Balancing means showing an equal number of atoms for each element on both sides of the equation. In chemical reactions, no mass is lost or gained. The same amount of matter is present before and after the reactions. So, the same number and kinds of atoms must be present on both sides of an equation. The equation below is not balanced.

$$C_3H_8 + O_2 \rightarrow CO_2 + H_2O$$

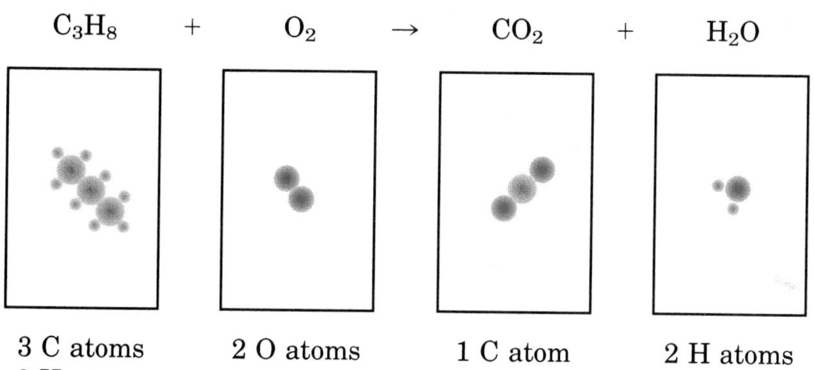

Figure 9.2 The equation $C_3H_8 + O_2 \rightarrow CO_2 + H_2O$ is not balanced. Equal numbers of carbon, hydrogen, and oxygen atoms are not present on both sides of the arrow.

There are three carbon atoms on the left, but only one on the right. To put the carbon in balance, place the coefficient 3 before the carbon dioxide on the right. In balancing an equation, *change only the coefficients. Never change the subscripts.* To do so would change the substance represented. Our equation is now

$$C_3H_8 \quad + \quad O_2 \quad \rightarrow \quad 3CO_2 \quad + \quad H_2O$$

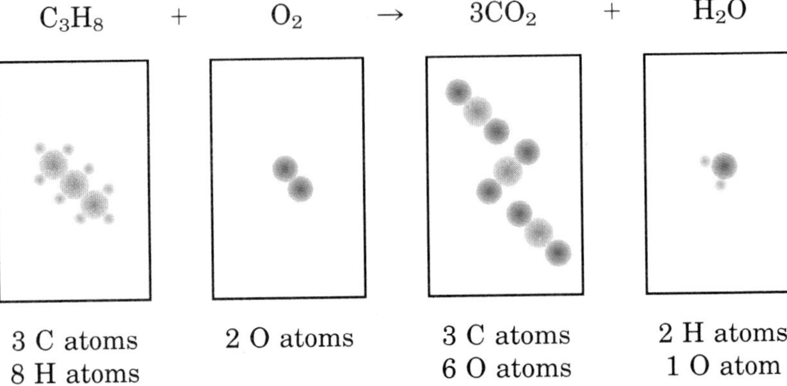

Figure 9.3 Adding two molecules of CO_2 to the right side of the equation balances the number of carbon atoms, but nothing else.

The carbon atoms are balanced, but the hydrogen atoms are not. There are eight hydrogen atoms on the left and only two on the right. By placing the coefficient 4 in front of the water, both the carbon and hydrogen atoms will be in balance. The equation is

$$C_3H_8 \quad + \quad O_2 \quad \rightarrow \quad 3CO_2 \quad + \quad 4H_2O$$

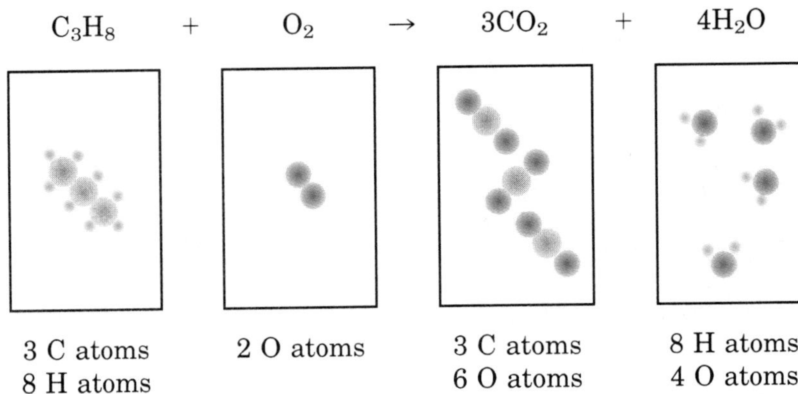

Figure 9.4 By adding three molecules of H_2O to the right side of the equation, the number of hydrogen atoms on the left and right become equal.

Only the oxygen remains to be balanced. There are two oxygen atoms on the left and ten oxygen atoms on the right. To balance the oxygen, place the coefficient 5 in front of the oxygen. The equation is now balanced.

$$C_3H_8 \quad + \quad 5O_2 \quad \rightarrow \quad 3CO_2 \quad + \quad 4H_2O$$

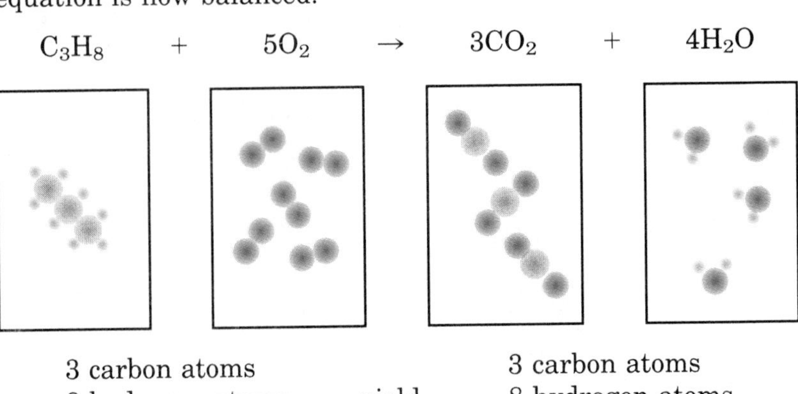

3 carbon atoms
8 hydrogen atoms
10 oxygen atoms

yield

3 carbon atoms
8 hydrogen atoms
10 oxygen atoms

Figure 9.5 The addition of four molecules of O_2 to the left side of the equation produces a balanced equation: three carbon atoms, eight hydrogen atoms, and ten oxygen atoms on each side of the equation.

In another reaction, the commercial production of hydrogen, steam reacts with a form of carbon to produce hydrogen and carbon monoxide.

$H_2O + C$	$\longrightarrow$	$H_2 + CO$
2 hydrogen atoms 1 carbon atom 1 oxygen atom	yield	2 hydrogen atoms 1 carbon atom 1 oxygen atom

Because we have equal numbers of hydrogen, carbon, and oxygen atoms on both sides of the equation, the equation is balanced.

Be careful not to confuse subscripts and coefficients in balancing equations. *Never change a subscript in an attempt to balance an equation.* Changing a subscript changes the substance. The resulting equation will then not represent the chemical change that actually takes place.

SAMPLE PROBLEM

Balancing Equations

Balance the equation:

$$LiAlH_4 + H_2O \rightarrow LiOH + Al(OH)_3 + H_2$$

Solving Process:

Looking at the equation, you can see that there is one lithium atom on each side and one aluminum atom on each side. Hydrogen appears in at least two places on each side, so you should balance that element last. In the equation as written, there is only one oxygen atom on the left while there are four oxygen atoms on the right. To balance the oxygen atoms, place a 4 as a coefficient before the water on the left. The equation now appears as

$$LiAlH_4 + 4H_2O \rightarrow LiOH + Al(OH)_3 + H_2$$

On the left there are now 12 hydrogen atoms, four from the lithium aluminum hydride and eight from the water. On the right there are only six hydrogen atoms, one in the lithium hydroxide, three in the aluminum hydroxide, and two in the free element, hydrogen. In balancing the hydrogen atoms on the right, you do not want to change the coefficients of the LiOH or the $Al(OH)_3$, because to do so would upset the balance of Li, Al, and O. You therefore balance the hydrogen atoms by changing the coefficient of the free hydrogen. By placing a 4 in front of the hydrogen, you provide for 12 hydrogen atoms on the right. The fully balanced equation then reads

$$LiAlH_4 + 4H_2O \rightarrow LiOH + Al(OH)_3 + 4H_2$$

You should do a final check of all the atoms to make sure everything is balanced.

PROBLEM SOLVING HINT

It is usually easier to balance the hydrogen and oxygen atoms last because they often appear in more than one place.

PRACTICE PROBLEMS

Write balanced equations for each of the following reactions.

1. $Cu + H_2O \rightarrow CuO + H_2$
2. $Al(NO_3)_3 + NaOH \rightarrow Al(OH)_3 + NaNO_3$
3. $KNO_3 \rightarrow KNO_2 + O_2$
4. $Fe + H_2SO_4 \rightarrow Fe_2(SO_4)_3 + H_2$
5. $O_2 + CS_2 \rightarrow CO_2 + SO_2$
6. $Cu + Cl_2 \rightarrow CuCl_2$
7. $Mg + N_2 \rightarrow Mg_3N_2$

Substitute symbols and formulas for names, and then write a balanced equation for each of the following reactions.

8. When copper(II) carbonate is heated, it forms copper(II) oxide and carbon dioxide gas.
9. Sodium reacts with water to produce sodium hydroxide and hydrogen gas.
10. Copper combines with sulfur to form copper(I) sulfide.
11. Silver nitrate reacts with sulfuric acid to produce silver sulfate and nitric acid.
12. Ethane, a component of natural gas, burns in oxygen to produce carbon dioxide and water.

Classifying Chemical Changes

There are hundreds of different kinds of chemical reactions. Many of these can be divided into five general types. Of these five, three are often used to make, or synthesize, new compounds.

Single Displacement. In this type of reaction, one element displaces another in a compound. For example, in the reaction

$$Cl_2(g) + 2KBr(aq) \rightarrow 2KCl(aq) + Br_2(l)$$

chlorine displaces bromine from potassium bromide.

In the reaction

$$2Al(cr) + Fe_2O_3(cr) \rightarrow 2Fe(cr) + Al_2O_3(cr)$$

aluminum displaces iron from iron(III) oxide. A **single displacement** reaction is recognized and predicted by its general form:

element + compound → element + compound

Double Displacement. There are hundreds of reactions in which the positive and negative portions of two compounds are interchanged.

$$PbCl_2(cr) + Li_2SO_4(aq) \rightarrow PbSO_4(cr) + 2LiCl(aq)$$
$$ZnBr_2(aq) + 2AgNO_3(aq) \rightarrow Zn(NO_3)_2(aq) + 2AgBr(cr)$$
$$BaCl_2(aq) + 2KIO_3(aq) \rightarrow Ba(IO_3)_2(cr) + 2KCl(aq)$$

MINUTE LAB

Double Displacement

Wear an apron and goggles. Mass 0.16 g barium nitrate and place it in a test tube with 5 cm^3 H_2O. Stopper and shake until dissolved. Mass 0.15 g sodium sulfate and place it in a second test tube containing 5 cm^3 H_2O. Stopper the test tube and shake until dissolved. Observe the tubes and record your observations. Mix the contents of the two test tubes. What did you observe? Use the solubility rules in Appendix Table A-7 to determine the name of the precipitate. Write the balanced chemical equation for the chemical reaction.

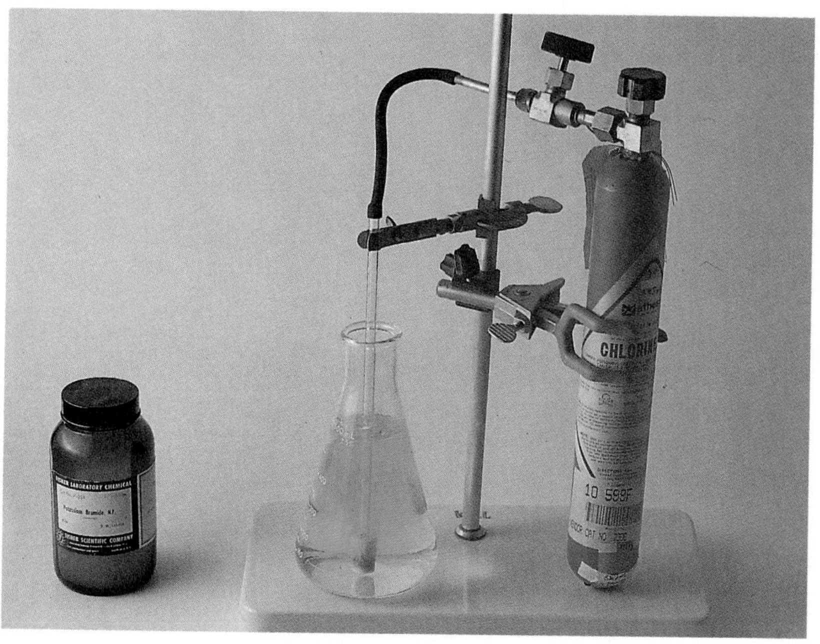

Figure 9.6 In the single displacement reaction (left), the element chlorine displaces bromine in a solution of potassium bromide, KBr. In the double displacement reaction (right), the positive and negative portions of cadmium nitrate and sodium sulfide are exchanged to form cadmium sulfide (orange) and sodium nitrate.

The form for a **double displacement** reaction is

compound + compound → compound + compound

Decomposition. Many substances will break up, or decompose, into simpler substances when energy is supplied.

$$CdCO_3(cr) \rightarrow CdO(cr) + CO_2(g)$$
$$Pb(OH)_2(cr) \rightarrow PbO(cr) + H_2O(g)$$
$$N_2O_4(g) \rightarrow 2NO_2(g)$$
$$PCl_5(cr) \rightarrow PCl_3(cr) + Cl_2(g)$$
$$H_2CO_3(aq) \rightarrow H_2O(l) + CO_2(g)$$
$$2KClO_3(cr) \rightarrow 2KCl(cr) + 3O_2(g)$$
$$2Ag_2O(cr) \rightarrow 4Ag(cr) + O_2(g)$$

Energy may be supplied in the form of heat, light, mechanical shock, or electricity. The general form for a **decomposition** reaction is

compound → two or more elements or compounds

Synthesis. In synthesis reactions, two or more substances combine to form one new substance.

$$NH_3(g) + HCl(g) \rightarrow NH_4Cl(cr)$$
$$CaO(cr) + SiO_2(cr) \rightarrow CaSiO_3(cr)$$
$$2H_2(g) + O_2(g) \rightarrow 2H_2O(g)$$

From the name, you might expect that synthesis reactions would be the most common method of preparing new compounds. However, these reactions are rarely as practical as one of the three preceding methods. The general form of a **synthesis** reaction is

element or compound + element or compound → compound

Figure 9.7 When mercury(II) oxide is heated, it decomposes to form liquid mercury and oxygen gas. Note the droplets of mercury on the sides of the test tube.

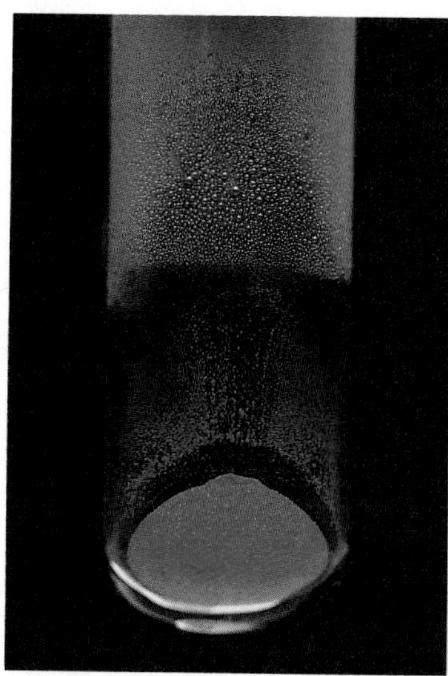

Figure 9.8 A cloud of ammonium chloride is formed when HCl and NH_3 combine in a synthesis reaction.

Combustion. Almost all organic compounds and some inorganic compounds will burn in air. Many will ignite quite readily and are considered *flammable*. When a compound burns in air, it is actually reacting with the oxygen in the air. As a result, chemists refer to combustion reactions as oxidation reactions. The products of the oxidation of a hydrocarbon under normal conditions are carbon dioxide and water vapor. As an example, consider the combustion of methane (CH_4), the main component of natural gas, and butane (C_4H_{10}), a fuel used in camping stoves.

$$CH_4 + 2O_2 \rightarrow CO_2 + 2H_2O$$
$$2C_4H_{10} + 13O_2 \rightarrow 8CO_2 + 10H_2O$$

The general form for **combustion** reactions is

hydrocarbon + oxygen → carbon dioxide + water

Not all reactions take one of these five general forms. Other classes of reactions will be considered later. Until then, we will deal chiefly with reactions of single or double displacement, decomposition, synthesis, or combustion (oxidation).

Figure 9.9 The butane in this camp stove serves as a source of energy when it combines with oxygen in a combustion reaction.

PRACTICE PROBLEMS

Classify each of the following reactions as single or double displacement, decomposition, synthesis, or combustion.

13. $CuO + H_2 \rightarrow Cu + H_2O$

14. $2H_2O_2 \rightarrow 2H_2O + O_2$

15. $2Ag + S \rightarrow Ag_2S$

16. $C_4H_8 + 6O_2 \rightarrow 4CO_2 + 4H_2O$

17. $2K + 2H_2O \rightarrow 2KOH + H_2$

18. $HCl + NaOH \rightarrow H_2O + NaCl$

Predict the products formed in each of the following reactions.

19. the single displacement reaction between aluminum, Al, and copper(II) nitrate, $Cu(NO_3)_2$

20. the synthesis of mercury(II) oxide from its elements

21. the double displacement reaction of sulfuric acid, H_2SO_4, and potassium hydroxide, KOH

22. the combustion of cyclopentane, C_5H_{10}

9.1 CONCEPT REVIEW

23. The illustrations below show three different types of reactions. What type of reaction is shown in each of the following?

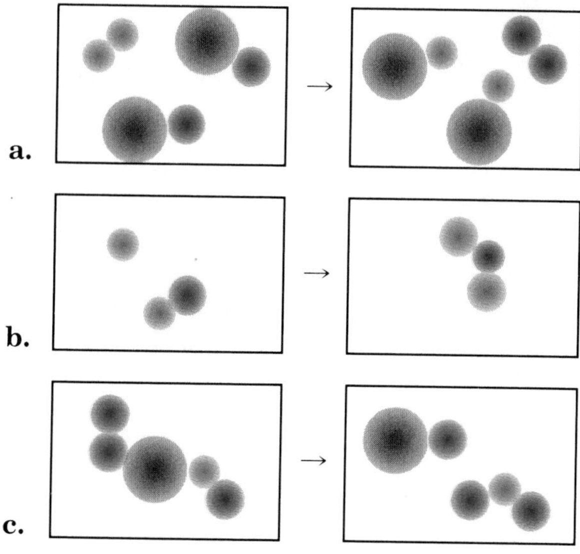

24. Write a balanced chemical equation to represent each of the following reactions.

a. antimony and chlorine gas react to form antimony trichloride

b. the decomposition of silver oxide to form metallic silver and oxygen gas

25. Four students try to balance the equation for the following reaction.

$$Cu_2S + O_2 \rightarrow Cu_2O + SO_2$$

a. The students obtain the following results. Tell what is wrong with each equation and why.

$$Cu_2S + 2O_2 \rightarrow Cu_2O_2 + SO_2$$
$$Cu_2S + O_2 \rightarrow Cu_2O + SO$$
$$Cu_2S + 2O_2 \rightarrow Cu_2O + SO_2$$
$$4Cu_2S + 6O_2 \rightarrow 4Cu_2O + 4SO_2$$

b. Write the correct balanced equation for the reaction.

26. Classify the following reactions as single displacement, double displacement, decomposition, synthesis, or combustion. Tell why you classified each reaction as you did.

a. $2Al(cr) + 3Cl_2(g) \rightarrow 2AlCl_3(cr)$

b. $Zn(cr) + 2HCl(aq) \rightarrow ZnCl_2(aq) + H_2(g)$

c. $Mg(ClO_3)_2(cr) \rightarrow MgCl_2(cr) + 3O_2(g)$

d. $3BaCl_2(aq) + 2H_3PO_4(aq) \rightarrow Ba_3(PO_4)_2(cr) + 6HCl(aq)$

27. Apply Why aren't mixtures representable in chemical equations?

CHEMISTRY AND

TECHNOLOGY

Developing Self-Fertilizing Plants

Nitrogen compounds are major nutrients for plants. Because approximately 78% of the air is nitrogen, you would think that there would be no problem providing plants with this nutrient. However, the major food crops of the world—rice, wheat, and corn—cannot utilize the nitrogen in the air directly. Through many industrial and natural chemical processes, nitrogen is converted into fertilizers in forms that plants can use.

The conversion of N_2 from the atmosphere into a form useful to plants is an important step in the nitrogen cycle, the interconversion of inorganic and organic nitrogen in nature. Both chemical and biological processes are involved. Commercial fertilizer production accounts for the conversion of about 3×10^7 metric tons of nitrogen each year. Other chemical processes, involving lightning, ultraviolet radiation, and combustion, account for another 4×10^7 metric tons annually. Biological sources, however, produce nearly twice as much usable nitrogen as these two sources combined. Much of this production is carried out by certain plants, such as peas and beans, whose seeds are in pods. These plants, called legumes, can use nitrogen directly from the air.

Legumes do not actually convert nitrogen themselves. Bacteria that live in nodules on the roots of the legumes can convert molecular nitrogen into ammonia, nitrites, and nitrates—compounds that plants can use. This process is called nitrogen fixation. In effect, these plants are self-fertilizing.

For years, farmers have grown legumes to help replenish nitrogen-poor soil. Farmers rotate legumes with grain crops or plow under the legumes as "green manure." In this way, the soil is enriched with the nitrogen products manufactured by the bacteria.

Researchers at the University of Sydney in Australia believe they have found a way to "infect" crops such as wheat with nitrogen-fixing bacteria. They found that low concentrations of the herbicide 2,4-D soften wheat cell walls and cause nodule-like swellings to form on the roots of wheat plants. Into these nodules, the researchers have been able to introduce a nitrogen-fixing bacterium that usually lives free in the soil. Comparison of treated wheat with untreated wheat showed that nitrogen was being fixed in the treated wheat. If the bacteria are able to remain for long periods of time in the nodules, farmers growing these altered crops will not have to rely on industrially produced fertilizers. These scientists may have come up with a means to improve production of major food crops and reduce environmental pollution without adding more synthetic fertilizer to our environment.

Thinking Critically

1. There are two kinds of bacteria that assist in nitrogen-fixing. One lives in root nodules of plants and the other is found free in the soil. Try to find out what these bacteria are and what chemical reactions are involved in nitrogen fixation.
2. Why is the Australian research important to the world's food shortage?

9.2 Stoichiometry

OBJECTIVES

- determine the mass of a reactant or product based on the mass of another reactant or product in a reaction.
- calculate the actual yield of a product as a percentage of the theoretical yield.
- determine the heat of reaction for a chemical reaction in which a specified amount of a substance is involved.

Stoichiometry is a word for that part of chemistry that deals with the amount of substances involved in chemical reactions, both as reactants and as products. Balanced equations represent the relationship between the number of particles that react and the number of particles produced. Because each type of particle has its own formula mass, there must be definite relationships between the masses that react and the masses of the products.

Mass-Mass Relationships

Balanced equations and the mass-mass relationships they convey help us answer several types of questions. How much of one reactant is needed to combine with a given amount of another reactant? How much product is produced from a specific amount of reactant? For instance, how much silver chloride can be produced from 17.0 g of silver nitrate? There are other, similar, questions. Fortunately, all these questions can be answered using the same procedure. The coefficients of a balanced equation give the relative amounts (in moles) of reactants and products. Calculations to find the masses of materials involved in reactions are called **mass-mass problems.**

SAMPLE PROBLEM

Mass-Mass

How many grams of silver chloride can be produced from the reaction of 17.0 g of silver nitrate with excess sodium chloride solution?

Solving Process:

Step 1. We are given silver nitrate and asked to find silver chloride. What do we know that would connect two different substances? A chemical equation would relate them in a quantitative way. Therefore, we write a balanced equation to represent the reaction that occurs. Silver nitrate is reacting with sodium chloride. Since we have compound plus compound, we predict that a double displacement reaction will occur.

$$AgNO_3(aq) + NaCl(aq) \rightarrow AgCl(cr) + NaNO_3(aq)$$

Step 2. We have thought of equations as written in terms of individual atoms and formula units. We can also consider the coefficients in the equation as indicating the number of *moles* of formula units that take part in the reaction. The equation above indicates that one formula unit of silver nitrate will produce one formula unit of silver chloride. It also indicates that one mole of silver nitrate formula units will produce one mole of silver chloride formula units. The problem states that

Figure 9.10 A white precipitate of silver chloride, AgCl, forms when solutions of silver nitrate, $AgNO_3$, and sodium chloride, NaCl, are mixed.

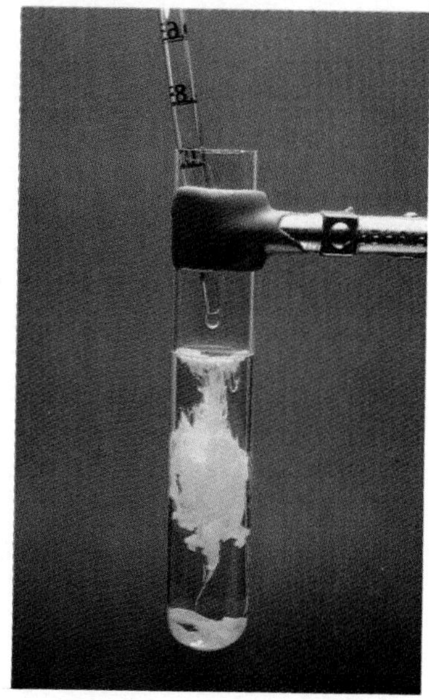

CAREERS IN CHEMISTRY

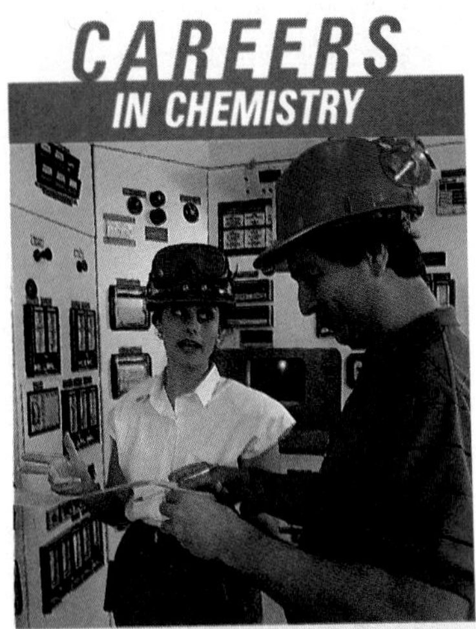

Chemical Engineer

The most important functions of chemical engineers are designing, constructing, and operating chemical plants. However, there are in the chemical enterprise many other aspects of interest to the chemical engineer. Chemical engineers need to be concerned with the economics of the chemical marketplace, and many engineers enter the field of sales. Quality control is another phase of chemical production under the supervision of chemical engineers, as is the maintenance of conditions suitable for preservation of our environment. If you are interested in chemistry but like to work with practical matters instead of more theoretical areas, you will find chemical engineering a fascinating career.

an excess of sodium chloride is used, which means that all the silver nitrate will react and there will be some sodium chloride left over.

Let us review where we are. We are given grams of silver nitrate. We are asked to find grams of silver chloride. Silver nitrate and silver chloride are related by the equation. The equation tells us that one mole of silver nitrate yields one mole of silver chloride.

Our solution begins by converting the grams of silver nitrate to moles. We use the table on the inside back cover to determine the molar mass of $AgNO_3$, which is 170 g.

$$17.0\text{ g }AgNO_3 \times \frac{1\text{ mol }AgNO_3}{170\text{ g }AgNO_3} \ldots$$

Step 3. We have now converted the silver nitrate to units that will enable us to relate it to the silver chloride. We use the equation to find the conversion factor to use in going from silver nitrate to silver chloride. The equation tells us that one mole of silver nitrate will produce one mole of silver chloride. Using this fact gives us a partial solution.

$$17.0\text{ g }AgNO_3 \times \frac{1\text{ mol }AgNO_3}{170\text{ g }AgNO_3} \times \frac{1\text{ mol }AgCl}{1\text{ mol }AgNO_3} \ldots$$

Step 4. We have now arrived at silver chloride, which is the substance we were asked to find. However, we were asked to find the answer in grams. To complete the problem, we must convert the moles of silver chloride to grams. Using the table inside the back cover, we can see that one mole of silver chloride is 144 g. The final solution then appears as:

$$17.0\,\cancel{\text{g }AgNO_3} \times \frac{\cancel{1\text{ mol }AgNO_3}}{170\,\cancel{\text{g }AgNO_3}} \times \frac{\cancel{1\text{ mol }AgCl}}{\cancel{1\text{ mol }AgNO_3}} \times \frac{144\text{ g }AgCl}{\cancel{1\text{ mol }AgCl}}$$

$$= 14.4\text{ g }AgCl$$

The problem above involved finding the mass of one substance from a given mass of another substance. All mass-mass problems can be solved in this way. This problem is an example of stoichiometry.

Let us review the process.

Step 1. *Write the balanced equation.*
The first step in the solution of any mass-mass problem is to write a balanced equation for the correct reaction. In this section, we will always assume that only one reaction occurs and that all of one reactant is used. In reality, several reactions may occur, the actual reaction may not be known, or all the reactant may not react.

Figure 9.11 Mass must be measured carefully so that accurate data will be available for solving mass-mass problems.

Step 2. *Find the number of moles of the given substance.* Express the mass of the given substance in moles by dividing the mass of the given substance by its molar mass.

$$\textit{grams of given substance} \times \frac{1\ \textit{mole}}{\textit{molar mass of given substance}} \cdots$$

Step 3. *Inspect the balanced equation to determine the ratio of moles of required substance to moles of given substance.* For example, look at the following equation.

$$2H_2(g) + O_2(g) \rightarrow 2H_2O(g)$$

In this reaction, 1 mole of oxygen reacts with 2 moles of hydrogen. From the same equation, 1 mole of oxygen will produce 2 moles of water. Also, 2 moles of water are produced by 2 moles of hydrogen. Once the equation is balanced, only the reactants and products directly involved in the problem should be in your calculations. Multiply the moles of given substance by the ratio:

$$\cdots \frac{\textit{moles of required substance}}{\textit{moles of given substance}} \cdots$$

Step 4. *Express the moles of required substance in terms of grams, then convert moles to grams.*

$$\cdots \frac{\textit{molar mass of required substance}}{1\ \textit{mole}} = \textit{grams of required substance}$$

Notice that, as you work through a problem of this kind, you first convert grams of given substance to moles, and then convert moles of required substance back to grams.

$$\left(\begin{matrix}\textit{start with}\\ \textit{grams given}\end{matrix}\right) \rightarrow \left(\begin{matrix}\textit{convert}\\ \textit{grams}\\ \textit{to moles}\end{matrix}\right) \rightarrow \left(\begin{matrix}\textit{use}\\ \textit{mole ratio}\end{matrix}\right) \rightarrow \left(\begin{matrix}\textit{convert}\\ \textit{moles}\\ \textit{to grams}\end{matrix}\right) \rightarrow \left(\begin{matrix}\textit{end with}\\ \textit{grams required}\end{matrix}\right)$$

This method is used because the balanced equation relates the number of moles of given substance to the number of moles of required substance. Now try the following two problems.

CHEMISTS AND THEIR WORK

St. Elmo Brady
(1884-1966)

St. Elmo Brady was the first African-American student to receive a doctorate in chemistry. He spent the rest of his life making certain that other African-American students would have the same opportunity. After finishing his degree at the University of Illinois in 1916, he went to Tuskegee Institute to develop what became the chemistry department there. He moved in 1920 to Howard University and accomplished the same task. In 1927 he moved to Fisk University and served 25 years building both undergraduate and graduate programs. After retirement, when he was nearly 70 years old, Brady was again asked to build a chemistry department, this time at Tougaloo, Mississippi. As a result of his efforts, thousands of students of all races have been able to study chemistry.

SAMPLE PROBLEMS

1. Mass-Mass

How many grams of Cu_2S could be produced from 9.90 g of CuCl reacting with an excess of H_2S gas?

Solving Process:

(1) We must write the balanced equation. CuCl and H_2S are reactants, Cu_2S is one product.

$$2CuCl(aq) + H_2S(g) \rightarrow Cu_2S(cr) + 2HCl(aq)$$

If we use the wrong reaction or do not balance the equation properly, we cannot get a correct answer.

(2) Find the number of moles of the given substance.

$$\begin{array}{rr} 1\text{Cu} \quad 1 \times 63.5 = & 63.5 \text{ g} \\ 1\text{Cl} \quad 1 \times 35.5 = & \underline{35.5 \text{ g}} \\ \text{molar mass of CuCl} = & 99.0 \text{ g} \end{array}$$

9.90 g CuCl	1 mol CuCl	. . .
	99.0 g CuCl	

(3) Determine the mole ratio of the required substance to the given substance. Notice that, although H_2S and HCl are part of the reaction, we do not consider them in this problem. They are not part of the calculations.

$$2CuCl + H_2S \rightarrow Cu_2S + 2HCl$$

9.90 g CuCl	1 mol CuCl	1 mol Cu_2S	. . .
	99.0 g CuCl	2 mol CuCl	

(4) We have now arrived at moles of Cu_2S. We were asked to find grams of Cu_2S, so we must convert moles of Cu_2S into grams of Cu_2S. We use the table inside the back cover.

$$\begin{array}{rr} 2\text{Cu} \quad 2 \times 63.5 = & 127 \text{ g} \\ 1\text{S} \quad 1 \times 32.1 = & \underline{32.1 \text{ g}} \\ \text{molar mass of } Cu_2S = & 159 \text{ g} \end{array}$$

9.90 ~~g CuCl~~	~~1 mol CuCl~~	~~1 mol Cu_2S~~	159 g Cu_2S
	99.0 ~~g CuCl~~	2 ~~mol CuCl~~	1 ~~mol Cu_2S~~

= 7.95 g of Cu_2S

Thus, we predict 9.90 g of CuCl will react to produce 7.95 g of Cu_2S. If the problem is set up correctly, all the factor units will divide out except the required units.

2. Mass-Mass: Another Example

How many grams of calcium hydroxide will be needed to react completely with 10.0 g of phosphoric acid? Note that we are asked to find the mass of one reactant that will react with a given mass of another reactant, H_3PO_4.

Solving Process:

(1) Write a balanced equation.

$$3Ca(OH)_2 + 2H_3PO_4 \rightarrow Ca_3(PO_4)_2 + 6H_2O$$

(2) Change 10.0 g H_3PO_4 to moles of H_3PO_4.

$$\frac{10.0 \text{ g } H_3PO_4}{} \left| \frac{1 \text{ mol } H_3PO_4}{98.0 \text{ g } H_3PO_4} \right. \cdots$$

(3) From the equation, 2 moles of H_3PO_4 will require 3 moles $Ca(OH)_2$.

$$\frac{10.0 \text{ g } H_3PO_4}{} \left| \frac{1 \text{ mol } H_3PO_4}{98.0 \text{ g } H_3PO_4} \right| \frac{3 \text{ mol } Ca(OH)_2}{2 \text{ mol } H_3PO_4} \cdots$$

(4) Change moles of calcium hydroxide into grams (mass) of calcium hydroxide.

$$\frac{10.0 \cancel{\text{ g } H_3PO_4}}{} \left| \frac{1 \cancel{\text{ mol } H_3PO_4}}{98.0 \cancel{\text{ g } H_3PO_4}} \right| \frac{3 \cancel{\text{ mol } Ca(OH)_2}}{2 \cancel{\text{ mol } H_3PO_4}} \left| \frac{74.1 \text{ g } Ca(OH)_2}{1 \cancel{\text{ mol } Ca(OH)_2}} \right.$$

$$= 11.3 \text{ g } Ca(OH)_2$$

PROBLEM SOLVING HINT

Use the unit(s) in your answer to check your work.

PRACTICE PROBLEMS

28. Glucose is used as a source of energy by the human body. The overall reaction in the body is

$$C_6H_{12}O_6 + 6O_2 \rightarrow 6CO_2 + 6H_2O$$

Calculate the number of grams of oxygen needed to oxidize 12.5 g of glucose to carbon dioxide and water.

29. Ammonia is synthesized from hydrogen and nitrogen according to the following equation.

$$N_2(g) + 3H_2(g) \rightarrow 2NH_3(g)$$

If an excess of nitrogen is reacted with 3.41 g of hydrogen, how many grams of ammonia can be produced?

30. Assume that in the decomposition of potassium chlorate, $KClO_3$, 80.5 g of O_2 form. How many grams of potassium chloride, the other product, would be formed?

31. In a single displacement reaction, 9.23 g of aluminum react with excess hydrochloric acid. How many grams of hydrogen will be produced?

32. The compound "cisplatin", $PtCl_2(NH_3)_2$, has been found to be effective in treating some types of cancer. It can be synthesized using the following reaction.

$$K_2PtCl_4(aq) + 2NH_3(aq) \rightarrow 2KCl(aq) + PtCl_2(NH_3)_2(aq)$$

a. How much "cisplatin" can be produced from 2.50 g K_2PtCl_4 and excess NH_3?

b. How much NH_3 would be needed?

FRONTIERS

Gas Stations

"Five dollars worth of gas, please. And put in some H_2, too!" You won't hear this at the local service station yet, but recent research into alternative automotive fuels may put a gas in the gas tank.

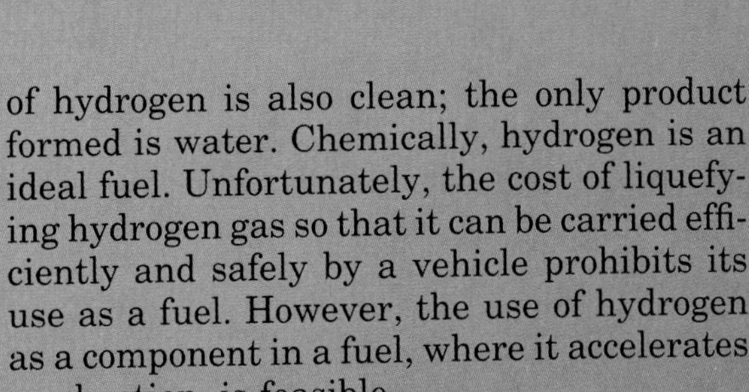

The truck shown here is modified to run on a mixture of compressed natural gas (CNG) and hydrogen gas. Because methane is the major component of natural gas, the mixture has been dubbed "Hythane" by its developer, Frank E. Lynch.

The abundance of natural gas, which is widely used for home heating, has spurred the development of compressed natural gas as an alternative automotive fuel. If the gas is compressed, a vehicle such as a delivery truck or commuter car can carry a sufficient supply of CNG conveniently for short-distance, urban use. However, because the combustion of CNG is a relatively slow reaction, incomplete combustion may take place in the engine. Unburned hydrocarbons, as well as products such as carbon monoxide, could be released to the atmosphere as pollutants.

The combustion of hydrogen, on the other hand, is rapid. That's why liquid hydrogen is used as a rocket propellant. The combustion of hydrogen is also clean; the only product formed is water. Chemically, hydrogen is an ideal fuel. Unfortunately, the cost of liquefying hydrogen gas so that it can be carried efficiently and safely by a vehicle prohibits its use as a fuel. However, the use of hydrogen as a component in a fuel, where it accelerates combustion, is feasible.

Initial tests indicate that the combustion of Hythane produces almost half as many unburned hydrocarbons but greater amounts of carbon monoxide as the combustion of CNG alone. Intrigued by these early results, researchers are testing the use of Hythane as a viable alternative to gasoline. Someday, the corner "gas" station may indeed sell gas.

Percentage Yield

The quantities we calculate in mass-mass problems are, of course, theoretical amounts. That is, we calculate the maximum possible yield for a particular product. In reality, the actual amounts of products are often much less. Chemists are frequently interested in comparing the actual amount of product from a reaction with the amount calculated by a mass-mass calculation. The comparison is usually expressed as a percentage and is called the percentage yield. The **percentage yield** of a product is the actual amount of product expressed as a percentage of the calculated theoretical yield of that product.

$$\text{Percentage yield} = \frac{\text{actual amount of product}}{\text{theoretical amount of product}} \times 100$$

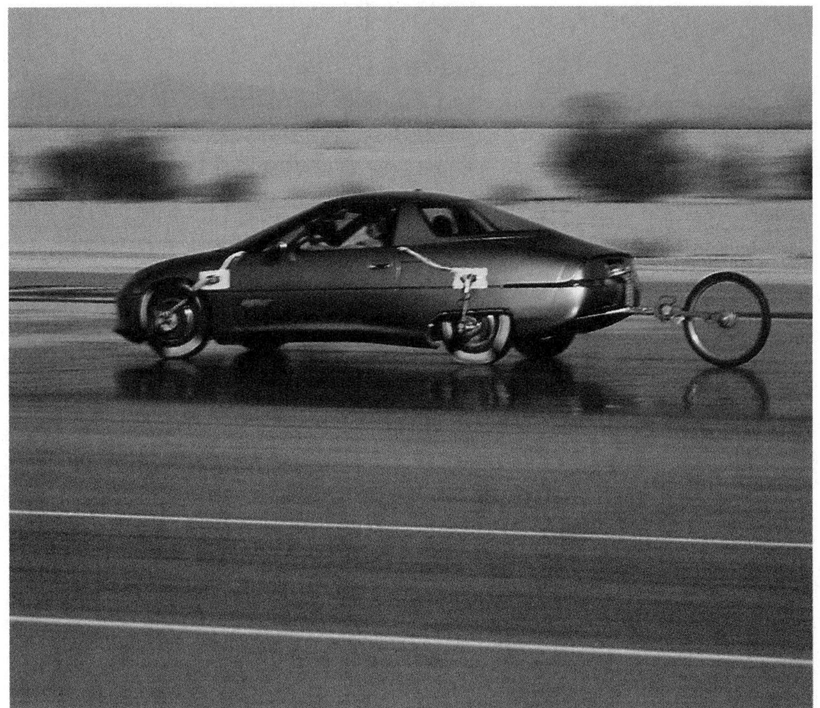

Figure 9.12 The actual miles per gallon in "average driving" that a new car delivers will usually vary from that indicated on the EPA sticker due to vehicle options, driving conditions, driving habits, and the vehicle's condition. In the same way, the actual yield of a reaction is usually less than the theoretical yield due to less than optimum conditions.

SAMPLE PROBLEM

Percentage Yield

Oil of wintergreen (methyl salicylate) is used in a variety of commercial products for its flavor and aroma. It is made by heating salicylic acid, $C_7H_6O_3$, with methanol, CH_3OH.

$$C_7H_6O_3 + CH_3OH \rightarrow C_8H_8O_3 + H_2O$$

A chemist starts with 1.75 g of salicylic acid and excess methanol and calculates the maximum possible yield to be 1.93 g. However, after the reaction is run, the chemist finds that the amount of methyl salicylate produced and isolated is only 1.42 g. What is the percentage yield of the process?

$$\frac{1.42 \text{ g}}{1.93 \text{ g}} \times 100 = 73.6\%$$

PRACTICE PROBLEMS

33. The actual amount of product in a reaction is 39.7 g although a mass-mass calculation predicted 65.6 g. What is the percentage yield of this product?

34. What is the percentage yield in the following reaction if 5.50 grams of hydrogen react with nitrogen to form 20.4 grams of ammonia?

$$N_2(g) + 3H_2(g) \rightarrow 2NH_3(g)$$

Mass-Energy

Balanced chemical equations can also be used to calculate the amount of energy absorbed or released during a chemical reaction. In this section, you will learn about heat absorbed or released. In Chapter 27, you will study more about energy.

When carbon in the form of coal is burned, heat is released.

$$C(cr) + O_2(g) \rightarrow CO_2(g) + \text{heat } (393.509 \text{ kJ})$$

According to this equation, one mole of carbon reacts with one mole of oxygen to produce one mole of carbon dioxide and 393.509 kJ of heat. The heat released (393.509 kJ) is called the heat of reaction. Heat in chemistry is represented by the letter q, so we will use q_r to represent the heat of reaction.

In the reaction above, the product must have less energy than the reactants because energy is given off. As you learned in Chapter 3, this reaction represents a system going downhill in energy. In that circumstance (a system that is losing energy), q would be negative.

The heat of reaction is based on the mole quantities shown in the balanced equation. If the mole quantities of reactants are different from those shown in the equation, a different amount of energy will be released or absorbed. If we were faced with a situation in which less (or more) than one mole of carbon is burned, we could calculate the value of q by using a problem solution similar to that of a mass-mass problem. In this case, the problem is a **mass-energy problem,** one in which the amount of energy absorbed or released during a reaction is calculated.

Figure 9.13 The heat that is produced by the burning of coal is used to run this steam engine.

SAMPLE PROBLEM

Mass-Energy

How much heat is generated by the reaction of 1.99 g of Na_2O_2 with water? The equation for the reaction is

$$2Na_2O_2 + 2H_2O \rightarrow 4NaOH + O_2 + 215.76 \text{ kJ}$$

Solving Process:

We are given the mass of sodium peroxide that reacts and asked to find the heat change involved. These two things, sodium peroxide and heat, are related by the equation. However, we must remember that the equation is in terms of moles. Therefore, the first step is to convert the mass of the sodium peroxide to moles.

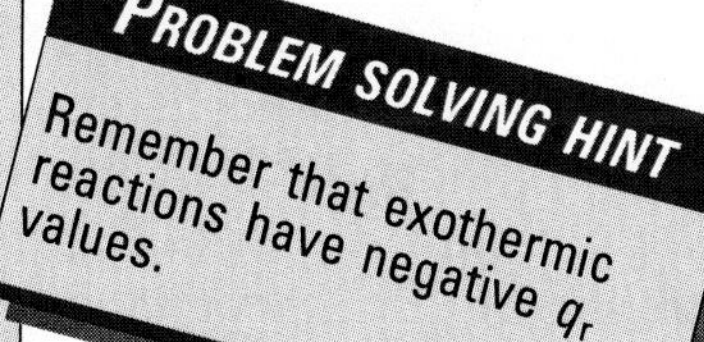

PROBLEM SOLVING HINT
Remember that exothermic reactions have negative q_r values.

$$\frac{1.99 \text{ g } Na_2O_2}{} \left| \frac{1 \text{ mol } Na_2O_2}{78.0 \text{ g } Na_2O_2} \right. \ldots$$

Now the ratio shown in the equation can be used to convert from sodium peroxide to heat. The equation tells us that 215.76 kJ of heat is released for every two moles of Na_2O_2.

$$\frac{1.99 \text{ g } Na_2O_2}{} \left| \frac{1 \text{ mol } Na_2O_2}{78.0 \text{ g } Na_2O_2} \right| \frac{215.76 \text{ kJ}}{2 \text{ mol } Na_2O_2} = 2.75 \text{ kJ}$$

$$q_r = -2.75 \text{ kJ}$$

The value of q_r is negative because energy was given off, and, therefore, the products had less energy than the reactants.

PRACTICE PROBLEMS

35. Compute the heat of reaction for the formation of 193 g of ammonium bromide from ammonia and hydrogen bromide.

$$NH_3 + HBr \rightarrow NH_4Br + 188.32 \text{ kJ}$$

36. Compute the heat of reaction for the decomposition of 0.772 g of cobalt(II) carbonate into cobalt(II) oxide and CO_2.

$$CoCO_3 + 81.6 \text{ kJ} \rightarrow CoO + CO_2$$

37. Compute the heat of reaction for the displacement of 0.0663 gram of bromine from sodium bromide by chlorine.

$$Cl_2 + 2NaBr \rightarrow 2NaCl + Br_2 + 100.18 \text{ kJ}$$

38. A chemist runs an experiment in which the following reaction occurs.

$$CrO_3(cr) + H_2O(l) \rightarrow H_2CrO_4(aq) + 5.4 \text{ kJ}$$

Careful measurement in a calorimeter of the change in temperature and the mass of water enables the chemist to determine the energy evolved. The calculations indicate the production of 6.18 kJ of energy. What mass of CrO_3 reacted?

Bridge to
BIOLOGY

Soaking Up the Rays

One of the most important reactions that takes place on Earth is photosynthesis. All life forms depend on a "fuel" source (food) to provide energy for their various life processes. Organisms, such as green plants, that can generate their own food are called autotrophs. Organisms that cannot generate their own food, and therefore must consume autotrophs, are called heterotrophs. Animals are heterotrophs. Thus, all life depends on the ability of autotrophs to produce food. Autotrophs, in turn, depend on sunlight as the source of energy used to make food. (There are a few deep-sea organisms that make use of another energy source.) Autotrophs convert about 10^{17} tons of carbon from CO_2 to food each year. The sunlight serves as the activation energy to start the first step in the photosynthetic process. A molecule of chlorophyll, whose structure is shown here, absorbs energy from the sun, and, in its energetic form, begins the series of reactions that result in the production of food, usually represented by glucose, $C_6H_{12}O_6$. The equation for the process of photosynthesis is

$$6CO_2 + 6H_2O + \text{light} \rightarrow C_6H_{12}O_6 + 6O_2$$

The importance of this process cannot be overemphasized. It takes place only because chlorophyll can absorb the energy in light.

Exploring Further

1. Does the photosynthesis reaction have a positive or negative heat of reaction? Explain, using the term *endothermic* or *exothermic* in your answer. Tell whether the reactants or products have greater energy.

2. Sunlight comes to us in quanta just as any other form of electromagnetic radiation does. Find out just what part of the chlorophyll molecule absorbs the quanta of sunlight. Find out what wavelengths of light are used in photosynthesis.

Chlorophyll a

R = $—C_{20}H_{39}$ (phytyl chain)

9.2 *CONCEPT REVIEW*

39. a. What mass of $Mg(OH)_2$ will react with 1.20 grams of HCl?

$$Mg(OH)_2 + 2HCl \rightarrow MgCl_2 + 2H_2O$$

b. What mass of $Mg(OH)_2$ will react with 2.40 grams of HCl in the same reaction? How can you determine the answer without completely redoing the calculations from above?

40. In what ways might the percentage yield of a reaction be affected if a chemist did not balance the equation before calculating the theoretical yield?

41. In the reaction $NH_3 + HBr \rightarrow NH_4Br + 188.32$ kJ, how much heat would be given off if 0.500 mole of HBr reacted?

42. Apply The heat of reaction for the combustion of fuels is an important consideration in the marketplace. Of what use is the heat of reaction for the combustion of fuels?

Chapter 9

REVIEW

Summary

9.1 Chemical Equations

1. Chemists use equations to describe the changes that substances undergo.
2. The physical state of substances in equations is shown by (g) for gas, (l) for liquid, (cr) for solid, and (aq) for a water solution of the substance.
3. Reactants are the starting substances in a reaction. Products are the substances resulting from a reaction.
4. A chemical equation represents changes that take place in a reaction. It also shows relative amounts of reactants and products.
5. Balancing an equation means adjusting coefficients so that there is the same number of atoms of each element on the left and right sides of the equation.
6. Five general types of reactions are single displacement, double displacement, synthesis, decomposition, and combustion reactions.
7. Under normal conditions, the oxidation of a hydrocarbon produces carbon dioxide and water.

9.2 Stoichiometry

8. The quantitative study of chemical changes is called stoichiometry.
9. A balanced equation indicates the relative numbers of moles of reactants and products in the reaction. The balanced equation is used as the basis for solving mass-mass problems.
10. The actual amount of a product obtained in a reaction is often expressed as a percentage of the theoretical amount.
11. Chemical equations can be used to calculate the amount of heat absorbed or given off in the chemical reaction of a specific amount of substance.

Key Terms

chemical reaction
reactant
product
single displacement
double displacement
decomposition
synthesis
combustion
stoichiometry
mass-mass problem
percentage yield
mass-energy problem

Review and Practice

43. Write balanced equations for the following reactions.
    a. $WO_3 + H_2 \rightarrow W + H_2O$
    b. $PdCl_2 + HNO_3 \rightarrow Pd(NO_3)_2 + HCl$
    c. $RbBr + AgCl \rightarrow AgBr + RbCl$
    d. $HfCl_3 + Al \rightarrow HfCl_2 + AlCl_3$
    e. $Zn + CrCl_3 \rightarrow CrCl_2 + ZnCl_2$
    f. $BaCO_3 + C + H_2O \rightarrow CO + Ba(OH)_2$
    g. $Cu + H_2SO_4 \rightarrow CuSO_4 + SO_2 + H_2O$
    h. $Pb(CH_3COO)_2(aq) + K_2CrO_4(aq) \rightarrow PbCrO_4(cr) + KCH_3COO(aq)$
    i. $RbCl(cr) + O_2(g) \rightarrow RbClO_4(cr)$
    j. $SiF_4(cr) + H_2O(l) \rightarrow H_2SiF_6(aq) + H_2SiO_3(cr)$
    k. $Sn(cr) + KOH(aq) \rightarrow K_2SnO_2(cr) + H_2(g)$
44. What mass of NH_4Cl can be produced from 7.87 grams of NH_3?
    $$NH_3 + HCl \rightarrow NH_4Cl$$
45. Substitute symbols and formulas for names and write balanced equations for each of the reactions described below.
    a. Ammonium nitrite decomposes to nitrogen gas and water.
    b. Sulfuric acid decomposes to water and sulfur trioxide.
    c. Ammonium nitrate decomposes to water and nitrogen(I) oxide.
46. Substitute symbols and formulas for names and write balanced equations for each of the following reactions.

a. Chromium displaces hydrogen from hydrochloric acid, with chromium(II) chloride as the other product.
b. Barium hydroxide reacts with carbon dioxide to form barium carbonate and water.
c. Calcium carbonate decomposes to calcium oxide and carbon dioxide gas.
d. Calcium carbide, CaC_2, reacts with water to produce calcium hydroxide and ethyne gas (acetylene), C_2H_2.
e. Hydrogen combines with sulfur to form hydrogen sulfide.
f. Pentane burns in oxygen to produce carbon dioxide and water.

47. Classify the following reactions as single displacement, double displacement, decomposition, synthesis, or combustion.
a. $4Na(cr) + O_2(g) \rightarrow 2Na_2O(cr)$
b. $Pb(NO_3)_2(aq) + Na_2CrO_4(aq) \rightarrow 2NaNO_3(aq) + PbCrO_4(cr)$
c. $NbI_3(cr) + I_2(cr) \rightarrow NbI_5(cr)$
d. $2Li(cr) + 2H_2O(l) \rightarrow 2LiOH(aq) + H_2(g)$
e. $2C_7H_{14}(l) + 21O_2(g) \rightarrow 14CO_2(g) + 14H_2O(g)$

48. Use the appropriate equations in Question 47 to solve the following mass-mass problems.
a. What mass of sodium oxide is produced by the reaction of 1.44 g of sodium with oxygen?
b. How much lead(II) nitrate will be needed to react with sodium chromate to produce 4.62 kg of lead(II) chromate?
c. What quantity of hydrogen gas is formed when 0.85 g of lithium reacts with water? How much lithium hydroxide will be formed?
d. What mass of water is given off when 192 kg of C_7H_{14} burn completely in air? How much oxygen will be used in the reaction?

49. Predict the products in the following reactions. Then write balanced equations.
a. decomposition of copper(II) oxide
b. copper and silver nitrate (displacement; copper(II) compound formed)
c. magnesium and oxygen (synthesis)
d. hydrochloric acid and silver nitrate (double displacement)
e. magnesium plus hydrochloric acid (displacement)
f. iron and oxygen (synthesis; iron(III) compound is formed)
g. iron and sulfur (synthesis; iron(II) compound is formed)
h. calcium hydroxide and sulfuric acid (double displacement)
i. zinc and sulfuric acid (single displacement)
j. benzene (C_6H_6) and oxygen (combustion reaction)
k. octane (C_8H_{18}) and oxygen (combustion reaction)

50. How many grams of $NaAlO_2$ can be obtained from 4.46 g of $AlCl_3$ according to the following reaction?

$$AlCl_3(aq) + 4NaOH(aq) \rightarrow NaAlO_2(aq) + 3NaCl(aq) + 2H_2O(l)$$

51. How many grams of hydrochloric acid are required to react completely with 61.8 grams of calcium hydroxide?

52. How many grams of hydrogen are produced when 4.72 grams of aluminum react with excess sulfuric acid?

53. How many grams of NH_3 are needed when 2.96 g of $Cr(NO_3)_3$ react according to the reaction: $Cr(NO_3)_3 + 6NH_3 \rightarrow Cr(NH_3)_6(NO_3)_3$?

54. A chemist carried out a reaction that should produce 21.8 g of a product, according to a mass-mass calculation. However, the chemist was able to recover only 13.9 g of the product. What percentage yield did the chemist get?

55. A calculation indicates that 82.2 g of a product should be obtained from a certain reaction. If a chemist actually gets 30.7 g, what is the percentage yield?

56. Chromium(III) hydroxide will dissolve in concentrated sodium hydroxide solution according to the following equation.

$$NaOH + Cr(OH)_3 \rightarrow NaCr(OH)_4$$

This process is one step in making high purity chromium chemicals. If you begin with 66.0 g $Cr(OH)_3$ and obtain 38.4 g of $NaCr(OH)_4$, what is your percentage yield?

57. Zinc oxide can be prepared industrially by treating zinc sulfide with oxygen. The by-product is sulfur dioxide, SO_2. An engineer expects to obtain a 78% yield of zinc oxide by this process. How much zinc sulfide should the chemical plant have on hand in order to prepare 2.0×10^4 kg of zinc oxide? Start by writing a balanced equation for the reaction.

58. Balance each of the following equations.

a. $PbCrO_4 + HCl + FeSO_4 \rightarrow PbCl_2 + Cr_2(SO_4)_3 + FeCl_3 + H_2O + Fe_2(SO_4)_3$

b. $IrCl_3(aq) + NaOH(aq) \rightarrow Ir_2O_3(cr) + HCl(aq) + NaCl(aq)$

c. $MoO_3(cr) + Zn(cr) + H_2SO_4(l) \rightarrow Mo_2O_3(cr) + ZnSO_4(aq) + H_2O(l)$

d. $Cu_2S(cr) + HNO_3(aq) \rightarrow Cu(NO_3)_2(aq) + CuSO_4(aq) + NO_2(g) + H_2O(l)$

e. $Ce(IO_3)_4(aq) + H_2C_2O_4(aq) \rightarrow Ce_2(C_2O_4)_3(aq) + I_2(aq) + CO_2(g) + H_2O(l)$

f. $KBr(cr) + H_2SO_4(aq) + MnO_2(cr) \rightarrow KHSO_4(aq) + MnSO_4(aq) + H_2O(l) + Br_2(l)$

59. In the decomposition of sodium chlorate, 31.7 g of O_2 are formed. How many grams of sodium chloride are produced?

60. The action of carbon monoxide on iron(III) oxide can be represented by the equation, $Fe_2O_3(cr) + 3CO(g) \rightarrow 2Fe(cr) + 3CO_2(g)$. What is the minimum amount of carbon monoxide used if 57.5 grams of iron were produced?

61. Claude-Louis Berthollet first prepared ethyne (acetylene) by sparking carbon electrodes in hydrogen gas.

$$2C + H_2 \rightarrow C_2H_2$$

How many grams of the carbon electrode will be consumed when 59.8 grams of acetylene are produced?

62. Compute the heat change for the reaction of 1.24 grams of NO according to the following equation.

$$2NO + O_2 \rightarrow 2NO_2 + 114.14 \text{ kJ}$$

63. Compute the heat change for the production of 17.1 grams of Fe_2O_3 according to the following equation.

$$4FeO + O_2 \rightarrow 2Fe_2O_3 + 560.4 \text{ kJ}$$

Concept Mastery

64. Applying Dalton's atomic theory, explain why chemical equations must be balanced.

65. Why is it that subscripts should not be changed in order to write a balanced chemical equation?

66. Why is it necessary to use a balanced chemical equation when solving a mass-mass problem?

67. Nitric acid, HNO_3, ranks thirteenth on the list of chemicals produced in greatest amount in the United States. It has extensive use in industry. Nitric acid can be produced by the action of water on nitrogen(IV) oxide. The balanced equation for the reaction is $3NO_2(g) + H_2O(l) \rightarrow 2HNO_3(aq) + NO(g)$. Write a sentence that summarizes the information given in this equation.

68. The reaction that describes the rusting of an old car body is $4Fe + 3O_2 \rightarrow 2Fe_2O_3$. Which of the diagrams on the next page illustrates this reaction?

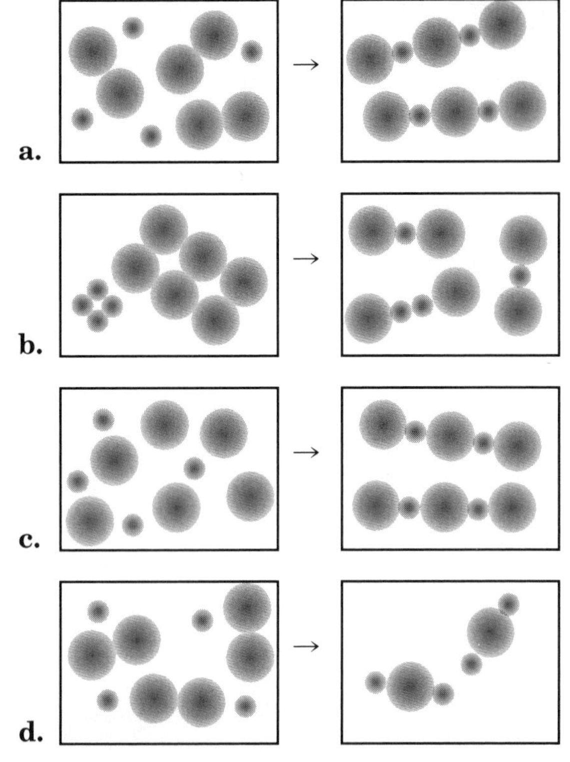

Application

69. Iron occurs in the United States in the form of oxide ores such as iron(III) oxide, Fe_2O_3. In a blast furnace the iron ore is reduced to iron metal by carbon monoxide, which becomes carbon dioxide. Write a balanced equation for the process.

70. Calcium chloride is used to melt ice or to keep it from forming on roads. This important compound can be produced by reacting calcium carbonate with hydrochloric acid. Two by-products are water and carbon dioxide. Write a balanced equation for the formation of calcium chloride by this reaction.

71. Cyclopropane, C_3H_6, was once used with oxygen as a general anesthetic. Write a balanced equation for the complete combustion of cyclopropane.

72. Aluminum hydroxide, $Al(OH)_3$, is a component in many antacids. It neutralizes the HCl in stomach acid in a double displacement reaction. Write a balanced equation for this reaction.

73. Ammonium nitrate, NH_4NO_3, is an important fertilizer and is also used in the manufacture of explosives and fireworks. It is produced by treating nitric acid, HNO_3, with ammonia gas, NH_3.
 a. Write a balanced equation for the reaction.
 b. What type of reaction is this?
 c. If 340 kg of ammonia gas are used with all the nitric acid necessary for reaction, how much ammonium nitrate can be formed?

74. In space vehicles, air purification for the crew is partly accomplished with the use of lithium peroxide, Li_2O_2. It reacts with waste CO_2 in the air according to the reaction $2Li_2O_2 + 2CO_2 \rightarrow 2Li_2CO_3 + O_2$. How many grams of oxygen are released by the reaction of 0.905 g CO_2?

75. Chlorine is prepared industrially by passing an electric current through brine, a concentrated solution of sodium chloride.

$$2NaCl(aq) + 2H_2O(l) \rightarrow 2NaOH(aq) + Cl_2(g) + H_2(g)$$

All three products are commercially valuable. If 1500 g of a brine solution that is 24% NaCl by mass is used, how many grams of each product can be produced?

76. Borax, $Na_2B_4O_7{\cdot}10H_2O$, is an important household cleaner. Its ore, kernite, has the formula $Na_2B_4O_7{\cdot}4H_2O$. If kernite is shipped and then treated with water to form borax at the destination, how much shipping mass is saved if 2500 kg of kernite are shipped?

77. Two components of smog are NO and NO_2. NO_2 is formed in car engines and emitted into the atmosphere. When NO_2 is exposed to sunlight it forms NO and O atoms. What type of reaction is this?

Critical Thinking

78. In the years following the French Revolution, Joseph Gay-Lussac made several important discoveries about how gases combine. For example, he noted that three volumes of hydrogen react with one volume of nitrogen to form two volumes of ammonia. Keep in mind that in those times, scientists did not know about atoms or the formulas of compounds.

a. Write a balanced chemical equation for this reaction.

b. How does this equation relate to Gay-Lussac's findings?

c. What can you infer about the contents of two equal volumes of reacting gases?

d. What word could be substituted for the word *volume* to describe this reaction on the simplest level?

e. What type of reaction did Gay-Lussac perform?

79. A chemical engineer was trying to project the amount of materials that a chemical plant would need for the second step in the production of sulfuric acid. The engineer used the "equation" $SO_2 + O_2 \rightarrow SO_3$. What problems could result from the use of this "equation"? Answer in terms of the quantities the engineer might order.

80. What factors can you think of that might cause a yield of less than 100 percent in a reaction?

Readings

Garcia, Arcesio. "A New Method to Balance Chemical Equations." *Journal of Chemical Education* 64, no. 3 (March 1987): 247–248.

Harjadi, W. "A Simpler Method of Chemical Reaction Balancing." *Journal of Chemical Education* 63, no. 11 (November 1986): 978–979.

Cumulative Review

1. What is the molarity of a 1.00×10^3-cm^3 solution containing 0.550 g of $Ni(IO_3)_2$?

2. Butyric acid, $CH_3CH_2CH_2COOH$, is formed when butter becomes rancid. Find the percentage composition of each element in this compound.

3. Tungsten is Swedish for "heavy stone." Its symbol, W, comes from wolfram, the German name for the element. It has a density of 19.3 g/cm^3. How many atoms are in a cubic centimeter of tungsten?

4. Convert 0.633 mol $BaSeO_4$ to formula units and to grams.

5. Convert 0.0731 mol $Sr(CN)_2 \cdot 4H_2O$ to grams.

6. Find the percentage composition of each element in $BaSO_4$.

7. What is the empirical formula of a compound with the composition 63.9% Cd, 11.8% P, and 24.3% O?

8. Compute the formula mass of Sr_3N_2.

9. What is the mass of 3.00 moles of $Zn_3(PO_4)_2$?

10. How would you prepare 100.0 cm^3 of a $0.100M$ solution of $ZrCl_4$?

11. Find the percentage composition of $Co(CH_3COO)_2$.

12. The compound styrene glycol is used in the manufacture of some plastics. Its percentage composition is 69.54% carbon, 7.30% hydrogen, and 23.16% oxygen. What is its empirical formula? If the molecular mass of styrene glycol is 138.2, what is its molecular formula?

CHAPTER Preview

CONTENTS

CHAPTER 10

Periodic Properties

Have you ever been given a list and been asked to tell which item doesn't fit with the others? To answer that question, you first need to find what the other items have in common. If the list consisted of hat, gloves, flowers, shoes, *and* jeans, *the answers would be easy to find. If the list were* Curie, Stravinski, Raoult, Boyle, *and* Gibbs, *the connection might not be so obvious.*

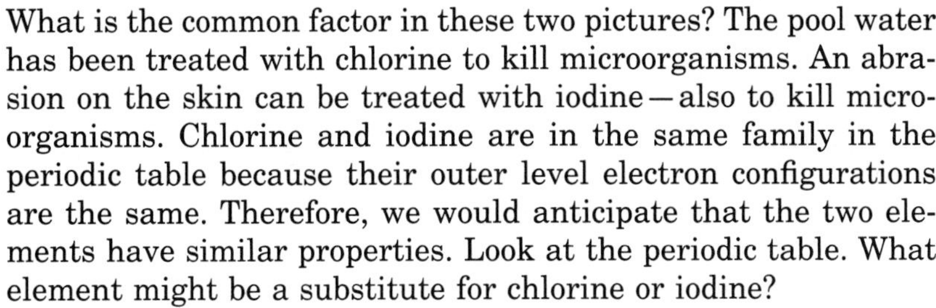

What is the common factor in these two pictures? The pool water has been treated with chlorine to kill microorganisms. An abrasion on the skin can be treated with iodine—also to kill microorganisms. Chlorine and iodine are in the same family in the periodic table because their outer level electron configurations are the same. Therefore, we would anticipate that the two elements have similar properties. Look at the periodic table. What element might be a substitute for chlorine or iodine?

EXTENSIONS AND CONNECTIONS

History Connection

All that Glitters, page 253

Bridge to Astronomy

Using Ionization Energy to Make Images: A view of how astronomers are using ionization properties of silicon to see dim objects, **page 260**

Everyday Chemistry

Coatings that Prevent Corrosion: A look at ways rust is prevented by coating iron, **page 262**

Frontiers

Metallic Hydrogen? Press On!: Predictions made from the periodic table are being used to investigate and maybe create a strange new material, **page 264**

OBJECTIVES

- use examples to explain the periodic properties of elements.
- state how atomic and ionic sizes change in groups and periods.
- predict oxidation numbers of elements.

10.1 Periodic Trends

The periodic table is a powerful tool of the chemist. The table is organized on the basis of the atomic structures of the elements. In this chapter, we will examine some properties whose variation depends upon electron configurations. These properties should be closely related to the positions of the elements in the periodic table. Remember, the elements do not have the properties *because* of their positions in the table. Rather, *both the position and the properties arise from the electron configurations of the atoms.*

Properties and Position

Elements in the same column have similar outer level electron configurations and the configurations change in a regular way from one column to the next as we scan across the table. Because the properties of the elements are determined by their electron configurations, we should be able to predict properties of most elements based on our knowledge of the behavior of a few.

We have already seen that as we scan across the table from left to right, we proceed from metallic elements, through metalloids, to nonmetals. When we drop down to the next period, the same pattern repeats. In other words, the properties are periodic. Similar properties occur at certain intervals of atomic number, as is stated in the modern periodic law. From the graph in Figure 10.1, you can see that density is a periodic property. The densities of the elements vary in a regular way when plotted against the atomic numbers of the elements. Metals have the highest densities and those nonmetals existing as gases at room temperature and one atmosphere of pressure have the lowest densities.

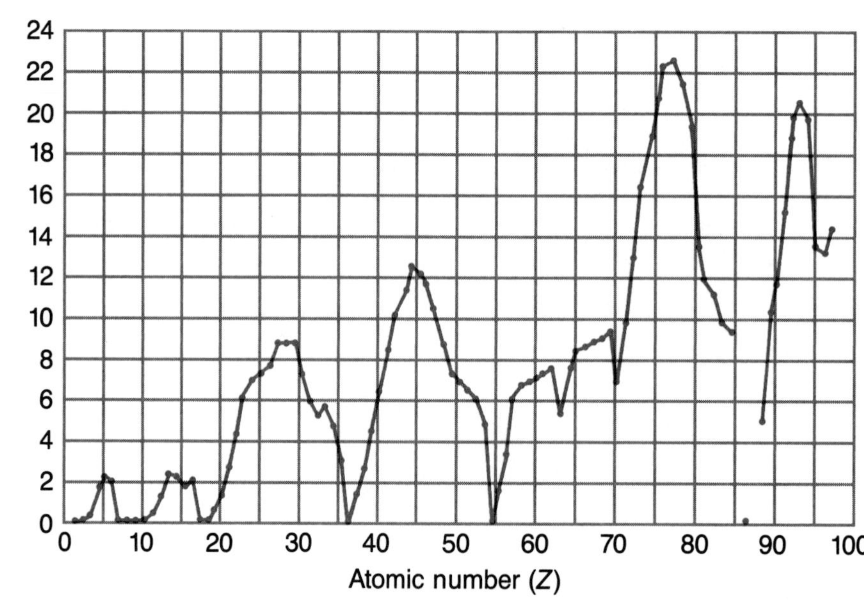

Figure 10.1 Density is a periodic property. When plotted against atomic numbers, the densities of the elements vary in a periodic way.

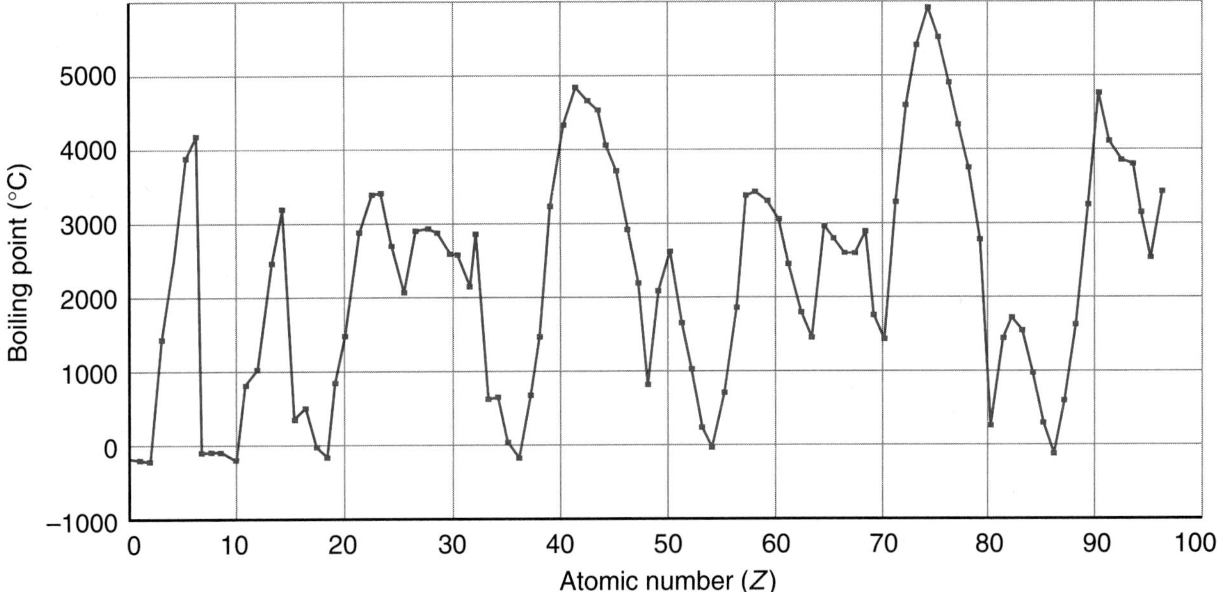

Figure 10.2 Boiling point is also a periodic property.

Radii of Atoms

As you look at the periodic table from top to bottom, each period represents a new, higher principal quantum number. *As the principal quantum number increases, the size of the electron cloud increases.* Therefore, the size of atoms in each group increases as you look down the table. Chemists discuss the size of atoms by referring to their radii. The radius of an atom without regard to surrounding atoms is the **atomic radius.** As you look across the periodic table, all the atoms in a period have the same principal quantum number. This fact might lead you to expect that all atoms in a period are of similar size. However, positive charge on the nucleus increases by one proton for each element in a period. As a result, the outer electron cloud is pulled in a little tighter. Consequently, one periodic property of atoms is that they generally decrease slightly in size from left to right across a period of the table. In summary, *atomic radii increase top to bottom and right to left in the periodic table.* In Table 10.1, you can see the general trends and the few exceptions to the rule.

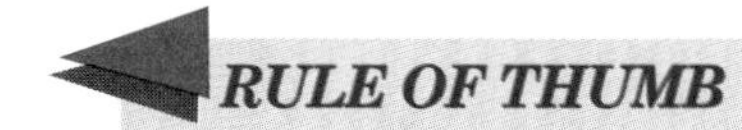

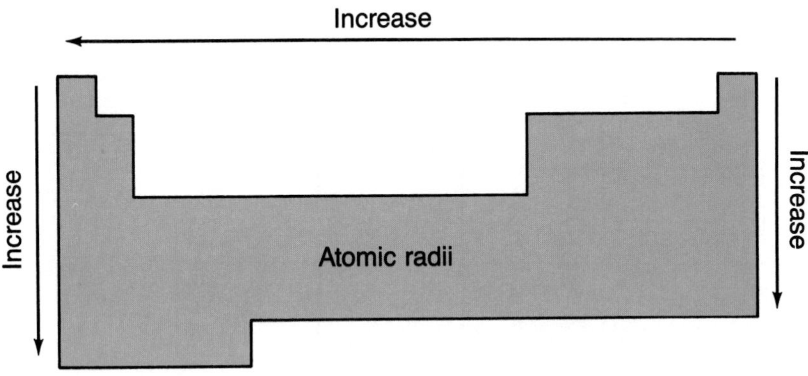

Figure 10.3 The general trend in atomic radii is an increase from right to left and top to bottom in the periodic table.

Table 10.1

Atomic and Ionic Radii (in picometers)

H 78																	He 122
Li 156 90 (1+)	Be 112 59 (2+)											B 85 41 (3+)	C 77 260 (4−)	N 71 132 (3−)	O 60 126 (2−)	F 69 119 (1−)	Ne 131
Na 186 116 (1+)	Mg 160 86 (2+)											Al 143 67.5 (3+)	Si 118 54 (4+)	P 109 212 (3−)	S 103 170 (2−)	Cl 91 167 (1−)	Ar 191
K 231 152 (1+)	Ca 197 114 (2+)	Sc 162 88.5 (3+)	Ti 147 74.5 (4+)	V 134 72 (4+)	Cr 128 75.5 (3+)	Mn 127 81 (2+)	Fe 126 69 (3+)	Co 125 79 (2+)	Ni 124 83 (2+)	Cu 128 87 (2+)	Zn 134 88 (2+)	Ga 134 76 (3+)	Ge 123 67 (4+)	As 121 72 (3+)	Se 117 184 (2−)	Br 119 182 (1−)	Kr 201
Rb 248 166 (1+)	Sr 215 132 (2+)	Y 180 104 (3+)	Zr 160 86 (4+)	Nb 146 78 (5+)	Mo 139 73 (6+)	Tc 136 78.5 (4+)	Ru 134 76 (4+)	Rh 134 80.5 (3+)	Pd 137 100 (2+)	Ag 144 129 (1+)	Cd 151 109 (2+)	In 167 94 (3+)	Sn 141 83 (4+)	Sb 161 90 (3+)	Te 138 207 (2−)	I 138 206 (1−)	Xe 218
Cs 262 181 (1+)	Ba 222 149 (2+)	Lu 174 100.1 (3+)	Hf 159 85 (4+)	Ta 146 78 (5+)	W 139 74 (6+)	Re 137 77 (4+)	Os 135 77 (4+)	Ir 136 76.5 (4+)	Pt 139 76.5 (4+)	Au 144 99 (3+)	Hg 151 116 (2+)	Tl 170 164 (1+)	Pb 175 133 (2+)	Bi 151 117 (3+)	Po 164 94 (4+)	At	Rn 232
Fr 280 180 (1+)	Ra 228 162 (2+)																

La 187 117.2 (3+)	Ce 182 115 (3+)	Pr 182 99 (3+)	Nd 181 98.3 (3+)	Pm 183 97 (3+)	Sm 180 95.8 (3+)	Eu 208 94.7 (3+)	Gd 180 93.8 (3+)	Tb 177 92.3 (3+)	Dy 178 91.2 (3+)	Ho 176 90.1 (3+)	Er 176 89 (3+)	Tm 176 88 (3+)	Yb 193 86.8 (3+)
Ac 203 112 (3+)	Th 179 108 (4+)	Pa 163 78 (5+)	U 156 103 (4+)	Np 155 75 (5+)	Pu 162 86 (4+)	Am 183 97.5 (3+)	Cm 174 97 (3+)	Bk 170 96 (3+)					

Sodium and chlorine are located at opposite ends of the third period. Sodium is found at the left side of the table and is a metal. Chlorine is on the right side of the table in Group 17 (VIIA) and is a nonmetal.

Both sodium and chlorine have partially filled third levels. The outer electrons that take part in reactions are separated from the positively charged nucleus by two inner energy levels. These two inner levels are filled (ten electrons). The chlorine nucleus contains seventeen protons; the sodium nucleus contains only eleven protons. The outer electrons of the chlorine atom are attracted by six more protons than are the outer electrons of the sodium atom. Therefore, the chlorine electrons are held more tightly, and the chlorine atom is smaller than the sodium atom. Sodium and chlorine atoms follow the general pattern shown in Table 10.1.

	1s	2s	2p	3s	3p
Sodium	(↑↓)	(↑↓)	(↑↓)(↑↓)(↑↓)	(↑)	
Chlorine	(↑↓)	(↑↓)	(↑↓)(↑↓)(↑↓)	(↑↓)	(↑↓)(↑↓)(↑)

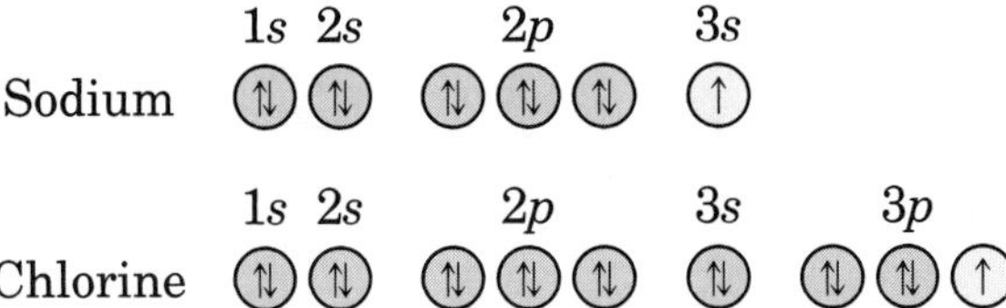

One obvious exception to the rule of thumb for size variation is the noble gas family. The sizes of the atoms for most of the elements have been determined by using a process called X-ray diffraction. In X-ray diffraction, X rays are passed through a crystal of the element. The resulting diffraction patterns reveal the arrangement, size, and spacing of the atoms. When most atoms form crystals, there is some interaction of the electrons in different atoms, pulling the atoms closer together and making the atoms smaller. In the noble gases, with a complete octet in the outer level of each atom, there is no electron interaction and the atoms remain far apart and larger.

HISTORY CONNECTION

All that Glitters . . .

An object submerged in water will displace its volume of water. Archimedes (287–212 B.C.) discovered the secret of density and water displacement while attempting to determine whether a king's crown was pure gold. He reasoned that a pure gold crown would displace less water than a crown of equal mass made of silver and gold. Because silver is less dense than gold, the same mass would require a greater volume than gold alone. When he placed the crown in water, he found that it displaced a greater amount than an equal mass of pure gold. The crown was not pure!

Radii of Ions

RULE OF THUMB

In general, when atoms unite to form compounds, the compound is more stable than the uncombined atoms. Consider a reaction between sodium and chlorine. The sodium atom holds its single outer 3*s* electron loosely. When chlorine and sodium atoms react, the chlorine removes the outer electron from the sodium atom. The resulting sodium ion has eleven protons, but only ten electrons, so it has a 1+ charge. There are two main reasons why the sodium ion is smaller than the sodium atom. First, the positively charged nucleus is now attracting fewer electrons. Second, with the loss of the $3s^1$ electron, the ion now has two energy levels whereas the atom had three levels. The sodium ion now has a new outer level ($2s^22p^6$) that is the same as the outer level of the noble gas, neon. It is important to remember that **noble gas configurations** are particularly stable because the noble gases have filled outer energy levels.

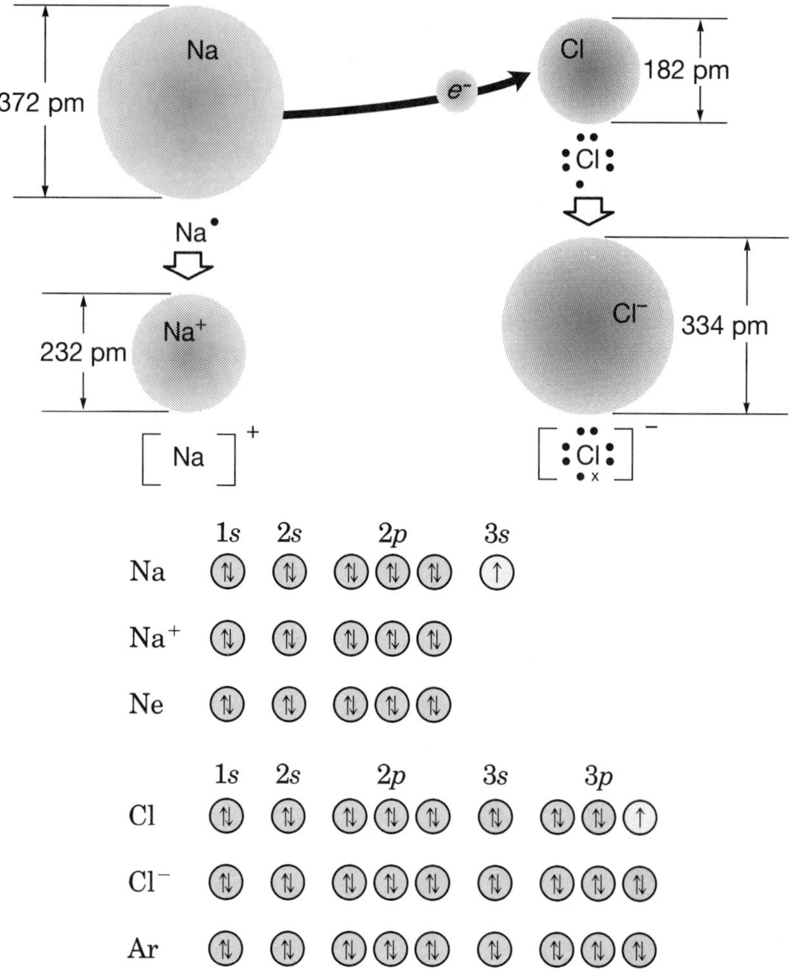

Figure 10.4 The sodium ion is smaller than the sodium atom because 11 protons are attracting only 10 electrons in the ion. The chloride ion is larger than the chlorine atom because 17 protons are attracting 18 electrons in the ion. The chlorine atom is smaller than the sodium atom because the electrons in the outer energy level of chlorine are attracted by 6 more protons than in the sodium atom.

The chlorine atom has gained an electron. The seventeen protons in its nucleus are now attracting eighteen electrons. The chloride ion thus has a 1− charge. Also, the chloride ion is larger than the chlorine atom. The ion now has the same outer level configuration as the noble gas, argon. The positively-charged sodium ions and the negatively-charged chloride ions now attract each other and form a crystal of sodium chloride, common table salt. This salt is more stable than the separate atoms or ions.

The compound formed from sodium ions and chloride ions does not consist of sodium chloride molecules. Instead, each cube of sodium chloride is a collection of equal numbers of Na^+ and Cl^- ions. When a salt is dissolved or melted, it will carry a current because the ions are free to move. However, a solid salt crystal will not carry an electric current because the individual ions are tightly bound in the crystal structure. The mobility of electric charge completes a circuit. Thus, the solution or molten substance is able to conduct a current. This property is a characteristic of compounds composed of ions.

Chemists discuss the size of ions by referring to their **ionic radii**. We have noted that the sodium ion is smaller than the sodium atom. Compared to its atom, the magnesium ion is even smaller. In losing two outer electrons, the unbalanced nuclear positive charge is larger than the negative charge on the electron cloud. The cloud shrinks in size. The nonmetals sulfur and chlorine form ions that are larger than their respective atoms. These elements gain electrons to form ions. The elements silicon and phosphorus do not gain or lose electrons readily. Although they may form ions, they tend to form compounds by sharing their outer electrons.

We can now look at some trends that will apply to any row of the periodic table. In general, *metallic ions, on the left and in the center of the table, are formed by the loss of electrons. They are smaller than the atoms from which they are formed. Nonmetallic ions are located on the right side of the table. They are formed by the gain of electrons and are larger than the atoms from which they are formed.* The metallic ions have an outer level that resembles that of the noble gas at the end of the preceding period. Nonmetallic ions have an outer level resembling that of the noble gas to the right in the same period.

RULE OF THUMB

PRACTICE PROBLEMS

1. From each of the following pairs of particles, select the particle that is larger in radius. Use the periodic table on the inside back cover.

a. Ar, Ne **b.** B, C **c.** N, P **d.** Cl, O **e.** Ca, Sc

2. From each of the following pairs of particles, select the particle that is smaller in radius. Use the periodic table on the inside back cover.

a. O, O^{2-} **b.** Mg, Mg^{2+} **c.** Te, Te^{2-} **d.** Ti, Ti^{4+}

Predicting Oxidation Numbers

Those electrons that are involved in the reaction of atoms with each other are the outer and highest energy electrons. You now know about electron configurations and the stability of the noble gas structures. Thus, it is possible for you to predict what oxidation numbers atoms will have.

Consider the metals in Group 1 (IA). Each atom has one electron in its outer level. The loss of this one electron will give these metals the same configuration as a noble gas. Group 1 (IA) metals have an oxidation number of 1+. What about the element hydrogen, which also is in Group 1 (IA)? Hydrogen can lose one electron and have an oxidation number of 1+. Note also that the hydrogen atom could attain the helium configuration by gaining one electron. If this change occurred, we would say that hydrogen has a 1− oxidation number. Hydrogen does have a 1− oxidation number in some compounds. In Group 2 (IIA), we expect the loss of the

Figure 10.5 Chromium shows several oxidation states, with the 3+ state being the most common. These solutions of $Cr(NO_3)_3$ (violet) and $CrCl_3$ (green) illustrate the 3+ oxidation state. Solutions of K_2CrO_4 (yellow) and $K_2Cr_2O_7$ (orange) illustrate the 6+ state.

two *s* electrons for the atom to achieve the same configuration as the prior noble gas element. That loss leads to a prediction of 2+ as the oxidation number for the alkaline earth metals. The elements in these two columns exhibit the oxidation numbers predicted for them.

Beginning with Group 3 (IIIB), we have atoms in which the highest energy electrons are not in the outer level. For instance, scandium has the configuration $1s^22s^22p^63s^23p^64s^23d^1$. Scandium's outer level is the fourth level containing two electrons. Its highest energy electron, however, is the one in the $3d$ sublevel. For the transition elements, it is possible to lose not only the outer level electrons but also some lower level electrons. The *d* electrons, because they are in an energy level one below the outer level, can be lost only after the electrons in the outer level have been lost and the outer level is empty. Further, these *d* electrons may be lost one at a time. The transition elements exhibit oxidation numbers varying from 1+ (representing loss of the outer electron or electrons) up to 8+. We would predict scandium to lose

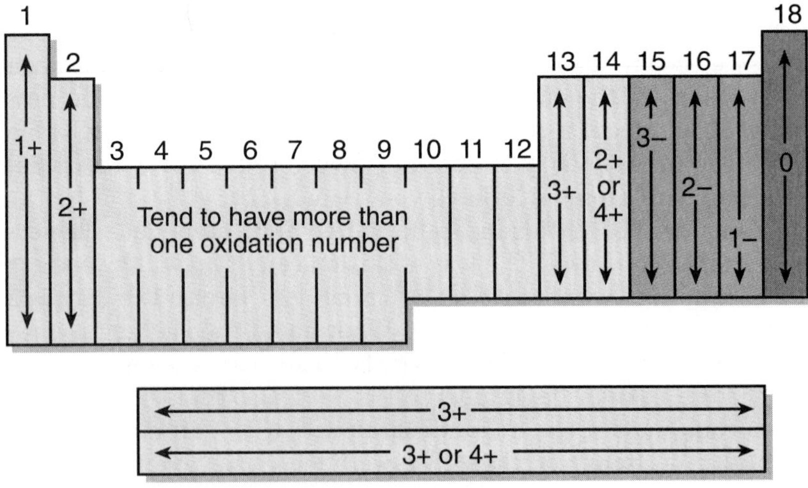

Figure 10.6 The oxidation number of an element can be predicted if one knows the element's location in the periodic table.

Figure 10.7 When exposed to the oxygen in air, iron will change from its 2+ oxidation state (blue-green) to its more stable 3+ oxidation state (reddish brown).

the two electrons in the outer 4*s* sublevel, giving it an oxidation number of 2+. It can also then lose the 3*d* electron to give it a 3+ oxidation number. In actual practice, the element shows only the 3+ oxidation number. Titanium, which has one more 3*d* electron than scandium has, should show 2+, 3+, and 4+, and it does. These oxidation numbers represent loss of $4s^2$, $4s^2$ and $3d^1$, and $4s^2$ and $3d^2$ electrons, respectively. As we continue across the fourth row, vanadium has a maximum oxidation number of 5+, chromium 6+, and manganese 7+. Iron, which has the configuration $1s^22s^22p^63s^23p^64s^23d^6$, has only 2+ and 3+ oxidation numbers. Recall that a half-full sublevel represents a particularly stable configuration. To take iron higher than 3+ would mean removing electrons from a half-full 3*d* sublevel. However, osmium, in the same group, does have an oxidation number of 8+.

With the exception of boron, Group 13 (IIIA) elements lose three electrons and have an oxidation number of 3+. Boron combines with other atoms solely by sharing electrons. Thallium, in addition to the 3+ oxidation number, exhibits a 1+ oxidation number. If we look at its configuration, we can understand why. The thallium configuration ends $6s^24f^{14}5d^{10}6p^1$. The large energy difference between the 6*s* and the 6*p* electrons makes it possible to lose only the 6*p* electron. That loss leads to an oxidation number of 1+. If stronger reaction methods are used, thallium also has an oxidation number of 3+. For the same reason, tin and lead in Group 14 (IVA) may have either a 2+ or 4+ oxidation number.

10.1 *CONCEPT REVIEW*

3. Place the following atoms in order of increasing size: B, Be, Mg, N, Na.
4. In each of the following pairs of particles, which particle is larger?
   a. Al, Al^{3+} **b.** P, P^{3-}
5. Predict oxidation numbers for each of the following elements: Ca, Cl, Cr, K, S.
6. **Apply** Would you predict iron to be more stable in the 2+ or 3+ oxidation state? Give a reason for your answer.

OBJECTIVES

- define ionization energy and electron affinity, and describe the factors that affect these properties.
- use multiple ionization energies to predict oxidation numbers of elements.

10.2 Reaction Tendencies

We know that when atoms form compounds, some atoms tend to give up electrons and become positive ions, while other atoms tend to gain electrons and become negative ions. The path of reactions and the properties of the products are largely dependent upon the tendencies of the reacting atoms to gain or lose electrons. We now want to examine the periodic nature of these atomic tendencies. In later chapters, we will be examining the attraction of an atom for electrons as the determining factor in the type of bond formed between atoms in a compound.

First Ionization Energy

The energy required to remove an electron from an atom is called its **ionization energy.** Our model of the atom was developed partly from determining the energy needed to remove the most loosely held electron from an atom. This energy is called the **first ionization energy** of that element. It is measured in kilojoules per mole (kJ/mol).

The first ionization energies of eighty-seven of the elements are graphed in Figure 10.8. Note that the first ionization energies, like many other properties of the elements, are periodic. In fact, the relative first ionization energies of two elements can be predicted by referring to their positions in the periodic table.

The first ionization energy tends to increase as atomic number increases in any horizontal row or period. In any column or group, there is a gradual decrease in first ionization energy as atomic

Figure 10.8 Ionization energy is a periodic property. Notice that the ionization energy values generally increase within each period.

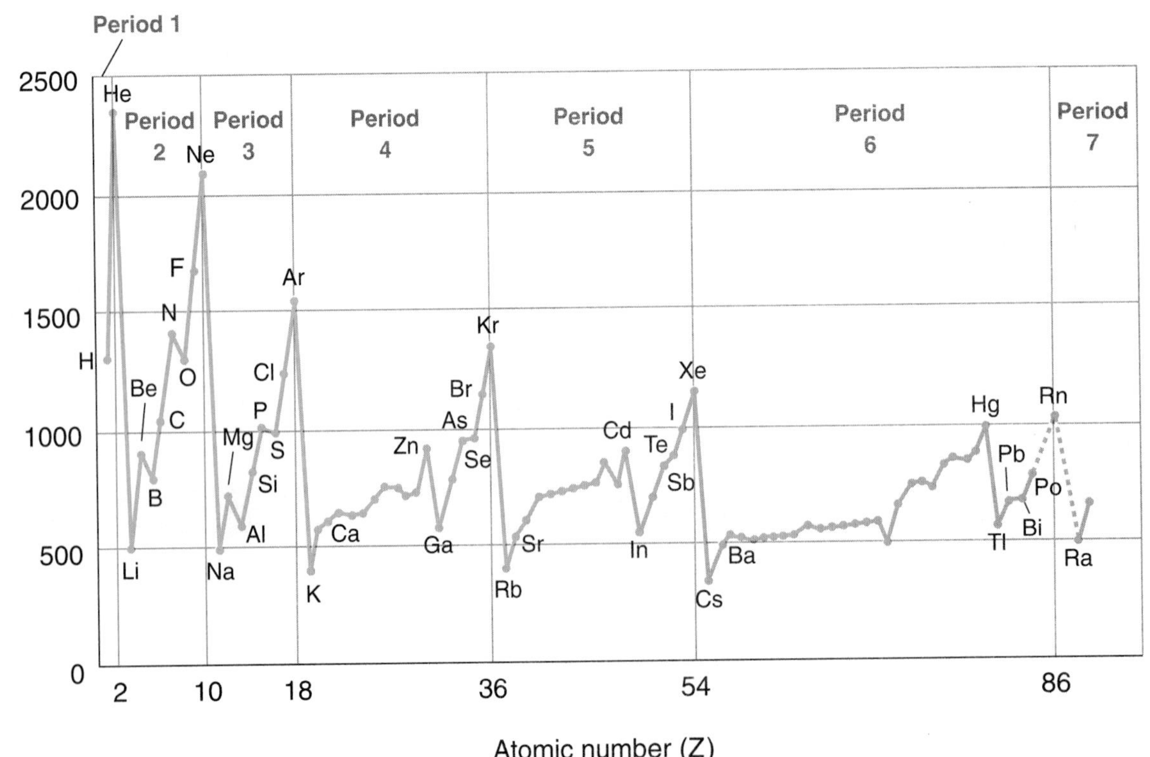

Table 10.2

First Ionization Energies (kilojoules per mole)							
H							**He**
1312.0							2372.3
Li	**Be**	**B**	**C**	**N**	**O**	**F**	**Ne**
520.2	899.5	800.6	1086.5	1402.3	1313.9	1681.0	2080.7
Na	**Mg**	**Al**	**Si**	**P**	**S**	**Cl**	**Ar**
495.9	737.8	577.5	786.5	1011.8	999.6	1255.5	1520.6
K	**Ca**	**Ga**	**Ge**	**As**	**Se**	**Br**	**Kr**
418.8	589.8	578.8	761.2	947	940.7	1139.9	1350.8
Rb	**Sr**	**In**	**Sn**	**Sb**	**Te**	**I**	**Xe**
403.0	549.5	558.2	708.4	834	869	1008.4	1170.4

number increases. Note, for example, the gradual decrease in first ionization energy in the alkali metal family, lithium through cesium. The same trend is seen in the noble gas family.

RULE OF THUMB

In general, elements can be classified as metals or nonmetals on the basis of first ionization energy. *A metal is characterized by a low first ionization energy.* Metals are located at the left side of the table. *An element with a high first ionization energy is a nonmetal.* Nonmetals are found at the right side of the table.

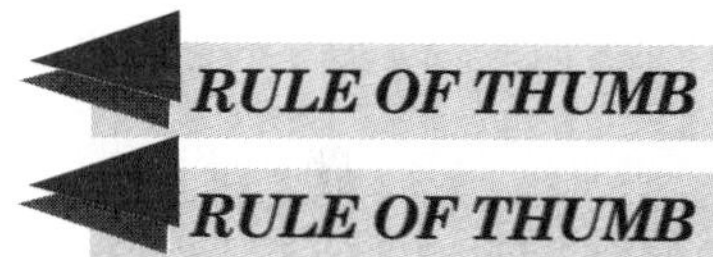

These patterns of first ionization energies provide strong evidence for the existence of energy levels in the atom. Our theories of structure are based on experimental evidence such as ionization energies and atomic spectra.

Look at Table 10.2. Notice that, in general, *first ionization energies decrease as you go down a column of the periodic table* (for instance, lithium, sodium, potassium). Both increased distance of the outer electrons from the nucleus and the **shielding effect,** in which inner electrons block the attraction of the nucleus for outer electrons, tend to lower ionization energy. Though it would seem that the increased nuclear charge of an element with a greater atomic number would increase first ionization energy, the lowering tendency caused by distance and the shielding effect is greater. Remember that the number of electrons in the outermost sublevel is the same for all elements in a particular group.

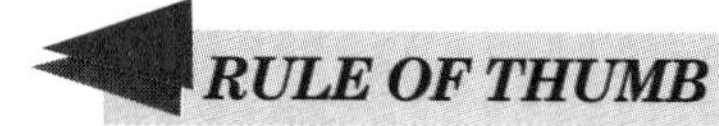

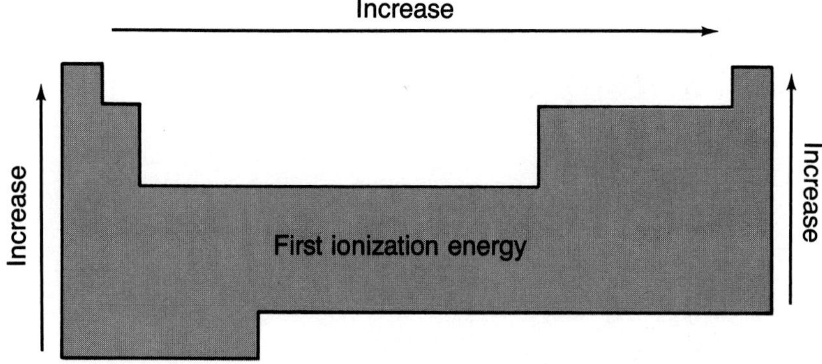

Figure 10.9 In general, the first ionization energies for the elements increase from left to right and from bottom to top of the periodic table.

In moving across a period of the periodic table, there is a general increase in first ionization energy as a result of increasing nuclear charge. However, there are some deviations from the expected trend of increasing first ionization energy. Look at the second row. There is a small decrease from beryllium ($1s^22s^2$) to boron ($1s^22s^22p^1$). In beryllium, the first ionization energy is determined by removing an *s* electron from a full *s* sublevel. In boron, it is determined by removing the lone *p* electron.

There is another slight decrease from nitrogen ($1s^22s^22p^3$) to oxygen ($1s^22s^22p^4$). Note that the nitrogen *p* sublevel is half-full. Recall from Chapter 6 that this is a state of special stability. Thus, a large amount of energy is needed to remove an electron from nitrogen's half-full *p* sublevel. Oxygen does not have the special stability of a half-full sublevel. Thus, oxygen has a lower first ionization energy than does nitrogen.

The patterns in ionization energy values can be explained by the same factors we discussed in Chapter 6 in connection with the periodic table. These factors are summarized in Table 10.3.

Bridge to
ASTRONOMY

Using Ionization Energy to Make Images

Astronomers are constantly trying to see dimmer objects in the universe. As additions to telescopes, they use devices that enhance their ability to see dim objects.

One such device, the charge-coupled device (CCD), makes use of the ionization properties of silicon. When a photon strikes a silicon atom, the atom becomes ionized. The number of free electrons generated varies directly with the radiant energy absorbed. These electrons are then detected electronically.

A silicon microchip can be divided into many small areas called pixels. The charge acquired by each pixel can be detected separately and the information assembled to produce images of dim objects. When cooled to temperatures as low as 100 K, the CCD can image even dimmer objects. The electronic circuitry can be designed to add images in much the same way as a time exposure is made on film. In addition, a CCD will record 70% of the photons that strike it, whereas film seldom records more than 1%. The CCD is reliable, and requires little power to operate. It is also sensitive to a wide range of wavelengths. In addition to its use in astronomy, the CCD can be used to measure and sort objects on an assembly line and read optical characters.

Exploring Further

1. Ionization energies often are reported in units of electron volts/atom. Find the conversion factor for kJ/mol to eV/atom. Using Table 10.2, write the first ionization energy of silicon in eV/atom.

2. Find out how the Hubble Space Telescope incorporates CCDs.

Table 10.3

Factors Affecting Ionization Energy
1. **Nuclear charge**—The larger the nuclear charge, the greater the ionization energy.
2. **Shielding effect**—The greater the shielding effect, the less the ionization energy.
3. **Radius**—The greater the distance between the nucleus and the outer electrons of an atom, the less the ionization energy.
4. **Sublevel**—An electron from a full or half-full sublevel requires additional energy to be removed.

▼ Multiple Ionization Energies

As additional electrons are lost from an atom it is possible to measure other ionization energies of an atom. These measurements give us the same evidence for atomic structure as first ionization energies give us. For example, the second ionization energy of aluminum is about three times as large as the first ionization energy, as shown in Table 10.4. The difference can be explained by the fact that the first ionization removes a *p* electron and the second removes an *s* electron from a full *s* sublevel. The third ionization energy is about one and one-half times as large as the second ionization energy. The second and third electrons are in the same sublevel. Yet, the third electron's ionization energy is greater because the positive nuclear charge remains constant as we remove electrons. As a result, the aluminum atom's remaining electrons are more tightly held. The fourth ionization energy is about four times as large as the third. Why is the jump so large between the third and fourth ionization energies? Look at the electron configuration. Aluminum has the configuration $1s^2 2s^2 2p^6 3s^2 3p^1$. The fourth electron would come from the second energy level that is both full and closer to the nucleus. The $2s^2 2p^6$ level with eight electrons is stable. Thus, a large amount of energy will be required to remove that fourth electron.

Table 10.4

Ionization Energies (kilojoules per mole)

Element	1st	2nd	3rd	4th	5th	6th
H	1312.0					
He	2372.3	5220				
Li	520.2	7300	11 750			
Be	899.5	1760	14 850	20 900		
B	800.6	2420	3 660	25 020	32 660	
C	1086.5	2390	4 620	6 220	37 820	46 990
Al	577.5	1810	2 750	11 580	14 820	18 360
Ga	578.8	1980	2 970	6 170	8 680	11 390

CHEMISTRY IN DEPTH

Smokeless Smoke

Put on an apron and goggles. Bring to class a battery-operated ionization-type smoke detector. Carefully burn a small 10 cm × 10 cm piece of newsprint in a metal pan in a sink while holding the detector about 30 cm above the flame. **CAUTION:** *Do not have the paper near other flammable objects.* How much smoke did you observe as the paper burned? Did the smoke detector sense the gases produced? Research how a smoke detector works and be prepared to discuss it in class the next day.

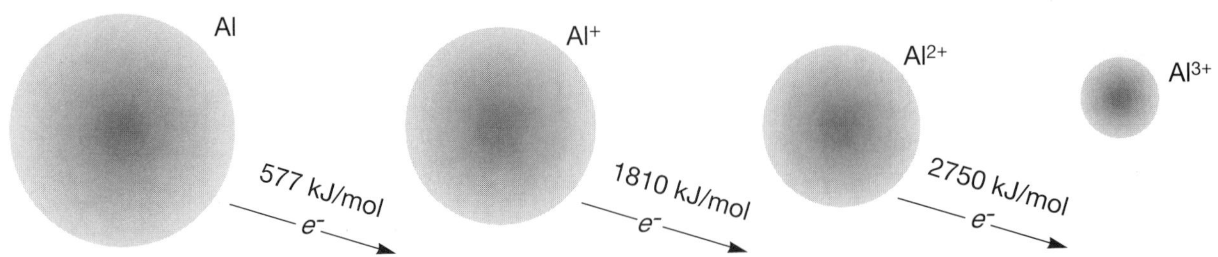

Figure 10.10 The second, third, and fourth ionization energies of aluminum are higher than the first because the same number of protons is attracting fewer electrons.

Look again at ionization energies for the aluminum atom. The first three are relatively small and close together when compared with later ionization energies. These energies help explain why aluminum has an oxidation number of 3+ in all of its compounds. This reasoning can be applied to any other atom. By knowing the ionization energies of the first six or so electrons, you can predict the most likely oxidation number or numbers by seeing where ionization energy increases greatly.

We have looked only at the aluminum atom, but the same reasoning can be used to explain similar data for other elements. This information can be applied as evidence for the theories of atomic structure.

EVERYDAY CHEMISTRY

Coatings that Prevent Corrosion

Corrosion is a popular term used to describe the reaction of some metals with their environment. For example, iron is an active metal and, assisted by some special properties of water, will unite with the oxygen of the air to form iron(III) oxide, or rust.

One possible means to prevent rust is to coat the iron or steel with a metal that does not react with its environment. Chromium plating trim on an automobile is an example of such a coating. Zinc plating is another method of protection. Under ordinary conditions, zinc reacts with water and carbon dioxide in air to produce a zinc carbonate coating that resists further corrosion. Building nails, for example, may be galvanized (coated with zinc) to prevent rusting before and after building is completed.

There are some forms of corrosion, like zinc carbonate, that are beneficial. For example, aluminum, when exposed to the air, forms a coating of aluminum oxide that prevents further corrosion from taking place. The patina on statues and sculpture and the green coating on many bronze or brass fixtures is $Cu_2CO_3(OH)_2$, also a protective covering.

Exploring Further

1. Investigate the processes by which chromium and zinc are electroplated on iron products. How do these processes differ?
2. What materials have replaced galvanized iron in buckets and downspouts? Why are these materials now used?

Table 10.5

Electron Affinities (kilojoules per mole)

H 72.766							**He** (−21)
Li 59.8	**Be** (−241)	**B** 23	**C** 122	**N** 0	**O** 141	**F** 328	**Ne** (−29)
Na 52.9	**Mg** (−230)	**Al** 44	**Si** 120	**P** 74	**S** 200.42	**Cl** 348.7	**Ar** (−34)
K 46.36	**Ca** (−156)	**Ga** 36	**Ge** 116	**As** 77	**Se** 194.91	**Br** 324.5	**Kr** (−39)
Rb 46.88	**Sr** (−167)	**In** 34	**Sn** 121	**Sb** 101	**Te** 190.16	**I** 295.3	**Xe** (−40)
Cs 45.5	**Ba** (−52)	**Tl** 48	**Pb** 101	**Bi** 101	**Po** (170)	**At** (270)	**Rn** (−41)

Parentheses indicate a calculated rather than an experimental value.

Figure 10.11 The reaction between aluminum, an element with a low electron affinity, and bromine, which has a high electron affinity, is shown here.

Electron Affinities

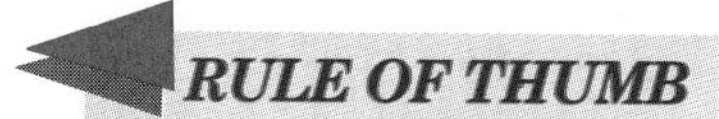

Now consider an atom's attraction for additional electrons. The attraction of an atom for an electron is called **electron affinity.** The same factors that affect ionization energy also affect electron affinity. In general, as electron affinity increases, an increase in ionization energy can be expected. *Metals have low electron affinities. Nonmetals have high electron affinities,* as shown in Table 10.5. Although not as regular as ionization energies, electron affinities still show periodic trends. Look at the column headed by hydrogen. The general trend as we go down the column is a decreasing tendency to gain electrons. We should expect this trend since the atoms farther down the column are larger. As a consequence, the nucleus is farther from the surface and attracts the outer electrons less strongly.

Look at the period beginning with lithium. The general trend as we go across is a greater attraction for electrons. The increased nuclear charge of each successive nucleus accounts for the trend. These elements with high electron affinities will tend to gain electrons and form negative ions.

How do we account for the exceptions of beryllium, nitrogen, and neon in the lithium period? The more stable an atom is, the less tendency it has to gain or lose electrons. The high negative value for beryllium is associated with the stability of the full 2*s* sublevel. If beryllium were to gain an electron, its configuration would be less stable than it previously was. Therefore, beryllium has a negative electron affinity. The failure of nitrogen to attract electrons more strongly is evidence of the stability of the half-full 2*p* sublevel. Neon has a stable, full octet of electrons in the outer level. Because this configuration is stable, neon will have no attraction for additional electrons. For a nonmetal, the greater the electron affinity, the greater the reactivity.

FRONTIERS

Metallic Hydrogen? Press On!

If you look at a periodic table of elements, you see that hydrogen is a Group 1(IA) element. Hydrogen and the metals that make up the rest of the Group 1(IA) elements share one metallic characteristic: low electron affinities. This property is just about the only metallic property hydrogen has under standard conditions. However, researchers, spurred by theoretical predictions, are searching for metallic hydrogen, a strange material that may form only at pressures millions of times greater than that of Earth's atmosphere.

Hydrogen exists as a transparent molecular gas and acts as an electrical insulator, both distinctly nonmetallic characteristics. But theoretical models predict that hydrogen will start gaining greater metallic characteristics at extremely high pressures. At pressures of about 2 million times that of atmospheric pressure, hydrogen should lose its electrical resistance and become a conductor—in other words, metallic hydrogen. At twice that pressure, about the pressure at the center of Earth, hydrogen acquires its most metallic characteristic—it becomes a superconductor, even at room temperature.

How can such a pressure be achieved? Place a small sample of hydrogen between the faces of a diamond press and squeeze it. Having done just that, researchers are now examining the first evidence that hydrogen does undergo a transition at pressures of about 2 million times that of Earth's atmospheric pressure. Hydrogen becomes opaque. The opacity indicates that the electron configuration of the entrapped hydrogen has changed, perhaps into that of metallic hydrogen. It is believed that much of the planet Jupiter is composed of this opaque metallic hydrogen.

The possible existence of metallic hydrogen is exciting news to nuclear fusion scientists. Some theoretical models predict that metallic hydrogen formed at these tremendous pressures could be stable at normal pressures. Such a material would make an ideal fuel for future fusion reactors.

10.2 CONCEPT REVIEW

7. Explain why magnesium has a first ionization energy higher than that of aluminum even though, generally, ionization energies increase from left to right in the periodic table.
8. Using Table 10.4, explain why lithium has an oxidation number of 1+ and beryllium has an oxidation number of 2+.
9. What is the meaning of a negative value for the electron affinity of an atom?
10. Why do atoms in Group 18 (VIIIA) have negative electron affinities?
11. **Apply** What can you suggest that would explain why antimony and bismuth have essentially identical electron affinities?

Chapter 10
REVIEW

Summary

10.1 Periodic Trends

1. Because properties of elements are based on electron configurations, many of these properties are predictable and repeat in periodic patterns.
2. Density is a property of elements that varies periodically. In general, densities of elements increase and then decrease as we move left to right across a period of elements.
3. Within a group of elements, the atomic radii of the atoms increase with increasing atomic number.
4. Within a period, the atomic radii of the atoms generally decrease with increasing atomic number.
5. Positive ions are smaller than the atoms from which they are produced.
6. Negative ions are larger than the atoms from which they are produced.
7. When atoms form ions they tend to take on noble gas configurations in the outer energy level.
8. Metals are found on the left side and center of the periodic table. Their atoms tend to lose electrons and thus have positive oxidation numbers.
9. Nonmetals are found on the right side of the periodic table. Their atoms tend to gain electrons and thus have negative oxidation numbers.
10. Oxidation numbers can be predicted from electron configurations by making use of the special stability of the full octet as well as the stability of full and half-full sublevels.

10.2 Reaction Tendencies

11. The relative ease with which an electron can be removed from an atom is related to the type of bond the atom is likely to form with another atom.
12. First ionization energy is the energy necessary to remove the first electron from an atom, leaving a positive ion.
13. First ionization energy is a periodic property. It tends to increase from left to right across a period and decrease from top to bottom through a group.
14. Metals are characterized by low ionization energy, while nonmetals generally have high ionization energy.
15. The factors that tend to lower ionization energy as we move down through a group of elements are increased distance of electrons from the nucleus and increased shielding effect from inner electrons.
16. Electron affinity is the attraction of an atom for an additional electron.
17. Metals have low electron affinities. Nonmetals have high electron affinities.
18. Atoms with filled or half-filled sublevels in their outer energy level tend to have lower electron affinities than neighboring atoms in a period.

Key Terms

atomic radius
noble gas configuration
ionic radius
ionization energy
first ionization energy
shielding effect
electron affinity

Review and Practice

12. What determines the characteristic properties of elements?
13. Why do some properties of elements repeat in periodic patterns?
14. How do atomic radii change from the top to the bottom of a column in the periodic table?

15. How do atomic radii change from left to right across a horizontal row of the periodic table? What is the main reason for this pattern?
16. In general, how is the radius of an atom related to the atom's attraction for outer-level electrons?
17. How does the size of a negative ion compare with the size of the atom from which it formed?
18. How does the size of a positive ion compare with the size of its atom?
19. From each of the following pairs of particles, select the particle that is larger in radius.
    a. V, Nb
    b. Cl^-, Br^-
    c. Rn, Fr
    d. Cs, At
    e. Cs^+, At^-
    f. Co, Co^{2+}
    g. P, P^{3-}
    h. Sn^{2+}, Sn
20. Compare metals and nonmetals in terms of (a) position on the periodic table and (b) the way in which they form ions.
21. How do ions of metallic elements compare in size with the atoms from which they are derived? Give reasons for the difference.
22. How do ions of nonmetallic elements compare in size with the atoms from which they are derived? Give reasons for the difference.
23. When oxygen forms the oxide ion, O^{2-}, it achieves a noble gas configuration.
    a. What is a noble gas configuration?
    b. Why is a noble gas configuration stable?
    c. The oxide ion has the configuration of what noble gas?
24. Why can we say that density is a periodic property of elements?
25. Describe the trend in densities from left to right across a period of elements.
26. When atoms lose electrons, how is a noble gas configuration achieved? Which elements are most likely to do this?
27. Why would we predict an oxidation number of 2+ for the transition elements? Do all these elements, in fact, exhibit oxidation numbers of 2+? Give reasons for your answer.
28. Predict oxidation numbers for each of the following elements.
    a. aluminum
    b. samarium
    c. arsenic
    d. polonium
    e. mercury
    f. titanium
    g. barium
    h. rubidium
    i. silver
    j. nickel
29. What is ionization energy?
30. Describe the trend in first ionization energy moving from left to right across a period of elements in the periodic table.
31. Which atom in each of the following pairs of atoms would have the lower first ionization energy?
    a. Al, B
    b. B, Tl
    c. F, N
    d. Mg, Na
    e. K, Ca
    f. Br, Cl
32. Compare metals and nonmetals in terms of first ionization energies.
33. How do first ionization energies change from top to bottom in a group of elements? List factors that influence this change and tell how each factor contributes to the change in ionization energy.
34. Characterize metals and nonmetals in terms of electron affinities.
35. The following are the endings of the electron configurations for several elements. For each, predict the possible oxidation numbers.
    a. $3s^23p^3$
    b. $4s^23d^2$
    c. $5s^24d^{10}5p^1$
    d. $6s^24f^{14}5d^{10}6p^2$
36. Write the electron configurations for tin and lead. Using these configurations, explain why tin and lead show oxidation numbers of both 2+ and 4+.
37. For each of the following pairs, which ion would you expect to be larger? Give reasons for your answers.
    a. As^{3+} and As^{3-}
    b. Tl^{3+} and Tl^+

38. What factors account for the small peaks and dips in ionization energy values moving from left to right across a period of elements? Use elements from the third period as examples in your answer.

39. Where on the periodic table are elements with the lowest densities? Which of the metal groups has the lowest densities?

40. Predict possible oxidation numbers for the following elements; state your reasoning. Check the actual oxidation numbers in Appendix Table A-3 and explain why your predictions may have deviated from those values.
   a. argon
   b. europium
   c. gallium
   d. uranium
   e. silicon
   f. antimony
   g. bromine
   h. cadmium
   i. cerium
   j. cobalt

41. Show how the octet rule would lead us to predict the formula Sr_3N_2 for the ionic compound strontium nitride. Use orbital notation to illustrate your answer.

Concept Mastery

42. What factors account for multiple oxidation numbers of a transition element? What factors tend to restrict the oxidation numbers an element can exhibit?

43. Using the data in Table 10.4, explain why the ionization energy changes so much for the third beryllium electron and for the fourth boron electron.

44. Explain the differences in the six ionization energies of carbon. See Table 10.4.

45. The electron affinities of magnesium and zinc are both negative. What structural features do their atoms possess that would lead to values less than zero for their electron affinities?

46. Both rubidium and silver have one electron in the outermost energy level (5*s*). Why do rubidium and silver differ in chemical properties?

47. The first ionization energies of hydrogen, helium, and lithium are 1312.0 kJ/mol, 2372.3 kJ/mol, and 520.2 kJ/mol, respectively. How do these first ionization values relate to the chemical properties of these three elements?

48. Why does Table 10.1 not show ionic radii for the noble gases?

49. Periodicity is not limited to a study of the elements. Many natural occurrences are periodic. Explain the concept of periodicity, using each of the following natural phenomena as examples.
   a. ocean tides
   b. the revolution of the moon around Earth
   c. foliage on a maple tree

50. Magnesium and calcium both form ionic compounds that are present in hard water.
   a. Draw an orbital notation diagram for both atoms and ions of magnesium and calcium.
   b. Compare the atomic and ionic radii for both elements.
   c. How are the principal quantum numbers of the outer electrons related to the atomic and ionic radii?

Application

51. In 1885, a rare metallic material, didymium, was found to consist of two metallic elements, praseodymium ($Z = 59$) and neodymium ($Z = 60$). Which of these elements would you expect to have a higher first ionization energy?

52. Selenium is named for the moon and has properties similar to those of tellurium, which was named for Earth. Which of these elements has the greater first ionization energy?

53. Graph the ionization energy against atomic radius for the elements lithium through chlorine. What conclusions can you make from the graph?

Critical Thinking

54. How would you generalize about a trend in ionization energies diagonally across the periodic table?

55. Why is it generally easier to predict oxidation numbers for elements in the "A" groups on the periodic table than for those in the "B" groups?

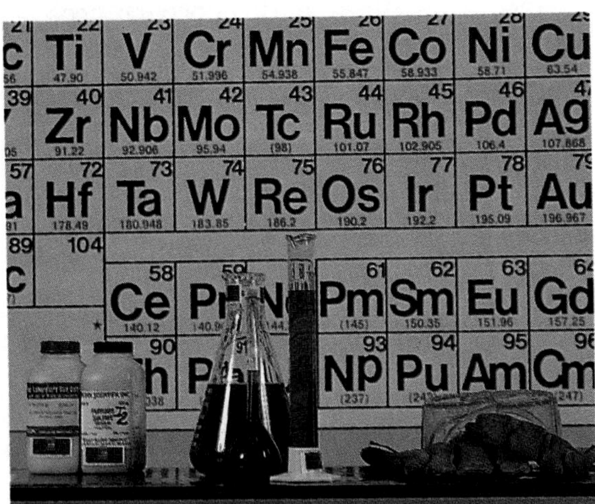

56. The elements in Groups 1 (IA) and 17 (VIIA) are found in nature only in compounds. In terms of their electron affinities and ionization energies, explain why this fact is true.

57. Considering only the first five periods, which two elements would you predict to combine with one another most readily? In your answer, take into account ionization energies and electron affinities of those elements.

58. Discuss two ways in which ionization energies provide evidence for the modern model of the atom. Consider trends both in first ionization energies and in the successive ionization energies.

Readings

Lloyd, D. R. "On the Lanthanide and 'Scandanide' Contraction." *Journal of Chemical Education* 63, no. 6 (June 1986): 502-503.

Mason, Joan. "Periodic Contractions Among the Elements." *Journal of Chemical Education* 65, no. 1 (January 1988): 17-20.

Mirone, Paolo. "How to Get More Ionization Energies in the Teaching of Atomic Structure. "*Journal of Chemical Education* 68, no. 2 (February 1991): 132-133.

Myers, R. Thomas. "The Periodicity of Electron Affinity." *Journal of Chemical Education* 67, no. 4 (April 1990): 307-308.

Rich, Ronald L., and Robert W. Suter. "Periodicity and Some Graphical Insights on the Tendency Toward Empty, Half-full, and Full Subshells." *Journal of Chemical Education* 65, no. 8 (August 1988): 702-704.

Strong, Judith A. "The Periodic Table and Electron Configurations." *Journal of Chemical Education* 63, no. 10 (October 1986): 834-846.

Thomsen, Dietrick E. "Partners for a Noble Element." *Science News* 132, no. 21 (November 21, 1987): 334.

Cumulative Review

1. Balance the following equations.

a. $Na + H_2O \rightarrow NaOH + H_2$

b. $Mg + HCl \rightarrow MgCl_2 + H_2$

c. $Sb + Cl_2 \rightarrow SbCl_3$

d. $Cl_2 + KBr \rightarrow KCl + Br_2$

e. $Na_2O_2 + H_2O \rightarrow NaOH + O_2$

2. According to the following equation, how many grams of $NaMnO_4$ can be prepared from 1.27 grams of $NaBiO_3$? What mass of $NaNO_3$ will be produced?

$2Mn(NO_3)_2 + 5NaBiO_3 + 14HNO_3 \rightarrow$
$2NaMnO_4 + 5Bi(NO_3)_3 + 3NaNO_3 + 7H_2O$

3. Balance the following equation.
$SnCl_2 + I_2 + HCl \rightarrow SnCl_4 + HI$
   a. Chemists do not balance equations because of rules. They balance an equation so that it will reflect the reality of what takes place during a chemical change. What information does the balanced version of the equation convey that the unbalanced equation does not?
   b. State the law that provides the basis for balancing chemical equations.
4. A solution of silver nitrate and a solution of potassium chloride are reacted according to the equation shown.
$AgNO_3(aq) + KCl(aq) \rightarrow AgCl(cr) + KNO_3(aq)$
   a. In what state is the AgCl produced?
   b. How many moles of KCl are present in 22.8 cm^3 of a 3.50*M* solution?
   c. Using your answer to part b, how many moles of $AgNO_3$ would react with that amount of KCl?
   d. If the amount of $AgNO_3$ you found in part c were dissolved in enough water to form 50.0 cm^3 of solution, what would be the molarity of the silver nitrate solution?
   e. Using the procedure described in this problem, suggest a method for determining the concentration (molarity) of substances in solution.
5. Sulfuric acid is the most widely used of all industrial chemicals. In its manufacture by the contact process, the last reaction dissolves $H_2S_2O_7$ in water to produce H_2SO_4. How much H_2SO_4 can be produced from 105.7 kg of $H_2S_2O_7$?
6. A gas camping stove is equipped with a tank containing 844 grams of propane, C_3H_8. What mass of oxygen is consumed in the complete combustion of this quantity of propane? How much carbon dioxide is given off?
7. Magnesium oxide reacts with water in the following reaction.

$$MgO + H_2O \rightarrow Mg(OH)_2$$

   What mass of magnesium hydroxide is obtained from 84.1 grams of magnesium oxide?
8. Two moles of H_2 and two moles of O_2 are in a closed container. A spark is introduced to cause H_2 and O_2 to unite to form water. Describe quantitatively what is in the container after the spark.
9. Acetylene, C_2H_2, is produced from the action of water on calcium carbide, CaC_2. A by-product is calcium hydroxide. How much acetylene can be formed from 128 g of calcium carbide?
10. Balance the following equation.

$$Cr(NO_3)_3 + NaOH \rightarrow Cr(OH)_3 + NaNO_3$$

11. What mass of silver oxide will be produced by the reaction of 2.37 g $AgNO_3$ according to the following equation?

$$2AgNO_3 + 2NaOH \rightarrow Ag_2O + 2NaNO_3 + H_2O$$

CHAPTER **Preview**

CONTENTS

CHAPTER 11

Typical Elements

We take the functioning of our bodies for granted when we perform physical feats like that shown opposite. The artificial hip shown in this X ray is one of hundreds of artificial body parts given to surgeons by modern technology that helps people perform everyday physical activities. The artificial hip is made of a titanium alloy that is compatible with living tissue. Could we use an element from Group 1 or Group 16 for this implant?

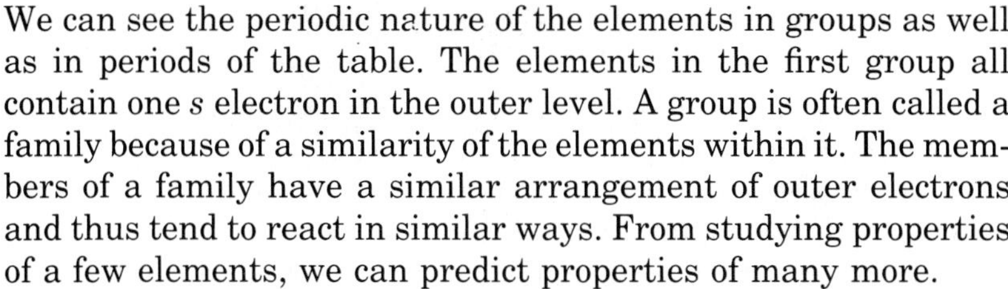

We can see the periodic nature of the elements in groups as well as in periods of the table. The elements in the first group all contain one *s* electron in the outer level. A group is often called a family because of a similarity of the elements within it. The members of a family have a similar arrangement of outer electrons and thus tend to react in similar ways. From studying properties of a few elements, we can predict properties of many more.

EXTENSIONS AND APPLICATIONS

OBJECTIVES

- define a family or group and explain what members of a chemical family have in common.
- list four ways in which hydrogen can bond and give an example of each.
- define the shielding effect and explain its importance to reactivity of atoms.
- list characteristics and give uses for representative elements in the alkali metal, alkaline earth metal, and aluminum groups.

11.1 Hydrogen and Main Group Metals

Groups of elements whose electron configurations end in s or p are classed as the main group elements. Their outer electrons are also their most energetic electrons. Those elements ending in s^1, s^2, or s^2p^1 tend to lose electrons and are therefore classified as metals. We don't think of hydrogen, a colorless gas, as a metal, but because it ends in s^1, we will still consider it in this section. Similarly, boron ends in s^2p^1 but reacts solely by sharing electrons, yet we will include it with the rest of Group 13.

Hydrogen

Hydrogen and the alkali metals have only one outer electron. Under the proper conditions, they can lose this electron and form positive ions. Unlike alkali metals, however, hydrogen can also gain or share electrons. Because of its unique properties, hydrogen is usually considered as a family by itself.

There are four ways in which a hydrogen atom can react. The first way a hydrogen atom can react is to lose its one electron to become a positive hydrogen ion. A **hydrogen ion,** H^+, is simply a bare proton. Remember that a proton is about one trillionth the size of an atom. Because of its small size, the hydrogen ion cannot exist as an independent species in the everyday world. This characteristic is important in understanding the behavior of acids, which will be considered later in this text.

The second way a hydrogen atom can react is by sharing its single outer electron. Recall that hydrogen gas is composed of diatomic molecules. Each molecule contains two atoms of hydrogen that share electrons with each other. Most nonmetals react with hydrogen to form compounds involving shared electrons. Examples of these compounds are HCl and H_2O. These compounds and their formation will be studied later in the text.

Figure 11.1 Iron displaces hydrogen from hydrochloric acid (left). The electrolysis of water produces both hydrogen and oxygen (right).

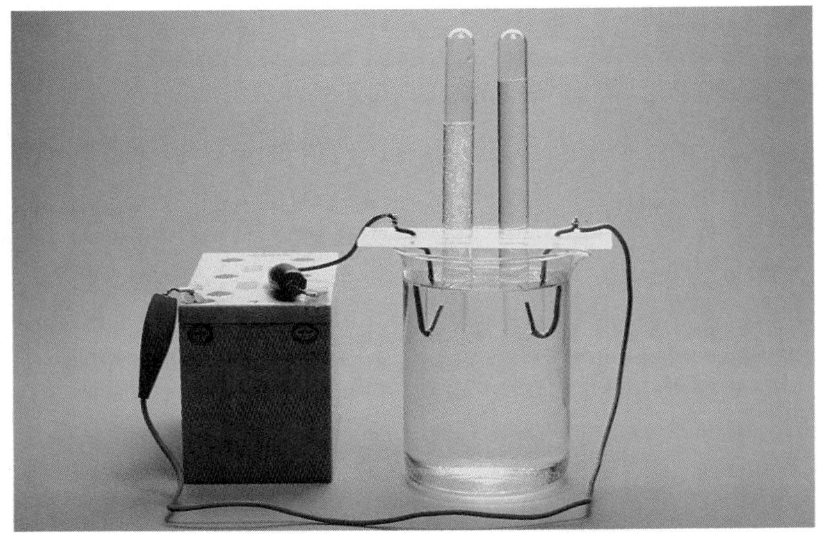

The third way hydrogen can react is to gain an electron. When this change occurs, the atom becomes a **hydride ion,** H^-. Such a reaction can take place only between hydrogen and atoms of elements that give up electrons easily. The most reactive metals, Groups 1 (IA) and 2 (IIA), form ionic hydrides. In these compounds, the radius of the hydride ion averages about 208 pm. In Table 10.1 on page 252, notice that the fluoride ionic radius is 119 pm. Thus, the hydride ion is larger than the fluoride ion. Such a large radius indicates that the single proton of the hydride ion has a very weak hold on the two electrons. We would expect, then, that the ionic hydrides would not be highly stable. Research confirms this conclusion. Ionic hydrides are found to be quite reactive compounds.

A fourth type of bonding involves the formation of bridges between two atoms by hydrogen atoms. The best examples of these compounds are found with the element boron and some of the transition metals. A study of such compounds is beyond the scope of this book. Because they are not common compounds, their behavior is a small fraction of the chemistry of hydrogen.

By sharing or gaining an electron, hydrogen attains the stable outer level configuration of helium. Hydrogen gas is used in the manufacture of other chemicals such as ammonia and methanol. One interesting use is the conversion of a vegetable oil such as corn oil into solid shortening, or margarine. The reaction that takes place involves the addition of hydrogen atoms to double bonds in the oil molecules, processes that will be described later in this textbook.

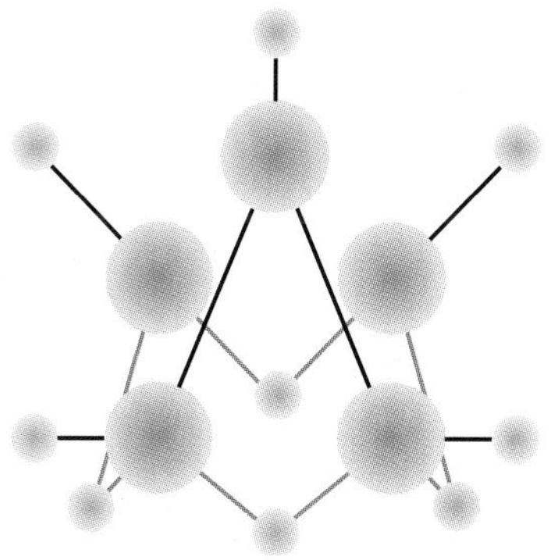

Figure 11.2 Hydrogen acts as a bridge between boron atoms. In this B_5H_9 molecule, there are four bridging bonds around the base and five "regular" B—H bonds.

Alkali Metals

The metals in Group 1 (IA) are reactive. If one member of a family forms a compound with an element or ion, we can predict that the other members will do the same. However, the members of a family are not the same in every way. The outer electron of lithium occupies a volume closer to the nucleus than the outer electron of sodium. The sodium atom has many more electrons between the outer electron and the nucleus than the lithium atom has. An increasing number of electrons between the outer level and the nucleus has a shielding effect. This effect blocks the attraction of the nucleus for the outer electrons. Larger atoms tend to lose their outer electrons more readily. This tendency is due to the increased distance of the electrons from the nucleus as well as the shielding effect.

As the atoms of the alkali metals increase in size, the nucleus increases in positive charge. However, the force of the increased positive charge is more than offset by the outer electron's distance from the nucleus and the shielding effect. As a result, we find that the alkali metals lose their outer electron more readily as we proceed down the group. This trend indicates that the most active metal would be francium in the lower left corner of the periodic

Figure 11.3 The alkali metal atoms increase in size with increasing atomic number. The peaks in the graph show that atomic radius is a periodic property.

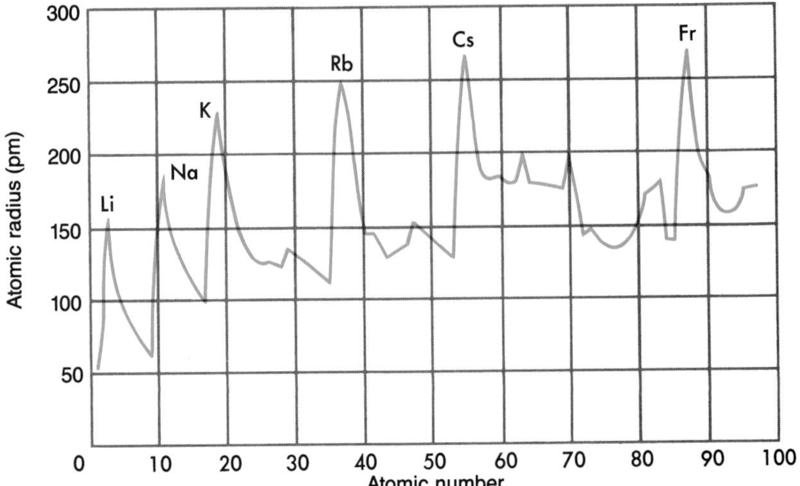

table. All of the elements in this group are active enough to displace hydrogen from its compounds. For example, when potassium metal reacts with water, potassium hydroxide and hydrogen gas are formed.

Sodium compounds are among the most important in the chemical industry. Millions of tons of sodium hydroxide are used each year in producing other chemicals, paper, and petroleum products. Sodium carbonate is also produced in millions of tons and used in manufacturing glass and other chemicals. Sodium sulfate is another substance used in manufacturing glass as well as paper and detergents. Sodium silicate is widely used in making soaps, detergents, paper, and pigments in addition to its use as a catalyst. A **catalyst** is a substance that speeds up a reaction. Sodium tripolyphosphate ($Na_5P_3O_{10}$) is used as a food additive, in softening water, and in making detergents.

Sodium compounds are important to the human body. They supply sodium ions, which are essential to the transmission of nerve impulses. Sodium ions and potassium ions are found in

Figure 11.4 Sodium carbonate is used to control large and small chemical spills involving acids.

different concentrations in different parts of the body. Table 11.1 shows some of the concentrations of these ions in the body. Note that the relative amounts of these ions vary throughout the body. Sodium and potassium are necessary elements in the diet. Their compounds are found in many foods. One sodium compound, sodium hydrogen carbonate, is used in baked goods. This compound is commonly known as baking soda.

Table 11.1

Distribution of Sodium and Potassium Ions in the Body (mg/100 g)

	K^+	Na^+
Whole blood	200	160
Plasma	20	330
Cells	440	85
Muscle tissue	250-400	6-160
Nerve tissue	530	312

There are two other 1+ ions that, because of their size, behave in a fashion similar to the alkali metal ions. These ions are the ammonium ion (NH_4^+) and the thallium(I) ion (Tl^+). Their compounds follow much the same patterns as the alkali metal compounds.

Lithium

The reactions of the alkali metals involve mainly the formation of 1+ ions. In general, the reactivity increases with increasing atomic number, with one exception. Lithium reacts more vigorously with nitrogen than any other alkali metal. Lithium is exceptional in other ways, too. Its ion has the same charge as the other alkali metals, but the unusual behavior is due to its smaller size. The ratio of charge to radius is often a good indication of the behavior of an ion. The charge/radius ratio of lithium more closely resembles the magnesium ion (Mg^{2+}) of Group 2 (IIA). In its behavior, the lithium ion (Li^+) resembles Mg^{2+} more closely than it resembles sodium (Na^+), the next member of its own family. This diagonal relationship is not unusual among the lighter elements. One example of this relationship is that lithium burns in air to form the oxide, Li_2O, as does magnesium to form MgO. The other alkali metals burn in oxygen to form the peroxide, M_2O_2, or the superoxide, MO_2 (where M = Na, K, Rb, or Cs). Another example of this relationship concerns the solubilities of compounds. The solubility of lithium compounds is similar to that of magnesium compounds, but not to that of sodium compounds.

Figure 11.5 Lithium metal is used to make batteries. The battery shown here is used in heart pacemakers.

The lithium atom also differs from the other alkali metal atoms in some physical properties. Unlike the other alkali metals that dissolve in each other in any proportion, lithium is insoluble in all but sodium. It will dissolve in sodium only above 380°C. In other respects, lithium metal is like the other members of its family. For example, it is a soft, silvery metal with a low melting

Figure 11.6 Sodium vapor lights are used on highways because they can be seen through fog.

point, as are the other alkali metals. All of the alkali metals will dissolve in liquid ammonia to give faintly blue solutions. These solutions conduct electricity.

The alkali metals form binary compounds with almost all nonmetals. In these compounds, nonmetals are in the form of negative ions. In solution, the lithium ion, because of its high charge/radius ratio, attracts water molecules more strongly than any other alkali metal ion.

Alkaline Earth Metals

The alkaline earth metals, Group 2 (IIA), are quite similar to the alkali metals except that they form the 2+ ion. Also, the alkaline earth metals are not as reactive as the alkali metals. Most alkaline earth metal compounds are ionic. These compounds are soluble in water, except for some hydroxides, carbonates, and sulfates.

Beryllium is used in making nonsparking tools, and magnesium is widely used in lightweight alloys. The other metals are too reactive to be used as free elements. The metals Ca, Sr, Ba, and Ra will displace hydrogen from water and other compounds.

Two calcium compounds find large markets. Lime (calcium oxide) is used to make steel, cement, and heat-resistant bricks. It is also applied to soils that are too acidic to farm without treatment. Some lakes that have become too acidic to support aquatic life (due to acid rain) have been treated with lime. Calcium chloride is used in a wide range of applications. One interesting use is in controlling road conditions through deicing in the winter and keeping down dust in the summer. It is also used in the paper and pulp industry.

Figure 11.7 Lime can be added to lakes to counteract the acid conditions caused by acid rain. Very large quantities of lime must be added to a lake of this size.

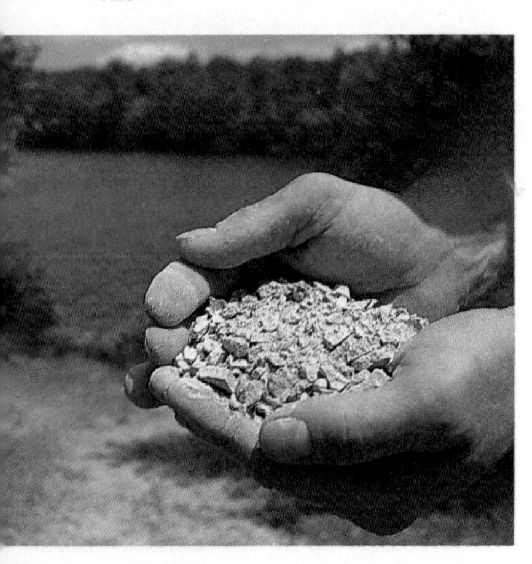

Both calcium and magnesium ions are important biologically. The amount of calcium ion present in various tissues affects a very large number of biochemical processes. One of the most important of these processes is the coagulation of blood. There are at least 18 substances involved in the clotting of blood, and these substances undergo at least 13 different reactions. We say "at least" because the process is not yet entirely understood by biochemists, and we count only those substances and reactions known to be involved. Calcium ions are definitely known to be reactants in four of the reactions, and it is strongly suspected that they are required for a fifth reaction. In addition to their role in blood clotting, calcium ions are a major constituent of bone. They also have a significant effect upon the release and absorption of chemicals, called hormones, used by the body as "messengers" from one organ to another.

Bridge to *BOTANY*

Magnesium in Plants

Plants produce their own food. In order to do so, they must have the necessary reactants and energy from sunlight. In addition, plants must contain chlorophyll, a complex compound that can capture the sun's energy and make it available to biochemical reactions. When plants make food, they are producing material for their own energy needs and for growth. Many plants are eaten by humans. In this process, we obtain the energy that was captured by the plant during photosynthesis. Humans also eat animals that eat plants. In this process, we are still obtaining energy from plants, although we do so indirectly.

At the center of the chlorophyll molecule is a magnesium ion. Thus, magnesium is an essential mineral for plants. If there is insufficient magnesium in the soil, plants cannot produce chlorophyll, and so they do not grow properly. Because chlorophyll is green, the lack of this compound can be detected visually. Plants that do not produce enough chlorophyll look pale or yellow. If the deficiency becomes severe, the plant may die because it cannot produce enough food to meet its energy requirements.

Exploring Further

1. From a drawing of the chlorophyll molecule on page 242, determine to what element(s) magnesium is attached in the chlorophyll molecule.
2. Use library materials to learn what properties of magnesium account for its being a part of a chlorophyll molecule.

Figure 11.8 The outer skin of airplanes is made of aluminum because this metal is light, strong, and flexible.

MINUTE LAB

Airplane Skin

The Federal Aviation Administration restricts the use of corrosive materials near the aluminum skins of airplanes. Check the strength of aluminum by trying to tear an empty aluminum soft drink can into two pieces. Grip the ends and twist. **CAUTION:** *Metal edges are sharp. Wear gloves, an apron, and goggles.* Using an L-shaped heavy wire, scratch the thin plastic coating on the inside of the can near the center. Place the can in an empty 600-cm^3 beaker. Add 6*M* HCl to the can so that the level is just above the scratch. **CAUTION:** *Acid is corrosive.* After a few minutes, the acid will have reacted with the can where the plastic liner was scratched. A thin black line will appear on the paint. Empty the acid into the beaker. Rinse the can with water to remove all traces of acid. Wearing gloves, try again to tear the can into two pieces. Explain your observations.

The Aluminum Group

Aluminum is the most plentiful metal in Earth's crust. Of the elements in Group 13 (IIIA), aluminum has the most practical uses. With three electrons in its outer level, aluminum is less metallic than the elements of Groups 1 (IA) and 2 (IIA). In forming compounds, it tends to share electrons rather than form ions. It is also less reactive than the metals of Groups 1 (IA) and 2 (IIA) metals. One aluminum compound, aluminum sulfate, is used in water purification, paper manufacture, and fabric dyeing. Large amounts of elemental aluminum are used to produce lightweight alloys for many items from soft drink cans to spacecraft.

When we think of electrical conductors, or wires, most of us think of copper. However, much long-distance wire is now made of aluminum. Although copper is a 50% better conductor than aluminum, it is also three times as dense as aluminum. As a result, an aluminum wire half the weight of a copper wire can conduct the same current. Many long-distance transmission lines from power plants are made of aluminum. Because aluminum's conductivity is dependent on its purity, only high quality aluminum can be used as a conductor. However, pure aluminum is a fairly soft metal. Consequently, long, thick aluminum wires would tend to sag under the force of their own weight. Engineers overcome this problem by wrapping the aluminum wire around a steel support cable. The combined weight and cost is still low enough to make aluminum wire competitive with copper.

11.1 CONCEPT REVIEW

1. Why do elements of a group have similar chemical properties?
2. List two of the four ways that hydrogen can form bonds. Give an example of each.
3. What effect does the shielding effect have on the stability of an alkali metal atom?
4. What substances are produced when the alkali metals react with water?
5. **Apply** Why is aluminum considered to be less metallic than the elements in Groups 1 and 2?

11.2 Nonmetals

The elements in Groups 14 through 18 tend to share or gain electrons when they react. There are a few exceptions. Tin, lead, and bismuth tend to lose electrons because their outer electrons are so far from the nucleus and are so well shielded. As far as chemists are aware, helium, neon, and argon form no compounds. Arsenic, antimony, silicon, and germanium have some metallic and some nonmetallic properties. In this section we will take a closer look at some of these elements.

OBJECTIVES

- define catenation and explain how it affects the ability of carbon to form compounds.
- explain the importance of nitrogen and phosphorus compounds to living things.
- state the relationship between the activities of elements and their locations in the periodic table.
- list characteristics and give uses for representative main group nonmetals.

Group 14 (IVA)

Group 14 (IVA) is the carbon group. The elements of Group 14 (IVA) have atoms with four electrons in the outer level. These elements generally react by sharing electrons. However, the tendency to lose electrons increases as the atomic number of Group 14 (IVA) elements increases. There are a few compounds in which carbon in the form of a **carbide ion** (C^{4-}) exists. Silicon and germanium, the next members of this family, are metalloids. They do not form 4− ions under any conditions. However, there are compounds in which silicon and germanium exist as a 4+ ion.

You know from Chapter 3 that the major part of carbon chemistry is classed as organic chemistry. In most **organic compounds,** electrons are shared between a carbon atom and one or more other carbon atoms. This characteristic causes the formation of long, chainlike molecules. The tendency to form chains and rings of similar atoms is called **catenation** (kat uh NAY shuhn). Of all the elements, only carbon exhibits catenation to any great extent. In general, those compounds that do not contain carbon are called **inorganic compounds.** There are several exceptions to this rule. A few carbon-containing substances that are considered inorganic are carbon itself, carbonic acid and its salts, carbides, cyanides, and the oxides and sulfides of carbon.

catena: (L) chain
catenation—chain-forming

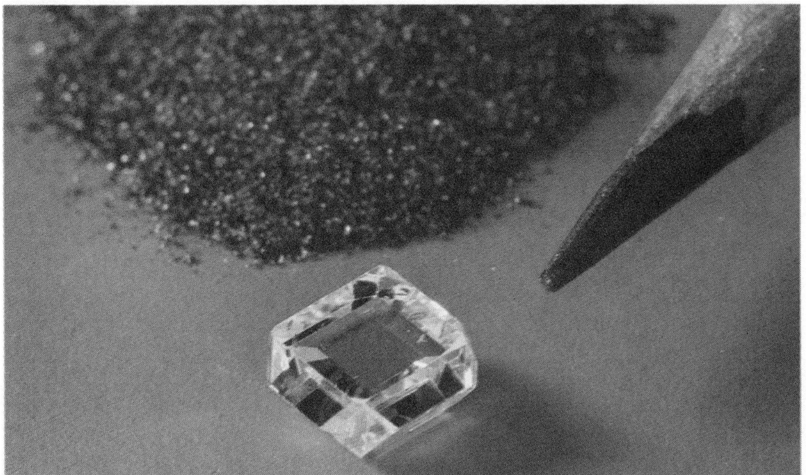

Figure 11.9 Diamond and graphite are two allotropes of carbon. Graphite is one component of pencil lead.

Word Origins

allos: (GK) another
tropos: (GK) turn, or way

allotrope—found in another way (or form)

Elemental carbon is found in nature in two different molecular forms, diamond and graphite. Different forms of the same element are called **allotropes.** In diamond, each carbon atom shares electrons with the four nearest carbon atoms. In graphite, the sharing is to the nearest three carbon atoms. The difference in structure results in different properties for these two forms of carbon, as will be described in Chapter 16.

Industrially, carbon is used in a form called "carbon black" made by the incomplete burning of natural gas or other fuel. Carbon black is often referred to as soot when it has accumulated in a place where it is unwanted. It is actually a microcrystalline form of graphite, and is used as a black pigment and wear-resistant additive in rubber tires. Carbon dioxide gas is a by-product of several chemical processes. It is collected, compressed, and sold as a liquid in steel cylinders. Customers use it for refrigeration, carbonating beverages, and producing other chemicals. Solid carbon dioxide is called "dry ice." Small crystals of dry ice are sometimes used in cloud-seeding to induce rain.

FRONTIERS

Welding Gems

At extreme pressures and temperatures deep within Earth, carbon atoms slowly crystallize into diamonds. To replicate this geologic process, scientists use large hydraulic presses to make synthetic diamonds from graphite. Under pressures 10 000 times that of the atmosphere and at a temperature of 2000°C, the graphite slowly changes structure and forms small diamond crystals. Similar methods are now standard for manufacturing synthetic diamonds for industrial use. However, the costly manufacturing process makes the diamonds expensive. Now scientists are looking within the flame of a welder's torch and not the depths of Earth for a new and cheaper way of producing diamonds.

Researchers have found microscopic crystals of diamonds in the by-products formed from burning ethyne. Ethyne, commonly called acetylene, is an organic compound used to fuel oxyacetylene welding torches. To produce the diamonds, an oxyacetylene torch is used to heat the surface of a molybdenum disk. The disk is cooled from beneath to keep it from melting. After a few minutes of heating at normal atmospheric pressure, microscopic diamond crystals form on the surface of the disk. The micrograph shows similar crystals magnified 50 000 times.

How the crystals form is still not understood. Researchers guess that a series of reactions take place between free carbon atoms and hydrocarbons in the gases produced by the torch. The reactions produce ringlike molecules of carbon vapor. As the reactions continue, the rings grow until they become unstable. Then they break apart into more stable structures, namely, diamond crystals. The crystals that are deposited on the disk continue to grow as they capture carbon atoms from the hydrocarbons in the surrounding gases.

Figure 11.10 The soot produced by burning wood in a fireplace must be removed periodically by a chimney sweep because it can become a fire hazard.

As you can see in Table 11.2, silicon is the second most plentiful element in Earth's crust after oxygen. Silicon is found in a large number of minerals such as quartz. Silicon shows only a slight tendency toward catenation, much less than carbon. The chemistry of silicon, like that of carbon, is characterized by electron sharing. In compounds called **silicates,** silicon is bound to oxygen, and each silicon atom is surrounded by four oxygen atoms.

Figure 11.11 Carbon compounds exhibit catenation, as shown by the structure of oleic acid (top). Silicon also exhibits this property, as shown by the structure of hexasilane (bottom).

$$CH_3-CH_2-CH_2-CH_2-CH_2-CH_2-CH_2-CH_2-CH=CH-CH_2-CH_2-CH_2-CH_2-CH_2-CH_2-CH_2-C(=O)-OH$$

Oleic Acid

$$H-SiH_2-SiH_2-SiH_2-SiH_2-SiH_2-SiH_2-H$$

Hexasilane

Silicon has many uses. Because it is a semiconductor, silicon is used in transistors, solar cells, and computer chips. Silicones are compounds of silicon. Some uses of silicones are in synthetic motor oils and in adhesives and automobile gaskets.

Tin and lead are distinctly metallic members of Group 14 (IVA). They are quite similar to each other except that for tin the 4+ state is more stable than the 2+, while the reverse is true for lead. These two metals are easily refined from their ores and have both been known since prehistoric times. In general, their uses are based on their lack of chemical reactivity. Tin and lead are common components of alloys, which are mixtures of metals. Examples of alloys are solder, which contains lead and tin, and bronze, which contains copper and tin.

Table 11.2

Abundances of Elements in Earth's Crust

Element	Percentage
Oxygen	45.5
Silicon	27.2
Aluminum	8.1
Iron	5.8
Calcium	4.66
Magnesium	2.76
Sodium	2.27
Potassium	1.84
Titanium	0.63
Hydrogen	0.152
Phosphorus	0.11
Manganese	0.1

HISTORY CONNECTION

Ancient Metallurgy in Peru

Copper metallurgy was prevalent on the northern Peruvian coast as early as 500 B.C. Some burial grounds contain copper and gold-plated copper objects. In the early centuries A.D., the Mochicas made the first copper-arsenic alloys, although they were not produced in quantity until about 900 A.D. The Incas about 1450 A.D. attempted to use copper-tin bronzes as the state metal, but they allowed continued production of the copper-arsenic alloys. The invasion of Peru by Pizzaro in the middle 1500s led to the introduction of European metallurgy and wiped out the Peruvian metallurgical tradition.

The Nitrogen and Phosphorus Group

Nitrogen and phosphorus are found in Group 15 (VA). Both have five outer electrons, which can be shared to form compounds. However, nitrogen and phosphorus differ considerably for adjacent members of the same family. Nitrogen occurs in all oxidation states ranging from 3− through 5+; phosphorus shows only 3−, 0, 3+, and 5+. Most nitrogen compounds are relatively unstable, tending to decompose to N_2. Conventional high explosives, for example, trinitrotoluene (TNT) and dynamite, utilize nitrogen compounds. Elemental nitrogen, N_2 gas, is one of the most stable substances known. It makes up 78 percent of Earth's atmosphere. The unreactive gas is used to surround reactive materials that would otherwise react with oxygen in the air. Liquid nitrogen is used to maintain very low temperatures.

Nitrogen compounds are produced from atmospheric nitrogen by nitrogen-fixing bacteria. These bacteria naturally convert molecular nitrogen to nitrogen compounds that can be used readily by plants. The plants use the nitrogen compounds to make amino acids, the essential components of proteins.

Synthetic nitrogen compounds are usually produced from atmospheric nitrogen. Huge quantities of liquid N_2, obtained from the air, are used in the Haber Process to produce ammonia, as discussed in Chapter 22. The most common use of ammonia is as a fertilizer. Much of the ammonia not used directly as a fertilizer is converted to other nitrogen compounds that are themselves fertilizers. An example is ammonium nitrate, which is also used in explosives. Some ammonia is converted to nitric acid. Nitric acid is widely used in the manufacture of fertilizer and explosives. Large quantities of ammonium sulfate are obtained by the steel industry's conversion of coal to coke. The $(NH_4)_2SO_4$ by-product is then used as a fertilizer.

Figure 11.12 Nitrogen-containing explosives such as dynamite (left) are used to demolish old buildings (right).

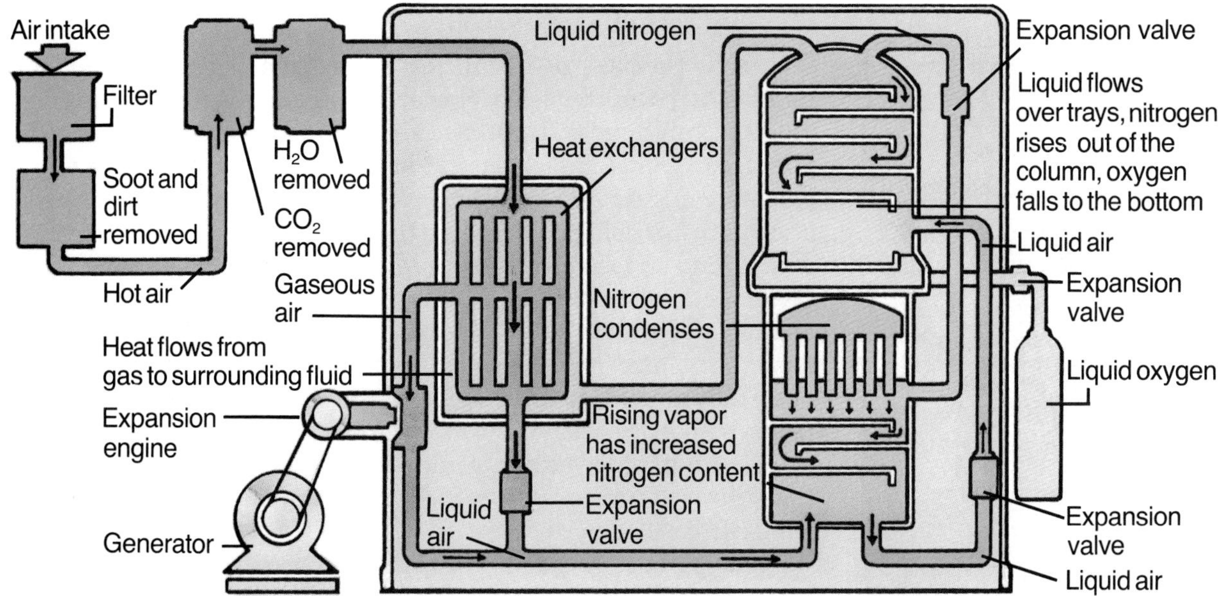

Figure 11.13 Nitrogen and other gases are obtained from liquid air. Air is liquefied through a complex process similar to that shown here.

Elemental phosphorus occurs as P_4 molecules and is solid at room temperature. The P_4 molecules "stack" in different ways to form several allotropes. One allotrope, white phosphorus, is so reactive that it ignites spontaneously on contacting air. Red phosphorus must be exposed to a flame to ignite. Black phosphorus, a third allotrope, is semiconducting. The principal source for phosphorus in nature is phosphate rock, $Ca_3(PO_4)_2$. Most phosphate rock is used in producing $Ca(H_2PO_4)_2$ and $CaHPO_4$ for use as fertilizer. Some phosphate rock is converted to phosphoric acid, which is used primarily in making fertilizer, but has other applications in industry.

Organic compounds containing both nitrogen and phosphorus are vital to living organisms. Utilization of energy by living systems involves a compound called adenosine triphosphate (ATP). The transfer of genetic information from generation to generation involves deoxyribonucleic acid (DNA). Each DNA molecule contains hundreds of nitrogen compounds attached to phosphate groups. Ribonucleic acid (RNA), used by cells in protein synthesis, also contains nitrogen compounds and phosphate groups.

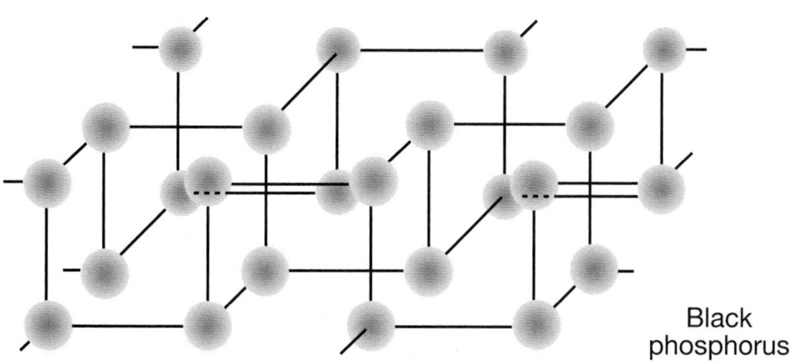

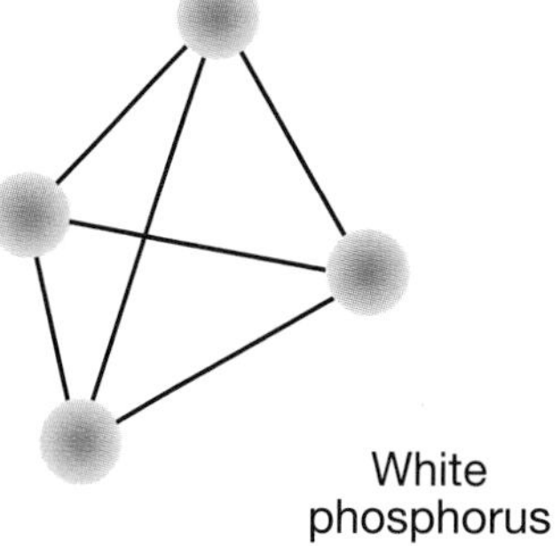

Figure 11.14 Black and white phosphorus are allotropes. Note the difference in the geometric arrangement of the phosphorus atoms.

amphoteros: (GK) both
amphoteric—can be both acid and base

Oxygen Family

Oxygen is the most plentiful element in Earth's crust and makes up about 21% of the atmosphere. Oxygen is a very reactive nonmetal, combining with all other elements except helium, neon, and argon. Since the oxygen atom has six electrons in its outer level, it can gain two electrons to achieve the octet configuration of neon. In so doing, it becomes the oxide ion, O^{2-}. With metals, which tend to lose electrons readily, oxygen forms ionic oxides. Oxygen can also react by sharing electrons with nonmetals. The behavior of oxides when dissolved in water depends on their structure. Ionic oxides generally react with water to produce basic solutions. However, oxides formed by the sharing of electrons tend to react with water to form acidic solutions. Some oxides can produce either acidic or basic solutions, depending on the other substances present. Such oxides are **amphoteric** (am foh TER ik) oxides.

Like carbon, oxygen has allotropes. Oxygen usually occurs in the form of diatomic oxygen molecules, O_2. The free oxygen you breathe from air is O_2. There is another allotrope of oxygen called ozone. Ozone is the triatomic form of oxygen, O_3, and is highly reactive. Ozone can be synthesized by subjecting O_2 to a silent electric discharge. In Europe, ozone is the main chemical used in water purification. Ozone is formed naturally in small amounts by lightning and, in the upper atmosphere, by ultraviolet radiation from the sun. It is also a major component of smog.

Pure oxygen is extracted from liquefied air, compressed, and sold in cylinders. Its largest uses are in the production of steel, artificial breathing atmospheres, rocket engines, and welding torches.

CAREERS IN CHEMISTRY

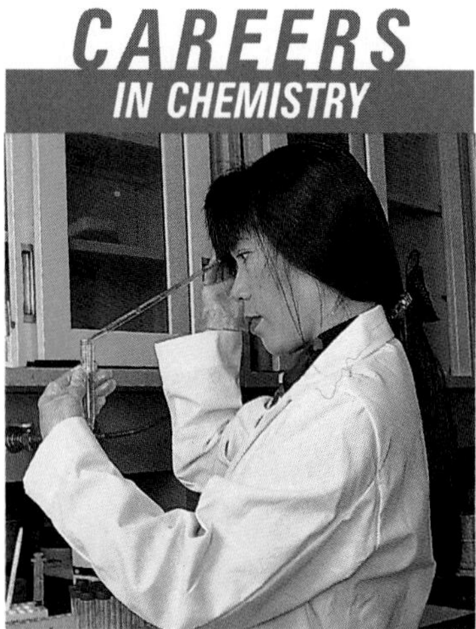

Forensic Chemist

As you know from numerous television shows, the laboratory plays a large part in crime detection and the analysis of evidence in preparation for a trial. Much of what takes place in a forensic laboratory is chemical in nature. Forensic chemists use modern instrumentation to analyze paint, blood, paper, hair, fabric, tissue, and other materials that could be used to solve a crime or as evidence in a trial.

Figure 11.15 Ozone is formed in the atmosphere by the reaction of lightning with oxygen. This ozone helps to shield Earth from dangerous radiation from the sun.

The chemistry of sulfur is similar to that of oxygen, especially in the behavior of the 2− ion. The S^{2-} ion shows characteristics similar to the oxide ion in solubility and acid-base behavior. Unlike oxygen, which forms O_3 as its longest chain, sulfur can form long chains of atoms attached to the S^{2-} ion. These ions, for example S_6^{2-}, are called polysulfide ions. Sulfur exhibits two important positive oxidation states represented by the oxides SO_2 and SO_3. When sulfur or sulfur compounds are burned in an ample supply of air, SO_2 is produced. Using a catalyst, SO_2 can be converted to SO_3. When SO_2 is dissolved in water, sulfurous acid (H_2SO_3) is produced. If SO_3 is combined with water, sulfuric acid (H_2SO_4) is produced. When these processes occur in the atmosphere, acid rain is formed. In industry, the combination of SO_3 and water is achieved by dissolving the SO_3 in H_2SO_4 and then adding water. Sulfuric acid is produced in huge quantities. Twice as much H_2SO_4 (by mass) is produced as the next most common chemical. The principal consumers of sulfuric acid in the United States are fertilizer, petroleum refining, steel, synthetic fiber production, paint, and pigment industries.

Figure 11.16 When sulfur burns in sufficient air, it forms sulfur dioxide gas. It is this gas that is responsible for the formation of acid rain.

The other members of the oxygen group are selenium, tellurium, and polonium. Selenium is a nonmetal. Tellurium and polonium are metalloids. Of these three elements, selenium is the most common.

Selenium is toxic, even in low doses. The Occupational Safety and Health Administration (OSHA) sets the permissible exposure limit at 200 μg Se/m^3. Yet, selenium is an essential part of our diet, in extremely small amounts. Red blood cells contain a vital compound that has four atoms of selenium per molecule. You will find, as you study chemistry and biology, that there are thousands of compounds with similar restrictions: a small amount of the substance is vital, but too much is toxic. The sixteenth century Swiss physician Paracelsus stated, "The dose makes the difference." When you read that a certain substance is harmful, you must know the amounts discussed.

VITAMINS	QUANTITY	% OF U.S. RDA	MINERALS	QUANTITY	% OF U.S. RDA
VITAMIN A	5000 I.U.	100	IRON (elemental)	18 mg	100
(AS BETA CAROTENE)			CALCIUM	130 mg	13
VITAMIN A (AS ACETATE)	1500 I.U.	30	PHOSPHORUS	100 mg	10
VITAMIN C	60 mg	100	IODINE	150 mcg	100
THIAMINE (B-1)	1.5 mg	100	MAGNESIUM	100 mg	25
RIBOFLAVIN (B-2)	1.7 mg	100	COPPER	2 mg	100
NIACIN	20 mg	100	ZINC	15 mg	100
VITAMIN D	400 I.U.	100	CHROMIUM	10 mcg	*
VITAMIN E	30 I.U.	100	SELENIUM	10 mcg	*
VITAMIN B-6	2 mg	100	MOLYBDENUM	10 mcg	*
FOLIC ACID	0.4 mg	100	MANGANESE	2.5 mg	*
VITAMIN B-12	6 mcg	100	POTASSIUM	37.5 mg	*
BIOTIN	30 mcg	10	CHLORIDE	34 mg	*
PANTOTHENIC ACID	10 mg	100			

*No U.S. RDA established.

CONTAINS IRON, WHICH CAN BE HARMFUL IN LARGE DOSES. CLOSE TIGHTLY AND KEEP OUT OF REACH OF CHILDREN. IN CASE OF OVERDOSE, CONTACT A PHYSICIAN OR POISON CONTROL CENTER IMMEDIATELY.

MILES

Figure 11.17 Many elements are essential to proper body function, in very small quantities. Vitamin tablets with mineral supplements can be used as a dietary supplement for these elements.

The Halogens

Group 17 (VIIA) contains fluorine, chlorine, bromine, iodine, and astatine. The elements of this group are called the halogen (salt-forming) family. In many chemical reactions, halogen atoms gain one electron. They become negatively charged ions with a stable outer level of eight electrons. As in other families already discussed, three factors determine the reactivity of the halogens. They are (1) the distance between the nucleus and the outer electrons, (2) the shielding effect of inner level electrons, and (3) the size of the positive charge on the nucleus. Fluorine atoms contain fewer inner-level electrons than the other halogens, so the shielding effect is the least for this element. The distance between the fluorine nucleus and its outer electrons is less than in the other halogens. Thus, the fluorine atom has the greatest tendency to attract other electrons. This attraction makes fluorine the most reactive nonmetal.

The astatine nucleus has the largest number of protons and the largest positive charge of the halogen elements. However, the increased charge on the nucleus is not enough to offset the distance and shielding effects. Thus, of all halogen atoms, the astatine nucleus has the least attraction for outer electrons.

In general, *on the right side of the periodic table, the nonmetallic elements become more active as we move from the bottom to the top. On the left side of the table, the metals become more active as we move from the top to the bottom.* The most active elements are located toward the upper right-hand and at the lower left-hand corners of the periodic table. Notice that fluorine is active because the atoms of fluorine have a great tendency to gain one electron and become negative ions. At the other extreme, the alkali metals are active because they hold the single outer electron loosely and it is easily removed. The groups between 1 (IA), and 17 (VIIA) vary between these two extremes.

The halogens usually react by forming negative ions or by sharing electrons. Fluorine is the most reactive of all the chemical elements. It reacts with all other elements except helium, neon, and argon. Fluorine is obtained from the mineral fluorspar, CaF_2.

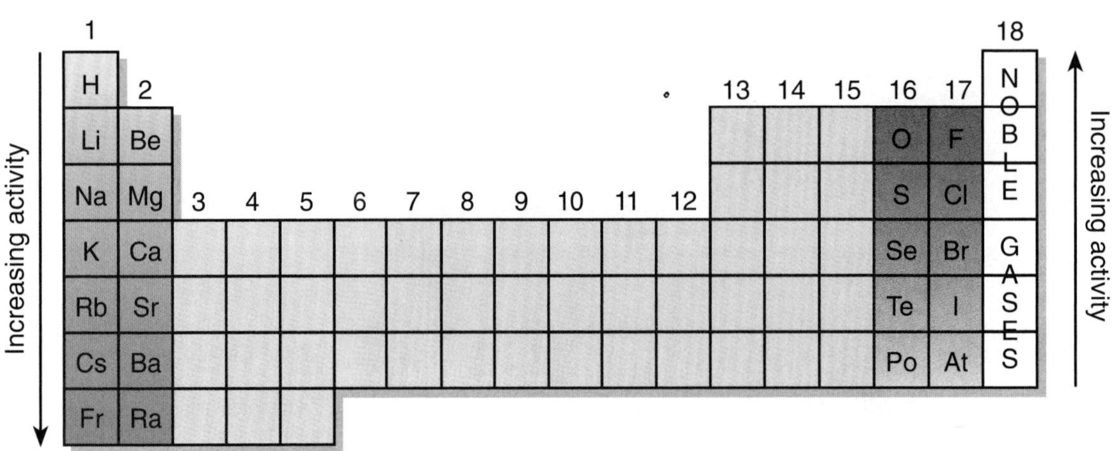

Figure 11.18 The activity of the elements on the two sides of the periodic table is indicated by the darkness of the tint in this illustration.

Like hydrogen, halogen atoms can form bridges between two other atoms. An example of such a compound is $BeCl_2$, shown in Figure 11.19. The halogens can also form compounds among themselves, as in ClF, ClF_5, BrF_5, IF_5, IF_7, BrCl, and ICl_3.

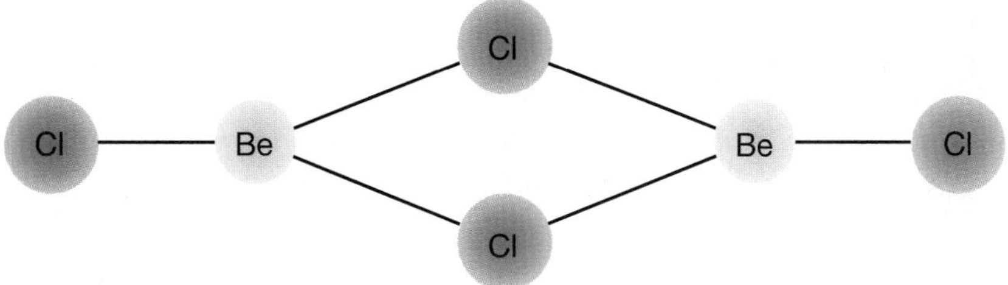

Figure 11.19 Beryllium chloride, $BeCl_2$, forms this molecule in which chlorine atoms form bridging bonds between two beryllium atoms.

Chlorine, though less abundant than fluorine in Earth's crust, is more commonly used in both the laboratory and industry. Chlorides of most elements are available commercially. These chlorides are quite often used in the laboratory as a source of positive metal ions bound with the chloride ions. The most familiar chloride is NaCl, common table salt.

Chlorine is produced primarily to produce other chemicals. Chlorine dioxide is used in water purification and as a bleaching agent in paper manufacturing. HCl is a by-product of many industrial processes. Its uses include steel manufacture, dye production, food processing, and oil well drilling.

The commercial preparation of chlorine led to one of the major water pollution crises of the early 1970s. Chlorine was produced by running an electric current through a solution of sodium chloride. Mercury, a toxic metal, was used to conduct current in the apparatus. The leakage of mercury into nearby water sources caused a public outcry. Industries had to remove the mercury from their waste before dumping it or switch to another method of producing chlorine. Chemists solved the problem by designing new types of electrochemical cells to produce chlorine.

Noble Gases

The noble gases, Group 18 (VIIIA), have very stable outer electron configurations. These elements are much less reactive than most other elements. For many years after their discovery, the noble gases were believed to be chemically unreactive and were commonly called inert. However, in 1962, the first "inert" gas compound was synthesized. Because these gases are not completely inert, they are usually referred to as the noble gases.

The first compound made with a noble gas was xenon hexafluoroplatinate ($XePtF_6$). Other compounds of xenon were soon produced. Once the techniques were known, the compounds could be made with increasing ease. Xenon difluoride (XeF_2) was first made by combining xenon and oxygen difluoride (OF_2) in a nickel tube at 300°C under pressure. The same compound can now be made directly from xenon and fluorine. In this process, fluorine

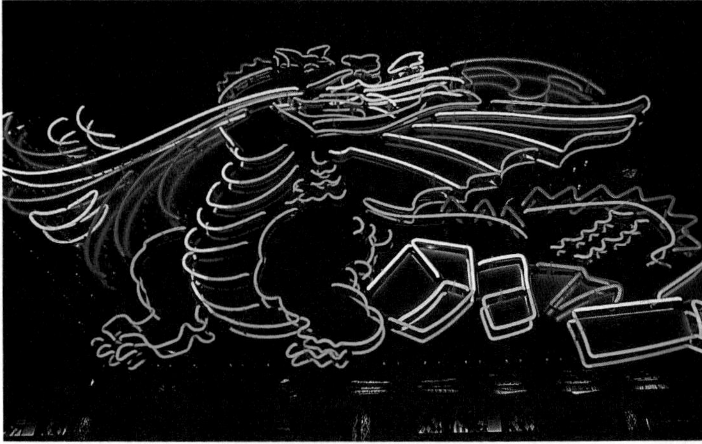

Figure 11.20 These gas discharge tubes (left) contain (from left to right) helium, neon, argon, krypton, and xenon. Because of the colored light they emit when a high voltage passes through them, the noble gases are used in "neon" lights for advertising and art (right).

and xenon are exposed to daylight in a glass container from which all the air has been removed. Compounds of xenon with oxygen and nitrogen have been synthesized. Krypton and radon compounds have also been produced.

Helium is the only element first discovered someplace other than on Earth. In 1868, the spectrum of an unknown element was discovered in sunlight. The element was then named helium after Helios, the Greek god of the sun. Helium is found in natural gas wells in the United States up to 7% by volume. This helium undoubtedly accumulated underground through the radioactive decay of certain elements emitting alpha particles. The boiling point of helium, −268.9°C, is the lowest of any known substance. Helium cannot be solidified at ordinary pressures, no matter how low the temperature. Liquid helium, because of its low boiling point, is used extensively in low temperature research. Helium is also used to fill balloons and in making artificial atmospheres such as those that deep-sea divers breathe.

Because they are inert, argon and helium are used to protect active metals during welding. Aluminum, for example, must be welded in an inactive atmosphere of argon or helium. If aluminum is welded in the air, the hot metal will catch on fire and burn in the air's oxygen. Argon is also used to fill light bulbs to protect the filament.

11.2 *CONCEPT REVIEW*

6. How does catenation affect the number of compounds carbon is able to form?

7. Fluorine differs from chlorine more than any other two adjacent halogens differ from each other. What explanation can you suggest for this phenomenon?

8. Why are nonmetallic elements at the top of a group more reactive than those at the bottom of the same group?

9. What three factors determine the reactivity of an element?

10. Apply Atmospheric nitrogen usually cannot be used by plants. Why, then, is it so important to plants?

CHEMISTRY AND

TECHNOLOGY

Recycling

Recycling has become an important environmental issue. There are two reasons for recycling. One is to reduce the amount of waste material that must be buried in landfills or incinerated. Landfills are unsightly, occupy space needed for more productive use, and may leak contaminants into water supplies. Incineration may introduce toxic gases and other pollutants into the atmosphere. The second reason for recycling is to slow down our consumption of resources that are nonrenewable or renewable only with difficulty. These resources include petroleum, natural gas, coal, ores, and radioactive minerals.

The principal materials being recycled at the present time are paper, glass, and aluminum. Paper is made primarily from trees and is the major component of trash. Trees, of course, are renewable, but replacing forests takes a long time. By recycling most of our paper (80-85%) we could greatly reduce the volume of trash entering landfills and save trees for other uses. Glass has been recycled for some time, because glass manufacturers have always introduced some broken glass into the glass-making process. Broken glass may make up from 10 to 80% of the glass furnace charge.

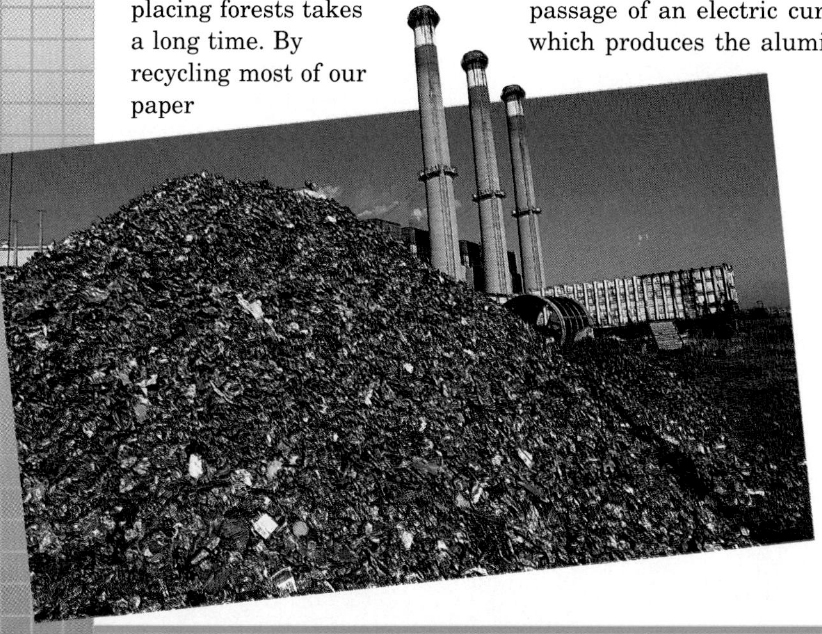

Aluminum is an ideal metal for making beverage cans. It is lightweight, and conducts heat rapidly so the food inside can be cooled quickly. Aluminum is produced from an ore called bauxite. Bauxite contains about 55% aluminum oxide, Al_2O_3. The bauxite is purified by dissolving the aluminum oxide in sodium hydroxide solution. The impurities are filtered off and the aluminum oxide precipitated by adding water. The aluminum oxide is then dissolved in cryolite, or sodium hexafluorosilicate, Na_3SiF_6. The solution is subjected to the passage of an electric current, which produces the aluminum metal. The production of aluminum from ore is a long, energy-consuming process.

One way to make the supply of aluminum ore last longer is to reuse the metal in objects that are no longer needed. Once a beverage can has been used, it cannot be refilled. However, it can be melted down and made into a new can. In addition to conserving the ore, recycling conserves energy. It takes only about one-third as much energy to make new cans from old as it does to make new cans from aluminum ore. Currently, 60% of aluminum cans are recycled in the United States.

Plastic products make up 9% of our trash by mass, but a whopping 30% by volume. With plastics, a new problem is encountered. There are many different kinds of plastics. Some of these are incompatible with others in a recycling procedure. Consequently, plastics must be separated by type before they can be reprocessed. Such a separation is labor intensive and, thus, expensive. Research into recycling plastics is currently being carried out in several locations in the United States.

Thinking Critically

1. Account for the difference in proportion of plastics in trash by mass (9%) and by volume (30%).
2. Find out why glass manufacturers introduce broken glass into the glass-making process. What properties of the finished product are determined by the amount of broken glass added?

OBJECTIVES

- define transition metals and list some of their uses.
- list representatives and some properties of lanthanoids and actinoids.

11.3 Transition Metals

Studying the transition metals is important for two reasons. First, these metals are the structural elements underlying much of our modern technology. For example, steel is an alloy of iron, a transition metal, and carbon. There are many specialty steels that include, in addition to iron, other transition metals such as cobalt, nickel, chromium, vanadium, tungsten and manganese.

The second reason for studying the transition metals is their fascinating chemistry. The involvement of the *d* sublevel electrons in their reactions leads to the formation of hundreds of thousands of compounds, many of which are important to biological systems.

Typical Transition Metals

The transition metals are those elements that have the highest energy electrons in *d* sublevels. In all groups except 12 (IIB), the *d* orbitals are only partially filled. These partially filled *d* orbitals make the chemical properties of the transition metals different from those of the main group metals. Remember that *d* electrons may be lost, one at a time, after the outer *s* electrons have been lost.

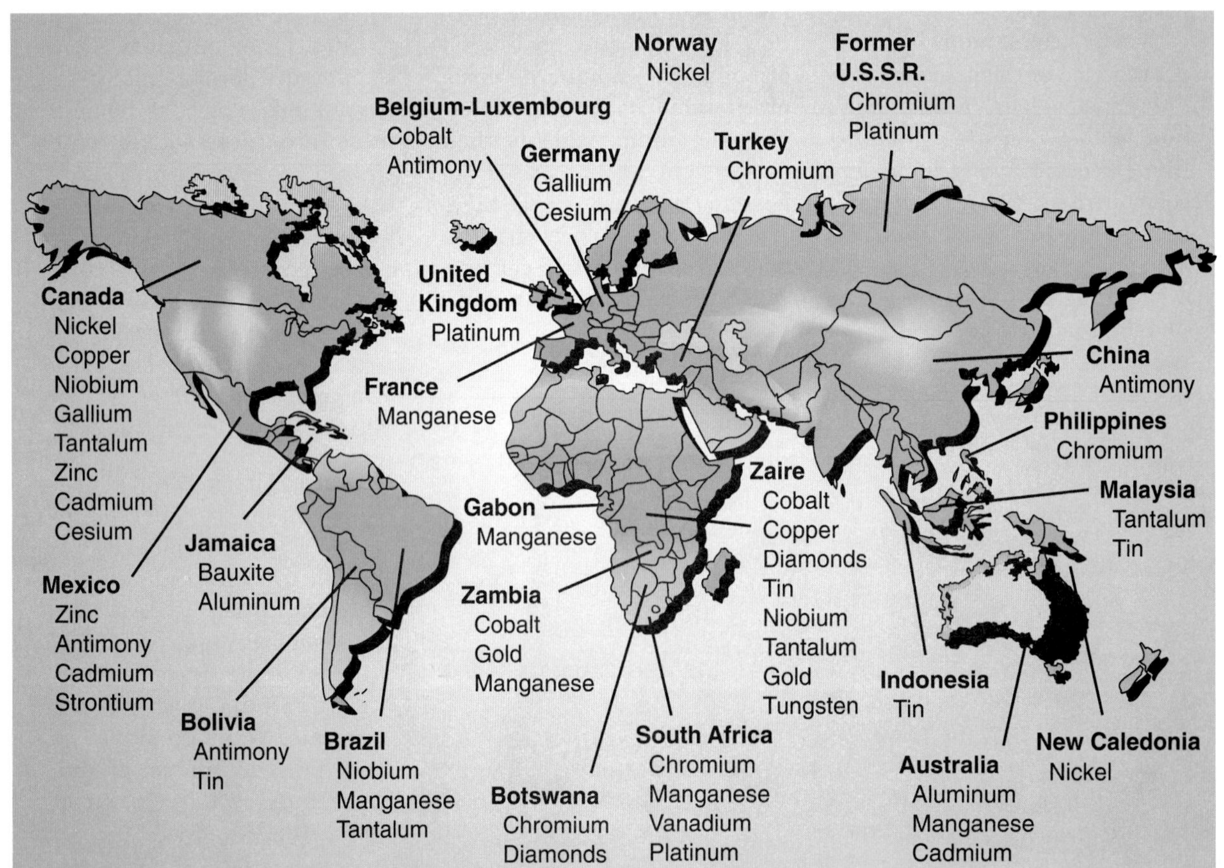

Figure 11.21 The map shows major sources of natural resources throughout the world. Note the number of transition metals that are included as important resources.

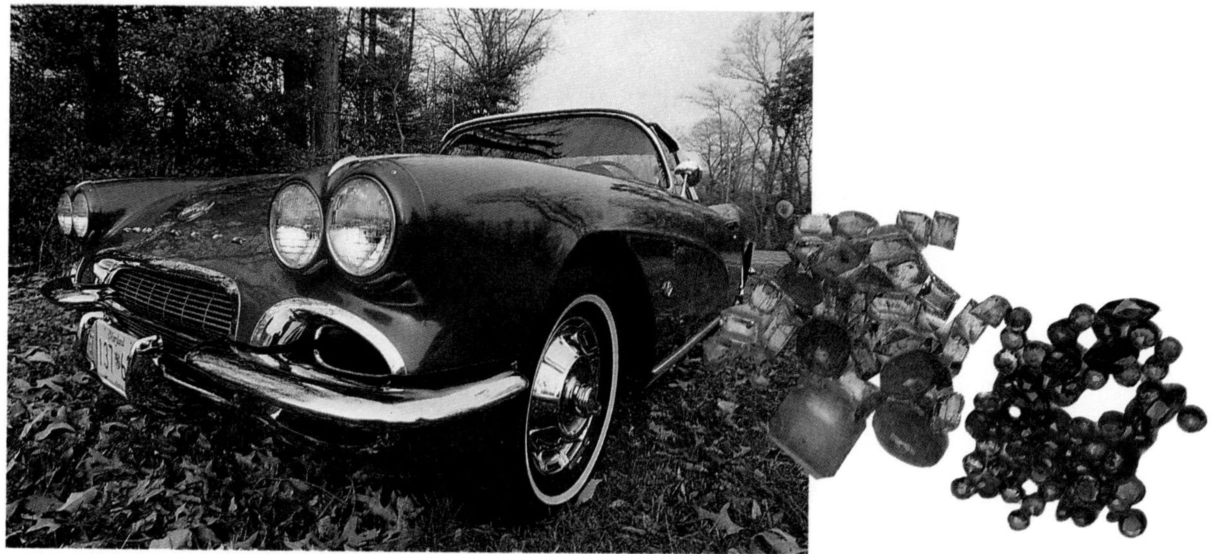

Figure 11.22 Chromium was used to protect steel parts of cars from corrosion (left). The colors of emeralds and rubies (right) are due to the presence of chromium impurities in the crystalline structure.

Transition metals have a wide variety of uses. Gold, silver, copper, and nickel are coinage metals. Catalytic converters contain platinum, palladium, and rhodium. Osmium is used to harden pen points. The radioactive isotope of cobalt, $^{60}_{27}Co$, is used in the treatment of cancer. Molybdenum is used in spark plugs. An important transition metal compound produced commercially is titanium(IV) oxide, TiO_2. The TiO_2 is used extensively as a white paint pigment. Other important oxides of transition metals include manganese dioxide, MnO_2, used in batteries, and iron and chromium oxides, used in audiotapes and videotapes.

Chromium

Chromium exhibits the typical properties of a transition metal. One property of chromium important to industry is its resistance to corrosion. Large quantities are used to make stainless steel and to "chromeplate" regular steel. In both cases, the chromium protects the iron in steel from corrosion.

Chromium usually reacts by losing three electrons to form the Cr^{3+} ion. Also of importance is the 6+ oxidation state. Recall that the outer electron configuration of chromium is $4s^1 3d^5$. This 6+ oxidation state is formed when chromium loses all five of its $3d$ electrons, as well as its outer level $4s$ electron. Two polyatomic ions are important examples of the 6+ state. These ions are the chromate ion, $CrO_4{}^{2-}$, and the dichromate ion, $Cr_2O_7{}^{2-}$.

Compounds containing chromium in the 6+ oxidation state are carcinogens; that is, they can cause cancer. Chromium also forms a Cr^{2+} ion. However, this ion is easily converted by oxygen in air to Cr^{3+}. Consequently, the 2+ state is not of great importance. The 4+ and 5+ states are very unstable, and so they have no practical uses. Many chromium compounds are highly colored. Small amounts of Cr^{3+} are found as impurities in the crystals of rubies and emeralds. These chromium ions cause the red color of rubies and the green color of emeralds.

ENVIRONMENT CONNECTION

Weeds Whack Wastes

Cadmium is a metal that contributes to toxic industrial waste along with copper and zinc. Now it has been discovered that the poisonous jimsonweed has the ability to absorb these metals from contaminated wastes. The weed manufactures a proteinlike material that absorbs and binds the metals within the plant cells in a nontoxic form. If enough of this proteinlike material can be produced synthetically, it could be used to remove metals from toxic wastes before they are thrown in a landfill.

Figure 11.23 Brass is an alloy of zinc and copper.

Zinc

Although zinc is classed as a transition element, its behavior differs slightly because of its full *d* sublevel. It exhibits only one oxidation state, 2+. The special stability of the full *d* sublevel leaves only the two outer 4*s* electrons available for reacting. Zinc is the second most important transition element (after iron) in biological systems. As an example, lack of zinc in the diet prevents the pancreas from producing some digestive enzymes. Over 25 zinc-containing proteins have been discovered.

Metallic zinc, like chromium, is corrosion resistant. It is used extensively as a coating to protect iron. The coating can be applied in three ways. When the iron is dipped in molten zinc, the process is called **galvanizing.** The coating is also applied electrically. The third method is to allow gaseous zinc to condense on the surface of iron. Another major use for metallic zinc is in the production of alloys. Especially important is its combination with copper to form brass.

Inner Transition Elements

The lanthanoids and actinoids are often called the **inner transition elements** because their highest energy electrons (*f* electrons) are inside the *d* sublevel and the outer level.

Neodymium, a Lanthanoid

The electron configuration of neodymium (nee oh DIHM ee um) ends $6s^2 4f^4$. From that configuration we would predict a 2+ oxidation state. However, Nd, as all the lanthanoids, shows 3+ as its most stable state. The 3+ ion is pale violet in solution. The metal itself is soft and quite reactive. It tarnishes when exposed to air. The metal can be mixed with other elements to form alloys with unusual conductivity and magnetic properties. The compound Nd_2O_3 is used in glass filters and in some lasers. Neodymium is

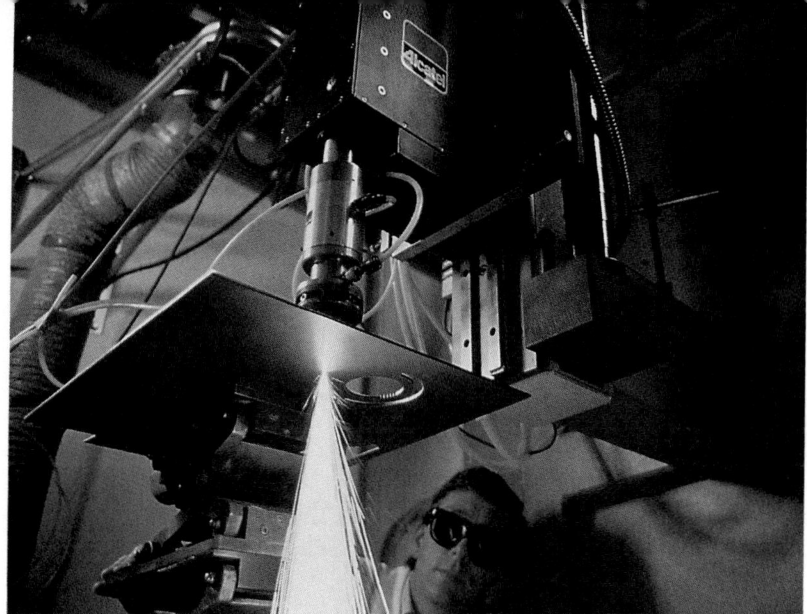

Figure 11.24 Neodymium is used in high-power pulsed lasers. These lasers are used to drill and weld metals and in experiments that attempt to produce nuclear fusion.

the second most abundant lanthanoid metal. Its compounds are separated from the other lanthanoids by a chromatographic technique, as explained further in Chapter 14. The metal is obtained from its fluoride by a single displacement reaction.

$$3Ca + 2NdF_3 \rightarrow 3CaF_2 + 2Nd$$

The lanthanoids have several interesting uses. Their compounds are used as the phosphors that give television picture tubes their colors. Lanthanoid compounds are also used to coat the insides of fluorescent light bulbs to alter the color given off.

Curium, an Actinoid

Curium's predicted electron configuration ends $7s^2 5f^8$. However, its actual configuration is $7s^2 5f^7 6d^1$. Plainly, the stability of the half-full *f* sublevel more than offsets the promotion of one electron to the 6*d*. The element exhibits a 3+ oxidation state in its compounds and the ion is pale yellow in solution. The element itself is a silvery, hard metal of medium density and high melting point. It does not occur in nature. All multigram quantities of curium have been produced by the slow neutron bombardment of the artificial element plutonium. Curium is reactive and highly toxic to the human body. Curium has potential use as an energy source if the heat generated by its nuclear decay can be converted to electric energy. This metal may be used as the energy source in the nuclear generators in satellites.

HISTORY CONNECTION

A Town That Chemists Will Remember

In 1788, near the Swedish town of Ytterby, a new mineral was discovered. Over the course of 100 years, this mineral yielded 16 new elements, most of them in the lanthanoid series. Four of these elements were named for the town: yttrium, ytterbium, erbium, and terbium. John Gadolin, a Finnish chemist, found the first element, yttrium; gadolinium, another of the group, was named for him.

11.3 CONCEPT REVIEW

11. What property makes zinc useful in galvanizing?
12. The lanthanoids were once called the "rare earths." What is meant by "earth"?
13. **Apply** What properties are desirable in coinage metals? Which transition metals would make the best coins?

Chapter 11

REVIEW

Summary

11.1 Hydrogen and Main Group Metals

1. Hydrogen is often considered a family by itself because of the unique ways in which it reacts.
2. The most active metals are listed toward the lower left-hand corner of the periodic table. The alkali metals, Group 1 (IA), are the most active metals. They react by losing one electron to form ions having a 1+ charge.
3. In large atoms, the shielding effect of inner-level electrons partially blocks the attraction of the nucleus for outer-level electrons. This effect, coupled with increased distance from the nucleus, results in increasing reactivity in larger alkali metal atoms.
4. Lithium behaves in many ways like other members of its family but, because of the small size of its ion, it acts in other ways like magnesium.
5. The alkaline earth metals, Group 2 (IIA), are second only to the alkali metals in reactivity. They form 2+ ions.
6. Aluminum, in Group 13 (IIIA), is less reactive than the alkaline earth metals, and forms 3+ ions. Aluminum is the most abundant metal in Earth's crust.

11.2 Nonmetals

7. The elements in Group 14 (IVA) have four electrons in the outer level. Carbon is a nonmetal; silicon and germanium are metalloids. Tin and lead, in the same family, are distinctly metallic.
8. An important property of carbon is its ability to form chains and rings of atoms. This ability is called catenation.
9. Nitrogen and phosphorus, in Group 15 (VA), form compounds by sharing electrons, but they differ from each other considerably.
10. An important use of nitrogen is in ammonia production. Compounds made from ammonia are used as fertilizers.
11. In Group 16 (VIA), oxygen is distinctly nonmetallic and reacts by gaining two electrons or by sharing electrons. Sulfur is similar to oxygen but also exhibits positive oxidation states.
12. Oxygen is the most plentiful element in Earth's crust. It forms compounds with all elements except certain noble gases. Many oxides react with water to form acids or bases.
13. The most active nonmetals are listed toward the upper right-hand corner of the periodic table. The halogens are the most active nonmetallic elements and usually react by gaining one electron.
14. In contrast to the alkali metals, increased shielding effect in the halogens makes the halogen elements less reactive moving downward through the family.
15. Fluorine is the most reactive of all elements. Chlorine is commonly produced by passing an electric current through a sodium chloride solution.
16. The noble gases, Group 18 (VIIIA), have very stable outer electron configurations and are much less reactive than most of the other elements.

11.3 Transition Metals

17. The transition elements are characterized by having partially filled *d* orbitals in all groups except 12 (IIB). These metals include major structural metals.
18. Chromium is a typical, corrosion-resistant transition metal, exhibiting more than one oxidation number.
19. Zinc is corrosion-resistant. Because of its full *d* sublevel, it does not exhibit more than one oxidation state.
20. Lanthanoids exhibit the 3+ oxidation state. Curium, an actinoid, also exhibits the 3+ oxidation state.

Key Terms

hydrogen ion
hydride ion
catalyst
carbide ion
organic compound
catenation
inorganic compound
allotrope
silicate
amphoteric
galvanizing
inner transition element

Review and Practice

14. What can you generalize about the electron structure of the elements in any given family?
15. Describe the four ways in which hydrogen can react.
16. What is the difference between a hydrogen ion and a hydride ion?
17. Lithium is not a typical alkali metal.
    - a. How does lithium differ chemically from other alkali metals?
    - b. How does lithium differ physically from other alkali metals?
    - c. What non-alkali metal does lithium resemble?
    - d. How is lithium similar to other alkali metals?
18. Marble, used in sculpture and architecture, is composed of almost pure calcium carbonate. Calcium is an alkaline earth metal. List all the alkaline earth metals and give additional uses for at least three of the elements or their compounds.
19. Classify each of the following as either organic or inorganic.
    - a. Na_4C
    - b. CH_4
    - c. CH_3COOH
    - d. $(NH_4)_2CO_3$
    - e. CO_2
    - f. C
    - g. C_3H_7Br
    - h. CS_2
    - i. H_2CO_3
20. Propane, C_3H_8, and butane, C_4H_{10}, are common organic fuels that are made up of chains of carbon atoms, as shown in the structures below.

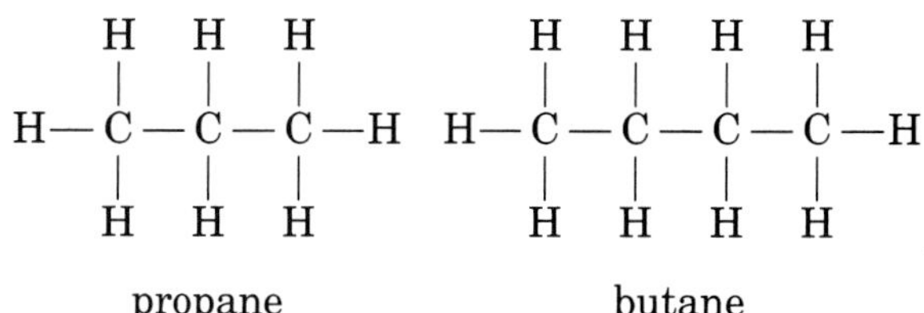

    - a. What is this tendency to form chains called?
    - b. How does this tendency in carbon compare with the same tendency in silicon?
    - c. Why would we call these compounds organic?
21. How do the structures of graphite and diamond differ? How do these materials illustrate allotropes?
22. What is the difference between silicon and silicates? What are three important products that use silicon?
23. What chemical property accounts for the uses of tin and lead, especially their uses in alloys?
24. Describe two ways that atmospheric nitrogen is changed to form useful compounds.
25. In 1669, Hennig Brandt noticed a new substance was produced when he heated urine and sand together in an attempt to make gold. Today we know the substance as phosphorus. In 1772, Daniel Rutherford noticed that a colorless gas—nitrogen—still remained after the oxygen and carbon dioxide were removed from an air sample. These two elements, phosphorus and nitrogen, have played an important part in chemistry.
    - a. How do the properties of nitrogen and phosphorus differ?
    - b. What properties do they both have?
    - c. How are both nitrogen and phosphorus important to living things?

d. List other uses for the element nitrogen and its compounds.
e. List other uses for phosphorus compounds.

26. How does the chemical behavior of oxygen differ when it combines with an active metal rather than a nonmetal?
27. What are the allotropes of oxygen?
28. Oxygen makes up one-fifth of the mass of the atmosphere, nine-tenths of the mass of Earth's water, and nearly half the mass of Earth's crust. Most of this oxygen is in the form of compounds resulting from the reactions between oxygen and other elements.
    a. Explain the two ways oxygen can bond to form oxides.
    b. If the resulting oxides are each dissolved in water, how do the solutions differ?
    c. Compare the chemical behavior of amphoteric oxides with that of other oxides.
29. Compare the chemical properties of oxygen and sulfur.
30. What is the meaning of the word *halogen* in this chapter?
31. Where are the halogens found on the periodic table?
32. In electronic terms, what usually happens when halogens react with active metals?
33. Which halogen is the most active? Which is the least active? What characteristics account for this difference in activity?
34. How does chemical activity vary among the elements on the left side of the periodic table? How does it vary among elements on the right side of the periodic table?
35. List several uses of chlorine in the laboratory and in industry.
36. Describe a method by which chlorine is produced industrially.
37. The noble gas helium was discovered on the sun by spectroscopy before it was discovered on Earth.
    a. List all the noble gases.
    b. What are their properties?
    c. Why are they no longer called inert gases?
    d. List several practical uses for the noble gases.
38. What is the major practical use of the transition elements?
39. Why is chromium an important metal in plating iron or steel and in making alloys with iron?
40. What are carcinogens?
41. Why is zinc important in the production of materials made from iron?
42. What oxidation state is common to all of the lanthanoids?
43. Why does the electron configuration of curium vary from the predicted configuration?
44. What is the importance of plutonium to the production of curium?
45. What characteristics of the lithium atom account for the differences in behavior that it exhibits in relation to the other alkali metals?
46. Use the periodic table to predict which element in each of the following pairs would be more active.
    a. cobalt, nickel
    b. copper, gallium
    c. mendelevium, nobelium
    d. chlorine, fluorine
    e. barium, radium
    f. titanium, zirconium
    g. sodium, potassium
    h. arsenic, cobalt
    i. beryllium, lithium
    j. oxygen, phosphorus
47. Use the periodic table to predict which element in each of the following pairs would be more active.
    a. actinium, thorium

b. americium, europium
c. erbium, fermium
d. arsenic, gallium
e. berkelium, californium
f. carbon, lithium
g. bromine, iodine
h. boron, carbon
i. potassium, rubidium
j. iodine, tellurium

48. For each of the following pairs of elements, determine the family name, and state two chemical and two physical properties that we could expect these elements to have.
   a. K, Na
   b. Ar, Ne
   c. I, Br
   d. Ba, Ca

49. One important form of oxygen is ozone.
   a. What is the formula for ozone?
   b. How can ozone be produced?
   c. What is a practical use for ozone?

Concept Mastery

50. Discuss the effects of shielding, atomic size, and nuclear charge on the relative reactivities of the alkali metals.

51. Lightning and the action of bacteria in the roots of legumes such as beans or peanuts are both said to be "nitrogen-fixing." What is meant by "nitrogen fixation?"

52. Why is phosphorus an essential nutrient in living systems?

53. How does the electronic structure of the transition elements differ from that in the main groups of the periodic table?

54. Many transition metals exhibit several oxidation states. Why does zinc exhibit only the 2+ oxidation state?

55. What is the shielding effect? How can it help you predict the difference in reactivity between calcium and barium?

56. Why are the alkali metals and most of the alkaline earth metals so little used in their uncombined forms?

57. Why do oxygen and lithium form the compound Li_2O when they combine, while calcium and oxygen form CaO?

58. Selenium is an example of a trace element that we must have in our diets. Why do we call it a trace element?

59. Why would you expect the noble gases to be relatively inert elements?

60. Which elements in the periodic table would you predict to have the least attraction for outer level electrons?

61. In some versions of the periodic table, hydrogen is placed at the top of both Group 1 (IA) and Group 17 (VIIA). Considering the chemistry of hydrogen, why might it fit in both places?

62. Write balanced equations for each of the following reactions.
   a. formation of each of the two oxides of sulfur
   b. reaction of each of these oxides with water

Application

63. Rubies and sapphires are two of the most valuable compounds of aluminum. They occur in nature and can also be synthesized. What are some other uses for aluminum or aluminum compounds?

64. Why does catenation make carbon compounds so important?

65. One property shared by most of the elements in Group 1 (IA) and Group 2 (IIA) is the ability to displace hydrogen from water. Write equations for the displacement of hydrogen from water by Li, Sr, K, and Ba.

66. Sodium compounds are used extensively in industry and in homes. Give at least one common use for each of the following compounds.
   a. $NaHCO_3$
   b. $Na_5P_3O_{10}$
   c. $NaOH$
   d. Na_2SO_4
   e. Na_2CO_3
   f. Na_2SiO_3

67. One of beryllium's uses is in alloys from which nonsparking tools are made. Suggest a situation in which nonsparking tools would be of particular value.

68. The amounts of important nutritive ingredients in fertilizer are usually given on the package as a series of three numbers, for instance 5-10-5. The last number is the percent potassium. Find out what the first two numbers designate. Write a short, general statement regarding what proportions of nutrients are useful in what sorts of situations.

Critical Thinking

69. Finely divided platinum is used as a catalyst in many industrial processes. What is the function of the platinum? Do you think the platinum will be used up in the process? Explain your answer.

70. The production of sulfuric acid is considered to be one of a country's major economic indicators. Why do you think this is so?

71. Is the compound SO_3 produced by the sharing or transfer of electrons?

72. The structures of graphite and diamond are shown here. These are two allotropes of carbon.
   a. What are allotropes?

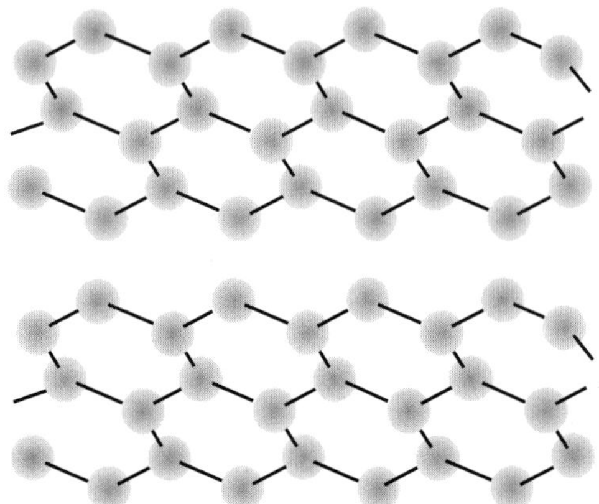

Graphite

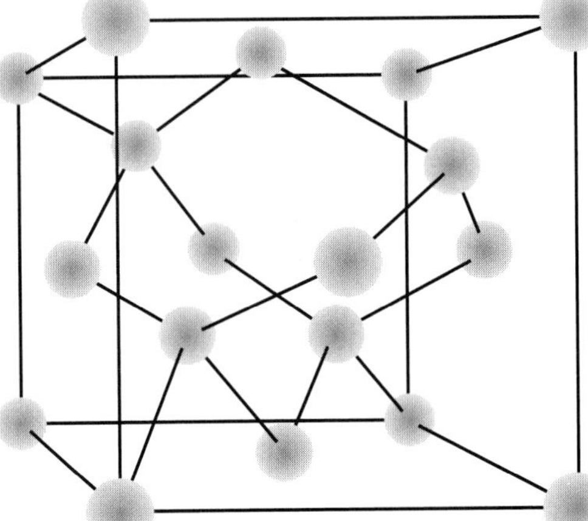

Diamond

   b. Study the structures and suggest reasons why graphite is slippery and is used as a lubricant, while diamond is not.
   c. Study the structures and suggest reasons why diamond is harder than graphite and is used as an abrasive coating.
   d. Use the structures to explain why graphite under great pressure turns into diamond.

Readings

Bachman, Peter K., and Russell Messier. "Diamond Thin Films." *Chemical and Engineering News* 67, no. 20 (May 15, 1989): 24-39.

Davenport, Derek A. "Going Against the Flow—The Isolation of Fluorine." *ChemMatters* 4, no. 4 (December 1986): 13-15.

Day, Richard. "Lead-Free Water." *Popular Science* 232, no. 1 (January 1988): 90-91, 116.

Dworetzky, Tom. "Gold Bugs." *Discover* 9, no. 3 (March 1988): 32.

Edwards, D. D. "Aluminum: A High Price for a Surrogate?" *Science News* 131, no. 16 (April 18, 1987): 245.

Raloff, Janet. "New Misgivings About Low Magnesium." *Science News* 133, no. 23 (June 4, 1988): 356.

Stolka, Milan. "Hard Copy Materials." *ChemTech* 19, no. 8 (August 1989): 487-495.

Cumulative Review

1. Complete and balance each of the following equations.
   a. $Zn + H_2SO_4 \rightarrow$
   b. $Mg + H_2O \rightarrow$
   c. $CuSO_4 + NaOH \rightarrow$
   d. $KI + Cl_2 \rightarrow$
2. How many grams of $AgNO_3$ are required to produce 38.7 grams of AgCNS according to the following reaction?

   $$AgNO_3 + KCNS \rightarrow AgCNS + KNO_3$$
3. Compute the atomic mass of bromine from the following data about its isotopes.

Isotopic mass	Abundance
78.918 332 u	50.54%
80.916 292 u	49.46%
4. Draw electron dot diagrams for yttrium, dysprosium, and indium.
5. Write balanced synthesis equations for the formation of each of the following.
   a. silver sulfide
   b. potassium bromide
   c. iron(III) oxide
   d. magnesium nitride
6. Look at the following orbital-filling diagrams, and decide which of them follow the arrow diagram and which do not. For those that are correct, write the electron configuration. For any others, correct the diagram. Do not change the total number of electrons present. Identify each element represented.

	1s	2s	2p		
a.	⇅	⇅	↑	○	○
b.	↑	↑	○	○	○
c.	⇅	⇅	⇅	○	○
d.	⇅	⇅	↑	↑	↑
e.	⇅	↑	↑	↑	○
7. Explain why the second ionization energy for lithium is more than four times the second ionization energy for beryllium.
8. How do atomic radii change, moving left to right across a period of elements? What factors account for this change?
9. Nitrogen gas can be produced by the decomposition of ammonium nitrite to give nitrogen and water. Write the balanced equation for this reaction. Calculate the mass of ammonium nitrite needed to produce 0.100 mole of nitrogen.
10. The density of magnesium is 1.738 g/cm^3. A mole of magnesium atoms has a mass of 24.305 g.
    a. What is the volume of a mole of magnesium atoms?
    b. What is the space taken up by one atom of magnesium?

CHAPTER **Preview**

CONTENTS

CHAPTER 12

Chemical Bonding

Are we looking at an optical illusion? Is the picture turned sideways? No. The objects in the photograph are bonded with methyl cyanoacrylate, or Super Glue. How does Super Glue hold things together? And how is methyl cyanoacrylate itself held together?

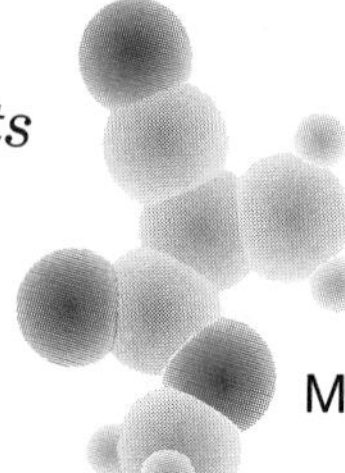

Methyl cyanoacrylate, $C_5H_5O_2N$

In past chapters, we have examined the structure and properties of atoms. This study has prepared us to learn how atoms attach together to form compounds. It is this attachment of atoms to each other that we refer to as a bond. In this chapter we will investigate three types of bonds, how they are formed, and how bonding affects the atoms that are bonded.

EXTENSIONS AND CONNECTIONS

Careers in Chemistry
Electrician, page 305

Home Economics Connection
Food Frequencies, page 307

Bridge to Metallurgy
Alloys: Metallic materials that aren't elements, page 310

Frontiers
Alien Alloys: A glimpse of high-tech alloys with remarkable properties, **page 312**

OBJECTIVES

- identify the type of bonding between two elements given their electronegativities.
- list factors that influence electronegativity and recognize it as a periodic property of elements.
- differentiate among properties of ionic, covalent, and metallic bonds.
- explain the use of infrared and microwave spectroscopy to determine the structure of molecules.

12.1 Bond Formation

We have experimental evidence that some atoms form bonds by transferring electrons. We also have evidence that atoms form bonds by sharing electrons. Therefore, we can expect that the bonds formed between atoms depend on the electron configurations of those atoms and on the attraction the atoms have for electrons. Because configuration and attraction are periodic properties, the bonding of atoms will vary in a systematic way.

Electronegativity

Both electron affinity and ionization energy are properties of isolated atoms. Chemists need a comparative scale relating the abilities of elements to attract electrons when their atoms are combined. The relative tendency of an atom to attract electrons to itself when it is bonded to another atom is called **electronegativity.** The elements are assigned electronegativities on the basis of many experimental tests.

Electronegativities of elements are influenced by the same factors that affect ionization energies and electron affinities. It is possible to construct an electronegativity scale using first ionization energies and electron affinities of the elements. Examine Table 12.1. Note that the trend in electronegativity follows the same trends as the ionization energies and electron affinities, as shown in Figure 12.2. *The most active metals* (lower left) *have the lowest electronegativities.* Fluorine, a nonmetal, has the highest electronegativity (4.10) of all the elements. Thus, in a bond involving fluorine, the bonding electrons are drawn closer to the fluorine atom than to the other atom. Electronegativities are not given for a number of the actinoids because not enough of these elements have been produced to measure their electronegativities accurately. Because only a few of the noble gases form any compounds and those few form only a small number of compounds, no electronegativities are given for the elements of Group 18 (VIIIA).

Figure 12.1 Potassium is a very active metal. In contact with water, it reacts vigorously, producing hydrogen.

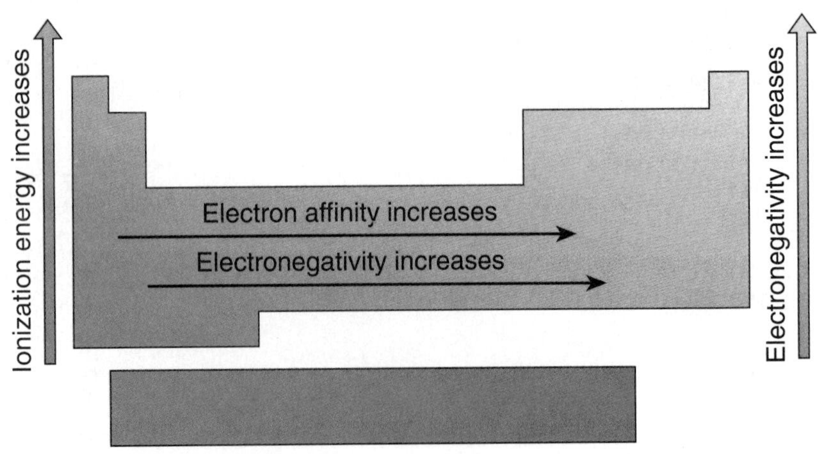

Figure 12.2 General trends for ionization energy, electron affinity, and electronegativity can be found in the periodic table.

Table 12.1

Electronegativities

1 **H** 2.20																	2 **He** —
3 **Li** 0.97	4 **Be** 1.47											5 **B** 2.01	6 **C** 2.50	7 **N** 3.07	8 **O** 3.50	9 **F** 4.10	10 **Ne** —
11 **Na** 1.01	12 **Mg** 1.23											13 **Al** 1.47	14 **Si** 1.74	15 **P** 2.06	16 **S** 2.44	17 **Cl** 2.83	18 **Ar** —
19 **K** 0.91	20 **Ca** 1.04	21 **Sc** 1.20	22 **Ti** 1.32	23 **V** 1.45	24 **Cr** 1.56	25 **Mn** 1.60	26 **Fe** 1.64	27 **Co** 1.70	28 **Ni** 1.75	29 **Cu** 1.75	30 **Zn** 1.66	31 **Ga** 1.82	32 **Ge** 2.02	33 **As** 2.20	34 **Se** 2.48	35 **Br** 2.74	36 **Kr** —
37 **Rb** 0.89	38 **Sr** 0.99	39 **Y** 1.11	40 **Zr** 1.22	41 **Nb** 1.23	42 **Mo** 1.30	43 **Tc** 1.36	44 **Ru** 1.42	45 **Rh** 1.45	46 **Pd** 1.35	47 **Ag** 1.42	48 **Cd** 1.46	49 **In** 1.49	50 **Sn** 1.72	51 **Sb** 1.82	52 **Te** 2.01	53 **I** 2.21	54 **Xe** —
55 **Cs** 0.86	56 **Ba** 0.97	71 **Lu** 1.14	72 **Hf** 1.23	73 **Ta** 1.33	74 **W** 1.40	75 **Re** 1.46	76 **Os** 1.52	77 **Ir** 1.55	78 **Pt** 1.44	79 **Au** 1.42	80 **Hg** 1.44	81 **Tl** 1.44	82 **Pb** 1.55	83 **Bi** 1.67	84 **Po** 1.76	85 **At** 1.96	86 **Rn** —
87 **Fr** 0.86	88 **Ra** 0.97	103 **Lr** —	104 **Unq** —	105 **Unp** —	106 **Unh** —	107 **Uns** —	108 **Uno** —	109 **Une** —									

57 **La** 1.08	58 **Ce** 1.08	59 **Pr** 1.07	60 **Nd** 1.07	61 **Pm** 1.07	62 **Sm** 1.07	63 **Eu** 1.01	64 **Gd** 1.11	65 **Tb** 1.10	66 **Dy** 1.10	67 **Ho** 1.10	68 **Er** 1.11	69 **Tm** 1.11	70 **Yb** 1.06
89 **Ac** 1.00	90 **Th** 1.01	91 **Pa** 1.14	92 **U** 1.30	93 **Np** 1.29	94 **Pu** 1.25	95 **Am** —	96 **Cm** —	97 **Bk** —	98 **Cf** —	99 **Es** —	100 **Fm** —	101 **Md** —	102 **No** —

Many chemical properties of the elements can be organized in terms of electronegativities. For example, bond strength is a measure of the energy that is needed to break the bonds between atoms in molecules of a compound. As seen in Table 12.2, the greater the strength of the bond between two atoms, the greater the difference in electronegativities.

Consider a reaction between two elements. Their relative attraction for electrons determines how they react. We can use the electronegativity scale to determine this attraction. Because electronegativity represents a comparison of the same property for each element, it is a dimensionless number.

Table 12.2

Bonds between Hydrogen and Halogens

Bond	Bond Strength (kJ/mol)	Electronegativity Difference
H–F	568.1	1.90
H–Cl	431.951	0.63
H–Br	366.25	0.54
H–I	298.32	0.01

PRACTICE PROBLEMS

Arrange the following elements in order of increasing attraction for electrons in a bond.

1. antimony, fluorine, indium, selenium
2. francium, gallium, germanium, phosphorus, zinc

Bond Character

Figure 12.3 When magnesium burns, it reacts with oxygen to form the ionic compound magnesium oxide.

Electrons are transferred between atoms when the difference in electronegativity between the atoms is quite high. If the electronegativity difference between two reacting atoms is small, we might expect a sharing of electrons. At what point in electronegativity difference does the changeover occur? The answer is not simple. For one thing, the electronegativity of an atom varies slightly depending upon the atom with which it is combining. Another factor is the number of other atoms with which the atom is combining. Therefore, a scale showing the percent of transfer of electrons (percent ionic character) has been constructed, Table 12.3. The amount of transfer depends on the electronegativity difference between two atoms.

When two atoms combine by a transfer of electrons, ions are produced. The attraction of the oppositely charged ions holds them together. When two elements combine by electron transfer, they form an ionic bond. If two elements combine by sharing electrons, they form a covalent bond.

From Table 12.3, we see that two atoms with an electronegativity difference of about 1.67 would form a bond that is 50% ionic and 50% covalent. For our purposes, we will consider an electronegativity difference of less than 1.67 as indicating a covalent bond. A difference of 1.67 or greater indicates an ionic bond.

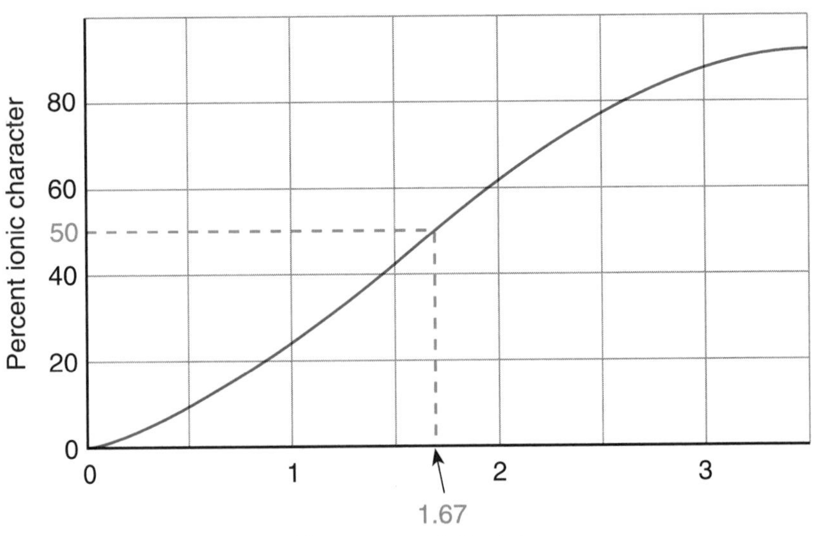

Figure 12.4 The percentage of ionic character of a bond between two atoms tends to increase as the difference in the electronegativities of the two elements increases.

Table 12.3

Character of Bonds										
Electronegativity Difference	0.00	0.65	0.94	1.19	1.43	1.67	1.91	2.19	2.54	3.03
Percent Ionic Character	0%	10%	20%	30%	40%	50%	60%	70%	80%	90%
Percent Covalent Character	100%	90%	80%	70%	60%	50%	40%	30%	20%	10%

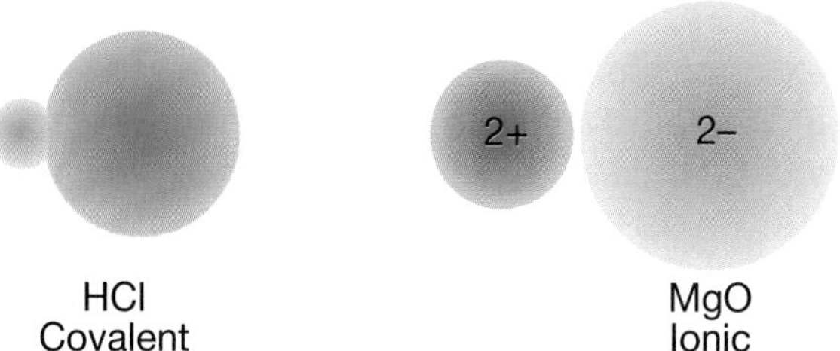

Figure 12.5 Covalent molecules, such as HCl, are held together by a mutual sharing of electrons as indicated by the overlap of the electron clouds. Ionic substances, such as MgO, are held together by an electrical force of attraction between oppositely charged ions.

Actually, unless the two atoms bonded are identical, all bonds have some ionic and covalent characteristics.

Think of magnesium reacting with oxygen. The difference in their electronegativities is 2.27 ($|3.50 - 1.23|$). We would predict the formation of an ionic bond between these two elements. On the other hand, consider boron and nitrogen. The difference in their electronegativities is 1.06 ($|3.07 - 2.01|$). We would predict that boron and nitrogen will form a covalent bond.

SAMPLE PROBLEM

Determining Bond Type

Decide whether an ionic bond or a covalent bond will form between the atoms for each of the following pairs of elements: B—P, Be—Si, C—Na, Li—O, Mg—N.

Solving Process:

The difference in electronegativity can be used to determine whether two atoms will form an ionic bond or a covalent bond. Therefore, use the values of the electronegativity found in Table 12.1 to find absolute value of the difference between the pairs of atoms listed above.

$$\begin{aligned} \text{B—P:} &\ |2.01 - 2.06| = 0.05 \\ \text{Be—Si:} &\ |1.47 - 1.74| = 0.27 \\ \text{C—Na:} &\ |2.50 - 1.01| = 1.49 \\ \text{Li—O:} &\ |0.97 - 3.50| = 2.53 \\ \text{Mg—N:} &\ |1.23 - 3.07| = 1.84 \end{aligned}$$

From Table 12.3, we can see that differences in electronegativities of 1.67 or greater lead to the formation of ionic bonds. Differences in electronegativities less than 1.67 lead to the formation of covalent bonds. Therefore the element pairs B—P, Be—Si, and C—Na will form covalent bonds and the element pairs Li—O and Mg—N will form ionic bonds.

CAREERS IN CHEMISTRY

Electrician

An electrician uses knowledge of the movement of electrons through metal in order to wire houses and businesses for many purposes, including lighting and heating. An electrician must have a high school diploma, take additional work in apprentice classes, and serve an apprenticeship. Electricians in most states are required to pass competency tests in order to gain a license. In high school, future electricians should take physics, chemistry, and vocational courses.

PRACTICE PROBLEMS

3. Classify the bonds between the following pairs of atoms as principally ionic or covalent.

a. Al—Si	**d.** Li—S	**g.** Ca—Cl
b. Ba—O	**e.** Ca—P	**h.** F—S
c. C—H	**f.** B—Na	**i.** Br—Rb

4. For each atom pair listed below, decide whether an ionic or a covalent bond would form between the elements.

a. hydrogen-iodine	**d.** chlorine-tellurium
b. astatine-beryllium	**e.** bromine-cerium
c. cobalt-fluorine	**f.** calcium-fluorine

Ionic Bonds

RULE OF THUMB

Sodium chloride is an excellent example of an ionic compound; that is, a compound with an ionic bond. *Ionic compounds are characterized by high melting points and the ability to conduct electricity in the molten state. They tend to be soluble in water and usually crystallize as sharply defined particles.*

Most of the properties of ionic compounds are best explained by assuming a complete transfer of electrons.

$$2Na^{\times} + :\ddot{\underset{..}{Cl}}:\ddot{\underset{..}{Cl}}: \rightarrow 2Na^{+} + 2:\ddot{\underset{..}{Cl}}\dot{\times}^{-}$$

If a chloride ion and a sodium ion are brought together, there will be an attractive force between them. If the ions are brought almost into contact, the force will be great enough to hold the two ions together. The electrostatic force that holds two ions together due to their differing charges is the **ionic bond.**

Elements can be assigned oxidation numbers for ionic bonding. Sulfur, for example, with six electrons in the outer level, will tend to gain two electrons. Thus, it attains the stable octet configuration. The oxidation number of sulfur for ionic bonding is 2−. The negative two is sulfur's electric charge after it gains two electrons.

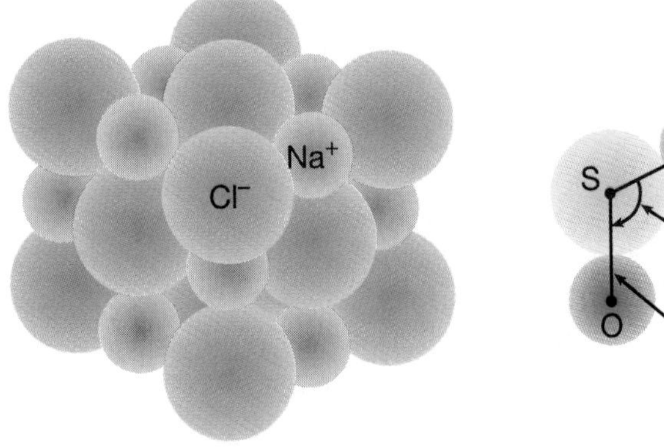

Figure 12.6 Sodium chloride exists as an array of sodium and chloride ions (left) while sulfur dioxide exists as molecules, each of which is composed of one sulfur and two oxygen atoms (right).

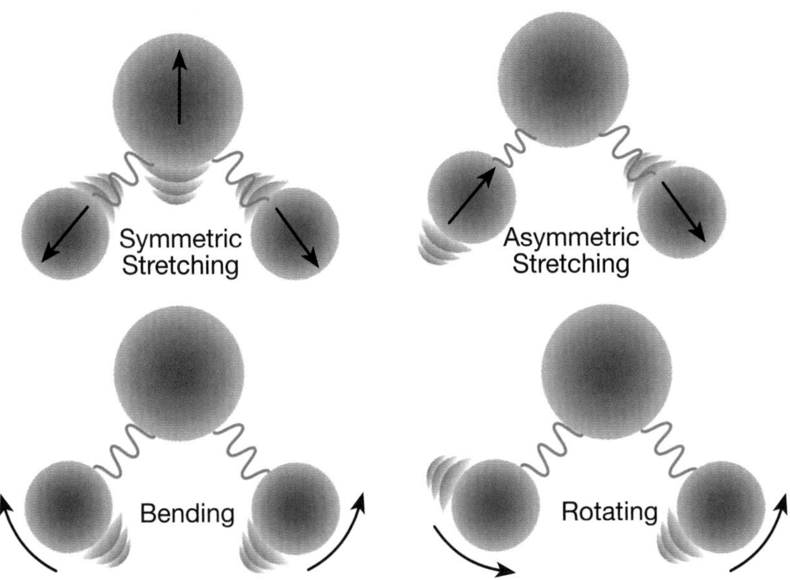

Figure 12.7 Molecules exhibit a variety of motions. Therefore, bond lengths and bond angles should be considered as average values.

Covalent Bonds

Atoms with the same or nearly the same electronegativities tend to react by sharing electrons. The shared pair or pairs of electrons constitute a **covalent bond.** Most covalent compounds are formed between atoms of nonmetals. *Covalent compounds typically have low melting points, do not conduct electricity, and are brittle.*

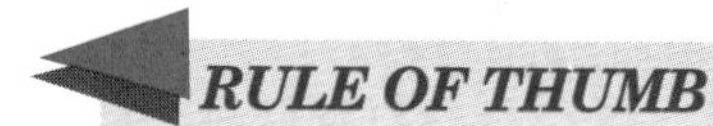

When two or more atoms bond covalently, the resulting particle is called a **molecule.** The line joining the nuclei of two bonded atoms in a molecule is called the **bond axis** as shown on the right of Figure 12.6. If one atom is bonded to each of two other atoms, the angle between the two bond axes is called the **bond angle.** The distance between nuclei along the bond axis is called the **bond length.** This length is not really fixed, because the bond acts much as if it were a stiff spring. The atoms vibrate as though the bond were alternately stretching and shrinking as shown in Figure 12.7.

Bonds also undergo bending, wagging, and rotational vibrations. These movements cause the bond angles and bond lengths to vary. The amplitudes of these vibrations are not large and the bond lengths and bond angles that we measure are, therefore, average values.

We have learned much of what we know about the structure of molecules from infrared and microwave spectroscopy. These methods of spectroscopy measure the amount of infrared or microwave radiation that a sample transmits or absorbs at different wavelengths. Infrared (IR) radiation has wavelengths of 700 nm to 50 000 nm and lies in a region of the electromagnetic spectrum between microwaves and visible light waves. The wavelengths of microwave radiation are typically between 50 000 nm and 30 cm, longer than infrared wavelengths.

HOME ECONOMICS CONNECTION

Food Frequencies

Microwave ovens operate by exposing food to frequencies (about 3 GHz) corresponding to the rotation of water molecules. Water molecules absorb this radiation and increase their rotational energy. The increased energy leads to a rise in the temperature of the food through molecular collisions.

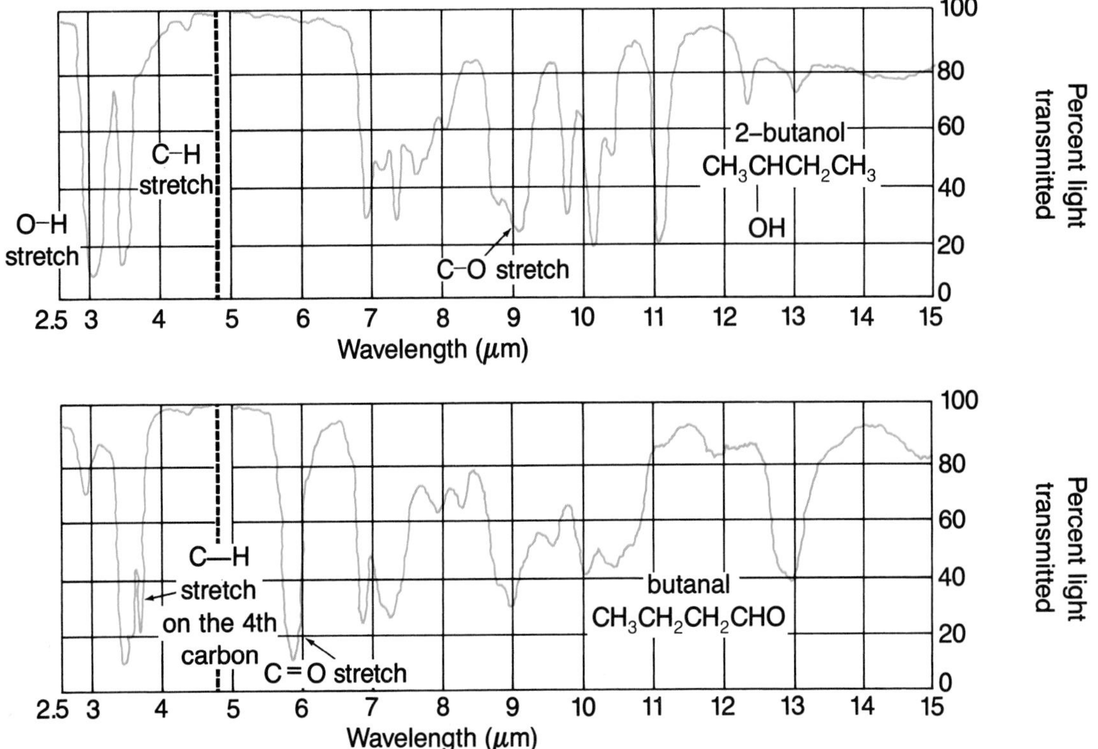

Figure 12.8 For the infrared spectra shown, note that the composition of both compounds is similar. However, the difference in the way the atoms are arranged causes different stretching motions in each molecule. Characteristic peaks in each spectrum can be used to identify the compound.

A molecular compound can often be identified by the infrared radiation it absorbs or transmits. Each molecular compound has a unique infrared spectrum. A sample of the compound is subjected to various wavelengths of IR radiation. The absorbance of IR radiation by the compound at each wavelength is measured and recorded graphically. Comparison with known spectra can be used to identify the compound, just as fingerprints can be used to reveal the identity of a person.

Look at the two spectra in Figure 12.8. The two compounds are not very different, yet the spectra are clearly distinguishable. As you can see, each compound absorbs IR radiation at specific wavelengths, and, therefore, at specific frequencies. The frequencies are determined by the structure of the molecule, the nature of the atoms bonded together in the molecule, and the effects of surrounding atoms. Radiation of wavelengths corresponding to those frequencies can be absorbed because the radiation has the same frequencies as the natural frequencies of vibration of the molecule. From measurements of these absorbances, we can compute bond strengths and determine a great deal about how and where particular atoms are bonded in the molecule.

Microwave radiation, on the other hand, affects the rotation of molecules. The way molecules rotate is determined principally by the mass of the molecule and the distribution of that mass. By substituting a different isotope of an atom in a molecule and seeing how that substitution changes its microwave spectrum, the location of that atom in the molecule can be determined.

Metallic Bonds

Metals are characterized by high electrical conductivity, luster (shininess), and malleability. What bonding arrangement could lead to these properties? Metals in the solid state form crystals in which each metal atom is surrounded by either eight or twelve neighboring metal atoms. Since metals have only 1, 2, or 3 outer-level electrons, they do not form normal covalent bonds with all their neighbors. Because all atoms of an element have the same attraction for their electrons, there is no tendency of metal atoms to form ions. Rather, metal crystals form when atoms crowd together and the outer-level orbitals from all those atoms overlap. The electrons can then move easily from one atom to the next. The electrons are said to be delocalized because they are not held in one "locality" as part of a specific ion or covalent bond. If an external electric field is applied, the electrons will flow through the metal, creating an electric current. Delocalized electrons interact readily with light, creating the luster of metals. When metals are pounded with a hammer, the atoms rearrange and the electrons rearrange into the orbitals of the atoms in the new positions. The delocalized electrons holding metallic atoms together constitute the **metallic bond.**

The properties of metals are determined by the number of outer electrons available. Group 1 (IA) metals have only one outer electron per atom. These metals are soft. Group 2 (IIA) metals have two outer electrons and are harder than Group 1 (IA) metals. In the transition elements, however, electrons from the partially filled *d* orbitals may take part in the metallic bond. Many of these metals are very hard.

Groups 3 (IIIB) through 6 (VIB) elements have three through six delocalized electrons. In the elements of Groups 7 (VIIB) through 10 (VIIIB), the number of delocalized electrons remains at six because not all of the *d* sublevel electrons of these elements are involved in the metallic bond. The number of delocalized electrons per atom begins to decrease with the metals of Groups 11

Figure 12.9 Some metals, such as iron, deform like putty during the forging process (left). Sodium, a Group 1 (IA) metal, is so soft it can be cut easily with a knife (right).

(IB) and 12 (IIB). Going from Groups 13 (IIIA) through 18 (VIIIA), the nonmetals, the metallic properties decrease more rapidly.

The strong metallic bond of our structural metals, such as iron, chromium, and nickel, makes them hard and strong. In general, the transition elements are the hardest and strongest elements. It is possible to strengthen some of the elements that have fewer delocalized electrons by combining them with other metals to form alloys. These alloys have properties different from those of pure elements.

Bridge to *METALLURGY*

Alloys

Many metallic materials are not pure elements. Brass, steel, and bronze are examples. These materials are alloys. An alloy is a metallic material that consists of two or more elements, usually metals.

Some pairs of metals are soluble in each other in all proportions. Alloys made from these pairs produce solid solutions, for example, copper-nickel. Some metal pairs will not dissolve completely in each other; thus, alloys of these pairs are heterogeneous mixtures, such as aluminum-silicon.

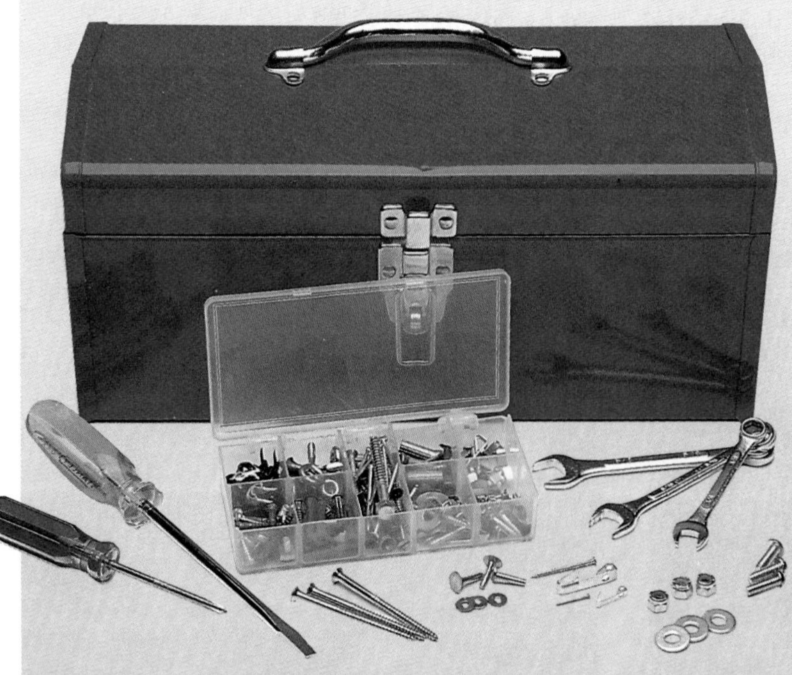

The solubility of one metal in another is determined mainly by the relative sizes of the atoms. Metals with atoms of similar size tend to be soluble in each other as are elements whose atoms are very much smaller than the other element.

Steel is an alloy of iron and a nonmetal, carbon, containing up to 2% carbon. Manufacturers often add other elements to steel to give it some special property. Iron is particularly subject to corrosion. By adding some chromium and nickel to the iron and carbon, we obtain stainless steel, which does not rust. Adding tungsten produces a steel that retains its hardness even at high temperatures. It is used to make cutting tools for metalworking. Manganese steels are very hard; they are used to make the jaws on rock crushing machinery and parts of bank vaults. Vanadium produces very tough steel that is used, among other things, to make the crankshafts in automobile engines.

Exploring Further

1. How are alloys classified?
2. Examine a phase diagram for an alloy and learn how to interpret the various areas of the diagram.

Table 12.4

Chemical Bond Summary					
INTERATOMIC BONDS	**Bond Type**	**Generally Formed between**	**Bond Formed by**	**Properties of Bond Type**	**Substances Utilizing Bond Type**
	Covalent	Atoms of nonmetallic elements of similar electronegativity	Sharing of electron pairs	Stable nonionizing molecules–not conductors of electricity in any phase.	OF_2, C_2H_6, $AsCl_3$, $GeCl_4$, C, SiC, Si
	Ionic	Atoms of metallic and nonmetallic elements of widely different electronegativities	Electrostatic attraction between ions resulting from transfer of electrons	Charged ions in gas, liquid, and solid. Solid is electrically nonconducting. Gas and liquid are conductors. High melting points.	NaCl, K_2O, BaS, LiH, CdF_2, $BaBr_2$, $ErCl_3$, CdO, Ca_3N_2
	Metallic	Atoms of metallic elements	Delocalized electron cloud around atoms of low electronegativity	Electrical conductors in all phases, lustrous, very high melting points.	Na, Au, Cu, Zn, Ac, Be, Gd, Fe, Dy

Polyatomic Ions

There are a large number of ionic compounds made of more than two elements. In these compounds, at least one of the ions consists of two or more atoms covalently bonded. However, the particle as a whole possesses an overall charge.

For example, consider the hydroxide ion (OH^-). $[:\ddot{O}:H]^-$

The oxygen atom is bonded covalently to the hydrogen atom. The hydrogen atom is stable with two electrons in its outer level. The hydrogen atom contributes only one electron to the octet of oxygen. The other electron required for oxygen to have a stable octet is the one that gives the 1− charge to the ion. Although the two atoms are bonded covalently, the combination still possesses charge. Such a group is called a polyatomic ion. Polyatomic ions form ionic bonds just as other ions do. Table 7.4 gives some of the more common polyatomic ions with their charges. Note that the charge of most polyatomic ions is negative. An important positive polyatomic ion is the ammonium ion, NH_4^+.

Figure 12.10 The sulfate ion contains S — O covalent bonds. The electron dot structure shows a stable octet for each atom. However, the particle as a whole has a negative charge.

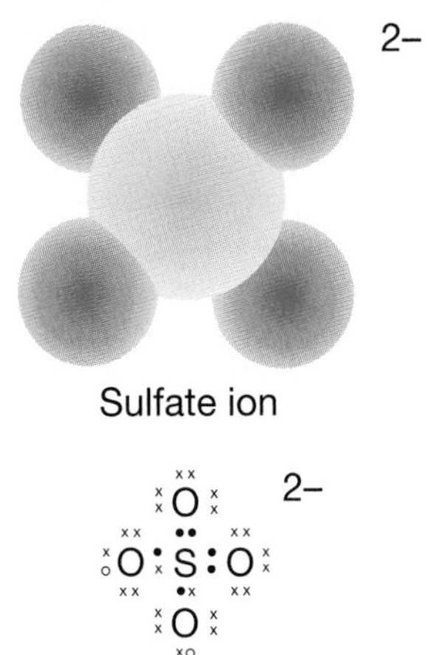

FRONTIERS

Alien Alloys

Wing components of this aircraft are made of an aluminum-lithium alloy, a material that is stronger than aluminum, but less dense.

Mixing molten metals has been a standard method of manufacturing alloys for centuries. This traditional method has prohibited manufacture of alloys using the more active metals, such as lithium. However, the aluminum industry, realizing the potential uses of Al-Li alloys in the aerospace industry, has designed furnaces and techniques to produce these alloys safely and economically.

Many of the new techniques involve changes in the cooling of the alloy. In standard methods, alloys are cooled slowly, allowing atoms to form regular, crystalline structures within the alloy. What would happen if cooling is accelerated to a rate equivalent to a temperature drop of 100 000 C° in the blink of an eye? The atoms would be frozen in position—producing glasslike materials. This rapid solidification has led to some unusual alloys. One such alloy is made from iron, boron, and silicon and has the magnetic properties of iron, the rigidity of glass, and doesn't corrode. Its use in electrical transformers may save several billion dollars yearly by reducing the waste of electrical energy.

Another technique chemically combines metals to form materials called intermetallic compounds. Unlike alloys, intermetallic compounds have properties that differ significantly from the metals they contain. One such intermetallic compound is aluminum-titanium. Unlike most alloys, this alloy maintains its strength at extremely high temperatures. Your future car may be powered by an efficient and lightweight aluminum-titanium engine.

12.1 CONCEPT REVIEW

5. Describe the periodic nature of electronegativity values of the elements.
6. Compare and contrast ionic, covalent, and metallic bonds.
7. How does the behavior of molecules differ with the absorption of infrared radiation and microwave radiation?
8. **Apply** Is the attraction of the sulfur atom for one of the bonding pairs of electrons in a molecule of SO_3 greater than or less than that of the sulfur atom in a molecule of SO_2? Explain.

12.2 Particle Sizes

OBJECTIVES

- differentiate among atomic radii, ionic radii, covalent radii, and van der Waals radii.
- discuss factors that affect the values of ionic radii and covalent radii.
- use covalent radii to calculate bond lengths.

In trying to make predictions about the reactivity or the properties of a substance, chemists frequently need to know how large the particles are that compose the substance. In Chapter 10, you learned about the concept of atomic radius and its periodic nature. In that chapter you also learned about the comparative sizes of ions and the atoms from which they were formed. Now we need to take a closer look at the sizes of particles in compounds.

Ionic Radii

We saw in Chapter 10 that a sodium ion is smaller than a sodium atom. We also found that a chloride ion is larger than a chlorine atom. Values for the radii of many ions have been determined, and some are listed in Table 10.1. These values are found from a combination of experimental data and simplifying assumptions. By adding the radii of two ions in a compound, we may find their **internuclear distance** in a crystal. How can ionic radii be determined? Consider compounds such as LiI or LiBr in which the negative ions are very much larger than the positive ions. In crystals of these compounds, we can assume that the negative ions are in contact. Therefore, the ionic radii of the negative ions are one-half the internuclear distances of these compounds. Once such radii are determined, the remainder can be determined from internuclear distances of other compounds. Remember that the radii are not fixed values. One reason for their variability is the "fuzziness" of the electron cloud. Another reason is the effect each ion has on its neighboring ions.

Covalent Radii

It is possible, by experiment, to determine the internuclear distance between two bonded atoms. For example, consider iodine(I) chloride, ICl. What are the radii of the iodine and chlorine atoms

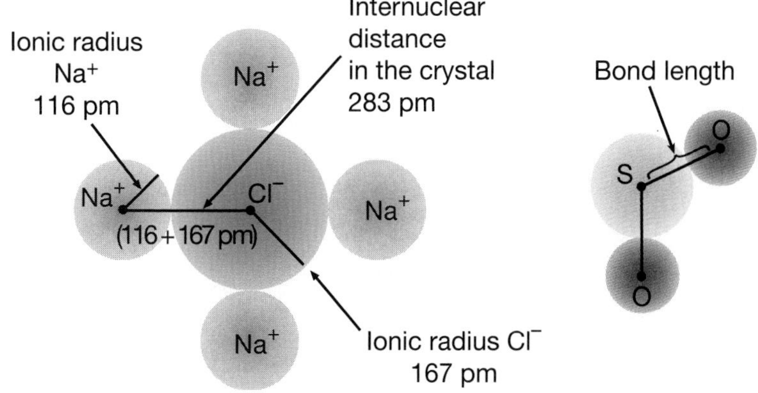

Figure 12.11 In ionic compounds such as sodium chloride (left), the internuclear distance is the sum of the radii of both ions. In a molecule such as sulfur dioxide (right), the bond length is the sum of the covalent radii of both atoms forming the bond.

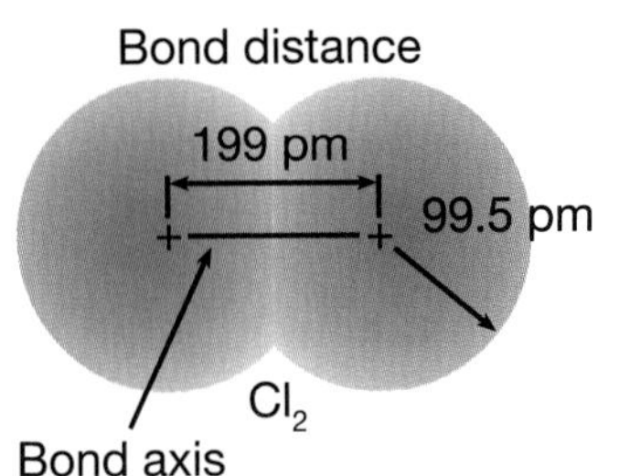

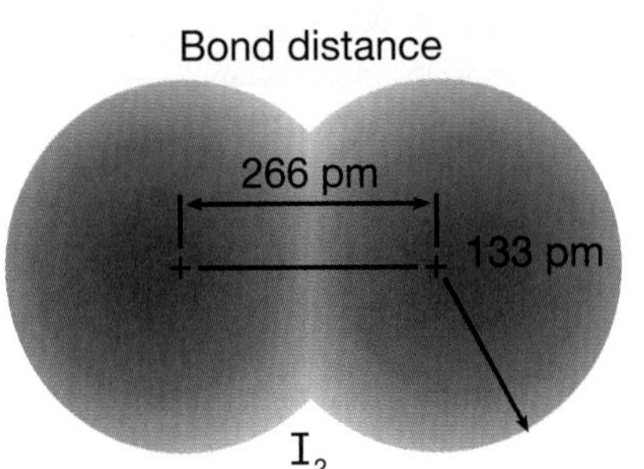

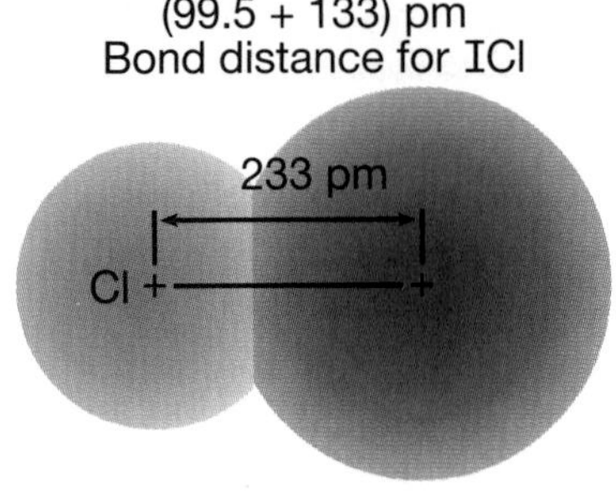

Figure 12.12 The internuclear distance is the sum of the covalent radii of each atom. Recall that the bond distances are not fixed values.

in this molecule? The internuclear distance in ICl is found to be 230 pm. The internuclear distance in Cl_2 is 199 pm, and in I_2 it is 266 pm as shown in Figure 12.12. One-half of each of these values might be taken as the radii of the chlorine and iodine atoms in a covalent bond: 99.5 pm and 133 pm. The sum of the iodine and chlorine radii would then be 233 pm. This sum is in good agreement with the observed bond distance in ICl. Covalent radii are only approximate. The value for hydrogen is less reliable than that for other atoms. Nevertheless, these radii are very useful in predicting bond lengths in molecules. Table 12.5 gives the covalent radii for some common atoms; Table 12.6 gives the bond lengths for some molecules. See for yourself how well the predicted bond lengths agree with the measured ones.

Remember that covalent radii are used to find the internuclear distance between atoms bonded to each other. Like electronegativities, covalent radii are average values. The radius of a particular atom is not constant. Its size is influenced by the other atom or atoms to which it is bonded.

Table 12.5

Covalent Radii (in picometers)*

Atom	Radius	Atom	Radius
Al	118	Li	134
As	120	Mg	130
B	82	N	70
Be	90	Na	154
Bi	150	O	73.5
Br	114	P	110
C	77.2	Pb	146
Ca	174	S	103
Cl	99.5	Sb	140
F	71.5	Sc	144
Ga	126	Se	119
Ge	122.3	Si	117.6
H	37.07	Sn	140.5
I	133	Te	142
K	196	Ti	152

*All digits given are significant.

Table 12.6

Experimental Bond Lengths (in picometers)

Molecule	Bond	Length
BCl_3	B—Cl	175
B_2H_6	B—H	132
$BeCl_2$	Be—Cl	177
Diamond	C—C	154.4
CH_4	C—H	109.4
CH_3I	C—H	109.6
	C—I	213.9
ClBr	Cl—Br	213.8
HF	H—F	91.7
H_2O	H—O	95.7
$LiCl_2(C_4H_8O_2)_2$	Li—O	195
NH_3	N—H	101.7
$OBe_4(CH_3COO)_6$	Be—O	166.6
OF_2	O—F	140.5
O_3	O—O	120.74
H_2SAlBr_3	S—Al	243
$(H_3Si)_2NN(SiH_3)_2$	Si—N	173

SAMPLE PROBLEM

Bond Length

Predict the length of the bond formed between an atom of arsenic and an atom of sulfur.

Solving Process:

The length of the bond is the sum of the covalent radii of the two elements. From Table 12.5, the covalent radius of arsenic is 120 pm and that of sulfur is 103 pm. Therefore, the bond length is the sum of 120 pm and 103 pm, or 223 pm.

PRACTICE PROBLEMS

9. Predict the length of the bond formed between each of the following pairs of atoms.

a. Al—Cl **b.** H—I **c.** N—P **d.** Se—S **e.** B—F

CHEMISTRY IN DEPTH

▼ Van der Waals Radii

A certain minimum distance is maintained between atoms that are not bonded to each other. This limitation exists because the electron cloud of one atom repels the electron cloud of other atoms.

In effect, colliding free atoms and molecules act as if they had a nearly rigid outer shell. This shell limits the closeness with which they may approach other atoms or molecules. As shown in Figure 12.13, bonded atoms are closer than atoms that are not bonded because the covalent bond consists of shared electrons. The radius of this imaginary rigid shell of a nonbonded atom is called the **van der Waals radius.** It is named for the Dutch physicist Johannes van der Waals. Chemists generally assume that an atom's rigid shell is at the point where the probability of finding electrons has dropped below 90 percent. ▲

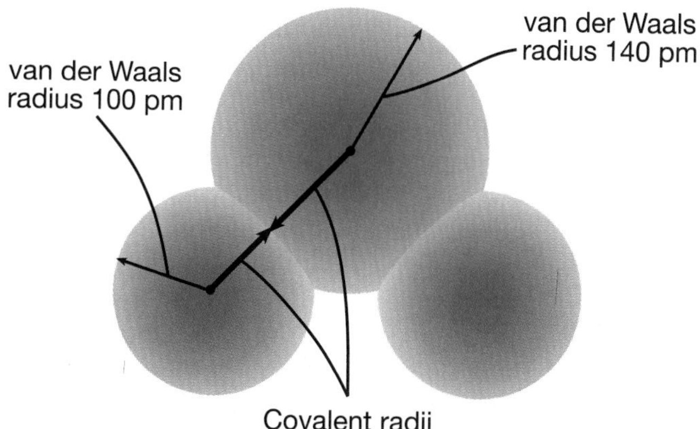

Figure 12.13 The van der Waals radius is the minimum distance between atoms on adjacent molecules. Note that the van der Waals radius will be larger than the covalent radius of a bonded atom.

Two Bonds Are Better Than One

Put on an apron and goggles. Place 5 cm^3 of white glue in a small, disposable plastic cup. Stir in about 1 cm^3 of a saturated solution of sodium borate, one drop at a time, until a sticky lump forms. Remove the lump from the cup and thoroughly rinse it under running tap water. Roll the lump into a ball. Can you bounce the ball? Set the ball on a flat surface and let it remain there undisturbed for 5 minutes. What happens to the ball?

White glue consists of long, chainlike molecules called polymers. When the glue and the sodium borate solution are mixed, the borate ions cross-link the polymers in the glue like rungs on a ladder. The borate ion can form cross-links because it has two bonding sites, one at each end of the ion. Conjecture on how cross-linking could account for the properties of the material you just made.

Table 12.7

van der Waals Radii (in picometers)

Atom	Radius	Atom	Radius
As	185	O	152
Br	185	P	180
C	170	Pb	202
Cl	175	S	180
F	147	Se	190
Ga	187	Si	210
H*	120	Sn	217
I	196	Te	206
N	155		

*The van der Waals radius for hydrogen when it is hydrogen-bonded (Section 17.2) is 100 pm.

Summary of Radii

Thus far, we have studied four radii—atomic, ionic, covalent, and van der Waals. How do these various measurements differ? How are these measurements the same? Atomic radii are measured on atoms in crystals.

Ionic radii differ from atomic radii because of the loss or gain of electrons. This difference was discussed in Chapter 10. Since most ionic radii are determined from ionic crystals, their values are consistent with the data. Covalent radii and van der Waals radii are quite variable due to the wide range of atoms to which the subject atom may be bonded. We would expect covalent radii to be less than an atomic radius. However, if an atom is bonded to more than one other atom, its electron cloud may be distorted. The distortion may make its covalent radius larger than the atomic radius. The same situation occurs with van der Waals radii. For both of these radii, the data given in the tables in this chapter represent average values for the atom bonded to its usual number of neighboring atoms. In every case, we can use the radii to predict the internuclear distance between atoms.

12.2 CONCEPT REVIEW

10. Explain how an atomic radius differs from each of the following: an ionic radius; a covalent radius; a van der Waals radius.

11. What is the minimum internuclear distance for a collision between a nitrogen molecule and a phosphorus molecule?

12. Why are the values of ionic and covalent radii approximate values?

13. Apply Select an atom from those listed in Table 12.7. Then choose two elements that will each covalently bond with the atom that you selected. Choose the two elements so that one decreases the van der Waals radius of the atom that you selected and the other increases its van der Waals radius. Explain how you made your two choices.

Chapter 12 REVIEW

Summary

12.1 Bond Formation

1. The relative tendency of a bonded atom to attract shared electrons to itself when it is bonded to another atom is called electronegativity.
2. Ionic bonds are formed by transfer of electrons between atoms with a large difference in electronegativity.
3. Ionic compounds are characterized by high melting points, solubility in water, and crystal formation.
4. Covalent bonds are formed by the sharing of electrons between atoms with either no difference or slight differences in electronegativity.
5. The bond axis is a line joining the nuclei of two bonded atoms. The length of the bond axis is called the bond length. The angle between two bond axes is called the bond angle.
6. A metallic bond is formed between atoms with few electrons in the outer level. These electrons circulate as delocalized electrons and allow metals to carry an electric current.
7. Polyatomic ions are composed of groups of atoms bonded covalently and possess an overall charge just as other ions.

12.2 Particle Sizes

8. The ionic radius is the best estimate chemists can make of the effective size of an ion.
9. Covalent radii can be used to predict the distance between bonded atoms.
10. Electron clouds repel each other strongly when two nonbonding atoms approach each other. The distance of closest approach for an atom is called the van der Waals radius of the atom.

Key Terms

electronegativity	bond angle
ionic bond	bond length
covalent bond	metallic bond
molecule	internuclear distance
bond axis	van der Waals radius

Review and Practice

14. Based on its location on the periodic table, predict which element in each of the following pairs has the greater electronegativity.
    - **a.** Rh — Ru
    - **b.** Ga — Ti
    - **c.** Cs — Sr
    - **d.** Al — Ti
    - **e.** Mo — Zr
    - **f.** Co — Re
15. Characterize the bond between the following pairs of elements as principally ionic or covalent.
    - **a.** Li and O
    - **b.** Se and F
    - **c.** Sr and Br
    - **d.** Ca and S
    - **e.** Cr and Cl
    - **f.** Mn and S
    - **g.** C and Br
    - **h.** Zn and I
16. What force holds ions together in ionic bonds? What causes this force?
17. Why is it difficult to give exact values for bond lengths and bond angles?
18. Using Table 12.5, predict the bond lengths indicated for the following substances.
    - **a.** F—F in F_2
    - **b.** C—Pb in $Pb(C_2H_5)_4$
    - **c.** Li—P in Li_3P
    - **d.** Rb—Si in Rb_4Si (Rb = 198 pm)
    - **e.** C—C in CH_3CH_3
    - **f.** N—O in N_2O_4
19. How does van der Waals radius differ from covalent radius?
20. List three characteristics of metals.
21. Why are the alkali metals and the alkaline earth metals very soft metals? How could they be strengthened?

22. Explain how $NaNO_3$ is an example of two different types of bonding.

23. Construct a graph of the number of delocalized electrons versus the atomic number for elements $Z = 21$ through $Z = 30$.

24. Three elements have electron configurations of $1s^22s^22p^4$, $1s^22s^22p^63s^1$, and $1s^22s^22p^63s^23p^5$. Their electronegativities (not in the same order) are 1.01, 2.83, and 3.50. The covalent radii in picometers are 73.5, 99.5, and 154. Identify each element and match with the correct electronegativity and covalent radius.

25. Use Tables 12.1 and 12.3 to estimate the percent ionic character and the percent covalent character of a bond between a magnesium atom and a phosphorus atom.

Concept Mastery

26. How is the chemical activity of elements related to their electronegativities?

27. Use the definition of electronegativity to explain why small differences in electronegativity between two atoms result in the formation of a covalent rather than an ionic bond between the atoms.

28. Considering your study of bond character and its relationship to electronegativity, how should you interpret the statement that $CaCl_2$ is an ionic compound while PCl_3 is a covalent compound?

29. How does the number of delocalized electrons in a metal affect its properties?

30. Explain how the nature of the metallic bond accounts for electrical conductivity of metals.

31. Without using the term electronegativity, formulate a statement to explain why sodium and chlorine atoms form an ionic bond, while carbon and chlorine atoms form a covalent bond.

32. Why is the term *ionic molecule* incorrect?

33. What factors influence electronegativity? Use your answer to explain the difference in electronegativities for fluorine and potassium.

34. Examine the way electronegativity changes as you look down through Group 1 (IA), the alkali metals, in Table 12.1. Explain how distance from the nucleus and the shielding effect may contribute to this change in electronegativity.

35. How does the oxidation number of an element relate to its charge in an ionic bond? How does the ion charge relate to the number of electrons transferred when the atom becomes an ion?

36. Many metals will stand a great deal of distortion without breaking. They can be bent, folded, rolled thin into sheets of foil, drawn through progressively smaller dies into thin wire, and can be beaten into various shapes. How does the electron structure of metals account for these properties?

Application

37. Each of the following diagrams indicates a type of radius. Tell whether each radius indicated is atomic, covalent, ionic, or van der Waals.

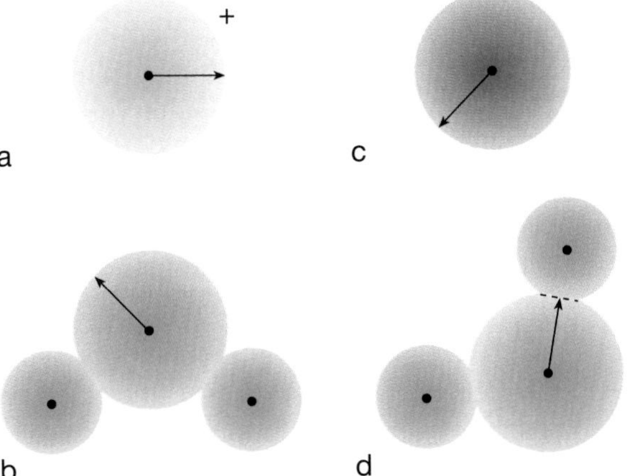

38. Think of ways different materials are used in building houses. For each of the following, give two examples of places where this type of substance, considering its properties, should be used? Why?
 a. a covalent compound
 b. an ionic compound
 c. a metal

39. Draw electron dot structures for the following polyatomic ions. Use dots to indicate the electrons from one type of atom, *x*'s to indicate electrons from the other type of atom, and *o*'s to indicate any electrons that are gained to stabilize the total electron configuration.
 a. NH_4^+ **b.** OH^- **c.** $C_2O_4^{2-}$ **d.** O_2^{2-}

42. Many pure metals are good conductors of electric current. However, the same metals are less conductive when in an unrefined or crude state. Explain this change in behavior.

43. The general trend in electronegativity from left to right across the periodic table is an increase. The trend down a column is a decrease. Explain each of the following observations.
 a. Zinc has an electronegativity of 1.66 and copper, which precedes it, a value of 1.75.
 b. Gallium has an electronegativity of 1.82 and aluminum, which precedes it, a value of 1.47.

Critical Thinking

40. Why are most ionic compounds brittle?

41. The metals copper, zinc, silver, cadmium, and gold are useful but are soft and relatively weak when pure—especially in comparison with metals that precede them in each period on the periodic table. Explain why this is fact. Suggest a way to add strength to these metals for practical use.

Readings

Borman, Stu. "New Tin-Based Propellane May Shed Light on Bonding Questions." *Chemical and Engineering News* 67, no. 32 (August 7, 1989): 29-31.

Imura, Toru. "Advanced Metallic Materials." *Chemtech* 19, no. 4 (April 1989): 234-237.

Cumulative Review

1. List ten elements having more than two possible oxidation states.

2. Briefly describe the development of classification for elements from Dobereiner to the modern periodic table.

3. Predict the oxidation number of the following elements, using only the periodic table as a guide.
 a. astatine, germanium, mercury, polonium, tin
 b. francium, hafnium, neodymium, rubidium, tellurium

4. Describe the structure of the modern periodic table.

5. What four factors affect the values obtained for ionization energies of an element?

6. Use the periodic table to predict oxidation numbers for the following elements.
 a. cesium **c.** niobium **e.** iodine
 b. zirconium **d.** selenium **f.** boron

7. Classify each of the elements in the previous problem as metal, metalloid, or nonmetal.

8. What are the family names of the elements in the following groups?
 a. 1 (IA) **c.** 17 (VIIA)
 b. 2 (IIA) **d.** 18 (VIIIA)

CHAPTER **P**review

CONTENTS

CHAPTER 13

Molecular Structure

The structure shown at the right is a model of a molecule of buckminsterfullerene, C_{60}. It is named for Buckminster Fuller, an American engineer and architect who invented the geodesic dome shown in the larger photograph. The domes designed by Fuller proved to be especially stable and able to withstand strong environmental forces.

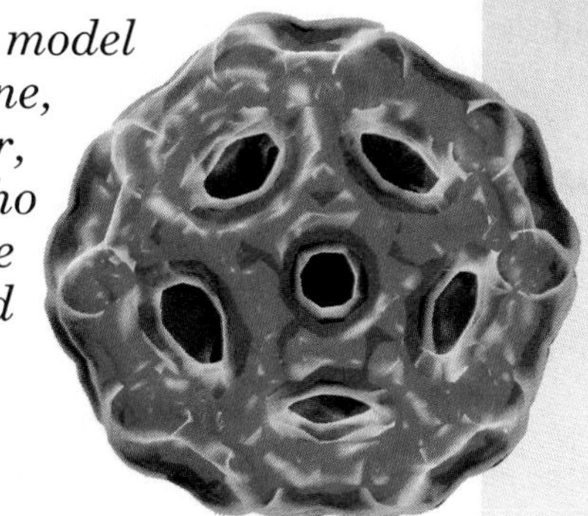

Like the geodesic dome, the structure of a C_{60} molecule is especially stable. In fact, there are a large number of even-numbered carbon atom clusters that are stable. These molecules are called, collectively, fullerenes. The C_{60} molecule is of special interest to scientists because it has many interesting and useful properties, such as magnetism and superconductivity, all of which are due to the structure of the molecule.

EXTENSIONS AND CONNECTIONS

OBJECTIVES

- use models to explain the structure of a given organic or inorganic molecule.
- describe hybrid orbitals and use hybridization theory to explain the bond angles in compounds.
- differentiate sigma and pi bonding and saturated and unsaturated carbon compounds.

13.1 Bonds in Space

A primary goal of studying chemistry is to learn how the macroscopic properties of matter are a consequence of its molecular structure. You have learned a great deal about the structure and properties of atoms. Their behavior is determined chiefly by their electron configurations. The behavior of molecules also depends on their structural characteristics. This section is devoted to examining the shapes of molecules, and what characteristics of their bonds produce those shapes.

Many consumer products are packaged in polystyrene foam, which is made of huge molecules of a substance named polystyrene. There are thousands of atoms in a molecule of polystyrene. How is it possible to study such a large molecule? Fortunately, atoms usually combine in large molecules in the same manner that they do in small molecules. By studying the relationship between structure and properties in small molecules, we can better understand the structure and properties of large molecules.

Electron Distribution

There are several ways of looking at the structure of molecules to account for their shape. We will consider two of these models. The first model takes into account the repulsive forces of electron pairs around an atom. The second model considers ways in which atomic orbitals can overlap to form orbitals around more than one nucleus. The electrons in these combined orbitals then serve to bind the atoms together.

In order to describe the shape of a molecule or polyatomic ion, it is useful to draw a Lewis electron dot diagram. In these diagrams, we arrange the outer electrons as dots around the atoms so that each atom ends up with a full outer level. *For all atoms that form covalent bonds, except hydrogen, eight electrons represent a full outer level.* Consider the water molecule. It is composed of one oxygen and two hydrogen atoms. The electron dot diagrams are ·Ö: and H× for these elements. All electrons are identical. We use different symbols for the electrons only to help us understand how we arrive at the final structure. By combining an oxygen and two hydrogens, we obtain the following electron dot diagram.

RULE OF THUMB

H:Ö:
 H

It is the only arrangement of electrons in which all three atoms can achieve a full outer level. Note that two pairs of electrons in the outer level of oxygen are involved in bonding the hydrogens. They are called **shared pairs.** The other two pairs of electrons are not involved in bonding. They are called **unshared pairs,** or lone pairs. Note that in counting electrons, a shared pair of electrons contributes to a full outer level for both atoms sharing the electron pair.

SAMPLE PROBLEM

Lewis Electron Dot Diagram

Draw the Lewis electron dot diagram for AsI_3.

Solving Process:

From the periodic table, we can see that arsenic has five electrons in its outer level and iodine has seven. Thus, arsenic tends to form three bonds and iodine one. The diagram is then

```
 ••   ××   ••
:I : As : I :
 ••   ×•   ••
     :I:
      ••
```

PRACTICE PROBLEMS

1. Draw Lewis electron dot diagrams for the following.
   a. H_2Te **c.** NI_3
   b. PF_3 **d.** CBr_4
2. How many unshared pairs of electrons are on the central atoms in 1c and 1d?

PROBLEM SOLVING HINT

The symbol used to represent electrons of a particular atom is arbitrary. However, each different element in a molecule should have a different electron symbol.

Electron Pair Repulsion

One way to account for the shape of molecules is to consider electron repulsion. Each bond and each unshared pair in the outer level of an atom form a charge cloud that repels all other charge clouds. In part, this repulsion is due to all electrons having the same charge. Another more important factor is the Pauli exclusion principle. Although two electrons of opposite spin may occupy the same orbital, electrons of the same spin may not do so. Repulsions resulting from the Pauli principle are much greater than electrostatic ones at small distances. Because of these repulsions, atoms cannot be compressed.

The repulsions between the charge clouds in the outer level of atoms determine the arrangement of the orbitals. The orbital arrangement, in turn, determines the shape of molecules. As a result, the following rule of thumb may be stated. *Electron pairs spread as far apart as possible to minimize repulsive forces.* Refer to Table 13.1 for diagrams of the bonding arrangement discussed here. If there are only two electron pairs in the outer level of the central atom, they will be on opposite sides of the nucleus. This arrangement is called linear. If there are three electron pairs, the axes of their charge clouds will be 120° apart. This arrangement is called trigonal planar and the electron pairs lie in the same plane as the nucleus of the central atom. If there are four electron pairs, the axes of the charge clouds will be farthest apart when they intersect at an angle of 109.5°. These axes will not all lie in the same plane but will form a tetrahedron. A tetrahedron is a figure having four faces, each of which is an equilateral triangle.

MINUTE  **LAB**

Like Charges Repel

As you read about electron pair repulsion theory, a balloon model of H_2O is helpful. Tie a string 50 cm long to each of two inflated balloons. Keeping the balloons separated, rub each balloon with a piece of wool cloth. Bring the strings together and observe how the balloons behave. Draw the electron dot diagram of water. Let the two balloons represent the two unshared pairs of electrons on the oxygen atom. Will the two shared pairs between oxygen and hydrogen be in the same plane as the unshared pairs? What is the bond angle between the H—O bonds?

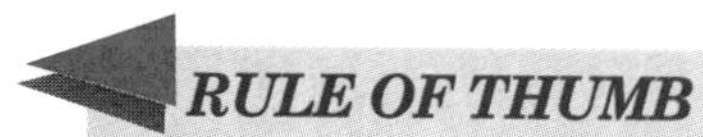

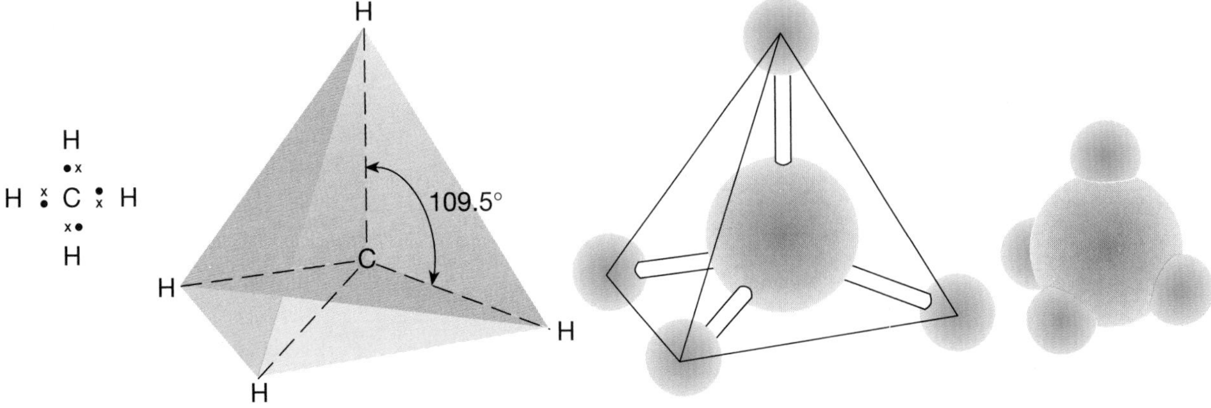

Figure 13.1 On the left, the Lewis dot diagram for methane shows a complete octet when four hydrogens bond to one carbon. The next diagram shows that each bond in methane points toward the vertex of a tetrahedron. The tetrahedral structure is also shown with a ball-and-stick model. The space-filling model on the right shows the relative volumes of the electron clouds.

The central atom is at the center and the axes extend out to the corners. See Figure 13.1.

The bonds and unshared electron pairs within a molecule determine the shape of the molecule. An unshared electron pair is acted upon by only one nucleus. Its charge cloud is shaped like a blunt pear with its stem end at the nucleus. A shared pair of electrons moves within the field of two nuclei. The cloud is therefore more slender because the attraction of the two nuclei restricts electron movement.

The electron pair repulsions in a molecule may not all be equal. The repulsion between two unshared pairs is greatest because they occupy the most space. The repulsion between two shared pairs is least because they occupy the least space. The repulsion between an unshared pair and a shared pair is an intermediate case.

unshared-unshared repulsion > unshared-shared repulsion > shared-shared repulsion

Let us consider the molecular shapes of the compounds CH_4, NH_3, and H_2O to illustrate this repulsion. See Figure 13.3. In each of these compounds, the central atom has four clouds around it. We expect the axes of all four charge clouds to point approximately in the direction of the corners of a tetrahedron.

In methane (CH_4) molecules, all clouds are shared pairs, so their sizes are equal and each bond angle is in fact 109.5°. The CH_4 molecule is therefore a perfect tetrahedron. In NH_3 molecules, there are one unshared pair and three shared pairs. The unshared pair occupies more space than any of the other three, so the bond clouds are pushed together and form an angle of 107° with each other. It is important to note that, although the electron clouds form a tetrahedron, one cloud is not involved in bonding. Therefore, the atoms composing the molecule form a trigonal pyramid. In H_2O molecules, two unshared pairs are present; both of these clouds are larger than the bond clouds. This additional cloud size results in a still greater reduction in the bond angle which is, in fact, 104.5°. Again, note that the electron clouds are tetrahedral but the molecule is "V" shaped, or bent. Note that in

Figure 13.2 Each hair on this student's head has a negative charge. Thus, the hairs repel each other.

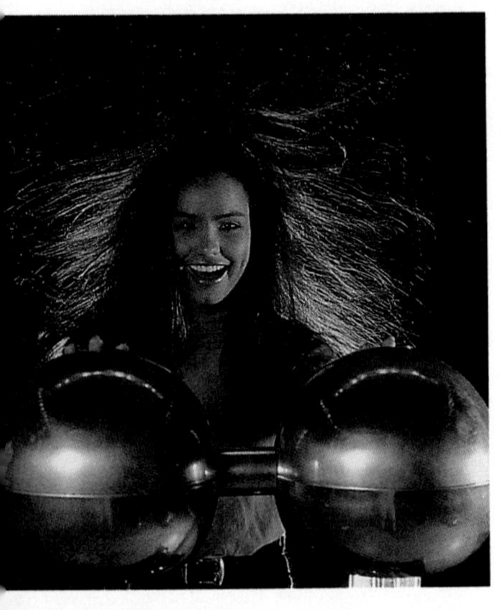

the three molecules discussed, each has four electron clouds. The differences in molecular shape result from the unequal space occupied by the unshared pairs and the bonds. The shapes for different numbers of clouds can be predicted in the same way and are listed in Table 13.1.

In most compounds, the outer level is considered full with four pairs or eight electrons. The outer level in some atoms can contain more than eight electrons (if the outer level is the third or higher level). A number of nonmetals, mainly the halogens, form compounds in which the outer level is expanded to 10, 12, or 14 electrons. Such an arrangement would also explain the formation of noble gas compounds. An example is xenon tetrafluoride, XeF_4. The structure of this compound is shown in Table 13.1. Xenon has eight electrons of its own in its outer level together with one electron from each of the four fluorine atoms.

SAMPLE PROBLEM

Bond Angles

Would the Cl—N—Cl bond angle in NCl_3 be greater than, less than, or equal to, 109.5°?

Solving Process:

First, draw an electron dot diagram for the molecule.

$$\begin{array}{c} \overset{\times\times}{\underset{\times}{\overset{\times}{}}\text{Cl}\underset{\times}{\overset{\times}{}}} \\ \overset{\bullet\times}{} \\ \underset{\times}{\overset{\times}{}}\overset{\times\times}{\underset{\times\times}{\text{Cl}}}\,\colon\,\underset{\bullet\bullet}{\text{N}}\,\colon\,\overset{\times\times}{\underset{\times\times}{\text{Cl}}}\underset{\times}{\overset{\times}{}} \end{array}$$

Note that there are four electron pairs in the outer level of the nitrogen atom. Because the four pairs would be expected to form a tetrahedron, the first approximation of a bond angle would be 109.5°. However, one of those pairs is unshared and has a greater repulsive effect. That force would push the three shared pairs closer together. Therefore, the Cl—N—Cl bond angle should be a little less than 109.5°.

PROBLEM SOLVING HINT

Drawing Lewis electron dot diagrams will help determine the number of electron pairs and whether these pairs are shared or unshared.

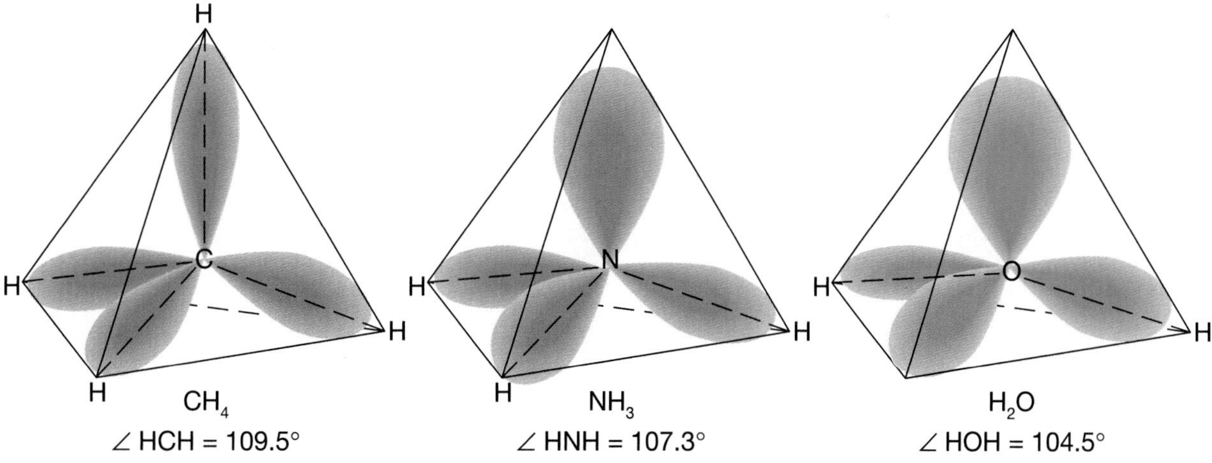

Figure 13.3 The central atom in each molecule is surrounded by four electron pairs. The influence of shared and unshared pairs accounts for the different shapes.

Table 13.1

Molecular Shapes				
Molecule	**Total Number of Electron Pairs**	**Number of Shared Pairs**	**Number of Unshared Pairs**	**Molecular Shape**
BeF_2	2	2	0	Linear
GaF_3	3	3	0	Trigonal planar
O_3	3	2	1	Bent
CH_4	4	4	0	Tetrahedral
NH_3	4	3	1	Trigonal pyramidal
H_2O	4	2	2	Bent
$NbBr_5$	5	5	0	Trigonal bipyramidal

Table 13.1 *continued*

Molecular Shapes				
Molecule	Total Number of Electron Pairs	Number of Shared Pairs	Number of Unshared Pairs	Molecular Shape
SF_4	5	4	1	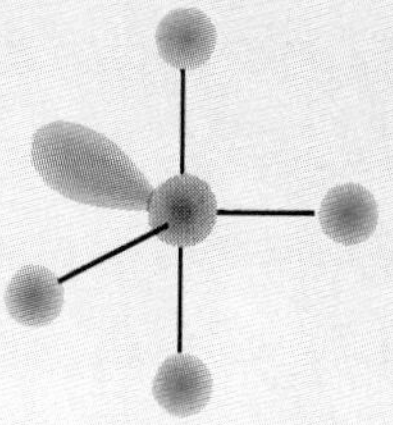 Irregular tetrahedron
BrF_3	5	3	2	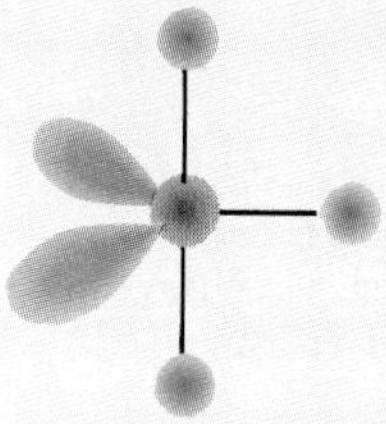 T-shaped planar
XeF_2	5	2	3	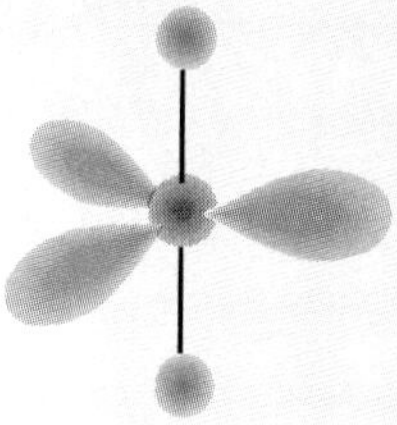 Linear
SF_6	6	6	0	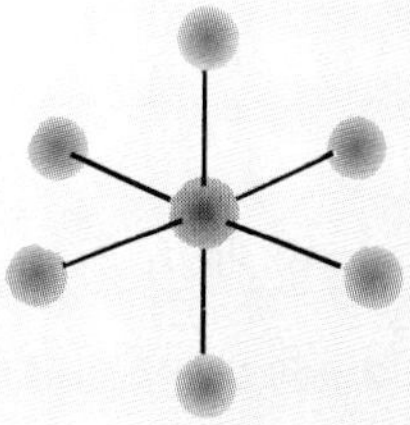 Octahedral
IF_5	6	5	1	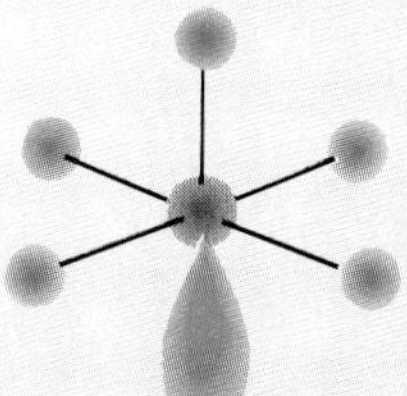 Square pyramidal
XeF_4	6	4	2	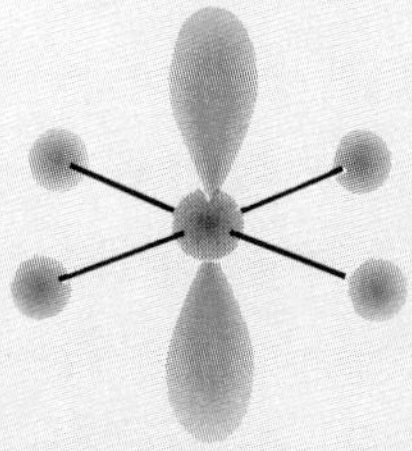 Square planar

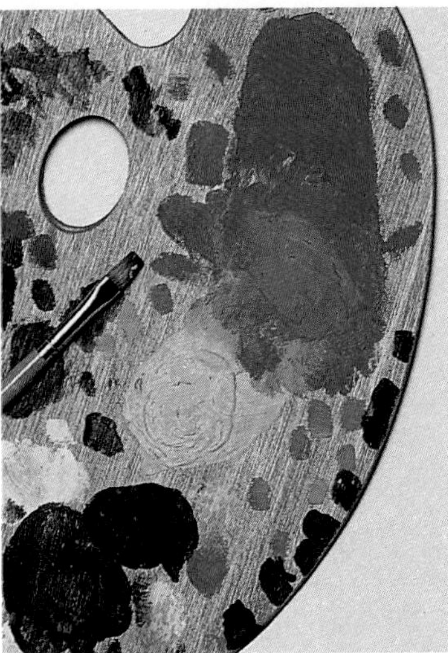

Figure 13.4 Forming hybrid orbitals is similar to mixing two colors of paint. The result is a combination of the two paints mixed and differs from each of them.

PRACTICE PROBLEMS

3. Predict whether the bond angle of each of the following is greater than, less than, or equal to 109.5°.
   a. F—N—F in NF_3
   b. F—O—F in OF_2
   c. F—Be—F in BeF_2
   d. F—As—F in AsF_5
   e. F—Te—F in TeF_4
   f. O—Xe—O in XeO_4

Hybrid Orbitals

Another model of molecular shape considers the different ways *s* and *p* orbitals can overlap when electrons are shared. This model is best illustrated using the element carbon. However, this model can also be applied to other atoms forming covalent bonds.

Consider the electron configuration of carbon, $2s^22p^2$. The electron dot diagram for carbon would be ∘C̈∘. Because the 2*s* orbital is full, we might expect that only the two unpaired electrons would be used to form bonds with other atoms, thus, X:C̈:X. We would predict the electron pairs to be distributed in a trigonal planar manner and the molecule itself to be V-shaped. The predicted bond angle would be slightly less than 120° because of the unshared pair. Instead, carbon actually forms four bonds distributed tetrahedrally. How is this structure possible?

Suppose we wish to make a gallon of fruit drink to serve at a birthday party. On hand we have one quart of limeade and three quarts of lemonade. However, we do not want to worry about who gets one flavor and who gets the other. Serving will be simpler if everyone could be served the same flavor. By mixing the limeade and the lemonade, we can obtain a gallon of single-flavored fruit drink. Note that we put together four nonidentical quarts and will get four identical quarts of mixture.

An analogous process takes place in the outer level of the carbon atom. The *s* orbital and the three *p* orbitals merge to form four orbitals that are identical. This process is call **hybridi-**

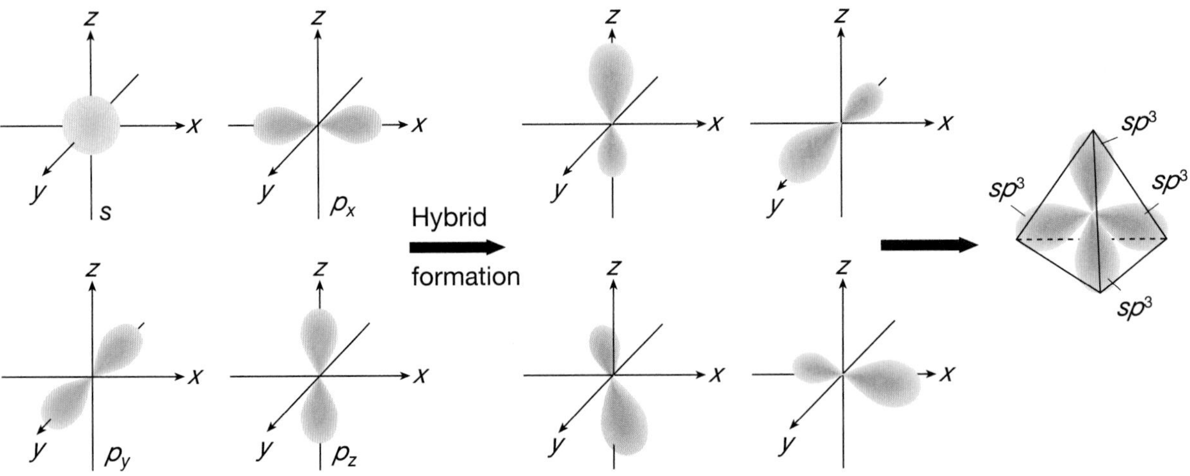

Figure 13.5 When carbon bonds, the *s* and three *p* orbitals merge to form four equivalent sp^3 hybrid orbitals. They are arranged in a tetrahedral shape.

zation and the orbitals formed are called **hybrid orbitals.** Note that four nonequivalent orbitals produced four equivalent orbitals.

Because one *s* and three *p* orbitals have merged, the hybrid orbitals are represented as sp^3 hybrids. These four orbitals are degenerate and contain one electron each. They are arranged in a regular tetrahedral shape. Each of these hybrid orbitals can then bond to another atom. If each of the hybrid orbitals bonds to an identical atom, the four bonds formed are equivalent.

Unhybridized Carbon
1*s* 2*s* 2*p*
carbon (⇅) (⇅) (↓) (↓) ( )
unhybridized orbitals
Hybridized Carbon
1*s* $2sp^3$
carbon (⇅) (↑) (↑) (↑) (↑)
4 hybrid orbitals

Geometry of Carbon Compounds

The bonding of four hydrogen atoms to one carbon atom forms methane. The bonds involve the overlap of the *s* orbital of each hydrogen atom with one of the sp^3 hybrid orbitals of a carbon atom. A three-dimensional representation of the formula of methane is shown in Figure 13.6. There is an angle of 109.5° between each carbon-hydrogen bond axis.

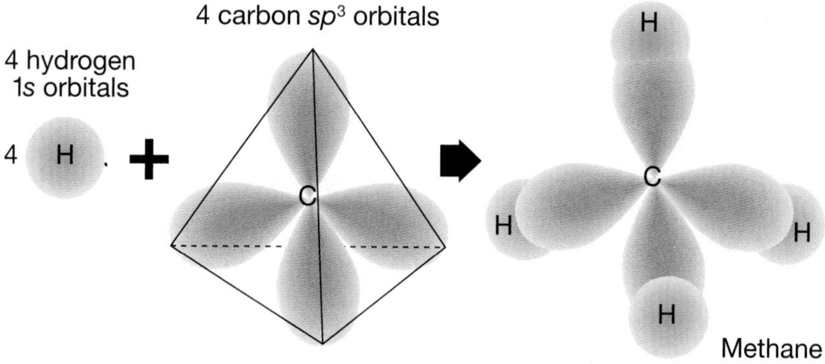

Figure 13.6 The methane structure can be explained by combining a hydrogen *s* orbital with each of the sp^3 hybrid orbitals of carbon.

Recall that carbon exhibits catenation. This structure occurs when two carbon atoms bond to each other by the overlap of an sp^3 orbital of one carbon with an sp^3 orbital of another carbon. The remaining three sp^3 orbitals of each carbon atom may bond with the *s* orbital of three hydrogen atoms. The compound formed in this way, C_2H_6, is called ethane. A three-dimensional structure of ethane is shown in Figure 13.7.

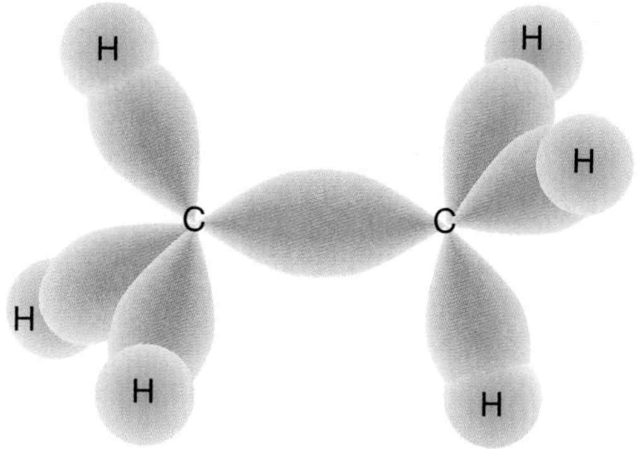

Figure 13.7 The shape of ethane is explained by combining two sp^3 hybrid carbon atoms. The single bond is a direct overlap of one hybrid orbital from each carbon atom.

Sigma and Pi Bonds

A covalent bond is formed when an orbital of one atom overlaps an orbital of another atom and they share the electron pair in the bond. For example, a bond may be formed by the overlap of two half-filled *s* orbitals. When two orbitals form a bond that lies directly on the bond axis, the bond is called a **sigma bond,** and is designated σ. See Figure 13.8. A sigma bond can also be formed by the overlap of an *s* orbital with a *p* orbital, the overlap of two *p* orbitals, the overlap of two hybrid orbitals, or the overlap of a hybrid orbital with an *s* orbital. In Figure 13.7, the *s* orbitals of hydrogen atoms are shown overlapping hybrid sp^3 orbitals of carbon. Also in Figure 13.7, the overlap of two sp^3 hybrid orbitals of carbon is shown.

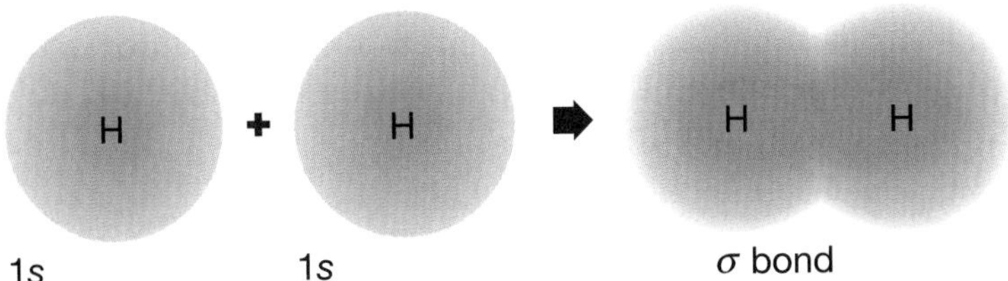

Figure 13.8 The overlap of two *s* orbitals is a sigma bond. The H-H bond in hydrogen gas is a sigma bond.

Because *p* orbitals are not spherical, when two half-filled *p* orbitals overlap, one of two types of bonds can form. If the two *p* orbitals overlap along an axis in an end-to-end fashion, a sigma bond forms. If the two *p* orbitals overlap sideways (parallel), they form a **pi bond,** designated π.

A molecule that shows both types of bonds is ethylene (ethene). In ethylene, one *s* orbital and two *p* orbitals of a carbon atom are considered to merge to form three sp^2 hybrid orbitals. These three sp^2 hybrid orbitals lie in the same plane at 120° bond angles. The third *p* orbital is not involved in this hybridization. It is perpendicular to the plane of the three sp^2 orbitals. See a model of this hybridization in Figure 13.9.

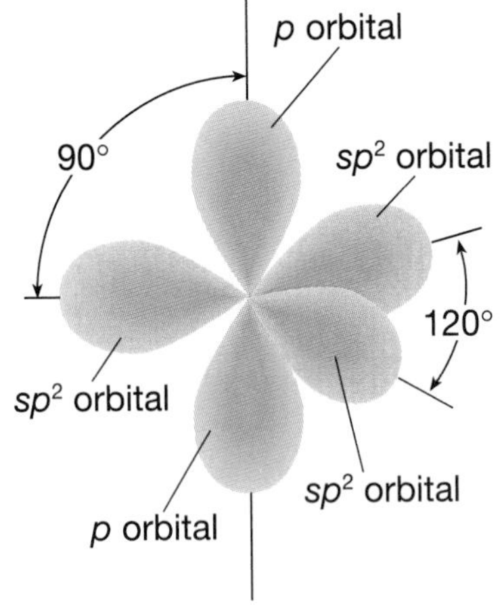

Figure 13.9 When one *s* orbital and two *p* orbitals merge to form three sp^2 orbitals, one *p* orbital is not hybridized. It is perpendicular to the sp^2 orbitals, which lie in the same plane at 120° angles from each other.

Now, if two sp^2 orbitals of adjacent carbon atoms overlap in an end-to-end fashion, a sigma bond is formed. Imagine that the two remaining sp^2 hybridized orbitals on each carbon atom overlap with an *s* orbital of two separate hydrogen atoms. Two other sigma bonds are formed, this time between sp^2 hybridized orbitals and *s* orbitals.

What happens to the remaining unhybridized *p* orbitals on the carbon atoms? These orbitals are now in position to overlap in a sideways fashion. They form a pi bond. Pi bonds are always formed by the sideways overlap of unhybridized *p* orbitals. The molecule formed by this series of bond formations, ethylene, is illustrated in Figure 13.10. Notice that ethylene contains one sigma bond and one pi bond between the carbon atoms. Thus, there is a double bond between the two carbon atoms, $H_2C{=}CH_2$. In a **double bond,** two pairs of electrons are shared between the bonding atoms. A double bond always consists of one sigma bond and one pi bond. The six atoms of ethylene lie in one plane.

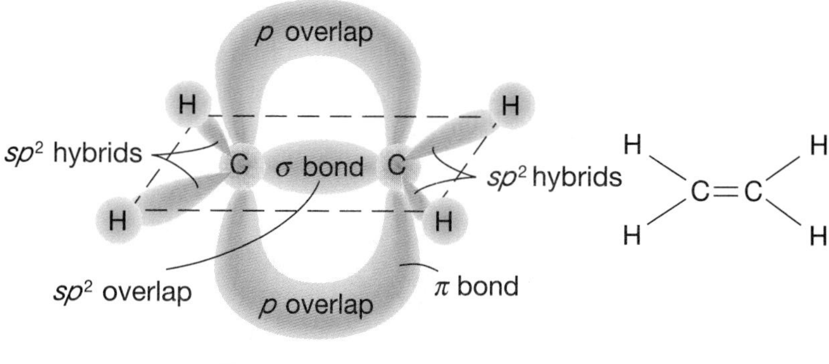

Figure 13.10 The shape of ethylene is explained by combining two sp^2 hybrid carbon atoms. The C-H sigma bonds all lie in the same plane. The unhybridized *p* orbitals of the carbon atoms combine to form the pi bond.

Now consider the bonding in the acetylene (ethyne) molecule, H—C≡C—H. In this molecule, the *s* orbital and one *p* orbital of each carbon atom form two *sp* hybrid orbitals. This particular hybridization leaves two *p* orbitals perpendicular to each other and to the *sp* hybrid orbitals. Two *sp* hybrid orbitals, one from each carbon, overlap to form one sigma bond. The two *p* orbitals from each atom overlap to form two pi bonds. Therefore, acetylene has a triple bond between its carbon atoms. The triple bond consists of one sigma bond and two pi bonds. In a **triple bond** three pairs of electrons are shared between the bonded atoms. Figure 13.11 summarizes the bonding that occurs in the acetylene molecule.

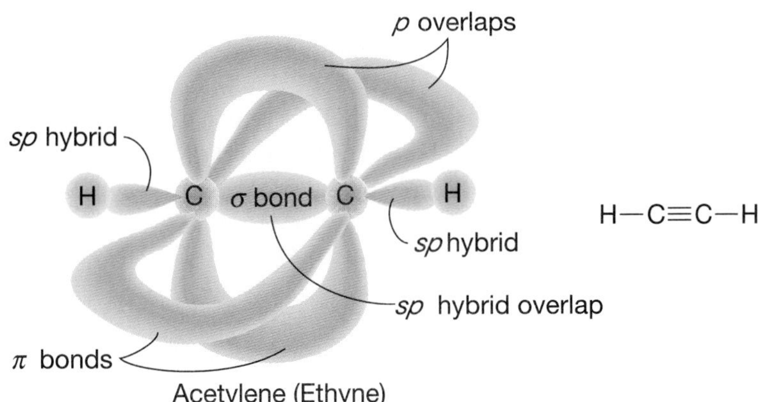

Figure 13.11 The shape of ethyne is explained by combining two *sp* hybrid carbon atoms. The two pi bonds are formed from the unhybridized *p* orbitals of both carbon atoms.

Both double and triple bonds are less flexible than single bonds are. Also, a multiple bond between two atoms is shorter and therefore stronger than a single bond between the same two atoms. Pi bonds are more easily broken than sigma bonds are because the electrons forming pi bonds are farther from the nuclei of the two atoms. As a result, molecules containing multiple bonds are usually more reactive than are similar molecules containing only single bonds. Compounds that contain double or triple bonds between carbon atoms are called **unsaturated compounds.**

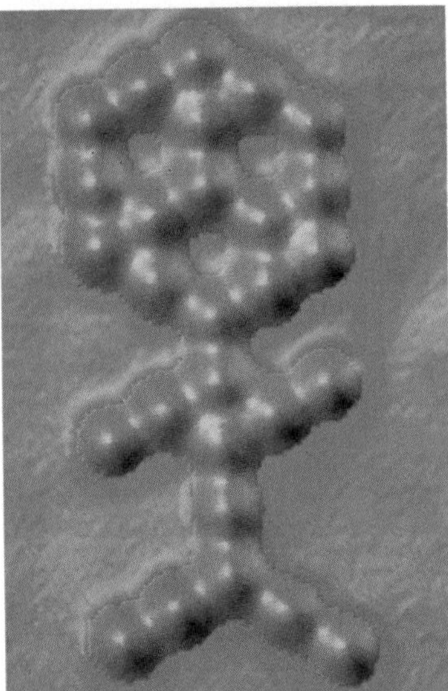

Figure 13.12 Twenty-eight carbon monoxide molecules on a platinum surface were used to create this figure, which is 5×10^3 pm tall.

Multiple Bond Molecular Shapes

The molecular formula for formaldehyde (methanal) is CH_2O. What is its electron dot diagram? If we allot the four electrons from carbon, one from each hydrogen, and six from the oxygen, the dot diagram ends up looking like this:

```
    H
    ˣ°   ••
  H ˣ C ° O :
    °     ••
```

Each hydrogen atom has two electrons, so the hydrogen outer levels are full. Likewise, the oxygen has eight electrons, and its outer level is full. However, carbon's outer level contains only six electrons. How can the electrons be adjusted so that carbon has eight, yet none are taken away from other atoms? If atoms share more than one pair of electrons, all atoms in the molecule can have full outer levels. Remember that both atoms get full credit for all shared pairs. For formaldehyde, carbon and oxygen share two pairs of electrons and the diagram becomes

```
    H
    ˣ°     ••
  H ˣ C ° : O :
    °
```

This diagram of formaldehyde is an example of a Lewis electron dot diagram of a molecule containing a double bond.

In the diatomic molecule, N_2, the two nitrogen atoms share three pairs of electrons. They are bound to each other by a triple bond. The electron dot diagram for the nitrogen molecule is

:N:::N:

How does the electron-pair repulsion theory predict the shapes of molecules containing multiple bonds? Recall that a double bond consists of four electrons occupying the space between the bonded atoms. The resulting cloud will occupy more space than a single bond. The triple bond occupies still more space than the double bond because six electrons occupy the space between the bonded atoms. How is molecular shape affected by the presence of multiple bonds? In the case of formaldehyde (methanal), there are three clouds around the carbon atom, two single bonds and one double bond. We anticipate that the clouds will assume a trigonal planar shape. The double bond will occupy somewhat more space. As a result, the H—C—H bond angle should be a little less than 120°, and the H—C=O bond angle a little more. Experiment shows a H—C—H bond angle of 116° and a H—C=O angle of 122°.

Figure 13.13 The methanal molecule contains a double bond. Thus, the H-C-H bond angle is smaller than predicted because a double bond occupies more space than a single bond does.

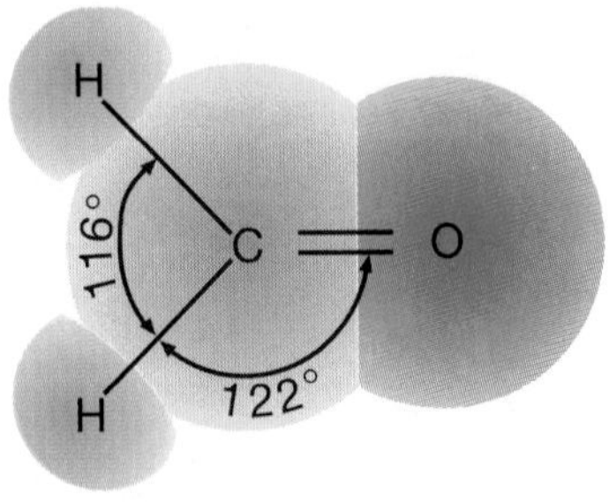

With two double bonds, each to a separate atom, we get a linear molecule. The organic compound, ketene, illustrates two double bonds as well as a double bond with two single bonds. The structure of the ketene molecule is

```
H
 \
  C=C=O
 /
H
```

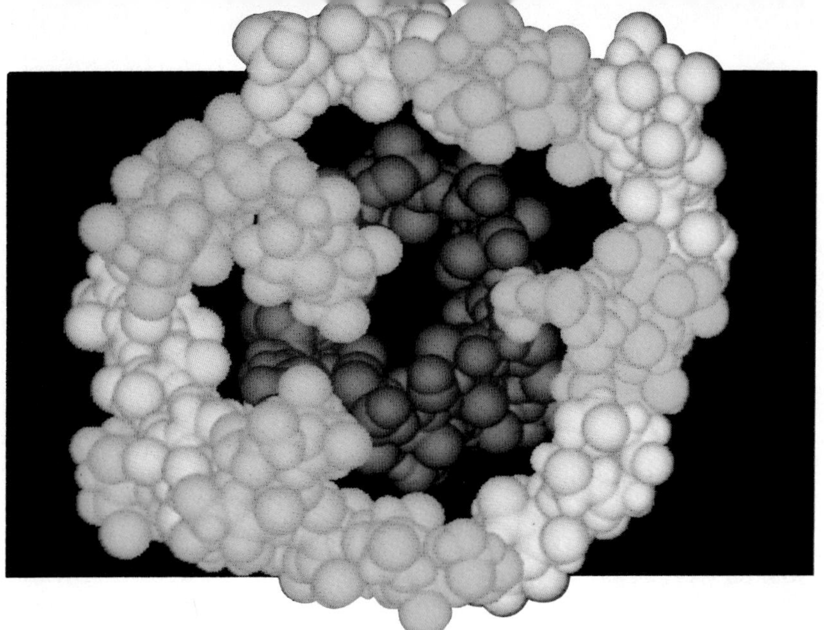

Figure 13.14 Computer simulations can create a three-dimensional model of a large molecule.

The C=C=O bond angle is 180°, while the H—C=C bond angle is slightly greater than 120°.

The one remaining case is that of eight electrons shared as a triple bond and a single bond. As you might expect, such an arrangement gives rise to a linear shape. Hydrogen cyanide illustrates this arrangement. It has the structure

$$H-C\equiv N$$

The H—C≡N bond angle is 180°. When individual molecules are investigated in detail, the experimental bond angles are not always exactly as we would predict. Bond angles are influenced by other factors that we will not study in this course.

▼ Benzene

CHEMISTRY IN DEPTH

One of the top 20 industrial chemicals in the United States is benzene, C_6H_6. It is used extensively in making drugs, dyes, and coatings as well as solvents. It is also highly toxic and can cause cancer. Each of the six carbon atoms in a benzene ring has three sp^2 hybrid orbitals and one p orbital. Sigma bonds are formed by the overlap of the sp^2 orbitals of six carbon atoms forming a ring of single bonds. The pi bonds of the benzene ring are formed by the sideways overlap of the p orbitals of the carbon atoms. However, note in Figure 13.15 that the unhybridized p orbitals can overlap to *both* sides. Because all six of these orbitals overlap to both sides, a circular orbital forms around the whole ring.

One of the characteristics of benzene is that the pi electrons can be shared all around the ring. Because the pi electrons are shared equally among all the carbon atoms and not confined to one atom or bond, they are delocalized. This **delocalization** of pi electrons among the carbon atoms in benzene results in greater stability of the compound.

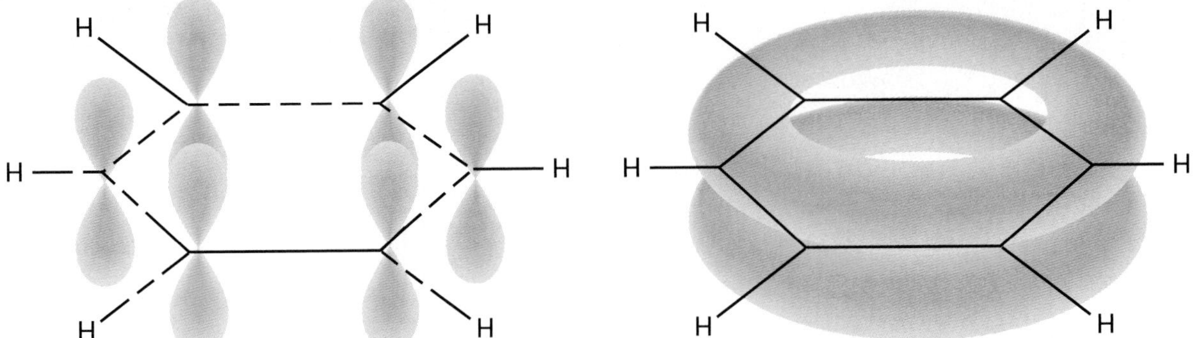

Figure 13.15 The benzene ring results from the overlap of six sp^2 hybrid carbon atoms. The unhybridized p orbitals form two pi clouds allowing electrons to be shared around the ring.

The following are representations of the benzene molecule. The solid lines in the drawing on the left represent sigma bonds formed between carbon sp^2 hybrid orbitals and hydrogen *s* orbitals, as well as between two adjacent carbon sp^2 hybrid orbitals. The circle represents the delocalized electrons in the pi bond formed from unhybridized carbon *p* orbitals. The drawing on the right is the simplified representation of the bonding in benzene. The vertices of the figure represent carbon atoms, and each straight line represents a bond formed between two carbon atoms.

Whenever multiple *p* orbital overlap can occur, the molecule is said to contain a **conjugated system.** Conjugated systems can occur in chains as well as in rings of atoms. 1,3-butadiene, $CH_2{=}CH{-}CH{=}CH_2$, is an example. Again, the conjugated system imparts a special stability to the molecule.

13.1 CONCEPT REVIEW

4. Draw a Lewis electron dot diagram for $CHCl_3$. Explain what each chemical symbol represents and why the electrons have the indicated locations.

5. Predict the shape and bond angles for the molecule in Problem 4.

6. A molecule has a trigonal planar shape. What hybridization would you expect to find on its central atom?

7. Draw a structural diagram for a propyne molecule, C_3H_4. Label each sigma bond and each pi bond.

8. Apply The structure of naphthalene can be represented by the structural formula ⌬⌬. Write the molecular formula for naphthalene, and compare its structure to that of benzene.

13.2 Molecular Arrangements

OBJECTIVES

- name and write formulas for simple organic compounds.
- define, explain, and give examples of isomerism.

Sometimes two or more compounds that are quite different have the same molecular formula. In order to clearly identify such compounds, we need to draw structural formulas and write distinctive names. The same principles apply to both organic and inorganic compounds. In this section we will deal with these variations in molecular shape.

Organic Names

In Chapter 7, it was mentioned that names for organic compounds carry a suffix describing how the atoms are bonded. The ending *-ane* was introduced at that time. You are now ready to expand your knowledge of organic compound names.

Compounds having only single bonds between carbon atoms are said to be **saturated compounds.** The ending *-ane* is used to name saturated compounds. For compounds containing a carbon-carbon double bond, the suffix *-ene* is used and for carbon-carbon triple bonds *-yne* is used. Recall that the name for $H_2C{=}CH_2$ is ethene, which indicates a double bond between the carbon atoms. The name for $H{-}C{\equiv}C{-}H$ is ethyne because there is a triple bond between carbon atoms. Ethene and ethyne are unsaturated compounds.

The compound known as propene, CH_3CHCH_2, has the following structure

```
     H
     |         ,H
 H — C — C = C
     |   |      \
     H   H       H
```

also written as $CH_3{-}CH{=}CH_2$.

Cyclopentene has the structure

```
     H     H
      \   /
 H      C
  \   /   \
    C       C — H
   / \      //
 H    C — C
    / |     \
  H   H       H
```

also written as .

Figure 13.16 shows other simplified diagrams used to represent cyclic compounds. It is understood that there is a carbon atom at the intersection of each pair of straight lines. We also know from Section 13.1 that carbon always forms four bonds. This bonding could be accomplished by four single bonds, a double and two single bonds, a single and a triple bond, or two double bonds. Therefore, when we use these simplified diagrams, we assume enough hydrogen atoms are bonded to each carbon to give that carbon four bonds.

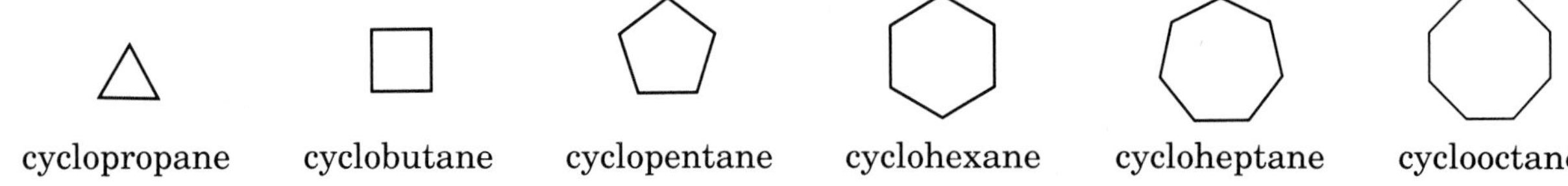

Figure 13.16 These simplified diagrams are used to represent the first six cycloalkanes.

PROBLEM SOLVING HINT

Prefixes are listed in Table 7.9, page 175.

PROBLEM SOLVING HINT

Note that the location of a multiple bond in a cyclic compound is arbitrary when only one multiple bond is present in the molecule.

SAMPLE PROBLEMS

1. Organic Nomenclature

Name the compound 

Solving Process:

There are three vertices in the figure; therefore, there are three carbon atoms in the compound and the stem of the name will be *prop*. The atoms are arranged in a ring, so the prefix *cyclo-* is used. Finally, there is a double bond, so the suffix is *-ene*. The name of the compound is *cyclopropene*.

2. Representing Organic Compounds

Draw cycloheptene.

Solving Process:

The stem of the name is *hept,* indicating seven carbon atoms. The prefix *cyclo-* means they are arranged in a ring. The suffix *-ene* means there is a double bond in the molecule.

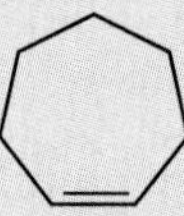

PRACTICE PROBLEMS

9. Name the following compounds.

a. $H-\overset{H}{\underset{H}{C}}-C\equiv C-H$ **b.**

10. Draw structural formulas for the following compounds.

a. cyclooctyne

b. cyclohexene

Word Origins

isos: (GK) equal, same
meros: (GK) part
isomer—made of the same parts

Isomers

The existence of two or more substances with the same molecular formula, but different structures, is called **isomerism** (i SOM eh rihz uhm). These structures are **isomers** (I soh murs). Because isomerism is common in organic chemistry, we will study the isomers of carbon compounds.

Consider the compound with the formula C_4H_{10}. There are two structures that can be written for this formula. Butane and methylpropane are examples of **structural isomers** or skeleton isomers, since it is the carbon chain that is altered. Methylpropane was at one time called isobutane since it is made of the same atoms as butane.

$$\begin{array}{ccccccccccc} & & H & & H & & H & & H & & \\ & & | & & | & & | & & | & & \\ H & - & C & - & C & - & C & - & C & - & H \\ & & | & & | & & | & & | & & \\ & & H & & H & & H & & H & & \end{array}$$

butane

$$\begin{array}{ccccccccc} & & & H & & & & & \\ & & & | & & & & & \\ & & H- & C & -H & & & & \\ & & H & | & & & H & & \\ & & | & | & & & | & & \\ H & - & C & - & C & - & C & - & H \\ & & | & & | & & | & & \\ & & H & & H & & H & & \end{array}$$

methylpropane

A second type of isomerism is positional isomerism. **Positional isomers** are two compounds that differ only in the position of something such as a double bond or an atom other than hydrogen and carbon. For example, consider the following two molecules.

$$CH_3-CH=CH-CH_3 \quad \text{and} \quad CH_2=CH-CH_2-CH_3$$

Both compounds are named butene, but plainly they are not the same compound. We name isomers such as these by specifying on which carbon atom the double bond begins. Number from the end, giving the double bond the lowest number. Thus, the compound $CH_3-CH=CH-CH_3$ is named 2-butene and the compound $CH_2=CH-CH_2-CH_3$ is named 1-butene.

When an atom or atoms different from hydrogen and carbon are introduced into an organic molecule, that part of the molecule is called a **functional group.** One of the most common functional groups is the alcohol group, $-OH$. For example, propanol is a three-carbon compound with single bonds and an alcohol group. However, the alcohol group could be placed on the end carbon or in the middle. The corresponding names reflect these locations.

$$\begin{array}{ccccccccc} & & H & & H & & H & & \\ & & | & & | & & | & & \\ H & - & C & - & C & - & C & - & OH \\ & & | & & | & & | & & \\ & & H & & H & & H & & \end{array}$$

1-propanol

$$\begin{array}{ccccccccc} & & H & & OH & & H & & \\ & & | & & | & & | & & \\ H & - & C & - & C & - & C & - & H \\ & & | & & | & & | & & \\ & & H & & H & & H & & \end{array}$$

2-propanol

Functional isomers have an element other than hydrogen or carbon being bonded in two different ways. In other words, the two isomers have the same atoms arranged in different functional groups. Ethanol is a liquid while methoxymethane is a gas at room temperature.

$$\begin{array}{ccccccc} & & H & & H & & \\ & & | & & | & & \\ H & - & C & - & C & - & OH \\ & & | & & | & & \\ & & H & & H & & \end{array}$$

ethanol

$$\begin{array}{ccccccccc} & & H & & & & H & & \\ & & | & & & & | & & \\ H & - & C & - & O & - & C & - & H \\ & & | & & & & | & & \\ & & H & & & & H & & \end{array}$$

methoxymethane

CHEMISTS AND THEIR WORK

Mary Lura Sherrill (1888-1968)

Leadership in educating young people in chemistry is an important quality for which some chemists have an exceptional talent. Mary Lura Sherrill was such a chemist. For 34 years, Sherrill taught chemistry at Mt. Holyoke College. Prior to that experience she had taught at Randolph-Macon Women's College, her alma mater, and the University of North Carolina. During World War II she also investigated antimalarial drugs for use by servicemen and women abroad. The American Chemical Society presented her with one of its top awards in 1947.

EVERYDAY CHEMISTRY

Gelatin

Nearly everyone has had the opportunity to see gelatin dissolve in water, cool in the refrigerator, and thicken. This process happens because of the structure and properties of the proteins that make up gelatin.

Gelatin is collagen that has undergone a structural change. Collagen, the protein found in the connective tissue of animals, is composed of three separate chains of amino acids intertwined. When heated in water, the weak bonds forming collagen are broken and the chains of proteins are unbraided.

Gelatin is generally dissolved in hot water. The dissolved gelatin is then added to other liquids and refrigerated. The weak bonds begin to form again, but they randomly form a large network that captures the liquid ingredients. The semisolid formed is called a gel. The more the gel is cooled, the more solid it becomes, because more weak bonds have had the opportunity to form.

Sometimes gels do not form or they are not firm enough. The best conditions for this particular gel formation are a pH of about 5 and addition of a small amount of sugar. Therefore, if fruits are added, those that are slightly acidic have a better chance of jelling.

Using pineapple in a gel illustrates the instability of gels. An enzyme in fresh pineapple will break the protein chains into small pieces that will not gel. Cooked or canned pineapple does not have this effect because the enzyme that destroys gelatin is itself destroyed in cooking.

Exploring Further

1. What are some fruits that are acidic enough to set up well in gelatin?
2. Look up sols and gels in a science handbook. How are they alike? How are they different?
3. Besides gelatin dessert, gelatin is also used to govern the consistency of many other products of the kitchen. What are some of these products?

A fourth type of isomerism can be seen in the two structures shown for butene. The formation of a π bond prevents the atoms on each end of the bond from rotating with respect to each other. Some compounds containing double bonds exhibit a kind of isomerism called geometric isomerism. **Geometric isomers** are composed of the same atoms bonded in the same order, but with a different arrangement of atoms around a double bond. We will use 2-butene (C_4H_8) to illustrate geometric isomers. Note that in the *cis* form of 2-butene, the $—CH_3$ groups and the hydrogen atoms are on the same side of the double bond. In the *trans* form, the $—CH_3$ groups and the hydrogen atoms are on opposite sides.

```
CH3         CH3          CH3          H
   \       /                \       /
    C == C                   C == C
   /       \                /       \
  H         H              H         CH3
 cis-2-butene              trans-2-butene
```

cis-2-butene   *trans*-2-butene

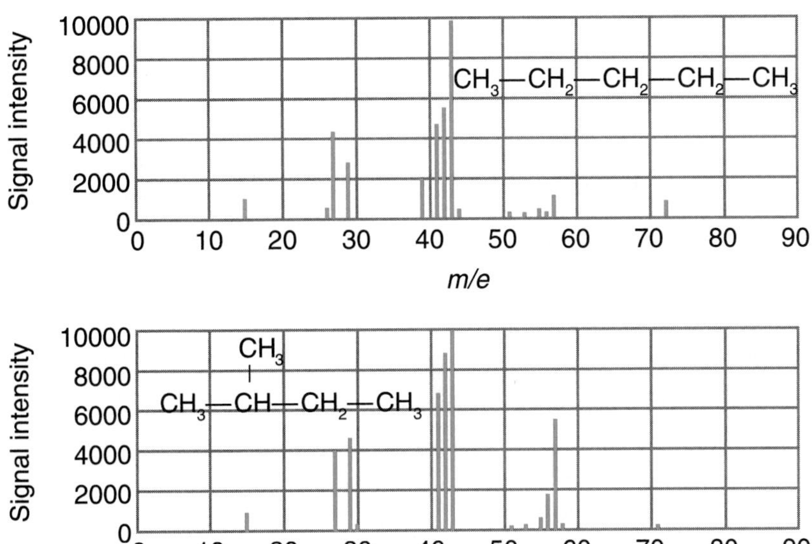

Figure 13.17 The mass spectrograms for the isomers pentane (top) and 2-methylbutane (bottom) are similar, but the difference in values for each peak can be used to identify each compound.

The mass spectrometer can be used to distinguish between isomers that have very similar properties. During the analysis, the sample being studied is ionized and divided into ion fragments. Each fragment has an *m/e* value (mass to charge ratio). In Figure 13.17, the mass spectrogram for pentane shows peaks at 15 different *m/e* values. Thus, the pentane molecule can be broken into 15 fragments that each have a different mass. Note the spectrogram for 2-methylbutane shows similar fragments. However, you can see that the relative intensities of the two isomers are significantly different.

Inorganic Compounds

The ground state electron configuration of carbon ends in $2s^2 2p^2$. Each of the four electrons was placed in a separate orbital. We treated the carbon atom as if it were in the state $2s^1 2p^3$, the configuration in which hybridization of orbitals could occur.

What about other atoms? For example, beryllium has two outer electrons, $2s^2$. We would expect *sp* hybridization, leading to linear bonding. Analysis bears out this prediction for molecules having a central atom ending in an s^2 configuration. For atoms with three outer electrons, we would predict sp^2 hybridization. Again, analysis confirms trigonal planar bond arrangements.

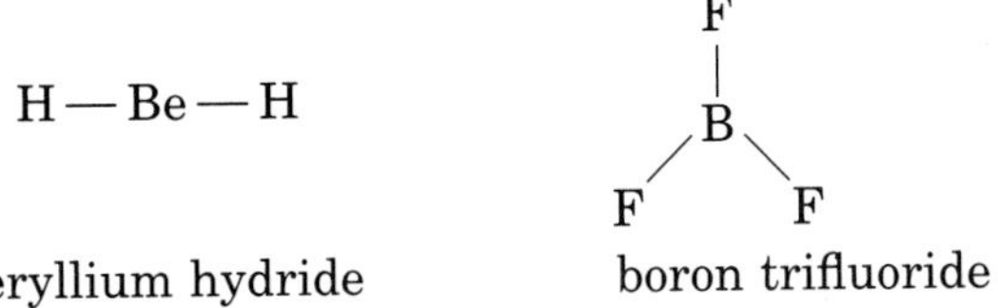

We can also apply other principles of molecular geometry and isomerism to inorganic compounds.

FRONTIERS

Super Soot

What has 32 sides, is hollow, and is the roundest molecule that can possibly exist? The answer, introduced in the chapter opener, is buckminsterfullerene, C_{60}, the best known member of a class of chemical substances called fullerenes. The structure of this molecule is so stable that experiments have revealed molecules of C_{60} bouncing back to shape after striking a steel plate at a speed of 7000 m/s. Because of the soccer ball-like shape and behavior of buckminsterfullerene molecules, tongue-tied and tired researchers have named them "buckyballs!"

Because of the hollow structure of the buckyball, substances may be trapped within the molecule. Theorists predict that C_{60} will not react with substances trapped within its molecular cage and envision buckyballs as molecular reaction chambers. Substances entrapped within the molecules would be forced to react with each other as the molecule is compressed. After decompression, the products would be extracted from the intact C_{60} molecule.

On the surface of a C_{60} molecule, things are different. Researchers have reacted buckminsterfullerene with fluorine and isolated $C_{60}F_{60}$. With the possibility of 60 bonding sites on each molecule, chemists anticipate sticking groups of atoms at various locations on the surface of the molecule. Such compounds could be designed to have specific chemical and physical properties.

Bond Summary

The substances in Table 13.2 illustrate the principles of molecular geometry. The angular values given are the actual values determined experimentally. Boron trichloride, BCl_3, has the shape we would expect from the electron-pair repulsion model. This compound is used in metallurgy to produce high purity metals including some used in the manufacture of computer chips.

Carbon dioxide, CO_2, is already familiar to you as a product of combustion and of respiration in living cells. It is also produced commercially for use in carbonated beverages, fire extinguishers, as the refrigerant dry ice, and as a propellant in aerosol cans containing such things as whipped cream.

Cyclohexane, C_6H_{12}, is a widely used solvent in many industries. It is also used by the chemical industry as a reagent in the synthesis of other chemicals.

Carbon tetrachloride, CCl_4, is no longer used extensively. At one time it was a common solvent and dry-cleaning fluid. However, it has been found that prolonged exposure to its vapors causes severe liver damage and may cause cancer. Carbon tetrachloride was used as a fire extinguishing liquid until it was discovered that, at high temperatures, it is converted to a very toxic gas called phosgene. Carbon tetrachloride is used in the chemical industry as a reagent in the synthesis of other chemicals.

Oxygen difluoride, OF_2, is also quite toxic. It is of interest mainly because it is one of the very few compounds in which oxygen exhibits a positive oxidation state.

Ammonia, NH_3, is produced commercially in huge quantities. It is used to make nitric acid, explosives, synthetic fibers, and hundreds of other materials. In addition, ammonia is used extensively in the production of fertilizers and as a refrigerant.

Ozone, O_3, is found in our atmosphere at an altitude of about 25 km. It protects us from the sun's harmful ultraviolet rays. Ozone is also a major component of smog. Industrially it is used as a disinfectant and as a bleaching agent. Small amounts of ozone are also used in the synthesis of organic compounds.

Methane, CH_4, is the principal constituent of natural gas and is therefore one of our major fuels. It is also widely used in the chemical industry in the synthesis of other organic compounds.

Acetylene (ethyne), C_2H_2, is another fuel that is important for welding and use in cutting torches. Acetylene is also a raw material for plastics and synthetic fibers.

The bond angles (96°) of trimethylarsine, $(CH_3)_3As$, seem to indicate that the bonding orbitals of the arsenic atom are *p* orbitals rather than the sp^3 hybrid orbitals we might have expected. Generally we find that the bonding orbitals of higher atomic mass elements are hybridized much less than in lighter elements. The reason may be that the heavier atoms can accommodate more bonded atoms around them because they are larger.

Table 13.2

Experimental Molecular Shapes		
Cl—B(—Cl)—Cl 120°	Cl—C(—Cl)(—Cl)—Cl 109.5°	H—C(—H)(—H)—H 109.5°
O=C=O 180°	F—Ö—F 103°	H—C≡C—H 180°
Cyclohexane (C_6H_{12}) 109.5°	H—Ö—H 104.5°	H_3C—As̈—CH_3, CH_3 96°
	H—N̈(—H)—H 107°	$[Cl_4Re—ReCl_4]^{2-}$ 87°, 104°
	O=Ö—O 117°	

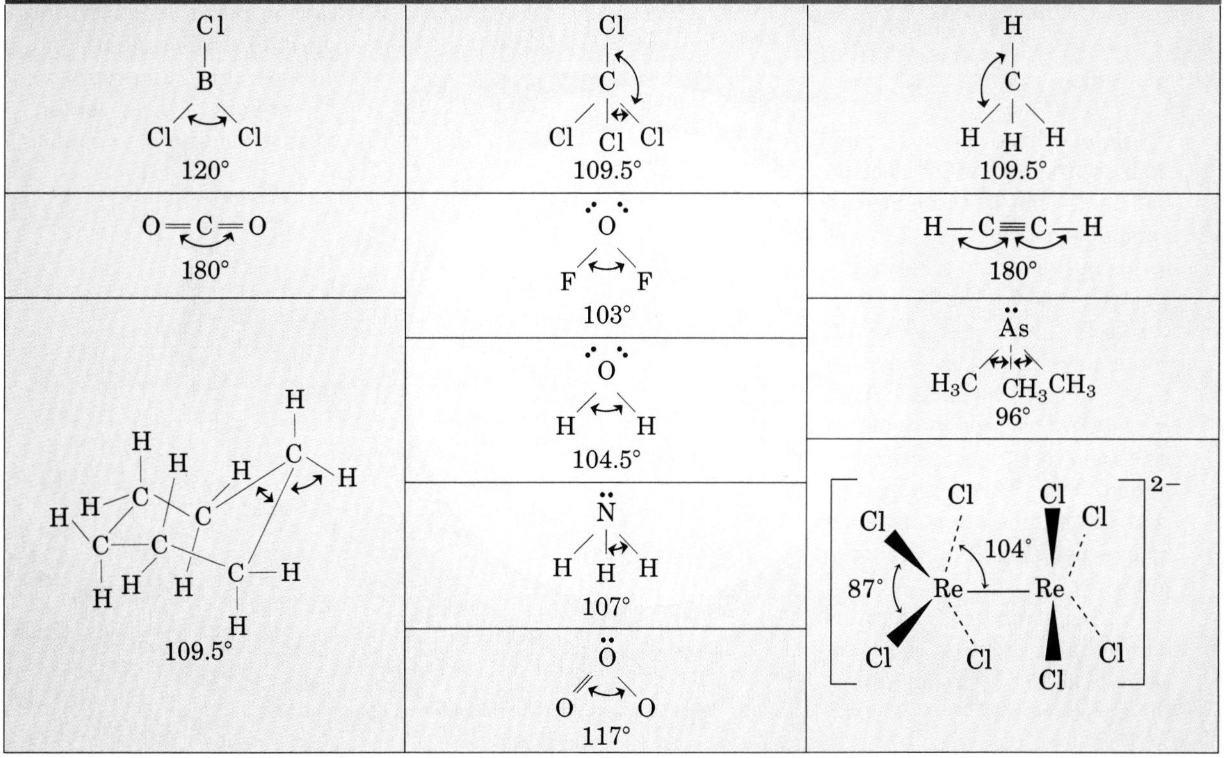

The Boranes

Boron and hydrogen form several compounds with unusual bonding arrangements. From the electron configurations of boron and hydrogen, we would predict boron hydride, BH_3, to be a planar molecule with 120° H—B—H bond angles. Instead, the two elements form a whole series of compounds called boranes. The simplest compound in the series is diborane(6), B_2H_6.

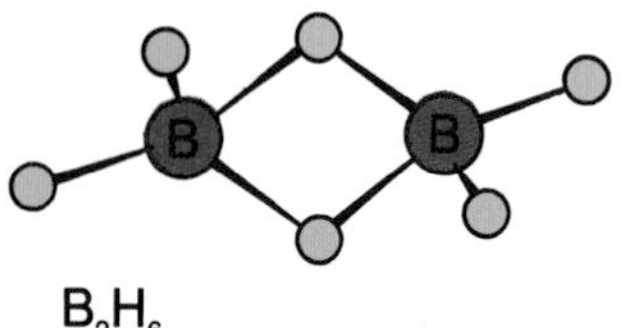

B_2H_6

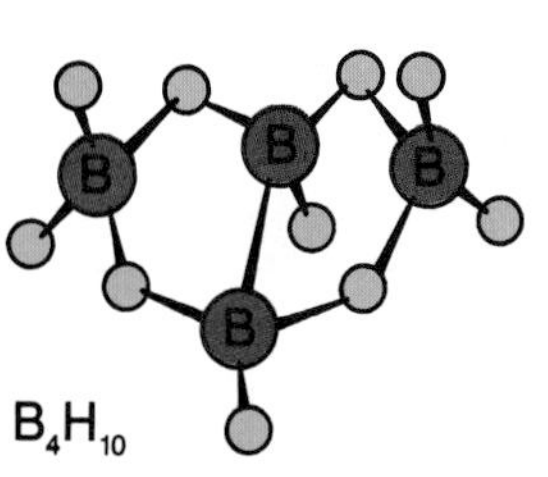

B_4H_{10}

All of these borane compounds are said to be electron deficient. That is, there are not enough electrons in the outer levels of the atoms to form any electron dot diagram to satisfy the octet rule. As a result, chemists have developed a three-center, two-electron bond model. In this model, a pair of electrons occupies an orbital spread over three atoms, so that the middle atom acts as a bridge between the other two atoms.

In the structural diagrams shown here, note that each contains bridging hydrogen atoms in three-center, two-electron bonds. In addition, all boranes having five or more boron atoms have at least one boron atom acting as a bridge between other boron atoms.

Most boranes are unstable. Several, including diborane(6), are spontaneously flammable. Diborane(6) is used as a reactant in making other chemicals. Diborane(6) has also been used on an experimental basis as a rocket fuel. The best rocket fuels are composed of low atomic mass elements. Lighter atoms travel faster than heavy atoms at the same temperature. The combustion of boranes produces higher-velocity gases in the rocket motor nozzle. Also, these compounds produce more energy than similar molecular mass organic fuels. For example, the energy produced by burning a mole of diborane(6), B_2H_6, is 2170 kJ, while one mole of ethane, C_2H_6, produces only 1560 kJ.

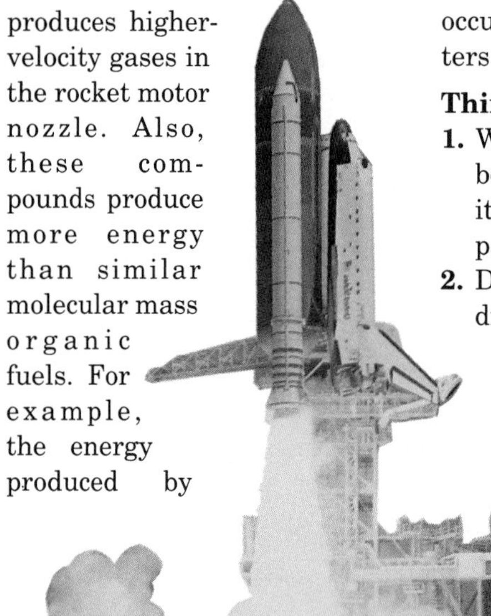

Carboranes are compounds in which one or more boron atoms have been replaced by carbon atoms. Their chemistry has been studied extensively in the hope that they too can be used as rocket fuels. Carborane polyesters have been used as ablative materials. An ablative material has the ability to carry away heat by gradually peeling away in layers. Space capsules use ablative materials in their heat shields. The peeling occurs when the capsule reenters Earth's atmosphere.

Thinking Critically

1. What is the structure of the borazine molecule? How does its electronic structure compare with benzene's?
2. Draw a Lewis electron dot diagram for diborane(6), and use this diagram to explain three-center, two-electron bonds.

Figure 13.18 As was mentioned on page 341, ammonia is used extensively to produce many useful commercial products such as these synthetic fabrics.

The octachlorodirhenate(III) ion, $[Re_2Cl_8]^{2-}$, is shown in Table 13.2 as an illustration of a metal-metal covalent bond, something that is not very common. In this ion, the chlorine atoms occupy the eight vertices of a cube.

In studying atomic structure chemists treat the electron as a particle, wave, and negative cloud. Several different approaches to bonding have also been studied and used to explain what can be observed. It is plain that chemists do not have a complete understanding of all factors in bonding. Therefore, more than one explanation is often needed to account for observations.

When faced with multiple explanations, scientists follow a basic rule. That rule is to try the simplest explanation first. If that method does not suffice, then the more complex ideas are applied until one is found to fit. The models of molecular structure presented in this chapter are arranged in order of increasing complexity. Electron-pair repulsion is the simplest model and should be applied first when attempting to explain the structure of a molecule.

13.2 CONCEPT REVIEW

11. Name these compounds.
  a. $CH_2{=}CH{-}CH_2{-}CH_2{-}CH_3$
  b. $CH_3{-}C{\equiv}C{-}CH_3$

12. Draw the structure of each of the following compounds.
  a. cyclopentene   **b.** nonane

13. Draw structures for two structural isomers, two positional isomers, and two geometric isomers of a molecule with the formula, C_5H_{10}.

14. Apply Are these structures examples of structural isomers? Justify your answer.

```
    H   H   H   H   H
    |   |   |   |   |
H — C — C — C — C — C — H
    |   |   |   |   |
    H   H   H   H   H
```

```
                  H
                  |
     H       HH — C — H
     |       |    |
H —  C   —   C  — C — H
     |       |    |
H —  C — HH       H
     |
     H
```

Chapter 13 REVIEW

Summary

13.1 Bonds in Space

1. The structure of a molecule or a polyatomic ion can be represented by a Lewis electron dot diagram, which shows the pattern of shared and unshared pairs of electrons.
2. The shape of a molecule can be predicted by taking into account the mutual repulsion of electron pairs due to the same electrostatic charge and the Pauli exclusion principle. Electron pairs spread as far apart as possible.
3. The shape of a molecule containing three or more atoms is determined by the number and type of electron clouds in outer levels of the atoms.
4. The actual bond angles may vary from predicted angles. Unshared electron pairs occupy the most space. Shared pairs occupy the least space.
5. The *s* and *p* orbitals of the same atom can combine to form hybrid orbitals. One *s* and three *p* orbitals can merge to form four tetrahedral sp^3 hybrid orbitals. One *s* and two *p* orbitals can merge to form three planar sp^2 hybrid orbitals. One *s* and one *p* orbital can merge to form two linear *sp* hybrid orbitals.
6. If two *s* orbitals or an *s* and a *p* orbital overlap, a sigma (σ) bond is formed. If two *p* orbitals overlap end to end, a sigma bond is formed. If two *p* orbitals overlap sideways, a pi (π) bond is formed. Pi electrons are farther from the nucleus.
7. Two atoms sometimes share more than one pair of electrons, forming double and triple bonds. This possibility must be considered when predicting geometry of a molecule.
8. Delocalization of pi electrons in conjugated chain and ring compounds such as benzene increases the stability of the structure.

13.2 Molecular Arrangements

9. Names of organic compounds carry a suffix to indicate how the atoms are bonded. The ending *-ane* is used for compounds in which all bonds are single bonds. The ending *-ene* is used for compounds with a carbon-carbon double bond. The ending *-yne* is used for compounds with a carbon-carbon triple bond.
10. Cyclic compounds may be represented using skeletal outlines with double and triple lines representing double and triple bonds respectively.
11. Isomers are compounds with the same molecular formula but with different arrangements of atoms and bonds.
12. Compounds may have several types of isomers including structural, geometric, positional, and functional.
13. Electron-pair repulsion is the simplest model used in predicting the structure of a molecule. Therefore, it should be applied before more complex explanations are tried.

Key Terms

shared pair	delocalization
unshared pair	conjugated system
hybridization	saturated compound
hybrid orbital	isomerism
sigma bond	isomer
pi bond	structural isomer
double bond	positional isomer
triple bond	functional group
unsaturated compound	functional isomer
	geometric isomer

Review and Practice

15. Draw Lewis electron dot structures for the following substances.
    a. H_2S **c.** HBr **e.** NH_3
    b. Br_2 **d.** CH_2Cl_2 **f.** $AlBr_3$

16. Using the ammonia molecule as an example, distinguish between shared pairs and unshared pairs of electrons.

17. Predict shapes for the following ions.
a. IO_4^- **c.** SiF_6^{2-} **e.** PO_4^{3-}
b. ClO_3^- **d.** SO_4^{2-} **f.** ClF_4^-

18. Predict the shapes of the following molecules. See Table 13.1.
a. H_2CO **c.** BF_3 **e.** S_2Cl_2
b. SeO_2 **d.** SF_6 **f.** SF_4

19. What is meant by hybrid in the term hybrid orbital?

20. Sketch a molecular diagram of butyne and one of pentyne, showing all bonds.

21. Name the following compounds.

a.

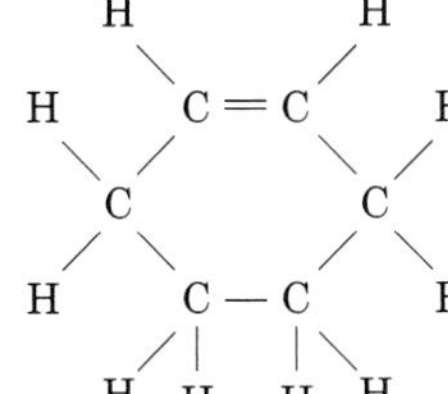

c.

```
   H   H   H
   |   |   |
H— C = C — C —H
           |
           H
```

b.

```
          H   H
          |   |
H—C≡C— C — C —H
          |   |
          H   H
```

d.

```
   H   H   H   H
   |   |   |   |
H— C — C — C — C —H
   |   |   |   |
   H   H   H   H
```

22. Name the following compounds.

a.

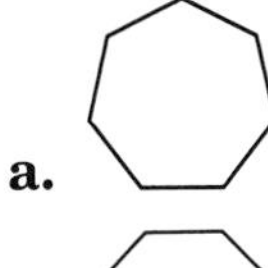

c.

b.

d.

23. Draw structural formulas for each of the following hydrocarbons.
a. cycloheptene **c.** cyclobutene
b. cyclobutane **d.** cyclononyne

24. Draw structural formulas for each of the following hydrocarbons.
a. propene **c.** heptane
b. octyne **d.** ethyne

25. Sketch a molecular diagram of butene, showing its orbitals and bonds.

26. Predict the bond angles indicated in the following compounds. Use Table 13.2.
a. H—Se—H in H_2Se
b. H—P—H in PH_3
c. C—N—C in $N(CH_3)_3$
d. Cl—P—Cl in PCl_3
e. F—C—F in CF_4
f. C—Pb—C in $Pb(C_2H_5)_4$

27. How many pairs of electrons are shared by two atoms with each of the following bonds between them?
a. single **b.** double **c.** triple

28. In the HCN molecule, the bonds are H—C and C≡N. Predict the shape of the molecule, the hybridization of the orbitals on the carbon atom, and the type (σ, π) of each bond.

29. A conjugated system stabilizes a molecule. Define "conjugated system" and give an example of one.

30. Why is benzene a particularly stable compound?

31. Define isomerism. Draw examples of two isomers of hydrocarbons containing five carbon atoms each.

32. Draw the structural isomers of the compound with the formula C_6H_{14}.

33. How do positional and functional isomers differ?

34. In what kinds of molecules do geometrical isomers occur? Draw examples using hydrocarbon molecules with five carbon atoms and one double bond.

35. What are functional isomers? Draw an example and include names.

36. Phosphorus pentachloride occurs as PCl_5 molecules in the gaseous state. As a solid, it is an ionic compound, $PCl_4^+ \; PCl_6^-$. Describe the shape and bonding in each of these species. See Table 13.1.

37. Draw two possible structural formulas for decyne and one for cyclononene.

Concept Mastery

38. How do forces resulting from the Pauli exclusion principle affect the shape of molecules? How do these forces compare with the electrostatic repulsions in a molecule? Taking both kinds of forces into account, state the general rule that applies to the shape of molecules when considering electron-pair repulsions.

39. What accounts for the difference in shape between the electron clouds of shared pairs and unshared pairs of electrons?

40. Why is the bond angle in NH_3 only 107° when the bond angle for BF_3 is 120°?

41. Explain why water has a bond angle of 104.5° instead of the predicted 109.5°.

42. In general, what effect does the presence of unshared pairs in an atom have on the orientation of bonds to the same atom?

43. Explain why the carbon atom forms four equal bonds instead of just two bonds involving electrons from *p* orbitals.

44. One might expect the bond angle for each C—H bond in methane to be 90°. Why is this prediction incorrect?

45. How are covalent bonds formed?

46. What is the major difference between sigma and pi bonds?

47. What kinds of orbital arrangements contribute to the bonding in ethene, $H_2C{=}CH_2$?

48. In what characteristics do multiple bonds differ from single bonds?

49. In general, how does the presence of double or triple bonds affect the geometry of adjacent single bonds?

50. What is meant by electrons being delocalized? In the case of benzene, describe how electrons are delocalized.

51. What is an unsaturated compound? Draw structural formulas of a saturated and an unsaturated organic hydrocarbon, each containing five carbon atoms.

52. How can isomers with very similar properties be distinguished from each other?

53. Explain the significance of the circles in the structural formula for naphthalene, $C_{10}H_8$.

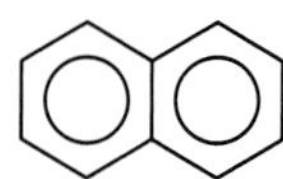

54. Butanol is an alcohol. One of its positional isomers is shown. Draw another positional isomer of butanol. Are there still other positional isomers? Explain your answer.

```
    H   H   H   H
    |   |   |   |
H — C — C — C — C — OH
    |   |   |   |
    H   H   H   H
```

55. Draw functional isomers of the following compounds.

a.

```
    H   O   H
    |   ‖   |
H — C — C — C — H
    |       |
    H       H
```

acetone

b.

```
H       H   H   H
 \      |   |   |
  N  —  C — C — C — H
 /      |   |   |
H       H   H   H
```

1-aminopropane

Application

56. Acetylene, ethyne, is widely used as fuel for welding torches and as a raw material for synthetic fibers and rubber. Sketch a molecular diagram of an ethyne molecule showing its bonds. Describe how these bonds are formed.

57. The largest moon of Saturn, Titan, has an atmosphere containing methane, ethane, ethene, propyne, hydrogen cyanide, and the compound diacetylene, C_4H_2, which contains two triple bonds. Draw the structural formula of each compound.

58. Fumaric acid, $C_2H_2(COOH)_2$, is used commercially in beverages. It occurs naturally in the metabolism of glucose in animal and plant cells. The structural formula of the *trans* isomer is shown.

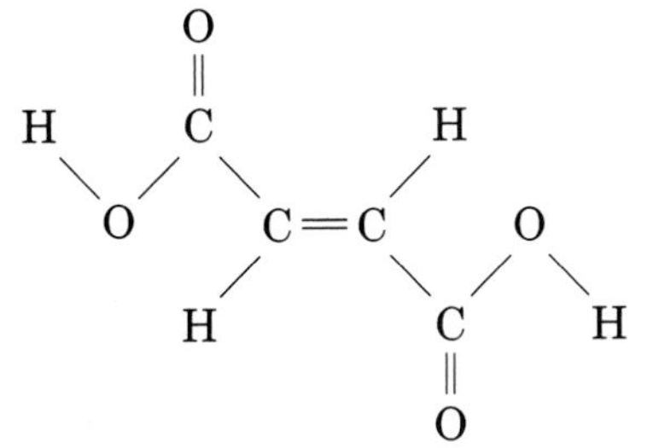

```
              O
              ‖
     H        C          H
      \      / \        /
       O         C=C         O
                /    \      / \
       H          C          H
                  ‖
                  O
```

Sketch the structural formula of the *cis* isomer, known as maleic acid.

59. In nutrition, scientists refer to fats as being saturated, unsaturated, or polyunsaturated. Considering that fats have long hydrocarbon chains, what do these terms mean when applied to fats?

64. Do the following two structures represent isomers, or are they representations of the same substance? Explain your reasoning.

```
     Cl              Cl
     |               |
     C               C
   / | \           / | \
  H  Cl  H        H  H  Cl
```

65. Explain why the general formula for a non-cyclic alkane such as pentane, C_5H_{12}, or hexane, C_6H_{14}, is C_nH_{2n+2}, where n is the number of carbon atoms in the chain. What are the general formulas for a cyclic alkane, a noncyclic alkene, and a noncyclic alkyne?

Critical Thinking

60. How does the conjugated bonding arrangement in benzene compare with the bonding in metals?

61. Why are there no compounds named methene or methyne?

62. Why are there no isomers for propane or ethane?

63. Explain why some noble gases such as Xe will form compounds and some such as Ne will not.

Readings

Al-Mousawi, Saleh M. "Molecular Shape Prediction and the Lone-Pair Electrons on the Central Atom." *Journal of Chemical Education* 67, no. 10 (October 1990): 861.

Curl, Robert F., and Richard E. Smalley. "Fullerenes." *Scientific American* 265, no. 4 (October 1991): 54–63.

Kluger, Jeffrey. "A Dream Come True." *Discover* 10, no. 1 (January 1989): 56.

Cumulative Review

1. Complete and balance the following equations for synthesis reactions.

a. $Al + O_2 \rightarrow$ **c.** $Na + I_2 \rightarrow$
b. $Mg + S_8 \rightarrow$ **d.** $Na + P_4 \rightarrow$

2. Classify the bonds between the following pairs of atoms as principally ionic or principally covalent.

a. barium and fluorine
b. bromine and rubidium
c. cesium and oxygen
d. iodine and antimony
e. nitrogen and sulfur
f. silicon and carbon

3. Explain why metals conduct electricity.

4. Sodium amide is used in preparing the dye indigo. This dye is used to color blue jeans. If the percentage composition of the compound is 58.9% sodium, 35.9% nitrogen, and 5.17% hydrogen, what is the empirical formula of the compound?

5. Arrange the following elements in order of increasing attractive force between the nucleus and the outer electrons.

a. boron **c.** manganese
b. francium **d.** zinc

CHAPTER **Preview**

CONTENTS

CHAPTER 14

Polar Molecules

The word polar *applies to bears, regions of Earth, magnets, and some molecules. These molecules are so named not because they are found exclusively in polar areas, but because they have two poles, just as Earth does.*

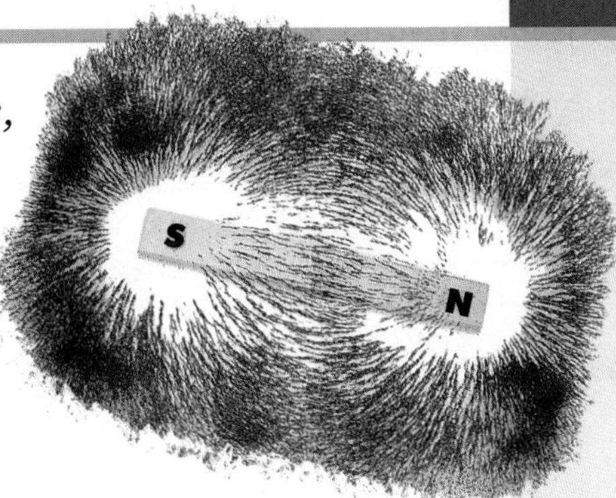

Most of the substances we encounter in our everyday life are made up of polar molecules. Water is, perhaps, the most important of all polar molecules. Because water molecules are polar, water is attracted by a charged rod, ice floats on liquid water, and water dissolves many substances. What does the word polar mean? What features cause a water molecule to be polar? What are the consequences of some molecules being polar? In this chapter we will investigate the structural features that cause a molecule to be polar and cause the behavior of polar molecules.

EXTENSIONS AND CONNECTIONS

OBJECTIVES

- distinguish between polar and nonpolar covalent bonds.
- use electronegativities to predict the comparative polarities of bonds.
- define *dipole* and compare the strengths of intermolecular forces based on dipole moments.
- define and describe the types of van der Waals forces and list the three factors contributing to them.

14.1 Molecular Attraction

Think about table salt and sugar. They look alike. They both dissolve in water. Yet they are quite different. Salt is made of ions. The sodium ions and chloride ions are oppositely charged and attract each other. Knowing the properties of ionic substances, we understand why salt is a solid. Sugar, on the other hand, is made of molecules. Each molecule of sugar contains twelve atoms of carbon, twenty-two atoms of hydrogen, and eleven atoms of oxygen bonded covalently. A sugar molecule is neutral, so what holds one sugar molecule to another? Why don't the sugar molecules just float away from each other and become a gas?

Substances composed of molecules exhibit a wide range of melting and boiling points. There must be, therefore, a wide range in the strength of forces holding molecules to each other. These forces are determined by the internal structure of a molecule. In this section, we will look at the way that internal structure affects the forces holding molecules to each other.

Polarity

Recall that electronegativity is an atom's ability to attract the electrons involved in bonding. Because no two elements have exactly the same electronegativities, in a covalent bond between different elements, one of the atoms attracts the shared pair more strongly than does the other. The resulting bond is said to be **polar covalent.** In the bond, the atom with higher electronegativity attracts the electrons more strongly, and that end of the bond will have a partial negative charge. The atom at the other end of the bond will have a partial positive charge.

Partial charges within a molecule are indicated by δ (delta). A water molecule, therefore, would be represented as follows:

$$\delta^{+}\mathrm{H}:\ddot{\underset{\cdot\cdot}{\mathrm{O}}}:\delta^{-} \qquad \text{bond angle } 104.5°$$
$$\mathrm{H}$$
$$\delta^{+}$$

Figure 14.1 CH_3Cl, HF, NH_3, and H_2O are polar molecules because the arrangement of the polar bonds is not symmetrical. The CCl_4 and CO_2 molecules are nonpolar because the polar bonds are symmetrically arranged. The arrows point toward the more electronegative atom.

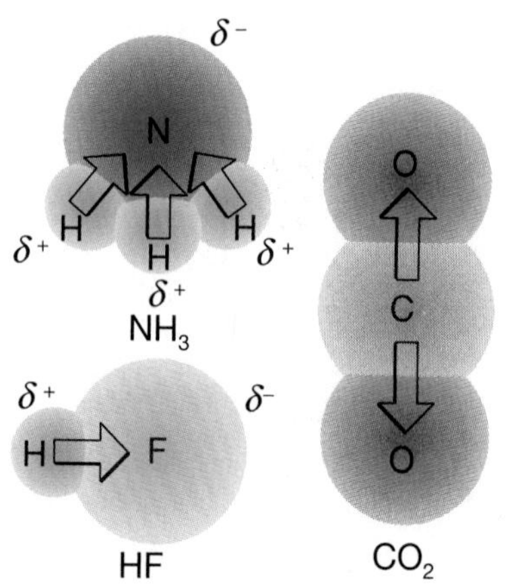

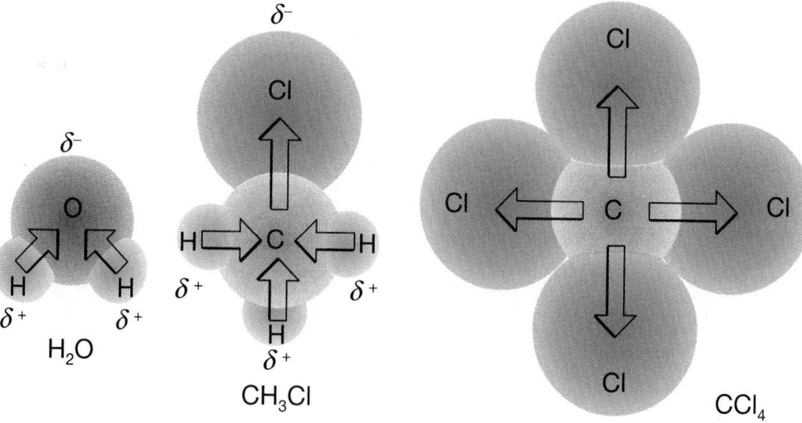

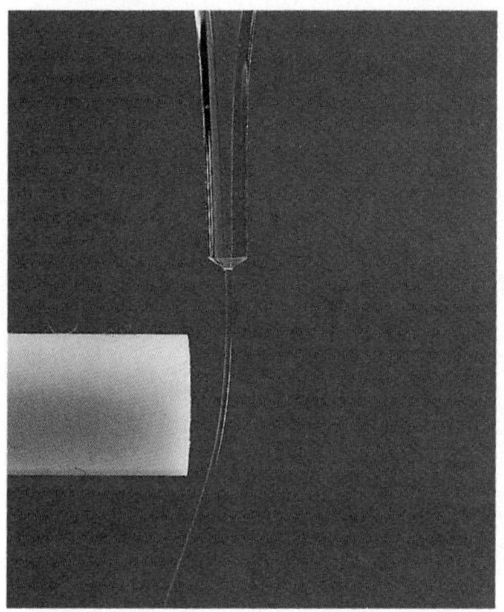
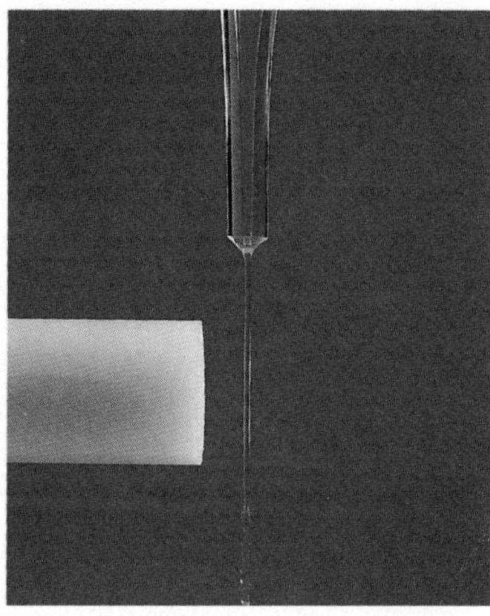

Figure 14.2 On the left, the polar molecules of water are attracted to a charged rod. The nonpolar molecules of carbon tetrachloride on the right are not affected by the rod.

Polar bonds may produce polar molecules. For example, look at Figure 14.1. The arrows show the direction in which electrons are attracted in each bond. In the ammonia and hydrogen fluoride molecules there is a concentration of negative charge near the nitrogen and fluorine atoms and positive charge near the hydrogen atoms. Consequently, ammonia and hydrogen fluoride are polar molecules. Now consider the carbon dioxide molecule. Each carbon-oxygen bond in this molecule is polar because oxygen has a greater electronegativity than carbon does. However, the polarities of the two bonds are in exactly opposite directions and therefore cancel each other. Even though the bonds are polar, the carbon dioxide molecule is nonpolar due to its linear geometry.

What about the water molecule? Is it polar or nonpolar? Both of the oxygen-hydrogen bonds are polar. In this case, however, the polar bonds do not cancel each other. Recall that the water molecule has an angular geometry. Therefore, there is a net concentration of negative charge near the oxygen atom and a slight positive charge near each hydrogen atom. For this reason, water is a polar molecule. In carbon tetrachloride, the four polar carbon-chlorine bonds are symmetrically distributed and therefore cancel each other. Carbon tetrachloride is not a polar molecule. In chloromethane, there is an asymmetrical distribution of charge. Therefore, chloromethane is a polar molecule. Polar bonds are a necessary but not a sufficient condition for polar molecules. In a polar molecule, the polar bonds cannot be symmetrically arranged.

Because it has both a positive and a negative pole, a polar molecule, such as water, is also said to be a **dipole,** or to have a dipole moment. A **dipole moment** is a measure of the strength of the dipole and is a property that results from the asymmetrical charge distribution in a polar molecule.

The dipole moment depends upon the size of the partial charges and the distance between them. In symbols,

$$\mu = Qd$$

where μ is the dipole moment, Q is the size of the partial charge in coulombs, and d is the distance in meters between the partial charges. Dipole moment is then expressed in coulomb·meters.* The higher the dipole moment, the stronger the intermolecular forces; and, consequently, the higher the melting point and boiling point for molecules of similar mass. Table 14.1 lists a number of common solvents in order of increasing polarity. The first three solvents are nonpolar.

*Older tables of dipole moments may be expressed in debyes (D) where 1 D = 3.338×10^{-30} C·m.

Table 14.1

Polarity of Solvents

Name	Formula	$\mu \times 10^{30}$ C·m
Cyclohexane	C_6H_{12}	0
Carbon tetrachloride	CCl_4	0
Benzene	C_6H_6	0
Toluene	$C_6H_5CH_3$	1.5
Ethoxyethane	$CH_3CH_2OCH_2CH_3$	4.07
Ammonia	NH_3	4.90
1-Butanol	$CH_3CH_2CH_2CH_2OH$	5.54
1-Propanol	$CH_3CH_2CH_2OH$	5.57
Ethanol	CH_3CH_2OH	5.64
Methanol	CH_3OH	5.64
Ethyl acetate	$CH_3CH_2OOCCH_3$	6.14
Water	HOH	6.14
Propanone	CH_3COCH_3	9.25

PRACTICE PROBLEMS

1. The following pairs of atoms are all covalently bonded. Arrange the pairs in order of decreasing polarity of the bonds. (See Table 12.1, page 303.)
   a. aluminum and phosphorus
   b. chlorine and carbon
   c. molybdenum and tellurium
   d. hydrogen and sulfur
   e. phosphorus and sulfur
   f. chlorine and silicon
2. Which of the following molecules are polar?
   a. HBr
   b. SO_2
   c. N_2
   d. BF_3

Figure 14.3 Liquids tend to form drops because their molecules have intermolecular attraction.

Weak Forces

Covalent compounds show a melting point range of over 3000 C°. How can we account for such a wide variation? The forces involved in some of these cases are called **van der Waals forces.** Johannes van der Waals was the first to account for these forces in calculations concerning gases. These forces are sometimes referred to as weak forces because they are much weaker than chemical bonds. Weak forces involve the attraction of the electrons of one atom for the protons of another.

It is important to note the difference between intramolecular forces and intermolecular forces. **Intramolecular forces** are forces within a molecule that hold atoms together, that is, covalent bonds. **Intermolecular forces** are forces between molecules that hold molecules to each other, that is, van der Waals forces.

The first van der Waals force that we will consider is the dipole-dipole force. With **dipole-dipole forces,** two molecules of the same or different substances that are both permanent dipoles, will be attracted to each other, Figure 14.4. Such would also be the case between two trichloromethane molecules, $CHCl_3$, or between a trichloromethane and an ammonia molecule.

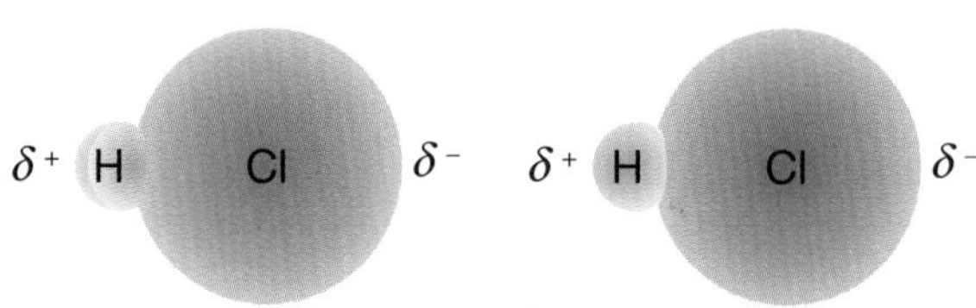

Figure 14.4 Dipole-dipole attraction exists between molecules that are permanent dipoles.

A dipole can also attract a molecule that is ordinarily not a dipole. When a dipole approaches a nonpolar molecule, its partial charge either attracts or repels the electrons of the other particle. For instance, if the negative end of the dipole approaches a nonpolar molecule, the electrons of the nonpolar molecule are repelled by the negative charge. The electron cloud of the nonpolar molecule is distorted by bulging away from the approaching dipole as shown in Figure 14.5. As a result, the nonpolar molecule is itself transformed into a dipole. We say it has become an **induced dipole.** Since it is now a dipole, it can be attracted to the permanent dipole. Interactions such as these are called **dipole-induced dipole forces.** An example of this force also occurs in a water solution of iodine. The I_2 molecules are nonpolar while the water molecules are highly polar.

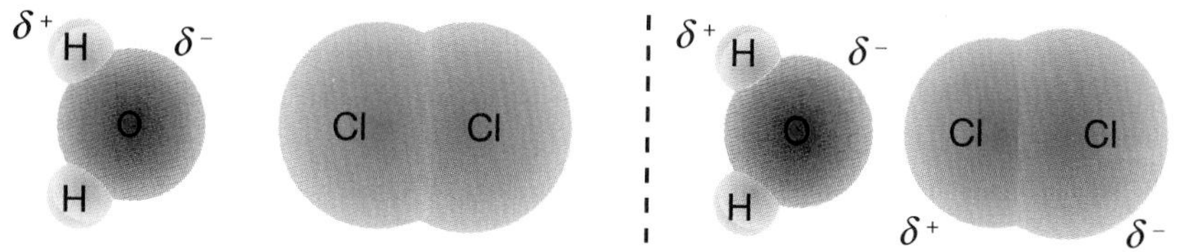

Figure 14.5 Dipole-induced dipole attraction occurs when a molecule that is a dipole causes a nonpolar molecule to become an induced dipole.

The attraction of two nonpolar molecules for each other must also be taken into account. For instance, there must be some force between hydrogen molecules; otherwise it would be impossible to form liquid hydrogen. Consider a hydrogen molecule. We usually assume that the electrons of the hydrogen molecule move uniformly about the nuclei of the two atoms. For an instant, both electrons may be at the same end of the molecule. Thus, the hydrogen molecule develops a temporary asymmetrical electron

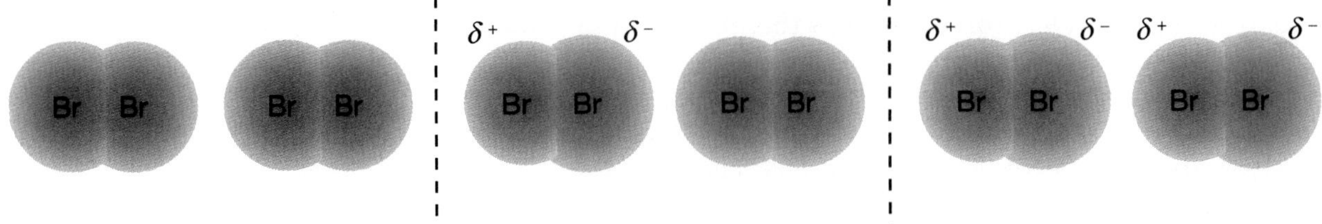

Figure 14.6 Dispersion forces exist between nonpolar molecules and are the result of the formation of temporary dipoles.

distribution. The end of the molecule where the electrons are concentrated develops a partial negative charge. The opposite end of the molecule develops a partial positive charge. In this way, a temporary dipole is set up. A **temporary dipole** can induce a dipole in the molecule next to it and an attractive force results. Another example is shown in Figure 14.6. The forces generated in this way are called **dispersion forces** or London forces after Fritz W. London, the physicist who first investigated them.

Of the factors contributing to van der Waals forces, dispersion forces are the most important. They are the only attractive forces that exist between nonpolar molecules. Even for most polar molecules, dispersion forces account for 85% or more of the van der Waals forces. Only in some special cases, such as those involving NH_3 or H_2O, do dipole-dipole interactions become more important than dispersion forces. We will examine these special cases in Chapter 17.

Many molecules will exhibit both dipole and dispersion interactions. However, we are only interested in the net result. The liquid and solid states of many compounds exist because of these intermolecular forces. Substances composed of nonpolar molecules are generally gases at room temperature or low-boiling liquids. Substances composed of polar molecules generally have

Table 14.2

Weak Forces Summary		
Type of Force	Dispersion forces	Dipole
Substances exhibiting force	Nonpolar molecules	Polar covalent molecules
Source of the force	Weak electric fluctuations which destroy spherical symmetry of electronic fields about atoms	Electric attraction between dipoles resulting from polar bonds
Properties due to the force	Low melting and boiling points	Higher boiling and melting points than nonpolar molecules of similar size
Examples	Ne (20 u; m.p. = −249°C) O_2 (32 u; m.p. = −219°C) F_2 (38 u; m.p. = −220°C) C_9H_{20} (128 u; m.p. = −54°C) Br_2 (160 u; m.p. = −7°C)	HF (20 u; m.p. = −84°C) CH_3OH (32 u; m.p. = −98°C) HCl (36.5 u; m.p. = −114°C) SeO_3 (127 u; m.p. = 118°C) ICl (162 u; m.p. = 27°C)

higher boiling points than nonpolar compounds have. Many substances composed of polar molecules are solids under normal conditions.

Weak intermolecular forces are effective only over short distances. They vary inversely as the sixth power of distance, $1/d^6$. Thus, if distance is doubled, the attractive force is $1/64$ as large.

Another weak force is the attractive force between an ion and the oppositely charged end of a dipole that comes near the ion. These ion-dipole forces are somewhat stronger than van der Waals forces, yet are not nearly as strong as chemical bonds.

14.1 *CONCEPT REVIEW*

3. Compare and contrast dipole-dipole forces, dipole-induced dipole forces, and dispersion forces.
4. What does the term *dipole* mean? Explain why *dipole* is a more descriptive term than *polar* to describe a polar molecule.
5. How does a polar molecule differ from a nonpolar molecule?
6. **Apply** Using Table 14.1, which solvent would be predicted to have a lower melting point, ethanol or toluene? Water or methanol?

FRONTIERS

Ion Trapping

The illustration is not showing the motion of a wobbly ice-skater but that of a single aluminum ion whizzing around in the electric field of an ion trap. Ion traps were once exotic experimental tools used to study the behavior of isolated ions. However, ion trap mass spectrometers may soon become standard tools used to measure minute amounts of chemicals.

The chamber of an ion trap confines individual ions in a quadripole electric field, which is the electric field associated with four point charges. The ion trap's electric field is generated by applying a high frequency potential difference to electrodes within the trap. In operation, gaseous chemical samples are ionized by reagents within the chamber. By applying a potential difference of precise frequency, ions of a specific charge-to-mass ratio are confined to certain trajectories in the electric field. The ions are literally trapped in different regions of the chamber. By slightly changing the frequency, these ions move into unstable trajectories, exit the trap, and are counted by an external detector. Because ions of a specific charge-to-mass ratio are ejected at a precise frequency, the number of ions of a particular molecular mass can be correlated to the frequency of the applied potential difference.

Researchers are using ion trap mass spectrometers to detect trace amounts of carcinogenic agricultural pesticides. Researchers have accurately measured concentrations of pesticide residue to as low as 50 parts per billion. These spectrometers may be used to analyze the safety of food crops.

OBJECTIVES

- define *complex ion, ligand, coordination number,* and *coordination compound.*
- name complex ions given their formulas, and write formulas for complex ions given their names.
- name coordination compounds given their formulas, and write formulas for coordination compounds given their names.

14.2 Coordination Chemistry

One important property of polar molecules is their behavior toward ions in solution. As an ionic crystal dissolves in water, surface ions become surrounded by polar H_2O molecules that adhere to the surface. The water molecule-ion clusters formed enter solution. The stability of these clusters is greatest when they have at their center a small ion of high charge. Because positive ions are usually smaller than negative ions, the clusters with which we are concerned have a positive ion at the center.

Ligands

When polar molecules or negative ions cluster around a central positive ion, a **complex ion** is formed. These polar molecules or negative ions are known as **ligands.** The number of points of attachment of the ligands around a central positive ion in a complex is called the **coordination number.** The most common coordination number found in complexes is six. These complex ions are described as octahedral. The ligands may be thought of as lying at the vertices of a regular octahedron with the central positive ion in the middle as shown in Figure 14.8. $[PtCl_6]^{2-}$ is an octahedral complex. Complex ions are widely used in analytical chemistry and as catalysts in industry.

Word Origins

ligare: (L) to bind

ligand—binds to the central ion in a complex

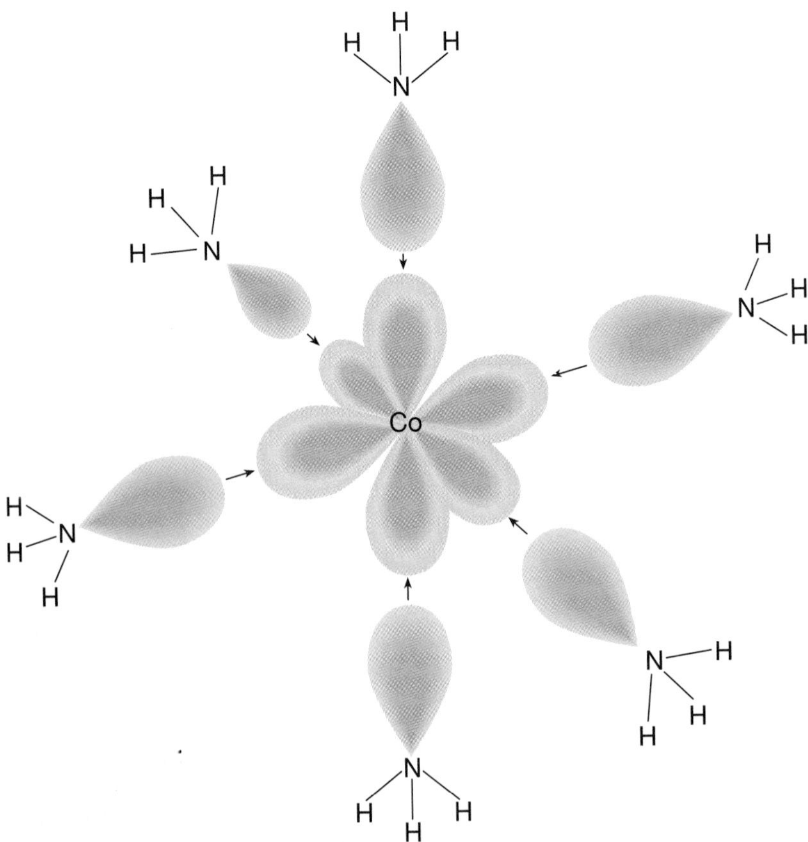

Figure 14.7 Unpaired electrons in NH_3 ligands are accepted by d^2sp^3 hybrid orbitals in a Co^{3+} ion.

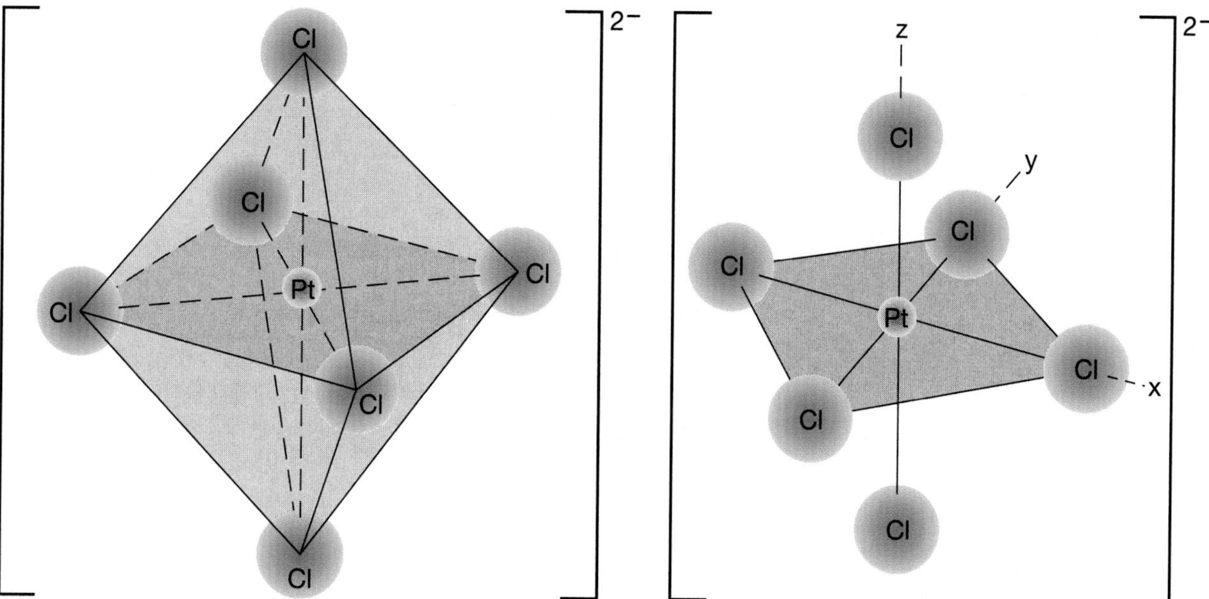

Figure 14.8 $[PtCl_6]^{2-}$ is a complex ion having an octahedral shape. Chloride ions are the ligands. On the right, a simpler representation of the octahedral shape is shown.

The coordination number four is also common. These complexes may be square-planar, with the ligands at the corners of a square and the central positive ion in the center. Others may be tetrahedral, with the ligands at the vertices of a regular tetrahedron and the central positive ion in the middle. Complexes of Ni^{2+}, Pd^{2+}, and Pt^{2+} are usually planar if the coordination number is four. An example of a square planar complex is $[NiBr_4]^{2-}$. A typical tetrahedral complex is $[CoI_4]^{2-}$, Figure 14.9.

The coordination number two is found in complexes of Ag^+, Au^+, and Hg^{2+}. Complex ions with coordination number two are always linear. The ligands are always located at the ends and the positive ion in the middle of a straight line. An example is $[Ag(CN)_2]^-$.

$$[NC—Ag—CN]^-$$

Ligands can be either molecules or negative ions. Molecular ligands are always polar and always have an unshared pair of electrons that is shared with the central ion. The most common

Figure 14.9 $[NiBr_4]^{2-}$ (left) forms a square planar complex, and $[CoI_4]^{2-}$ (right) exists as a tetrahedron. Both Ni^{2+} and Co^{2+} have a coordination number of four in these complexes.

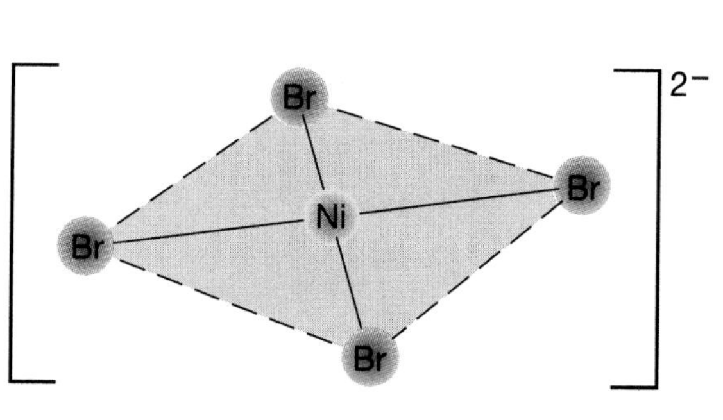

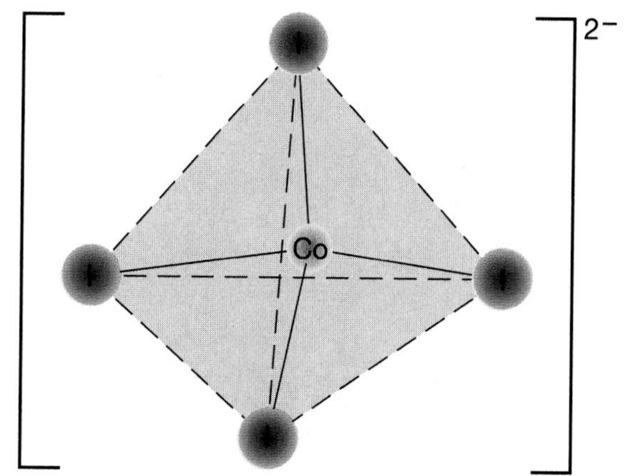

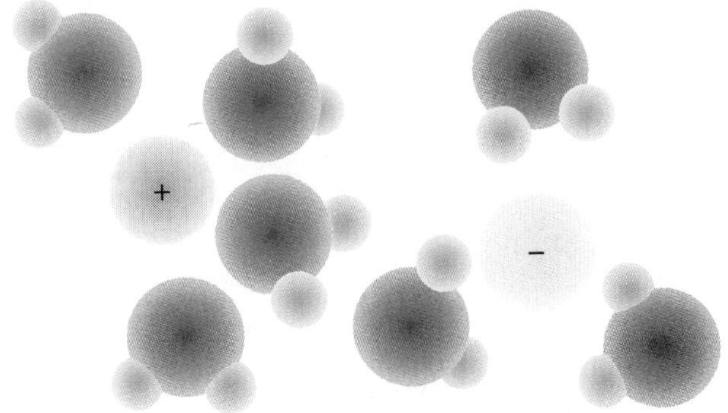

Figure 14.10 When an ionic substance dissolves in water, the water molecules cluster around the ions, forming complex ions.

ligand is water. The hydrated compounds mentioned in Section 8.2 are composed of positive ions surrounded by water ligands and the negative ions. Ammonia, NH_3, is also a common ligand. Many negative ions can also act as ligands in complexes. Some of the most important are the following: fluoride, F^-; chloride, Cl^-; bromide, Br^-; iodide, I^-; cyanide, CN^-; thiocyanate, SCN^-; and oxalate, $C_2O_4^{2-}$.

The oxalate ion has two of its oxygen atoms attached to the positive ion. Such a ligand is shown in Figure 14.11, and is called didentate ("two-toothed"). A didentate ligand attaches at two points. Therefore, two didentate ligands can form a tetrahedral complex. Three didentate ligands can form an octahedral complex. Tridentate and tetradentate ligands are also known.

One widely used hexadentate ligand is the ethylenediaminetetraacetate ion (EDTA), usually produced in the form of one of its salts. The calcium disodium salt, called calcium sodium edetate, is used in medicine. EDTA is also found as a cleansing agent in detergents and shampoos, as a preservative in foods, and as a cleansing agent in metal polishes. It is used in purifying vegetable oil. EDTA has even been used to clean surfaces contaminated by radioactive materials.

MATH CONNECTION

Platonic Solids

The regular octahedron made of eight equilateral triangles is one of only five Platonic solids. The Platonic solids are those solids whose faces are congruent regular polygons. Others that are common in coordination chemistry and in crystals are the hexahedron (cube) with six square faces and the tetrahedron with four equilateral triangles. The remaining two solids are the dodecahedron with twelve pentagons as faces and the icosahedron with twenty equilateral triangular faces.

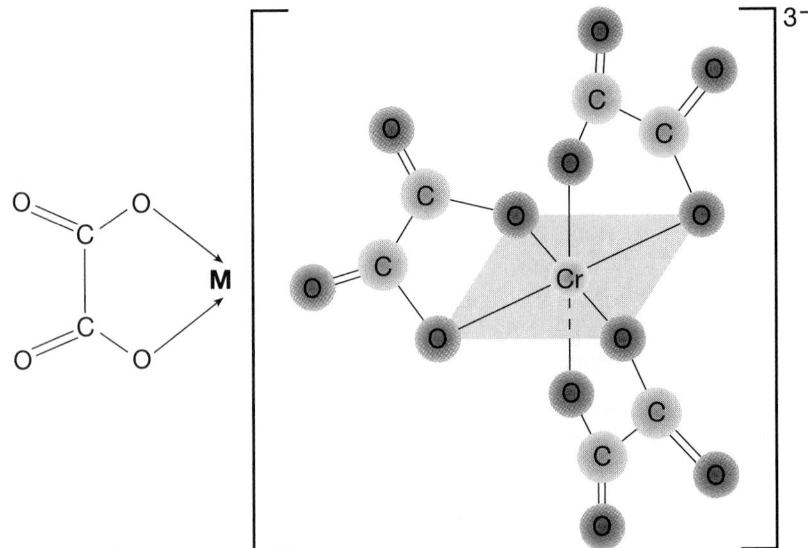

Figure 14.11 The oxalate ion (left) is a didentate ligand. In $[Cr(C_2O_4)_3]^{3-}$ (right), three oxalate ions form an octahedral complex with Cr^{3+}.

Bridge to
MEDICINE

Removing Toxic Metals from the Body

The human body does not have built-in processes for eliminating most transition metal atoms. Small amounts of several of these metals, such as chromium, copper, iron, manganese, and molybdenum, are essential to health. However, most transition metals, and large amounts of all of them, are toxic. In cases where a human has ingested large amounts of a transition metal, physicians use polydentate ligands to form complexes with the metal atoms. These complexes prevent the metal atoms from interfering with bodily processes. In addition, in most cases the body is able to eliminate the complex, thereby getting rid of the toxic atom. For example, calcium sodium edetate is used to complex lead. Dimercaprol complexes arsenic, gold, and mercury. Penicillamine complexes copper.

There are some problems that may arise from using some of these complexing agents. For example, the lanthanoid metals are more toxic when complexed by organic ligands than they are as the free metals.

Exploring Further

1. If most metals are toxic, what metals can be used to make artificial body parts?
2. What problems might result from using silver and mercury in dental amalgams?
3. Find out why lead poisoning is a particular problem for young children. Why is lead poisoning less of a problem than it was thirty years ago?

▼ Names and Formulas of Complex Ions

CHEMISTRY IN DEPTH

In naming a complex ion, the ligands are named first, followed by the name of the central ion. The names used to identify the ligands are listed in Table 14.3. The name for each ligand is modified by using a prefix indicating how many of that ligand appear in the formula. The prefixes used are *di-*, *tri-*, *tetra-*, *penta-*, and *hexa-*. If no prefix is used, it is assumed that the ligand appears only once in the formula. If more than one kind of ligand appears in the formula, the names are listed alphabetically without regard to numerical prefixes. For example, pentaammine would precede chloro because we disregard the *penta,* and the *a* of ammine comes before the *c* of chloro.

The central ion is named in the usual manner if the whole complex ion has a positive charge. If, however, the whole complex ion has a negative charge, the ending of the central ion is changed to *-ate*. In the case of the elements listed in Table 14.4, the Latin stems are used in negatively charged complex ions. If the central ion has more than one possible oxidation number, a Roman numeral indicating the correct oxidation number is included in parentheses following the name of the complex ion.

Figure 14.12 From left to right are solutions containing the colorless $[Cu(CN)_2]^-$ ion and colorful $[Cu(H_2O)_4]^{2+}$, $[Cu(en)_2]^{2+}$, and $[CuBr_4]^{2-}$ ions. The abbreviation *en* represents the ethylenediamine ligand.

Table 14.3

Some Common Ligands			
Neutral Molecules		**Anions**	
Ligand	**Name**	**Ligand**	**Name**
CO	carbonyl	Br^-	bromo
NH_3	ammine	CN^-	cyano
NO	nitrosyl	Cl^-	chloro
H_2O	aqua	F^-	fluoro
		I^-	iodo
		OH^-	hydroxo
		$C_2O_4^{2-}$	(oxalato)*
		O^{2-}	oxo
		S^{2-}	thio
		SCN^-	thiocyanato
		$S_2O_3^{2-}$	thiosulfato

*Names of organic ligands are always shown in parentheses.

Table 14.4

Latin Derived Names Used in Metal Complexes	
Metal	**Latin Form**
copper	cuprate
gold	aurate
iron	ferrate
lead	plumbate
silver	argentate
tin	stannate

PROBLEM SOLVING HINT

In writing the formula of a complex ion, the brackets do not include the charge on the ion.

SAMPLE PROBLEMS

1. Naming a Positive Complex Ion

Name the complex ion $[CrCl(NH_3)_5]^{2+}$.

Solving Process:

The two ligands present in this complex ion are *chloro* (the chloride ion) and *ammine* (the ammonia molecule). Ammine precedes chloro alphabetically. Because there are five ammonia molecules, ammine is prefixed with *penta-*. The ligands are, then, pentaamminechloro. The central ion, chromium, is named with its ending unchanged because the complex ion, as a whole, is positive. However, chromium has more than one oxidation number. We must, therefore, follow the name with a Roman numeral, in this case (III). The complete name is written as one word and followed by the word *ion*.

pentaamminechlorochromium(III) ion

2. Naming a Negative Complex Ion

Name the complex ion, $[IrCl_6]^{3-}$.

Solving Process:

The ligands are six chloride ions named *hexachloro*. The central ion is iridium. The complex ion is negative, so the name iridium is changed to *iridate*. Iridium here is in the oxidation state (III). The full name of the complex is

hexachloroiridate(III) ion.

PRACTICE PROBLEMS

7. Name the following complex ions.
   - **a.** $[Cd(NH_3)_2]^{2+}$
   - **b.** $[Cr(H_2O)_4]^{2+}$
   - **c.** $[Co(NH_3)_6]^{2+}$
   - **d.** $[Cu(NH_3)_4]^{2+}$
   - **e.** $[PtBr(NH_3)_3]^{+}$
8. Name the following complex ions.
   - **a.** $[PtI_6]^{2-}$
   - **b.** $[GeF_6]^{2-}$
   - **c.** $[Sb(OH)_6]^{-}$
   - **d.** $[PdCl_4]^{2-}$
   - **e.** $[Fe(CN)_6]^{4-}$
9. Name the following complex ions.
   - **a.** $[SbF_6]^{-}$
   - **b.** $[BF_4]^{-}$
   - **c.** $[AlF_6]^{3-}$
   - **d.** $[CoCl(H_2O)(NH_3)_4]^{2+}$
   - **e.** $[Cd(NH_3)_4]^{2+}$

Figure 14.13 The colors of many gemstones are due to small quantities of transition metal ions. For example, Cr in the mineral beryl produces an emerald, and Fe in beryl produces aquamarine. Cr in corundum produces a ruby, and corundum containing Fe or Ti is a sapphire.

As you may have already noticed, formulas for complex ions are written in an order different from that of the names. In the formulas, the symbol for the central ion is placed first, followed by the symbols or formulas for the ligands. Negative ligands are listed first, followed by neutral molecules. Within those two groups, the ligands are listed in alphabetical order of their symbols. Formulas for complex ions are always enclosed in brackets. Polyatomic ions or molecules acting as ligands are always enclosed in parentheses even if only occurring once.

Figure 14.14 Nickel forms a variety of complex ions. Left to right are $[Ni(H_2O)_6]^{2+}$, $[Ni(NH_3)_6]^{2+}$, and a strawberry red precipitate of a nickel and dimethylglyoxime complex.

Below are listed several other complex ions and their names. Use these examples as a check on your understanding of names and formulas for complex ions.

$[SiF_6]^{2-}$	hexafluorosilicate(IV) ion
$[Zn(NH_3)_2]^{2+}$	diamminezinc ion
$[Ir(H_2O)(NH_3)_5]^{3+}$	pentaammineaquairidium(III) ion

SAMPLE PROBLEMS

1. Writing a Positive Complex Ion Formula

Write the formula for the diamminepalladium(II) ion.

Solving Process:

The central ion is palladium, Pd. The ligands are two ammonia molecules, NH_3. The charge on the palladium ion is its oxidation number, 2+. The ammonia molecules are neutral, so they do not contribute to the charge of the complex ion. The complex ion, then, has a charge of 2+.

$$[Pd(NH_3)_2]^{2+}$$

2. Writing a Negative Complex Ion Formula

Write the formula for the carbonylpentacyanoferrate(II) ion.

Solving Process:

The central ion is iron. See Table 14.4. The two ligands are CO and CN^-. Because cyano is charged and carbonyl is not, cyano will precede carbonyl in the formula. The charge on the iron ion is 2+, on the carbonyl it is 0, and on the five cyanides, 5− (1− each). The formula, then, is

$$[Fe(CN)_5(CO)]^{3-}$$

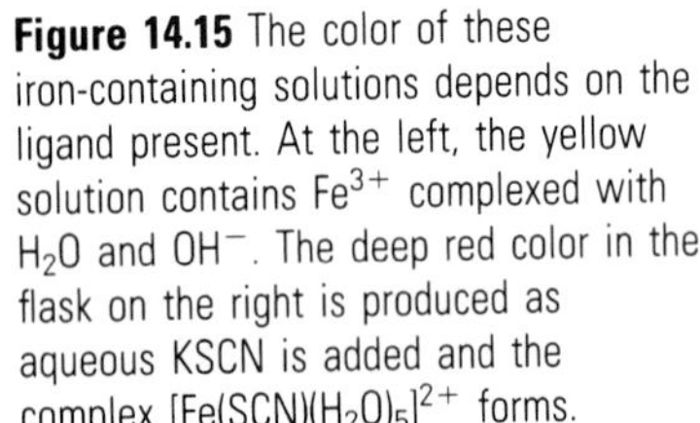

Figure 14.15 The color of these iron-containing solutions depends on the ligand present. At the left, the yellow solution contains Fe^{3+} complexed with H_2O and OH^-. The deep red color in the flask on the right is produced as aqueous KSCN is added and the complex $[Fe(SCN)(H_2O)_5]^{2+}$ forms.

PRACTICE PROBLEMS

10. Write formulas for the following complex ions.
  a. tetraammineplatinum(II) ion
  b. hexaamminecobalt(III) ion
  c. hexaaquairidium(III) ion
  d. tetraamminepalladium(II) ion

11. Write formulas for the following complex ions.
  a. hexachloropalladate(IV) ion
  b. amminetrichloroplatinate(II) ion
  c. tetraiodoaurate(III) ion
  d. tetracyanoaurate(III) ion

12. Write formulas for the following complex ions.
  a. octacyanotungstate(V) ion
  b. hexahydroxoantimonate(V) ion
  c. pentacarbonylnitrosyliron(II) ion
  d. tetracarbonylcobalt(I) ion

Figure 14.16 Colorless complex ions are formed by ions with empty or full *d* orbitals. The colored complex ions are formed when *d* orbitals are partially filled. Shown here from left to right are solutions containing the ions Sc^{3+}, Cr^{3+}, Co^{2+}, Ni^{2+}, Cu^{2+}, and Zn^{2+}. All ions are complexed with six water molecules.

Coordination Compounds

Sometimes, the charge of the central ion in a complex ion is just matched by the charges of the ligands. A neutral compound has then been formed. Occasionally, a complex will form in which neither the ligands nor the central atom has a charge. In that case, the name of the central atom is followed by zero (0). Complex ions can also form ionic compounds just as any monatomic ion can. The charges of all particles present must, of course, add up to zero. Coordination compounds are compounds formed in any of the ways just listed.

SAMPLE PROBLEMS

1. Naming a Coordination Compound

Name the following coordination compound:

$$[Ni(NH_3)_6]Br_2$$

Solving Process:

Determine the name of the complex ion. The complex ion is hexaamminenickel(II). Then name the compound as you would any other ionic compound. The complete name is hexaammine-nickel(II) bromide.

2. Writing a Formula for a Coordination Compound

Write the formula for the coordination compound, hexacarbonylchromium(0).

Solving Process:

The central atom is chromium without a charge. The ligands are carbonyl groups, also without charge. Thus, the formula is $[Cr(CO)_6]$.

CHEMISTS AND THEIR WORK

Jokichi Takamine
(1854-1922)

It is not often that industrial chemists receive the same fame that research chemists do. However, Jokichi Takamine, a Japanese industrial chemist, had many "firsts" in the field of chemistry. He was the first to make a successful separation of the starch-hydrolyzing enzyme called diastase. As a result of this accomplishment, he made enzyme production commercially feasible. At the age of 30, while in the United States, he learned of the use of superphosphate as a fertilizer. He returned to his native country and soon started an artificial fertilizer company to use the process in Japan. In 1901, he was the first to isolate pure adrenalin.

PRACTICE PROBLEMS

13. Name the following coordination compounds.

a. $[Ni(H_2O)_2(NH_3)_4](NO_3)_2$
b. $[CoCl(NH_3)_5]Cl_2$
c. $K_3[Fe(CN)_6]$
d. $K_2[Sn(OH)_6]$
e. $[Ag(NH_3)_2]ReO_4$

14. Write formulas for the following coordination compounds.

a. hexaaquacobalt(II) iodide
b. hexaamminecobalt(II) chloride
c. potassium tetracyanonickelate(II)
d. diamminedichloropalladium(II)
e. pentacarbonylruthenium(0)

Below are several more examples of coordination compounds with their names. Use these examples as a check on your understanding of naming and writing formulas for coordination compounds.

$Na_2[Sn(OH)_6]$	sodium hexahydroxostannate(IV)
$Ag_3[Fe(CN)_6]$	silver hexacyanoferrate(III)
$[V(CO)_6]$	hexacarbonylvanadium(0)

Bonding in Complexes

Almost any positive ion might be expected to form complexes. In practice, the complexes of the metals of the first two groups in the periodic table have little stability. By far the most important and most interesting complexes are those of the transition metals. The transition metals have partially filled *d* orbitals that can become involved in bonding. These positive ions are small

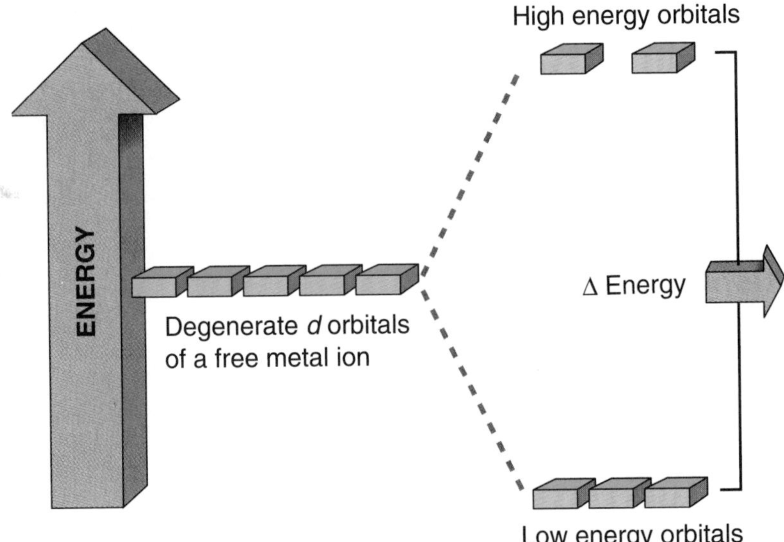

Figure 14.17 The intense colors of many complex ions are the result of the splitting of the *d* orbitals. Energy is absorbed as electrons move from the lower to higher energy levels.

because of the strong attractive force of the nucleus for the inner electrons. They also have high oxidation numbers. The combination of small size and high charge results in a high charge density on the central ion. Such a high charge density is particularly favorable to complex ion formation.

In an isolated ion of one of the metals of the first transition series, the five 3*d* orbitals are degenerate. That is, they have the same energy. No energy is absorbed if an electron is transferred from one of the 3*d* orbitals to another. However, in the complex, the ligands affect the energies of the different 3*d* orbitals. In an octahedral complex, the *d* orbitals are split into a higher energy group of two orbitals and a lower energy group of three orbitals. Many complex ions have intense colors. The blue color that is typical of solutions of copper(II) compounds is an example. This blue color is due to the $[Cu(H_2O)_6]^{2+}$ ion. The colors of complexes arise from electron transitions between the split *d* orbitals. The split represents only a small energy gap. As the electrons move from the lower to the higher energy group, they absorb light energy. The phthalocyanine blue and green dyes and paint pigments are tetradentate ligand complexes of copper. Their intense color is due to the splitting of the *d* orbitals of the copper ion. These dyes find wide use in inks.

Ligands are either negative ions or polar molecules, while the central atom is a positive ion. These facts suggest that the bonding forces in a complex are electrostatic, just as they are in a salt crystal. In fact, there are strong similarities between the structures of salts and the structures of complex ions. On the other hand, the ligands have an unshared pair of electrons that they are capable of donating. The central ion always has unoccupied orbitals into which electron pairs might be placed. This combination suggests that the bonds are covalent. The name **coordinate covalent bond** has been given to covalent bonds in which both of the electrons in the shared pair come from the same atom. The chemistry of coordinate covalent bonds is exactly like that of any other covalent bond. The bonds of most complex ions have characteristics of both covalent and ionic bonding types. The covalent character dominates.

PRINTING CONNECTION

Ink Composition

Some printing inks consist of a pigment (color) dispersed in a vehicle such as linseed oil or kerosene. Other oils and solvents are also used as vehicles depending on the particular kind of printing machinery to be used. Additives are also often blended in the vehicle. They may stabilize the pigment if it is subject to decomposition. They may also alter the finish of the ink to give a glossy appearance or a matte effect.

14.2 CONCEPT REVIEW

15. If an ion has a positive charge, would you expect to find it in the central position in a complex, as a ligand, or either one?
16. How is it possible for a ligand to be a neutral molecule?
17. Name the following.
    a. $[MoI(CO)_5]^-$
    b. $[Fe(C_2O_4)_3]^{3-}$
    c. $[TiCl(H_2O)_5]^{2+}$
18. Write formulas for the following.
    a. hexacarbonylmolybdenum(0)
    b. dicyanocuprate(I) ion
19. **Apply** In terms of energy levels of electrons, explain the deep red color of the complex ion $[Fe(SCN)(H_2O)_5]^{2+}$.

OBJECTIVES

- define *chromatography, mobile phase,* and *stationary phase.*
- define, describe, and name uses for the different types of chromatography.

14.3 Chromatography

One of the routine tasks facing analytical chemists and scientists engaged in chemical research is the separation of different substances from mixtures. You have already been introduced to two such operations, distillation and crystallization. Another separation method is chromatography, which will be described in this section.

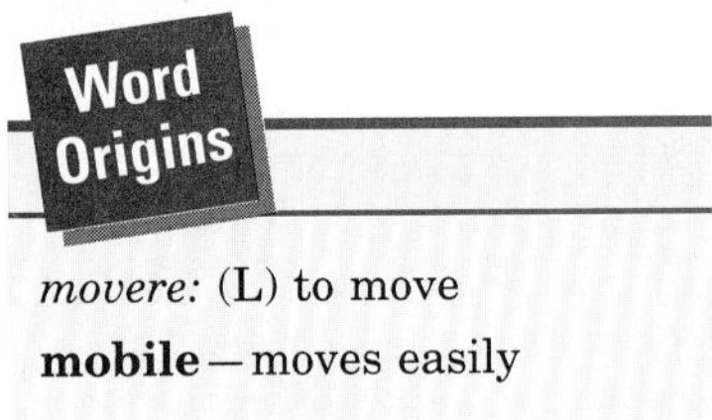

movere: (L) to move
mobile – moves easily

Fractionation

Substances separated from a mixture are called fractions because they are parts of the whole. The overall separation of parts from a whole by any process may be called **fractionation. Chromatography** is a method of fractionation based on polarity. We will examine several different chromatographic techniques.

In chromatography, a mobile phase containing a mixture of substances is allowed to pass over a stationary phase that has an attraction for polar materials. The **mobile phase** consists of the mixture to be separated dissolved in a liquid or gas. The **stationary phase** consists of either a solid or a liquid adhering to the surface of a solid. The substances in the mixture will travel at different rates due to varying polarity. There are several polarity factors determining the rate at which each component of the mixture migrates. A polar substance will have attraction for the solvent as well as an attraction for the stationary phase.

The stationary phase will attract some substances more strongly than others. The slowest migrating substance will be the one with the greatest attraction for the stationary phase. The fastest migrating substance will be the one with the least attraction for the stationary phase. Thus, substances can be separated.

Figure 14.18 The packing components for a typical chromatography column are shown.

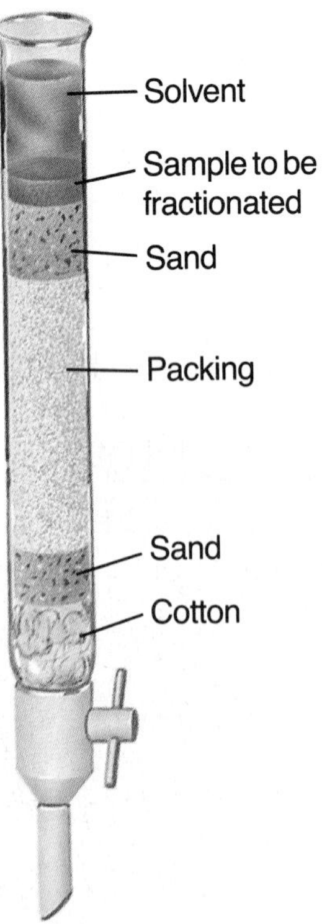

Chromatography

Column chromatography is used for extremely delicate separations. Complex substances such as vitamins, proteins, and hormones, not easily separated by other methods, can be separated by column chromatography. In column chromatography, a glass or plastic column, typically 1 cm in diameter and 50 cm long, is used to carry out the separation. The column is packed with a stationary phase such as aluminum oxide. Other packings used are calcium carbonate, magnesium carbonate, sodium carbonate, activated charcoal, clays, gels, and many organic compounds. The size of the particles of the stationary phase will be in the range of 150–200 μm.

The mixture to be fractionated is dissolved in a solvent. This solution is the mobile phase, which is added to the top of the column. The various components of the mixture are attracted to the stationary phase at the top of the tube. Then fresh solvent is poured onto the top of the column and allowed to percolate through the column. The solvent may be water, ammonia, an acid, an alcohol, or some other organic or inorganic solution.

Each substance in the mobile phase migrates down the tube at a different rate. The rate of travel of a particular substance depends upon the attraction of the substance for the stationary phase and its attraction for the solvent. If the substance has a high attraction for the stationary phase it will move down the column very slowly. A substance with less attraction for the stationary phase moves faster. The greater the concentration of solvent the faster a substance will move through the column. However, it is usually the case that very high concentrations of the solvent do not result in efficient separations. Solvent concentration, then, is usually a compromise between speed and effective separation.

After separation, it may be necessary to recover the material in each separate zone for identification. One method involves forcing the zones out the bottom of the tube, one at a time. The substance in each zone is dissolved and then recovered by evaporation. A second method uses solvents of increasing polarity until each zone comes out the bottom along with the solvent. Identification is then made by any number of methods.

chroma: (GK) color
graphein: (GK) to write

chromatography—writing with color

EVERYDAY CHEMISTRY

New Tips on the Laundry

Sebaceous glands are found near the hair follicles and secrete fats and oils, cellular debris, and keratin. These fats and oils rub off on clothing and cause dirt to adhere more readily. Researchers have some high-technology tools such as neutron bombardment, radiotracer analysis, and thin-layer chromotography to help them find out what happens after you put soiled clothing into the laundry basket.

Apparently, a series of reactions occurs. First, oxygen reactions make the oils more polar and thus more easily dissolved by soaps and detergents. However, as the oils stay in the clothing, more reactions take place which bind them even more closely.

One of the oils, squalene, gradually forms long chains that bond with fibers and turn them yellow. Polyesters sometimes resist the yellowing effects of the oil, but cottons are very difficult to lighten.

If you launder often, not allowing the oils to get to the second stage, you are more likely to be happy with the results of your laundry.

Exploring Further

1. Explain the action of polar molecules in helping detergents or soaps do their work.
2. Find out the names of the other oils that are secreted by the sebaceous glands. In your search, find out what sebum is.

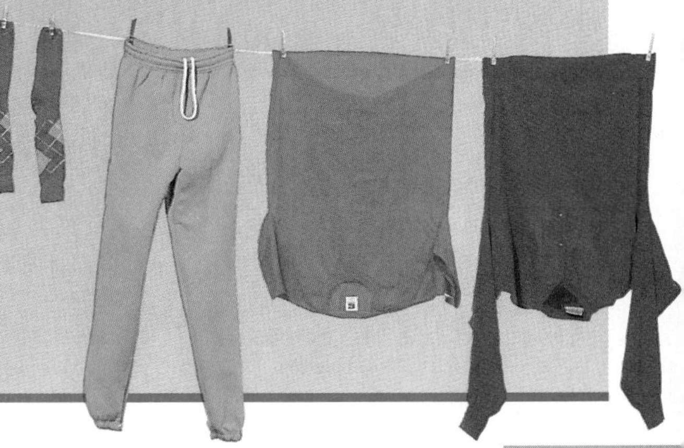

MINUTE LAB

The Ink Blot Test

Put on an apron and goggles. Flatten a piece of filter paper of the type used in coffee makers. Use scissors to cut four long strips, each 2 cm wide, from the circle. Using a different pen or marker for each, draw a line 1.5 cm from the bottom of each of the four paper strips. Add water to a large beaker until it is about 1 cm deep. Arrange the four filter paper strips around the edge of the beaker so the bottom edge is just immersed in the water. The ink line should be above the water level. Fold the top of the paper strip over the edge of the beaker to hold the strip in place.

After five minutes, remove the paper strips and hang them up to dry. How do you know that some of the inks are a mixture of different dyes? What evidence do you have to prove that some of the inks are insoluble in water? Water is a polar solvent. What type of substance, polar or nonpolar, would water be unable to separate? Did the dye of any ink go as far up the filter paper as the water traveled? What would you guess about the ink being polar or nonpolar? Identify the ink and the paper as being either the mobile or stationary phase.

High performance liquid chromatography (HPLC) is a technique designed to overcome the speed limitations of conventional column chromatography. If the particle size in the stationary phase is reduced to the range 5-10 μm, the surface area of the particles in the stationary phase is increased, and much better separations can be made. However, the small particle size severely restricts the rate of flow of the mobile phase. Consequently, high pressures are required to force the mobile phase through the stationary phase at a reasonable rate. Typically, pressures as much as several hundred times normal atmospheric pressure are employed. The apparatus needed for HPLC is considerably more expensive than that for conventional column chromatography. The pumps used must not only achieve high pressures, but must produce a stream of mobile phase at a very steady rate without any pulsing. Columns may be 1 to 5 mm in diameter and 10 to 30 cm long.

Ion chromatography is another specialized application of column chromatography. In this case, the column is packed with a material called an ion exchange resin. The resin is a complex material made of extremely large molecules called polymers (Chapter 30). Attached to the polymers are ions that have a weak attraction for the resin. When the mobile phase passes over the resin, ions in the mixture have a greater attraction for the resin, and they displace the weakly held ions. The ions with the strongest attraction for the resin "stick" first. Ions with less attraction stick farther down the column and a separation is thus achieved. Different resins are available, so that by using an appropriate resin, either positive ions or negative ions may be separated.

Paper chromatography is a form of chromatography in which the separations are carried out on paper rather than in glass columns. Strips of paper are placed in a closed container in which the atmosphere is saturated with water vapor or solvent vapor. A drop of the solution to be separated is placed at the top of

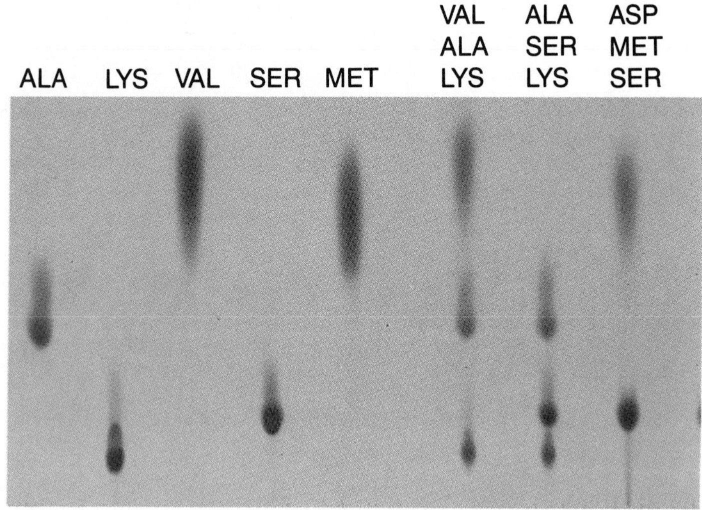

Figure 14.19 Thin layer chromatography may be used to separate amino acids. The letters at the top are abbreviations of different amino acids. Note that each amino acid separates out at a certain level on the chromatogram.

the paper. The paper is then overlapped into a tray of solvent at the top of the container. The solvent moves down the paper by capillary action, separating the constituents of the drop. In an alternate method, the paper may be placed in solvent at the bottom of the box. In this case, the drop of the solution would be placed at the bottom of the paper and the solvent would ascend the paper. In either case, any separations are seen as a series of colored spots on the paper strip. If the separated fractions are colorless, they can be sprayed with solutions that will produce colored compounds. Some of these compounds may fluoresce under ultraviolet light.

Paper chromatography is simple, fast, and has a high resolving power. One great difficulty with using paper chromatography separation arises out of its extremely small scale. Also, quantitative determinations are difficult. In addition, a control is needed to determine which spots on the paper belong to which compound. Even with these difficulties, the method is still extremely useful.

Thin layer chromatography combines some of the techniques and principles used in both column and paper chromatography. A glass or plastic plate is coated with a very thin layer of stationary phase, as is used in column chromatography. A spot of an unknown mixture is applied to the plate as a spot is applied to the paper in paper chromatography. The glass plate is then placed in an atmosphere of solvent vapor and solvent, as is done in paper chromatography. The procedure from this point is just the same as that for paper chromatography. The thin layer technique is used frequently in separating mixtures of different biological materials.

Figure 14.20 Ion exchange resins are often used in home water softeners.

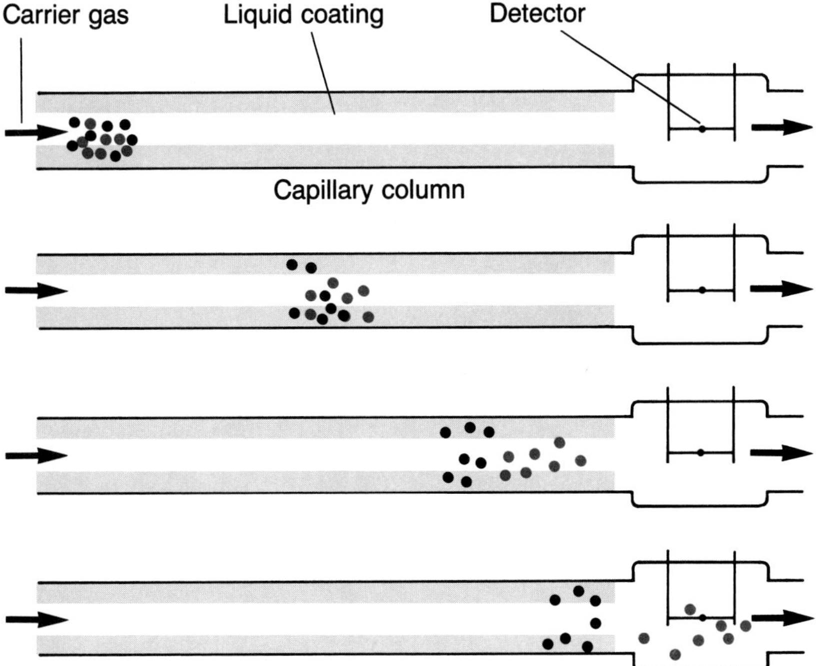

Figure 14.21 In gas chromatography a mixture is driven through a column by an unreactive carrier gas. The column is packed with a solid coated with a liquid that gradually adsorbs various components of the mixture to different degrees, thus separating them. Each component then passes over the detector and its presence is recorded.

Drug Detection

One important use of chromatography is in the detection of drugs used by athletes. In many sports at various levels, competitors are routinely tested for use of drugs by analyzing samples of their urine before and after a contest. These samples are analyzed using both gas and column chromatography. Sometimes the chromatographic technique is combined with mass spectrometry for a more detailed analysis. The types of drugs tested in this way include stimulants, narcotics, anabolic steroids, relaxants, and diuretics.

Stimulants are used to enhance performance by pushing the human body to its limits, and sometimes beyond. Narcotics are used to dull the pain of injuries. Anabolic steroids enable athletes to increase their muscle mass at an accelerated rate. Why would an athlete use a relaxant? Pistol, rifle, and archery competitors may use relaxants to steady their aim by slowing down their breathing and pulse rates. Diuretics can be employed by wrestlers and boxers who must not exceed certain weight limits. Diuretics provide rapid weight loss. Not only does use of these drugs result in unfair competition, it also endangers the health, and in some cases the life, of the users.

Athletes are not the only people who abuse their bodies by taking drugs. Some so-called "recreational" drugs are used by people in many walks of life and from a wide variety of economic backgrounds. Many government agencies and large corporations now require a drug test of current and prospective employees. The method of testing is again an analysis of urine. The urine is tested for the presence of the drugs themselves or for compounds, called metabolites, produced by the body when the drugs are used. Testing typically analyzes for amphetamine, methamphetamine ("ice"), cocaine ("crack"), cannabinoids (from marijuana), opiates (codeine, morphine, and heroin), and phencyclidine (PCP or "angel dust").

A combination of different analytical methods is commonly used for testing. Initially, the presence of the drug or metabolite is established using antibodies or antigens in an immunochemical process. If evidence of the presence of the drug or metabolite is found, chromatography is used to concentrate the suspected substance. Then, mass spectrometry is used to confirm the substance and give some indication of quantity. The quantity of drug or metabolite in the urine is difficult to correlate with the amount used since urinary concentration varies with diet, sex, age, and other factors.

Thinking Critically

1. Certain types of steroids, such as cortisone, are common, legal medications used by people with diseases such as arthritis and asthma. Find out how testing allows for, or doesn't allow for, these legal medications.
2. Race horses commonly undergo similar drug testing. Explain why this testing might be necessary.

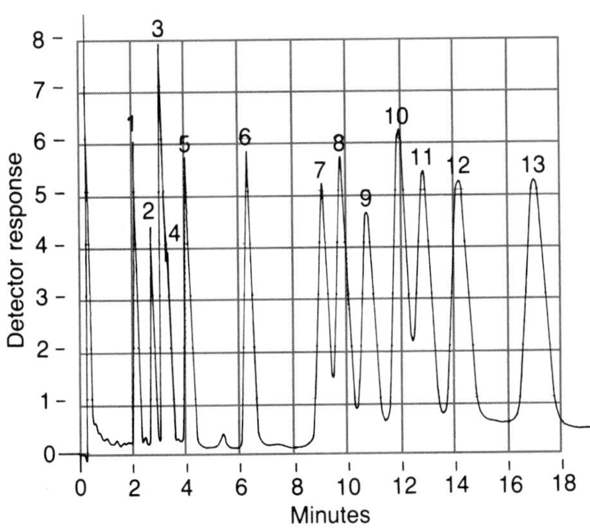

Figure 14.22 This drawing of a gas chromatogram shows the pesticides present in a sample of run-off water.

Gas chromatography is a chromatographic technique for the analysis of volatile liquids and mixtures of gases or vapors. The gases to be analyzed are carried along by an inert gas such as helium in the mobile phase. The gases are fractionated on a stationary phase by a method similar to column chromatography. They are then carried by the inert helium through a tube fitted with an electrocouple. The varying amounts of separated substances in the helium produce varying amounts of current. These variations are recorded and interpreted by computer. The amount of gas present in each fraction can be determined from the area under the curve.

Gas chromatography has many practical uses. Gasoline mixtures that contain over 100 different compounds can be separated and characterized by gas chromatography. A common test performed by environmental scientists is analysis of agricultural run-off for organic pesticide residues. Figure 14.22 shows such a chromatogram.

HOME ECONOMICS CONNECTION

Hard Water

Many homes in areas with hard water are equipped with water softeners. *Hard* water is water containing dissolved substances that react with soap, forming a precipitate, or soap scum. Water softeners contain ion exchange resins quite similar to those used in ion chromatography. The resins contain sodium ions and chloride ions. As the hard water passes through the water softener, sodium ions are exchanged for any other metallic ions in the water, and chloride ions are exchanged for other negative ions. The water is now considered *softened* because there are no substances present that will form precipitates with soap.

14.3 CONCEPT REVIEW

20. Chromatographic methods are sometimes classified as two-dimensional or three-dimensional. List each of the techniques introduced in this section under one of those two headings.

21. The mobile phase and the stationary phase perform different functions in chromatography. Differentiate between these functions.

22. Which technique would you suggest using if you had to separate and recover large quantities of a polar organic material?

23. Apply Home water softeners frequently use an ion exchange resin to replace ions such as Ca^{2+} and SO_4^{2-} with Na^+ and Cl^-. If the resulting water is used for drinking, sodium ions are not desirable in the water either. Use the principles involved in ion chromatography to explain how the undesirable ions in hard water could be removed without adding other metallic ions.

Chapter 14
REVIEW

Summary

14.1 Molecular Attraction

1. A polar bond is one in which a shared pair of electrons is attracted to one atom more strongly than to the other atom.
2. A molecule containing polar bonds can itself be polar or nonpolar depending on its bond arrangement. A molecule is polar if the polar bonds are arranged asymmetrically and nonpolar if the polar bonds are arranged symmetrically.
3. Van der Waals forces are the net result of dipole-dipole, dipole-induced dipole, and dispersion forces.

14.2 Coordination Chemistry

4. A complex ion is composed of molecular or negative-ion ligands clustered around a central positive ion.
5. The number of points of attachment of ligands around the central ion in a complex is called the coordination number.
6. A coordination compound consists of a complex ion and other ions or a neutral central atom and neutral ligands combined to form a compound.
7. Transition metals, because of their partially filled *d* orbitals, characteristically form complex ions.
8. The bonds of complex ions have both ionic and covalent bonding characteristics.

14.3 Chromatography

9. Chromatography is a method of separating a mixture of substances into identifiable chemical fractions based on differences in polarity of the individual fractions.

Key Terms

polar covalent
dipole
dipole moment
van der Waals force
intramolecular force
intermolecular force
dipole-dipole force
induced dipole
dipole-induced dipole force
temporary dipole
dispersion force
complex ion
ligand
coordination number
coordinate covalent bond
fractionation
chromatography
mobile phase
stationary phase
column chromatography
high performance liquid chromatography (HPLC)
ion chromatography
paper chromatography
thin layer chromatography
gas chromatography

Review and Practice

24. The following pairs of atoms are all covalently bonded. Arrange the pairs in order of decreasing polarity of the bonds using Table 12.1, page 303.
    a. phosphorus and sodium
    b. fluorine and nitrogen
    c. nitrogen and oxygen
    d. arsenic and iodine
    e. chlorine and tellurium
    f. antimony and sulfur
25. What forces hold molecular substances in the liquid and solid states?
26. What are dipole-dipole forces? Give an example.
27. Water has a bond angle of 104.5° between the two O—H bonds. Sketch a diagram of a water molecule, indicating the positive and negative ends of the molecule.
28. Classify each of the following as either an intramolecular force or an intermolecular

force. Explain each answer.
a. van der Waals forces
b. covalent bonds

29. Describe two ways that a nonpolar molecule can become dipolar. Give an example of each.

30. Chloroform, $CHCl_3$, is a polar molecule. When considering intermolecular forces between $CHCl_3$ molecules, explain why dispersion forces may be more important than dipole-dipole forces.

31. Define central ion, ligand, and coordination number. Give an example of each.

32. What factors determine coordination number?

33. What is the coordination number in each of the following ions?
a. $[PtCl_6]^{2-}$
b. $[PdCl_4]^{2-}$
c. $[Sb(OH)_6]^-$
d. $[Cu(CN)_2]^-$

34. What is the coordination number in each of the following ions?
a. $[Fe(CN)_5(CO)]^{3-}$
b. $[CoCl(H_2O)(NH_3)_4]^{2+}$
c. $[FeCl_2(C_2O_4)]^{2-}$
d. $[Fe(H_2O)_5(NO)]^{2+}$
e. $[CrCl_2(H_2O)_4]^+$
f. $[Co(CO_3)(NH_3)_4]^+$

35. What is a didentate ligand? Give an example.

36. Predict and sketch the shape of the $[Al(H_2O)_6]^{3+}$ complex.

37. Sketch the shape of the complex ion $[PdCl_4]^{2-}$.

38. What is the oxidation number of the central metal in each item of Problem 33?

39. What is the oxidation number of the central metal in each item of Problem 34?

40. Determine the oxidation number of the central metal in each of the following complex ions.
a. $[CrCl_2(NH_3)_4]^+$
b. $[Co(C_2O_4)_3]^{3-}$
c. $[Mn(CN)_6]^{2-}$
d. $[CrBr_4(C_2O_4)]^{3-}$

41. Name the following complex ions.
a. $[GaF_5]^{2-}$
b. $[SbS_4]^{3-}$
c. $[Au(CN)_2]^-$
d. $[CrCl_2(H_2O)_4]^+$

42. Write the formula for each of the following complex ions.
a. tetraamminecopper(II)
b. hexaaquanickel(II)
c. tetracyanonickelate(II)
d. pentaaquachlorochromium(III)

43. Name the following coordination compounds.
a. $[VCl_2(H_2O)_4]$
b. $[CrBr_2(NH_3)_4]NO_3$
c. $Na_2[Ni(CN)_4]$
d. $K_3[Rh(CN)_6]$

44. Write the formulas for the following coordination compounds.
a. sodium tetrafluorocobaltate(II)
b. hexaaquavanadium(III) bromide
c. potassium tetrachloroferrate(III)
d. cesium trioxalatoferrate(III)

45. Why do transition metals make effective central ions in complexes?

46. How do complex ions form compounds?

47. What is fractionation?

48. What is thin layer chromatography? How is it used?

49. What is a monodentate ligand? Give an example.

50. Define tridentate ligand and tetradentate ligand.

51. Write the name for
$[Pt(NH_3)_4][CoCl_4(NH_3)_2]_2$.

52. Write both names and formulas for the complex ions or coordination compounds made up of the following ions and molecules. Be sure to show the correct charges for ions.
a. one chloride ion, one ruthenium(III) ion, and five ammonia molecules
b. four ammonia molecules, two chloride ions, and one platinum(IV) ion
c. one manganese(II) ion and six water molecules
d. one rhodium(III) ion and six cyanide ions
e. six carbon monoxide molecules and one chromium atom

f. two chloride ions, four ammonia molecules, and one iridium(III) ion
g. one thiocyanate ion, one iron(III) ion, and five water molecules

Concept Mastery

53. Covalent bonds may be polar, as in H_2O, or nonpolar, as in I_2. Explain the differences and similarities between polar and nonpolar covalent bonds.
54. What is a dipole? Explain how some molecules with polar bonds are dipoles while other molecules are not. Explain why water is a dipole and carbon dioxide is not.
55. Explain how a nonpolar molecule can contain polar bonds. Give an example.
56. Describe the nature of a coordinate covalent bond. What conditions lead to formation of a coordinate covalent bond?
57. Describe the process of chromatography in terms of the mobile phase and stationary phase.
58. How are column and paper chromatography similar? How are they different?
59. Why would paper chromatography be impractical in an industry that must separate the components of large quantities of a mixture?
60. Carbon dioxide has a dipole moment of 0. The dipole moment of sulfur dioxide is 1.6×10^{-30} C·m. Draw structural diagrams for both molecules and use them to explain the difference in polarity.
61. Compare the structures of complex ions with those of covalent compounds.
62. What is the difference between ordinary covalent bonds such as the H—N bonds in ammonia molecules and coordinate covalent bonds such as the Co—N bond in the $[Co(NH_3)_6]^{2+}$ ion?
63. Would the use of chromatography be more effective in separating a mixture of polar substances or a mixture of nonpolar substances? Explain your answer.
64. Helium is the most common carrier gas used in gas chromatography. Why do you think helium is used rather than oxygen, methane, or another gas?

Application

65. Ethoxyethane was commonly used as a surgical anesthetic until safer, less flammable, materials were available. Methanol, called wood alcohol, is used as an industrial solvent although it is poisonous. Using Table 14.1, compare the following properties of these solvents.
    a. polarity
    b. melting point
    c. boiling point
66. $[PtCl_2(NH_3)_2]$ has been found to be a powerful anti-cancer drug. Write the name of this coordination compound.
67. Forming a complex ion in solution is often easier than dissolving the metal salt. For example, AgBr, the photosensitive substance in photographic film, will not dissolve in water. However, when $Na_2S_2O_3$ is added, $[Ag(S_2O_3)_2]^{3-}$ is formed and the AgBr is now dissolved. What is the name of the $[Ag(S_2O_3)_2]^{3-}$ ion?
68. When complex ions and their compounds were first discovered, they were frequently named after the scientists who discovered them. $NH_4[Cr(SCN)_4(NH_3)_2]\cdot H_2O$, for example, was named Rienecke's salt. What is the present name of the $[Cr(SCN)_4(NH_3)_2]^-$ ion?
69. Some complexes were originally given names based on their colors. For example, $[CoCl(NH_3)_5]Cl_2$ was named purpureocobaltic chloride because the compound is purple. What is the present name of this compound?
70. Carbon monoxide is a deadly gas even in relatively low concentrations in air. Use reference books to find out how carbon monoxide acts in the human body and use your knowledge from this chapter to suggest a reason for its danger.

Critical Thinking

71. Determine whether each of the following would more likely be formed by polar or nonpolar molecules.
- **a.** a solid at room temperature
- **b.** a liquid with a high boiling point
- **c.** a gas at room temperature
- **d.** a liquid with a low boiling point

72. Why is gas chromatography primarily used to separate mixtures of organic compounds?

73. Considering what you have learned about forces between atoms and molecules, why do you think all of the elements in Group 18 (VIIIA) exist as gases at room temperature?

74. Purple dye could result from a mixture of red and blue dyes. How could it be determined whether the color was obtained by using one purple dye or a mixture of red and blue?

75. Suppose you rub a balloon on a wool sweater. The balloon becomes negatively charged by gaining electrons during the rubbing. Next, you place the balloon on a wall and it stays in place. Evidently the wall now has a positive charge because there is a force holding the negatively charged balloon to the wall. Propose an explanation for the wall becoming positively charged.

76. The noble gas elements, Group 18 (VIIIA), have low boiling points and freezing points that are very close to their boiling points. Using your knowledge of bonding and forces, formulate a statement that explains this observation.

77. The dipole moment of compound X_2A equals 1.84×10^{-30} C·m. Compound X_3D has a dipole moment of 1.50×10^{-30} C·m. Which compound is more polar? If they have approximately the same formula mass, which compound would you predict to have the higher melting point?

78. Even though cyanide compounds are poisonous, they are used in the reclamation of silver from film and in other processes in the refining of precious metals. Suggest a reason that these compounds are useful despite their toxicity.

Readings

Bertholf, Roger L. "Confirmatory Testing for Drugs of Abuse in Urine by Gas Chromatography/Mass Spectrometry." *The Chemist* 67, no. 5 (May 1990): 7–20

McClure, M. W. "Saving Arnold." *ChemMatters* 6, no. 4 (December 1988): 4–6.

Rouvray, Dennis H. "Predicting Chemistry from Topology." *Scientific American* 255, no. 3 (September 1986): 40–47.

Cumulative Review

1. What shape would you predict for the NI_3 molecule? For CCl_4?

2. Describe the shapes of *s* and *p* orbitals.

3. Explain why carbon, which has a predicted outer configuration of $2s^22p^2$, can form four equivalent bonds.

4. Why is benzene more stable than a compound with three single and three double bonds?

5. What shape would you predict for H_2Se? Why?

6. Draw as many isomers of C_5H_{10} as you can. Hint: There are 12 isomers. All contain a C=C double bond or are cyclic.

7. Predict the N—C=O bond angle in $(NH_2)_2CO$.

CHAPTER **Preview**

CONTENTS

CHAPTER 15

Kinetic Theory

Up! Up! and Away! A hot-air balloon ride is great fun and, oh, so quiet! The only sound is the intermittent pshhhh as the flame springs from the nozzle of the propane heater warming the air. Just a few minutes ago, the balloon was being inflated by air forced into it by a fan. But the inflated balloon didn't rise until the air inside it was heated. Does hot air lift the balloon? If so, how does heating air change its properties? Why do hot-air balloons rise?

To answer the question just posed we must begin with the notion that the molecules of air are constantly moving. All the particles that make up matter are in constant motion. It is the motion of the particles in a substance that determines if it is a solid, liquid, or gas. We will study motion in this chapter.

EXTENSIONS AND CONNECTIONS

OBJECTIVES

- list and explain the basic assumptions of the kinetic theory.
- relate pressure to molecular motion.
- differentiate among an open-arm manometer, a closed-arm manometer, and a barometer.
- calculate the pressure of gases in both open-arm and closed-arm manometers using appropriate units.

15.1 Pressure

Think about using an air pump to inflate a bicycle tire or a basketball. When you push down on the pump handle, you get the sensation that the air is pushing back. It is. In the middle of the seventeenth century, the English inventor Robert Hooke developed an explanation of such behavior. He predicted that there were particles in nature that were in constant motion. The particles are the molecules of the air. Their motion causes them to bump into the walls of their container and exert pressure. The resistance you feel in the handle of an air pump is caused by air pressure exerted on the surface of the plunger by the air trapped inside the pump. In this section we will examine pressure and other consequences of molecular motion.

Molecules in Motion

The kinetic theory is the name we give today to Hooke's explanation of gas behavior. The **kinetic theory** explains the effects of temperature and pressure on matter. First, let us consider some basic assumptions of the theory. *All matter is composed of small particles* (atoms, ions, or molecules). *These small particles are in constant motion.* Finally, *all collisions between particles are perfectly elastic.* "Perfectly elastic" means that there is no change in the total kinetic energy of two particles before and after their collision. It may be difficult to imagine that all particles in a large structure such as the space shuttle are in constant motion. However, as we shall see, many of the properties of matter are the result of such motion.

At room temperature (25°C), the average speed of an oxygen molecule is 443 meters per second. This speed is equivalent to just over 1700 kilometers per hour. At this speed, the molecules collide with each other frequently. We can determine the average number of collisions a molecule undergoes in a unit of time. We do

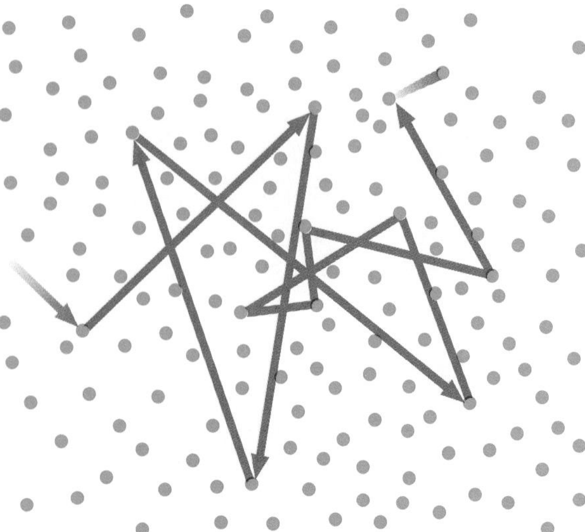

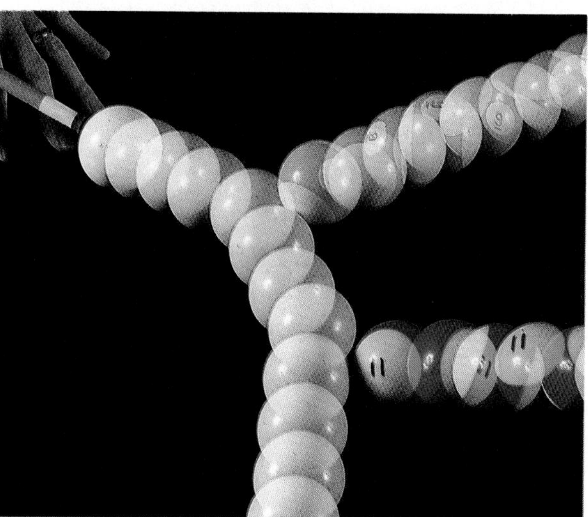

Figure 15.1 A particle of a gas travels in straight lines between collisions (left) that are perfectly elastic; that is, kinetic energy is conserved, as in the collision between two billiard balls (right).

so by finding the average distance a molecule travels before colliding with another molecule. This distance is called the **mean free path** of the molecule. For oxygen at 25°C and standard atmospheric pressure, the mean free path is 106 nm. The diameter of the oxygen molecule is about 0.339 nm. Thus, an oxygen molecule at 25°C will travel, on the average, a distance of about 314 times its own diameter between collisions.

Each molecule will have a little more than four and a half billion collisions per second. We give these figures for oxygen as examples of the speed, the distance of travel, and the number of molecular collisions in a gas. These factors vary with the temperature, the number of particles in a given volume, and the mass of the particles composing the gas.

Pressure

Besides colliding with each other, gas molecules collide with the walls of the container in which the gas is confined. When a gas molecule collides with the wall of a container, it exerts a force on the container. It is the force of collision and the number of collisions with the walls of a container that cause gas **pressure,** Figure 15.2. This pressure is measured in terms of the force per unit area, as illustrated in Figure 15.3.

The molecules and atoms of the gases present in the air are constantly hitting the surface of Earth and everything on it. As a result, you and everything surrounding you are subject to a certain pressure from the molecules of the air. Air pressure varies from place to place and from time to time in a particular place. In order to facilitate communications, scientists have agreed on a standard of pressure as representing an average air pressure at sea level. The standard has been set as 101.325 kilopascals (kPa) and is known as standard atmospheric pressure. One **pascal** is a pressure of one newton per square meter (N/m^2). In this course, you will generally use units of kilopascals.

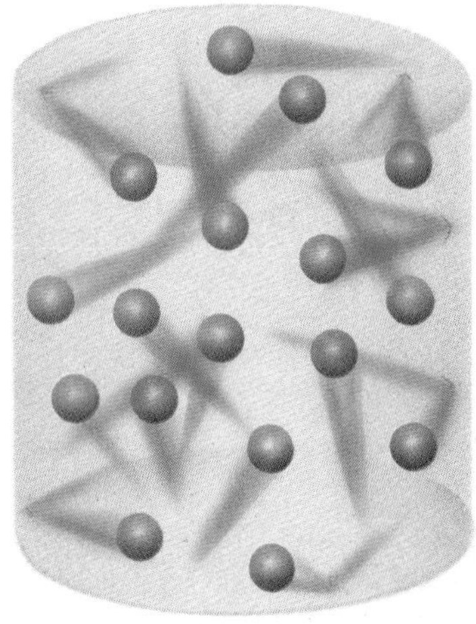

Figure 15.2 The pressure exerted by a gas on its container is the same in every direction and is caused by the gas molecules colliding with the container walls.

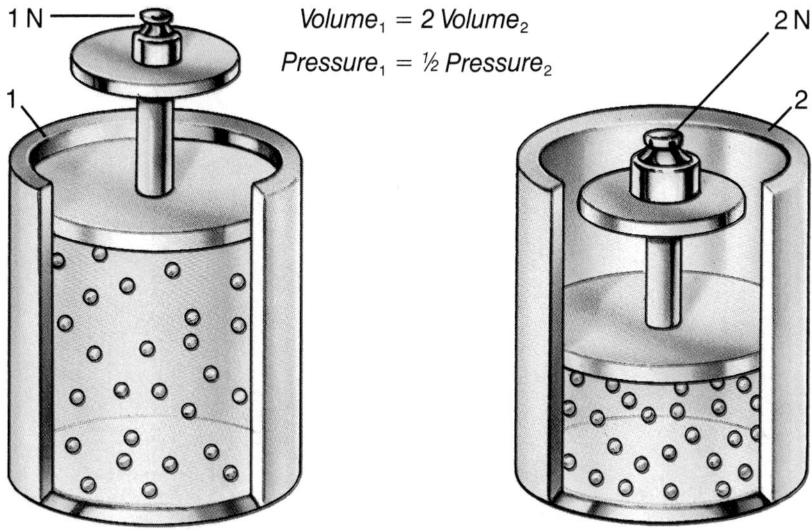

Figure 15.3 If the volume of a gas is halved, the number of collisions with the walls of the container doubles. Thus, the pressure in container 2 is twice that of container 1.

Word Origins

manos: (GK) thin
baros: (GK) heavy
metron: (GK) measure

manometer—instrument to measure thinness of air
barometer—instrument to measure heaviness of air

Measuring Pressure

In measuring gas pressure, an instrument called a **manometer** (muh NAHM uht ur) is used. Two types of manometers are shown in Figure 15.4. In the "open" type, the atmosphere exerts pressure on the column of liquid in the open arm of the U-tube. The gas being studied exerts pressure on the other arm. The difference in liquid level between the two arms is a measure of the difference in pressure between the atmosphere and the contained gas. If you know the density of the liquid in the manometer, you can calculate the pressure difference between the gas and the atmosphere.

The "closed" type of manometer has a vacuum above the liquid in one arm. The operation of a closed-arm manometer is independent of atmospheric pressure. A closed-arm manometer used to measure atmospheric pressure is called a **barometer.** Most barometers are manufactured with a scale calibrated to read the height of a column of mercury in millimeters. By definition, standard atmospheric pressure will support a column of mercury 760 mm high. Because standard atmospheric pressure is defined as 101.325 kPa, we can state that 101.325 kPa = 760 mm Hg. By dividing both sides of the equation by 101.325 we find that 1 kPa = 7.501 mm Hg.

Closed-arm manometers can be used to measure actual or "absolute" gas pressure. It is also possible to calculate the absolute pressure of a gas using an open manometer. However, a barometer must be available to measure the atmospheric pressure on the outside arm of the open manometer. The following Sample Problem shows some typical calculations involving gas pressure.

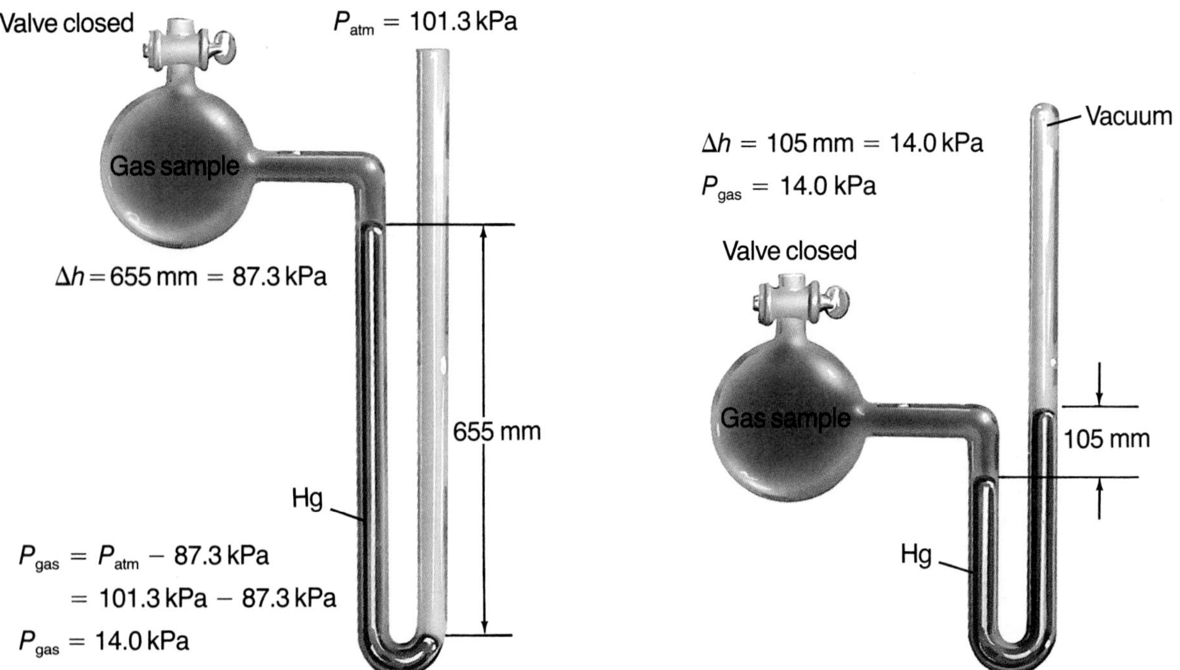

Figure 15.4 Gas pressure is measured by an open-arm manometer (left) and a closed-arm manometer (right).

EVERYDAY CHEMISTRY

Air Pressure Shapes the Tire

Automobile tires must meet demanding standards if they are to perform safely. They must generate sufficient friction with the road surface to avoid slipping. They must cushion the ride for passengers as the car passes over rough surfaces. They must also respond to turning without deforming to such an extent that they fail. The modern automobile tire accomplishes these sometimes contradictory tasks by two means. One means is the construction of fabric coated with rubber. The other means is the inflation of the tire with air. Air molecules are forced into the tire with an air pump. The concentration of air molecules inside the tire is greater than that outside. Therefore, the pressure exerted by those molecules will be greater than atmospheric pressure. The construction of the tire is such that this increased pressure forces the tire against the rim of the wheel. The construction of the tire prevents it from expanding, or ballooning. The pressure exerted by the air molecules helps maintain the shape of the tire, especially in cornering. When a tire hits a bump, for example, its volume is suddenly decreased. The decreased volume increases the pressure on the air inside the tire. The air molecules, in turn, exert increased pressure on the interior walls of the tire and quickly restore the tire to its correct volume and shape.

Exploring Further

1. Why do tire manufacturers suggest that you measure the pressure of a tire *before* the automobile is driven?
2. How do the tires on drag-strip race cars differ from standard automobile tires? How do these tires function?

SAMPLE PROBLEM

Determining the Pressure of a Gas

The open manometer in Figure 15.4 is filled with mercury. The difference between mercury levels in the two arms is 6 mm. What is the total pressure, in kilopascals, of the gas in the container if the atmospheric pressure is 101.3 kPa?

Solving Process:

The mercury is higher in the arm connected to the gas. Thus, the pressure exerted by the gas must be less than that of the atmosphere. As a result, we must subtract the pressure of the mercury from atmospheric pressure to get the gas pressure. Before subtracting, however, we must convert the 6 mm difference in height to kilopascals.

$$\frac{6\ \cancel{\text{mm}}}{} \left| \frac{1\ \text{kPa}}{7.501\ \cancel{\text{mm}}} \right. = 0.8\ \text{kPa}$$

Now we can subtract the two pressures.

$$101.3\ \text{kPa} - 0.8\ \text{kPa} = 100.5\ \text{kPa}$$

PROBLEM SOLVING HINT

The greater height of the liquid in one arm of a manometer is a measure of the greater pressure of the gas in the opposite arm.

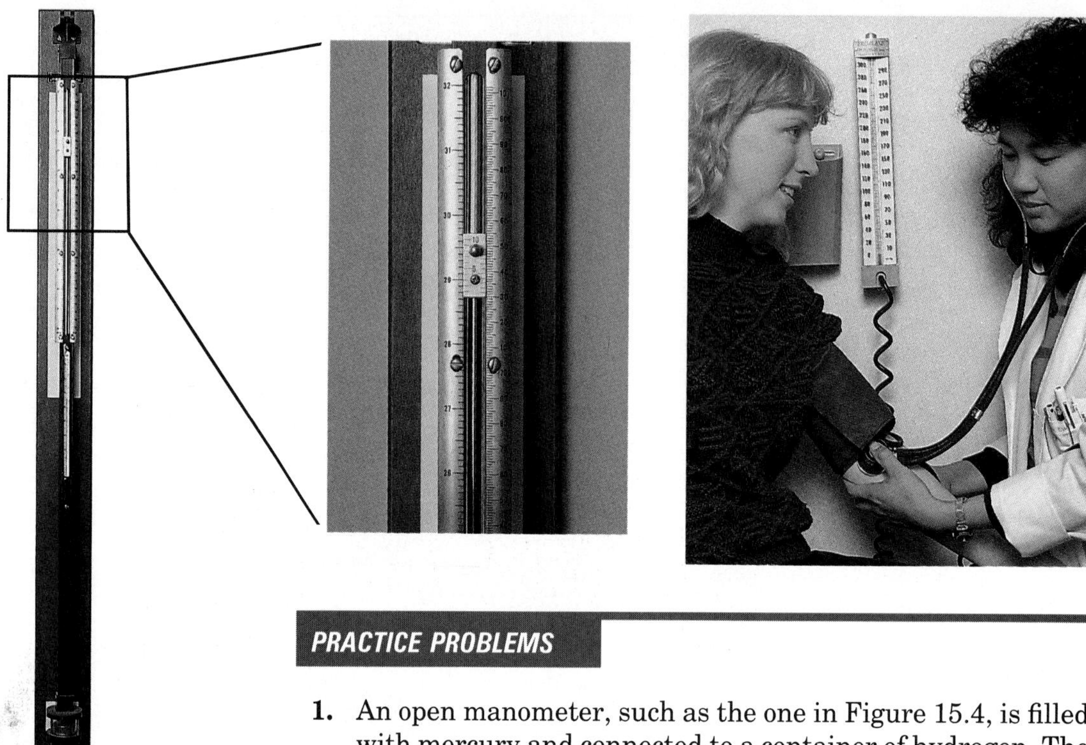

Figure 15.5 The barometer on the left is used to measure air pressure in the laboratory. A sphygmomanometer, shown on the right, is used to measure blood pressure.

PRACTICE PROBLEMS

1. An open manometer, such as the one in Figure 15.4, is filled with mercury and connected to a container of hydrogen. The mercury level is 62 mm higher in the arm of the tube connected to the gas. Atmospheric pressure is 97.7 kPa. What is the pressure of the hydrogen in kilopascals?
2. A closed manometer, like the one in Figure 15.4, is filled with mercury and connected to a container of nitrogen. The difference in the height of mercury in the two arms is 691 mm. What is the pressure of the nitrogen in kilopascals?
3. An open manometer connected to a tank of argon has a mercury level 38 mm higher in the atmospheric arm. If atmospheric pressure is 96.3 kPa, what is the pressure of the argon in kilopascals?
4. A closed manometer is filled with mercury and connected to a container of helium. The difference in the height of mercury in the two arms is 86.0 mm. What is the pressure, in kilopascals, of the helium?

15.1 CONCEPT REVIEW

5. List the three basic assumptions of the kinetic theory.
6. How can the kinetic theory be used to explain the pressure of a gas?
7. Compare and contrast an open-arm manometer, a closed-arm manometer, and a barometer.
8. **Apply** Normal human blood pressure is sometimes expressed as "120/80." The value is actually a ratio of two pressures, each measured in millimeters of mercury. Express these pressures in kPa.

15.2 Motion and Physical States

OBJECTIVES

- relate temperature and energy transfer to molecular motion.
- determine the relative velocities of gas molecules at the same temperature.
- differentiate among the four states of matter.
- describe characteristics of substances in each of the three common states of matter in terms of the kinetic theory and bonding in the substances.

A very simple definition of each physical state describes properties of matter in that state. A solid has a definite shape and a definite volume. A liquid has a definite volume but assumes the shape of its container. A gas has neither a definite shape nor volume, but expands to fill its container completely. Why does matter behave differently in each state? In this section we will answer that question by looking at how the motions of the particles in solids, liquids, and gases differ. We will also look at the behavior of particles in a plasma, the fourth state of matter.

Kinetic Energy and Temperature

The average speed of the particles in a gas depends only on the temperature and the mass of the particles. How are temperature and particle mass related?

Kinetic energy is the energy an object possesses because of its motion. The kinetic energy of an object is equal to $mv^2/2$, where m is its mass and v is its velocity. Consider a sample of matter. The kinetic energy of each of its particles depends on the mass of the particle and its velocity. As you recall, the particles in matter are constantly colliding with each other. At any instant we would expect the particles in the sample to have a wide range of velocities and, therefore, a wide range of kinetic energies. If we could measure the kinetic energy of each of the particles and divide that sum by the number of particles, we could arrive at the average kinetic energy of the particles. But such a task is impossible. There is another way of indicating the average kinetic energy of the particles of a sample of matter. We can specify its temperature. The temperature of a substance is a measure of the average kinetic energy of its particles. According to the kinetic theory, the particles of a substance are constantly undergoing elastic collision with each other. As you recall, in an elastic collision, the amount of kinetic energy lost by one particle is gained by the other. Because kinetic energy is conserved, the average kinetic energy of the particles remains the same.

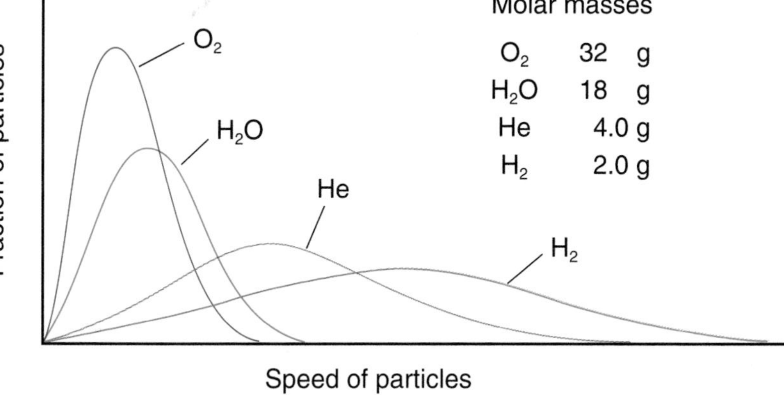

Figure 15.6 At the same temperature, the average speed of the particles of a gaseous substance depends on the molar mass of the substance.

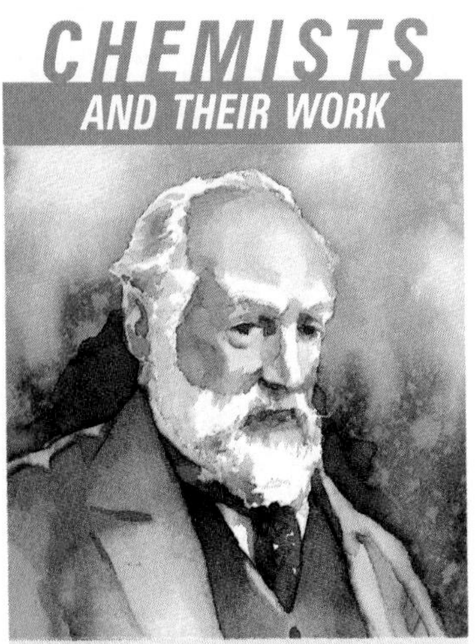

CHEMISTS AND THEIR WORK

William Thomson, Lord Kelvin (1824-1907)

William Thomson was a child prodigy. He entered the University of Glasgow at the age of 11 and graduated second in his class. After further study at Cambridge University and in Paris, he then returned to the University of Glasgow as professor of natural philosophy (science).

During his active career Thomson investigated such topics as Earth's age, the relationship of work and heat, the absolute temperature scale, and entropy. He also made many contributions to the theoretical and practical development of telegraphic communications, often using instruments that he invented. In 1886, William Thomson was knighted as Lord Kelvin for his contribution to the trans-Atlantic cable.

Because the mass of each particle of a substance remains constant, a decrease in the temperature of that substance means the particles of the substance are moving more slowly. On the other hand, an increase in temperature of a substance means that the particles of that substance are moving faster.

If samples of two substances are at the same temperature, their particles must have the same average kinetic energy. This equality does not mean that the particles in each substance have the same average velocities. As you recall, a particle's kinetic energy depends on its mass and velocity. If the two substances are at the same temperature, the less massive particles that make up one substance will have higher average velocities than the more massive particles of the other substance.

According to kinetic theory alone, it should be possible to lower the temperature to a point where all molecular motion ceases. The temperature at which all molecular motion should cease is known as **absolute zero.** It is −273.15°C. This value is usually rounded to −273°C.

To make a temperature scale based on absolute zero, scientists have agreed on a system known as the absolute, or **Kelvin scale.** The zero point of the Kelvin scale is absolute zero. The divisions, or degrees, are the same size as those of the Celsius scale. Therefore,

$$K = °C + 273$$

Thus, 25°C is 298 K, or 298 kelvins, and −25°C is 248 kelvins. The kelvin is the SI unit of temperature.

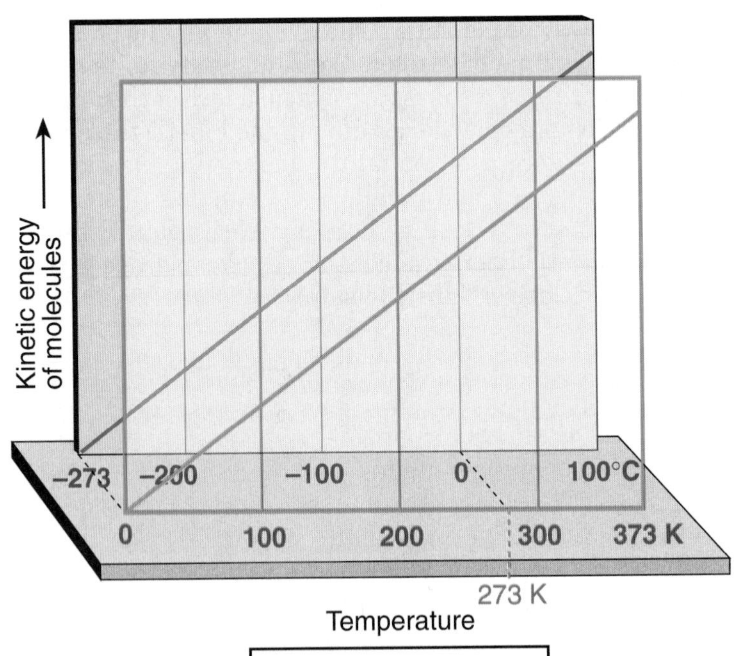

Figure 15.7 Theoretically, absolute zero (−273°C) is the temperature at which molecular motion ceases; that is, the average kinetic energy of the molecules is zero.

Temperature can be used to determine the direction of flow of energy. Energy spontaneously flows from a warmer object to a cooler one. Kinetic theory explains the flow of energy in terms of particle collisions. Because the warm object is at a higher temperature than the cool object, the average kinetic energy of its particles is greater than that of the particles in the cool object. Because the objects are in contact, the particles of the warm object can collide with those of the cool object. Gradually, the particles in the warm object will transfer kinetic energy to the particles in the cool object through collisions. As a result, the particles in the cool object gain kinetic energy. Eventually the average kinetic energy of the particles in both objects will be equal and the two objects will have the same temperature. For example, you can feel that a bowl of hot soup warms the air around it. Left undisturbed, the soup and the bowl will eventually cool to room temperature. As you recall from Chapter 3, the amount of energy transferred between two objects at different temperatures is called heat. Heat, like other forms of energy, is measured in joules.

MINUTE LAB

Too Much in an Elevator

Have you ever detected the pungent aroma of perfume or after-shave lotion in an elevator? Did someone just splash on too much of the scent? Let's investigate the possibility that the proper amount was used but another variable was responsible for the strong aroma.

Place a few drops of perfume or after-shave lotion in a heat-resistant petri dish. Set the dish on crushed ice. Have your lab partner sit near the dish and note the time required to detect the aroma of the sample. Thoroughly wash and dry the dish. Fan away any scent in the air. Repeat the investigation. This time gently warm the sample on a hot plate set on low. Turn off the plate when the sample is warmed. Describe how warming could account for any difference in the times noted by your partner by considering the motion of the molecules in the sample. What factors could affect the scent worn by someone standing in an elevator?

PRACTICE PROBLEMS

9. Convert the following Kelvin temperatures to Celsius temperatures.
a. 86 K **b.** 191 K **c.** 533 K **d.** 321 K **e.** 894 K

10. Express the following Celsius temperatures as Kelvin temperatures.
a. 23°C **b.** 58°C **c.** −90°C **d.** 18°C **e.** 25°C

11. Convert the following Kelvin temperatures to Celsius temperatures.
a. 872 K **b.** 690 K **c.** 384 K **d.** 20 K **e.** 60 K

12. At a temperature of 25°C, which of the following gas molecules moves fastest?
a. N_2 **b.** F_2 **c.** CO_2 **d.** O_2

States of Matter

Matter exists in four states—solid, liquid, gas, and plasma. Thus far, our discussion of the kinetic theory has been mostly about gases. However, kinetic theory can also be used to explain the behavior of solids and liquids. Plasmas are considered to be a special case and will be discussed later.

Gas particles are independent of each other and move in straight lines. Change of direction occurs only when one particle collides with another, or when a particle collides with the walls of the container. Gas particles, then, travel in a completely random manner. Because they travel until they collide with a neighbor or with the walls of their container, gases assume the shape and volume of their container.

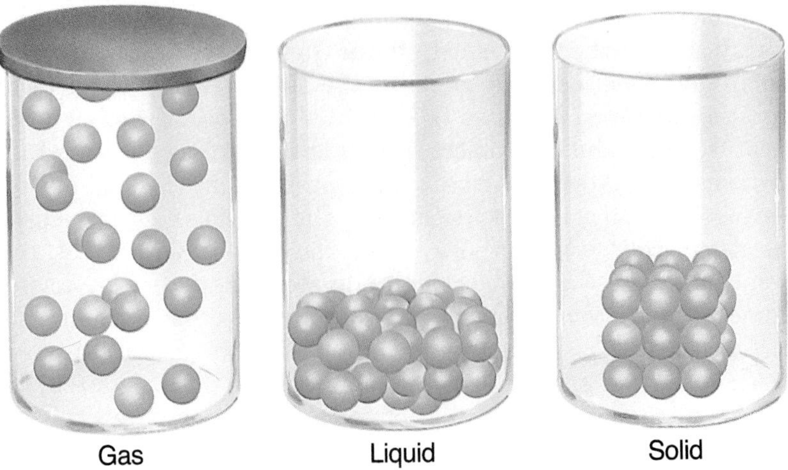

Figure 15.8 According to the kinetic theory, particles of a gas travel in a completely random manner, particles of a liquid appear to vibrate around moving points, and particles of a solid appear to vibrate around fixed points.

The particles of liquids and solids are much closer than those of gases. Unlike a gas, the particles of liquids and solids do not act independently of each other. The motion of their particles is more limited than that of the particles of a gas. In a liquid, the distance between particles is great enough to allow the particles to slip by each other. Thus, liquids, although they have a definite volume, assume the shapes of their containers.

The particles of a liquid have what appears to be a vibratory type of motion. Actually, each particle is traveling a straight-line path between collisions with near neighbors. The point around which the vibration seemingly occurs often shifts as one particle slips past another.

In solids, a particle occupies a relatively fixed position in relation to the surrounding particles. A particle of a solid appears to vibrate around a fixed point. Again, the particle is actually traveling a straight-line path between collisions with very near neighbors. For example, a molecule of oxygen gas at 25°C travels an average distance equal to 314 times its own diameter before colliding with another molecule. In a solid, however, the particles are closely packed and travel a distance equal to only a fraction of their diameters before colliding. Unlike liquids, solids have their particles arranged in a definite pattern. Solids, therefore, have both a definite shape and a definite volume.

The physical state of a substance at room temperature and standard atmospheric pressure depends mostly on the bonding in the substance. Ionic compounds have strong electric charges holding the ions together and exist as solids. Molecular substances are attracted to each other by van der Waals forces. Molecular compounds of high molecular mass tend to be solids. Nonpolar molecules of low molecular mass tend to be gases. Greater molecular mass and greater polarity both tend to make substances form a condensed state, either liquid or solid.

The arrangement of particles is important to the chemist. In the following chapters we will discuss some of the special characteristics of solids, liquids, and gases.

▼ Plasma

CHEMISTRY IN DEPTH

When matter is heated to very high temperatures (>5000°C), the collisions between particles are so violent that electrons are knocked away from atoms. Such a state of matter, composed of electrons and positive ions, is called a **plasma.**

Most of the universe is made of plasma. Stars are in a plasma state. Outer space is not empty. It is composed of extremely thin plasma. The Van Allen radiation belts that surround Earth are made of plasma. Matter in a neon tube or in a cyclotron is in the form of plasma. Many lasers, such as the CO_2 laser, which is used in industry to cut metals and clothing into patterns, have plasma tubes. Scientists are working on the character of plasmas because a nuclear reaction called fusion occurs in plasmas at high temperatures. If the fusion reaction can be controlled, it promises to be an energy source second only to the sun.

Word Origins

plasso: (GK) form

plasma – form of matter

FRONTIERS

Ready! Aim! Plasma!

Ions implanted on the surface of this metal gear harden the surface without subjecting the gear to high temperatures that could deform the metal. Strengthening metals and ceramics is one of the uses of ion implantation, a technique that allows engineers to alter the properties of a material by placing specific ions on or beneath its surface.

One method of implanting metallic ions uses plasma guns. The experimental procedure takes place in a vacuum and implants metallic ions – in this case, yttrium ions – on a silicon wafer. Yttrium ions are used as electrodes to which an intense electrical impulse is applied. The yttrium at the tip of the electrode immediately vaporizes and forms a cloud of plasma that streams through a small annular ring. The positively charged ring repels the positively charged metallic ions of the plasma and accelerates them toward the silicon wafer about 5 cm away. As the yttrium electrode is repeatedly pulsed, the surface of the silicon wafer is washed with yttrium ions, which form a thin metallic film.

If the silicon is negatively charged by an electrical impulse in tandem with the metallic electrode, the yttrium ions embed themselves farther beneath the surface of the silicon. By altering the size and duration of the electrical pulses of the yttrium electrode and that of the silicon wafer, the depth of the ion-implantation layer can be precisely controlled.

Using two plasma guns, researchers have demonstrated that two different metallic ions can be deposited simultaneously or in alternating layers. Ion-implantation techniques may allow future materials scientists to alter existing materials atom by atom or tailor-make new materials with specific chemical and physical properties.

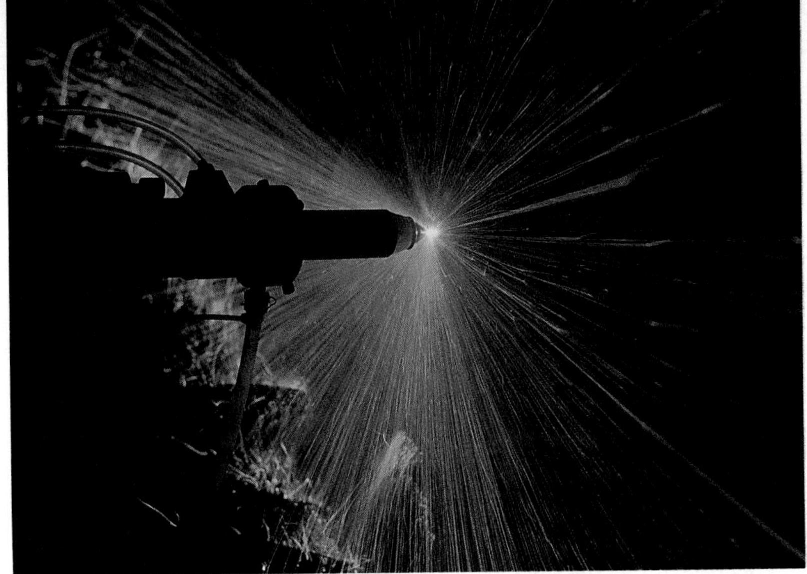

Figure 15.9 CO_2 lasers like this one are used to cut metals, plastics, and cloth into patterns.

Although a plasma consists of charged particles, it is electrically neutral because, as a whole, it has the same number of negative and positive charges. The particles of a plasma move at high speeds due to its high temperature. Because a plasma is composed of rapidly moving charged particles, it acts as a conducting fluid and, thus, has a magnetic field. A plasma is greatly affected by electric and magnetic fields.

The study of plasmas is called magnetohydrodynamics (MHD). This study involves confining plasmas. The difficulty with confining a plasma, which would be required for a self-sustaining nuclear fusion reaction, is two-fold. First, the plasma would quickly vaporize its container. Second, and more important, the plasma would tend to lose energy to the container rapidly so that it would quickly cool to temperatures below that needed for reaction. Consequently, a shaped magnetic field interacting with the magnetic field of the plasma is used to confine the plasma. Certain confined plasmas could undergo nuclear fusion and be used as future sources of energy. Scientists are also researching the use of confined plasmas as advanced propulsion units for space vehicles.

▲

15.2 CONCEPT REVIEW

13. Why is the word *average* always used in describing the kinetic energy or velocity of the particles in a sample of a substance?

14. A cylinder of helium gas and a cylinder of neon gas are at room temperature. What is the ratio of the average speeds of the atoms in the two containers if the ratio of their atomic masses is approximately 1:5?

15. Compare the intermolecular forces in two substances of the same molecular mass at the same temperature if one substance is a solid and the other is a liquid.

16. How does plasma differ from the other states of matter?

17. Apply Explain why the handle of a metal spoon placed in a dish of ice cream gets cold.

CHEMISTRY AND
TECHNOLOGY

The Versatile Medical Laser

The medical use of lasers dates from 1961. From the beginning, lasers were used because they produce intense, focused light. At the points where the beams are absorbed by tissues, intense heating takes place.

Lasers can be used for internal surgery, destroying target tissue without harming external tissue. The intense heat produced in a target tissue may be conducted to other parts of the body and inadvertently cause damage. By using short bursts of light of about 1-ms duration instead of continuous light, the heating effects can be more effectively controlled.

Different wavelengths of light are used to target different tissues. A YAG laser (erbium-yttrium-aluminum-garnet) will produce a 0.2-ns pulse of light with a wavelength of 2900 nm. The light can clean calcification from bone. A xenon-chlorine laser, on the other hand, produces 10-ns pulses of light with a wavelength of 308 nm. This light can vaporize bone without heat damage to the other tissue.

If the laser is used with optical fibers, surgery can be performed in places in the body previously unreachable. Lung disorders, for example, can be treated by channeling the laser light through optical fibers in a hole in the chest. Arteries can be threaded with bundles of optical fibers and the laser light can be transmitted along them to destroy blockages.

Some lasers produce such high-energy photons that the photons can break molecular bonds. Researchers found that by breaking bonds, tissue could be removed from the cornea of the eye. Using a 193-nm wavelength argon laser in 10-ns bursts, 0.2 microns of tissue could be removed. Through this process a cornea could be curved into a shape to correct sight defects.

Finally, lasers can drive chemical reactions. There is some hope that laser photochemistry can be a treatment for cancer. In this experimental procedure, patients are injected with a dye that concentrates in cancerous tissue. Optical fibers transmitting blue-violet light of a krypton laser are then placed in the patient's veins. The exact location of the cancerous tissue can be determined visually because the dye fluoresces when illuminated by the laser light. The tissue is then illuminated by light of a second laser which further excites the dye molecules. The excited dye molecules in turn excite molecular oxygen which becomes reactive and destroys the tumor.

Lasers have proved to be extremely valuable medical tools. Their ultimate uses are limited only by the creativity of the scientists who use them.

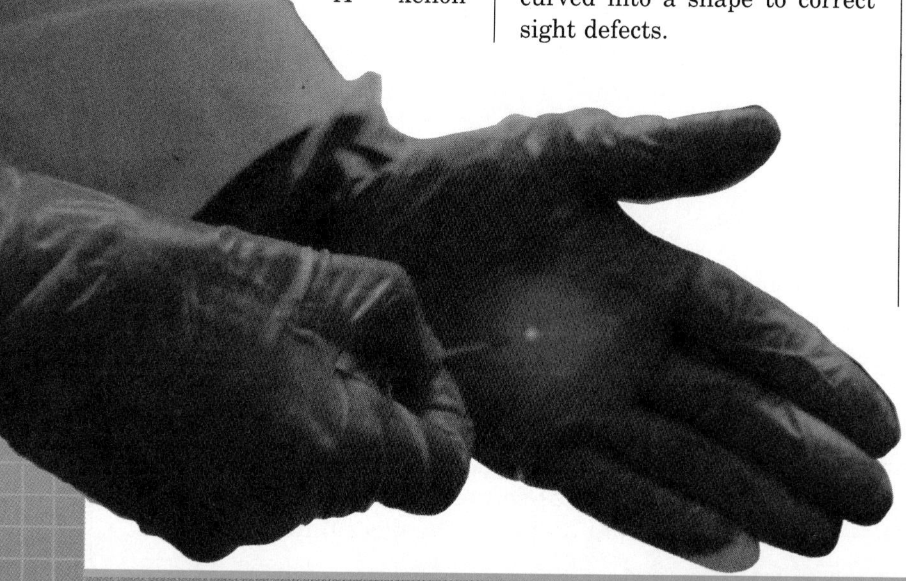

Thinking Critically

1. In which parts of the electromagnetic spectrum would you expect to find the light from the YAG and xenon-chlorine lasers?
2. How is photon energy related to the wavelength of light?
3. Some lasers will have pulse rates in the picoseconds. How do picoseconds relate to nanoseconds?
4. There are possible side effects of mutation and adjacent tissue damage when laser light is used. Should lasers be used in spite of the possible side effects?

Chapter 15
REVIEW

Summary

15.1 Pressure

1. The kinetic theory explains the effects of temperature and pressure on matter. This theory assumes that (1) all matter is composed of small particles, (2) these particles are in constant motion, and (3) collisions between these particles are perfectly elastic.
2. At 25°C, an oxygen molecule travels at just over 1700 kilometers per hour in straight lines and has more than 4.5×10^9 collisions per second.
3. A gas exerts pressure on its container because the gas molecules are constantly colliding with the walls of the container. Each collision exerts a force on the walls of the container.
4. A manometer is an instrument used to measure gas pressure.
5. Standard atmospheric pressure is defined as 101.325 kilopascals and will support a column of mercury 760 mm high. Therefore 1 kPa = 7.501 mm Hg.

15.2 Motion and Physical States

6. Temperature is a measure of the average kinetic energy of the particles of a substance.
7. Kinetic energy is the energy an object possesses because of its motion. It is related to the mass and the velocity of the object by the equation $KE = mv^2/2$.
8. A change of one kelvin is equivalent to a change of one Celsius degree (K = °C + 273). The Kelvin zero point is absolute zero.
9. Energy flows from an object of higher temperature to one of lower temperature until both reach the same temperature.
10. Heat is the amount of energy transferred from a warmer object to a cooler object. Heat is measured in joules.
11. The particles of a gas travel in a straight-line path between collisions. Each collision causes a change in the velocity of the particles.
12. The particles of a liquid appear to be vibrating around a point that is moving with respect to neighboring particles.
13. The particles of a solid appear to be vibrating around a point that is fixed with respect to neighboring particles.
14. A plasma consists of electrons and positive ions in random motion. It is strongly affected by electric and magnetic fields.

Key Terms

kinetic theory
mean free path
pressure
pascal
manometer
barometer
absolute zero
Kelvin scale
plasma

Review and Practice

18. Make an observation that is explained by the kinetic theory. Use the three basic assumptions of the kinetic theory to account for the observation.
19. What is meant when we say that a collision is perfectly elastic?
20. What is meant by the mean free path of a particle?
21. Define *pascal*. Why are kilopascal units usually used instead of pascals?
22. Explain how a gas exerts pressure on its container.
23. For what purpose is a manometer used?
24. What is the value of standard atmospheric pressure in kPa?
25. An open manometer like the one in Figure 15.4 is used to measure the pressure of a gas sample. The mercury level is 12 mm higher in the gas sample arm. What

is the pressure, in kilopascals, of the gas in the container if the air pressure is 98.7 kilopascals?

26. In a closed manometer, assume that the height of the levels differs by 522 mm Hg. What is the pressure, in kilopascals, of the gas in the container?

27. An open manometer is connected to a container of carbon dioxide. The mercury level is 24 mm higher in the arm open to the atmosphere. If the pressure of the atmosphere is 100.3 kPa, what is the pressure of the carbon dioxide?

28. If the heights of the two levels of mercury in a barometer differ by 800.0 mm, what is the atmospheric pressure?

29. How does an increase in the average kinetic energy of a substance affect its temperature?

30. In terms of the kinetic theory, what is the significance of absolute zero?

31. How can the kinetic energy of a moving object be calculated?

32. Convert the following temperatures as indicated.

a. 606 K to °C

b. 751°C to K

c. 18°C to K

d. −14°C to K

33. Describe the Kelvin temperature scale. Explain its similarities to and differences from the Celsius temperature scale.

34. The diagrams shown represent models of the three common states of matter. Tell which state each diagram represents. Explain your choice for each diagram.

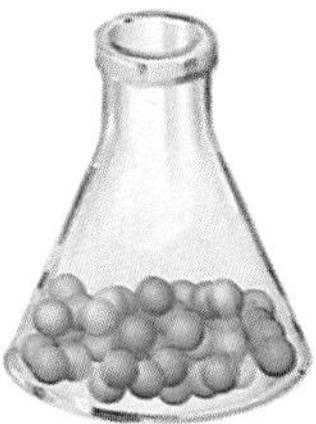

35. What is the primary factor that determines whether a substance is a solid, a liquid, or a gas at room temperature?

36. What is plasma? Describe the characteristics of the plasma state.

37. Describe how scientists hope to use plasma in production of energy.

38. What is magnetohydrodynamics?

39. Which has the greater kinetic energy: an object with a mass of 10 g traveling at 2 m/s or an object with a mass of 2 g traveling at 10 m/s?

40. What is the difference between a barometer and a closed manometer?

41. Compute the kinetic energy of an auto with a mass of 1500 kg traveling at 50 km/h.

Concept Mastery

42. Why is it necessary to know the atmospheric pressure in order to determine the pressure of a gas when measured with an open manometer?

43. Examine the diagram of the manometer. What kind of manometer is shown here? What is the pressure of the gas in the sample bulb? If this type of manometer were to be used to measure pressures several times that of atmospheric pressure, how would the manometer have to be different from the one shown?

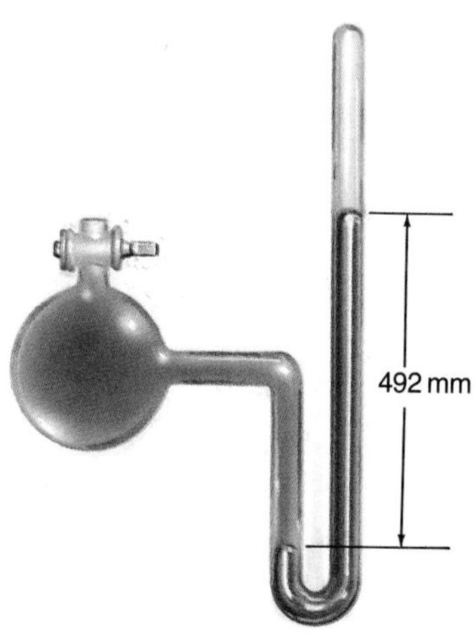

44. Explain what a barometer is and how a mercury barometer works.
45. In an open manometer, if the atmospheric pressure decreases, in which direction would the mercury in the open arm move?
46. In a closed manometer, why is there no need to state in which arm the mercury is higher?
47. You have one container of Cl_2 gas and an identical container of CH_4 (methane) gas. Both are at 18°C. What is the difference in the average kinetic energy of the molecules? What is the difference in the average velocity of the molecules?
48. Describe the differences in particle motion among the states of matter.
49. Why are most ionic compounds solids at room temperature?
50. In terms of the kinetic energy of molecules, explain why your coat is warm when you take it off.
51. Examine the diagram of the manometer shown. What is the pressure difference, in kilopascals, between the atmosphere and the gas in the bulb? What is the pressure of the gas in the bulb if the air pressure is 102.3 kPa? Describe how this manometer would appear if the pressure of the gas were 102.3 kPa also.

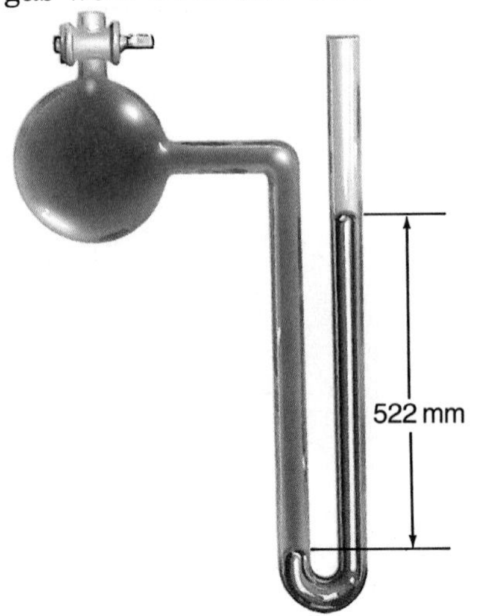

52. Consider a system composed of fizzing soda being poured over ice. Describe the molecular motion of the soda, ice, and the carbon dioxide above the liquid.
53. Hot tea is poured into a cool cup and allowed to stand for five minutes. Describe and explain the changes that are likely to occur in this system over the five-minute period.
54. Which are the condensed states of matter? Why are these states described as condensed?
55. A gas sample occupying 300 cm^3 is moved to a container with a volume of 200 cm^3. Describe any changes in each of the following values.
    a. average kinetic energy of the gas molecules
    b. the pressure of the gas
    c. the average speed of the gas molecules
    d. the number of collisions of a gas molecule with other gas molecules
56. What is wrong with saying, "This container is half full of hydrogen chloride gas"? Is there anything wrong with saying, "Here is a bottle full of argon gas"?
57. The mercury level in the sample arm of an open manometer is 525 mm higher than in the open arm when the atmospheric pressure is 99.0 kPa. What is the pressure of the gas? If the apparatus is placed in a vacuum, what would the difference in levels become?
58. An open manometer is filled with mercury and the closed end is filled with ammonia, NH_3. Hydrogen chloride gas, HCl, is slowly added to the ammonia, and solid NH_4Cl begins to form. What happens to the level of mercury in the open arm?
59. Water has a molecular mass of about 18 u and is a liquid at room temperature and standard pressure. Carbon dioxide has a molecular mass of about 44 u but is a gas at the same condition of temperature and

pressure. Compare the two in terms of average kinetic energy, average molecular velocity, and mean free path. How do you account for the fact that CO_2 is a gas even though its molecules have more than twice the mass of H_2O molecules?

Application

60. Compute the average kinetic energy of an O_2 molecule at 25°C. The mass of one mole of O_2 divided by the Avogadro constant equals the mass of one O_2 molecule.
61. Use your answer from Problem 60 to calculate the average velocity of a hydrogen molecule at 25°C.
62. Assume there are two identical rubber rafts at the same temperature, inflated to exactly the same volume. One was filled by an air compressor and contains mostly N_2. The other was inflated with the use of a container of compressed carbon dioxide, CO_2. There is an identical pinhole leak in each. Which raft will deflate faster?
63. A sphygmomanometer is used to determine blood pressure. What is measured by a sphygmomanometer? What do these measurements mean in terms of blood flow and blood pressure?
64. Barometric pressure is frequently given in weather forecasts as inches of mercury. If 1 inch = 2.54 cm, and the forecast reports a barometric pressure of 29.7 inches of mercury, what is the atmospheric pressure in kilopascals?

Critical Thinking

65. A container of oxygen has a volume of 2.00 dm^3. A quantity of nitrogen is added to the same container. Will the pressure in the container change? Explain.
66. Suppose you have two vials, one containing ammonia and the other containing chlorine. When they are opened across the room, which would you expect to smell first? Why?
67. A gas sample occupies a volume of 900 cm^3. If the volume of that same gas sample is changed to 300 cm^3, explain how the pressure will be affected.
68. In terms of elasticity, explain why it is harder to hit a home run on a slow pitch than it is to hit a home run on a fast pitch.
69. Scientists make use of the kinetic theory to separate $^{235}_{92}U$ from $^{238}_{92}U$. Both isotopes are changed to UF_6, which is a gas at just above room temperature. How might the compounds of these two isotopes be separated?

Readings

Castleman, Jr., A.W., and R.G. Keesee. "Gas-Phase Clusters: Spanning the States of Matter." *Science* 241, (July 1, 1988): 36-42.

Cumulative Review

1. Write electron configurations for phosphorus, calcium, molybdenum, and lead.
2. Describe the general trend of first ionization energies of elements across a period. Explain why this value changes in the way it does.
3. Which of the following bonds has a greater polarity? Explain.
   a. Ru—Si **b.** Be—Br
4. Arrange the following elements in order of increasing attraction for electrons in a bond: Ni, F, Ti, C, Sc.

CHAPTER Preview

CONTENTS

CHAPTER

16

Solids

The photo on the left is the dazzling interior of a geode—a crystal rock garden in a stone. What was once a liquid-filled cavity such as the shell of a clam has changed slowly into an array of shimmering mineral crystals. These minerals have many of the properties that characterize solids. As you will see, this is because all solids are composed of crystals.

The study of crystals is an extension of the study of kinetic theory. As you recall, solids are composed of particles that seemingly vibrate about fixed positions. What determines the arrangement of these positions? Does the arrangement account for the properties of a solid? As you know, different solids have different properties. Do these solids differ in the arrangement of their particles? To answer these questions we must look at solids on the atomic level.

EXTENSIONS AND CONNECTIONS

OBJECTIVES

- describe characteristics of all solid substances.
- distinguish among cubic, body-centered cubic, and face-centered cubic cells.
- explain the relationship of melting point to bonding type and to crystal type.
- distinguish isomorphous and polymorphous crystals.

16.1 Crystal Structure

Have you ever examined table salt under a magnifying glass? If so, you have seen that the crystals appear to be little cubes. The lengths of the edges may vary, but the angles between the surfaces are always exactly 90°.

The systematic study of crystals began with Nicolaus Steno, in 1669. He observed that corresponding angles between faces on different crystals of the same substance were always the same. This fact was true regardless of the size or source of the crystals.

Steno's observation has been extended to all intensive properties (density, melting point, face angles, and other similar properties). For example, a single crystal of the mineral beryl, $Be_3Al_2Si_6O_{18}$, with a mass of approximately 40 metric tons was unearthed in New Hampshire. This huge crystal was found to be identical in intensive properties to all other beryl crystals. The size or mass of the crystal is not important. It can now be stated that *the extensive properties of crystals vary while the intensive properties remain the same.* As you recall, extensive properties include mass, volume, and length.

Crystals

The study of the solid state is really a study of crystals. All true solid substances are crystalline. Apparent exceptions to this statement can be explained in either of two ways. In some cases, substances we think of as solids are not solids at all. In other cases, the crystals are so small that the solid does not appear crystalline to the unaided eye. The relationship of properties to structure is important to chemists in the study of crystals as well as in other aspects of chemistry. Once the relationship is established, the chemist can use that knowledge to predict the properties of new substances.

Figure 16.1 All edges of a salt crystal meet at angles of 90° whether the crystal is large or small.

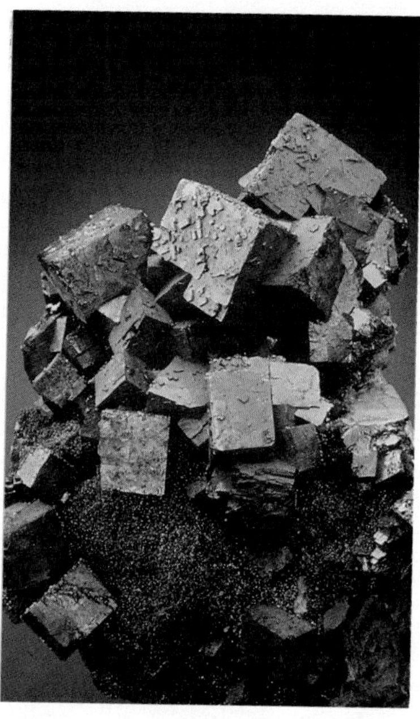

Figure 16.2 Two minerals that form cubic crystals are fluorite, CaF_2 (left); and galena, PbS (right).

All crystals of a certain substance are made of similar small units. These units are then repeated over and over again as the crystal grows. The units that compose a crystal are too small to be seen. Yet before any methods existed for studying crystal structure, scientists suggested that crystals form by repetition of identical units. Consider the patterns on wallpaper or drapery fabrics. These patterns are applied by rollers that repeat the design with each rotation. In a crystal, the forces of chemical bonding play the same role that the roller does in printing. They cause the basic pattern to be repeated. However, the "design" of a crystal must be composed of atoms, molecules, or ions instead of ink. Unlike the two-dimensional units of wallpaper, the units in crystals are three-dimensional.

Figure 16.3 These minerals represent different crystal systems. Molybdenite, MoS_2 (left), has a hexagonal structure; manganite, MnO(OH) (center), has a monoclinic structure; and amazonite, $KAlSi_3O_8$ (right) is triclinic.

A **crystal** is defined as a rigid body in which the particles are arranged in units which form a repeating pattern. The arrangement of these units is determined by the bonds between the particles. Therefore, the bonding in the crystal partially determines the properties of the crystal.

There is a relationship between the repeating units and the external shape of the crystal. Long ago, crystallographers (krihs tuh LAHG ruh furs) classified crystals on the basis of their external shapes into seven "crystal systems," Table 16.1.

Table 16.1

Seven Crystal Systems			
	Lengths of the Unit Cell Axes	**Angles between the Unit Cell Axes**	**Crystal System**
	all equal	all = 90°	cubic (Figure 16.2)
	2 equal 1 unequal	all = 90°	tetragonal (Figure 16.10, left)
	3 equal 1 unequal	1 = 90° 3 = 60°	hexagonal (Figure 16.3, left)
	all equal	all ≠ 90°	rhombohedral (Figure 16.15)
	all unequal	all = 90°	orthorhombic (Figure 16.13)
	all unequal	2 = 90°, 1 ≠ 90°	monoclinic (Figure 16.3, center)
	all unequal	all ≠ 90°	triclinic (Figure 16.3, right)

Unit Cells

Each substance that crystallizes does so according to a particular geometric arrangement. The simplest repeating unit in this arrangement is called the **unit cell.** It is possible to have more than one kind of unit cell with the same shape. For example, a unit cell in a tetragonal crystal may be simple tetragonal or it may be body-centered.

Consider the different kinds of three-dimensional unit cells. Fourteen such cells are possible, Table 16.2.

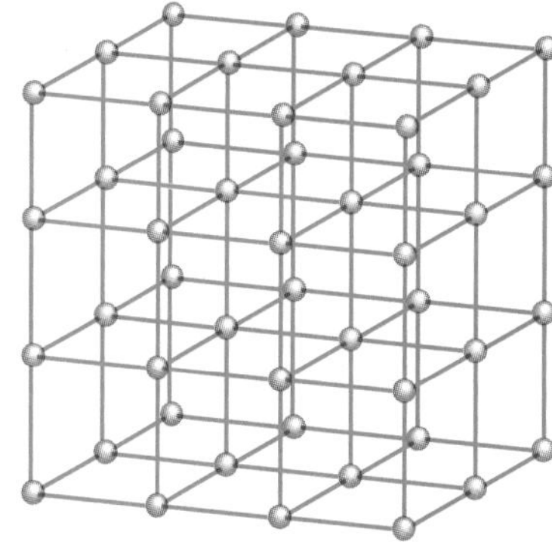

Figure 16.4 A simple cubic cell (top) is shown expanded into its space lattice (bottom). Unit cells and space lattices are mental models.

Table 16.2

Unit Cells

Crystal System	Unit Cell
Cubic	simple, body-centered, face-centered
Tetragonal	simple, body-centered
Orthorhombic	simple, single face-centered, body-centered, face-centered
Monoclinic	simple, single face-centered
Triclinic	simple
Rhombohedral	simple
Hexagonal	simple

Three of the simplest unit cells are **simple cubic, body-centered cubic (bcc),** and **face-centered cubic (fcc).** These unit cells are illustrated in Figure 16.5. Note the packing arrangement of the particles in each unit cell. In the simple cubic cell, each particle has six immediate neighbors. In the body-centered cell, each particle has eight immediate neighbors; and in the face-centered unit cell, each has twelve. The three-dimensional arrangement of unit cells repeated over and over in a definite geometric arrangement is called a **space lattice.**

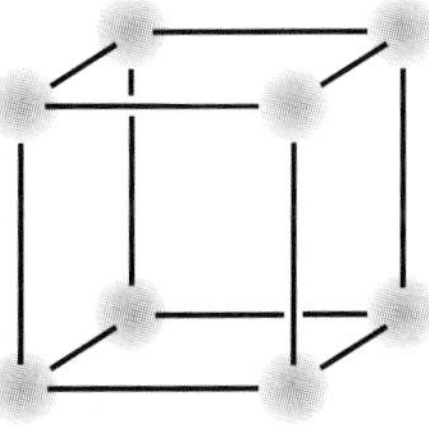

Simple cubic

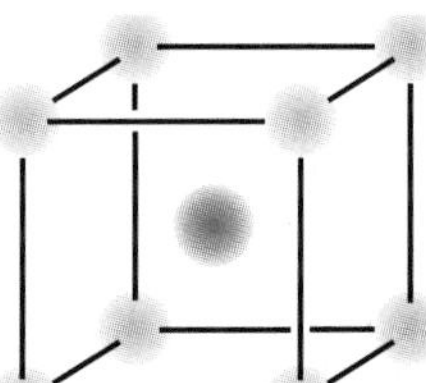

Body-centered cubic

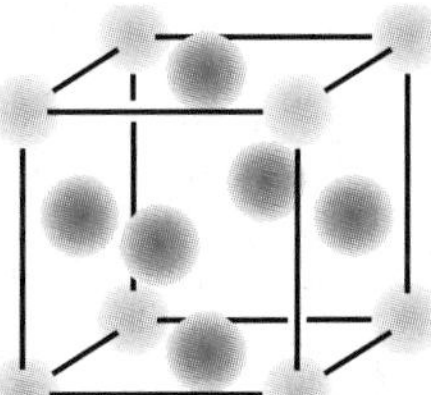

Face-centered cubic

Figure 16.5 The three unit cells of the cubic system shown at the top have different packing arrangements within the same shape. The same arrangements shown in the bottom row indicate that the particles in crystals are much closer than those used to illustrate unit cells.

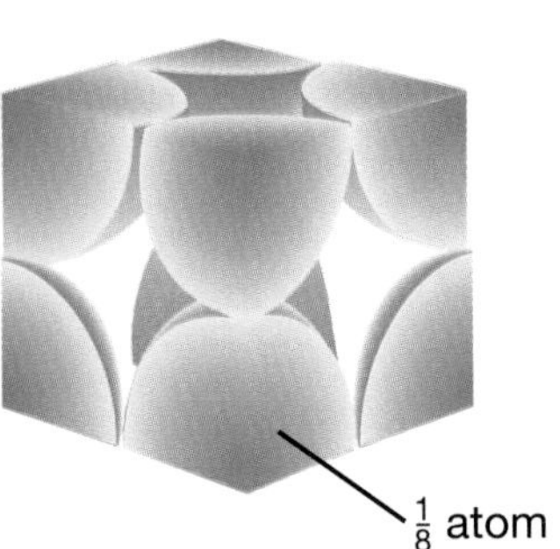

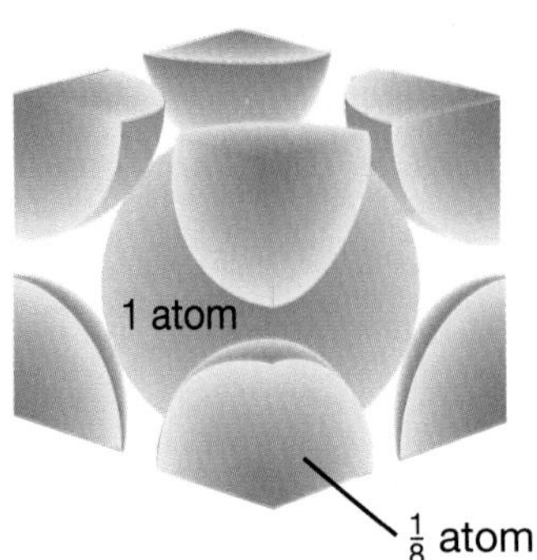

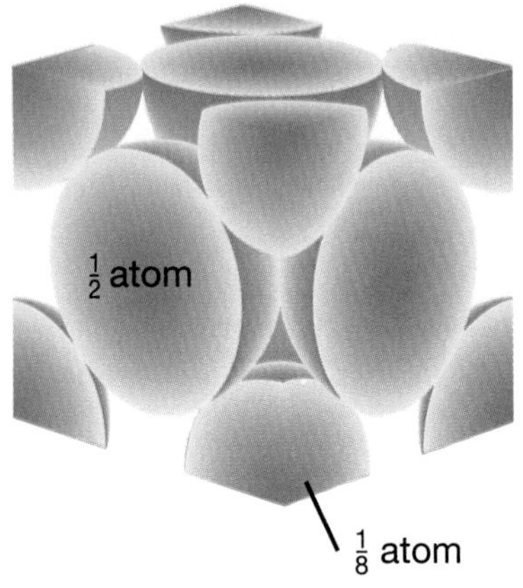

Figure 16.6 This representation of the cubic unit cells shows how atoms are shared between unit cells. The contents of the unit cells shown are 1, 2, and 4 atoms, respectively.

It should also be pointed out that space lattices and unit cells have no real physical existence. They are mental models that help demonstrate crystal structures.

Compound Unit Cells

Sodium chloride crystallizes in a structure similar to that pictured in Figure 16.7. How would you classify this unit cell? It apparently is some form of cubic, but which one? Look more closely at the arrangment of the ions. Each Cl^- ion is surrounded by six Na^+ ions. Each Na^+ ion in turn is surrounded by six Cl^- ions. If you consider either arrangment alone, you can see that the unit of repetition is face-centered cubic. Thus, the unit cell can be considered face-centered cubic, even though more than one kind of particle is present.

The particular crystal structure of an ionic compound such as sodium chloride is determined principally by the ratio of the radii of the ions. Since negative ions are generally larger than positive ions, the positive ions must be large enough to keep the negative

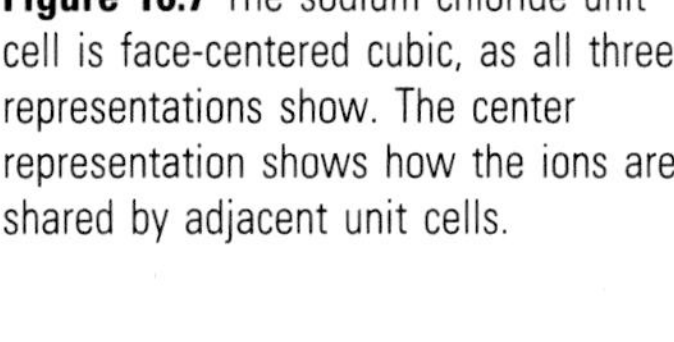

Figure 16.7 The sodium chloride unit cell is face-centered cubic, as all three representations show. The center representation shows how the ions are shared by adjacent unit cells.

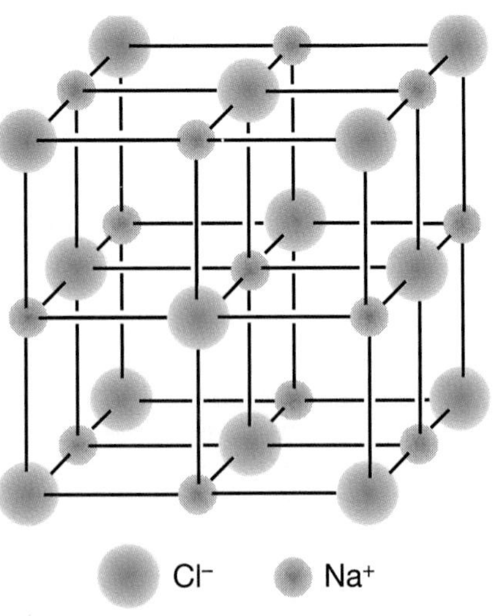

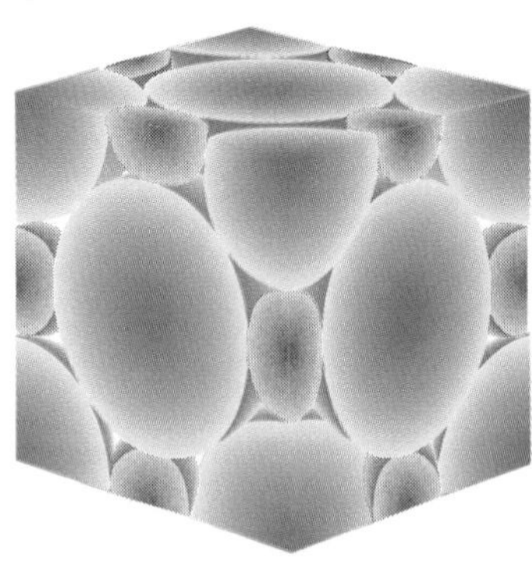

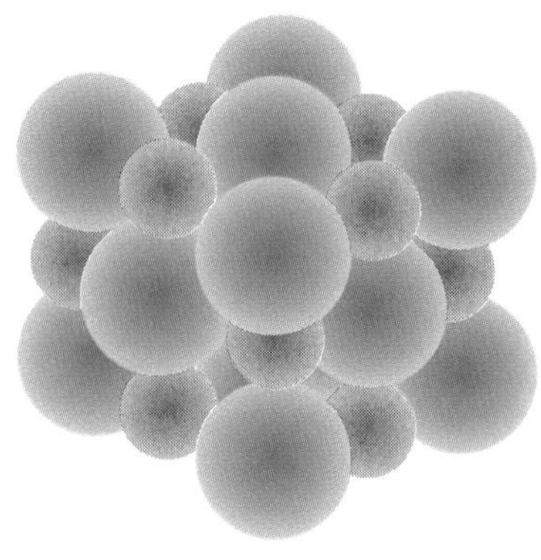

ions from coming into contact with each other, but small enough not to come into contact with neighboring positive ions. The radius ratio of Na^+ to Cl^- is 0.69 and that of Cs^+ to Cl^- is 1.08. This difference is the reason NaCl has a face-centered cubic structure and CsCl is simple cubic.

Simple salts are those formed by the chemical combination of the elements of Group 1 (IA) (the alkali metals) and the elements of Group 17 (VIIA) (the halogens). These salts, except for those of Cs, always have structures based on the face-centered cubic lattice. Salts of cesium differ because of the large size of the Cs^+ ion. Sodium chloride is typical of this class of compounds. The same model would serve for other members of the group and also for many other binary compounds, like MgO and CaO.

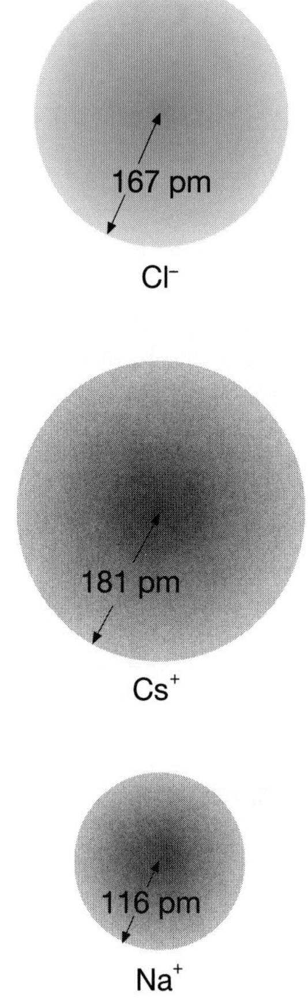

Figure 16.8 The size difference between the sodium ion and the cesium ion accounts for the difference in unit cell arrangements in NaCl and CsCl.

Closest Packing

The crystal structures of elements are usually simple. Imagine placing a group of spheres as close as possible on a table and holding them so that they cannot roll apart. Then place another layer on top of the first one in an equally close arrangement. If we continue with more layers, a close-packed structure results. This structure is the kind found in the majority of metals. It is difficult to visualize the lattice in such a structure, since the lattice is really only a system of imaginary lines. However, the experienced eye will detect that the lattice is either hexagonal or face-centered cubic in the close-packed arrangement. They are often called **hexagonal closest packing** (hcp) and **cubic closest packing**

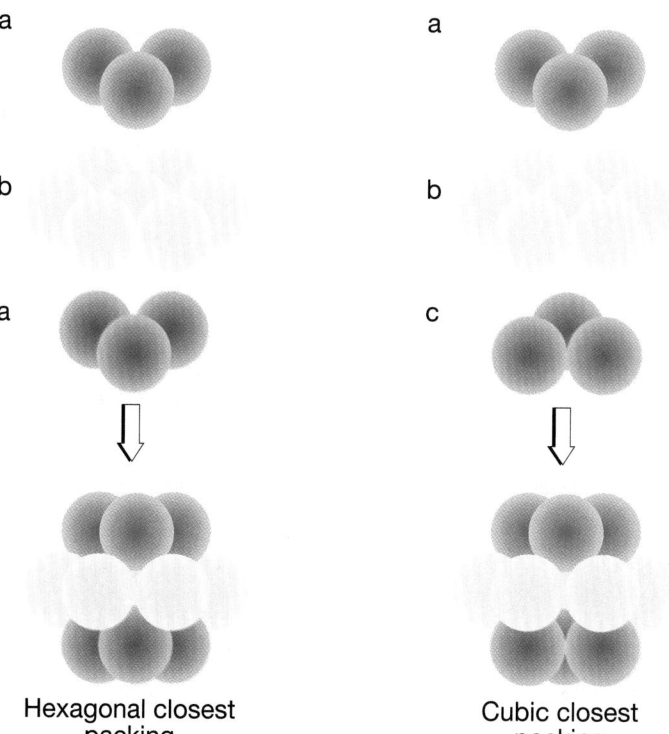

Figure 16.9 When spheres are closest packed so that spheres in the third layer are directly over those in the first layer *(aba),* the arrangement is hexagonal-closest packed (left). When the spheres are packed in the *abc* arrangement, the arrangement of spheres in the third layer is rotated 60° with respect to those in the first layer. The result is a cubic-closest packed structure (right).

INDUSTRIAL ARTS CONNECTION

Shop Crystals

Industrial shop tools may contain diamond and silicon carbide crystals to increase their hardness. Engravers use diamond-tipped scribes to incise lines on the hard surfaces of metal and glass. Sandpaper and grinding wheels found in most wood and metal shops are made of silicon carbide, an abrasive known as emery. When a grinding wheel becomes worn, a dressing tool can be used to return the wheel to its proper shape. If you look at a dressing tool, you will see that its surface is encrusted with diamond dust.

(ccp), Figure 16.9. Note that cubic closest packing and face-centered cubic are two names for the same arrangement. Hexagonal closest packing is the one most frequently found in metals at room temperature. Another structure, found particularly in the metals of Group 1 (IA) of the periodic table, is based on a body-centered cubic lattice. This structure is not quite so closely packed as the other two. In a body-centered cubic structure, while the openings between the atoms are smaller, there are more openings.

Iron has a body-centered cubic structure at ordinary temperatures. At higher temperatures, it has a face-centered cubic structure. This fact is of great practical importance. Iron, in the form of steel, always contains a small amount of carbon. Carbon atoms are smaller than iron atoms, and at high temperatures they fit into the open spaces in the face-centered structure. When the iron cools, it changes to the body-centered cubic form. In that form, the carbon atoms cannot fit into the smaller spaces. Either the iron lattice is distorted by the oversize carbon atoms, or the carbon separates out of the iron as iron carbide, Fe_3C.

Iron and Fe_3C crystals exist in many sizes and shapes. The final structure of the crystal is determined by the percentage of carbon and the rate of cooling. These differences in crystal structure result in the great versatility of steel as an industrial material. They also account for the fact that the properties of steel can be changed greatly by heat treatment.

Elementary Crystals

Almost all metals are packed in one of three kinds of unit cells. These three are body-centered cubic, face-centered cubic, or hexagonal closest packed. Figure 16.11 lists the most stable unit cell arrangements for metals at room temperature. The packing of many metals, like iron, changes as the temperature rises. The

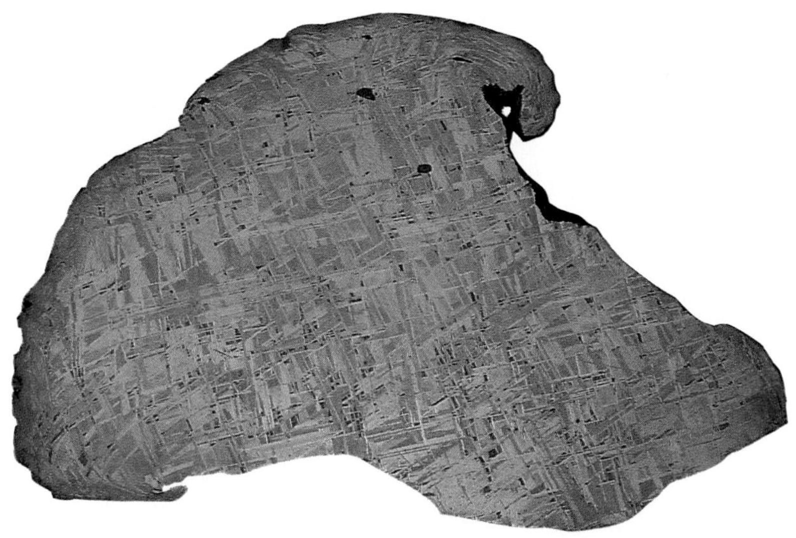

Figure 16.10 Chalcopyrite, $CuFeS_2$ (left) is an example of a mineral that has a tetragonal crystal structure. The nickel-iron meteorite (right) shows a granular pattern resulting from heating and recrystallization.

CRYSTALLINE FORMS OF METALS

1	2	3	4	5	6	7	8	9	10	11	12	13	14	15	16	17	18
H																	He
Li	Be											B	C	N	O	F	Ne
Na	Mg											Al	Si	P	S	Cl	Ar
K	Ca	Sc	Ti	V	Cr	Mn	Fe	Co	Ni	Cu	Zn	Ga	Ge	As	Se	Br	Kr
Rb	Sr	Y	Zr	Nb	Mo	Tc	Ru	Rh	Pd	Ag	Cd	In	Sn	Sb	Te	I	Xe
Cs	Ba	Lu	Hf	Ta	W	Re	Os	Ir	Pt	Au	Hg	Tl	Pb	Bi	Po	At	Rn
Fr	Ra	Lr	Rf	Ha													

La	Ce	Pr	Nd	Pm	Sm	Eu	Gd	Tb	Dy	Ho	Er	Tm	Yb
Ac	Th	Pa	U	Np	Pu	Am	Cm	Bk	Cf	Es	Fm	Md	No

Body-centered cubic

Face-centered cubic

Hexagonal closest packed

Figure 16.11 The stable unit cell packing arrangements for metals are shown in this periodic chart.

general, though not universal, trend is toward the hexagonal closest packed arrangement.

There are two crystal forms of carbon—diamond and graphite. Their structures are illustrated in Figure 16.12. You can see that in graphite, the atoms within each layer have a hexagonal arrangement and are held close together by strong covalent bonds. The layers themselves are relatively far apart and are held together only by weak van der Waals forces.

In diamond, each carbon atom is bonded covalently to four other carbon atoms. The four atoms are at the vertices of a tetrahedron. This arrangement forms a face-centered cubic lattice.

Figure 16.12 The difference in the crystalline structures of graphite and diamond accounts for their very different physical properties.

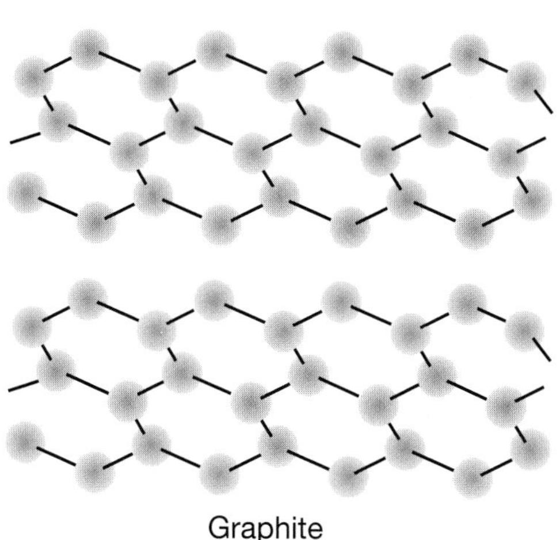

Graphite

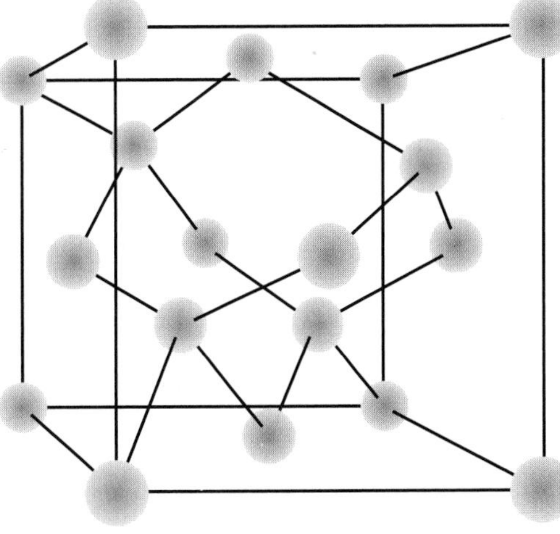

Diamond

Figure 16.13 The orthorhombic sulfur crystals (left) consist of stacked S_8 molecules, which exist in a puckered-ring formation (right).

MINUTE LAB

Hard as Diamonds

Diamond is a hard material. Its structure is rigid because each carbon atom is bonded to four other carbon atoms. Four atoms form the vertices of a tetrahedron with the fifth atom at its center. You can better visualize a tetrahedron by building a model. Cut six equal-length pieces from drinking straws. Use thin wire or string to connect three pieces into a triangle. Form another triangle by connecting two pieces to the base of the first triangle. Then connect the top vertices of the two triangles with the remaining piece. Inspect the tetrahedron. Is the tetrahedron rigid?

Identify the position of the five carbon atoms in your tetrahedron. Locate similar tetrahedral structures within the diamond crystal shown in Figure 16.12. Explain how the tetrahedral structures give rigidity to diamond.

Since each carbon atom is bonded to all its nearest neighbors, a diamond crystal is a single molecule.

Nonmetallic elements have relatively low melting points. Elements like sulfur or iodine form crystals in which the lattice positions are occupied by molecules. The molecules are attached to each other by weak van der Waals forces.

In sulfur crystals, which are orthorhombic, the S_8 molecule contains eight atoms arranged as shown in Figure 16.13. The atoms within the molecule are much closer to each other (208 pm) than they are to the atoms of neighboring molecules (370 pm). The sulfur lattice is orthorhombic.

Network Crystals

In order to melt most molecular solids, enough energy must be added to overcome the van der Waals forces that hold the molecules in the crystal lattice. Observations in the laboratory show that molecular compounds of low molecular mass have very low melting points ranging from −272°C to about 400°C. However, covalently bonded substances, such as diamond, have melting points in the range 1000°C to 4000°C. Another example is silicon carbide, which melts at about 2700°C. How can these very high melting points be explained?

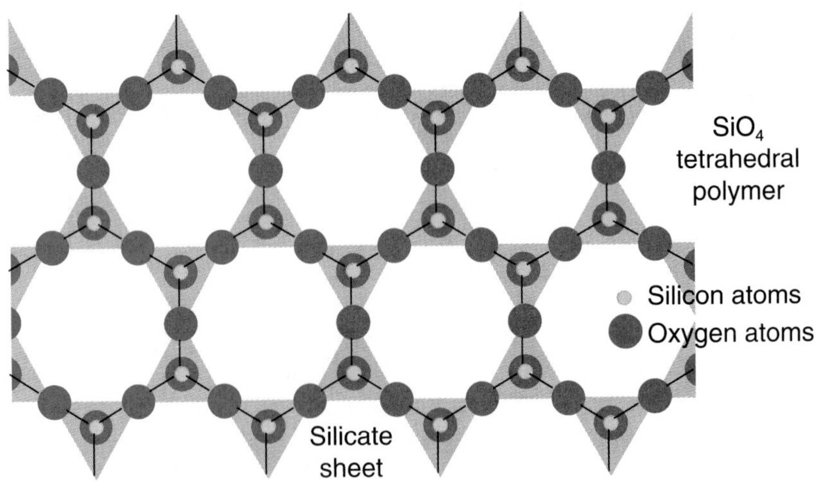

Figure 16.14 Silicate compounds generally have high melting points. The macromolecule shown here consists of repeating SiO_4 units. This substance melts at a high temperature because covalent bonds must be broken to liquefy the crystal.

In silicon carbide, each carbon atom is surrounded by four silicon atoms to which it is covalently bonded. Each of these silicon atoms, in turn, is surrounded by four carbon atoms to which it is covalently bonded. Thus, each atom in the crystal network is bonded to its four nearest neighbors. Silicon carbide has the same structure as diamond except every other atom is silicon. We may think of the entire crystal as one giant molecule. In fact, this type of structure is often called a **network crystal** or a **macromolecule.** There are many substances composed of macromolecules. All of these substances have very high melting points. In order to melt the substance, you must break covalent bonds, which are about ten times stronger than van der Waals forces.

Table 16.3

Characteristics of Crystals

Crystal Type	Bonding	Melting Points	Examples
Ionic	Electrostatic forces	300-1000°C	NaCl, LiBr
Metallic	Delocalized electrons	100-3500°C	Na, W
Macromolecular	Covalent	2000-4000°C	C, Si
Molecular	van der Waals forces	−260-400°C	CH_4, C_6H_6

FRONTIERS

Quirky Quasicrystals

The twelve-sided grain shown here is an alloy of aluminum, copper, and iron. Experiments indicate that the grains of this alloy are not composed of only one type of unit cell as are crystals. Thus, this alloy is not a crystalline solid. However, its orderedness indicated it is not an amorphous solid. Instead, it is called a quasicrystalline solid.

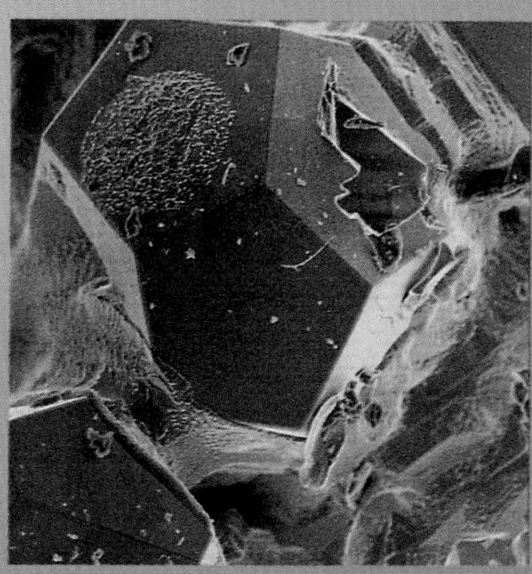

Some quasicrystalline solids have unique properties. For example, one such solid is a metallic glass that contains large amounts of aluminum. It has the hardness of a glass and the low density of aluminum. These properties might make it ideal in the future for fabricating lightweight components for airplanes and automobiles.

How quasicrystals form and grow is not understood. Some theorists hypothesize that a quasicrystal is actually two crystals that have penetrated each other. Others predict that quasicrystals form from two different unit cells.

Figure 16.15 Calcite is the rhombohedral form of calcium carbonate.

Isomorphism and Polymorphism

There are many solid compounds and only 32 combinations of unit cells and crystal systems in which the compounds may crystallize. Therefore, it is not surprising that many different substances have the same crystalline structure. Crystals of different solids with the same structure and shape are **isomorphous.** The most likely compounds to form isomorphous crystals are those which have the same type of chemical formula and differ by only one atom from the same group in the periodic table. For example, NaCl and KCl are isomorphous, as are NaCl and NaBr. There are, however, compounds of considerably different chemical formulas that have isomorphous crystals. The compounds have isomorphous crystals because the compounds have essentially the same radius ratio. Such compounds include NaCl, PbS, ScN, NH_4I, and KCN.

It is also possible that the same substance may crystallize into two or more different patterns. A single substance having two or more crystalline shapes is said to be **polymorphous.** Calcium carbonate, $CaCO_3$, has two crystalline forms: calcite, which is rhombohedral; and aragonite, which is orthorhombic. Surprisingly, calcite is isomorphous with crystalline $NaNO_3$, while aragonite is isomorphous with crystalline KNO_3. Aragonite is the normal form for $CaCO_3$. If aragonite is subjected to a high temperature, it rearranges to calcite. Another polymorphous mineral whose crystal formations are dependent upon temperature is FeS_2. At low temperatures, iron disulfide crystallizes in the orthorhombic form as marcasite, while at high temperatures it crystallizes in the cubic form as pyrite.

Pressure, as well as temperature, can affect crystal formation. You already know that carbon is polymorphous. Carbon crystallizes as graphite at low pressures. At high pressures, carbon crystallizes as diamond. Another mineral whose crystal structure is affected by pressure is Al_2O_3. At low pressures it crystallizes in the orthorhombic form as andalusite, while at high pressures it crystallizes in the triclinic form as kyanite.

Word Origins

isos: (GK) equal, same
polys: (GK) many
morphe: (GK) form, shape

isomorphous—having the same shape

polymorphous—many shapes

16.1 *CONCEPT REVIEW*

1. What characteristic is common to all solid substances?
2. Describe the three most common types of packing found in crystals of metallic compounds.
3. What factors account for the hardness and high melting points of covalently bonded substances like silicon carbide?
4. How does a network crystal differ from an ionic crystal and a molecular crystal?
5. Name and describe the crystal structures of the two polymorphic forms of carbon.
6. How does the existence of unit cells account for Steno's observations?
7. **Apply** How does the crystal structure of graphite enable it to be used as a common lubricant for locks?

16.2 Special Structures

Now that you have acquired some knowledge of crystal structure, you can look at some special structures that lead to practical applications. Imperfect crystals, hydrated crystals, and those substances that appear to be solid but are not have all found important uses in our technological society. We will also examine a class of substances that have properties of both liquids and solids.

OBJECTIVES

- identify and explain the types of crystal defects.
- describe the chemistry of semiconductors.
- distinguish between hydrated ions and anhydrous substances.
- describe the structure and properties of crystals, liquid crystals, and amorphous solids.

Crystal Defects

A perfect crystal is rare. Most crystals contain defects of one or more types. We will cover two basic types of defects.

The first type of defect occurs within the unit cell structure. Look at a plane of a simple crystal such as sodium chloride. The positive and negative ions alternate in such a crystal lattice. However, it is entirely possible that one of the ions may be missing from its proper position and occupy a space where no ion usually occurs. This change causes an imperfect crystal. Another possibility is that an ion may be missing completely from its position in the lattice. If a defect of this type occurs, for every positive ion missing there is a negative ion missing. This arrangement preserves the electrical neutrality of the crystal. It is sometimes possible for foreign ions, atoms, electrons, or molecules to occupy these spaces vacated by the normal ions of the crystal.

The second basic type of defect concerns the manner in which the unit cells are joined. These defects are called **dislocations.** In some crystals, an extra layer of atoms extends part of the way into a crystal. The resulting crystal is said to have an edge dislocation. It is also possible for the particles to be slightly out of position. This defect is due to unequal growth while the crystal forms. Such a defect is termed a screw dislocation.

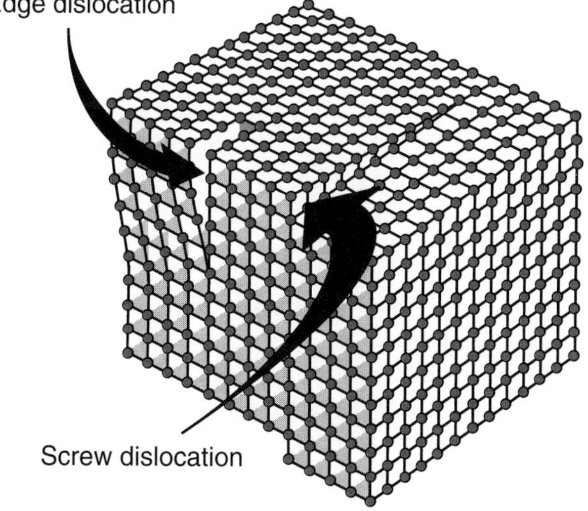

Figure 16.16 Edge and screw dislocations are defects in the way unit cells are joined and cause structural defects in the crystals.

CHEMISTRY IN DEPTH

Semiconductors

Sometimes defects in crystals are useful. For example, in the manufacture of semiconductor material for transistors, almost perfect crystals are **doped.** That is, impurities are added deliberately. Silicon and germanium, Group 14 (IVA), are the most common elements used in transistors. A pure crystal of either of these elements is an extremely poor electrical conductor. However, if a small amount of another element is added, current will readily flow through the resulting *doped* crystal.

A closer look at the bonding of silicon or germanium reveals why these elements behave as semiconductors. Both silicon and germanium have four electrons in the outer level, and crystallize in a structure similar to that of diamond. Thus, all four electrons are involved in bonding. Under the proper circumstances, an electron can be excited sufficiently to break out of the bond and move freely through the crystal. Thus, these substances are called semiconductors.

Arsenic atoms have five electrons available for bonding and gallium atoms have three. If arsenic atoms are introduced into a crystal of germanium, extra electrons are present. When a voltage is applied, the extra electrons in the lattice will move.

Figure 16.17 Arsenic-doped crystals of germanium conduct electricity because they contain extra electrons that can move freely through the crystal.

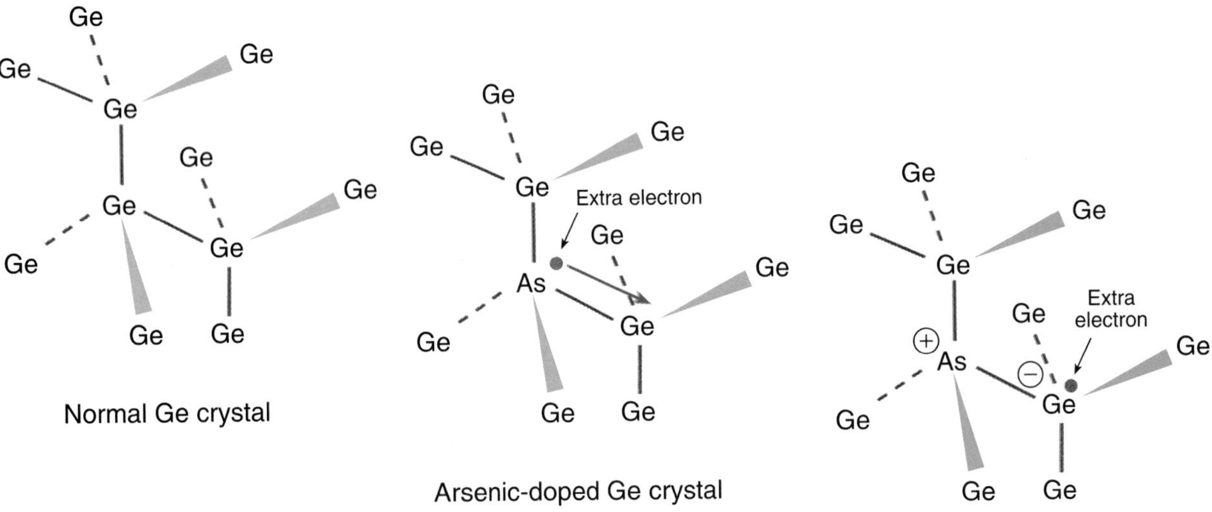

On the other hand, gallium atoms have only three electrons in the outer level. So, when gallium atoms are introduced into a crystal of germanium, the crystal will be short of electrons as shown in Figure 16.18. The resulting electron-deficient lattice, however, will also conduct electricity. It does so by moving electrons into the "holes" created by the gallium atom. Note that in both types of doping, the crystal is still electrically neutral.

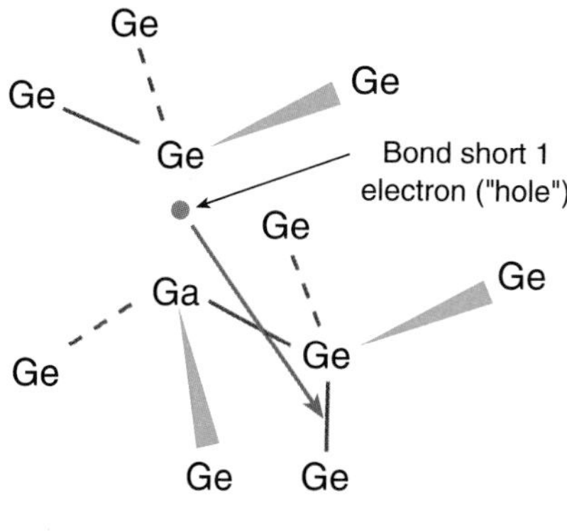

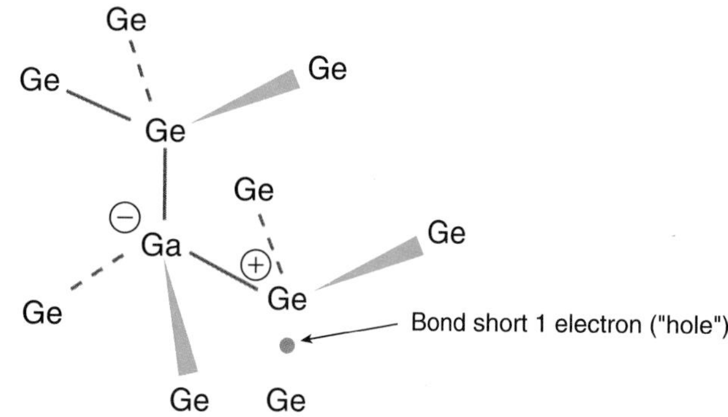

Figure 16.18 Gallium-doped crystals of germanium can also conduct electricity because they contain holes into which electrons can move.

After their development in the late 1940s, transistors were used in place of vacuum tubes in electronic circuits because of their small size, long life, and resistance to shock. Today's integrated circuits, such as silicon microprocessors, contain thousands of transistors. Computers, radios, heart pacemakers, and microwave ovens are a few of the many products that semiconductor devices have either made possible or improved.

In order to produce an integrated circuit module, a chip, which is a thin wafer of silicon, is cut from a very pure crystal. Extremely pure silicon is produced by the process of zone refining. A cylindrical crystal of silicon is held vertically and a heating coil placed around it. The coil melts a narrow band of the crystal. As the coil is passed down the crystal, the impurities remain in the moving liquid zone, leaving behind a more pure crystal. After several passes of the heating coil, a crystal of silicon with not more than one alien atom per ten billion silicon atoms can be produced. A wafer of this pure silicon is exposed to oxygen or water vapor at about 1000°C. The surface layers of silicon atoms are oxidized to silicon dioxide, which forms a protective layer. The oxide surface is coated with a photosensitive material called a photoresist. The coated wafer is then exposed to ultraviolet light through a "mask" containing the desired pattern of areas to be doped. The exposed areas of photoresist are removed by solvent and the oxide coating is then removed by hydrofluoric acid. After removing the remainder of the photoresist, the chip is exposed to vapor of the appropriate Group 13 (IIIA) or Group 15 (VA) element for doping. These gaseous atoms diffuse into the surface only where it is unprotected by oxide, that is, in the pattern of the mask. By repeating this process several times, complex circuits can be built on one silicon surface.

Figure 16.19 This group of integrated circuit chips is made from ultra pure silicon.

CHEMISTRY AND

SOCIETY

Space and Defense Spin-offs

Critics of the space program and national defense sometimes question the large sums of money spent by the government on these programs. On the other hand, advocates cite many worthwhile objectives. However, both sides often fail to recognize the spin-offs that have become part of our everyday lives.

The need to lessen the weight and volume of instruments in rockets accelerated the development of miniaturized electronic circuits, especially the computer chip. Chemists contributed to this progress by developing methods to analyze and produce the pure silicon used to manufacture the chips. Today's personal computer and many electronic gadgets are spin-offs of the space program.

Did you know that "black-boxes" on aircraft are coated with fireproofing materials first developed for the space program? These boxes record instrument readings and communications while the plane is in flight. The recordings are used to analyze the causes of plane crashes and so must be shielded from impact and fire.

Similarly, the heat shields of spacecraft are coverings that protect the craft from the intense heat produced when it reenters Earth's atmosphere. The shields are coated with ablātive materials specially developed by chemists for the Mercury, Gemini, and Apollo space programs. When heated, the surfaces of ablative materials simply peel away, because the heat changes the molecular structures. As the surfaces peel, they carry away heat. Coating the heat shields with these materials prevents the spacecraft from absorbing too much heat. Today, we find similar ablatives in coatings on reactant pellets used in fusion research.

Alloys, such as beryllium-copper, and adhesives, such as polyamide adhesive, were first developed for the space program. Now they are finding daily use. The Be-Cu alloy is used to fabricate highly reliable springs for mechanical devices. The polyamide adhesive is used to bond components of electronic circuits.

The need for instruments to measure and monitor the health of astronauts thousands of miles in space spurred advancements in medical technology. Today, we can see these advancements in instruments that can measure the elasticity of bone and the tremor of a muscle. Likewise, sensors developed to monitor the pulse, respiration rate, blood pressure, and temperature of astronauts are routinely used in modern hospitals.

The Defense Department requires instant, reliable communications and accurate weather forecasts in all parts of the world. Such demands fostered the launching of the first communications and weather satellites. The so-called "spy-in-the-sky" satellites, used to obtain intelligence data for the military establishment, have led to many civilian applications. For the surveillance satellites to function, a substantial improvement in the optical systems of their cameras was needed. Chemists helped to develop new optical glass for these systems. Today we recognize these improvements in our cameras, binoculars, and camcorders.

Analyzing the Issue

1. How has miniaturization affected the communications industry?
2. Prepare for a classroom debate on the economic benefits of spin-offs versus the initial costs of their development. Try to assess how expenditures that lead to these spin-offs may have affected other national programs.

Hydrated Crystals

If solids crystallize from water solutions, molecules of water may be incorporated into the crystal structure. For some ionic substances, the attraction for water molecules is so high that the water molecules become chemically bonded to the ions. Ions that are chemically bonded to water molecules are called **hydrated ions.** Crystals containing hydrated ions are called hydrated crystals. As we saw in Chapter 14, it is also possible for ions to be surrounded by, or coordinated with, solvent molecules other than water.

Many common chemical compounds are normally hydrated. It is possible to remove the water molecules from some hydrated crystals. It can be done by raising the temperature or lowering the pressure, or both. The resulting compound, without the water molecules, is said to be anhydrous. **Anhydrous** means without water.

The ions of some anhydrous substances have such a strong attraction for water molecules that the dehydrated crystal will recapture and hold water molecules from the air. Such a substance is called a **hygroscopic** (hi gruh SKAHP ihk) substance. Sodium carbonate, Na_2CO_3, is hygroscopic.

As you recall from Chapter 8, in formulas of hydrated compounds, the water of hydration is indicated by a dot following the formula for the anhydrous compound. For example, if plaster of paris is mixed with water, solid material called gypsum is quickly formed. The hydration is written as $2CaSO_4{\cdot}H_2O + 3H_2O \rightarrow 2[CaSO_4{\cdot}2H_2O]$. Some substances are so hygroscopic that they take up enough water from the air to dissolve and form a liquid solution. These substances are said to be **deliquescent** (del ih KWES uhnt). Sodium hydroxide and calcium chloride are deliquescent.

Special care is required in storing hygroscopic substances. You may have noticed that some reagents in the laboratory stockroom have formed liquids or hardened in the bottle. This action can be reduced by making sure bottles are closed tightly.

Word Origins

a-: (GK) negates what comes after
hydor: (GK) water
anhydrous—without water

hygros: (GK) wet
skopos: (GK) to look
hygroscopic—looks wet

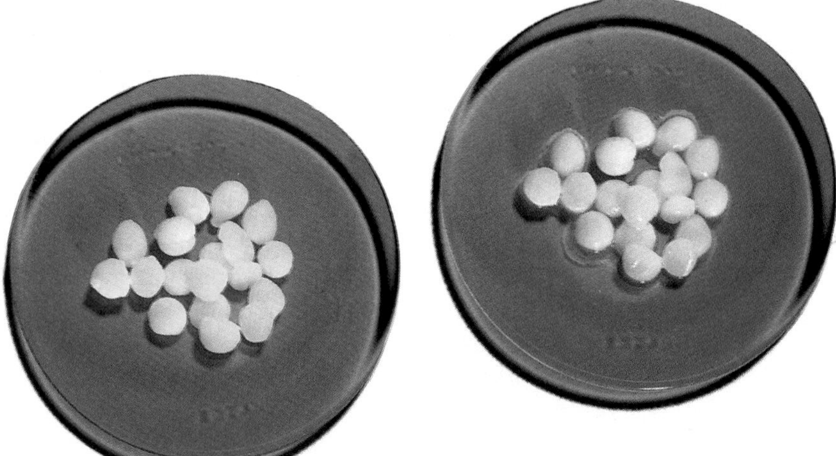

Figure 16.20 Sodium hydroxide (left) has such a strong attraction for water molecules that it begins to dissolve (right) if exposed to water vapor in the air.

EVERYDAY CHEMISTRY

Using Anhydrous Materials as Desiccants

Most electronic devices, and even some electrical appliances, are packaged with a small packet of colorless crystals. The crystals are silica gel, an anhydrous material. Some anhydrous materials, such as silica gel, gain water molecules so easily that they can be used to remove water from other substances. When an anhydrous material is used in this manner, it is called a desiccant. The purpose of including a desiccant in the packaging of an electronic device is to absorb water vapor from the air. Maintaining a dry atmosphere around the device helps prevent corrosion of metal parts. Just as important is the prevention of the condensation of water vapor in the wiring of the device. Even a tiny quantity of liquid water on a circuit board can create a short circuit, preventing proper operation of the device, or destroying part of the circuit. Silica gel is manufactured by precipitation from a solution of sodium silicate. The precipitate is a jelly-like mass (hence the name), which is washed to remove impurities. The result is a form of hydrated silica, reasonably represented by the formula H_2SiO_3. The hydrated material is then heated to dehydrate it. The resulting silica (SiO_2) can absorb water equivalent to 40% of its own mass.

Exploring Further

1. What is the function of desiccants?
2. What mass of water can a mole of silica absorb?
3. Phosphorus pentoxide, P_2O_5, has a much stronger attraction for water than SiO_2. Why isn't P_2O_5 used as a desiccant in consumer products?

CHEMISTRY IN DEPTH

Liquid Crystals

As we have seen, true solids are crystalline. Their constituent particles are arranged in a highly ordered manner. The arrangement is *ordered* because the particles are spaced in a regular, repeating fashion in all three dimensions. In Chapter 15, we saw that the structure of liquids is much less regular. Particles can actually change their positions in a liquid. The structure of a liquid is less ordered than the structure of a solid. The disorder of the liquid extends to all three dimensions. In general, when a pure solid is heated, it has a sharp melting point at a specific temperature. At that temperature the solid changes to a liquid, and the order in all three dimensions is destroyed.

However, some solid materials can lose their crystalline order in only one or two dimensions at the melting point. At a specific higher temperature, the remaining order will also be destroyed. Between these two transition temperatures, these materials retain some degree of order. These substances are called **liquid crystals.** If they retain two-dimensional order, they are called *smectic* substances. If they retain only one-dimensional order, they are called *nematic* substances. A third class of liquid crystals (with two-dimensional order) will not be considered here.

The structure leading to such properties has been determined. Liquid crystals are formed by long, rodlike molecules arranged parallel to each other. When the attractive force between layers of the molecules is overcome by energy, a smectic material results, with the layers remaining intact. If energy overcomes both the attraction between layers and end-to-end attraction, only parallel orientation remains. We then have a nematic material because only one-dimensional order remains.

Substances in liquid crystal form are said to be *mesomorphic*. Mesomorphic materials exhibit anisotropy. Anisotropic materials show different properties in different directions. As you know, the "lead" in pencils is made of graphite. Graphite smudges easily because the layers of carbon atoms in a graphite crystal can slide over each other and be separated from the crystal. Graphite is anisotropic because of its bonding structure. Liquid crystals have optical and electrical anisotropy. Interestingly, along with several other biological systems, both muscle fibers and nerve fibers exhibit liquid crystal properties.

Some liquid crystals become transparent when subjected to a pulse of high-frequency current. Then, if subjected to a low-frequency pulse, they become opaque. This property has led to the widespread use of liquid crystals for digital displays in watches and calculators and the video screens of miniature televisions.

Figure 16.21 Some insects, like this beetle, appear different in different light because of liquid crystals in their wings.

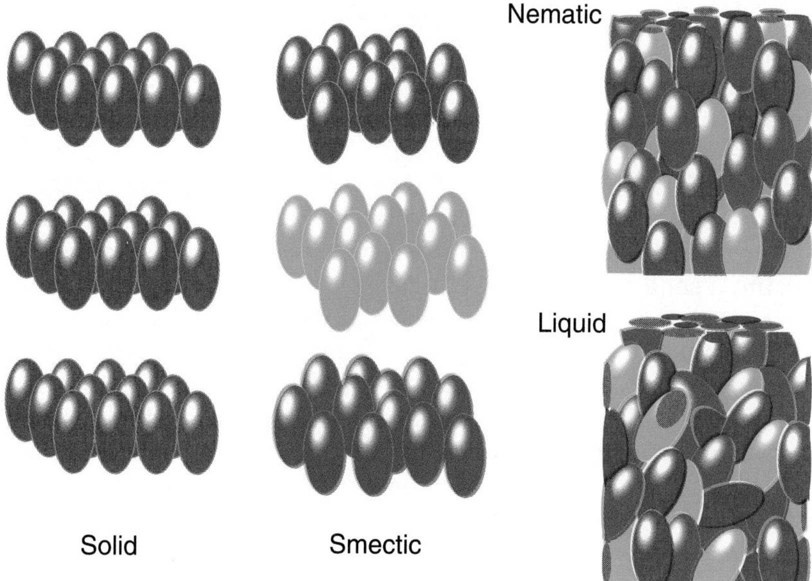

Figure 16.22 Solids, liquid crystals, and liquids have decreasing degrees of order.

EVERYDAY CHEMISTRY

Liquid Crystal Thermometers

The forces that align the molecules in liquid crystals are weak. Consequently, mechanical stress, electromagnetic fields, and chemical substances can easily affect liquid crystals. Certain liquid crystals are also affected by temperature. At most temperatures, the liquid crystals are transparent because they transmit light. However, the properties of these crystals change within narrow temperature ranges. In one temperature range, some crystals will begin to reflect certain wavelengths of white light and transmit the remaining wavelengths. In another temperature range, other crystals will transmit and reflect other colors of white light. As a result, the crystals will become colored as the temperature changes. This change in properties is used to indicate temperature.

In your classroom or at home, you may have an aquarium equipped with a liquid crystal thermometer. In the thermometer, several different liquid crystals are present, each crystal being sensitive to a different temperature. Because the various liquid crystals are applied in the proper order, the strip functions as a thermometer.

Exploring Further

1. How do liquid crystal thermometers differ from expansion-type thermometers?
2. Research the use of liquid crystals in the medical procedure called thermography.

Amorphous Materials

There are many substances that appear to be solids, but are not crystalline. Examination of their structures reveals a disordered arrangement of particles. These materials are said to be **amorphous,** or without crystalline form. Crystals are characterized by the repetition of the same arrangement over and over. They possess *long-range order.* Liquids, on the other hand, have a generally random arrangement, with only occasional regular arrangements among just a few particles. They possess, at best, only *short-range order.*

Glass is an excellent example of an amorphous material. Glass does not have a sharply defined melting point as ice has. Sharp melting points are characteristic of solids. At the melting point, a solid loses its long-range order and liquefies. When glass is heated, it does not reach a point at which it suddenly becomes liquid. Instead, it gradually softens more and more over a wide temperature range. The hotter it gets, the more easily the glass flows. Upon cooling, glass does not reach a specific temperature at which it turns into a solid. As it cools, it flows more and more slowly. We say its viscosity increases. The resistance of a liquid to flow is called its **viscosity** (vis KAHS uh tee). Glass and cold molasses are good examples of viscous materials. Water and alco-

a-: (GK) negates what comes after
morphe: (GK) form, shape
amorphous—without shape

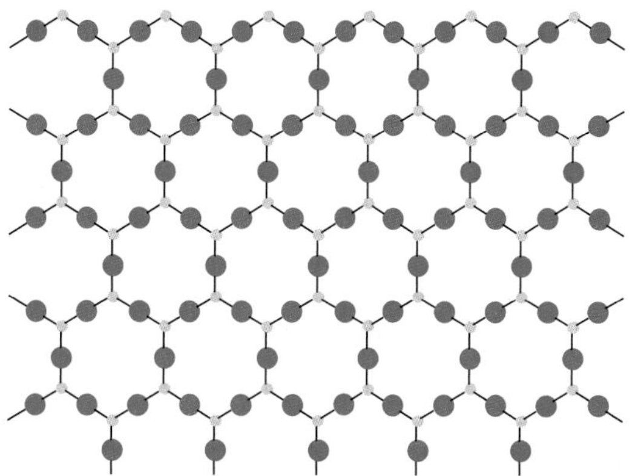

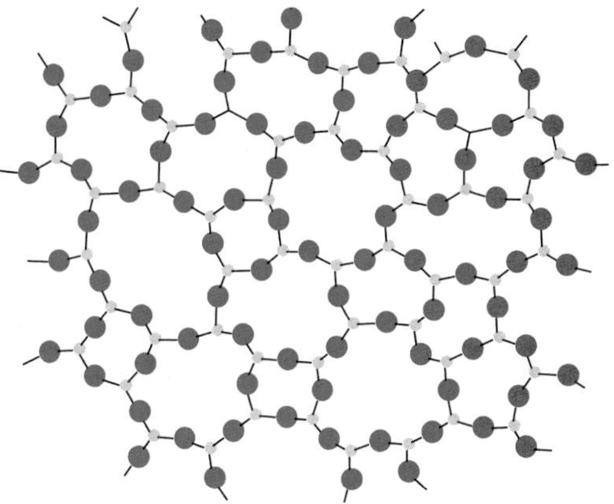

hol are good examples of liquids with low viscosities, or nonviscous liquids. Amorphous materials are said to be supercooled liquids. Supercooled liquids may become so viscous that they exhibit many characteristics usually associated with solids: rigidity, hardness, elasticity. Yet the application of an external force over a long period of time causes the material to flow and become permanently deformed. Butter is another example of an amorphous material, or supercooled liquid. For equations in the chapters that follow, true crystalline solids will retain the symbol (cr), and amorphous materials will be designated by (amor).

Figure 16.23 This two-dimensional representation of crystalline quartz (left) shows long-range order. The quartz glass (right), on the other hand, lacks order.

One form of sulfur contains long chains. It is an amorphous form and is called plastic sulfur. A few hours after it is prepared, it changes back into the stable orthorhombic form. In most amorphous substances the amorphous form is unstable. Some materials, such as glass, remain in the amorphous form for long periods instead of changing to a more stable crystalline form as sulfur does. Substances that can occur in a long-lasting amorphous form are said to be **metastable.** Although a metastable form is not the most stable form, a substance in this form is not likely to change unless subjected to some outside disturbance. Glass is a metastable substance. It normally occurs in the amorphous form, but even glass may be crystallized under the proper conditions. The crystallizing times for many metastable substances have been calculated and range from centuries to millions of years.

meta: (GK) near, between
stare: (L) to stand
metastable—stable if undisturbed

16.2 CONCEPT REVIEW

8. Perfect crystals are rare. Explain two basic ways in which there could be a defect within the unit cell structure.
9. How can a crystal of silicon conduct electrical current when doped with gallium, which has less than the necessary number of electrons to form crystal lattice bonds?
10. How are anhydrous compounds usually prepared?
11. How do liquid crystals differ from ordinary liquids?
12. How do amorphous materials differ from crystalline materials?
13. **Apply** Calcium chloride, a deliquescent substance, is sometimes applied to unpaved roads to control dust. What properties of calcium chloride account for its use to reduce dust?

Chapter 16 REVIEW

Summary

16.1 Crystal Structure

1. Steno's law states that the corresponding angles between faces on different crystals of the same substance are always the same.
2. All true solid substances are composed of crystals.
3. A crystal is a rigid body in which particles are arranged in a repeating pattern. The smallest unit of the repeating pattern in a crystal is called the unit cell.
4. The repetition of the unit cell in a crystal forms an arrangement of particles called a space lattice.
5. Pure metallic elements generally form crystals in a type of closest-packing arrangement.
6. Molecular substances form crystals in which the molecules are held together by weak van der Waals forces. These substances have low melting points.
7. Some substances form macromolecular crystals that are characterized by having very high melting points.
8. Crystals that have the same structure and shape but are of different substances are said to be isomorphous. A single substance that has two or more crystalline shapes is said to be polymorphous.

16.2 Special Structures

9. Crystal defects are imperfections in the regular repetition of the unit cell arrangement. Defects can be a result of displaced or missing particles. Other defects are edge dislocations and screw dislocations.
10. Semiconductors are modified by adding impurities to crystals of substances such as silicon and germanium. This doping process creates a crystal with more electrons than needed for bonding or holes where electrons may go.
11. A hygroscopic substance has such a strong attraction for water molecules that it will capture water molecules from the air. A deliquescent substance is so hygroscopic that it takes enough water from the air to dissolve itself.
12. Some crystals contain hydrated ions, which are ions that have water molecules chemically bonded to them.
13. Liquid crystals are formed by substances that retain some degree of order at their melting point. If they retain two-dimensional order they are called smectic. If they retain order in only one dimension they are called nematic.
14. Substances that seem to be solids but do not have a crystalline form are called amorphous substances.
15. The resistance of a liquid to flow is called its viscosity.

Key Terms

crystal	isomorphous
unit cell	polymorphous
simple cubic	dislocation
body-centered cubic	doped
face-centered cubic	hydrated ion
space lattice	anhydrous
hexagonal closest packing	hygroscopic
	deliquescent
cubic closest packing	liquid crystal
	amorphous
network crystal	viscosity
macromolecule	metastable

Review and Practice

14. What is a crystal? What determines the properties of a crystal?
15. Why is it stated that the study of the solid state is the study of crystals?
16. How do simple cubic, body-centered cubic, and face-centered cubic unit cells differ in the physical arrangement of the particles?

17. Why is the body-centered cubic lattice, which is common for alkali metals, not considered a "most closely packed" form?
18. Compare the physical properties of diamond and graphite and explain how bonding affects those properties.
19. Cite reasons why nonmetallic elements have low melting points.
20. Why do most molecular compounds generally have low melting points?
21. List and explain properties that are common to most network crystals.
22. What is a macromolecule? What is another name for this type of structure?
23. Describe the lattice structure of a crystal of a molecular compound. Explain how the forces holding the crystal together differ from the forces holding an ionic, a metallic, and a network crystal together. Compare the properties of these four types of crystals and explain the differences.
24. What does the term *isomorphous* mean? How does an isomorphous substance differ from a polymorphous substance?
25. The compound $CaCO_3$ is polymorphous. One form, calcite, occurs naturally as a rhombohedral crystal. The other form, aragonite, occurs naturally as an orthorhombic crystal. Describe each of these two crystal forms.
26. Name two types of crystal defects.
27. How is selective doping accomplished in the production of integrated circuits?
28. Sometimes crystal forms of compounds can vary as their water of hydration varies. For example, $FeSO_4$ is an off-white monoclinic crystal, $FeSO_4{\cdot}5H_2O$ is a white triclinic crystal, and $FeSO_4{\cdot}7H_2O$ is a green monoclinic crystal. Describe a monoclinic and a triclinic crystal.
29. What are hydrated ions?
30. How is an anhydrous substance produced?
31. What happens when hygroscopic and deliquescent substances are exposed to moist air?
32. What are liquid crystals? How do they differ from ordinary liquids?
33. What is the difference between smectic and nematic substances?
34. How do amorphous materials differ from crystalline substances?
35. What is the principal characteristic of amorphous materials?
36. What are the characteristics of a metastable substance? Give an example.
37. The Great Salt Lake in Utah contains an estimated four million tons of dissolved lithium chloride. What crystal structure does solid lithium chloride have?
38. In terms of its structure, explain why diamond can scratch all other materials.
39. Graphite is described as anisotropic. How does graphite exhibit this property?

Concept Mastery

40. What is the relationship between a unit cell and a space lattice?
41. From the diagram of the NaCl lattice, show why NaCl is the simplest formula.
42. Describe the two crystal forms that are most closely packed. Which form is more frequently found in metals at room temperature?
43. Describe how the process of zone refining removes impurities from silicon cylinders that are to be made into integrated circuits. Why is pure silicon necessary?
44. Why do people commonly refer to glass as solid?
45. Why do you think that metallic elements usually have simple crystal structures that involve closest packing?
46. How might you produce a crystalline hydrate from an anhydrous substance?

47. Compare the meanings of the terms long-range order and short-range order in describing the structures of solid and liquid substances.

48. Why is it not accurate to refer to silicon carbide as a molecular substance even though its bonds are covalent?

49. The diagram shown here represents the unit cell structure of cesium chloride, CsCl. What type of unit cell is it? Why do you think CsCl has this type of unit cell when all other salts of Group 1 (IA) ions and Group 17 (VIIA) ions are face-centered cubic?

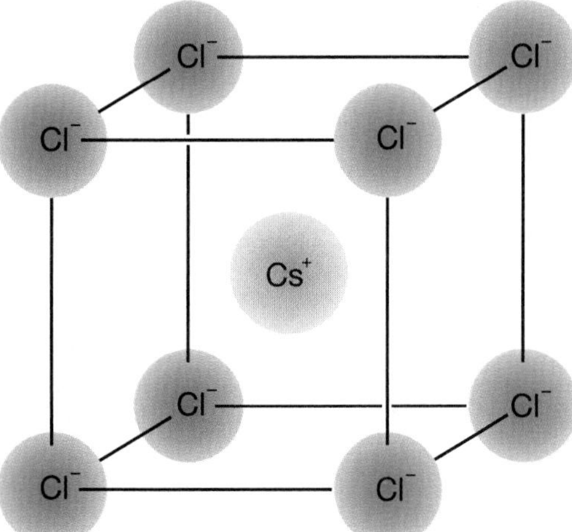

50. The crystal form for $Fe(TaO_3)_2$ is light brown and tetragonal. Iron(II) iodide, FeI_2, is gray and hexagonal. List the differences in tetragonal and hexagonal crystal forms.

51. Explain how doping with arsenic changes the properties of silicon and germanium. What happens when gallium is used instead of arsenic?

Application

52. List two applications for liquid crystals. State the properties that make them useful in each application.

53. One important property of motor oil is its viscosity. What is meant by viscosity? Why is it an important consideration in choosing a motor oil for a car?

54. Arrange the following in order of increasing viscosity at room temperature.
a. honey **b.** rubbing alcohol **c.** salad oil

55. A box of washing soda, $Na_2CO_3{\cdot}10H_2O$, was labeled "453.6 g" but was found to contain a mass of only 410.2 g. What could account for the difference?

56. Semiconductor devices have been produced using diamonds. Explain how diamonds could be semiconductors.

57. The opal is a hydrated gemstone. Explain why this makes the opal more easily broken than many other gemstones.

58. A technician preparing a calcium chloride solution spills some of the crystals on the tabletop but forgets to clean it up. A day later, she notices drops of liquid on the table. What is the probable explanation for the presence of the liquid?

Critical Thinking

59. Break-resistant glass is produced by replacing sodium ions on the surface of the regular glass with potassium ions, which are larger than the sodium ions. How would this make the glass more resistant to breaking?

60. In an experiment, two steel paper clips are heated until they are red-hot. One is cooled slowly, and the other is cooled quickly by dipping it in cold water. After cooling, both are bent in the same way several times. Which one would break first? Why? How does this explain why steel is often "tempered"?

61. Sodium chloride, table salt, is not deliquescent, but magnesium chloride is. Magnesium chloride is a common impurity in table salt. What undesirable effect might this have on a shaker full of table salt? What might be done to eliminate the problem?

Chapter 16

REVIEW

62. How does Steno's observation of crystals support the atomic theory?

63. Certain kinds of glass tableware are referred to as "crystal" ware. Why is this term probably improperly applied?

64. Refer to the simple cubic unit cell shown in Figure 16.6. If the crystal is of an element, the mass of one unit cell of a simple cubic crystal equals the mass of one atom because there are eight corners of the crystal and one-eighth of the atom on each corner is included in that unit cell.
   a. The mass of one unit cell of a face-centered cubic crystal equals the mass of how many atoms?
   b. How many atoms are in one unit cell of a body-centered cubic crystal?

Readings

Cara, Robert J. "Superconductors Beyond 1-2-3." *Scientific American* 263, no. 2 (August 1990): 42-49.

Fine, Gerald J. "Glass and Glassmaking." *Journal of Chemical Education* 68, no. 9 (September 1991): 765-768.

Vogel, Shawna. "Snow Clones." *Discover* 10, no. 2 (February 1989): 52-54.

Willson, C. Grant, and Murrae J. Boden. "Resist Materials for Microelectronics." *CHEMTECH* 19, no. 2 (February 1989): 102-111.

Wools, A. A. "Coordination and Radius Ratio." *Journal of Chemical Education* 66, no. 6 (June 1989): 509.

Cumulative Review

1. Find the percentage of water in a crystal of $CuSO_4 \cdot 5H_2O$.

2. A hydrate of nickel(II) chlorate having a mass of 9.88 g was heated to drive off the water. The anhydrous material had a mass of 6.68 g. What is the empirical formula of the hydrate?

3. If an open manometer shows a mercury level 64 mm higher in the arm connected to the confined gas, what is the pressure of that gas? The pressure of the air is 92.7 kPa.

4. If a closed manometer shows a difference in mercury levels of 421 mm, what is the pressure on the open end expressed in kilopascals?

5. How does plasma differ from the other three states of matter?

6. What kinds of substances are usually separated by gas chromatography?

7. Perform the following calculations. Express your results in scientific notation.
   a. $\dfrac{(3.1 \times 10^6\ \text{kg/m}^3)\ (6.0 \times 10^{-2}\ \text{m}^3)}{(6.0 \times 10^3\ \text{m}^2)\ (2.4 \times 10^4\ \text{m})}$
   b. $\dfrac{(2.31 \times 10^4\ \text{m}^2)\ (6.80 \times 10^5\ \text{m})}{(1.06 \times 10^7\ \text{m})\ (2.89 \times 10^{-4}\ \text{m})}$

8. How many grams of lead(II) sulfate are in a mole of the compound?

9. Name each of the following.
   a. P_2O_5
   b. $Ca_3(AsO_4)_2$
   c. $Fe(CN)_6{}^{4-}$
   d. CH_4

10. What type of chemical reaction is each of the following?
   a. $Fe_2O_3(cr) + 2Al(cr) \rightarrow Al_2O_3(cr) + 2Fe(cr)$
   b. $2Fe(cr) + 3Cl_2(g) \rightarrow 2FeCl_3(cr)$
   c. $CaCl_2(aq) + K_2CO_3(aq) \rightarrow CaCO_3(cr) + 2KCl(aq)$
   d. $2C_2H_6(g) + 7O_2(g) \rightarrow 4CO_2(g) + 6H_2O(g)$

11. Balance each of the following equations.
   a. $F_2 + H_2O \rightarrow HF + O_3$
   b. $BaSO_4 + C \rightarrow BaS + CO$
   c. $PH_3 + NO_2 \rightarrow H_3PO_4 + N_2$
   d. $HBr + Mg \rightarrow MgBr_2 + H_2$

12. A compound that has a molecular mass of 176 u is found to have an empirical formula of $C_3H_4O_3$. What is its molecular formula?

CHAPTER **Preview**

CONTENTS

CHAPTER

17

Liquids

Since the advent of the space age, Earth has acquired some new nicknames. One, evident in the photograph on this page, is the "blue planet." The blue color is due to water, which covers 70.8% of Earth's surface. As our space probes have brought us views of other planets and their satellites, we have realized that we inhabit a sphere unique in our solar system.

It is impossible for us to think of life without thinking of liquid water. Why is water so important? First, water is a liquid and will flow. Second, our bodies are largely water. Finally, water has some unusual properties even for a liquid. In this chapter you will examine the properties of liquids and the special properties of water that make it so vital to our lives.

EXTENSIONS AND CONNECTIONS

OBJECTIVES

- explain the properties of liquids and changes of state in terms of the kinetic theory.
- use Le Chatelier's principle to explain reversible changes of state in a closed system.
- determine the relationship between energy and change of state and perform related calculations.

17.1 Changes of State

Water exists in three states: solid, liquid, and gas. Each of these states is stable under certain conditions. If large changes of the conditions occur, the motion of water molecules is altered. This causes the state to change. The same principles apply to all other substances. To understand the role water and other liquids play, we need to investigate the processes of change of state.

Melting and Freezing

Almost all solids and liquids expand when the temperature is raised. According to the kinetic theory, when the temperature of a solid is raised, the velocity of the particles increases. As the temperature increases, the particles collide with each other more often and with a greater force. Thus, they move farther apart. If the temperature of a solid is raised sufficiently, the particles will move far enough apart for some of them to slip over one another. The ordered arrangement of the solid state breaks down. When such a change takes place, we say the solid has melted.

In the liquid state, there will be a temperature (and pressure) at which particles travel so slowly that they can no longer slip past one another. The particles settle into an ordered arrangement and form a solid. Then, we say the liquid freezes. All pure liquids have a definite freezing point and all pure solids have a definite melting point. Melting and freezing take place at the same temperature for a pure substance.

Vapor Equilibrium

The average kinetic energy of atoms or molecules in a gas is a constant for all substances at a given temperature. This average kinetic energy can be calculated for any particular temperature. If we were to measure the kinetic energy of individual particles in a gas, we would find that few had exactly the predicted kinetic energy. Some particles would have more and some would have less kinetic energy than the average, as shown in Figure 17.1.

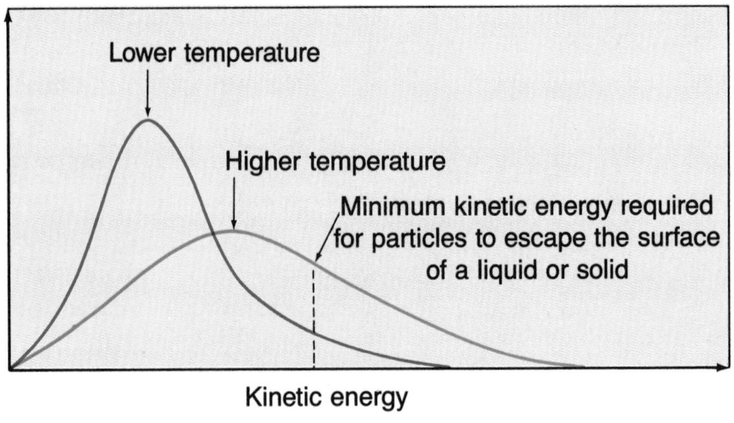

Figure 17.1 At a given temperature, the particles of a substance have a range of kinetic energies. At a higher temperature (green curve), the average kinetic energy of the particles is greater than at a lower temperature (red curve). More particles can escape the surface at the higher temperature, as seen by the steam condensing over the beaker.

Equilibrium in a closed container of fluid

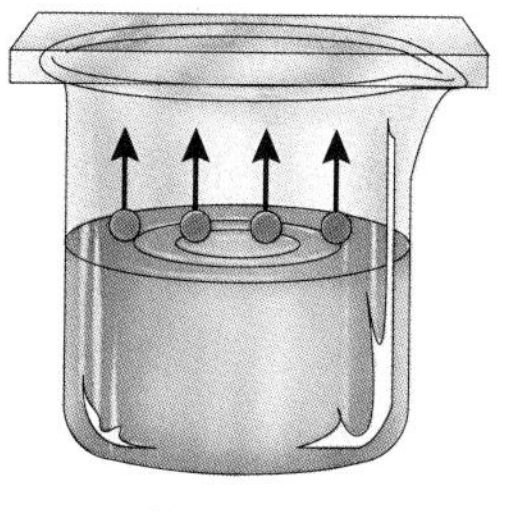

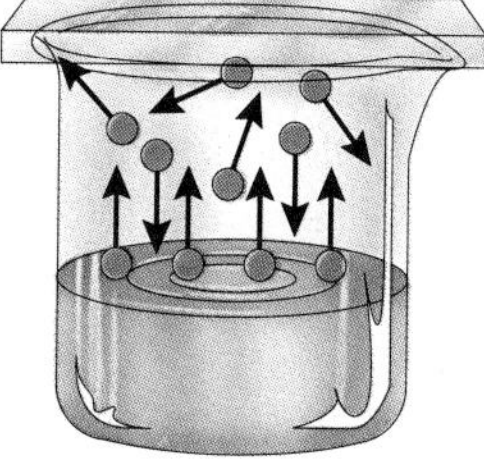

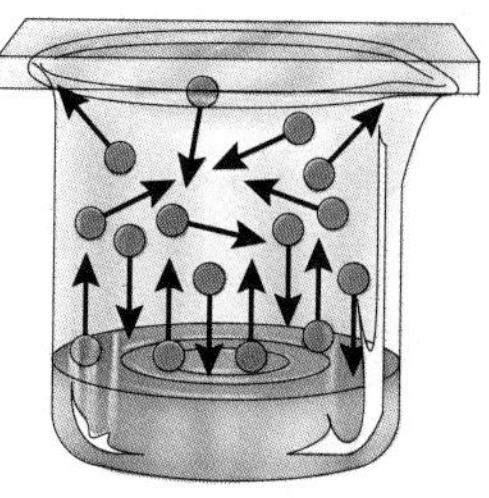

Figure 17.2 When the container is sealed initially, the particle movement is away from the liquid surface. The system reaches dynamic equilibrium when the number of molecules leaving the liquid surface equals the number reentering the liquid.

Most would have a kinetic energy close to the calculated amount. However, we would sometimes find a molecule with a kinetic energy considerably above or below the average.

All that has been said about the collisions of particles in a gas is also true of particles in a solid or a liquid. A molecule in a liquid, because of several rapid collisions with other molecules, might gain kinetic energy considerably above the average value. Imagine that molecule on the surface of the liquid. If it gains enough kinetic energy to overcome the attractive force of nearby molecules, it may escape from the liquid surface. The same process may also occur at the surface of a solid. The molecules that escape from the surface of a solid or a liquid form a vapor. This vapor is made of molecules or atoms of the substance in the gaseous state. A gas and a vapor are the same. We usually use the word *gas* for those substances that are gaseous at room temperature. The word **vapor** is used for the gaseous state of substances that are liquids or solids at room temperature.

A molecule of a solid or liquid that has escaped the surface behaves as a gaseous molecule. It is possible for this molecule to collide with the surface of the liquid it left. If its kinetic energy is sufficiently low at the time of such a collision, the molecule may be captured and again be a part of the liquid. However, if the container holding the liquid is open, there is little chance that the molecule will return to the surface it left.

If the solid or liquid is in a closed container, then there is an increased chance that the molecule will return to the surface. In fact, a point will be reached where just as many molecules return to the surface as leave the surface, as shown in Figure 17.2. There will be a constant number of molecules in the solid or liquid phase, and a constant number of molecules in the vapor phase. Such a situation in which there is no net change is known as an **equilibrium** condition. The situation we just discussed is a dynamic equilibrium because molecules are continuously escaping from and returning to the surface. In a **dynamic equilibrium,** two opposing changes are occurring at the same rate. Thus, the overall result remains constant. When a substance is in equilibrium with its vapor, the gaseous phase of the system is said to be **saturated** with the vapor of that substance. A gaseous phase that is saturated holds all the vapor it can at that temperature and pressure.

SPORTS CONNECTION

Speed Skating

Speed skaters use skates with longer, narrower blades than ice hockey players or figure skaters. They do not have to make sudden stops or abrupt changes in direction. In addition, their skate stroke is almost completely straight ahead, while the skaters from the other sports often must describe an arc with their skate blades.

The physical change from liquid to vapor is represented in equation form as

$$X_{(l)} \rightarrow X_{(g)}$$

where X represents any vaporizable substance, such as water. The opposite process can be represented by

$$X_{(l)} \leftarrow X_{(g)}$$

The two equations can be combined

$$X_{(l)} \leftrightarrows X_{(g)}$$

This kind of change is called a **reversible change.** A reversible process has reached equilibrium when the changes are occurring at the same rate in both directions.

Le Chatelier's Principle

The vapor phase exerts a pressure that depends on the temperature. In a closed container, a liquid and its vapor will reach equilibrium at a specific pressure for any particular temperature. If the temperature is increased, the vapor pressure will increase. Thus, the equilibrium shifts. This shifting of the equilibrium was observed and described by the French scientist Le Chatelier in 1884. Le Chatelier's principle is expressed as follows: *If stress is applied to a system at equilibrium, the system will tend to readjust so that the stress is reduced.* The stress may be a change in temperature, pressure, concentration, or other external force. Le Chatelier's principle applies to any system in equilibrium.

We will learn how this principle is applied by examining the water-ice system. The freezing of water and the melting of ice constitute a reversible system that can come to equilibrium. It can be represented by the equation

$$H_2O(l) \leftrightarrows H_2O(cr)$$

It would not be wise to place a bottle of water in the freezer. Water expands when it freezes and can burst the bottle. Water is one of the few substances that expands when it freezes. A quantity of water will occupy more space as a solid than as a liquid.

$$\underset{\text{less space}}{H_2O(l)} \leftrightarrows \underset{\text{more space}}{H_2O(cr)}$$

You have seen that when you press on something, it tends to become smaller; that is, it takes up less space. That knowledge can be applied to the behavior of water under pressure.

For example, an ice skater can skim over the ice with little effort. The skate blades are not really touching the ice but are traveling on a thin film of liquid water. The presence of this water can be explained by Le Chatelier's principle.

Both pressure and temperature must be considered in this example. We will discuss pressure first. The entire weight of the skater is directed onto the ice through the blades on the skates.

The surface area of the blades in contact with the ice is probably less than 15 cm^2. The entire weight (force) of the skater is concentrated on this small area. Remember that pressure is force/area. Thus, the blades exert a great pressure on the ice at the points of contact. A piece of ice (solid) occupies a greater volume than an equal mass of water (liquid). Increased pressure causes a reduction in volume, and the ice changes to water. The change of ice to water will tend to reduce the pressure (stress) caused by the blades. As a result, the blades travel on a film of liquid water. Applying pressure to a water-ice equilibrium tends to favor the formation of liquid water because it takes up less space.

Temperature also plays an important role in skating. Friction between the ice and the blades of the skates warms the blades. This increased temperature (stress) will be relieved by the flow of heat into the ice. Increased temperature causes an increase in molecular motion and the ice melts. The low energy ice molecules are changed into more energetic water molecules. The blades will be cooled by this change. Both the increased pressure and the increased temperature enable the skater to glide over the surface of the ice with very little friction.

Measuring Vapor Pressure

Many techniques are available for measuring vapor pressure. Figure 17.3 shows two methods of finding the vapor pressures of substances. The apparatus in Figure 17.3a is especially useful for finding the vapor pressures of solids at elevated temperatures. A carefully controlled constant temperature bath is used to maintain the temperature of a substance. When the substance and its vapor reach equilibrium, the vapor pressure and temperature can be measured with a manometer and thermometer, respectively. Often, the vapor pressure will be read at temperatures over a wide range. Using these data, the changing behavior of a substance with temperature is obtained. Such information is valuable in industrial processes involving the substance tested.

Figure 17.3 Two devices used for measuring vapor pressure are shown. In each, the vaporized liquid exerts a force (pressure) on the mercury in the tube.

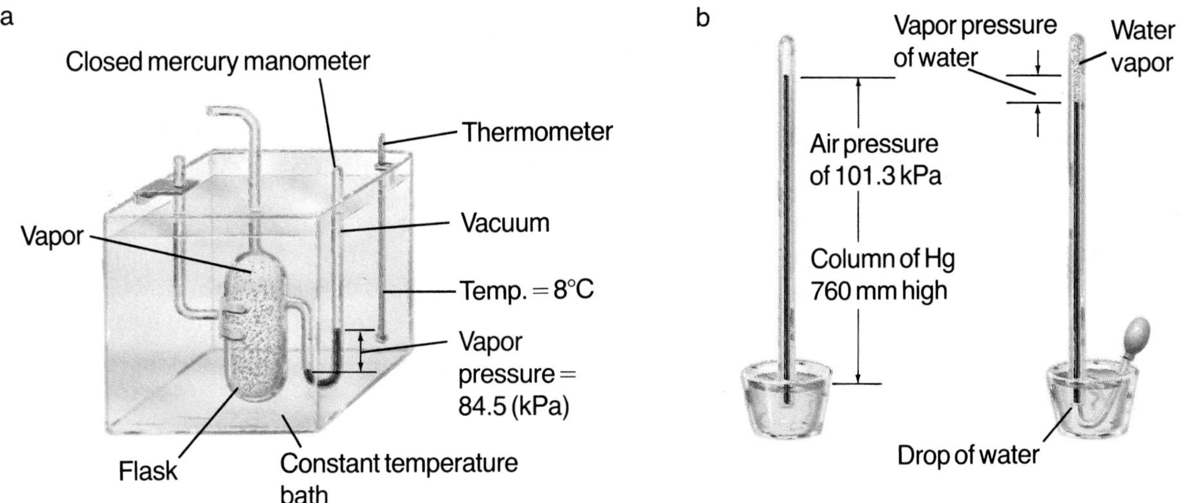

Table 17.1

Vapor Pressures of Some Substances at 25°C			
Substance	**Vapor Pressure (kPa)**	**Substance**	**Vapor Pressure (kPa)**
Mercury (Hg)	0.000 246 0	Bromine (Br_2)	28.720
Turpentine ($C_{10}H_{16}$)	0.588 4	Acetone (CH_3COCH_3)	30.786
Water (H_2O)	3.167 2	Carbon disulfide (CS_2)	48.113
Carbon tetrachloride (CCl_4)	15.250	Sulfur dioxide (SO_2)	392.23

The left side of Figure 17.3b shows an ordinary barometer. The space above the mercury column in the closed tube is essentially a vacuum because mercury has a very low vapor pressure. Thus, there are no molecules bombarding the surface of the mercury. The pressure on the enclosed column is zero. If we introduce a drop of a liquid to the column, as shown on the right side of Figure 17.3b, the liquid will float to the top of the mercury because the mercury is more dense. When the liquid reaches the vacuum above the mercury, evaporation takes place. Eventually, an equilibrium will be established between the liquid and its vapor. Now there are gas molecules bombarding the top of the mercury column, which will push it down. The pressure exerted on the mercury column from inside the tube is the vapor pressure of the liquid and is equal to the change in level of the mercury column. This method is accurate but not very practical. For one thing, the barometer is now contaminated. For another, it is difficult to change or control the temperature in such an apparatus.

Table 17.1 gives the vapor pressures of some substances near room temperature. *Substances with low vapor pressures have strong intermolecular forces. Those with high vapor pressures have weak intermolecular forces. Ionic compounds do not exert a significant vapor pressure because the interionic forces are too strong to be overcome.*

Solids Changing State

Melting Point

In a mixture of solid and liquid states in a closed container, there will be a dynamic equilibrium between the molecules of the solid and the liquid. Remember that each state is also in equilibrium with its vapor. Because there is only one vapor, the solid and liquid have the same vapor pressure, Figure 17.4. In fact, **melting point** is defined as the temperature at which the vapor pressure of the solid and the vapor pressure of the liquid are equal.

The melting point temperature of a substance depends on the intermolecular forces in the substance. *Substances with weak intermolecular forces have lower melting points than substances with strong intermolecular forces.* Thus, nonpolar substances with low molar masses have lower melting points than polar substances with similar molar masses.

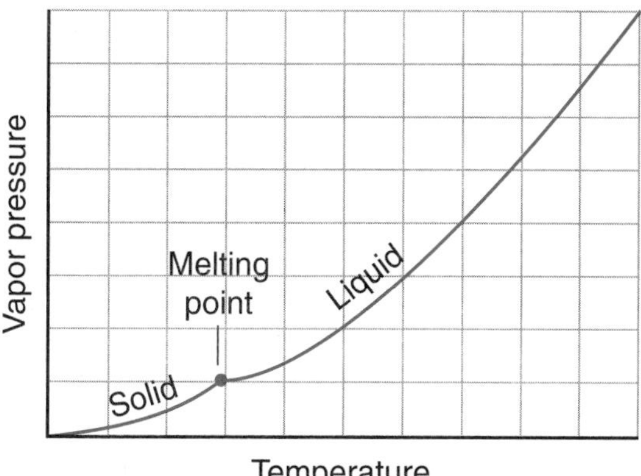

Figure 17.4 The graph shows the melting point of a substance to be that temperature where the vapor pressures of the solid and liquid phases are equal.

Sublimation

Some solids have a vapor pressure high enough at room temperature to vaporize rapidly if not kept in a closed container. Such substances will change directly from a solid to a gas, without passing through the liquid state. This process is known as **sublimation.** Dry ice (solid CO_2) and moth crystals are two examples of substances that sublime.

Word Origins

sublimare: (L) to uplift, elevate, refine

sublime—to change directly from a solid to a gas

Boiling Point

A liquid and its vapor can be at equilibrium only in a closed container. There is little chance that the molecules leaving the surface of a liquid will return to the liquid phase if the liquid container is open to the air. When a liquid is exposed to the air, it may gradually disappear. The disappearance is due to the constant escape of molecules from its surface. The liquid is said to evaporate.

As the temperature of a liquid is increased, the vapor pressure of that liquid increases because the kinetic energy of the molecules increases. Eventually, the kinetic energy of the molecules becomes large enough to overcome the internal pressure of the liquid. The internal pressure is caused by the pressure of the atmosphere on the liquid surface. When this pressure is overcome, the molecules are colliding violently enough to push each

Figure 17.5 Dry ice (left) and iodine (right) are two substances that sublime.

other apart. They are pushed far enough apart, in fact, to form bubbles of gas within the body of the liquid. These bubbles rise to the surface of the liquid because the vapor of which they are composed is less dense than the surrounding liquid. At this point, the liquid is boiling. The **normal boiling point** is the temperature at which the vapor pressure is equal to standard atmospheric pressure, or 101.325 kPa. Boiling point is a function of pressure. At lower pressures, the boiling point is lower.

Note carefully the difference between evaporation and boiling. Evaporation occurs only at the surface. Boiling, on the other hand, takes place throughout the body of a liquid.

volare: (L) to fly
volatile—tending to fly away (evaporate)

Adding energy to a liquid at its boiling point will cause the liquid to change rapidly to a gas by boiling. In a like manner, if we remove energy from a gas at the boiling point of the substance, the gas will change to a liquid. The boiling point of a liquid is also the condensation point of the vapor state of the liquid.

Different liquids boil at different temperatures. A liquid that boils at a low temperature and evaporates rapidly at room temperature is said to be **volatile.** Examples of volatile liquids are alcohols and ethers of low molar masses. Liquids that boil at high temperatures and evaporate slowly at room temperature, such as molasses and glycerol, are nonvolatile. Volatile substances have high vapor pressures at ordinary temperatures; nonvolatile substances have low vapor pressures at similar temperatures.

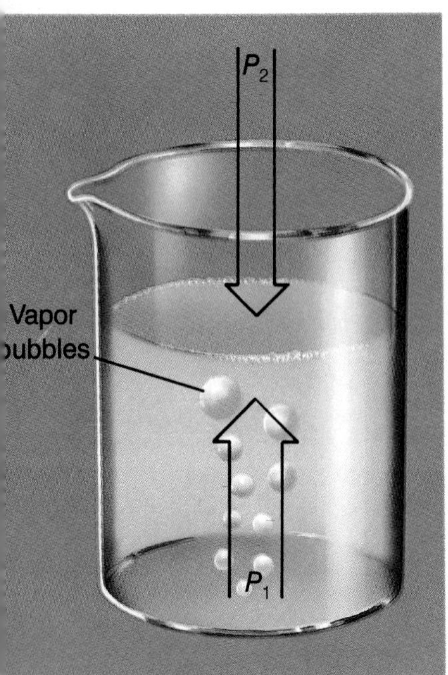

Figure 17.6 A liquid will boil when its vapor pressure (P_1) becomes large enough to overcome the internal pressure of the liquid, which is equal to the atmospheric pressure. At that point, bubbles of vapor form and can remain stable until they reach the liquid's surface.

SAMPLE PROBLEMS

1. Reading Vapor Pressure Graphs

What is the vapor pressure of CCl_4 at 60°C?

Solving Process:

Find 60°C on the *x* axis of Figure 17.7. Follow that line vertically to the red curve showing the vapor pressure of CCl_4. Move horizontally to the *y* axis and read the pressure, 57 kPa.

2. Finding a Boiling Point

What is the boiling point of water at a pressure of 90.0 kPa?

Solving Process:

Locate the pressure of 90.0 kPa on the *y* axis of Figure 17.7. Move horizontally to the blue line representing the vapor pressure of water. Read the boiling point, 97°C, on the *x* axis.

PRACTICE PROBLEMS

1. Using Figure 17.7, find boiling points of the following.
   a. $CHCl_3$ at 70.0 kPa
   b. H_2O at 100.0 kPa
   c. CCl_4 at 80.0 kPa
   d. H_2O at 11.0 kPa
2. Which of the three substances in Figure 17.7 will boil at 30°C if the pressure is lowered from 40 kPa to 30 kPa?

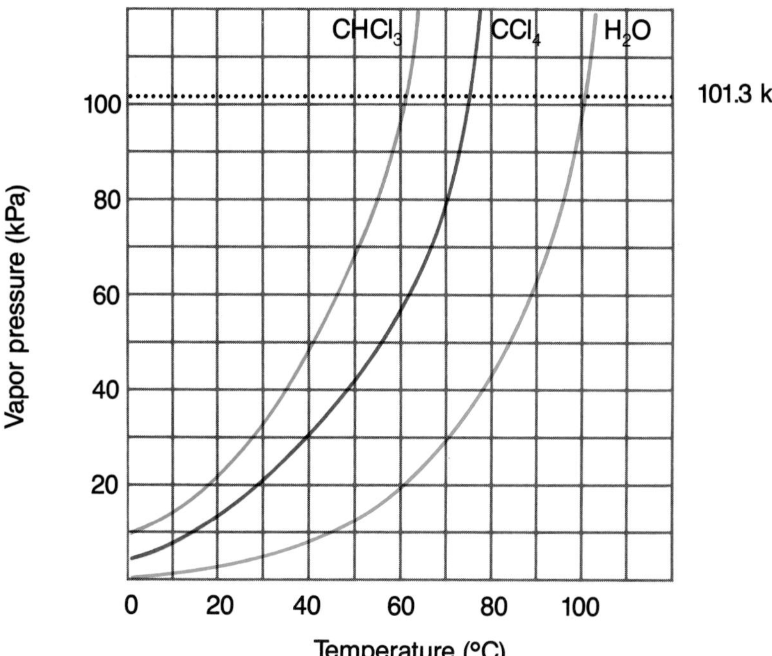

Figure 17.7 The temperature at which a substance boils is lowered when atmospheric pressure is lowered. At any temperature, the vapor pressure of CCl_4 and $CHCl_3$ is greater than that of H_2O.

Liquefaction of Gases

We have considered the condensation of gases to liquids only for substances that are normally solids or liquids at room temperature. What about substances that are normally gases at room temperature? Can they be condensed? Yes, under the correct conditions. The condensation of substances that are normally gases is called **liquefaction** (lik wuh FAK shuhn). A gas must be below a certain temperature before it can be liquefied. Cooling reduces the kinetic energy of the molecules to the point where the van der Waals attraction is sufficient to bind the molecules together. It is also necessary to compress some gases before they will liquefy. The van der Waals forces are effective for only short distances. Compression forces the molecules of these gases close enough for the van der Waals forces to take effect.

For every gas, there is a temperature above which no amount of pressure will result in liquefying the gas. The point is called the critical temperature (T_c) of the gas. Another way of expressing this principle is to say that the critical temperature represents the maximum temperature at which the liquid can exist. Above that temperature, the molecules are simply traveling too fast to stick together no matter how highly compressed they may be. The pressure needed to liquefy the gas at the critical temperature is called the critical pressure. There is really nothing critical about the critical pressure; it is simply the equilibrium pressure at the critical temperature. Less pressure can be used to liquefy the gas if its temperature is lowered.

Many gases have critical temperatures above normal room temperature. Sulfur dioxide, for example, has a critical temperature of 430.65 K. It can be liquefied by increased pressure alone, if

MINUTE LAB

Cold Water Boils

Put on an apron and goggles. Obtain a 50-60 cm^3 plastic syringe from your teacher. Draw 30 cm^3 of water at room temperature into the syringe. Holding the syringe with the tip up, gently press on the plunger until no air remains in the syringe. Seal the tip where a needle normally would be by covering the opening tightly with your finger. Use your other hand to pull on the plunger to enlarge the volume of space occupied by the water. **CAUTION:** *Do not release the plunger quickly – the end of the syringe will break.* Allow the plunger to return slowly to its original position. Repeat several times, being certain to remove any air first. What did you observe? What happened to the size of the bubbles that formed as the plunger was pulled down? When you held the plunger in a down position, did bubbles break loose from the bottom and rise to the surface? Use Figure 17.7 to predict the pressure in the syringe at 30°C.

CAREERS IN CHEMISTRY

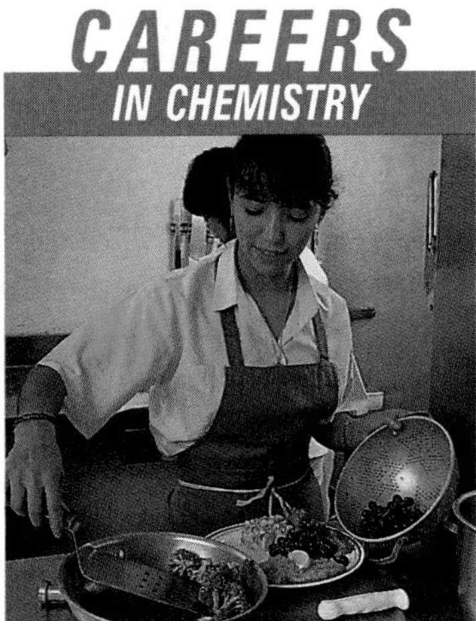

Chef

A chef in a fine restaurant or hotel must balance nutrients with taste and appearance. This process requires a knowledge of the chemical reactions that take place among foods. Beginning cooks need a high school education geared toward science and food preparation. However, a chef must have advanced training at a culinary arts school, a community college, or in apprenticeship with a chef. As more people decide to eat out, the demand for chefs will increase.

Table 17.2

Critical Temperature and Pressure		
Substance	**Critical temperature (K)**	**Critical pressure (kPa)**
Water (H_2O)	647.14	21.76×10^3
Sulfur dioxide (SO_2)	430.65	7.89×10^3
Carbon dioxide (CO_2)	304.19	7.381×10^3
Oxygen (O_2)	154.77	5.080×10^3
Nitrogen (N_2)	126.26	3.398×10^3
Hydrogen (H_2)	33.24	1.297×10^3

the temperature does not exceed 430.65 K. (See Table 17.2.) The critical temperature of a gas is an indication of the strength of the attractive forces between its atoms or molecules. The low critical temperature of hydrogen indicates that there are weak forces between its molecules. The high critical temperature of water indicates the existence of strong attractive forces between molecules.

Table 17.3 summarizes a number of properties that depend on intermolecular attraction. If the intermolecular forces are strong, then few of the individual molecules will have enough energy to escape from the liquid. As a result, the vapor pressure will be low at room temperature. If the vapor pressure is low, the substance will not evaporate readily and is considered to be nonvolatile. The temperature would have to be raised considerably to increase the vapor pressure to the boiling point. Finally, with strong intermolecular forces, the molecules of a substance would

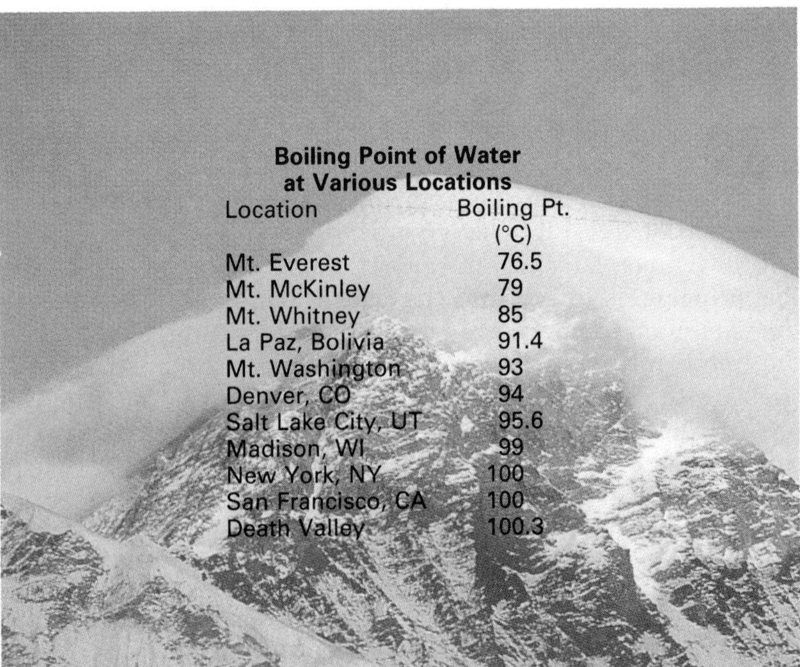

Boiling Point of Water at Various Locations

Location	Boiling Pt. (°C)
Mt. Everest	76.5
Mt. McKinley	79
Mt. Whitney	85
La Paz, Bolivia	91.4
Mt. Washington	93
Denver, CO	94
Salt Lake City, UT	95.6
Madison, WI	99
New York, NY	100
San Francisco, CA	100
Death Valley	100.3

Figure 17.8 It takes longer to cook food at high elevations than at sea level because water boils at a lower temperature and the food does not get as hot as it would at sea level.

not have to be slowed much in order for them to stick together to form a liquid; that is, the critical temperature for this substance would be high.

Table 17.3

Intermolecular Forces	
Strong intermolecular forces lead to:	**Weak intermolecular forces lead to:**
Nonvolatile substances	Volatile substances
High boiling points	Low boiling points
Low evaporation rates	High evaporation rates
High critical temperatures	Low critical temperatures
Low vapor pressures at room temperature	High vapor pressures at room temperature

EVERYDAY CHEMISTRY

The Human Cooling System

Any time the human body needs to lose heat, such as during exertion or hot weather, the sweat glands go into action. In doing so, they utilize the process of cooling through evaporation. Humans sweat at other times, too. Even relatively sedentary people may give off as much as a dm^3 of perspiration a day. It evaporates almost unnoticed and keeps the body within one or two degrees of normal throughout the day.

Body temperature is regulated by the hypothalamus, part of the brain. When warm blood and/or heat receptors in the skin signal the hypothalamus that some temperature regulation is needed, the sweat glands are activated. The person perspires. Perspiration is secreted by sweat glands located in the skin, and is composed of water, urea, and some salts. Water molecules on the skin are in motion and those with the most kinetic energy escape. This loss lowers the average kinetic energy of the remaining molecules. These molecules gain heat from the body, and again some of them escape. Again the average kinetic energy of the remaining molecules is lowered. The body is cooled as perspiration uses body heat to evaporate the water molecules. As the body cools it eventually stops sweating.

Exploring Further

1. Exercise therapists recommend that after vigorous exercise, a person cool down for a few minutes before showering. Why do you think they make this recommendation?
2. Why does using a fan appear to cool a person who is sweating?

CHEMISTRY IN DEPTH

Phase Diagrams

Much of the information we have discussed can be shown in a **phase diagram,** which shows how the states of matter in a system are affected by changes in temperature and pressure. Figure 17.9 is a phase diagram for water. Each colored area represents all of the conditions under which water in that phase can exist. The solid-vapor line, *AB*, gives the temperatures and pressures at which solid water and water vapor are in equilibrium. This line represents the equilibrium vapor pressure of ice at temperatures from −79°C to point *B*. The liquid-vapor line, *BD*, gives the temperatures and the pressures at which liquid water and water vapor are in equilibrium. This line represents the equilibrium vapor pressure of liquid water at temperatures from point *B* to 374°C. Point *B* is called the **triple point** because all three states are in equilibrium at this temperature and pressure (0.01°C and 0.611 kPa). Point *D* represents the critical point; above this point, there is no vapor pressure curve. Only the gaseous state exists at pressures and temperatures above this point.

T_m is the normal melting point, which occurs where *BC*, the solid-liquid equilibrium line, is cut by the standard atmospheric pressure line. T_b, the normal boiling point, is that temperature at which the liquid-vapor equilibrium curve is cut by the pressure line of 101.325 kPa. Only the solid-vapor, *AB*, and liquid-vapor, *BD*, lines represent vapor pressure information. The line *BC* simply indicates the pressure-temperature conditions under which the solid and liquid can be in equilibrium.

The solid-liquid equilibrium line, *BC*, for water has a negative

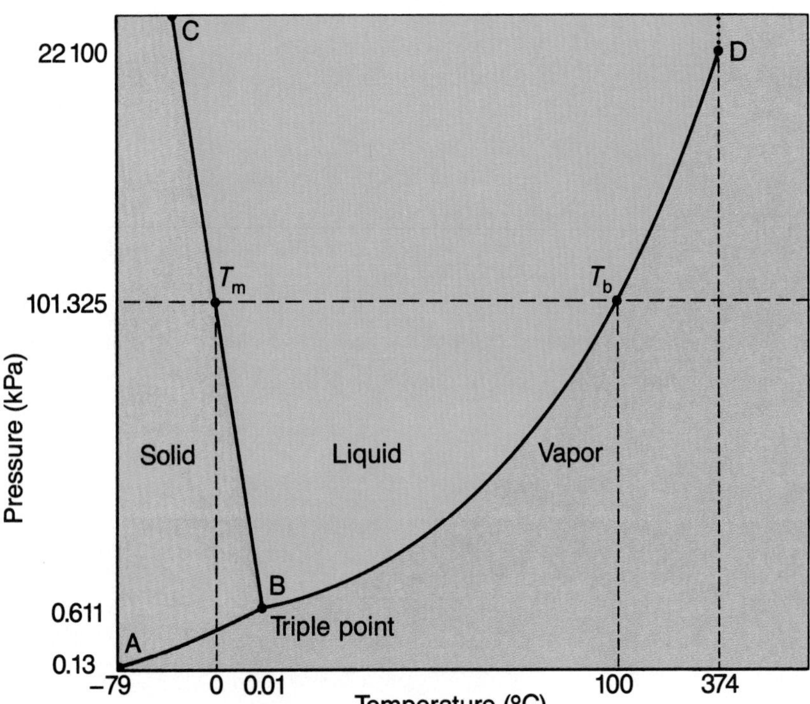

Figure 17.9 The phase diagram for water shows the relationships among pressure, temperature, and physical state. The diagram is distorted to emphasize its features. Note the nonlinear scales.

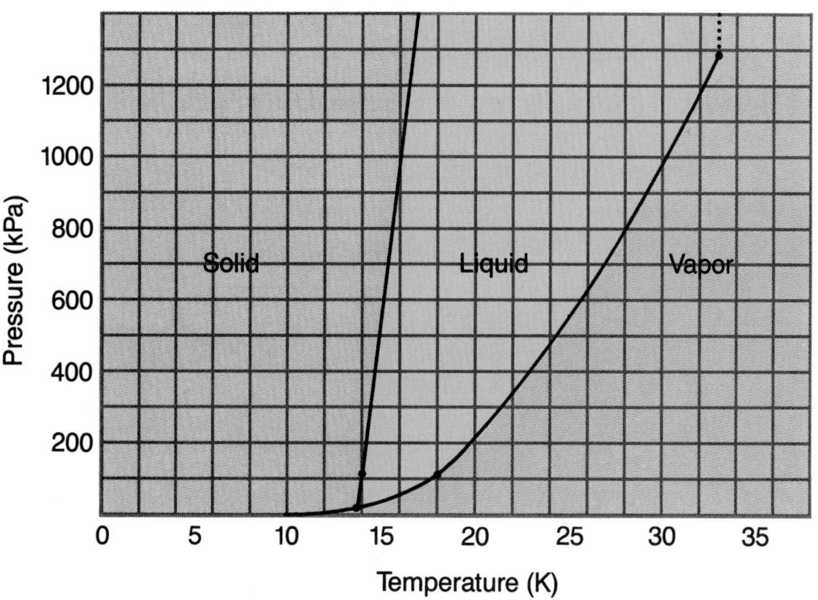

Figure 17.10 This phase diagram shows the effects of temperature and pressure on the physical state of the element hydrogen.

slope, indicating that a rise in pressure lowers the freezing point. Recall that water expands when it freezes. Most substances contract when they freeze and their solid-liquid equilibrium lines have positive slopes, such as that for hydrogen in Figure 17.10.

FRONTIERS

Flow of Silicate Magma

The lava flow of a volcano is external evidence of the movement of molten material, called magma, within Earth's crust. Lava itself is magma that has reached the surface of Earth. Magmas are mostly solutions of silicates—compounds of silicon and oxygen. The basic structure of silicates is usually tetrahedral. Each silicon atom is bonded to four oxygen atoms, with each oxygen atom located at an apex of a tetrahedron. Researchers have found that silicon in liquid silicates can bond with a fifth oxygen atom.

Research has shown that liquid silicates containing traces of five-bonded silicon atoms may disrupt the liquid structure of the silicate, decreasing its viscosity and causing it to flow more easily.

Geologists are interested in these new forms of silicon because they may influence magma flow. More viscous magma tends to dissolve greater amounts of gases than does less viscous magma. As the concentration of gases in magma increases, the tendency of the magma to explode in a volcanic eruption increases.

SAMPLE PROBLEM

Reading Data From a Phase Diagram

Find the triple point temperature, triple point pressure, critical pressure, critical temperature, normal boiling point, and normal melting point of substance Q from the phase diagram given.

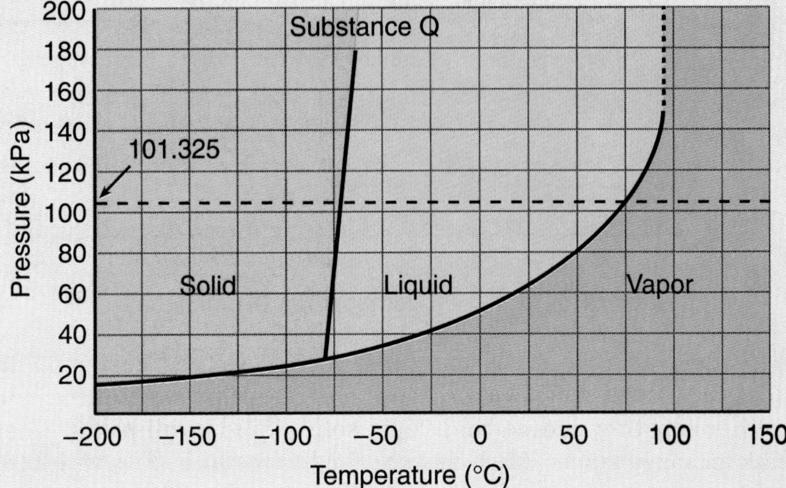

Figure 17.11 The graph shows the vapor diagram for Q, a hypothetical substance.

Solving Process:

Find the point in Figure 17.11 where all three states are in equilibrium. This point is the triple point. Read the triple point temperature, −80°C, from the x axis. Read the triple point pressure, 23.0 kPa, from the y axis. Follow the liquid-vapor equilibrium line up to the critical point (above which there is only vapor). Read the coordinates of this point: critical pressure = 148.0 kPa and critical temperature = 95°C. Find the horizontal line corresponding to normal atmospheric pressure (101.325 kPa). Follow it across to the solid-liquid equilibrium line and read the normal melting point of −74°C. Continue across the normal atmospheric pressure line until you get to the normal boiling point, 75°C, at the liquid-vapor equilibrium line.

PRACTICE PROBLEMS

3. Using Figure 17.10, determine the following for hydrogen.
   - **a.** critical temperature
   - **b.** critical pressure
   - **c.** triple point temperature
   - **d.** triple point pressure
   - **e.** normal melting point
   - **f.** normal boiling point
4. Using Figure 17.9, determine the state of matter that exists for water under the following conditions.
   - **a.** 50°C and 0.1 kPa
   - **b.** −30°C and 50.0 kPa
   - **c.** 105°C and 1000 kPa
   - **d.** 30°C and 100 kPa

CHEMISTS AND THEIR WORK

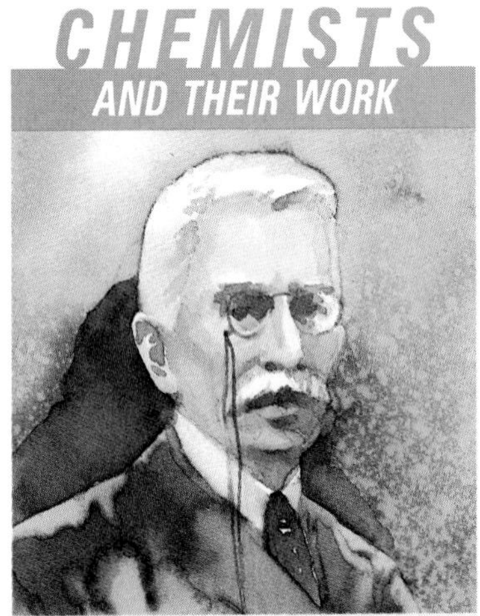

Henri Louis Le Chatelier (1850-1936)

Following in his father's footsteps, Henri Le Chatelier studied mining engineering. He obtained a post as professor at the School of Mines in France. There he did research in a number of areas including cement, ceramics, and the chemistry of flames. His objective in studying flames was to understand fire well enough to engineer preventive measures for fires in mines. During his research he invented two new ways of measuring extremely high temperatures. It was during his study of heat that he formulated the principle now known by his name. His work fit so well with related work done previously by Norwegian and American chemists that the principle was accepted and put to use immediately.

Energy and Change of State

We have seen that the loss or gain of energy from a system has an effect upon the equilibrium that exists between states. It is important for the chemist to be able to treat these energy changes quantitatively in order to completely describe a change that has taken place in a system.

When energy is added to a solid substance, its temperature increases until its melting point is reached. Upon the addition of more energy, the substance begins to melt. The temperature, however, remains the same until all of the substance has melted. Before the melting point is reached, the added energy increases the kinetic energy of the molecules; that is, the temperature is raised. At the actual melting point, the added energy changes the positions of the particles. In other words, the potential energy of the particles is increased as the physical state changes. The energy required to melt one gram of a specific substance at its melting point is called the energy of melting or the **enthalpy of fusion,** ΔH_{fus}, of that substance. A similar phenomenon takes place at the boiling point. The energy required to vaporize one gram of a substance at its boiling point is called the **enthalpy of vaporization,** ΔH_{vap}, of the substance. For example, the enthalpy of vaporization of water is 2260 J/g, and the enthalpy of fusion of ice is 334 J/g.

Enthalpy is an energy quantity that will be defined in Chapter 27. For processes that take place at constant pressure, enthalpy is equal to the heat transferred. Because almost all the work you do in the laboratory is in open vessels, you are working at a constant pressure, atmospheric pressure. Consequently, you can think of enthalpy as heat transferred until you study the more specific definition of enthalpy in Chapter 27.

Word Origins

en: (GK) in
thalpein: (GK) to warm
enthalpy—internal warmth; energy in or out

SAMPLE PROBLEMS

1. Enthalpy of Fusion

How much heat is required to melt 5.67 g of iron(II) oxide, FeO, if its enthalpy of fusion is 32.2 kJ/mol?

Solving Process:
Since $q = m \cdot \Delta H_{fus}$,

$$q = \frac{5.67 \cancel{\text{g FeO}}}{} \bigg| \frac{1 \cancel{\text{mol FeO}}}{71.8 \cancel{\text{g FeO}}} \bigg| \frac{32.2 \text{ kJ}}{1 \cancel{\text{mol FeO}}}$$

$$q = 2.54 \text{ kJ}$$

2. Enthalpies of Fusion and Vaporization

How much energy is necessary to convert 10.0 g of ice at −10.0°C to steam at 150°C?

Solving Process:

It is often helpful in problems of this type to draw a graph to indicate the steps in changing the ice to steam. See Figure 17.12.

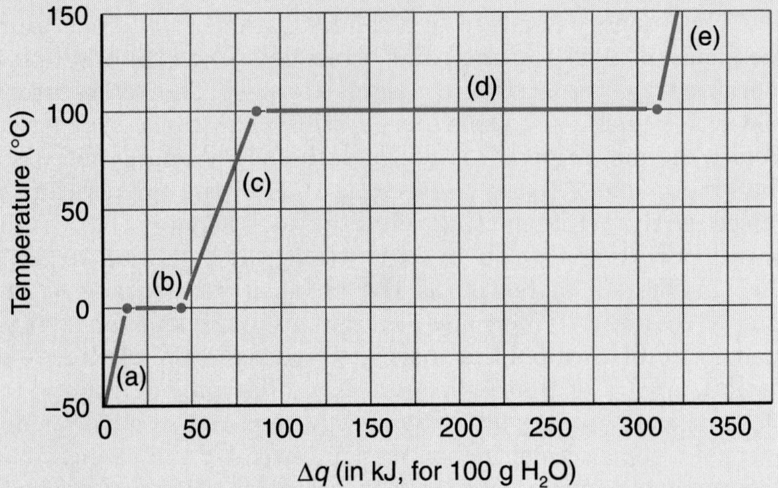

Figure 17.12 The graph shows the energy changes involved in heating ice at −50°C to steam at 150°C.

PROBLEM SOLVING HINT

Review the discussion of specific heat in Chapter 3 before you solve the sample problem.

(a) The ice must be warmed to its melting point, 0.0°C. The energy that must be absorbed for the ice to reach 0.0°C is calculated by

$$q = m(\Delta T)C_p$$

The specific heat, C_p, of ice is 2.06 J/g·C°.

$$q = \frac{10.0\ \cancel{g} \mid 10.0\ \cancel{C°} \mid 2.06\ \text{J}}{\mid \mid \cancel{g}\cdot\cancel{C°}} = 206\ \text{J}$$

(b) The ice must be melted. The enthalpy of fusion of ice is 334 J/g. The change of state energy is calculated as follows:

$$q = mass \times enthalpy\ of\ fusion = m \times \Delta H_{\text{fus}}$$

$$q = \frac{10.0\ \cancel{g} \mid 334\ \text{J}}{\mid \cancel{g}} = 3340\ \text{J}$$

(c) The water must now be heated from 0.0°C to its boiling point, 100.0°C. The C_p of water is 4.18 J/g·C°.

$$q = \frac{10.0\ \cancel{g} \mid 100.0\ \cancel{C°} \mid 4.18\ \text{J}}{\mid \mid \cancel{g}\cdot\cancel{C°}} = 4180\ \text{J}$$

(d) The water must now be vaporized. The enthalpy of vaporization of water is 2260 J/g. The change of state energy is calculated as follows:

$$q = mass \times enthalpy\ of\ vaporization = m \times \Delta H_{\text{vap}}$$

$$q = \frac{10.0\ \cancel{g} \mid 2260\ \text{J}}{\mid \cancel{g}} = 22\,600\ \text{J}$$

(e) Finally, the steam must be heated from 100.0°C to 150.0°C. The C_p of steam is 2.02 J/g·C°.

$$q = \frac{10.0\ \cancel{g} \mid 50.0\ \cancel{C^\circ} \mid 2.02\ J}{\mid \mid \cancel{g}\cdot\cancel{C^\circ}} = 1010\ J$$

The total heat that must be absorbed is the sum of the five steps.

206 J + 3340 J + 4180 J + 22 600 J + 1010 J = 31 300 J

PRACTICE PROBLEMS

5. You have a 46.0-g sample of H_2O at a temperature of −58.0°C. How many joules of energy are necessary to:
   a. heat the ice to 0°C?
   b. melt the ice?
   c. heat the water from 0.0°C to 100.0°C?
   d. boil the water?
   e. heat the steam from 100.0°C to 114°C?
6. How much energy is needed to melt 25.4 g of I_2? The enthalpy of fusion of I_2 is 61.7 J/g.
7. How much energy will be needed to melt 4.24 g of Pd? The enthalpy of fusion of Pd is 162 J/g.
8. Using data from Appendix Table A-3, calculate the heat needed to raise the temperature of 5.58 kg of iron from 20.0°C to 1000.0°C.
9. Using information from the sample problem on pages 435–437, calculate the heat required to change 70.0 g of ice at −64°C to steam at 522°C.
10. Using data from Appendix Table A-3, calculate the energy released when a 28.9 g piece of copper is cooled from its melting point to 25.0°C.

17.1 CONCEPT REVIEW

11. Describe freezing and boiling in terms of the kinetic theory of particle motion.
12. In a closed system, a liquid and its vapor can reach equilibrium. Is it necessary for there to be equal numbers of molecules in the liquid and vapor states? What must be equal?
13. Apply Le Chatelier's principle to the equilibrium between a liquid and its vapor by assuming a decrease in temperature. What would happen to the vapor pressure? Why?
14. How much energy must be removed to lower the temperature of 244 cm^3 of cola from 25°C to 0°C? Assume the density of the cola is 1.02 g/cm^3 and that the specific heat of the cola is 4.16 J/g · C°. How many grams of ice would have to be melted to accomplish the removal of that amount of energy? Assume the ice is at 0°C when placed in the glass.
15. **Apply** Why is ice more slippery to walk on when the temperature is 0°C than when the temperature is −10°C?

OBJECTIVES

- use polarity to explain hydrogen bonding.
- explain the unique properties of water in terms of its molecular structure.
- explain surface tension and capillary rise on the basis of unbalanced surface forces.

17.2 Special Properties

You already know that water has an unusual property—it expands when it freezes. Water has a number of other properties that would not be expected of such a small molecule of low mass. For example, molecules of CO_2, O_2, Cl_2, C_2H_6, and N_2 all have greater mass than H_2O. Yet, all are gases at room temperature. What properties does water possess that cause it to be a relatively nonvolatile liquid at room temperature? In this section, you will learn about the structure that leads to these properties.

Hydrogen Bonding

For many substances, changes of state can be predicted from atomic and molecular structure, which affects interatomic and intermolecular forces. However, in a number of substances, the observed melting and boiling points differ from the predicted ones. Many substances that do not behave as expected have two things in common. First, their molecules contain hydrogen. Second, the hydrogen is covalently bonded to a highly electronegative atom, which has almost complete possession of the electron pair shared with the hydrogen atom. This leaves the hydrogen atom with a strong partial positive charge. In fact, at the point of attachment of the hydrogen atom, there is a nearly bare hydrogen nucleus, or proton. A molecule containing both a highly electronegative atom and a hydrogen atom is therefore highly polar. The only elements electronegative enough to cause bonded hydrogen to behave in this manner are nitrogen, oxygen, and fluorine. There are compounds other than water, such as ammonia, that are highly polar and exhibit unexpected properties. Figure 17.13 represents a comparison of the hydrogen compounds of two different families of elements.

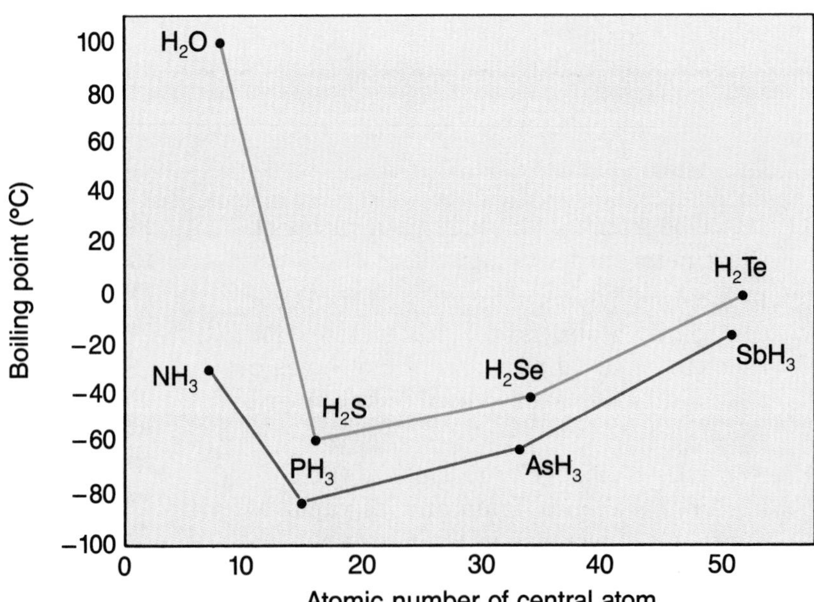

Figure 17.13 Substances that contain hydrogen bonded to an atom of high electronegativity have boiling points that are higher than would be predicted. Notice the effects of hydrogen bonding on the boiling points of H_2O and NH_3.

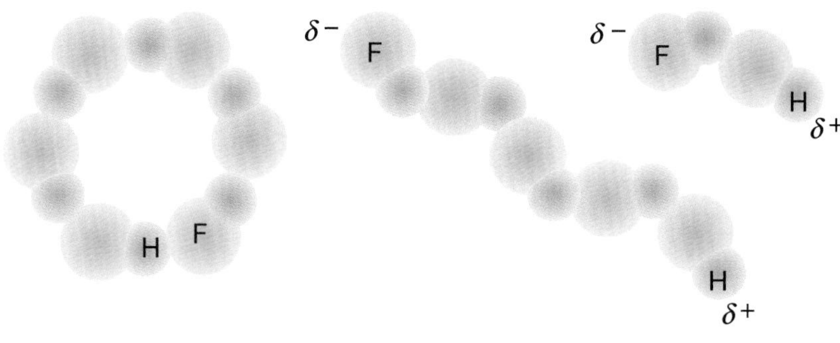

Figure 17.14 The high electronegativity of fluorine leaves hydrogen with a partial positive charge in HF. As a result, the HF molecules form chains linked by hydrogen bonds.

If an actual H^+ ion existed, it would consist of a bare proton. Consider the size of a proton compared with the size of the next largest ion with a 1+ charge, Li^+. The H^+ ion would have a full positive charge in about one trillionth of the space! It simply would not exist near other particles without interacting with them. Consequently, hydrogen is always covalently bonded, even with the most electronegative elements.

In a molecule containing hydrogen that is bonded to a highly electronegative element, the proton is not completely bare. However, the partial charge on the hydrogen end of the molecule is much more concentrated than that at the positive end of an average dipole. Hydrogen is the only element to exhibit this property. All other positive ions have inner levels of electrons shielding their nuclei. In a substance composed of polar molecules containing hydrogen, the hydrogen atom is attracted to the negative portion of other molecules. Because the hydrogen atom has been reduced to a proton with almost no electron cloud density, the attractive force is strong. This attractive force is called the **hydrogen bond.** A hydrogen bond is not as strong as an actual chemical bond but it can hold the two molecules firmly together.

Because of the special properties of a hydrogen bond, the attractive force between hydrogen-bonded molecules is much greater than the attractive force between other dipoles with the same electronegativity difference. Hydrogen bonding is a subdivision of the large class of interactions called dipole attractions. It is considered apart from other dipole attractions because it has so much influence on the properties of substances.

Figure 17.15 Hydrogen bonding occurs between the bases in DNA molecules. These bonds are formed between hydrogen atoms and either oxygen or nitrogen atoms.

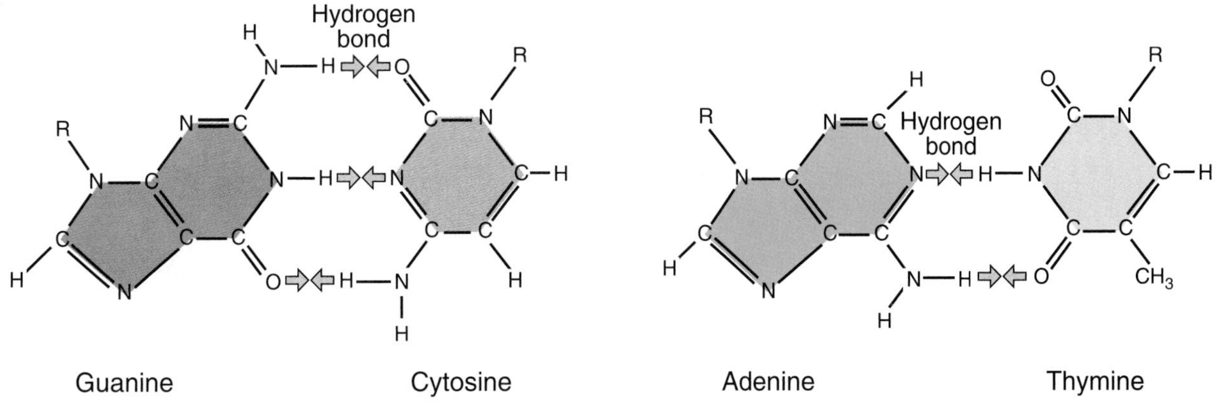

Hydrogen bond

Guanine Cytosine

Hydrogen bond

Adenine Thymine

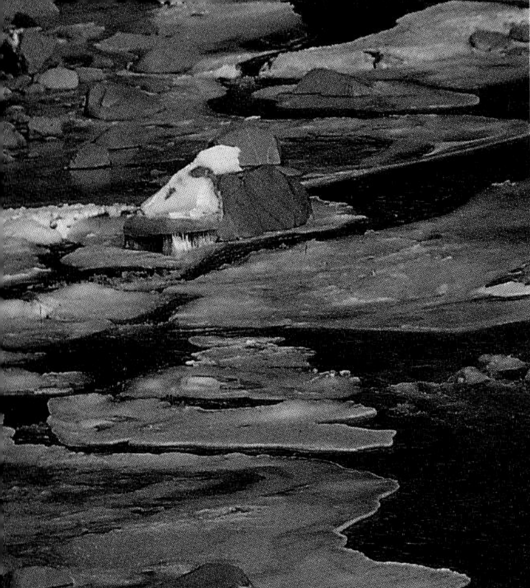

Figure 17.16 Because of hydrogen bonding, water molecules in ice are farther apart than in liquid water, causing ice to be less dense than liquid water. Therefore, ice floats.

Hydrogen Bonding in Water

The effects of hydrogen bonding can be seen in the properties of water. For example, when water is frozen, each molecule of water is hydrogen bonded to four other water molecules, Figure 17.17. The two hydrogen atoms that are part of the central water molecule are attracted to the oxygen atoms of two other water molecules. Hydrogen atoms from two other water molecules are attracted to the oxygen atom of the central water molecule. This open crystalline structure occupies a large amount of space.

When ice melts, many, but not all, of the hydrogen bonds are broken. As some of the bonds are broken, the lattice collapses. The water molecules move closer. The same number of molecules now occupy less space. Thus, liquid water is more dense than ice. As water is heated above 0°C, more hydrogen bonds are broken, and the molecules continue to move closer. At 3.98°C, most of the hydrogen bonds have been broken. Above 3.98°C, the water expands with increased temperature. Above this temperature, the density of water starts to decrease. Now, we can understand why water has its maximum density at 3.98°C. Figure 17.18 shows changes in density and volume as water melts or freezes.

Because ice is less dense than water, ice floats on lakes and rivers. This layer of ice provides insulation that helps to prevent these bodies of water from freezing solid. Thus, some aquatic organisms can live through freezing temperatures.

Surface Tension and Capillary Rise

Obtain a needle and a glass of water. With tweezers, place the needle carefully on the surface of the water. Be sure there is no soap on your hands, the tweezers, or the needle. With a little practice, you will be able to suspend the needle on the surface of the water. Why doesn't the needle sink? Have you ever poured a drink into a glass so that the surface of the liquid was higher than the rim of the glass?

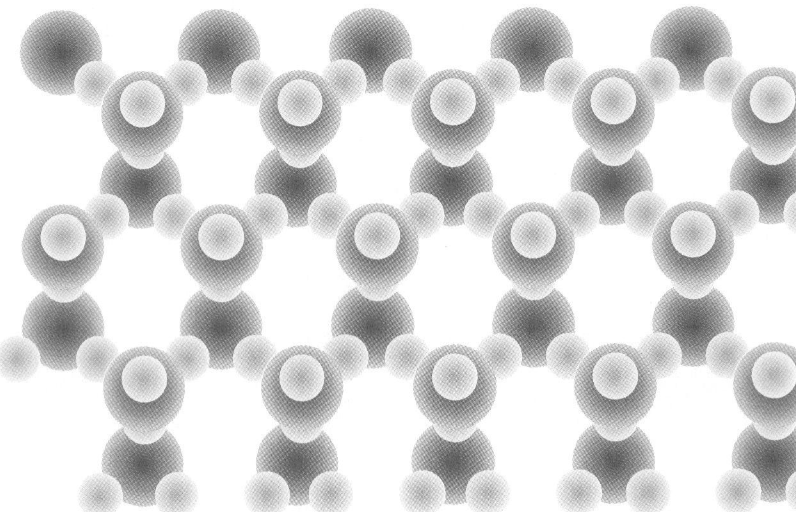

Figure 17.17 Hydrogen bonding causes water to expand as the temperature falls below 3.98°C. This expansion is seen in this molecular model of ice.

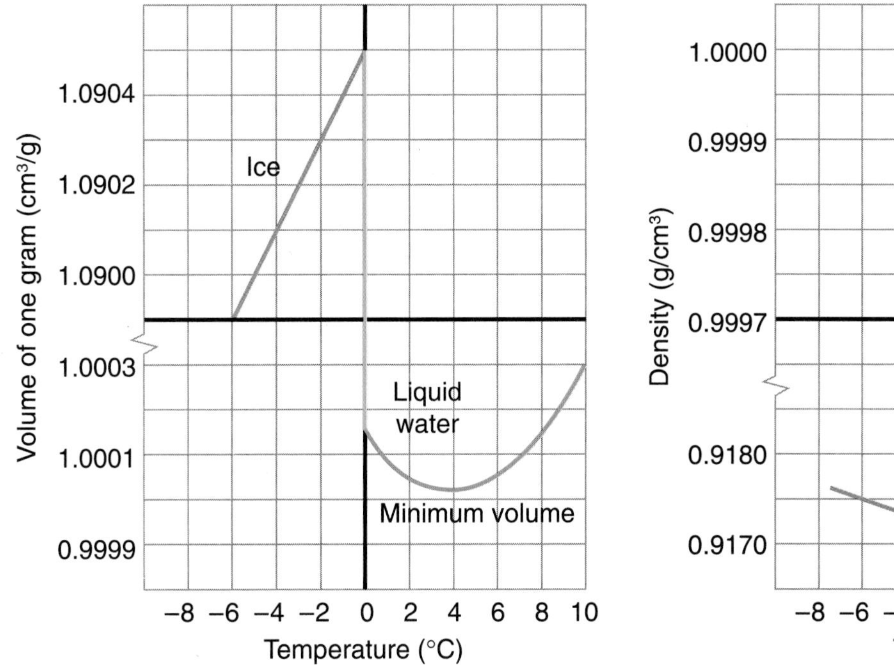

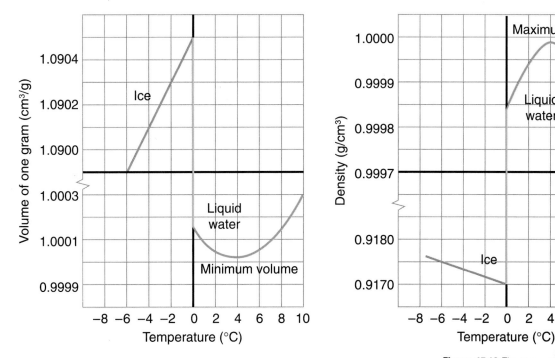

Figure 17.18 The temperature at which water reaches its minimum volume (left graph) corresponds to the maximum density of water (right graph).

The particles at the surface have special properties because they are subjected to unbalanced forces, as shown in Figure 17.19. These unbalanced forces help explain the **surface tension,** or apparent elasticity of the surface. You know that jarring the overfilled glass destroys the forces at the surface and the liquid overflows.

Particles in the interior of a liquid are subject to attractive forces in all directions. Particles at the surface experience attractive forces laterally and toward the interior of the liquid. Thus, the particles at the surface experience a net force directed perpendicularly into the liquid. The net force not only accounts for the surface tension, but also helps explain why liquids form spheres when dropped. Because a sphere has the least surface area for any given volume, liquids tend to assume a spherical shape when dropped.

Figure 17.19 Molecules in the interior of a liquid are attracted in all directions. Molecules at the surface have a net inward attraction that results in surface tension suspending the needle on the surface.

capillus: (L) hair

capillary—a tube that has a hair-thin internal diameter

The unbalanced force also accounts for the phenomenon known as capillary rise. **Capillary rise,** or capillary action, is the rise of a liquid in a tube of small diameter. If there is an attractive force between a liquid and the solid wall of the capillary tube, the liquid will rise in the tube. Capillary rise for water is shown in Figure 17.21 (left). The attractive force between the water molecules and the glass relieves the unbalanced force on the surface water molecules. The water will continue to rise in the capillary tube until the various forces (attraction between glass and water, attraction between water molecules, and the force of gravity on the water column) are balanced. Capillary rise is one method used to measure surface tension. The amount of surface tension is related to the density of the liquid, the radius of the capillary tube, the force of gravity on the liquid column, and the height to which the liquid rises.

EVERYDAY CHEMISTRY

Capillary Action and Types of Paper

Paper is manufactured almost entirely from vegetable fibers. These fibers may come from rags, grass, straw, or waste paper, but primarily they come from wood. The fibers are prepared as a pulp in water. The pulp is then formed into sheets, and the water is drained away, leaving a mat of fibers.

So that the paper will take ink without it running, the surface of the paper may then be treated to produce various specialty papers, such as the paper this book is printed on or your notebook paper. Untreated papers are used for paper bags and paper boxes.

Another use for untreated paper is in the making of paper towels. If you dip the corner of a paper towel in water, you can watch the water gradually move up the paper. This movement is a good example of capillary action. The spaces between the individual fibers act just like a capillary tube. The attraction of the water for the fibers causes it to move up the spaces between fibers.

Try writing with a felt-tip marker on a paper towel. Note how the ink spreads through capillary action on the fibers. Many inks will spread so much that the writing becomes illegible. If you try the same thing with a coated paper, such as notebook paper, you will note little or no capillary action. The reason no movement takes place is that the coating materials have filled the spaces between the fibers.

Exploring Further

1. Determine how watermarks are put into paper.
2. Learn how the technique of paper chromatography depends on capillary action to separate components of a mixture.

Figure 17.20 Surface tension holds a water strider on the surface of water (left). Falling drops of a liquid are spherical because particles at the surface of the drop have a net force inward (right).

The surface tension of both water and mercury is high at room temperature, though that of mercury is higher. Water will rise quite readily in a capillary tube while mercury is depressed, as shown in Figure 17.21 (center). Mercury will not "wet" the glass of the tube. That is, there is not enough attractive force between the mercury and the glass to overcome the attractive forces that exist between the atoms of mercury. Compare the meniscus of water with that of mercury in Figure 17.21 (right). Can you think of an explanation for the difference in behavior?

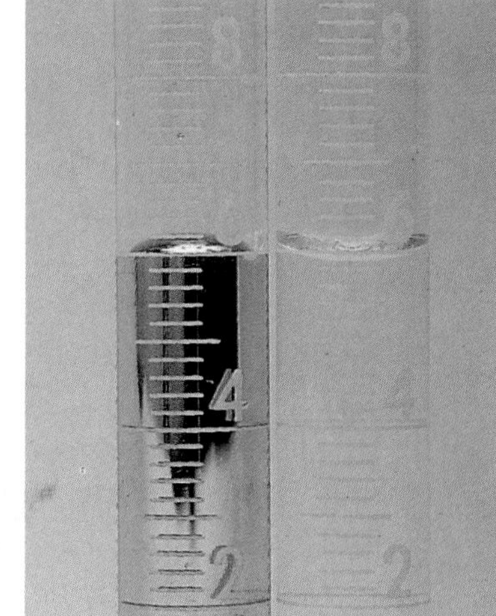

Figure 17.21 Water rises readily in a capillary tube (left). Mercury, on the other hand, is depressed in the capillary tube (center). Compare the concave meniscus of water with the convex meniscus of mercury in glass graduated cylinders (right).

17.2 *CONCEPT REVIEW*

16. Why is the partial charge on a hydrogen atom bonded to a highly electronegative element more concentrated than for any other atom bonded to that same element?

17. Explain why water expands when it freezes.

18. Water has a greater capillary rise than does ethanol under identical circumstances. Explain why this is true.

19. Apply Why do paper towels absorb water throughout when only one corner is immersed in the water?

Summary

17.1 Changes of State

1. When the temperature of a solid is raised, the particles increase in velocity until they slip over one another, and the solid melts.
2. Molecules in a liquid can gain enough energy to overcome attractive forces and break away from the liquid surface to become molecules of vapor.
3. In a gas-liquid dynamic equilibrium, the rate at which particles move from the liquid to the gas phase is equal to the rate at which particles move from the gas phase to the liquid phase. Thus, the number of particles in both the liquid and gaseous phases remains constant.
4. The vapor pressure of a substance is the pressure exerted by the gaseous phase in equilibrium with the liquid or solid phase, at a given temperature.
5. Le Chatelier's principle states that if stress is applied to a system at equilibrium, the system readjusts so that the stress is reduced.
6. Substances that have strong intermolecular forces have lower vapor pressures than substances that have weak intermolecular forces at the same temperature.
7. Ionic compounds are not volatile and do not exert a significant vapor pressure because their interionic forces are too strong.
8. At the melting/freezing point of any substance, the vapor pressure of the liquid and the vapor pressure of the solid are equal.
9. Sublimation is the change of state in which a substance passes directly from the solid to the gaseous state.
10. Evaporation is the process whereby molecules escape from the surface of a liquid or solid.
11. A liquid boils when its vapor pressure is equal to the internal pressure of the liquid. The normal boiling point of a liquid is the temperature at which the vapor pressure is equal to standard atmospheric pressure.
12. Every gas has a critical temperature (T_c) above which pressure will not liquefy the gas. The critical pressure (P_c) is the pressure that will produce liquefaction at T_c.
13. A phase diagram is a graph showing the relationship among temperature, pressure, and physical state.
14. The triple point is the temperature and pressure at which all three phases of a substance are in equilibrium.
15. Adding energy to a substance may change its kinetic energy or its potential energy. If the kinetic energy is increased, the temperature of the substance will rise. If the potential energy is raised, the substance will change its state.
16. The enthalpy of fusion is the amount of energy required to melt one gram of a solid substance at its melting point. The enthalpy of vaporization is the amount of energy required to vaporize one gram of a liquid at its boiling point.

17.2 Special Properties

17. Compounds containing hydrogen bonded to fluorine, oxygen, or nitrogen are highly polar and exhibit unusual properties. These properties occur because of the formation of hydrogen bonds.
18. Ice is less dense than liquid water, which is most dense at 3.98°C. The expansion of water as it freezes into ice occurs because hydrogen bonding pulls the molecules into an open crystalline structure that occupies more space than the liquid.
19. Unbalanced forces account for the surface tension of liquids. Capillary rise of liquids in tubes with small diameters is due to surface tension.

Chapter 17 REVIEW

Key Terms

vapor	liquefaction
equilibrium	phase diagram
dynamic equilibrium	triple point
saturated	enthalpy of fusion
reversible change	enthalpy of vaporization
melting point	hydrogen bond
sublimation	surface tension
normal boiling point	capillary rise
volatile	

Review and Practice

20. Name each of the following changes of state.
a. liquid to gas
b. solid to gas
c. gas to liquid
d. solid to liquid

21. Describe what happens when a substance melts.

22. Differentiate between liquids and solids in terms of particle spacing, arrangement, and motion.

23. How do the freezing points and melting points of pure substances compare?

24. Describe how it is possible for a molecule to leave a liquid and enter the gas phase even though the temperature of the liquid is below the boiling point.

25. What is a vapor? How does a vapor differ from what is commonly called a gas?

26. Describe equilibrium in terms of a liquid-vapor system. Why is this situation referred to as a dynamic equilibrium?

27. Describe the effect of applying pressure to ice that is near its melting point. Explain why pressure has this effect.

28. What is the relationship between vapor pressure and intermolecular forces? Explain why this relationship holds true.

29. What is the relationship between melting point and vapor pressure?

30. Describe the process of sublimation. Give an example of a substance that sublimes readily.

31. State the definition of boiling point in terms of vapor pressure.

32. How are boiling and evaporation alike? How are they different?

33. How are boiling point and condensation point related?

34. What is the difference between volatile and nonvolatile substances? Give one example each of a volatile solid, a volatile liquid, a nonvolatile solid, and a nonvolatile liquid.

35. Using the following data, plot the vapor pressure vs. the temperature for magnesium and lithium.

Vapor Pressure (kPa)	Temperature (K)	
	Mg	**Li**
0.2	891	1040
1	979	1145
2	1025	1200
5	1085	1275
10	1140	1345
20	1205	1415
50	1295	1530
101.3	1366	1620

36. Using the graph from the previous question, determine the boiling point of each of the following under the conditions listed.
a. Li at 80 kPa
b. Mg at 60 kPa
c. Mg at 30 kPa
d. Li at 70 kPa

37. What is liquefaction? What conditions are necessary for liquefaction of a gas?

38. Define critical temperature and critical pressure. Give two examples of each.

39. Explain why hydrogen has a very low critical temperature while water has a very high critical temperature.

40. From Table 17.2, determine which substances could be liquefied at room temperature (25°C) by applying pressure.

41. What is the triple point for a substance?

42. Using the phase diagram shown, determine the boiling point, melting point, triple point, and the critical temperature and pressure for substance X.

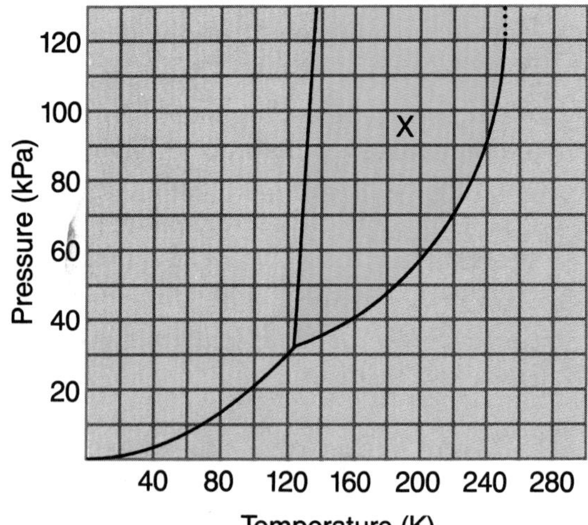

43. Using Figure 17.10, determine the state of matter that exists for hydrogen under the following conditions.
  a. 20 K and 750 kPa
  b. 7 K and 255 kPa
  c. 16 K and 1250 kPa
  d. 15 K and 250 kPa
  e. 22 K and standard pressure

44. What happens as heat is added to a solid substance below its melting point? What happens as the melting point is reached? For both questions, answer in terms of temperature and give a kinetic description as well.

45. Define enthalpy of fusion and enthalpy of vaporization for a substance.

46. Calculate the energy in joules required to melt 86.4 g of gallium at its melting point. Use information from Appendix Table A-3.

47. Calculate how much energy it takes to convert 400.0 g of ice at −25°C to steam at 275°C.

48. What is the final temperature when 750 J of energy is added to 9.0 g of ice at 0.0°C?

49. What conditions must be met in order for hydrogen bonding to take place in a collection of molecules?

50. Explain hydrogen bonding in terms of polarity.

51. Why do molecules with hydrogen bonded to nitrogen, oxygen, or fluorine have unusual properties when compared with other molecules having atoms with similar electronegativity differences?

52. Why does water decrease in density when it changes from a liquid to a solid?

53. Why do liquids exhibit surface tension?

54. Explain differences in behavior between water and mercury in a capillary tube.

55. At room temperature, what happens to a solid that has a high vapor pressure? What is this process called?

56. Describe how it is possible for a molecule to leave the surface of a solid. (How do solids evaporate?)

57. An ice cube at 0.0°C is heated until it melts and the water reaches 20.0°C. The amount of heat needed for this is 5.000×10^3 J. What is the mass of the ice cube?

58. From the following data and that of Appendix Table A-3, calculate the heat required to raise 45.0 g of cesium metal from 24.0°C to 880.0°C. The specific heat of solid Cs is 0.2421 J/g·C°; specific heat of liquid Cs is 0.2349 J/g·C°; specific heat of gaseous Cs is 0.1564 J/g·C°.

59. Bromine, Br_2, has a critical temperature of 315°C and a critical pressure of 1.03×10^4 kPa. Its triple point is at −7.3°C and 5.87 kPa. Its normal boiling point is 59.35°C, and its normal melting point is −7.25°C. Draw a phase diagram.

Concept Mastery

60. What is Le Chatelier's principle? Describe the equilibrium that exists in an inflated balloon. How does pressing on a

balloon to decrease its size demonstrate Le Chatelier's principle?

61. If an H_2O sample is at 100°C, is the sample liquid water or steam? Explain.

62. Ethyl acetate is a sweet-smelling substance often used in organic chemistry as a solvent or a reactant. When it is in an open container, its odor soon is evident at some distance. From this description, what can you infer about its volatility, boiling point, evaporation rate, critical temperature, and vapor pressure?

63. At constant pressure, how does the quantity of heat necessary to change a liquid to a gas compare with the quantity of heat that must be removed in order to condense that gas?

64. Describe what happens to the crystal lattice as ice melts.

65. Why would you expect the boiling point of HF to be higher than that of HBr?

66. Why can many substances be identified by their melting or boiling points?

67. At any given temperature, the molecules in an object can have a large range of kinetic energies. How, then, is it possible to measure the temperature of the object? What does that temperature represent?

68. If there were a 20-g sample of water in a container at 80°C, how could the water be made to boil without raising the temperature?

69. Place the following hydrogen halides in order of expected vapor pressure, lowest to highest, at room temperature.
a. HBr **b.** HI **c.** HF **d.** HCl

70. On the basis of molecular mass, where would HF be expected to fit in the list in the previous problem? However, HF does not follow the pattern expected. How can you explain this?

71. What happens to a boiling substance when the heat source is turned up? Will the temperature increase? Explain.

72. A cup contains 10 cm^3 of a liquid. Explain how each of the following factors affects its rate of evaporation.
a. temperature
b. surface area
c. intermolecular forces

73. Figure 17.10 shows the phase diagram for hydrogen. What effect will increasing the pressure have on the freezing point of hydrogen? The boiling point?

74. A flask contains a mixture of acetone and water. How would knowledge of the boiling points of the liquids enable them to be separated?

75. A well-stirred mixture of 100 grams of ice and 100 grams of water is at equilibrium in an insulated container. Fifteen grams of water at 40°C are added. Will the temperature of the water change? Explain your reasoning.

76. A student is cooling a liquid to its freezing point and below. She records the temperature every minute. The readings in Celsius degrees are 169, 151, 140, 128, 118, 113, 113, 113, 113, 109, 100, 92, and 87. Explain the pattern of these temperature readings.

77. Use a dictionary to look up the general meaning of the word *saturated*. Use this definition to characterize the vapor phase of a liquid-vapor equilibrium.

78. Why would it be unlikely that the liquid-vapor equilibrium line in a phase diagram for a substance would ever have a negative slope?

Application

79. Explain why wet laundry hung on an outside clothesline when the temperature is less than 0°C will freeze, but then dry.

80. Explain the "freezer burn" that occurs on food stored in a freezer for long periods. How can it be prevented?

81. Many boxed baking mixes give high altitude directions. Why are these directions necessary? Would you expect the high altitude baking temperature to be greater than, less than, or equal to the baking temperature at sea level?

82. Freeze drying is a process in which food products and other items are dried after being frozen. Often, this is done in a "vacuum" chamber in which a very low air pressure can be maintained. Explain why the vacuum chamber might be required.

83. People frequently run dehumidifiers in their homes in the summer and humidifiers in the winter. How can the action of these devices help keep us comfortable? Why might these devices reduce fuel bills for air conditioning and heating?

84. In a weather report, if the relative humidity is 100%, the air is saturated with water vapor. Explain this saturated system in terms of dynamic equilibrium.

85. Would the boiling point of water at each of the following locations be greater than, less than, or equal to 100°C? Why?
- **a.** top of Mt. Whitney
- **b.** on a New York City sidewalk
- **c.** Death Valley
- **d.** Miami Beach

86. The human body has its own cooling system. In terms of the kinetic energy of the molecules involved, explain why the evaporation of perspiration cools the body. If the day is warm, why are you more uncomfortable if the humidity is high?

87. Before a person donates blood, a small sample of the blood is taken for testing. This sample is usually obtained by pricking a finger to get a drop of blood. A small glass tube is touched to the drop, and the blood goes up into the tube. Explain why the blood rises in the tube.

88. Walking or driving on ice that is near the melting point is generally more dangerous than when the ice is 10°C colder. At colder temperatures, ice begins to exhibit a "dry" quality as you walk on it. Formulate an explanation for this phenomenon. Which do you think is better for skating, ice just below the melting point or ice that is 20°C colder? Give reasons for your choice.

89. In past centuries, tall structures called shot towers were used to manufacture lead shot. Kettles of molten lead were poured slowly down through the center of the tower from a platform at the top. The shot cooled as it fell. It was gathered at the bottom and graded into sizes by passing it through a series of sieves. Explain why shot towers were able to produce uniformly shaped shot.

Critical Thinking

90. Use information from Appendix Table A-3 to determine whether attractive forces are stronger in magnesium or in aluminum. On what information did you base your judgment?

91. Why might a burn from steam be more severe than a burn from boiling water?

92. Many mixtures of organic compounds can be separated by distillation. Often, the process of vacuum distillation is used. In this process, the distillation apparatus is attached to a vacuum pump, which maintains the system inside at a very low pressure. Vacuum distillation has two main applications. The first is separating mixtures of components having high boiling points. The second is the separation of compounds that decompose at high temperatures. Explain how vacuum distillation could be beneficial in each of these situations.

93. A student needs 245 grams of ice for an experiment involving heat transfer. She decides to make the ice by measuring 245 grams of liquid water into a shallow bowl placed in a freezer. A week later, she is ready to conduct her experiment. Before beginning, she decides to check the mass of the ice. To her surprise, she finds the mass to be only 239 grams. Can you explain what happened to the rest of the water and suggest a way the student could avoid this problem in the future?

94. Hot molten rock wells up at fissures deep in the oceans. Boiling has been observed at these sites. Would the boiling occur at higher, lower, or the same temperature as boiling at the ocean's surface? Give reasons for your answer. No bubbles of steam ever make it to the ocean surface to give evidence of the boiling far below. How can you account for this fact?

95. For any substance, the enthalpy of vaporization is larger than the enthalpy of fusion. Suggest an explanation for this observation.

Readings

Aubert, James H., *et al.* "Aqueous Foams." *Scientific American* 254, no. 5 (May 1986): 74-82.

Berry, R. Stephen. "When the Melting and Freezing Points are Not the Same." *Scientific American* 263, no. 2 (August 1990): 68-74.

Davis, Gode. "Self-Cooling Cans." *Popular Science* 230, no. 4 (April 1987): 53.

Morris, Daniel Luzon. "Cooking With Steam." *ChemMatters* 5, no. 1 (February 1987): 17-19.

Ring, Terry A. "Making Powders." *CHEMTECH* 18, no. 1 (January 1988): 60-64.

Rowell, Charles F. "Flash Point!" *ChemMatters* 4, no. 4 (December 1986): 10-11.

Salem, Lionel. *Marvels of the Molecule.* New York: VCH Publishers, 1987.

Cumulative Review

1. What is an exothermic reaction? Why is energy often required to start an exothermic reaction? What is this energy called?

2. Balance the equation below, then answer the following questions.

$$As + Cl_2 \rightarrow AsCl_3$$

a. What type of reaction is this?
b. How much chlorine will be needed to react with 3.44 g of arsenic?

3. How do atomic radii and ionic radii differ for (a) the alkali metals, and (b) the halogens? Explain differences between the two families.

4. Describe the nature of ionic bonds. List two pairs of elements that are likely to form ionic bonds.

5. How is it possible for a molecule to have polar bonds but not be a dipole? Give an example of a substance that fits this description.

6. What kinds of forces exist among molecules of nonpolar substances? How do these forces affect the melting points and boiling points of these substances?

7. What is a crystal? Why do we study crystals in order to learn about solids?

8. How many particles are found in a simple cubic unit cell?

9. What are the most common crystal structures found in metals?

10. How does doping affect the physical properties of silicon and germanium?

11. How do gases exert pressure?

12. Compare the distance between molecules in a gas at room temperature with the size of the molecules themselves.

13. An open manometer shows a mercury level 97.1 mm higher in the arm connected to the confined gas. The air pressure is 98.5 kPa. What is the pressure of the confined gas?

CHAPTER **Preview**

CONTENTS

CHAPTER 18

Gases

*The scuba (**s**elf-**c**ontained **u**nder-water **b**reathing **a**pparatus) allows divers to descend to depths of as much as 100 meters. How are divers able to breathe underwater?*

In the atmosphere, the pressure of the air entering the lungs is the same as the pressure on the chest. When a diver is beneath the water's surface, the water exerts an additional pressure on the chest. Below the 0.5 meter dive mark, the pressure on the chest is too great to allow breathing of atmospheric air. A scuba removes the pressure differential problem by supplying a breathing mixture at an elevated pressure. Using the pressure regulator shown here, a diver adjusts the pressure of his or her breathing gas to equal that of the surrounding water. In this chapter, you will study the behavior of gases and how various conditions change the characteristics of the gases.

EXTENSIONS AND CONNECTIONS

Careers in Chemistry

Meteorologist, page 455

Sports Connection

Ballooning and Charles's Law, page 458

Frontiers

Stirling Shiver: Engines that use helium, page 462

Chemists and Their Work

Jacques Alexandre César Charles, page 466

Everyday Chemistry

Compressed Gas and Conveniences: How refrigerators and air conditioners work, page 468

Chemistry and Society

Losing the Ozone Layer: Earth's atmosphere is in trouble, page 470

OBJECTIVES

- explain the concept of an ideal gas.
- describe the conditions of STP.
- relate the laws of Boyle, Dalton, and Charles and perform calculations using these laws.

18.1 Variable Conditions

Look at the photograph of the scuba diver again. Notice particularly the size of the tank containing the breathing mixture. It has a volume of about 30 dm^3. Normal human respiration requires roughly 0.5 dm^3 of breathing mixture for each breath. How can the tank hold enough mixture for the diver to stay down for an hour or more? The answer is that the mixture in the tank is under pressure. Why does pressure ensure a sufficient supply of mixture? This section of Chapter 18 is devoted to examining how various conditions affect gases.

Kinetic Theory of Gases

You already know many characteristics of gases because air is composed of gases. When we speak of a cubic centimeter of a solid or a hundred cubic centimeters of a liquid, we are referring to a definite amount of matter at a given temperature and pressure. Both solids and liquids expand and contract with pressure and temperature changes. However, the change is usually small enough to be ignored. This statement is not true for gases. When the temperature of a gas is raised or lowered, the change in its volume is significant. A gas has no particular volume. Instead, a given amount of gas occupies the entire volume of its container. According to the kinetic theory, a gas is made of very small particles that are in constant random motion. Gas particles of a substance are not held in a fixed position as they are in the solid form of the substance. Also, the gas particles are not held close together by van der Waals forces as they are in the liquid form of the substance. Instead, gas particles are free to spread far apart from each other.

Gas molecules are much smaller than the distances between molecules. They can be treated as **point masses,** which have no volume or diameter because they are so small and so far apart. A

Figure 18.1 A gas, such as the helium in these balloons, has a volume and exerts a pressure.

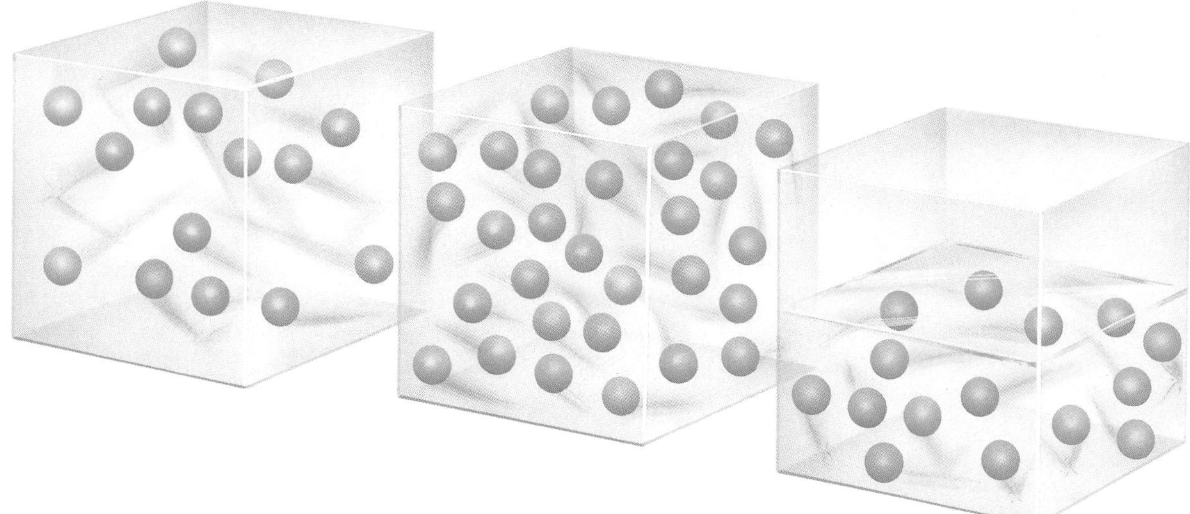

Figure 18.2 The force of the collisions between the gas particles and the sides of the container represents the pressure of the gas. The pressure in a given container (left) is greater if the number of particles increases (center) or if the volume decreases (right).

gas composed of point masses does not actually exist, but it is a useful way of thinking about gases. An **ideal gas** is composed of molecules with mass but with no volume and no mutual attraction between the molecules. In the latter part of this chapter, you will study more about real gases and how they differ in behavior from ideal gases.

The volume of a gas, the number of gas particles in that volume, the pressure of the gas, and the temperature of the gas are variables that depend on one another. Therefore, in discussing quantities of gases, it is necessary to specify not only the volume but also the pressure and the temperature. **Standard atmospheric pressure** is 101.325 kilopascals and **standard temperature** is 0°C. We indicate that a gas has been measured at standard conditions by the capital letters **STP** (standard temperature and standard atmospheric pressure). In scientific literature, gas volumes are reported as so many m^3, dm^3, or cm^3 at STP. The system is convenient, but what can we do if we must actually measure the volume of a gas in the laboratory when the pressure is 98.7 kilopascals and the temperature is 22°C? In this chapter, we will describe how a measured gas volume can be adjusted mathematically to the volume the gas will occupy at STP.

We have seen that a gas exerts pressure on the walls of its container because gas molecules collide with the walls. The pressure exerted by a gas then depends on three factors. The three factors are the number of molecules, the volume they are in, and the average kinetic energy of the molecules. A change in any of these factors changes the pressure exerted by the gas. If the number of molecules in a constant volume increases, the pressure increases. If the number of molecules and the volume remain constant but the kinetic energy of the molecules increases, the pressure increases. The kinetic energy depends on the temperature. If the temperature increases, the average kinetic energy will also increase.

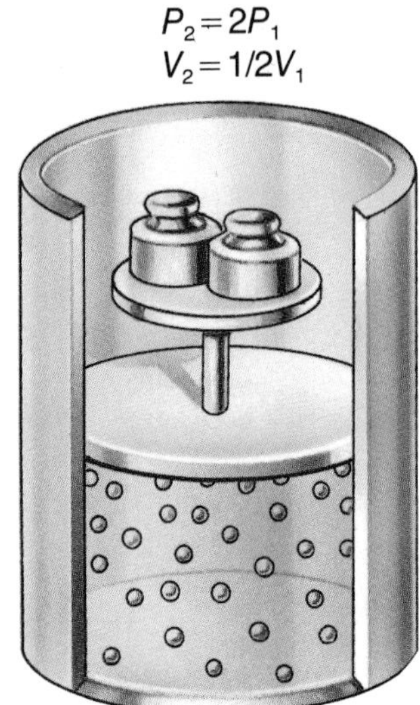

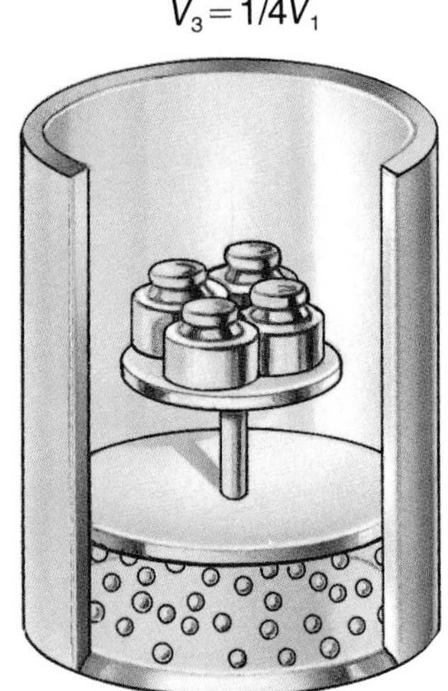

Figure 18.3 At constant temperature, an increase in pressure on a gas decreases the volume of that gas. The number of molecules in each cylinder remains constant.

Boyle's Law

Let us consider first the relationship between the pressure and volume of a gas when both the number of molecules and their average kinetic energy are constant; that is, they do not change when the pressure and volume change. Consider the container of gas with a movable piston in the top as shown at the left in Figure 18.3. If the piston is lowered until it is half the original distance from the bottom, there is only half as much space as before. The same number of molecules occupy half the volume. The molecules hit the walls of the container twice as often and with the same force per collision. So, the pressure is twice as high when the volume is half as much. When the volume is one fourth the original volume, the pressure is four times as high. We conclude that, at constant temperature, pressure increases as volume decreases. More precisely, the pressure varies inversely with volume. The product of pressure and volume is then a constant.

The British chemist Robert Boyle arrived at this principle by experiment 300 years ago. The relationship is called Boyle's law. **Boyle's law** states: *If the amount and the temperature of a gas remain constant, the pressure exerted by the gas varies inversely as the volume.* When we put the relationship into mathematical form, we obtain

$$P \propto \frac{1}{V}$$

where the symbol $\propto$ is a sign that means "varies with."

Then $$P = \frac{k}{V} \quad \text{or} \quad PV = k$$

In this equation P is the pressure and V is the volume. The k is a constant that takes into account the number of molecules and the temperature. Thus, pressure varies inversely with a change in volume.

Air Pressure vs. Gravity

Put on goggles and an apron. Pour water into a small fruit-juice glass until it overflows. Place a 3″ × 5″ index card over the mouth of the glass. Hold the card across the top of the glass while you invert the glass over a sink. Remove your hand from the card. What did you observe? Explain. Which force appears to be stronger, gravity or air pressure? What is the pressure of one standard atmosphere?

Applying Boyle's Law

If we know the volume a gas occupies at one pressure, we can use Boyle's law to calculate the volume it will occupy at a different pressure. For example, it is difficult to experiment with gases under standard conditions. Experiments are often carried out at room temperature and pressure. Since the temperature and atmospheric pressure vary from day to day, experimental results cannot be compared easily. It is desirable, therefore, to adjust all results mathematically to standard conditions. If P_1 is measured pressure and V_1 is measured volume, P_2 is standard atmospheric pressure and V_2 is volume at standard atmospheric pressure, then

$$V_1 = \frac{k}{P_1} \quad \text{and} \quad V_2 = \frac{k}{P_2}$$

In the first equation, $k = V_1P_1$. Since k is a constant, we may substitute V_1P_1 for k in the second equation:

$$V_2 = \frac{V_1P_1}{P_2} \quad \text{or} \quad V_2 = \frac{V_1 \mid P_1}{\mid P_2}$$

We have derived this relationship by using Boyle's law. Note that the original volume is simply multiplied by the ratio of the two pressures to find the new volume. In dealing with changes in pressure, there will always be possible two pressure ratios, one of which will be greater than one and the other less than one. Which ratio to use is decided by applying Boyle's law to the problem. For instance, if a gas is changing from 87.1 kPa to 101.3 kPa, the pressure is increasing. Boyle's law enables you to say immediately that the volume will decrease. The new volume after the pressure change will be less than the volume before the change. Since the new volume is less, you must multiply the old volume by a ratio whose value is less than one. The two possible ratios are 87.1/101.3 and 101.3/87.1. You would choose 87.1/101.3 because multiplying the old volume by a ratio whose value is less than one will produce a new volume smaller than the original volume.

The same process can be used to change the volume of a gas to correspond to a pressure other than standard. For instance, if we wish to change the volume of a gas measured at 16.0 kPa to the volume it would occupy at 8.8 kPa, the mathematical operations would be as follows. Correcting for a pressure change from 16.0 kPa to 8.8 kPa is equivalent to expanding the gas. The new volume would be greater, and the ratio of pressures must be greater than 1. The proper ratio is 16.0 kPa/8.8 kPa.

In a similar manner, you can compute a new pressure for a gas if you know its original volume, original pressure, and new volume. In this case, you would multiply the original pressure by a ratio of volumes. The same decision concerning which ratio to use would still apply. If the volume increased, for instance, then Boyle's law could be used to predict that the pressure decreased. If the pressure decreased, you would multiply the old pressure by

CAREERS IN CHEMISTRY

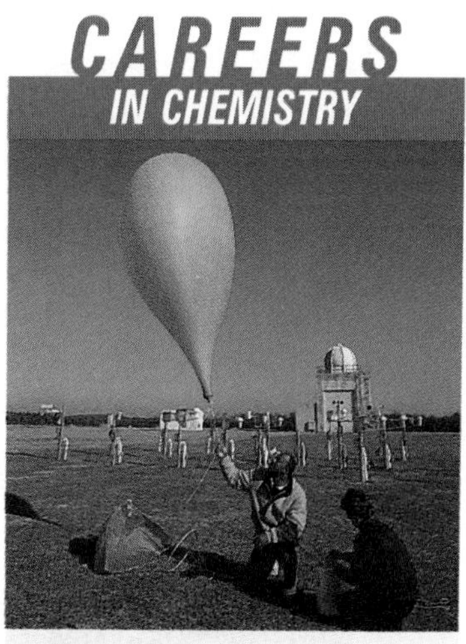

Meteorologist

Most people think of meteorologists as weather forecasters. While forecasting is certainly one of the important functions of meteorologists, their work actually covers a much broader field. Meteorologists study all aspects of Earth's atmosphere. In doing so, they also become involved in the study of the atmospheres of other components of the solar system. The advent of spacecraft that can send us pictures of atmospheres on other bodies has helped meteorologists understand our atmosphere much better. In studying atmospheric phenomena, the meteorologist must understand the chemical reactions that take place and how those changes affect weather and climate. The problems we are experiencing with the ozone layer illustrate how important chemistry is to the study of atmospheric behavior.

a ratio whose value was less than one. Do not fall into the habit of "plugging" numbers into equations. Visualize the change to be made in the volume or pressure, and then multiply by the appropriate ratio.

SAMPLE PROBLEM

Pressure Correction

A gas is collected in a 242-cm^3 container. The pressure of the gas in the container is measured and determined to be 87.6 kPa. What is the volume of this gas at standard atmospheric pressure? (Assume that the temperature remains constant.)

Solving Process:

Standard atmospheric pressure is 101.325 kPa. A change to standard atmospheric pressure would compress the gas. Therefore, the gas would occupy a smaller volume. If the volume is to decrease, then the ratio of pressures by which the original volume is to be multiplied must be less than 1. The two possible ratios by which the original volume could be multiplied are

$$\frac{101.3 \text{ kPa}}{87.6 \text{ kPa}} \quad \text{and} \quad \frac{87.6 \text{ kPa}}{101.3 \text{ kPa}}$$

The latter value is the proper one in this case because it is less than 1 and decreases the volume. The corrected volume is

$$\frac{242 \text{ cm}^3}{} \left| \frac{87.6 \not{\text{kPa}}}{101.3 \not{\text{kPa}}} \right. = 209 \text{ cm}^3$$

PROBLEM SOLVING HINT

Determine whether the gas must be expanded or compressed to STP. This change determines which ratio to use.

PRACTICE PROBLEMS

1. If 4.41 dm^3 of gas are collected at a pressure of 94.2 kPa, what volume will the same gas occupy at standard atmospheric pressure, assuming the temperature remains the same?
2. If some oxygen gas at 101 kPa and 25°C is allowed to expand from 5.0 dm^3 to 10.0 dm^3 without changing the temperature, what pressure will the oxygen gas exert?
3. Correct the following volumes of gas from the indicated pressures to standard atmospheric pressure. (Use 101.3 kPa.)
   a. 844 cm^3 at 98.5 kPa
   b. 273 cm^3 at 59.4 kPa
   c. 116 m^3 at 90.0 kPa
   d. 77.0 m^3 at 105.9 kPa
4. Make the indicated corrections in the following gas volumes. Assume constant temperature.
   a. 338 cm^3 at 86.1 kPa to 104.0 kPa
   b. 0.873 m^3 at 94.3 kPa to 102.3 kPa
   c. 31.5 cm^3 at 97.8 kPa to 82.3 kPa
   d. 524 cm^3 at 110.0 kPa to 104.5 kPa

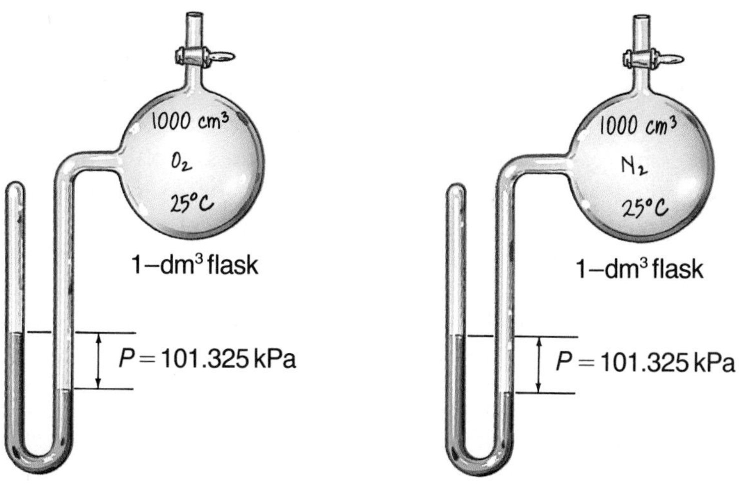

Figure 18.4 When 1000 cm^3 O_2 (left) and 1000 cm^3 N_2 (center) are mixed (right), the pressure of the mixture is the sum of the individual pressures. Note that all three flasks have the same volume and temperature.

Dalton's Law of Partial Pressure

How much pressure is exerted by a particular gas in a mixture of gases? Because more than one gas is occupying the container, each gas contributes to the total pressure. We say that each gas in the mixture exerts its partial pressure. John Dalton was the first to form a hypothesis about partial pressures. After experimenting with gases, he concluded that each gas exerts the same pressure it would if it were present alone at the same temperature. When a gas is one of a mixture, the pressure it exerts is called its partial pressure. **Dalton's law** of partial pressure states: *The total pressure in a container is the sum of the partial pressures of all the gases in the container.*

$$P_{total} = P_1 + P_2 + \ldots + P_n$$

P_1, P_2, and P_n are partial pressures. Gases in a single container are all at the same temperature and have the same volume. Therefore, the difference in their partial pressures is due only to the difference in the numbers of molecules present.

We can mix 1000 cm^3 of O_2 and 1000 cm^3 of N_2, each at room temperature and 101.325 kilopascals. The volume of the mixture is then adjusted to 1000 cm^3 with no change in temperature. The

Figure 18.5 At very high altitudes, mountain climbers must use a supplementary oxygen supply because the partial pressure of oxygen is too low for normal breathing.

pressure exerted by this mixture will be 202.650 kilopascals. However, the pressure exerted by the oxygen will still be 101.325 kilopascals (one half of the pressure). Also, the pressure exerted by the nitrogen will be 101.325 kilopascals (one half of the pressure). Air is an example of such a mixture. The air contains nitrogen, oxygen, argon, carbon dioxide, and other gases in small amounts. The total pressure of the atmosphere at standard conditions is 101.325 kilopascals. If 78% of the molecules present are nitrogen molecules, then 78% of the pressure is due to nitrogen. The partial pressure of nitrogen in the air at standard conditions is, then, 0.78×101.325 kilopascals or 79 kilopascals.

SPORTS CONNECTION

Ballooning and Charles's Law

Perhaps unknowingly, hot air balloonists make use of Charles's law when they "fill" their balloons and glide away on the breezes. A propane heater is used to heat the air in the balloon envelope. As the air is heated, the gas expands, becoming less dense. Because the density of air inside the envelope is less than the density outside the envelope, the balloon is buoyant—like a cork in water—and literally floats in the air. As the air cools, it contracts and becomes more dense and less buoyant. Thus, the balloonist can control altitude by heating or cooling the gas to increase or decrease buoyancy.

Table 18.1

Composition of Air (dry)

Gas	Percentage composition
nitrogen	78.084
oxygen	20.948
argon	0.934
carbon dioxide	0.031 5
neon	0.001 818
helium	0.000 524
krypton	0.000 114
xenon	0.000 008 7

One way to collect a sample of a gas is by bubbling it through water. This procedure is known as collecting a gas by water displacement. It is useful for collecting many gases, but the gas must be practically insoluble in water. When the collection is finished, water vapor is present in the container along with the gas collected. The pressure in the container actually is the sum of the partial pressures of the gas and the water vapor. We know that each of the gases exerts the same pressure it would if it were present alone in the container. Therefore, if we subtract the value for water vapor pressure from the total pressure, the result will be the pressure of the collected gas. The vapor pressure of water at various temperatures has been measured. We need only to consult Table 18.2 to determine the partial pressure of water.

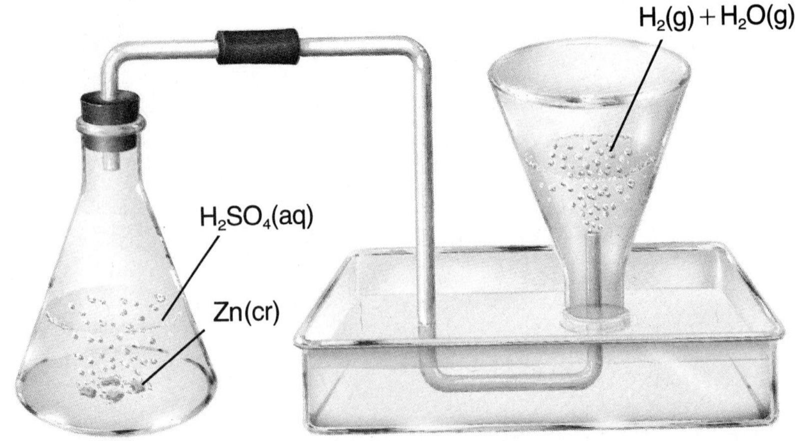

Figure 18.6 Hydrogen can be produced by the reaction of zinc with sulfuric acid. The gas is then collected over water in a setup similar to this one. The hydrogen gas collected this way also contains water vapor, $H_2O(g)$.

Table 18.2

Vapor Pressure of Water					
Temperature (°C)	Pressure (kPa)	Temperature (°C)	Pressure (kPa)	Temperature (°C)	Pressure (kPa)
0	0.6	20	2.3	30	4.2
3	0.8	21	2.5	32	4.8
5	0.9	22	2.6	35	5.6
8	1.1	23	2.8	40	7.4
10	1.2	24	3.0	50	12.3
12	1.4	25	3.2	60	19.9
14	1.6	26	3.4	70	31.2
16	1.8	27	3.6	80	47.3
18	2.1	28	3.8	90	70.1
19	2.2	29	4.0	100	101.3

SAMPLE PROBLEM

Volume of a Dry Gas

A quantity of gas is collected over water at 8°C in a 353-cm^3 vessel. The manometer indicates a pressure of 84.5 kPa. What volume would the dry gas occupy at standard atmospheric pressure and 8°C?

Solving Process:

(a) We must determine what part of the total pressure is due to water vapor. Table 18.2 indicates that at 8°C, water has a vapor pressure of 1.1 kPa. To find the pressure of the collected gas, use Dalton's law.

$$P_{gas} = P_{total} - P_{water}$$
$$= 84.5 \text{ kPa} - 1.1 \text{ kPa}$$
$$= 83.4 \text{ kPa}$$

(b) Since this pressure is less than standard, the gas would have to be compressed to change it to standard. The pressure ratio by which the volume is to be multiplied must be less than 1. The correct volume is

$$353 \text{ cm}^3 \times \frac{83.4 \cancel{\text{kPa}}}{101.3 \cancel{\text{kPa}}} = 291 \text{ cm}^3$$

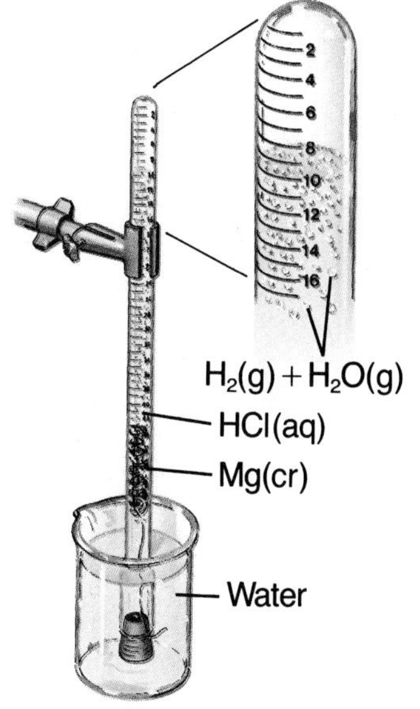

Figure 18.7 The total pressure of a gas collected over water is the sum of the pressure exerted by the gas and the pressure exerted by water vapor. In this reaction, hydrogen gas is produced by the reaction of magnesium with hydrochloric acid.

PRACTICE PROBLEMS

5. A gas is collected over water and occupies a volume of 596 cm^3 at 43°C. The atmospheric pressure is 101.1 kPa. What volume will the dry gas occupy at 43°C and standard atmospheric pressure? The vapor pressure of water at 43°C is 8.6 kPa.

6. The following gas volumes were collected over water under the indicated conditions. Correct each given volume to the

Figure 18.8 Charles's law can be demonstrated when air-filled balloons are placed in liquid nitrogen (at 77 K). The volume of the air is greatly reduced at this temperature. When the balloons are poured out of the nitrogen and warmed to room temperature, they reinflate to their original volume.

volume that the dry gas would occupy at standard atmospheric pressure and the indicated temperature. Assume that the temperature remains constant.

a. 888 cm^3 at 14°C and 93.3 kPa
b. 30.0 cm^3 at 16°C and 77.5 kPa
c. 34.0 m^3 at 18°C and 82.4 kPa
d. 384 cm^3 at 12°C and 78.3 kPa
e. 8.23 m^3 at 27°C and 87.3 kPa

Charles's Law

Jacques Charles, a French physicist, noticed a simple relationship between the volume of a gas and the temperature. He found that the volume of any gas doubled when the temperature increased from 0°C to 273°C (at constant pressure). For each Celsius degree increase in temperature, the volume of the gas increased by 1/273 of its volume at 0°C. Similarly, Charles found that a gas will decrease by 1/273 of its 0°C volume for each Celsius degree decrease in temperature. This finding would suggest that at −273°C, a gas would have no volume, or would disappear. This temperature is called absolute zero. However, all gases become liquid before they are cooled to this low temperature, and Charles's relationship does not hold for liquids or solids.

Charles's experimental information led to the formation of the absolute or Kelvin temperature scale. Thus far, we have always defined the Kelvin scale in terms of the Celsius scale. Now we are in a position to define the Kelvin scale directly. The zero point of the Kelvin scale is absolute zero. The other reference point in defining the Kelvin scale is the triple point of water, 0.01°C, which is defined as 273.16 K.

We can now describe the volume-temperature relationship that Charles found in terms of the Kelvin scale. *The volume of a quantity of gas, held at a constant pressure, varies directly with the Kelvin temperature.* This is called **Charles's law.** In calculations involving gases, the temperatures given in Celsius degrees must be converted to Kelvins where K = °C + 273.15.

Applying Charles's Law

Charles's law states that the volume of a definite amount of gas varies directly as the absolute temperature: $V = k'T$. For a volume of gas V_1, $V_1 = k'T_1$. After a temperature change with pressure constant, the new volume would be $V_2 = k'T_2$. Since $k' = V_1/T_1$, substituting gives

$$V_2 = \frac{V_1}{T_1} T_2 \quad \text{or} \quad V_2 = \frac{V_1}{} \left| \frac{T_2}{T_1} \right.$$

To correct the volume for a change in temperature, you must multiply the original volume by a ratio. For temperature changes, the ratio must be expressed in kelvins.

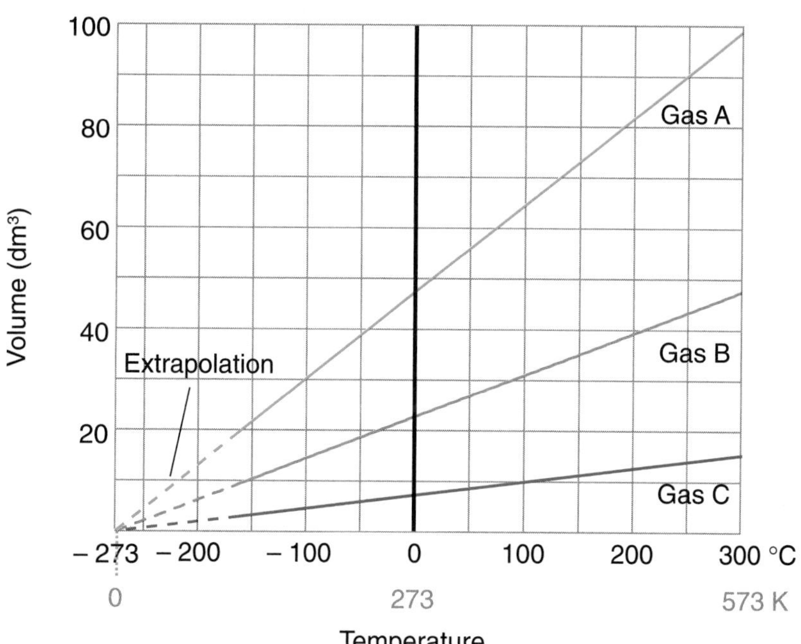

Figure 18.9 At constant pressure, the volume of a gas varies directly with temperature. Extrapolation of the volume-temperature line shows that, at absolute zero, the volume of a gas is theoretically zero. Note that the two temperature scales are comparable.

SAMPLE PROBLEM

Temperature Correction

A 225-cm^3 volume of gas is collected at 58°C. What volume would this sample of gas occupy at standard temperature? Assume a constant pressure.

Solving Process:

The temperature decreases and the pressure remains constant. Charles's law states that if the temperature of a gas decreases at constant pressure, the volume will decrease. Therefore, the original volume must be multiplied by a fraction less than 1. Convert both the initial (58°C) and the final (0°C) temperatures to the Kelvin scale (331 K and 273 K). The two possible temperature ratios are the following.

$$\frac{331\text{ K}}{273\text{ K}} \quad \text{and} \quad \frac{273\text{ K}}{331\text{ K}}$$

The correct ratio is 273 K/331 K because it is less than 1. The corrected gas volume is

$$\frac{225\text{ cm}^3}{} \left| \frac{273\ \cancel{\text{K}}}{331\ \cancel{\text{K}}} \right. = 186\text{ cm}^3$$

PRACTICE PROBLEMS

7. A gas occupies a volume of 60.0 cm^3 at 36°C. What volume will the same gas occupy at standard temperature if the pressure remains constant?

MINUTE LAB

Charles's Law in a Freezer

At home, inflate a small balloon and tie it closed. Tape one end of a piece of string to the side of the balloon at its widest point. Wrap the string around the balloon tightly but without stretching the string or deforming the balloon. Tape the string to the balloon where it meets the previously taped end. Use a marker to mark the string where the two ends meet so you can later measure the circumference. Use the marker to mark where the string lies on the balloon also. Place the balloon in a freezer and wait fifteen minutes. When you check the cold balloon, be prepared to pull the string tight and mark it where the two ends meet so you can later determine the circumference of the cold balloon. What did you observe about the balloon's volume when you opened the freezer? State a relationship between temperature and volume when the pressure remains constant. Is this a direct or an inverse mathematical relationship?

8. Correct the following volumes of gases for a change from the temperature indicated to standard temperature (pressure is constant).
   a. 617 cm^3 at 9°C
   b. 609 cm^3 at 83°C
   c. 942 cm^3 at 22°C
   d. 7.12 m^3 at 988 K
9. Correct the following volumes of gases for the temperature changes indicated (P is constant).
   a. 2.90 m^3 at 226 K to 23°C
   b. 7.91 m^3 at 52°C to 538 K
   c. 667 cm^3 at 431 K to 41°C
   d. 4.82 m^3 at 22°C to 31°C

18.1 *CONCEPT REVIEW*

10. Which of these variables are held constant in Boyle's law: number of particles, volume, temperature, pressure?
11. What are the properties of an ideal gas?
12. What is meant by STP? Why is it important when discussing quantities of gases?
13. **Apply** If a scuba diver is to remain submerged for 1 hour, what pressure must be applied to force sufficient air into the tank to be used? Assume 0.500 dm^3 of air per breath at standard atmospheric pressure, a respiration rate of 38 breaths per minute, and a tank capacity of 30 dm^3.

FRONTIERS

Stirling Shiver

On the hottest days, the coolest room in any office building will be the computer room. Computers work best with cool computer chips. The noisiest room is the one in which the air conditioners are located. Thanks to an invention from an eighteenth century Scottish clergyman, there may be a quiet change coming.

In 1816, Robert Stirling was granted a patent for an external combustion engine—the Stirling engine. In an internal combustion engine, the heat source is the rapidly-burning gasoline or diesel fuel inside a cylinder of the engine. The heat causes the entrapped gases to expand and push down on a piston. In the Stirling engine, the heat source is external to the engine. The heat is transferred through a plate to an inert gas. The heat causes the entrapped gas to expand and push down on a piston. The Stirling engine can become a cooling mechanism by reversing the process; that is, by *pulling* the piston upward. The trapped gas beneath the piston will expand and cool rapidly. The cooled gas can then be used as a refrigerant.

Researchers are presently designing cryogenic coolers based on the Stirling engine. These coolers, which use helium as the entrapped gas, have been used to reach temperatures close to −200°C. Helium is environmentally safer than the chlorofluorocarbons presently used as refrigerants.

18.2 Additional Considerations of Gases

OBJECTIVES

- solve problems involving the change of more than one condition for gases.
- explain Graham's law and solve problems using it.
- differentiate between an ideal gas and a real gas.

You now know how pressure, volume, temperature, and number of particles affect gas characteristics. However, you have dealt with these variables only two at a time, with the other variables held constant. What would happen to the volume of a gas if both pressure and temperature changed? What happens when a gas is released from its container? Thus far, you have worked only with the ideal gas. How do real gases behave? In this section, we will look into the answers to these questions.

Combined Gas Law

As you have seen, the volume of a gas is affected by both temperature and pressure. It is possible to correct a volume to a new set of conditions. For example, laboratory experiments are almost always made at temperatures and pressures other than standard. It is then necessary to correct the laboratory volumes of gases for both temperature and pressure. The correction is made by multiplying the original volume by two ratios, one for temperature and the other for pressure. Because multiplication is commutative, we can use the ratios in either order. We may think of the process as correcting the volume for a pressure change while the temperature is held constant. Then, we correct for the temperature change while the pressure is held constant. The two changes do not have any effect on each other.

SAMPLE PROBLEM

Volume Correction to STP

The volume of a gas measured at 75.6 kPa pressure and 60.0°C is to be corrected to correspond to the volume it would occupy at STP. The measured volume of the gas is 10.0 cm^3.

Solving Process:

The pressure must be increased from 75.6 kPa to 101.3 kPa. Thus, the volume must decrease, which means the pressure ratio must be less than one. The correct pressure ratio is 75.6 kPa/101.3 kPa. The temperature must be decreased from 333 K to 273 K. This change will also decrease the volume. Therefore, the correct temperature ratio is 273 K/333 K. The problem then becomes

$$\frac{10.0 \text{ cm}^3}{} \left| \frac{75.6 \text{ kPa}}{101.3 \text{ kPa}} \right| \frac{273 \text{ K}}{333 \text{ K}} = 6.12 \text{ cm}^3 \text{ at STP.}$$

PROBLEM SOLVING HINT

Remember that STP = 101.325 kPa and 0°C and that temperature must be in kelvins, where K = °C + 273.

PRACTICE PROBLEMS

14. Freon-12, dichlorodifluoromethane (CCl_2F_2), was a widely-used refrigerant. Consider a 2.23-dm^3 sample of gaseous CCl_2F_2 at a pressure of 4.85 kPa and a temperature of

−1.36°C. Calculate the volume of the gas at a pressure of 1.38 kPa and a temperature of 5.5°C.

15. Correct the volumes of the following gases as indicated.

a. 7.51 m^3 at 5°C and 59.9 kPa to STP

b. 351 cm^3 at 19°C and 82.5 kPa to 36°C and 94.5 kPa

c. 7.03 m^3 at 31°C and 111 kPa to STP

d. 955 cm^3 at 58°C and 108.0 kPa to 76°C and 123.0 kPa

e. 960.0 cm^3 at 71°C and 107.2 kPa to 13°C and 59.3 kPa

Diffusion and Graham's Law

One assumption of the kinetic theory is that gas molecules travel in straight lines. Because a molecule often collides with other molecules, its actual path is a series of straight lines connected end to end in no particular pattern. However, if the concentration of a particular substance is greater in one area of a container than in another, its molecules will gradually spread out. This random scattering of gas molecules is called **diffusion.** As gas molecules diffuse, they become more and more evenly distributed throughout their container.

Look at Figure 18.10. At the left, some bromine, a liquid at room temperature, has been placed in the bottom of the cylinder. At 25°C, bromine is quite volatile. As it evaporates, the concentration of Br_2 molecules is highest just above the liquid surface. However, as time passes, the Br_2 molecules diffuse to fill the entire cylinder.

All gases do not diffuse at the same rate. The rate of diffusion varies directly as the velocity of the molecules. At the same temperature, molecules of small mass diffuse faster than molecules of large mass because they travel faster. They also will pass through a small hole (effuse) more rapidly than molecules of high mass.

Figure 18.10 This series of photographs shows the diffusion of bromine vapor with time.

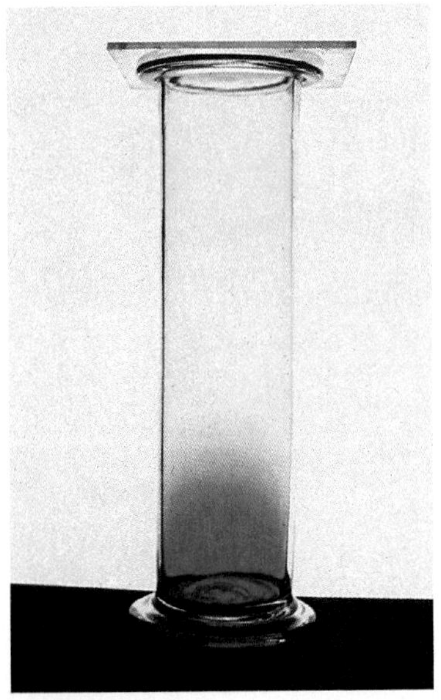

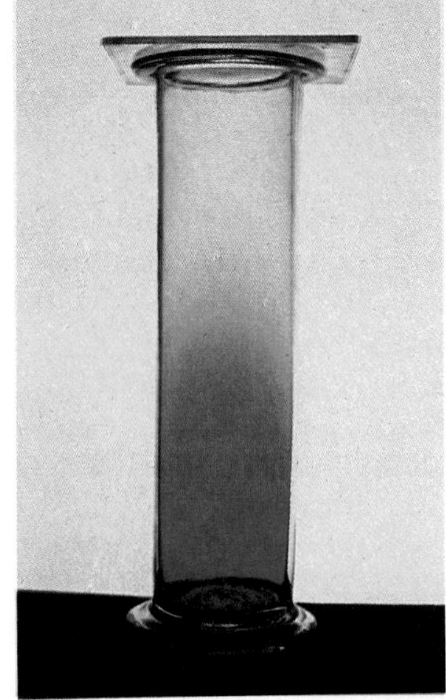

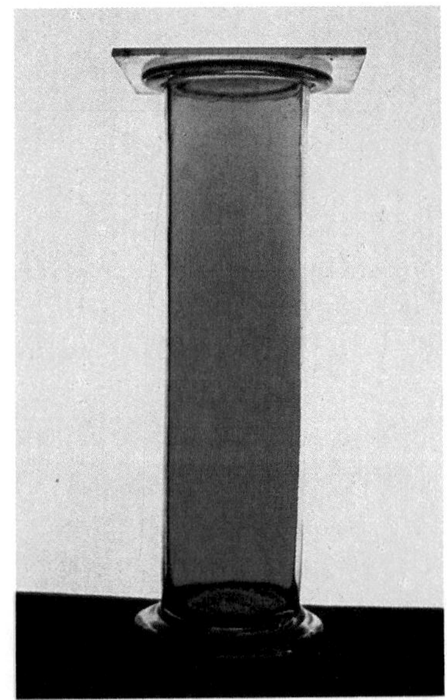

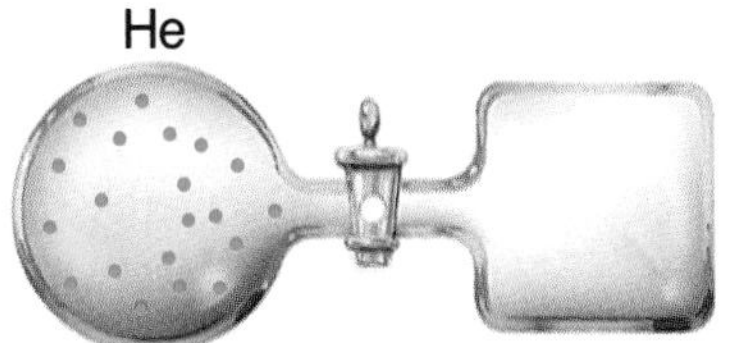

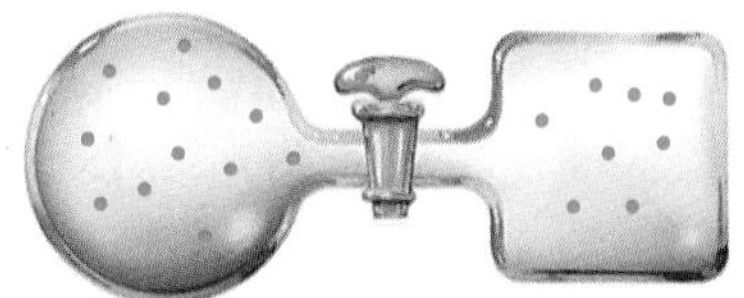

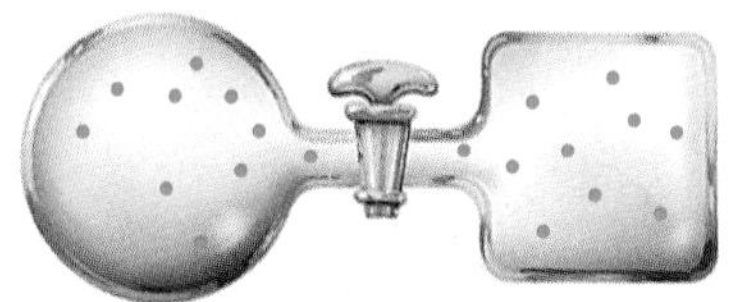

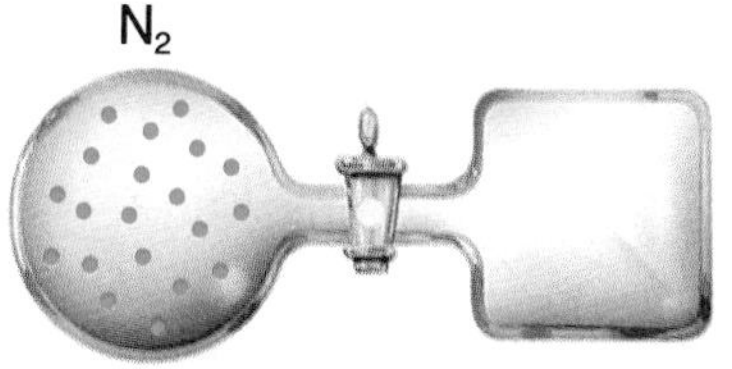

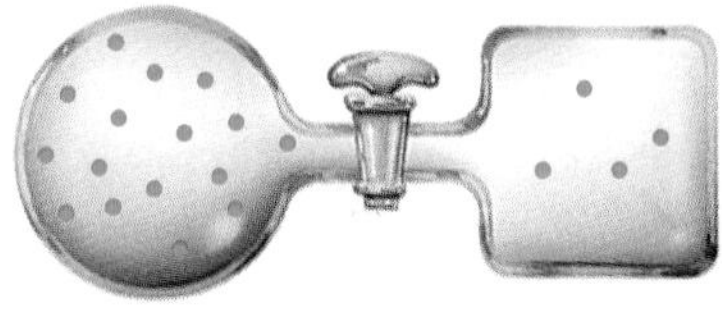

Figure 18.11 Helium (top), which is lighter, diffuses faster than nitrogen (bottom) when the stopcock is first opened. After a period of time, the gases will have diffused to an equal degree.

In Chapter 15 we found that if two substances are at the same temperature, their average kinetic energies must be the same. Thus,

$$KE_1 = KE_2 \text{ and } KE = \tfrac{1}{2}\, mv^2$$

then

$$\tfrac{1}{2}\, m_1 v_1^2 = \tfrac{1}{2}\, m_2 v_2^2$$

$$m_1 v_1^2 = m_2 v_2^2 \quad \text{or} \quad \frac{v_1^2}{v_2^2} = \frac{m_2}{m_1}$$

which is equivalent to

$$\frac{v_1}{v_2} = \sqrt{\frac{m_2}{m_1}}$$

From this equation, we see that the relative rates of diffusion of two gases vary inversely as the square roots of their molecular masses. This relationship is true only when the temperature is the same for each. This principle was first formulated by a Scottish chemist, Thomas Graham, and is known as Graham's law. **Graham's law** states: *The relative rates at which two gases under identical conditions of temperature and pressure will diffuse vary inversely as the square roots of the molecular masses of the gases.*

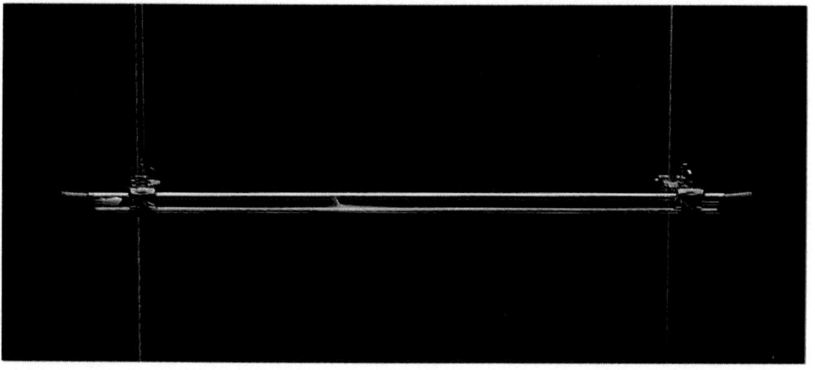

Figure 18.12 When ammonia (right) and HCl (left) are placed in cotton plugs at the ends of a long tube, a white band of ammonium chloride is formed in the tube. Because ammonia has a smaller mass, it diffuses faster than HCl and the band is formed closer to the HCl.

Figure 18.13 At a uranium processing plant, uranium hexafluoride gas is passed through a series of diffusion chambers to increase the concentration of $^{235}_{92}U$. In its natural state, uranium is about 0.7% $^{235}_{92}U$; it must be enriched to about 3% for use as fuel in a nuclear power plant.

CHEMISTS AND THEIR WORK

Jacques Alexandre César Charles (1746-1823)

Although he was a professor at the University of Paris, Charles also held a minor government post which provided an additional income. His investigation of gases was not confined to the law we know by his name. After the earliest balloon flights began, Charles realized that hydrogen would be the best gas for flying balloons. He built a balloon using hydrogen, made several flights in it himself, and managed on one flight to soar to a height exceeding one mile. In addition to his work with gases, Charles was instrumental in the spread through France of Benjamin Franklin's ideas about electricity.

During World War II, the United States undertook the development of nuclear fission weapons. One of the problems that had to be solved was the separation of uranium of mass number 235 from natural uranium, which contains mostly $^{238}_{92}U$. The method used most often was gaseous diffusion using the compound uranium hexafluoride, UF_6. In this process, isotopes can be separated by diffusion because they have different atomic masses. The difference in mass between $^{235}_{92}UF_6$ and $^{238}_{92}UF_6$ is very small. The square root of the ratio of their masses is 1.0043/1. Consequently, the gas had to be passed through thousands of stages before sufficient enrichment of $^{235}_{92}U$ was achieved.

SAMPLE PROBLEM

Gas Diffusion

Find the relative rate of diffusion for the gases krypton and bromine.

Solving Process:

First, remember that bromine is diatomic, so that its diffusing particle is actually the Br_2 molecule. The relative rates of diffusion are then given by the equation

$$\frac{v_{Kr}}{v_{Br_2}} = \sqrt{\frac{m_{Br_2}}{m_{Kr}}}$$

Substituting the formula masses for the symbols, we obtain

$$\frac{v_{Kr}}{v_{Br_2}} = \sqrt{\frac{159.8}{83.8}}$$

and the relative rate of diffusion is 1.38. Since krypton is the lighter gas, it will diffuse 1.38 times as fast as bromine.

PRACTICE PROBLEMS

16. Compute the relative rate of diffusion of helium to argon.
17. Compute the relative rate of diffusion of argon to radon.
18. What is the ratio of the velocity of hydrogen molecules to that of oxygen molecules when both gases are at the same temperature? Remember that both elements are diatomic.
19. What will be the ratio of the velocity of helium atoms to the velocity of radon atoms when both gases are at the same temperature?
20. At a certain temperature, the velocity of oxygen molecules is 0.0760 m/s. What is the velocity of helium atoms at this temperature?

▼ Deviations from Ideal Behavior

CHEMISTRY IN DEPTH

So far in this chapter, we have made two assumptions when discussing gases. The first is that gas molecules have no volume. The second is that gas molecules have no attraction for each other. These assumptions are true only for an ideal gas. In a **real gas,** the particles occupy a finite volume and have an attraction for each other. However, for many real gases at low pressure, the molecules closely approach the behavior of ideal gas molecules.

At low pressures, the molecules of both ideal and real gases are far apart. The volume occupied by the molecules is small when compared with the total gas volume. Most of the total volume is empty space. As the pressure is increased, the gas molecules are forced closer. Ideal gas molecules still remain relatively far apart because they have no volume, but real gas molecules begin to occupy a significant portion of the total volume. A further increase in pressure of a real gas does not always cause the predicted decrease in volume. If the molecules are slowed down enough, the van der Waals forces will have an effect.

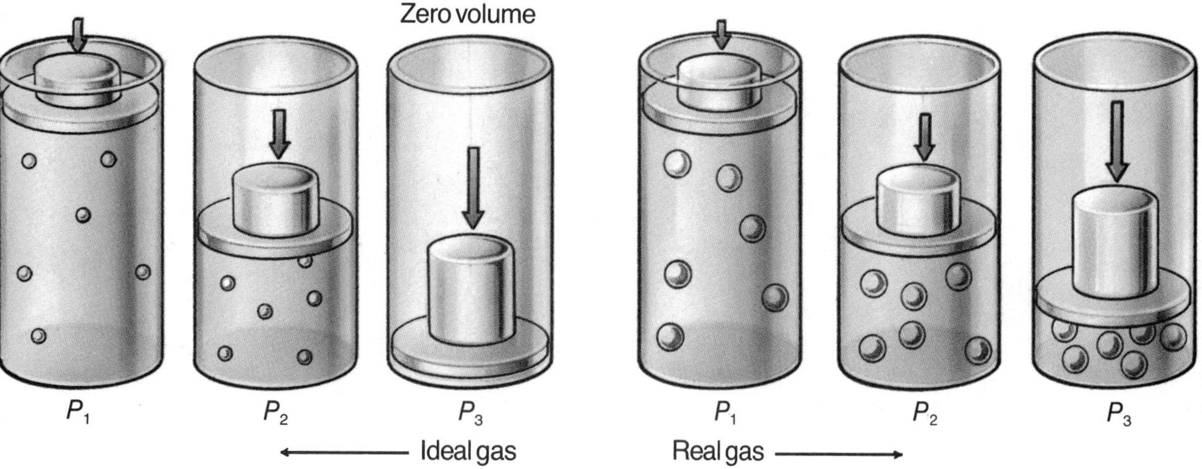

Figure 18.14 An increase in pressure increases the number of molecules per unit volume. Thus, the density of real and ideal gases increases. However, real gases liquefy rather than go to zero volume.

Table 18.3

Critical Temperature	
Gas	**Critical Temperature (K)**
He	5.19
H_2	33.24
Ne	44.44
N_2	126.26
O_2	154.77
CO_2	304.19
SO_2	430.65
H_2O	647.14

Our second assumption, that gas particles exert no attractive forces on each other, is, therefore, not true. If the gases are made up of polar molecules such as ammonia or water, the attractive forces are large and the behavior of the gases is significantly different from that predicted for ideal gases. Even between the atoms of noble gases such as argon or helium there are weak forces of attraction, the dispersion forces.

For most common gases, the ideal gas laws are accurate to 1% at normal laboratory temperatures and pressures. It will be assumed, for convenience, that these gases exhibit ideal gas properties. Generally, the lower the critical temperature of a gas, the more closely the gas obeys the ideal gas laws. Using this knowledge, we can estimate that helium more nearly approaches ideal behavior than any of the other gases listed in Table 18.3.

EVERYDAY CHEMISTRY

Compressed Gas and Conveniences

Air conditioners and refrigerators operate on the Joule-Thomson effect. The gas in a refrigerator is compressed in a pump operated by an electric motor. Compression of the gas raises its temperature. The hot gas is passed through a coil of finned tubing over which air from the room is blown. The air is moved by a fan operated by the electric motor. The air cools the compressed gas. The energy removed by the air can be felt as heat. Put your hand near the vent where the refrigerator meets the floor. When the refrigerator runs, you will feel the warm air being blown into the room. The cooled gas is then allowed to escape at low pressure through a small hole into tubing. As the gas expands rapidly, its temperature drops quickly. The cold gas flows through the tubing in the walls of the refrigerator, removing the heat from the refrigerator, and cooling the food. In an air conditioner, outside air is used to cool the gas. Room air is passed over the tubing containing the cold gas. The cooled air is then returned to the room. The energy removed from the cooled gas in an air conditioner can also be detected as heat. If a room air conditioner is running, put your hand in the path of the air exhaust and feel the warm air.

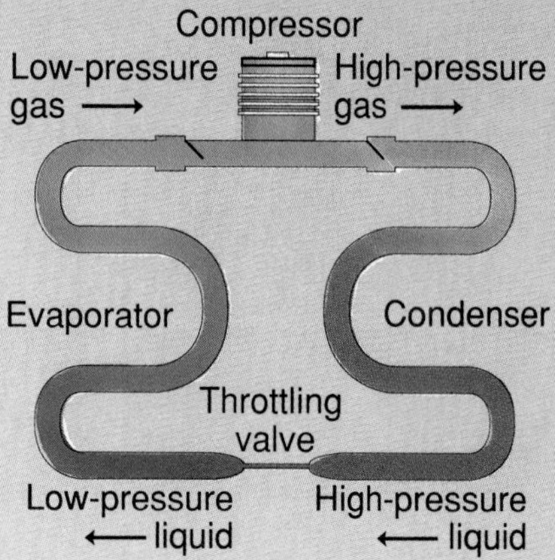

Exploring Further

1. Heat pumps can either cool air or heat it. How do they work?

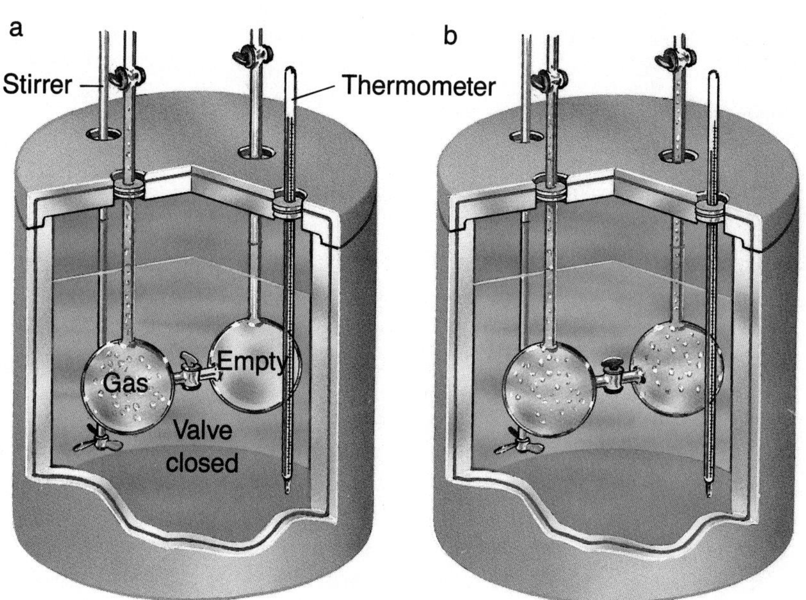

Figure 18.15 When the valve between the flasks in (a) is opened, gas rushes into the empty flask. Because the system is insulated, the temperature in (b) will be lower than the temperature in (a).

There is a property of real gases that depends upon the attractive forces existing between molecules. If a highly compressed gas is allowed to escape through a small opening, its temperature decreases. This phenomenon is known as the Joule-Thomson effect, after the two scientists who first investigated it. In order for the gas to expand, the molecules of the gas must do some work to overcome the attractive forces between them. The energy used to do this work comes from the kinetic energy of the molecules. As their kinetic energy decreases, the temperature falls. Consider the apparatus shown in Figure 18.15. The system shown is completely insulated so that no heat exchange can take place with the surroundings. Such an isolated system is known as an **adiabatic** (ay dee uh BAT ihk) **system.** The temperature of the gas in Figure 18.15b after the expansion will be lower than the temperature of the same gas in Figure 18.15a before the expansion. Refrigeration systems depend on adiabatic cooling. The Joule-Thomson effect is also observed in the operation of aerosol cans. As the product and the propellant are released through the nozzle of the can, the can and its contents become cooler.

Word Origins

a: (GK) negates what comes after it
dia: (GK) through, across
bainein: (GK) to pass; go
adiabatic – (energy) cannot pass through

18.2 CONCEPT REVIEW

21. If a sample of gas occupies 4.40 dm^3 at 19°C and 91.0 kPa, what volume will the same gas occupy at standard temperature and pressure?

22. Compute the relative rate of diffusion of the gases chlorine and xenon at the same temperature.

23. Would the weak forces between molecules of a gas tend to make real gases less compressible or more compressible than the ideal gas?

24. Apply To produce a new perfume, would you choose substances with small or large molecular masses? Why?

Losing the Ozone Layer

Recent research has revealed that, over the last two decades, the ozone layer in Earth's atmosphere has been thinning at an annual rate of as much as 2.5%. The ozone layer over the continent of Antarctica thins to about half its normal thickness during the southern hemisphere winter (September-October). The principal cause of this depletion of the ozone layer appears to be reactions involving chlorofluorocarbons (CFCs). CFCs are made of the elements carbon, chlorine, and fluorine. These compounds have been widely used as propellants in aerosol cans, as solvents, as refrigerants in air conditioners, refrigerators, and freezers, and as the foaming agent in the manufacture of plastic materials such as polyurethane foam.

The most common CFC has the formula CCl_2F_2. When struck by ultraviolet light high in the atmosphere, the molecule releases a chlorine atom.

$$CCl_2F_2 + h\nu \rightarrow CClF_2 + Cl$$

where $h\nu$ represents ultraviolet light.

The chlorine atom reacts with ozone, O_3, converting it to oxygen, O_2.

$$Cl + O_3 \rightarrow ClO + O_2$$

The following sequence of reactions then takes place.

(1) $ClO + ClO + M \rightarrow (ClO)_2 + M$
(2) $(ClO)_2 + h\nu \rightarrow Cl + ClO_2$
(3) $ClO_2 + M \rightarrow Cl + O_2 + M$
(4) $Cl + O_3 \rightarrow ClO + O_2$
(5) $Cl + O_3 \rightarrow ClO + O_2$

In these equations, M represents a third molecule that can carry off some excess energy resulting from the reaction. This molecule is usually part of an aggregate of molecules, typically an ice crystal. Ice crystals also provide a surface on which these reactions can take place. Note that this series of reactions is cyclic; that is, the product of equation (5) is the reactant of equation (1). Thus, the process continues indefinitely. If you add the five equations, the net result is the conversion of two ozone molecules to three oxygen molecules.

$$2O_3 \rightarrow 3O_2$$

The chemistry of the ozone layer is far more complex than is presented here, but scientists believe that 80% of the damage to the ozone layer is due to these reactions involving CFCs. Other halogen-containing compounds in commercial use that are dangerous to the ozone layer are $CFCl_3$, $CF_2ClCFCl_2$, CF_2ClCF_2Cl, CF_2ClCF_3, CF_3Br, CF_2BrCl, and CCl_4. Other gases suspected of being destructive are NO, NO_2, and $ClONO_2$. The United States has already banned the use of CFCs as propellants in aerosol cans. In September, 1987, the major industrial nations of the world agreed to a gradual reduction in the use of CFCs. However, the CFCs already in the atmosphere will persist for many years. Even if we were to stop all CFC use now, it has been calculated that we would be well into the 22nd century before ozone levels returned to those of the 1960s. Thus, the problem is not solved.

Analyzing the Issue

1. Find out what other gases could be used as propellants and foaming agents. What are their advantages and disadvantages?
2. Why will CFCs persist in the atmosphere even after their use is discontinued?

Chapter 18 REVIEW

Summary

18.1 Variable Conditions

1. A gas is composed of very small particles in constant random motion. The distance between particles is very large compared with the size of the particles.
2. An ideal gas is an imaginary gas that has particles with zero volume and no mutual attraction.
3. The volume of a gas depends not only on the number of particles but also on temperature and pressure.
4. Standard temperature and standard atmospheric pressure (STP) are 0°C and 101.325 kilopascals. Gas volumes are usually reported in m^3, dm^3, or cm^3 at STP.
5. Gas pressure depends on the number of molecules per unit of volume and the average kinetic energy of the molecules.
6. Boyle's law states that, at constant temperature, the volume of a gas varies inversely as its pressure ($V = k/P$).
7. Dalton's law states that the total pressure in a container is the sum of the partial pressures of the individual gases in the container.
8. Charles's law states that, at constant pressure, the volume of a gas varies directly with the absolute temperature ($V = kT$).
9. The absolute temperature scale is defined by two points. One point is absolute zero (0 K). The other point is the triple point of water (273.16 K, or 0.01°C). To convert Celsius degrees to kelvins, K = °C + 273.15.

18.2 Additional Considerations of Gases

10. When applying the combined gas laws, the corrections for temperature and pressure have no effect upon each other.
11. Diffusion is the process by which gas molecules scatter randomly to become evenly distributed throughout the entire space in which they are confined.
12. Because their particles have higher velocity, gases made up of light molecules diffuse or effuse faster than gases made up of heavier molecules at the same temperature.
13. Graham's law states that, under constant temperature and pressure, the relative diffusion and effusion rates of two gases vary inversely as the square roots of their molecular masses. Mathematically expressed, Graham's law is $v_1/v_2 = \sqrt{m_2/m_1}$.
14. The particles of real gases, unlike ideal gases, have both volume and mutual attraction. At high pressures and low temperatures, these two factors take effect.
15. At normal laboratory temperatures and pressures, most common gases behave nearly as ideal gases. The lower the critical temperature of a gas, the more nearly it behaves as an ideal gas.

Key Terms

point mass
ideal gas
standard atmospheric pressure
standard temperature
STP
Boyle's law
Dalton's law
Charles's law
diffusion
Graham's law
real gas
adiabatic system

Review and Practice

25. What are point masses?
26. Of these variables—volume, temperature, number of particles, and pressure, which are held constant in Dalton's law of partial pressure?

27. What is standard temperature in °C? What is standard atmospheric pressure in kPa? What abbreviation is used to specify these standard conditions?
28. On what factors does the pressure of a gas depend?
29. State Boyle's law. Explain in kinetic terms why Boyle's law holds true.
30. Correct the following volumes of gas from the indicated pressures to standard atmospheric pressure. Assume constant temperature.
   a. 817 cm^3 at 80.8 kPa
   b. 50.0 m^3 at 55.1 kPa
   c. 13.7 dm^3 at 87.1 kPa
   d. 641 cm^3 at 93.1 kPa
   e. 405 cm^3 at 229 kPa
31. Make the indicated corrections in the following gas volumes. Assume constant temperature.
   a. 0.600 m^3 at 110.0 kPa to 62.4 kPa
   b. 380.0 cm^3 at 66.0 kPa to 42.1 kPa
   c. 0.338 dm^3 at 102.4 kPa to 47.3 kPa
   d. 0.123 m^3 at 104.1 kPa to 117.7 kPa
32. To what volume would you have to change 85.0 dm^3 of gas at 104.4 kPa in order to decrease its pressure to 21.0 kPa? Assume no temperature change occurs.
33. A cylinder of compressed gas has a volume of 85.0 dm^3 and a pressure of 984 kPa. What volume would the gas occupy if allowed to escape into a balloon at a pressure of 125 kPa? Assume temperature does not change.
34. A cylinder contains 4.30 dm^3 of a gas at a pressure of 105.0 kPa. Keeping the temperature constant, a piston is moved in the cylinder until the volume of the cylinder is 2.8 dm^3. What is the pressure at this volume?
35. State Dalton's law of partial pressures.
36. A sample of O_2 gas is collected over water.
   a. What is meant by the vapor pressure of water?
   b. How does the vapor pressure of water affect the observed pressure of a gas collected over water?
   c. How can you determine the actual pressure of the oxygen obtained? Why must you know the temperature at which the gas was collected?
37. The following volumes of gases were collected over water under the indicated conditions. Correct each volume to the volume that the dry gas would occupy at standard pressure. Assume temperature does not change.
   a. 871 cm^3 at 12°C and 84.1 kPa
   b. 0.317 dm^3 at 26°C and 115.7 kPa
   c. 7.83 m^3 at 20°C and 107 kPa
   d. 964 cm^3 at 29°C and 111.5 kPa
38. A chemist collects 8.00 cm^3 of gas over water at STP. What volume would the dry gas occupy at STP?
39. State Charles's law.
40. Determine the volume of each of the following gases at standard temperature. Assume the pressure is constant.
   a. 5.93 cm^3 at 492 K
   b. 2.27 m^3 at 9°C
   c. 819 cm^3 at 21°C
   d. 4.67 dm^3 at 287 K
41. Correct the following volumes of gases for the temperature change indicated. Assume the pressure is constant.
   a. 2.97 m^3 at 72°C to 502 K
   b. 19.0 cm^3 at 56.0 K to 53°C
   c. 5.18 dm^3 at 76°C to 6°C
   d. 882 cm^3 at 42°C to 455 K
42. Maintaining constant pressure, the volume of a gas is increased from 15.0 dm^3 to 30.0 dm^3 by heating it. If the original temperature was 20.0°C, what is the new temperature?
43. For the following gases, correct the volumes to the conditions indicated.
   a. 654 cm^3 at 6°C and 65.3 kPa to 4°C and 108.7 kPa
   b. 2.13 m^3 at 95°C and 103 kPa to STP

c. $4.76\ dm^3$ at 6°C and 124.5 kPa to 25°C and 99.4 kPa
d. $61.4\ cm^3$ at 67°C and 96.8 kPa to 0°C and 52.2 kPa

44. At STP, the volume of a gas is $325\ cm^3$. What volume does it occupy at 20.0°C and 93.3 kPa?

45. If a helium-filled balloon has a volume of $3.40\ dm^3$ at 25.0°C and 120.0 kPa, what is its volume at STP?

46. Find the volume of a gas at STP if it measures $806\ cm^3$ at 26°C and 103.0 kPa.

47. A gas has a volume of $3.04 \times 10^{-3}\ m^3$ at 12°C and a pressure of 99.7 kPa. What pressure will cause the gas to have a volume of $3.25 \times 10^{-3}\ m^3$ at 25°C?

48. How do diffusion and effusion differ?

49. Use an example to explain Graham's law.

50. For each of the following pairs of gases, state which will diffuse more rapidly under the same conditions of temperature and pressure.
a. CO_2 and Br_2 **c.** N_2 and HCl
b. NO_2 and C_3H_8 **d.** CCl_2F_2 and SO_3

51. At room temperature, what is the ratio of the average kinetic energies of He and O_2? What is the ratio of the average velocity of a helium atom to the average velocity of an oxygen molecule?

52. Calculate the relative rates of diffusion of methane and sulfur dioxide.

53. Under what conditions do real gases behave like ideal gases?

54. What factors cause real gases to behave in a nonideal way?

55. Relate the critical temperature of a gas to how well it obeys the ideal gas law. Which gas, CO_2 or SO_2, comes closer to behaving ideally?

56. What is the Joule-Thomson effect?

57. State the definition of *adiabatic system* and give an example of one.

58. A gas of volume *V* is placed in a container. Determine the new volume if conditions are changed in each of the following ways. Express your answer in terms of *V*; for instance, $2V$, $\frac{1}{4}V$, and so on.
a. The pressure is doubled.
b. The Kelvin temperature is doubled.
c. The pressure is reduced to ¼ the original pressure.
d. The absolute temperature is reduced to ⅔ the original temperature.
e. The pressure and absolute temperature are both doubled.
f. The pressure is doubled, and the absolute temperature is halved.
g. The absolute temperature is reduced to ¾ the original absolute temperature, and the pressure is reduced to ½ the original pressure.
h. The absolute temperature is increased to 1¼ times the original absolute temperature and the pressure is decreased to ¼ times the original pressure.

59. A chemist collected $56.1\ cm^3$ of gas in an open manometer. The next day, the chemist noted that the volume had changed to $57.9\ cm^3$ and the barometer reading was 99.4 kPa. The temperature had not changed. What had been the barometer reading on the previous day when the gas was collected?

60. Assume $2.00\ dm^3$ of helium, $4.50\ dm^3$ of oxygen, and $3.00\ dm^3$ of argon at 18°C and 100.0 kPa are all pumped into a $0.500\text{-}dm^3$ container. What will be the final pressure in the container? Assume no temperature change. A solid substance that reacts with oxygen is introduced into the container and all of the oxygen reacts with it. Assume the substance takes up negligible volume. What will be the pressure inside the container?

61. Theoretically, a gas would have zero volume at absolute zero. Why would it never actually achieve this volume?

62. If the gas pressure in an aerosol can is 151.6 kPa at 25°C, what is the pressure inside the can if it is heated to 300°C?

63. A tank for compressed gas has a maximum safe pressure limit of 825 kPa. The pressure gauge reads 388 kPa when the temperature is 24°C. What is the highest temperature the tank can withstand safely?

64. Which variables—volume, temperature, number of particles, and pressure—are held constant in Charles's law?

65. A chemist collects 372 cm^3 of gas over water at 90°C and 111.0 kPa. What volume would the dry gas occupy at 2°C and 98.0 kPa?

66. At 20°C and 93.0 kPa, 30.0 cm^3 of a gas are collected over water. What volume would the dry gas occupy at STP?

Concept Mastery

67. In terms of particle spacing and freedom of motion, explain how gases differ from solids and liquids.

68. Why can gases be compressed more easily than can solids or liquids?

69. Why does the phrase "a cubic decimeter of methane" have little meaning?

70. A characteristic of gases is that they spread throughout the space available to them. Explain, in kinetic terms, how this change occurs.

71. Why do different gases at the same temperature diffuse at different rates?

72. Imagine a gas trapped in a leakproof tube having a vertically moving piston. The compressed gas just supports the piston. Additional weight is added to the piston. Analyze this situation using Boyle's law. Explain the result of the added weight in kinetic terms.

73. Hydrogen and oxygen are easily collected over water. However, it is not advisable to try to collect carbon dioxide or hydrogen chloride by this method. Can you suggest a reason why?

74. Nitrogen and an unknown gas, at equal pressure and temperature, separately effuse through an equal number of identical pinhole openings. It takes 84 seconds for one liter of the unknown gas to effuse, and 32 seconds for an equal volume of nitrogen to effuse. How does the molecular mass of the unknown gas compare with that of the nitrogen? Explain.

75. Explain Charles's law in kinetic terms.

Application

76. A "vacuum line" is not a complete vacuum but contains some gas under very low pressure. A device called a McLeod gauge measures the pressure in a vacuum line by compressing a large volume of gas from the system until it has a much smaller volume and a pressure high enough to be measured. If a 640 cm^3 sample from a line of unknown pressure were compressed to a volume of 0.048 cm^3 with a pressure of 90.0 kPa, what was the original pressure in the line? Assume constant temperature.

77. Yeast is used to make bread dough rise. The yeast ferments sugar and produces bubbles of carbon dioxide that become trapped in the dough. When the bread is baked, the yeast is killed, yet the bread continues to rise as it bakes. Explain how this change takes place.

78. A natural gas tank is constructed so that the pressure remains constant. On a cold day when the temperature was 10°C, the volume of gas in the tank was determined to be 3.04×10^3 m^3. What would the volume be on a warm day when the temperature is 35°C?

79. An internal combustion engine in a car operates by injecting fuel vapor and air into a cylinder, compressing the gas with

a piston, and igniting the pressurized gas. Suppose an engine cylinder has a volume of 420 cm^3 when the gas is injected. The gas has a temperature of 32°C and a pressure of 92.0 kPa. At the time of firing, the cylinder volume has changed to 49.4 cm^3, and the gas is at a temperature of 400.0°C. What is the pressure of the gas at that time?

80. Near the top of Mt. Everest, 8840 m above sea level, the atmospheric pressure is 34.0 kPa.
   a. At what temperature will water boil at this altitude? Use Table 18.2.
   b. How would you cook potatoes under these conditions?

81. Explain why there is a "Do not incinerate!" warning on aerosol cans.

82. Explain how using a bicycle pump to inflate a tire demonstrates Boyle's law.

Critical Thinking

83. People sometimes playfully discharge compressed carbon dioxide fire extinguishers. This action is very dangerous because the carbon dioxide can freeze skin and cause severe frostbite. Explain this danger, considering that the gas inside the fire extinguisher is at room temperature.

84. Heating, ventilating, and air conditioning systems in large buildings often use air-mixing rooms in which outdoor air is combined with indoor air before the mixture is heated or cooled. Engineers need to know the average temperature of the air in this room in order to adjust the equipment. To arrive at an average temperature, an averaging thermometer is used. This device consists of a long flexible tube filled with gas. The tube can be routed through several areas of the room. The tube is sealed at one end and the other end has a pressure sensor that is calibrated to read in °C. Using what you have learned from this chapter, explain how such a device can work. What would need to be done to the thermometer before using it?

Readings

Gushee, David E. "Global Climate Change." *CHEMTECH* 19, no. 8 (August 1989): 470-479.

Oman, Henry. "Controlling CO_2 Buildup in the Atmosphere." *CHEMTECH* 18, no. 2 (February 1988): 116-119.

Salem, Lionel. *Marvels of the Molecule.* New York: VCH Publishers, 1987.

Cumulative Review

1. What requirements must be met by the escaping molecules when a liquid at room temperature evaporates?

2. State Le Chatelier's principle. Describe a situation to which Le Chatelier's principle applies.

3. What happens in sublimation?

4. How can a gas be liquefied?

5. Describe the dynamic equilibria existing in a sealed bottle of a soft drink held at a constant temperature. There are two equilibria to be considered.

6. Relate the physical properties of graphite and diamond to their crystal structure.

7. Describe melting point and boiling point in terms of vapor pressure.

8. How many moles of $CaCl_2$ are found in 146 grams of $CaCl_2$?

9. Compare solids, liquids, and gases in terms of mean free path.

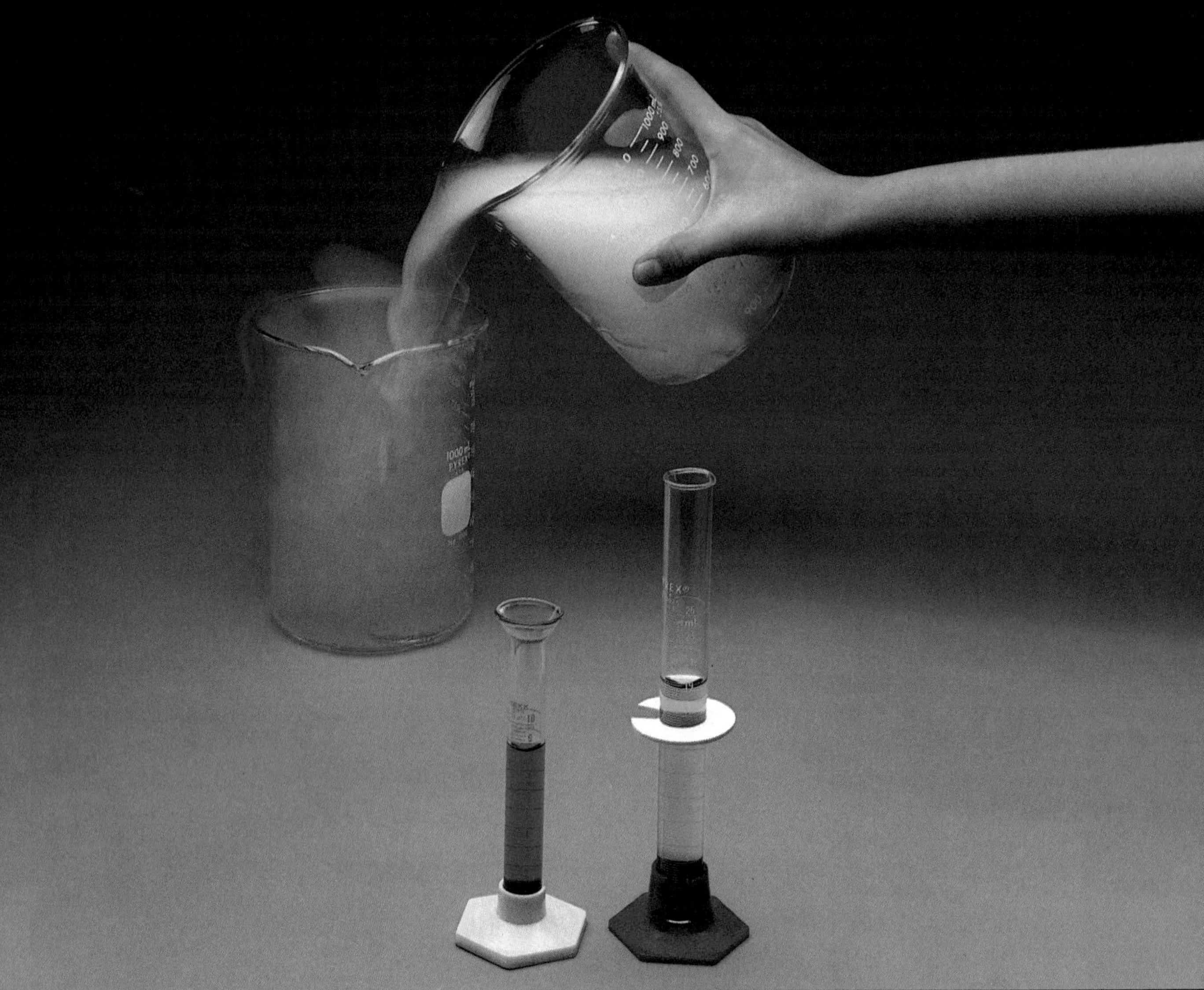

CHAPTER Preview

CONTENTS

CHAPTER 19

Gases and the Mole

Have you ever received a helium-filled balloon? The helium is less dense than air and, therefore, has the buoyancy needed to rise. Other gases are also less dense than air. Some gases, however, are more dense than air. Carbon dioxide, CO_2, is shown in the photo on the left as it is produced from dry ice. CO_2 can be poured because it is more dense than air. Although CO_2 is colorless, its path can be seen here because the cold CO_2 condenses water from the air.

How can you account for differences in gas density? In this chapter you will learn that equal volumes of gas have equal numbers of particles. You will learn about the relationship between the volume, number of particles, and mass of a gas. Finally, you will see how this relationship can be used to predict how a volume of a gas is related to the mass of another substance or volume of another gas in a chemical reaction.

EXTENSIONS AND CONNECTIONS

OBJECTIVES

- state Avogadro's principle.
- define molar volume.
- explain and use the ideal gas equation.
- compute the molecular mass of a gas from its mass, temperature, pressure, and volume.

19.1 Avogadro's Principle

In Chapter 18, we examined the effect of temperature and pressure on the volume of a constant mass of gas. Different gases, however, have molecules and atoms of different masses. In this chapter, we will look at the effect of the number of particles on the other gas variables, particularly volume. We will also find out how to obtain mass measurements of gases at various temperatures and pressures.

Molar Volume

Suppose we place two different gases at exactly the same temperature and pressure in separate containers that have exactly the same volume. At a given temperature, the average kinetic energy of all gas molecules will be the same regardless of size or mass. Molecules with greater mass will travel slower, lighter molecules will travel more rapidly. However, the average kinetic energy, $mv^2/2$, is the same for all. At equal kinetic energies, any differences in pressure are due to the presence of different numbers of molecules of each gas. At the same pressure, there is an equal number of molecules in the two containers. *At equal temperatures and equal pressures, equal volumes of gases contain the same number of molecules.* This statement is called **Avogadro's principle,** after Amadeo Avogadro.

Let n represent the number of moles of a gas, and let V represent the volume. Then, for two gases under similar conditions, Avogadro's principle states: if $V_1 = V_2$, then $n_1 = n_2$. Conversely, if the number of moles of two gases under similar conditions is equal, then their volumes are equal. Thus, we can conclude that 1 mole of any gas at STP will occupy the same volume as 1 mole of any other gas at STP. For example, 1 mole of oxygen has a mass of 32.0 g, and at STP 1000 cm^3 of oxygen has a mass of 1.43 g.

Figure 19.1 One mole of each of the three gases shown occupies 22.4 dm^3 and contains 6.02×10^{23} particles. Note, however, that the masses differ.

Therefore, a mole of oxygen at STP will occupy

$$\frac{32.0\ \cancel{g}}{1\ \text{mol}}\left|\frac{1000\ \cancel{cm^3}}{1.43\ \cancel{g}}\right|\frac{1\ dm^3}{1000\ \cancel{cm^3}} = 22.4\ dm^3/mol$$

One mole of hydrogen gas has a mass of 2.016 g, and 1000 cm^3 of hydrogen has a mass of 0.0899 g at STP. Thus, 1 mole occupies

$$\frac{2.016\ \cancel{g}}{1\ \text{mol}}\left|\frac{1000\ \cancel{cm^3}}{0.0899\ \cancel{g}}\right|\frac{1\ dm^3}{1000\ \cancel{cm^3}} = 22.4\ dm^3/mol$$

The volume occupied by 1 mole of any gas under standard conditions is 22.4 dm^3. This volume is called the **molar volume** of the gas at STP.

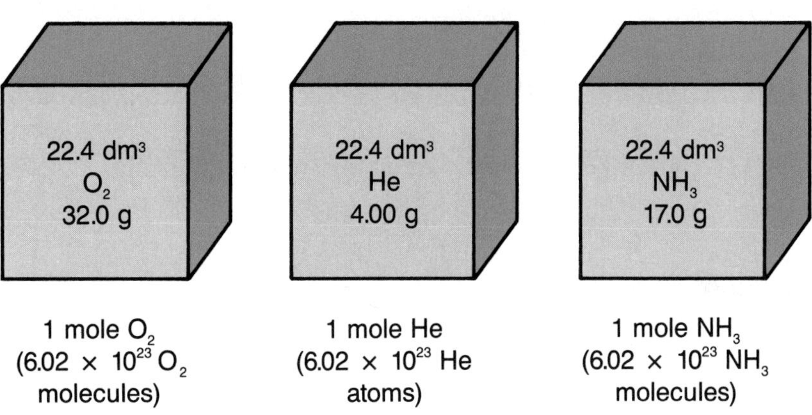

Figure 19.2 One mole of any gas occupies a volume of 22.4 dm^3 under standard conditions.

Ideal Gas Equation

We are now in a position to combine all four variables concerned with the physical characteristics of gases. These variables are pressure, temperature, volume, and number of particles. Charles's law states that volume varies directly as the absolute temperature. Boyle's law states that volume varies inversely as the pressure. These two laws may be combined as

$$V = k''\left(\frac{T}{P}\right),\ \text{or}\ PV = k''T$$

The constant, k'', depends upon the number of particles present. Therefore, it will change if we add or remove molecules. Because n represents the number of molecules measured in moles, we can write the equation using the symbols n and R to replace k''.

$$k'' = nR$$

therefore,

$$PV = nRT$$

The equation $PV = nRT$ is called the **ideal gas equation.** The value of the new constant R can be obtained by substituting into the equation a set of known values of n, P, V, and T. We know that standard atmospheric pressure P is 101.325 kPa, molar volume V is 22.4 dm^3, standard temperature T is 273 K, and the number of moles n is 1.

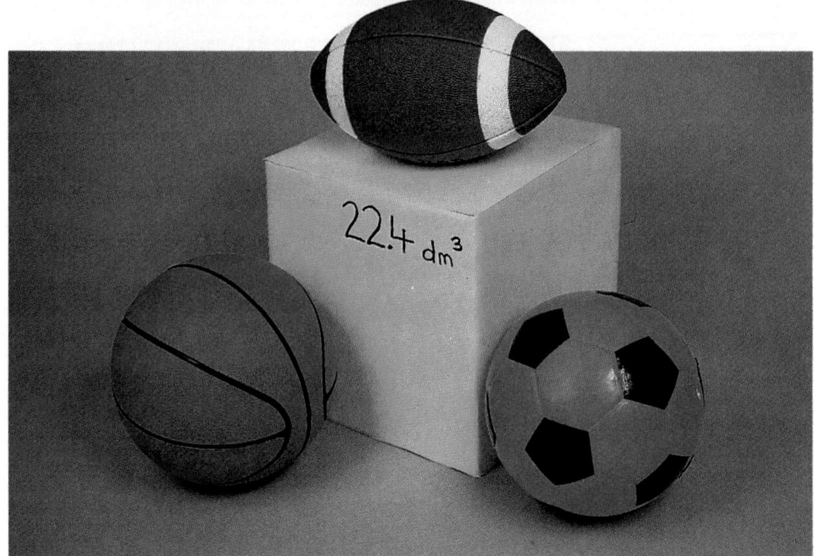

Figure 19.3 The cube has a volume of 22.4 dm^3, the volume occupied by 1 mole of an ideal gas at STP. The balls are shown for comparison.

$$(101.325 \text{ kPa})(22.4 \text{ dm}^3) = (1 \text{ mol})(R)(273 \text{ K})$$

$$R = \frac{(101.325 \text{ kPa})(22.4 \text{ dm}^3)}{(1 \text{ mol})(273 \text{ K})}$$

$$R = 8.31 \text{ dm}^3\cdot\text{kPa/(mol}\cdot\text{K)}$$

This value for R is constant and is used in calculations involving the ideal gas equation. We can determine the number of moles, n, in a quantity of a substance by dividing its mass by its formula mass.

PROBLEM SOLVING HINT

Remember that temperature in the ideal gas equation is always in kelvins (K).

SAMPLE PROBLEMS

1. Ideal Gas Equation—Finding Pressure

What pressure is exerted by 4.50 moles of gas in a 198 dm^3 container at a temperature of 8°C?

Solving Process:

The unknown variable is pressure, so the first step in solving the problem is to solve the ideal gas equation for pressure.

$$PV = nRT, \text{ so } P = nRT/V$$

Substitute the known values and calculate P.

$$P = \frac{4.50 \cancel{\text{mol}}}{198 \cancel{\text{dm}^3}} \left| \frac{8.31 \cancel{\text{dm}^3}\cdot\text{kPa}}{\cancel{\text{mol}}\cdot\cancel{\text{K}}} \right| \frac{281 \cancel{\text{K}}}{} = 53.1 \text{ kPa}$$

2. Ideal Gas Equation—Finding Volume

What volume would be occupied by 3.22 moles of gas at a temperature of 35°C and a pressure of 93.2 kPa?

Solving Process:

The unknown variable is volume so the first step in the solution is to solve the ideal gas equation for volume.

$$PV = nRT, \text{ so } V = nRT/P$$

Substitute known values and calculate V.

$$V = \frac{3.22 \cancel{\text{mol}}}{93.2 \cancel{\text{kPa}}} \left| \frac{8.31 \text{ dm}^3 \cdot \cancel{\text{kPa}}}{\cancel{\text{mol}} \cdot \cancel{\text{K}}} \right| \frac{308 \cancel{\text{K}}}{} = 88.4 \text{ dm}^3$$

3. Ideal Gas Equation—Finding Number of Moles

How many moles of gas can be contained in a 1.44 dm^3 flask at 32°C and 93.5 kPa?

Solving Process:

The unknown variable is the number of moles, so the first step in solving the problem is to solve the ideal gas equation for n.

$$PV = nRT, \text{ so } n = PV/(RT)$$

Substitute known values and calculate n.

$$n = \frac{93.5 \cancel{\text{kPa}}}{305 \cancel{\text{K}}} \left| \frac{1.44 \cancel{\text{dm}^3}}{} \right| \frac{\text{mol} \cdot \cancel{\text{K}}}{8.31 \cancel{\text{dm}^3} \cdot \cancel{\text{kPa}}} = 0.0531 \text{ mol}$$

4. Ideal Gas Equation—Finding Temperature

What temperature in °C is a gas if 2.31 mol of it occupy 61.0 dm^3 at a pressure of 94.6 kPa?

Solving Process:

The unknown variable is the temperature, so the first step in solving the problem is to solve the ideal gas equation for T.

$$PV = nRT, \text{ so } T = PV/(nR)$$

Substitute known values and calculate T.

$$T = \frac{94.6 \text{ kPa}}{2.31 \text{ mol}} \left| \frac{61.0 \text{ dm}^3}{} \right| \frac{\text{mol} \cdot \text{K}}{8.31 \text{ dm}^3 \cdot \text{kPa}} = 301 \text{ K}$$

The calculated value is in K and the problem requires an answer in °C. Therefore, 301 − 273 = 28°C.

PRACTICE PROBLEMS

1. What pressure is exerted by 0.622 mol of gas contained in a 9.22-dm^3 vessel at 16°C?
2. How many moles of gas occupy a 486-cm^3 flask at 11°C and 66.7 kPa pressure?
3. What volume is occupied by 0.684 mol of gas at 99.1 kPa and 9°C?
4. At what temperature is a gas if 0.0851 mol of it is contained in a 604-cm^3 vessel at 100.4 kPa?
5. What pressure is exerted by 0.003 06 mol of gas in a 25.9-cm^3 container at 9°C?

How to Bag a Mole

Put on an apron and goggles. **CAUTION:** *Dry ice is very cold and will damage skin. Wear heavy gloves when handling dry ice.* Place 1.75 g of crushed CO_2(cr), dry ice, into a one-quart zipper-close plastic bag. Seal the end closed. Allow the solid CO_2 to completely sublime and inflate the bag. Place a 2 dm^3 container inside a 4 dm^3 container. Completely fill the smaller container with water. Submerge the gas-filled bag in the water. Measure the volume of the water that overflows into the larger container. This water volume will approximate the volume of the gas in the bag. Determine the air temperature and pressure in your laboratory today. Use $PV = nRT$ to calculate the volume of CO_2 gas that should be produced from 1.75 g of dry ice at the laboratory temperature and pressure. Account for the difference between the two volumes.

Bridge to ENGINEERING

Heat Engines

Consider a gas in a cylinder containing a piston, as shown in the figure. If the internal pressure exceeds the external pressure (P_{ex}), the gas expands, doing work. Work is represented by w, a positive quantity, which is a symbol for work done *on* the gas. If the gas *does work*, w will be a negative quantity.

The actual amount of work done by a gas depends on the conditions under which expansion takes place. Specifically, w depends on whether the volume change occurred in an adiabatic system or in an isothermal (constant temperature) system. It also depends upon the difference between the external and internal pressures during the change.

Chemists define a reversible expansion of a gas as an ideal change. Reversible changes do not occur in practice, but they are convenient to use to find the maximum amount of work in a change. In actual practice we deal with irreversible changes that always involve less than the ideal maximum work. The most practical change is the irreversible change against a constant pressure. This change is the type we find in an engine. In this case

$$w = -P_{ex}\Delta V$$

Exploring Further

1. In the mathematical expressions for the work done in an isothermal reversible change, one form uses V_2/V_1 and another uses P_1/P_2. Why are initial and final conditions reversed in the two expressions?
2. Find out how horsepower rating is obtained for one type of heat engine.

Molecular Mass Determination

The ideal gas equation can be used to solve a variety of problems. One type, which we will illustrate here, is the calculation of the molecular mass of a gas from laboratory measurements. Such calculations are of importance to the chemist in determining the formulas and structures of unknown compounds.

The number of moles n of a substance is equal to mass m divided by the molecular mass M; $n = \frac{m}{M}$. Therefore, the ideal gas equation may be written

$$PV = \frac{mRT}{M}, \text{ or } M = \frac{mRT}{PV}$$

This modified form of the ideal gas equation may be used in many other types of problems. For example, the equation can be modified to solve for the density of a gas. Because $D = m/V$, then

$$D = \frac{m}{V} = \frac{PM}{RT}$$

SAMPLE PROBLEM

Molecular Mass from Gas Measurements

Suppose we measure the mass of the vapor of an unknown compound contained in a 273-cm^3 gas bulb. We find that the bulb contains 0.750 g of gas at 97.2 kPa pressure and 61°C. What is the molecular mass of the gas?

Solving Process:

Since the unknown quantity is the molecular mass, we will use the modified ideal gas equation

$$M = \frac{mRT}{PV}$$

Before we can substitute the known values into the ideal gas equation, °C must be converted to K. We get the following expression

$$\underset{m}{\frac{0.750\text{ g}}{\underset{P}{97.2\ \cancel{\text{kPa}}}}}\ \Bigg|\ \overset{R}{\frac{8.31\ \cancel{\text{dm}^3}\cdot\cancel{\text{kPa}}}{\text{mol}\cdot\cancel{\text{K}}}}\ \Bigg|\ \overset{T}{\frac{334\ \cancel{\text{K}}}{\underset{V}{273\ \cancel{\text{cm}^3}}}}\ \Bigg|\ \frac{1000\ \cancel{\text{cm}^3}}{\underset{\text{conversion to dm}^3}{1\ \cancel{\text{dm}^3}}}$$

$$= 78.4\text{ g/mol}$$

The result is, therefore, 78.4 g/mol. Note that all other units in the problem divide out.

PROBLEM SOLVING HINT

The units remaining at the end of the problem serve as a check on the problem setup. Answers other than g/mol indicate an error in the setup.

PRACTICE PROBLEMS

6. What is the molecular mass of a gas if 150.0 cm^3 have a mass of 0.922 g at 99°C and 107.0 kPa?
7. What is the molecular mass of a gas if 3.59 g of it occupy 4.34 dm^3 at 99.2 kPa and 31°C?
8. What is the molecular mass of a gas if 0.858 g of it occupies 150.0 cm^3 at 106.3 kPa and 2°C?
9. What is the molecular mass of a gas if 8.11 g of it occupy 2.38 dm^3 at 109.1 kPa and 10.0°C?

19.1 CONCEPT REVIEW

10. State Avogadro's principle.
11. What is meant by the term *molar volume*? What are its numerical value and units?
12. State the ideal gas equation.
13. Find the number of moles of an ideal gas contained in a 4.70-dm^3 tank at 4°C and 91.5 kPa.
14. **Apply** A chemist was asked to find the molecular mass of a gas. She massed the gas and found that 2.73 g of the gas occupied 315 cm^3 at 24°C and 94.2 kPa. What is the molecular mass of the gas?

CHEMISTRY AND
TECHNOLOGY

Gases from Air

Several of the gases comprising air have extensive industrial and commercial uses. Oxygen, nitrogen, neon, argon, krypton, and xenon are these useful gases, and they are all produced by the fractional distillation of liquid air. Before air can be liquefied, it is filtered to remove soot and other small dirt particles that would clog the equipment used later in the process. The air is then compressed to five or six times normal atmospheric pressure and cooled. Some of the water and carbon dioxide solidify during the cooling process. The remainder of the water and carbon dioxide, as well as impurities such as hydrocarbons, are removed by a special filter. The filtered air is then compressed to about 10 megapascals (MPa). As a result of being compressed, the temperature of the air rises. This energy is removed in a device called a heat exchanger. A heat exchanger consists of a series of tubes passing through a cylindrical vessel. One fluid passes through the tubes. Another cooler fluid flows through the vessel surrounding the tubes. Heat flows from the hotter fluid to the cooler fluid. The compressed air is cooled and then further compressed to about 15 MPa.

Some of the cold gas is used to run one or more compressors. In the process of doing that work, the gas uses some of its internal energy. Thus, its temperature drops even more. Most of the cold, compressed gas is allowed to expand through a valve. A large part of the gas is liquefied through the Joule-Thomson effect.

The liquid and extremely cold gases then enter a two-stage distillation column. In the column, the air is separated into high-purity nitrogen, oxygen, and other fractions. The high-purity products are sold and low-purity nitrogen gas is used as a coolant in the heat exchangers. The other liquid fraction passes to a specialized distillation unit where neon, argon, krypton, and xenon are produced. During distillation, each component is removed at its boiling point.

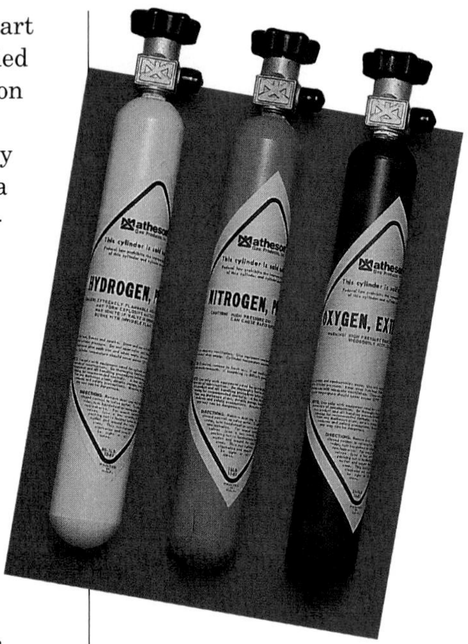

Oxygen is the most important gas obtained from liquid air, with nitrogen ranking second. The major industrial use for oxygen is in the steelmaking industry, where it is used to burn off impurities in open-hearth or basic oxygen furnaces. Oxygen is also consumed by the chemical industry in the production of artificial atmospheres, for the oxidizing of rocket fuels, and for processing wastewater. Nitrogen is used mostly for the production of ammonia. Nitrogen is also used in creating an inert atmosphere where the presence of oxygen might cause problems. It is also used as a refrigerant and for quick-freezing foods. Argon is another air-derived gas used to produce inert atmospheres. In the welding of very active metals, for instance, nitrogen may be reactive and argon is used to protect the metal during welding. Aluminum, titanium, and zirconium are often protected in this manner. Argon is also used extensively to fill light bulbs. The major use for neon, krypton, and xenon is the filling of advertising signs, although new applications are being found, especially in the field of lighting. It should be noted that helium is not recovered from the air but is obtained from natural gas wells where it makes up as much as 2% of the natural gas produced.

Thinking Critically

1. What is the purpose of the compression in the liquefying process?
2. A small amount of liquid oxygen is produced by the pressure-swing adsorption system. How does that system work?

19.2 Gas Stoichiometry

OBJECTIVES

- solve gas volume-mass, mass-gas volume, and volume-volume problems.
- identify the limiting reactant and be able to solve problems based upon it.

One of the most frequent calculations performed by chemists and chemical engineers concerns the quantitative relationships between and among reactants and products in chemical reactions. In earlier work you learned about mass-mass relationships. Working with gases, however, we frequently encounter quantities of gas expressed in volume units. In this section you will learn how to deal with problems involving gas volumes in quantitative chemical calculations.

Mass-Gas Volume Relationships

In Section 9.2, we discussed a method of finding the mass of one substance from a specific mass of another substance. It is usually awkward to measure the mass of a gas. It is easier to measure the volume under existing conditions and convert this measurement to the volume under standard conditions. One mole of gas molecules occupies 22.4 dm^3 (STP). This knowledge enables us to determine the volume of gas in a reaction by using the balanced equation for the reaction.

All mass-gas volume problems in this book can be solved in a manner similar to that shown in the following sample problem. Try to keep in mind the following four steps:

Step 1. *Write a balanced equation.*
Step 2. *Find the number of moles of the given substance.*
Step 3. *Use the ratio of the moles of the given substance to the moles of required substance to find the moles of gas.*
Step 4. *Express moles of gas in terms of volume of gas.*

Remember that 1 mole of gas occupies 22.4 dm^3 at STP.

SAMPLE PROBLEM

Mass-Gas Volume

What volume of hydrogen at STP can be produced when 6.54 g of zinc reacts with hydrochloric acid?

Solving Process:

(a) Write a balanced equation for the reaction.

$$2HCl(aq) + Zn(cr) \rightarrow H_2(g) + ZnCl_2(aq)$$

(b) Express the mass (6.54 g) of zinc in moles.

$$\frac{6.54 \text{ g Zn}}{} \Big| \frac{1 \text{ mol Zn}}{65.4 \text{ g Zn}} \cdots$$

(c) From the balanced equation, determine the mole ratio. Note that 1 mole of zinc yields 1 mole of hydrogen gas.

$$2HCl(aq) + Zn(cr) \rightarrow H_2(g) + ZnCl_2(aq)$$
$$1 \text{ mole Zn} \rightarrow 1 \text{ mole } H_2$$

Figure 19.4 The reaction of zinc metal with hydrochloric acid produces hydrogen gas.

SPORTS CONNECTION

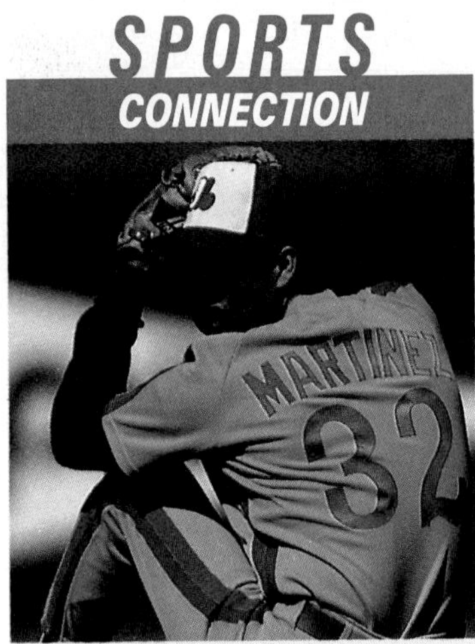

Humid Homers

Are you more likely to hit a home run in humid weather or in dry weather? By measuring the mass of 22.4 dm^3 of air, you can determine the "average molecular mass" of the mixture of gases composing air. Such a measurement will show that the average molecular mass of air is about 28. The molecular mass of water vapor is 18. Therefore, humid air is less dense than dry air. As a result, the lighter water molecules provide less resistance to the passage of a ball through the air. The difference, however, is extremely small. A well-hit ball will travel only about 2 to 3 cm farther in humid air.

Find the moles of hydrogen produced.

$$\frac{6.54 \text{ g Zn}}{} \left| \frac{1 \text{ mol Zn}}{65.4 \text{ g Zn}} \right| \frac{1 \text{ mol } H_2}{1 \text{ mol Zn}} \cdots$$

(d) Use molar volume to express the volume of hydrogen in terms of dm^3 of hydrogen.

$$\frac{6.54 \cancel{\text{ g Zn}}}{} \left| \frac{1 \cancel{\text{ mol Zn}}}{65.4 \cancel{\text{ g Zn}}} \right| \frac{1 \cancel{\text{ mol}} H_2}{1 \cancel{\text{ mol Zn}}} \left| \frac{22.4 \text{ dm}^3}{1 \cancel{\text{ mol}}} \right. = 2.24 \text{ dm}^3 \ H_2$$

We conclude that 2.24 dm^3 of hydrogen at STP are produced when 6.54 g of zinc react completely with hydrochloric acid.

PRACTICE PROBLEMS

15. An excess of hydrogen reacts with 14.0 grams of nitrogen. How many cm^3 of ammonia are produced at STP?

16. How many cm^3 of hydrogen at STP are produced from 28.0 grams of zinc reacting with an excess of sulfuric acid?

Gas Volume-Mass Relationships

It is also possible to determine the mass of a substance in a reaction when the volume of a gaseous substance is known. We use a procedure similar to that used in mass-gas volume relationships.

We begin with gas volume and find the mass of another substance. Previously, we started with the mass of a substance and found the volume of gas. However, we are still concerned with the mole relationships. The procedure in the sample problem below is as follows.

Step 1. *Write a balanced equation.*
Step 2. *Change volume of gas to moles of gas.*

Figure 19.5 If the volume of hydrogen produced by this reaction between potassium and excess water is measured, the mass of the potassium can be found.

Step 3. *Determine the ratio of moles of the given substance to moles of required substance.*

Step 4. *Express moles of required substance as grams of required substance.*

After you have worked enough problems to become familiar with these procedures, you should be able to vary your approach to suit the problem.

Figure 19.6 When powdered antimony is placed in an atmosphere of chlorine, antimony (III) chloride is produced in a vigorous reaction.

SAMPLE PROBLEM

Gas Volume-Mass

How many grams of NaCl can be produced by the reaction of 112 cm^3 of chlorine at STP with an excess of sodium?

Solving Process:

(a) Determine the balanced equation.

$$2Na(cr) + Cl_2(g) \rightarrow 2NaCl(cr)$$

(b) Express the volume of chlorine as moles of chlorine at STP.

$112\ cm^3\ Cl_2$	$1\ dm^3$	$1\ mol$	
	$1000\ cm^3$	$22.4\ dm^3$	. . .

(c) Determine the mole ratio.

$$2Na(cr) + Cl_2\,(g) \rightarrow 2NaCl(cr)$$
$$1\text{ mole } Cl_2 \rightarrow 2\text{ moles NaCl}$$

$112\ cm^3\ Cl_2$	$1\ dm^3$	$1\ mol$	$2\ mol\ NaCl$	
	$1000\ cm^3$	$22.4\ dm^3$	$1\ mol\ Cl_2$	. . .

(d) Convert moles of NaCl to grams of NaCl.

Na	$1 \times 23.0\ g = 23.0\ g$
Cl	$1 \times 35.5\ g = 35.5\ g$
molar mass of NaCl	$= 58.5\ g$

$112\ \cancel{cm^3\ Cl_2}$	$1\ \cancel{dm^3}$	$1\ \cancel{mol}$	$2\ \cancel{mol\ NaCl}$	$58.5\ g\ NaCl$
	$1000\ \cancel{cm^3}$	$22.4\ \cancel{dm^3}$	$1\ \cancel{mol\ Cl_2}$	$1\ \cancel{mol\ NaCl}$

$$= 0.585\ g\ NaCl$$

We conclude that 112 cm^3 of Cl_2, plus enough sodium to react completely with the Cl_2, will yield 0.585 g of NaCl.

PROBLEM SOLVING HINT

Remember that the coefficients in chemical equations express the number of moles of each substance.

PRACTICE PROBLEMS

17. Bromine reacts with 5.60×10^3 cm^3 of hydrogen to yield what mass of hydrogen bromide at STP?

18. How many grams of solid antimony(III) chloride can be produced from 3570 cm^3 of chlorine gas at STP reacting with an excess of antimony?

Volume-Volume Relationships

The equation for the complete burning of methane is

$$CH_4(g) + 2O_2(g) \rightarrow CO_2(g) + 2H_2O(g)$$

Notice that all reactants and all products of the reaction are gases. Gas is more easily measured by volume than by mass. Therefore, we will solve problems involving gases by using Avogadro's principle instead of converting to moles. According to Avogadro's principle, equal numbers of molecules of a gas occupy equal volumes at the same temperature and pressure. Therefore, the ratio of the combining volumes of gases is the same as the ratio of the combining moles. Thus, you can use the coefficients of the balanced equation as the ratios of the combining volumes.

SAMPLE PROBLEM

Volume-Volume

How many dm^3 of oxygen are required to burn 1.00 dm^3 of methane? (All of these substances are gases measured at the same temperature and pressure.)

Solving Process:

(a) Write a balanced equation.

$$CH_4(g) + 2O_2(g) \rightarrow CO_2(g) + 2H_2O(g)$$

(b) Determine the ratio of moles from the equation.

$$CH_4 + 2O_2 \rightarrow CO_2 + 2H_2O$$

We see that 1 mole CH_4 reacts with 2 moles O_2.

(c) This mole ratio is the same as the ratio of volumes, so one dm^3 of CH_4 reacts with 2 dm^3 of O_2.

$$\frac{1.00\ \cancel{dm^3\ CH_4}}{} \left| \frac{2\ dm^3\ O_2}{1\ \cancel{dm^3\ CH_4}} \right. = 2.00\ dm^3\ O_2$$

We conclude that 1.00 dm^3 of methane will be completely burned by 2.00 dm^3 of oxygen.

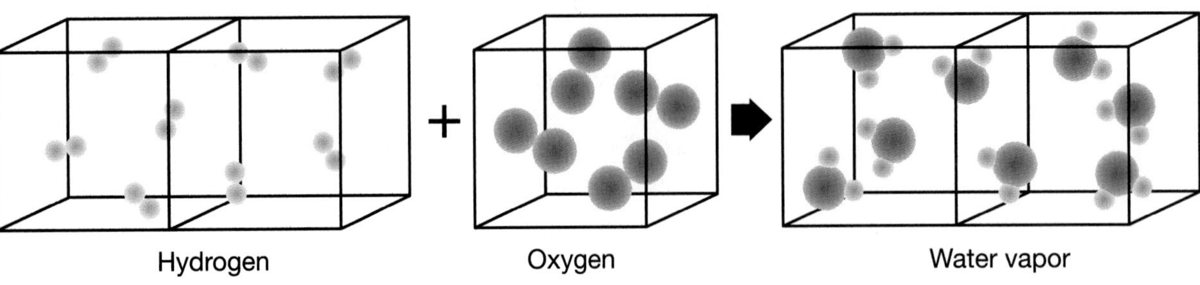

Figure 19.7 Just as two molecules of hydrogen react with one molecule of oxygen to form two molecules of water, so two volumes of hydrogen react with one volume of oxygen to produce two volumes of water vapor.

Figure 19.8 The methane in natural gas burns as it reacts with oxygen in air to produce carbon dioxide and water vapor.

PRACTICE PROBLEMS

19. What volume of oxygen is required to burn 401 cm^3 of butane, C_4H_{10}, completely? (All substances are gases measured at the same temperature and pressure.)

20. What volume of bromine gas is produced if 75.2 dm^3 of Cl_2 react with excess HBr? (All substances are gases measured at the same temperature and pressure.)

$$Cl_2(g) + 2HBr(g) \rightarrow Br_2(g) + 2HCl(g)$$

21. Heptane, C_7H_{16}, is a typical component of gasoline. When gasoline is injected into the cylinder of an automobile engine, it is vaporized. What volume of oxygen is required to burn 917 cm^3 of heptane vapor when both gases are at the same temperature and pressure? What volume of air (20.9% O_2) will supply the necessary oxygen?

22. What volume of O_2 is required to oxidize 499 cm^3 of NO to NO_2? (Assume STP.)

23. What volume of hydrogen is produced when 941 m^3 of C_6H_6 are produced? (All substances are gases measured at the same temperature and pressure.)

$$C_6H_{14}(g) \rightarrow C_6H_6(g) + 4H_2(g)$$

Limiting Reactants

Suppose 4.00 dm^3 of hydrogen and 1.00 dm^3 of oxygen are placed in a container and ignited by means of a spark. An explosion occurs and water is formed.

$$2H_2(g) + O_2(g) \rightarrow 2H_2O(g)$$

We know that two volumes of hydrogen are all that can combine with one volume of oxygen, so 2.00 dm^3 of hydrogen are left

Figure 19.9 Limiting factors are common in everyday life. For example, when the student runs out of either dough or candies, no more cookies of this type can be made.

unreacted. To take another example, suppose we drop nine moles of sodium into a vessel containing four moles of chlorine. If we warm the container slightly, the sodium will burn with a bright yellow flame, and crystals of sodium chloride will be formed.

$$2Na(cr) + Cl_2(g) \rightarrow 2NaCl(cr)$$

We know that one mole of Cl_2 will react completely with two moles of Na, so one mole of sodium will remain unreacted. For these two reactions we say that the hydrogen and sodium are in "excess" and that the oxygen and chlorine are "limiting reactants." In a chemical reaction, the **limiting reactant** is the one that is completely consumed in the reaction. It is not present in sufficient quantity to react with all of any other reactant. The reactants that are left are said to be in excess and are called **excess reactants.** The amount of product is therefore determined by the limiting reactant.

Figure 19.10 In this reaction, 8 molecules of H_2 will react with 4 molecules of O_2 to produce 8 molecules of H_2O. Four oxygen molecules are left unreacted. The hydrogen is the limiting reactant; oxygen is in excess.

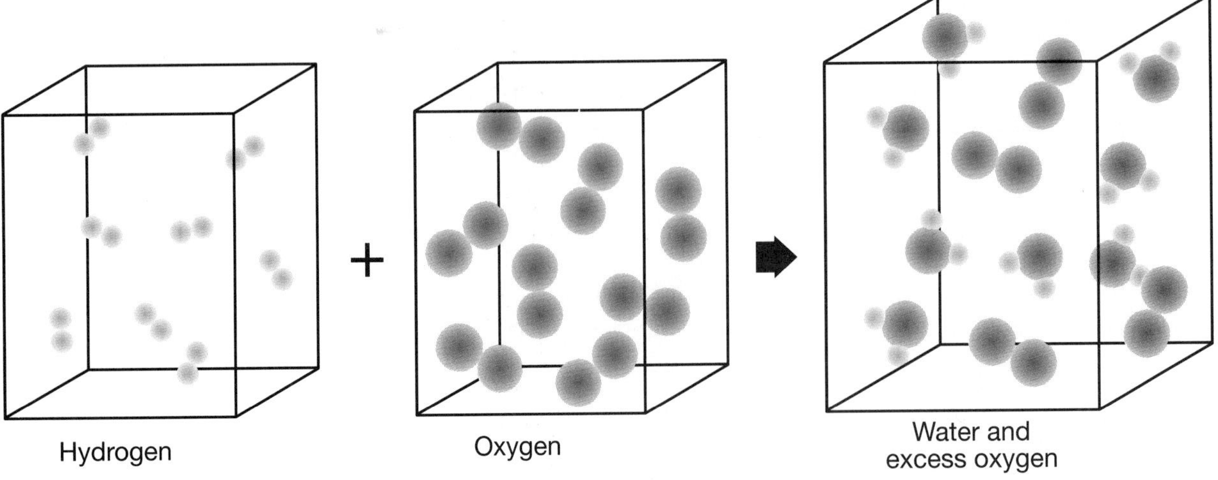

SAMPLE PROBLEMS

1. Limiting Reactants

How many grams of CO_2 are formed if 10.0 g of carbon are burned in 20.0 dm^3 of oxygen? (Assume STP.)

Solving Process:

(a) Write a balanced equation.

$$C(amor) + O_2(g) \rightarrow CO_2(g)$$

(b) Change both quantities to moles.

$$\frac{10.0\ \cancel{g\ C}}{} \left| \frac{1\ mol\ C}{12.0\ \cancel{g\ C}} \right. = 0.833\ mol\ C$$

$$\frac{20.0\ \cancel{dm^3}\ O_2}{} \left| \frac{1\ mol}{22.4\ \cancel{dm^3}} \right. = 0.893\ mol\ O_2$$

(c) Compare the amount of CO_2 produced by complete reaction of each of the reactants. The equation indicates that

$$1\ mole\ C + 1\ mole\ O_2 \rightarrow 1\ mole\ CO_2$$

Therefore, 0.833 mol C would produce 0.833 mol CO_2, and 0.893 mol O_2 would produce 0.893 mol CO_2. Because carbon would produce fewer moles of carbon dioxide, the carbon limits the reaction. Some oxygen (0.060 mole) is left unreacted. We call carbon the limiting reactant.

(d) Complete the problem on the basis of the limiting reactant.

$$\frac{0.833\ \cancel{mol\ C}}{} \left| \frac{1\ \cancel{mol\ CO_2}}{1\ \cancel{mol\ C}} \right| \frac{44.0\ g\ CO_2}{1\ \cancel{mol\ CO_2}} = 36.7\ g\ CO_2$$

We conclude that 10.0 g of carbon will react with excess O_2 to form 36.7 g of CO_2.

Figure 19.11 A fire is a chemical reaction between oxygen and combustible substances in fuel. In a forest fire, a firebreak makes the fuel the limiting reactant. When a campfire is covered with dirt, oxygen becomes the limiting reactant.

CHEMISTS AND THEIR WORK

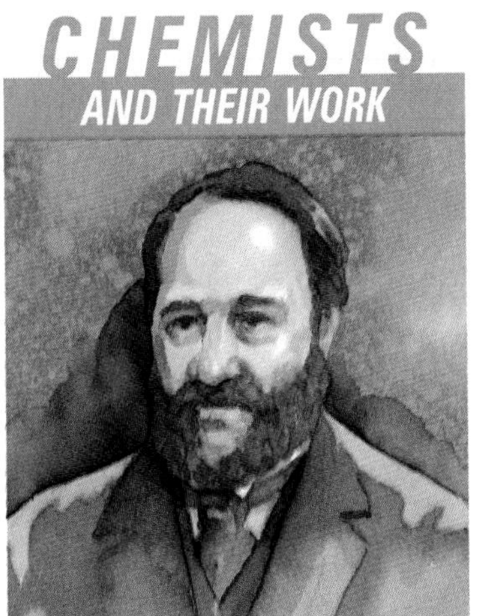

James Prescott Joule (1818-1889)

James Prescott Joule was an ardent measurer of natural phenomena, and most of his measurements had to do with heat. Even as a teenager, he published papers on the heat generated by electric motors. Eventually, he was able to produce specific relationships between an electric current in a circuit or mechanical work performed and the heat generated. His work was not immediately accepted because he was not in an academic position and was almost entirely self-educated. It was largely through the effort of William Thomson (Lord Kelvin) that Joule's work finally received the recognition it deserved. The unit of energy (and work) was named for him. His work led directly to the formulation of the law of conservation of energy. Together with Lord Kelvin, Joule formulated the Joule-Thomson effect. In addition, he made other scientific discoveries and eventually was to receive many honors from the scientific community.

SAMPLE PROBLEMS (continued)

2. Limiting Reactants—Another Example

How many grams of aluminum sulfide can form from the reaction of 9.00 g of aluminum with 8.00 g of sulfur?

Solving Process:

(a) $$2Al(cr) + 3S(l) \rightarrow Al_2S_3(cr)$$

(b) $$\frac{9.00 \text{ g Al}}{} \left| \frac{1 \text{ mol Al}}{27.0 \text{ g Al}} \right. = 0.333 \text{ mol Al}$$

$$\frac{8.00 \text{ g S}}{} \left| \frac{1 \text{ mol S}}{32.1 \text{ g S}} \right. = 0.249 \text{ mol S}$$

(c) 2 moles Al yield 1 mole Al_2S_3.

$$\frac{0.333 \text{ mol Al}}{} \left| \frac{1 \text{ mol } Al_2S_3}{2 \text{ mol Al}} \right. = 0.167 \text{ mol } Al_2S_3$$

3 moles S yield 1 mole Al_2S_3

$$\frac{0.249 \text{ mol S}}{} \left| \frac{1 \text{ mol } Al_2S_3}{3 \text{ mol S}} \right. = 0.0830 \text{ mol } Al_2S_3$$

(d) We know that 0.0830 mole of Al_2S_3 is less than 0.167 mole of Al_2S_3, so sulfur is the limiting reactant. Complete the problem on the basis of the limiting reactant.

$$\frac{0.0830 \text{ mol } Al_2S_3}{} \left| \frac{150.0 \text{ g } Al_2S_3}{1 \text{ mol } Al_2S_3} \right. = 12.5 \text{ g } Al_2S_3$$

Thus, using sulfur as the limiting reactant, 0.0830 times the formula mass of aluminum sulfide will give the mass of 12.5 g of aluminum sulfide produced from 8.00 g sulfur and an excess of aluminum. An alternate method for determining the limiting reactant is as follows.

3. Limiting Reactants—An Alternate Method

How many grams of aluminum sulfide can form from the reaction of 9.00 g of aluminum with 8.00 g of sulfur?

Solving Process:

(a) $$2Al(cr) + 3S(l) \rightarrow Al_2S_3(cr)$$

(b) According to the equation, 2 moles Al require 3 moles S. Thus, 9.00 g of Al, converted to moles, require

$$\frac{9.00 \text{ g Al}}{} \left| \frac{1 \text{ mol Al}}{27.0 \text{ g Al}} \right| \frac{3 \text{ mol S}}{2 \text{ mol Al}} = 0.500 \text{ mol S}$$

Since we have only

$$\frac{8.00 \text{ g S}}{} \left| \frac{1 \text{ mol S}}{32.1 \text{ g S}} \right. = 0.249 \text{ mol S}$$

sulfur is the limiting reactant. If we had more than 0.500 mole of sulfur, aluminum would have been the limiting reactant.

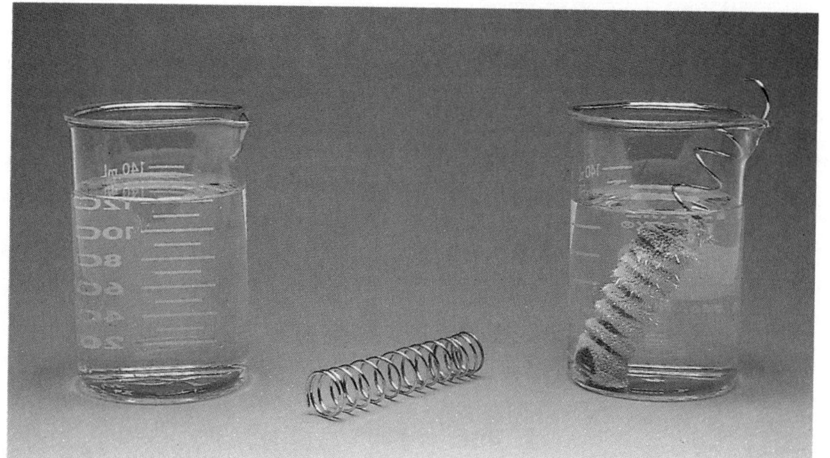

Figure 19.12 This copper wire has been placed in a silver nitrate solution. After the reaction has stopped, some copper wire remains. Silver nitrate was the limiting reactant.

PRACTICE PROBLEMS

24. $2NaBr(aq) + 2H_2SO_4(aq) + MnO_2(cr) \rightarrow Br_2(l) + MnSO_4(aq) + 2H_2O(l) + Na_2SO_4(aq)$
What mass of bromine could be produced from 2.10 g of NaBr and 9.42 g of H_2SO_4?

25. $2Ca_3(PO_4)_2(cr) + 6SiO_2(cr) + 10C(amor) \rightarrow P_4(cr) + 6CaSiO_3(cr) + 10CO(g)$
What volume, in cubic decimeters, of carbon monoxide gas at STP is produced from 4.14 g $Ca_3(PO_4)_2$ and 1.20 g SiO_2?

26. $H_2S(g) + I_2(aq) \rightarrow 2HI(aq) + S(cr)$
What mass of sulfur is produced by 4.11 g of I_2 and 317 cm^3 of H_2S at STP?

27. $2Cl_2(g) + HgO(cr) \rightarrow HgCl_2(cr) + Cl_2O(g)$
What volume, in cubic centimeters, of Cl_2O gas can be produced from 116 cm^3 of Cl_2 gas at STP and 7.62 g of solid HgO?

19.2 *CONCEPT REVIEW*

28. $2K_2MnO_4(aq) + Cl_2(g) \rightarrow 2KMnO_4(aq) + 2KCl(aq)$
What volume of chlorine gas at STP is required to completely react with 4.80 g of K_2MnO_4?

29. $Ca(cr) + 2HCl(aq) \rightarrow CaCl_2(aq) + H_2(g)$
What mass of calcium is required to produce 8.32 dm^3 of hydrogen by this reaction?

30. $H_2(g) + Cl_2(g) \rightarrow 2HCl(g)$
What volume of hydrogen chloride can be produced from 1.60 dm^3 of hydrogen?

31. $Cr(OH)_3(cr) + NaOH(aq) \rightarrow NaCrO_2(aq) + 2H_2O(l)$
What mass of $NaCrO_2$ can be obtained from the reaction of 7.40 g $Cr(OH)_3$ with 7.60 g NaOH?

32. Apply Why do we not have to convert to moles in the solution of a volume-volume problem?

Chapter 19
REVIEW

Summary

19.1 Avogadro's Principle

1. According to Avogadro's principle, at equal temperatures and pressures, equal volumes of gases contain equal numbers of molecules.
2. The molar volume of a gas is the volume occupied by 1 mole of the gas. At STP, molar volume of a gas is 22.4 dm^3.
3. The ideal gas equation is $PV = nRT$, where P is pressure, V is volume, n is the number of moles of gas, R is a constant, and T is the kelvin temperature.
4. The molecular mass of a gas may be determined by using a modified form of the ideal gas equation.

19.2 Gas Stoichiometry

5. The relationship between moles of a gas and its volume at STP allows us to relate volumes of gaseous reactants or products to quantities of other reactants or products in chemical reactions.
6. A limiting reactant is a reactant that is completely consumed in a reaction and thereby limits the amount of product. Any remaining reactants are said to be in excess.

Key Terms

Avogadro's principle
molar volume
ideal gas equation
limiting reactant
excess reactant

Review and Practice

33. State Avogadro's principle.
34. Define molar volume.
35. What is the molar volume of a gas at STP?
36. Calculate the number of moles of gas at STP in each of the following volumes.
    a. 168 dm^3
    b. 0.422 dm^3
    c. 3.50×10^3 cm^3
    d. 0.48 cm^3
    e. 1.55 m^3
    f. 0.000 089 m^3
37. Calculate the volume occupied at STP by each of the following quantities of gas.
    a. 3.00 mol of H_2
    b. 0.005 45 mol of methane
    c. 2.99×10^{23} molecules of chlorine
38. State the ideal gas equation and give the meaning of each symbol.
39. What pressure is exerted by 0.400 mol of gas in a 10.0-dm^3 vessel at 27°C?
40. How many moles of methane are in a cylinder having a volume of 104 m^3 if the pressure is 389 kPa at 25°C?
41. An engineer wishes to design a container that will hold 12.0 mol of ethane at a pressure no greater than 5.00×10^2 kPa and a temperature of 52°C. What is the minimum volume the container can have?
42. At what temperature will a 10.9-dm^3 container holding 0.877 mol of hydrogen sulfide have a pressure of 2.00×10^2 kPa?
43. What is the molecular mass of a gas if a volume of 5.00×10^2 cm^3 has a mass of 1.00 g at −23°C and 105.0 kPa?
44. Calculate the molecular mass of a gas if 0.132 g occupies a volume of 41.5 cm^3 at 24.9°C and 98.4 kPa.
45. How many grams of sodium chloride could be produced from the reaction of 23.0 g of sodium with 22.4 dm^3 of chlorine at STP?
46. Determine how many dm^3 of CO_2 at STP can be produced from 15.7 g of Fe_2O_3 according to the following reaction.

$$Fe_2O_3(cr) + 3CO(g) \rightarrow 2Fe(cr) + 3CO_2(g)$$

47. Joseph Priestley first prepared pure oxygen by decomposing mercury(II) oxide. At STP, what volume of oxygen is produced from 4.66 g of mercury(II) oxide?

48. The complete burning of ethane, C_2H_6, produces CO_2 and water vapor as the only products. Assume all substances are gases measured at the same temperature and pressure.
   a. How many dm^3 of ethane would produce 7.07 dm^3 of CO_2?
   b. How many dm^3 of O_2 would be required for the reaction?

49. Carbon disulfide will burn completely to produce CO_2 and SO_2. Assume all substances are gases measured at the same temperature and pressure.
   a. How many dm^3 of SO_2 gas at STP can be produced from 2.22 dm^3 of CS_2 vapor?
   b. How many moles of CO_2 would be produced in the reaction?

50. State the definitions of limiting reactants and excess reactants.

51. If 5.75 g of nickel are mixed with 5.22 g of sulfur, how many grams of nickel(II) sulfide will be produced by their reaction?

52. If 4.44 grams of CaO are mixed with 7.77 grams of water, how many grams of calcium hydroxide form?

53. Balance the following equation.

$$ZnS(cr) + O_2(g) \rightarrow ZnO(cr) + SO_2(g)$$

   a. What mass of ZnO could be produced from 418 g ZnS and 185 dm^3 of O_2?
   b. What volume, at STP, of SO_2 would be produced from 4.66 g ZnS and 1250 cm^3 of O_2?

54. Cyclopropane, C_3H_6, reacts with oxygen to produce water and carbon dioxide.
   a. What volume of carbon dioxide gas at 25°C and standard atmospheric pressure will be produced when 8.4 g of cyclopropane react with an excess of oxygen?
   b. If there are 5.5 dm^3 of oxygen available for this reaction, what is the limiting reactant? What reactant is in excess?

55. Hydrogen gas was prepared in 1776 by Cavendish by passing steam through a red-hot iron rifle barrel. This process also produced Fe_3O_4.
   a. Write a balanced equation for the reaction.
   b. At STP, what volume of H_2 could be produced from 3.00 g of water?
   c. What mass of iron would react in the production of 73.6 dm^3 of hydrogen measured at STP?

56. Laughing gas, an anesthetic, has the formula N_2O. What mass of laughing gas occupies a volume of 0.300 dm^3 at STP?

57. Suppose that 9.50×10^2 dm^3 of helium measured at STP are pumped into a sealed container having a volume of 78.0 dm^3 and a mass of 36.8 kg.
   a. What will be the combined mass of the container and the helium?
   b. What pressure will the helium exert in the container at 39°C?

58. What volume of chlorine gas at 24°C and 99.2 kPa is required to react with 2.51 g of silver in the following equation?

$$2Ag(cr) + Cl_2(g) \rightarrow 2AgCl(cr)$$

59. What mass of mercury(II) chloride will react with 0.567 dm^3 of ammonia at 27°C and 102.7 kPa according to the following equation?

$$HgCl_2(aq) + 2NH_3(aq) \rightarrow Hg(NH_2)Cl(cr) + NH_4Cl(aq)$$

60. What volume of oxygen at 26°C and 102.5 kPa is required to burn 684 m^3 of methane at 101°C and 107.5 kPa?

Concept Mastery

61. Containers A and B have equal volumes and both contain gases. At the same temperature, the pressure in container B is four times that in container A. What can you say about the number of gas molecules in containers A and B?

62. Assume that in each of the following situations, there is a limiting reactant and an excess reactant. State what the limiting reactant and the excess reactant are most likely to be.
   a. Rust, iron(III) oxide, forms on the door panel of an automobile.
   b. A laboratory burner produces a sooty, yellow flame.
   c. Iron filings are heated together with sulfur to produce a dark material that shows no evidence of magnetism.
   d. Charcoal briquets burn in a grill.

63. A 4.0-dm^3 sample of a colorless gas is decomposed to produce 2.0 dm^3 of nitrogen and 6.0 dm^3 of fluorine measured at STP. What is the formula for the colorless gas?

64. At STP, the density of sulfur dioxide is 2.86 g/dm^3, and the density of carbon dioxide is 1.96 g/dm^3.
   a. Calculate the molar volume of each gas.
   b. How do the molar volumes of these gaseous compounds compare with the molar volume of an elemental gas? Explain.

65. A closed container at STP contains 0.25 g nitrogen, 1.5 g argon, and 0.050 g hydrogen. Assuming ideal gas behavior, what is the volume of the container?

Application

66. While resting, the average human male uses 0.200 dm^3 of oxygen per hour at STP for each kg of body mass. Assume that all this oxygen is used to produce energy by oxidizing glucose in the body according to the following reaction.

$$C_6H_{12}O_6(aq) + 6O_2(aq) \rightarrow 6CO_2(aq) + 6H_2O(l)$$

What is the mass of glucose required per hour for a resting male having a mass of 60.0 kg? What volume, at STP, of carbon dioxide would be produced?

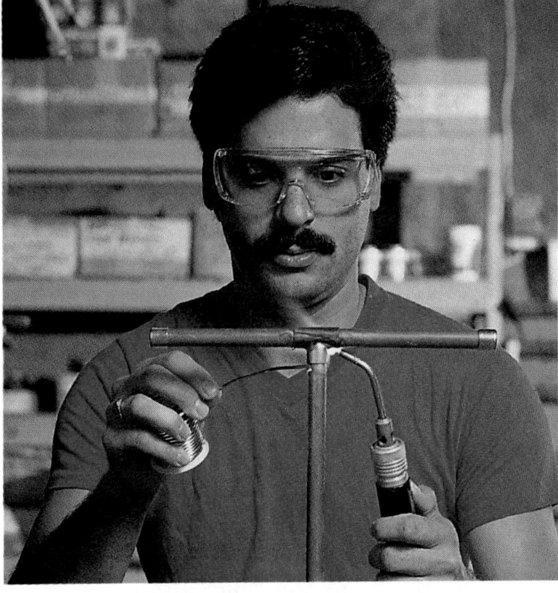

67. Propane torches are commonly used to provide heat for simple plumbing repairs. Propane, C_3H_8, burns completely to produce carbon dioxide and water. At STP, what volume of oxygen is needed to burn completely 2.53 dm^3 of propane?

68. A gas cylinder containing nitrogen has a leaky valve. The volume of the cylinder is 32.4 dm^3. A technician records that the pressure gauge reads 97.0 kPa on a day when the laboratory temperature is 23°C. After four days, the gauge reads 81.4 kPa at 19°C.
   a. How many moles of nitrogen leaked from the cylinder?
   b. How much mass did the cylinder lose?
   c. If the cylinder had contained helium under the same conditions, would it have leaked faster, slower, or at the same rate? Explain.

69. Dry ice is solid carbon dioxide. It sublimes to give carbon dioxide gas. If a 386.4-g sample of dry ice is placed in a tank with a volume of 2.500 dm^3, what is the pressure when all the dry ice has sublimed? The temperature is a constant 32°C.

70. Methanol (wood alcohol), CH_3OH, is manufactured by the following reaction.

$$CO(g) + 2H_2(g) \xrightarrow[\substack{260°C, \\ 3.0 \times 10^3 \text{ kPa}}]{\text{Cu catalyst}} CH_3OH(g)$$

Assuming ideal gas behavior, what is the ratio of the total volume of reactants to the volume of product?

Critical Thinking

71. Chemical analysis of cyclopentane results in an empirical formula of CH_2. Devise an experiment that would show that the formula for cyclopentane is C_5H_{10}, and not CH_2.

72. If a student performs an experiment to determine the molecular mass of a gaseous compound using the modified ideal gas equation, $M = mRT/(PV)$, and forgets to correct for the fact that the gas was collected over water, would the results be high or low? Explain your answer.

Readings

Ember, Lois R., et al. "The Changing Atmosphere." *Chemical and Engineering News* 64, no. 47 (November 1986): 14-64.

Grove, Noel. "An Atmosphere of Uncertainty." *National Geographic* 171, no. 4 (April 1987): 536.

Cumulative Review

1. If a sample of gas occupies 642 cm^3 at 93.9 kPa, what volume will it occupy at 109.0 kPa if the temperature remains constant?

2. If a sample of gas collected over water at 50°C occupies 62.5 cm^3 at a total pressure of 114.9 kPa, what volume will the dry gas occupy at 25°C and 97.8 kPa?

3. If a sample of gas occupies 286 cm^3 at 98.1 kPa and 42°C, what volume will it occupy at 42.2 kPa and 60.0°C?

4. Define an ideal gas.

5. In kinetic terms, explain how a gas exerts pressure on its container. If no other conditions are changed, how does the addition of more gas affect the pressure? Explain why either doubling the amount of gas or compressing the gas by half would have the same effect on pressure.

6. Where are the most chemically reactive metals and nonmetals located on the periodic table? Explain why elements in these positions have high reactivity.

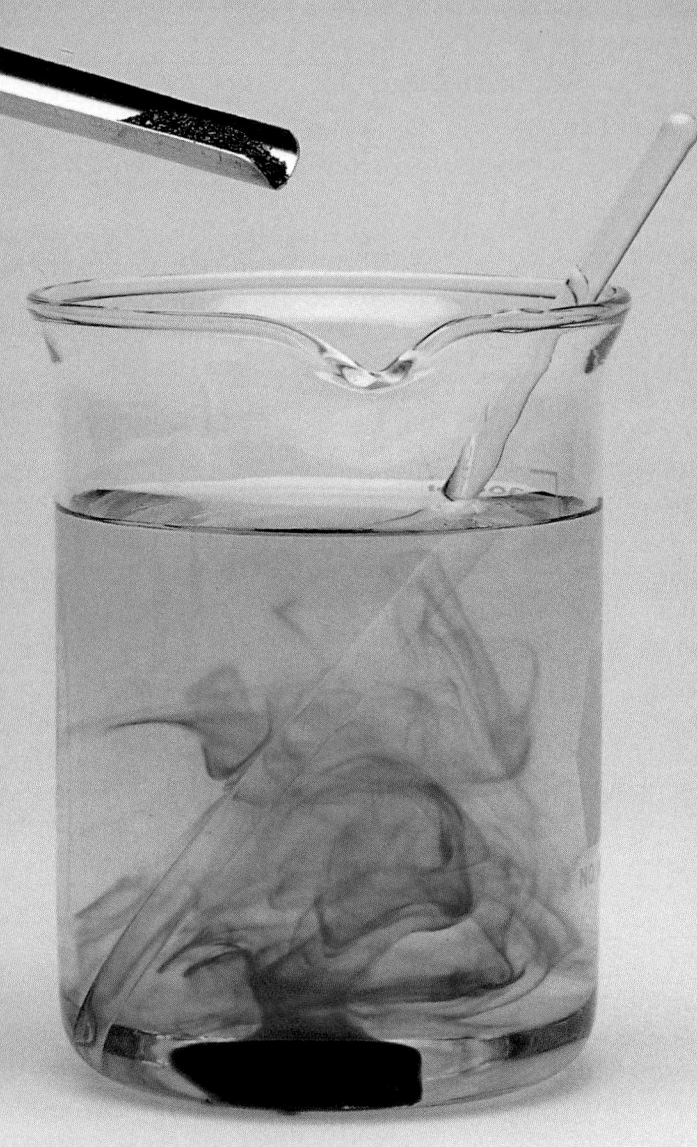

CHAPTER **Preview**

CONTENTS

Solutions and Colloids

CHAPTER 20

The beaker contains water into which some potassium permanganate, $KMnO_4$, has been added. From your experience, you can probably tell that the potassium permanganate is dissolving in the water because the water is turning purple. What is happening in the beaker? Why will some substances but not others dissolve in water? You will investigate these questions as you begin to study how the process of dissolving occurs.

A solution is a homogeneous mixture. It is as uniformly mixed as a mixture can be. The paint also looks closely mixed. However, it is a heterogeneous mixture, not a solution. The difference between the two mixtures is not apparent. Therefore, we must look more closely at the properties and characteristics of these two types of mixtures.

EXTENSIONS AND CONNECTIONS

OBJECTIVES

- describe and explain the processes of solvation, dissociation, and dissolving.
- discuss factors affecting the solubility of one substance in another.
- relate enthalpy of solution to endothermic and exothermic dissolving processes.
- differentiate among and solve problems involving molarity, molality, mole fraction, and mass percent.

20.1 Solutions

We will begin our study of solutions by looking at the physical process of dissolving. One important aspect of solutions is the relative amounts of the substances present in the solution. These amounts can be described by the concept of concentration.

The Dissolving Process

What do we mean by solution? In Chapter 3, we referred to homogeneous matter as being the same throughout. It is often made of only one substance — a compound or an element — but may also be a mixture of several substances. Such a homogeneous mixture is called a solution. Most solutions consist of a solid dissolved in a liquid. The particles in a true solution are molecules, atoms, or ions that will pass easily through the pores of filter paper. Solutions cannot be separated into their components by filtration.

The substance that occurs to the greater extent in a solution is said to do the dissolving and is called the **solvent.** The less abundant substance is said to be dissolved and is called the **solute.** The most common solvent is water. Using water as a typical solvent, let us look at the mechanism of solvent action. Water molecules are very polar. Because they are polar, they are attracted to other polar molecules and to ions. Table salt, NaCl, is an ionic compound made of sodium and chloride ions. If a salt crystal is put into water, the polar water molecules are attracted to ions on the crystal surfaces. The water molecules gradually surround and isolate the surface ions. The ions become hydrated, Figure 20.1. The attraction between the hydrated sodium and chloride ions and the remaining crystal ions becomes so small that the hydrated ions are no longer held by the crystal. They

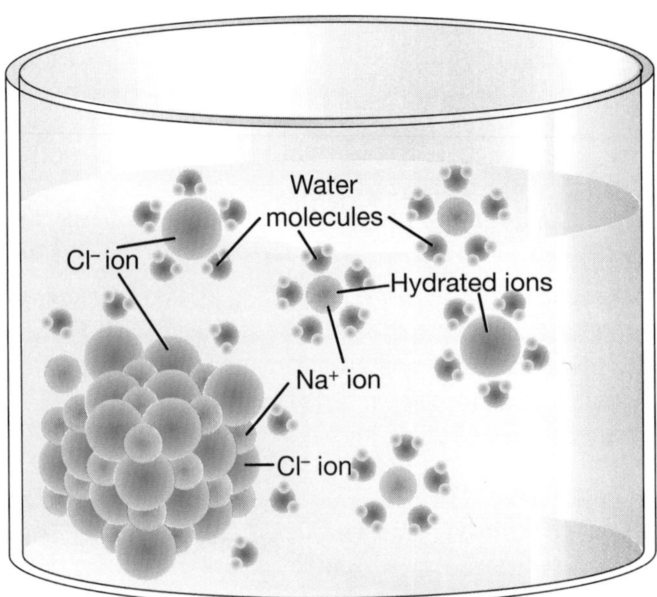

Figure 20.1 Table salt dissolves in water because polar water molecules gradually surround and isolate sodium and chloride ions.

gradually move away from the crystal into solution. This separation of ions from each other is called **dissociation.** The surrounding of solute particles by solvent particles is called **solvation.**

Whenever you are considering a solution of an ionic compound, it is important to keep the following in mind. When the ions are dissociated, each ion species in the solution acts as though it were present alone. Thus, a solution of sodium chloride acts as a solution of sodium ions and chloride ions. There is no characteristic behavior of "sodium chloride" in solution because there really is no sodium chloride in solution. There is simply a solution containing both sodium ions and chloride ions uniformly mixed with water molecules.

Solvent-Solute Combinations

Four simple solution situations can be considered. They are listed in Table 20.1. Not all possible combinations of substances will fit into these four rigid categories. We will include ionic substances with polar substances in our discussion.

Table 20.1

Solvent-Solute Combinations

Solvent Type	Solute Type	Is Solution Likely?
Polar	Polar	yes
Polar	Nonpolar	no
Nonpolar	Polar	no
Nonpolar	Nonpolar	yes

(1) Polar solvent-polar solute.
The mechanism of solution involving a polar solvent and a polar solute is the one we have already described for salt and water. The polar solvent particles solvate the polar solute particles. If the solvent is water, this process is called **hydration.** The solvent particles attach themselves due to the polar attraction. The intracrystalline forces are reduced so much that the surface particles are carried away by the solvent particles.

(2) Polar solvent-nonpolar solute.
Because the solvent particles are polar, they are attracted to each other. However, the solute particles in this case are nonpolar and have little attraction for particles of the solvent. Thus, solvation does not occur, and solution to any extent is unlikely, as we see if we try to dissolve wax in water.

(3) Nonpolar solvent-polar solute.
Reasoning similar to that of the second case applies here. The solvent particles are nonpolar and thus have little attraction for the solute particles. In addition, the solute particles in this case are polar and are attracted to each other. Again, because there is no solvation, solution to any extent is unlikely as we see if we try to dissolve salt in hexane.

CHEMISTS AND THEIR WORK

Agnes Pockels (1862-1935)
Agnes Pockels was the daughter of an Austrian army officer. Although she attended high school for some time, she was largely self-taught. In her late teens, she made her first observations on the changes in surface tension of water contaminated by oil. Her first paper was printed nearly ten years later. Most of her experimentation was done at home with homemade apparatus. Among other makeshift devices, she developed the Pockels' trough, which was used to measure surface tension.

At 69, Pockels won the Laura-Leonard prize for her work on the properties of surface layers and films. An extension of her work in surface chemistry led to the Nobel Prize for Irving Langmuir in 1932.

Table 20.2

Ionic Dissolving Ability		
	Name	**Formula**
increasing ability ↓	Cyclohexane	C_6H_{12}
	Carbon tetrachloride	CCl_4
	Benzene	C_6H_6
	Toluene	C_7H_8
	Ethoxyethane	$C_4H_{10}O$
	Ethyl acetate	$C_4H_8O_2$
	1-Butanol	$C_4H_{10}O$
	1-Propanol	C_3H_8O
	Propanone	C_3H_6O
	Ethanol	C_2H_6O
	Methanol	CH_4O
	Water	H_2O

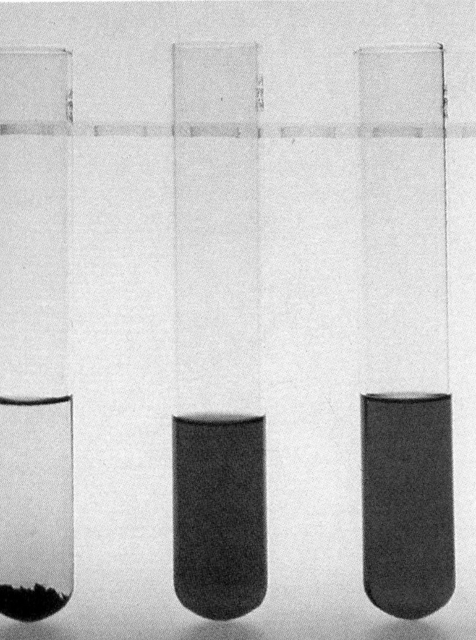

Figure 20.2 Iodine, a nonpolar solid, dissolves (from left to right) in water, propanone, and carbon tetrachloride. The extent of solubility is greatest in the nonpolar solvent, CCl_4.

(4) Nonpolar solvent-nonpolar solute.
Only van der Waals forces exist among the nonpolar solvent particles. The same is true for the nonpolar solute particles. Thus, all particles in the solution are subject only to van der Waals forces, and solution can occur. Random motion of solute molecules will cause some of them to leave the surface of the solute. As time passes, the particles of nonpolar solutes become randomly dispersed. Thus, wax will dissolve in hexane. There can be solvation in such cases, but the forces are far weaker than those in solutions involving polar compounds.

Not all nonpolar substances, however, are soluble in each other. Let us consider the most common type of solution: a solid dissolved in a liquid. The solubility of a nonpolar solid in a nonpolar liquid depends on two factors: its melting point and its enthalpy of fusion. When the solid dissolves, a liquid solution results. The solid is undergoing a phase change. Solids with low melting points and low enthalpies of fusion will be more soluble than those with high melting points and high enthalpies of fusion. This difference is due to stronger attractive forces within the crystals of high melting point substances. These are the forces that must be overcome during solution.

Table 20.2 lists a number of common solvents in order of increasing ability to dissolve highly polar and ionic materials. Compare the order here with that in Table 14.1.

Gases, Liquids, and Solids in Solution

Since there are three common physical states of matter (solid, liquid, and gas), there are nine possible combinations of solvent-solute pairs. These combinations are given in Table 20.3. Note that because all gases expand to fill the volume of their containers, all mixtures of gases are solutions.

Figure 20.3 The brass used to make instruments is an alloy and consists of a solution of copper and zinc. At higher concentrations of zinc, the alloy becomes a heterogeneous mixture.

Table 20.3

Possible Solution Combinations		
Solvent	**Solute**	**Common Example**
gas	gas	helium-oxygen (deep-sea diver's gas)
gas	liquid	air-water (humidity)
gas	solid	air-naphthalene (mothballs)
liquid	gas	water-carbon dioxide (soft drink)
liquid	liquid	water-acetic acid (vinegar)
liquid	solid	water-salt (seawater)
solid	gas	palladium-hydrogen (gas stove lighter)
solid	liquid	silver-mercury (dental amalgam)
solid	solid	gold-silver (ring)

Word Origins

miscere: (L) to mix

miscibility—the ability to be mixed

The property of mutual solubility of two liquids is called **miscibility.** If two liquids are mutually soluble in all proportions, they are said to be completely miscible. Ethylene glycol (automobile antifreeze) and water are two such liquids. Both water and ethylene glycol molecules are polar. Water and carbon tetrachloride, however, do not appreciably dissolve in each other and are, therefore, immiscible. Carbon tetrachloride molecules are nonpolar. Two liquids such as diethyl ether ($C_4H_{10}O$) and water, which dissolve in each other to some extent but not completely, are referred to as partially miscible.

A number of metals (such as gold and silver) are mutually soluble and form solid-solid solutions. Such solid metal-metal solutions constitute one type of alloy.

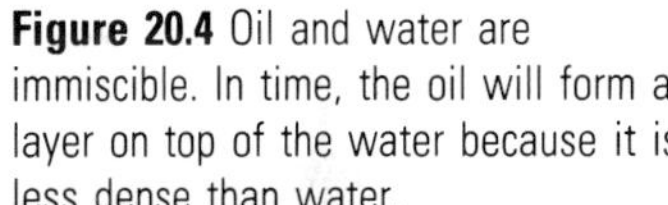

Figure 20.4 Oil and water are immiscible. In time, the oil will form a layer on top of the water because it is less dense than water.

Solution Equilibrium

When crystals are first placed in a solvent, many particles may leave the surface and go into solution. As the number of solute particles in solution increases, some of the dissolved particles return to the surface of the crystal. Eventually the number of particles leaving the crystal surface equals the number returning to the surface. This point is called **solution equilibrium.**

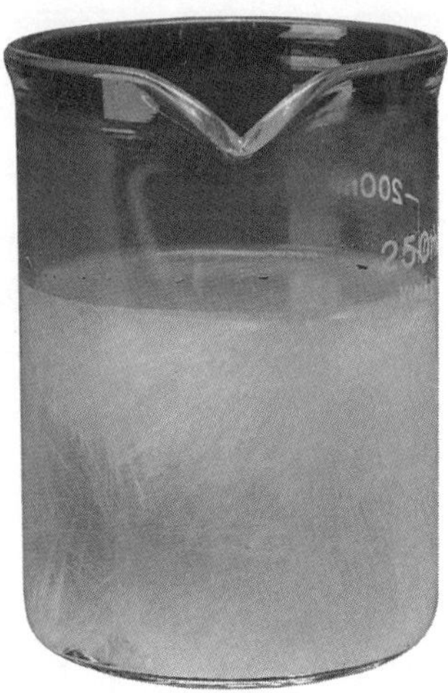

Figure 20.5 When a small crystal of sodium acetate is added to a supersaturated solution of sodium acetate (left), crystal growth begins immediately (center) and quickly spreads throughout the solution (right).

A solution in which an undissolved substance is in equilibrium with the dissolved substance is called a **saturated solution.** A solution containing less than the saturated amount of solute for that temperature is an **unsaturated solution.**

At a specific temperature, the maximum amount of solute that will dissolve in a given quantity of solvent at saturation is a fixed amount. For instance, at 20°C, a maximum of 64.2 grams of nickel(II) chloride will dissolve in 100 cm^3 of water. This relationship is called the **solubility** of the substance at 20°C. When 64.2 grams of $NiCl_2$ dissolve in 100 cm^3 of water at 20°C, the solution is saturated. The solubility of a substance can be changed by altering the temperature.

Larger amounts of solute can usually be dissolved in a solvent at a temperature higher than room temperature. If the hot solution is then cooled, an unstable solution is formed. This solution contains more solute than a saturated solution can normally hold. The solution is called a **supersaturated solution.** Supersaturation is possible because solids will not crystallize unless there is an angular surface upon which to start crystallization. A container that has a smooth interior and which contains a dust-free solution has no such surfaces. However, a supersaturated solution will crystallize almost instantly if a crystal of the solute is introduced. How could you find out if a solution is saturated, unsaturated, or supersaturated?

Precipitation Reactions

Precipitation reactions are an important tool of the analytical chemist. The differing solubilities of substances can be used to separate them from mixtures. If a solution contains two different ionic solutes, they can be separated by choosing a reactant that will precipitate only one ion. For example, consider a mixture of sodium nitrate and barium chloride. A chemist wishes to know what percentage of the mixture is $BaCl_2$. The chemist first finds the mass of the mixture before dissolving it in water. Once dissolved, the barium ions can be precipitated with sulfate ions from a soluble sulfate such as sulfuric acid. The chemical equation for the precipitation reaction is

$$BaCl_2(aq) + H_2SO_4(aq) \rightarrow BaSO_4(cr) + 2HCl(aq)$$

The three other ions present in solution do not form insoluble compounds in this reaction. Thus, this reaction separates the barium as a precipitate from the solution of the original mixture.

On the other hand, the chloride ions could be precipitated from salt solution with silver ions from a soluble silver salt such as silver nitrate. (See Appendix Table A-7.) The equation is

$$NaCl(aq) + AgNO_3(aq) \rightarrow AgCl(cr) + NaNO_3(aq)$$

The product is collected, dried, and the mass obtained. By solving a mass-mass problem, the chemist can find the amount and, thereby, the percentage of $BaCl_2$ in the original mixture.

SAMPLE PROBLEM

Precipitation

A mixture consists of cadmium sulfate and cesium iodide. Find the percentage of cesium iodide in the mixture.

Solving Process:

Determine the mass of the mixture. Then, dissolve the mixture in water. The solution now consists of four ions: Cd^{2+}, Cs^{+}, SO_4^{2-}, and I^{-}. To separate one ion, it must be precipitated by a substance that will not react with other ions in solution. By consulting Appendix Table A-7, we find that Cd^{2+} may be precipitated by a Group 1 hydroxide, carbonate, or sulfide, none of which will react with Cs^{+}. Alternatively, barium nitrate or barium acetate could be added to give a precipitate of barium sulfate. The iodide ion is unaffected. Either precipitate then would be filtered, dried, and massed. By solving a mass-mass problem, the quantity of the original $CdSO_4$ or CsI could be computed. From this mass value the percentage composition of the original mixture could be calculated.

PRACTICE PROBLEMS

1. For each of the following substances, select a reagent that can be used to precipitate the positive ion.
   a. $CuBr_2$ **b.** $BaCl_2$ **c.** $Ca(OH)_2$ **d.** NiF_2
2. For each of the substances in Practice Problem 1, select a reagent that can be used to precipitate the negative ion.
3. Predict the result of each of the following reactions.
   a. $HF(aq) + Hg_2(NO_3)_2(aq) \rightarrow$
   b. $LiI(aq) + AgNO_3(aq) \rightarrow$
   c. $K_2S(aq) + Co(CH_3COO)_2(aq) \rightarrow$
   d. $NaOH(aq) + Cr_2(SO_4)_3(aq) \rightarrow$

Figure 20.6 The rate of dissolving can be increased by stirring (left), by grinding the solute (center), and by heating (right).

Dissolving Effects

The rate of solution of a solid in a liquid is affected by the surface area of the crystal that is exposed to fresh solvent. When the area of exposed surface is increased, more solute particles are subjected to solvation. The surface area can be increased by breaking the crystal to be dissolved into very small particles. The rate of dissolving can also be increased by stirring the mixture as the solute is dissolving. Without stirring, the solvent in contact with the solid soon becomes saturated. By stirring, the saturated solvent is moved away from the surface of the solid solute. Fresh solvent can then come into contact with the solid surface.

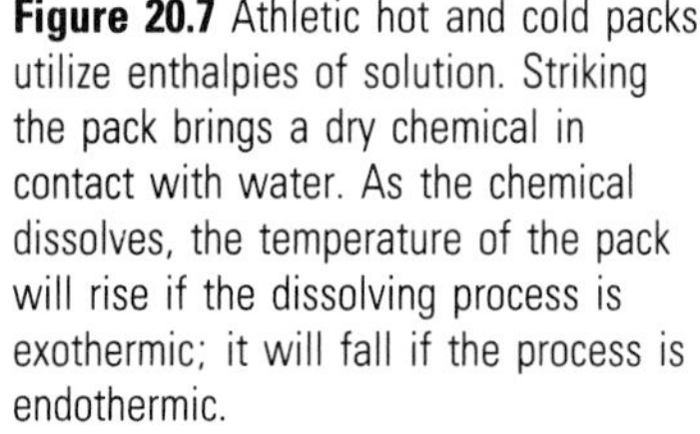

Figure 20.7 Athletic hot and cold packs utilize enthalpies of solution. Striking the pack brings a dry chemical in contact with water. As the chemical dissolves, the temperature of the pack will rise if the dissolving process is exothermic; it will fall if the process is endothermic.

Solution rate is also a function of the kinetic energy of both the solute and solvent particles. The kinetic energy of a system is increased by heating it. The faster the solvent particles are moving, the more rapidly they will circulate among the crystal particles. This motion exposes fresh solute surface, thus increasing solution rate. Finally, with increased kinetic energy, particles are more easily removed from the crystal.

The energy change that always occurs when one substance dissolves in another is called the **enthalpy of solution** (ΔH_{sol}). Consider the dissolving of a solid in a liquid. The solution process is usually endothermic. Because the energy of the system has increased, the enthalpy of solution has a positive value ($\Delta H_{sol} > 0$). A positive enthalpy of solution should be expected when we realize that during the dissolving process the solid becomes a component of a liquid solution. We would expect that a change from a solid to a component of a liquid would be an endothermic process. Occasionally, the dissolving of a solid in a liquid is an exothermic process, that is, $\Delta H_{sol} < 0$. The dissolving of sodium hydroxide in water is a well-known example. This example and other exceptions take place because the formation of certain hydrated ions is an exothermic process.

When a gas dissolves in water, the change of the solute is from a gas to a component of a liquid solution. The enthalpy of solution of a gas dissolving in water is negative ($\Delta H_{sol} < 0$) because the change of the solute is an exothermic process.

Using our knowledge of enthalpies of solutions, we can predict the effect of temperature on solubility. Most solids have positive enthalpies of solution ($\Delta H_{sol} > 0$). They are more soluble in hot water than in cold water. However, gases have negative enthalpies of solution ($\Delta H_{sol} < 0$). They are more soluble in cold water. The bubbles of gas that appear in water as it is heated below its boiling point are bubbles of air.

Pressure has little effect on solutions unless the solute is a gas. The amount of gas that dissolves in a given amount of solvent is greater at high pressure than it is at low pressure. *The mass of a gas that will dissolve in a liquid at a given temperature varies directly with the partial pressure of that gas.* This statement is Henry's law, named in honor of William Henry, the English chemist who first discovered this relationship.

The carbonation of soft drinks is a good example of Henry's law. After filling, the bottles and cans are placed in a chamber in which the partial pressure of carbon dioxide is much higher than that of the atmosphere. After some of the excess carbon dioxide has dissolved in the soft drink, the bottles and cans are sealed. When you open the can or bottle, the partial pressure of carbon dioxide above the solution drops, and the drink fizzes as the carbon dioxide bubbles out of solution.

Figure 20.8 When a soft drink is opened, the partial pressure of CO_2 over the liquid is decreased and the excess dissolved CO_2 bubbles out of solution.

Bridge to *OCEANOGRAPHY*

Underwater Diving

The most dangerous gas in the compressed air breathed by a diver is nitrogen. High concentrations of dissolved nitrogen in the blood may lead to nitrogen narcosis, a condition in which a diver feels euphoric, becomes disoriented, and may panic.

When a diver ascends, the pressure decreases and the nitrogen comes out of solution. If the diver rises too fast, the dissolved nitrogen does not have a chance to travel to the lungs to be exhaled. Instead, bubbles of nitrogen form in the blood vessels and nervous tissue of the body. These bubbles can lead to paralysis, convulsions, and the "bends." The bends is so named because nitrogen gas bubbles frequently collect at bends in the blood vessels, such as in joints. The pressure created there is extremely painful and prevents the diver from straightening the joint.

A diver who has ascended too quickly can be treated in a decompression chamber in which the pressure is quickly raised to the pressure at which the diver was working and then gradually decreased to normal.

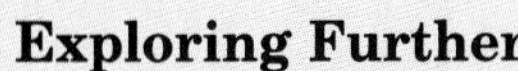

Exploring Further

1. Divers sometimes breathe mixtures of oxygen and helium because helium is virtually insoluble in blood. Find out how much oxygen is typically included in deep-diving mixtures.
2. Research how oxygen becomes toxic at high partial pressures.

MINUTE LAB

Chill Out

Put on an apron and goggles. Add 30 cm^3 of water to a small beaker that contains 10 g NH_4Cl. Feel the bottom of the beaker and record your observations. Add 30 cm^3 of water to a second beaker that contains 10 g of $CaCl_2$. Feel the bottom of the beaker and record your observations. Which dissolving process had a positive enthalpy of solution? Which process was exothermic? Suggest practical applications for dissolving processes that have positive enthalpies of solution and those that have negative enthalpies of solution.

Concentration

In Chapter 8 you learned about a concentration unit called molarity. You found that a one-molar (1*M*) solution contains 1 mole of solute in 1 dm^3 of solution. If 1 mole of sodium chloride is dissolved in enough water to make 1 dm^3 of solution, the solution is a 1*M* solution of sodium chloride. Sodium chloride is in the form of dissociated ions in solution. Therefore, the solution can also be said to be 1*M* in sodium ions and 1*M* in chloride ions.

Molarity is the most common concentration unit in chemistry. With it, measurement of the amount of solute in a volume of solution is fast and convenient. If the solution measured has a known molarity, a measurement of volume is also a measurement of a number of particles. Each unit of volume contains a known number of ions or molecules. Multiplying the concentration in molarity (moles per dm^3) by the volume (dm^3) will give you the number of particles (in moles).

Whenever the concentration of solute is known with a high degree of certainty, a solution is called a standard solution. Standard solutions are frequently used in analytical chemistry.

PRACTICE PROBLEMS

Calculate the molarity of the ions in Practice Problems 4 and 5.

4. Br^- for 193 g $MgBr_2$ in 5.00×10^2 cm^3 solution

5. Ca^{2+} for 8.28 g $Ca(C_5H_9O_2)_2$ in 2.50×10^2 cm^3 solution

6. How many moles of Ca^{2+} are in 2.00 dm^3 of 0.523*M* $CaCl_2$ solution?

Sometimes, it is convenient to express concentration in terms of moles of solute per kilogram of solvent, or **molality.** A solution that contains 1 mole of solute in each 1000 grams of solvent is called a one-molal (1*m*) solution. This solution differs from a one-molar (1*M*) solution, which contains 1 mole of solute in 1 dm^3 of solution. Molality is most useful in studying the colligative properties of solutions, which will be discussed in Chapter 21.

Another method of describing concentration used frequently in organic chemistry is mole fraction. The **mole fraction** shows the comparison of moles of solute to moles of solution. It is the ratio of the number of moles of one solution component (either the solute or the solvent) to the total number of moles of all components in the solution.

Biologists frequently express the concentration of solutions in mass percent. A 5% solution of sodium hydroxide, for example, contains 5 g NaOH in each 100 g of solution.

A less frequently used concentration quantity is normality, *N*. Normalities can be found several ways. The simplest method of calculating normality is to multiply the molarity of the solution by the total charge of the positive ions in the compound.

SAMPLE PROBLEMS

1. Molality

If 52.0 g of K_2CO_3 are dissolved in 518 g of H_2O, what is the molality of the resulting solution?

Solving Process:

Formula mass of K_2CO_3 = 2(39.1 g) + 12.0 g + 3(16.0 g) = 138 g

Divide solute by solvent and convert to the proper units.

$$\frac{52.0 \text{ g } \cancel{K_2CO_3}}{518 \text{ g } \cancel{H_2O}} \left| \frac{1 \text{ mol } K_2CO_3}{138 \text{ g } \cancel{K_2CO_3}} \right| \frac{1000 \text{ g } \cancel{H_2O}}{1 \text{ kg } H_2O} = 0.727m$$

2. Mole Fraction

What is the mole fraction of alcohol in a solution made of 2.00 moles of ethanol and 8.00 moles of water?

Solving Process:

The mole fraction of ethanol in the 10.0 moles of solution is

$$\frac{2.00 \text{ mol ethanol}}{10.00 \text{ mol solution}} = 0.200$$

Any size sample of this solution will have an ethanol mole fraction of 0.200. The sum of the mole fractions of all components of a solution must equal 1.

3. Mass Percent

Find the mass percent of 142 g of H_2O_2 in 331 g of H_2O.

Solving Process:

$$\frac{142 \cancel{g}}{142 \cancel{g} + 331 \cancel{g}} \times 100\% = 30.0\% \ H_2O_2$$

PRACTICE PROBLEMS

Calculate the molality of each of the following solutions.

7. 199 g of $NiBr_2$ in 5.00×10^2 g of water

8. 92.3 g of KF in 1.00×10^3 g of water

9. 0.059 g KF in 0.272 g water

Calculate the mole fraction for each component in the following.

10. 12.3 g of C_4H_4O in 1.00×10^2 g of C_2H_6O

11. 156 g of $C_{12}H_{22}O_{11}$ in 3.00×10^2 g H_2O

12. 75.6 g of $C_{10}H_8$ in 6.00×10^2 g $C_4H_{10}O$

Calculate the mass percent concentration for the following.

13. the solutions in Practice Problems 7 and 8

14. the solutions in Practice Problems 9 and 10

15. the solutions in Practice Problems 11 and 12

CAREERS IN CHEMISTRY

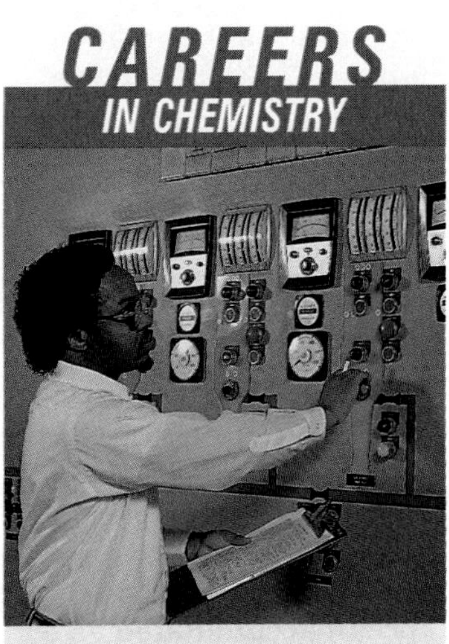

Water Treatment Worker

Water treatment workers are responsible for both the physical and chemical processes that make water safe for human use. They must be familiar with the processes of disinfection, sedimentation, and filtration to assure that water is free of harmful bacteria, toxic chemicals, and turbidity. Water treatment workers should take chemistry and biology courses in high school and may be required to take additional training provided by the plant or at a college.

Bridge to
OCEANOGRAPHY

Seawater

After swimming in the ocean, if you let the water evaporate from your skin, you will notice a fine deposit of salt crystals on your skin. Seawater is a solution.

Where did the ocean come from? Scientists theorize that Earth was quite hot at its early stage of formation. Any water present at that time would have vaporized and been lost to space. One theory is that our present ocean accumulated as a result of water emerging from the interior of Earth, principally as a result of volcanic action. The effects of erosion (weathering) produced a flow of water that contained dissolved minerals, and this solution accumulated to form the ocean.

The minerals in the sea are constantly forming sediment on the seafloor. This sediment is the result of chemical reactions within seawater and living organisms in the sea that use the solutes. When these organisms die, their remains sink to the bottom. It is generally assumed by scientists today that seawater represents a state of equilibrium between the addition of new salts and the precipitation of solutes.

Seawater is a commercial source of two elements: bromine and magnesium. Sand, gravel, and diamonds are obtained from the seafloor. The seafloor is also drilled for petroleum.

The most prolific life forms are found around Antarctica. Recall that gases have a negative enthalpy of solution. Cold water will dissolve more gas than warm water. The ocean around Antarctica, therefore, is rich in O_2 and supports much sealife.

Exploring Further

1. What is the current explanation of the origins of the oceans?
2. What evidence is there to indicate that seawater is a solution?
3. Why would diamonds be found on the seafloor in shallow water?

20.1 CONCEPT REVIEW

16. How do the processes of dissociation and solvation enter into the dissolving process?
17. How does the polar nature of a solute affect its solubility in polar and nonpolar solvents?
18. What does a positive enthalpy of solution indicate about the dissolving process?
19. **Apply** Why is stirring necessary to dissolve sugar added to cold lemonade?

20.2 Colloids

OBJECTIVES

- distinguish among colloids, solutions, and suspensions.
- classify colloids.
- describe properties of colloids and explain how these properties depend upon particle size.

There are mixtures that have some, but not all, of the characteristics of solutions. For example, these mixtures can pass through filter paper just as solutions can. However, unlike solutions, these materials cannot move across membranes such as those that enclose living cells.

Colloids and Phases

In 1861, Thomas Graham, an English chemist, tested the passage of different substances through a parchment membrane. He found that one group of substances passed readily through the membrane, and another group did not pass through it at all. He called the first group **crystalloids** and the second group colloids. The name *colloid* means "gluelike"; glue was one of the substances that did not pass readily through the membrane. Graham thought the ability or inability to pass through the membrane was due to elementary particle size. It was later discovered that any substance could be used to produce a colloid. Included were those substances Graham had classified as crystalloids. **Colloids** are now defined as mixtures composed of two phases of matter, the dispersed phase and the continuous phase. They are an intermediate class between suspensions and solutions. Colloid particles (dispersed phase) are larger than the single atoms, ions, or molecules of solutions. They are smaller than the particles of suspensions, which can be seen through a microscope and which settle out of suspension on standing.

kolla: (GK) glue

colloid—gluelike material

Colloids are classified based on the states of their dispersed and continuous phases. Liquids and solids dispersed in gases are aerosols. Fog and smoke are examples of aerosols. Foams are gases dispersed in liquids or solids. Whipped cream is a liquid

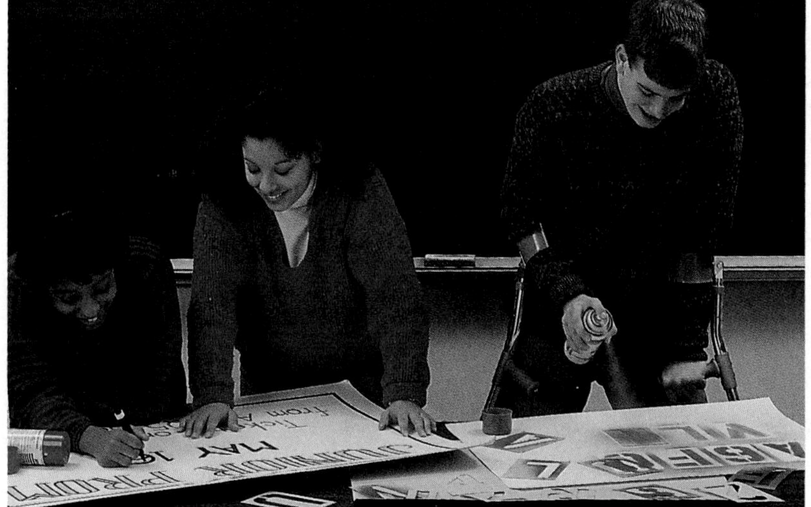

Figure 20.9 Spray paints are examples of a class of colloids called aerosols.

foam; marshmallows are solid foam. Emulsions are colloids in which liquids are dispersed in other liquids or in solids. Mayonnaise is a liquid emulsion; cheese is a solid emulsion. Sols are dispersions of solids in liquids or other solids. Jellies and paints are sols; pearls and opals are solid sols.

Colloidal Sizes

If a finely ground substance is placed in water, one of three things will happen. First, it may form a true solution that is simply a dispersion of atoms, molecules, or ions of the substance into a solvent. Particles in a solution do not exceed ~ 1 nm in size.

Second, the particles may remain larger than ~ 100 nm. These particles are large enough to be seen with a microscope and gradually fall to the bottom of the container. Because the particles are temporarily suspended and settle out upon standing, this mixture is called a **suspension.**

RULE OF THUMB

Particles from ~ 1 to ~ 100 nm in size usually remain dispersed throughout the medium. Such a mixture is called a colloid. Colloids represent a transition between solutions and suspensions. However, they are considered heterogeneous, with the medium as the continuous phase and the dispersed substance as a separate phase.

Actually, substances show unusual properties even when only one of the three dimensions of the particles is less than 100 nm. **Colloid chemistry** is the study of the properties of matter whose particles are colloidal in size in at least one dimension.

Table 20.4

Comparison of Solutions, Suspensions, and Colloids

Type	Particle size	Permanence
Solution	<1 nm	permanent
Colloid	>1 nm but < 100 nm	permanent
Suspension	>100 nm	settle out

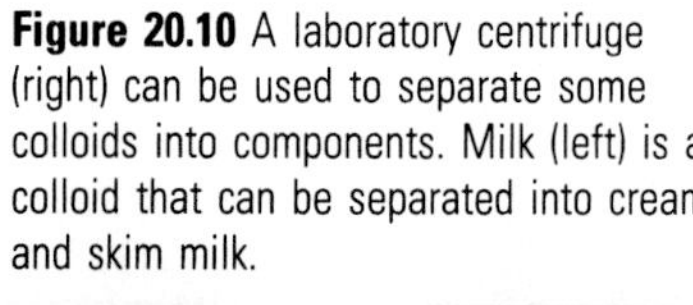

Figure 20.10 A laboratory centrifuge (right) can be used to separate some colloids into components. Milk (left) is a colloid that can be separated into cream and skim milk.

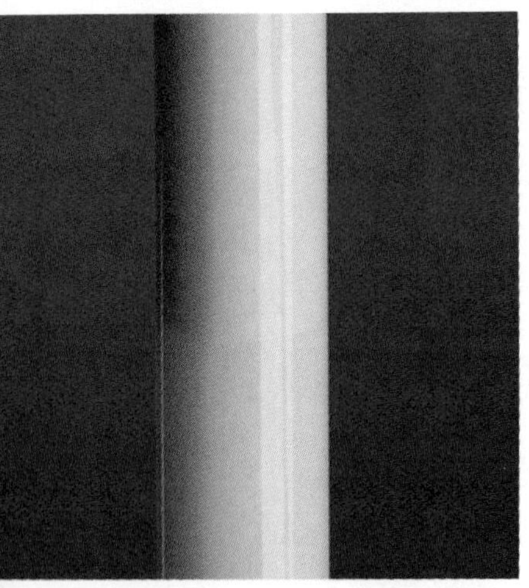

Figure 20.11 A light beam passes through a solution, but a colloid scatters the light (left). The Tyndall effect can also be seen in air (right) when suspended dust and water droplets scatter light.

Properties of Colloids

If a beam of light is allowed to pass through a true solution, some of the light will be absorbed, and some will be transmitted. The particles in solution are not large enough to scatter the light. However, if light is passed through a colloid, the light is scattered by the larger, colloidal particles. The beam becomes visible from the side. This effect is called the **Tyndall effect.** You may have seen this effect in the beam of a searchlight in the night air. In this situation, the light is scattered by suspended water droplets in air. You may also have observed this effect as the sunbeam coming through a hole in the blinds. Here, suspended dust particles in the air scatter the sunlight.

As you know, colloidal particles are too small to be seen with an ordinary microscope. In 1912, Richard Zsigmondy, a German professor of chemistry, designed the ultramicroscope. Through ultramicroscopy, in which the Tyndall effect is used, it is possible to "see" colloidal particles.

If you observe colloids under an ultramicroscope, you will notice that they have another interesting property. The particles of a colloid are in continuous random motion. This motion is called **Brownian motion.** The constant bombardment of the colloid particles by the smaller molecules of the medium is the source of the motion. The motion is the result of the collision of

Table 20.5

Properties of Solutions, Colloids, and Suspensions		
Solutions	**Colloids**	**Suspensions**
Do not settle out	Do not settle out	Settle out on standing
Pass unchanged through ordinary filter paper	Pass unchanged through ordinary filter paper	Separated by ordinary filter paper
Pass unchanged through membrane	Separated by a membrane	Separated by a membrane
Do not scatter light	Scatter light	Scatter light
Affect colligative properties	Do not affect colligative properties	Do not affect colligative properties

many molecules with the particle. It is as if a large crust of bread in a pond is moved back and forth as first one fish and then another hits the bread from opposite sides. Brownian motion is named in honor of Robert Brown, a biologist. Brown first noticed it while observing the motion of particles in a suspension of pollen grains in water.

Word Origins

ad: (L) to
sorbere: (L) to soak up
adsorption—the soaking up of substances

As a result of electric forces, solid and liquid surfaces tend to attract and hold substances with which they come into contact. This phenomenon is called **adsorption.** The stationary phase in chromatography is an adsorbing material. Colloidal particles, because of their small size, have an extremely large ratio of surface to mass. Consider a cube that measures 1 cm on an edge. It has a volume of 1 cm^3 and a surface area of 6 cm^2. If that cube is cut into 1000 smaller cubes of equal size, the surface area of that same amount of matter becomes 60 cm^2. If we subdivide these cubes even further, until they measure 10 nm on edge, the surface area increases to 6 000 000 cm^2. Such a large surface area makes colloidal particles excellent adsorbing materials, or adsorbents. Dispersed particles have the property of adsorbing charge on the surface.

Word Origins

elektron: (GK) electricity
phoreus: (GK) carrier
electrophoresis—carrying by electric charges

If a colloid is subjected to an electric field, a migration of the particles can be observed. The positive particles are attracted to the cathode. The negative particles are attracted to the anode. This migration, called **electrophoresis,** is evidence that colloidal particles are charged. The separation of amino acids and peptides obtained in protein analysis is accomplished rapidly using electrophoresis. The process is also common in nucleic acid research.

Semipermeable membranes can be used to separate ions and colloidal particles. The ions pass through the membrane but the colloidal particles, because they are larger, do not. This type of separation is utilized in kidney dialysis.

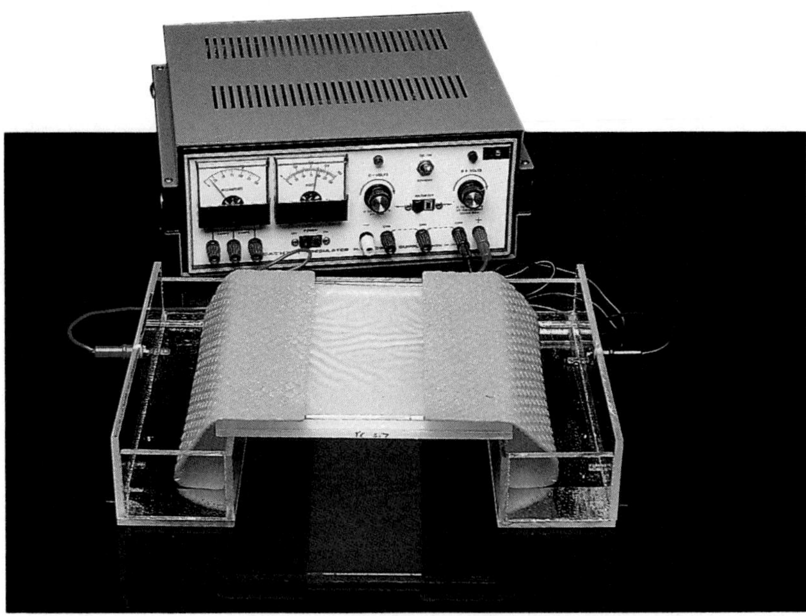

Figure 20.12 An electrophoresis apparatus is used in separating mixtures containing charged particles. Negatively-charged particles move toward the anode. Positively-charged particles move toward the cathode.

EVERYDAY CHEMISTRY

Preparing Preserves with Pectin

If you enjoy preserves such as jams and jellies, you can thank a long molecule named pectin for your pleasure. Pectin is found naturally in grapes, berries, green apples, and citrus fruits. It is a long, chainlike molecule made up of smaller structures similar to those of sugars. Pectin is what causes preserves to gel.

To make preserves from fruits that contain natural pectin, the fruits are first boiled to release the pectin from the walls of the cells. Pectin is added to fruits that do not have it naturally. Within the boiling juice, the negatively-charged pectin molecules repel each other. The charge on the molecules also causes them to be surrounded by water molecules. Fruits like grapes or green apples are acidic and neutralize the charge on the pectin molecule. Neutralization of fruits that are not naturally acidic is achieved by the addition of lemon juice. Sugar is added and the solution is boiled until the sugar concentration reaches 60-65%. The concentrated sugar solution attracts water molecules from the pectin. The pectin molecules then bond to each other, trapping the other ingredients to form a gel similar to gelatin.

The amount of pectin added to preserves is important. Too much pectin causes preserves to be rubbery; too little pectin makes them runny. Similarly, if the fruit lacks sufficient acid or too little sugar is added, no gel will form. The result will be syrup.

Making preserves used to be an annual ritual in many homes, and allowed fruits to be enjoyed throughout the year. Getting the right balance of the critical ingredients was an art. The chemistry of the process may not have been understood, but the products were delicious!

Exploring Further

1. Find out the difference between jams and jellies.
2. Peaches and apricots are not acidic and do not contain pectin. How can preserves be made from them?
3. Concentrated sugar solutions are hygroscopic. What is the mechanism by which sugar attracts water?

20.2 CONCEPT REVIEW

20. Distinguish between the dispersed phase and the continuous phase in a colloid.
21. How do colloids differ from solutions and suspensions?
22. How is the Tyndall effect dependent upon particle size?
23. **Apply** List the products found in your kitchen that you can classify as aerosols, foams, emulsions, gels, or sols.

Chapter 20 REVIEW

Summary

20.1 Solutions

1. A true solution is a homogeneous mixture of molecules, atoms, or ions that cannot be separated by filtration.
2. When ionic compounds are dissolved in water, the ions become hydrated and dissociate.
3. Polar solvents tend to dissolve polar solutes; nonpolar solvents tend to dissolve nonpolar solutes.
4. Two liquids that are mutually soluble in all proportions are completely miscible.
5. A solution has reached solution equilibrium when the rates of particles leaving and returning to the solution are equal. When solution equilibrium is reached, the solution is said to be saturated.
6. The rate of solution is increased by increasing the surface area of crystal exposed, stirring to prevent the buildup of saturated solution around solute crystals, and increasing the kinetic energy of solute and solvent.
7. Enthalpy of solution is the energy change that occurs when one substance is dissolved in another.
8. Henry's law: The mass of a gas that will dissolve in a liquid at a given temperature varies directly with the partial pressure of that gas.
9. Molarity, molality, mole fraction, and mass percent are common concentration units.

20.2 Colloids

10. A colloid is composed of two phases, the dispersed phase and the continuous phase.
11. In terms of particle size, colloids are intermediate between solutions and suspensions. Colloidal particles range between ~ 1 and ~ 100 nm in at least one dimension.
12. Colloids can scatter light (Tyndall effect), undergo constant random motion (Brownian motion) as a result of bombardments by the particles of the suspending medium, and act as excellent adsorbing materials.

Key Terms

solvent	enthalpy of solution
solute	molality
dissociation	mole fraction
solvation	crystalloid
hydration	colloid
miscibility	suspension
solution equilibrium	colloid chemistry
saturated solution	Tyndall effect
unsaturated solution	Brownian motion
solubility	adsorption
supersaturated solution	electrophoresis

Review and Practice

24. Define solute and solvent.
25. What is the most common solvent? Why is it the most common?
26. Describe the process by which water dissolves an ionic compound.
27. Why is a solution likely to form when a polar solute and a polar solvent are combined? Why is a solution unlikely to form when a polar solute and a nonpolar solvent are combined or when a nonpolar solute and a polar solvent are combined?
28. What factors determine the solubility of a nonpolar solid in a nonpolar liquid?
29. "Oil and water do not mix" is an old adage. What term describes two liquids that are not mutually soluble?
30. What is a saturated solution? Describe the equilibrium process that takes place in a saturated solution.
31. How is solubility specified?

32. Using Appendix Table A-7, select a reagent that will precipitate only one of the positive ions in each of the following mixtures.
  a. $AgNO_3$ and $NaNO_3$
  b. $FeCl_3$ and NaCl
  c. $BaCl_2$ and KNO_3
  d. KCl and $CuSO_4$

33. Predict the results of the following reactions.
  a. Na_2CO_3 (aq) + $FeSO_4$ (aq) →
  b. K_2CrO_4 (aq) + $Ba(NO_3)_2$ (aq) →
  c. $Na_2C_2O_4$ (aq) + $CaCl_2$ (aq) →
  d. K_3PO_4 (aq) + $Zn(CH_3COO)_2$ (aq) →

34. Discuss three actions that can increase the rate of solution of a solute in a solvent. Explain how each of these actions works to increase rate of solution.

35. Explain why ΔH_{sol} is negative for some solids, such as sodium hydroxide, dissolved in water.

36. Relate temperature conditions to the solubilities of solids and gases in liquids.

37. State Henry's law. Explain how Henry's law accounts for what happens when a bottle of soda is opened.

38. What effect would an increase in pressure have on the solubility of a solid in a liquid? What effect would it have on the solubility of a gas in a liquid?

39. Describe the content of a 1*M* solution of a molecular substance such as glucose, $C_6H_{12}O_6$.

40. Calculate the molarity of the following for the given volume of solution.
  a. 31.1 g $Al_2(SO_4)_3$ in 1.00×10^3 cm^3
  b. 48.4 g $CaCl_2$ in 1.00×10^2 cm^3
  c. 313.5 g $LiClO_3$ in 2.50×10^2 cm^3

41. Determine the concentration of each type of ion in 0.100*M* solutions of each of the following compounds.
  a. KBr **c.** $CuCl_2$
  b. Rb_2S **d.** $Mg(NO_3)_2$

42. Calculate the concentration of Ca^{2+} ions in a solution of 21.0 g of $CaCl_2$ in 5.00×10^2 cm^3 of solution. Determine the concentration of Cl^- ions in the same solution.

43. How many moles of CH_3COO^- ions are in 46.0 cm^3 of a 0.250*M* solution of $Pb(CH_3COO)_2$?

44. Calculate the molality of the following solutions.
  a. 20.0 g of NH_4Cl in 4.00×10^2 g of water
  b. 145 g of CH_3COCH_3 in 0.320 kg of water

45. Calculate the molality of the following solutions.
  a. 98.0 g RbBr in 824 g water
  b. 85.2 g $SnBr_2$ in 1.40×10^2 g water
  c. 10.0 g $AgClO_3$ in 201 g water
  d. 0.059 g KF in 0.272 g water

46. Calculate the mole fraction for each component in the following solutions.
  a. 67.4 g of C_9H_7N in 2.00×10^2 g C_2H_6O
  b. 5.48 g of $C_5H_{10}O_5$ and 3.15 g of CH_6N_4O in 21.2 g H_2O

47. The mole fraction of benzene in a solution with cyclohexane is 0.125. What does this statement tell you about the solution? How much benzene would have to be combined with 2.00 mol of cyclohexane to give this mole fraction?

48. Find the mass percent concentration of 12.0 g of glucose, $C_6H_{12}O_6$, in 2.50×10^2 g of water.

49. Compute the masses of solute needed to make the following solutions.
  a. 1.00 dm^3 of 0.780*M* $Sc(NO_3)_3$
  b. 2.00 dm^3 of 0.179*M* $Er_2(SO_4)_3$
  c. 1.00×10^2 cm^3 of 0.626*M* VBr_3
  d. 2.50×10^2 cm^3 of 0.0965*M* $DyCl_3$

50. Consider the mixture of sodium nitrate and barium chloride mentioned on page 504. The mass of the mixture was measured, then the mixture was dissolved in water. Excess sulfuric acid was added. The precipitate was dried and its mass was measured.

a. What was the precipitate? See Appendix Table A-7.
b. If the precipitate had a mass of 25.3 g, what mass of barium chloride was in the original mixture?
c. If the original mixture had a mass of 56.2 g, what percentage was barium chloride?

51. Compute the mass of solute that will yield the following solutions.
a. $Fe_2(C_2O_4)_3$ to be added to 1.00×10^3 g of water for a 0.851*m* solution
b. $VOBr_3$ to be added to 1.00×10^3 g water for a 0.534*m* solution
c. $C_7H_4O_2Br_2$ to be added to 2.00×10^2 g of C_2H_6O so that the mole fraction of the solvent is 0.510
d. $C_{14}H_{16}N_2$ to be added to 1.00×10^3 g of $C_4H_{10}O$ so that the mole fraction of the solute is 0.363

52. A solution is prepared containing 84.0 g of $LiNO_3 \cdot 3H_2O$ in 3.00×10^2 g of water. What is the molality of Li^+ ions? Note that you must account for H_2O added in the hydrate.

53. A chemist has a 5.00-g mixture of silver and potassium nitrates. To isolate the silver, HCl is added. The dry AgCl precipitate has a mass of 3.50 g. What was the percentage of $AgNO_3$ in the original mixture?

54. How did Graham differentiate between a colloid and a crystalloid?

55. Distinguish between colloids and true solutions.

56. Describe the structure of a colloid.

57. How does a colloid differ from a suspension?

58. Compare the sizes of particles in solutions, colloids, and suspensions.

59. What causes the Tyndall effect? Where can the Tyndall effect be observed?

60. Describe Brownian motion. How is it caused?

61. What is adsorption? Why do colloidal materials make good adsorbing materials?

Concept Mastery

62. What composition and properties characterize a true solution?

63. How do hydration and solvation differ?

64. Explain why some nonpolar substances are insoluble in each other.

65. Under what conditions can a supersaturated solution form?

66. What is the relationship between ΔH_{sol} and solubility for solids and gases?

67. Sodium hypochlorite, NaClO, is the active ingredient in liquid chlorine bleach. How does a 0.500*M* NaClO solution differ from a 0.500*m* NaClO solution?

68. Using Figure 20.1, explain why the orientation of water molecules surrounding sodium ions differs from that of water molecules surrounding chloride ions.

69. The ionic compound calcium chloride is commonly used to melt ice on roads. Why is a solution of this compound not truly a "calcium chloride solution?"

70. Iodine crystals are relatively insoluble in water. However, they do dissolve in alcohol, benzene, and several other organic solvents. Explain these solubilities.

71. Two liquid substances are stirred together and then allowed to stand. After standing, the liquids have formed layers with a clearly defined interface between the two phases. Does this fact mean the liquids are completely immiscible? Explain.

72. What is the usefulness of solutions whose concentrations are given in molarity rather than in mass percent?

73. Devise an experiment to find out if a sample of sucrose solution is unsaturated, saturated, or supersaturated.

74. Devise a method to make a supersaturated solution of $Na_2S_2O_3$.

75. Describe the process of electrophoresis and explain why the dispersed particles in a colloid exhibit this behavior.

Application

76. Ammonium sulfate and ammonium carbonate are both ionic compounds and are important in the production of fertilizer. Explain why both compounds are soluble in water but insoluble in cyclohexane.

77. NaOH is commonly used as a drain cleaner. How would the fact that it has a negative ΔH_{sol} be a help in using NaOH to clear a kitchen drain?

78. Ethylene glycol is used as automobile antifreeze and has a freezing point of −13.2°C and a boiling point of 198°C at standard pressure. Ethylene nitrate has a freezing point of −22.3°C and a boiling point of 197°C at standard pressure, yet it would not be usable mixed with water in a car radiator. In terms of solubility, hypothesize why this observation is true.

79. Proteins compose enzymes, connective tissue, and many other important components of our bodies. Proteins are very complex chains of amino acids cross-linked by covalent and hydrogen bonding into specific shapes and can have molecular masses of millions of atomic mass units. Could a protein that has one dimension less than 100 nm and another dimension much larger than that act as a dispersed particle in a colloid? Explain.

Critical Thinking

80. Predict the solubility of the first substance in the second on the basis of comparative polarities.
   a. RbF in ethanol
   b. CuS in water
   c. ethanol in water
   d. NCl_3 in C_6H_6

81. Predict the solubility of the following in water on the basis of comparative polarities.
   a. CuF_2 **c.** Rb_2S **e.** CsI
   b. $ScCl_3$ **d.** ThS_2 **f.** LiBr

82. Laboratory glassware can be dried by rinsing the glassware with distilled water, then ethanol, then ether, which evaporates very quickly. Why is the ethanol-rinsing step necessary?

Reading

Rona, Peter A. "Mineral Deposits from Sea-Floor Hot Springs." *Scientific American* 254, no. 1 (Jaunary 1986): 84-92.

Cumulative Review

1. A roll of wire has a mass of 57.2 grams per meter of length. What would be the length of 1.40×10^3 kilograms of wire?

2. Balance the following equations.
   a. $C_3H_8 + O_2 \rightarrow CO_2 + H_2O$
   b. $FeBr_3 + Na_2S \rightarrow Fe_2S_3 + NaBr$

3. Calculate the empirical formula of a compound whose percentage composition is 0.58% H, 44.89% K. 36.74% O, and 17.79% P.

4. How does atomic radius change as you look across a period of elements? Explain why the radius changes in this way.

5. If a gas occupies 629 cm^3 at 24°C and 99.0 kPa, what volume will it occupy at STP?

6. What is the molecular mass of a gas if 2.20 g occupy 2.13 dm^3 at 20.4°C and 107.1 kPa?

CHAPTER **Preview**

CONTENTS

CHAPTER 21

Colligative Properties

Petroleum is the source of an enormous number of the consumer products we take for granted every day. Petroleum is a complex mixture of substances. In order to utilize these substances, they must be separated into useful fractions. The method that chemists have developed for this separation is distillation. The tower in the large photograph is a petroleum distilling tower.

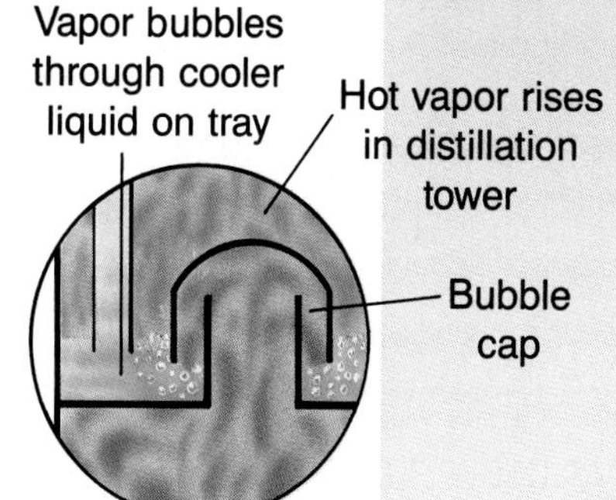

In this chapter you will study some of the properties of solutions that are affected by solutes and how they can be used in distillation processes. The drawing on this page shows a cross section of a small part of a distillation column. The column is divided into sections by horizontal plates. Short sections of pipe are welded to holes in the plates and caps are suspended above the pipes. Vapors rise through the column and condensed liquids flow downward. Eventually, a steady state is reached, and each plate contains a different fraction of petroleum.

EXTENSIONS AND CONNECTIONS

Chemists and Their Work

Hazle J. Shorter, page 524

Frontiers

More than One, Less than Many: How freezing points are affected by the number of atoms, **page 528**

Chemistry and Technology

Drinking Water from Seawater: Some ways to make drinkable water from seawater, **page 536**

OBJECTIVES

- state Raoult's law and use it to calculate the vapor pressure of a solution.
- explain how the process of fractional distillation can be used to separate components of a solution.
- identify the effect of solute particles on the boiling point and freezing point of a solvent.

21.1 Vapor Pressure Changes

Automobile owners add antifreeze/summer coolant to the water in their automobile radiators. From the name of the product, it is obvious that its purpose is to prevent the water from freezing and to help in cooling the engine in hot weather. How does the antifreeze prevent freezing or overheating? It is not that the antifreeze itself has such a low freezing point or high boiling point. The best protection is afforded by a mixture that is roughly 50% antifreeze and 50% water. In this section we will investigate the behavior of solutions in freezing and boiling. We will look at those properties of solvents that are changed by solutes.

Word Origins

kolligativ: (GK) glue together

colligative—properties of substances (solute and solvent) together

Raoult's Law

Properties of a solution that are different from those of the pure solvent are vapor pressure, freezing point, boiling point, and rate of diffusion of particles through a membrane. These properties, which are determined by the *number* of particles in solution rather than by the *type* of particle, are **colligative properties.**

Consider a solute dissolved in a liquid solvent. Some of the solute particles take up space on the liquid surface normally occupied by solvent particles. As a result, these solute particles decrease the opportunity for solvent particles to escape from the liquid surface by evaporation. Because fewer solvent particles escape into the vapor phase, a decrease in vapor pressure occurs. Thus, if the solute is nonvolatile, the vapor pressure of a solution is always less than that of the pure solvent at the same temperature. The lowering of the vapor pressure of the solvent varies directly as the mole fraction of dissolved solute. *Any nonvolatile solute at a specific concentration lowers the vapor pressure of a solvent by an amount that is characteristic of that solvent.* The characteristics of the solute are not involved. Ionic compounds and molecular compounds with high melting points are typical nonvolatile solutes. To determine the vapor pressure of a solution, we must correct the vapor pressure of the pure solvent for the presence of the solute. The equation used is

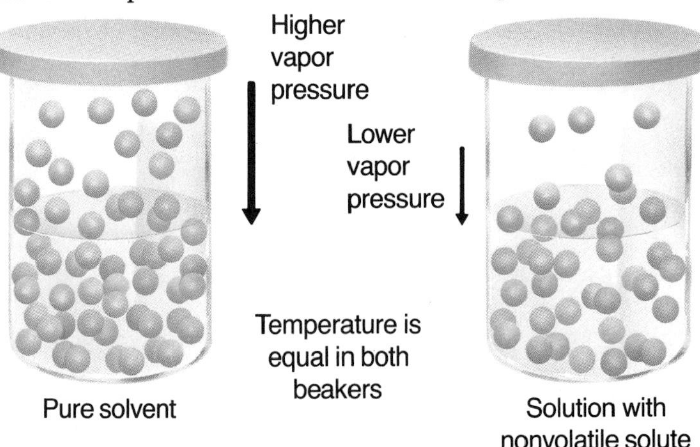

Figure 21.1 The presence of a nonvolatile solute partially blocks the escape of solvent molecules from the liquid and, thus, lowers the vapor pressure of the solvent.

$$vapor\ pressure_{(solution)} = mole\ fraction_{(solvent)} \times vapor\ pressure_{(solvent)}$$

$$P_{solution} = x_{solvent}P_{solvent}$$

This expression is a mathematical statement of **Raoult's law,** named for Francis Raoult, a French chemist. He first stated the principle that the *vapor pressure of a solution varies directly as the mole fraction of solvent.* Thus, a solution of water and glucose in which the mole fraction of water is 0.5 will have a vapor pressure that is 0.5 times the vapor pressure of water alone. The difference between the vapor pressure of the solvent and the vapor pressure of the solution, $P_{solvent} - P_{solution}$, is the lowering of the vapor pressure.

Raoult's law describes an ideal solution. An **ideal solution** is one in which all intermolecular attractions are the same. In other words, the solute-solute, solvent-solvent, and solute-solvent attractions are all essentially the same. All the subsections of this chapter except the last apply strictly to ideal solutions.

SAMPLE PROBLEM

Raoult's Law: Nonvolatile Solute

Sugar is extremely soluble in water and is not a volatile substance. What would the vapor pressure of water be at 70°C if 1.00×10^2 g of the water have dissolved 2.00×10^2 g of sucrose, $C_{12}H_{22}O_{11}$?

Solving Process:

In order to compute the vapor pressure of the water in the solution we must know the mole fraction of the water. Therefore, the first step in solving the problem is to convert both solute and solvent quantities to moles.

$$100\ \text{g}\ H_2O \times \frac{1\ \text{mol}\ H_2O}{18.0\ \text{g}\ H_2O} = 5.56\ \text{mol}\ H_2O$$

$$200\ \text{g}\ C_{12}H_{22}O_{11} \times \frac{1\ \text{mol}\ C_{12}H_{22}O_{11}}{342\ \text{g}\ C_{12}H_{22}O_{11}} = 0.585\ \text{mol}\ C_{12}H_{22}O_{11}$$

Total moles of solution = 5.56 + 0.585 = 6.14 mol solution.
Mole fraction of water = 5.56/6.14 = 0.905.
The vapor pressure of water at 70°C (Table 18.2, page 459) is 31.2 kPa.
Using Raoult's Law, the vapor pressure of the solution = (0.905)(31.2) = 28.3 kPa

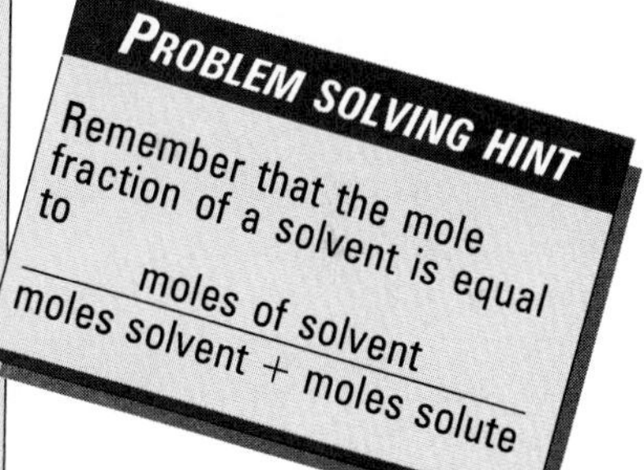

PROBLEM SOLVING HINT
Remember that the mole fraction of a solvent is equal to
moles of solvent
moles solvent + moles solute

PRACTICE PROBLEMS

1. Find the vapor pressure of a water solution in which the mole fraction of $HgCl_2$ (a nonvolatile, nonionizing solute) is 0.163 at 25°C. Use the vapor pressure of water in Table 18.2, page 459.

2. Ethylene glycol is commonly used in automobile antifreeze. If, in a mixture of ethylene glycol and water, the mole fraction of ethylene glycol is 0.100, what is the vapor pressure of water over the mixture at 100°C, the normal boiling point of water?

CHEMISTRY IN DEPTH

Raoult's Law Calculations

In the case of a volatile solute, Raoult's law is often inadequate to predict the behavior of a real solution. However, there are many solutions whose behavior approaches the ideal closely enough to be treated as such. Each volatile component of an ideal solution has a vapor pressure that can be determined by Raoult's law.

CHEMISTS AND THEIR WORK

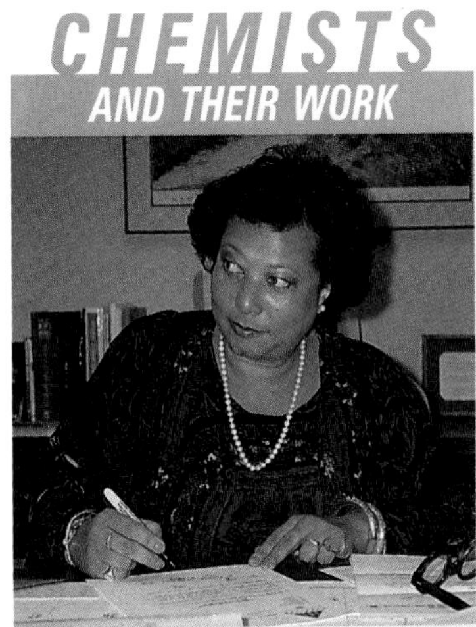

Hazle J. Shorter
(1942–)

Hazle Shorter is an example of a person whose interest in chemistry led her first into another field and then back to chemistry. She received her B.S. in chemistry from Manhattanville College of the Sacred Heart, but her interest in the biological aspects of chemistry led her into medicine. She received her M.D. from Howard Medical School and is presently Director of Medical Communications Worldwide for the DuPont/Merck Pharmaceutical Company. Shorter is responsible for all medical communications to physicians and health care professionals concerning new medicinal and pharmacological products.

SAMPLE PROBLEM

Raoult's Law: Volatile Solute

Consider a solution composed of 1.00 mole of benzene, C_6H_6, and 1.00 mole of toluene, $C_6H_5CH_3$. The mole fraction of each volatile component in the solution is 0.500. Thus, the number of benzene molecules in the solution equals the number of toluene molecules. At 25°C, benzene has a vapor pressure of 12.7 kPa, and toluene has a vapor pressure of 3.79 kPa. What is the vapor pressure of the resulting solution?

Solving Process:

According to Dalton's law, the vapor pressure is the sum of the individual pressures of benzene and toluene. The vapor pressure of benzene in the solution is

$$(0.500)(12.7 \text{ kPa}) = 6.35 \text{ kPa}$$

The vapor pressure of toluene in the solution is

$$(0.500)(3.79 \text{ kPa}) = 1.90 \text{ kPa}$$

Thus, the vapor pressure of the resulting solution is 8.25 kPa (6.35 kPa + 1.90 kPa).

Note especially that the vapor phase is much richer in benzene, the more volatile component, than is the liquid phase. The vapors of the two substances are in the same volume and at the same temperature. Thus, the ratio of their pressures must be equal to the ratio of the number of vapor molecules of each. This ratio is 6.35/1.90 or 3.34/1.

PRACTICE PROBLEMS

3. Find the total vapor pressure of a solution of ethanal (mole fraction 0.300, vapor pressure at 18°C = 86.3 kPa) in methanol at 18°C. (Vapor pressure of methanol at 18°C = 11.6 kPa.)

Fractional Distillation

We take advantage of the difference in vapor pressures of two components of a solution in the process of fractional distillation. Recall from Chapter 3 that distillation is a physical process of separating substances that have different boiling points.

In order to understand how the distillation process takes place, we need to look at the boiling and condensing behavior of a solution. We will use a solution consisting of benzene and toluene as an example. Solutions of different proportions of these substances will be allowed to come to equilibrium with their vapor at the boiling point. The composition of the liquid phase and of the vapor phase will then be measured. The data obtained will be plotted on a graph. Such a graph is shown in Figure 21.2. The y coordinate of each point on the green curve represents the boiling point of a solution whose composition is given by the x coordinate of the point. In the same way, the y coordinate of each point on the orange curve represents the condensing point of a vapor whose composition is given by the x coordinate. If a solution and its vapor are in equilibrium, they must both be at the same temperature. For instance, in Figure 21.2, liquid of composition A would be in equilibrium with vapor B. From the boiling points of the pure substances, we can see that benzene (BP = 80.1°C) is more volatile than toluene (BP = 110.6°C). As a consequence, the vapor in equilibrium with a solution of any composition will always be enriched in the more volatile component, in this case benzene.

Consider boiling a solution of benzene and toluene in which the mole fraction of benzene is 0.3. It will boil at the temperature represented by point *A* in Figure 21.2. The vapor in equilibrium with that solution will have the composition represented by point *B*, which is at the same temperature. It is richer in benzene than the original solution because benzene is more volatile than toluene. If we condense this vapor, we will obtain a solution that is represented by point *C*. If we then boil the solution of composition

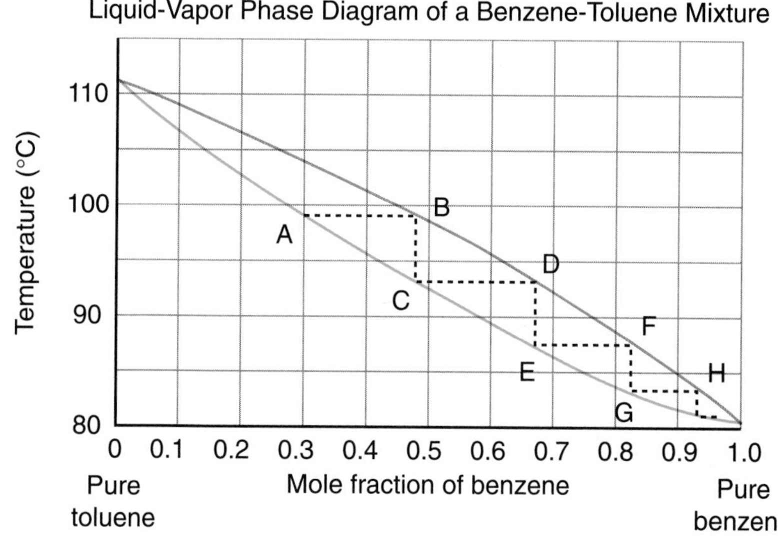

Figure 21.2 The lower curve represents the boiling point and composition of a benzene-toluene mixture of varying proportions. The upper curve represents the vapor concentrations. The stepwise line indicates how a benzene-toluene mixture can be separated into its components by fractional distillation.

Figure 21.3 The packing material in this distillation column (left) increases the surface area of the column. Thus, this column has the separating capabilities of one many times its height. Fractionating towers (right) are used in industry to separate petroleum into its components.

C, it will produce a vapor of composition D. This vapor is even richer in benzene. If this vapor is condensed, the resulting solution will boil at point E to produce vapor (point F) whose mole fraction of benzene is greater than 0.8. This process can be continued until nearly pure benzene is obtained as vapor, and almost pure toluene remains behind.

It is possible to construct a distillation apparatus in which each step takes place in a separate section of the equipment. In the laboratory, a fractional distillation apparatus, Figure 21.3 (left), is often used in separating volatile liquids. A column is packed with stainless steel rings, glass beads, or broken ceramic pieces. These packings offer a surface on which the liquid and vapor can come to equilibrium. The top of the column is cooler because it is farther from the heat source at the bottom. The liquid and vapor in equilibrium at the top are of different composition from the liquid and vapor in equilibrium at the bottom. Each layer in the column acts as a separate stage for the distillation.

Table 21.1

Petroleum Distillation Fractions	
Name	**Boiling point range (°C)**
Gasoline	60-280
Jet fuel	190-450
Kerosene	350-550
Diesel fuel	430-700
Fuel oil	550-800
Lubricating oil	600-1000

In industry, a fractionating tower like that in Figure 21.3 (right) is used for distillation on a commercial scale. The industrial tower uses bubble plate columns, as shown on page 521. As the different steps of the distillation are accomplished, petroleum is separated into useful products. The products are drawn off at different levels of the tower according to their boiling points. Some of the common products of the fractional distillation of petroleum and their boiling point ranges are given in Table 21.1. These products do not have a specific boiling point because they represent a mixture or a solution of various hydrocarbon molecules with boiling point ranges that fall between the values shown in the table.

Figure 21.4 Because sodium chloride lowers the freezing point of water, it is mixed with ice to maintain lower temperatures while freezing homemade ice cream.

Boiling Point and Freezing Point

The presence of nonvolatile solute particles in a solvent causes the boiling point of a solution to be raised. The boiling point of a liquid is the temperature at which the vapor pressure of the liquid equals the atmospheric pressure. In a solution, then, a higher temperature is needed to put enough solvent particles into the vapor phase to equal atmospheric pressure. The boiling point of a solution is, therefore, higher than that of the pure solvent. Figure 21.5 shows the elevated boiling point of a solvent such as water when a nonvolatile solute is added. Note that the curve is shifted to the right.

How does the addition of a nonvolatile solute affect the freezing point of a solution? The freezing point is the temperature at which the vapor pressures of the solid and liquid are equal. Because the addition of solute particles lowers the vapor pressure of the liquid, the vapor pressures of the solid and liquid will be equal at a lower temperature. Solutions, then, will freeze at a lower temperature than the pure solvent alone. The freezing point is said to be depressed. You can see this decrease in the freezing point of a solution in the graph in Figure 21.5.

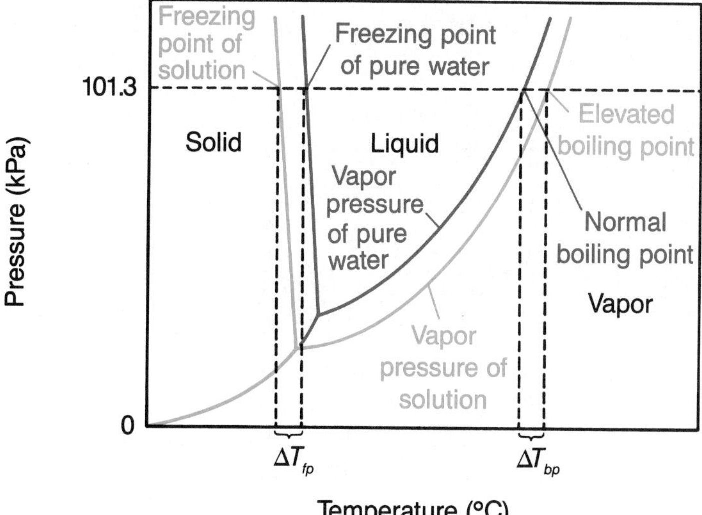

Figure 21.5 The addition of nonvolatile solute particles to a solvent such as water lowers the vapor pressure of the solvent and causes it to freeze at a lower temperature and to boil at a higher temperature.

Because a solute lowers the freezing point of water, sodium chloride is mixed with ice in making homemade ice cream. NaCl and $CaCl_2$ are used to deice sidewalks and streets. However, if the outside temperatures fall below the freezing points of the resulting salt solutions, ice will form anyway. That is why these salts are ineffective as deicers at very low temperatures.

In summary, *the addition of a nonvolatile solute to a liquid causes both a boiling point elevation and a freezing point depression.* Both boiling point elevation and freezing point depression occur because the vapor pressure of the solvent is lowered by the solute. These changes depend only on the concentration of the solute particles, and not upon the composition of the solute.

FRONTIERS

More than One, Less than Many

An iceberg melts at 0°C, as does a snowflake. But what happens when there are only several thousand molecules of water in an ice crystal? Scientists are asking whether the physical properties of a small amount of a substance are different from those of a larger amount of the same substance.

Researchers are studying clusters, the name given to small amounts of a substance on the order of 10^1 to 10^4 atoms. A cluster can be distinguished from a bulk sample because a cluster has a much greater fraction of its atoms occupying its surface. For example, in a cluster of 55 argon atoms, 42 atoms reside on the surface.

Researchers have found that clusters of 1500 sodium atoms begin to display the organization of a solid. Below that number, some scientists hypothesize that the cluster behaves as a liquid. As an atom joins a cluster, it loses one or more of its electrons to the other atoms in the cluster. The electrons of the atoms in the cluster occupy common energy levels. After the cluster grows to a certain size, it begins to grow geometrically, as a crystal of a solid does. Other scientists hypothesize that the smaller cluster can have solid characteristics at much smaller sizes. They hypothesize that liquid characteristics of smaller clusters can be accounted for by considering the cluster to have the interior of a solid while its surface is that of a liquid.

Even more intriguing is the fact that computer simulations indicate that the melting point and the freezing point of a substance may differ. Scientists working in this field must answer many questions before the picture is clear.

21.1 CONCEPT REVIEW

4. Explain the effect of a solute on the boiling point of a solvent. Use the terms *vapor pressure, atmospheric pressure,* and *solute particles* in your answer.
5. State Raoult's law.
6. Explain how a fractional distillation apparatus works.
7. **Apply** Find another industry, besides petroleum, in which fractional distillation is an essential step.

21.2 Quantitative Changes

OBJECTIVES

- calculate the effect of a solute on the boiling point, freezing point, and osmotic pressure of a solution.
- determine the molecular mass of a solute from freezing point, boiling point, or osmotic pressure data.
- explain the concept of osmotic pressure.
- relate the concept of ion activity to nonideal solutions.

One of the most important measurements made by chemists in research is the molecular mass of a compound. In Chapter 19 you learned how to make such a measurement on a gaseous compound or the vapor of a substance. For many substances, however, the temperature required to change a substance to vapor will result in decomposition of the substance. Other methods useful for determining molecular masses are the measurement of the effect the substance has on the boiling point, freezing point, or osmotic pressure of a solvent. In this section you will learn how to use these techniques for the measurement of molecular mass.

Calculating Freezing and Boiling Points

It has been found experimentally that 1 mole of nonvolatile solute particles will raise the boiling temperature of 1 kg of water by 0.515 C°. The same concentration of solute will lower the freezing point of 1 kg of water by 1.853 C°. These two figures are the molal boiling point constant, K_{bp}, and molal freezing point constant, K_{fp}, for water. The corresponding constants for some other common solvents are given in Appendix Table A-8. A 1m solution of sugar in water contains 1 mol of solute particles per 1 kg of water. However, a 1m solution of table salt contains 2 mol of solute particles (1 mol of Na^+ and 1 mol of Cl^- ions). A 1m solution of calcium chloride contains 3 mol of solute particles per 1 kg of water (1 mol of Ca^{2+} and 2 mol of Cl^- ions). The 1m sugar solution freezes at 1.853 C° below the freezing point of pure water. However, the 1m salt solution freezes at 2(1.853 C°), or 3.706 C°, below the freezing point of pure water. The 1m solution of calcium chloride freezes at 3(1.853 C°), or 5.559 C°, below the freezing point of pure water. The multiple lowering of the freezing point and elevation of the boiling point by ionic substances supports the theory of dissociation.

Figure 21.6 $CaCl_2$ is more effective at melting ice than NaCl because it produces three ions (Ca^{2+} + $2Cl^-$) and thus lowers the freezing point more than equimolar amounts of NaCl.

SAMPLE PROBLEM

Freezing Point and Boiling Point Changes

If 85.0 grams of sugar are dissolved in 392 grams of water, what will be the boiling point and freezing point of the resulting solution? The molecular formula of sugar is $C_{12}H_{22}O_{11}$.

Solving Process:

(a) Determine the number of moles of solute. The mass of one mole of $C_{12}H_{22}O_{11}$ equals 342 g.

Since there are 85.0 g of sugar present, the number of moles of sugar in 392 g of water is found by

$$\frac{85.0 \text{ g } \cancel{C_{12}H_{22}O_{11}}}{} \left| \frac{1 \text{ mol } C_{12}H_{22}O_{11}}{342 \text{ g } \cancel{C_{12}H_{22}O_{11}}} \right. = 0.249 \text{ mol of } C_{12}H_{22}O_{11}$$

(b) Convert this quantity to mol/kg of water. The result is the molality of the solution.

$$\frac{0.249 \text{ mol } C_{12}H_{22}O_{11}}{392 \text{ } \cancel{\text{g water}}} \left| \frac{1000 \text{ } \cancel{\text{g water}}}{1 \text{ kg water}} \right. = 0.635m \text{ solution}$$

(c) Determine the boiling point elevation. The boiling point is raised 0.515 C° for each mole of sugar added to 1 kg of water. Therefore, the boiling point is

$$100°C + (0.635m)(0.515 \text{ C}°/m) = (100 + 0.327)°C = 100.327°C$$

(d) Determine the freezing point depression. The freezing point is lowered 1.853 C° for each mole of sugar added to 1 kg of water. Therefore, the freezing point is

$$0°C - (0.635m)(1.853 \text{ C}°/m) = (0 - 1.18)°C = -1.18°C$$

PROBLEM SOLVING HINT

When determining significant digits, consider the 100°C and 0°C points to be exact because the Celsius scale is defined by these points.

MINUTE LAB

Zero Degrees And Counting

Put on an apron and goggles. Fill two 400 cm^3 beakers with crushed ice and add 50 cm^3 of cold tap water to each. Place a thermometer and a stirring rod in each beaker. Stir with the stirring rods, and wait until both beakers are at constant temperature. **CAUTION:** *Don't use the thermometers as stirring rods. They break easily.* Record the thermometer readings after they reach constant temperature. Then add 75 g of NaCl (fine rock salt) to one of the beakers. Stir both beakers (some of the salt will dissolve) and wait several minutes until the temperature is constant before recording the readings. What did you observe happening to the temperature in the beaker that had the salt added? How do you explain the observed temperature change?

Let us review what we have done. To determine the change in the freezing point, we multiplied the freezing point constant (K_{fp}) by the molality (m). We used a similar process for determining the boiling point elevation. We can write two equations representing the mathematical steps we have performed.

$$\Delta T_{fp} = mK_{fp} \qquad \Delta T_{bp} = mK_{bp}$$

SAMPLE PROBLEM

ΔT_{fp} and ΔT_{bp} for an Ionic Solute

If 26.4 grams of nickel(II) bromide are dissolved in 224 grams of water, what will be the boiling point and freezing point of the resulting solution? (Assume 100% dissociation and no interaction between ions.)

Solving Process:

(a) Determine the number of moles of solute. The formula mass of one mole of $NiBr_2$ is 58.7 g + 2(79.9 g) = 219 g. The number of moles of $NiBr_2$ is

$$\frac{26.4 \text{ g } NiBr_2}{} \left| \frac{1 \text{ mol } NiBr_2}{219 \text{ g } NiBr_2} \right. = 0.121 \text{ mol } NiBr_2$$

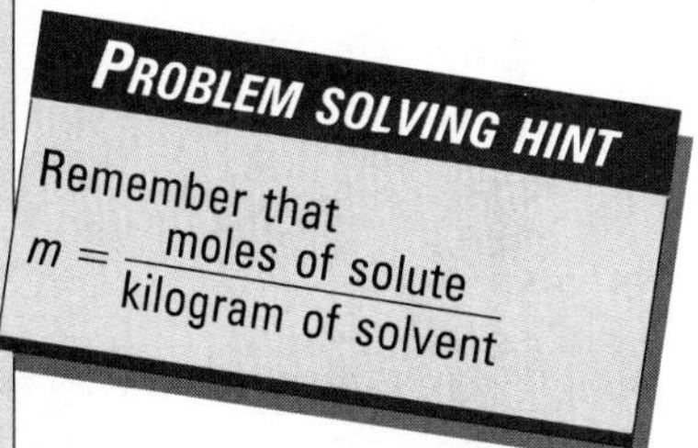

(b) Determine the molality of the solution. Because the mass of water is 224 g, the molality equals

$$\frac{0.121 \text{ mol}}{224 \text{ g water}} \left| \frac{1000 \text{ g water}}{1 \text{ kg water}} \right. = 0.540m \text{ } NiBr_2$$

However, the molality in total particles is three times this number 3(0.540) or 1.62m because

$$NiBr_2(cr) \rightarrow Ni^{2+}(aq) + 2Br^-(aq)$$

(c) Determine the boiling point.

$$\begin{aligned} \text{Boiling point} &= 100°C + \Delta T_{bp} = 100°C + mK_{bp} \\ &= 100°C + (1.62m)(0.515 \text{ C°}/m) \\ &= 100.834°C \end{aligned}$$

(d) Determine the freezing point.

$$\begin{aligned} \text{Freezing point} &= 0°C - \Delta T_{fp} = 0°C - mK_{fp} \\ &= 0°C - (1.62m)(1.853 \text{ C°}/m) = -3.00°C \end{aligned}$$

Figure 21.7 In the Antarctic seas, there are species of fish that have antifreeze proteins in their blood.

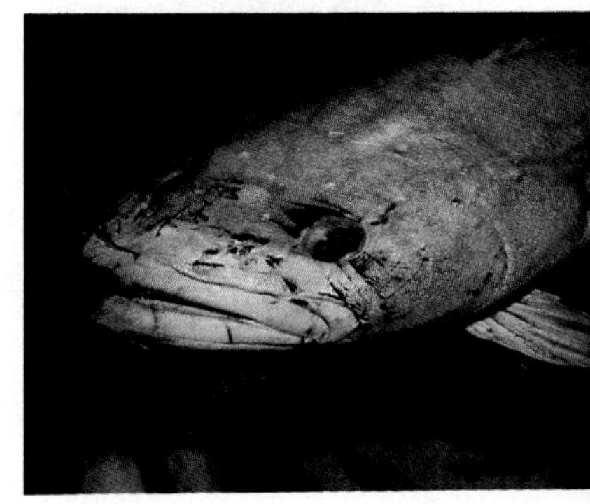

PRACTICE PROBLEMS

Compute the boiling and freezing points of these solutions.

8. 25.5 g $C_7H_{11}NO_7S$ (4-nitro-2-toluenesulfonic acid dihydrate) in 1.00 × 10^2 g H_2O (nonionizing solute)

9. 1.00 × 10^2 g $C_{10}H_8O_6S_2$ (1,5-naphthalenedisulfonic acid) in 1.00 × 10^2 g H_2O (nonionizing solute)

10. 21.6 grams $NiSO_4$ in 1.00 × 10^2 g H_2O (assume 100% ionization)

11. 77.0 grams $Mg(ClO_4)_2$ in 2.00 × 10^2 g H_2O (assume 100% ionization)

12. 41.3 grams $C_{15}H_9NO_4$ (2-methyl-1-nitroanthraquinone) in 1.00 × 10^2 grams $C_6H_5NO_2$ (nitrobenzene)

▼ Experimental Determination of Molecular Mass

CHEMISTRY IN DEPTH

Molecular mass of a solute may be determined by using boiling point elevation or freezing point depression. A known mass of the solute is added to a known mass of a solvent and the shift in the boiling or freezing point is measured. This change can be used to calculate the number of moles of solute this mass represents.

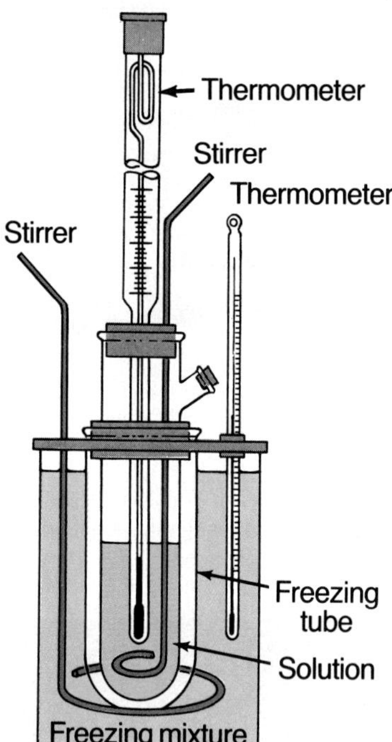

Figure 21.8 This apparatus is used in molecular mass determinations. The freezing point depression of the solution in the tube is carefully measured and the number of moles of solute calculated.

SAMPLE PROBLEM

Molecular Mass Determination

If 99.0 grams of nonionizing solute are dissolved in 669 grams of water, and the freezing point of the resulting solution is −0.960°C, what is the molecular mass of the solute?

Solving Process:

(a) Determine the molality of the solution.

$$\Delta T_{fp} = mK_{fp} \qquad m = \frac{\Delta T_{fp}}{K_{fp}}$$

$$m = \frac{0.960\ \cancel{C^\circ}}{} \left| \frac{1m}{1.853\ \cancel{C^\circ}} \right. = 0.518\ \text{mol/kg}$$

(b) Calculate the molecular mass.

$$\frac{99.0\ \text{g}}{669\ \cancel{\text{g}}} \left| \frac{1000\ \cancel{\text{g}}}{1\ \cancel{\text{kg}}} \right| \frac{1\ \cancel{\text{kg}}}{0.518\ \text{mol}} = 286\ \text{g/mol}$$

concentration | convert to g/kg | convert to g/mol

PRACTICE PROBLEMS

Calculate the molecular mass of the nonionic solutes. Use Appendix Table A-8 if necessary.

13. 8.02 grams of solute in 861 grams of water lower the freezing point to −0.430°C.

14. 64.3 grams of solute in 3.90×10^2 grams of water raise the boiling point to 100.680°C.

15. 20.8 grams of solute in 128 grams of acetic acid lower the freezing point to 13.5°C.

16. 10.4 grams of solute in 164 grams of phenol lower the freezing point to 36.3°C.

17. 2.53 grams of solute in 63.5 grams of nitrobenzene lower the freezing point to 3.40°C.

Osmosis and Osmotic Pressure

Suppose a membrane separates pure water from a solution of sugar in water. Suppose only the water molecules can pass through the membrane. A membrane that allows only certain types of particles to pass through is called a **semipermeable membrane.** The water moves through the membrane from the pure solvent into the solution more rapidly than it does from the solution into the pure solvent. The movement of solvent through a membrane from an area of higher solvent concentration (lower solute concentration) to an area of lower solvent concentration (higher solute concentration) is called osmosis. Osmosis applies to any solvent system; water is the most familiar example and the

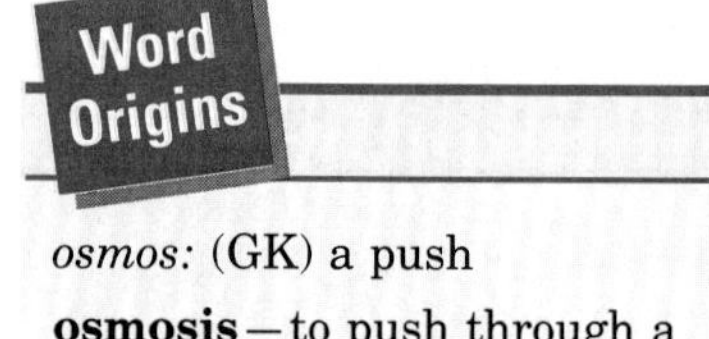

osmos: (GK) a push
osmosis—to push through a membrane

system that most often concerns biologists. During osmosis, solvent molecules at first pass more rapidly into the solution than in the opposite direction. As time goes on, however, the solution becomes more and more dilute, and the rate at which solvent molecules pass back into the pure solvent increases.

The net movement of solvent particles into the solution results in an increase in volume on the solution side of the membrane. The increasing volume of solution begins to exert a backward pressure on the membrane, and eventually the pressure balances the pressure of the water moving through the membrane from the solvent side. Thus, the rate of flow in both directions becomes equal, and dynamic equilibrium is reached. At equilibrium, solvent molecules continue to move through the membrane in both directions, but there is no net change in their concentration. At equilibrium, the pressure exerted by the increased volume is called the **osmotic pressure** of the solution. It is equivalent to the pressure that must be applied to prevent the net flow of solvent through a membrane from the pure solvent into a solution of the same solvent.

It is important to remember that osmotic pressure results from two things. First, the difference in concentration of solvent (such as water) on two sides of a membrane is due to the presence of a greater concentration of solute on one side. Second, the solute particles cannot pass through the membrane, whereas the solvent particles can.

Osmotic pressure is a colligative property that is of great importance in living systems because it is the mechanism by which all the cells of a living organism take in water. A living cell can be considered as an aqueous solution enclosed in a semipermeable membrane, the cell membrane. The fluid that surrounds the cell must have the same osmotic pressure as the fluid within the cell. If it does not, water flows into the cell, causing it to swell and burst, or flows out, causing the cell to shrink or dehydrate. These considerations are important in the administration of intravenous (IV) fluids. As shown in Figure 21.11, if the IV fluid is less concentrated than the blood, fluid will flow into red blood

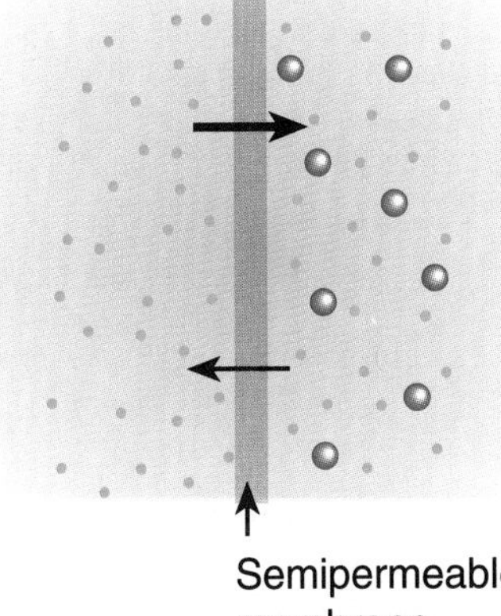

Figure 21.9 Osmosis is the unequal flow of water molecules through a semipermeable membrane when one side of the membrane has a higher concentration of a solute that cannot pass through the membrane.

Figure 21.10 There is a net transfer of solvent molecules into the solution (a) by osmosis until the height of the solution column (b) is sufficient to exert enough pressure to stop the osmosis. Osmotic pressure can be measured as the force applied to the piston (c) to stop the osmotic flow through the membrane.

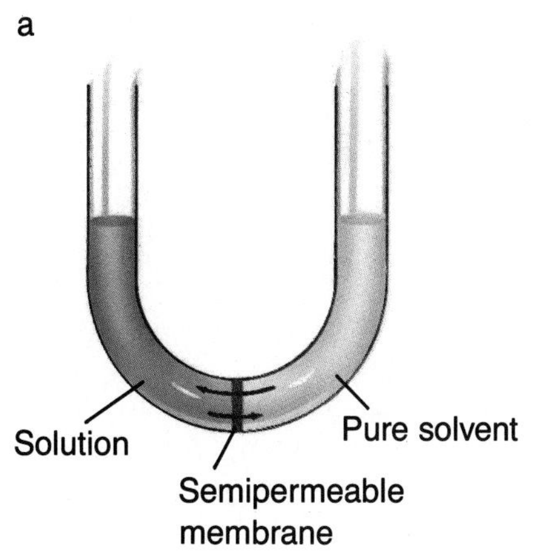

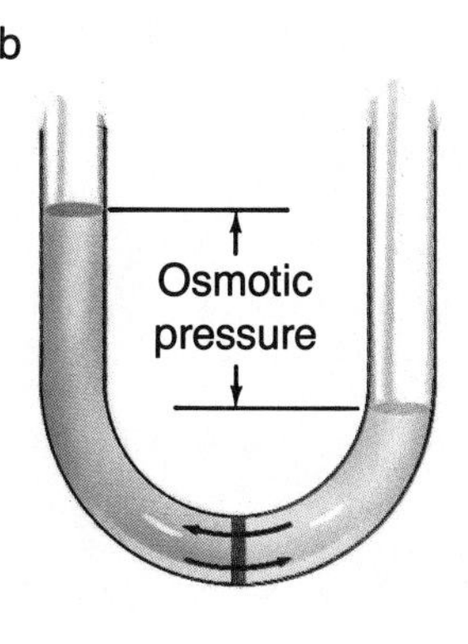

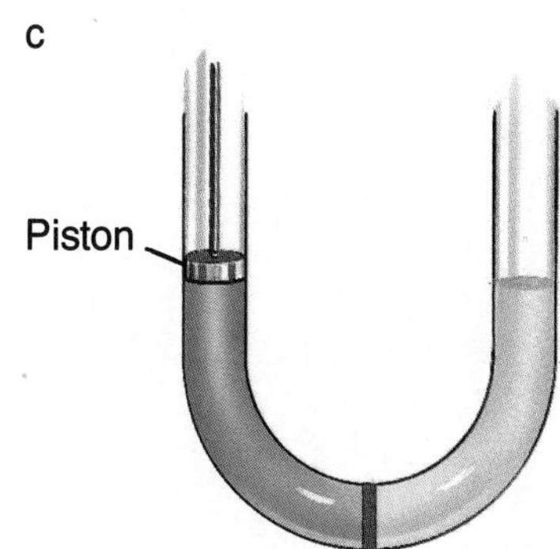

Figure 21.11 When a red blood cell (a) is surrounded by a solution of low concentration, fluid will flow into the cell and cause it to burst (b). If the cell is surrounded by a solution of high concentration, fluid will flow out of the cell, and the cell will shrivel up (c).

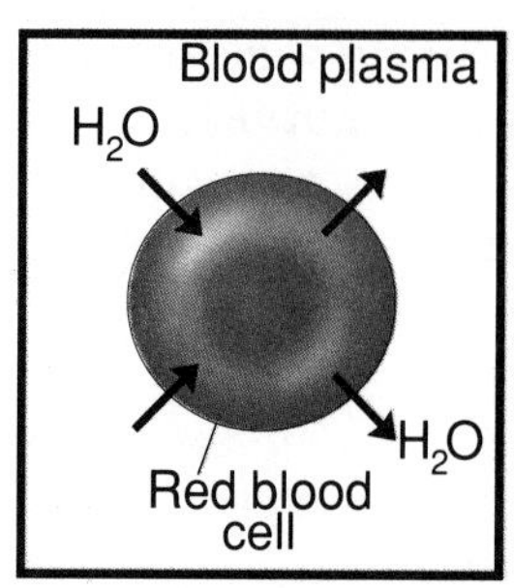

a Osmosis

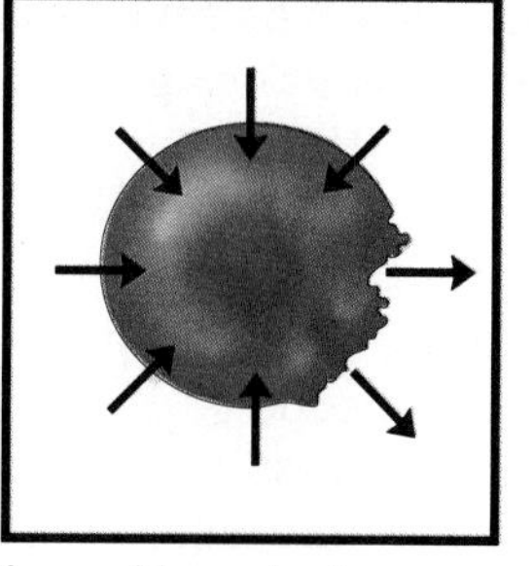

b Hemolysis

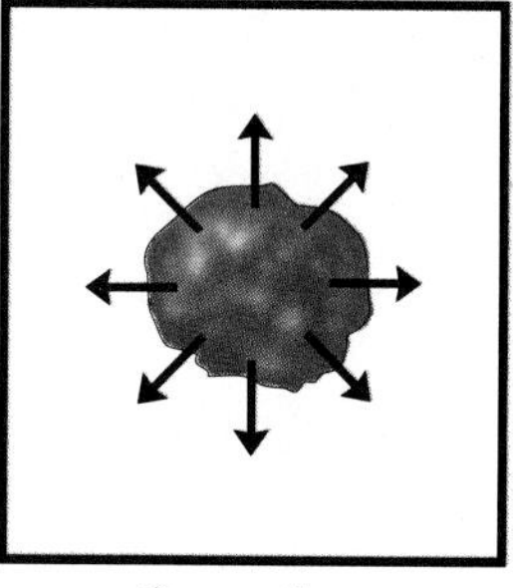

c Crenation

cells, causing them to burst, a process known as hemolysis. If the IV fluid has a greater concentration than that of the blood, the red blood cells shrivel as fluid flows out.

Osmosis has many interesting applications. A pickle is simply a cucumber that has been placed in brine, lost water by osmosis, and shriveled. Osmosis is involved in the preservation of fruit with sugar and of meat with salt. Because of the high sugar or salt concentrations, bacteria on the candied fruit and salted meat become dehydrated and die. The movement of water from soil into plant roots and then into the upper parts of the plant is at least partly due to osmosis.

CHEMISTRY IN DEPTH

Osmotic Pressure Calculations

Because osmotic pressure is a colligative property, it can be expressed in an equation such as

$$\Pi = mK_{\text{osm}}$$

where Π is the osmotic pressure. However, the constant for osmosis, K_{osm}, is highly temperature dependent. As a result, osmotic pressure is expressed in the form

$$\Pi = MRT$$

In this equation, M is the molarity of the solution, R is the ideal gas constant, and T is the Kelvin temperature.

Figure 21.12 After several hours in a strong NaCl solution, a carrot (right) shows the effects of osmosis. Water has flowed out of the carrot cells into the salt solution, leaving the carrot limp. The carrot in tap water (left) is not affected.

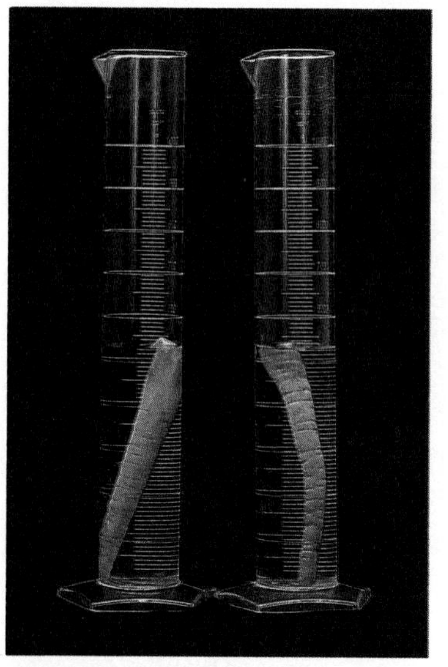

SAMPLE PROBLEM

Osmotic Pressure

What osmotic pressure is exerted by a 1.82M solution at 18°C?

Solving Process:

$$\Pi = MRT$$

Since the units of R are $\text{dm}^3\cdot\text{kPa/mol}\cdot\text{K}$, we will express our molarity as $1.82\ \text{mol}/1.00\ \text{dm}^3$.

$$\Pi = \frac{1.82\ \cancel{\text{mol}}}{1.00\ \cancel{\text{dm}^3}} \left| \frac{8.31\ \cancel{\text{dm}^3}\cdot\text{kPa}}{\cancel{\text{mol}\cdot\text{K}}} \right| \frac{291\ \cancel{\text{K}}}{} = 4.40 \times 10^3\ \text{kPa}$$

PRACTICE PROBLEMS

18. The average osmotic pressure of blood is 780 kPa at 25°C. What concentration of glucose, $C_6H_{12}O_6$, in water will have the same osmotic pressure?
19. The osmotic pressure of a hemoglobin solution is 0.40 kPa at 25°C. The solution contains 0.300 g of hemoglobin in 30.0 cm^3 of solution. Find the molecular mass of hemoglobin.
20. Calculate the osmotic pressure at a temperature of 25°C of a solution with concentration 0.490*M*.

Dilute solution

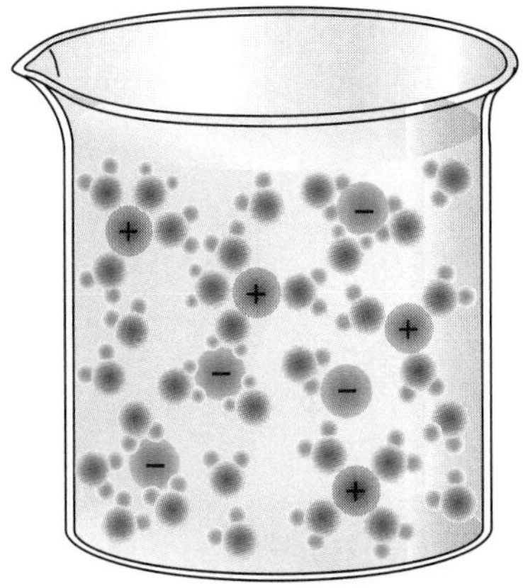

Concentrated solution

Figure 21.13 Attractive forces between ions are greater in concentrated solutions (bottom) than in dilute solutions (top), so the freezing point depression and boiling point elevation in concentrated solutions are not as great as would be expected.

Nonideal Solutions

Chemists generally make the assumption that ionic compounds are completely dissociated in water solution. However, data obtained from colligative property experiments seem to contradict the assumption. The same indication comes from measurements of the solubility of certain compounds. The reason for the deviation of these solutions from ideal behavior is the attractive force between oppositely charged ions. All real solutions deviate slightly from Raoult's law. Up to now we have assumed that all solutions are ideal, that is, they obey Raoult's law precisely.

Let us look at a specific case of deviation from Raoult's law. Two moles of sodium chloride should produce two moles each of sodium ions and chloride ions. Therefore, you would expect that a 2*m* solution of NaCl would lower the freezing point of the water by 4(1.853 C°) or 7.412 C°. Salt actually lowers the freezing point only about three-quarters of this amount. We can explain this deviation by assuming that the solute is completely dissociated into ions but that these ions interact.

The actual ion effectiveness in freezing point depression and boiling point elevation is known as the **activity** of the ion. In dealing with a nonideal solution, chemists use activities in place of concentrations. As the solution becomes more concentrated, each ion individually becomes less effective through increased interaction with its neighboring ions.

21.2 CONCEPT REVIEW

21. Find the boiling point and freezing point of a solution of 100.0 g of α-carotene (a nonvolatile solute) in 294 g of benzene.
22. Find the molecular mass of a substance, 4.80 g of which in 150.0 g of water forms a solution that freezes at −0.319°C.
23. Describe the process of osmosis and give an example of its importance in humans.
24. Why do real solutions of ionic compounds differ slightly from Raoult's law?
25. **Apply** Calcium chloride does not form an ideal solution in water. Would you need to add more or less calcium chloride to a road to melt ice if it were an ideal solute? Explain.

CHEMISTRY AND

TECHNOLOGY

Drinking Water from Seawater

A natural resource we often take for granted is water we drink. We cannot drink the water from the ocean because it has a high concentration of dissolved ions. If you were to drink seawater, the osmotic pressure in your intestines would dehydrate you so severely that death would result eventually.

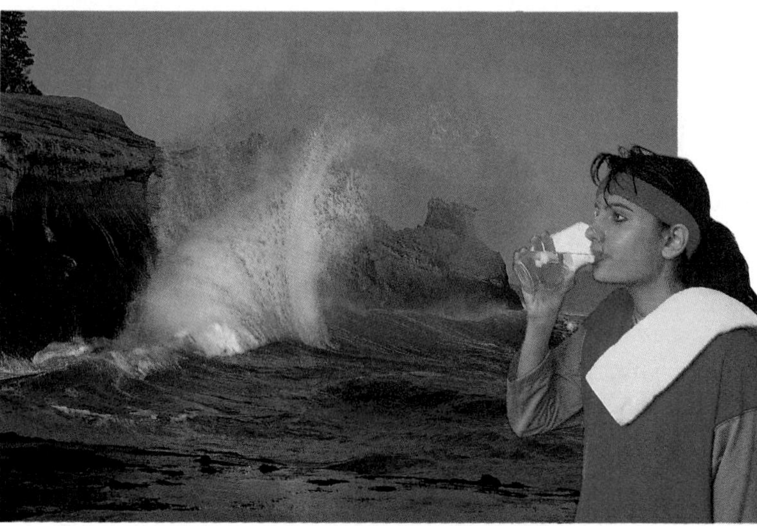

The water we use for drinking is called potable water. Because of population growth, waste, and pollution, our supplies of potable water are being depleted. One recourse available is to convert seawater into potable water by removing the dissolved salts, a process called desalination.

One method of desalination is to distill the water. Unfortunately, distillation requires the expenditure of large quantities of energy. The source of this energy is the burning of fossil fuels, which are also endangered resources. Even countries that have enormous quantities of fossil fuels are finding the production of potable water by distillation a very expensive process.

A more energy-efficient process, electrodialysis, involves passing the seawater across membranes in an electric field. The water on one side of the membrane is depleted in ions while the water on the other side is enriched. After several passes, the depleted water is sufficiently pure to be drinkable. Each pass results in a lowering of salt content of between 20 and 60%. Seawater is not considered potable until at least 98.6% of its salt has been removed.

Another popular technology for new potable water plants involves osmosis. Not only is it energy efficient, it costs 20–30% less than other processes. You have learned that the diffusion of pure solvent through a semipermeable membrane into a solution creates a difference in pressure across the membrane. If we now apply a pressure greater than the osmotic pressure to the solution side of the membrane, pure solvent will be forced through the membrane. The membranes may be in the form of fine tubing or in the form of a spiral sheet. This process is called reverse osmosis because the applied pressure forces the solvent to move in the opposite direction from that of osmosis.

Other processes are being tested. One is the freezing of seawater. When seawater is frozen, much of the salt is squeezed out. The ice can then be melted and refrozen. These steps can be repeated until the water is drinkable. Still another approach is to pass the seawater through substances called ion exchange resins. Here, hydrogen ions are substituted for other positive ions and hydroxide ions are substituted for other negative ions. The hydrogen ions and hydroxide ions then combine to form water. With our present technology, however, neither ion exchange nor freezing appears to be economically practical.

Thinking Critically

1. Why might you expect desalination by freezing to be an expensive process?
2. Find out how ion exchange resins work. What uses do they have in addition to desalination?

Chapter 21 REVIEW

Summary

21.1 Vapor Pressure Changes

1. Amounts of solutes affect the colligative properties of a solution. These properties include vapor pressure, boiling point, freezing point, and osmotic pressure of a solution.
2. A nonvolatile solute lowers the vapor pressure of a solvent by an amount that depends on the mole fraction of the solvent and the nature of the solvent.
3. The vapor pressure of a solution is the sum of the vapor pressure of each pure component multiplied by its mole fraction in the solution.
4. Many substances can be separated by taking advantage of the difference in their vapor pressures. The process used is called fractional distillation.
5. The addition of a nonvolatile solute to a solvent results in both an elevation in the boiling point and a depression in the freezing point because the vapor pressure of the solvent has been lowered by the presence of the solute.

21.2 Quantitative Changes

6. Changes in boiling and freezing points may be calculated from the relations $\Delta T_{fp} = mK_{fp}$ and $\Delta T_{bp} = mK_{bp}$.
7. Molecular mass of a solute may be determined by dissolving a known mass of the substance in a known mass of solvent and measuring the resulting shift in freezing or boiling point.
8. Osmotic pressure is pressure built up between the two sides of a semipermeable membrane as a result of unequal passing of solvent particles through the membrane.
9. Osmotic pressure can be computed from $\Pi = MRT$, where M is molarity, T is Kelvin temperature, and R is the ideal gas constant.
10. Ideal solutions obey Raoult's law. Some ionic solutions deviate because of attractive forces between ions of opposite charges.

Key Terms

colligative property
Raoult's law
ideal solution
semipermeable membrane
osmotic pressure
activity

Review and Practice

26. List four colligative properties. What determines the colligative properties of a solution?
27. What sorts of substances are categorized as nonvolatile solutes?
28. What is the vapor pressure of a solution composed of 67.0 grams of maltose (malt sugar, $C_{12}H_{22}O_{11}$), a nonvolatile solute, in 2.00×10^2 grams of water at 60.0°C?
29. What is the vapor pressure at 24°C of a solution of 8.27 grams of benzene, C_6H_6, (vapor pressure = 12.1 kPa) and 3.87 grams of chloroform, $CHCl_3$, (vapor pressure = 25.2 kPa)?
30. Give the definition of boiling point and freezing point.
31. What effect does the addition of a nonvolatile solute have on the boiling point of a solution? Why?
32. What effect does the addition of a nonvolatile solute have on the freezing point of a solution? Why?
33. State the definitions and give the numerical values of K_{fp} and K_{bp} for water.
34. Calculate the boiling point and freezing point of each of the following solutions of molecular substances.
    a. 97.5 g of $C_{12}H_{22}O_{11}$ (sucrose) in 185 g of water

b. 14.0 g of $C_{10}H_8$ (naphthalene) in 25.0 g of C_6H_6 (benzene)
c. 5.00×10^2 g $C_{20}H_{27}NO_{11} \cdot 3H_2O$ (amygdalin trihydrate) in 5.00×10^2 g of water
d. 2.50×10^2 grams of $C_7H_4BrNO_4$ (3-bromo-2-nitrobenzoic acid) in 5.00×10^2 g of C_6H_6 (benzene)
e. 60.0 g of C_9H_{18} (propylcyclohexane) in 1.00×10^2 g of acetic acid

35. Use the following liquid-vapor phase diagram for a solution of two volatile liquids, A and B, to answer the questions.
a. What are the boiling points of pure A and pure B?
b. If the mole fraction of A is 0.3, what is the boiling point of the solution?
c. What mole fraction of A will the vapor contain at this temperature?
d. If the vapor at this temperature is condensed, what will be the boiling point of the resulting solution?
e. If the resulting solution is boiled, what will be the mole fractions of A and B in the vapor?

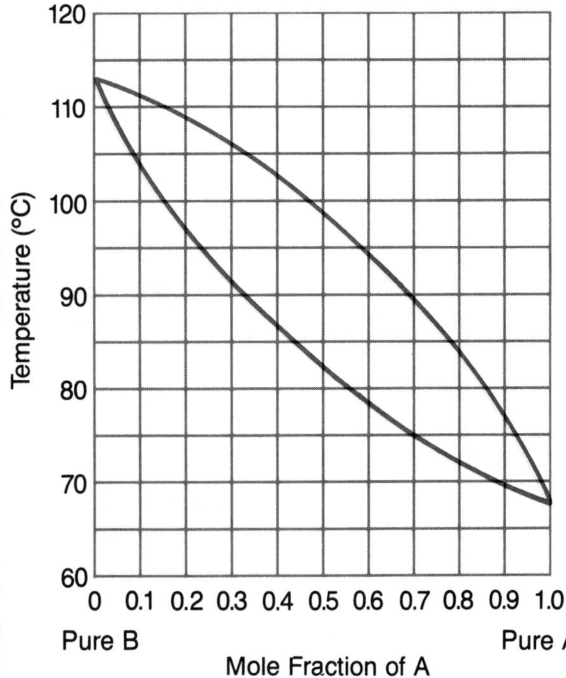

f. What is the composition of a solution of A and B that boils at 100°C? What is the composition of the vapor above that solution?

36. What is the molecular mass of a nonionic solute if 5.60 grams dissolved in 104 grams of water lower the freezing point to −0.603°C?

37. Suppose 1.500 g of a compound are dissolved in 35.00 g of camphor. The freezing point of pure camphor is 178.75°C, the freezing point of the solution is 164.4°C, and the K_{fp} for camphor is 37.7 C°·kg/mol. Assuming no ionization, what is the molecular mass of the compound?

38. What is a semipermeable membrane?

39. What causes osmotic pressure? Give an example of where it is important in the human body.

40. What osmotic pressure is exerted by a solution that is 4.04*M* at 26°C?

41. What is the molarity of a solution with an osmotic pressure of 1.14×10^4 kPa at 37°C?

42. Why are ice-melting compounds such as NaCl or $CaCl_2$ not effective if the ground temperature is extremely cold?

43. A sample of seawater taken from the Atlantic Ocean freezes at −2.14°C, and a sample taken from the Arctic Ocean freezes at −1.96°C. What is the total molality of ionic solutes present in each?

44. How many moles of the ionic compound NaBr dissolved in 1.00 kg of water would lead to the same vapor pressure as would 0.246 mole of glucose, $C_6H_{12}O_6$?

45. What mass of the ionic compound $SrCl_2$ dissolved in 1.00 kg of water would lead to the same vapor pressure lowering of water as would 87.3 g of fructose, $C_6H_{12}O_6$?

46. Find the osmotic pressure of a solution containing 8.10 grams of solute of molar mass 1310 grams in 1.00 dm^3 of water at 22°C.

47. If 18.6 g of a solute with molar mass 8940 g are dissolved in enough water to make 1.00 dm^3 of solution at 25°C, what is the osmotic pressure of the solution?

48. When 96.0 g of a solute are dissolved in enough water to make 1.00 dm^3 of solution at 25°C, the osmotic pressure of the resulting solution is 15.8 kPa. What is the molecular mass of the solute?

49. When 69.7 grams of a solute with molar mass 2790 grams are dissolved in enough water to make 1.00 dm^3 of solution at 20.0°C, what is the osmotic pressure of the solution?

50. When 2.00×10^2 grams of a solute are dissolved in enough water to make 1.00 dm^3 of solution at 21.0°C, the osmotic pressure of the resulting solution is 1.00×10^2 kPa. What is the molecular mass of the solute?

Concept Mastery

51. Explain why the vapor pressure of a solvent decreases as a nonvolatile solute is added to the solution.

52. What are the characteristics of an ideal solution?

53. What is the relationship between Raoult's law and ideal solutions?

54. What characteristics must the components of a solution have in order to be separated by fractional distillation?

55. Use the liquid-vapor phase diagram for benzene and toluene in Figure 21.2 on page 525 to answer the following questions.
    - **a.** What is the boiling point of toluene?
    - **b.** What is the boiling point of the solution when the mole fraction of benzene is 0.30?
    - **c.** What is the mole fraction of toluene in solution when the mixture boils at 85°C?
    - **d.** Compare the mole fraction of benzene in the solution at 95°C with the mole fraction of benzene in the vapor phase at the same temperature. Explain the difference.
    - **e.** If you mixed 2.00 moles of toluene with 0.50 mole of benzene and heated the solution, at what temperature would it begin to boil at standard atmospheric pressure? Describe the composition of the vapor at the onset of boiling.
    - **f.** If the horizontal axis of the graph were changed to read in mole fraction of toluene, how would the curves appear?

56. List the following water solutions in order of their predicted freezing points, lowest to highest. Assume complete ionization.
    - **a.** 0.10m NaCl
    - **b.** 0.15m HCl
    - **c.** 0.05m $CaCl_2$
    - **d.** 0.10m $(NH_4)_3PO_4$

57. Describe ionic activity, then explain ionic activity using a CaI_2 solution as an example. Assume CaI_2 does not form an ideal solution.

58. Chemists sometimes determine the melting point of a substance in order to check its purity. Would the melting point of a substance containing impurities be higher or lower than that of the pure substance? Explain.

59. The Dead Sea contains dissolved minerals, including approximately 10% $MgCl_2$. Explain why the vapor pressure of pure water at 20°C is higher than the vapor pressure of a sample of equal volume and temperature from the Dead Sea.

60. Assuming the water vapor and carbon dioxide are removed, suggest a method to separate liquid air into O_2, N_2, and Ar.

61. Explain why a plant will wilt if placed in a solution of sugar or salt that is more concentrated than that inside the plant. What will happen if the wilted plant is then placed in pure water?

Application

62. Drugs injected into the bloodstream by an IV (intravenous) injection are in a 0.9% solution of NaCl, which contains the same concentration of dissolved particles as the blood.
    a. What would happen to the blood cells if a more concentrated solution were injected?
    b. What would happen to the blood cells if a more dilute solution were used?
63. A mechanic wishes to prepare an ethylene glycol solution in 20.0 kg of water to use as an antifreeze. The resulting solution should have a freezing point of −30.0°C. Ethylene glycol has a molecular mass of 62.0 g/mol. What mass of ethylene glycol should be used? The vapor pressure of ethylene glycol is negligible.
64. Destructive distillation is the heating of a substance in a closed container without oxygen. Methanol, wood alcohol, was first prepared by the destructive distillation of wood. The fractional distillation of the resulting liquid yielded acetic acid (b.p. = 117.90°C), acetone (b.p. = 56.2°C), and methanol (b.p. = 65.0°C).
    a. Which substance would be removed from the liquid first?
    b. Which would remain after the other two were removed?
    c. Which two would be the most difficult to separate from each other? Why?
65. The ionic compounds calcium chloride and sodium chloride, as well as the nonionic compound urea, NH_2CONH_2, are commonly used to melt street ice.
    a. Per mole, which compound would be most effective?
    b. Per kg, which is most effective?

Critical Thinking

66. The vapor pressure of a $1m$ NaCl solution equals that of a $1m$ KI solution. Explain why the vapor pressure of these solutions is the same.
67. Columns used for fractional distillation are often heated by a steam jacket. The temperature inside the jacket is controlled in order to provide a temperature gradient. Would the temperature be controlled so the column becomes hotter or cooler going upward in the column? Why?
68. Most automobile manufacturers recommend that an antifreeze (ethylene glycol) and water mixture be kept in the car's cooling system in summer as well as in winter. Why is this practice advisable?

Readings

Berry, R. Stephen. "When the Melting and Freezing Points Are Not the Same." *Scientific American* (August 1990): 68–74.

Friedman, R. "Seawater to Drink." *Technology Review 92* (Aug./Sept. 1989): 14–15.

Storey, Kenneth B. and Janet M. Storey. "Frozen and Alive." *Scientific American* (December 1990): 92–97.

Cumulative Review

1. What is the molality of a solution made from 1.00×10^2 grams of water and 20.0 grams of $Fe_2(C_2O_4)_3$?
2. What mass of Li_2S must be dissolved in 2.50×10^2 cm^3 of water to prepare a $0.560m$ solution?
3. What is the mole fraction of acetone, $(CH_3)_2CO$, in a solution of 50.0 grams of acetone and 50.0 grams of water?
4. What is the molarity of a solution made from 2.73 g of KNO_3 dissolved in 500.0 cm^3 of solution?

Chapter 21 REVIEW

5. What is the mass percent of $Ba(NO_3)_2$ in a solution formed by dissolving 20.6 g of $Ba(NO_3)_2$ in 293 g of water?

6. If the solubility of $KMnO_4$ at 20°C is 6.38 g in 100 cm^3 of water, how much $KMnO_4$ will have to be dissolved in 5.00×10^2 cm^3 of water at 20°C to produce a saturated solution?

7. You have a 32.0-g sample of ice at a temperature of −43.0°C. How many joules of energy are necessary to heat the ice to 0°C? How many joules are needed to melt the ice? What total amount of energy is required to change the ice at −43°C to steam at 130°C?

8. Find the relative rates of diffusion for Ne and Kr.

9. What is the molecular mass of a gas if a 151-cm^3 sample has a mass of 0.873 g at 90.0°C and 105 kPa?

10. Adrenaline (epinephrine) is nonionic, and is a hormone that triggers the release of extra glucose into the bloodstream when the body is stressed. Adrenaline is composed of 59.0% C, 7.15% H, 7.65% N, and 26.2% O.
   a. What is the empirical formula for the hormone adrenaline?
   b. A 0.985-g sample of adrenaline is dissolved in 36.0 g of CCl_4. There is a resulting elevation of 0.750 C° in the boiling point of the solution. What are the molecular mass and molecular formula for adrenaline? The molal boiling-point elevation constant, K_{bp}, for CCl_4 is 5.02 C°·kg/mol.

11. A cubic crystal measures 1.0 cm on a side.
   a. Calculate the volume of the crystal.
   b. Calculate the surface area of the crystal.
   c. The crystal is crushed into smaller cubic crystals averaging 0.1 cm on an edge. How many crystals would result? What would be the total surface area?
   d. The crystals are crushed again to give a fine powder made of cubic crystals that average 1×10^{-6} m on a side. What would be the total surface area now? Express your answer in m^2 and km^2.

12. Balance the following equations.
   a. $C_6H_{12}O_6 \rightarrow C_2H_5OH + CO_2$
   b. $H_3PO_3 \rightarrow H_3PO_4 + PH_3$
   c. $As_2S_3 + (NH_4)_2S \rightarrow (NH_4)_3AsS_3$
   d. $BaO + N_2 + C \rightarrow Ba(CN)_2 + CO$
   e. $SiO_2 + HF \rightarrow SiF_4 + H_2O$
   f. $PI_3 + H_2O \rightarrow H_3PO_3 + HI$

13. A 0.527-gram sample of copper was reacted completely with excess sulfur. After removing the sulfur residue, the mass of the resulting compound was found to be 0.659 gram. Determine the empirical formula of the compound.

14. What mass of sulfuric acid may be obtained by reacting 10.0 kg of SO_3 with water? What volume at STP of SO_3 would have to be generated to give this mass?

15. When 6.423 grams of a hydrate of calcium sulfate are heated until all water is driven off, the mass of the resulting anhydrous calcium sulfate is 5.082 grams. What is the formula of the hydrate?

16. In addition to producing heat during solution and reacting with grease to form soluble materials, crystal-type drain cleaners contain a small amount of aluminum along with the main ingredient, sodium hydroxide. The aluminum reacts with sodium hydroxide according to the following equation.

$$2Al + 2NaOH + H_2O \rightarrow 2NaAlO_2 + 3H_2$$

If a 500-gram can of drain cleaner actually contains 455 grams of sodium hydroxide and 45.0 grams of aluminum shavings, what volume of hydrogen at STP will be produced in the reaction? Suggest a reason that this action would help clear drains.

CHAPTER Preview

CONTENTS

Reaction Rate and Chemical Equilibrium

CHAPTER 22

Striking a kitchen match isn't a very mysterious way of starting a chemical reaction. As you probably guessed, the heat generated by the friction of the match head in contact with a rough surface started the combustion reaction. You now know that heat was the activation energy of this reaction. Why is activation energy necessary for a reaction? What does activation energy activate?

Some reactions take place until one of the reactants is used up. Other reactions run for a while and then appear to stop even though unused reactants remain. These reactions have reached equilibrium. In this chapter, you will learn about reaction rates and systems that reach equilibrium.

EXTENSIONS AND CONNECTIONS

OBJECTIVES

- distinguish between thermodynamic stability and kinetic stability.
- list and describe the factors that influence the rate of reaction.
- distinguish among heterogeneous catalyst, homogeneous catalyst, and inhibitor.
- describe and determine reaction mechanisms for simple reactions.

22.1 Reaction Rates

Different reactions take place at different rates. The rate at which a reaction occurs is important information for chemists and chemical engineers. If a reaction is very slow, the chemist may not wish to make use of it in a research project. Likewise, the chemical engineer probably will not be able to design a profitable process around a reaction which is very slow. Consequently, chemists study reaction rates and factors affecting these rates.

There is a legend concerning a warrior who saved the life of the king's daughter. In gratitude, the king granted the warrior whatever he wanted as a reward. The warrior noticed a chessboard nearby. He asked that he be given one grain of wheat on the first square of the board, two grains on the second, four grains on the third, and so on, until the last square on the board was covered. The king, without thinking, quickly agreed. There are 64 squares on a chessboard. The last square would require 2^{63} grains of wheat. The total to cover the board would be 3.7×10^{19} grains. That is enough wheat to fill the Grand Canyon over 200 meters deep! Amounts quickly increase when they are doubled repeatedly. The king agreed to give more than he owned.

There are many reactions in which the rate doubles for each 10 C° rise in temperature. You can see that a significant increase in temperature could mean a tremendous increase in rate, just as the number of grains of wheat required to cover squares on the chessboard became huge after a number of doublings.

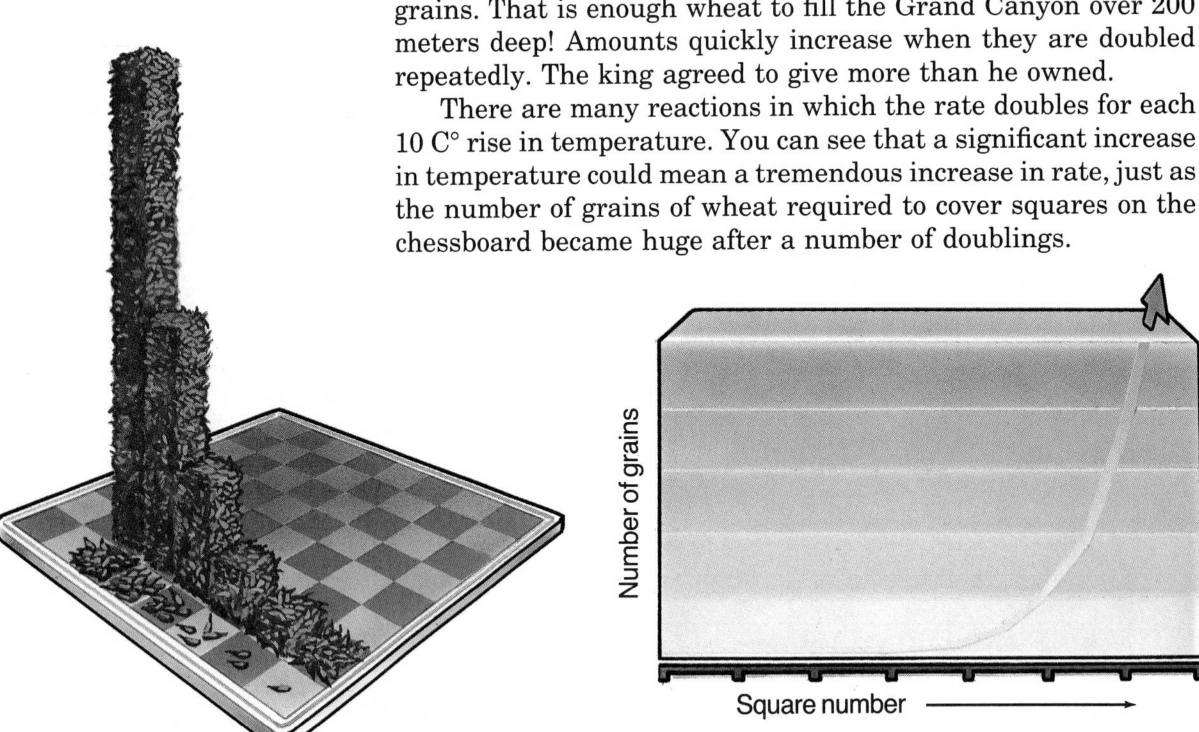

Figure 22.1 The effect of doubling the amount of wheat on each successive square of a checkerboard (left) is shown in the dramatic rise of the graph (right).

Stability of Compounds

Consider the reaction

$$2TiO \rightarrow 2Ti + O_2$$

The overall energy change for this reaction is positive; the reaction does not take place spontaneously. Titanium(II) oxide does not spontaneously decompose at room temperature. It is said to be **thermodynamically stable.**

When the overall energy change for a reaction is negative, a reaction will proceed spontaneously. However, some spontaneous reactions take place so slowly that hundreds of years may pass before any observable change occurs. For example, the overall energy change for the combustion of glucose, a common sugar,

$$C_6H_{12}O_6(cr) + 6O_2(g) \rightarrow 6CO_2(g) + 6H_2O(l)$$

is −2870 kJ. At room temperature, the reaction proceeds so slowly that we can never detect a change. Thus, sugar is said to be **kinetically stable.** Note that a compound is stable in terms of its tendency to decompose if the overall energy change for that reaction is positive. A system may have an overall energy change which is negative and still be stable if the rate of change is imperceptible.

Reversible Reactions and Equilibrium

Consider what happens when we add an ice cube to a beaker of water. The temperature of the water drops. Energy is absorbed as the ice cube gets smaller and solid ice is converted into liquid water. We may represent this process by an equation.

$$\textit{solid} + \textit{energy} \rightarrow \textit{liquid}$$

If we use a thermos bottle instead of a beaker and our ice cube is large enough, the temperature of the water will drop to 0°C. After the temperature reaches 0°C, we observe no further melting of the ice. If we now add a piece of metal that has been chilled to −20°C, we may be surprised to find that the ice cube grows larger! The process can go either way. At 0°C an equilibrium exists.

$$\textit{solid} + \textit{energy} \rightleftarrows \textit{liquid}$$

The relative amounts of solid and liquid can be changed by adding or removing a small amount of energy without changing the temperature. We say that the water-ice mixture is in equilibrium. Many chemical reactions also result in an equilibrium mixture.

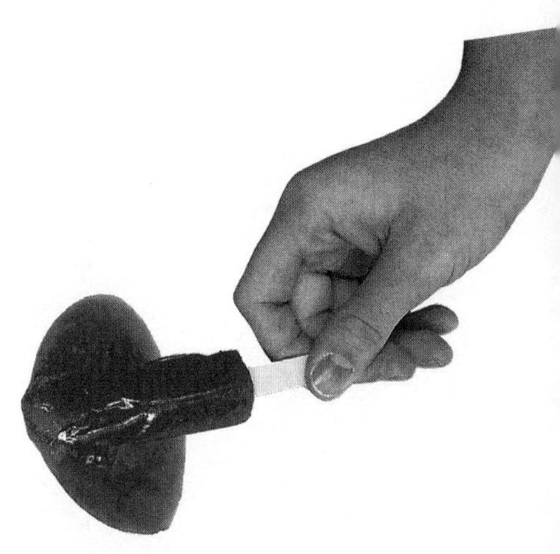

Figure 22.2 The melting and refreezing of a frozen fruit stick indicate that melting and freezing are reversible physical changes.

The study of chemical reactions is the study of the breaking and the forming of chemical bonds. The formation of chemical bonds is a complex subject. So far we have discussed only the simplest chemical reactions, those that go to completion. *A reaction goes to completion if it proceeds until all of one of the reactants is used completely and the reaction stops.* Reactions of this kind go from reactants to products. Not all reactions go to completion. Consider the following reaction.

$$H_2(g) + I_2(g) \rightarrow 2HI(g)$$

The arrow means the reaction is read from left to right, but this equation is only partially correct. The bond between the hydrogen and iodine in the hydrogen iodide molecule is a weak bond. Therefore, hydrogen iodide breaks easily into hydrogen gas and iodine vapor. The following equation represents this reaction.

$$H_2(g) + I_2(g) \leftarrow 2HI(g)$$

Notice the direction in which the arrow points. This reaction is read from right to left. We now combine the two equations.

$$H_2(g) + I_2(g) \rightleftarrows 2HI(g) \quad \text{(reversible reaction)}$$

The first reaction is said to go from left to right. The second reaction is said to go from right to left. The combined equation represents a **reversible reaction.** Reversible reactions eventually reach equilibrium.

Reaction Rate

Suppose the product of a reversible reaction decomposes faster than the reactants form the product. Then there will always be more reactant than product. Here is an example. HI decomposes to H_2 and I_2 more rapidly than H_2 unites with I_2 to form HI. There will always be more hydrogen and iodine than hydrogen iodide. Consider a flask containing hydrogen, iodine, and hydrogen

Figure 22.3 Calcium carbonate stalagmites and stalactites form as a result of a reversible reaction between limestone and groundwater containing carbon dioxide.

$$CaCO_3(cr) + CO_2(g) + H_2O(l) \rightleftarrows Ca^{2+}(aq) + 2\ HCO_3^-(aq)$$

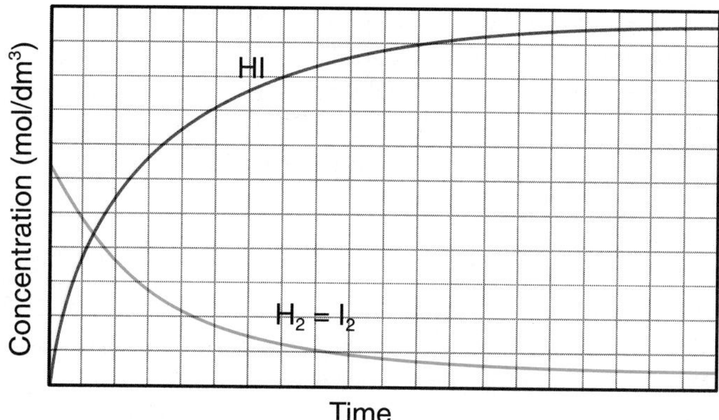

Figure 22.4 The rate for the reaction $H_2 + I_2 \rightleftarrows 2HI$ can be expressed as the rate of appearance of the product, HI, or the rate at which the reactants, H_2 and I_2, disappear.

iodide. Hydrogen and iodine are combining to form hydrogen iodide. The rate of appearance of hydrogen iodide is defined as the rate of the reaction from left to right.

$$H_2(g) + I_2(g) \rightarrow 2HI(g) \qquad \text{(rate of appearance of HI)}$$

But the hydrogen iodide is decomposing rapidly, more rapidly than H_2 and I_2 can combine to produce it. The rate of disappearance of hydrogen iodide is defined to be the reaction rate of the reaction read from right to left.

$$H_2(g) + I_2(g) \leftarrow 2HI(g) \qquad \text{(rate of disappearance of HI)}$$

Reaction rate is usually defined in terms of the rate of disappearance of one of the reactants. It can also be defined as the rate of appearance of one of the products. The usual units for reaction rates are (mol/dm^3)/s. What we are actually measuring is the rate of change of concentration in a constant volume process. If we know the two reaction rates, we can predict whether product or reactant will be in the higher concentration at equilibrium. In the next sections, we will consider some factors that affect reaction rates.

Nature of Reactants

The nature of the reactants involved in a reaction will determine the kind of reaction that occurs. Reactions with bond rearrangement or electron transfer generally take longer than reactions without these changes. Ionic reactions (such as double displacement and neutralization reactions) occur almost instantaneously. They are rapid because ions of one charge are attracted by ions of opposite charge and collide often.

In a double displacement reaction between ions, no electron transfer is involved. Reactions between neutral molecules are slower than ionic reactions because electron transfer and bond rearrangement must occur. As pointed out in Chapter 15, most molecular collisions are elastic. The molecules simply rebound and move away unchanged. However, some collisions do have enough energy to cause changes in the electron clouds of the colliding molecules. When the change occurs, the colliding molecules

CAREERS IN CHEMISTRY

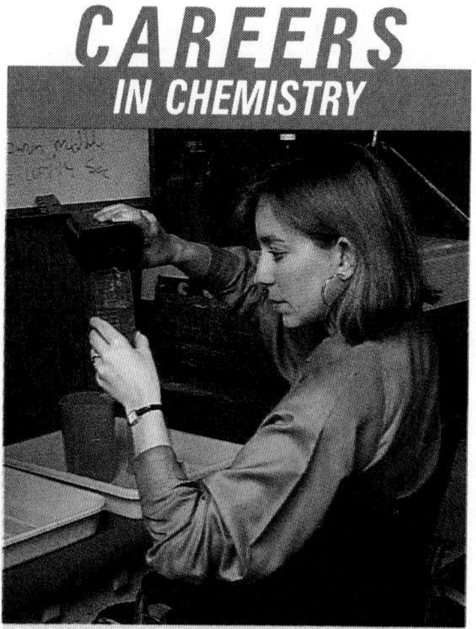

Photographer

Being a photographer is more than just aiming a camera and shooting a picture. Most of the hard work—developing the film and printing the picture—comes after the picture is taken. The photographer needs to know the function of the developer, the stop bath or acid rinse, and the fixer in developing black and white photos, as well as the roles of the developer, bleach, and stabilizer in color photography. These are all chemical processes. Prospective photographers should have a knowledge of chemistry and physics. Many photographers are self-taught, but technical schools and colleges offer formal training.

may form an **activated complex.** The energy required to form the activated complex is known as the activation energy. If the activation energy is high, few collisions have enough energy to form the activated complex. As a result, the reaction may be so slow that it cannot be detected. Consider the following reaction of bromoethane with hydroxide ion.

$$CH_3CH_2Br + OH^- \rightarrow CH_3CH_2OH + Br^-$$

The activated complex is $CH_3CH_2(OH)Br^-$. From Figure 22.5, we can see that the activation energy is 89.5 kJ per mole of CH_3CH_2Br. We can also see that the enthalpy change for the reaction is −77.2 kJ. We can now see why some substances are kinetically stable even though many of their reactions may have negative overall energy changes. These reactions have very high activation energies. Consequently, unless very strong reaction conditions are used, the reactions do not take place.

For example, hydrogen and oxygen can be kept in the same container at room temperature for years without reacting to form water. Although the molecules collide, the activation energy will not be reached. If, however, the mixture is heated to 800°C, or a flame or spark is introduced, a violent reaction occurs. The heat, flame, or spark furnishes the activation energy.

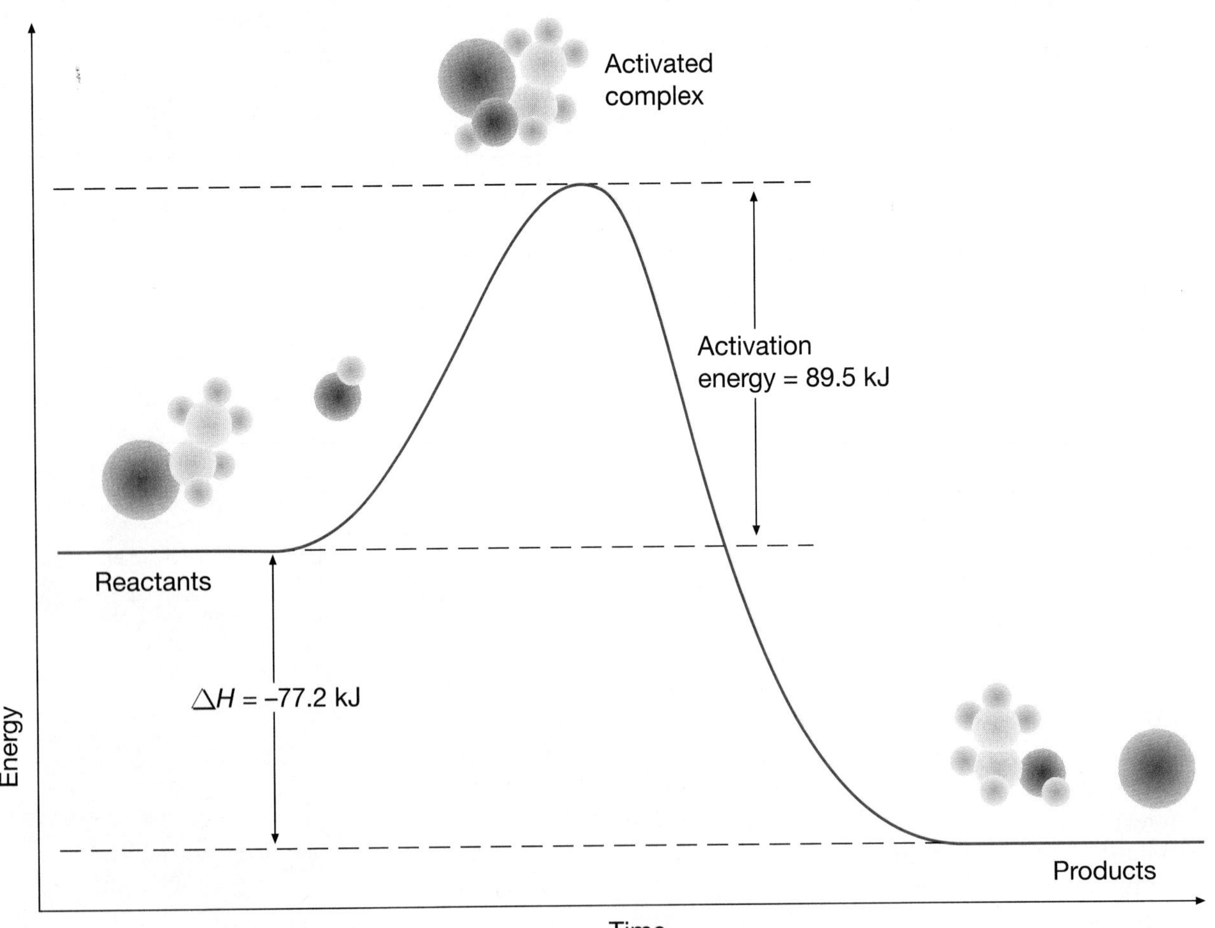

Figure 22.5 Activation energy is the energy that must be attained in order for a collision between the reactants to result in the formation of an activated complex.

FRONTIERS

Laser-Zapped Ions

Researchers have demonstrated that the ionization rate of hydrogen chloride molecules can be controlled by laser excitation. The scientists consider this achievement as the first step towards controlling chemical reactions by directly controlling molecules.

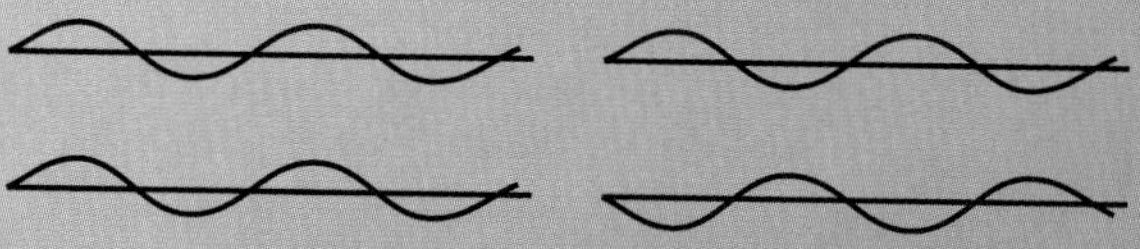

Researchers produce HCl^+ ions by exciting electrons in HCl molecules using laser beams of two different wavelengths of light. The important factor in this experiment is that the rate at which the ionization energy delivered to the HCl can be controlled by altering the phase relationship between the two laser beams. The diagram above represents light waves of the same wavelength. The light waves on the left are in phase. Such light would lead to constructive interference. In this case the photon density would be sufficient for absorption and ionization of the molecules. The light waves on the right are out of phase such that they would undergo destructive interference and would not lead to ionization.

The phase relationship between the two beams of laser light can be controlled by passing the two beams through hydrogen or argon gas. By changing the pressure of the gas, one beam is delayed relative to the other. Thus, by controlling the amount of delay, the amount of interference between the beams and ultimately the ionization rate of the HCl molecules are controlled.

Concentration

Another factor affecting reaction rate is concentration. Concentration refers to the quantity of matter that exists in a unit volume. For instance, in Chapter 8 we discussed concentration in mol/dm^3 of solution. We referred to this concentration as the molarity of the solution. For a reaction to take place, the particles must collide. If the number of particles per unit volume is increased, the chance of their colliding is also increased. Reconsider the equation

$$H_2(g) + I_2(g) \rightarrow 2HI(g)$$

If we keep the concentration of the hydrogen molecules the same, we would expect doubling the concentration of iodine to double the number of collisions between iodine and hydrogen molecules. In turn, the reaction rate would double. Actual experiment confirms that the rate of reaction varies directly as the concentration of iodine. Written in equation form, it is

$$rate_1 = k_1[I_2]$$

MINUTE LAB

More is Better

Put on an apron and goggles. Generate oxygen gas by adding 25 cm^3 of 6% H_2O_2 slowly to a 250-cm^3 beaker containing 0.3 g MnO_2. **CAUTION:** *H_2O_2 causes burns.* Cover the beaker with a glass plate. Ignite the tip of a wooden splint by holding it with tongs in the flame of a laboratory burner. Remove the splint from the flame. Gently blow on the burning splint to extinguish it, taking care to allow the tip of the splint to continue to smolder. Quickly remove the glass plate from the beaker and lower the smoldering splint into it. **CAUTION:** *Do not drop the splint into the beaker.* Record your observations. Extinguish the splint in water.

Based on your observations of the smoldering splint in air and in the oxygen-filled beaker, write a summary statement that describes the relationship between the rate of a reaction and the concentration of reactants. Why might an athlete who trains at a high altitude have an advantage competing in an event held at a lower altitude?

where the brackets around the I_2 ([ ]) mean concentration expressed in mol/dm^3.

What if the concentration of iodine remains constant, and the concentration of hydrogen is allowed to vary? We can say that the number of collisions and, therefore, the reaction rate, varies directly as the concentration of hydrogen molecules. We write

$$rate_2 = k_2[H_2]$$

If we allow the concentration of both iodine and hydrogen to vary, what will happen? If we double the concentration of hydrogen and also double the concentration of iodine, there will be four times as many molecules. What will the reaction rate be? There will be four times as many molecules per unit volume and four times as many collisions. The reaction rate is found to be quadrupled. We conclude, then, that *the reaction rate varies directly as the product of the concentrations of the reactants (hydrogen and iodine).* We write

$$rate = k[H_2][I_2]$$

In this case, the constant k depends upon the size, speed, and kind of molecule involved in the reaction. Each reaction has only one value of k for a given temperature; this k is called the **specific rate constant** of the reaction. It should be pointed out here that the actual mechanism of this reaction involves the breaking of the I—I bond before collision. However, it can be demonstrated mathematically that the same rate expression results.

Consider the reaction $H_2O_2 + 2HI \rightarrow 2H_2O + I_2$. The rate expression for the reaction is written

$$rate = k[H_2O_2][HI]$$

Even though two HI molecules appear in the equation, only one appears in the rate expression. The only way to be sure of the rate expression is to use experimental data, as will be shown later.

An increase in the pressure on a gas at constant temperature results in a decrease in the volume occupied by the molecules. Since there are more molecules per unit volume, there has been an increase in concentration. Increasing the pressure on a gas, then, will also increase reaction rate.

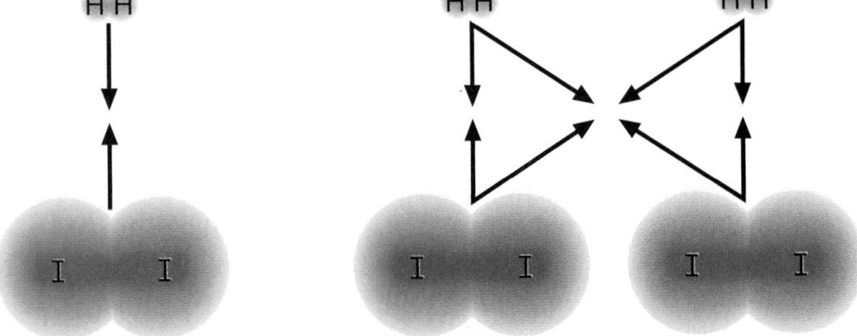

Figure 22.6 An increase in the concentration (number of particles) of the reactants increases the number of effective collisions that form products.

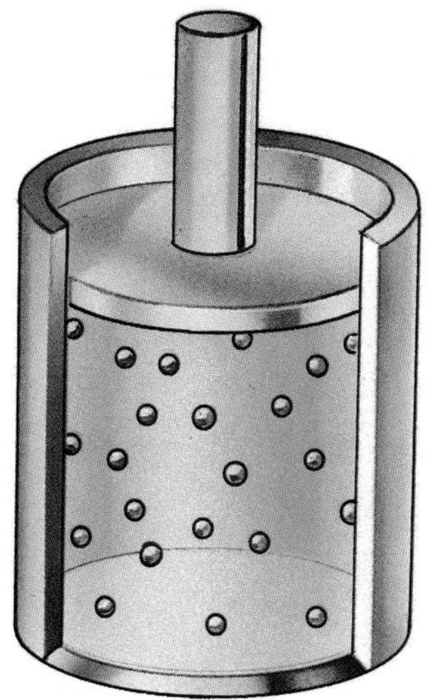

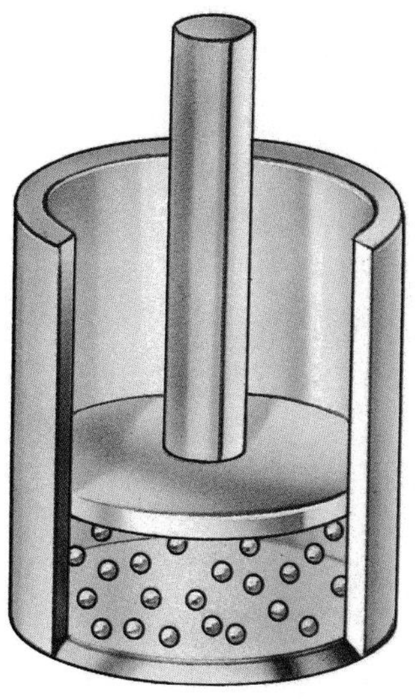

Figure 22.7 Increasing the pressure on a gas increases its concentration.

Reactions such as the preceding in which the reactants are in the same phase are called **homogeneous reactions.** Chemical reactions that take place at the interface between two phases are called **heterogeneous reactions.** An example of a heterogeneous reaction is zinc (a solid) reacting with sulfuric acid (a liquid). The reaction takes place on the surface of the zinc that is the interface between the two phases. If more surface is exposed, the reaction will take place more rapidly. Increasing the surface area increases the number of surface molecules in the same space. Exposing more surface, in effect, increases concentration. Increasing surface area, then, increases the rate of a reaction.

Temperature

Reaction rate is determined by the frequency of collision between molecules and increases as the frequency of collision increases. According to the kinetic theory, the speed and, therefore, the kinetic energy of molecules increases as the temperature increases. Increased speed means that more collisions will occur and the reaction rate will increase. However, the increase in reaction rate depends less on the increase in the number of collisions than it does on the increase in the number of molecules that have the activation energy. Molecules must collide with a kinetic energy sufficient to form the activated complex. Otherwise a collision will not lead to reaction. Figure 22.8 shows the energy distribution of molecules of a substance at two different temperatures. On the graph, the area under each curve indicates the number of molecules present. At temperature T_1, few molecules have attained activation energy. At temperature T_2, ($T_2 > T_1$), the average kinetic energy of molecules is greater. Thus, many more molecules have reached the activation energy. The same number of molecules is present at this higher temperature. However, the fraction of molecules that have attained the activation energy is greater at the higher temperature, T_2.

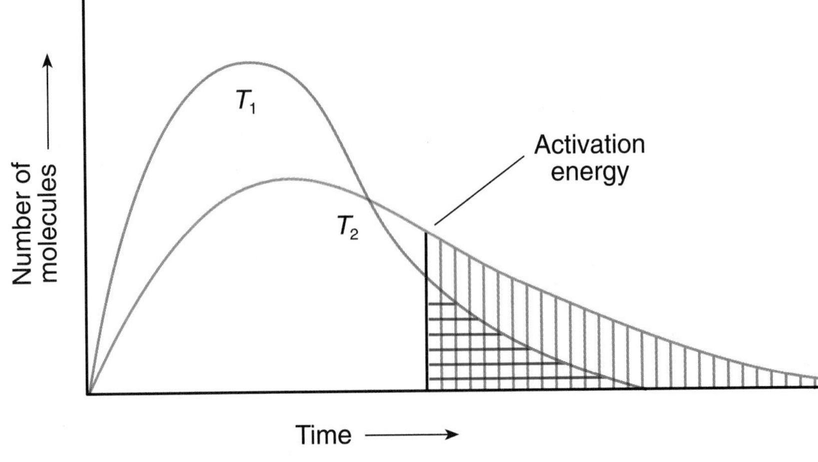

Figure 22.8 As the temperature of a molecular substance is raised from T_1 to T_2, more molecules will have the activation energy, that is, the minimum energy needed to react. Thus, the rate of reaction will be greater at temperature T_2.

MATH CONNECTION

Increasing Surface Area

The increase in the surface area of a crystalline solid by pulverizing it is easily seen through mathematics. Suppose we have a single cubic salt crystal 10 mm on a side. Its surface area is 6×10^2 mm^2 or 600 mm^2. If we were able to break that crystal into 1000 cubes each 1 mm on a side, the total area becomes $1000 \times 6 \times 1$ mm^2, or 6000 mm^2. As you can see, this is ten times as much surface. Breaking each of the small cubes into 1000 cubes 0.1 mm on a side would create a total surface area of $1000 \times (1000 \times 6 \times 0.01$ mm^2), or 60 000 mm^2. This is 100 times as much area as the original cube. With the increase of the surface area by a factor of 100, it is apparent why the rate of a reaction involving a solid would increase significantly when the solid is pulverized.

Figure 22.9 is a graph that shows the energy changes involved in the reaction of hydrogen with oxygen. This graph can be thought of as a map of the potential energy possessed by the atoms and molecules taking part in a reaction. The reactants are the gases H_2 and O_2, which are considered to have no potential energy (0 kJ). As two molecules approach, the kinetic energy of motion is transformed into the potential energy of repulsion of electron clouds. As more and more kinetic energy is transformed into potential energy, the line of the graph, which represents potential energy, rises. If the molecules have enough kinetic energy to approach close enough to react chemically, they form an activated complex. Note that the potential energy graph peaks at the formation of the activated complex. From here, the activated complex must come apart. The activated complex can fall back on the left side and break into H_2 and O_2 molecules, or it can fall down on the right side and come apart to produce products. The chance of falling either way is equal. If the activated complex does react, energy will be released and water will be formed. The effect

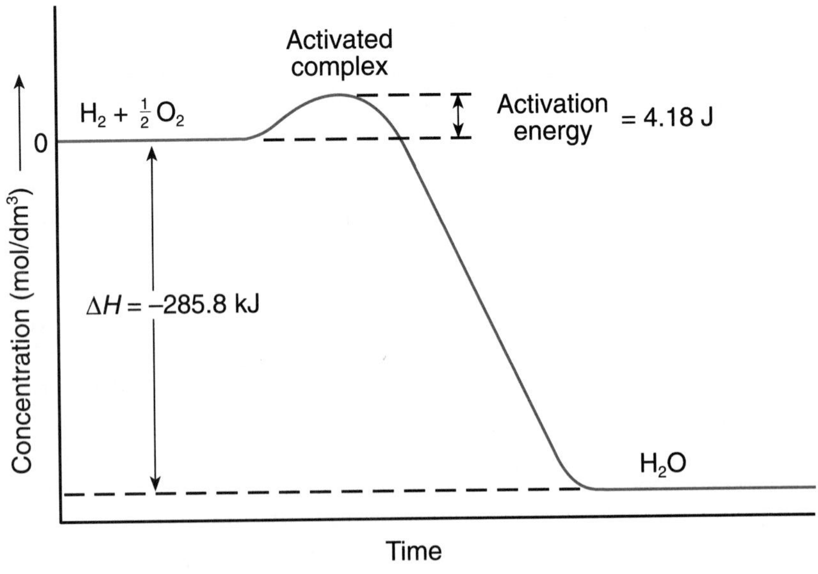

Figure 22.9 As H_2 and O_2 molecules approach each other, their kinetic energy changes to potential energy. If sufficient energy is present, they form an activated complex. The formation of this complex begins a series of steps that produces water.

of raising the temperature is to produce more activated complexes through an increased number of collisions and more energetic collisions. Thus, with the increase in number of activated complexes, the number that react will also increase.

An increase in temperature will increase the rate of any reaction. More activated complexes are formed because the number of collisions having the required activation energy increases.

kata: (GK) down, very
lyein: (GK) to loosen, break up

catalysis—breaking up (reacting) more intensely

Catalysis

A substance that increases a reaction rate without being permanently changed is called a catalyst. **Catalysis** is the process of increasing rates of reaction by the presence of a catalyst. The catalyst appears to be chemically unaffected by the reaction. It changes the reaction mechanism in such a way that the activation energy required is less than in the uncatalyzed reaction. We will discuss two kinds of catalysts: the heterogeneous (or contact) catalyst, and the homogeneous catalyst.

The reaction of sulfur dioxide gas with oxygen gas

$$2SO_2(g) + O_2(g) \rightarrow 2SO_3(g)$$

is extremely slow at room temperature. If these two gases are brought into contact in the presence of solid vanadium(V) oxide, V_2O_5, the reaction is rapid. Vanadium(V) oxide is the catalyst in this reaction. Notice that vanadium(V) oxide is a solid and the reactants are gases. The V_2O_5 lowers the required activation energy by providing a surface on which the activated complex can form.

A reaction in which the reactants and catalysts are not in the same phase is a heterogeneous reaction. The catalyst is called a

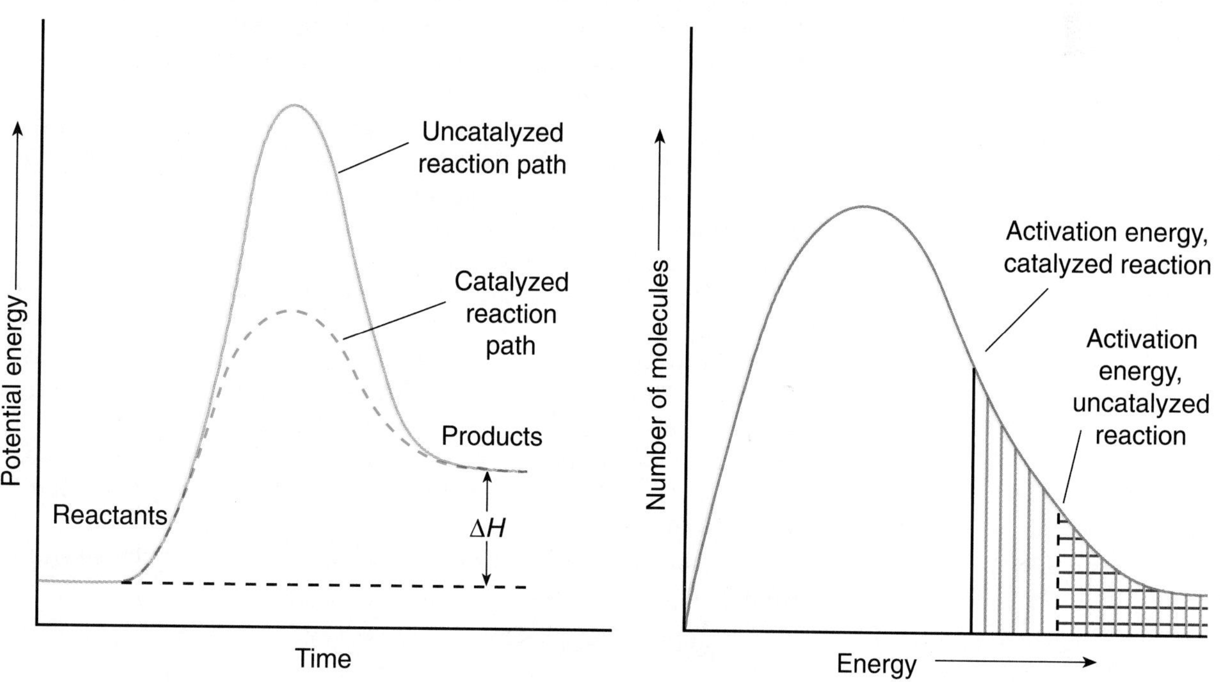

Figure 22.10 A catalyst speeds up a reaction by providing a new reaction pathway at a lower energy (left). The catalyzed reaction is faster because a larger number of molecules possess the lowered activation energy (right).

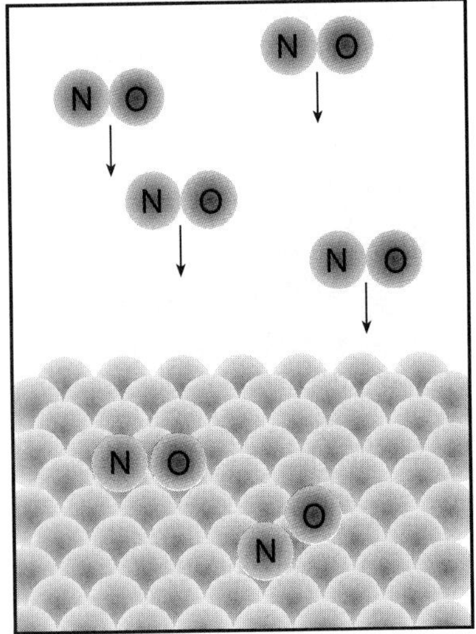

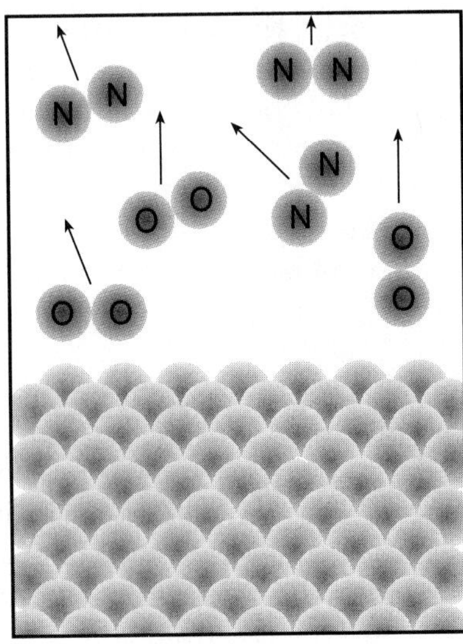

Figure 22.11 One of the reactions that takes place in a catalytic converter is the decomposition of nitrogen(II) oxide to nitrogen and oxygen gas (top). The platinum-rhodium catalyst provides a surface on which the NO decomposes. The catalyst surface is large because it coats the surface of a honeycomb structure seen in the cutaway view of a catalytic converter (bottom).

heterogeneous catalyst. This kind of catalyst has a surface on which the substances can react. Platinum and other finely divided metals and metallic oxides are common examples of this kind of catalyst. Most heterogeneous catalysts work by adsorbing one of the reactants. Adsorption is the adherence of one substance to the surface of another. In the process of adsorbing a molecule, such as O_2, the catalytic surface attracts the O_2 molecule. This attraction weakens the O—O bond to the point where the other reactant can break the O—O bond. The reaction then proceeds. Catalytic converters on automobile exhaust systems use a contact catalyst.

A **homogeneous catalyst** exists in the same phase as the reactants. This kind of catalyst does enter into the reaction, but is returned unchanged in a final step of the reaction mechanism. It forms an intermediate compound or compounds that react more readily than the uncatalyzed reactants. They react more readily because they require less activation energy. As an example of homogeneous catalysis, consider the hydrolysis of sucrose (table sugar). The reaction is

$$\underset{\text{Sucrose}}{C_{12}H_{22}O_{11}(aq)} + H_2O(l) \rightarrow \underset{\text{Glucose}}{C_6H_{12}O_6(aq)} + \underset{\text{Fructose}}{C_6H_{12}O_6(aq)}$$

INGREDIENTS: WHOLE WHEAT, SUGAR, SORBITOL, GELATIN,
VITAMINS AND MINERALS: NIACINAMIDE, ZINC (OXIDE), IRON, VITAMIN B6 (PYRIDOXINE HYDROCHLORIDE), VITAMIN B2 (RIBOFLAVIN), VITAMIN B1 (THIAMIN HYDROCHLORIDE), FOLIC ACID, AND VITAMIN B12.
TO KEEP THIS CEREAL FRESH, BHT HAS BEEN ADDED TO THE PACKAGING.

Figure 22.12 BHT, butylated hydroxytoluene, is an inhibitor that is added to products to preserve their freshness because it acts as an antioxidant.

The reaction is normally very slow. If, however, the solution is made acidic, the presence of the acid causes the reaction to proceed readily. In the reaction, all substances are in aqueous solution (the same phase). Thus, the reaction is a homogeneous one, and the acid is a homogeneous catalyst.

Catalysts are used a great deal in industry, as well as in the chemical laboratory. Other substances, called **inhibitors,** are also used to affect reaction rates. These substances do not "slow up" a reaction. Rather they "tie up" a reactant or catalytic substance in a complex, so that it will not react. Preservatives used in foods and medical preparations are included to avoid spoilage. These substances are examples of inhibitors.

Word Origins

in: (L) in
habere: (L) to have, hold

inhibitor—that which holds a substance from reacting

Reaction Mechanism

CHEMISTRY IN DEPTH

Most reactions occur in a series of steps. Each step normally involves the collision of only two particles. Steps involving three or more particles are unlikely. There is little chance of three or more particles colliding with the proper geometry and energy to cause a reaction.

If a reaction consists of several steps such as the following

$$A \rightarrow B$$
$$B \rightarrow C$$
$$C \rightarrow \textit{final product}$$

one of the steps will be slower than all the others. This step is called the **rate determining step.** Only this step will affect the rate. The series of reaction steps that must occur for a reaction to go to completion is called the **reaction mechanism.**

At a given temperature, the rate of a reaction varies directly as the product of the concentrations of the reactants in the slowest step. For the reaction $H_2(g) + I_2(g) \rightarrow 2HI(g)$, the rate expression was $rate = k[H_2][I_2]$. For the general reaction $A + B \rightarrow C$, the

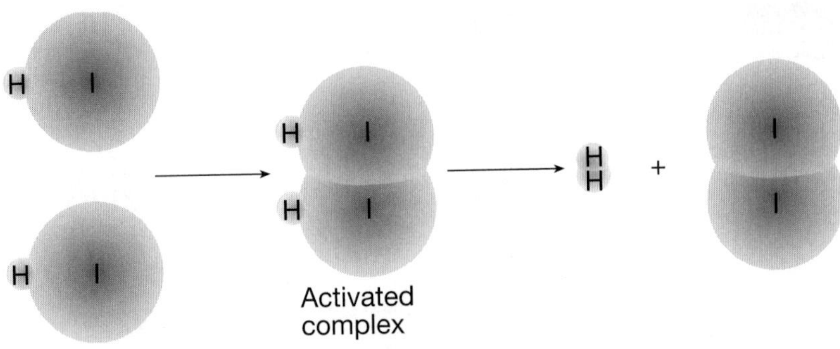

Figure 22.13 The probable mechanism for the decomposition of HI is a one-step reaction in which two HI molecules collide, forming H_2 and I_2.

rate expression would be *rate* = $k[A][B]$. Hydrogen iodide decomposes into hydrogen and iodine. The equation is $2HI(g) \rightarrow H_2(g) + I_2(g)$. This reaction is $HI + HI \rightarrow H_2 + I_2$. The rate expression (if it is a one-step reaction) is *rate* = $k[HI][HI]$, or *rate* = $k[HI]^2$.

How do we know if a reaction is a single step? The only way to obtain information is by experiment. The observation of reaction rates gives insight into the mechanisms of reactions. Consider

$$C_2H_4Br_2(l) + 3I^-(aq) \rightarrow C_2H_4(l) + 2Br^-(aq) + I_3^-(aq)$$

taking place in one step as indicated by the equation. Four particles ($C_2H_4Br_2$ and $3I^-$) would have to collide all at once in the right orientation to react. Such a collision is highly unlikely.

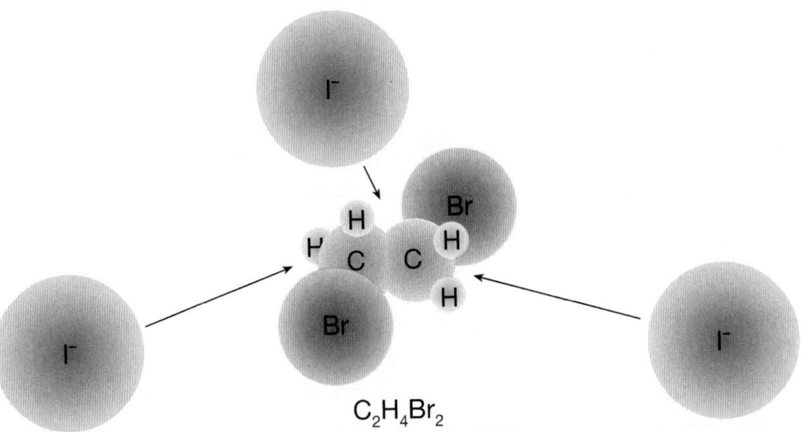

Figure 22.14 The probability of four particles colliding at once in the correct orientation is not very great.

Instead, the reaction has been shown to obey the rate expression, *rate* = $k[C_2H_4Br_2][I^-]$. The rate data tell us that we need to consider multiple-step mechanisms. The rate determining step, according to the data, involves only one I^- with the $C_2H_4Br_2$. There are a large number of possible mechanisms that would agree with the observed rate expression. Consider just one.

$$C_2H_4Br_2 + I^- \rightarrow C_2H_4Br^- + IBr \quad \text{(slow)}$$
$$C_2H_4Br^- \rightarrow C_2H_4 + Br^- \quad \text{(fast)}$$
$$IBr + I^- \rightarrow Br^- + I_2 \quad \text{(fast)}$$
$$I_2 + I^- \rightarrow I_3^- \quad \text{(fast)}$$

Because the slow step requires one I^- for every $C_2H_4Br_2$ molecule, both substances have an equal effect on the rate.

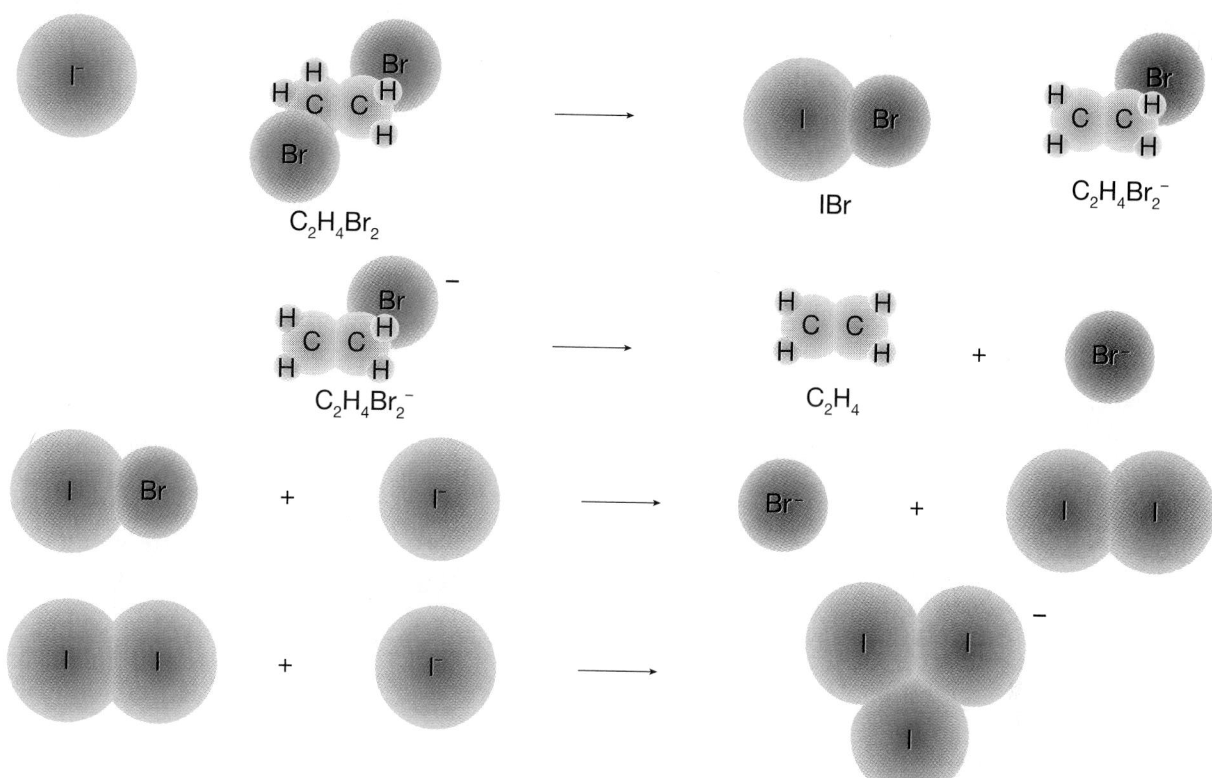

Figure 22.15 Shown here is one of several mechanisms that would account for the reaction $C_2H_4Br_2(l) + 3I^-(aq) \rightarrow C_2H_4(l) + 2Br^-(aq) + I_3^-(aq)$.

Review the rate expression for this reaction. The exponents of the concentration factors are spoken of as the order of the expression. Thus, the reaction is first order with respect to $C_2H_4Br_2$ and first order with respect to I^-. Adding the exponents for all concentrations in the expression gives the overall order. The reaction then is second order.

SAMPLE PROBLEM

Rate Expression

What is the rate expression for the following reaction? What is the value of the rate constant for the reaction?

$$H_2O_2 + 2HI \rightarrow 2H_2O + I_2$$

Trial	$[H_2O_2]$	[HI]	Rate
1	0.1*M*	0.1*M*	0.0076 (mol/dm³)/s
2	0.1*M*	0.2*M*	0.0152 (mol/dm³)/s
3	0.2*M*	0.1*M*	0.0152 (mol/dm³)/s

Solving Process:

(a) By comparing trials 1 and 2, we can see that doubling the [HI] doubles the rate. Thus, there is a direct relationship: *rate* ∝ [HI].

MINUTE LAB

Faster and Faster

Put on an apron and goggles. Label a 250-cm^3 beaker A and add to it 100 cm^3 of water. Dissolve 10 g of potassium sodium tartrate, $KNaC_4H_4O_6{\cdot}4H_2O$, in the water in beaker A. Label two small beakers B and C. Add 40 cm^3 of 6% H_2O_2 to beaker B. **CAUTION:** *H_2O_2 causes burns.* In beaker C dissolve 1 g $CoCl_2{\cdot}6H_2O$ in 15 cm^3 water.

Place beaker A on a hot plate and heat the solution to 85°C. Remove the beaker from the hot plate and place it on a tray. Add to it the contents of beaker B. Record your observation of the rate of reaction. Add the contents of beaker C to beaker A. **CAUTION:** *Beaker A may overflow.* Again, record your observation of the reaction rate. What evidence do you have that the contents of beaker C acted as a homogeneous catalyst?

(b) By comparing trials 1 and 3, we see that doubling the $[H_2O_2]$ also doubles the rate and we again have a direct relationship: $rate \propto [H_2O_2]$.

(c) The rate expression is, then,

$$rate = k[H_2O_2][HI]$$

(d) The value of k is given by

$$k = \frac{rate}{[H_2O_2][HI]} = \frac{0.0076\ (\text{mol/dm}^3)/\text{s}}{(0.1\ \text{mol/dm}^3)(0.1\ \text{mol/dm}^3)}$$

$$= 0.76\ \text{dm}^3/\text{mol·s}$$

PRACTICE PROBLEMS

1. Assume that NO(g) and H_2(g) react according to the rate expression: $rate = k[NO]^2[H_2]$. How does the rate change if
a. the concentration of H_2 is doubled?
b. the volume of the enclosing vessel is suddenly halved?
c. the temperature is decreased?

2. For the reaction $H_2(g) + I_2(g) \rightarrow 2HI(g)$, the following data were obtained.

Experiment	Initial $[H_2]$	Initial $[I_2]$	Initial rate of formation of HI
1	1.0M	1.0M	0.20 $(\text{mol/dm}^3)/\text{s}$
2	1.0M	2.0M	0.40 $(\text{mol/dm}^3)/\text{s}$
3	2.0M	2.0M	0.80 $(\text{mol/dm}^3)/\text{s}$

a. Write the rate expression for this reaction.
b. What would be the initial rate of formation of the hydrogen iodide if the initial concentrations of the hydrogen gas and the iodine gas were each 0.50M?

3. The reaction $CH_3COCH_3 + I_2 \rightarrow CH_3COCH_2I + HI$ is run under carefully controlled conditions in the presence of an excess of acid. Write the rate expression for the reaction using the following data.

Initial concentration		Initial Rate
$[CH_3COCH_3]$	$[I_2]$	
0.100M	0.100M	1.16×10^{-7} $(\text{mol/dm}^3)/\text{s}$
0.0500M	0.100M	5.79×10^{-8} $(\text{mol/dm}^3)/\text{s}$
0.0500M	0.500M	5.78×10^{-8} $(\text{mol/dm}^3)/\text{s}$

4. It has been determined that the reaction in problem 3 is also first order with respect to the acid catalyst. Propose a mechanism for the slow step consistent with the data.

Bridge to
BIOLOGY

Enzymes in the Body

Every living organism is a vast chemical factory. Living organisms are made of cells that are constantly building cellular structures and breaking down large molecular structures as waste in the process called metabolism. The metabolism of an organism is the sum of all the chemical reactions taking place throughout the organism.

Reactions of metabolism do not usually take place spontaneously under normal conditions of temperature and pressure found in living organisms. Cell metabolism is possible by use of biochemical catalysts called enzymes, complex protein molecules.

Originally, biochemists thought that the enzyme molecule and the reactant molecule, called the substrate, had to fit together like a lock and key. More recent research, however, has shown that the attraction of polar parts of each molecule, enzyme and substrate, for polar parts of the other, can cause each molecule to change shape slightly so that they fit together quite closely. The alteration in the shape of the substrate makes it susceptible to attack by another molecule.

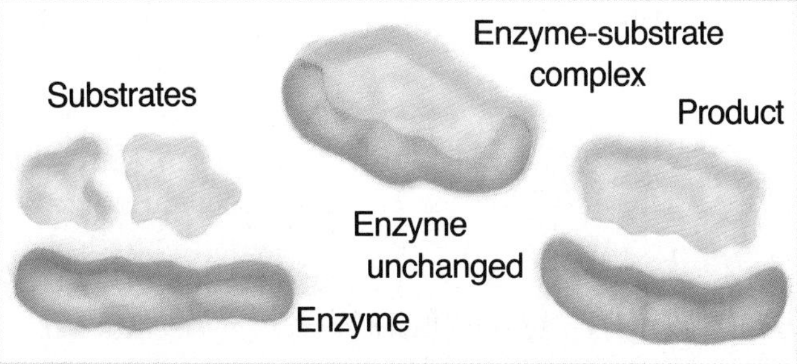

Enzymes are, in general, quite specific. That is, they will participate with only one certain substrate in the catalysis of only one certain reaction. Once the reaction has taken place, the substrate is released and the enzyme is ready to take on another substrate molecule. The names of enzymes usually end in the suffix *-ase*. The remainder of the name indicates what function the enzyme performs in the metabolism of the cell. For example, glucose-6-phosphatase is an enzyme that catalyzes the addition of a phosphate group to the glucose molecule. Because virtually all metabolic reactions are reversible, glucose-6-phosphatase also catalyzes the removal of a phosphate group from glucose-6-phosphate.

The system of substrate, enzyme, and product usually remains in equilibrium, and so obeys Le Chatelier's principle.

Exploring Further

1. Certain metabolic diseases are caused by the lack of an enzyme due to a genetic defect. One such condition is phenylketonuria. Find out what enzyme is missing and how the disorder is treated.
2. What controls the amounts of various enzymes found in cells?

22.1 CONCEPT REVIEW

5. How does thermodynamic stability differ from kinetic stability?
6. What factors can affect the rate of a chemical reaction?
7. How does the action of a catalyst differ from that of an inhibitor?
8. What is a reaction mechanism?
9. **Apply** Why might reaction rates be of interest to physicians?

OBJECTIVES

- determine an equilibrium constant expression for a system at equilibrium.
- use Le Chatelier's principle to explain the effects of changes in concentration, pressure, and temperature on an equilibrium system.
- relate relative amounts of product and reactant to the equilibrium constant.
- calculate equilibrium constants and concentrations of reactants or products for a reaction.

22.2 Chemical Equilibrium

Equilibrium is often defined as a state of balance. Consider two children of equal weight at opposite ends of a seesaw. By slightly adjusting their positions, the children could balance the seesaw so that it is motionless; neither end would rise or fall. A balanced seesaw is an example of static equilibrium because nothing is moving. In earlier chapters, you learned about dynamic equilibrium—a situation in which change was taking place even though an overall balance was maintained. You learned about the equilibrium of melting and freezing ice, and also about the equilibrium of dissolving and crystallizing solute. In those cases, however, only physicial changes took place while the balance was maintained. Now you will be introduced to dynamic equilibria involving chemical changes. Chemical changes are taking place, but in such a way that there is no overall change in the composition of the system.

Equilibrium Constant

We have seen that the reaction of H_2 and I_2 to form HI is an equilibrium reaction. As the reaction proceeds, H_2 and I_2 are used and the concentration of the reactants decreases. Thus, fewer collisions between H_2 and I_2 molecules occur per unit time. The reaction rate of the hydrogen and iodine reaction decreases. The reverse reaction, the collision of two HI molecules to form H_2 and I_2, does not occur initially because the concentration of HI is zero.

However, as the concentration of HI increases, the number of collisions between HI molecules increases. Therefore, the reverse reaction, the decomposition of hydrogen iodide, also steadily increases. The rates of the forward and reverse reactions of the hydrogen iodide reaction are

$$\textit{rate of forward reaction} = k_f[I_2][H_2]$$
$$\textit{rate of reverse reaction} = k_r[HI]^2$$

Figure 22.16 If this harried teen bails out the water at the same rate at which it enters the boat, he can maintain the boat in a state of dynamic equilibrium. If not, he's in big trouble.

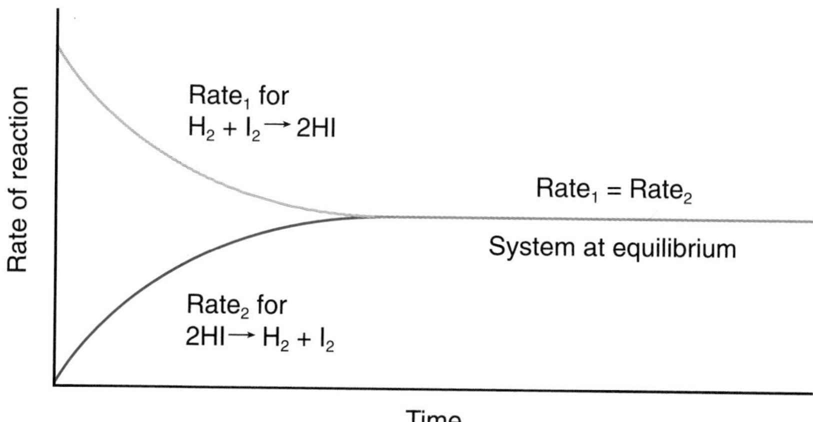

Figure 22.17 Equilibrium is established when rates of the forward and reverse reactions are equal.

Equilibrium is attained when the rates of the two opposing reactions are equal. Thus, at equilibrium, the two rates are equal,

$$k_f[I_2][H_2] = k_r[HI]^2$$

Dividing both sides by $k_r[I_2][H_2]$

$$\frac{k_f[I_2][H_2]}{k_r[I_2][H_2]} = \frac{k_r[HI]^2}{k_r[I_2][H_2]}$$

we obtain

$$\frac{k_f}{k_r} = \frac{[HI]^2}{[I_2][H_2]}$$

Since both k_f and k_r are constants, the ratio k_f/k_r is also a constant. This new constant is the equilibrium constant, K_{eq}.

$$K_{eq} = \frac{[HI]^2}{[I_2][H_2]}$$

Note that the equilibrium constant K_{eq} is capitalized to distinguish it from the specific rate constant k.

Must we know the rate expressions for the forward and reverse reactions for every equilibrium condition? In 1867, C.M. Guldberg and P. Waage, two Norwegian chemists, found that the numerator of the equilibrium expression contained the concentrations of the products of the reaction, and the denominator the concentrations of the reactants of the reaction. They also found that the exponents in the equilibrium expression were the same as coefficients from the chemical equation. Thus, it is possible to determine the equilibrium constant without knowing the reaction mechanism.

For the general equation $mA + nB \rightarrow sP + rQ$, the equilibrium constant is

$$K_{eq} = \frac{[P]^s[Q]^r}{[A]^m[B]^n}$$

Thus, K_{eq} establishes a relationship between the concentrations of the reactants and products of a reaction. If K_{eq} is very small (much less than 1), equilibrium will be established before much product is formed.

If K_{eq} is large (much greater than 1), the reaction will approach completion. For example, a K_{eq} of 1×10^{-8} would mean that equilibrium will be established before much product is formed. A K_{eq} of 1×10^{8} would mean that equilibrium will be established only after a great deal of product is formed. An industrial chemist would want to use a reaction with a large K_{eq} in order to obtain large amounts of product.

Consider the following example of the reaction between carbon dioxide and hydrogen.

$$CO_2(g) + H_2(g) \rightleftarrows CO(g) + H_2O(g)$$

At 1120°C, a measurement at equilibrium shows that the concentration of each substance is 0.01*M*, except H_2O, which is 0.02*M*. Substitution of the observed concentrations into the equilibrium expression results in a calculation of $K_{eq} = 2$.

$$K_{eq} = \frac{[CO][H_2O]}{[CO_2][H_2]} = \frac{[0.01][0.02]}{[0.01][0.01]} = 2$$

Thus, the products are favored. On the other hand, consider the reaction: $PCl_5(g) \rightleftarrows PCl_3(g) + Cl_2(g)$. Measurement of its equilibrium constant at 200°C indicates a K_{eq} of 0.457. In this case, the reactant is favored.

For any reaction, K_{eq} remains constant only if the temperature remains constant. Each reaction has a unique K_{eq} for every temperature. Above, we noted that the K_{eq} for $CO_2 + H_2 \rightleftarrows CO + H_2O$ is 2 at 1120°C. If the temperature is lowered to 500°C, the same reaction has a $K_{eq} = 5.5$.

The concentrations of solids and pure liquids are constants that can be determined from their densities. The concentrations of gases and solutes, on the other hand, must be obtained for the specific conditions under consideration.

SAMPLE PROBLEM

Equilibrium Constants

What is the equilibrium constant for the following reaction if the final concentrations are $CH_3COOH = 0.302M$, $CH_3CH_2OH = 0.428M$, $H_2O = 0.654M$, and $CH_3CH_2OOCCH_3 = 0.655M$?

$$CH_3COOH + CH_3CH_2OH \rightleftarrows CH_3CH_2OOCCH_3 + H_2O$$

Solving Process:

$$K_{eq} = \frac{[CH_3CH_2OOCCH_3][H_2O]}{[CH_3COOH][CH_3CH_2OH]} = \frac{[0.655][0.654]}{[0.302][0.428]} = 3.31$$

PRACTICE PROBLEMS

10. Write equilibrium expressions for the following reactions.
  a. $NH_2COONH_4 \rightleftarrows CO_2 + 2NH_3$
  b. $4HCl + O_2 \rightleftarrows 2Cl_2 + 2H_2O$

c. $NH_4HS \rightleftarrows NH_3 + H_2S$
d. $CuSO_4 \cdot 5H_2O \rightleftarrows CuSO_4 + 5H_2O$

11. At a given temperature, the reaction (all gases) $CO + H_2O \rightleftarrows H_2 + CO_2$ produces the following concentrations: $CO = 0.200M$; $H_2O = 0.500M$; $H_2 = 0.32M$; $CO_2 = 0.42M$. Find the K_{eq} at that temperature.
12. Hydrogen sulfide decomposes according to the equation: $2H_2S(g) \rightleftarrows 2H_2(g) + S_2(g)$. At 1065°C, measurement of an equilibrium mixture of these three gases shows the following concentrations: $H_2S = 7.06 \times 10^{-3}M$, $H_2 = 2.22 \times 10^{-3}M$, and $S_2 = 1.11 \times 10^{-3}M$. What is the value of K_{eq} for this equation?

SAMPLE PROBLEM

Equilibrium Concentration

What is the equilibrium concentration of SO_3 in the following reaction if the concentration of SO_2 and O_2 are each $0.0500M$ and $K_{eq} = 85.0$? The equation for the reaction is

$$2SO_2 + O_2 \rightleftarrows 2SO_3$$

Solving Process:

$$K_{eq} = \frac{[SO_3]^2}{[SO_2]^2[O_2]}$$

$$85.0 = \frac{[x]^2}{[0.0500]^2[0.0500]}$$

$$x^2 = 0.0106$$

$$x = 0.103M$$

MATH CONNECTION

Calculators and Roots

Calculators equipped with $\sqrt[x]{y}$ keys can calculate universal roots. To obtain the xth root of the number y, enter the value of y, press the $\sqrt[x]{y}$ key, enter the value of x, and press the = key. Other calculators may be equipped with universal power keys designated as y^x. In such cases, you must use the algebraic identity, $\sqrt[x]{y} \equiv y^{1/x}$ to calculate a root. To obtain the xth root of a number y using the y^x key, enter the value of y, press the y^x key, enter x, press the $1/x$ key followed by the = key. Some calculators may have universal roots and power keys designated as $\sqrt[y]{x}$ and x^y, respectively. When using these calculators, the order of entering the values of x and y must be reversed to the order stated above.

Figure 22.18 Quicklime is produced in this kiln from the reaction
$CaCO_3(s) \rightleftarrows CaO(s) + CO_2(g)$
The carbon dioxide gas is removed from the kiln to shift the equilibrium to the right so that the reaction is carried to completion.

PRACTICE PROBLEMS

13. At a given temperature, the K_{eq} for the gas phase reaction, $2HI(g) \rightleftarrows H_2(g) + I_2(g)$, is 1.40×10^{-2}. If the concentrations of both H_2 and I_2 at equilibrium are $2.00 \times 10^{-4}M$, find [HI].
14. If the temperature in the reaction in Practice Problem 11 is changed, the K_{eq} becomes 2.40. By removing some H_2 and CO_2 and adding H_2O all concentrations except CO are adjusted to the values given in Practice Problem 11. What is the new CO concentration?
15. At 60.2°C, the equilibrium constant for the reaction, $N_2O_4(g) \rightleftarrows 2NO_2(g)$, is 8.75×10^{-2}. At this temperature, a vessel contains nitrogen tetroxide, N_2O_4, at a concentration of $1.72 \times 10^{-2}M$ at equilibrium. What concentration of nitrogen dioxide, NO_2, does it contain?

Le Chatelier's Principle

The conditions affecting equilibrium are temperature, pressure, and concentration of reactants and products. If a system is in equilibrium and a condition is changed, then the system will shift toward restoring the equilibrium. As you learned in Chapter 17, this statement is called Le Chatelier's principle. Le Chatelier first described the effect of stress (change of conditions) upon reversible systems, such as water and melting ice, at equilibrium. His principle holds for reaction equilibrium as well. If stress is put on a reversible reaction at equilibrium, the system will shift in such a way to relieve the stress. Let us see how the principle applies.

The reaction for the preparation of ammonia by the Haber process is

$$N_2(g) + 3H_2(g) \rightleftarrows 2NH_3(g) + \textit{energy}$$

Let us consider the effect of changes in concentration, pressure, and temperature on the equilibrium.

If the concentration of either of the reactants is increased, the number of collisions between reactant particles will increase. The result is an increase of the reaction rate toward the right. As the amount of NH_3 increases, the rate of the reverse reaction will also increase. However, the net result to the system as a whole is to shift the reaction toward the right. That is, more product is produced.

Remember that the ideal gas equation can be written as $n/V = P/RT$. The reactants and products are in the same container so the volume and temperature are the same for both. Thus, pressure will determine the concentration, n/V. If the pressure is increased in the Haber process, more product is formed. Consider the situation if the pressure on the reaction system is doubled.

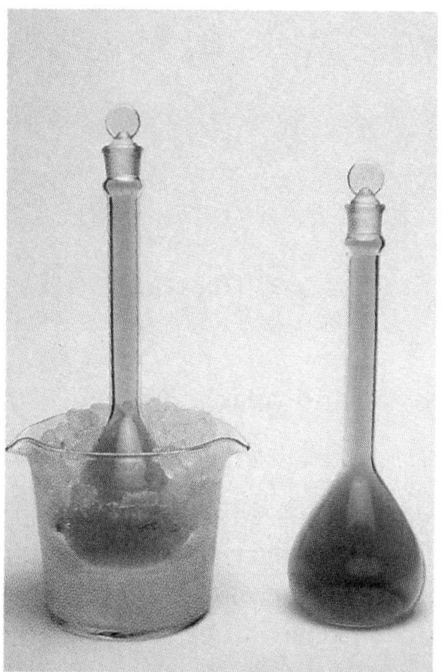

Figure 22.19 These flasks contain brown $NO_2(g)$ and colorless $N_2O_4(g)$ in equilibrium: $2NO_2(g) \rightleftharpoons N_2O_4(g)$. In the left photograph, the flasks are at 0°C and 20°C. In the right photograph, the flasks are at 0°C and 80°C. Increasing the temperature favors the formation of the brown $NO_2(g)$.

The concentrations of nitrogen, hydrogen, and ammonia are all doubled. The equilibrium expression for this reaction shows that the concentration of ammonia is squared.

$$K_{eq} = \frac{[NH_3]^2}{[N_2][H_2]^3}$$

Squaring a doubled concentration means the reverse reaction must then speed up by a factor of 4. On the other hand, the concentration of hydrogen is cubed. Cubing a doubled concentration would increase the rate by a factor of 8. However, the H_2 concentration is multiplied by the concentration of nitrogen, which has also been doubled. Doubling the pressure should increase the rate of the forward reaction by $2^3 \times 2$, or 16 times! The net result is clearly an increase in the formation of product.

In the reaction $H_2(g) + Cl_2(g) \rightarrow 2HCl(g)$, all substances are also gases. The number of moles of reactants is the same as the number of moles of product. Because concentration varies directly with the pressure, a change in the pressure on this reaction will not shift the equilibrium. The rate in each direction would be affected in the same way. An increase in pressure will always drive a reaction in the direction of the smaller number of moles of gas. Pressure, of course, has an effect only on the gases in a reaction. In reactions involving more than one phase, pressure changes will only affect the concentration of gases. A reaction taking place in solution or one involving only pure liquids or solids would be unaffected by pressure.

Both the forward and reverse reactions at equilibrium are speeded by an increase in temperature. However, their rates are increased by different amounts. Also, the value of the equilibrium constant itself is changed by a change in temperature. How can you predict the shift in an equilibrium subjected to a temperature change? One easy way is to consider energy as either a reactant or a product.

CHEMISTS AND THEIR WORK

Fritz Haber
(1868-1934)

Fritz Haber received his doctorate degree from the University of Berlin in 1891. His research interests included the study of flames, the reaction of gases when heated, and methods of producing nitrogen compounds from atmospheric nitrogen. He received the 1919 Nobel Prize in chemistry for developing the Haber process. However, because the process was used by Germany to make explosives during World War II, many scientists denounced his award. Haber was forced to leave his homeland during Hitler's regime because he was Jewish.

Figure 22.20 The lenses of solar darkening sunglasses contain small, trapped crystals of colorless silver chloride, which decomposes in light to form atoms of silver and chlorine, $AgCl + light \rightarrow Ag + Cl$. As light intensity increases, more opaque silver atoms form, and the lenses darken. The reaction is reversible. So, as light intensity decreases, the reverse reaction occurs and the lenses lighten.

Energy is produced when hydrogen and nitrogen react.

$$N_2(g) + 3H_2(g) \rightleftarrows 2NH_3(g) + \textit{energy}$$

If energy is considered to be a product, the addition of energy would increase the concentration of the product. The equilibrium would be shifted to the left to reduce this stress. In the Haber process, the reverse reaction (the decomposition of ammonia) is favored by the addition of energy.

In this equilibrium reaction, however, equilibrium is often attained before enough product is produced to make the process economical. In such circumstances, the equilibrium can be shifted to give a higher yield of product. The conditions that produce the highest yield are called the **optimum conditions.**

In the Haber process, there are five optimum conditions. (1) High concentrations of hydrogen and nitrogen should be maintained. (2) Ammonia should be removed as it is formed so that its concentration never increases. (3) A temperature is used that is high enough to maintain a reasonable rate but low enough not to favor the reverse reaction. (4) A catalyst is used to lower the required activation energy. (5) High pressure is maintained throughout the process. Each condition increases yield by shifting the equilibrium to favor formation of product.

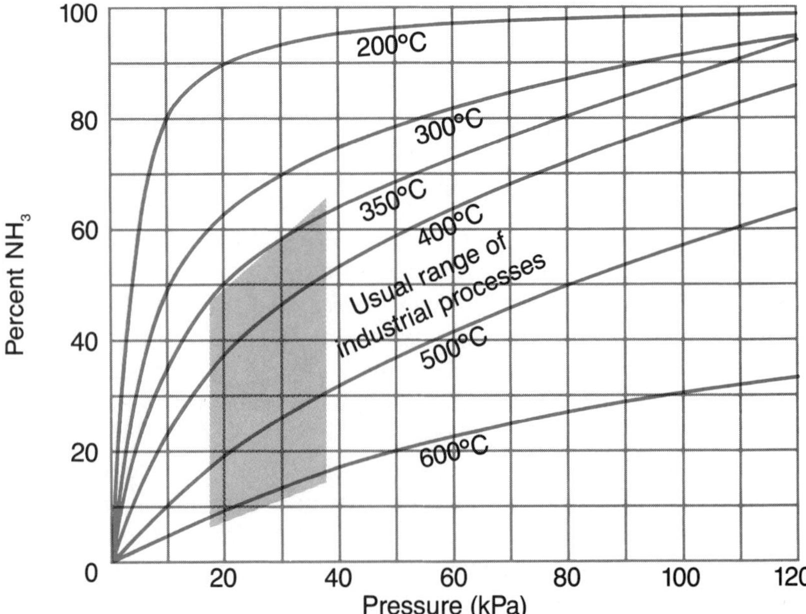

Figure 22.21 This graph shows that by controlling the pressure and the temperature of the reaction system in the Haber process, the yield of ammonia can be increased.

Bridge to
AGRICULTURE

Ammonia as a Fertilizer

The continued increase in the world's population is putting a strain on available food resources. In the United States, we are fortunate to have a surplus of food; we are able to export food to many parts of the world where supplies are insufficient to maintain life. Fertilizers are one class of agrichemicals that are widely used to improve the yield of crops from a given parcel of land. Plants need a number of elements to grow properly, but the principal ones are nitrogen, phosphorus, and potassium. Most of the nitrogen that is used as fertilizer comes from ammonia, NH_3.

In the United States we produce more than 33 billion pounds of ammonia each year! About 85% of that ammonia is used in fertilizers such as ammonium sulfate, ammonium nitrate, and urea. A common process for producing ammonia is the Haber process, named for Fritz Haber, a German scientist who developed the process in the early part of this century. As we have already seen in an earlier section, that reaction is heavily influenced by the conditions under which it is run. The process is usually carried out at pressures of 15 to 30 MPa and a temperature near 500°C. Even higher pressures would be necessary if it were not for the use of a catalyst that is a mixture of iron and oxides of other elements. The Haber process is now so efficient that the cost of ammonia is mainly that of the hydrogen used in making it.

Exploring Further

1. List and give the chemical formulas of ammonia compounds used as fertilizers.
2. If catalysts were not used, why would higher pressures and temperatures be necessary to produce ammonia by the Haber process?
3. How is nitrogen important to plant life?

22.2 CONCEPT REVIEW

16. Write the equilibrium constant equation for the following reaction.

$$H_2Cr_2O_7(aq) + 4NaOH(aq) \rightarrow 2Na_2CrO_4(aq) + 3H_2O(l)$$

17. What would be the effect on the $H_2Cr_2O_7$ concentration if you added sodium hydroxide to the system in the question above when the system is at equilibrium?

18. If an equilibrium constant for a system has a value of 5.62×10^{49}, would you expect to find mostly reactant or mostly product after the system has come to equilibrium?

19. **Apply** If you wished to maximize the products in the following equilibrium, would you run the reaction at a high pressure or at a low pressure?

$$CaCO_3(cr) \rightarrow CaO(cr) + CO_2(g)$$

Chapter 22 REVIEW

Summary

22.1 Reaction Rates

1. A process may be spontaneous but proceed so slowly that no change is apparent. The reactants are said to be kinetically stable in this situation.
2. A reversible reaction is a reaction in which products react to form the original reactants.
3. Reaction rate is the rate of disappearance of one of the reactants of a reaction or the rate of appearance of one of the products.
4. Even though the overall energy change may be negative, many reactions require an initial input of energy to form an activated complex. This energy is called the activation energy.
5. Increasing the concentration of a reactant increases the rate of reaction by increasing the number of collisions.
6. Increasing the pressure of a gas or the surface area of a heterogeneous reactant increases reaction rate.
7. Increasing the temperature increases the rate of reaction because collisions are more frequent and more of the collisions involve sufficient energy to form the activated complex.
8. A catalyst is a substance that causes an increase in reaction rate without being permanently changed.
9. A catalyst changes the reaction mechanism so that the activation energy required is lower.
10. A homogeneous catalyst is one that is in the same phase as the reactants.
11. Most reactions take place in a series of steps called the reaction mechanism. The slowest step in the mechanism determines the rate of the overall reaction.

22.2 Chemical Equilibrium

12. The equilibrium constant, K_{eq}, is the ratio of the forward rate constant, k_f, to the reverse rate constant, k_r.
13. If the equilibrium constant for a process has a large value, the process goes substantially to completion. Conversely, if the constant is small, little product is formed when equilibrium is reached.

Key Terms

thermodynamically stable
kinetically stable
reversible reaction
reaction rate
activated complex
specific rate constant
homogeneous reaction
heterogeneous reaction
catalysis
heterogeneous catalyst
homogeneous catalyst
inhibitor
rate determining step
reaction mechanism
optimum conditions

Review and Practice

20. How is reaction rate defined? In what units are reaction rates expressed?
21. Explain how concentration of reactants affects reaction rate.
22. What does the term $[HNO_3]$ mean? In what units is it expressed?
23. Write a reaction rate expression for the following single-step reaction.

$$N_2O_4 + H_2O \rightarrow HNO_2 + HNO_3$$

24. What effect on reaction rate is caused by increasing the pressure of a gaseous reactant? Why is this effect similar to that achieved by increasing concentration?
25. How does temperature affect the rate at which activated complexes are formed? Describe the two main mechanisms that affect this rate.
26. What does a catalyst do? In general, how does a catalyst carry out its function?

27. Platinum acts as a heterogeneous catalyst in many hydrogenation reactions because it adsorbs large quantities of hydrogen. What is meant by adsorption? How does a heterogeneous catalyst function?

28. How does a homogeneous catalyst differ from a heterogeneous catalyst?

29. Define *reaction mechanism.*

30. For each of the following rate expressions, determine the order of each reactant listed and the overall order of the reaction.

a. *rate* = $k[NO_2][NO_2]$

b. *rate* = $k[NO_2]^2[O_2]$

31. NO will react with bromine according to the reaction $2NO + Br_2 \rightarrow 2NOBr$. Experimentally, the following data were obtained.

Trial	Initial [NO]	Initial $[Br_2]$	Initial rate of formation of NOBr
1	1.0*M*	1.0*M*	0.80 $(mol/dm^3)/s$
2	1.0*M*	2.0*M*	1.60 $(mol/dm^3)/s$
3	2.0*M*	2.0*M*	6.40 $(mol/dm^3)/s$

a. Write the rate expression for this reaction.

b. Calculate the rate constant, *k*.

c. What would be the initial rate of formation of NOBr if the initial concentrations of NO and Br_2 were 0.60*M* and 0.25*M* respectively?

d. If the initial concentration of NO were 1.4*M*, what initial concentration of Br_2 would produce NOBr at an initial rate of 0.75 $(mol/dm^3)/s$?

e. Note that concentrations and rates are all specified "initial." How would rate differ if measured later in the reaction? Why would it differ?

32. Balance the equation and write the equilibrium constant expression, K_{eq}, for each of the following reactions. All reactants and products are gases.

a. $NH_3 + O_2 \rightleftarrows NO + H_2O$

b. $COCl_2 \rightleftarrows CO + Cl_2$

c. $H_2O + CO \rightleftarrows H_2 + CO_2$

d. $N_2O_4 \rightleftarrows NO_2$

33. Determine the equilibrium constant of the following chemical reaction using the data given.

$$2SO_2 + O_2 \rightleftarrows 2SO_3$$

At equilibrium at 295°C, a 2.00-dm^3 flask was found to contain 0.35 mole of SO_2, 0.70 mole of O_2, and 1.40 moles SO_3.

34. At 2000°C, nitrogen and oxygen react according to the following equation.

$$N_2 + O_2 \rightleftarrows 2NO$$

The equilibrium constant for this reaction at 2000°C is 1.2×10^{-4}. At equilibrium, the concentrations of N_2 and O_2 are found to be 0.166*M* and 0.145*M* respectively. What is the concentration of NO?

35. List and describe the four factors that influence the rate of a reaction.

36. The following reaction goes to completion. $2NO(g) + H_2(g) \rightarrow$
$N_2O(g) + H_2O(g) + 340\ 270$ J

a. What effect would an increase in temperature have on its rate?

b. Assume the reaction is a single-step reaction. What would be the effect on the reaction rate if the hydrogen gas concentration were doubled? If the pressure on the reaction were doubled?

37. In the following hypothetical reaction, which will be in greater concentration at equilibrium, D_2, E_2, or DE?

$D_2 + E_2 \rightarrow 2DE$ *rate* = 0.26 $(mol/dm^3)/s$
$2DE \rightarrow D_2 + E_2$ *rate* = 0.15 $(mol/dm^3)/s$

38. On the same set of axes, show an energy diagram of a reaction, both catalyzed and uncatalyzed.

39. $NaCl + AgNO_3 \rightarrow AgCl + NaNO_3$ is an ionic reaction. Why are ionic reactions more rapid than other reactions?

40. When steam is passed over iron filings, iron(II) oxide, iron(III) oxide, and hydrogen are formed. If this hydrogen is collected and passed over the heated iron oxides, water and iron will again be formed. Write two balanced equations (one for each iron oxide) showing these reversible reactions.

41. If K_{eq} for $2A + B \rightleftarrows 2C$ is 8.0, set up the expression used to calculate the concentration of C at equilibrium. Calculate the equilibrium concentration of C if the starting conditions were 0.50 mol each of A and B in a 10.0-dm^3 container.

42. Ammonia combines with oxygen to produce water vapor and nitrogen.

a. Write a balanced equation for the equilibrium reaction.

b. Write the expression for the equilibrium constant.

c. At a certain temperature, the concentration of each substance is 1.0*M*. Calculate K_{eq} for that temperature.

d. Calculate K_{eq} if $[NH_3]$ is 3.0*M*, $[O_2]$ is 2.0*M*, $[H_2O]$ is 4.0*M*, and $[N_2]$ is 2.0*M*.

Concept Mastery

43. What is meant when we say that iron(III) oxide is thermodynamically stable?

44. Describe what is happening in a reversible reaction as it begins, as it progresses, and after it reaches equilibrium.

45. How does an activated complex form?

46. In terms of activation energy, account for the kinetic stability of a substance.

47. A certain reaction takes place in three steps as follows.

$$X \xrightarrow{\text{fast}} Y;\ Y \xrightarrow{\text{slow}} Z;\ Z \xrightarrow{\text{fast}} \text{final product}$$

a. Which is the rate determining step?

b. If you were to examine the contents of the reaction vessel after the reaction had begun but before completion, what relative amounts of each substance would you expect to find?

48. The rate expression for the reaction $2NO(g) + O_2(g) \rightarrow 2NO_2(g)$ is $rate = k[NO]^2[O_2]$. Predict the effect on the reaction when each of the following is increased.

a. [NO] **c.** pressure
b. $[O_2]$ **d.** temperature

49. In the last section, study the discussion of the equilibrium established in the Haber process.

a. Explain the effect of increasing the concentration of reactants.

b. Explain the effect of increasing the pressure on the system.

c. Explain the effect of increasing the temperature of the system.

d. What would be the effect of removing the ammonia formed during the reaction?

50. Use the following reaction and Le Chatelier's principle to explain the effects of changes in each of the following on a system in equilibrium.

$$PCl_5(g) \rightleftarrows PCl_3(g) + Cl_2(g) + \text{energy}$$

a. concentration of reactant

b. pressure in the system

c. temperature

51. Relate thermodynamic stability of substances to the change in energy when the substance decomposes.

52. How do the relative rates of the forward and reverse reactions in a reversible reaction affect the quantities of reactants and products at equilibrium?

53. Why is it possible for a reaction to have a large negative value for the change in energy when it decomposes, yet apparently not take place?

Application

54. Titanium can be obtained from its ore by the following reactions.

$$TiO_2 + C + 2Cl_2 \rightarrow TiCl_4 + CO_2$$
$$2Mg + TiCl_4 \rightarrow Ti + 2MgCl_2$$

Why would grinding the C and TiO_2 into powder increase the rate for the first reaction?

55. Why must a fire in a fireplace be started with paper and kindling? Why not light the logs directly with a match?

56. In the chemical industry, reactions are generally carried out either by batch or continuous-flow processes. In a batch process, reactants are placed in a vessel and allowed to react. The vessel is then emptied, and products are isolated and purified. In a continuous-flow process, reactants are continuously pumped into a reaction chamber, and products are continuously removed. Explain which process is better adapted to reversible reactions that reach equilibria?

Critical Thinking

57. Refer to Figure 22.9 on page 552.

a. From what source do hydrogen and oxygen obtain the energy required to form an activated complex?

b. After forming an activated complex, the complex may fall either toward product or back to reactants. How can you explain the fact that the reaction goes substantially to completion?

58. How would an increase in pressure affect the equilibrium of a reaction in which the products occupied more volume than the reactants?

59. Many exothermic reactions have relatively high activation energies. In order to start them, energy must be supplied from an outside source. Why is it usually unnecessary to keep on supplying outside energy to the reaction?

60. In the reaction $NO_2 + CO \rightarrow NO + CO_2$, it has been experimentally determined that the rate expression is second order in NO_2 and zero order in CO. Propose a possible mechanism for the reaction.

Readings

Gordus, Adon A. "Chemical Equilibrium." *Journal of Chemical Equilibrium* 68 no. 2 (February 1991): 138-140; and 68 no. 3 (March 1991): 215-217.

Teichner, S.J. "Aerogels." *CHEMTECH* 21 no. 6 (June 1991): 372-377.

Cumulative Review

1. Describe the process of solvation.

2. Describe three procedures that can be used to increase the rate of solution of a solid solute in a solvent.

3. What is the molarity of a solution that contains 16.0 g HIO_3 in 1.00×10^2 cm^3 of solution?

4. What effect does a nonvolatile solute have on the boiling point of a solvent? Why does it have this effect?

5. What is the vapor pressure at 25°C of a solution of 98.1 g of sucrose, $C_{12}H_{22}O_{11}$, dissolved in 1.00×10^2 g of water?

CHAPTER **Preview**

CONTENTS

CHAPTER

23

Acids, Bases, and Salts

Have you ever tried to pick up something when your hands were soapy? Soap makes your hands slick because soap creates a basic solution, and bases feel slick. If we were to taste soap, it tastes bitter, another property of a base. Lemonade, however, tastes sour because it contains an acid. Sour taste is a physical property of acids.

Of the top 100 chemical products produced in the United States, seven are acids, three are bases, and twelve are a type of compound known as salts. Many of these substances are consumed in quantities of billions of pounds per year. Most of that consumption is by industry in the production of consumer products. The role played by these substances makes them important groups of compounds to understand. In this chapter, you will learn what these three groups are, their relationship to each other, how they react, and how they may be used.

EXTENSIONS AND CONNECTIONS

Chemistry and Society

Chemical Economics: A view of how chemical processes affect the economics of production, page 585

Everyday Chemistry

Baking: An Acid-Base Reaction: How this reaction makes batter rise, page 589

Chemists and Their Work

Mary Engle Pennington, page 591

OBJECTIVES

- distinguish the definitions of acids and bases as outlined in the theories of Arrhenius, Brønsted-Lowry, and Lewis.
- name acids and bases.
- define acidic and basic anhydrides and write formulas for them.
- define and give examples of strong and weak acids and bases.

23.1 Acids and Bases

Acids and bases have a particular relationship to one another. Therefore, we will look at these two different groups of compounds together. There are a number of ways of defining acids and bases. In this section, three such definitions will be presented. The definitions of acids and bases have changed over the years because of the changing ways chemists have dealt with substances. Originally, most chemistry was carried out in aqueous solution. Consequently, the first definitions of acids and bases included their behavior in water. As chemists began using other solvents, the definitions were broadened.

Arrhenius Theory

It was discovered long ago that most acids, bases, and salts, when dissolved in water, conduct an electric current. Because solutions of these substances conduct a current, these substances are called **electrolytes.**

In 1887, the Swedish chemist Svante Arrhenius published a paper that discussed acids and bases as electrolytes. He knew that aqueous solutions containing acids or bases conducted an electric current. Arrhenius concluded that these substances released charged particles when dissolved. He called these charged particles ions. His theory was that *acids were substances that ionized in water solution to produce hydrogen ions, H^+, or free protons.* He also believed that *bases were substances that ionized to produce hydroxide ions, OH^-, in water solution.* The following are examples.

$$HCl(g) \rightarrow H^+(aq) + Cl^-(aq)$$
$$NaOH(cr) \rightarrow Na^+(aq) + OH^-(aq)$$

We now know that in the case of NaOH, the ions already exist in the solid and merely dissociate in solution. Note the difference between dissociation and ionization. In dissociation, ions separate. In ionization, neutral molecules react with water to form charged ions.

Figure 23.1 Some sea slugs defend themselves by ejecting a stream of sulfuric acid, H_2SO_4. Sulfuric acid ionizes completely in water.

Brønsted-Lowry Theory

As the knowledge of catalysts and nonaqueous solutions increased, it became necessary to redefine the terms acid and base. In 1923, an English scientist, T. M. Lowry, and a Danish scientist, J. N. Brønsted, independently proposed new definitions. They stated that *in a chemical reaction, any substance that donates a proton is an acid* and *any substance that accepts a proton is a base.* Remember that H^+ is just a proton. For example, when hydrogen chloride gas is dissolved in water, ions are formed.

$$\underset{\text{acid}}{HCl(g)} + \underset{\text{base}}{H_2O(l)} \rightarrow H_3O^+(aq) + Cl^-(aq)$$

In this reaction, hydrogen chloride is an acid, and water is a base. Notice that the hydrogen ion (H^+) from the acid has combined with a water molecule to form the polyatomic ion H_3O^+, which is called the **hydronium** (hi DROH nee uhm) **ion.** There is strong evidence that the hydrogen ion is never found free as H^+. The bare proton is so strongly attracted by the electrons of surrounding water molecules that H_3O^+ forms immediately. Consider the opposite reaction.

$$\underset{\text{acid}}{H_3O^+(aq)} + \underset{\text{base}}{Cl^-(aq)} \rightarrow HCl(g) + H_2O(l)$$

In this reaction, the H_3O^+ ion is an acid. It acts as an acid because it donates a proton to the chloride ion, which is a base. The hydronium ion is said to be the conjugate acid of the base, water. The chloride ion is called the conjugate base of the acid, hydrochloric acid. In general, any acid-base reaction is described as:

acid + base → conjugate base + conjugate acid

The **conjugate base** of an acid is the particle that remains after a proton has been released by the acid. The **conjugate acid** of a base is formed when the base acquires a proton from the acid. Table 23.1 contains some bases and their conjugate acids.

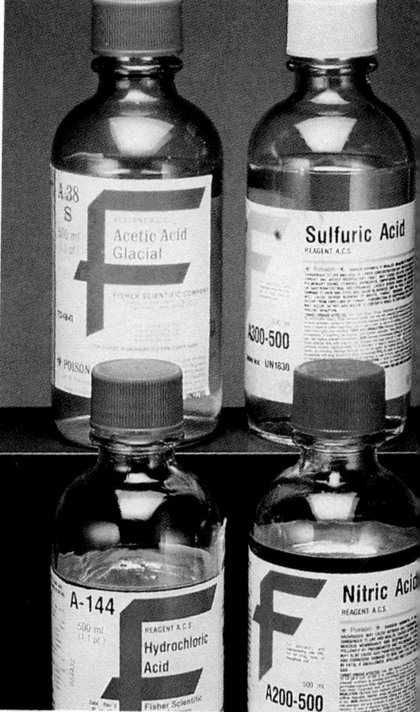

Figure 23.2 Four common laboratory acids are hydrochloric acid, HCl, sulfuric acid, H_2SO_4, nitric acid, HNO_3, and acetic acid, CH_3COOH.

Table 23.1

Bases and Their Conjugate Acids

Base	Name	Conjugate acid	Name
CH_3COO^-	acetate ion	CH_3COOH	acetic acid
NH_3	ammonia	NH_4^+	ammonium ion
CO_3^{2-}	carbonate ion	HCO_3^-	hydrogen carbonate ion
CN^-	cyanide ion	HCN	hydrocyanic acid
$H_2PO_4^-$	dihydrogen phosphate ion	H_3PO_4	phosphoric acid
H_2NNH_2	hydrazine	$H_2NNH_3^+$	hydrazinium ion
HSO_4^-	hydrogen sulfate ion	H_2SO_4	sulfuric acid
OH^-	hydroxide ion	H_2O	water
NO_3^-	nitrate ion	HNO_3	nitric acid
ClO_4^-	perchlorate ion	$HClO_4$	perchloric acid
S^{2-}	sulfide ion	HS^-	hydrogen sulfide ion
H_2O	water	H_3O^+	hydronium ion

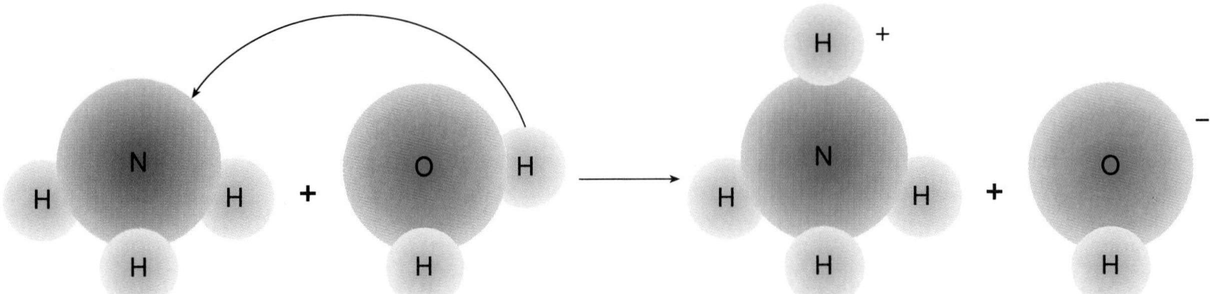

Figure 23.3 When ammonia is added to water, a water molecule donates a proton to the ammonia molecule. Water acts as an acid, and ammonia acts as a base.

Consider what happens when ammonia is added to water.

$$\underset{\text{base}}{NH_3(g)} + \underset{\text{acid}}{H_2O(l)} \rightarrow \underset{\text{conjugate acid}}{NH_4^+(aq)} + \underset{\text{conjugate base}}{OH^-(aq)}$$

In this reaction, water acts as an acid because it donates a proton to the ammonia molecule. The ammonium ion is the conjugate acid of ammonia, a base, which receives a proton from water. The hydroxide ion is the conjugate base of the acid, water.

PRACTICE PROBLEMS

1. Identify the acid, base, conjugate acid, and conjugate base in the following reactions.
   a. $HNO_3(aq) + NaOH(aq) \rightarrow H_2O(l) + NaNO_3(aq)$
   b. $NaHCO_3(aq) + HCl(aq) \rightarrow NaCl(aq) + H_2CO_3(aq)$
2. What is the conjugate base of each of the following?
   a. H_2SO_3 **c.** NH_3
   b. H_2CO_3 **d.** HF

CHEMISTRY IN DEPTH

Lewis Theory

In 1923, the same year that Brønsted and Lowry proposed their acid-base theory, another idea appeared. Gilbert Newton Lewis, an American chemist, proposed even broader definitions of acids and bases. However, Lewis focused on electron behavior rather than proton transfer. He defined *an acid as an electron-pair acceptor and a base as an electron-pair donor.* These definitions are more general than Brønsted's. They apply to solutions and reactions that do not even involve hydrogen or hydrogen ions.

Consider the reaction that occurs between ammonia and boron trifluoride.

$$BF_3(g) + NH_3(g) \rightarrow F_3BNH_3(g)$$

The electronic structures of boron trifluoride and ammonia are

$$\begin{array}{c} F \\ F:\overset{..}{\underset{..}{B}} \\ F \end{array} \quad \text{and} \quad \begin{array}{c} H \\ :\overset{..}{\underset{..}{N}}:H \\ H \end{array}$$

Note that boron has an empty orbital and can accept two more electrons in its outer level. Since boron trifluoride can accept an electron pair, it is a Lewis acid. Now consider the structure of ammonia. Note that the nitrogen atom has an unshared electron pair that can be donated to the boron atom. Ammonia is a Lewis base because it can donate an electron pair. If we use dots to represent the electrons involved in the reaction, the equation can be written

$$\underset{\text{Lewis base}}{H_3N\!:} + \underset{\text{Lewis acid}}{BF_3} \rightarrow \underset{\text{Addition product}}{H_3N\!:\!BF_3}$$

Consider again the reaction of ammonia gas and water.

$$\underset{\text{Lewis base}}{H\!:\!\overset{\displaystyle H}{\underset{\displaystyle H}{\ddot{\underset{\cdot\cdot}{N}}}}\!:} + \underset{\text{Lewis acid}}{H—O—H} \rightarrow \left[H\!:\!\overset{\displaystyle H}{\underset{\displaystyle H}{\ddot{\underset{\cdot\cdot}{N}}}}\!:\!H\right]^+ + OH^-$$

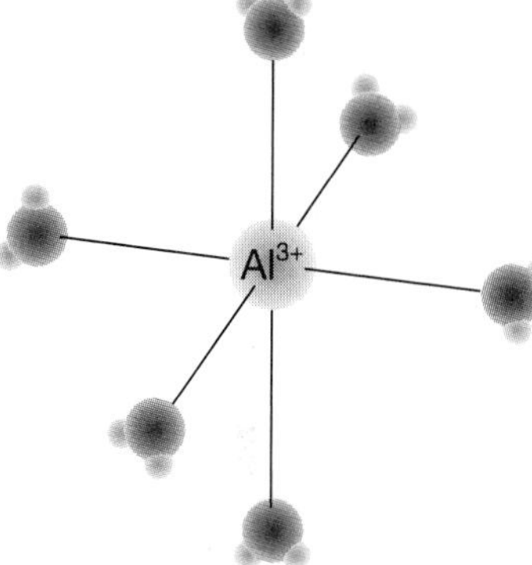

Figure 23.4 The Al^{3+} ion acts as a Lewis acid by accepting electron pairs from water and forms the octahedral complex, $[Al(H_2O)_6]^{3+}$.

The ammonia donates an electron pair and is the Lewis base. The hydrogen atom attached to the oxygen of the water molecule acts as the Lewis acid. Notice that ammonia is a base in the Arrhenius theory, the Brønsted-Lowry theory, and the Lewis theory.

The formation of complex ions can be viewed in terms of the Lewis acid-base theory. Recall from Chapter 14 that central ions have empty orbitals and can therefore act as electron pair acceptors. These ions are Lewis acids. Ligands, on the other hand, have unshared electron pairs that they can donate. Ligands are Lewis bases. For example, the aluminum ion, Al^{3+}, has all outer orbitals empty, and water,

$$:\ddot{O}—H \\ \quad | \\ \quad H$$

has unshared electron pairs. It is to be expected, then, that the reaction between Al^{3+} and water where aluminum has coordination number 6,

$$Al^{3+} + 6H_2O \Rightarrow [Al(H_2O)_6]^{3+},$$

proceeds vigorously to the right.

Table 23.2

Summary of Acid-Base Theories		
Theory	**Acid Definition**	**Base Definition**
Arrhenius	Any substance that releases H^+ ion in water solution	Any substance that releases OH^- ions in water solution
Brønsted-Lowry	Any substance that donates a proton	Any substance that accepts a proton
Lewis	Any substance that can accept an electron pair	Any substance that can donate an electron pair

Figure 23.5 Nitric acid, HNO_3, reacts with copper, producing nitrogen dioxide gas, NO_2.

PRACTICE PROBLEMS

3. Classify each of the following substances as either a Lewis acid or a Lewis base.
   a. Cl^- **c.** Na^+
   b. CO_3^{2-} **d.** Br^-

Write equations for the formation of complex ions in the following reactions:

4. $Ag^+ + NH_3 \rightarrow$ (coordination number = 2)
5. $Cu^{2+} + NH_3 \rightarrow$ (coordination number = 4)
6. $Co^{3+} + NH_3 \rightarrow$ (coordination number = 6)
7. $Fe^{3+} + CN^- \rightarrow$ (coordination number = 6)
8. $Zn^{2+} + OH^- \rightarrow$ (coordination number = 4)

Naming Binary Acids

Binary acids are acids containing only two elements. If you look at Table 23.3, you will notice that the prefix is *hydro-* and the suffix is *-ic* in the names of binary acids.

To name a binary acid, we determine what stem to use with the prefix *hydro-* and the suffix *-ic* by finding the stem of the name of the element that combines with hydrogen to form the acid. For instance, in HCl, chlorine will have the stem *-chlor-*, and, in HF, fluorine will have the stem *-fluor-*. To this stem, the prefix *hydro-* and the suffix *-ic* are added. One example of a non-binary acid named according to these rules is hydrocyanic acid, HCN. An exception to the naming system for a binary acid is seen in the naming of hydroazoic acid, HN_3, where the root $-azo-$ is used for nitrogen.

Table 23.3

Naming Binary Acids

Binary Compound + Water	Prefix	Stem	Suffix	Name
Hydrogen chloride gas dissolved in water	Hydro-	-chlor-	-ic	Hydrochloric acid
Hydrogen iodide gas dissolved in water	Hydro-	-iod-	-ic	Hydroiodic acid
Hydrogen sulfide gas dissolved in water	Hydro-	-sulfur-	-ic	Hydrosulfuric acid

Naming Ternary Acids and Bases

Ternary acids are acids that contain three elements. The ternary acids we will be working with have oxygen as the third element. We find the stem by determining what element is combined with oxygen and hydrogen in the acid molecule. We determine the prefix, if there is one, and the suffix by the number of oxygen atoms in each molecule.

Generally, the most common form of a ternary acid is given the suffix *-ic*. No prefix is used. Examples of common ternary acids are sulfur*ic* acid, H_2SO_4, chlor*ic* acid, $HClO_3$, and nitr*ic* acid, HNO_3.

If a second acid is formed containing the same three elements, but having one fewer oxygen atom, this acid is given the suffix *-ous*. Again, there is no prefix. Examples of these acids are sulfu*rous* acid, H_2SO_3, chlor*ous* acid, $HClO_2$, and nitr*ous* acid, HNO_2.

If a third acid containing still fewer oxygen atoms is formed, it is given the prefix *hypo-* and the suffix *-ous*. An example is *hypo*chlor*ous* acid, HClO.

Acids containing one more oxygen atom than the common form contains are named by adding the prefix *per-* to the common name; for example, $HClO_4$ is *per*chloric acid. See the examples in Table 23.4.

Table 23.4

Naming Ternary Acids					
Compound	**Number of Oxygen Atoms**	**Prefix**	**Stem**	**Suffix**	**Name of Acid**
H_2SO_4	4	no prefix	sulfur-	-ic	sulfuric
H_2SO_3	3	no prefix	sulfur-	-ous	sulfurous
$HClO_4$	4	per-	-chlor-	-ic	perchloric
$HClO_3$	3	no prefix	chlor-	-ic	chloric
$HClO_2$	2	no prefix	chlor-	-ous	chlorous
HClO	1	hypo-	-chlor-	-ous	hypochlorous

It is not possible, without previous knowledge, to know which form of an acid is most common. If the name of one form is known, the other ternary acids containing the same elements can be named. Bromine forms only two acids with hydrogen and oxygen: HBrO and $HBrO_3$. Instead of being named bromous and bromic acids, they are named hypobromous and bromic acids. The exception occurs because they contain the same number of oxygen atoms as hypochlorous and chloric acids. The pattern of the chlorine acids is followed for the bromine acids because both elements are in the same group in the periodic table. The same pattern is followed in naming the ternary acids of the other halogens. In addition to those acids previously mentioned, the most common acid containing phosphorus, oxygen, and hydrogen is H_3PO_4. The most common inorganic acid containing carbon, oxygen, and hydrogen is H_2CO_3. As you can see, it is useful to memorize the names of common acids.

Arrhenius bases are composed of metallic, or positively charged, ions and the negatively charged hydroxide ion. These bases are named by adding the word *hydroxide* to the name of the positive ion. Examples are sodium hydroxide, NaOH, and calcium hydroxide, $Ca(OH)_2$.

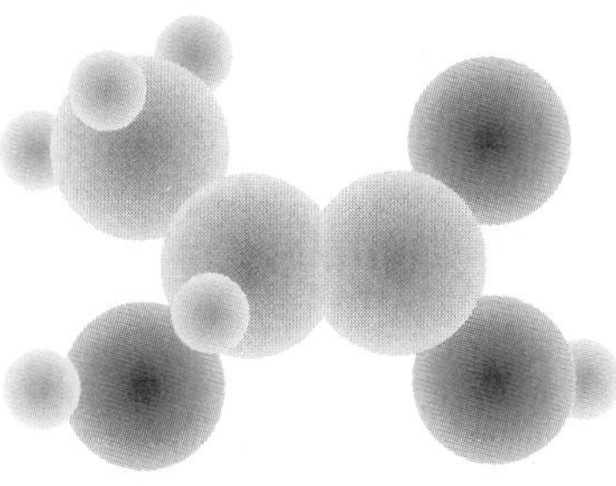

Figure 23.6 Lactic acid is a carboxylic acid. It is a solid at room temperature and is widespread in nature. The sour taste of yogurt, the tartness of sauerkraut, and the acid taste of perspiration are all due to lactic acid.

Naming Organic Acids and Bases

There are several kinds of organic acids. The most important type is the carboxylic acid, characterized by the carboxyl group

$$-\overset{\displaystyle O}{\overset{\|}{C}}-OH \quad \text{or} \quad -COOH$$

Carboxylic acids are named by adding the ending *-oic acid* to the name of the hydrocarbon from which the acid is derived. For example, HCOOH is methanoic acid, which is also known as formic acid. Similarly, $CH_3CH_2CH_2CH_2CH_2COOH$ is hexanoic acid and CH_3CH_2COOH is propanoic acid.

Organic bases contain nitrogen with its unshared pair of electrons. The most common organic bases contain the amine group

$$-\overset{\displaystyle H}{\overset{|}{N}}-H \quad \text{or} \quad -NH_2$$

Amines are named by adding the ending *-amine* to the name of the hydrocarbon from which they were derived. For example, $CH_3CH_2NH_2$ is ethanamine.

PRACTICE PROBLEMS

9. Name each of the following binary acids.
- **a.** HBr(aq)
- **b.** HF(aq)

10. Name each of the following ternary acids.
- **a.** H_2SeO_3
- **b.** HNO_2
- **c.** H_3PO_4
- **d.** H_3AsO_3
- **e.** HIO_3

11. Write formulas for each of the following acids.
- **a.** carbonic acid
- **b.** nitric acid
- **c.** arsenic acid
- **d.** selenic acid
- **e.** hypoiodous acid

12. Name each of the following bases.
- **a.** $Ca(OH)_2$
- **b.** KOH
- **c.** $Al(OH)_3$
- **d.** CH_3NH_2
- **e.** RbOH

13. Write formulas for each of the following bases.
- **a.** cesium hydroxide
- **b.** propanamine
- **c.** butanamine
- **d.** lithium hydroxide

Acid-Base Behavior

Consider a compound having the formula HO*X*. If the element *X* is highly electronegative, it will have a strong attraction for the electrons it is sharing with the oxygen. As these electrons are pulled toward *X*, the oxygen, in turn, will pull strongly on the electrons it is sharing with the hydrogen. The hydrogen ion, or proton, will then be easily lost. In this case, HO*X* is behaving as an acid. For example, consider hypochlorous acid, HOCl. The combination of oxygen and chlorine, both of which have a high electronegativity, creates a greater attraction for the electrons shared with hydrogen than oxygen would alone. The hydrogen is then easily lost.

$$H \rightarrow O - Cl$$

If the element *X* has a relatively low electronegativity, the oxygen will tend to pull the shared electrons away from *X*. The hydrogen will remain joined to the oxygen. Since in this case the formation of the hydroxide ion, OH^-, is likely, HO*X* is behaving as a base. For example, in lithium hydroxide, oxygen has such a high electronegativity it has captured the outer electron of lithium, which has a relatively low electronegativity. Oxygen, then, has a lesser attraction for the electrons shared with the hydrogen. As a result, the lithium and hydroxide ions dissociate to produce a base in water.

$$Li \rightarrow O - H$$

We know that nonmetals have high electronegativities and metals have low electronegativities. We can conclude, then, that nonmetals will tend to form acids, and metals will tend to form bases.

Some substances can react as either an acid or a base. If one of these substances is in the presence of a proton donor, then it reacts as a base. In the presence of a proton acceptor, it acts as an acid. Such a substance is said to be **amphoteric.** Water is the most common amphoteric substance.

$$\underset{\text{proton donor}}{HCl} + \underset{\text{base}}{H_2O} \rightarrow H_3O^+ + Cl^-$$

$$\underset{\text{proton acceptor}}{NH_3} + \underset{\text{acid}}{H_2O} \rightarrow NH_4^+ + OH^-$$

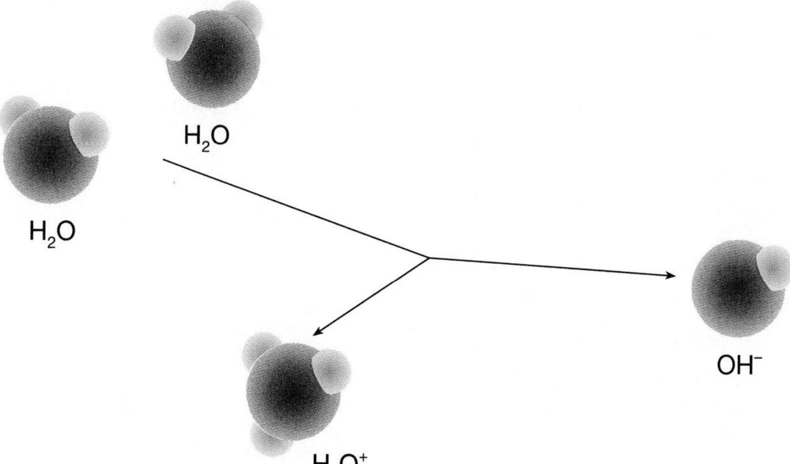

Figure 23.7 The reaction between two water molecules shows the amphoteric nature of water. The reaction produces a hydronium ion and a hydroxide ion.

Acidic and Basic Anhydrides

Anhydrous means without water, so anhydrides are acids or bases that have had water removed. When sulfur dioxide is dissolved in water, sulfurous acid is formed. Any oxide that will produce an acid when dissolved in water is called an **acidic anhydride**.

$$\underset{\text{acidic anhydride}}{SO_2(g)} + \underset{\text{water}}{H_2O(l)} \rightarrow \underset{\text{acid}}{H_2SO_3(aq)}$$

When sulfur-containing fuels such as coal and petroleum are burned, SO_2 is introduced into the atmosphere. Then, when water vapor condenses and rain falls, the SO_2 reacts with the water to produce sulfurous acid.

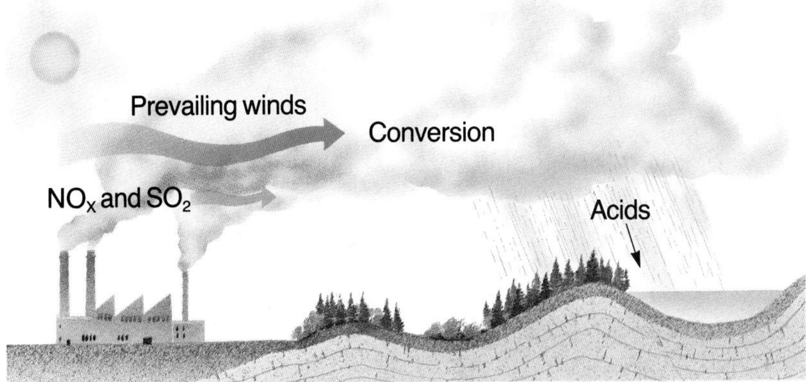

Figure 23.8 Acid rain is produced when sulfur oxides and nitrogen oxides in the air dissolve in water.

If sodium oxide is added to water, sodium hydroxide, a base, is formed. Any oxide that will produce a base when dissolved in water is called a **basic anhydride.**

$$\underset{\text{basic anhydride}}{Na_2O(cr)} + \underset{\text{water}}{H_2O(l)} \rightarrow \underset{\text{base}}{2NaOH(aq)}$$

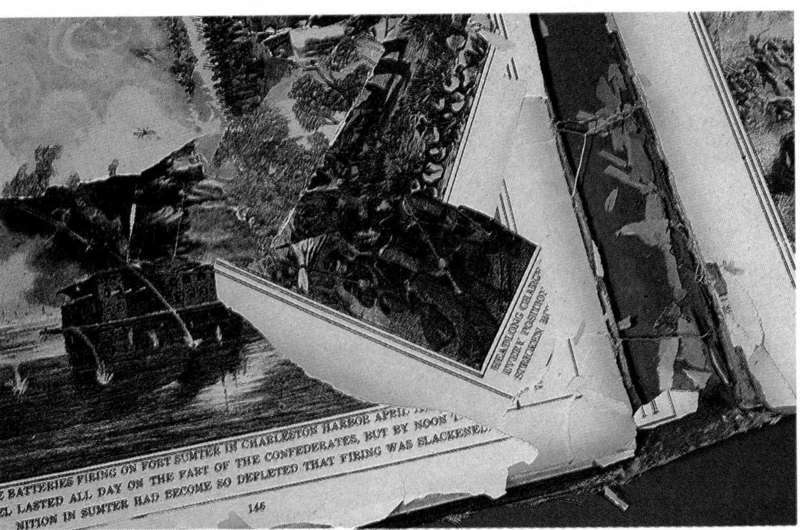

Figure 23.9 Many valuable documents are disintegrating because acidic anhydrides were used as filler by paper manufacturers.

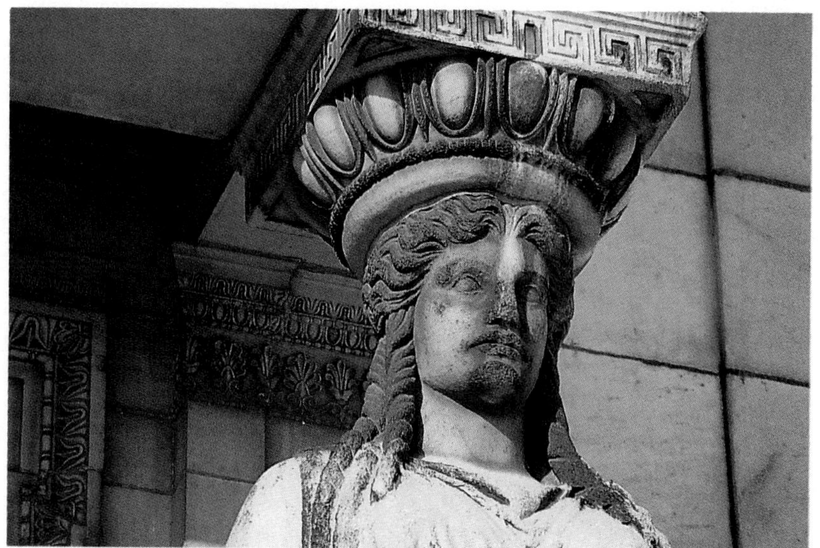

Figure 23.10 The effects of acid rain caused by the dissolving of acidic anhydrides present in air can cause serious damage to structures made from materials such as limestone.

Organic acids may also form anhydrides. An example is acetic anhydride. It is formed by removing a water molecule from two acetic acid molecules.

$$CH_3{-}\overset{\overset{\displaystyle O}{\|}}{C}{-}O{-}H + H{-}O{-}\overset{\overset{\displaystyle O}{\|}}{C}{-}CH_3 \rightarrow CH_3{-}\overset{\overset{\displaystyle O}{\|}}{C}{-}O{-}\overset{\overset{\displaystyle O}{\|}}{C}{-}CH_3 + H_2O$$

SAMPLE PROBLEM

Determining the Formula of an Anhydride

Write the formula for the anhydride of nitric acid, HNO_3.

Solving Process:

The word anhydride means *without water*, so we need to remove H_2O from HNO_3. Because there is only one H atom in each HNO_3 molecule, we need to use two molecules.

$$\left.\begin{matrix} HNO_3 \\ HNO_3 \end{matrix}\right\} - H_2O \text{ gives } N_2O_5$$

Thus, removing water from HNO_3 leaves the anhydride N_2O_5.

PRACTICE PROBLEMS

14. Predict the acidic or basic nature of the following anhydrides.

a. Li_2O **b.** MgO **c.** CO_2 **d.** K_2O **e.** SeO_2

15. Write formulas for the anhydrides of the following.

a. $Ba(OH)_2$ **b.** HIO_4 **c.** H_6TeO_6 **d.** $Al(OH)_3$ **e.** $Zn(OH)_2$

MINUTE LAB

Let's Bag Air Pollution

Put on an apron and goggles. You can see how gases responsible for acid rain diffuse around and from their source. Place drops of varying sizes of 0.04% bromocresol green indicator in the bottom half of a plastic petri dish so that the drops are about 1 cm apart. Leave the center of the petri dish empty. Place a zipper-close plastic bag on a piece of white paper. Carefully place the bottom half of the petri dish, with the drops of indicator, inside the plastic bag. In the center of the petri dish, place one drop of 0.5 M KNO_2 and two drops of 2.0 M H_2SO_4. **CAUTION:** *NO_2 gas is corrosive; KNO_2 and H_2SO_4 are skin irritants.* Seal the bag. Observe, comparing the various color changes that take place in large and small drops, and drops near to and far from the source. Record your observations every 15 seconds. To clean up, open the bag inside an operating fume hood. Rinse out the petri dish. The NO_2 gas produced is soluble in the water of the indicator drops and reacts according to the equation: $2NO_2 + H_2O \rightarrow HNO_3 + HNO_2$. Name the two acids that form.

Strengths of Acids and Bases

Not all acids and bases are completely ionized in water solution. An acid such as hydrochloric that is considered to ionize completely into positive and negative ions is called a **strong acid**. A base such as sodium hydroxide that is completely dissociated into positive and negative ions is called a **strong base**.

Some acids and bases ionize only slightly in solution and are called **weak acids** and **weak bases.** The most important base of this kind is ammonia. In water solution, this base ionizes only partially into NH_4^+ and OH^-. The major portion of the ammonia molecules remains unreacted. Ammonia is a weak base. Acetic acid also ionizes only slightly in water solution. Acetic acid is a weak acid.

Do not equate the solubility of an acid or base with its strength. For example, some bases, such as $Ca(OH)_2$ or $Mg(OH)_2$, are not considered soluble in water, but what small amounts of the bases that do dissolve dissociate completely.

Table 23.5

Relative Strengths of Some Acids and Bases

Compound	Formula	Relative Strength
Hydrochloric acid	HCl	strong acid ↑
Phosphorous acid	H_3PO_3	
Phosphoric acid	H_3PO_4	
Hydrofluoric acid	HF	
Hydroselenic acid	H_2Se	
Acetic acid	CH_3COOH	
Carbonic acid	H_2CO_3	
Hydrosulfuric acid	H_2S	neutral solution
Hypochlorite ion	ClO^-	
Cyanide ion	CN^-	
Ammonia	NH_3	
Carbonate ion	CO_3^{2-}	
Aluminum hydroxide	$Al(OH)_3$	
Phosphate ion	PO_4^{3-}	
Silicate ion	SiO_3^{2-}	↓
Hydroxide ion	OH^-	strong base

23.1 CONCEPT REVIEW

16. Describe the differences in the three acid-base definitions of this chapter.

17. What are conjugate acids and conjugate bases?

18. What are acidic and basic anhydrides?

19. Name the following substances:
a. $HCl(aq)$ **b.** H_2SO_4 **c.** KOH

20. **Apply** What kind of element would you expect to find in position *X* of the compound HOX if the compound is determined to be amphoteric?

Chemical Economics

Just as a chemist must balance equations, so must a chemical company balance its account books if it is to stay in business. Chemical firms take in raw materials and produce a product with more monetary value than the raw materials have. The increase in value is due to the cost of the process, including the company's investment and the company profit.

The two main factors that determine the price of a product are cost of materials and cost of processing. Gold is expensive because the raw material, gold-bearing rock, is quite scarce. Other elements that are expensive because of their scarcity include bismuth, iridium, osmium, palladium, platinum, rhenium, rhodium, ruthenium, and tellurium. In addition, most of the radioactive elements beyond bismuth in the periodic table are unavailable from Earth's crust and must be produced synthetically. Titanium ore is plentiful, yet titanium metal is expensive because the process used to produce it is costly. The most expensive materials are those that are both scarce and costly to produce.

Chemists and chemical engineers can do nothing about the scarcity of raw materials, but they can, and do, work continuously to create new processing methods that will lower the cost of making a product. One interesting method of cutting costs that has become important in recent years is cogeneration. Many of the processes in the chemical industry produce heat. In the past, this heat has been transferred to cooling water. Thus, that energy was essentially wasted. Now that energy is used to operate a plant in which electricity is generated. The electricity can then be sold to a local electric utility. The income from the sale of the electricity is used to offset some of the cost of manufacturing the product.

Some economic factors affecting the chemical industry are those common to all industries: equipment wears out, competition becomes keener, and labor-management conflicts must be resolved. In recent years, the United States chemical industry has had to meet increased competition from foreign firms, both at home and abroad. Research is one way chemists meet competition. Chemical engineers contribute in a number of ways. Optimizing the usage of equipment is one such way. As the technology of a particular manufacturing process is improved, a company must make a decision about replacement. That is, which path is better: continuing to use the old equipment in an expensive process until it wears out, or spending the money to replace the old equipment with the latest design now in order to recognize an immediate saving in the process?

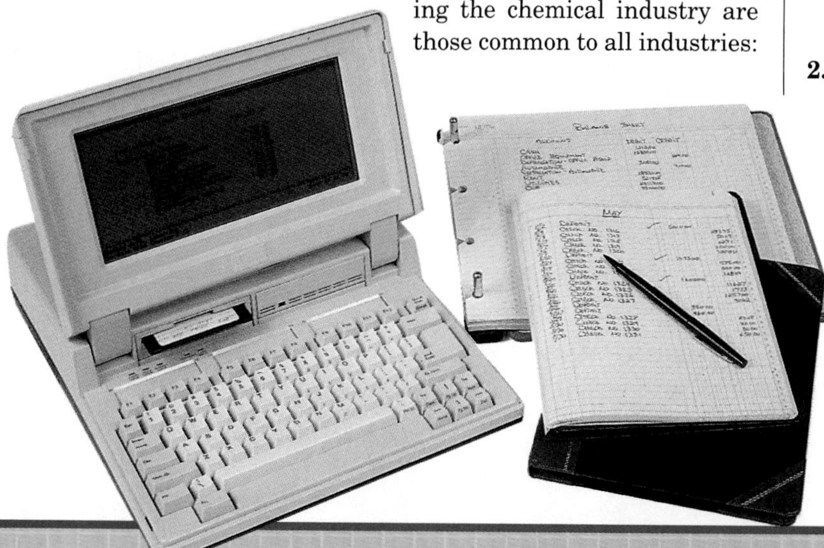

Analyzing the Issue

1. If you were a plant engineer, what factors would you consider when deciding whether or not to build a cogeneration unit?
2. A plant manager must decide whether to keep old equipment, which results in an expensive process, or replace the old equipment with expensive new equipment. The new equipment costs much less to run. When might it be a wise decision to use the old equipment? When might it be more economical to use the new equipment?

OBJECTIVES

- explain the concept of neutralization and the composition of a salt and be able to name salts.
- write net ionic equations.
- derive and use ionization constants.
- compute the percent ionization of a weak electrolyte.

23.2 Salts and Solutions

Now that you have seen the relationship of acids and bases, you are ready to examine the third group of inorganic substances, the salts. When you see the word "salt," you most likely think of common table salt, NaCl. There are, however, thousands of salts. You may have already worked with many of them in the laboratory and in the questions and problems in this book. Salts bear a close relationship to both acids and bases, as you will learn in this section.

Definition of a Salt

An Arrhenius acid is composed of positive hydrogen ions combined with negative nonmetallic ions. Metallic bases are composed of negative hydroxide ions combined with positive metallic ions. An Arrhenius acid reacts with an Arrhenius base to form a salt and water. A **salt** is a crystalline compound composed of the negative ion of an acid and the positive ion of a base. The water is formed from the hydrogen ion of the acid and the hydroxide ion of the base. If the water is evaporated, the negative ions of the acid will unite with the positive ions of the base to form another compound called a salt. If such a reaction results in the removal of all hydrogen and hydroxide ions in solution, the resulting salt will be neither an acid nor a base. We say that the acid and base have neutralized each other. The reaction of an acid and a base is called a **neutralization reaction**. For example, if equivalent amounts of chloric acid and sodium hydroxide react, the salt sodium chlorate and water are formed.

$$\underset{\text{acid}}{HClO_3(aq)} + \underset{\text{base}}{NaOH(aq)} \rightarrow \underset{\text{water}}{H_2O(l)} + \underset{\text{salt}}{NaClO_3(aq)}$$

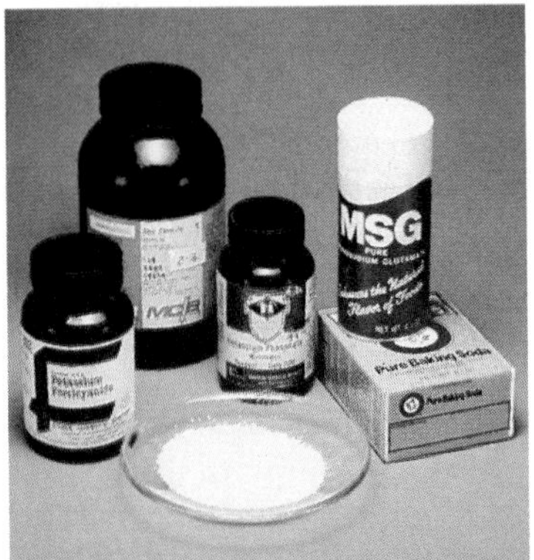

Figure 23.11 A wide variety of salts are used in chemical laboratories and in everyday life.

Figure 23.12 Basic minerals, such as limestone, react with acidic solutions, such as rainwater, to form solutions of dissolved salts. These cave formations result when water evaporates from the salt solutions.

Salts may also result from the reactions of acidic or basic anhydrides with a corresponding base, acid, or anhydride. Note that water is not a product of a reaction between two corresponding anhydrides.

$Na_2O + H_2SO_4 \rightarrow Na_2SO_4 + H_2O$ (basic anhydride + acid)
$2NaOH + SO_3 \rightarrow Na_2SO_4 + H_2O$ (base + acidic anhydride)
$Na_2O + SO_3 \rightarrow Na_2SO_4$ (basic anhydride + acidic anhydride)

Although a salt is formed by neutralization, solutions of some salts in water are not neutral. It is possible to obtain salts that are acidic or basic. For example, if sodium hydroxide reacts with sulfuric acid in a 1:1 mole ratio, the product, sodium hydrogen sulfate, is called an acidic salt.

$$H_2SO_4(aq) + NaOH(aq) \rightarrow H_2O(l) + NaHSO_4(aq)$$

$NaHSO_4$ still contains an ionizable hydrogen atom. In a similar manner, partially neutralized bases form basic salts. Acidic salts such as $NaHSO_4$ and basic salts such as $Cu_2(OH)_2CO_3$ are not neutral in solution.

There is a relationship between the name of an acid and the name of the salt it forms. Binary acids (prefix *hydro-*, and suffix *-ic*) form salts ending in *-ide*. As an example, hydrochloric acid forms chloride salts. Ternary acids form salts in which *-ic* acids form *-ate* salts and *-ous* acids form *-ite* salts. Prefixes from ternary acid names remain in the salt names.

In naming acidic and basic salts, each ion is named separately. Hydrogen is generally named immediately before the names of any negative ions and hydroxide immediately after the names of any positive ions. Thus, $NaHC_2O_4$ is sodium hydrogen oxalate and $Pb_2(OH)_2CO_3$ is lead(II) hydroxide carbonate. Table 23.6 lists names and formulas of several acids and their negative ions.

Table 23.6

Acid and Ion Names

Acid		Ion
HF hydrofluoric acid	→	F^- fluoride
$HMnO_4$ permanganic acid	→	MnO_4^- permanganate
H_2SO_4 sulfuric acid	→	SO_4^{2-} sulfate
HNO_2 nitrous acid	→	NO_2^- nitrite
$HClO$ hypochlorous acid	→	ClO^- hypochlorite
H_2CO_3 carbonic acid	→	HCO_3^- hydrogen carbonate

PRACTICE PROBLEMS

21. Name the following salts.
    a. $NaHSO_4$
    b. $KHC_4H_4O_6$
    c. NaH_2PO_4
    d. $NaHS$
    e. $Al(OH)SiO_3$
22. Write formulas for each of the following salts.
    a. sodium hydrogen carbonate
    b. sodium monohydrogen phosphate
    c. ammonium hydrogen sulfide
    d. potassium hydrogen sulfate
    e. tin(II) hydroxide nitrate

Net Ionic Equations

For reactions taking place in water, it is customary for chemists to write equations in the ionic form. In this method, only those ions taking part in the reaction are written. Other ions present in the solution but not involved in the reaction are known as **spectator ions** and are not included in the equation. For example, in the reaction of HCl(aq) with NaOH(aq), the sodium and chloride are spectator ions because they are unchanged in the reaction.

$$H^+(aq) + Cl^-(aq) + Na^+(aq) + OH^-(aq) \rightarrow H_2O(l) + Cl^-(aq) + Na^+(aq)$$

Figure 23.13 The net ionic equation for the precipitation of lead(II) iodide by reacting lead(II) nitrate and sodium iodide solutions is $Pb^{2+}(aq) + 2I^-(aq) \rightarrow PbI_2(cr)$. Sodium and nitrate are spectator ions.

In writing net ionic equations, dissolved salts are considered to be in their ionic form. We write insoluble salts in their undissociated form. The following rule must be observed when writing net ionic equations. *Substances occurring in a reaction in molecular form are written as molecules. Those substances occurring as ions are written as ions.*

Weak acids and weak bases are written in molecular form, but strong acids and strong bases should be written in the ionized form. An acid may contain more than one ionizable hydrogen atom. Such an acid is called a **polyprotic acid**. Examples of polyprotic acids are sulfuric acid, H_2SO_4, phosphoric acid, H_3PO_4, and carbonic acid, H_2CO_3. Below are listed some rules of thumb for deciding whether to use ions or molecules in writing net ionic equations. These rules are not applicable in all cases but work well in most reactions. If your equation must be exact, you should use a chemical handbook. The handbook will help you determine whether substances are to be written as ions or molecules.

Rule 1. *Binary acids:* HCl, HBr, and HI are strong; all other binary acids and HCN are weak. Strong acids are written in ionic form; weak acids are written in molecular form.

Rule 2. *Ternary acids:* If the number of oxygen atoms in the inorganic acid molecule exceeds the number of hydrogen atoms by two or more, the acid is strong. We will consider all organic carboxylic acids as weak.

Strong: $HClO_3$, $HClO_4$, H_2SO_4, HNO_3, H_2SeO_4
Weak: $HClO$, H_3AsO_4, H_2CO_3, H_4SiO_4, HNO_2

Rule 3. *Polyprotic acids:* In the second and subsequent ionizations the acids are always weak, whether or not the original acid is strong or weak.
Rule 4. *Bases:* Hydroxides of the Groups 1 (IA) and 2 (IIA) elements except beryllium are strong bases. All others including ammonia, hydroxylamine, and organic bases are weak.
Rule 5. *Salts:* Salts are written in ionic form if soluble, and in undissociated form if insoluble. Use the solubility rules in Appendix Table A-7.

Ionic: $K^+ + Cl^-$, $Zn^{2+} + 2NO_3^-$
Undissociated: $AgBr$, $BaSO_4$

Rule 6. *Oxides:* Oxides are always written in molecular or undissociated form.
Rule 7. *Gases:* Gases are always written in molecular form.

EVERYDAY CHEMISTRY

Baking: An Acid-Base Reaction

Baking soda and baking powder give rise to acid-base neutralization reactions that result in light and delicate pastries. How does this change happen?

Baking soda is sodium hydrogen carbonate, $NaHCO_3$, which is basic in water solution. When it is used with a batter that contains acidic ingredients, such as yogurt, sour cream, lemon juice, cream of tartar, or vinegar, it will react with this acid to produce carbon dioxide gas. This gas is trapped in the batter and causes the batter to rise as it is baking.

If the batter does not contain acidic ingredients or if a more delicate texture is desired, then baking powder is used. Baking powder is a mixture of baking soda, a dry acid such as cream of tartar or tartaric acid, and starch to keep the ingredients dry. When baking powder is added to a batter that contains water, the base and the acid react.

Sometimes double-action baking powder is used. The double-acting baking powder has two acids mixed with the baking soda. The first, such as tartaric acid, reacts with the baking soda at room temperature and forms small bubbles as soon as it is added to the batter. After baking for a while, the second acid, usually sodium aluminum sulfate, reacts with the remaining baking soda while the batter is in the oven. This double action results in a finer textured cake, for example, and more cheers for the cook.

Exploring Further

1. What do you think happens if too much or too little baking soda is used in a recipe?
2. Write the equations for the reactions of baking soda with acetic, citric, lactic, and tartaric acids.

SAMPLE PROBLEMS

1. Net Ionic Equations—Weak Acid

Convert the following balanced equation to an equation in net ionic form.

$$H_2SiO_3 + 2NaOH \rightarrow Na_2SiO_3 + 2H_2O$$

Solving Process:

According to Rule 2, since there are three oxygens and two hydrogens in silicic acid, it must be weak. Rule 4 tells us that NaOH is strong. The sodium silicate salt is soluble, so we apply Rule 5. Water is an oxide, and Rule 6 applies. Using these rules the ionic equation is written as

$$H_2SiO_3 + 2Na^+ + 2OH^- \rightarrow 2Na^+ + SiO_3{}^{2-} + 2H_2O$$

To make the equation a net ionic equation, we remove (by subtracting from both sides) those species that appear on both sides. For our reaction, two sodium ions appear on each side and are removed. Thus, the Na^+ ions are spectator ions. The net ionic equation is then:

$$H_2SiO_3 + 2OH^- \rightarrow SiO_3{}^{2-} + 2H_2O$$

2. Net Ionic Equations—Strong Acid

Convert the following balanced equation to an equation in net ionic form.

$$2HCl + Ba(OH)_2 \rightarrow BaCl_2 + 2H_2O$$

Solving Process:

Rule 1: HCl is a strong acid.
Rule 4: $Ba(OH)_2$ is a strong base.
Rule 5: $BaCl_2$ is soluble.
Rule 6: H_2O is molecular.
Ionic equation:

$$2H^+ + 2Cl^- + Ba^{2+} + 2OH^- \rightarrow Ba^{2+} + 2Cl^- + 2H_2O$$

Net Ionic Equation: $2H^+ + 2OH^- \rightarrow 2H_2O$
Final form: $H^+ + OH^- \rightarrow H_2O$

PROBLEM SOLVING HINT

Note that the net ionic equation should be divided through by 2 in order to put it in its lowest terms.

PRACTICE PROBLEMS

Reduce the following balanced equations to net ionic form.

23. $2AgNO_3(aq) + H_2SO_4(aq) \rightarrow Ag_2SO_4(cr) + 2HNO_3(aq)$

24. $H_4SiO_4(aq) + 4NaOH(aq) \rightarrow Na_4SiO_4(aq) + 4H_2O(l)$

25. $4HCl(aq) + 2Cr(NO_3)_2(aq) + 2HgCl_2(aq) \rightarrow$
$2CrCl_3(aq) + Hg_2Cl_2(cr) + 4HNO_3(aq)$

26. $2Mn(NO_3)_2(aq) + 5NaBiO_3(cr) + 14HNO_3(aq) \rightarrow$
$2NaMnO_4(aq) + 5Bi(NO_3)_3(aq) + 7H_2O(l) + 3NaNO_3(aq)$

27. $2CuSO_4(aq) + 2NH_4CNS(aq) + H_2SO_3(aq) + H_2O(l) \rightarrow$
$2CuCNS(cr) + (NH_4)_2SO_4(aq) + 2H_2SO_4(aq)$

Ionization Constant

Acetic acid is a weak acid and ionizes only slightly. The equation for the ionization of acetic acid at equilibrium is

$$CH_3COOH(l) + H_2O(l) \rightleftarrows CH_3COO^-(aq) + H_3O^+(aq)$$

The equilibrium constant for this reaction is

$$K_{eq} = \frac{[CH_3COO^-][H_3O^+]}{[CH_3COOH][H_2O]}$$

The CH_3COO^- and H_3O^+ ion concentrations are small, and the concentration of CH_3COOH is almost unaffected by the ionization. When acetic acid ionizes, hydrogen ions attach to a water molecule and form the hydronium ion (H_3O^+). However, acetic acid is a weak acid. Thus, few hydrogen ions are formed. The concentration of water remains essentially constant (55.6 mol/dm^3). Thus, we can multiply the concentration of water by the equilibrium constant and obtain the equation:

$$K_{eq}[H_2O] = \frac{[H_3O^+][CH_3COO^-]}{[CH_3COOH]}$$

Because $[H_2O]$ is constant, the product of the equilibrium constant and the concentration of water ($K_{eq}[H_2O]$) produces a new constant. This new constant is called the **ionization constant**, and is given the symbol K_a. For any weak acid ($HA + H_2O \rightleftarrows H_3O^+ + A^-$), the ionization constant is

$$K_a = \frac{[H_3O^+][A^-]}{[HA]}$$

In a similar way, we can write ionization constant expressions for weak bases (K_b). Ammonia is a weak base that reacts with water as follows:

$$NH_3 + H_2O \rightleftarrows NH_4^+ + OH^-$$

The ionization constant expression for ammonia is

$$K_b = \frac{[NH_4^+][OH^-]}{[NH_3]}$$

CHEMISTS AND THEIR WORK

Mary Engle Pennington (1872-1952)

Mary Engle Pennington became interested in medical chemistry when she was 12 years old. In 1890 she enrolled in the University of Pennsylvania and completed the requirements for a B.S. degree in two years. She was not awarded the degree because she was a woman. Ironically, she received her PhD in Chemistry in 1895. She began doing bacteriological analyses for local doctors and eventually became chief of the Food Research Laboratory in the United States Department of Agriculture, devising ways of preventing food spoilage.

Figure 23.14 Muriatic acid, an industrial grade of hydrochloric acid, can be used to clean excess mortar from bricks or ceramic tiles.

SAMPLE PROBLEM

Ionization of a Weak Acid

What is the hydronium ion concentration of a 0.100M solution of formic acid, HCOOH? Formic acid has a K_a equal to 1.77×10^{-4}.

Solving Process:

$$HCOOH(l) + H_2O(l) \rightleftarrows H_3O^+(aq) + HCOO^-(aq)$$

$$K_a = \frac{[H_3O^+][HCOO^-]}{[HCOOH]} = 1.77 \times 10^{-4}$$

The balanced equation shows that $[H_3O^+]$ is equal to $[HCOO^-]$ in this example. If we let $x = [H_3O^+]$ or $[HCOO^-]$, then the concentration of formic acid [HCOOH] is $(0.100 - x)$. Because x is so small when compared with the concentration of formic acid, the term $(0.100 - x)$ is approximately equal to 0.100M. Therefore,

$$1.77 \times 10^{-4} = \frac{x^2}{0.100}$$

$$\text{and } x^2 = (0.100)(1.77 \times 10^{-4}) = 1.77 \times 10^{-5}$$

$$x = 4.21 \times 10^{-3}M$$

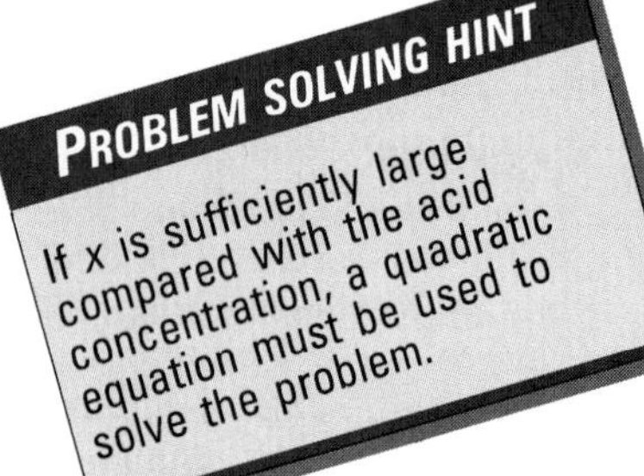

PROBLEM SOLVING HINT

If x is sufficiently large compared with the acid concentration, a quadratic equation must be used to solve the problem.

PRACTICE PROBLEMS

28. What is the hydronium ion concentration in 0.0200M HClO? $K_a = 2.88 \times 10^{-8}$

29. What is K_b for the base N_2H_4 (hydrazine) if a 0.500M solution of hydrazine has the following concentrations at equilibrium? $[N_2H_4] = 0.499M$, $[OH^-] = 1.23 \times 10^{-3}M$, $[N_2H_5^+] = 1.23 \times 10^{-3}M$

Figure 23.15 In the Amazon basin, stinging ants are abundant. These ants give off formic acid. This acid can occur in such quantity that it becomes a pollutant in streams. The name of the acid is derived from the Latin term for ant—*Formica*.

Percent of Ionization

When a weak acid or weak base is dissolved in water, it ionizes only slightly. It is often desirable in such cases to know just how much of a substance is ionized. This amount is usually expressed in terms of percent, and is called the **percent of ionization**. For example, at room temperature, we found the hydronium ion concentration of 0.100M formic acid, HCOOH, to be $4.21 \times 10^{-3}M$. We know that

$$HCOOH + H_2O \rightleftarrows H_3O^+ + HCOO^-$$

Thus, we can find the percent ionization of formic acid by dividing either $[H_3O^+]$ or $[HCOO^-]$ by the original [HCOOH] and then multiplying by 100 percent. Either $[H_3O^+]$ or $[HCOO^-]$ can be used in the ratio, because the concentrations of the two ions are equal.

SAMPLE PROBLEM

Percent of Ionization

Find the percent of ionization of a $0.100M$ solution of formic acid if the hydronium ion concentration is $4.21 \times 10^{-3}M$.

Solving Process:

$$\text{Percent of ionization} = \frac{[\textit{amount ionized}]}{[\textit{original acid}]} \times 100\%$$

$$= \frac{4.21 \times 10^{-3}M}{0.100M} \times 100\% = 4.21\%$$

PRACTICE PROBLEMS

30. The ionization constant of acetic acid, CH_3COOH, is 1.75×10^{-5}. Find the percent of ionization of $0.200M$ acetic acid.

31. A solution of $1.00M$ HA in water ionizes 2.00%. Find K_a.

CHEMISTRY IN DEPTH

Common Ion Effect

There are times when a chemist may wish to change the concentration of a specific ion in a solution. Such changes are often made by adding a new substance to the solution.

Acetic acid ionizes in a water solution to form both acetate and hydronium ions. What will happen if we add some sodium acetate to the solution? Sodium acetate is a soluble salt and dissociates completely into acetate (CH_3COO^-) and sodium ions.

$$NaCH_3COO(cr) \rightarrow Na^+(aq) + CH_3COO^-(aq)$$

Acetic acid ionizes into acetate and hydronium ions.

$$H_2O(l) + CH_3COOH(l) \rightleftarrows H_3O^+(aq) + CH_3COO^-(aq)$$

K_a for acetic acid is

$$\frac{[CH_3COO^-][H_3O^+]}{[CH_3COOH]} = 1.75 \times 10^{-5}$$

K_a is a constant and does not change unless the temperature changes. If sodium acetate is added to an acetic acid solution, acetate ion concentration is increased and a shift in the equilibrium occurs. Because there are more particles of CH_3COO^-, there will be more collisions between CH_3COO^- and H_3O^+. Thus, the rate of reaction will increase toward acetic acid. Some of the excess acetate ions unite with hydronium ions to form molecular acetic acid and water. This reaction results in the removal of hydronium ions from solution, thus decreasing hydronium ion concentration. The acetic acid concentration increases slightly. A new equilibrium is established with more acetate ions and fewer hydronium ions. K_a remains unchanged.

The acetate ion is common to both acetic acid and sodium acetate. The effect of the acetate ion on the acetic acid in solution is called the **common ion effect**. The addition of a common ion increases the concentration of one of the products of the ionization. Thus, the reaction shifts toward the opposite side in accordance with Le Chatelier's principle. By adding acetate ions in the form of sodium acetate, we have placed a stress on the system, a surplus of acetate ions. The system shifts to relieve the stress by reacting hydronium ions with acetate ions. In the process, acetate and hydronium ions are consumed and acetic acid is produced.

SAMPLE PROBLEM

Common Ion Effect

What is the hydronium ion concentration in a solution that is $0.10M$ in hydrazoic acid, HN_3, and $0.050M$ in sodium azide, NaN_3? $K_a = 1.9 \times 10^{-5}$

Solving Process:

First, write the equation for the equilibrium system.

$$HN_3 + H_2O \rightleftarrows H_3O^+ + N_3^-$$

Then write the expression for the equilibrium constant.

$$K_a = \frac{[H_3O^+][N_3^-]}{[HN_3]}$$

The hydronium ion is the quantity we seek, so we will let x represent that concentration. If x hydronium ions are produced, then x acid molecules ionized and x azide ions were produced from the acid. However, azide ions were also produced from the soluble sodium salt. Since sodium azide is soluble, we assume it dissociates completely into $Na^+(aq)$ and $N_3^-(aq)$. In that case, the salt contributes a concentration of $0.050M$ to the azide ion concentration. The total azide ion concentration would be $0.050M + x$. If x acid molecules ionized and we started with $0.10M$, then at equilibrium the molecular acid concentration must be $0.10M - x$.

$$1.9 \times 10^{-5} = \frac{[x][x + 0.050]}{[0.10 - x]}$$

Since the K_a for hydrazoic acid is very small, we can assume that the tiny amount that ionizes will be very small compared with the original amount. That assumption also means that the amount of azide ion produced will also be very small compared with the amount obtained from the salt. Consequently, we can neglect the x in $x + 0.050$ and in $0.10 - x$.

$$1.9 \times 10^{-5} = \frac{[0.050\,x]}{[0.10]}$$

$$x = 3.8 \times 10^{-5}M$$

PRACTICE PROBLEMS

32. What is the hydronium ion concentration in a solution $0.150M$ in HNO_2 and $0.0300M$ in $NaNO_2$? Refer to Appendix Table A-9 for the K_a.

33. What is the hydronium ion concentration in a solution of $0.403M$ in HBrO and $0.000\ 195M$ in NaBrO? Refer to Appendix Table A-9 for the K_a.

34. Given a solution of the weak acid H_2CO_3, how could the hydronium ion concentration of this solution be reduced? How could the carbonate ion concentration of this same solution be reduced?

Polyprotic Acids

A polyprotic acid has been defined as an acid containing more than one ionizable hydrogen atom. Sulfuric acid is an example of a polyprotic acid. In sulfuric acid, the hydrogen atoms leave the molecule one at a time. Sulfuric acid is a strong acid because the first hydrogen atom ionizes completely. The remaining hydrogen sulfate ion ionizes only slightly, and is considered a weak acid. The reaction can be represented as follows.

$$H_2SO_4(l) + H_2O(l) \rightarrow H_3O^+(aq) + HSO_4^-(aq) \quad \text{ionizes readily (strong acid)}$$

$$HSO_4^-(aq) + H_2O(l) \rightleftarrows H_3O^+(aq) + SO_4^{-2}(aq) \quad \text{ionizes slightly (weak acid)}$$

Phosphoric acid is a triprotic acid (an acid containing three ionizable hydrogen atoms); it ionizes as follows.

$$H_3PO_4(cr) + H_2O(l) \rightleftarrows H_3O^+(aq) + H_2PO_4^-(aq)$$
$$K_a = 7.08 \times 10^{-3}$$

Figure 23.16 Sulfuric acid, H_2SO_4, is a polyprotic acid. It is produced in quantity greater than that of any other chemical manufactured in the United States. It is produced in factories such as the one shown.

Figure 23.17 Citrus fruits are rich in citric acid. Citric acid is a polyprotic acid containing three ionizable hydrogen atoms. It is an organic acid that has three carboxyl groups.

$$H_2PO_4^-(aq) + H_2O(l) \rightleftarrows H_3O^+(aq) + HPO_4^{2-}(aq)$$
$$K_a = 6.31 \times 10^{-8}$$

$$HPO_4^{2-}(aq) + H_2O(l) \rightleftarrows H_3O^+(aq) + PO_4^{3-}(aq)$$
$$K_a = 4.17 \times 10^{-13}$$

Note that in each successive step, K_a is smaller, because the acid ionized is weaker than the acid ionized in the previous step. There are two factors making each successive step of the ionization weaker for all polyprotic acids. First, to use phosphoric acid as an example, you can see that in the first step, a positive ion (H_3O^+) is separating from a negative ion ($H_2PO_4^-$). In the second step, a positive ion (H_3O^+) is separating from a negative ion of charge 2− (HPO_4^{2-}). This second step will be less likely to occur, because HPO_4^{2-} has a greater attraction for the H_3O^+ than $H_2PO_4^-$ does. For the third step, the separation of PO_4^{3-} and (H_3O^+) is even less likely. The second reason for successive steps being weaker is that the hydronium ion produced by the first step acts as a common ion for the second and subsequent steps. Recall that a common ion reduces the amount of ionization, as predicted by Le Chatelier's principle.

Calculations involving diprotic acids are complex. There are three unknown quantities: hydronium ion, protonated anion, and anion. There is a large difference between the extent of first and second ionizations. Thus, it is usually possible to ignore the hydronium ion contributed by the second ionization. The same assumption is also applicable to tri- or tetraprotic acids.

PRACTICE PROBLEMS

35. Select the polyprotic acids from the following acids.
- **a.** HCOOH
- **b.** H_2SeO_3
- **c.** CH_3COOH
- **d.** H_3BO_3
- **e.** H_2CO_3
- **f.** HOOCCOOH

36. Calculate the ratio of the first ionization constant to the second ionization constant for each of the polyprotic acids listed in Table A-9 of the Appendix. Do you see any pattern?

23.2 CONCEPT REVIEW

37. How does a salt differ from an acid or a base?

38. What is the basic rule for writing net ionic equations? Write a net ionic equation for the reaction of hydrobromic acid and potassium hydroxide.

39. Write an ionization constant expression for the following reaction:
$H_2Se + H_2O \rightarrow H_3O^+ + HSe^-$

40. The ionization constant for propanoic acid, CH_3CH_2COOH, is 1.34×10^{-5}. Find the percent of ionization of 0.100*M* propanoic acid. Why is the computation of percent of ionization limited to weak electrolytes?

41. Apply Oxalic acid, HOOCCOOH, can be used as a rust remover. List two types of compounds that would reduce the $[H_3O^+]$ when added to an oxalic acid solution.

Chapter 23 REVIEW

Summary

23.1 Acids and Bases

1. According to the Arrhenius theory, acids produce H^+ ions in water solution, while bases produce OH^-, hydroxide ions.
2. The Brønsted-Lowry theory defines an acid as any substance that donates a proton in a reaction and a base as any substance that accepts a proton.
3. According to the theory proposed by G.N. Lewis, an acid accepts an electron pair, while a base donates an electron pair during the formation of a bond.
4. Ternary acids contain three elements. Binary acids contain hydrogen and a nonmetal. Ternary acid names depend on the number of oxygen atoms compared to the most common acid containing those three elements.
5. For a compound HO*X*, if *X* attracts electrons strongly, the compound will act as an acid. If *X* has low electronegativity, HO*X* will tend to act as a base.
6. Metallic oxides tend to act as basic anhydrides, forming bases when added to water. Nonmetallic oxides tend to act as acidic anhydrides.
7. Strong acids and bases are those in which ionization or dissociation processes go to completion in water solution.

23.2 Salts and Solutions

8. A salt is a crystalline compound composed of the negative ion of an acid and the positive ion of a base.
9. In ionic reactions in solution, not all ions actively take part in the reaction. In net ionic equations, only the reacting species are shown.
10. The ionization of a weak acid is an equilibrium process. The degree to which the process proceeds is represented by an ionization constant.
11. Adding ions that are the same as one of those produced by the ionization of a weak electrolyte to a solution of the electrolyte suppresses its ionization.
12. Polyprotic acids contain more than one ionizable hydrogen atom. Each successive ionization occurs to a lesser extent.

Key Terms

electrolyte
hydronium ion
conjugate base
conjugate acid
binary acid
ternary acid
amphoteric
anhydrous
acidic anhydride
basic anhydride
strong acid
strong base
weak acid
weak base
salt
neutralization reaction
spectator ion
polyprotic acid
ionization constant
percent of ionization
common ion effect

Review and Practice

42. According to Arrhenius, what are the characteristics of acids and bases?
43. How do hydronium ions form? Why do they form in aqueous acid solution?
44. What are conjugate acids and bases? What are the conjugate acid and conjugate base in the ionization of hydrogen chloride in water?
45. For each of the following reactions, label the acid, base, conjugate acid, and conjugate base.
    a. $NH_3(g) + H_3O^+(aq) \rightarrow NH_4^+(aq) + H_2O(l)$
    b. $CH_3OH(l) + NH_2^-(aq) \rightarrow CH_3O^-(aq) + NH_3(g)$
    c. $OH^-(aq) + H_3O^+(aq) \rightarrow H_2O(l) + H_2O(l)$
    d. $NH_2^-(aq) + H_2O(l) \rightarrow NH_3(g) + OH^-(aq)$
46. How do binary acids and ternary acids differ?

REVIEW

47. Draw an electron-dot diagram for each of the following substances. Decide if the substance would be a Lewis acid or a Lewis base.
a. $AlCl_3$ **c.** PH_3 **e.** Zn^{2+}
b. SO_3 **d.** Xe **f.** CO_3^{2-}

48. There are three common theories of acids and bases. Explain how each of the following theories defines acids and bases.
a. Arrhenius **c.** Lewis
b. Brønsted-Lowry

49. What element is always present in binary acids? What two elements are usually present in ternary acids?

50. Define acidic anhydride and basic anhydride. Give an example of each.

51. Write formulas for the acids or bases formed from the following anhydrides.
a. Na_2O **c.** N_2O_5 **e.** TeO_2
b. CaO **d.** Rb_2O **f.** Cs_2O

52. Write formulas for the anhydrides of the following substances.
a. $Sc(OH)_3$ **c.** HIO_3 **e.** $Cd(OH)_2$
b. CsOH **d.** $Ga(OH)_3$ **f.** KOH

53. What properties cause a substance to be termed amphoteric?

54. Why do we call an aqueous solution of HCl strong and a solution of acetic acid weak?

55. In writing net ionic equations, why are insoluble salts not written as ions?

56. Calculate K_a for a 0.100M HF solution if $[H_3O^+]$ is $8.13 \times 10^{-3}M$.

57. What is the benzoate ion concentration in a 0.0718M solution of benzoic acid, which is C_6H_5COOH? See Appendix Table A-9 for K_a.

58. What is the hydronium ion concentration in a solution that is 0.100M in formic acid, HCOOH, and 0.020M in potassium formate? Refer to Appendix Table A-9.

59. What is a polyprotic acid? Describe how a polyprotic acid such as H_2SO_4 ionizes when placed in water.

60. Name the following acids.
a. HI(aq) **c.** H_2Se(aq)
b. H_2TeO_3(aq) **d.** HIO_3(aq)
(Note: H_2SO_3 is sulfurous acid.)

61. Write formulas for the following acids.
a. phosphorous acid
b. hypophosphorous acid
c. hydrotelluric acid
d. periodic acid
e. iodic acid
f. hydrochloric acid

62. Write an expression for the ionization constant for each of the following weak acids.
a. $HClO_2$ **c.** H_2CO_3 (first ionization)
b. HIO **d.** C_6H_5COOH

63. Give the name of the sodium salt that would result from a neutralization reaction involving each of the following acids and NaOH.
a. $HBrO_3$ **d.** HCOOH **e.** HF
b. H_3PO_4 (Assume loss of 1 proton.)
c. NH_3 (acting as an acid)

64. What is $[H_3O^+]$ in a 0.884M solution of oxalic acid, HOOCCOOH?

65. What is $[H_3O^+]$ in a 0.0500M phenol solution? $K_a = 1.02 \times 10^{-10}$
$$C_6H_5OH + H_2O \rightleftarrows H_3O^+ + C_6H_5O^-$$

Concept Mastery

66. For the compound HOX, how will the compound in which X is a metal differ in water solution from the compound in which X is a nonmetal? Explain this difference.

67. Why is the K_a smaller for each succeeding ionization of phosphoric acid?

68. Illustrate and explain the amphoteric nature of water by comparing HNO_3 in solution to the combination of CaO and water. Write equations for both processes.

69. Draw structures for butanoic acid and pentanamine. Why are organic amines considered to be bases?

Application

70. Give the names and formulas of the salts obtained from complete neutralization reactions between the following acid-base pairs.

a. sodium hydroxide and phosphoric acid

b. potassium hydroxide and boric acid, H_3BO_3

c. cadmium hydroxide and hydrobromic acid

d. lithium hydroxide and silicic acid, H_4SiO_4

e. barium hydroxide and sulfurous acid

71. Reduce the following complete equations to net ionic form.

a. $6Cr(NO_3)_2 + 3CuSO_4 \rightarrow 3Cu + 4Cr(NO_3)_3 + Cr_2(SO_4)_3$

b. $H_2SO_4 \rightarrow H_2O + SO_3$

c. $P_4O_{10} + 6H_2O \rightarrow 4H_3PO_4$

d. $4CuCNS + 7KIO_3 + 14HCl \rightarrow 4HCN + 4CuSO_4 + 7ICl + 7KCl + 5H_2O$

Critical Thinking

72. Can substances considered acids and bases under the Arrhenius and Brønsted-Lowry theories also be considered acids and bases under the Lewis theory? Explain, using ammonia as an example.

73. What is the effect of dissolving potassium sulfite in a 0.100*M* solution of H_2SO_3? Would you expect the acidity to increase or decrease? Describe a mechanism that explains your answer. How does Le Chatelier's principle apply to this process?

Readings

Dagani, Ron. "Water Cluster Cradles H_3O^+ Ion in Stable Cagelike Structure." *Chemical and Engineering News* 69 no. 14 (8 April, 1991): 47-48.

deLange, A.M., and J.H. Potgieter. "Acid and Base Dissociation Constants of Water and Its Associated Ions." *Journal of Chemical Education,* 68 no. 4 (April 1991): 304-305.

Gordus, Adon A. "Chemical Equilibrium: V." *Journal of Chemical Education* 68, no. 7 (July 1991): 566-568.

Piacenti, Franco. "Conservation of Monumental Buildings," *Chemistry International* 13 no. 4. (July 1991): 147-152.

Cumulative Review

1. What is meant by a reversible reaction?

2. In what units is the concentration expressed in rate expressions?

3. What two effects does an increase in temperature have on reaction rate?

4. Describe the difference that exists between homogeneous reactions and heterogeneous reactions.

5. What effect would an increase in pressure have on the following equilibrium?

$$CH_3N_2CH_3(g) \rightleftarrows C_2H_6(g) + N_2(g)$$

6. Would you expect methane gas, CH_4, to dissolve in water? Explain your answer.

7. Differentiate among unsaturated, saturated, and supersaturated solutions.

8. What is the difference between concentrated and dilute solutions?

9. When NH_4NO_3 is dissolved in water, the solution gets cold. Is the enthalpy of solution of NH_4NO_3 in water positive, or negative?

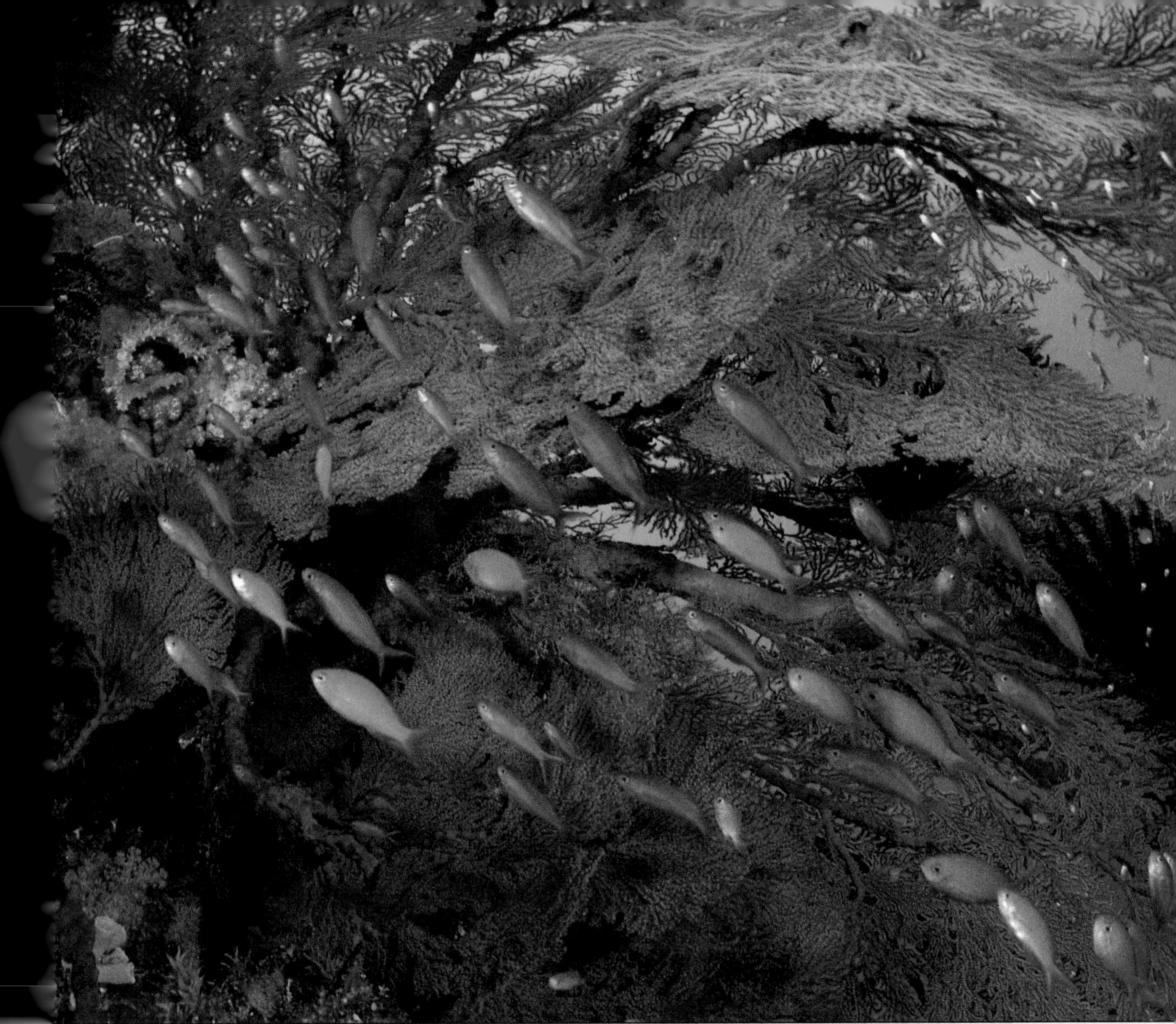

CHAPTER **Preview**

CONTENTS

Solutions of Electrolytes

CHAPTER 24

The world's seas contain many unusual creatures. One group, the tunicates, squirts jets of water when disturbed—hence, their common name: sea squirts. Sea squirts, shown here, take up vanadium from sea water, where it occurs to the slight extent of 0.3 to 3 parts per million. The vanadium is found in blood cells of the sea squirt at concentrations of about 3700 parts per million. In other words, the organism can concentrate vanadium by a factor of over 1000.

Tunicates are so named because they build an envelope, or tunic, around themselves. Vanadium acts as a catalyst in the synthesis of tiny rods called microfibrils. The tunic is then spread over these microfibrils. Vanadium is found in sea water as ions. Recall that ions in solution are electrolytes. In this chapter, you will learn about the behavior of many electrolytes in solution.

EXTENSIONS AND CONNECTIONS

OBJECTIVES

- explain the concept of solubility product and solve problems using the solubility product constant.
- discuss the auto-ionization of water and solve problems using the ion product constant for water.
- explain how the pH scale is used for measuring solution acidity.
- describe the processes of hydrolysis and buffering.

24.1 Water Equilibria

Have you ever seen a swimming pool attendant check the pH of the pool water? A balance must be found between conditions that will prevent the spread of bacteria in the water and conditions that are safe for human occupancy. In this section we will look at the meaning of pH as well as other characteristics of water-based solutions.

Solubility Product Constant

The pH is a measure of the concentration of hydrogen ions in solution. It may also be important to know about other ion concentrations. For example, it is possible to have many Na^+ and Cl^- ions in the same volume of solution. However, some pairs of ions form insoluble compounds, so it is not possible to have a high concentration of both of these ions in solution. In that case, their solubility product is used to calculate the amount of each ion that can exist in solution.

Silver bromide is an ionic compound that is only slightly soluble in water. When an ionic compound is in a solution so concentrated that the solid is in equilibrium with its ions, the solution is saturated. The equilibrium equation for a saturated solution of silver bromide is

$$AgBr(cr) \rightleftarrows Ag^+(aq) + Br^-(aq)$$

The equilibrium constant (K_{eq}) for this equilibrium system is

$$K_{eq} = \frac{[Ag^+][Br^-]}{[AgBr]}$$

However, because we are dealing with a solid substance, the AgBr concentration is constant. Both sides of the equation can be multiplied by [AgBr], giving a new expression.

$$K_{eq}[AgBr] = \frac{[Ag^+][Br^-][AgBr]}{[AgBr]}$$

$$K_{eq}[AgBr] = [Ag^+][Br^-]$$

The term $K_{eq}[AgBr]$ is a constant. This new constant is called the **solubility product constant** K_{sp}.

$$K_{sp} = [Ag^+][Br^-]$$

At room temperature, 25°C, K_{sp} of silver bromide is 5.01×10^{-13}. Consider some silver bromide allowed to stand in water. The silver bromide dissolves until an equilibrium exists between the undissolved solid and the ions in solution. The product of the Ag^+ and Br^- ion concentrations will then be 5.01×10^{-13}.

$$[Ag^+][Br^-] = 5.01 \times 10^{-13}$$

Using the following equation, we see that for every silver ion there is one bromide ion.

$$\text{AgBr(cr)} \rightleftarrows \text{Ag}^+\text{(aq)} + \text{Br}^-\text{(aq)}$$

$$[\text{Ag}^+] = [\text{Br}^-], \text{ or } [\text{Ag}^+][\text{Ag}^+] = 5.01 \times 10^{-13}$$

therefore,

$$[\text{Ag}^+]^2 = 5.01 \times 10^{-13}$$

$$[\text{Ag}^+] = 7.08 \times 10^{-7}M$$

In a saturated solution containing only silver bromide and water, the concentration of silver ions is 7.08×10^{-7} mol/dm^3. The concentration of bromide ions is also 7.08×10^{-7} mol/dm^3. Suppose, however, that some potassium bromide solution is added. What will happen? KBr dissociates completely to K^+ and Br^- ions. The increased number of bromide ions collide more frequently with silver ions. The equilibrium is thus shifted toward the solid silver bromide. When equilibrium is again established, the concentration of silver ion, $[Ag^+]$, has decreased. At the same time, the concentration of bromide ion, $[Br^-]$, has increased. Solid silver bromide has precipitated out, and the K_{sp}, 5.01×10^{-13}, has remained the same. This reaction is an example of the common ion effect. The addition of a common ion removes silver ion from solution and causes the equilibrium to shift toward the solid silver bromide. We could also say the addition of a common ion decreases the solubility of a substance in solution.

SAMPLE PROBLEMS

1. Common Ion

What will be the silver ion concentration in a saturated solution of silver bromide if 0.100 mol of KBr is added to 1.00 dm^3 of the solution? Assume no increase in the solution volume.

Solving Process:

Potassium bromide is a soluble salt. Thus, it will contribute a bromide ion concentration of $0.100M$. The total bromide ion

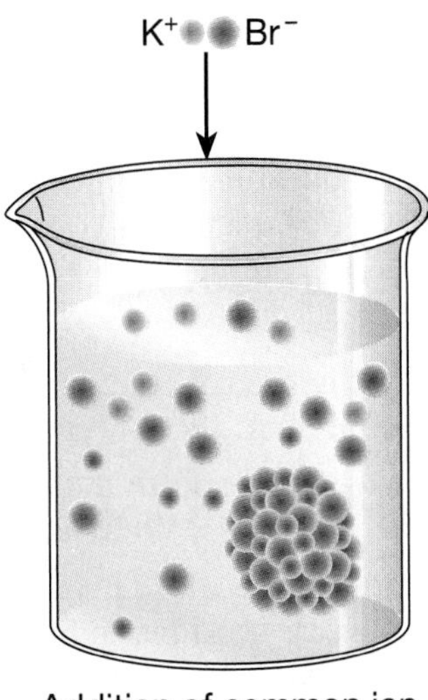

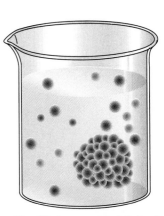

AgBr in KBr solution

Equilibrium reestablished

Figure 24.1 The addition of the common ion Br^- to the AgBr equilibrium system decreases the Ag^+ concentration as more AgBr precipitates.

concentration will consist of 0.100M from the KBr plus some unknown amount from the AgBr. We will designate the amount of bromide ion from the AgBr as y. If the AgBr produces a bromide ion concentration of y, it must also produce a silver ion concentration of y, because $AgBr \rightarrow Ag^+ + Br^-$. Substituting these values in the solubility product expression gives us

PROBLEM SOLVING HINT

Review scientific notation in Chapter 8.

$$5.01 \times 10^{-13} = y(0.100 + y)$$

We know that y will be small compared to 0.100 because silver bromide is almost insoluble. Thus, $0.100 + y$ is essentially equal to 0.100.

$$5.01 \times 10^{-13} = 0.100y$$
$$y = 5.01 \times 10^{-12}M$$

2. Ion Concentration

What is the iodate concentration in a saturated solution of copper(II) iodate? The K_{sp} of $Cu(IO_3)_2$ is 7.41×10^{-8}.

$$Cu(IO_3)_2 \rightleftarrows Cu^{2+} + 2IO_3^-$$

Solving Process:

For every copper ion, there are two iodate ions. If $[Cu^{2+}] = x$, then $[IO_3^-] = 2x$.

$$K_{sp} = [Cu^{2+}][IO_3^-]^2$$
$$7.41 \times 10^{-8} = x(2x)^2$$
$$7.41 \times 10^{-8} = x(4x^2)$$
$$7.41 \times 10^{-8} = 4x^3$$
$$x^3 = 1.85 \times 10^{-8}$$
$$[Cu^{2+}] = x = 2.65 \times 10^{-3}M$$
$$[IO_3^-] = 2x = 5.30 \times 10^{-3}M$$

Figure 24.2 When photographic paper, which is coated with silver bromide, is dipped into solutions during the developing and printing processes, the AgBr does not dissolve and wash off the paper. This is due to the very small K_{sp} of AgBr.

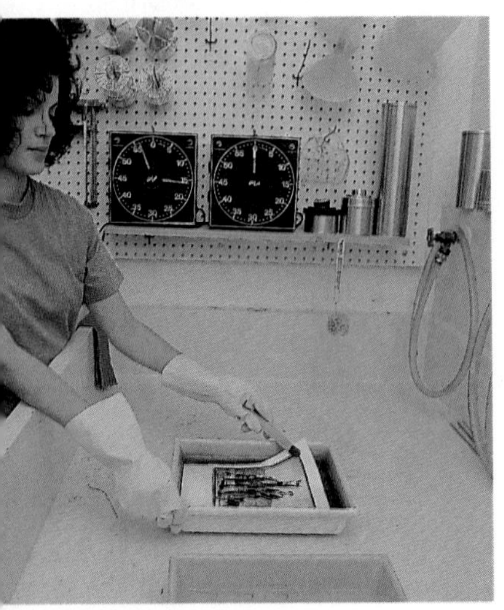

PRACTICE PROBLEMS

1. Write the solubility product expression for the following.
   a. $PbI_2(cr) \rightleftarrows Pb^{2+}(aq) + 2I^-(aq)$
   b. $Cu_3(PO_4)_2(cr) \rightleftarrows 3Cu^{2+}(aq) + 2PO_4^{3-}(aq)$
2. The solubility product constant of silver iodide is 8.32×10^{-17}. What is $[Ag^+]$ in a solution at equilibrium?
3. If $[D^+]$ is $2.00 \times 10^{-5}M$ at equilibrium, what is the K_{sp} for D_2A?
4. What is the concentration of Be^{2+} in a saturated solution of $Be(OH)_2$? $K_{sp} = 1.58 \times 10^{-22}$
5. A saturated solution of PbI_2 has a lead ion concentration of $1.21 \times 10^{-3}M$. What is K_{sp} for PbI_2?
6. What will be the $[Tl^+]$ in a saturated solution of TlBr if 0.050 mol NaBr is added to 500.0 cm^3 of solution? Assume there is no increase in volume. Use Appendix Table A-10.

SAMPLE PROBLEM

Predicting Precipitation

Suppose 5.00×10^2 cm^3 of $0.0200M$ NaCl are added to 5.00×10^2 cm^3 of $0.0100M$ $AgNO_3$. If the K_{sp} of AgCl is 1.78×10^{-10}, will a precipitate of AgCl form?

Solving Process:

The total volume of solution will be 1.000×10^3 cm^3 because 5.00×10^2 cm^3 of each solution are used. Therefore, the concentrations of both the NaCl solution and the $AgNO_3$ solution will be halved.

$$[NaCl] = \tfrac{1}{2} \times 0.0200M = 0.0100M$$

$$[AgNO_3] = \tfrac{1}{2} \times 0.0100M = 0.005\ 00M$$

Because both NaCl and $AgNO_3$ are soluble and dissociate completely

$$K_{sp} = [Ag^+][Cl^-]$$
$$= (0.005\ 00)(0.0100) = 0.000\ 0500 = 5.00 \times 10^{-5}$$

However, the ion product cannot exceed the K_{sp}, 1.78×10^{-10}. Therefore, AgCl will precipitate out of solution until

$$[Ag^+][Cl^-] = 1.78 \times 10^{-10}.$$

PRACTICE PROBLEMS

7. If 250.0 cm^3 of $0.100M$ $FeCl_3$ solution are added to 250.0 cm^3 of $0.010M$ NaOH solution, will a precipitate form? Show all of your calculations. Use Appendix Table A-10.
8. Will a precipitate form if 0.500 dm^3 of $0.100M$ LiCl solution is added to 0.500 dm^3 of $0.100M$ Na_2CO_3? Show all of your calculations. Use Appendix Table A-10.

Ionization of Water

Conductivity experiments have shown that water ionizes according to the equation

$$H_2O(l) + H_2O(l) \rightleftarrows H_3O^+(aq) + OH^-(aq)$$

An electric current is carried in a solution by the ions in the solution. Because pure water is a poor conductor of electricity, it must contain few ions. In pure water, the concentration of H_3O^+ is equal to the concentration of OH^-. Each molecule of H_2O that ionizes produces one ion each of H_3O^+ and OH^-.

Because water ionizes, it should be possible to find an equilibrium constant for this reaction.

$$K_{eq} = \frac{[H_3O^+][OH^-]}{[H_2O][H_2O]}$$

Figure 24.3 Pure water does not ionize enough to light the bulb in this circuit, but a current *does* flow, as shown on the ammeter.

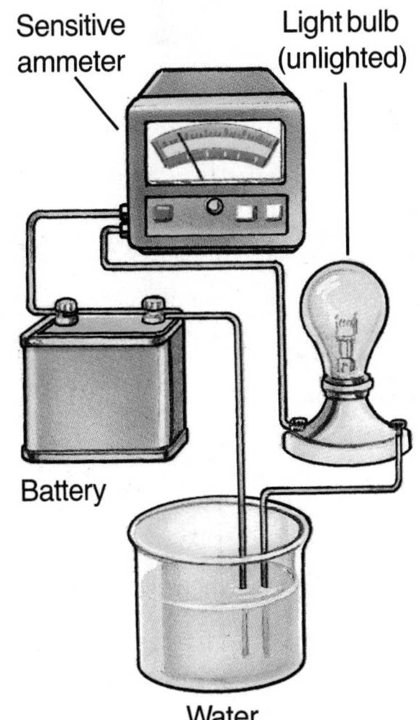

HEALTH CONNECTION

Keeping the pH Right

The pH of swimming pool water must be maintained in a narrow range, between 7.4 and 7.6, to prevent problems. Pool water acidity must be kept high enough to prevent the growth of algae and bacteria and low enough to avoid irritating the eyes and skin of swimmers. Algae can clog pool filters and cause odors. Bacteria cause illness and eye infections. If the acidity is too low, the disinfectants, notably sodium hypochlorite, will be less effective. An acidity that is too high will cause the pool plaster to become pitted. Chemicals used to raise the acidity are HCl and $NaHSO_4$. These chemicals neutralize the excess OH^- ions. Sodium carbonate is used to neutralize excess H_3O^+ ions and lower the acidity.

Conductivity experiments have indicated that pure water contains 1×10^{-7} mole of H_3O^+ (and the same amount of OH^-) per cubic decimeter at room temperature (25°C). Therefore

$$K_{eq} = \frac{(1.00 \times 10^{-7})(1.00 \times 10^{-7})}{[H_2O]^2}$$

We can find K_{eq} if we can arrive at some value for $[H_2O]$. Because water ionizes so slightly, we can approximate $[H_2O]$ in pure water by assuming that no ionization occurs. We assume that pure water contains only molecular H_2O.

One mole of water has a mass of 18.0 g. One dm^3 of pure water has a mass of 1000 g. The concentration of water in pure water is then

$$[H_2O] = \frac{1 \text{ mol}}{18.0 \not{g}} \left| \frac{1000 \not{g}}{1 \text{ dm}^3} \right. = 55.6 \text{ mol/dm}^3$$

We can now find K_{eq} for water.

$$K_{eq} = \frac{(1.00 \times 10^{-7})(1.00 \times 10^{-7})}{(55.6)^2}$$

Because the value 55.6 mol/dm^3 is relatively constant, we can multiply both sides of the equation by $(55.6)^2$. The new constant is

$$K_{eq}(55.6)^2 = [H_3O^+][OH^-] = 1.00 \times 10^{-14}$$

We call this new constant $K_{eq}(55.6)^2$ the **ion product constant of water,** K_w.

$$K_w = [H_3O^+][OH^-] = 1.00 \times 10^{-14}$$

K_w is a constant for all dilute aqueous solutions at room temperature (25°C). Although the concentrations of H_3O^+ and OH^- may change when substances are added to water, the product of $[H_3O^+]$ and $[OH^-]$ remains the same.

SAMPLE PROBLEM

Concentration of Hydroxide Ion in Solution

What is the $[OH^-]$ in a water solution with $[H_3O^+] = 1.00 \times 10^{-5}M$?

Solving Process:

If an acid, HA, is added to water,

$$HA + H_2O \rightarrow H_3O^+ + A^-$$

an excess of hydronium ion is produced. Collisions between H_3O^+ and OH^- also increase.

$$H_3O^+ + OH^- \rightarrow H_2O + H_2O$$

As the $[H_3O^+]$ increases, the $[OH^-]$ decreases; K_w remains 1.00×10^{-14}. If the $[H_3O^+]$ is increased to $1.00 \times 10^{-5}M$ by the addition of acid, the $[OH^-]$ must decrease.

$$[H_3O^+][OH^-] = 1.00 \times 10^{-14} \text{ and } [OH^-] = \frac{1.00 \times 10^{-14}}{[H_3O^+]}$$

$$[OH^-] = \frac{1.00 \times 10^{-14}}{1.00 \times 10^{-5}} = 1.00 \times 10^{-9}M$$

If a base is added to water, the equilibrium shifts in the opposite direction, and the solution becomes more basic.

$$OH^-(aq) + H_3O^+(aq) \rightarrow H_2O(l) + H_2O(l)$$

H_3O^+ is removed from solution, and the ion product constant remains 1.00×10^{-14}.

PRACTICE PROBLEMS

9. What is the hydroxide ion concentration in a solution with hydronium ion concentration $6.80 \times 10^{-10}M$?

10. What is the hydronium ion concentration in a solution with hydroxide ion concentration $5.67 \times 10^{-3}M$?

pH Scale

The ionization of water is so slight that it is almost never considered in the actual production or use of acids and bases. Why, then, was it introduced? Knowledge of the ion product constant for water has enabled chemists to develop a simple acidity scale, called the pH scale. The scale can be used to indicate the basicity as well as the acidity of any water solution. The **pH scale** is a measure of hydronium ion concentration and, thus, of acidity.

The concentration of H_3O^+ is expressed in powers of 10, from 10^{-14} to 10^0. For convenience, scientists choose to express the $[H_3O^+]$ as **pH,** which is the negative log of $[H_3O^+]$ and always gives a positive value.

$$pH = -\log[H_3O^+]$$

For example, if $[H_3O^+] = 10^{-7}M$, the negative log of $[H_3O^+] = +7$. Thus, the pH equals 7, indicating a neutral solution.

Figure 24.4 The pH scale is a measure of hydronium ion concentration. The scale is logarithmic and has values from 0 to 14. The pH of some common substances is shown.

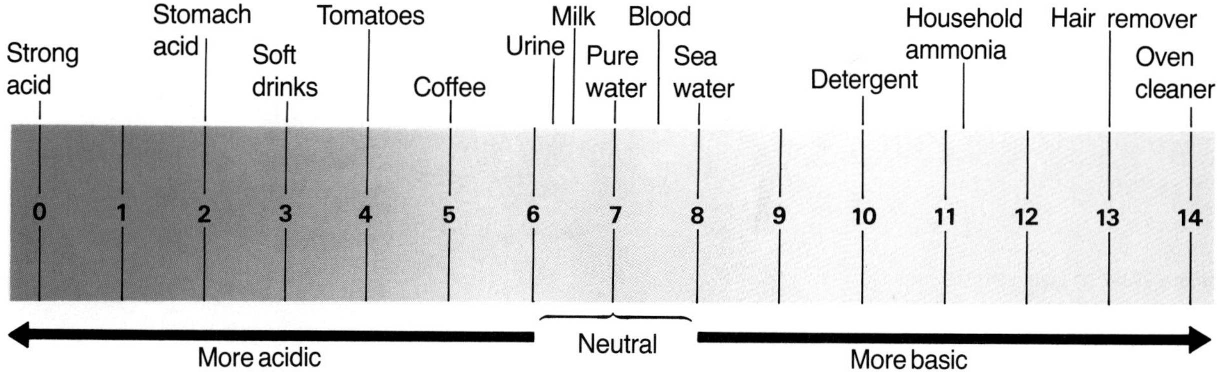

PROBLEM SOLVING HINT

Predict whether the solution will be acidic or basic before you work on the problem. After you have performed the calculation, check to see if your answer is in the correct range, based on your prediction.

SAMPLE PROBLEM

pH Determination

One-tenth mole of HCl is added to enough water to make 1 dm^3 of solution. What is the pH of the solution? (Assume that the HCl is 100% ionized.)

Solving Process:

$$pH = -\log[H_3O^+]$$
$$[H_3O^+] = 0.1M = 1 \times 10^{-1} \text{ mol/dm}^3$$
$$pH = -\log[H_3O^+] = -\log(1 \times 10^{-1})$$
$$pH = -(\log 1 + \log 10^{-1}) = -[0 + (-1)] = 1$$

PRACTICE PROBLEMS

11. Find the pH of solutions with the following H_3O^+ concentrations.

a. $1.00 \times 10^{-3}M$ **c.** $6.59 \times 10^{-10}M$ **e.** $9.47 \times 10^{-8}M$
b. $1.00 \times 10^{-6}M$ **d.** $7.01 \times 10^{-6}M$ **f.** $6.89 \times 10^{-14}M$

12. Find the H_3O^+ concentration of the following solutions.

a. pH = 3.000 **c.** pH = 6.607 **e.** pH = 6.149
b. pH = 10.000 **d.** pH = 2.523 **f.** pH = 7.662

Note that as the hydronium ion concentration increases and a neutral solution is made more acidic, the pH goes from 7 toward 0. If the pH of a solution falls between 7 and 14, the solution is basic. Figure 24.6 shows the pH of several common substances.

Suppose we wish to know the pH of a solution that is basic or of which we know only the hydroxide ion concentration. We can use the ion product constant of water to find the relationship between pH and hydroxide ion concentration. If we take the logarithm of both sides of the ion product of water, we get

$$\log([H_3O^+][OH^-]) = \log 1.00 \times 10^{-14}$$
$$\log[H_3O^+] + \log[OH^-] = -14$$

Figure 24.5 Plants are sensitive to the pH of soil. Hydrangeas produce blue flowers in acid soil (left) and pink flowers in basic soil (right).

Multiplying both sides of the equation by -1, we have

$$-\log[H_3O^+] + (-\log[OH^-]) = 14$$

If we now designate $-\log[OH^-]$ as pOH, and substitute, then

$$pH + pOH = 14$$

SAMPLE PROBLEMS

1. pH of Solution

Find the pH of a solution with an $[OH^-]$ of $1.00 \times 10^{-9}M$.

Solving Process:

$$[H_3O^+] = 10^{-14}/[OH^-] = 10^{-14}/10^{-9} = 10^{-5}M$$
$$pH = -\log[H_3O^+] = -\log(1.00 \times 10^{-5})$$
$$= -(\log 1.00 + \log 10^{-5}) = -[0 + (-5)] = 5$$

PROBLEM SOLVING HINT
Use the log key of a scientific calculator to determine the log and/or pH.

2. H_3O^+ Concentration

Find the concentration of H_3O^+ if the pOH of a solution is 11.700.

Solving Process:

$$pH = 14.000 - pOH = 14.000 - 11.700 = 2.300$$
$$pH = -\log[H_3O^+]$$
$$-2.300 = \log[H_3O^+]$$
$$0.700 - 3 = \log[H_3O^+]$$
$$5.01 \times 10^{-3}M = [H_3O^+]$$

PRACTICE PROBLEMS

13. Find the pH of each of the following solutions.
- **a.** pOH = 2.00 **c.** pOH = 1.263 **e.** pOH = 9.714
- **b.** pOH = 7.00 **d.** pOH = 4.976 **f.** pOH = 3.004

14. Determine the pH of solutions with the following $[OH^-]$.
- **a.** $1.00 \times 10^{-4}M$ **d.** $3.45 \times 10^{-8}M$
- **b.** $1.00 \times 10^{-6}M$ **e.** $4.97 \times 10^{-10}M$
- **c.** $2.64 \times 10^{-13}M$ **f.** $2.93 \times 10^{-2}M$

15. Find the $[H_3O^+]$ of each of the solutions in Problem 13.

Figure 24.6 The pH of some common substances is shown here. The intensity of the red or blue color is an indication of the degree of acidity or basicity, respectively.

pH of Some Common Substances

Substance	pH
0.1*M* HCl	1
Stomach contents	2
Vinegar	2.9
Soft drinks	3
Grapes	4
Beer	4.5
Pumpkin pulp	5
Bread	5.5
Intestinal contents	6.5
Urine	6.6
Bile	6.9
Saliva	7
Blood	7.4
Eggs	7.8
0,1*M* NH_3(aq)	11.1
0.1*M* NaOH	13

Hydrolysis

A salt is composed of positive ions from a base and negative ions from an acid. When a salt dissolves in water, it releases ions having an equal number of positive and negative charges. Some salts form neutral solutions, but others react with water to form acidic or basic solutions. The reaction of a salt with water to form an acidic or basic solution is called **hydrolysis.**

CAREERS IN CHEMISTRY

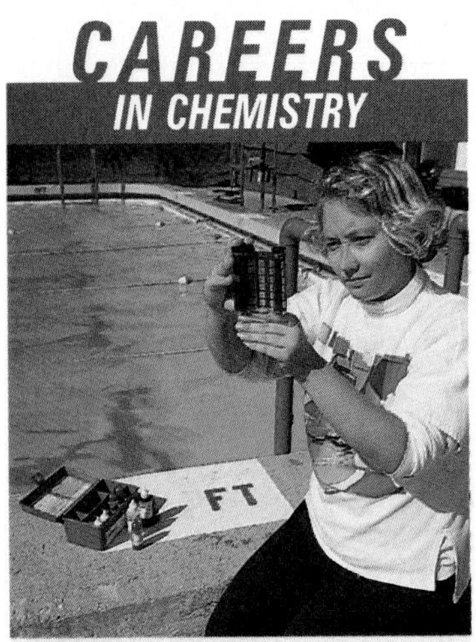

Swimming Pool Technician

A swimming pool technician may work for a large pool service company or for schools, colleges, and clubs that have large aquatic programs. The technician has the responsibility to make certain that the pool is chemically safe and free of bacteria and algae. The technician checks and balances pH and mineral salts and uses chemical compounds to alleviate imbalances. Swimming pool technicians need on-the-job training, but high school chemistry is important in understanding the interactions among the various chemicals used. With the increasing number of private and public pools, the outlook for jobs in this area is good.

If potassium chloride is dissolved in water, a neutral solution results. Each ion from the salt K^+ and Cl^- is hydrated with no apparent reaction (except hydration). Water ionizes very slightly to form H_3O^+ and OH^- ions. The solution contains ions from the salt and ions from the water. In solution, the positive potassium ion could unite with the negative chloride ion or the negative hydroxide ion. However, potassium chloride is a salt that dissociates completely in water. Potassium hydroxide is a strong base that also dissociates completely. The same reasoning applies to Cl^- and H_3O^+. No reaction occurs. The four ions remain in solution as ions, and the solution is neutral. The ions produced by the salt of a strong acid and strong base do not react with water. There is no change in the concentrations of H_3O^+ or OH^-. Therefore, such a solution has a pH of 7. No hydrolysis occurs.

If we test a solution of the salt aluminum chloride, we will find that the solution is acidic. With the exception of the metals of Groups 1 and 2, metallic hydroxides are all weak bases and have low solubilities. The positive ions of such metals are strongly hydrated and give up a proton from the associated water molecule. When $AlCl_3$ dissolves in water, the Al^{3+} ions become hydrated.

$$AlCl_3(cr) + 6H_2O(l) \rightarrow Al(H_2O)_6{}^{3+}(aq) + 3Cl^-(aq)$$

The hydrated aluminum ions then undergo hydrolysis.

$$Al(H_2O)_6{}^{3+}(aq) + H_2O(l) \rightleftarrows Al(OH)(H_2O)_5{}^{2+}(aq) + H_3O^+(aq)$$

Because HCl is a strong acid, the H_3O^+ ions do not combine with the Cl^- ions, and the solution is acidic.

If we test a solution of sodium carbonate, we will find it to be basic. Sodium carbonate forms Na^+ and $CO_3{}^{2-}$ as it dissolves in water. The $CO_3{}^{2-}$ ions react with water molecules.

$$CO_3{}^{2-}(aq) + H_2O(l) \rightleftarrows HCO_3{}^-(aq) + OH^-(aq)$$

This reaction produces a large excess of hydroxide ions. Hydrogen carbonate ion, $HCO_3{}^-$, is an exceedingly weak acid, so that the reverse reaction, the hydrolysis of hydrogen carbonate ion, proceeds only to a very slight extent.

$$HCO_3{}^-(aq) + H_2O(l) \rightleftarrows H_3O^+(aq) + CO_3{}^{2-}(aq)$$

More OH^- ions form than H_3O^+ ions, so the solution is basic.

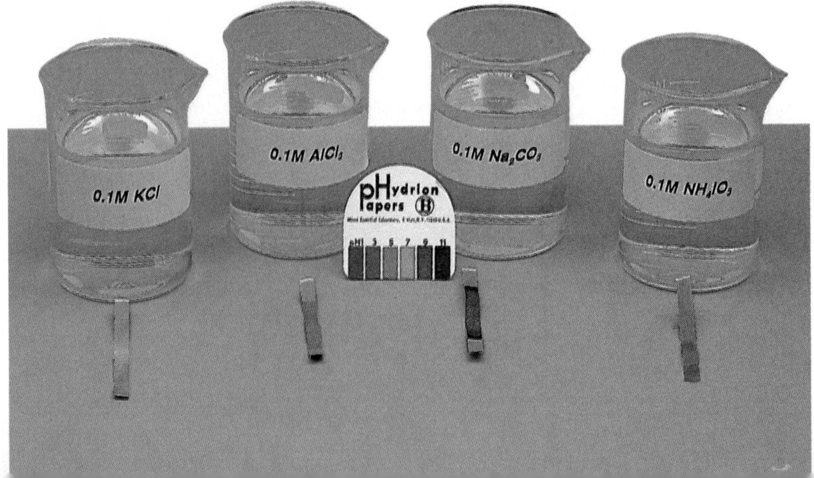

Figure 24.7 pH paper shows that salt solutions can be acidic, basic, or neutral depending upon the strength of the acid and base that formed the salt.

If the salt ammonium acetate is dissolved in water, ammonia and acetic acid are formed. Ammonia is a weak, slightly ionized base. Acetic acid is a weak, slightly ionized acid. Both H_3O^+ and OH^- ions are removed from solution. This reaction results in a neutral solution because acetic acid and ammonia are of the same degree of weakness. For other salts of two weak ions, you must know their relative degrees of weakness in order to predict the acidity of the solution. For example, a solution of NH_4CN is basic. The acid, HCN, is far weaker than the base NH_3. A solution of NH_4IO_3 is acidic because HIO_3 is a stronger acid than NH_3 is a base.

Table 24.1

Formation of Salt Solutions

Positive ion from	Negative ion from	Resulting solution
Strong base	Strong acid	Neutral
Strong base	Weak acid	Basic
Weak base	Strong acid	Acidic
Weak base	Weak acid	Varies

Table 24.1 summarizes the four kinds of salt solutions that can be formed. You can refer to the set of rules on pages 588 and 589 of Chapter 23 to determine the strength of each acid and base.

SAMPLE PROBLEM

Hydrolysis

Would a solution of $Bi(NO_3)_3$ be acidic, basic, or neutral?

Solving Process:

Bismuth ion (Bi^{3+}) is derived from $Bi(OH)_3$, a weak base (Rule 4, page 589). Nitrate ion (NO_3^-) is derived from nitric acid, HNO_3, a strong acid (Rule 2, page 588). The bismuth ion will undergo hydrolysis and the nitrate ion will not.

$$Bi^{3+} + 2H_2O \rightarrow Bi(OH)^{2+} + H_3O^+$$

H_3O^+ is produced, and the solution will be acidic.

PRACTICE PROBLEMS

16. Will mixing equal quantities of $1M$ solutions of the following give an acidic, basic, or neutral solution?
- **a.** strong acid and strong base
- **b.** weak acid and strong base
- **c.** strong acid and weak base

17. Using the rules in Chapter 23, predict the acidic, basic, or neutral character of the solutions of the following salts.

a. $CrBr_3$ **c.** $NiSO_4$ **e.** $MgC_4H_4O_6$
b. NH_4ClO_4 **d.** GaI_3 **f.** K_2CO_3

CHEMISTS AND THEIR WORK

Mae C. Jemison
(1956-)

Mae Jemison is a physician, a chemical engineer, and an astronaut. She served in the Peace Corps in Africa, managing the health care delivery system for the Peace Corps and the United States Embassy. She returned to general practice in 1985, while taking advanced engineering courses at night and doing research on artificial organ design. Her diversity and hard work were rewarded in 1987 when she was selected from 2000 applicants to be one of 15 members of the astronaut training program. She is currently assigned as a mission specialist on a space lab.

MINUTE LAB

Buffer Preparation

A buffer system can absorb moderate amounts of acid or base without a significant change in pH. Put on an apron and goggles. To prepare a system, add 3 cm^3 of 17.4*M* acetic acid to 47 cm^3 distilled H_2O. **CAUTION:** *Acid is corrosive.* Divide the diluted acid equally between two 100 cm^3 beakers. Dissolve 4 g of sodium acetate in the diluted acid in one of the beakers. Stir and label it "buffer." Label three large test tubes *acid, buffer,* and *water.* Add equal amounts of the diluted acetic acid, the buffer, and distilled water separately to the respectively labeled test tubes. Add a few drops of universal indicator to each test tube. Place a stirring rod in each test tube. Add a filled dropper of 1*M* NaOH to each test tube. **CAUTION:** *NaOH is caustic.* Stir and observe the color. Fill the dropper and repeat many times and record your observations. Which of the three test tubes served as a control in this experiment? How was the definition of a buffer verified? Do you think that buffered aspirin would be better for your stomach than unbuffered aspirin? Explain.

Buffers

Many of the fluids in your body must be maintained within a very narrow pH range if you are to remain healthy. There are also many instances in chemistry when the maintenance of a certain pH is important. In both cases, the end is accomplished in the same way, by the creation of a buffer system. A **buffer system** is a solution that can absorb moderate amounts of acid or base without a significant change in its pH. A buffer provides ions that will react with H_3O^+ or OH^- if they are introduced into the solution. Because the added H_3O^+ or OH^- ions are thereby neutralized, the pH of the system remains nearly constant.

Consider the hydrolysis of ammonia in water.

$$NH_3(g) + H_2O(l) \rightleftarrows NH_4^+(aq) + OH^-(aq)$$

If we look at the reverse reaction in this equilibrium, we can see that NH_4^+ ions will react with a base. What happens when we dissolve ammonium ions (from NH_4Cl, for example) in water? The ions are hydrolyzed, and an equilibrium is produced.

$$NH_4^+(aq) + H_2O(l) \rightleftarrows NH_3(aq) + H_3O^+(aq)$$

From the reverse reaction, we can see that ammonia molecules will react with acids. By preparing a solution containing both ammonium ions and ammonia molecules, we would have our desired buffer solution. The NH_3 molecules would react with any added H_3O^+, and the NH_4^+ ions would react with any OH^-.

Buffer solutions are prepared by using a weak acid with one of its salts or a weak base with one of its salts. Stated in general terms, the reactions would appear as follows.

For the weak acid HA and its salt NaA that dissociates to Na^+ and A^-,

$$HA + OH^- \rightarrow H_2O + A^-$$

$$A^- + H_3O^+ \rightarrow HA + H_2O$$

The weak acid, HA, will react with added OH^-. The negative ion, A^-, from the salt will react with added H_3O^+.

For the weak base MOH and its salt MA that dissociates to M^+ and A^-,

$$MOH + H_3O^+ \rightarrow M^+ + 2H_2O$$

$$M^+ + OH^- \rightarrow MOH$$

The weak base, MOH, will react with added H_3O^+. The positive ion from the salt, M^+, will react with added OH^-.

Buffers are most efficient at neutralizing added acids or bases when the concentrations of HA and A^- (or MOH and M^+) are equal. By choosing the correct weak acid (or base) we can prepare a buffer solution of almost any pH value. Note that there is a common ion between the weak acid or base and its salt. The behavior of a buffer solution can always be predicted on the basis of Le Chatelier's principle and the common ion effect.

Blood is buffered principally by the hydrogen carbonate ion, HCO_3^-.

$$HCO_3^- + H_3O^+ \rightleftharpoons H_2CO_3 + H_2O$$
$$HCO_3^- + OH^- \rightarrow H_2O + CO_3^{2-}$$

When the H_2CO_3 reaches the lungs, it decomposes to form CO_2.

$$H_2CO_3 \rightleftharpoons H_2O + CO_2$$

CO_2 is then exhaled. There is also buffering in the blood by the dihydrogen phosphate ion, $H_2PO_4^-$.

Bridge to MEDICINE

Buffering and Hyperventilation

Athletes sometimes upset the buffering of their blood through a process called hyperventilation. In the excitement and anxiety of a contest, they breathe more rapidly and deeply than they need to. When hyperventilation occurs, a person expels more CO_2 than necessary, upsetting the carbonic acid equilibrium. According to Le Chatelier's principle, as the CO_2 is exhaled, more H_2CO_3 decomposes to replace the CO_2 in the reaction $H_2CO_3 \rightleftharpoons H_2O + CO_2$. As the H_2CO_3 is used, the equilibrium between carbonic acid and the hydrogen carbonate ion is upset and HCO_3^- is consumed in replacing the H_2CO_3.

$$HCO_3^- + H_3O^+ \rightleftharpoons H_2CO_3 + H_2O$$

Eventually, the HCO_3^- concentration drops to the point at which it is insufficient to maintain the blood pH at a safe level. Because using HCO_3^- also consumes H_3O^+, the blood pH is raised (becomes more basic). One response of the body to this condition is a constriction of the cerebral blood vessels. As the blood flow to the brain is reduced, the individual becomes dizzy and can lapse into unconsciousness. At that point, the body's reflex mechanisms usually restore normal breathing and the blood pH returns to normal.

Exploring Further

1. How is a paper bag used to treat hyperventilation?
2. Would hyperventilation have any effect on the $H_2PO_4^-$ ion that helps to buffer the blood?

24.1 *CONCEPT REVIEW*

18. What is a solubility product constant?

19. Copper(II) arsenate, $Cu_3(AsO_4)_2$, has an extremely low solubility ($K_{sp} = 7.6 \times 10^{-36}$). What would happen to the concentration of AsO_4^{3-} ions in a saturated solution of copper(II) arsenate if some solid $Cu(NO_3)_2$ is added?

20. What is the value of the ion product constant for water? What is its significance?

21. Describe the pH scale.

22. What happens when a salt undergoes hydrolysis?

23. Apply Why is it necessary for blood to contain a buffering system?

CHEMISTRY AND

SOCIETY

Acid Rain

The nonmetallic oxides produced by the burning of fossil fuels such as coal and petroleum are all acidic anhydrides. Thus, when these gases come in contact with water vapor or water droplets in the atmosphere, they react to produce these acids.

$$SO_2 + H_2O \rightarrow H_2SO_3$$
$$SO_3 + H_2O \rightarrow H_2SO_4$$
$$2NO_2 + H_2O \rightarrow HNO_2 + HNO_3$$
$$CO_2 + H_2O \rightarrow H_2CO_3$$

Rain formed from these water droplets is acidic. It is considered normal for rainfall to be slightly acidic. Nonpolluted air contains CO_2, and rain in equilibrium with atmospheric CO_2 has a pH of 5.65. However, there are times when the rainfall in some areas is highly acidic. Rain with a pH below 5.3 is considered sufficiently acidic to cause concern. Acid rain has a number of harmful effects. One effect is the increase in the rate of deterioration of stone. Buildings constructed of stone are affected over time. The art work and carvings that decorate buildings and public areas are damaged. The fine detail on statuary is destroyed. Acid rain dissolves some compounds, such as carbonates, in the stone while the remaining substances become a powdery coating. In addition, the acid rain eats into the stone leaving tiny cracks. When water fills these cracks and freezes, its expansion causes cracking and breaking of the stone.

Acid rain also has adverse effects on living organisms. Rain falling in a region from the Ohio River Valley to upper New York State has been measured with pH values as low as 4.1. Acid rain has increased the acidity of some lakes in the northeastern United States and Canada to such an extent that fish can no longer live in them. Fish and shellfish cannot tolerate acidities below pH 4.8. Acid rain also tends to dissolve vital minerals in the soil. These minerals are then washed away in run-off. Crops grown in these depleted soils give poor yields, if they grow at all. When the acidic run-off reaches waterways, it further interferes with fish growth and development.

Correcting these problems by treating statues with preservatives and lakes with basic compounds such as lime is not really the answer. Although the source of the problem is known, much research on the chemistry of the atmosphere remains to be done in order to understand how to combat the problem. Removing the offending oxides from exhausts and using alternate energy sources are much preferred courses of action at the present time. Fortunately, the electric power industry in the United States has begun to attack the problem. The industry has succeeded in reducing the sulfur emissions from fuel burning power plants by 50% over the last decade. This reduction has been accomplished through several avenues. One of the most important has been the switch to low sulfur fuels. Another is the scrubbing of stack gases before they are released to the atmosphere. In this process, the stack gases percolate through a solution that absorbs the oxides of sulfur. The solution is renewed frequently, and waste sulfur can be recovered from the spent solution.

Analyzing the Issue

1. What natural processes contribute to acid rain?
2. Air pollution can be carried by wind and cause acid rain to fall far from the source of the pollution. Who should be responsible for cleaning up the pollution? What if the source of the pollution cannot be identified?

24.2 Titration

OBJECTIVES

- state the principles and uses of indicators.
- explain the process of titration and perform calculations using the data from titrations.

It is frequently necessary to measure the concentration of a solution. One technique for such measurements is the laboratory procedure of titration. In titrating acids and bases, the use of colored materials called indicators is involved. In this section you will learn about indicators and titration.

Indicators

In the laboratory, how do we know whether we are using an acidic, basic, or neutral solution? We could taste the solution to determine whether it is bitter or sour. However, this method is dangerous as well as inexact because many solutions are toxic or corrosive. We could experiment to determine what reactions the solution undergoes, but the experiment might take too much time. A chemist generally uses a **pH meter** to determine the degree of acidity in a solution. This electronic device indicates the pH of a solution directly when its electrodes are immersed in the solution. We will investigate the operation of the pH meter further in Chapter 26.

There are times when a pH meter is not available, or its use is not convenient. At those times, substances called indicators are used. **Indicators** are weak organic bases and acids whose colors differ from the colors of their conjugate acids or bases. The examples below show how the pH of a solution affects the color when an indicator is added. A number of indicators and their color changes are listed in Appendix Table A-11.

Assume the formula for an indicator is HIn. When in water, the equilibrium established is

$$\underset{\text{yellow}}{\text{HIn}} + \text{H}_2\text{O} \rightleftarrows \text{H}_3\text{O}^+ + \underset{\text{red}}{\text{In}^-}$$

Note that the molecular acid form, HIn, has one color, yellow, while the ionized form, In^-, the conjugate base, has a different color, red. Consider what happens when this indicator is placed in an acidic solution. The solution is acidic because of an excess of hydronium ions. This excess will shift the indicator equilibrium (above) to the left and the color observed will be yellow. On the other hand, if the indicator is placed in a basic solution, the hydroxide ions from the base will react with the hydronium ions to form water. The shortage of hydronium ions will then shift the indicator equilibrium to the right and we will observe a red color. By choosing an indicator whose equilibrium shifts at the proper pH, we can follow the course of many reactions. For example, consider an indicator that is blue above pH 6 and yellow below pH 5. If we place that indicator in a solution and the solution turns yellow, then we know that the pH of the solution is less than 5. If we then begin to add a base, the solution will gradually change to green and then to blue as the pH changes from less than 5, through 5 and 6, to more than 6.

Figure 24.8 The indicator methyl red is shown (back to front) at pH 2, pH 5, and pH 10.

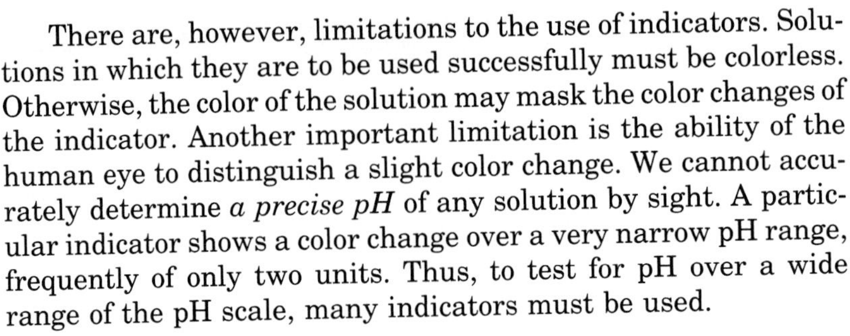

There are, however, limitations to the use of indicators. Solutions in which they are to be used successfully must be colorless. Otherwise, the color of the solution may mask the color changes of the indicator. Another important limitation is the ability of the human eye to distinguish a slight color change. We cannot accurately determine *a precise pH* of any solution by sight. A particular indicator shows a color change over a very narrow pH range, frequently of only two units. Thus, to test for pH over a wide range of the pH scale, many indicators must be used.

Word Origins

titulus: (L) inscription or title (Only nobles of high rank had titles. The word was transformed in French to *titre,* meaning rank.)

titration—determination of rank (concentration) of a solution.

Titration

Chemists often need to find the concentration of a solution. **Titration** is an analytical method in which a standard solution is used to determine the concentration of another solution. A **standard solution** is one for which the concentration is known.

If an acid is added to a base, a neutralization reaction occurs. The acid unites with the base to form a salt and water.

$$\textit{acid} + \textit{base} \longrightarrow \textit{salt} + \textit{water}$$

For example, acetic acid can be neutralized by sodium hydroxide to form sodium acetate and water. By adding an indicator, we can see at what point complete neutralization occurs.

To carry out a titration, a buret is filled with a standard solution. A small amount of indicator is added to a measured amount of the solution of unknown concentration. The buret is opened and the standard solution is allowed to flow (with stirring) into the solution to be titrated. Eventually a color change occurs with the addition of a single final drop, indicating the endpoint of the reaction. At the endpoint, an amount of standard solution has been added that just completely reacts with the solution titrated. The endpoint may also be detected using a pH meter. If the pH meter is connected to a recorder, a graph of pH versus cm^3 of titrating solution added is obtained. The

Figure 24.9 In this titration, a standard NaOH solution in the buret has been added to an acid solution in the flask until the phenolphthalein indicator turned from colorless to pink, indicating the endpoint of the titration.

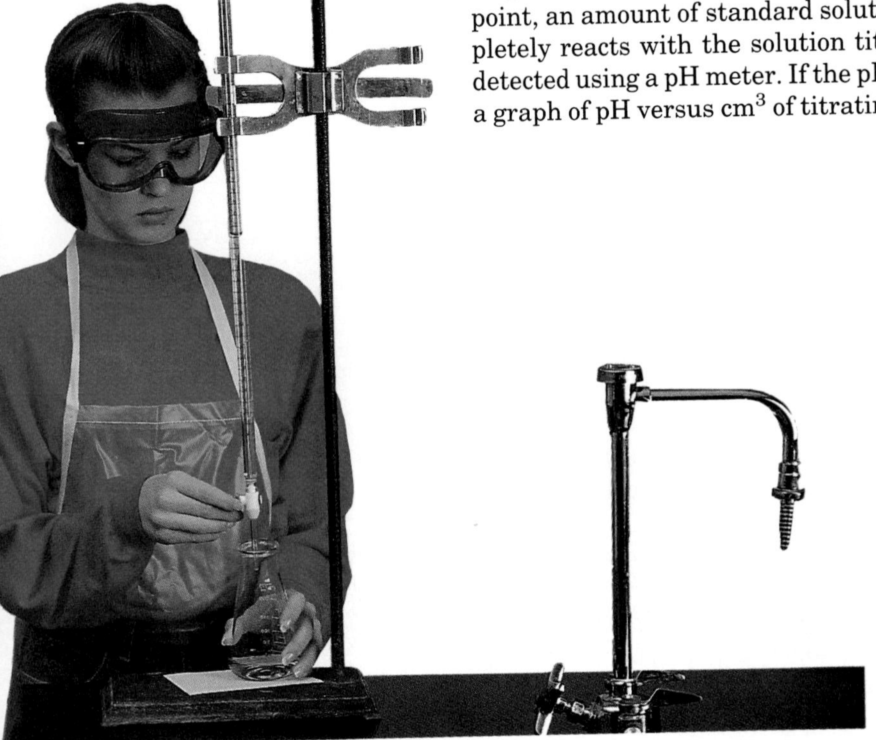

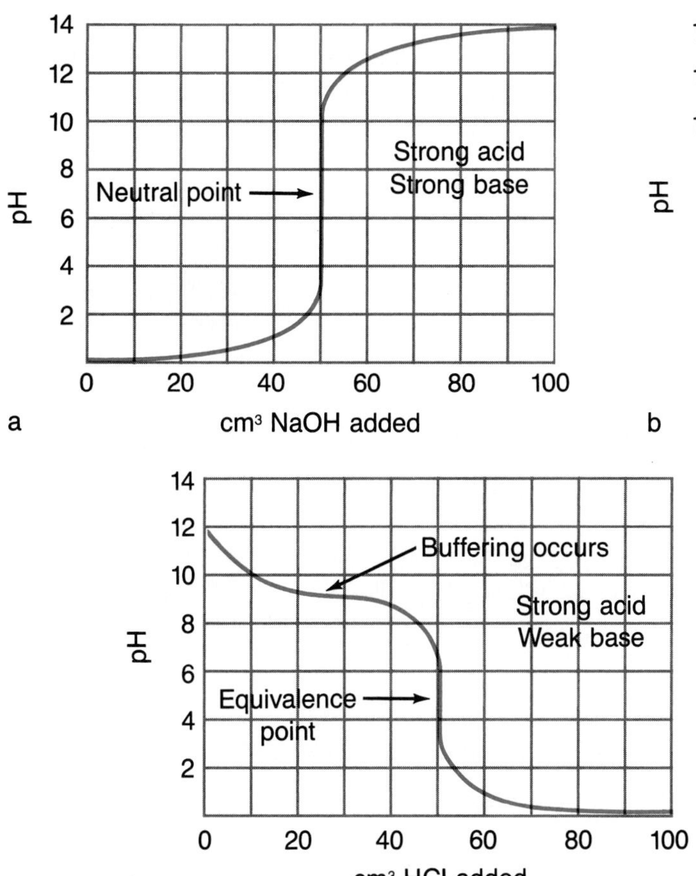

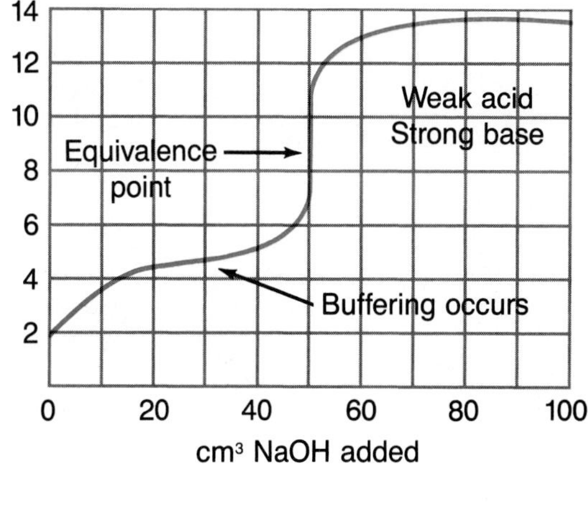

Figure 24.10 These graphs show the change in pH throughout titration of a strong acid with a strong base (a), a weak acid with a strong base (b), and a strong acid with a weak base (c). The curve in (a) is for the titration of 50.0 cm^3 of 0.100*M* HCl with 0.100*M* NaOH. The point at pH 2 when 47 cm^3 of NaOH have been added means that after we have added 47 cm^3 of NaOH to 50.0 cm^3 of HCl, the pH has increased to 2. When 52 cm^3 of NaOH have been added, the pH has increased to 11.

endpoint corresponds to the middle of that portion of the graph showing a very large change in pH with the addition of a small amount of titrating solution. The graph in Figure 24.10a shows the titration of a strong acid by a strong base. Weak electrolytes show a slightly different shape, shifted toward the pH of the stronger reactant.

The moles of standard solution that were used in the titration can be calculated by multiplying the volume of standard solution used by its molarity.

$$\underset{\text{standard solution}}{moles} = \underset{\text{standard solution}}{volume} \times \underset{\text{standard solution}}{molarity}$$

The moles in the titrated solution of unknown concentration are then found using the coefficients in the balanced equation. By dividing the moles of the titrated solution by the volume of that solution, the concentration of the titrated solution can be found.

$$\underset{\text{titrated solution}}{molarity\ (M)} = \frac{\overset{\text{titrated solution}}{moles}}{\underset{\text{titrated solution}}{volume}}$$

Suppose we wanted to find the molarity of vinegar. Vinegar is acidic because it is dilute acetic acid. We use a standard NaOH solution that we know to be 0.500*M*. We will use phenolphthalein as an indicator. Two burets, including the tips, are filled with solution, one with vinegar and one with standard NaOH, and the initial volumes

Figure 24.11 The endpoint of this titration has been reached when the addition of a single drop of a basic solution changes the color of the indicator (phenolphthalein is shown here).

recorded. A carefully measured amount (20-25 cm^3) of the vinegar solution is allowed to run into a beaker that contains a few drops of phenolphthalein. Phenolphthalein is colorless in acid solution.

While stirring constantly, the basic solution is allowed to run slowly into the beaker. The acid is neutralized when one drop of NaOH and then one drop of vinegar solution alternately changes the color. We then record the volume used for each solution.

PROBLEM SOLVING HINT
Remember that 1000 cm³ = 1 dm³.

SAMPLE PROBLEMS

1. Titration

Suppose we used 15.0 cm^3 of 0.500*M* NaOH and 25.0 cm^3 of vinegar solution of unknown concentration. What is the molarity of the vinegar solution?

Solving Process:

(a) The balanced equation for the neutralization is

$NaOH(aq) + CH_3COOH(aq) \rightarrow NaCH_3COO(aq) + H_2O(l)$

(b) Because the concentration of the base is given, determine the moles of NaOH.

$$\frac{15.0\ \cancel{cm^3\ soln}}{} \left| \frac{1\ \cancel{dm^3}}{1000\ \cancel{cm^3}} \right| \frac{0.500\ mol\ NaOH}{1.00\ \cancel{dm^3\ soln}}$$

$= 0.007\ 50$ mol NaOH

(c) Because the coefficients from the equation are all 1's, equal moles of NaOH and CH_3COOH react.

$$0.007\ 50\ mol\ NaOH = 0.007\ 50\ mol\ CH_3COOH$$

(d) Determine the molarity of the acid.

$$\frac{0.007\ 50\ mol\ CH_3COOH}{25.0\ \cancel{cm^3\ soln}} \left| \frac{1000\ \cancel{cm^3}}{1\ dm^3} \right. = 0.300M\ CH_3COOH$$

2. Neutralization

How many cm^3 of 0.200*M* KOH will exactly neutralize 15.0 cm^3 of 0.400*M* H_2SO_4?

Solving Process:

(a) $2KOH(aq) + H_2SO_4(aq) \rightarrow K_2SO_4(aq) + 2H_2O(l)$

(b) The number of moles of H_2SO_4 is

$$\frac{15.0\ \cancel{cm^3}\ soln}{} \left| \frac{1\ \cancel{dm^3}}{1000\ \cancel{cm^3}} \right| \frac{0.400\ mol\ H_2SO_4}{1.00\ \cancel{dm^3\ soln}}$$

$= 0.006\ 00$ mol H_2SO_4

(c) Because 2 moles KOH are required for 1 mole H_2SO_4, 0.0120 ($2 \times 0.006\ 00$) mole of KOH will be required.

(d) Thus,

$$\frac{0.0120\ \cancel{mol\ KOH}}{} \left| \frac{1.00\ \cancel{dm^3\ soln}}{0.200\ \cancel{mol\ KOH}} \right| \frac{1000\ cm^3}{1\ \cancel{dm^3}}$$

$= 60.0$ cm^3 KOH soln

PRACTICE PROBLEMS

24. How many cm^3 of $0.0947M$ NaOH are needed to neutralize 21.4 cm^3 of $0.106M$ HCl?

25. If 26.4 cm^3 of LiOH solution are required to neutralize 21.7 cm^3 of $0.500M$ HBr, what is the concentration of the basic solution?

26. If 23.4 cm^3 of $0.551M$ NaOH is used to titrate 50.0 cm^3 of HCl to the endpoint, what is the concentration of the HCl solution?

Bridge to *BIOCHEMISTRY*

Sodium, Potassium, and Nerves

Sodium and potassium ions are essential to the proper functioning of the human nervous system. In the nervous system, messages are sent throughout the body as pulses of electric current. Nerve cells have long cylindrical fibers called axons, which transmit the nerve impulses. The cell membrane surrounding these axons does not conduct electricity. How, then, does the impulse move along the axon?

A biochemical "pump" maintains a much higher concentration of sodium ions outside the axon than inside. At the same time, the potassium ion concentration inside the cell is much higher than it is outside. Thus, a potential difference builds up across the cell membrane. When a nerve cell is stimulated, the membrane suddenly allows sodium ions to pass into the cell. The change in potential at that point on the axon "triggers" the next axon section to allow entry of sodium ions.

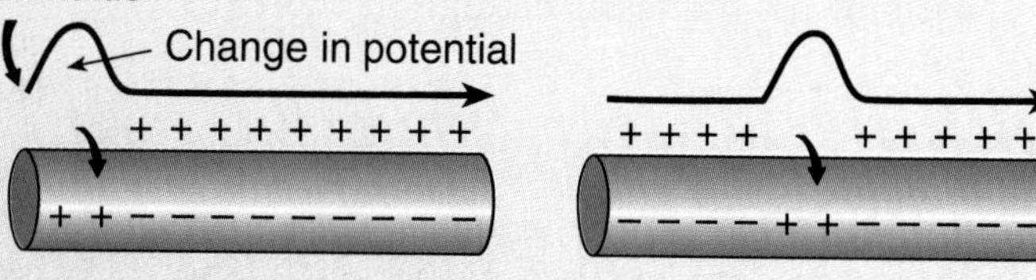

Exploring Further

1. Why do you think the ion ratio that is important is Na^+/K^+ rather than, say, Na^+/Mg^{2+}?
2. Find out what part of the nervous system is attacked by multiple sclerosis.

24.2 CONCEPT REVIEW

27. Explain how an indicator works.

28. Could methyl orange be used as an indicator in a titration of an acid with a base that produced a yellow colored salt? Explain, using Appendix Table A-11.

29. Why is it important, when doing a titration, to choose an indicator that changes color at a pH close to that of the expected endpoint?

30. **Apply** What could a student do (besides starting over) if he or she added too much standard solution during a titration and went past the endpoint?

Chapter 24

REVIEW

Summary

24.1 Water Equilibria

1. The solubility product constant, K_{sp}, for an ionic substance is equal to the product of the concentrations of the dissociated ions found in solution.
2. The ion product constant for water, K_w, is 1.00×10^{-14} at 25°C and is equal to the product of $[H_3O^+]$ and $[OH^-]$.
3. The pH of a solution is equal to $-\log [H_3O^+]$. The pH of pure water and neutral solutions is 7. Acid solutions have a pH below 7. Basic solutions have a pH above 7.
4. Some salts hydrolyze in water to form either an acidic or a basic solution.
5. Buffer solutions are capable of absorbing moderate amounts of acid and/or base without a significant change in their pH value.

24.2 Titration

6. Indicators are weak organic acids and bases whose colors differ from their conjugate bases and acids. They may be used to indicate the pH of a solution.
7. Titration is a laboratory procedure used to find the concentration of a substance in solution. In titration, a solution of a substance of unknown concentration is reacted with a standard solution of another substance.

Key Terms

solubility product constant
ion product constant of water
pH scale
pH
hydrolysis
buffer system
pH meter
indicator
titration
standard solution

Review and Practice

31. Calcium carbonate is only slightly soluble in water.
    - **a.** Write an equilibrium equation for the dissociation of $CaCO_3$ in water.
    - **b.** Write expressions for K_{eq} and K_{sp} for the equilibrium in a saturated solution of $CaCO_3$.
32. The K_{sp} of $CdCO_3$ is 2.5×10^{-14}. What is the concentration of Cd^{2+} ions in a saturated aqueous solution at equilibrium?
33. The concentration of sulfide ions in a saturated solution of silver sulfide is $3.4 \times 10^{-17}M$. Calculate the K_{sp} for silver sulfide.
34. What is the value for the ion product constant for water?
35. What is the hydronium ion concentration of a solution in which the hydroxide ion concentration is $2.77 \times 10^{-10}M$? What is the pH of the solution?
36. Determine the hydroxide ion concentration in a solution whose hydronium ion concentration is $4.49 \times 10^{-9}M$. What is the pH of the solution?
37. What is the pH of 0.0001*M* NaOH?
38. What is the pH of freshly pressed apple juice, if the H_3O^+ concentration is $1.74 \times 10^{-4}M$?
39. What kinds of compounds function as indicators?
40. What are two limitations of indicators?
41. What volume of 0.196*M* LiOH is required to neutralize 25.0 cm³ of 0.413*M* HBr?

$$LiOH + HBr \rightarrow LiBr + H_2O$$

42. If 75.0 cm³ of 0.823*M* $HClO_4$ require 95.5 cm³ of $Ba(OH)_2$ for complete neutralization, what is the concentration of the $Ba(OH)_2$ solution?

$$Ba(OH)_2 + 2HClO_4 \rightarrow Ba(ClO_4)_2 + 2H_2O$$

43. What is the concentration of a solution of NaOH if 21.2 cm^3 of a 0.0800*M* solution of HCl are needed to neutralize 25.0 cm^3 of the base?

44. If 86.2 cm^3 of 0.765*M* sodium hydroxide neutralize 50.0 cm^3 of hydrochloric acid solution, what is the concentration of the acid?

45. If 40.8 cm^3 of 0.106*M* sulfuric acid neutralize 50.0 cm^3 of potassium hydroxide solution, find the concentration of the base.

46. Find the hydroxide ion concentration of a solution with a pOH of 6.13.

47. Using Figure 24.6, calculate pOH for each of the following.
   a. stomach contents **d.** saliva
   b. beer **e.** eggs
   c. pumpkin pulp **f.** bile

48. Calculate the $[OH^-]$ for lemon juice, pH = 2.3, and baking soda solution, pH = 8.5.

49. The K_{sp} of $Mg(OH)_2$ is 1.20×10^{-11}. What is the pH of a saturated solution of magnesium hydroxide at equilibrium?

50. What is the pH of a 0.300*M* solution of HCN in which the CN^- concentration has been adjusted to 0.001 00*M*?

51. What is the pH of a solution that is 0.0100*M* in chloroacetic acid, $CH_2ClCOOH$, and 0.001 50*M* in $NaCH_2ClCOO$?

Concept Mastery

52. Describe the mechanism by which the addition of KBr to a saturated solution of AgBr reduces the concentration of silver ions.

53. Describe the two types of salts that produce a neutral solution when dissolved in water.

54. Predict whether each of the following salts in solution would form an acidic, a basic, or a neutral solution. Write the ionic equation showing the interaction with water for those that do not form neutral solutions.
   a. NaCl **c.** $AlBr_3$
   b. K_2CO_3 **d.** $Ca(NO_3)_2$

55. How does a buffer system function? What are the usual components of a buffer system?

56. What is the relationship between the common ion effect and buffer solutions?

57. What is a titration? Why is a standard solution necessary for a titration?

58. When the pH changes by 3 units, for example, from 2 to 5, by what factor does the $[H_3O^+]$ change?

59. How does the pOH scale differ from the pH scale?

60. The following equilibria in solution form a buffering system.

$$CH_3COOH + H_2O \rightleftarrows CH_3COO^- + H_3O^+$$
$$CH_3COO^- + H_2O \rightleftarrows CH_3COOH + OH^-$$

   a. Describe what happens when a small amount of a base is added to the system.
   b. Describe what happens when a small amount of an acid is added to the system.
   c. Why is it necessary to have an acetate salt in solution along with the acetic acid in order to form an effective buffering system?

61. Sketch the general shape you would expect for the titration graph of each of the following acid-base pairs.
   a. KOH and HCl
   b. HBr and NH_3
   c. $Be(OH)_2$ and H_2SO_4
   d. HCN and LiOH

62. If 0.050 mol Na_2SO_4 is added to 500.0 cm^3 of a saturated solution of $BaSO_4$, what will be the concentration of Ba^{2+}? K_{sp} for $BaSO_4$ is 1.10×10^{-10}.

63. The K_{sp} of MnS is 2.50×10^{-13}. What concentration of sulfide ion is needed in a $0.100M$ solution of $Mn(NO_3)_2$ to begin precipitation of MnS?

Application

64. A chemist wishes to precipitate as much Ag^+ as possible in the form of AgCl from a solution of $AgNO_3$. Suggest a method to accomplish this goal.

65. Soaps are salts formed by neutralization of fatty acids, which are weak acids, by NaOH or KOH. Will a soap solution be acidic, neutral, or basic?

66. A student is titrating 25.0 cm^3 of a solution of NaOH with a $0.247M$ solution of HCl. Accidentally, the student adds too much HCl. The amount of HCl solution added is 34.38 cm^3. Using a buret filled with the unknown NaOH solution, the student finds that the addition of 4.77 cm^3 of base results in an endpoint. What is the concentration of the NaOH?

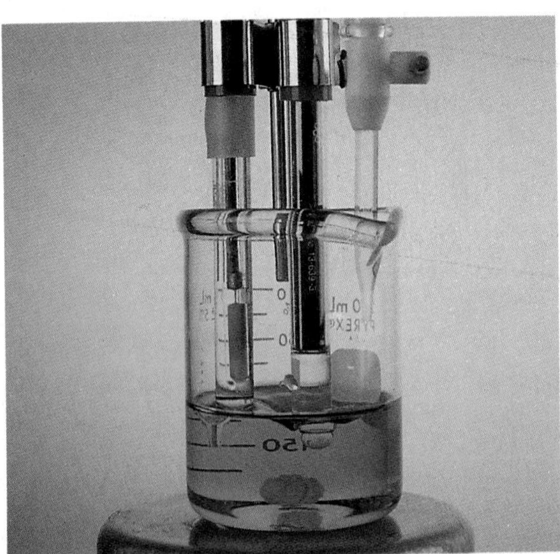

67. Using Appendix Table A-11, suggest an effective indicator for each of the titrations whose endpoints occur at the following pH values.
a. 9.9 **b.** 4.1 **c.** 6.3

68. The pH of a sample of water from a swimming pool was measured. The water turned methyl orange to the color yellow and propyl red to the color pink. What are the maximum and minimum pH values possible for the water sample? See Appendix Table A-11.

69. In each of the following, if you were carrying out a titration with the pairs of aqueous reactants, would you expect an endpoint at an acidic, basic, or neutral pH?
a. NH_3 and HCl
b. C_6H_5COOH and KOH
c. HCN and $Ca(OH)_2$
d. TlOH and HBr
e. $NaHSO_3$ and NaOH

70. Hemoglobin is the oxygen carrier in the blood. The following equation shows the reaction of hemoglobin with oxygen. Hb is hemoglobin and HbO_2 represents oxygenated hemoglobin.

$$HbH^+(aq) + H_2O + O_2(aq) \rightleftarrows HbO_2(aq) + H_3O^+(aq)$$

a. As blood passes through the capillaries of the lungs, the $[O_2]$ is increased. How does this affect the equilibrium of the reaction?
b. If acidic metabolic products increase in the blood, what happens to the pH of the blood?
c. How would this change in pH affect the ability of the blood to transport oxygen?

Critical Thinking

71. A solution contains a mixture of zinc, copper, and cadmium ions in equal concentrations. A solution containing sulfide ions is added slowly. Arrange the metal ions in the order in which they will begin to precipitate as sulfides.

72. Without doing any calculations, state whether the pH of an HCN solution will be greater than or less than that of an

equimolar HBrO solution. Explain your answer. Refer to Appendix Table A-9.

73. If a solution of NaCl is added slowly to a solution that is 1.00*M* in Ag^+ and 1.00*M* in Pb^{2+}, which salt, AgCl or $PbCl_2$, will precipitate first? See Appendix Table A-10. Could you use this method to effectively separate lead and silver as their chlorides? Explain.

74. You need a standard solution of a base in order to titrate an acid. However, all you have on hand is a standard solution of a different acid and a bottle of the crystalline base. Devise a method to prepare a standard base solution.

75. If 0.1 cm^3 of 1.0*M* HCl is added to 1 dm^3 of pure water, the pH changes from 7 to 4. If the same amount of HCl is added to 1 dm^3 of seawater, the pH changes only to 6.4. Suggest an explanation for this phenomenon.

76. Examine the titration curves in Figure 24.10 on page 617. For which titrations would the choice of an indicator be most critical? Explain why it is critical.

Readings

Charola, A. Elena. "Acid Rain Effects on Stone Monuments." *Journal of Chemical Education* 64, no. 5 (May 1987): 436-437.

Corcoran, Elizabeth. "Cleaning Up Coal." *Scientific American* 264, no. 5 (May 1991): 106-116.

Starkey, Ronald, et al. "Who Knows the K_a Values of Water and the Hydronium Ion?" *Journal of Chemical Education* 63, no. 6 (June 1986): 473-474.

Cumulative Review

1. What is the molarity of 1.00 dm^3 of a solution containing 4.66 g $Hg(CN)_2$?

2. What is the molality of a solution made by dissolving 0.944 g K_3AsO_4 in 10.0 g water?

3. What is the mole fraction of solute in a solution made by dissolving 1.00×10^2 g $Rb_2C_4H_4O_6$ in 500.0 g H_2O?

4. What is the vapor pressure, at 30°C, of a solution of 39.5 g $C_6H_{12}O_6$ in 1.00×10^2 g H_2O?

5. What are the freezing and boiling points of a solution of 42.2 grams of $C_6H_{12}O_6$ in 2.00×10^2 grams H_2O?

6. Calculate the molecular mass of a substance if 22.2 g of the substance dissolved in 2.50×10^2 g H_2O lowers the freezing point to −1.83°C.

7. Write the rate expression for the following reaction, assuming it is a single-step reaction.

$$PCl_3 + Cl_2 \rightarrow PCl_5$$

8. Write the equilibrium constant expression for the following reaction.

$$HCO_3^- + OH^- \rightleftarrows H_2O + CO_3^{2-}$$

9. In the following equation, label the acid, base, conjugate acid, and conjugate base.

$$NH_3 + HClO \rightarrow NH_4^+ + ClO^-$$

10. Using Appendix Table A-9, calculate the percent of HCOOH molecules that are ionized in a 0.10*M* HCOOH solution, then in a 0.01*M* solution. Explain why your answers support the results that would be expected from Le Chatelier's principle.

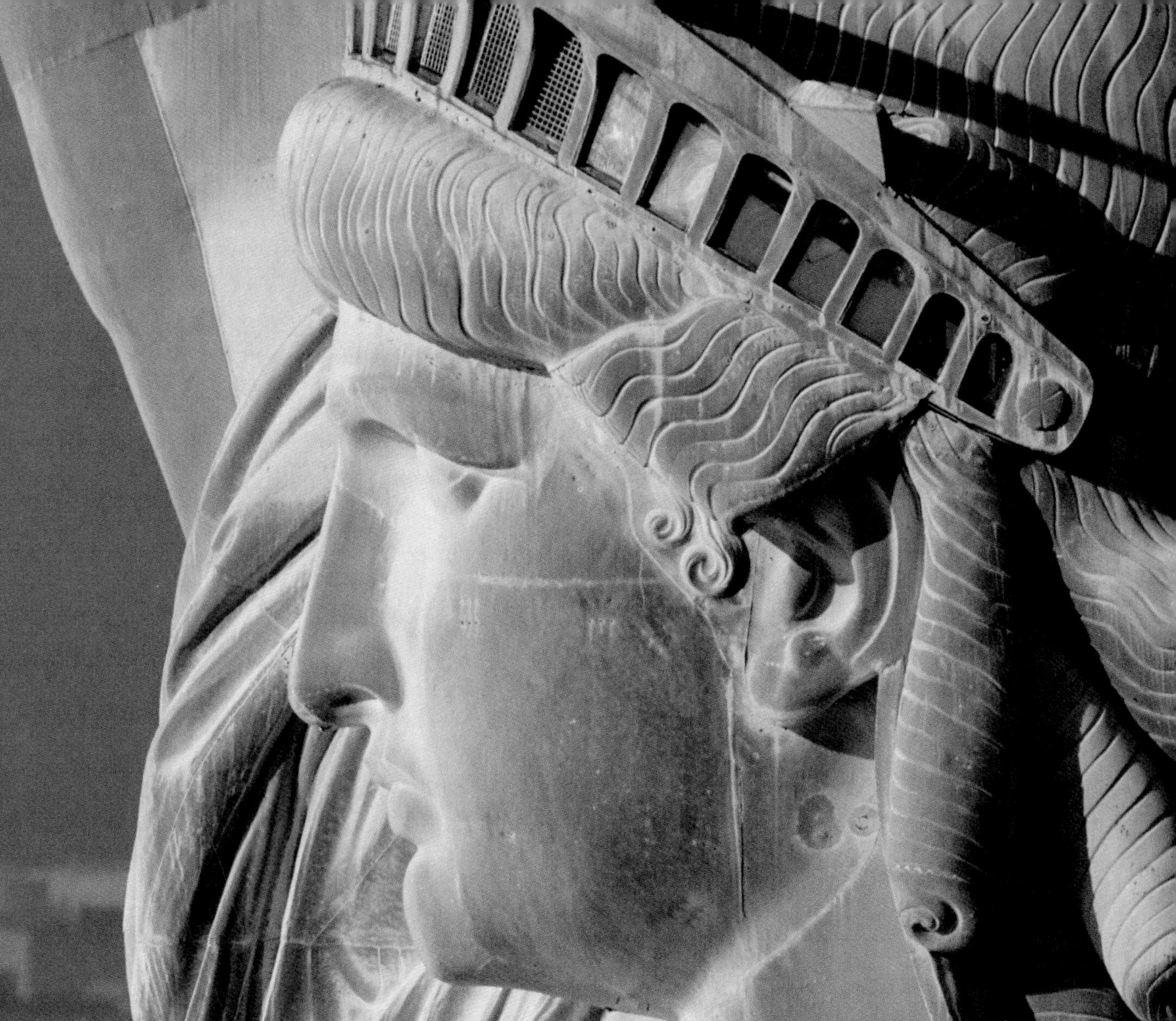

CHAPTER Preview

CONTENTS

Oxidation-Reduction

CHAPTER 25

The Statue of Liberty has become a symbol of freedom. Manufactured from copper sheets, it was erected in 1886 in New York harbor, where the light in the torch was considered an aid to navigation. More importantly, the statue became one of the first sights seen by immigrants seeking a better life in the United States and came to symbolize freedom and friendship.

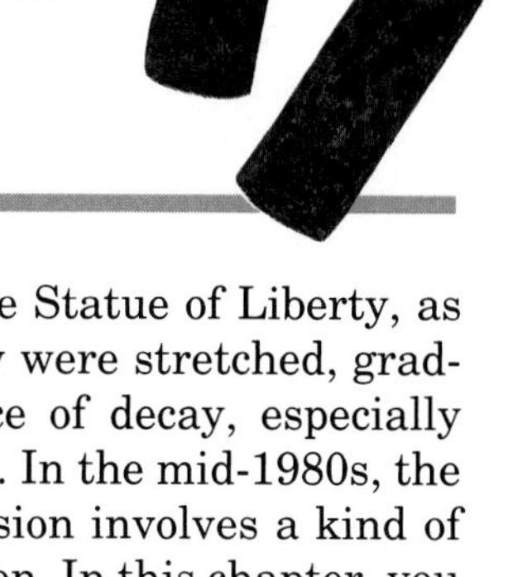

Over the decades, the copper sheets of the Statue of Liberty, as well as the steel skeleton over which they were stretched, gradually corroded. Inspectors found evidence of decay, especially where rivets penetrated the copper sheets. In the mid-1980s, the statue was restored. The process of corrosion involves a kind of chemical reaction called oxidation-reduction. In this chapter, you will learn about these reactions and practice balancing the equations that represent them.

EXTENSIONS AND CONNECTIONS

OBJECTIVES

- compare the process of oxidation with the process of reduction.
- explain what constitutes an oxidizing agent and a reducing agent.
- describe how to assign oxidation numbers to atoms in compounds.
- state how to identify oxidation-reduction reactions.

25.1 Oxidation and Reduction Processes

When you strike a match or light the gas in a gas stove, you might say that you are burning fuel. The usual meaning of "burning" is the combining of fuel with oxygen in air. But how does the fuel in the solid rocket of a space shuttle burn? After the first few seconds at lift-off, the shuttle is above the oxygen-containing atmosphere. How can the fuel continue to burn? In this section we will look at the type of reaction that allows burning to occur even in the absence of atmospheric oxygen.

Types of Reactions

There are various ways to classify chemical reactions. In one method, reactions are classified into two different types. In the first type, ions or molecules react with no apparent change in the electronic structure of the particles. In the second type, ions or atoms undergo changes of electronic structure. Electrons may be transferred from one particle to another, or the way atoms share electrons may be changed. This second type of reaction, involving electron changes, is called an **oxidation-reduction reaction** or redox reaction. The terms oxidation-reduction and redox are used interchangeably.

The distinction between redox reactions and all other reactions is an important one. Table 25.1 compares a double displacement reaction with a simple redox reaction. Notice in the double displacement reaction that all the substances taking part are ionic. There is no change in the charges on any of the ions. Therefore, there is no transfer of electrons between atoms. Notice that a precipitate ($BaSO_4$) is formed. The production of a precipitate is nearly always a result of a non-redox reaction. Most acid-base reactions are also the non-redox type.

The second reaction shown in Table 25.1 is quite different. On the left side of the equation, the chlorine molecule has no charge.

Figure 25.1 "Cold light" is produced by oxidation-reduction in this laboratory reaction involving luminol (left) and in fireflies (right). Fireflies use light flashes to attract mates.

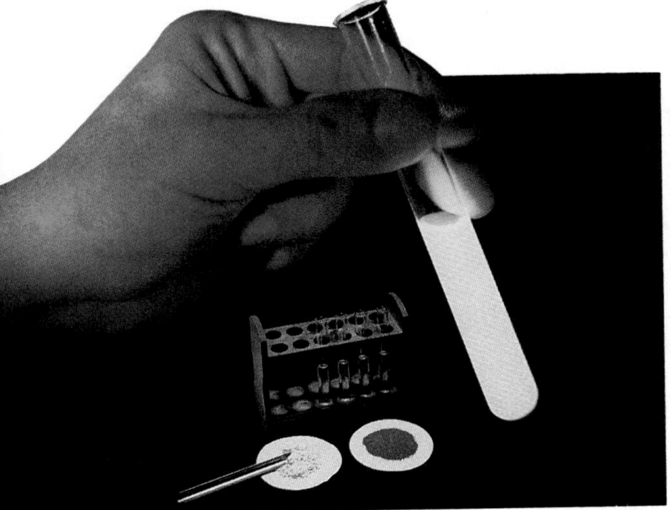

On the right side of the equation, two Cl^- ions are present. Each chlorine atom has gained one electron to form a Cl^- ion. The Br^- ions on the left side of the equation have each lost one electron to form a bromine molecule. Therefore, two electrons were transferred from bromide ions to chlorine atoms. This transfer is an example of a redox reaction.

Table 25.1

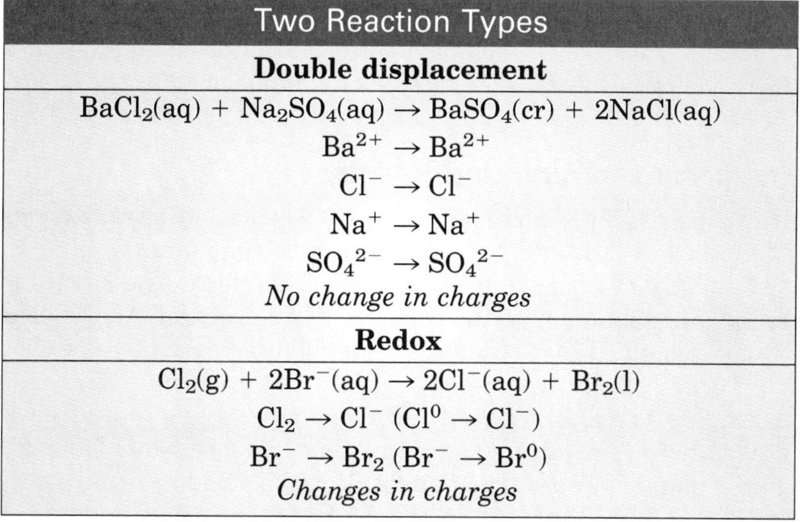

Two Reaction Types
Double displacement
$BaCl_2(aq) + Na_2SO_4(aq) \rightarrow BaSO_4(cr) + 2NaCl(aq)$ $Ba^{2+} \rightarrow Ba^{2+}$ $Cl^- \rightarrow Cl^-$ $Na^+ \rightarrow Na^+$ $SO_4^{2-} \rightarrow SO_4^{2-}$ *No change in charges*
Redox
$Cl_2(g) + 2Br^-(aq) \rightarrow 2Cl^-(aq) + Br_2(l)$ $Cl_2 \rightarrow Cl^-$ $(Cl^0 \rightarrow Cl^-)$ $Br^- \rightarrow Br_2$ $(Br^- \rightarrow Br^0)$ *Changes in charges*

Many important chemical changes involve oxidation-reduction reactions. The corrosion of metals, most forms of energy production, and many life processes take place by the transfer of electrons. In the rest of this section we will see many other ways in which oxidation-reduction reactions are important.

Oxidation and Reduction

The term *oxidation* was first applied to the combining of oxygen with other elements. Iron rusts and carbon burns. Both are oxidation reactions. Chemists later recognized that other nonmetallic elements unite with substances in a manner similar to that of

Figure 25.2 The rusting of iron is an oxidation reaction.

oxygen. Hydrogen, antimony, and sodium all burn in chlorine, and iron burns in fluorine. In each case, the element that is burning is losing electrons. Because these reactions are similar, **oxidation** is defined as the loss of electrons from an atom or ion.

The term *reduction* was originally limited to the type of reaction in which oxide ores were "reduced" to their metals. In many of these reactions, oxygen is removed, and the free element is produced. However, the free element can be produced in similar reactions from compounds that are not oxides. The similarity led chemists to formulate a more generalized definition of reduction. **Reduction** is the gain of electrons by atoms or ions.

Oxidizing and Reducing Agents

In an oxidation-reduction reaction, electrons are transferred. Electrons are lost and gained at the same time and the number lost must equal the number gained. Therefore, oxidation and reduction must occur at the same time in a reaction. All the electrons exchanged in a redox reaction must be accounted for.

The substance in a redox reaction that gives up electrons is called the **reducing agent.** The reducing agent contains the atoms that are oxidized (the atoms that lose electrons). Zinc is a good example of a reducing agent. It can be oxidized to the zinc ion, Zn^{2+}. The substance in the redox reaction that gains electrons is called the **oxidizing agent.** It contains the atoms that are reduced (the atoms that gain electrons). Chlorine, Cl_2, is a good example of an oxidizing agent. It can be reduced to the chloride ion, Cl^-.

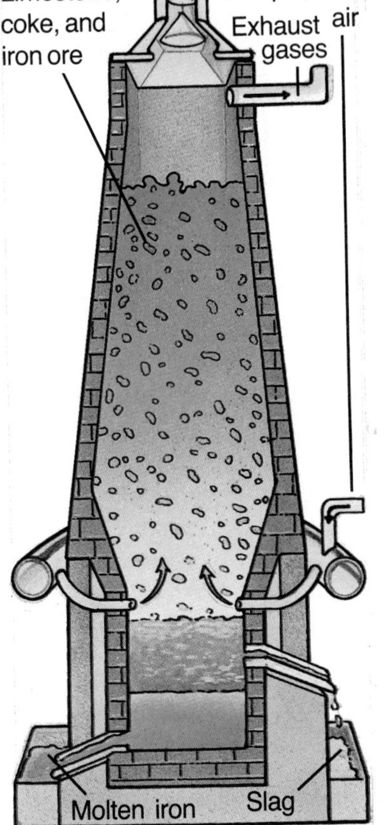

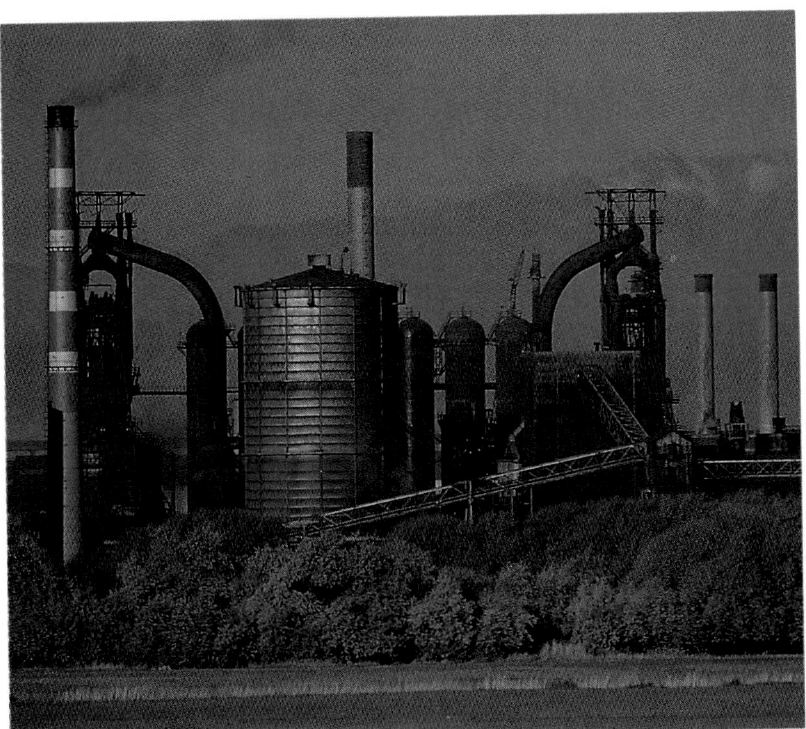

Figure 25.3 A blast furnace is used in the reduction of iron ore compounds to elemental iron. The molten iron is then poured out of the bottom of the furnace Shown in the photograph are two blast furnaces with the associated equipment needed to operate them.

$$\underset{\substack{\text{substance}\\\text{oxidized;}\\\text{reducing}\\\text{agent}}}{Zn(cr)} + \underset{\substack{\text{substance}\\\text{reduced;}\\\text{oxidizing}\\\text{agent}}}{Cl_2(aq)} \rightarrow Zn^{2+}(aq) + 2Cl^-(aq)$$

To summarize, oxidation is the loss of electrons and reduction is the gain of electrons. The oxidizing agent is the substance containing the element that is reduced. The reducing agent contains the element that is oxidized. In the reaction

$$Fe_2O_3(cr) + 3CO(g) \rightarrow 2Fe(l) + 3CO_2(g)$$

iron is reduced, so Fe_2O_3 is the oxidizing agent. Carbon is oxidized, so CO is the reducing agent.

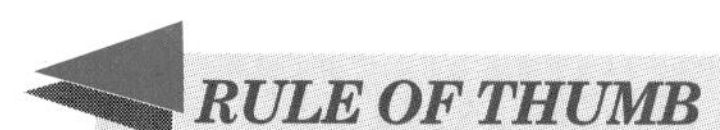

If a substance gives up electrons readily, it is said to be a strong reducing agent. Its oxidized form, however, is normally a poor oxidizing agent. If a substance gains electrons readily, it is said to be a strong oxidizing agent. Its reduced form is a weak reducing agent.

Bridge to METALLURGY

Ore to Metal in a Blast Furnace

The production of iron in a blast furnace involves a series of oxidation-reduction reactions. An enclosed conveyer belt carries iron ore (Fe_2O_3), coke (C), and limestone ($CaCO_3$) to the top of the furnace. Tall cylinders called stoves extract heat from the hot exhaust gases. This energy is then used to raise the temperature of the incoming air. A blast of this hot air (about 900°C) is admitted near the bottom of the furnace. It causes the coke to burn and produce temperatures at the bottom of the furnace of about 2000°C.

$$C(cr) + O_2(g) \rightarrow CO_2(g) + \text{heat}$$

$$CO_2(g) + C(cr) \rightarrow 2CO(g)$$

The carbon monoxide produced in this way is used in the reduction of the Fe_2O_3. The main reaction in the blast furnace is

$$Fe_2O_3(cr) + 3CO(g) \rightarrow 2Fe(l) + 3CO_2(g)$$

The CO_2 gas produced by this reaction, and that produced by the decomposition of limestone at a temperature of about 1870°C,

$$CaCO_3(cr) \rightarrow CaO(cr) + CO_2(g)$$

is then used in another redox reaction. The carbon dioxide gas reacts with the unburned coke to produce more carbon monoxide. The principal contaminant of the ore is dirt mostly in the form of SiO_2. Lime, CaO, and the silicon dioxide, SiO_2, react to form slag. The molten iron and the slag are tapped off separately near the bottom of the furnace. The hot exhaust gases, containing some combustible substances, are burned in the stoves.

Exploring Further

1. What is the reducing agent in the conversion of iron ore to iron?
2. How is coke produced?

Table 25.2

Oxidation-Reduction Reactions						
Reaction	Element Oxidized	Electrons Lost per Atom	Element Reduced	Electrons Gained per Atom	Oxidizing Agent	Reducing Agent
$4Fe + 3O_2 \rightarrow 2Fe_2O_3$	Fe	3	O	2	O_2	Fe
$C + O_2 \rightarrow CO_2$	C	4	O	2	O_2	C
$H_2 + Cl_2 \rightarrow 2HCl$	H	1	Cl	1	Cl_2	H_2
$Fe_2O_3 + 3CO \rightarrow 2Fe + 3CO_2$	C	2	Fe	3	Fe_2O_3	CO
$CuO + H_2 \rightarrow Cu + H_2O$	H	1	Cu	2	CuO	H_2
$3Cu^{2+} + 2Fe \rightarrow 3Cu + 2Fe^{3+}$	Fe	3	Cu	2	Cu^{2+}	Fe

Oxidation Numbers

How is it possible to determine whether a redox reaction has taken place and which substances have been oxidized or reduced? We do so by checking to see whether any electrons have been transferred between atoms during a chemical reaction. This checking is done by looking at the oxidation numbers of the atoms taking part in the reaction. Any change in the oxidation number of a particular atom indicates that an oxidation-reduction reaction has taken place.

We have already seen in Chapters 7 and 10 how to determine the oxidation numbers of monatomic ions. The oxidation number of a monatomic ion is equal to the charge on the ion. In the next section we will discuss other rules for assigning oxidation numbers to species other than monatomic ions.

Assigning Oxidation Numbers

The oxidation number of an atom is the apparent charge that atom has when it is assigned electrons, in accordance with certain rules. To determine the apparent charge, you may find it helpful to draw an electron dot structure. However, the structure you draw will not give you the complete answer; it only helps you to visualize an atom, ion, or molecule.

Figure 25.4 Copper exists in two common oxidation states, 1+ and 2+, as shown by these two hydrated chlorides. Copper(I) chloride, or cuprous chloride, is green. Copper(II) chloride, or cupric chloride, is greenish-blue.

Suppose you want to know the oxidation number of the sodium atom. The electron dot symbol for sodium is Na•. The atom is not charged. The number of electrons is equal to the number of protons. The apparent charge of the sodium atom is 0; therefore, its oxidation number is 0.

The sodium ion, however, is indicated in this manner, Na^+, showing that there is one more proton than electron. Since its apparent charge is 1+, its oxidation number is 1+.

Certain conventions have also been established for assigning oxidation numbers to atoms in molecules and polyatomic ions. When electrons are shared, they are assigned to the more electronegative element. If two identical atoms share electrons, the electrons are split between the two atoms. For example, consider free chlorine, Cl_2, whose electronic structure is

$$:\ddot{\underset{\cdot\cdot}{Cl}}\!:\!\underset{\times\times}{\overset{\times\times}{Cl}}\!\times$$

Since each chlorine atom has the same electronegativity, the two chlorine atoms share the electrons equally. Thus, each is assigned seven electrons in the outer level, giving a net charge of 0 for each. Free chlorine, Cl_2, is given an oxidation number of 0.

Chlorine in hydrogen chloride

$$H\!:\!\underset{\times\times}{\overset{\times\times}{Cl}}\!\times$$

has an oxidation number different from 0. The chlorine atom is more electronegative than the hydrogen atom. All the electrons shared are, therefore, assigned to the chlorine atom. Thus, the chlorine atom will have 18 electrons and 17 protons, and a resulting apparent charge of 1−. Its oxidation number is 1−. Hydrogen has had its electron assigned to chlorine, and will have one less electron than proton. The hydrogen atom's apparent charge will be 1+ and its oxidation state will be 1+.

Consider the electron dot diagram for sulfuric acid, H_2SO_4.

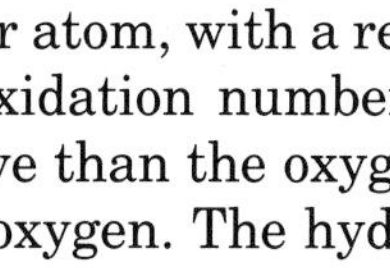

$$\begin{array}{c} O \\ H:\underset{\times\times}{\overset{\times\times}{O}}:\underset{\cdot\cdot}{\overset{\cdot\cdot}{S}}:\underset{\times\times}{\overset{\times\times}{O}}:H \\ \times\underset{\times\times}{O}\times \end{array}$$

It can be seen that the oxygen atoms share electrons with both sulfur and hydrogen atoms. Since oxygen is more electronegative than sulfur, the shared electrons are assigned to each oxygen atom. Thus, the sulfur atom is assigned six fewer electrons than it has protons. The sulfur atom, with a resulting apparent charge of 6+, is assigned an oxidation number of 6+. Each hydrogen atom, less electronegative than the oxygen atoms, will also have its electron assigned to oxygen. The hydrogen oxidation number in this compound is 1+. Each oxygen atom has assigned to it all the shared electrons from either the sulfur or hydrogen atoms, or both. This assignment gives each oxygen atom a 2− oxidation number (ten electrons, eight protons). The total of the oxidation numbers of all the atoms in a compound must be zero. In H_2SO_4, one sulfur atom has an oxidation number 6+, four oxygen atoms each have the oxidation number 2−, and two hydrogen atoms

MINUTE LAB

Alchemy: Iron to Copper

Put on an apron and goggles. Use steel wool to clean an iron nail. Dissolve 1.5 g $CuSO_4 \cdot 5H_2O$ in 50 cm^3 of distilled H_2O in a 100 cm^3 beaker. **CAUTION:** *$CuSO_4$ is both toxic and a skin irritant.* Why does the solution have a blue color and what is its relationship to the Cu^{2+} ion? Dip the end of the nail into the solution for about 15 seconds. Examine the surface of the nail when you remove it from the solution. Record your observations. Repeat for another 15 seconds and examine again. Is the coating metallic? If so, which metal is it? Is the coating plated in the usual sense? (See Chapter 26 for a discussion on electroplating.) What is the oxidation number of a metal? How did a red coating on the nail come from a blue solution?

each have 1+. Therefore, (6+) + 4 (2−) + 2 (1+) = 0, and the apparent charge of the compound is zero.

In sulfur dichloride, SCl_2, sulfur has a different oxidation state. Consider the electron dot structure for SCl_2.

```
   ××
  ×Cl×
   •×   ××
  :S ×Cl×
   ••   ××
```

The sulfur atom shares only four electrons with chlorine. Chlorine is more electronegative than sulfur. Thus the shared electrons are assigned to the chlorine. This assignment gives sulfur an oxidation state of 2+ (14 electrons, 16 protons). Each chlorine then has an oxidation state of 1− (18 electrons, 17 protons).

The oxidation number of an atom may change from compound to compound. Therefore, an electron dot structure must be made for each new compound and the oxidation number determined from this electronic structure. Drawing electron dot structures takes time, however. Fortunately, there is an easier way. General rules have been made to enable you to determine oxidation numbers more easily. These rules are summarized in Table 25.3.

Table 25.3

Oxidation Number Rules

Rule 1. *The oxidation number of any free element is 0.* This statement is true for all atomic and molecular structures: monatomic, diatomic, or polyatomic.

Rule 2. *The oxidation number of a monatomic ion* (Na^+, Ca^{2+}, Al^{3+}, Cl^-) *is equal to the charge on the ion.* Some atoms have several possible oxidation numbers. For example, iron can be either 2+ or 3+; tin, 2+ or 4+.

Rule 3. *The oxidation number of each hydrogen atom in most of its compounds is 1+.* There are some exceptions. In compounds such as lithium hydride (LiH), hydrogen, being the more electronegative atom, has an oxidation number of 1−.

Rule 4. *The oxidation number of each oxygen atom in most of its compounds is 2−* (H_2O). In peroxides, (Na_2O_2, H_2O_2), each oxygen is assigned an oxidation number of 1−.

Rule 5. *The sum of the oxidation numbers of all the atoms in a particle must equal the apparent charge of that particle.* In $SO_4{}^{2-}$, sulfur has an oxidation number of 6+ and each oxygen has an oxidation number of 2−, for a net charge of 2−.

Rule 6. *In compounds, the elements of Group 1 (IA), Group 2 (IIA), and aluminum have positive oxidation numbers of 1+, 2+, and 3+, respectively.*

CHEMISTS AND THEIR WORK

Jane Haldimond Marcet
(1769-1845)

Science writer Jane Haldimond Marcet had a profound effect on the chemists of her day and those that followed. Her book, *Conversations in Chemistry,* was first published in the early 1800s and was continually revised through sixteen editions to reflect new scientific breakthroughs. By the 1860s, nearly 200 000 copies had been sold in England and the United States. The book, written in dialogue form, provided experiments to illustrate chemical concepts—a method that today would be labeled a discovery approach. Marcet's work provided the chemical foundations for a young bookbindery clerk named Michael Faraday. Later, as a famous scientist, Faraday referred to *Conversations in Chemistry* as "an anchor in chemical knowledge."

SAMPLE PROBLEM

Oxidation Numbers

What are the oxidation numbers of the elements in Na_2SO_4?

Solving Process:

According to Rule 6, the oxidation number of sodium is 1+. According to Rule 4, the oxidation number of oxygen is 2−. According to Rule 5, the total of all oxidation numbers in the formula unit is 0. Letting x = oxidation number of sulfur, we have

$$2(1+) + x + 4(2-) = 0$$
$$x = 6+$$

PROBLEM SOLVING HINT

When you solve an equation, be sure to add the terms to *both* sides so the equation remains balanced.

PRACTICE PROBLEMS

1. Give the oxidation number for each indicated atom.
   - **a.** S in Na_2SO_3
   - **b.** Mn in $KMnO_4$
   - **c.** N in $Ca(NO_3)_2$
   - **d.** C in Na_2CO_3
   - **e.** N in NO_2
   - **f.** S in HSO_4^-
   - **g.** S in $H_2S_2O_7$
   - **h.** S in Al_2S_3
   - **i.** Mn in $MnCl_2$
   - **j.** C in $C_{12}H_{22}O_{11}$

Identifying Oxidation-Reduction Reactions

Oxidation numbers can be used to determine whether oxidation and reduction (electron transfer) occur in a specific reaction. Even the simplest reaction may be a redox reaction. Let us see how it is possible to determine whether a reaction is actually a redox reaction.

The direct combination of sodium and chlorine to produce sodium chloride is a simple example.

$$2Na(cr) + Cl_2(g) \rightarrow 2NaCl(cr)$$

As a reactant, each sodium atom has an oxidation number of 0. In the product, the oxidation number of each sodium atom is 1+. Similarly, each chlorine atom as a reactant has an oxidation number of 0. As a product, each chlorine atom has an oxidation number of 1−. Since a change of oxidation number has occurred, an oxidation-reduction reaction has taken place.

$$2Na(cr) + Cl_2(g) \rightarrow 2Na^+Cl^-(cr)$$

The change in oxidation number can result only from a shift of electrons between atoms. This shift of electrons alters the apparent charge, that is, the oxidation number.

A gain of electrons means the substance is reduced. It also means that the oxidation number is algebraically lowered. In contrast, a loss of electrons means the substance is oxidized. When an atom is oxidized, its oxidation number increases.

Figure 25.5 Sodium metal reacts vigorously in an atmosphere of chlorine. Sodium is oxidized; chlorine is reduced.

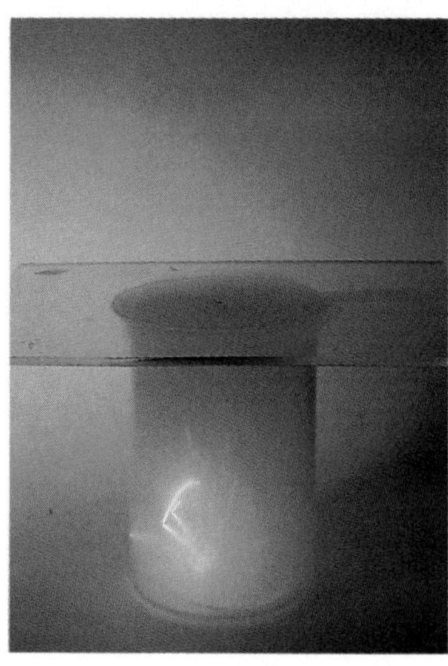

The equation for a reaction can be used to determine whether the reaction is a redox reaction. It can also be used to find the substance oxidized, the substance reduced, and the oxidizing and reducing agents. Since the oxidation number of sodium in the equation

$$2Na(cr) + Cl_2(g) \rightarrow 2NaCl(cr)$$

changed from 0 to 1+, sodium is oxidized, making sodium the reducing agent. A reducing agent always loses electrons to another substance and is, therefore, always oxidized as it reduces the other substance.

Solving problems involving reduced or oxidized elements and reducing and oxidizing agents is much easier if we write the equation in net ionic form. Redox reactions are also easier to balance in net ionic form. Before proceeding, review the rules in Chapter 23, pages 588–589, for writing net ionic equations.

PROBLEM SOLVING HINT

Use this mnemonic device to help you remember the definitions of oxidation and reduction: LEO goes GER. Lose Electrons Oxidize; Gain Electrons Reduce.

SAMPLE PROBLEM

Identifying Oxidizing and Reducing Agents

Which substance is the reducing agent and which is the oxidizing agent in the following reaction?

$$16H^+(aq) + 2MnO_4^-(aq) + 5C_2O_4^{2-}(aq) \rightarrow$$
$$2Mn^{2+}(aq) + 8H_2O(l) + 10CO_2(g)$$

Solving Process:

In this reaction, manganese is converted from 7+ to 2+. This conversion is accomplished by the gain of five electrons. Therefore, MnO_4^- is the oxidizing agent.

$$\underset{\text{oxidizing agent}}{MnO_4^-} \rightarrow Mn^{2+} \quad \text{reduction}$$

The carbon loses one electron, going from 3+ to 4+, and is converted to carbon dioxide. Therefore, carbon is oxidized, and $C_2O_4^{2-}$ is the reducing agent.

$$\underset{\text{reducing agent}}{C_2O_4^{2-}} \rightarrow CO_2 \quad \text{oxidation}$$

Figure 25.6 Liquid oxygen and liquid hydrogen mix in a space shuttle's main tank. They react to produce some of the energy that launches the space vehicle.

PRACTICE PROBLEMS

2. Identify the oxidizing and reducing agents in each of the following reactions.
   a. $Mg(cr) + Cu(NO_3)_2(aq) \rightarrow Mg(NO_3)_2(aq) + Cu(cr)$
   b. $SnCl_4(l) + Fe(cr) \rightarrow SnCl_2(cr) + FeCl_2(cr)$
   c. $2Na(cr) + S(cr) \rightarrow Na_2S(cr)$
3. What element is oxidized in each of the reactions in Problem 2? How many electrons does each of these elements lose?

SAMPLE PROBLEM

Redox

For the following reaction, tell what is oxidized, what is reduced, and identify the oxidizing and reducing agents.

$$3As_2S_3(cr) + 10NO_3^-(aq) + 10H^+(aq) + 4H_2O(l) \rightarrow 6H_3AsO_4(aq) + 9S(cr) + 10NO(g)$$

Solving Process:

Nitrogen is reduced ($5+ \rightarrow 2+$) and sulfur is oxidized ($2- \rightarrow 0$). The nitrate ion (NO_3^-) is the oxidizing agent because it contains nitrogen, which is reduced. The arsenic(III) sulfide (As_2S_3) is the reducing agent; it contains sulfur, which is oxidized. The oxidation number of arsenic does not change.

PRACTICE PROBLEMS

Some of the following unbalanced reactions are oxidation-reduction reactions, and some are not. In each case: **(a)** *Is the reaction redox?* **(b)** *If yes, name the element reduced, the element oxidized, the oxidizing agent, and the reducing agent.*

4. $H_2(g) + N_2(g) \rightarrow NH_3(g)$

5. $C(cr) + H_2O(g) \rightarrow CO(g) + H_2(g)$

6. $AgNO_3(aq) + FeCl_3(aq) \rightarrow AgCl(cr) + Fe(NO_3)_3(aq)$

7. $H_2CO_3(aq) \rightarrow H_2O(l) + CO_2(g)$

8. $MgSO_4(aq) + Ca(OH)_2(aq) \rightarrow Mg(OH)_2(aq) + CaSO_4(cr)$

9. $H_2O_2(aq) + PbS(cr) \rightarrow PbSO_4(cr) + H_2O(l)$

10. $KCl(cr) + H_2SO_4(aq) \rightarrow KHSO_4(aq) + HCl(g)$

11. $HNO_3(aq) + H_3PO_3(aq) \rightarrow NO(g) + H_3PO_4(aq) + H_2O(l)$

12. $HNO_3(aq) + I_2(cr) \rightarrow HIO_3(aq) + NO_2(g) + H_2O(l)$

13. $Na_2S(aq) + AgNO_3(aq) \rightarrow Ag_2S(cr) + NaNO_3(aq)$

14. $H^+(aq) + NO_3^-(aq) + Fe^{2+}(aq) \rightarrow H_2O(l) + NO(g) + Fe^{3+}(aq)$

15. $FeBr_2(aq) + Br_2(l) \rightarrow FeBr_3(aq)$

16. $S_2O_3^{2-}(aq) + I_2(cr) \rightarrow S_4O_6^{2-}(aq) + I^-(aq)$

17. $H_2O_2(aq) + MnO_4^-(aq) \rightarrow O_2(g) + Mn^{2+}(aq)$

25.1 CONCEPT REVIEW

18. How do oxidation and reduction differ?

19. What is an oxidizing agent? a reducing agent?

20. How are shared electrons assigned to the atoms in molecules and polyatomic ions?

21. How can you identify a redox reaction?

22. Apply Some of the energy needed to launch a space shuttle is provided by a redox reaction. Liquid oxygen and liquid hydrogen react, when combined, to form water vapor. Write an equation for this reaction. Circle the oxidizing agent and underline the reducing agent.

Making Light of Toxic Waste

Of the major areas of toxic-waste cleanup explored over the past few years, two successful practices have relied to some degree on light. One is the cleanup of volatile chlorinated hydrocarbon wastes and the other is the cleanup of ground water contamination.

Liquid or gaseous waste that contains volatile chlorinated hydrocarbons can be converted into useful materials by a reductive photo-dechlorination (RPD) process. This process uses ultraviolet light and a reducing atmosphere to carry out the conversion. RPD uses moderate temperatues and does not produce carbon soot or compounds of high molecular mass. From hazardous compounds such as vinyl chloride, 1,1-dichloroethylene, and dichloromethane, the process has been able to produce hydrogen chloride, ethylene, acetylene, and ethane. When the value of the recovered new compounds is considered, the RPD process pays for itself.

Underground disposal of toxic chemicals, routine leaks from landfills, and rusted storage tanks have all contributed to groundwater contamination. These hazardous substances in groundwater can also be broken down by light. Research scientists have added a photocatalyst to the water and then pumped it through long narrow glass tubes exposed to sunlight. The sunlight activates the catalyst, which breaks down the toxic materials into nontoxic substances. The process is able to break down the contamination from the industrial cleaner trichloroethylene by using titanium dioxide as a catalyst. This catalyst reacts with the water to create hydroxide ions. These convert the trichloroethylene into water, carbon dioxide, and dilute hydrochloric acid. In addition to the chlorinated hydrocarbons, the researchers believe that the process will work also with benzene, dyes, and pesticides.

Thinking Critically

1. Discuss how either of the described processes could be adapted for industrial use. Do they use materials that are in plentiful supply and are reasonably priced?
2. What is a "reducing atmosphere"?
3. Consider the reaction using the titanium dioxide catalyst. Try to write an equation realizing that the catalyst only activates the reaction.
4. The use of microbes has also evolved as a promising treatment technology. Try to find some literature on a cleanup process that uses microbes.

25.2 Balancing Redox Equations

OBJECTIVES

- explain the concept of half-reactions.
- determine how to balance redox equations by the half-reaction method.

Oxidation-reduction reactions are represented by chemical equations just as any other chemical changes are. Therefore, redox equations must also be balanced. However, trying to balance a redox reaction by trial and error can be time consuming and very frustrating. Fortunately, there is a systematic way to approach this task. Remember as you learn this technique that chemical equations can be added, subtracted, or multiplied.

Balancing a Redox Equation

The half-reaction method of balancing redox equations involves separating the reaction into two half-reactions. One of these half-reactions represents the oxidation that is taking place, and the other half-reaction represents the reduction. The number of electrons gained in the process of reduction must equal the number of electrons lost in oxidation. You may add half-reaction equations involving equal numbers of electrons to obtain a balanced redox equation. The first rule to keep in mind while writing half-reactions is: *Always write the formulas of molecules and ions as they actually occur.* For example, in the reaction involving nitric acid and phosphorous acid

$$HNO_3(aq) + H_3PO_3(aq) \rightarrow NO(g) + H_3PO_4(aq) + H_2O(l)$$

there are no actual N^{5+}, N^{2+}, P^{3+}, or P^{5+} ions. The substances must be represented in the form in which they actually occur using the rules for writing net ionic equations on pages 588–589. In this reaction, the actual ions and molecules are NO_3^-, NO, H_3PO_3, and H_3PO_4. There is one exception to this rule. We will represent H_3O^+ by H^+ to simplify balancing. Always keep in mind that H^+ represents H_3O^+.

SAMPLE PROBLEM

Redox

Find the proper ions and molecules to balance the equation.

$$K_2Cr_2O_7(aq) + FeSO_4(aq) \rightarrow$$
$$Cr_2(SO_4)_3(aq) + K_2SO_4(aq) + Fe_2(SO_4)_3(aq)$$

Solving Process:

Check the oxidation number of each element before and after the reaction. Chromium and iron are the two elements changing oxidation number, Cr from 6+ to 3+ and iron from 2+ to 3+. The chromium is reduced and the iron is oxidized. All reactants and products are soluble salts and strong electrolytes, so they are shown in ionic form.

$$Cr_2O_7^{2-} + Fe^{2+} \rightarrow Cr^{3+} + Fe^{3+}$$

PRACTICE PROBLEMS

23. For each of the following unbalanced equations, choose the substances acting as oxidizing agents and reducing agents.

a. $K_2Cr_2O_7(aq) + Na_2SO_3(aq) + HCl(aq) \rightarrow Cr_2(SO_4)_3(aq) + KCl(aq) + NaCl(aq)$

b. $MnO_2(cr) + KI(aq) + HCl(aq) \rightarrow MnCl_2(aq) + KCl(aq) + I_2(cr)$

c. $MnO_2(cr) + H_2O_2(aq) + HCl(aq) \rightarrow MnCl_2(aq) + H_2O(l) + O_2(g)$

d. $K_2MnO_4(aq) + HCl(aq) \rightarrow MnCl_2(aq) + KCl(aq) + Cl_2(g)$

e. $Al(cr) + HCl(aq) \rightarrow AlCl_3(aq) + H_2(g)$

Each half-reaction and redox reaction can be balanced three ways: by electrons, by total charge, and by atoms. Any two of these three ways are sufficient to give an equation overall balance. The third method may be used as a check. In the following illustrations, electrons and atoms are used to obtain the balanced equation. Then charge balance is used as a check.

Using the Half-Reaction Method

Consider the nitric acid-phosphorous acid reaction.

$$HNO_3(aq) + H_3PO_3(aq) \rightarrow NO(g) + H_3PO_4(aq) + H_2O(l)$$

Balancing the Reduction Half-Reaction

Begin to balance the equation by writing the skeleton half-reaction for the reduction process.

$$NO_3^- \rightarrow NO$$

Nitrogen is reduced from oxidation number 5+ to 2+ and, therefore, each nitrogen atom gains three electrons. By incorporating the electrons in the skeleton equation, the half-reaction is balanced with respect to electrons.

$$NO_3^- + 3e^- \rightarrow NO$$

Figure 25.7 Lightning provides the energy for the oxidation-reduction reaction in which nitrogen gas and oxygen gas react to form nitrogen(II) oxide, NO. It is estimated that some 40 million metric tons of N_2 are oxidized to NO by lightning every year.

Now the half-reaction must be balanced with respect to atoms. Note that one nitrogen atom appears on each side of the equation, and the equation is balanced with respect to nitrogen. For oxygen, however, there are three atoms on the left and only one on the right. It is, therefore, necessary to add two oxygen atoms to the right side of the half-reaction. From the information about the reaction taking place, you know that no oxygen is generated. However, there must be some species present that contains oxygen. Here is a second important rule for balancing redox reactions: *In aqueous solutions, H^+, H_2O, and OH^- are available.* The nature of the reactants determines which will participate in the reaction. In an acid solution, H^+ and H_2O are always available; in a basic solution, OH^- and H_2O are always available. In the reaction now being considered, two acids are involved, and the available substances are H^+ and H_2O. Thus, the two oxygen atoms to be added to the right-hand side of the reduction half-reaction must be present in the form of water.

$$NO_3^- + 3e^- \rightarrow NO + 2H_2O$$

The half-reaction is now balanced with respect to nitrogen and oxygen atoms. However, with the introduction of hydrogen on the right side, four hydrogen atoms must be placed on the left. In the acidic solution, hydrogen is available in the form of hydrogen ions. Hydrogen must be added to the equation in ionic form.

$$NO_3^- + 3e^- + 4H^+ \rightarrow NO + 2H_2O$$

The half-reaction is now balanced with respect to electrons and to atoms. Let us check the balance by total charge. On the left there is one nitrate ion with a 1− charge, three electrons with 1− charges, and four hydrogen ions with 1+ charges. The sum of these charges shows a net left-hand charge of 0. On the right, nitrogen(II) oxide and water are both neutral molecules and the total charge is 0. The half-reaction is then balanced with respect to total charge.

Balancing the Oxidation Half-Reaction

The same procedure can be repeated with the oxidation half-reaction.

$$H_3PO_3 \rightarrow H_3PO_4$$

The phosphorus atom changes in oxidation state from 3+ in H_3PO_3 to 5+ in H_3PO_4 as a result of losing two electrons. The half-reaction balanced with respect to electrons is then

$$H_3PO_3 \rightarrow H_3PO_4 + 2e^-$$

The half-reaction is already balanced with respect to hydrogen and phosphorus. Adding one oxygen atom to the left side of the equation should balance it with respect to atoms. The oxygen must be added in the form of water, making the half-reaction

$$H_3PO_3 + H_2O \rightarrow H_3PO_4 + 2e^-$$

HISTORY CONNECTION

Politics and Chemistry – Do They Mix?

The writing of the first chemical equation is generally credited to Antoine Lavoisier (1743–1794), a French chemist. Lavoisier was a brilliant experimenter who proved the law of conservation of matter and interpreted the true nature of burning. He had many interests besides science, such as economics, education, social welfare, and agriculture. Lavoisier was responsible for improving conditions in prisons and hospitals. He started a model farm and experimental stations to demonstrate the advantages of scientific agriculture. As a result, farming methods in France improved.

As a member of the company that collected taxes for the king, Lavoisier was considered a member of the ruling class. Although he was a reformer and political liberal, and took an active role in the French Revolution, his membership in the unpopular tax-collecting agency made him a target of suspicion. In spite of his many services to his country, he was forced to leave his home and laboratory. During the Reign of Terror that followed the revolution, Lavoisier was arrested and condemned to death in a trial that lasted less than a day. That same day, he was guillotined and his body thrown into a common grave. The history of chemistry was changed forever because this brilliant chemist and reformer took a government position.

Figure 25.8 This blue flame is characteristic of hydrogen. It is produced as hydrogen is oxidized by oxygen in the air. Water is the product.

Now the hydrogen is out of balance, and two hydrogen atoms must be added to the right side in the form of H^+.

$$H_3PO_3 + H_2O \rightarrow H_3PO_4 + 2e^- + 2H^+$$

In checking charges, notice that each side is neutral.

Combining the Half-Reactions

The two half-reactions are now balanced, but three electrons have been shown to be gained and only two electrons lost. You know that the number of electrons lost must equal the number of electrons gained. Before adding the two half-reactions, it is necessary to have the same number of electrons in each half-reaction. In balancing oxidation-reduction reactions, you must find the least common multiple of the number of electrons lost and gained. In this example, the least common multiple of electrons is 3×2 or 6. The reduction half-reaction must then be adjusted so that six electrons are gained by the substance being reduced. Similarly the oxidation half-reaction is adjusted so that six electrons are lost by the substance being oxidized. By multiplying the reduction half-reaction by 2 and the oxidation half-reaction by 3, the following equations are obtained.

$$2(NO_3^- + 3e^- + 4H^+ \rightarrow NO + 2H_2O)$$
$$3(H_3PO_3 + H_2O \rightarrow H_3PO_4 + 2e^- + 2H^+)$$

$$2NO_3^- + 6e^- + 8H^+ \rightarrow 2NO + 4H_2O$$
$$3H_3PO_3 + 3H_2O \rightarrow 3H_3PO_4 + 6e^- + 6H^+$$

Adding the two equations, the following is obtained.

$$2NO_3^- + 6e^- + 8H^+ + 3H_3PO_3 + 3H_2O \rightarrow$$
$$2NO + 4H_2O + 3H_3PO_4 + 6e^- + 6H^+$$

Note that electrons, hydrogen ions, and water molecules appear on both sides of the equation. By subtracting those quantities that appear on both sides of the equation, the equation may be simplified.

$$2NO_3^- + 2H^+ + 3H_3PO_3 \rightarrow 2NO + H_2O + 3H_3PO_4$$

If any electrons remain on either side, a mistake has been made. Note that this equation is balanced with respect to electrons, atoms, and total charge.

SAMPLE PROBLEM

Balancing Redox Reactions

Silver will react with nitric acid to produce silver nitrate, nitrogen(II) oxide, and water. Write and balance the equation.

Solving Process:

The equation for this reaction, in *acidic* solution, is

$$Ag(cr) + HNO_3(aq) \rightarrow AgNO_3(aq) + NO(g) + H_2O(l)$$

Balancing the reduction half-reaction proceeds as follows.

Skeleton	$NO_3^- \rightarrow NO$
Electrons	$NO_3^- + 3e^- \rightarrow NO$
Oxygen	$NO_3^- + 3e^- \rightarrow NO + 2H_2O$
Hydrogen	$NO_3^- + 3e^- + 4H^+ \rightarrow NO + 2H_2O$
Check charge	$0 = 0$

Balancing the oxidation half-reaction proceeds as follows.

Skeleton	$Ag \rightarrow Ag^+$
Electrons	$Ag \rightarrow Ag^+ + e^-$
Check charge	$0 = 0$

The least common multiple for the electrons is three. Therefore, the oxidation half-reaction can be multiplied by three to obtain the correct number of electrons.

$$3Ag \rightarrow 3Ag^+ + 3e^-$$

Adding the two half-reactions and simplifying produces

$$3Ag + NO_3^- + 4H^+ \rightarrow 3Ag^+ + NO + 2H_2O$$

Figure 25.9 The tarnish on this silver spoon is silver sulfide. It results from the oxidation of silver by sulfur-containing compounds in food and in the atmosphere.

PRACTICE PROBLEMS

24. Balance the following reactions using the half-reaction method. All reactions are carried out in acidic solution.

a. $Cu(cr) + AgNO_3(aq) \rightarrow Cu(NO_3)_2(aq) + Ag(cr)$

b. $HNO_3(aq) + H_2S(g) \rightarrow S(cr) + NO(g)$

c. $Pb(cr) + PbO_2(cr) + H_2SO_4(aq) \rightarrow PbSO_4(cr)$

▼ A More Complex Redox Reaction

CHEMISTRY IN DEPTH

Suppose at least one of the substances in a reaction appears as two or more atoms per formula unit of reactant. What happens then? Consider the following reaction.

$$I_2(cr) + HClO(aq) + H_2O(l) \rightarrow HIO_3(aq) + HCl(aq) \quad (acidic)$$

The oxidation number of iodine changes from 0 to 5+, and the oxidation number of chlorine changes from 1+ to 1−. However, iodine is a diatomic molecule, so both atoms are oxidized. If one atom must lose five electrons to go from 0 to 5+, two atoms must lose ten electrons. Each chlorine is reduced from 1+ to 1− and must gain two electrons for each atom. The balancing of the oxidation half-reaction then proceeds as follows.

Skeleton	$I_2 \rightarrow 2IO_3^-$
Electrons	$I_2 \rightarrow 2IO_3^- + 10e^-$
Oxygen	$I_2 + 6H_2O \rightarrow 2IO_3^- + 10e^-$
Hydrogen	$I_2 + 6H_2O \rightarrow 2IO_3^- + 10e^- + 12H^+$
Check charge	$0 = 0$

The reduction half-reaction proceeds as follows.

Skeleton	$HClO \rightarrow Cl^-$
Electrons	$HClO + 2e^- \rightarrow Cl^-$
Oxygen	$HClO + 2e^- \rightarrow Cl^- + H_2O$
Hydrogen	$HClO + 2e^- + H^+ \rightarrow Cl^- + H_2O$
Check charge	$1- = 1-$

To obtain the overall redox equation, proceed as follows.

Electron least common multiple = 10

The reduction half-reaction must be multiplied by 5 to obtain the correct number of electrons.

$$I_2 + 6H_2O \rightarrow 2IO_3^- + 10e^- + 12H^+$$
$$5HClO + 10e^- + 5H^+ \rightarrow 5Cl^- + 5H_2O$$

Adding and simplifying

$$I_2(cr) + 5HClO(aq) + H_2O(l) \rightarrow 2IO_3^-(aq) + 7H^+(aq) + 5Cl^-(aq)$$

EVERYDAY CHEMISTRY

Photography

Photographic film consists of tiny grains of silver bromide embedded in a thin layer of gelatin. These grains contain many silver ions and bromide ions. When a picture is taken, light from the scene being photographed passes through the camera's lens and shutter, and strikes the light-sensitive silver bromide on the film. Photons of light cause an electron on each bromide ion that is hit to be ejected from the ion. The electrons then attach to the silver ions, reducing them to metallic silver. The greatest number of silver ions is reduced in the areas that have been exposed to the greatest amount of light. An image is formed on the film, but it cannot yet be seen. To develop the image, the film is placed in a solution of a reducing agent, or developer. This substance, usually hydroquinone ($C_6H_4(OH)_2$), reduces all the silver ions in any grain that contains a metallic silver atom.

$$2AgBr(cr) + C_6H_4(OH)_2(aq) \rightarrow 2Ag(cr) + C_6H_4O_2 + 2HBr(aq)$$

The reducing agent does not react with silver bromide that has not been exposed to light. Because metallic silver is dark and silver bromide is light, an image having light and dark areas is produced.

After the film has been developed, it is placed in a solution of a fixer, which contains thiosulfate ions. The thiosulfate forms a soluble complex with unreduced silver ions, and they are dissolved away. The fixed film then goes through a wash to remove any remaining developer or fixer. However, the image that is formed is "negative"; that is, light areas in the scene are dark on the film, and vice versa.

Exploring Further

1. What would happen to a photograph if a fixer were not used?
2. Find out how color film is produced.

A Redox Reaction in Basic Solution

Although most redox reactions occur in acidic solution, it is important to realize that some can take place in basic solution. Balancing oxidation-reduction reactions in basic solutions can proceed in a similar manner to those in acidic solutions. However, the mechanics of balancing oxygen and hydrogen using OH^- and H_2O can become very tedious. *An easier way is to balance the equation as though it were acidic. Then, sufficient OH^- is added to each side to neutralize any H^+. Finally, any H_2O appearing on both sides is subtracted.*

Consider the oxidation of oxalate ion to carbonate ion by permanganate ion, which in turn is reduced to manganese(IV) oxide. This is a redox reaction that takes place in base.

$$C_2O_4^{2-}(aq) + MnO_4^-(aq) + OH^-(aq) \rightarrow MnO_2(cr) + CO_3^{2-}(aq) + H_2O(l)$$

The reduction half-reaction

Skeleton	$MnO_4^- \rightarrow MnO_2$
Electrons	$MnO_4^- + 3e^- \rightarrow MnO_2$
Oxygen	$MnO_4^- + 3e^- \rightarrow MnO_2 + 2H_2O$
Hydrogen	$MnO_4^- + 3e^- + 4H^+ \rightarrow MnO_2 + 2H_2O$
Hydroxide	$MnO_4^- + 3e^- + 4H^+ + 4OH^- \rightarrow MnO_2 + 2H_2O + 4OH^-$
Neutralize	$MnO_4^- + 3e^- + 4H_2O \rightarrow MnO_2 + 2H_2O + 4OH^-$
Subtract water	$MnO_4^- + 3e^- + 2H_2O \rightarrow MnO_2 + 4OH^-$
Check charge	$4- = 4-$

The oxidation half-reaction

Skeleton	$C_2O_4^{2-} \rightarrow 2CO_3^{2-}$
Electrons	$C_2O_4^{2-} \rightarrow 2CO_3^{2-} + 2e^-$
Oxygen	$2H_2O + C_2O_4^{2-} \rightarrow 2CO_3^{2-} + 2e^-$
Hydrogen	$2H_2O + C_2O_4^{2-} \rightarrow 2CO_3^{2-} + 2e^- + 4H^+$
Hydroxide	$4OH^- + 2H_2O + C_2O_4^{2-} \rightarrow 2CO_3^{2-} + 2e^- + 4H^+ + 4OH^-$
Neutralize	$4OH^- + 2H_2O + C_2O_4^{2-} \rightarrow 2CO_3^{2-} + 2e^- + 4H_2O$
Subtract water	$4OH^- + C_2O_4^{2-} \rightarrow 2CO_3^{2-} + 2e^- + 2H_2O$
Check charge	$6- = 6-$

For the overall redox reaction the least common multiple of electrons is 6. Multiply the manganese half-reaction by 2 and the oxalate half-reaction by 3.

$$2MnO_4^- + 6e^- + 4H_2O \rightarrow 2MnO_2 + 8OH^-$$

$$12OH^- + 3C_2O_4^{2-} \rightarrow 6CO_3^{2-} + 6e^- + 6H_2O$$

Add the half-reactions and simplify. The result is the overall redox reaction.

$$2MnO_4^-(aq) + 4OH^-(aq) + 3C_2O_4^{2-}(aq) \rightarrow 2MnO_2(cr) + 6CO_3^{2-}(aq) + 2H_2O(l)$$

▲

Summary of Balancing Redox Reactions by the Half-Reaction Method

Because it is very difficult to balance most redox equations by trial and error, it is more convenient to use the half-reaction method. The half-reaction method consists of five basic steps.

1. Write the skeleton equations for the oxidation and reduction half-reactions.
2. For each half-reaction in turn perform the following steps:

In Acidic Solution	**In Basic Solution**
a. Balance the half-reaction with respect to electrons.	**a.** Balance as if acidic.
b. Balance all of the atoms except hydrogen and oxygen.	**b.** Add sufficient OH^- to each side to neutralize any H^+.
c. Balance oxygen atoms by adding H_2O.	**c.** Neutralize by combining H^+ and OH^- to form H_2O.
d. Balance the hydrogen-atoms by adding H^+.	**d.** Subtract any H_2O appearing on both sides of the equation.
e. Check to see that the charges are balanced.	**e.** Check to see that the charges are balanced.

3. Balance the total number of electrons in the two half-reactions by finding the least common multiple of electrons lost and gained. Multiply the appropriate half-reaction(s) by the appropriate factor(s).
4. Add the two half-reaction equations and simplify.
5. Perform a final check on the equation to make sure that the number of atoms and the charge are balanced.

Figure 25.10 When a zinc plate is placed in a lead acetate solution, crystals of lead form on the plate. Lead acetate is the oxidizing agent.

Figure 25.11 The copper in this roof has reacted with carbon dioxide and water in the air to form a hydrated copper carbonate, which is green.

PRACTICE PROBLEMS

Balance the following equations using the half-reaction method. All reactions occur in acidic solution unless indicated.

25. $MnO_4^-(aq) + H_2SO_3(aq) + H^+(aq) \rightarrow Mn^{2+}(aq) + HSO_4^-(aq) + H_2O(l)$
26. $Cr_2O_7^{2-}(aq) + H^+(aq) + I^-(aq) \rightarrow Cr^{3+}(aq) + I_2(cr) + H_2O(l)$
27. $NH_3(g) + O_2(g) \rightarrow NO(g) + H_2O(g)$ (*basic*)
28. $As_2O_3(cr) + H^+(aq) + NO_3^-(aq) + H_2O(l) \rightarrow H_3AsO_4(aq) + NO(g)$
29. $I_2(cr) + H_2SO_3(aq) + H_2O(l) \rightarrow I^-(aq) + HSO_4^-(aq) + H^+(aq)$
30. $H_3AsO_4(aq) + Zn(cr) \rightarrow AsH_3(g) + Zn^{2+}(aq)$
31. $MnO_4^{2-}(aq) + H^+(aq) \rightarrow MnO_4^-(aq) + MnO_2(cr)$
32. $MnO_4^-(aq) + SO_2(g) \rightarrow Mn^{2+}(aq) + SO_4^{2-}(aq) + H^+(aq)$
33. $NO_2(g) + OH^-(aq) \rightarrow NO_2^-(aq) + NO_3^-(aq)$ (*basic*)
34. $HgS(cr) + Cl^-(aq) + NO_3^-(aq) \rightarrow HgCl_4^{2-}(aq) + S(cr) + NO(g)$

25.2 CONCEPT REVIEW

35. What is a half-reaction?
36. Balance the following equation, using the half-reaction method.

$$Cr_2O_7^{2-} + SO_3^{2-} \rightarrow Cr^{3+} + SO_4^{2-}$$
(in acid)

37. **Apply** Silver metal can be produced by placing zinc metal in a solution containing Ag^+ ions. Write equations for the two half-reactions involved. Label each half-reaction as oxidation or reduction.

Chapter 25 REVIEW

Summary

25.1 Oxidation and Reduction Processes

1. A redox reaction is a type of reaction that involves the transfer of electrons from one atom or ion to another.
2. Oxidation is that part of a redox reaction in which electrons apparently are removed from an atom or ion.
3. Reduction is that part of a redox reaction in which electrons apparently are added to an atom or ion.
4. An oxidation reaction does not take place without a corresponding reduction. If electrons are lost, something must be present to accept them.
5. Any substance that loses electrons in a redox reaction is called a reducing agent. Any substance that gains electrons in a redox reaction is an oxidizing agent.
6. If a substance is easily oxidized (loses electrons), it is called a strong reducing agent because it causes other substances to be reduced.
7. If a substance is easily reduced (gain electrons), it is called a strong oxidizing agent because it causes other substances to be oxidized.
8. It is possible to determine whether or not a redox reaction has taken place by finding changes in oxidation numbers between reactants and products. A change indicates that electrons have been transferred.
9. Drawing electron dot structures and comparing electronegativities can help in assigning oxidation numbers to atoms.

25.2 Balancing Redox Equations

10. Redox reactions can be balanced more easily by the half-reaction method than by trial and error. One half-reaction is written to show the oxidation taking place. The other shows the reduction.
11. In all chemical reactions, charge, atoms, and electrons are conserved. Any two of these principles can be used to balance an equation. The third principle may be used as a check to see whether the reaction is properly balanced.

Key Terms

oxidation-reduction reaction
oxidation
reduction
reducing agent
oxidizing agent

Review and Practice

38. How do oxidation-reduction reactions differ from other chemical reactions?
39. Bromine reacts with iodide ions according to the following equation.
    $Br_2(aq) + 2I^-(aq) \rightarrow 2Br^-(aq) + I_2(aq)$
    a. Describe the net electron transfer in this reaction.
    b. What is oxidized in this process?
    c. What is reduced in this process?
40. What occurs in the process of oxidation? Why is the process $Mg + Cl_2 \rightarrow MgCl_2$ considered an oxidation process? What is reduced in the reaction?
41. What occurs in the process of reduction? Why is the process $2NaBr \rightarrow 2Na + Br_2$ considered a reduction process? What is oxidized in the reaction?
42. Why is it impossible to characterize the reactions in the previous two questions simply as oxidation or as reduction?
43. Distinguish between oxidizing agents and reducing agents.
44. Identify the oxidizing and reducing agents in the reactions shown in Problems 40 and 41.
45. Determine the oxidation numbers for the element named in each case.
    a. copper in a copper penny

b. copper in the ionic compound $CuSO_4$
c. hydrogen in NaH_2PO_4
d. oxygen in N_2O_5
e. nitrogen in NH_4^+
f. rubidium in any compound

46. Determine the oxidation numbers for atoms of the underlined elements.

a. $Al\underline{P}O_4$ **d.** $Ba(\underline{Cl}O)_2$
b. $(NH_4)_3\underline{As}O_4$ **e.** $Ba\underline{Si}O_3$
c. $(NH_4)_2\underline{Si}F_6$ **f.** $Pb(\underline{I}O_3)_2$

47. For each of the following reactions, tell whether it is a redox reaction, and if it is, name the element oxidized, the element reduced, the oxidizing agent, and the reducing agent.

a. $Fe(OH)_2 + H_2S \rightarrow FeS + 2H_2O$
b. $SnS_2 + 2NH_4HS \rightarrow (NH_4)_2SnS_3 + H_2S$
c. $H_2S + Br_2 \rightarrow S + 2HBr$
d. $3CuS + 8HNO_3 \rightarrow 3Cu(NO_3)_2 + 3S + 2NO + 4H_2O$

48. In the half-reaction method of balancing redox equations, what does each half-reaction represent?

49. Balance the following equations after putting them in net ionic form.

a. $Cu(cr) + HNO_3(aq) \rightarrow Cu(NO_3)_2(aq) + NO(g) + H_2O(l)$
b. $Fe(NO_3)_2(aq) + HNO_3(aq) \rightarrow Fe(NO_3)_3(aq) + NO(g) + H_2O(l)$
c. $Zn(cr) + HNO_3(aq) \rightarrow Zn(NO_3)_2(aq) + NO_2(g) + H_2O(l)$
d. $Sb(cr) + H_2SO_4(aq) \rightarrow Sb_2(SO_4)_3(aq) + SO_2(g) + H_2O(l)$
e. $H_2S(g) + H_2SO_3(aq) \rightarrow S(cr) + H_2O(l)$
f. $HCl(aq) + HNO_3(aq) \rightarrow HClO(aq) + NO(g)$
g. $Ag(cr) + HClO_3(aq) + HCl(aq) \rightarrow AgCl(cr) + H_2O(l)$
h. $KI(aq) + O_2(g) + HI(aq) \rightarrow KI_3(aq) + H_2O(l)$
i. $HNO_3(aq) + H_2SO_4(aq) + Hg(l) \rightarrow Hg_2SO_4(cr) + NO(g) + H_2O(l)$
j. $CO(g) + I_2O_5(g) \rightarrow CO_2(g) + I_2(g)$

50. Alkali metals displace hydrogen from water according to the following general reaction.

$$2M + 2H_2O \rightarrow 2MOH + H_2$$

a. What element is oxidized?
b. What element is reduced?
c. What is the reducing agent?
d. What is the oxidizing agent?
e. Write half-reactions and a balanced net ionic equation for the reaction.

51. $KClO_3$ and HCl will react in aqueous solution to produce KCl, chlorine gas, and water.

a. What element is oxidized?
b. What element is reduced?
c. What substance is the oxidizing agent?
d. What substance is the reducing agent?
e. Using half-reactions, write a balanced net ionic equation for the reaction.

52. Find the oxidation numbers of the underlined elements in the following substances.

a. $\underline{Bi}PO_4$
b. $Co\underline{Se}O_4 = (Co = 2+)$

Concept Mastery

53. How does the process of balancing redox equations for reactions that take place in acidic solution differ from the process used when the reactions take place in basic solution?

54. Of all common elements, list two that would be powerful reducing agents. List two that would be powerful oxidizing agents. Give reasons for your choices.

55. Consider the formation of aluminum chloride from aluminum and chlorine.

$$2Al + 3Cl_2 \rightarrow 2Al^{3+} + 6Cl^-$$

a. Using *x*'s to represent electrons initially in Al atoms and *o*'s to represent electrons initially in Cl atoms, sketch electron dot diagrams for all reactants and products to show how the redox reaction takes place.

b. Why is Al given the positive oxidation number and Cl given the negative oxidation number in aluminum chloride?

56. Assuming maximum appropriate values for oxidation numbers, write formulas for hypothetical binary compounds of the following pairs of elements.

a. Ba and N
b. Cr and O
c. Ca and P
d. Sr and H
e. Fr and C
f. Li and O
g. Sc and F
h. Ti and Cl
i. Fe and S

57. Draw the electron dot diagram for the thiosulfate ion and assign oxidation numbers to each atom. What is unusual about this ion?

Application

58. One way copper metal is obtained is by allowing a solution containing copper(II) ions to trickle over scrap iron. Write equations for the two half-reactions involved. Assume Fe becomes Fe^{2+}. Label each half-reaction as an oxidation or reduction half-reaction.

59. The burning of a propane torch is an example of a redox reaction.

$$C_3H_8 + 5O_2 \rightarrow 3CO_2 + 4H_2O$$

a. What element is oxidized?
b. What element is reduced?
c. What bonding characteristics of carbon make it difficult to assign oxidation numbers to carbon in an organic compound?

60. For the following, write balanced equations; then circle the oxidizing agent and underline the reducing agent.

a. Aluminum can be etched. Areas that are not to be etched are painted for protection. Then, HCl is applied to the exposed metal, producing aluminum chloride and hydrogen gas.

b. Some plants, such as beans, change atmospheric nitrogen to usable nitrogen in the soil. This change must be a continuous process, because nitrates in the soil will react with glucose, $C_6H_{12}O_6$, from plant matter in the soil to form carbon dioxide, nitrogen gas, and water.

c. Copper(II) sulfate is used as a swimming pool algicide. It can be prepared by pouring hot sulfuric acid over copper. Sulfur dioxide and water are also produced.

61. Hydrogen peroxide is commonly used as a bleaching agent and an antiseptic. H_2O_2 molecules react with each other to produce oxygen gas and water.

a. Write an equation for the production of oxygen and water from hydrogen peroxide and balance the equation by half-reactions.
b. What is the oxidizing agent in the reaction?
c. What is the reducing agent in the reaction?
d. What element is reduced?
e. What element is oxidized?

Critical Thinking

62. Explain why solutions of good reducing agents, such as photographic developing solutions, have a limited shelf life.

63. What would happen if an aluminum spoon were used to stir an iron(II) chloride solution? What would happen if an iron spoon were used to stir an aluminum nitrate solution? Write an equation for the reaction that takes place and balance it by half-reactions. (Hint: See Table 12.1 on page 303.)

Chapter 25 REVIEW

Readings

Brimhall, George. "The Genesis of Ores." *Scientific American* 264. no. 5 (May 1991): 84-91.

Kauffman, Joel M. "Simple Method for Determination of Oxidation Numbers of Atoms in Compounds." *Journal of Chemical Education* 63, no. 6 (June 1986): 474-475.

Miller, James A., and George A. Fisk. "Combustion Chemistry." *Chemical and Engineering News* 65, no. 35 (August 31, 1987): 22-46.

Schobert, Harold H. "The Geochemistry of Coal." *Journal of Chemical Education* 66, no. 3 (March 1989): 242-244.

Zino, Kenneth. "Performance Sentinels." *Popular Science* 230, no. 6 (June 1988): 70-71.

Cumulative Review

1. The solubility product for $Cd_3(PO_4)_2$ is 2.50×10^{-33}. What is the concentration of Cd^{2+} in a saturated solution of $Cd_3(PO_4)_2$?

2. Which step in a reaction mechanism determines the overall reaction rate?

3. What substance is produced from the anhydride SrO and water?

4. Write the following equation in net ionic form.

$$NaH_2PO_4 + HCl \rightarrow NaCl + H_3PO_4$$

5. A solution of 0.100*M* propanoic acid, CH_3CH_2COOH, has an H_3O^+ concentration of $1.16 \times 10^{-3}M$.

a. What is the K_a for the acid?

b. What is the percent ionization of propanoic acid?

6. What is the H_3O^+ concentration in a solution that is 0.100*M* in HF and 0.0100*M* in NaF? What is the pH?

7. When iodine reacts with concentrated nitric acid, HIO_3 and NO_2 gases are formed. At STP, how much iodine is needed to produce 30.0 dm^3 of NO_2?

8. About 90% of our iodine is obtained from the iodide salts in the brine that comes up in California oil fields. Chlorine gas is passed through the brine and displaces the iodine in iodide compounds. How many grams of iodine will be produced from 284 g of NaI?

9. How many grams of K_2SO_3 can be oxidized to K_2SO_4 by 7.90 g of $KMnO_4$, which will be reduced to MnO_2?

CHAPTER **Preview**

CONTENTS

CHAPTER 26

Electrochemistry

Corrosion of a metal object can be prevented by electroplating, coating the object with a corrosion-resistant metal. Electroplating is a process in which a thin layer of the corrosion-resistant metal is deposited on the object through the use of an electric current. Both the shiny chromium trim on this car and the chromium on the trophy are resistant to corrosion.

The object to be plated is usually a metal, because a conductive surface is required for the deposition of the coating. The metal coating can be an alloy or a single element. The conditions of depositing are so controlled that the coating becomes an integral part of the metal, not just a separate outer level of atoms.

In this chapter you will learn about the reaction of a substance to an electric current and the generation of an electric current by substances.

EXTENSIONS AND CONNECTIONS

OBJECTIVES
- describe the construction and functioning of electrolytic and voltaic cells.
- compare electrolytic and electronic conduction.
- describe the movement of anions and cations and their reactions in cells.

26.1 Cells

You press the button on a flashlight and light is produced. The light results from the passage of an electric current through the flashlight bulb. Where does the electricity come from? The obvious answer is the battery. What, then, is a flashlight battery? How does it function? How does this type of battery compare with an automobile battery? In this section you will study the principles needed to answer these questions.

Complete Circuits and Conductivity

If you close the switch of the apparatus on the left in Figure 26.1, the bulb does not light because the circuit is not complete. If you complete the circuit by placing a conducting material across the two electrodes, as in the apparatus on the right, the bulb will light. This is an easy way of observing which substances are conductors of electricity. If you touch the two electrodes to a piece of copper or other metal, the bulb lights. Metals are conductors of electricity. Copper is one of the best conductors; for this reason it is used in electric equipment. If you touch the electrodes to a piece of glass or a sulfur crystal, the bulb does not light. Most nonmetallic solids, including salts, are nonconductors. If you fill the beaker with pure water, benzene, alcohol, sugar solution, or other solution of a nonionizing substance, the bulb does not light. These liquids are all nonconductors. However, if you add a small amount of an electrolyte such as sodium chloride (or any other soluble salt), hydrochloric acid, or sodium hydroxide, the bulb lights. These solutions conduct electricity. What materials conduct electricity in solution? Why do they conduct electricity?

A **galvanometer** is an instrument for detecting electric current. **Electric current** is the flow or movement of charged particles. Let us take pieces of two different metals and connect them to the terminals of a galvanometer. The two pieces of metal are

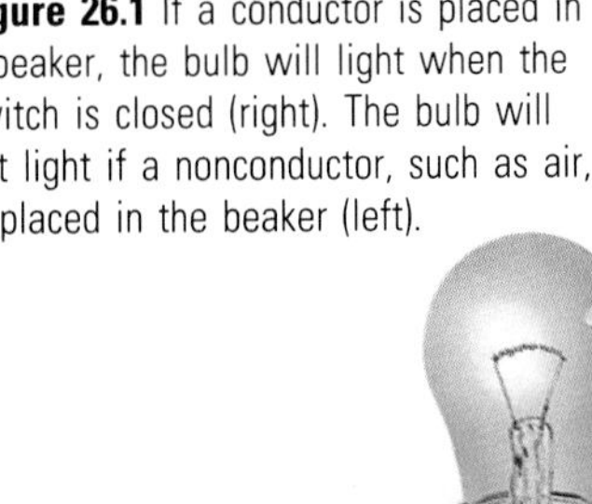

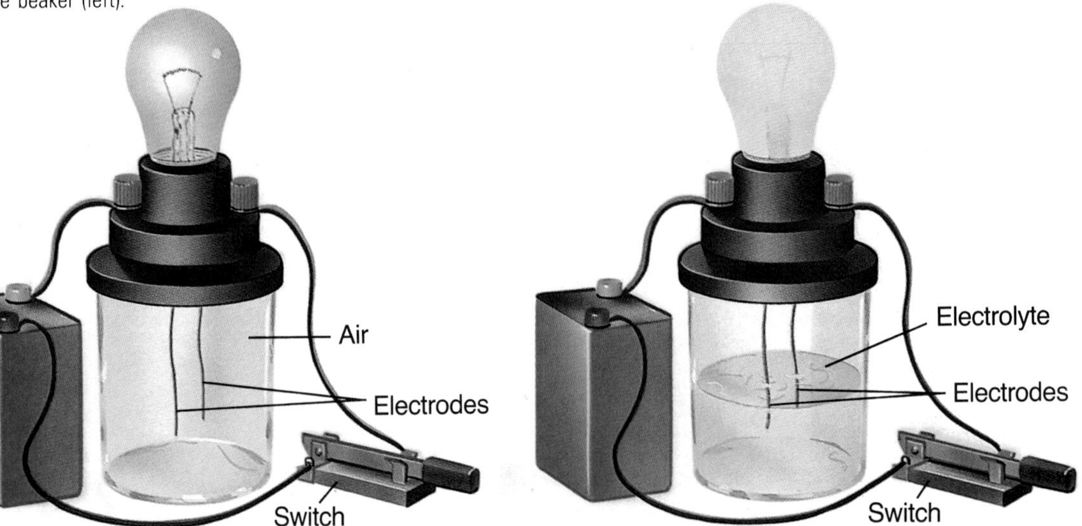

Figure 26.1 If a conductor is placed in a beaker, the bulb will light when the switch is closed (right). The bulb will not light if a nonconductor, such as air, is placed in the beaker (left).

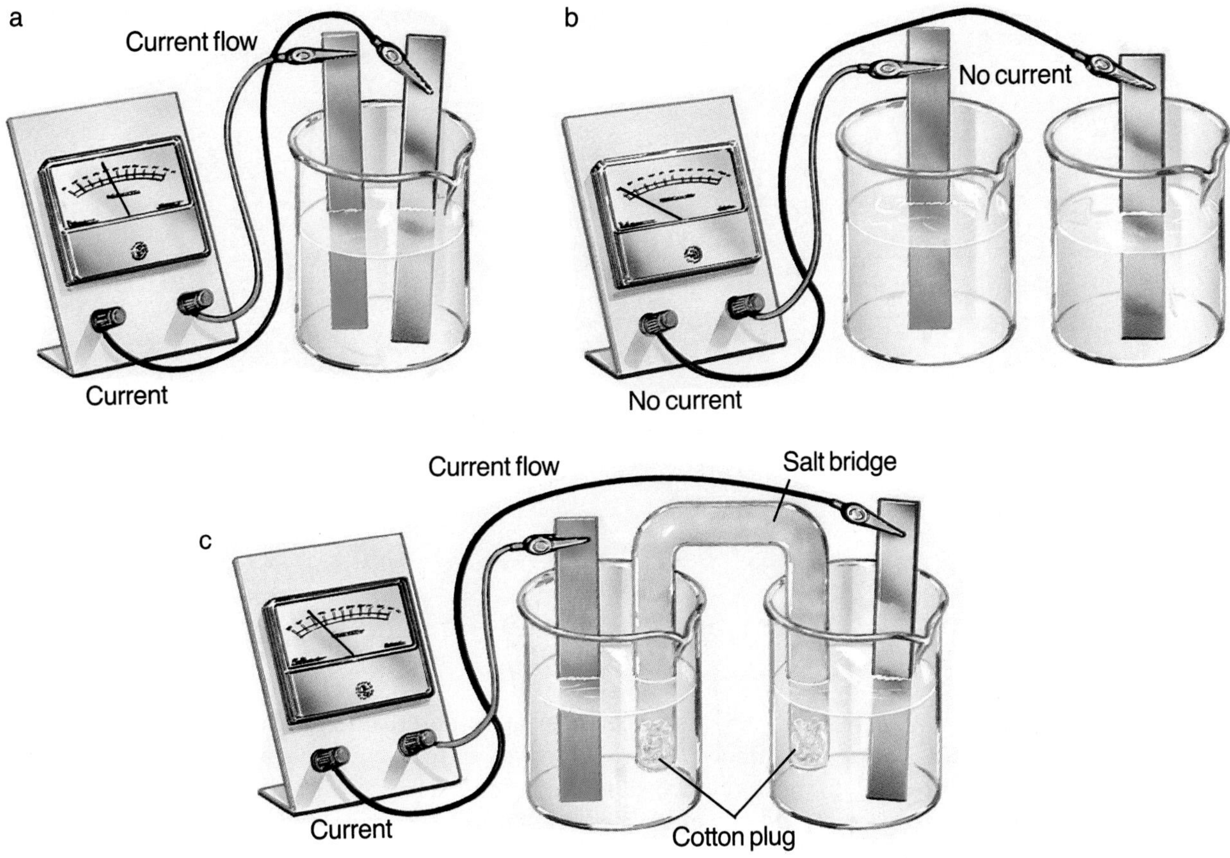

Figure 26.2 Two different metals are placed in a salt solution and connected to the terminals of a galvanometer. The galvanometer registers current flow in (a) and (c). In (b), the circuit is incomplete and thus no current flows.

kept from making contact with each other and put into a salt solution. The galvanometer needle registers a flow of current. How is the current produced? If we place the two metal plates into separate beakers containing salt solutions, as shown in Figure 26.2b, no current flows. However, when we add a salt bridge to the same system, a current is produced. The **salt bridge** is a U-tube containing an ionic substance in solution. The function of the salt bridge is to complete the circuit between the two separate solutions without mixing them. This apparatus produces a current just as if the electrodes were in the same solution.

Metallic Conduction and Potential Difference

Metals, in general, are excellent conductors of electricity. This statement is true whether the metal is in the solid state or the liquid state. Mercury is used in some scientific apparatus because it is a liquid at room temperature and has excellent electric conductivity. In Chapter 12, you read that the outer electrons of metal atoms are delocalized. If the potential energy of the electrons in the metal is raised, these electrons will flow to a point where their potential energy is lower. Therefore, conduction takes place by the movement of electrons through a metal. This kind of conduction is known as **metallic conduction** or **electronic conduction.**

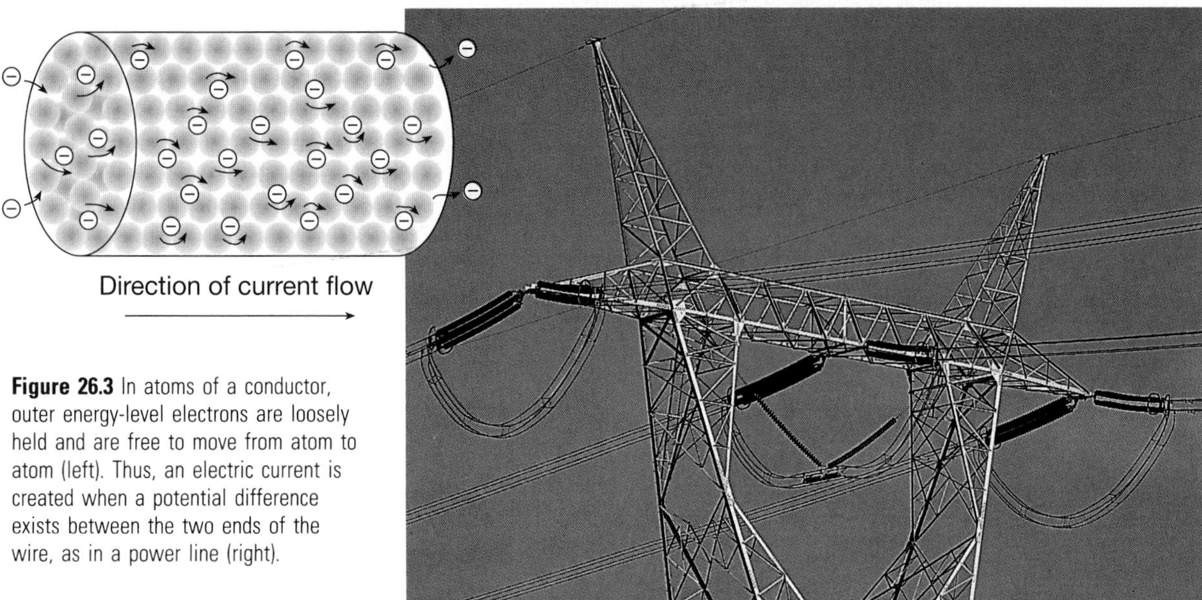

Figure 26.3 In atoms of a conductor, outer energy-level electrons are loosely held and are free to move from atom to atom (left). Thus, an electric current is created when a potential difference exists between the two ends of the wire, as in a power line (right).

LANGUAGE CONNECTION

Scientists' Names Live On

The names of some famous scientists live on long after they die. For example, the galvanometer, an instrument that detects an electric current, was named for the Italian anatomist Luigi Galvani. He discovered that muscles in contact with two dissimilar metals twitched as they would when subjected to an electric discharge.

The unit *volt* is from the name of Italian physicist Allesandro Giuseppe Antonio Anastasio Volta, who built the first voltaic cells. Many words, such as *voltage* and *voltaic,* are derived from the word *volt.*

The unit *ampere* is from the name of the French physicist Andre Marie Ampere. He worked with electric currents in wires and the magnetic forces they generated.

A difference in electric potential can be brought about in a number of ways. One way is to use a generator. Suppose we connect the two ends of a long wire to a generator. The generator adds electrons at one end of the wire and removes them at the other end. As a result, a potential energy difference is created at the ends of the wire, and current flows. Energy is required to create a potential difference and work is done on the system when a current is made to flow through a wire.

Metals are excellent conductors of electricity because their electrons are free to move when a small potential difference is created. In nonconducting substances, the outer electrons are tightly held. Very large potential differences are required to move electrons in these nonconducting substances that we call insulators. However, even the best insulators break down and conduct a current if the potential difference is high enough.

Electric potential difference is measured in units called **volts.** The voltage (potential difference) produced by a generator or battery can be thought of as electric pressure. In fact, it is often easier to consider a generator or battery as an electron pump. The current, or number of electrons that flow through the circuit per second, is measured in SI units called **amperes.**

Electrolytic Conduction

Solid acids, bases, and salts (for example, oxalic acid, sodium hydroxide, and sodium chloride) are nonconductors. However, when any one of these substances is dissolved in water, the resulting solution is a conductor. Those substances that are conductors in solution are known as electrolytes. In fact, any substance that produces ions in solution is an electrolyte. Conduction that takes place by the migration of ions is called **electrolytic conduction.** Electrolytic conduction is possible because ions move freely in aqueous solution. Salts are ionic even in the solid state, but a salt must either dissolve or melt in order for the ions to separate, or dissociate, from each other and become free to move. Acids and bases may be either ionic or molecular substances. However, when they dissolve in water, ions are formed.

Electrolytic conduction takes place in a device called an electrolytic cell. If we connect the cell to a direct current, one of the electrodes will be negative and the other electrode will be positive. It is customary to call the negative electrode the cathode and the positive electrode the anode. If we immerse the electrodes in an ionic solution, positive ions in the solution will be attracted to the cathode. For this reason, positive ions are called **cations.** Negative ions will be attracted toward the anode. Consequently, they are called **anions.** The movement of ions through a solution results in an electric current just as the movement of electrons in a metal results in a current.

Electrode Reactions

What happens to a moving ion when it reaches the electrode to which it is attracted? Consider molten sodium chloride, a system that contains only positive sodium ions and negative chloride ions, and no other particles. Electrodes that are inert, that is, electrodes that do not react chemically with sodium or chloride ions, will be used. The positive sodium ions, or cations, are attracted to the cathode. The cathode is made negative by the action of a generator which, in effect, pumps electrons into it. The electrons in the cathode are at a state of high potential energy. The sodium ion has a positive charge. It has an attraction for electrons. An

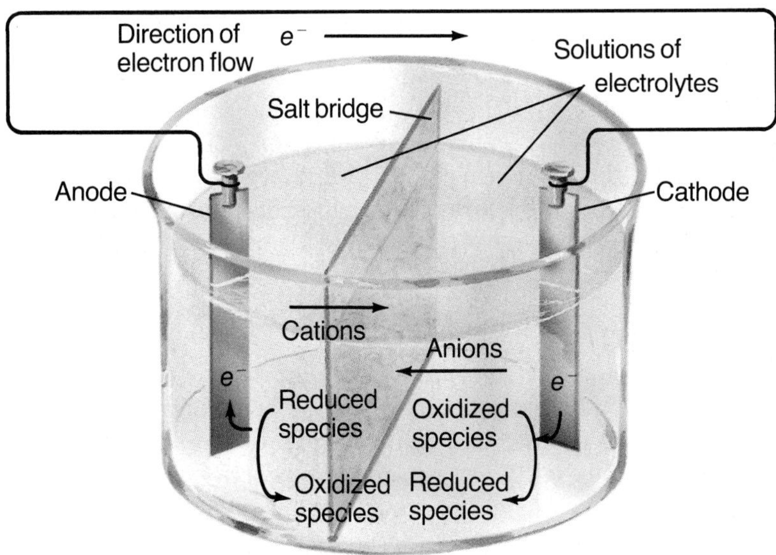

Figure 26.4 An electrolytic process involves electron transfer at the interface of a solution and an electrode. At the cathode, the species in solution receives electrons and is reduced. At the anode, the solution species releases electrons and is oxidized. Electrons in the external circuit flow from the anode to the cathode.

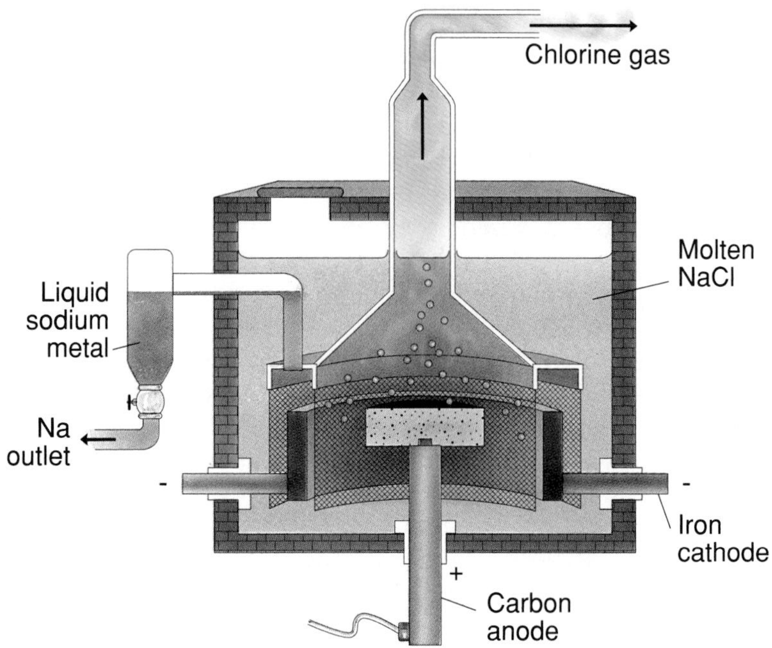

Figure 26.5 A Downs cell is used in the industrial production of sodium metal and chlorine gas. The molten NaCl is actually a mixture. Other salts are added to lower the melting point. Thus, corrosion is reduced and power costs are less.

electron in a sodium atom would have a lower potential energy than an electron on the cathode. Thus, electrons move from the cathode (high potential energy) to the sodium ions (lower potential energy). At the cathode, each sodium ion will be converted into a sodium atom by the addition of an electron. This change is a chemical reaction and can be shown by an equation.

$$Na^{+} + e^{-} \rightarrow Na$$

Note that this chemical change represents a gain of electrons. In Chapter 25 you read that electron gain is called reduction. Reduction always occurs at the cathode. In this case, the sodium ion is reduced to sodium metal.

Now consider what happens at the anode. The anode has a positive charge, and negative ions are attracted to it. The anode is positive because the generator is, in effect, pumping electrons out of it. Thus, electrons in the anode may be said to exist at a state of low potential energy. Because the chloride ion has a negative charge, its outer electrons have a higher potential. Electrons move from a state of higher potential energy to a state of lower potential energy. Therefore, when chloride ions reach the electron-deficient anode, they give up electrons to it. The chemical change that occurs at the anode can be shown by the following equation.

$$2Cl^{-} \rightarrow Cl_2 + 2e^{-}$$

Note that in this reaction chloride ions lose electrons to become chlorine atoms that then combine to form Cl_2 molecules. The $Cl_2(g)$ bubbles off at the anode. This reaction results in a loss of electrons. A loss of electrons is oxidation. Oxidation always occurs at the anode.

The oxidation and reduction processes are shown by separate equations because they take place at different points. Recall that these processes do not occur independently. The reduction process cannot occur without the oxidation process going on at the same time. The generator does not produce the electrons; it pumps the electrons toward the cathode and away from the anode. The role of the generator is to raise the potential energy of the electrons on the cathode and to reduce the potential energy of the electrons on the anode.

These electrode reactions are called half-reactions. The anode reaction involves a loss of electrons and the cathode reaction involves a gain of electrons. Therefore, the number of electrons must balance when we add the two half-reactions. The cathode reaction must be multiplied by 2. The overall reaction for the electrolysis of molten sodium chloride is

$$2Na^+(l) + 2Cl^-(l) \rightarrow 2Na(l) + Cl_2(g)$$

The process by which an electric current produces a chemical change is called **electrolysis.** Electrolysis of molten sodium chloride is the commercial process used to produce metallic sodium. Chlorine gas is produced as a by-product during this process.

FRONTIERS

Biosensors

A sample of blood is withdrawn from a patient and inserted into an analyzer. By the time the doctor has keyed in the patient's name, an analysis has been performed. These new medical devices are possible because of microelectronic biosensors.

The first biosensor was made in the early 1960s to measure blood glucose. It consisted of a circuit containing a cell and two electrodes, which were immersed in a solution of an electrolyte. The solution was separated from a blood sample by a membrane through which oxygen could diffuse, and the membrane was coated with glucose oxidase, an enzyme that catalyzes the oxidation of glucose by oxygen. The current in the circuit depended on the rate at which dissolved oxygen gas diffused into the sensor. The higher the glucose concentration in the blood, the more oxygen was consumed by the enzyme reaction. Thus, less oxygen diffused into the sensor and the current flowing in the circuit was less.

Modern biosensors can be miniaturized to fit within a hypodermic needle. Researchers are devising methods of using animal nerve cells in biosensors that may one day be used to monitor very small concentrations of substances in the environment.

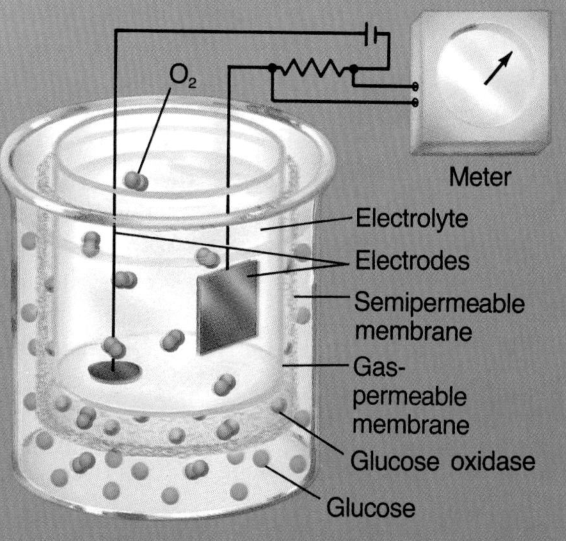

Figure 26.6 In this photo, molten aluminum is being drawn off from the bottom of the electrolytic cell, where it accumulates during the process.

Electrolysis is a common method of preparing active metals. The metal produced in the largest quantity by electrolysis is aluminum. The electrolytic cells are lined with carbon, which acts as a cathode. The anodes are also made of carbon. Aluminum oxide is the compound from which the aluminum is to be obtained. However, aluminum oxide has a very high melting point, about 2000°C. Consequently, aluminum producers add cryolite, Na_3AlF_6, which melts at about 1000°C and does not interfere with the electrolysis. The aluminum oxide dissolves in the molten cryolite at 1000°C, and the electrolysis proceeds. The molten aluminum sinks to the bottom of the cell, where it is tapped off. The oxygen produced reacts with the carbon anodes to form CO_2. As a result, the anodes must be replaced periodically.

Electrolysis of a Salt Solution

Look now at what happens in the electrolysis of a concentrated *solution* of sodium chloride in water. In this electrolysis, water molecules are present in addition to sodium ions and chloride ions. At the cathode, electrons could pass from the electrode to either a sodium ion or to a water molecule. Because an electron has a lower potential energy on a water molecule, electrons at the cathode are transferred to water molecules rather than to sodium ions. When a water molecule acquires an extra electron, the electron is accepted by hydrogen, forming a hydrogen atom. This change leaves a hydroxide ion. The hydrogen atoms produced combine to form H_2 molecules. This hydrogen gas bubbles from the solution at the cathode. At the anode, on the other hand, there is the possibility of the electrode gaining electrons from either water molecules or chloride ions. In a *concentrated* sodium chloride solution, electrons transferring from chloride ions to the anode undergo a greater reduction in energy than from water molecules to the anode. Thus, chlorine gas is produced at the anode, just as in the molten salt cell. What remains in the cell after the hydrogen and chlorine gases have left? Water molecules remain around the anode. The solution surrounding the cathode acquires a high concentration of sodium ions and hydroxide ions. If this solution is drained off and evaporated, solid sodium hydroxide remains. Commercial sodium hydroxide is produced this way. Chlorine gas and hydrogen gas are also produced.

The half-reactions are

$$2Cl^- \rightarrow Cl_2 + 2e^-$$

$$2H_2O + 2e^- \rightarrow 2OH^- + H_2$$

The equation for the overall reaction may be written

$$2Cl^-(aq) + 2H_2O(l) \rightarrow Cl_2(g) + H_2(g) + 2OH^-(aq)$$

It should be noted that sodium metal, chlorine gas, hydrogen gas, and sodium hydroxide are all produced commercially by using rock salt. Rock salt is a very cheap raw material. The costly part of this operation is the electric power required.

In the processes we have just described, both kinds of electric conduction are involved. In the generator and the cables leading to the electrodes, metallic conduction is taking place. The electric path (circuit) in electrolysis is completed through the liquid between the electrodes. This part of the process is electrolytic conduction. Both kinds of conduction involve the movement of electric charge. For electrolytic conduction, it is necessary that ions be present and be free to move. Ions move freely in solution or as molten salts. Solid salts do not conduct electricity.

Voltaic Cells

The processes that have been described are examples in which electric energy is used to cause chemical changes. These processes take place in electrolytic cells. When you use a flashlight, however, you are using a chemical process to produce electricity. In

Figure 26.7 Rock salt is used in the commercial preparation of sodium, chlorine, hydrogen, and sodium hydroxide. The rock salt is obtained from mines and brought to the surface, where it is stored until it is transported.

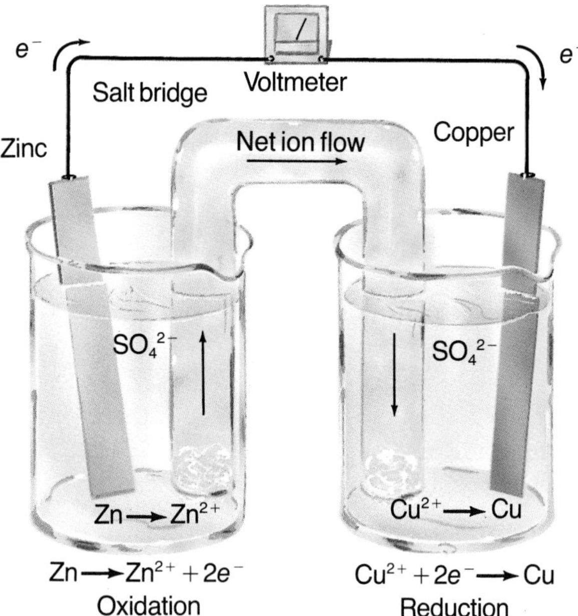

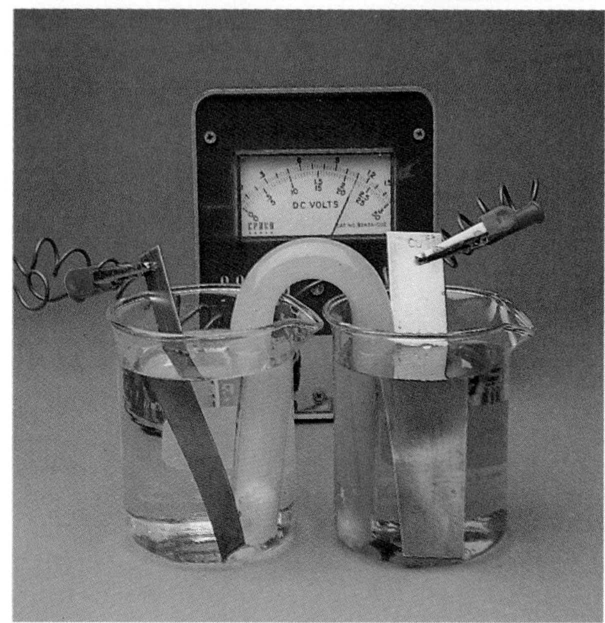

Figure 26.8 These voltaic (or galvanic) cells consist of zinc metal in a solution of zinc ions connected with a salt bridge and an external circuit to copper metal in a solution of copper ions.

other words, you are converting chemical potential energy into electric energy. A device that makes such a conversion is known as a **voltaic cell.** An example of a simple voltaic cell is shown in Figure 26.8. Note that there is no external source of electrons (no generator). In this cell, there are two compartments separated by a porous barrier or salt bridge that prevents the two solutions from mixing by diffusion. However, the barrier permits the migration of ions from one side of the cell to the other. The cathode in the beaker on the right consists of a strip of copper immersed in a copper(II) sulfate solution. The anode in the beaker on the left consists of a strip of zinc immersed in a zinc sulfate solution. If the two electrodes are connected through a precision voltmeter, an electric current flows, and the voltmeter reads 1.103 volts. Electrons flow from the zinc anode through the voltmeter to the copper cathode. Chemical changes occur at the electrodes. Zinc is a more active metal than copper because its outer electrons have a higher potential energy than outer electrons in copper atoms. In other words, zinc has a greater tendency to give up electrons than does copper. The anode half-reaction that occurs at the zinc electrode is

$$Zn \rightarrow Zn^{2+} + 2e^-$$

Because this change involves loss of electrons, it is an oxidation reaction. Zinc metal is oxidized at the anode (the zinc bar in the zinc half-cell in Figure 26.8). The potential energy of electrons on the copper electrode is higher than the potential energy the electrons would have on the copper ions in the solution. The cathode transfers electrons to the copper ions and copper metal plates out on the cathode. The cathode reaction is

$$Cu^{2+} + 2e^- \rightarrow Cu$$

Because electrons are produced on the zinc anode, it has a negative charge. Electrons leave the copper cathode, so it has a positive charge. Electrons flow from the zinc anode through the wire to the copper cathode. At the copper cathode, electrons are transferred to copper ions. This process involves a gain of electrons and is a reduction reaction. Note that the operation of the cell involves both oxidation and reduction, but the two changes do not occur at the same electrode. Rather, the electrons travel through the external circuit (the voltmeter) and do electrical work.

Note that these changes at the electrons occur without producing an electric current. Consider a piece of zinc placed in a copper sulfate solution as shown in Figure 26.9. Zinc metal is converted to Zn^{2+} ions, and Cu^{2+} ions are converted to copper metal. Energy is also produced but is released in the form of heat that is not available for practical use. No current is produced because the electrons are transferred directly from Zn atoms to Cu^{2+} ions.

ENGINEERING CONNECTION

Nightlighting from the Sun

A photovoltaic cell is a voltaic cell that is activated by radiation, usually light. This type of cell is finding increasing use in safety applications. Many freeways now have emergency telephones powered by sunlight through the voltaic cell. Some outdoor security lighting uses a solar cell to collect energy during the day. When the sun sets, the battery powers the lighting. These sun-powered lights are used along walkways, alleys, and porches to provide light where electric power is either inaccessible or prohibitive in cost.

Structure of Voltaic Cells

A voltaic cell, then, is a device used to produce electric energy from an oxidation-reduction reaction. The main feature of the voltaic cell, as shown in Figure 26.8, is the porous barrier or salt bridge that separates the two solutions and keeps them from mixing freely. If this barrier is not porous enough to allow ions to migrate through it, the cell will not operate. The anode compartment acquires an excess of positive zinc ions. In order to maintain neutrality, it must have negative ions to balance the positive zinc ions. At the same time, the cathode compartment uses copper ions. In order to maintain neutrality, it must lose negative ions. Sulfate ions move through the porous barrier from the cathode compartment to the anode compartment. In that way electric neutrality is maintained in the two compartments. The cell continues to operate as long as there is a potential energy difference between the half-cells. The two compartments in the cell are called half-cells because the reactions that occur in them are half-reactions.

Figure 26.9 When a zinc strip is placed in blue $CuSO_4$ solution (left), a redox reaction occurs. After a few days, the blue color disappears as Cu^{2+} is reduced to copper metal. Most of the zinc metal has been oxidized to Zn^{2+} which is present in the colorless solution (right).

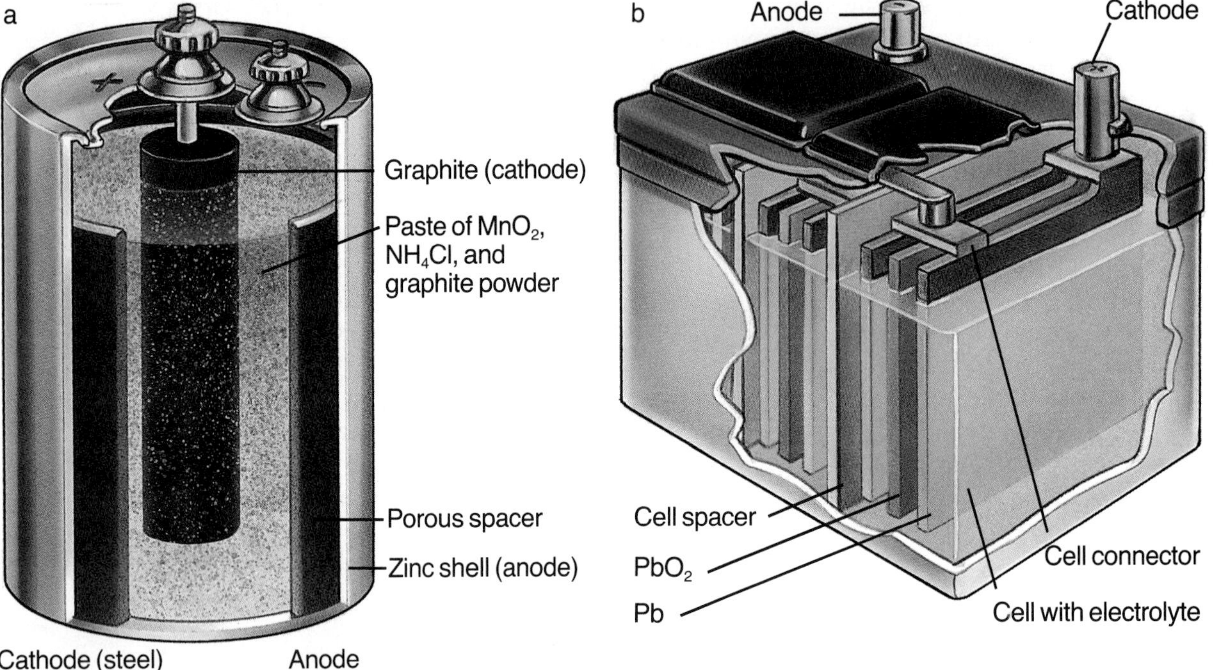

Figure 26.10 A dry cell (a), an automobile battery (b), and a mercury battery (c) are all voltaic cells. The automobile battery contains an electrolyte (H_2SO_4) in solution. The dry cell and mercury cell each have an electrolyte in paste form.

The two most commonly used voltaic cells are the dry cell and the lead storage cell. A flashlight battery is an example of a dry cell. An automobile battery is a lead storage cell. Mercury cells are increasingly being used in calculators and watches.

Chemists have a shorthand method of representing cell reactions. The oxidation half-cell is written first. The reduced and oxidized species are separated by a vertical line. The zinc half-cell from the zinc-copper cell would be written

$$Zn|Zn^{2+}$$

The reduction half-cell is written in the reverse order.

$$Cu^{2+}|Cu$$

The two half-cells are separated by two vertical lines representing the salt bridge. The entire zinc-copper cell is thus shown as

$$Zn|Zn^{2+}||Cu^{2+}|Cu$$

Fuel Cells

In the most common method of generating electric current, a fuel is burned in air and the heat produced is used to boil water. The steam is then passed into a turbine, which is used to turn a generator. Each step in this series has an efficiency of less than 100%. As a consequence, only about 20% of the potential energy in the fuel is finally realized as electric energy. If the intermediate steps between fuel and electricity could be eliminated, the efficiency would be greatly enhanced. A device that accomplishes just that end is the fuel cell. A fuel cell acts much like a battery. However, fuel cells do not have the very limited life that batteries

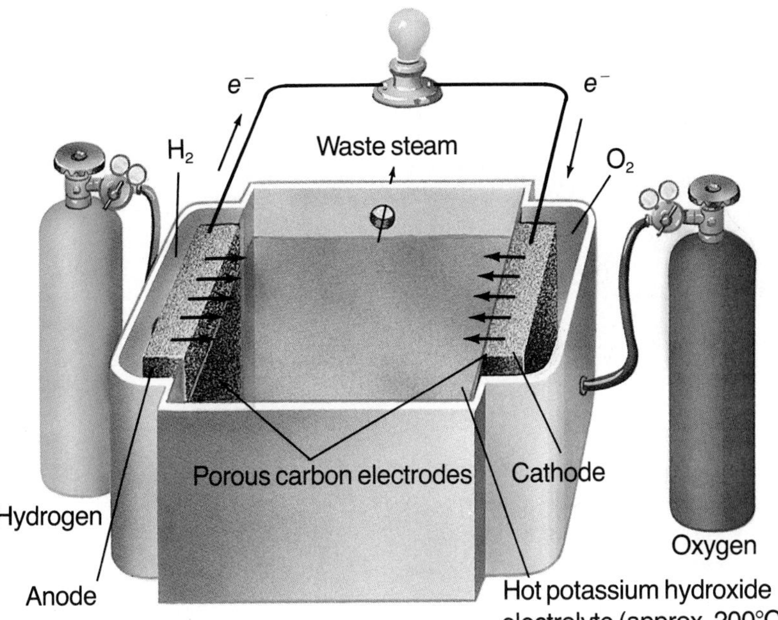

Figure 26.11 A hydrogen-oxygen fuel cell like this one was used in Apollo spacecraft. Instead of producing toxic wastes, as combustion engines do, this fuel cell produces drinkable water when the waste steam is condensed.

have, because there is a continuous feed of fuel and oxygen or air to the cell. The fuel is fed to an electrode on one side of an electrolyte solution, and the oxygen is fed to an electrode on the other side. The electrolyte must be capable of transferring electrons between the electrodes. In addition to the cell itself, an installation must have pumps, fuel reservoirs, and control circuits. The main feature of the fuel cell process is that it separates the oxidation of fuel molecules from the reduction of oxygen. Fuel cells are used in demanding installations such as manned spacecraft, but at present are too expensive for common applications.

26.1 *CONCEPT REVIEW*

1. Describe the difference between metallic conduction and electrolytic conduction.
2. Define cathode and anode in terms of oxidation and reduction.
3. What is the difference between an electrolytic cell and a voltaic cell?
4. The two types of electrodes in a lead storage battery are made of porous lead and compressed insoluble lead(IV) oxide. These electrodes are immersed in concentrated sulfuric acid. The equation for the overall reaction in this cell is

$$Pb(cr) + PbO_2(cr) + 2H_2SO_4(aq) \rightarrow 2PbSO_4(cr) + 2H_2O(l)$$

   a. Identify the reactions taking place at the anode and the cathode and write the shorthand representation for this cell.
   b. What substance loses electrons in the reaction? What substance gains electrons in the reaction?
5. **Apply** Describe what would happen if a sample of molten magnesium bromide were subjected to an electric current.

CHEMISTRY AND

SOCIETY

Cadmium Batteries: Recharging and Recycling

Cadmium has been designated as a hazardous waste, a toxic chemical that must be disposed of carefully. That designation creates a problem for millions of people who regularly need to discard cadmium in the form of rechargeable nickel-cadmium (Ni-Cd) batteries.

Cadmium batteries are not the first batteries to cause waste problems. Lead is a toxic waste, and large quantities of lead are found in automobile batteries. Manufacturers now recycle these batteries. Mercury, another hazardous material, coats the electrodes in alkaline batteries. In the 1980s, mercury accounted for one percent of the weight of a cell. By 1992, it was only 0.025 percent of the weight.

How does cadmium get into waste dumps? Most Ni-Cd batteries are sealed in cordless appliances such as toothbrushes, power tools, and vacuum cleaners. When the appliances are worn out or beyond reasonable repair, they are discarded. The consumer does not realize that he or she has contributed to toxic waste. In landfills, the batteries can release cadmium as they lie buried. Eventually, the cadmium can be leached into the water supply. If the batteries are burned with garbage, the ash can become a hazardous waste.

Some states have attempted to add a surcharge on batteries in the same way they have legislated a deposit on bottles and cans. Other states plan to start community recycling projects. The major problem is that cadmium has been designated a hazardous waste and the spent batteries cannot be transported without costly permits. Manufacturers are unwilling to recycle because of the cost and the small volume of returns. Exemptions from requiring transportation permits have been made for lead batteries being transported to recycling, and manufacturers believe some exemption for cadmium would make the recycling cost-effective.

An alternative plan is to find a substitute for the Ni-Cd cell. A nickel-nickel hydride battery has been tested. Other metal hydrides seem to be promising alternatives for toys and photoflash units, but they do not sustain the constant rate of discharge that larger appliances require. So even though some rechargeable batteries may be replaced with batteries of nontoxic metals, the short-term outlook is for continued use of the Ni-Cd battery—but, with increased awareness of the toxic waste problems its disposal creates.

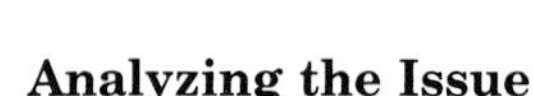

Analyzing the Issue

1. Contact the Health Department in your city or county to determine what steps they have taken to deal with the cadmium problem.
2. Debate with your class the question of labor-saving and energy-saving devices against the need to keep the environment as free of toxic waste as possible.
3. What kind of a campaign could you start to make consumers aware of the dangers in disposing of cordless tools and appliances that run on Ni-Cd batteries?

26.2 Quantitative Electrochemistry

OBJECTIVES

- describe the derivation and uses of a table of standard reduction potentials.
- state the Nernst equation and show how it relates the voltage of a cell to the conditions under which the cell is operating.
- explain the working of a pH meter.
- use Faraday's laws to calculate the amounts of substances reacting in electric cells.

It is often advantageous to know how much electric potential is present in a voltaic cell, or how much metal reacts in an electrolytic cell. In this section you will learn how electric potentials are calculated for voltaic cells as well as how to predict the products of the electrolysis of solutions. You will also learn how to compute the amounts of substances involved in the reactions that take place in both voltaic and electrolytic cells.

Redox Potentials

There is no way in which the potential energy of a single half-cell can be measured. However, the difference in potential between two half-cells in a voltaic cell can be measured by means of a voltmeter. This potential difference is a measure of the relative tendency of two substances to take on electrons. For example, consider the zinc-copper cell that gives a voltmeter reading of 1.103 volts. If we arbitrarily assigned a potential of zero to the copper half-cell, we would say the zinc half-cell has a potential of −1.103 volts. The negative sign indicates that zinc ions are less likely to take on electrons than the copper ions. On the other hand, if we were to assign a potential of zero to the zinc half-cell, we would say that the copper half-cell has a potential of +1.103 volts. In principle, any half-cell could be used for the reference half-cell. That is, any half-cell could be the one assigned zero potential. In practice, the zero potential is assigned to a standard hydrogen half-cell, which can be used in any voltaic cell. The half-reaction is

$$2H^+(aq) + 2e^- \rightleftarrows H_2(g)$$

The hydrogen half-cell, Figure 26.12, consists of a sheet of platinum with a specially treated surface. The platinum is immersed in a one molar (1*M*) ideal solution of H^+ ions. Hydrogen gas, from a cylinder, is bubbled into the solution around the plat-

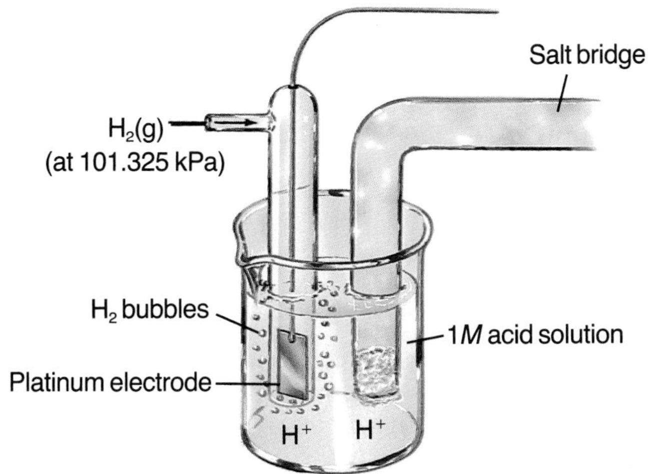

Figure 26.12 The apparatus shown is used in the reduction of hydrogen. The hydrogen reduction reaction is assigned a potential of zero so that the half-cell potentials of other substances can be measured by comparison.

CAREERS IN CHEMISTRY

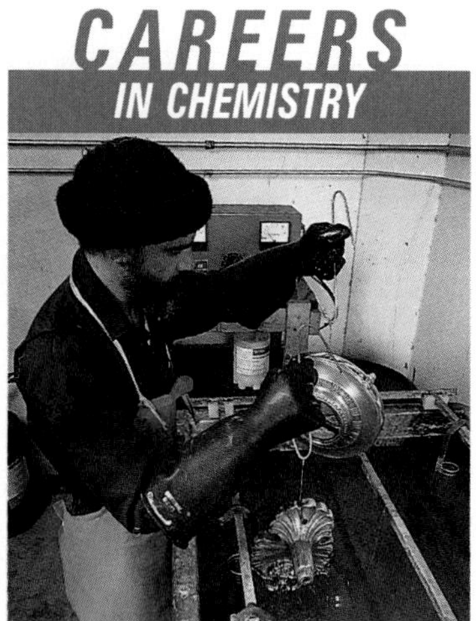

Anodizer

An anodizer has responsibility for the electroplating of metals. Among his or her duties are cleaning the materials to be plated, preparing the correct bath and the plating material, attaching the terminals correctly, making certain the current is accurate and sufficient, and inspecting the final product for full coverage. An anodizer is so named because the metal that is plating out is placed at the anode. The object to be plated is at the cathode. High school chemistry provides a useful starting place for the anodizer. Most of the training is on the job.

inum at a pressure of 101.325 kPa. The H_2 molecules are adsorbed onto the platinum and form the electrode. The hydrogen half-cell is an ideal that can only be approached in practice.

Standard Electrode Potentials

The hydrogen half-cell cannot operate by itself. It must be connected to another half-cell so that a complete voltaic cell exists. If we use the hydrogen half-cell with the zinc half-cell, our voltmeter reads −0.7626 volt. We assign this potential to the zinc half-cell. If we use the hydrogen half-cell with the copper half-cell, the voltmeter reads +0.340 volt. This potential is assigned to the copper half-cell. In this way, potentials can be experimentally determined for almost all oxidation-reduction half-reactions. Some of these half-cell potentials are given in Table 26.1. Potentials are dependent on temperature, pressure, and concentration. The values given in the table are for a temperature of 25°C, a pressure of 101.325 kPa, and a 1*M* ionic concentration. A table of half-cell potentials is of great practical importance. It enables us to predict the direction for a large number of chemical reactions, the maximum voltage that can be produced by a particular voltaic cell, and the products of an electrolytic cell.

In Table 26.1, the substance on the left side of the arrow in each case is an oxidizing agent. It is an electron acceptor. The oxidizing agent with the highest positive potential is the strongest oxidizing agent. The oxidizing agents in this table are arranged in the order of increasing strength from top to bottom. Think of this series of substances as a list arranged in order of ability to attract electrons in competition with other oxidizing agents. The substance on the right side of the arrow in each case is a reducing agent. The strongest reducing agent is at the top of the table and the strength of the reducing agents decreases toward the bottom of the table. Let us see how this table can be used to predict the course of a particular reaction. Remember that electrons must flow from an electron donor (reducing agent) to an electron acceptor (oxidizing agent).

The half-reactions in Table 26.1 are all written as reductions. If we reverse a reaction and write it as an oxidation, the voltage remains the same in magnitude, but is opposite in sign.

$$Li^+ + e^- \rightarrow Li \qquad -3.040 \text{ volts}$$

but

$$Li \rightarrow Li^+ + e^- \qquad +3.040 \text{ volts}$$

Consider what will happen if a strip of copper is immersed in a silver nitrate solution. A possible reaction could be written as

$$Cu(cr) + 2Ag^+(aq) \rightarrow 2Ag(cr) + Cu^{2+}(aq)$$

If this reaction is to occur, copper must give up electrons to silver ions. Table 26.1 shows that silver ions have a greater attraction for electrons (more positive voltage) than copper ions.

Table 26.1

Standard Reduction Potentials (at 25°C, 101.325 kPa, 1*M*)			
Half-Reaction	**E° (Volts)**	**Half-Reaction**	**E° (Volts)**
$Li^+ + e^- \rightarrow Li$	−3.040	$SO_4^{2-} + 4H^+ + 2e^- \rightarrow H_2SO_3 + H_2O$	0.158
$K^+ + e^- \rightarrow K$	−2.924	$Cu^{2+} + e^- \rightarrow Cu^+$	0.159
$Rb^+ + e^- \rightarrow Rb$	−2.294	$HAsO_2 + 3H^+ + 3e^- \rightarrow As + 2H_2O$	0.248
$Cs^+ + e^- \rightarrow Cs$	−2.923	$UO_2^{2+} + 4H^+ + 2e^- \rightarrow U^{4+} + 2H_2O$	0.27
$Ba^{2+} + 2e^- \rightarrow Ba$	−2.92	$Bi^{3+} + 3e^- \rightarrow Bi$	0.317 2
$Sr^{2+} + 2e- \rightarrow Sr$	−2.89	$Cu^{2+} + 2e^- \rightarrow Cu$	0.340
$Ca^{2+} + 2e^- \rightarrow Ca$	−2.84	$O_2 + 2H_2O + 4e^- \rightarrow 4OH^-$	0.401
$Na^+ + e^- \rightarrow Na$	−2.713	$Cu^+ + e^- \rightarrow Cu$	0.520
$La^{3+} + 3e^- \rightarrow La$	−2.37	$I_2 + 2e^- \rightarrow 2I^-$	0.535 5
$Mg^{2+} + 2e^- \rightarrow Mg$	−2.356	$H_3AsO_4 + 2H^+ + 2e^- \rightarrow HAsO_2 + 2H_2O$	0.560
$Ce^{3+} + 3e^- \rightarrow Ce$	−2.34	$O_2 + 2H^+ + 2e^- \rightarrow H_2O_2$	0.695
$Nd^{3+} + 3e^- \rightarrow Nd$	−2.32	$Rh^{3+} + 3e^- \rightarrow Rh$	0.7
$H_2 + 2e^- \rightarrow 2H^-$	−2.25	$Tl^{3+} + 3e^- \rightarrow Tl$	0.72
$Sc^{3+} + 3e^- \rightarrow Sc$	−2.03	$Fe^{3+} + e^- \rightarrow Fe^{2+}$	0.771
$Be^{2+} + 2e^- \rightarrow Be$	−1.97	$NO_3^- + 2H^+ + e^- \rightarrow NO_2 + H_2O$	0.775
$Al^{3+} + 3e^- \rightarrow Al$	−1.676	$Hg_2^{2+} + 2e^- \rightarrow 2Hg$	0.796 0
$U^{3+} + 3e^- \rightarrow U$	−1.66	$Ag^+ + e^- \rightarrow Ag$	0.799 1
$Ti^{2+} + 2e^- \rightarrow Ti$	−1.63	$O_2 + 4H^+(10^{-7}M) + 4e^- \rightarrow 2H_2O$	0.815
$Hf^{4+} + 4e^- \rightarrow Hf$	−1.56	$AmO_2^+ + 4H^+ + e^- \rightarrow Am^{4+} + 2H_2O$	0.82
$No^{3+} + 3e^- \rightarrow No$	−1.2	$NO_3^- + 2H^+ + 2e^- \rightarrow NO_2^- + H_2O$	0.835
$Mn^{2+} + 2e^- \rightarrow Mn$	−1.18	$OsO_4 + 8H^+ + 8e^- \rightarrow Os + 4H_2O$	0.84
$Cr^{2+} + 2e^- \rightarrow Cr$	−0.90	$Hg^{2+} + 2e^- \rightarrow Hg$	0.853 5
$2H_2O + 2e^- \rightarrow H_2 + 2OH^-$	−0.828	$2Hg^{2+} + 2e^- \rightarrow Hg_2^{2+}$	0.911 0
$Zn^{2+} + 2e^- \rightarrow Zn$	−0.762 6	$Pd^{2+} + 2e^- \rightarrow Pd$	0.915
$Cr^{3+} + 3e^- \rightarrow Cr$	−0.74	$NO_3^- + 4H^+ + 3e^- \rightarrow NO(g) + 2H_2O$	0.957
$Ga^{3+} + 3e^- \rightarrow Ga$	−0.529	$Br_2 + 2e^- \rightarrow 2Br^-$	1.065 2
$U^{4+} + e^- \rightarrow U^{3+}$	−0.52	$SeO_4^{2-} + 4H^+ + 2e^- \rightarrow H_2SeO_3 + H_2O$	1.151
$2CO_2 + 2H^+ + 2e^- \rightarrow H_2C_2O_4$	−0.475	$Ir^{3+} + 3e^- \rightarrow Ir$	1.156
$S + 2e^- \rightarrow S^{2-}$	−0.447	$Pt^2 + 2e^- \rightarrow Pt$	1.188
$Fe^{2+} + 2e^- \rightarrow Fe$	−0.44	$O_2 + 4H^+ + 4e^- \rightarrow 2H_2O$	1.229
$Cr^{3+} + e^- \rightarrow Cr^{2+}$	−0.424	$Tl^{3+} + 2e^- \rightarrow Tl^+$	1.25
$2H_2O + 2e^- \rightarrow H_2 + 2OH^-(10^{-7}M)$	−0.414	$Pd^{4+} + 2e^- \rightarrow Pd^{2+}$	1.263
$Cd^{2+} + 2e^- \rightarrow Cd$	−0.402 5	$Cl_2 + 2e^- \rightarrow 2Cl^-$	1.358 28
$Ti^{3+} + e^- \rightarrow Ti^{2+}$	−0.37	$Au^{3+} + 2e^- \rightarrow Au^+$	1.36
$PbI_2 + 2e^- \rightarrow Pb + 2I^-$	−0.365	$Cr_2O_7^{2-} + 14H^+ + 6e^- \rightarrow 2Cr^{3+} + 7H_2O$	1.36
$PbSO_4 + 2e^- \rightarrow Pb + SO_4^{2-}$	−0.350 5	$MnO_4^- + 8H^+ + 5e^- \rightarrow Mn^{2+} + 4H_2O$	1.51
$In^{3+} + 3e^- \rightarrow In$	−0.338 2	$Au^{3+} + 3e^- \rightarrow Au$	1.52
$Tl^+ + e^- \rightarrow Tl$	−0.336 3	$H_5IO_6 + H^+ + 2e^- \rightarrow IO_3^- + 3H_2O$	1.603
$Co^{2+} + 2e^- \rightarrow Co$	−0.277	$2HBrO + 2H^+ + 2e^- \rightarrow Br_2 + 2H_2O$	1.604
$H_3PO_4 + 2H^+ + 2e^- \rightarrow H_3PO_3 + H_2O$	−0.276	$PbO_2 + SO_4^{2-} + 4H^+ + 2e^- \rightarrow PbSO_4 + 2H_2O$	1.698
$Ni^{2+} + 2e^- \rightarrow Ni$	−0.257	$H_2O_2 + 2H^+ + 2e^- \rightarrow 2H_2O$	1.763
$Sn^{2+} + 2e^- \rightarrow Sn$	−0.136	$Au^+ + e^- \rightarrow Au$	1.83
$Pb^{2+} + 2e^- \rightarrow Pb$	−0.125 1	$Co^{3+} + e^- \rightarrow Co^{2+}$	1.92
$Hg_2I_2 + 2e^- \rightarrow 2Hg + 2I^-$	−0.040 5	$S_2O_8^{2-} + 2e^- \rightarrow 2SO_4^{2-}$	1.96
$Fe^{3+} + 3e^- \rightarrow Fe$	−0.04	$O_3 + 2H^+ + 2e^- \rightarrow O_2 + H_2O$	2.075
$2H^+ + 2e^- \rightarrow H_2$	0.000 0	$F_2 + 2e^- \rightarrow 2F^-$	2.87
$Sn^{4+} + 2e^- \rightarrow Sn^{2+}$	0.154	$F_2 + 2H^+ + 2e^- \rightarrow 2HF$	3.053

← Weak Oxidizing Agents/Strong Reducing Agents

Strong Oxidizing Agents/Weak Reducing Agents →

a b c

Figure 26.13 A feathery deposit of silver metal occurs on copper metal immersed in silver nitrate solution (a, b) as copper metal is oxidized. No reaction occurs between copper metal and zinc sulfate solution (c).

$$Cu^{2+} + 2e^- \rightarrow Cu \qquad +0.340 \text{ volt}$$
$$Ag^+ + e^- \rightarrow Ag \qquad +0.7991 \text{ volt}$$

Therefore, this reaction does occur spontaneously. In fact, it may be done in the laboratory by immersing a coil of copper wire in a silver nitrate solution. When silver ions are converted to silver atoms, a beautiful feathery deposit of metallic silver appears on the copper coil. Notice that in the balanced equation for this reaction the number of silver atoms is doubled. This does not affect the value of the standard reduction potential for the reduction of silver. The standard reduction potential of a substance, like density, is an intensive property and therefore does not depend upon the amount of material present.

You can see that any metal in the table can displace any of the metals below it in the table. In other words, *any metal ion in the table can take electrons away from any metal above it. For nonmetals, any nonmetal can donate electrons to any nonmetal below it in the table.*

PROBLEM SOLVING HINT

Remember that a spontaneous reaction must have a positive voltage.

SAMPLE PROBLEM

Predicting Reactions

A strip of copper metal is immersed in a solution containing zinc ions. A possible reaction would be the following.

$$Cu(cr) + Zn^{2+}(aq) \rightarrow Cu^{2+}(aq) + Zn(cr)$$

Does this reaction occur spontaneously as written?

Solving Process:

If this reaction is to occur, copper atoms must give up electrons to zinc ions. From Table 26.1, we obtain the following half-reactions.

$$Zn^{2+} + 2e^- \rightarrow Zn \qquad -0.7626 \text{ volt}$$
$$Cu^{2+} + 2e^- \rightarrow Cu \qquad +0.340 \text{ volt}$$

Copper ions have a greater attraction (more positive voltage) for electrons than zinc ions. Therefore, we can predict that the copper atoms cannot give electrons to the zinc ions. The reaction, as it is written, is not spontaneous. If the copper half-reaction is reversed, and the two half-reactions are added, the voltage for the overall reaction will be negative. A negative voltage for a redox reaction indicates that the reaction is not spontaneous as written and that the reverse reaction will occur.

PRACTICE PROBLEMS

Use Table 26.1 to predict whether the following reactions will occur spontaneously as written.

6. $Al^{3+}(aq) + Ni(cr) \rightarrow Ni^{2+}(aq) + Al(cr)$

7. $Ag^{+}(aq) + Co(cr) \rightarrow Co^{2+}(aq) + Ag(cr)$

8. $Sn^{2+}(aq) + Pb^{2+}(aq) \rightarrow Sn^{4+}(aq) + Pb(cr)$

9. $Au^{3+}(aq) + Fe(cr) \rightarrow Au(cr) + Fe^{2+}(aq)$

Cell Potential

Suppose we wish to find what voltage we could obtain from a cell using $Mg|Mg^{2+}$ and $Cu|Cu^{2+}$. In this cell, Mg is oxidized and Cu^{2+} is reduced. From Table 26.1, we see that

$$Mg^{2+} + 2e^{-} \rightarrow Mg \qquad -2.356 \text{ volts}$$

Thus, the oxidation reaction is

$$Mg \rightarrow Mg^{2+} + 2e^{-} \qquad +2.356 \text{ volts}$$

The reduction reaction is

$$Cu^{2+} + 2e^{-} \rightarrow Cu \qquad +0.340 \text{ volt}$$

The sum of the two potentials is +2.696 volts. This quantity is the voltage of the cell if the ion concentration in each compartment is one molar, and the temperature is 25°C. The direction of electron flow in the external circuit can be determined from the fact that the magnesium atom is losing electrons. Magnesium (the anode) gives up electrons, which flow through the external circuit to the copper electrode. There, the reduction reaction occurs. Would it be possible to construct a cell in which copper metal is used as the anode and magnesium ion is reduced at the cathode? Remember, a negative voltage indicates that the reverse reaction occurs.

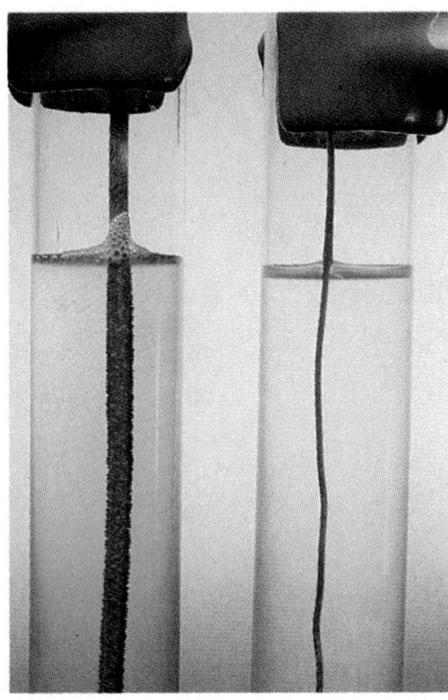

Figure 26.14 Magnesium metal is oxidized by copper(II) ions and copper(II) ions are reduced to copper metal in the test tube on the left. Copper metal does not react with magnesium sulfate solution, as shown by the test tube on the right.

The reduction potentials in Table 26.1 are expressed in volts. This voltage indicates how strong a tendency each half-reaction has to gain electrons. Lithium, at the top of the table, with a voltage of −3.040 has a weak tendency to gain an electron, and fluorine at the bottom of the table has a strong tendency to gain an electron.

SAMPLE PROBLEM

Cell Voltage

What voltage should be produced by the following cell?

$$Fe|Fe^{2+}||Br_2|Br^-$$

Solving Process:

The table of standard reduction potentials lists the voltage for the two half-reactions as

$$Fe^{2+} + 2e^- \rightarrow Fe \qquad -0.44 \text{ V}$$

$$Br_2 + 2e^- \rightarrow 2Br^- \qquad 1.065\ 2 \text{ V}$$

Since we have iron being oxidized to iron(II) ions, the voltage for that half-cell would be +0.44 V. The sum of the two half-cell voltages gives us the cell potential.

$$0.44 + 1.065\ 2 = 1.51 \text{ V}$$

PROBLEM SOLVING HINT

Remember that if a reaction yields a negative voltage, the reverse reaction yields a positive voltage.

PRACTICE PROBLEMS

10. Predict the voltages produced by the following cells. Use Table 26.1.

a. $Zn|Zn^{2+}||Fe^{2+}|Fe$

b. $Mn|Mn^{2+}||Br_2|Br^-$

c. $H_2C_2O_4|CO_2||MnO_4^-|Mn^{2+}$

d. $Ni|Ni^{2+}||Hg_2{}^{2+}|Hg$

e. $Cu|Cu^{2+}||Ag^+|Ag$

f. $Pb|Pb^{2+}||Cl_2|Cl^-$

CHEMISTRY IN DEPTH

Products of Electrolysis

In Section 26.1, we saw that there could be competition for reaction at the electrodes in an electrolytic cell. In the example described, the competition was between sodium ions and water molecules and between water molecules and chloride ions.

To predict which reactions will actually occur at an electrode, we use the table of standard reduction potentials. The energy required to cause an electrode reaction to occur varies directly with the negative voltage of that reaction. The reaction that requires the least amount of energy is associated with the least negative (most positive) voltage. Since most electrolysis reactions take place in water solution, we need to know the voltage required to oxidize and reduce water. In investigating the voltages involved in these reactions, we must remember that water dissociates slightly into H^+ and OH^-. In a neutral solution, both of these ions are found in a concentration of $10^{-7}M$. The reactions of interest to us are the reduction of water,

$$2H_2O + 2e^- \leftrightarrows H_2 + 2OH^- \qquad -0.414 \text{ volt } (10^{-7}M\ OH^-)$$

and the oxidation of water,

$$2H_2O \leftrightarrows 4e^- + 4H^+ + O_2 \qquad -0.815 \text{ volt } (10^{-7}M\ H^+)$$

SAMPLE PROBLEM

Predicting Products of Electrolysis

Predict the products of the electrolysis of aqueous potassium iodide.

Solving Process:

The possible reduction reactions are

$$K^+ + e^- \rightarrow K \qquad -2.924\ V$$

$$2H_2O + 2e^- \rightarrow H_2 + 2OH^- \qquad -0.414\ V$$

The least negative value is the reduction of water, so the product at the cathode is hydrogen gas. The possible oxidations are

$$2I^- \rightarrow I_2 + 2e^- \qquad -0.5355\ V$$

$$2H_2O \rightarrow 4e^- + 4H^+ + O_2 \qquad -0.815\ V$$

The least negative value is the oxidation of iodide ion, so the product at the anode is iodine.

PROBLEM SOLVING HINT

The reaction that requires the least amount of energy is associated with the least negative (most positive) voltage.

PRACTICE PROBLEMS

11. Predict the products of the electrolysis of the following 1*M* aqueous solutions.

a. Na_2SO_4 **c.** CoF_2 **e.** LiBr
b. $CuCl_2$ **d.** $Pb(NO_3)_2$ **f.** NaCl

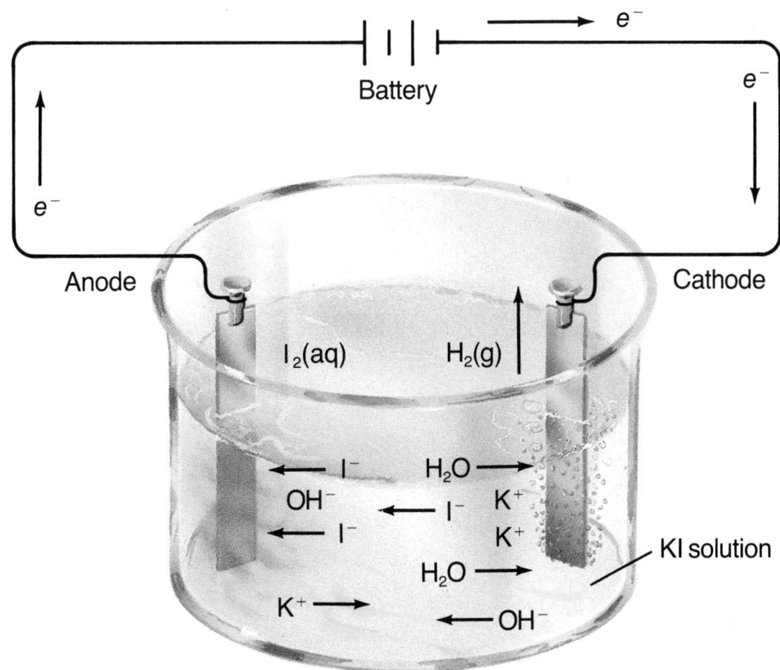

Figure 26.15 In the electrolysis of an aqueous KI solution, H_2O is reduced to H_2 gas at the cathode. Iodide ions are oxidized to I_2 at the anode.

Effect of Conditions on Cells and Energy

In the latter part of the nineteenth century, German chemist H.W. Nernst found a relationship between the voltage of a cell and the conditions under which it was operating. The voltage of a cell under standard conditions is written $E°$. The values in Table 26.1 are $E°$ values for 1*M* solutions at 25°C and 101.325 kPa pressure. If we now call the voltage of a cell at other than standard conditions E, we can use the relationship worked out by Nernst.

$$E = E° - \frac{RT}{nF} \, ln \, \frac{(concentration\ of\ products)}{(concentration\ of\ reactants)}$$

Let us look at each of the factors in the equation. The value of R is the universal gas constant that was developed in the ideal gas equation in Chapter 19. However, in applying R to electric measurements it must be expressed in convenient units. These units will be different from the $dm^3 \cdot kPa/mol \cdot K$ used in gas calculations. Conversion to J/mol·K gives a value of 8.31. The value of T is the absolute temperature in K. The value of n is the number of electrons transferred in the balanced equation representing the change taking place in the cell. The value of F is a conversion factor from volts to joules per mole and is equal to 96 485. The symbol *ln* stands for the natural logarithm. It can be replaced by (2.30 log) when using common logarithms. If we combine $(R/F)2.30$ with the standard temperature, 298 K, we obtain 0.05916. We may then write the Nernst equation in a simpler form.

$$E = E° - \frac{0.05916}{n} \log \frac{(concentration\ of\ products)}{(concentration\ of\ reactants)}$$

at standard temperature

A table of logarithms is given in Appendix B. Let us now apply this equation to a sample problem.

PROBLEM SOLVING HINT

Refer to Appendix B to review logarithms or use the log key of a scientific calculator.

SAMPLE PROBLEM

Nonstandard Concentrations

What would be the voltage of a cell using $Ni|Ni^{2+}$ (2.50*M*) for one half-cell and $Ag|Ag^{+}$ (0.100*M*) for the other half-cell? Assume pressure and temperature to be at standard values.

Solving Process:

(a) Determine the equation for the total cell reaction. Compare the oxidizing abilities of Ni^{2+} and Ag^{+} in Table 26.1. Silver has the greater tendency to gain electrons. On that basis, nickel is oxidized and silver is reduced.

$$Ni(cr) + 2Ag^{+}(aq) \rightarrow Ni^{2+}(aq) + 2Ag(cr)$$

(b) Determine the values to be substituted into the Nernst equation.

$$E = E° - \frac{0.059\ 16}{n} \log \frac{[Ni^{2+}]}{[Ag^{+}]^2}$$

Solids have constant concentrations and, therefore, do not appear in the equation.

From Table 26.1, $E° = \overset{\text{Ni oxidation}}{+0.257 \text{ volt}} + \overset{\text{Ag}^+ \text{ reduction}}{0.799\,1 \text{ volt}} = 1.056$ volts

$n = 2$ (the number of electrons transferred)

$$E = 1.056 \text{ V} - \frac{0.059\,16}{2} \log \frac{[2.50]}{[0.100][0.100]}$$

$$= 1.056 - 0.029\,58(\log 250)$$
$$= 1.056 - 0.029\,58(2.398)$$
$$= 1.056 - 0.070\,93 = 0.985 \text{ volt}$$

Note that the concentration of Ag^+ is squared in the expression because the coefficient of Ag^+ is 2. The concentrations of reactants and products are handled in the Nernst equation exactly as they are in an equilibrium constant expression.

PRACTICE PROBLEMS

12. What voltage could you expect (ideally) from the following cell at 25°C?
$Pb|Pb^{2+}(0.485M)||Sn^{4+}(0.652M)|Sn^{2+}(0.346M)$
13. What voltage could you expect (ideally) from the following cell at 25°C?
$Ni|Ni^{2+}(1.00M)||Cu^{2+}(0.0100M)|Cu$
14. What voltage could you expect (ideally) from the following cell at 25°C?
$Fe|Fe^{2+}(0.720M)||Ag^{+}(0.785M)|Ag$

A Shocking Experience

Have you ever experienced a sharp pain when you began chewing a piece of gum that still had some of the aluminum foil wrapper remaining on it? What is the cause of that pain? If you or your lab partner have metal fillings in your teeth, bite down on a small, clean piece of aluminum foil. When the foil comes in contact with the silver-mercury or silver-tin amalgam in the filling, a voltage is produced. Did you experience a mild, sharp pain in the tooth? Was there a metallic taste in your mouth? The $Al^{3+}|Al$ reduction half-reaction has an $E°$ of -1.676 V. The $Hg^{2+}|Ag_2Hg_3$ reduction half-reaction has an $E°$ of 0.85 V. The aluminum ion reduction reaction has the less positive voltage, so aluminum metal foil will be oxidized to soluble Al^{3+} ions. The presence of these ions in your mouth accounts for the metallic taste. Write the equation for this oxidation half-reaction. A positive voltage would account for the mild, sharp pain. Calculate the voltage of this cell. Is it positive? The fluid in the mouth served as the salt bridge.

pH Meter

The Nernst equation can be applied to half-cells as well as to an entire cell. Consider the hydrogen half-cell at normal temperature and pressure.

$$H^+ + e^- \rightarrow \tfrac{1}{2}H_2$$

$$E = E° - 0.05916 \log \frac{1}{[H^+]}$$

Using $\log 1/x = -\log x$, we can rearrange the above equation.

$$E = E° + 0.05916 \log [H^+]$$

We know that $E°$ for the hydrogen half-cell is 0.0000 volt. We also know that $-\log [H^+] = \text{pH}$. Substituting these values we get

$$E = -0.05916 \times \text{pH}$$

You can readily see from the Nernst equation that any reaction involving hydrogen ions, H^+ (actually hydronium ions, H_3O^+), would have a potential dependent on the concentration of

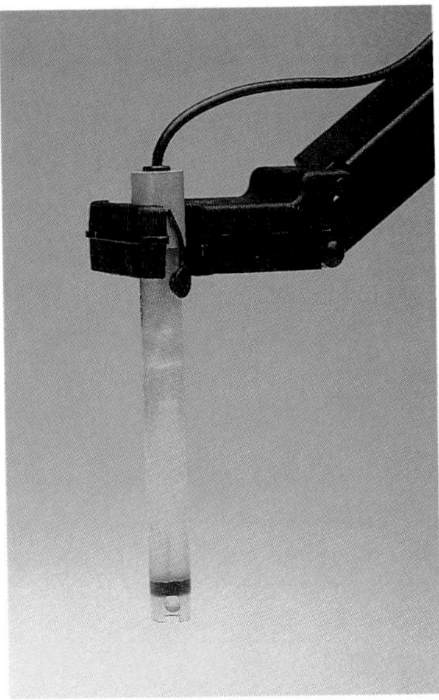

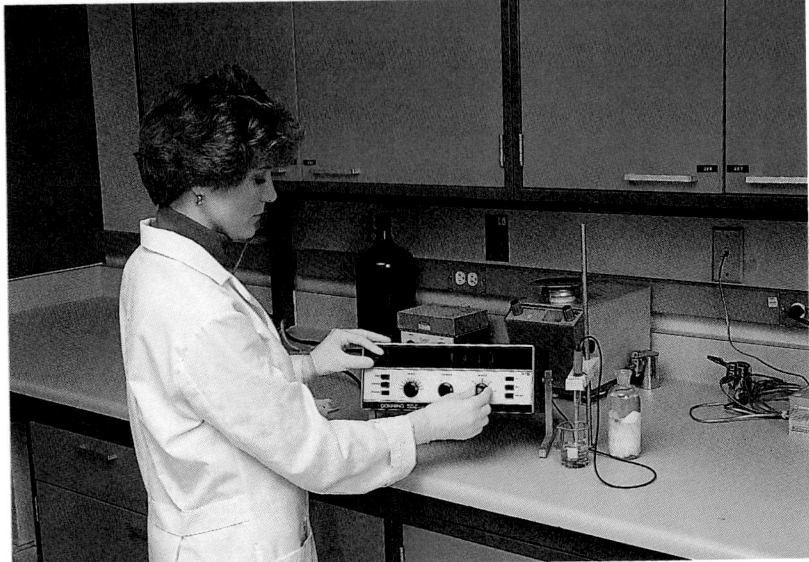

Figure 26.16 The calomel electrode shown on the left is part of the pH meter (right) used to measure electrically the hydrogen ion concentration of a solution. The voltage difference between electrodes is measured in pH units. A pH meter may use a single probe in which the electrodes are combined for convenience, or two separate electrodes.

H^+. This principle is used in a pH meter. You learned in Chapter 24 that a pH meter is used to measure the hydrogen ion concentration, $[H^+]$. By combining a reference electrode with a hydrogen electrode, we can measure the pH of a solution electrically. The reference electrode usually chosen is the saturated calomel (Hg_2Cl_2) electrode because it has a known constant voltage. This electrode is immersed in the solution whose pH is to be determined. The potential developed by the electrode is then measured with a millivoltmeter. Thus, you can see that the voltage developed by the electrode is a linear function of the pH. It is this relationship that enables us to calibrate the pH meter directly in units of pH rather than in millivolts. This calibration saves the chemist the trouble of converting from volts to pH units.

Figure 26.17 Gold is refined by electrolysis. To plate out one gram of gold requires about 1470 coulombs. Gold is plated out on thin gold foil cathodes, which are then melted and poured into bars.

Faraday's Laws

The equations for the electrode half-reactions can be balanced like any other chemical equation. They have the same quantitative significance. Consider the equation for the formation of sodium metal in the electrolysis of molten sodium chloride.

$$Na^+ + e^- \rightarrow Na$$

This equation indicates that 1 mole of electrons is required to produce 1 mole of sodium atoms. However, electricity is not usually measured in moles. The more common unit for electricity is the coulomb. One coulomb (C) is the quantity of electricity produced by a current of one ampere flowing for one second, 1 ampere·second (A·s). A charge of 96 485 coulombs is the equivalent of 1 mole of electrons. This quantity allows us to relate charge to the amount of substance oxidized or reduced in an electrochemical reaction.

$$96\,485 \text{ coulombs} = 1 \text{ mole of electrons}$$

We could produce one mole of sodium by using a current of 10 amperes for 9648.5 seconds. Any combination of amperes and seconds that gives a product of 96 485 A·s will yield one mole of sodium.

In the anode reaction for the electrolysis of sodium chloride, one chlorine molecule requires the release of two electrons. Thus, the production of one mole of chlorine gas will require two moles of electrons.

The principles that we have just discussed are expressed in more concise form as Faraday's laws. They were proposed by Michael Faraday.

Since the amount of electricity produced is equal to the current in amperes multiplied by the time in seconds, we may say:

$$\textit{coulombs} = \textit{amperes} \times \textit{seconds}$$

Further, we can write the following expression.

$$\textit{moles of electrons} = \frac{\text{ampere}}{} \left| \frac{\text{second}}{} \right| \frac{1 \text{ coulomb}}{1 \text{ ampere·second}} \left| \frac{1 \text{ mole } e^-}{96\,485 \text{ coulomb}} \right.$$

From a balanced equation, we obtain the relationship between the formula mass of a substance and the moles of electrons.

$$\frac{\textit{mass of substance}}{\textit{per mole of electrons}} = \frac{\textit{formula mass of substance}}{\textit{moles of electrons transferred}}$$

Combining the two expressions, we get the mathematical statement of Faraday's laws.

$$\textit{mass of substance} = \frac{\text{coulombs}}{} \left| \frac{1 \text{ mole } e^-}{96\,485 \text{ coulombs}} \right| \frac{\text{formula mass}}{\text{mole } e^-}$$

Note that the coefficient of the substance in the equation must be taken into account.

SAMPLE PROBLEMS

1. Faraday's Law Calculation

What mass of copper will be deposited by a current of 7.89 amperes flowing for a period of 1200 seconds?

Solving Process:

The cathode reaction is

$$Cu^{2+}(aq) + 2e^- \rightarrow Cu(cr)$$

and therefore 2 moles of electrons plate out 63.5 g Cu(cr). By combining the expressions, we obtain

$$\frac{7.89 \cancel{A}}{} \left| \frac{1200 \cancel{s}}{} \right| \frac{1 \cancel{C}}{\cancel{A·s}} \left| \frac{1 \cancel{\text{mol } e^-}}{96\,485 \cancel{C}} \right| \frac{1 \cancel{\text{mol Cu}}}{2 \cancel{\text{mol } e^-}} \left| \frac{63.5 \text{ g Cu}}{1 \cancel{\text{mol Cu}}} \right.$$

$$= 3.1 \text{ g Cu}$$

PROBLEM SOLVING HINT
Two moles of electrons are needed to reduce one mole of Cu^{2+} to Cu.

2. Faraday's Law Calculation—Another Example

What mass of Cr^{3+} ion is produced by a current of 0.713 ampere flowing for 12 800 seconds? The equation is

$$14H_3O^+(aq) + 6Fe^{2+}(aq) + Cr_2O_7^{2-}(aq) \rightarrow 6Fe^{3+}(aq) + 2Cr^{3+}(aq) + 21H_2O(l)$$

Solving Process:

You must verify that the moles of electrons involved in the equation as written are equal to six ($Cr_2O_7^{2-} \rightarrow 2Cr^{3+}$). Then, by substituting in the Faraday's law equation, the following expression is obtained.

$$\frac{0.713 \cancel{A}}{} \left| \frac{12\,800 \cancel{s}}{} \right| \frac{1 \cancel{C}}{\cancel{A \cdot s}} \left| \frac{1 \cancel{mol\ e^-}}{96\,485 \cancel{C}} \right| \frac{2 \cancel{mol\ Cr^{3+}}}{6 \cancel{mol\ e^-}} \left| \frac{52.0\ g\ Cr^{3+}}{1 \cancel{mol\ Cr^{3+}}} \right. = 1.64\ g\ Cr^{3+}$$

CHEMISTS AND THEIR WORK

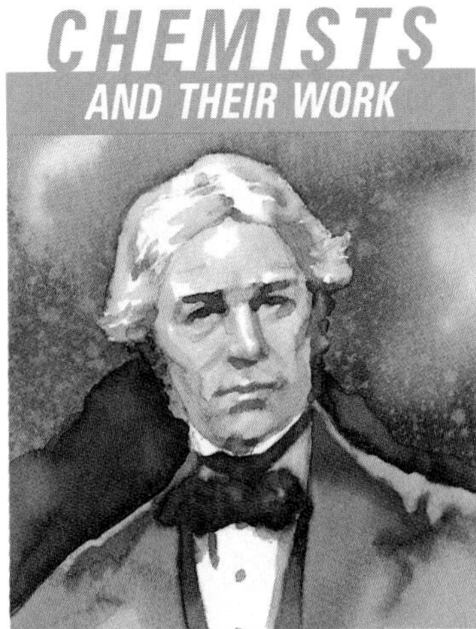

Michael Faraday
(1791-1867)

Michael Faraday began his scientific career in 1813 as an apprentice to a chemist. In 1825, he became the director of the laboratory where he had been apprenticed, and in 1833, he became professor of chemistry at the Royal Institution in England. His discoveries and inventions are numerous, such as liquefying gases, producing temperatures lower than 0°F, and developing the laws of electrolysis named after him. The designs of modern electric motors are based on his work with electricity, and he invented the transformer. Faraday also became a popular lecturer on science topics.

After 1839, Faraday did little work. His mental health had deteriorated, possibly due to his constant exposure to mercury in his experiments.

PRACTICE PROBLEMS

15. How many grams of silver will be deposited by a current of 1.00 A flowing for 9650 s?

16. A current of 5.00 A flows through a cell for 10.0 min. How many grams of silver could be deposited during this time? Write the cathode reaction.

17. What current must be used to plate 1.75 mole of copper on an electrode in 6.24 min?

18. What period of time, in hours, was required if a current of 5.00 A was passed through a salt (sodium chloride) solution and 1.00 mole of chlorine was produced?

19. How many minutes would be necessary to deposit 0.375 g of calcium from a cell with a current of 3.93 A?

One of the chief uses of Faraday calculations is in the electroplating industry. In this industry, objects are coated with a metal for various purposes such as protection against corrosion, added strength, or decorative effect. In an electrolytic cell, the object to be plated is made the cathode. The anode is made of the metal to be deposited on the object, and the solution contains a soluble salt of the metal to be deposited. A potential sufficient to force electrons through the circuit is applied. At the cathode (object to be coated), ions of the plating metal are reduced to atoms on the surface of the object. At the anode, atoms of the plating metal are oxidized to ions which then migrate through the solution. One of the most important electroplating operations is the chrome plating of steel automobile parts. The parts not only look nice and shine beautifully, but are also protected from corrosion. ▲

EVERYDAY CHEMISTRY

The Chemistry of Corrosion

Corrosion, the chemical reactions of metals with their environment, causes billions of dollars of damage to metal structures each year. The corrosion of iron has a large impact on the economy, because much of the iron currently being produced is used to replace or repair iron objects that have been damaged or destroyed by corrosion.

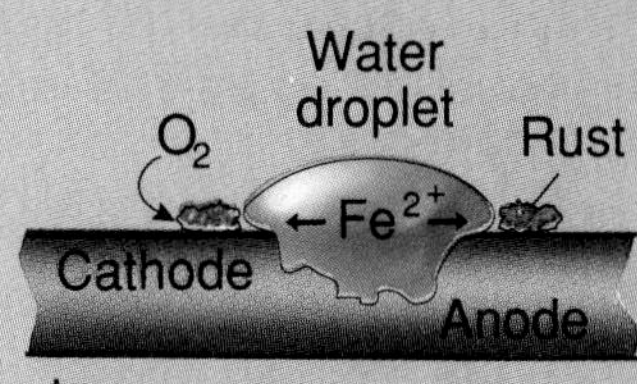

Corrosion processes, including rusting, are electrochemical processes. Water and oxygen are needed for corrosion to occur, and the presence of hydrogen ions speeds up the reaction. The iron is oxidized and, thus, acts as an anode.

$$Fe \rightarrow Fe^{2+} + 2e^-$$

Oxygen in the air reacts with water or water vapor to form hydroxide ions. Because the oxygen gains electrons in this reaction, it is reduced. A piece of soot or other debris from the air acts as an inert cathode at which the following reduction takes place.

$$O_2 + 2H_2O + 4e^- \rightarrow 4OH^-$$

Carbon dioxide from the air can dissolve in the water, forming carbonic acid, which acts as an electrolyte. The essential components of a voltaic cell are present—an oxidation and a reduction reaction and an electrolyte (a solution of H_3O^+ and HCO_3^-). The Fe^{2+} ions can be further oxidized by the oxygen in the air to Fe^{3+} ions, before or after combining with the OH^- ions. The $Fe(OH)_3$ formed soon becomes rust, $Fe_2O_3{\cdot}xH_2O$, a hydrated iron (III) oxide.

If you have ever had to polish silver or brass, you have been introduced to another form of corrosion, tarnish. The term tarnish is usually applied to metals other than iron. Silver reacts with contaminants in the air, mainly hydrogen sulfide, to form silver sulfide, a black deposit on the silver.

The crystal structure of a metal is an important factor in the rate and extent of corrosion. Aluminum and magnesium usually react quickly when exposed to air or water, but the thin, closely packed film of oxide that forms on the surface protects the remaining metal from further corrosion. The rust that forms on the surface of iron, however, is so porous and flaky that the corrosion reaction can readily occur in the underlying metal. This flaking is why iron will rust away completely and must be replaced.

Exploring Further

1. In general, corrosion reactions take place because a more active metal displaces a less active one. Find out what metals are more reactive than iron. How can they be used to prevent corrosion of iron?
2. One home method of cleaning tarnished silver is to heat it on a bed of aluminum foil in a solution of an electrolyte. Write a balanced equation for this reaction.

26.2 CONCEPT REVIEW

20. Explain how the table of reduction potentials is constructed.
21. What conditions affect the voltage of a half-reaction?
22. How does a pH meter work?
23. What two factors affect the mass of a metal that will be deposited in an electroplating operation?
24. **Apply** What voltage would you predict for the following cell? $2Ga + 3Cl_2 \rightarrow 2GaCl_3$

Chapter 26

REVIEW

Summary

26.1 Cells

1. An electric current is the flow or movement of charged particles.
2. A material may be tested for conductivity by providing a means for it to complete a circuit. The presence of current flow may be detected or measured by a device such as a light bulb or galvanometer.
3. Metallic conduction, also called electronic conduction, takes place by the flow of electrons through a metal.
4. Electric current flows whenever two points having potential difference are connected by a conductor. Electric potential difference is measured in volts, V.
5. The rate of electron flow, electric current, is measured in amperes, A.
6. Electrolytic conduction takes place by the migration of ions.
7. Positive ions are called cations. Negative ions are called anions.
8. Reactions in which electrons are gained (reductions) occur at the cathode.
9. Reactions in which electrons are lost (oxidation reactions) occur at the anode.
10. Electrolysis is the process by which an electric current produces a chemical change.
11. In an electrolytic process, the sum of the oxidation and reduction half-reactions occurring at the electrodes represents the total change occurring in the process.
12. A device that can convert chemical energy into electric energy is called a voltaic cell. In a voltaic cell, oxidation and reduction of differing substances occur at the electrodes.
13. A voltaic cell must have a barrier that is porous enough for ions to migrate through but will not allow the solutions to mix.

26.2 Quantitative Electrochemistry

14. The potentials of half-cell reactions are measured against the potential of the hydrogen half-cell, which is assigned a value of zero. Half-cell potential is an indication of the relative tendency of a substance to take on electrons.
15. The standard electrode potentials are a group of half-reactions arranged in order of their reduction potentials under standard conditions.
16. Standard electrode potentials can be used to predict the direction of a reaction, the maximum voltage that can be produced by a particular voltaic cell, and the products of an electrolytic cell.
17. A reaction in a voltaic cell will occur spontaneously if the voltage for the overall reaction is positive.
18. The potential of a voltaic cell is the sum of the potentials of the half-cell reactions.
19. In an electrolytic cell, when more than one half-reaction can occur at an electrode, the reaction with the least negative voltage requires the least energy and, therefore, takes place.
20. The Nernst equation relates the voltage of a cell to the conditions under which the cell is operating.
21. The relationship between hydrogen ion concentration and the voltage of the hydrogen half-cell can be used as the basis for the electrical measurement of pH.
22. One coulomb, C, is the quantity of electricity produced by a current of one ampere flowing for one second. One mole of electrons is equal to 96 485 coulombs.

Key Terms

galvanometer
electric current
salt bridge
metallic conduction
electronic conduction
volt

Chapter 26 REVIEW

ampere
electrolytic conduction
cation
anion
electrolysis
voltaic cell

Review and Practice

25. Describe a general method to determine whether or not a material conducts electric current.
26. A galvanometer is used to detect electric current. What is an electric current?
27. What is the purpose of using a salt bridge in chemistry?
28. What condition must be present in order to cause an electric current to flow from one end of a conductor to another?
29. Under what conditions will an insulator conduct a current?
30. The volt and the ampere are two electrical units. What does each measure?
31. What is an electrolyte? What kinds of substances are electrolytes?
32. What are the two types of electrodes in electrolytic cells and what are their characteristics?
33. What names are given to positive and negative ions?
34. What is electrolysis? What is the function of a generator in electrolysis?
35. At which electrode does oxidation occur? Where does reduction occur?
36. Answer the following for both the anode and the cathode in the electrolysis of a concentrated sodium chloride solution.
    a. What two reactions are possible?
    b. Which reaction actually occurs?
    c. Why does this reaction occur in preference to the other?
37. What is the function of a voltaic cell?
38. Why is it necessary to have a reference half-cell? Why can't we use just a voltmeter to determine half-cell potential?
39. Describe the materials needed and the conditions required in a hydrogen half-cell. Explain how it is used to determine potential of other half-cells.
40. How does the standard reduction potential of a substance or ion relate to its strength as an oxidizing agent? As a reducing agent? Using Table 26.1, give two examples each of very strong oxidizing agents and reducing agents.
41. How does the voltage of a redox reaction relate to its spontaneity?
42. Use Table 26.1 to determine which of the following reactions will proceed spontaneously. For the reactions that do proceed, complete the net ionic equation.
    a. $Na + Cl_2 \rightarrow ?$
    b. $Fe^{2+} + Cu \rightarrow ?$
    c. $Cu + H_2 \rightarrow ?$
    d. $Cu^{2+} + Ag^+ \rightarrow ?$
    e. $Zn + Pb^{2+} \rightarrow ?$
    f. $Fe + Pb^{2+} \rightarrow ?$
43. For each of the reactions in Problem 42 that proceeds, determine the cell potential and the direction of external current flow between substances.
44. Use Table 26.1 to calculate potentials for the following cells.
    a. $Pb|Pb^{2+}||Fe^{3+}|Fe^{2+}$
    b. $Co|Co^{2+}||HBrO|Br_2$
    c. $Cr|Cr^{3+}||NO_3^-|NO_2^-$
45. Calculate the voltaic cell potential for the forward reaction in the following equilibrium.

$$Fe + Ni^{2+} \rightleftarrows Fe^{2+} + Ni$$

Is the forward reaction spontaneous? Why or why not?

46. What maximum voltage could you expect from a cell that consisted of $Al|Al^{3+}$ and $Ag|Ag^+$ half-cells? Write the electrode reactions and the overall reaction.
47. When two or more reactions are possible at an electrode, what factor determines which reaction will take place?

48. What voltage could you expect from the following cell at 25°C?

$Mg|Mg^{2+}(0.170M)||Cl_2|Cl^-(0.100M)$

49. What voltage would be obtained at 25°C from the following cell?

$Cd|Cd^{2+}(0.500M)||Fe^{3+}(0.100M)|Fe^{2+}$ $(0.200M)$

50. How is the potential of the hydrogen half-cell related to the pH of the solution in which it is immersed?

51. What is the potential of the hydrogen half-cell in a solution of pH 9.5?

52. What is the potential of the hydrogen half-cell in a solution in which $[H_3O^+]$ is $4.0 \times 10^{-2}M$?

53. How long, in hours, must a current of 20.0 A be passed through a NaCl solution to produce 40.0 g of NaOH?

54. What is a coulomb? What is the relationship between coulombs and moles of electrons?

55. A current of 4.5 amperes flows through an electrolysis apparatus for 6.0 hours. How many coulombs passed through the circuit? How many moles of electrons were transferred?

56. A silver ion solution is subjected to a current of 5.00 A for 2.00 hours.
   a. Write the cathode reaction.
   b. How much silver would be plated out?

57. What feature of redox reactions allows them to be driven by electric current and to produce electric current in voltaic cells?

58. From each of the following pairs, select the better reducing agent.
   a. Cd, Fe
   b. Ca, Mg
   c. Pb, Zn
   d. Br^-, Cu

59. From each of the following pairs, select the better oxidizing agent.
   a. Ca^{2+}, Li^+
   b. Fe^{3+}, $Hg_2{}^{2+}$
   c. Cl_2, Hg^{2+}
   d. Cu^+, Cu^{2+}

60. Combine each of the following pairs of half-reactions into a balanced ionic equation that will proceed spontaneously in the forward direction. Determine the potential that would result from the reaction taking place under standard conditions.
   a. $Be^{2+} + 2e^- \rightarrow Be$
   $I_2 + 2e^- \rightarrow 2I^-$
   b. $Hg^{2+} + 2e^- \rightarrow Hg$
   $Fe^{2+} + 2e^- \rightarrow Fe$
   c. $Pb^{2+} + 2e^- \rightarrow Pb$
   $Pd^{4+} + 2e^- \rightarrow Pd^{2+}$
   d. $H_3PO_4 + 2H^+ + 2e^- \rightarrow H_3PO_3 + H_2O$
   $H_2O_2 + 2H^+ + 2e^- \rightarrow 2H_2O$

61. The electrolysis of 10.0 g H_2O produces what volumes in dm^3 of O_2 and H_2 at STP? Give each electrode reaction.

62. Use the Nernst equation to compute the voltage of a cell in which one half-cell is made with Zn and $1.00M$ Zn^{2+} and the other half-cell is Zn and $0.100M$ Zn^{2+} at standard conditions.

63. Compute the voltage of a cell in which one half-cell is the hydrogen half-cell with hydrogen at a pressure of 203 kPa and $1.00M$ H^+. The other half-cell is Zn and Zn^{2+} at $1.00M$. All other conditions are standard.

Concept Mastery

64. Why are metals conductors of electricity? Why is the term electronic conduction applied to conduction through metals?

65. What processes will cause an ionic solid to conduct an electric current? Explain why it will not conduct in the solid state.

66. Assume an electric current is passed through molten KBr. Describe the reaction that takes place at each electrode. Explain how the current causes a chemical change that would not otherwise occur in the salt.

67. For the zinc-copper voltaic cell, why does the current in the wire flow from zinc to copper rather than in the opposite direction?

68. Write the two half-reactions for each of the following voltaic cells. Identify the anode and cathode reactions. What is the direction of current flow in the external circuit?
   a. $Mg|Mg^{2+}||Hg^{2+}|Hg$
   b. $Cd|Cd^{2+}||Br_2|Br^-$

69. In a voltaic cell, why is it important that there be a barrier between one electrode with its associated substances and the other electrode with its associated substances?

Application

70. Predict the products of electrolysis of 1*M* aqueous solutions of each of the following substances.
   a. $CaCl_2$
   b. K_2SO_4
   c. $Hg(NO_3)_2$
   d. Na_2S

71. What is produced at the anode and the cathode during the electrolysis of an aqueous solution of copper(II) bromide?

72. A certain process for plating gold from a solution containing Au^{3+} ions requires that a small current be passed for a long period of time. In order to plate 4.00 g of gold in 10.0 days, what current would have to be maintained in the electrolysis apparatus?

73. In the Alcoa process for producing aluminum, an aluminum ore containing aluminum oxide, Al_2O_3, is reacted to produce aluminum chloride, which is then electrolyzed in the molten state.
   a. Write the half-reaction that occurs at each electrode.
   b. Write the net ionic equation for the electrolysis of aluminum chloride and identify the product produced at each electrode.
   c. If a current of 50.0 A is passed through the melt for 2.00 hours, what mass of aluminum is produced?
   d. If a current of 0.500 A were used to produce 1.00 kg of aluminum, how long would the electrolysis take?
   e. In the reaction of part c, how many coulombs and electrons were used?

74. Tin and zinc are both used to coat iron to prevent rust and corrosion. Both are effective as long as the coating is intact. Once the coating is scratched, cracked, or broken, oxygen and water containing dissolved ions are able to come in contact with both the iron and tin or the iron and zinc at the same time. Which metal, tin or zinc, is better able to preserve iron under these conditions? To make your decision, consult Table 26.1 and consider the following possible anode reactions.

$$Fe(cr) \rightarrow Fe^{2+}(aq) + 2e^-$$
$$Zn(cr) \rightarrow Zn^{2+}(aq) + 2e^-$$
$$Sn(cr) \rightarrow Sn^{2+}(aq) + 2e^-$$

The cathode reaction is the reduction of oxygen.

$$O_2(g) + 4H^+(aq) + 4e^- \rightarrow 2H_2O(l)$$

Give the reasoning for your choice of zinc or tin as the better protective coating.

75. In the lead-acid storage battery, such as the one shown in Figure 26.10b, the anodes are lead and the cathodes consist of lead(IV) oxide packed into a conductive metal screen or grid. The electrolyte is an aqueous solution of sulfuric acid, which ionizes strongly in water to produce H^+ (H_3O^+) and HSO_4^- ions. When the battery is generating current, the following half-reactions occur at the electrodes.

Cathode reaction
$$PbO_2(cr) + HSO_4^-(aq) + 3H^+(aq) + 2e^- \rightarrow PbSO_4(cr) + 2H_2O(l)$$

Anode reaction
$$Pb(cr) + HSO_4^-(aq) \rightarrow PbSO_4(cr) + H^+(aq) + 2e^-$$

a. Write a complete net ionic equation for the cell reaction of a storage battery.
b. What are the products of the reaction? At which electrodes are they produced?
c. Examine the net equation and determine what happens to the sulfuric acid as the cell discharges.
d. You can determine how well-charged a battery is by measuring the density of the electrolyte solution. Would you expect the density to increase or decrease as the battery becomes discharged? Density of pure H_2SO_4 is 1.84 g/cm^3.
e. Write the net ionic equation for the reaction that occurs during the charging of a lead-acid battery. What is oxidized and what is reduced during the reaction?
f. In order for a battery to be rechargeable, the products of the discharge reaction must remain on or near the electrodes. Thus, they are available for conversion back to the original reactants by the charging current. What characteristic of the main reaction product makes a lead-acid battery rechargeable?

76. To plate a spoon with silver, it can be placed as one electrode in a solution containing silver ions. The other electrode is composed of silver. In what direction must current flow within the cell? Will the spoon be the anode or the cathode?

77. A common dry cell, shown in Figure 26.10a, is an inexpensive type of "flashlight battery". The cathode half-reaction is the following.

$$2NH_4^+(aq) + 2MnO_2(cr) + 2e^- \rightarrow Mn_2O_3(cr) + 2NH_3(aq) + H_2O(l)$$

What material makes up the anode? Write the half-reaction that occurs at the anode.

78. Ordinary dry cells cannot be recharged. The nickel-cadmium battery, however, is rechargeable. While in operation, cadmium reacts with hydroxide ion at the anode to produce $Cd(OH)_2$. At the cathode, nickel(IV) oxide reacts with water to form nickel(II) hydroxide and hydroxide ions.

a. Write balanced half-reactions for the electrode reactions of this battery.
b. Write the net ionic equation of the reaction that takes place during the recharging of the battery.

Critical Thinking

79. Of acids, hydroxide bases, and salts, which is least likely to conduct in the pure liquid (molten) state? Explain.

80. Why do you think that platinum, despite its cost, is often used for making electrodes for electrolytic reactions?

81. Instead of using a salt bridge or another porous barrier, why not just connect two half-cells with a piece of copper wire?

82. Do the oxidation and reduction reactions of a voltaic cell take place when there is no external conductor connecting the anode with the cathode? Explain.

83. Explain why dry cells cannot really be dry.

84. Why are aluminum nails used to attach aluminum gutter to houses? What might happen if iron nails were used?

Readings

Amato, Ivan. "Biotrodes: Food for Thought." *Science News* 131, no. 6 (February 7, 1987): 92-93.

Ashley, Steven. "Multi-Fuel Cells." *Popular Science* 231, no. 5 (November 1987): 94-95.

Capulong, Eduardo R.C. "Nickel-Hydrogen: A Better Battery?" *Popular Science* 231, no. 5 (November 1987): 67.

Dagani, Ron. "New Class of Superconductors Pushing Temperatures Higher." *Chemical and Engineering News* 66, no. 20 (May 16, 1988): 24-29.

Dworetzky, Tom. "New Life for a Cell: The Ceramic Solution." *Discover* 8, no. 7 (July 1987): 16.

Ellis, Arthur B. "Superconductors." *Journal of Chemical Education* 64, no. 10 (October 1987): 836-841.

Hesse, Joseph J. "Tainted Water." *ChemMatters* 6, no. 1 (February 1988): 13-15.

Lindstrom, Olle. "That Incredible Fuel Cell." *CHEMTECH* 18, no. 8 (August 1988): 490-497; no. 9 (September 1988): 553-559.

McGeough, J.A., and M.B. Barker. "Electrochemical Machining." *CHEMTECH* 21, no. 9 (September 1991): 536-542.

Scott, David. "High-Energy Battery." *Popular Science* 230, no. 3 (March 1987): 60.

Wilt, Rachel. "Leaf Jewelry." *ChemMatters* 5, no. 4 (December 1987): 14-15.

Cumulative Review

1. What is the modern definition of oxidation?

2. What does an oxidizing agent do?

3. Show oxidation numbers for the indicated atoms in the following.
- **a.** U in UF_6
- **b.** Si in $Al_2(SiF_6)_3$
- **c.** Cl in $Pb(ClO_4)_2$ (lead is 2+)
- **d.** As in $Fe_3(AsO_4)_2$ (iron is 2+)

4. Balance the following redox equations by half-reactions.
- **a.** $Fe^{2+} + Cr_2O_7^{2-} \rightarrow Fe^{3+} + Cr^{3+}$ (in acid solution)
- **b.** $S^{2-} + H_2O_2 \rightarrow SO_4^{2-} + H_2O$
- **c.** $IO_3^- + I^- + H_3O^+ \rightarrow H_2O + I_2$

5. Define a Brønsted-Lowry base.

6. What is the name of $H_2Te(aq)$?

7. What is the name of HClO?

8. What kind of elements form oxides that combine with water to form bases?

9. What is the ionization constant for the dihydrogen diphosphate ion, $H_2P_2O_7^{2-}$, if a 0.100*M* solution of the ion has a H_3O^+ concentration of $1.58 \times 10^{-4}M$? What is the percentage ionization?

10. Which would you expect to produce a more acidic solution, $H_4IO_6^-$ or $H_3IO_6^{2-}$?

11. What is the OH^- concentration in a solution with H_3O^+ concentration of $3.61 \times 10^{-6}M$? What is the pH of the solution?

12. A platinum electrode has a mass of 7.601 g before being plated with nickel, and 9.186 g after the nickel is deposited. If the nickel was obtained from an alloy sample with mass 4.525 g, what percentage of the alloy is nickel?

13. A chemist must analyze a sample of brass with mass 2.60 g for copper content. Brass is an alloy of copper and zinc. It is usually contaminated with tin, lead, and iron. The zinc and iron do not interfere with the electroanalysis. The alloy is dissolved in nitric acid, and the tin present is precipitated as SnO_2. The solution is then treated with sulfuric acid, precipitating lead as $PbSO_4$. The resulting solution is prepared for electrolysis to determine the copper content. The mass of the cathode is found to be 26.041 grams. After electrolysis, the cathode is dried and again measured. The new mass is 28.118 grams. What is the percentage of copper in the brass?

CHAPTER **Preview**

CONTENTS

CHEMISTRY IN DEPTH

Energy and Disorder

CHAPTER 27

When you take a ride on a roller coaster, a chain driven by an engine slowly pulls the car to the very top of the first hill. Then, the cars are released. There seems to be no stopping the exciting downhill plunge as the potential energy of the cars is converted to kinetic energy.

Many chemical reactions follow a path similar to that of a roller coaster. An initial input of energy starts the reaction, which may proceed through many up-and-down steps before reaching final products. Each step must provide sufficient activation energy for the next. For example, when you make a fire, the energy to light a match comes from friction. The heat produced by the burning match lights a roll of newspaper, providing even more heat to start wood sticks, which then ignite logs in a fireplace. In this chapter, you'll see how chemists look at energy changes.

OBJECTIVES

- state two reasons that reactions occur.
- calculate changes in internal energy.
- state the reasons that enthalpy changes occur in chemical reactions.
- calculate enthalpies of formation and use them to calculate enthalpies of reaction.

27.1 Introductory Thermodynamics

If you play a portable radio and tape player for an hour, it will use about 200 kJ of energy. In comparison, people use a lot more energy to stay warm on cold days and cool on hot days, to move about by car or plane, or to cook food. When energy is used to run factories, heat and cool stores, plow fields, and transport goods by truck and train, it turns out that an average of 4000 kJ of energy are used by each person in the United States per hour around the clock. Energy consumption in the United States is presently about 8.5×10^{18} joules per year. In order to supply this need, chemists and engineers are constantly investigating more efficient methods of providing usable energy and using that energy. There are limits, however, to how efficient these processes can be made. In this section we will study some of the laws that describe the transfer of energy, including those laws limiting our ability to convert energy from one form to another.

Why Reactions Occur

RULE OF THUMB

RULE OF THUMB

Exothermic reactions generally take place spontaneously. On the other hand, endothermic reactions are generally not spontaneous. In everyday life, we can see that changes in nature are usually "downhill." *That is, a system in nature tends to go from a state of higher energy to one of lower energy.* A ball will roll down a hill spontaneously, but not up. *Also, natural processes tend to go from an orderly state to a disorderly one.* Consider a box containing 36 marbles arranged in three layers—a layer of 12 red, a layer of 12 blue, and a layer of 12 white marbles, as shown in Figure 27.1. Now, suppose you empty the box of marbles into a bag, shake the bag, and pour the marbles back into the box. The colors will then be randomly arranged.

There are some exceptions to these two general rules. For example, it is possible to form products that are at a higher energy or in a more ordered state than the reactants. Keep in mind that unless otherwise stated we will be concerned with reactions taking place at constant temperature and constant pres-

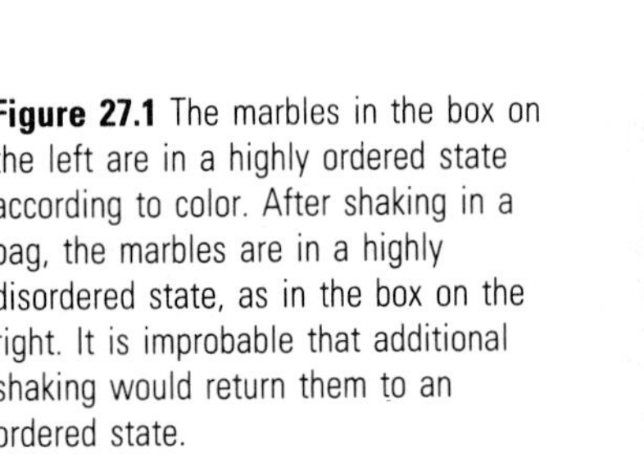

Figure 27.1 The marbles in the box on the left are in a highly ordered state according to color. After shaking in a bag, the marbles are in a highly disordered state, as in the box on the right. It is improbable that additional shaking would return them to an ordered state.

sure. Reactions taking place at constant temperature are called **isothermal processes.** Reactions that take place at constant pressure are called **isobaric processes.** Scientists call studies concerning the flow of energy **thermodynamics.**

Word Origins

isos: (GK) equal, same
therme: (GK) heat
baros: (GK) weight

isothermal—of equal heat (constant temperature)
isobaric—of equal weight (constant pressure)

therme: (GK) heat
dynamikos: (GK) power, strength

thermodynamics—heat power, study of the flow of energy

Internal Energy

Every system has some internal energy. The **internal energy** is symbolized by U. Since we will be interested only in changes in the internal energy of a system, ΔU, we do not have to know absolute values of U for systems.

There are two ways of transferring energy to a system, by heating the system or by doing work on it. A system may also transfer energy to its surroundings by giving off heat or by doing work on the surroundings. The change in energy of a system undergoing a process can be represented by the equation

$$\Delta U = q + w$$

In this equation, ΔU is positive when energy is added to the system, q represents the amount of heat *absorbed by* the system, and w represents the amount of work *done on* the system. Thus, q has a *positive* sign if heat flows into the system and a *negative* sign if heat flows out of the system. The sign of w is negative when the system does work on the surroundings, as when a battery does work to start a car. The sign of work is positive when the surroundings do work on the system, as when a piston compresses a gas in a cylinder.

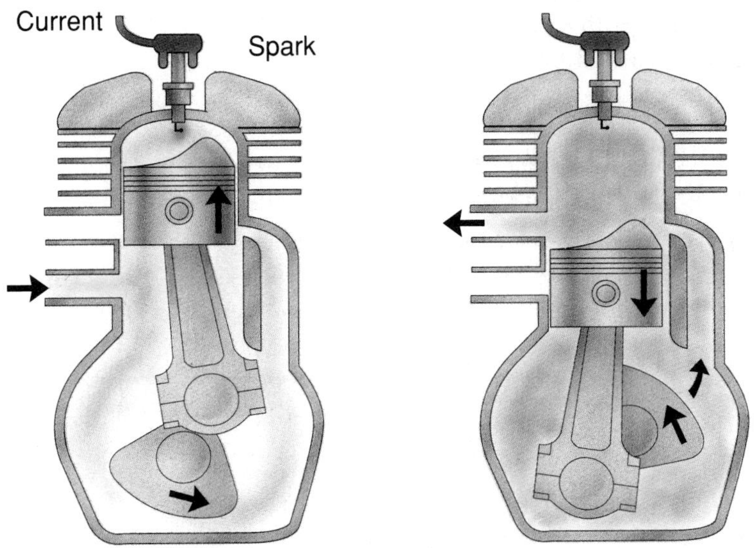

Figure 27.2 Many small gasoline engines, such as those used on chain saws, are two-stroke engines. In the left drawing, the surroundings do work on the system by compressing the gases as the piston moves upward. As the gases burn and expand, the system does work on the surroundings by pushing the piston downward to complete the cycle.

There are several ways of doing work, but the most important to a chemist is the expansion or compression of the system, that is, pressure-volume work. If the system expands, it is doing work on the surroundings, and energy is being transferred from the system. Other kinds of work include electric, magnetic, elastic, and surface tension.

SAMPLE PROBLEMS

1. Change in Internal Energy

A system receives 466 kJ of heat from its surroundings and the surroundings do 56.0 kJ of work on the system. What is the change in the system's internal energy?

Solving Process:

The change in internal energy is represented by

$$\Delta U = q + w$$

Since heat is received by the system, q is positive. Also, since work is done on the system, w is positive. Therefore,

$$\Delta U = 466 + 56.0 = 522 \text{ kJ}$$

2. Change in Internal Energy

What would be the change in internal energy in the previous example if the work had been done by the system instead of on the system?

Solving Process:

Since the work was done by the system, w is negative.

$$\Delta U = 466 - 56.0 = 4.10 \times 10^2 \text{ kJ}$$

PROBLEM SOLVING HINT

Remember that both q and w represent energy put into the system. If the system gives off heat, q is negative; if the system does work, w is negative.

PRACTICE PROBLEMS

1. A certain system absorbs 350 joules of heat and has 230 joules of work done on it. What is the value of ΔU?
2. For a certain change, the value of ΔU is -2.37 kJ. During the change, the system absorbs 650 J. How much work did the system do?
3. A system receives 419 kJ of heat from its surroundings and does 389 kJ of work on the surroundings. What is the change in its internal energy?
4. A system gives off 196 kJ of heat to the surroundings and the surroundings do 4.20×10^2 kJ of work on the system. What is the change in the internal energy of the system?

State Functions

If you measure the distance from the center of Baltimore to the center of Philadelphia on a map of the eastern United States, you will find that the distance is 90 ± 1 miles. However, if you drive from one city to the other by the most direct interstate highway, you will discover that you actually travel 98 miles. On an older highway, you will drive 103 miles. Although you may travel many different paths, the actual distance between the two cities stays constant.

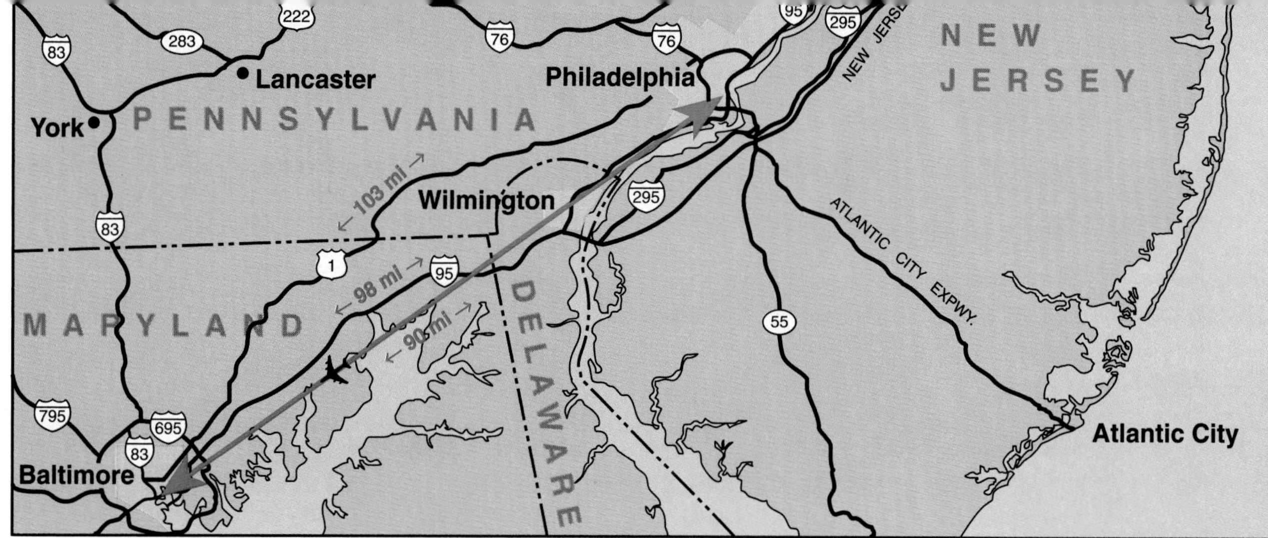

Figure 27.3 If you drive from Baltimore to Philadelphia on I-95, you will travel about 98 miles. If you take U.S. Route 1, you will travel about 103 miles. No matter which route you take, you will have changed your position by 90 miles, the direct distance between the two cities.

Several quantities in thermodynamics are similar to the distance between two cities. They do not vary and do not depend on the path taken from initial conditions to final conditions. For instance, when the temperature of a system is changed, the change is represented by ΔT. No matter how the change is made,

$$\Delta T = T_2 - T_1$$

where T_2 is the final temperature and T_1 is the initial temperature. The same can be said for volume and pressure.

$$\Delta V = V_2 - V_1 \quad \text{and} \quad \Delta P = P_2 - P_1$$

Such characteristics as T, V, and P are called state functions. A **state function** is one whose value depends only on the current state of the system. The amount of change in a state function depends only on the initial and final states, not on the path followed in getting from the initial state to the final state. Internal energy, U, is also a state function. However, q and w, heat and work, depend on the path by which a system gets from an initial state to a final state. They are analogous to the actual mileage traveled from Baltimore to Philadelphia.

Enthalpy

In Chapter 17, you studied the energy required to melt a solid and boil a liquid under the terms *enthalpy of fusion* and *enthalpy of vaporization*. In Chapter 20, you learned about the energy changes that accompany the dissolving of substances, under the name *enthalpy of solution*. Now, you will learn more about enthalpy and see that it includes more than just heat transfer.

If we rearrange the internal energy equation we get

$$q = \Delta U - w$$

In the laboratory, most work is pressure-volume work. At constant pressure, we have

$$w = -P\Delta V$$

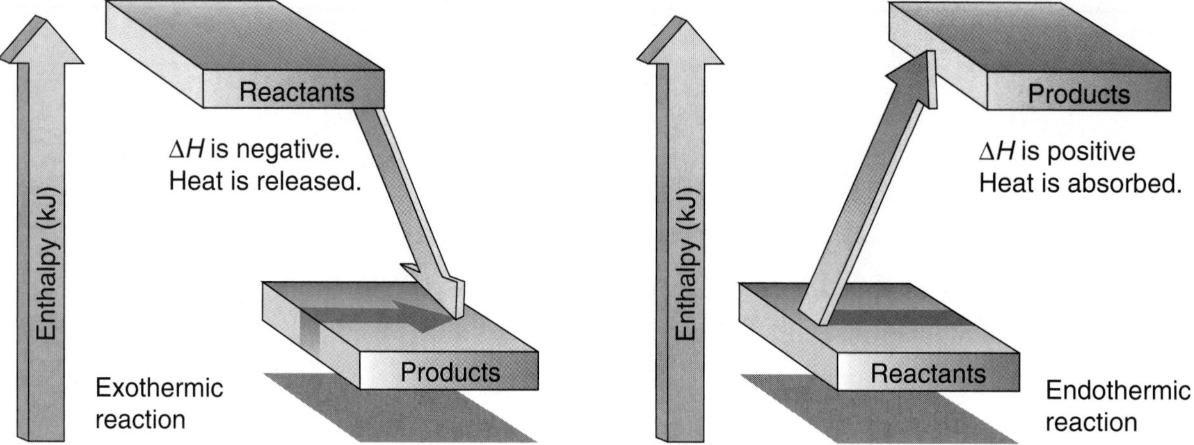

Figure 27.4 The products in an exothermic reaction (left) are lower in enthalpy than the reactants. The products of an endothermic reaction (right) are higher in enthalpy than the reactants.

endon: (GK) in, within
thalpein: (GK) to warm
enthalpy—internal warmth, energy in (or out)

Substituting, we obtain

$$q_p = \Delta U + P\Delta V$$

which can be rewritten as

$$q_p = \Delta(U + PV)$$

The subscript p represents a constant pressure process. The quantity $U + PV$ occurs frequently in energy considerations and is given the name **enthalpy** and the symbol H. Thus,

$$q_p = \Delta H$$

Enthalpy is a state function, so

$$\Delta H = H_2 - H_1$$

In an exothermic reaction the products have less enthalpy than the reactants, so $\Delta H < 0$. In an endothermic reaction $\Delta H > 0$.

You can think of the exothermic change as a downhill change. Because the products have less energy than the reactants, the system went down the "energy hill." Conversely, if the reaction is endothermic, the system went up the "energy hill." Still another way of considering these changes is in terms of the "height" of the energy in the system. If energy is produced, the height of the energy is reduced ($\Delta H < 0$). If energy is absorbed, the height of the energy content is increased ($\Delta H > 0$).

The change in enthalpy that occurs when a chemical reaction takes place is due primarily to the energy required to break the chemical bonds in the reactants and the energy produced by forming the chemical bonds in the products. A balanced chemical equation can be used to represent the energy absorbed or released during a chemical reaction.

When carbon (as coal) is burned, energy is released.

$$C(cr) + O_2(g) \rightarrow CO_2(g) + \text{energy} \quad (393.5 \text{ kJ})$$

One mole of carbon reacts with one mole of oxygen to produce one mole of carbon dioxide and 393.5 kJ of energy. The energy released is called the enthalpy of reaction and is represented by ΔH. In this case, $\Delta H = -393.5$ kJ.

FRONTIERS

Cool Sweats and Hot Socks

In summer you might wear a T-shirt and in winter a sweater because one keeps you cool and the other keeps you warm. As you know, the cooling and warming don't occur because the materials in the garments absorb or produce heat. The T-shirt keeps you cool by allowing perspiration to evaporate and the sweater keeps you warm by insulating your body. But wouldn't it be nice to have clothing designed to keep you comfortable as the thermometer either rose or fell?

Fabrics that have been treated with polyethylene glycol, PEG, can generate or absorb heat, depending on air temperature. Presently these materials are being used to produce thermal underwear and stockings, but ski wear using PEG is also being developed. The unusual properties of the clothing result from physical changes in PEG that has been blended with the clothing fabric. PEG is a polymer, a long, chainlike molecule made up of repeating units. The molecular mass of the PEG used to treat the fabric is about 1000.

PEG polymers can be visualized as long coiled springs that change tension as the temperature changes. At room temperature, the polymers are loosely coiled. As the temperature falls, the polymers deform into more compressed shapes. In doing so, they release heat. The deformation temperature of a PEG depends upon its molecular mass. Thus, clothing that can release heat at specific temperatures can be designed.

When temperatures rise, the deformed PEGs absorb heat and uncoil. In doing so, the polymers act as cooling mechanisms. Researchers hope to exploit this property by developing PEG-treated fabrics that cool the skin by directly absorbing its heat. The day of air-conditioned clothing may not be that far in the future.

Standard States

The height of a hill cannot be measured in absolute terms. We can describe its level above the surrounding plain or we can say it is a certain number of meters above sea level. The figure we use depends upon the standard of reference used. In the same way, we cannot measure enthalpy in absolute terms. We can only measure changes in enthalpy. Our reference will be substances in their standard states. By standard states we mean the enthalpy they have at 298.15 K (25°C) and 100.000 kPa. Note that standard conditions in thermodynamics are *not* the same as those for gas laws. We specify both the temperature and the pressure, since a change in either can have an effect on the enthalpy of a substance. *In measuring enthalpy, we arbitrarily set the enthalpy of the free elements equal to zero.* A free element is one that is not in a compound.

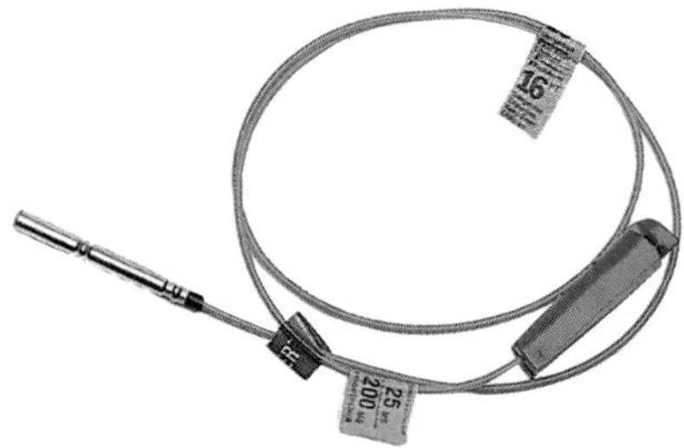

Figure 27.5 Mercury fulminate, used in blasting caps (left) is thermodynamically unstable. It detonates easily, producing large amounts of energy. Mercury fulminate is too unstable to be used in large quantities, but is used to supply activation energy for the detonation of dynamite (right).

Enthalpy of Formation

In reaction of carbon and oxygen, we know that carbon and oxygen are assigned enthalpies of zero because they are free elements. Therefore, the standard molar enthalpy for the formation of carbon dioxide must be −393.5 kJ/mol. The change in enthalpy when one mole of a compound is produced from the free elements in their standard states is known as the standard **enthalpy of formation.** This quantity is expressed in units of kilojoules per mole. A negative sign represents an exothermic reaction. Thus the compound has less enthalpy than the elements from which it was formed. By the same reasoning, a compound produced by an endothermic reaction would have a positive enthalpy of formation. Compounds like CO_2 with large negative enthalpies of formation are thermodynamically stable. **Thermodynamic stability** depends on the amount of energy that would be required to decompose the compound. One mole of CO_2 would require 393.5 kJ of energy to decompose it to the elements carbon and oxygen. On the other hand, mercury fulminate, $Hg(OCN)_2$, produces 268 kJ when one mole decomposes. It is explosive and is used in making detonator caps. Thus, CO_2 has higher thermodynamic stability than $Hg(OCN)_2$.

Appendix Table A-6 lists the enthalpies of formation for some compounds. The symbol used for enthalpy of formation is ΔH_f°. The superscript "°" is used to indicate that the values given are those at 100.000 kPa and 25°C. The subscript "$_f$" designates the value as enthalpy of formation.

Calculation of Enthalpy of Reaction

Let us apply the law of conservation of energy to a reaction. To do so, we will make use of the mathematical symbol, Σ. This Greek uppercase sigma is used to represent a sum. Thus, ΣQ is the sum of all values of Q. Now, consider the reaction *reactants* → *products*. The enthalpy of the products, $\Sigma \Delta H_f^\circ{}_{(\text{products})}$, must equal the enthalpy of the reactants, $\Sigma \Delta H_f^\circ{}_{(\text{reactants})}$, plus any change in enthalpy (ΔH_r°) during the reaction.

$$\Sigma\Delta H_{f\,(products)}^{\circ} = \Sigma\Delta H_{f\,(reactants)}^{\circ} + \Delta H_r^{\circ}$$

Solving for ΔH_r° we get

$$\Delta H_r^{\circ} = \Sigma\Delta H_{f\,(products)}^{\circ} - \Sigma\Delta H_{f\,(reactants)}^{\circ}$$

If the enthalpy of formation of each reactant and product is known, we can calculate the amount of energy produced or absorbed. We can then predict whether a reaction will be exothermic or endothermic.

SAMPLE PROBLEM

Enthalpy Change

Calculate the enthalpy change in the following reaction.

carbon monoxide + oxygen → carbon dioxide

Solving Process:

First, write a balanced equation. Include all the reactants and products.

$$2CO(g) + O_2(g) \rightarrow 2CO_2(g)$$

Each formula unit represents one mole. Remember that free elements have zero enthalpy by definition. Using the table of enthalpies of formation, Appendix Table A-6, the total enthalpy of the reactants is

$$\Sigma\Delta H_{f\,(reactants)}^{\circ} = \frac{2\ \cancel{\text{mol CO}}}{} \left| \frac{-110.5\ \text{kJ}}{\cancel{\text{mol CO}}} \right. + 0\ \text{kJ} = -221.0\ \text{kJ}$$

The total enthalpy of the product ($2CO_2$) is

$$\Sigma\Delta H_{f\,(products)}^{\circ} = \frac{2\ \cancel{\text{mol CO}_2}}{} \left| \frac{-393.5\ \text{kJ}}{\cancel{\text{mol CO}_2}} \right. = -787.0\ \text{kJ}$$

The difference between the enthalpy of the reactants and the enthalpy of the product is

$$\Delta H_r^{\circ} = \Sigma\Delta H_{f\,(products)}^{\circ} - \Sigma\Delta H_{f\,(reactants)}^{\circ}$$

$$\Delta H_r^{\circ} = -787.0\ \text{kJ} - (-221.0\ \text{kJ}) = -566.0\ \text{kJ}$$

This difference between the enthalpy of the products and the reactants (−566.0 kJ) is released as the enthalpy of reaction.

$$2CO(g) + O_2(g) \rightarrow 2CO_2(g) + \textit{enthalpy of reaction}$$

PROBLEM SOLVING HINT

Be careful with signs. Most enthalpies of formation are negative. Likewise, enthalpies of reaction for most spontaneous processes are negative.

PRACTICE PROBLEMS

5. Compute ΔH_r° for the following reaction.
$2NO(g) + O_2(g) \rightarrow 2NO_2(g)$
6. Compute ΔH_r° for the following reaction.
$4FeO(cr) + O_2(g) \rightarrow 2Fe_2O_3(cr)$

Figure 27.6 Shown here is an illustration of Hess's law. When Os and O_2 react to form gaseous OsO_4, the same change in enthalpy occurs whether the reaction goes in one step (left) or in two steps (right).

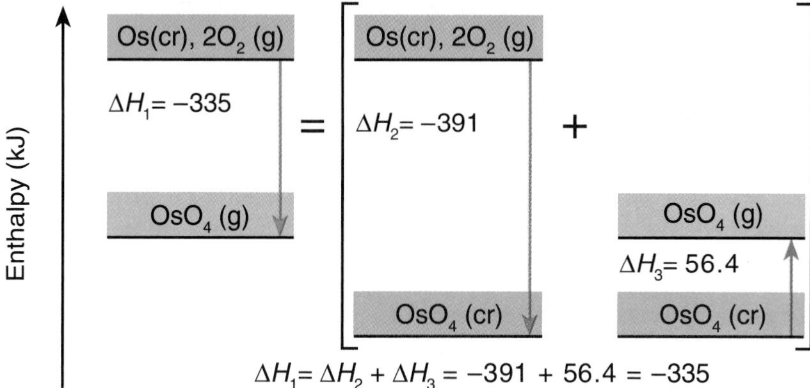

Hess's Law

Consider a reaction A → C that can be broken into two steps, A → B and B → C.

$$\Delta H_{r(1)}{}^{\circ} = \Delta H_f^{\circ}B - \Delta H_f^{\circ}A \qquad \Delta H_{r(2)}{}^{\circ} = \Delta H_f^{\circ}C - \Delta H_f^{\circ}B$$

The enthalpy change for the overall change of A to C is $\Delta H^{\circ} = \Delta H_{r(1)}{}^{\circ} + \Delta H_{r(2)}{}^{\circ}$ because enthalpy is a state function. The principle just illustrated is known as **Hess's law:** *the enthalpy change for a reaction is the sum of the enthalpy changes for a series of reactions that add up to the overall reaction.*

For the reaction in the previous sample problem, we can make the steps

$$C + O_2 \rightarrow CO_2 \qquad \Delta H = -393.5 \text{ kJ}$$
$$2CO \rightarrow CO_2 + C \qquad \Delta H = -172.5 \text{ kJ}$$

Adding, we get $2CO + O_2 \rightarrow 2CO_2$, $\Delta H = -566.0$ kJ, which is the same result as before.

PRACTICE PROBLEMS

7. Barium oxide reacts with sulfuric acid as follows.

$$BaO(cr) + H_2SO_4(l) \rightarrow BaSO_4(cr) + H_2O(l)$$

Calculate the enthalpy of the reaction from these data:

$$SO_3(g) + H_2O(l) \rightarrow H_2SO_4(l) \qquad \Delta H_r{}^{\circ} = -78.2 \text{ kJ}$$
$$BaO(cr) + SO_3(g) \rightarrow BaSO_4(cr) \qquad \Delta H_r{}^{\circ} = -213.4 \text{ kJ}$$

27.1 CONCEPT REVIEW

8. List two ways to change the internal energy of a system.

9. In a reaction whose enthalpy change is positive, compare the energy involved in breaking the bonds of reactants with that involved in forming bonds of products.

10. State two reasons that reactions occur.

11. Apply Does the burning of propane in an outdoor grill have a positive or negative enthalpy of reaction? Explain.

CHEMISTRY AND
SOCIETY

Environmental Quality

The news media today are full of articles expressing concern about the quality of our environment. In some localities the air is polluted. In other areas, waterways are contaminated. The very existence of some species of plants and animals is threatened. Yet, there are points of encouragement.

As an example of what has been done and what still remains to be done, consider the Great Lakes region. This natural drainage system was in serious trouble when both U.S. and Canadian clean-up efforts began in 1972. A bilateral agreement between the United States and Canada established goals for reducing pollutants and a mechanism for the joint monitoring of compliance. The first target was phosphorus pollution, which causes algal blooms and subsequent depletion of the oxygen in the water when the algae die. Sewage treatment, low-phosphorus detergents, and restriction of phosphorus discharge from industry have reduced the phosphorus by 80 percent.

After the initial cleanup of the lakes, some less-obvious problems emerged. One was farm runoff of fertilizer, wastes, and pesticides. However, urban runoff can be worse because lawns and golf courses often get a more intensive fertilizer and pesticide treatment than do farmers' fields. The agricultural runoff can be controlled through no-till planting, careful use of fertilizer, and animal waste management.

Only in the last few years has it been recognized that many contaminants reach the lakes through the air. Automobile exhaust contributes nitrogen in the form of NO_x, which represents a mixture of nitrogen oxides. Contaminated sewage sludge that is spread on fields as fertilizer or burned in municipal incinerators also contributes toxic substances to the atmosphere. Treatment of toxic waste and contaminated groundwater by aeration results in airborne contaminants. Contaminated groundwater discharging into the lakes contributes pollutants. The sediments in the lakes have accumulated pollutants and can give up these pollutants over a long period of time, thus complicating cleanup efforts.

Enormous progress has been made in cleaning up the whole Great Lakes region. At one time, Lake Erie was considered "dead." Today, commercial fishing once again goes on. Even so, the Great Lakes cleanup has a long way to go.

Analyzing the Issue.

1. What is thermal pollution? What kinds of processes contribute to thermal pollution? What steps do industries take to avoid thermal pollution?
2. Find out how phosphorus pollution depletes the oxygen in bodies of water.
3. Is air or water pollution a problem in your community? If so, research the problem, its cause, and any measures being taken to clean up the pollution. Organize a debate on the economic costs of the cleanup versus the health benefits obtained.

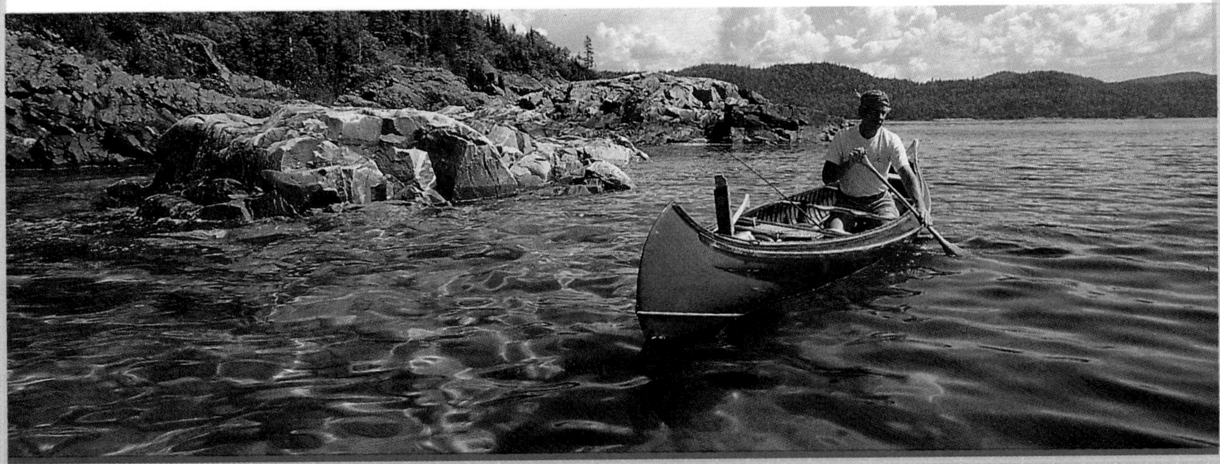

OBJECTIVES

- describe and give examples of changes in entropy.
- relate Gibbs free energy to the spontaneity of reactions and to equilibrium.
- perform calculations involving Gibbs free energy and entropy, equilibrium constants, or cell voltage.

27.2 Driving Chemical Reactions

Earlier in this chapter, it was mentioned that nature tends to run downhill in both energy and order. What happens in a change in which energy is produced but the system becomes more orderly? Or, what happens when a change is endothermic but produces disorder? In this section you will learn how tendencies toward lower energy and greater disorder combine to determine what really happens in a change.

Entropy

Highly exothermic reactions tend to take place spontaneously. However, weak exothermic reactions and some endothermic reactions can also occur spontaneously. Sometimes these reactions will proceed under stronger reaction conditions, such as a temperature increase.

Consider the reaction of steam with very hot carbon to form carbon monoxide and hydrogen.

$$C(cr) + H_2O(g) + \text{energy} \rightarrow H_2(g) + CO(g)$$

The products have a higher enthalpy than the reactants. Therefore, since energy is absorbed in this process, ΔH is positive. It has been determined experimentally that if 1 mole of carbon reacts with 1 mole of steam, then $\Delta H = 131$ kJ. Most spontaneous reactions seem to have negative ΔH values. Since ΔH is positive in this reaction, some additional factor must be involved.

This additional factor is the degree of disorder or **entropy.** We have seen in Chapter 16 that there is a very orderly arrangement of atoms in crystalline solids. In liquids, there is less order. Two liquids dissolved in each other make a more disordered system than the two liquids separated. Gases lack any orderly arrangement. A gas at high temperature is more disordered than one at a low temperature. The degree of disorder, or entropy, is represented by the symbol S. Entropy is a state function.

Word Origins

endon: (GK) in, within
tropein: (GK) to turn around, change
entropy—internal turning, disorder

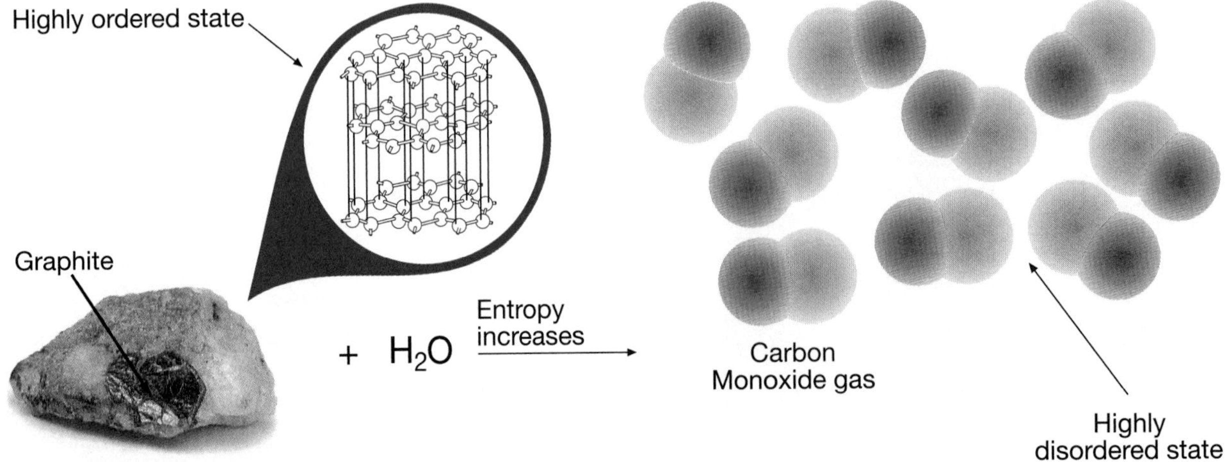

Figure 27.7 In the formation of CO gas from graphite and steam, the degree of disorder increases because the carbon atoms move from a highly ordered crystalline arrangement in the graphite to a highly disordered state as gaseous carbon monoxide. Thus, there is an increase in entropy.

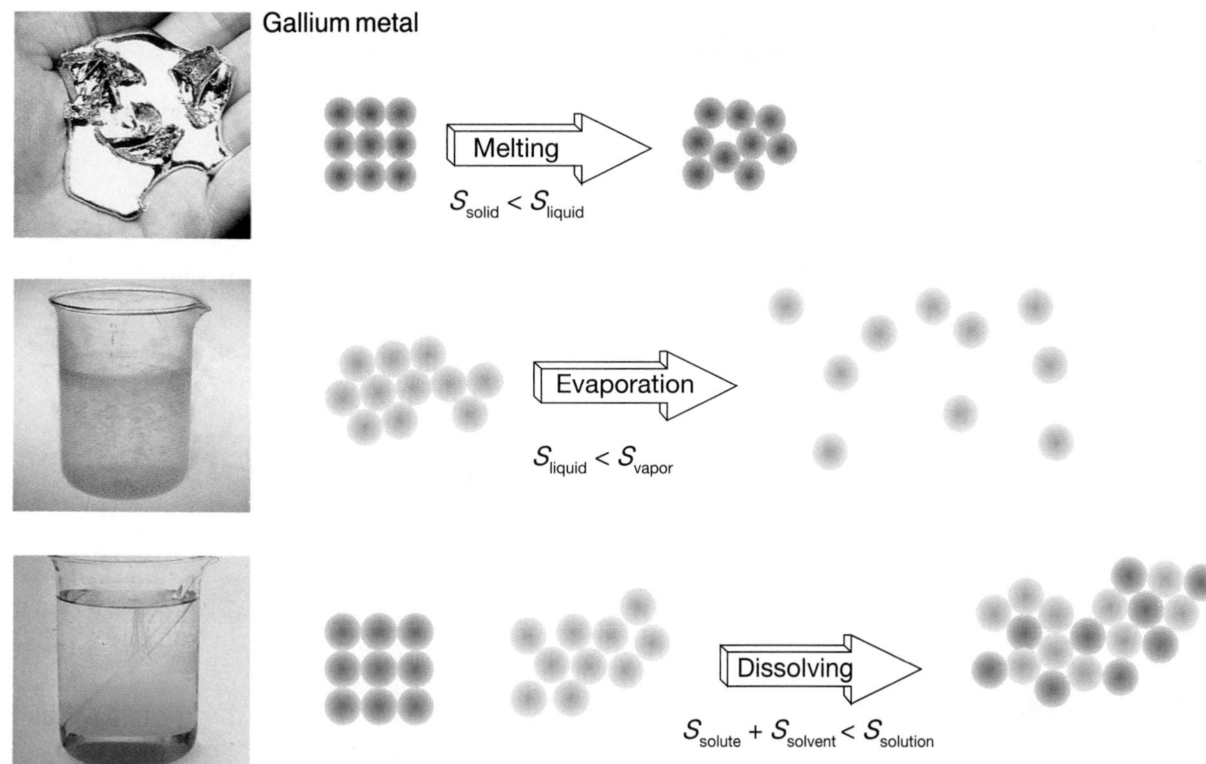

Figure 27.8 Each of the three physical processes shown here—melting of gallium, evaporation, and dissolving—results in a change to a state of greater disorder, that is, increased entropy. Thus ΔS is positive for all three processes.

Let's look at some examples of entropy changes. In the reaction $2H_2O \rightarrow 2H_2 + O_2$, there are more particles after the change than before. Consequently, the products represent a more disordered system. Another way of considering entropy is the probability of finding a particular object in a specific place. Imagine the change $H_2O(l) \rightarrow H_2O(g)$. What happens to the chances of finding a water molecule in a specific place when that water changes from liquid to vapor? In the liquid state, water molecules have no systematic arrangement. However, the water molecules are close together, so we might expect to find a water molecule almost anywhere we looked in a container of liquid water. On the other hand, we know that the molecules of a vapor are quite far apart. Thus, the chances of finding a water molecule at a specific location in a container of water vapor are much less than when the H_2O was a liquid. Thus, the entropy of the gas is greater than that of the liquid.

Look at one final example. As we raise the temperature of a bar of iron from room temperature to 100°C, the iron atoms gain energy and move more rapidly. As a result, although their organization in a crystal lattice remains, the increased motion means that an atom is less likely to be found in a particular spot as often as before. Entropy is increased when the bar is heated.

Change in entropy is symbolized as ΔS. A positive value for ΔS means an increase in the degree of disorder. That is to say, the system becomes less ordered. Such a change (positive ΔS) occurs when a solid is converted to a liquid or a gas. When liquid or gas is converted to a solid, ΔS is negative.

MINUTE LAB

Heat Shrink

Put on an apron and goggles. Hang a 9-cm rubber band from a pencil extending over the edge of a desk. Form a hook from a paper clip and use it to attach a mass to stretch the rubber band until it is about 20 cm in length. Use a meterstick to measure the distance from the bottom of the mass to the floor. Predict what will happen when the rubber band is heated. Use a portable electric hair dryer to heat the rubber band. How many millimeters does the mass move and in which direction? Allow the rubber band to cool and check to see if it returns to its original position. The polymer chains in the rubber band are straighter and more ordered when stretched. They are twisted and disordered when the band is relaxed. What is the sign of ΔS when the rubber band is stretched? Why does it have this sign?

PRACTICE PROBLEMS

12. Will the entropy change for each of the following be positive or negative?

a. Sugar dissolves in tea.

b. Frost forms on a window pane.

c. Air is pumped into a tire.

d. Acetone evaporates from nail polish remover.

13. Will the entropy change for each of the following be positive or negative?

a. $CaCO_3(cr) \rightarrow CaO(cr) + CO_2(g)$

b. $N_2(g) + 3H_2(g) \rightarrow 2NH_3(g)$

Gibbs Free Energy

Suppose a reaction is exothermic but involves an increase in order. The enthalpy change tends to make the reaction spontaneous, whereas the entropy change tends to prevent reaction. How can we decide whether such a reaction will occur at all? The same problem occurs with an endothermic reaction that produces disorder. Here, the entropy change favors a spontaneous reaction, but the enthalpy change does not. The solution to this problem was developed by the American chemist J. Willard Gibbs. He introduced a quantity called free energy that indicates whether a reaction will occur or not.

Gibbs free energy is a state function represented by G. Thus, ΔG represents the change in the Gibbs free energy. **Gibbs free energy** is defined in terms of enthalpy and entropy in the following equations

$$G = H - TS$$
$$\Delta G = \Delta H - T\Delta S$$

where T is the temperature in kelvins (absolute temperature).

Figure 27.9 The reaction that occurred in the tragic burning of the zeppelin *Hindenburg* is the burning of hydrogen.

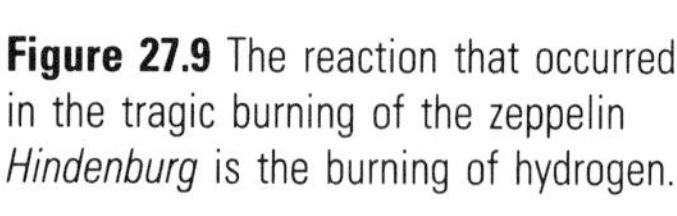

$$2H_2(g) + O_2(g) \rightarrow 2H_2O(g)$$

This is an example of a reaction that is spontaneous even though two gas molecules are formed from three, resulting in a decrease in entropy. Because ΔH has a large negative value, ΔG is negative, and the reaction is spontaneous.

It can be shown, both by theory and by experiment, that in a spontaneous change, ΔG is always negative. If $\Delta G < 0$ for a reaction, the reaction is **exergonic.** If $\Delta G > 0$ for a reaction, the reaction is **endergonic.** If a reaction takes place at low temperatures and involves little change in entropy, then the term $T\Delta S$ will be negligible. In such a reaction, ΔG is largely a function of ΔH, the change in enthalpy. Thus, most reactions occurring spontaneously at room temperature have a negative ΔH.

Highly endothermic reactions can occur only if $T\Delta S$ is large. Thus, the temperature is high, or there is a large increase in entropy. In the endothermic reaction of carbon with steam, both of these conditions occur. ΔS is positive because the ordered arrangement of carbon in the solid is converted to the disordered arrangement in CO gas. T is high because the reaction only takes place at red heat (600–900°C) or higher. If the temperature decreases, the reaction stops and goes into reverse.

If ΔH and ΔS have the same sign, there will be some temperature at which ΔH and $T\Delta S$ will be numerically equal, and ΔG will be exactly zero. This state is the thermodynamic definition of a system in equilibrium. At equilibrium, the value of the Gibbs free energy G is at a minimum for the system.

To summarize, if ΔG for a particular reaction is negative, the reaction is spontaneous. If ΔG is positive, the reaction is not spontaneous. If ΔG is zero, the reaction is at equilibrium. These criteria are based on the enthalpy and entropy changes for reactions as shown in the equation for the Gibbs free energy. Changes in nature tend toward a low energy state (large negative values for ΔH) and a state of high randomness (large positive values for ΔS). Most changes occur because of some combination of these tendencies. See Table 27.1.

All spontaneous processes proceed toward equilibrium. For example, a ball rolls down a hill and not up. The bottom of the hill is where the ball has the least potential energy. Chemical potential energy, technically called free energy, is least when a system is at equilibrium.

An interesting result of the entropy contribution to the Gibbs free energy equation is that molecules like H_2, O_2, and N_2, which are stable on Earth, do not exist on the sun and stars. To see why they do not, consider the case of N_2. In order to decompose one mole of N_2 molecules, much energy must be supplied.

$$N_2 + \textit{energy} \rightarrow 2N \qquad \textit{energy} = \Delta H = +941 \text{ kJ}$$

Table 27.1

Reaction Spontaneity and Values of ΔH and ΔS

ΔH	ΔS	ΔG	Comments on Reaction
−	+	−	Always spontaneous
+	+	+ or −	Spontaneous at high temperatures
−	−	+ or −	Spontaneous at low temperatures
+	−	+	Never spontaneous
ΔH =	$T\Delta S$	0	At equilibrium

Word Origins

exo: (GK) out, outside
endon: (GK) in, within
ergon: (GK) work

endergonic—taking in work (energy)

exergonic—giving out work (energy)

Figure 27.10 Near the stars, nitrogen exists as discrete atoms rather than diatomic molecules. The high temperatures favor the reaction $N_2 \rightarrow 2N$, because $T\Delta S$ has a much higher value than it does at the temperatures on Earth.

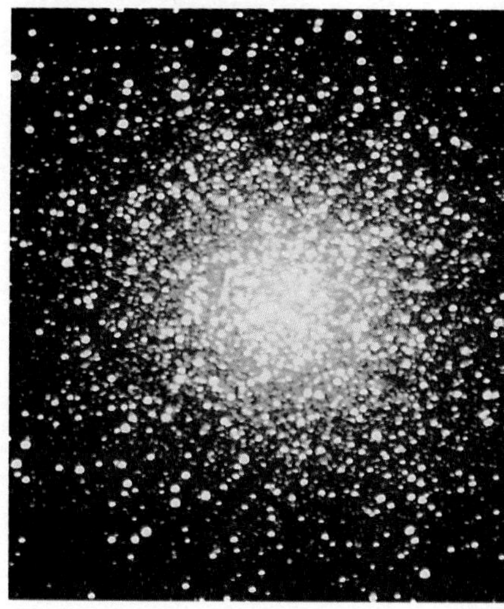

Because ΔH has such a large positive value, at ordinary Earth temperatures, N_2 is a very stable molecule. This stability is a direct result of $T\Delta S$ being small in comparison to the large positive ΔH (which makes ΔG positive). A gas composed of separate nitrogen atoms has a greater entropy than one made of N_2 molecules. The pairing of nitrogen atoms is a kind of order. Therefore, the decomposition of these molecules represents an increase in entropy (a positive ΔS). If N_2 molecules are exposed to higher temperatures (like those near the sun), the value of $T\Delta S$ is greater than 941 kJ and thus ΔG is negative. As a result, nitrogen exists near the sun only as discrete atoms.

All three quantities, enthalpy, entropy, and free energy, depend on temperature. However, we will work only at 298.15 K and 100.00 kPa. Thus, we will always be working with substances in their standard states.

EVERYDAY CHEMISTRY

Chemistry and Human Life

What can chemists do about the problems facing the human population of our planet? Fuel is a central concern of most people. Fuels supply the energy used by people for cooking, manufacturing clothing, building homes, and all the other items that make life easier. Presently, the source of energy for most of the world is the store of coal, petroleum, and natural gas found in Earth's crust. These resources will eventually run out. There are many other possible sources of energy, and the development of these sources requires the work of chemists.

Energy from the sun may be used in several ways. One is through the process of photosynthesis in plants. As chemists learn more about the photosynthetic pathway, they will develop chemical systems that can imitate those in the plants. The use of photocells, which can convert sunlight directly to electricity, is already a reality. Eventually, the efficiency of photo cells will be improved enough to make this an economical method of generating electricity. A third method of utilizing sunlight is to develop a chemical system that can absorb sunlight and use the energy to decompose water to hydrogen and oxygen. The hydrogen can then be burned in the oxygen to release heat as an energy source.

Exploring Further

1. Any method for harnessing solar energy must incorporate ways to provide a continuous supply of energy at night, on cloudy days, and in winter. Do research to find out how this problem is overcome in the use of photovoltaic cells and with solar production of hydrogen.
2. Many methods of obtaining useful energy from heat involve boiling water to steam at one location, piping the steam to another location, and condensing the steam to reclaim the heat. What can you say about the changes in enthalpy and entropy involved in boiling water and condensing the steam?

Gibbs Free Energy Calculations

Appendix Table A-6 lists the standard Gibbs free energies of formation ($\Delta G_f°$), enthalpies of formation ($\Delta H_f°$), and entropies ($S°$) of some substances. The superscript "°" shows these values have been obtained for standard states. The subscript "f" shows they are values for the formation of one mole of the compound from the elements. We already know that the enthalpy change for a reaction is found by

$$\Delta H_r° = \Sigma\Delta H_f°_{(products)} - \Sigma\Delta H_f°_{(reactants)}$$

Similarly, we may compute the Gibbs free energy and entropy changes.

$$\Delta G_r° = \Sigma\Delta G_f°_{(products)} - \Sigma\Delta G_f°_{(reactants)}$$
$$\Delta S_r° = \Sigma S°_{(products)} - \Sigma S°_{(reactants)}$$

SAMPLE PROBLEM

State Function Changes

What is the change in entropy and the change in free energy for the reaction between methane and oxygen under standard measurement conditions? The equation for the reaction is $CH_4(g) + 2O_2(g) \rightarrow CO_2(g) + 2H_2O(l)$.

Solving Process:

(a) Organize the data you will use from Appendix Table A-6 into a table.

Value	CH_4	O_2	CO_2	H_2O
$\Delta G_f°$(kJ/mol)	−50.72	0	−394.359	−237.129
$S°$(J/mol·K)	186.264	205.138	213.74	69.91

(b) To find $\Delta G°$, use the following equations. It is important to have a balanced chemical equation.

$$\Delta G_r° = \Sigma\Delta G_f°_{(products)} - \Sigma\Delta G_f°_{(reactants)}$$
$$= [CO_2 + 2H_2O] - [2O_2 + CH_4]$$

Multiply each $\Delta G_f°$ by the number of moles from the balanced equation. Substitute these values into the equation used to determine $\Delta G_r°$.

$$\frac{1\ \cancel{mol\ CO_2}}{} \left| \frac{(-394.359\ kJ)}{\cancel{mol\ CO_2}} \right. = -394.359\ kJ$$

$$\frac{2\ \cancel{mol\ H_2O}}{} \left| \frac{(-237.129\ kJ)}{\cancel{mol\ H_2O}} \right. = -474.258\ kJ$$

$$\frac{2\ \cancel{mol\ O_2}}{} \left| \frac{0\ kJ}{\cancel{mol\ O_2}} \right. = 0\ kJ$$

PROBLEM SOLVING HINT

Working with state functions involves handling many negative values, so be careful with signs. Check carefully to make sure that chemical equations are balanced.

$$\frac{1\ \cancel{\text{mol CH}_4}}{} \left| \frac{(-50.72\ \text{kJ})}{\cancel{\text{mol CH}_4}} \right. = -50.72\ \text{kJ}$$

$\Delta G_r° = [(-394.359\ \text{kJ}) + (-474.258\ \text{kJ})] -$
$[(0\ \text{kJ}) + (-50.72\ \text{kJ})]$

$\Delta G_r° = -817.90\ \text{kJ}$

(c) To find $\Delta S°$:

$$\Delta S_r° = \Sigma S°_{(\text{products})} - \Sigma S°_{(\text{reactants})}$$
$$= [CO_2 + 2H_2O] - [2O_2 + CH_4]$$

Multiply each $S°$ by the number of moles from the balanced equation.

$$\frac{1\ \cancel{\text{mol CO}_2}}{} \left| \frac{213.74\ \text{J}}{\cancel{\text{mol}}\cdot\text{K}\ \cancel{\text{CO}_2}} \right. = 213.74\ \text{J/K}$$

$$\frac{2\ \cancel{\text{mol H}_2\text{O}}}{} \left| \frac{69.91\ \text{J}}{\cancel{\text{mol}}\cdot\text{K}\ \cancel{\text{H}_2\text{O}}} \right. = 139.82\ \text{J/K}$$

$$\frac{2\ \cancel{\text{mol O}_2}}{} \left| \frac{205.138\ \text{J}}{\cancel{\text{mol}}\cdot\text{K}\ \cancel{\text{O}_2}} \right. = 410.276\ \text{J/K}$$

$$\frac{1\ \cancel{\text{mol CH}_4}}{} \left| \frac{186.264\ \text{J}}{\cancel{\text{mol}}\cdot\text{K}\ \cancel{\text{CH}_4}} \right. = 186.264\ \text{J/K}$$

$\Delta S° = [(213.74\ \text{J/K}) + (139.82\ \text{J/K})]$
$- [(410.276\ \text{J/K}) + (186.264\ \text{J/K})]$

$\Delta S° = -242.98\ \text{J/K}$

PRACTICE PROBLEMS

14. Compute the Gibbs free energy change for the following reaction. Is the reaction spontaneous?

$$BaCl_2(aq) + H_2SO_4(aq) \rightarrow BaSO_4(cr) + 2HCl(aq)$$

15. Compute $\Delta S_r°$ for the following reaction.

$$Ca(cr) + 2H_2O(l) \rightarrow Ca(OH)_2(cr) + H_2(g)$$

16. Compute the Gibbs free energy change for the following reaction. Is the reaction spontaneous?

$$2HF(g) \rightarrow H_2(g) + F_2(g)$$

17. Hydrogenation is the addition of hydrogen to carbon-carbon double bonds in organic compounds. For example, some vegetable oils are hydrogenated to make them solid at room temperature for use in products such as margarine. Using the data given, determine the change in entropy for the hydrogenation of octene. $S°$ for octene = 462.8 J/mol·K; $S°$ for octane = 463.6 J/mol·K

$$C_8H_{16}(g) + H_2(g) \rightarrow C_8H_{18}(g)$$

Figure 27.11 The process of hydrogenation is used to produce margarine and solid shortening from liquid vegetable oil.

Gibbs Free Energy and Equilibrium

We saw in Chapter 22 that in reactions where K_{eq} is much greater than 1, the reactants are converted almost entirely to products. On the other hand, reactions where K_{eq} is much less than 1 would not proceed to any significant extent. In this chapter, we have seen that reactions with ΔG greater than zero do not occur and that those with ΔG less than zero do occur. As you might expect, there is a connection between Gibbs free energy and the equilibrium constant. We will not derive the relationship here, as it involves complex mathematics. Simply stated, Gibbs free energy and equilibria are related by the expression

$$\Delta G = -2.30RT(\log K_{eq})$$

In the expression, R is the universal gas constant, T is the absolute temperature, and *log* is a logarithm in base 10.

SAMPLE PROBLEMS

1. Gibbs Free Energy Change

The equilibrium constant for the reaction of CO_2 and H_2 to form CO and H_2O at 1120°C is 2. What is ΔG for this reaction? Use $\Delta G = -2.30RT(\log K_{eq})$, $R = 0.008\ 31$ kJ/mol·K, and a calculator or log tables.

Solving Process:

$$\begin{aligned}\Delta G &= -2.30RT(\log K_{eq})\\ &= (-2.30)(0.008\ 31\ \text{kJ/mol·K})(1393\ \text{K})(\log 2)\\ &= (-2.30)(0.008\ 31\ \text{kJ/mol·K})(1393\ \text{K})(0.301)\\ &= -8.01\ \text{kJ/mol}\end{aligned}$$

2. Equilibrium Constant

The Gibbs free energy change for the reaction $2Cu_2O + 2NO \rightleftarrows 4CuO + N_2$ is -389 kJ/mol at 25°C. What is K_{eq} for this reaction?

Solving Process:

Solving the equation relating Gibbs free energy to equilibrium for K_{eq}, we find that

$$\log K_{eq} = \frac{\Delta G}{-2.30\ RT}$$

$$\log K_{eq} = \frac{-389\ \text{kJ/mol}}{(-2.30)(0.008\ 31\ \text{kJ/mol·K})(298\ \text{K})} = 68.297$$

Using a calculator or log tables, we find the antilog and convert this value to

$$K_{eq} = 1.98 \times 10^{68}$$

The equilibrium constant is so large that we can safely assume that all the reactants are converted to products.

PROBLEM SOLVING HINT

In thermodynamics, absolute temperatures are nearly always required. Change all Celsius temperatures to Kelvin before doing anything else. Then, you won't use the incorrect value by accident.

CHEMISTS AND THEIR WORK

Jennie R. Patrick
(1949-)

Jennie R. Patrick was the first American black woman to earn a doctorate in chemical engineering. After receiving a B.S. degree in chemical engineering from the University of California at Berkeley, she completed her Ph.D. at Massachusetts Institute of Technology. Dr. Patrick's area of research is high pressure thermodynamic phenomena, particularly as they relate to supercritical fluids. Dr. Patrick is interested in her community and supports activities that help African-American youth to succeed in both school and society. She now works at Southern Company Services, Inc., in Birmingham, Alabama, where she is helping to address future energy concerns in the electric utility industry.

PRACTICE PROBLEMS

18. Calculate K_{eq} at 25.0°C for the reaction $CO + H_2 \rightleftarrows H_2CO$ if $\Delta G = +28.95$ kJ.

19. At 627°C, K_{eq} for the reaction $2SO_3 \rightleftarrows 2SO_2 + O_2$ is 3.16×10^{-4}. Compute ΔG for this reaction.

20. For the decomposition of PCl_5 at 200.0°C, the equilibrium constant is 0.457. What is the Gibbs free energy change for this reaction?

Energy and Electric Cells

The Gibbs free energy change for a reaction is related to another quantity in addition to the equilibrium constant. Perhaps you have already noticed that the expression $\Delta G° = -2.30\,RT\,(\log K)$ is similar to the Nernst equation discussed in Chapter 26. The Nernst equation may be written as

$$E = E° - \frac{2.30RT}{nF} \log K$$

At equilibrium, there is no net change, so E would equal zero. Solving the equation for $(2.30RT)\log K$ and substituting 0 for E, we obtain

$$2.30RT \log K = nFE°$$

Substituting the equivalent expression in the Gibbs free energy equation we get

$$\Delta G° = -nFE°$$

These relationships are important to a chemist. Voltages are measured more easily and with greater accuracy than are concentrations and changes in enthalpy. Once the cell voltages have been measured, they can be used to calculate the Gibbs free energy change for the reaction. With other information, enthalpy and entropy changes can be found. Measurement of voltages may also be used to determine the equilibrium constant for reactions. When measuring $E°$, we must remember that "°" stands for "in standard states," which are ideal conditions that can only be approached in the laboratory. As you may recall from Section 21.2, the effectiveness of an ion is called its activity. Because, in most cases, the activities are unknown, the voltage developed by a cell is measured at several electrolyte concentrations. The data are than graphed, and the voltage values extrapolated to find $E°$.

We have seen that Gibbs free energy, standard cell potentials, and equilibrium constants are related. Any of these factors can be used as a predictor of reaction spontaneity. A reaction tends to be spontaneous if K_{eq} is considerably larger than 1, if $E°$ is positive, and if ΔG is less than zero.

SAMPLE PROBLEM

Free Energy-Voltage

The following cell was set up with all substances in their standard states and with standard conditions

$$Zn \mid Zn^{2+} \parallel I_2 \mid I^-$$

Calculate the free energy change for the overall reaction.

Solving Process:

From Table 26.1 we obtain the voltages for the two half-reactions. The reduction half-reactions involved are:

$$Zn^{2+} + 2e^- \rightarrow Zn \qquad -0.7626 \text{ V}$$
$$I_2 + 2e^- \rightarrow 2I^- \qquad 0.5355 \text{ V}$$

Recall from Chapter 26 that the cell voltage is found from adding the individual half-reaction voltages and that reversing a half-reaction reverses the sign of the voltage for that half-cell. Thus, the voltage generated by the cell in question is $-(-0.7626 \text{ V}) + 0.5355 \text{ V} = 1.2981 \text{ V}$. In the balanced oxidation-reduction equation, two electrons are transferred.

$$Zn + I_2 \rightarrow Zn^{2+} + 2I^-$$

Voltage and free energy are related by the equation

$$\Delta G° = -nFE°$$

Substituting the values given in the problem, we get

$$\Delta G° = (-2 \text{ mol } e^-)\left(96\,485 \frac{\text{J}}{\text{mol } e^- \cdot \text{V}}\right)(1.2981 \text{ V})$$

$$\Delta G° = -2.5049 \times 10^5 \text{ J} = -250.49 \text{ kJ}$$

PROBLEM SOLVING HINT

When you look up the potential for a half reaction, don't forget to reverse the sign if the reaction is in the direction opposite to that shown in Table 26.1.

PRACTICE PROBLEMS

21. What is the Gibbs free energy change for the following reaction at standard conditions?

$$2U^{3+} + Hg^{2+} \rightarrow 2U^{4+} + Hg$$

22. What is the Gibbs free energy change for the following reaction at standard conditions?

$$Zn + 2Hg^{2+} \rightarrow Zn^{2+} + Hg_2^{2+}$$

27.2 CONCEPT REVIEW

23. What is entropy? Discuss a physical change in which entropy increases.

24. How is it possible for an endothermic reaction ($\Delta H > O$) to be spontaneous?

25. Apply Describe the process that takes place in a flashlight battery in terms of enthalpy, entropy, and free energy.

Chapter 27 REVIEW

Summary

27.1 Introductory Thermodynamics

1. Systems in nature tend to move toward states of low energy and high disorder.
2. The change in internal energy of a system is the sum of the heat absorbed from the surroundings and the work done on the system as given by the equation, $\Delta U = q + w$.
3. A state function is a quantity whose value depends only on the current state of a system. U, P, V, and T are examples of state functions.
4. The change in a state function is independent of the path and depends only on the initial and final states.
5. The change in enthalpy of a reacting system is the energy absorbed or produced during a chemical reaction and is represented by ΔH.
6. For an exothermic reaction, ΔH is negative because the products are at a lower enthalpy than the reactants. For an endothermic reaction, ΔH is positive.
7. The standard enthalpy of formation, ΔH_f°, of a compound is the energy absorbed or released when one mole of the compound is formed from its elements.
8. A thermodynamically stable compound has a large negative enthalpy of formation.
9. Enthalpy change, ΔH_r°, for a reaction is the difference between the enthalpies of the products and the reactants.
10. Hess's law states that the change in enthalpy for a reaction consisting of a series of steps is the sum of the enthalpy changes for all the steps.

27.2 Driving Chemical Reactions

11. Entropy is the degree of disorder of a system. Entropy change is designated by ΔS.
12. When a system increases in disorder, ΔS is positive. In the series of state changes for a substance, solid → liquid → gas, entropy increases.
13. When the enthalpy change and entropy change differ in tendency, the net effect is found from the equation $\Delta G = \Delta H - T\Delta S$. In this equation, G is Gibbs free energy, H is enthalpy, T is the absolute temperature, and S is entropy.
14. The Gibbs free energy change for a reaction is related to the equilibrium constant for that reaction by the expression $\Delta G = -2.30RT(\log K_{eq})$.
15. Combining the expression for Gibbs free energy with the Nernst equation yields an expression that relates Gibbs free energy to cell potential. Thus, cell potential, which is measured easily, can be used to predict reaction spontaneity.

Key Terms

isothermal process
isobaric process
thermodynamics
internal energy
state function
enthalpy
enthalpy of formation
thermodynamic stability
Hess's law
entropy
Gibbs free energy
exergonic
endergonic

Review and Practice

26. In terms of order, how do most natural processes proceed?
27. Give three examples of state functions.
28. Define internal energy. What is its relationship to heat and work?
29. List four ways energy may be transferred between a system and its surroundings.
30. For a certain process, both q and w are negative. What do these values mean in

terms of heat transfer and work done? How does the internal energy of the system change?

31. A system received 622 kJ of energy from its surroundings and did 813 kJ of work on the surroundings. What was the change in its internal energy?

32. In a process, the internal energy of a system decreases 10.4 kJ. The system absorbs 62.5 kJ of energy from its surroundings. How much work did the system do?

33. Potassium bromide, KBr, has a standard enthalpy of formation of −393.798 kJ/mol. Explain what this statement means.

34. Calculate the enthalpy change for each of the following reactions. All components are at standard states. Solutions are 1*M*.

a. $Zn(cr) + H_2SO_4(aq) \rightarrow H_2(g) + ZnSO_4(aq)$

b. $Fe_2O_3(cr) + 3CO(g) \rightarrow 3CO_2(g) + 2Fe(cr)$

35. State Hess's law. Use an example to explain it.

36. Define entropy. Why is entropy a state function?

37. In terms of energy change and entropy change, what are the general trends of reactions in nature?

38. Give the name of the quantity represented by each of the following.

a. G **c.** q **e.** U

b. H **d.** S **f.** w

39. What are endergonic and exergonic reactions? Which type of reaction occurs spontaneously?

40. Relate Gibbs free energy in terms of entropy, enthalpy, and temperature. Write the equation, then explain it in words.

41. What is meant by thermodynamic equilibrium? What is the value of ΔG at thermodynamic equilibrium?

42. Using Appendix Table A-6, compute $\Delta G_r°$ for each of the following reactions. Decide whether the reaction would occur spontaneously or not. In addition, compute $\Delta H_r°$ and $\Delta S_r°$ for each reaction.

a. $PbBr_2(cr) + Cl_2(g) \rightarrow PbCl_2(cr) + Br_2(l)$

b. $H_2O(l) \rightarrow H_2O(g)$

c. $2C_2H_6(g) + 7O_2(g) \rightarrow 4CO_2(g) + 6H_2O(l)$

d. $Cu_2S(cr) + S(cr) \rightarrow 2CuS(cr)$

e. $CuS(cr) + 2O_2(g) \rightarrow CuSO_4(cr)$

43. The following equation shows the combustion of ethane.

$$2C_2H_6(g) + 7O_2(g) \rightarrow 4CO_2(g) + 6H_2O(g)$$

Use Hess's law to calculate the enthalpy of combustion for ethane, C_2H_6, from these data.

$$C_2H_4(g) + 3O_2(g) \rightarrow 2CO_2(g) + 2H_2O(g) \quad \Delta H_r = -1323 \text{ kJ/mol}$$

$$C_2H_4(g) + H_2(g) \rightarrow C_2H_6(g) \quad \Delta H_r = -137 \text{ kJ/mol}$$

$$H_2(g) + \tfrac{1}{2}O_2(g) \rightarrow H_2O(g) \quad \Delta H_f° = -242 \text{ kJ/mol}$$

44. For the reaction $N_2 + 3H_2 \rightleftarrows 2NH_3$, at 25°C the equilibrium constant, K_{eq}, is 6.00×10^5. Calculate ΔG for the reaction at that temperature.

45. The Gibbs free energy change for the reaction $2CO_2 \rightleftarrows 2CO + O_2$ at a temperature of 1.000×10^{3}°C is 338 kJ. What is the equilibrium constant for the reaction at that temperature?

46. Calculate the change in Gibbs free energy for the following reactions.

a. $2Ce + 3Cu^{2+} \rightarrow 2Ce^{3+} + 3Cu$

b. $S^{2-} + I_2 \rightarrow S + 2I^-$

47. Tin(IV) ions and iron react according to the following reaction.

$$Sn^{4+} + Fe \rightarrow Fe^{2+} + Sn^{2+}$$

a. Calculate the Gibbs free energy change for the reaction.

b. Determine the equilibrium constant, K_{eq}, for the reaction at 25°C.

Concept Mastery

48. In general, how does the spontaneity of a reaction relate to whether the reaction is exothermic or endothermic?

49. Relate change in enthalpy to whether a process is exothermic or endothermic.

50. What sign will the enthalpy change of a system have if internal energy increases and constant-pressure work is done by the system on the surroundings?

51. Why must standard states be established for substances that are to be studied in thermodynamics?

52. How is the enthalpy of formation of a compound related to its thermodynamic stability?

53. In the following situations, tell whether ΔH and $T\Delta S$ each have positive or negative values.
   a. A reaction is exothermic and involves a decrease in order.
   b. A reaction is endothermic and involves a decrease in order.
   c. A reaction is exothermic and involves an increase in order.
   d. A reaction is endothermic and involves an increase in order.

   In which of these situations is it unclear whether the reaction will occur spontaneously? How can it be determined in these cases whether the reaction in question will be spontaneous?

54. List two factors that allow highly endothermic reactions to occur.

55. Why are diatomic molecules of elements stable under Earth conditions but not under the conditions that exist on the sun and stars?

56. How can a great increase in entropy help drive a highly endothermic reaction? Think about what would have to occur in order for the reaction to reverse.

57. What relationship exists between the spontaneity of a reaction and values for cell potential, change in Gibbs free energy, and equilibrium constant for that reaction?

Application

58. Platinum is usually found in nature in the elemental state. It is used in jewelry and in scientific equipment. Platinum, along with all other elements, has an enthalpy of zero. Why is the enthalpy of any free element zero?

59. A blacksmith heats an iron bar from 41°C to 1400°C, hammers it to shape, and cools it to 75°C. What is ΔT for the iron bar in this process?

60. A student describes the burning of a lantern as an endothermic reaction because heat was needed to light it. Is this statement a valid description? Explain.

61. Using Appendix Table A-6 and the equation $H_2 \rightarrow 2H$, calculate the temperature at which hydrogen exists as separate atoms in equilibrium with H_2.

62. Using Appendix Table A-6, at what temperature will the following reaction exhibit thermodynamic equilibrium?

$$2Ag + Cl_2 \rightarrow 2AgCl$$

63. Under standard conditions, what voltage should be obtained from a cell employing a reaction with a Gibbs free energy change of −64 kilojoules? The balanced equation for the reaction shows one electron transferred.

Critical Thinking

64. Will the change in entropy for each of the following be positive or negative?
   a. Dry ice sublimes.
   b. An ice cube melts.
   c. Raindrops form in a cloud.
   d. Carbohydrates are broken down in the cells to release carbon dioxide and water.

e. A solution of sodium chloride and a solution of silver nitrate are combined; crystals of silver chloride settle to the bottom of the container.
f. Helium under pressure in a cylinder is used to inflate a balloon.

65. Give an illustration of an increase in disorder that is different from the marble example used in the text.

66. In each of the following processes, determine whether q and w will have positive or negative values.
a. A sample of propane burns in air.
b. Gas from ignited fuel expands inside an engine cylinder.
c. A beaker of warm solution is put into ice water to cool.
d. Nitrogen gas is pumped into a tank.

67. Explain diffusion of gases by using the concept of entropy.

68. Explain why ΔS is positive for dissolving a solid in a liquid and negative for dissolving a gas in a liquid.

69. The solution of most solids is endothermic yet spontaneous. What factor causes the solution process for solids to be spontaneous?

Readings

Hauserman, W.B. "Thermodynamics of Resource Recycling." *Journal of Chemical Education* 65, no. 12 (December 1988): 1045-1047.

Marsella, Gail. "Fireside Dreams." *ChemMatters* 6, no. 4 (December 1988): 13-15.

Marsella, Gail. "Hot and Cold Packs." *ChemMatters* 5, no.1 (February 1987): 7-11.

Stover, Dawn. "Sidewalk Cooler." *Popular Science* 232, no. 6 (June 1988): 70-71.

Tykodl, R. J. "Thermodynamics and Reactions in the Dry Way." *Journal of Chemical Education* 63, no. 2 (February 1986): 107-111.

Cumulative Review

1. If a sample of gas occupies 163 cm^3 at 52.4 kPa, what volume will it occupy at 68.7 kPa? Temperature is constant.

2. If a sample of gas occupies 258 cm^3 at 97°C, what volume will it occupy at 77°C and constant pressure?

3. If a gas occupies 94.5 dm^3 at 46°C and 87.5 kPa, what volume will it occupy at 18°C and 104.6 kPa?

4. Which gas would diffuse faster, CO_2 or UF_6?

5. In what two ways do real gases differ from an ideal gas?

6. What is Avogadro's principle?

7. What volume will be occupied by 0.161 mole of gas at 43.3 kPa and 52°C?

8. What voltage could you expect to obtain from a voltaic cell having a reaction with a free energy change of −26.6 kJ? One electron is transferred.

9. What is the modern definition of reduction?

10. What is the oxidation number of P in Ag_3PO_4?

11. What is the oxidation number of chlorine in $Hg(ClO_3)_2$? Mercury is 2+.

12. How can an oxidation-reduction reaction be identified?

CHAPTER Preview

CONTENTS

CHAPTER 28

Nuclear Chemistry

The picture on the opposite page is a Tokamak, a device for harnessing the same source of energy as exists in the sun—nuclear fusion. Should scientists succeed in perfecting this device, people on Earth could stop worrying about energy supplies. It has been estimated that the "fuel" for a Tokamak is so plentiful that it would last about 5 billion years into the future.

The source of the sun's energy is nuclear reactions in its core. Nuclear reactions involve the nucleus of one or more atoms changing. Chemical and physical changes obey the laws of conservation of energy and conservation of mass. Nuclear changes obey the law of conservation of mass-energy, a result of Einstein's demonstration of the equivalence of mass and energy.

EXTENSIONS AND CONNECTIONS

OBJECTIVES

- describe the operation of particle accelerators and fission reactors.
- describe the probing of the nucleus through the use of accelerators and reactors.
- apply three tests for relative stability of nuclides.
- explain the concept of half-life and write nuclear equations.

28.1 Nuclear Structure and Stability

Susan was very ill. Her doctor was puzzled by the combination of her symptoms. A preliminary examination indicated a kidney problem. He referred Susan to a specialist in kidney function and diseases. He used a complex of a radioactive nuclide, $^{99m}_{43}Tc$, as the diagnostic agent. The results showed that Susan's medical problem could be cured with medication. What is $^{99m}_{43}Tc$? How can it be used to diagnose disease? In this chapter we will investigate the phenomena that can be used to answer such questions.

Investigating Nuclear Structure

Most of what we know about nuclear structure has been learned by bombarding atomic nuclei with high-energy particles and observing results. One way of bombarding atomic nuclei is using accelerators to speed up a stream of charged particles with which to bombard nuclei. Another way is the use of a nuclear reactor to expose nuclei to a random bombardment by neutrons.

When a particle collides with a nucleus, one possible result is that the particle may simply bounce off the nucleus. However, during such a collision, some of the energy of the bombarding particle may be transferred to the nucleus. The nucleus is now excited, or metastable, and can undergo internal change. This internal change can result in the emission of detectable energy. For example, the $^{99m}_{43}Tc$ nuclide used in the diagnosis of Susan's illness was in the excited state, as was indicated by the *m* (for metastable) in its symbol. The specialist was able to diagnose Susan's illness by studying the exact location of the energy that was being given off by the $^{99m}_{43}Tc$.

Another possible result of bombardment is the shattering of the nucleus into many parts that can be detected and analyzed. A third possibility is the merging of the bombarding particle and the nucleus to form a new nucleus. There are also other possibilities. The final result depends on three factors: the relative stability of the target nucleus, the kind of bombarding particle, and the energy of the bombarding particle.

Figure 28.1 Accelerators can be either linear or circular. Drift tubes in a linear accelerator are shown.

Accelerators

An **accelerator** is a linear or circular device that is used to increase the velocity of charged particles. When the charged particle has been given a very high velocity and, thus, a very high energy, it is aimed at the target material. The interaction that results from the collision helps scientists understand nuclear structure.

In a linear accelerator, a particle is introduced using a device similar to the mass spectrometer ion-beam generator. The charged particle in motion is "pushed" or accelerated by an electromagnetic wave down the "pipe" forming the core of the accelerator. The wave, however, is traveling at the speed of light, and the particle is traveling more slowly. If the wave should overtake

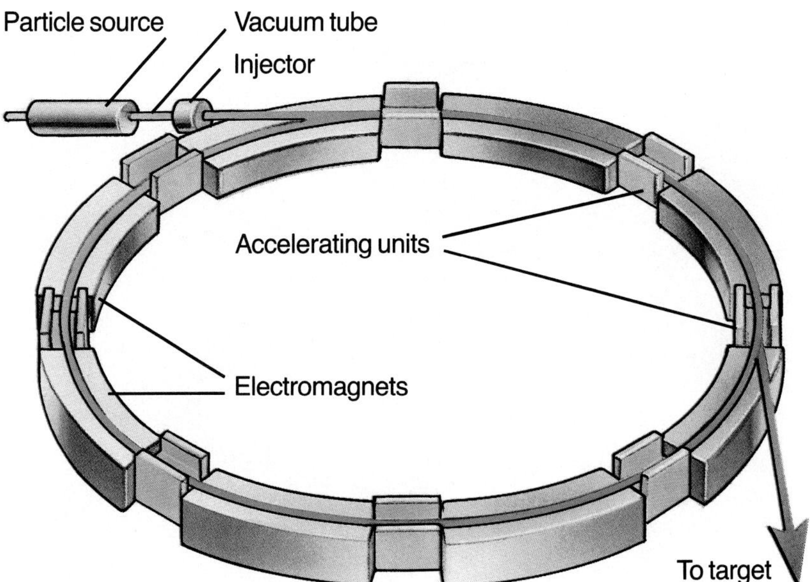

Figure 28.2 A synchrotron uses magnets to accelerate charged particles in a circular path.

the particle, the particle would then decelerate. Therefore, there are grounded tubes in the pipe called **drift tubes.** The particle drifts unaffected through the tube until the next accelerating part of the radio wave is in the correct position. Careful design of the drift tubes as well as control of the radio wave frequency is important to assure maximum acceleration.

In a circular accelerator, the pipe through which the particles move is in the shape of a circle, or slightly modified circle, Figure 28.2. Particles are accelerated by electric fields in several locations around the ring. The path of the particles is confined to the ring by huge magnets. As the particles are accelerated, the magnetic field must be increased to keep the particles in the ring. Careful synchronization of acceleration and increasing magnetic fields is important. Hence, the device is called a **synchrotron.**

The greater the energy of collision between the accelerated particles and the target, the more the nuclear scientists can learn about the structure and the behavior of the nucleus. Recently, some experiments have involved two rings whose beams collide head-on. An accelerator is used to produce a stream of high-energy particles that are stored in one ring. The accelerator can then generate a second stream to collide with the particles in the storage ring. The energy of collision is then the sum of the energies of the two streams and exceeds the energy of the collision of one stream with a stationary target.

At Waxahachie, Texas, construction is underway for the largest such accelerator. The device is called the Superconducting Super Collider. It will be capable of accelerating opposing currents of protons and colliding them. There will be more than ten thousand magnets in the 84-kilometer circumference of the ring. The total energy of both protons when they collide will be about 40 teravolts (TeV). That much energy is equivalent to adding 3.86×10^{18} joules of energy to a mole of a substance.

Word Origins

syn: (GK) with
chronos: (GK) time
tron: (GK) instrument

synchrotron—instrument with electric and magnetic fields at the same time

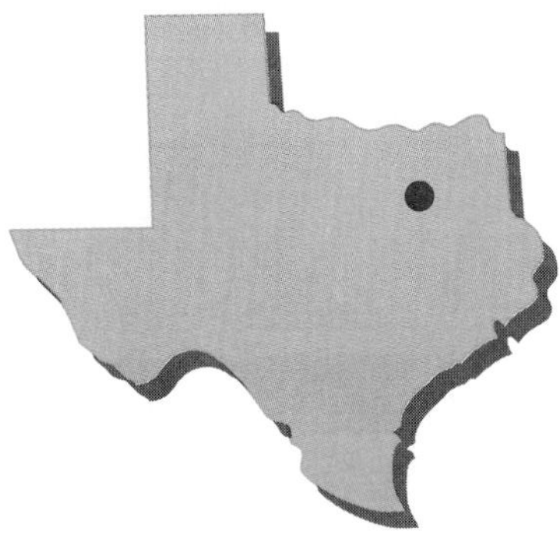

Figure 28.3 On the right, the central part of a 17-meter long dipole magnet to be used in the Superconducting Super Collider is being assembled. The map on the left shows the location of this huge synchrotron, which is 84 km in circumference.

Fission

Scientists have studied several types of nuclear changes. One such change is fission. **Fission** is the breaking apart of a very heavy nucleus into two approximately equal parts. Fissioning nuclei are most likely to split into two fragments having a mass ratio of three to two. A few synthetically produced heavy nuclei will fission spontaneously, as will a few naturally occurring nuclei such as uranium-235. Other heavy nuclei can be made unstable by exposure to slow-moving neutrons. Because neutrons have no charge, they easily penetrate to the nucleus and cause changes in the energy and the structure of the target nucleus.

findere: (L) to split
fission – splitting of atoms

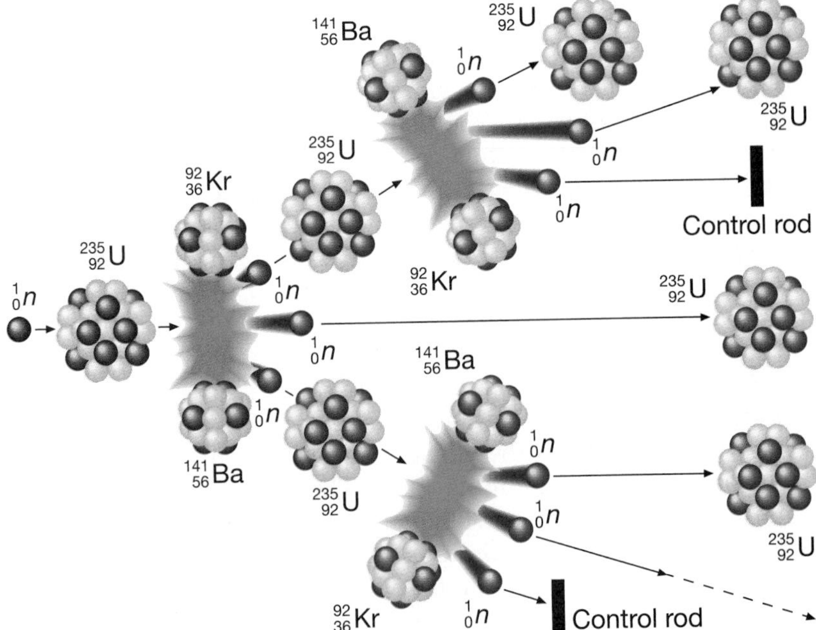

Figure 28.4 The fission process in a reactor consists of a chain of fission reactions. Control rods are used to break the chain and slow or stop the reaction.

When some heavy nuclei are struck by neutrons, the neutrons are absorbed. The nucleus then splits into two heavy fragments and two or three new neutrons are emitted. These newly produced neutrons, in turn, can strike other nuclei, causing them to split and produce more neutrons. This process continues until a very large number of the nuclei have reacted. Note that neutrons are reactants as well as products in this process. A nuclear reaction of this type is called a chain reaction. When not controlled, this type of chain reaction can "run away" and result in a nuclear explosion that we refer to as a nuclear bomb, often incorrectly called an atomic bomb.

Nuclear Reactors

We have seen that a large amount of energy is released in the fission process. By controlling the chain reaction, energy released during fission can be put to a practical use. A **nuclear reactor** is a device for controlling nuclear fission. In a nuclear reactor, a fissionable element is used as a fuel. Some fissionable isotopes require such high-energy neutrons to fission that they cannot sustain a chain reaction. Other isotopes can sustain a chain reaction with slow neutrons. These nuclides are used as fuel in reactors. Reactors can be designed to produce energy for electric power generation plants or for propulsion units in ships and submarines.

Reactors can also be used as bombarding devices. Because there are so many neutrons being emitted in a reactor, placing a material into a reactor will expose the nuclei of the material to an intense bombardment by neutrons. Thus, you see how a reactor can be used as a tool to investigate nuclear structure and properties.

The chief fuels for nuclear reactors are $^{235}_{92}U$ and $^{239}_{94}Pu$ (itself a product of a reactor). It is also possible to convert $^{232}_{90}Th$ to $^{233}_{92}U$ in a reactor and use the uranium isotope as a fuel. The neutrons being produced during fission are traveling too fast to initiate the fission process efficiently in other atoms. Therefore, most reactors contain a material called a **moderator** that slows down the neutrons through collisions but does not absorb them. Water and graphite are good moderators.

The reaction rate is controlled by rods composed of efficient neutron-absorbing materials such as boron, cadmium, or gadolinium. With the rods fully inserted, most neutrons are absorbed by the rods. As the rods are withdrawn from the reactor core, fewer neutrons are absorbed, and more neutrons continue the chain reaction. The power level of the reactor is controlled by the extent to which rods are inserted in the reactor core. The energy produced by the fissioning atoms is in the form of the kinetic energy of the particles produced. As these particles collide with the materials in the reactor, their kinetic energy is gradually transferred to these other materials and the temperature of the whole system increases. It is necessary, then, to have a coolant to keep the

HISTORY CONNECTION

The Manhattan Project

In 1939, following the first "splitting" of the atom, Albert Einstein wrote to President Franklin Roosevelt expressing the feelings of many physicists that there were military possibilities in a sustained nuclear chain reaction. In 1942, the Manhattan Engineer District of the Corps of Engineers was organized to produce a nuclear bomb. This project was called the Manhattan Project. On December 2, 1942, a group headed by Enrico Fermi at the University of Chicago produced the first human-made nuclear chain reaction. The first nuclear bomb explosion took place near Los Alamos, New Mexico, on July 16, 1946.

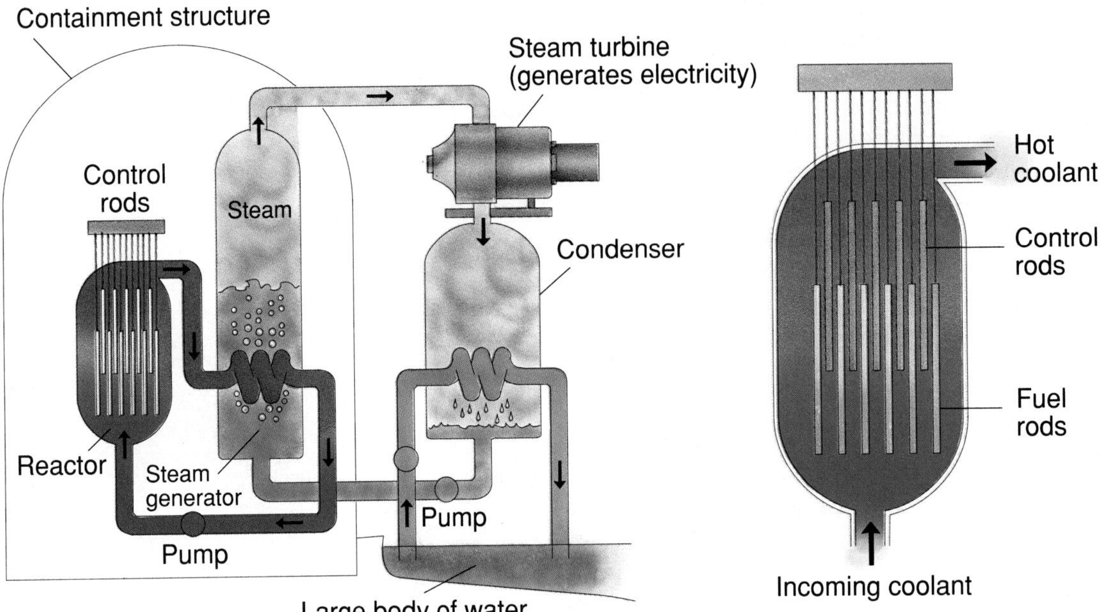

Figure 28.5 The diagram on the left shows the parts of a pressurized water nuclear reactor. The diagram on the right is a schematic of the core of the reactor.

reactor from overheating. Several substances have been used as coolants, with water being the most common. Liquid metals, such as sodium and sodium-potassium alloys, and gases, such as helium, have been tried. The energy gained by the coolant is used to perform useful work, such as powering an electric generator.

Because of the intense radioactivity the whole reactor system is encased in protective materials. The **containment vessel** keeps radioactive materials from escaping into the environment. It also shields personnel in the plant from the radiation produced in the reactor.

On April 26, 1986, the water coolant levels in Reactor No. 4 at Chernobyl, Ukraine, were not properly managed. The nuclear reactor overheated, and the cooling water turned to steam. The fuel rods ruptured and spilled fuel into the coolant. As a result of the tremendous steam pressure that was generated, an explosion occurred. Reactor No. 4 did not have a containment vessel. The top of the reactor and the roof of the reactor building were blown off, and radioactive materials were blown high into the atmosphere. The core of the reactor itself underwent a meltdown.

Note that this explosion was a result of physical and chemical changes. It was *not* a nuclear explosion. Because of the nature of their structure, nuclear power reactors cannot detonate in a nuclear explosion. They can, however, blow up as a result of the buildup of tremendous pressures and temperatures generated by a runaway nuclear reaction. The reactor at Chernobyl underwent an explosion caused by a runaway of this type.

The demand for energy in the United States has surpassed our ability to produce it from present sources. Worldwide resources of fossil fuels such as petroleum and coal may last only a few decades. We have begun to utilize fission reactors as a source of electric power. However, they represent only a fraction of the total power production facilities currently in operation. Fission reactors do not represent an unlimited source of energy. However, they do offer a source of energy for the immediate future. As with other energy sources, there are problems associated with nuclear power. For example, what is to be done with the highly radioactive waste material? Will strict safety measures be maintained at reactor sites? European countries are aggressively pursuing nuclear power as a means of satisfying energy demands. The future of fission power reactors as an alternative and clean energy source will have to be carefully considered.

Nuclear Equations

Nuclear reactions can result in the change of one element into another. If a reaction changes the number of protons in the nucleus, an atom with a different atomic number is obtained. This change is called **transmutation.** A transmutation can be represented by a nuclear equation.

The most common isotope of natural uranium is $^{238}_{92}U$. By a natural process, uranium will transmute to $^{234}_{90}Th$. The $^{238}_{92}U$ nuclide decays by emitting an alpha particle. The resulting nuclide contains two fewer protons and two fewer neutrons than

Figure 28.6 The nuclear decay of $^{238}_{92}U$ takes place in a series of steps. Note that each new product has a different half-life.

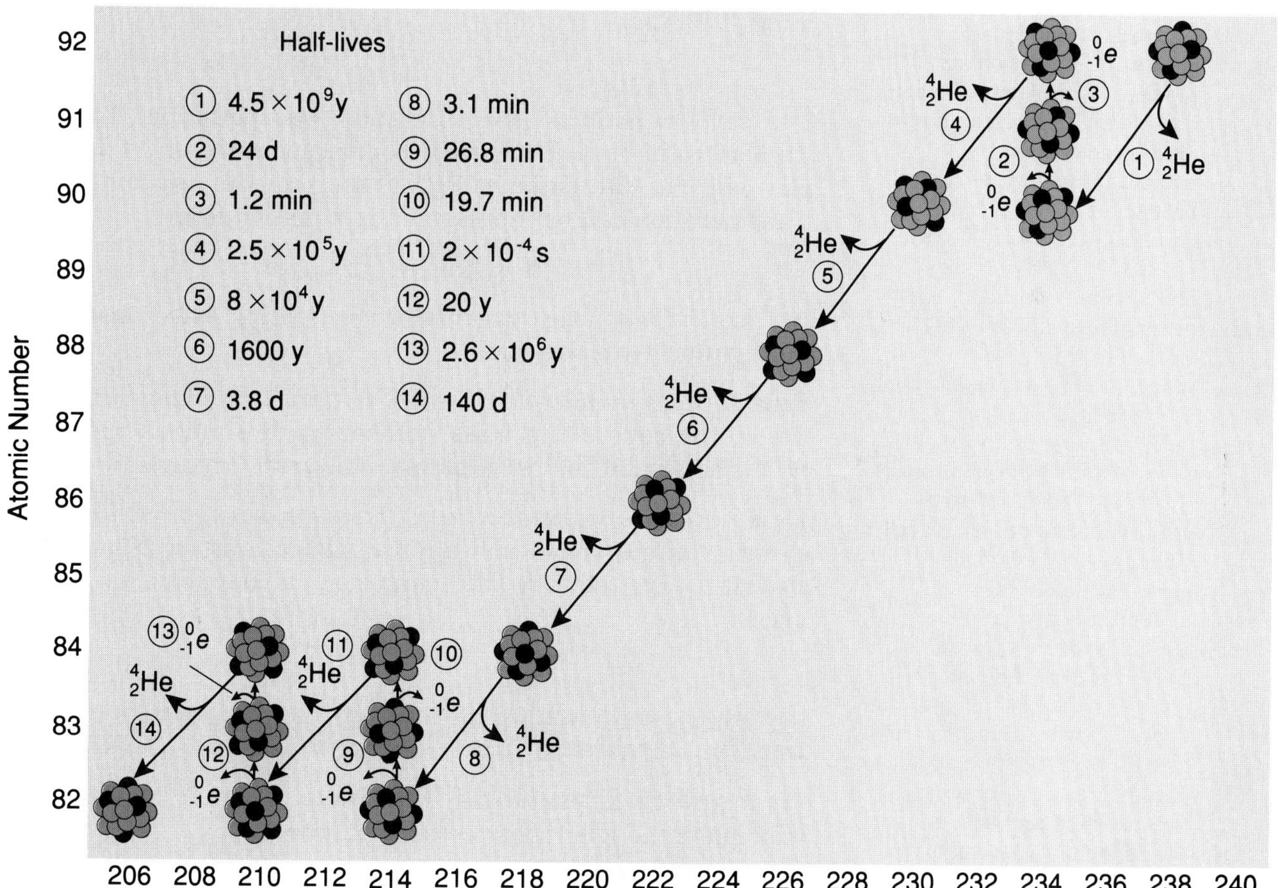

$^{238}_{92}U$ does. Thus, it has an atomic number of 90 and a mass number of 234. This new nuclide is a thorium atom, $^{234}_{90}Th$.

$$^{238}_{92}U \rightarrow {}^{234}_{90}Th + {}^{4}_{2}He$$

Other typical alpha emission reactions are:

$$^{211}_{84}Po \rightarrow {}^{207}_{82}Pb + {}^{4}_{2}He$$
$$^{219}_{86}Rn \rightarrow {}^{215}_{84}Po + {}^{4}_{2}He$$

The nuclide $^{234}_{90}Th$ decays by emitting a beta particle. Because the mass of an electron is negligible, the new nuclide also has the mass number 234, but has one more positive charge than before. Its new atomic number is 91. The new atom is $^{234}_{91}Pa$.

$$^{234}_{90}Th \rightarrow {}^{0}_{-1}e + {}^{234}_{91}Pa$$

Other typical beta emission reactions are:

$$^{210}_{83}Bi \rightarrow {}^{210}_{84}Po + {}^{0}_{-1}e$$
$$^{223}_{87}Fr \rightarrow {}^{223}_{88}Ra + {}^{0}_{-1}e$$

The series of disintegrations that begins with $^{238}_{92}U$ ends with $^{206}_{82}Pb$, which is a stable nuclide as shown in Figure 28.6.

Let's look at a transmutation that involves *K*-capture. The nuclide $^{100}_{46}Pd$ decays by *K*-capture. Its atomic number is decreased by one and its mass number remains the same. The new atom is $^{100}_{45}Rh$.

$$^{100}_{46}Pd + {}^{0}_{-1}e \rightarrow {}^{100}_{45}Rh$$

The earliest artificial transmutation was performed by Lord Rutherford in 1911. Rutherford bombarded nitrogen-14 with alpha particles. He obtained $^{17}_{8}O$ and protons as products. This transmutation can be represented in equation form.

$$^{14}_{7}N + {}^{4}_{2}He \rightarrow {}^{17}_{8}O + {}^{1}_{1}H$$

Use the following two rules when writing equations representing nuclear changes.

Rule 1. *Mass number is conserved in a nuclear change.* In other words, the sum of the mass numbers before the change must equal the sum of the mass numbers after the change.

Rule 2. *Electric charge is conserved in a nuclear change.* The total electric charges before and after a change must be equal. This electric charge refers to the charges on subatomic particles and nuclei.

CHEMISTS AND THEIR WORK

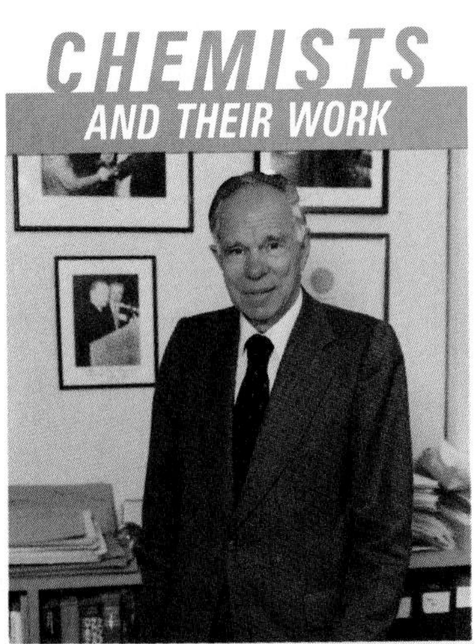

Glenn Theodore Seaborg (1912-)

Imagine patenting an element! That action is precisely what Glenn Seaborg did. He and his research team discovered several transuranium elements and patented two of them, assigning the patents to the United States government. His discovery of these elements was a result of the work he did in developing the first nuclear weapons during World War II. Seaborg had been educated and worked at the University of California in Berkeley. You can see his affinity for the university in the names of two of the elements he discovered, berkelium and californium. In 1951 Seaborg was awarded a Nobel Prize in chemistry. He has served as chairman of the United States Atomic Energy Commission (the forerunner of the Nuclear Regulatory Commission) and president of the American Association for the Advancement of Science.

SAMPLE PROBLEMS

1. Nuclear Equation—The New Atom Produced

Complete the following nuclear equation.

$$^{18}_{9}F \rightarrow {}^{0}_{+1}e + ?$$

Solving Process:

(a) Find the mass number of the unknown product.

$$\text{mass no. of } {}^{0}_{+1}e + \text{mass no. of ?} = 18$$
$$\text{mass no. of ?} = 18 - \text{mass no. of } {}^{0}_{+1}e$$
$$= 18 - 0$$
$$= 18$$

(b) Find the charge of the unknown product.

$$\text{charge of } {}^{0}_{+1}e + \text{charge of ?} = 9$$
$$\text{charge of ?} = 9 - \text{charge of } {}^{0}_{+1}e$$
$$= 9 - 1$$
$$= 8+$$

(c) Determine the identity of the unknown product and complete the nuclear equation. Turn to the periodic table and find which nuclide has 8 protons. This nuclide is oxygen. The symbol for an oxygen atom with a mass number of 18 is ${}^{18}_{8}O$. Thus, the completed nuclear equation is

$${}^{18}_{9}F \rightarrow {}^{0}_{+1}e + {}^{18}_{8}O$$

2. Nuclear Equation—*K*-Capture

Complete the following nuclear equation.

$${}^{83}_{37}Rb + ? \rightarrow {}^{83}_{36}Kr$$

Solving Process:

(a) Find the mass of the unknown reactant.

$$83 - 83 = 0$$

(b) Find the charge of the unknown reactant.

$$36 - 37 = -1$$

(c) Determine the identity of the unknown reactant and complete the nuclear equation. The unknown reactant has a charge of -1 and a mass of zero. The only particle that fits this description is an electron. The symbol for an electron is ${}^{0}_{-1}e$. Thus the completed nuclear equation is

$${}^{83}_{37}Rb + {}^{0}_{-1}e \rightarrow {}^{83}_{36}Kr$$

PROBLEM SOLVING HINT

Mass number and electric charge are both conserved in a nuclear equation.

PRACTICE PROBLEMS

Complete the following equations.

1. ${}^{27}_{12}Mg$ decays by beta-minus emission
2. ${}^{49}_{24}Cr$ decays by beta-plus emission
3. ${}^{76}_{36}Kr$ decays by *K*-capture
4. ${}^{213}_{88}Ra$ decays by alpha emission
5. ${}^{231}_{90}Th$ decays by alpha emission

Figure 28.7 One method proposed to dispose of radioactive wastes is to pour a mixture of the wastes and molten glass into molds made of graphite. After the glass hardens the cylinders formed would be placed in containers and stored underground.

CHEMISTRY IN DEPTH

Stability of Nuclides

Not all isotopes of an element are equally stable. What do we mean by a stable isotope? A stable isotope is one whose nucleus will not spontaneously decay. An unstable nucleus is one that undergoes spontaneous change of some kind. Some nuclei, such as $^{99m}_{43}Tc$, decay by giving off a quantum of gamma radiation. Various other unstable nuclei decay by emitting a particle with or without an accompanying quantum. There is also the possibility of the nucleus capturing an electron from the innermost energy level. This process is called *K*-capture because an old name for the innermost energy level was the *K*-level. It is possible to estimate which nuclides will be the most stable by applying the three following rules.

RULE OF THUMB

Rule 1. *The greater the binding energy per nucleon the more stable the nucleus.* The binding energy is the energy needed to separate the nucleus into individual protons and neutrons.

Consider the $^{16}_{8}O$ nuclide. It contains eight protons, eight electrons, and eight neutrons. We can think of it as eight hydrogen (protium) atoms, each of which contains one proton and one electron, and eight neutrons. Each protium atom has a mass of 1.007 825 2 u. Each neutron has a mass of 1.008 665 2 u. Thus, the total mass of an $^{16}_{8}O$ atom should be 16.131 923 2 u. However, the actual mass of the $^{16}_{8}O$ atom is 15.994 915 0 u. The difference between the calculated mass and the actual mass is called the **mass defect.**

$$\text{mass of } 8\ ^{1}_{1}\text{H atoms} + \text{mass of } 8\ ^{1}_{0}n = \text{expected mass of } ^{16}_{8}\text{O atom}$$

$$\begin{aligned} 8(1.007\,825\,2\text{ u}) + 8(1.008\,665\,2\text{ u}) &= 16.131\,923\,2\text{ u} \\ \text{actual mass of } ^{16}_{8}\text{O atom} &= \underline{15.994\,915\,0\text{ u}} \\ \text{mass defect} &= 0.137\,008\,2\text{ u} \end{aligned}$$

For an $^{16}_{8}O$ atom, the mass defect is 0.137 008 2 u. This mass has been converted to energy and released in the formation of the oxygen nucleus. Thus, it is also the energy that must be put back into the nucleus to separate the nucleons.

We can convert this mass defect of 0.137 008 2 u into its energy equivalent using the equation

$$E = mc^2$$

$$E = \frac{0.137\,008\,2\text{ u}}{} \left| \frac{1.660\,40 \times 10^{-27}\text{ kg}}{1\text{ u}} \right| \frac{(2.997\,93 \times 10^{8}\text{ m})^2}{(\text{s})^2}$$

$$= 2.044\,57 \times 10^{-11}\ \frac{\text{kg·m}^2}{\text{s}^2}$$

Recall that one joule is the energy required to maintain a force of one newton through a distance of one meter. A newton is equivalent to kg·m/s^2. Thus,

$$E = 2.044\,57 \times 10^{-11}\text{ J}$$

CHEMISTS AND THEIR WORK

James Chadwick
(1891-1974)

Capitalizing on the research and theories of Rutherford, Bothe, and Joliot-Curie, James Chadwick produced experimental evidence of the existence of the neutron. The discovery of the neutron explained a number of puzzling features about atomic structure, including the existence of isotopes and the masses of atoms. Chadwick spent some of the time during World War II in the United States, working on a combined British and United States effort to develop a nuclear weapon.

The energy described by this equation is called the **binding energy.** If we divide the total binding energy by the total number of nucleons in the oxygen atom, we obtain the binding energy per nucleon.

Total nucleons = 8 protons + 8 neutrons = 16 nucleons

$$\text{Energy per nucleon} = \frac{2.044\,57 \times 10^{-11}\ \text{J}}{1.6 \times 10^{1}\ \text{nucleons}}$$

$$= 1.277\,85 \times 10^{-12}\ \frac{\text{J}}{\text{nucleon}}$$

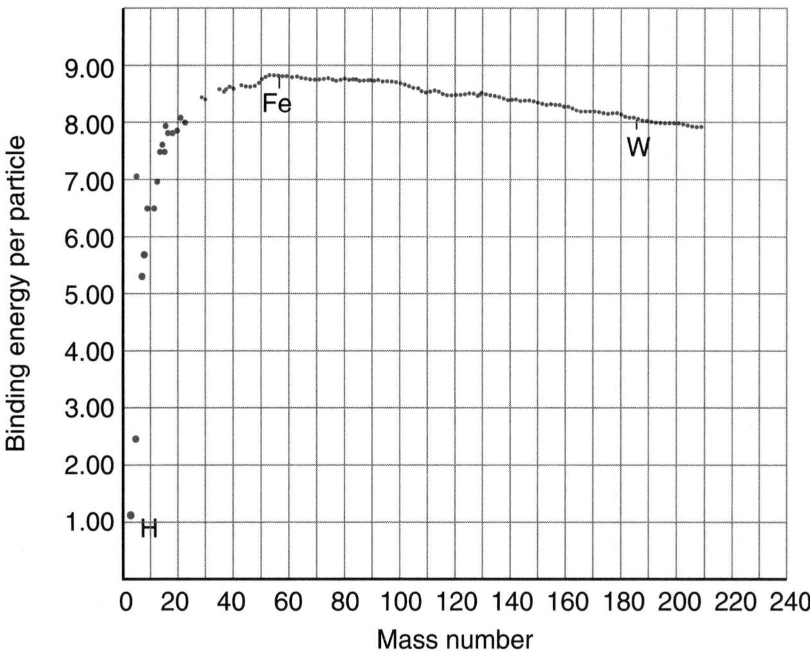

Figure 28.8 Iron (Fe) and its nearest neighbors have the highest binding energy per nucleon. Elements with mass numbers greater or less than iron have lower binding energies per nucleon.

The greater the binding energy per nucleon, the greater the stability of the nucleus. In Figure 28.8, the binding energy per nucleon is graphed against the mass number of known nuclides. Note that energy will be released in two reaction types involving the nucleus. It is released when one large nucleus splits to form two medium-sized nuclei as in a nuclear reactor. It is also released when two small nuclei join to form a medium-sized nucleus. In both cases, the medium-sized nuclei have greater binding energies per nucleon than the nuclei from which they were produced.

Rule 2. *Nuclei of low atomic numbers with a 1:1 neutron-proton ratio are very stable.* In Figure 28.9, the ratio of neutrons to protons is plotted for the known stable nuclei. For low atomic numbers, the ratio has a value very close to one. However, as the atomic number increases, the value of the neutron-proton ratio steadily increases. The closer the value of the neutron-proton ratio of a nuclide is to the broken line in this figure, the more stable the nuclide is.

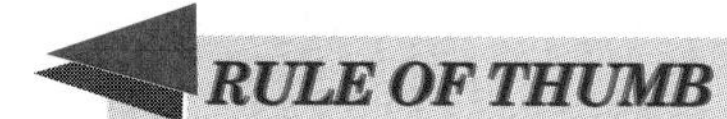

Figure 28.9 Stable nuclei are represented by the shaded blocks. The nuclei in region A emit neutrons or beta particles. The nuclei in region B emit alpha particles. Those nuclei in C emit positrons or capture electrons.

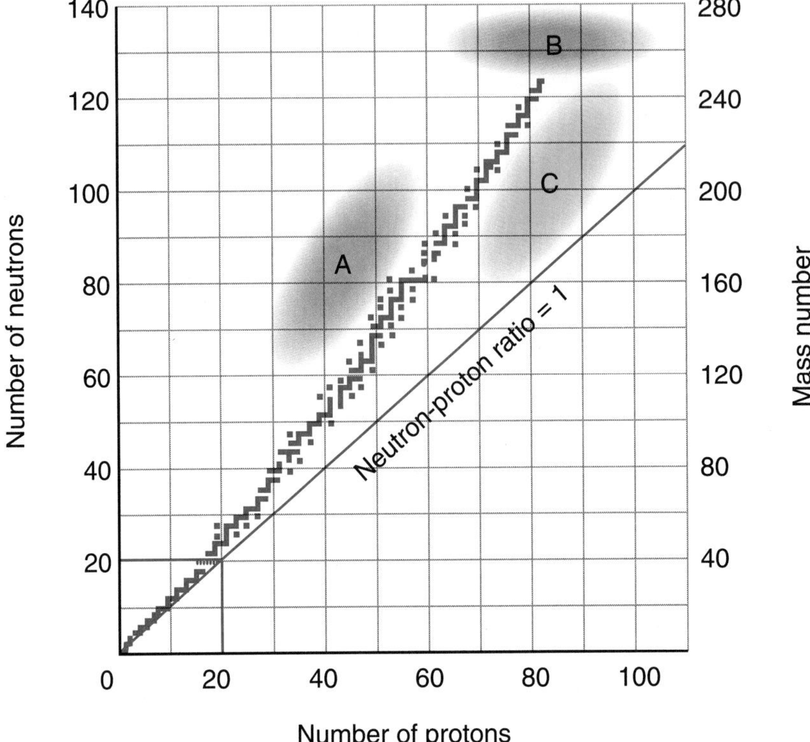

RULE OF THUMB

Rule 3. *The most stable nuclei tend to contain an even number of both protons and neutrons.* Of the known stable nuclei, 57.8 percent have an even number of protons and an even number of neutrons. Those nuclei with an even number of one kind of nucleon but an odd number of the other are slightly less stable. Thus, 19.8 percent of stable nuclei have an even number of neutrons but an odd number of protons; and 20.9 percent of stable nuclei have an even number of protons and an odd number of neutrons. Only 1.5 percent of stable nuclei have both an odd number of neutrons and an odd number of protons.

In Figure 28.9, nuclei falling within the regions *A, B,* and *C* are all unstable. Those nuclei lying in region *A* have excess neutrons and become more stable either by emitting neutrons or, more commonly, beta particles. For example,

$$^{131}_{53}I \rightarrow ^{131}_{54}Xe + ^{0}_{-1}e$$

Nuclei falling within the *B* region are too large for stability and are usually alpha emitters.

$$^{227}_{92}U \rightarrow ^{223}_{90}Th + ^{4}_{2}He$$

When a nucleus in the *B* region emits an alpha particle, its composition moves parallel to the 1:1 ratio line toward the *A* region. A nucleus in the *C* region has excess protons and can become stable by positron emission or by *K*-electron capture. Either the loss of a positron (β^+) or the capture of an electron (β^-) results in a new atom. This atom has the same mass number as the original atom but has an atomic number with a value one unit lower.

$$^{129}_{55}Cs + {}^{0}_{-1}e \rightarrow {}^{129}_{54}Xe$$
$$^{145}_{64}Gd \rightarrow {}^{145}_{63}Eu + {}^{0}_{1}e$$

Note that in the heavy-element radioactivity series, several alpha emissions are always followed by beta emission. The beta emission moves the nucleus back toward the stable region. ▲

Figure 28.10 Carbon-14, an isotope of carbon with a half-life of 5730 years, can be used to date previously living artifacts that are less than 20 000 years old.

Half-Life

The rate at which many radioactive nuclides decay has been determined experimentally. The rate of disintegrations is measured in a unit called a becquerel, Bq. One becquerel is one disintegration per second. An older unit for rate was the curie, Ci, equal to 3.7×10^{10} Bq.

The number of atoms that disintegrate in a unit of time varies directly as the number of atoms present. The length of time it takes for one-half of the atoms to disintegrate has been chosen as a standard for comparison purposes. This time interval is called **half-life.** For example, the half-life of $^{131}_{56}Ba$ is 12 days. If we start with a given number n of atoms of $^{131}_{56}Ba$, then at the end of 12 days, $n/2$ atoms will have changed into another element or isotope. At that time, we will have $n/2$ $^{131}_{56}Ba$ atoms left. At the end of the next 12 days, half of the remaining atoms will have disintegrated and we will have $n/4$ $^{131}_{56}Ba$ atoms left. In 12 more days, half of these atoms will have disintegrated and $n/8$ atoms of $^{131}_{56}Ba$ will remain. How many atoms will remain at the end of another 12 days?

Table 28.1

Half-Life and Decay Mode of Selected Nuclides

Nuclide	Half-Life	Decay Mode	Nuclide	Half-Life	Decay Mode
$^{3}_{1}H$	12.26 years	β^-	$^{100}_{46}Pd$	4.0 days	K-capture and γ
$^{6}_{2}He$	0.797 second	β^-	$^{129}_{55}Cs$	32.1 hours	K-capture and γ
$^{14}_{6}C$	5730 years	β^-	$^{149}_{61}Pm$	53.1 hours	β^- and γ
$^{19}_{8}O$	29.1 seconds	β^- and γ	$^{145}_{64}Gd$	25 minutes	β^+ and γ
$^{20}_{9}F$	11.56 seconds	β^- and γ	$^{183}_{76}Os$	12.0 hours	K-capture and γ
$^{26}_{14}Si$	2.1 seconds	β^+ and γ	$^{212}_{82}Pb$	10.64 hours	β^- and γ
$^{39}_{17}Cl$	55.5 minutes	β^- and γ	$^{194}_{84}Po$	0.5 second	α
$^{49}_{21}Sc$	57.5 minutes	β^- and γ	$^{210}_{84}Po$	138.40 days	α and γ
$^{63}_{27}Co$	52 seconds	β^-	$^{226}_{88}Ra$	1602 years	α and γ
$^{71}_{30}Zn$	2.4 minutes	β^- and γ	$^{227}_{92}U$	1.3 minutes	α
$^{69}_{32}Ge$	36 hours	β^+ and γ	$^{235}_{92}U$	7.1×10^8 years	α and γ
$^{81}_{33}As$	33 seconds	β^-	$^{238}_{92}U$	4.51×10^9 years	α
$^{82}_{35}Br$	35.34 hours	β^- and γ	$^{236}_{94}Pu$	2.85 years	α and γ
$^{87}_{37}Rb$	4.8×10^{10} years	β^-	$^{242}_{94}Pu$	3.79×10^5 years	α
$^{91}_{42}Mo$	15.49 minutes	β^+ and γ	$^{250}_{96}Cm$	1.7×10^4 years	Spontaneous fission

These half-life figures are determined experimentally for a large number of atoms of an individual nuclide. They predict the behavior of large numbers of atoms. At present, it is not possible to predict the exact instant when an individual atom will decay.

ARCHEOLOGY CONNECTION

Carbon-14 Dating

The radioactive nuclide $^{14}_{6}C$ is formed at a constant rate in the upper atmosphere. The $^{14}_{6}C$ is also decaying at a steady rate. Therefore, its concentration in the atmosphere remains constant. All $^{14}_{6}C$ is converted to CO_2 immediately after formation. This CO_2 is used in the photosynthetic process and becomes a part of the plant. Animals eat the plant or plant-eating animals and take in $^{14}_{6}C$. Once an organism dies, however, no more $^{14}_{6}C$ is taken in, and that $^{14}_{6}C$ already present continues to decay, with a half-life of 5730 years. Thus, by comparing the $^{14}_{6}C$ content of the once-living archeological specimen with its concentration in living organisms, a date of death for the artifact can be found. After four half-lives, the $^{14}_{6}C$ activity is so low that it becomes impossible to separate from background radiation. Consequently, $^{14}_{6}C$ dating is only good for specimens up to 20 000 years old.

SAMPLE PROBLEM

Half-Life

If you start with 2.97×10^{22} atoms of $^{91}_{42}Mo$, how many atoms will remain after 62.0 minutes? The half-life of $^{91}_{42}Mo$ is 15.49 minutes.

Solving Process:

Divide 62.0 min by 15.49 min to find the number of half-lives.

$$\text{Number of half-lives} = \frac{62.0\ \cancel{\text{min}}}{15.49\ \cancel{\text{min}}} = 4.00$$

The $^{91}_{42}Mo$ will go through four half-life decay cycles in 62.0 minutes.

$$(1/2)^4 = 1/16$$

One-sixteenth of the atoms will remain. Multiplying 1/16 by the number of atoms you started with gives

$$2.97 \times 10^{22}\ \text{atoms}\ (1/16) = 1.86 \times 10^{21}\ \text{atoms}$$

Thus, after 62.0 minutes, 1.86×10^{21} atoms of $^{91}_{42}Mo$ remain.

PRACTICE PROBLEMS

6. What part of a sample of $^{69}_{32}Ge$ will remain after 15 days?
7. If there are 5.32×10^{9} atoms of $^{129}_{55}Cs$, how much time will pass before the amount remaining is 5.20×10^{6} atoms?
8. If you start with 5.80×10^{28} atoms of $^{242}_{94}Pu$, how many will remain after 3.03×10^{6} years?

28.1 CONCEPT REVIEW

9. What are the ways in which scientists investigate the structure and properties of the nucleus? Describe the action of a linear particle accelerator as it is used in this investigation.
10. What three rules allow us to estimate relative nuclide stability? Would $^{212}_{82}Pb$ be predicted to be stable or unstable?
11. Explain the functions of the moderator and the containment vessel in a nuclear reactor.
12. Complete the following nuclear reaction: $^{231}_{91}Pa \rightarrow ^{227}_{89}Ac + ?$
13. **Apply** A sample of $^{129}_{55}Cs$ contains 2.54×10^{18} atoms. How many of these atoms will remain after 96.3 hours?

28.2 Nuclear Applications

OBJECTIVES

- explain how synthetic elements are produced.
- list the biological effects of radiation and the units used to measure them.
- describe the uses for radioactive nuclides.
- describe fusion reactions and the ongoing research being devoted to controlling the fusion process.

There is more to nuclear chemistry than huge accelerators and fission reactors. There are practical applications of radioactive nuclides in the chemistry laboratory, the hospital, the geologist's laboratory, and the archeologist's "dig." We must also be aware that radioactive nuclides, improperly handled, can pose a threat to health. In this section you will learn about some of the applications and dangers of radioactive nuclides.

Synthetic Elements

Elements with atomic numbers greater than 92 are called the transuranium elements. All of the synthesized transuranium elements have been produced by converting a lighter element into a heavier one. Such a change requires an increase in the number of protons in the nucleus. One of the processes of synthetic transmutation occurs as follows. A nuclear reactor produces a high concentration of neutrons that are "packed" into the nucleus of $^{239}_{94}\text{Pu}$. As the mass number builds, a beta particle is emitted. When beta emission occurs, a neutron is lost, and a proton is gained. This process produces an element that has an atomic number greater than that of the original element. The process can be written

$$^{239}_{94}\text{Pu} + {}^{1}_{0}n \rightarrow {}^{240}_{94}\text{Pu}$$
$$^{240}_{94}\text{Pu} + {}^{1}_{0}n \rightarrow {}^{241}_{94}\text{Pu}$$
$$^{241}_{94}\text{Pu} \rightarrow {}^{241}_{95}\text{Am} + {}^{0}_{-1}e$$

The nuclide $^{241}_{95}\text{Am}$ in turn can be used as a target to produce another element with a higher atomic number. The element with the highest atomic number reached in this manner is $^{256}_{100}\text{Fm}$.

A second method of synthesizing transuranium elements makes use of nuclear explosions that produce vast numbers of neutrons. Some of these neutrons are captured by uranium

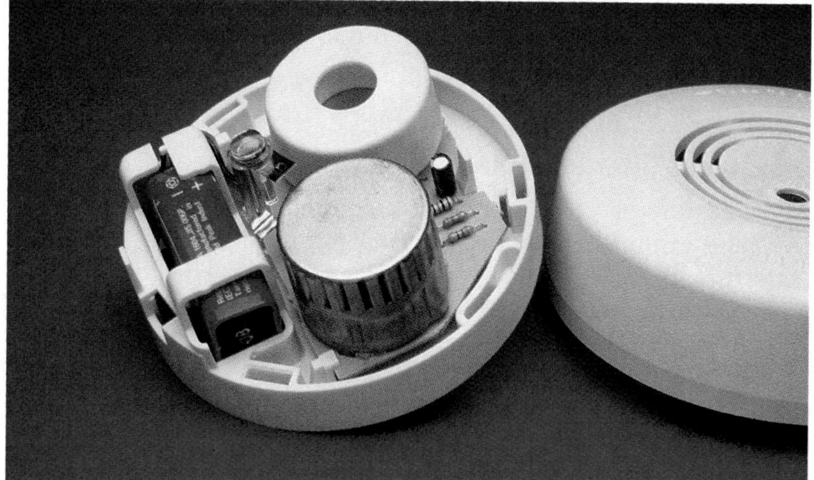

Figure 28.11 Ionization-type smoke detectors use $^{241}_{95}\text{Am}$ to ionize the air. The ions conduct an electric current, indicated by the arrows in the diagram on the right. The current is monitored by a current sensor that triggers the alarm when the current is interrupted by particles in smoke.

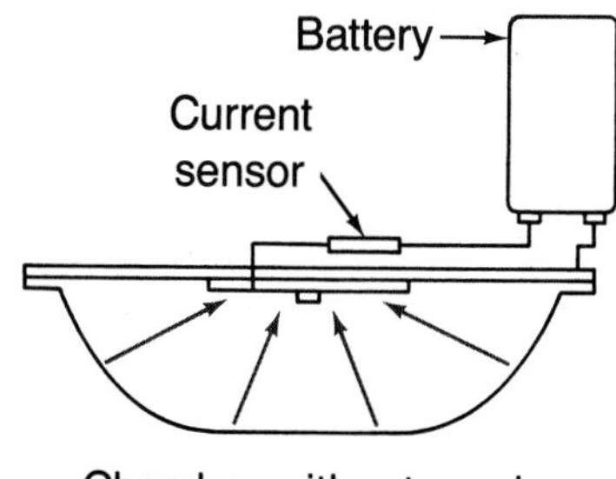

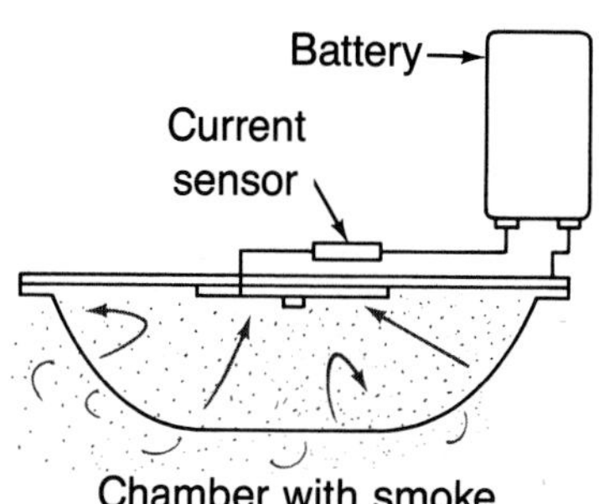

atoms. Successive electron emissions produce new elements. The element with the highest atomic number produced in this way is also $^{256}_{100}Fm$.

Elements with atomic numbers greater than 100 have been produced using other elements to bombard target elements. The nuclide $^{256}_{101}Md$ was created by bombarding $^{254}_{99}Es$ with alpha particles.

$$^{254}_{99}Es + ^{4}_{2}He \rightarrow ^{256}_{101}Md + 2^{1}_{0}n$$

Nobelium, atomic number 102, was created by using carbon nuclei and curium.

$$^{12}_{6}C + ^{244}_{96}Cm \rightarrow ^{254}_{102}No + 2^{1}_{0}n$$

The production of lawrencium, atomic number 103, can use neon and californium or einsteinium.

$$^{22}_{10}Ne + ^{254}_{99}Es \rightarrow ^{262}_{103}Lr + ^{12}_{6}C + 2^{1}_{0}n$$

One way that element 104 can be produced is by the bombardment of plutonium by neon. An alternate method uses carbon and californium.

$$^{12}_{6}C + ^{249}_{98}Cf \rightarrow ^{257}_{104}Unq + 4^{1}_{0}n$$

Element 105 can be produced by the bombardment of californium with nitrogen.

$$^{15}_{7}N + ^{249}_{98}Cf \rightarrow ^{260}_{105}Unp + 4^{1}_{0}n$$

Other nuclides have also been produced.

$$^{209}_{83}Bi + ^{58}_{26}Fe \rightarrow ^{267}_{109}Une$$

A heavy-ion accelerator in California is being prepared to accelerate particles as heavy as bromine nuclei.

It is highly possible that elements with even greater atomic numbers can be produced. The elements produced thus far are characterized by low yields and extremely short half-lives. Only a few atoms of elements 103–109 were first prepared, and these had half-lives of seconds. However, the latest lawrencium isotope produced, $^{262}_{103}Lr$, has a half-life of 216 minutes. Nuclear scientists believe that elements with atomic numbers as high as 126 might be produced.

Too Close for Comfort

A color TV emits X rays. Investigate what happens to the intensity of the radiation you absorb when you move farther from the source. Cover a tabletop with newspaper. Cover the reflector of a flashlight with dark paper or tempera paint. In a darkened room, shine the flashlight on the tabletop from a height of 10 cm. Mark the limits of the lighted area. Repeat this procedure from 20 cm and then 30 cm above the table. What do you notice about the intensity of the illuminated area as the light moves farther from the paper's surface? Calculate the area illuminated during each of the three trials. Compute a ratio of areas to the nearest whole number. The same amount of energy continues to fall on the paper as the light moves farther away, but the energy is spread over a larger area. If the intensity was 1 unit at 10 cm, what was it at 20 cm? At 30 cm?

Predict how much X-ray energy you would receive if you were 4 meters from the TV instead of 1 meter. Test your hypothesis using the flashlight 40 cm above the table.

Biological Effects of Radiation

All radiation (particles and electromagnetic waves) has an effect on living organisms. If the radiation has enough energy, it can penetrate living cells and disrupt their function. The disruption is particularly dangerous if a nucleic acid molecule is affected. Nucleic acids make up the genetic material of cells.

The amount of radiation, called kerma, is measured in a unit called the **gray,** Gy. One gray is equivalent to the transfer of exactly one joule of energy to one kilogram of living tissue. An older unit used to measure the same quantity is the rad, which is equal to 0.01 gray. The damage done biologically is better indicated by the absorbed dose, which is measured in the unit **sievert,** Sv. A sievert is equal to a gray multiplied by factors that determine how much of the energy transferred was actually absorbed by the tissue. An older measure of absorbed dose is the rem, which is equal to 0.01 sievert.

We are always exposed to some radiation from rocks containing radioactive elements, cosmic rays, and radioactive atoms naturally present in foods and water. These sources expose the average person to about one millisievert each year. About 0.02 mSv per year has been added to the exposure by testing nuclear weapons and operating nuclear power plants.

The radiation from radioactive sources varies in its effect on humans. All three types of radiation (α, β, and γ) occur in a range of energies. However, we may make the following general observations. Alpha particles are the least penetrating; a thin cotton garment will stop them. However, if an alpha particle gets inside the body, for example, when we breathe or eat contaminated material, it will do the most damage. Remember that an alpha particle has a large mass and charge compared with the other types of radiation. Gamma rays are the most penetrating.

Radiation exposure must sometimes be balanced against any other factors in a value judgment. For example, an X ray of a leg would produce about 0.2 mSv. However, if your leg is broken, you may decide to have your leg x-rayed. The danger of an incorrectly set bone outweighs the slight additional exposure to radiation.

Uses of Radioactive Nuclides

Because radioactive elements are easily detected by their radiation, they can be used as **tracers.** Tracers have a number of practical applications in chemistry. In quantitative analysis, a small amount of a radioactive nuclide of the element sought is introduced into the sample. The proportion of the unstable nuclide recovered in the analytical process is measured. That ratio, along with the actual total amount of substance recovered, can be used to compute the original quantity of unknown.

CAREERS IN CHEMISTRY

Science Writer

Science writers explore and report on major scientific and technological developments, explaining science discoveries and breakthroughs in everyday language. They call attention to issues that have important implications for society and spread information outside the scientific community. Sometimes, science writers are scientists, but more often they are people with excellent writing skills and a strong interest in science. Science writers are usually employed by the media. After a writer develops a strong reputation, he or she may be offered a contract for a book.

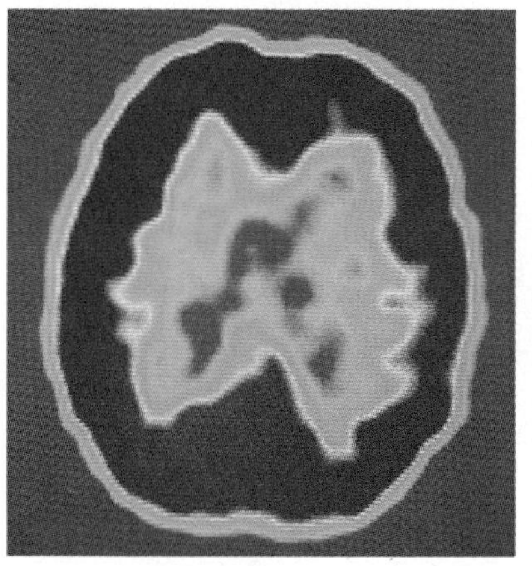

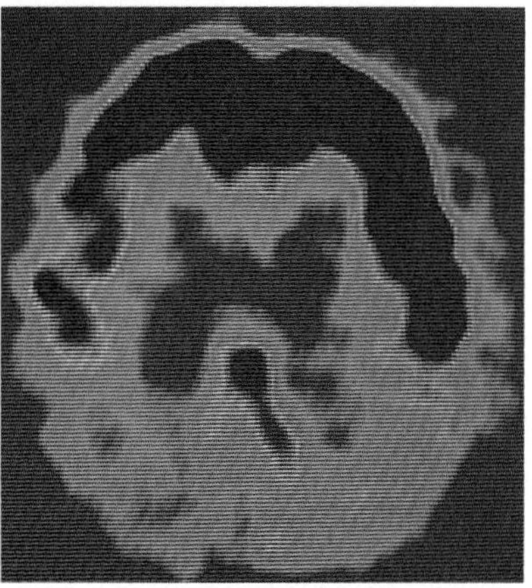

Figure 28.12 Radioisotopes can be used to find brain problems. The left scan shows a healthy brain. The right scan shows the brain of a patient with Alzheimer's disease.

CHEMISTRY AND

SOCIETY

Food Irradiation

Food spoils for many reasons. These reasons include the presence of parasites, insects, bacteria, and internal enzymes causing decomposition. Bacteria and other microorganisms pose a much greater threat to health than does any other cause, including contaminants such as inorganic toxins, pesticides, and food additives. People who eat spoiled food may get food poisoning, which

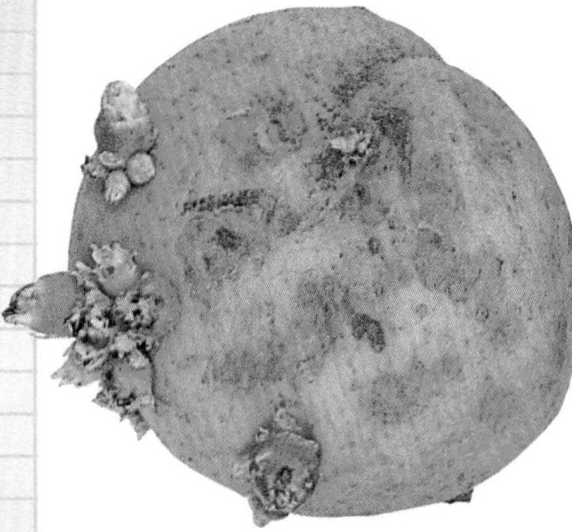

causes several thousand deaths each year in the United States. Much of this suffering could be prevented by a method of food preservation called radiation sterilization. Radiation sterilization prevents spoilage due to parasites, insects, and bacteria. The process involves exposing food to gamma radiation from $^{60}_{27}Co$.

Gamma radiation from $^{60}_{27}Co$ disrupts the metabolism of organisms, especially bacteria in the food, to such an extent that they die, or at least cannot reproduce. The radiation is so effective that it is used in sterilizing surgical instruments in some hospital operating rooms. The radiation does not change the natural flavor and consistency of the food. Depending on the dose level, irradiated food may last for weeks to years without refrigeration, thus prolonging its shelf life. The radiation involves only one-fifth the energy used in sterilization by heat, and raises the temperature of the food by only 3 C°. At present, dosage in the United States is limited by the Food and Drug Administration (FDA) to 1000 grays. That level preserves fruits and vegetables best. Meats require considerably higher doses. So far, the FDA has approved irradiation as a method of preservation for only a few foods. For example, the FDA has approved the irradiation of spices imported from countries with poor sanitation records.

People in favor of this method of food preservation believe there are many benefits. Chemical preservatives could be eliminated in some foods. Irradiation prevents potatoes and onions from sprouting and can slow down the ripening of some fruits. It can also destroy fungi which grow on foods in long-term storage. People who favor food preservation by irradiation think that it may help provide food for hungry people around the world.

People opposed to food irradiation are those who are concerned about the use of any nuclear reactions. Also, many are concerned about chemical changes in the food that result from the food irradiation. The radiation produces unique radiological products (URP) in the food. The current FDA ruling is that the URPs cannot exceed one part per million. Some people believe that even that level of URPs is harmful.

Research in the preservation of food with gamma radiation is ongoing. There is still much to learn, and there are many decisions to be made.

Analyzing the Issue

1. After further investigation of URPs, decide how you would vote on a referendum to approve radiation sterilization.
2. What are some of the negative effects of chemical preservatives that would not exist if food were preserved by irradiation?

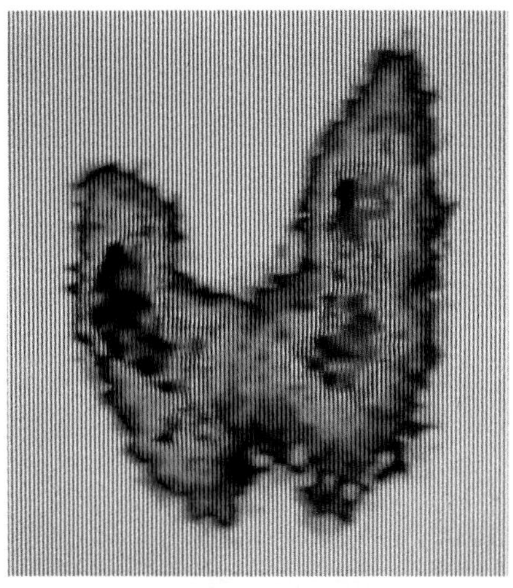
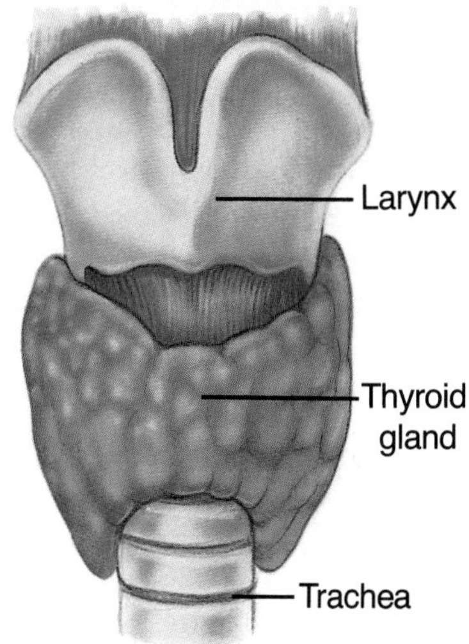

Figure 28.13 Iodine is a substance that concentrates in the thyroid gland. Because it does, radioactive $^{131}_{53}I$ is used to diagnose and treat disorders of the thyroid.

Radioactive nuclides have also been used extensively to study the mechanisms of reactions. For example, the study of photosynthesis is greatly assisted by the use of radioactive nuclides.

Naturally occurring radioactive nuclides can be used for dating certain objects. Using the half-life of a nuclide, we can date when organisms died up to about 20 000 years ago. By a similar technique, we can date the time of formation of rocks as far back as the origin of Earth. Radioactive nuclides of argon and uranium are used to date rocks; $^{238}_{92}U$ has been used to estimate the age of Earth. Radiocarbon dating, using the ratio of $^{14}_{6}C$ to $^{12}_{6}C$ in the artifact and the half life of $^{14}_{6}C$, is the method most often used for dating ancient artifacts made of cloth or wood.

Suppose we want to know what part of the human body utilizes a certain substance in our diet. Using a radioactive nuclide of an element in the substance, we would prepare some of the substance. We would have to use a very low level of radiation. After a person eats or drinks the substance, we would examine the person's body with radiation detectors. The detectors would be set to observe only radiation from the nuclide being used in the diagnosis. The part of the body activating the detectors would be where the substance is concentrated.

Fusion

The peak of the binding energy curve in Figure 28.8 occurs near iron, $Z = 26$. We can see how both fission and the emission of small particles by atoms of high atomic number lead to more stable atoms. Stability could also be gained by combining small nuclei into larger ones. A nuclear reaction in which two or more small nuclei combine to form one larger nucleus is called a **fusion reaction.** Note also that the slope of the binding energy curve is greater on the low atomic number side than it is on the high atomic number side. We should therefore expect fusion reactions to produce much greater amounts of energy per particle than fission reactions. This prediction is supported by observation.

fundere: (L) to melt

fusion — melting together of nuclei

Figure 28.14 The magnetic field coils of a fusion reactor are shown in this interior view.

Scientists have harnessed the fission reaction in a controlled process on a small scale for limited power production. When we harness the fusion reaction for power production, we will have a solution to energy problems for some time to come. Some of the difficult problems associated with fission reactions as a power source will also be eliminated. The availability of "fuel" for nuclear fusion reactions is much greater than for fission reactions. Also, fusion reactions useful for power production do not produce as much radioactive waste as do fission reactions. The most likely reactions for fusion generators are the following.

$$^{2}_{1}H + ^{2}_{1}H \rightarrow ^{3}_{2}He + ^{1}_{0}n$$
$$^{2}_{1}H + ^{2}_{1}H \rightarrow ^{3}_{1}H + ^{1}_{1}H$$
$$^{2}_{1}H + ^{3}_{1}H \rightarrow ^{4}_{2}He + ^{1}_{0}n$$

The last of these reactions is the most promising from the standpoint of power production. However, it does have drawbacks. Tritium ($^{3}_{1}H$) is a radioactive nuclide that occurs naturally in only tiny amounts. It can be produced by exposing $^{6}_{3}Li$ to neutrons.

$$^{6}_{3}Li + ^{1}_{0}n \rightarrow ^{4}_{2}He + ^{3}_{1}H$$

Another disadvantage is that the reactor itself will become radioactive as it is exposed to these radiations.

On the other hand, the reactions involving only deuterium ($^{2}_{1}H$) are free of radioactivity. Also, the supply of deuterium could provide Earth with energy for as long as 10^{12} years. Extraction of deuterium from water supplies would require only a small fraction of the power output of a fusion reactor.

Present research could produce practical fusion installations within your lifetime. If certain technical problems are solved, these reactors could appear in the near future. Let us look at some of these problems.

Figure 28.15 As a pulse of ions is fired in Sandia National Laboratories' Particle Beam Fusion Accelerator II, electrical discharges are visible on the surface of the water that covers the accelerator.

Stellar Nucleosynthesis

About 75 percent of the matter in the universe is hydrogen and almost 25 percent is helium. All other elements make up about 1 percent of the matter in the universe. Why are there such massive amounts of these two light elements? When a star forms, a cloud of hydrogen, helium, and dust contracts through the mutual gravitational attraction the particles have for one another. As these particles fall inward in the cloud, they pick up speed and their temperatures are thus increased. Eventually, the core of the star is hot enough (10^7 K) to start a fusion reaction among hydrogen atoms.

$$^{1}_{1}\text{H} + ^{1}_{1}\text{H} \rightarrow ^{2}_{1}\text{H} + ^{\ 0}_{+1}e + \upsilon \text{ (neutrino)}$$
$$^{2}_{1}\text{H} + ^{1}_{1}\text{H} \rightarrow ^{3}_{2}\text{He}$$
$$^{3}_{2}\text{He} + ^{3}_{2}\text{He} \rightarrow ^{4}_{2}\text{He} + 2^{1}_{1}\text{H}$$

As the hydrogen fuel is exhausted, the star continues to collapse. This collapse increases the density of the star and raises the temperature to a range (10^8 K) where helium fusion reactions begin.

$$^{4}_{2}\text{He} + ^{4}_{2}\text{He} \rightarrow ^{8}_{4}\text{Be}\ \textit{(unstable)}$$
$$^{8}_{4}\text{Be} + ^{4}_{2}\text{He} \rightarrow ^{12}_{6}\text{C}$$
$$^{12}_{6}\text{C} + ^{4}_{2}\text{He} \rightarrow ^{16}_{8}\text{O}$$

Similar processes can continue in massive stars until iron nuclei are formed. Nuclei heavier than iron consume energy when formed. As a consequence, the star starts a rapid collapse, iron accumulates in the core, and the nuclear reactions in the core cease. The collapse of such a large amount of matter produces an incredible amount of energy. The collapse finally ends in a catastrophic explosion of the entire star, called a supernova. All naturally occurring heavier-than-iron elements have been produced in supernova explosions.

Exploring Further

1. Stars produce larger quantities of elements with even atomic numbers. Why?
2. Why do larger stars burn up faster?

Fusion Reactors

In order for two atomic nuclei to undergo fusion, they must come almost into contact. Recall that the nuclear force extends only about a distance equal to the diameter of the nucleus. However, since all nuclei are positively charged, they tend to repel each other strongly. In plasma fusion technology, fusion reactions occur when the reactants are held in the temperature range of $10^{8\circ}$C to $10^{9\circ}$C for a suitable time. The energy requirements to achieve these temperatures are quite high. The length of time required depends upon how closely packed the particles are during that time. At these high temperatures, matter is in the form of plasma. One major problem in designing a plasma fusion reactor is containing the fuel at that temperature. Containers made from materials cannot be used. The plasma would lose energy to them so rapidly that fusion temperatures would never be reached. Because plasmas are charged particles, they are affected by magnetic fields. By properly shaping a magnetic field, a plasma may be contained. As fusion occurs and the temperature rises, the pressure of a plasma causes it to expand. With charged particles in the plasma moving around at tremendous velocities, the plasma itself generates electric currents. Both of these effects make leakage from the magnetic "bottle" a difficult problem to solve. All of these problems are topics of present research. Who can predict what role developments in this fascinating field will play in shaping our future!

FRONTIERS

Cold Fusion Confusion

In the spring of 1989, the results of an electrolysis experiment indicated the first room-temperature nuclear fusion reaction. The announcement of cold fusion, as it is called, to the news media and the scientific community produced instant response. Media reporters realized that cold fusion could produce unlimited and cheap electric power and announced the dawning of a new energy age. The response from the scientific community ranged from attempts to confirm the results to dismissal of the results as theoretically impossible. Most scientists took a wait-and-see attitude, letting researchers verify and validate the results.

According to the original research, the electrolysis of heavy water, water made from deuterium, produced more energy than expected. The energy was accounted for by citing preliminary evidence that deuterium underwent fusion within the lattice of the palladium electrode. Cold fusion—and confusion about it—were born.

Researchers trying to repeat the experiment have had poor results. Some have detected less energy than originally reported. Others have detected neutrons or gamma rays that might account for fusion but in amounts too small or unsynchronized to be energy released by standard fusion processes. Still others have declared the original research flawed and have dismissed cold fusion. Most of the scientific community is highly skeptical.

28.2 CONCEPT REVIEW

14. What is the effect of radiation on living tissue?

15. Describe the use of a radioactive nuclide as a biological tracer.

16. Explain how carbon-14 is used in dating objects.

17. List two advantages and two disadvantages of using a fusion reaction for energy production purposes.

18. Apply Give an example of a fusion reaction that could theoretically be used to produce an atom of ${}^{275}_{113}\text{Uut}$.

Chapter 28
REVIEW

Summary

28.1 Nuclear Structure and Stability

1. Nuclear structure can be investigated by bombarding nuclei with high-energy particles from devices such as particle accelerators or nuclear reactors.
2. After having been struck by a particle, a nucleus may undergo one of several changes. Some of these changes are to become metastable (excited), break into many parts, or absorb the particle.
3. In linear accelerators, charged particles are accelerated by electromagnetic waves. In circular accelerators or synchrotrons, charged particles are accelerated by electric fields and held in a circular path by electromagnets.
4. Fission is the breaking apart of an unstable nucleus into two roughly equal parts.
5. Some heavy nuclei can be made to fission by absorbing slow neutrons. The neutrons emitted in the reaction can, in turn, cause other nuclei to fission, leading to a chain reaction.
6. A nuclear reactor is a device for containing and controlling a fission reaction so that it can be put to practical use.
7. Moderators such as water or graphite are used in reactors to slow down neutrons so that collisions with nuclei produce fission.
8. A coolant material is required to keep nuclear reactors from overheating. The energy absorbed by the coolant can be used to do useful work.
9. If a reaction changes the number of protons in a nucleus, a different element results. The changing of one element into another is called transmutation.
10. In completing equations for nuclear changes, both mass number and electric charge must be conserved.
11. Three relationships can be used to predict the stability of nuclides.
    a. The greater the binding energy per nucleon is, the more stable is the nucleus.
    b. Nuclei of low atomic number having a 1:1 neutron-proton ratio are stable.
    c. The most stable nuclei tend to contain an even number of both protons and neutrons.
12. The calculated mass defect of an atom indicates the transformation of mass into energy.
13. Binding energy is the energy needed to separate the nucleus into individual protons and neutrons.
14. The half-life of a radioactive element is the time it takes for one-half of the atoms of a sample of the substance to decay.

28.2 Nuclear Applications

15. New elements may be synthesized by bombarding nuclei with neutrons or with the nuclei of other elements.
16. All radiation has an effect on living organisms. The food and water we consume and our environment consistently emit small amounts of naturally occurring radiation.
17. Amount of radiation is measured in grays. Damage done to living tissue is indicated by absorbed dose, which is measured in sieverts.
18. Radioactive nuclides are useful as tracers in chemical analysis and in the study of reaction mechanisms. Some naturally occurring nuclides are useful in archeological and geological dating.
19. Fusion is the combining of two or more small nuclei into one larger nucleus.
20. In hot-plasma technology, fusion reactants must be heated to between 10^8°C and 10^9°C. A shaped magnetic field is used to contain a plasma at these extremely high temperatures.

Chapter 28 REVIEW

Key Terms

accelerator
drift tube
synchrotron
fission
nuclear reactor
moderator
containment vessel
transmutation
mass defect
binding energy
half-life
gray
sievert
tracer
fusion reaction

Review and Practice

19. What are the two principal ways of bombarding a nucleus?
20. In the symbol $^{99m}_{43}\text{Tc}$, what do each of the numbers and the *m* represent?
21. List three possible results of a collision of a particle with a nucleus.
22. By what means are particles accelerated in a linear accelerator?
23. Explain why drift tubes are necessary in a linear accelerator.
24. By what means are particles accelerated in a circular accelerator?
25. What is the purpose of electromagnets in a circular accelerator? Why must the magnets be variable?
26. Explain why circular accelerators are called synchrotrons.
27. What happens to an atom during fission? What kinds of nuclei fission spontaneously? How can fission be induced in other elements?
28. What is the source of neutrons that keep a nuclear chain reaction going?
29. What is a nuclear reactor?
30. What is the purpose of the moderator in a nuclear reactor? Name two substances commonly used as moderators.
31. How are control rods used to control reaction rate in a reactor?
32. What type of element should be used as a control rod in a reactor? Give three examples of elements that are frequently used in control rods.
33. Why is a coolant necessary in a reactor?
34. What purposes are served by a containment vessel?
35. What are two problems involved in the operation of a nuclear reactor?
36. List the three general processes by which a naturally radioactive nucleus is able to decay.
37. What is meant by a stable nucleus?
38. How is the ratio of neutrons to protons in a nucleus related to stability?
39. What happens during *K*-electron capture? How is the nucleus affected by this process?
40. What is measured using the unit becquerel? What is the value of one becquerel?
41. If the half-life of $^{90}_{38}\text{Sr}$ is 28.8 years, how many atoms of $^{90}_{38}\text{Sr}$ will be left after 86.4 years if there were initially 1.00 mol of atoms?
42. How long will it take 6.00×10^{20} atoms of $^{71}_{30}\text{Zn}$ to disintegrate to 1.88×10^{19} atoms?
43. One of the cobalt nuclides, $^{60}_{27}\text{Co}$, has a half-life of 5.3 years. If an initial sample contains 10.0 g of $^{60}_{27}\text{Co}$, what mass will be left after 21.2 years?
44. What net change in a nucleus results in transmutation?
45. How is a nucleus changed by emission of a beta particle?
46. When the nuclide $^{210}_{84}\text{Po}$ loses an alpha particle, what happens to the atomic number of the nucleus? What happens to the mass number?
47. What two quantities are conserved in a nuclear change?
48. Complete the following equations.
    a. $^{3}_{1}\text{H} \rightarrow ? + {}^{0}_{-1}e$
    b. $^{61}_{30}\text{Zn} \rightarrow ? + {}^{0}_{+1}e$
    c. $^{9}_{3}\text{Li} \rightarrow {}^{9}_{4}\text{Be} + ?$

d. $^{240}_{96}Cm \rightarrow ? + ^{4}_{2}He$
e. $^{199}_{84}Po + ^{0}_{-1}e \rightarrow ?$ (*K*-capture)

49. Write a balanced nuclear equation for each of these changes.
a. alpha emission by $^{242}_{94}Pu$
b. beta minus emission by $^{30}_{13}Al$
c. electron capture by $^{55}_{26}Fe$
d. positron emission by $^{93}_{44}Ru$

50. Write the symbols, including atomic number and mass number, for the radioactive nuclides that would give each of these products. Use these symbols to write a balanced nuclear equation for each change.
a. $^{253}_{100}Fm$ from alpha emission
b. $^{80}_{37}Rb$ from electron capture
c. $^{140}_{61}Pm$ from positron emission
d. $^{211}_{83}Bi$ from beta minus emission

51. List three methods of synthesizing transuranium elements.

52. How can transuranium elements be synthesized by nuclear explosions?

53. Describe the method generally used in the synthesis of elements with atomic numbers greater than 100.

54. Write an equation for production of $^{257}_{104}Unq$ by bombardment of $^{249}_{98}Cf$ by another element.

55. List three natural sources of radiation from the environment.

56. Compare the effects of internal and external gamma and alpha radiation on the human body.

57. What are three advantages of fusion over fission reactions?

58. What feature of hot-plasma fusion reactions causes the greatest difficulty in putting these reactions to practical use? How are magnets used to overcome this problem?

59. Beginning with *n* atoms of an unstable nuclide, how many atoms will remain after 3 half-lives? After 5 half-lives?

60. Identify the mode of nuclear decay represented by each of the following. Particles or radiation given off are not shown.
a. $^{x}_{y}A \rightarrow ^{x-4}_{y-2}X$
b. $^{x}_{y}A \rightarrow ^{x}_{y-1}X$
(Two modes are possible.)
c. $^{x}_{y}A \rightarrow ^{x}_{y+1}X$
d. $^{x}_{y}A \rightarrow ^{2x/3}_{2y/3}X + ^{x/3}_{y/3}Q$ (approximately)

61. Compute the mass defect, binding energy, and binding energy per nucleon for iron if the actual mass of a $^{56}_{26}Fe$ nucleus is 55.920 66 u.

62. If the binding energy per nucleon is 1.13 $\times$ 10^{-12} J for $^{4}_{2}He$, calculate the mass defect.

63. Unnilquadium, element 104, was prepared by bombarding $^{249}_{98}Cf$ with $^{12}_{6}C$ nuclei in the Heavy Ion Linear Accelerator. Assuming that the product of the reaction had a mass number of 257, write an equation for this change. Write an equation for the bombardment of the same target with $^{13}_{6}C$ to form a product of mass number 259.

64. Compute the relative stabilities of the two isotopes $^{148}_{60}Nd$ (mass 147.916 93 u) and $^{150}_{60}Nd$ (mass 149.920 91 u) on the basis of the binding energy per particle for each nuclide.

65. At noon on Monday, you measure the mass of a sample of $^{183}_{76}Os$ and find the mass is 802 g. What mass of $^{183}_{76}Os$ will remain at noon on the following Monday? Refer to Table 28.1.

Concept Mastery

66. Why are accelerators essential to the study of atomic nuclei?

67. What advantage does an accelerator with two colliding streams have over a single-stream accelerator that has a stationary target?

68. Why must a chain reaction be controlled in order to be useful?

69. Describe how the chain reaction of a nuclear reactor produces energy to do useful work, such as powering generators.

70. Define mass defect. What is the cause of mass defect?

71. What is binding energy? Relate binding energy to mass defect. Relate binding energy to nuclear stability.

72. Explain what is meant by the statement that the half-life of $^{26}_{14}$Si is 2.1 seconds.

73. Describe how packing a nucleus with neutrons can lead to the formation of a nucleus of a different element.

74. What effect of radiation on living tissue is most dangerous?

75. Define *tracer*. Describe how one is used in quantitative analysis.

76. How can exposure to neutrons induce fission in a nucleus?

Application

77. The carbon nuclide, $^{14}_{6}$C, is used in dating organic materials. However, the accuracy of radiocarbon dating diminishes rapidly in material that is older than about 20 000 years. Suggest an explanation for the inaccuracy.

78. The nuclides $^{85}_{36}$Kr and $^{90}_{38}$Sr are both products of nuclear fission and are released in minute amounts into the atmosphere during the reprocessing of atomic fuels. Explain why $^{85}_{36}$Kr may attack skin and lungs but does not accumulate in the body, while $^{90}_{38}$Sr may accumulate in the bones, possibly causing bone cancer and leukemia.

79. The carbon in a bone from an ancient campsite is found to contain only one-eighth the proportion of $^{14}_{6}$C that occurs in organisms living today. Approximately how old is the bone?

80. How many neutrons must be absorbed by a $^{238}_{92}$U nucleus to be changed to $^{243}_{95}$Am? Assume that every other neutron absorbed undergoes β^- decay, beginning with the first neutron. Write the nuclear equations.

81. The naturally occurring radioactive nuclide $^{232}_{90}$Th decays through a series of steps similar to those shown for $^{238}_{92}$U in Figure 28.6. Without writing equations for the individual steps, determine the nuclide that is the final product if the steps consist of the emission of six alpha particles and four beta particles. Check your answer by writing the equations for the individual steps if the order of emission is α, β, β, α, α, α, α, β, β, α.

Critical Thinking

82. What property must a particle have in order to be accelerated in a linear accelerator or a synchrotron?

83. Explain why neutrons can penetrate easily to the nucleus. Could a proton penetrate as easily? Explain.

84. Why is the commonly used term *atomic bomb* inaccurate?

85. Why do you think that water is the most popular cooling material used in nuclear reactors?

86. The fission of large nuclei and the fusion of small nuclei both result in medium-sized nuclei. Why do both processes give off energy?

87. Radiation is most hazardous to living organisms if it affects nucleic acid. Why would this effect be particularly dangerous to organisms?

Readings

Atwood, Charles H. "Chernobyl—What Happened." *Journal of Chemical Education* 65, no. 12 (December 1988): 1037-1042.

Freeman, David H. "Fission in the Fusion Camp." *Discover* 10, no. 12 (December 1989): 32–42.

Hanson, David J. "Radon Tagged as Cancer Hazard by Most Studies, Researchers." *Chemical and Engineering News* 67, no. 6 (February 6, 1989): 7-13.

Ishihara, Takehiko. "Radioactive Waste Disposal." *Oceanus* 30, no. 1 (Spring 1987): 61-65.

Kaufmann, George B. "Beyond Uranium." *Chemical and Engineering News* 68, no. 7 (November 19, 1990): 18-29.

Kluger, J. "The Residue of Nuclear Hubris." *Discover* 10, no. 1 (January 1989): 10-11.

Navratil, James D., *et al.* "The Most Useful Actinide Isotope: Americium-241." *Journal of Chemical Education* 67, no. 1 (January 1990): 15-16.

Wood, Clair G. "Carbon-14 Dating." *ChemMatters* 7, no. 1 (February 1989): 12-15.

Cumulative Review

1. What happens to a positive ion at the cathode of an electrolytic cell?

2. Why is it necessary to separate the half-cells when constructing a voltaic cell?

3. Will Ni react with Zn^{2+}? Explain. Use the table of standard reduction potentials.

4. What voltage do you expect from a cell with $Co \mid Co^{2+}$ and $Pb \mid Pb^{2+}$ half-cells?

5. How many grams of Sn would be deposited from a Sn^{2+} solution by a current of 0.506 ampere flowing for 22.0 minutes?

6. What will be the products of the electrolysis of a water solution of $CuBr_2$?

7. What voltage could you expect from the following cell at 25°C?

$$Pb \mid Pb^{2+}(0.282M) \mid F_2 \mid F^-(0.0400M)$$

8. Compute the enthalpy change for the following reaction.

$$\Delta H_f° CuSO_4(aq) = -839.662 \text{ kJ/mol}$$

$$Cu(cr) + 2H_2SO_4(aq) \rightarrow CuSO_4(aq) + SO_2(g) + 2H_2O(l)$$

9. Compute the free energy change for the following reaction.

$$P_4O_{10}(cr) + 6H_2O(l) \rightarrow 4H_3PO_4(l)$$

$$\Delta H_r° = -454.3 \text{ kJ and } \Delta S° = 3.76 \text{ J/K at } 25°C$$

10. How do standard conditions in thermodynamics differ from standard temperature and standard atmospheric pressure used in working with the gas laws?

11. Define the quantities q and w.

12. What mass of aluminum is required to react with 4.90 dm^3 of chlorine at STP according to the following equation?

$$2Al + 3Cl_2 \rightarrow 2AlCl_3$$

13. What mass of sodium chloride can be obtained from the reaction of 49.5 g HCl with 91.8 g NaOH?

$$HCl + NaOH \rightarrow NaCl + H_2O$$

14. What mass of sodium aluminate is obtained from the reaction of 134 g $Al(OH)_3$ with 635 g NaOH according to the following equation?

$$Al(OH)_3 + 3NaOH \rightarrow Na_3AlO_3 + 3H_2O$$

15. How much energy is needed to heat 8.16 g of ice at −16°C to steam at 102°C?

16. A gas collected over water at 21°C and 103 kPa is dried. What volume does it occupy when dry if it occupied 37.0 dm^3 when wet?

17. What is the molecular mass of a gas if 2.20 g of it occupy 2.13 dm^3 at 20.0°C and 107.1 kPa?

18. How many dm^3 of hydrogen at STP can be obtained from 21.0 g of Zn?

$$Zn + 2HCl \rightarrow ZnCl_2 + H_2$$

CHAPTER Preview

CONTENTS

Classes of Organic Compounds

CHAPTER 29

You may not know it, but organic compounds play a major role in your life. Your clothing is made of natural and synthetic organic materials such as cotton, wool, rayon, and polyester. Your shoes may be made from natural or synthetic leather, both organic materials. The tires on your bicycle contain other organic material, such as rubber and nylon. Even the road you ride on may be covered with macadam, an organic material. If all organic materials were removed from the photographs, there would be little remaining.

Of the more than 10 million substances known to scientists, almost nine million are organic substances. Understanding organic chemistry would be impossible if we were not able to classify these substances.

EXTENSIONS AND CONNECTIONS

Frontiers

Paddy Gas: Reevaluating a source of natural methane gas, **page 744**

Mechanics Connection

Ethyne's Flame, page 746

Chemists and Their Work

Emma Perry Carr, page 748

Bridge to Medicine

Nuclear Magnetic Resonance: How the structures of organic molecules are determined by studying the magnetic fields of atomic nuclei, **page 753**

Everyday Chemistry

Dyes: A look at the compounds used to color your life, **page 760**

OBJECTIVES

- differentiate between aromatic and aliphatic hydrocarbons, saturated and unsaturated hydrocarbons, and chain and cyclic hydrocarbons.
- identify, name, and write structural formulas of alkanes, alkenes, and alkynes.
- identify, name, and write structural formulas of branched, cyclic, and aromatic hydrocarbons.

Word Origins

alephein: (GK) to anoint

aliphatic — oil (traditionally used to anoint)

29.1 Hydrocarbons

This morning you probably put on some piece of clothing made from an artificial fiber, such as rayon, nylon, Orlon, or polyester. You may have ridden to school on a bus or in a car that has synthetic rubber tires. When you bought a pack of notebook paper, it was covered with a plastic wrap. At lunch your drink may have been in a plastic cup. When you comb your hair, you may use a plastic comb. The body of your ballpoint pen is a plastic. All of these materials — synthetic fibers, synthetic rubber, plastics — are produced from petroleum. Petroleum is almost entirely composed of hydrocarbons. The study of organic chemistry is best started by examining these compounds.

Classification of Hydrocarbons

Recall that hydrocarbons are composed of only carbon and hydrogen. Almost all other organic compounds can be considered as derivatives of the simple hydrocarbons. Hydrocarbons are classified as aromatic or aliphatic. **Aromatic hydrocarbons** contain one or more benzene rings. All other hydrocarbons are **aliphatic** (al ih FAT ihk) **hydrocarbons.**

Aliphatic compounds may have carbon atoms bonded in rings or chains. The chain compounds may be further classified on the basis of the individual carbon-carbon bonds. A chain compound in which all carbon-carbon bonds are single bonds is called an **alkane** (AL kayn). This type of compound is also called a **saturated hydrocarbon.** Thus, additional atoms can be bonded to the original atoms in the compound only by breaking the compound into two or more fragments. The types of hydrocarbons are summarized in Table 29.1.

Alkanes

The alkanes are the least complex hydrocarbons. Recall that the first four members of the alkane series and their formulas are methane, CH_4; ethane, C_2H_6; propane, C_3H_8; and butane, C_4H_{10}. After butane, the members of the alkane series are named using the Greek or Latin prefix for the number of carbon atoms. The word ending characteristic of this family is *-ane.*

Each alkane differs from the next by a $-CH_2-$ group. You can think of this series as being formed by removing a hydrogen atom from one of the carbon atoms, adding a $-CH_2-$ group, and replacing the hydrogen. For example:

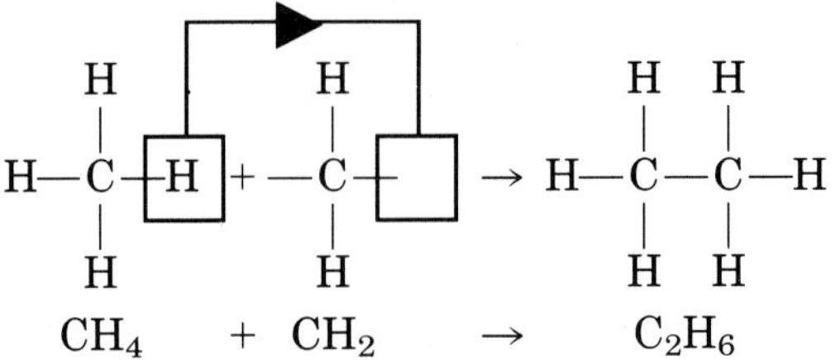

Table 29.1

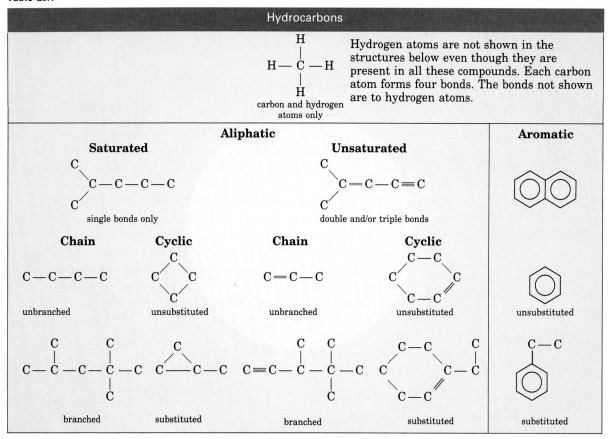

Hydrocarbons				
H–C(H)(H)–H carbon and hydrogen atoms only	Hydrogen atoms are not shown in the structures below even though they are present in all these compounds. Each carbon atom forms four bonds. The bonds not shown are to hydrogen atoms.			
Aliphatic				**Aromatic**
Saturated (C)(C)C—C—C—C single bonds only		**Unsaturated** (C)(C)C=C—C≡C double and/or triple bonds		(naphthalene ring structure)
Chain	**Cyclic**	**Chain**	**Cyclic**	
C—C—C—C unbranched	(four-membered carbon ring) unsubstituted	C=C—C unbranched	(six-membered carbon ring with one double bond) unsubstituted	(benzene ring) unsubstituted
C—C(C)—C—C(C)(C)—C branched	(three-membered carbon ring)—C—C substituted	C≡C—C(C)—C(C)(C)—C branched	(six-membered carbon ring with one double bond)—C—C substituted	(benzene ring)—C—C substituted

However, the equation on the previous page does not represent the actual preparation of an alkane.

Compounds whose structures differ from each other by a specific structural unit is called a **homologous** (hoh MAHL uh guhs) **series.** A member of a homologous series is called a homolog. In the case of the alkanes, the specific structural unit is $—CH_2—$. A general formula can be written for all of the members of a homologous series such as the alkanes. For the alkanes, the formula is C_nH_{2n+2}, where n is the number of carbon atoms in the compound.

The simplest alkane is methane. Methane contains one carbon atom and four hydrogen atoms. The structural formula for methane is shown in Table 29.1. As you can see, methane consists of the specific alkane structural unit $—CH_2—$ bonded to two hydrogen atoms. The structural formulas of the next three compounds in the alkane series are

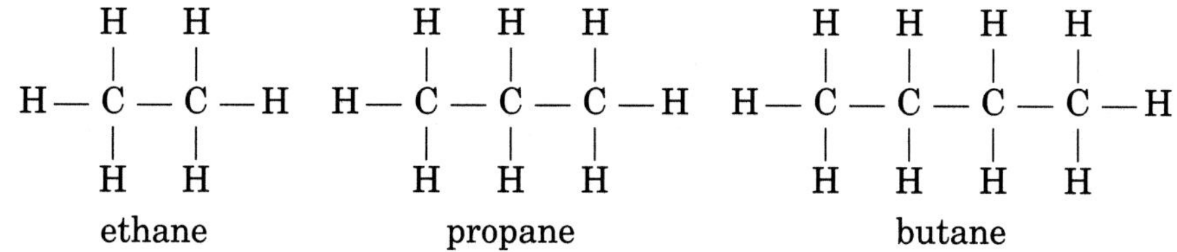

omios: (GK) same, like
logos: (GK) word
homologous—compounds that agree with a general formula

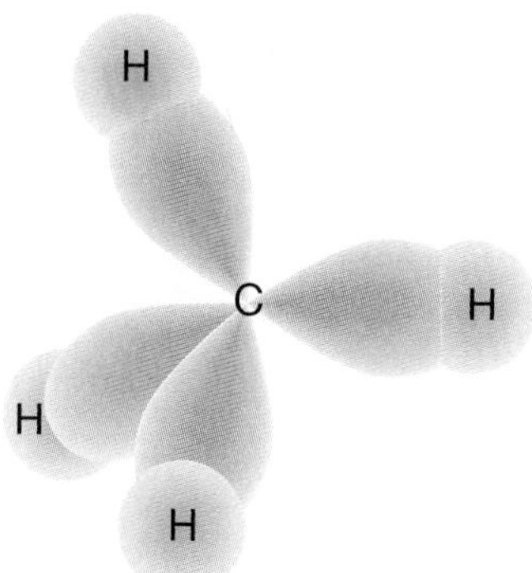

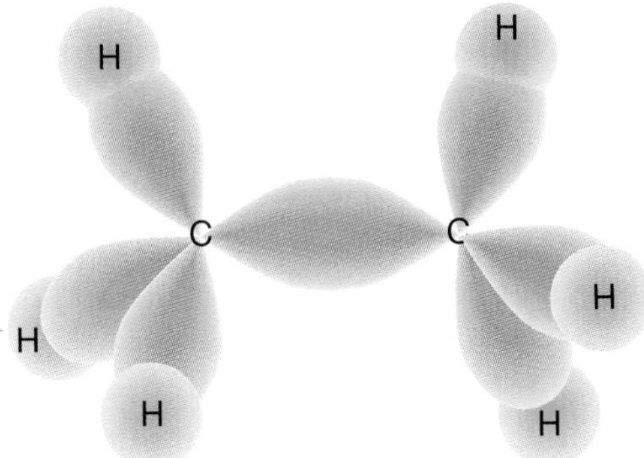

Figure 29.1 Methane, CH_4, and ethane, C_2H_6, are the first two members of the alkane series of hydrocarbons.

If one hydrogen atom, together with its associated electron, is removed from a hydrocarbon molecule, a **radical** is left. Free radicals are extremely reactive in most instances and exist for only a fraction of a second. Their existence is often merely implied from the study of reaction mechanisms. There are some radicals, however, that are so stabilized by delocalization that they can exist for significant amounts of time.

$$
\begin{array}{c}
\text{H} \\ | \\ \text{H}-\text{C}-\text{H} \\ | \\ \text{H} \\ \text{methane}
\end{array}
\rightarrow
\begin{array}{c}
\text{H} \\ | \\ \text{H}-\text{C}\cdot \\ | \\ \text{H} \\ \text{methyl radical}
\end{array}
\qquad
\begin{array}{c}
\text{H}\quad\text{H}\quad\text{H} \\ |\qquad|\qquad| \\ \text{H}-\text{C}-\text{C}-\text{C}-\text{H} \\ |\qquad|\qquad| \\ \text{H}\quad\text{H}\quad\text{H} \\ \text{propane}
\end{array}
\rightarrow
\begin{array}{c}
\text{H}\quad\text{H}\quad\text{H} \\ |\qquad|\qquad| \\ \text{H}-\text{C}-\text{C}-\text{C}\cdot \\ |\qquad|\qquad| \\ \text{H}\quad\text{H}\quad\text{H} \\ \text{propyl radical}
\end{array}
$$

Radicals are named by substituting the ending *-yl* for the normal *-ane* ending of the parent compound.

There are general trends in physical properties within homologous series. For instance, as the molecular mass of the compounds in a series increases, the boiling point increases as shown in Table 29.2. Alkanes are soluble in nonpolar solvents. As you would expect from such similar structures, the chemical properties of homologs are also very similar.

Naming Branched Alkanes

For convenience in naming organic compounds, carbon atoms in a structural formula are given position numbers. In an unbranched chain molecule, the numbering of carbon atoms can begin at either end of the chain as shown for butane.

$$
\begin{array}{c}
\text{H}\quad\text{H}\quad\text{H}\quad\text{H} \\ |\qquad|\qquad|\qquad| \\ \text{H}-\text{C}^1-\text{C}^2-\text{C}^3-\text{C}^4-\text{H} \\ |\qquad|\qquad|\qquad| \\ \text{H}\quad\text{H}\quad\text{H}\quad\text{H} \\ \text{butane}
\end{array}
\qquad
\begin{array}{c}
\text{H}\quad\text{H}\quad\text{H}\quad\text{H} \\ |\qquad|\qquad|\qquad| \\ \text{H}-\text{C}^4-\text{C}^3-\text{C}^2-\text{C}^1-\text{H} \\ |\qquad|\qquad|\qquad| \\ \text{H}\quad\text{H}\quad\text{H}\quad\text{H} \\ \text{butane}
\end{array}
$$

Some alkanes are branched chains of carbon atoms. In naming branched alkanes, we must first find the longest chain of carbon atoms. This chain is used as the basis of the compound name

Table 29.2

Alkanes					
Name of Alkane	**Formula**	**Melting Point, °C**	**Boiling Point, °C**	**Name and Formula of Radical**	
Methane	CH_4	−183	−162	Methyl	CH_3-
Ethane	C_2H_6	−172	−89	Ethyl	C_2H_5-
Propane	C_3H_8	−188	−42	Propyl	C_3H_7-
Butane	C_4H_{10}	−138	−1	Butyl	C_4H_9-
Pentane	C_5H_{12}	−130	36	Pentyl	$C_5H_{11}-$
Hexane	C_6H_{14}	−95	69	Hexyl	$C_6H_{13}-$
Heptane	C_7H_{16}	−91	98	Heptyl	$C_7H_{15}-$
Octane	C_8H_{18}	−57	126	Octyl	$C_8H_{17}-$
Nonane	C_9H_{20}	−54	151	Nonyl	$C_9H_{19}-$
Decane	$C_{10}H_{22}$	−30	174	Decyl	$C_{10}H_{21}-$
—					
Dodecane	$C_{12}H_{26}$	−10	216	Dodecyl	$C_{12}H_{25}-$
—					
Hexadecane	$C_{16}H_{34}$	18	287	Hexadecyl	$C_{16}H_{33}-$
Heptadecane	$C_{17}H_{36}$	22	302	Heptadecyl	$C_{17}H_{35}-$
Eicosane	$C_{20}H_{42}$	36	344	Eicosyl	$C_{20}H_{41}-$

and is called the parent chain. The carbon atoms of the parent chain do not necessarily occur in a row, as shown below.

```
      H   H       H       H
      |   |       |       |
  H — C¹— C²———— C³———— C — H
      |   |       |       |
      H   H   H — C⁴— H   H
                  |
              H — C⁵— H
                  |
              H — C⁶— H
                  |
                  H
```

The longest chain contains six carbon atoms. Thus, the parent chain is hexane, C_6H_{14}. The carbon atoms of the longest chain are given position numbers beginning at the end of the parent chain that is closer to the branch. A group of atoms such as the CH_3- group that is attached to the main chain is called a **branch** or **substituent.** We use the word substituent because the branch has been substituted for a hydrogen atom on the parent chain. The branch is named as a radical. We indicate, by number, the position of the carbon atom of the parent chain to which the branch (radical) is attached. In our example, the branch is attached to the third carbon and is a methyl radical. The parent chain is hexane. The name of this compound is 3-methylhexane. The name is written with a hyphen between the substituent position number and the substituent name. The substituent and the parent are written as one word. Numbering of the carbon atoms of the parent alkane chain always begins at the end that will give the lowest position numbers to the substituents.

How do we name an alkane that has more than one branch? Consider the following example:

```
                                H
                                |
                           H— C —H
                                |
      H    H    H    H— C —H    H    H
      |    |    |          |         |    |
H— C⁶— C⁵— C⁴———— C³———— C²— C¹—H
      |    |    |          |         |    |
      H    H    H    H— C —H    H    H
                                |
                                H
```

The parent chain is hexane. Both an ethyl group and a methyl group are attached to the parent chain. They are attached to the third carbon. The name of this alkane is 3-ethyl-3-methylhexane. Note that the radicals appear in the name of the compound in alphabetical order. Why is the name of this compound not 4-ethyl-4-methylhexane?

FRONTIERS

Paddy Gas

Estimating the amounts of hydrocarbons emitted into the atmosphere globally every year has challenged environmental scientists. Their estimates require sorting through and analyzing vast amounts of data collected by scientists doing on-site research. Such on-site research in China has now given scientists new information about the amount of methane produced from the decay of organic matter in rice fields.

Scientists had estimated that one-seventh of the amount of methane emitted into the atmosphere each year from human activity comes from the cultivation of rice paddies. These estimates were based on the methane emission rates collected from rice fields in the United States and Europe. Recent research indicates that the emission rates of western rice fields may be far lower than those of Chinese fields. The methane emission rates were calculated from 13 000 individual readings made during growing seasons in 1988 and 1989. Methane samples were obtained from 24 sites in the Szechuan province of China. The samples were taken about every other day during the approximately 120-day growing season. The gas samples were analyzed using gas chromatography.

The average methane emission rate of the Chinese rice paddy fields is 58 milligrams per square meter per hour (58 $mg/m^2 \cdot h$). These values were much higher than the range of 4-16 $mg/m^2 \cdot h$ reported from western rice fields. Calculations based on the higher emission rates indicate that one-fifth of the methane emitted globally could be accounted for by the rice fields of China alone. Now environmental scientists can improve their estimates of hydrocarbon emissions.

If there are two or more substituent groups that are alike, it is convenient to use prefixes such as *di-*, *tri-*, and *tetra-* instead of writing each substituent separately. A comma is placed between the position numbers of the substituents that are alike. For example, the name of the following compound is 2,3-dimethylhexane. Why is the name not 4,5-dimethylhexane?

```
              H
              |
   H  H — C — H  H          H
   |      |      |          |
H— C¹ ——— C² ——— C³ ——————— C — H
   |      |      |          |
   H      H  H — C⁴ — H     H
                 |
             H — C⁵ — H
                 |
             H — C⁶ — H
                 |
                 H
```

Note that *trimethyl-* means three one-carbon branches, while *propyl-* means one three-carbon branch. In naming a compound that has multiple branches, check that the number of branches and the number of locations agree. That is, if there are two location numbers, there must be two substituents; three locations require three substituents.

PRACTICE PROBLEMS

1. Choose the correct name for this compound.

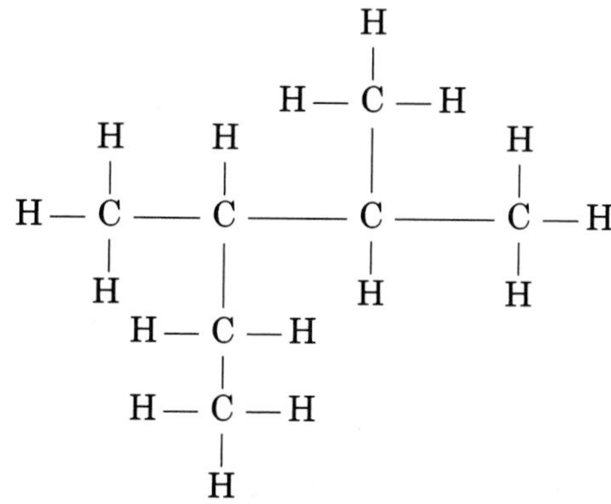

a. 2-ethyl-3-methylbutane
b. 2-methyl-3-ethylbutane
c. 2,3-dimethylpentane
d. 2-methylhexane

2. Name each of the following.

```
                 H
                 |
             H — C — H
   H    H        |          H
   |    |        |          |
a. H— C — C ———— C ———————— C — H
      |   |      |          |
      H   H      |          H
             H — C — H
                 |
                 H
```

```
      H   H        H          H
      |   |        |          |
b. H— C — C ——————— C ——————— C — H
      |   |        |          |
      H   H    H — C — H      H
                   |
               H — C — H
                   |
                   H
```

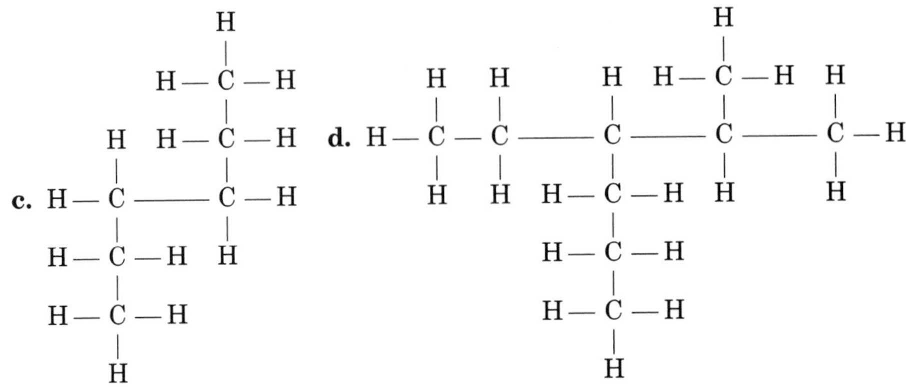

3. Write structural formulas for the following.
   a. methylpropane
   b. 2,3,4-trimethyloctane
   c. 3-ethyl-2-methylhexane
   d. 4-propyloctane

Isomers of Alkanes

It is important for you at this point to review the information on isomers in Chapter 13. Most organic compounds have isomers. Because they are composed of the same number and kind of atoms, isomers have similar properties. In general, however, unbranched compounds have higher melting and boiling points than their highly branched isomers. There is no known way of predicting exactly how many isomers most compounds can form. Each of these two structures is an isomer of butane.

```
                                            H
                                            |
                                         H—C—H
    H   H   H   H                   H       |       H
    |   |   |   |                   |       |       |
 H—C—C—C—C—H             H—C———C———C—H
    |   |   |   |                   |       |       |
    H   H   H   H                   H       H       H
        butane                     2-methylpropane
```

Pentane, C_5H_{12}, the next member of the alkane family, has three isomers. Hexane, C_6H_{14}, has five isomers. Heptane, C_7H_{16}, has nine. It is estimated that triacontane, $C_{30}H_{62}$, has over four billion isomers. Isomers are named according to the longest chain and not according to the total number of carbon atoms in the molecule. Thus, the second structure of butane is a compound named 2-methylpropane.

Our work will be easier if we modify the structural formulas to a condensed form. There can be only one bond between carbon and hydrogen atoms. In the condensed form of a structural formula, carbon atoms are still written separately. However, the hydrogen atoms that are attached to a carbon atom are grouped with that carbon atom. Thus, the structural formulas of the isomers of pentane, Figure 29.2, may be written as follows.

MECHANICS CONNECTION

Ethyne's Flame

Acetylene is used in welding and cutting metal. The gas is usually burned with pure oxygen as the oxidizing agent. The heat of combustion of the acetylene creates a small pool of molten metal where the flame hits the surface. By drawing the flame across a piece of metal, a cut may be made as the pieces separate. On the other hand, if two pieces of metal are pushed together as the flame is applied, they may be joined by the molten metal.

$CH_3-CH_2-CH_2-CH_2-CH_3$

pentane

CH_3
|
$CH_3-CH_2-CH-CH_3$

2-methylbutane

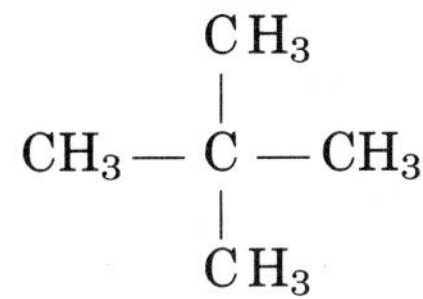

2,2-dimethylpropane

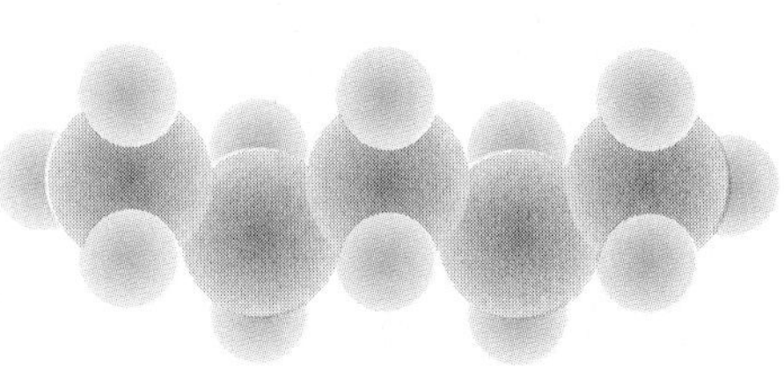

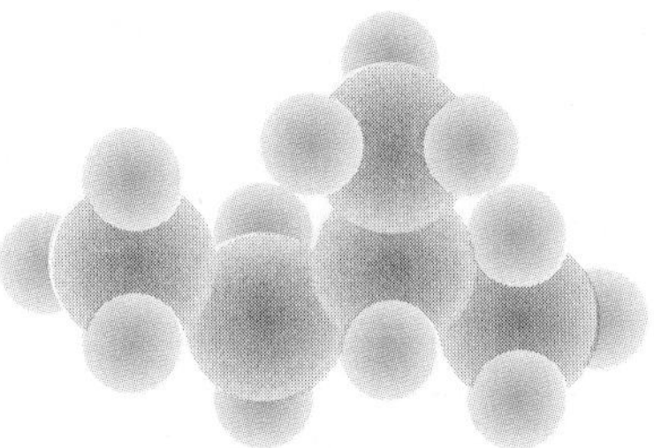

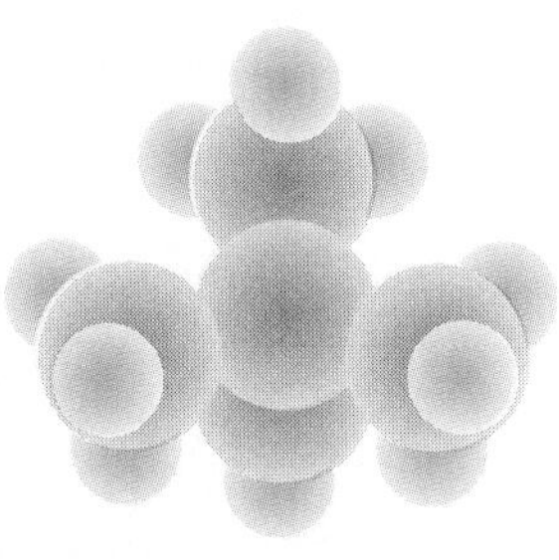

Figure 29.2 Note from the models of the isomers of pentane that the C—C bonds are not at 90° or 180° as shown by the structural formulas.

PRACTICE PROBLEMS

4. Name each of the following.

a.
$\quad\quad\quad CH_3 \quad\ CH_3$
$\quad\quad\quad | \quad\quad\quad |$
$CH_3-CH-CH-CH_2-CH_3$

b.
CH_3
|
$CH_2-CH_2-CH_2-CH_2-CH_2-CH_2-CH_3$

c.
$\quad\quad\quad\quad\quad\quad\quad\quad\quad\quad\quad\quad\quad\quad CH_3$
$\quad\quad\quad\quad\quad\quad\quad\quad\quad\quad\quad\quad\quad\quad |$
$CH_3-CH_2-CH_2-CH_2-CH-C-CH_3$
$\quad\quad\quad\quad\quad\quad\quad\quad\quad\quad\quad\quad | \quad\quad |$
$\quad\quad\quad\quad\quad\quad\quad\quad\quad\quad\quad CH_3 \quad CH_3$

d.
$\quad\quad\quad CH_3 \quad CH_3 \quad CH_3$
$\quad\quad\quad | \quad\quad\quad | \quad\quad\quad |$
$CH_3-CH-CH-CH-CH_2-CH_3$

e.
$\quad\quad\quad CH_3$
$\quad\quad\quad |$
$CH_3-C-CH_2-CH_3$
$\quad\quad\quad |$
$\quad\quad\quad CH_3$

5. Draw the structural formula (condensed form) for each of the following.
a. 2-methylheptane
b. tetramethylbutane
c. 2,2,4-trimethylpentane
d. 3-ethyl-2-methylpentane
e. 3-ethylhexane

6. Draw the structural formula for each of the isomers of hexane.

7. Name each isomer in Practice Problem 6.

kyklos: (GK) round
cyclic—round arrangement

Cycloalkanes

The cyclic forms of saturated hydrocarbons are called cycloalkanes. Recall from Chapter 13 that stylized symbols are used to represent the cyclic hydrocarbons. Cycloalkanes have the general formula C_nH_{2n} and occur in petroleum in such forms as cyclopentane and cyclohexane. In naming cycloalkanes, the ring carbons are numbered so as to give the substituents the lowest set of numbers. Notice how the following compounds are named.

CH_3 (on cyclopropane) — methylcyclopropane

CH_3, CH_3 (on cyclohexane) — 1,3-dimethylcyclohexane

$-CH_2-CH_3$ (on cyclopentane) — ethylcyclopentane

The normal bond angles of a single-bonded carbon atom are 109.5°. The interior angle of an equilateral triangle is 60°, as is the C—C—C bond angle in cyclopropane. Because the angles between the carbon atoms are less than 109.5°, bonding electrons are forced close and exert a force on each other that causes weakening of the bonds. These bonds are called "strained" bonds and lead to a less stable molecule than a similar chain hydrocarbon. Other cycloalkanes also exhibit strained bonds.

Cyclohexane is an industrially important cycloalkane. It is produced from benzene and is used as a raw material in making one type of nylon.

CHEMISTS AND THEIR WORK

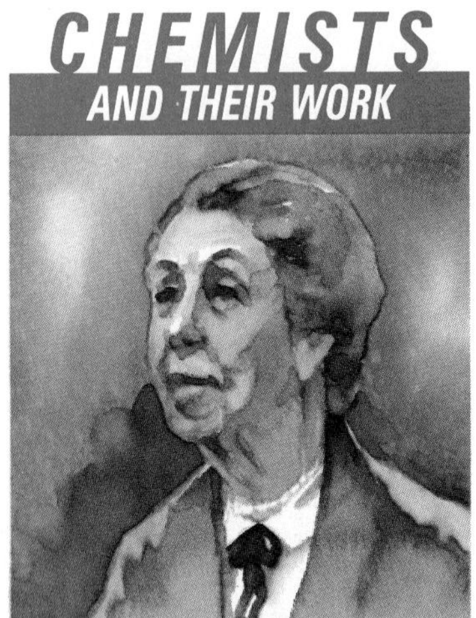

Emma Perry Carr
(1880-1972)
Emma Perry Carr received a Ph.D. in physical chemistry from the University of Chicago. She then joined the chemistry faculty of Mount Holyoke. In three years she became head of the college's chemistry department and remained so for 33 years. Carr directed her students in the synthesis of complex organic compounds and the analysis of their structure using spectroscopy, a fairly modern technique at that time. She made significant contributions to determining the structure of the carbon-carbon double bond. In 1937 she received the first Garvan medal, which honors an American woman for distinguished service in chemistry.

Alkenes

Hydrocarbons that contain multiple bonds are called **unsaturated hydrocarbons.** Unsaturated hydrocarbons with double bonds between carbon atoms are called **alkenes** (AL keens) or olefins (OH leh fihns).

H:C::C:H (H above and below each C)

carbon-carbon double bond

Alkenes are another homologous series of hydrocarbons. The alkene series has the general formula C_nH_{2n}. As with the alkanes, there are general trends in the physical and chemical properties of alkenes as the molecular mass increases.

Because the carbon atoms in an alkene are held together by two pairs of electrons, they are closer than two carbon atoms held by a single bond. The double bond is stronger than a single bond. Because the second pair of electrons (the π bond) is farther from the two nuclei than the sigma bond, the double bond is not twice as strong as a single bond. In addition, the greater π bond-nuclei distance makes a double bond more reactive than a single bond. The less tightly held π electrons are more easily attacked by a reactant. The arrangement of the π electron cloud off the axis prevents free rotation of the atoms on either end of the bond. This rigidity allows for the possibility of geometric isomers as shown on pages 336 to 338 in Chapter 13.

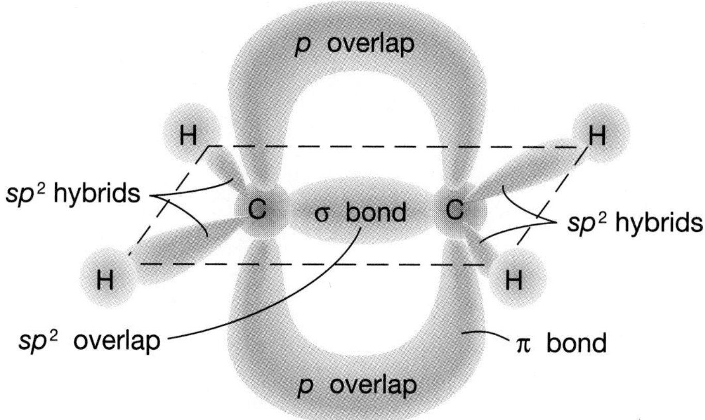

Ethylene (Ethene)

Figure 29.3 The π bond in ethene makes ethene more reactive than the corresponding alkane, ethane. The π electron cloud also prevents rotation around the double bond axis.

Naming Alkenes

With the introduction of double bonds, a new way of forming positional isomers is introduced. Butene in an unbranched chain can still have two isomers, as shown here.

```
H        H   H   H               H   H   H   H
 \       |   |   |               |   |   |   |
  C¹ ═ C² ─ C³ ─ C⁴─H        H─ C¹ ─ C² ═ C³ ─ C⁴─H
 /           |   |               |           |
H            H   H               H           H
      1-butene                       2-butene
```

Compounds with double bonds, which exist in isomeric form, are named by using a position number for the double bond. The number comes from the carbon atom on which the double bond begins and is always placed before the name of the parent compound.

Alkenes are numbered so that the lowest position number is assigned to the first carbon atom to which the double bond is attached. The parent compound is named from the longest continuous chain containing a double bond. Thus, the name of the following alkene is 3-propyl-1-heptene.

```
                                ¹CH₂
                                 ‖
                                ²CH
                                 |
⁷CH₃ ─⁶CH₂ ─⁵CH₂ ─⁴CH₂ ─³CH ─ CH₂ ─ CH₂ ─ CH₃
                3-propyl-1-heptene
```

In some compounds, the double bond also makes possible geometric isomerism. This type of isomerism was also discussed in Chapter 13.

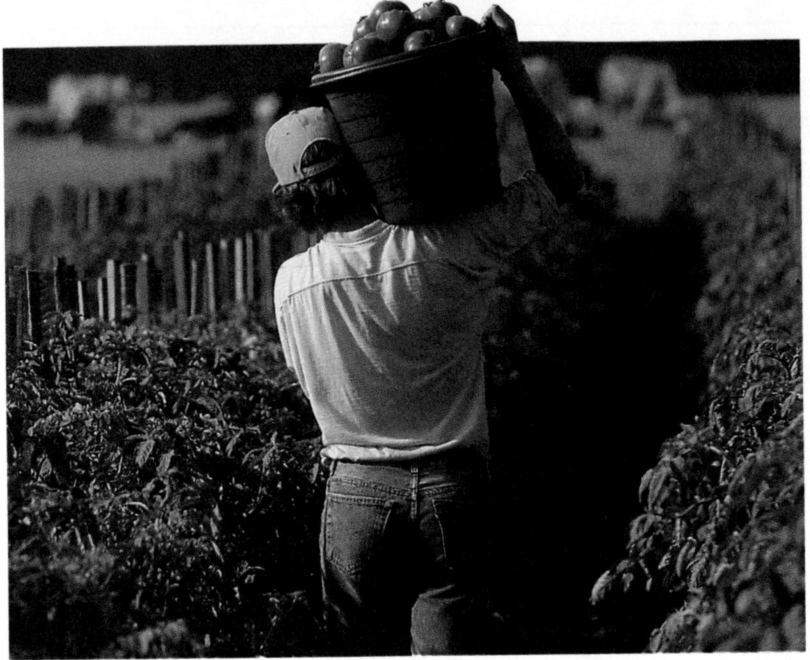

Figure 29.4 Ethene (ethylene) is a plant hormone essential to the growth and maturation of fruits and vegetables. Tomatoes treated with ethene ripen at the same time for harvesting.

Two alkenes are commercially important, ethene (common name *ethylene*) and propene (common name *propylene*). In fact, ethene is the number one organic chemical in industry. More than 17 million tons are produced annually! It is obtained chiefly as one product of the refining of petroleum (Chapter 30). It is used to produce plastics, antifreeze, synthetic fibers, and solvents. Propene is also a by-product of petroleum refining and is used to manufacture plastics and synthetic fibers. Double bonds may also be found in cyclic compounds. A typical example of a cycloalkene is cyclohexene.

$$\begin{array}{c} \text{H} \quad\quad \text{H} \quad \text{H} \\ \diagdown \quad\quad | \quad\quad | \\ \text{C}=\text{C}-\text{C}-\text{H} \\ \diagup \quad\quad\quad\quad | \\ \text{H} \quad\quad\quad\quad\;\; \text{H} \end{array}$$

propene

cyclohexene

A single molecule may contain more than one double bond. In that case the *-ene* ending must be preceded by a prefix indicating the number of double bonds in the molecule. Thus, the compound $CH_2{=}CH{-}CH{=}CH_2$ is named 1,3-buta*di*ene. 1,3-butadiene is produced from petroleum in large quantities to be used in the production of synthetic rubber.

Alkynes

A third homologous series of hydrocarbons consisting of molecules containing triple bonds between carbon atoms is called **alkynes** (AL kyns).

$$\text{H:C:::C:H}$$

carbon-carbon triple bond

Alkynes constitute a homologous series with the general formula C_nH_{2n-2}. They are important raw materials for industries producing synthetic materials such as plastics and fibers. Chemically, alkynes are very reactive. The alkynes are named just as the alkenes, except the ending *-yne* replaces *-ene*. *Acetylene* is the common name for ethyne, the first member of this series. Ethyne is commercially the most important member of the alkyne family. The first two members of the alkyne family are ethyne, C_2H_2, and propyne, C_3H_4.

$$\mathrm{H-C\equiv C-H} \qquad \mathrm{H-C\equiv C-\overset{\displaystyle H}{\underset{\displaystyle H}{\overset{|}{\underset{|}{C}}}}-H}$$

ethyne (acetylene) propyne

In naming alkynes, the numbering system for location of the triple bond and the substituent groups follows the same pattern as was used for naming the alkenes. For example, the name of the following compound is 4,4-dimethyl-2-pentyne.

$$\overset{1}{\mathrm{CH_3}}-\overset{2}{\mathrm{C}}\equiv\overset{3}{\mathrm{C}}-\overset{\displaystyle \mathrm{CH_3}}{\underset{\displaystyle \mathrm{CH_3}}{\overset{|\,4}{\underset{|}{\mathrm{C}}}}}-\overset{5}{\mathrm{CH_3}}$$

4,4-dimethyl-2-pentyne

If a compound contains both double and triple bonds, the double bonds take precedence in numbering; that is, the double bond is given the lower number and is named first.

Aromatic Hydrocarbons

To an organic chemist, one of the most important organic compounds is benzene, a cyclic hydrocarbon.

The benzene ring is diagrammed as ⌬ .

In this structural representation of C_6H_6, it is assumed that there is a carbon atom at each corner with one hydrogen atom attached.

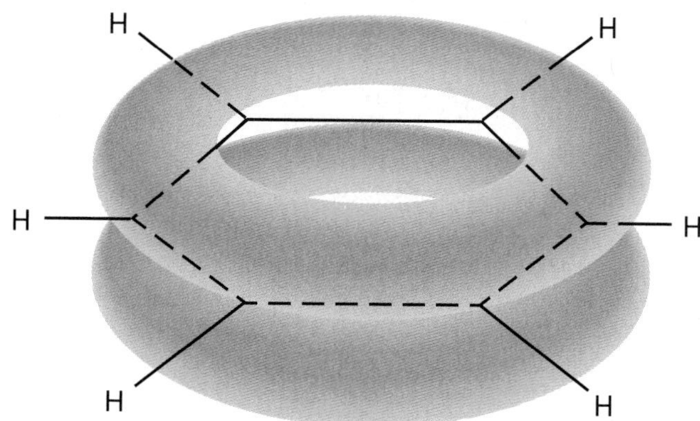

Figure 29.5 Aromatic hydrocarbons have a benzene ring or similar structure. The common characteristics of these compounds arise from the delocalization of the π electrons as shown here.

Note from Chapter 13 that the benzene ring has a system of delocalized electrons. Therefore, it possesses great stability. The benzene ring diagram should not be confused with the symbol for cyclohexane. Cyclohexane, C_6H_{12}, is an alkane composed of six carbon atoms bonded only by single bonds.

Thousands of compounds are derived from benzene, and their study constitutes a whole branch of organic chemistry. Most of these compounds have rather distinctive odors. Thus, they are called aromatic compounds. Aromatic compounds are organic compounds that have a benzene ring or a similar structure. These compounds are normally named as derivatives of benzene. Aromatic compounds occur in small quantities in some petroleum reserves. They occur to a large extent in coal tar obtained from the distillation of coal. Some compounds consist of a fused system of several rings. These compounds have properties similar to benzene. An example of a fused ring compound is naphthalene.

naphthalene

The radical formed by removing a hydrogen atom from a benzene ring is called the phenyl radical. The symbol for the phenyl radical is

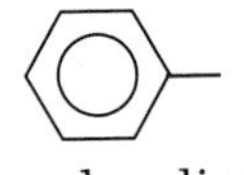

phenyl radical

Some examples of benzene compounds are

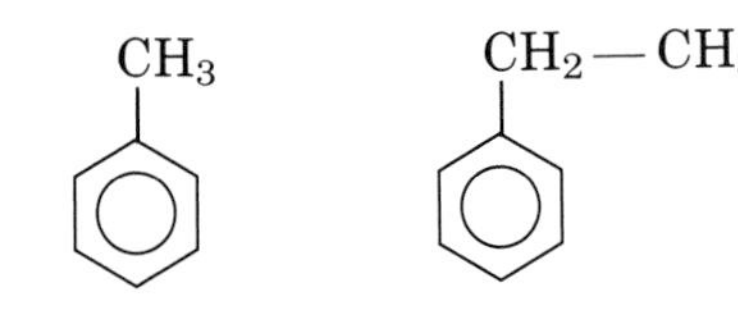

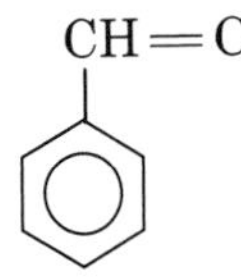

$CH_3—CH—CH_3$

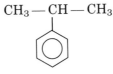

CH_3

CH_3

methylbenzene (toluene) | ethylbenzene | phenylethene (styrene) | 2-phenylpropane (cumene) | 1,4-dimethylbenzene (*p*-xylene)

Benzene, toluene, and xylene are synthesized from petroleum. Ethylbenzene and cumene are made from benzene, while styrene is made from ethylbenzene. All of these compounds are important to our economy. Benzene is used in making plastics and fibers. Toluene is used to improve the quality of gasoline, as well as in the production of explosives and other chemicals. The explosive TNT is 2,4,6-trinitrotoluene. Ethylbenzene is converted almost entirely to styrene. Styrene is a vital component of synthetic rubber, plastics, and paints. The compound *p*-xylene (p = *para*) is a raw material for polyester fibers. Cumene is used in making plastics. A very large part of our synthetic materials industry is based upon aromatic compounds containing a benzene ring.

Bridge to
MEDICINE

Nuclear Magnetic Resonance

Both protons and neutrons have the property of spin and that motion creates a magnetic field. Just as electrons pair in orbitals, like nucleons also pair. If two neutrons pair, their spins will be opposite and their magnetic fields will cancel each other. However, for nuclei with unpaired nucleons, the nucleus as a whole should possess a magnetic field.

If a nucleus with a magnetic field is placed in an external magnetic field, two energy states are possible for the nucleus. The two fields are either aligned or opposed. The opposed fields are a higher energy state. The energy required to "flip" the nucleus from one state to the other is affected by nearby atoms in the molecule. The energy can be measured by a process called nuclear magnetic resonance, NMR, spectroscopy.

About two-thirds of naturally occurring nuclei have magnetic fields. An NMR spectrum of an organic molecule shows several energy peaks. These peaks correspond to hydrogen atoms (protons) in different locations in a molecule.

NMR spectroscopy has been adapted to a medical diagnostic technique called magnetic resonance imaging, MRI. By having the detector scan the body while varying the magnetic field, a three-dimensional image can be obtained. MRI can be used to detect tumors, to monitor chemical processes in the body, and to locate malformations in organs.

Exploring Further

1. What are other applications of NMR spectroscopy?
2. What natural nuclides could be used by biochemists in NMR spectroscopy?

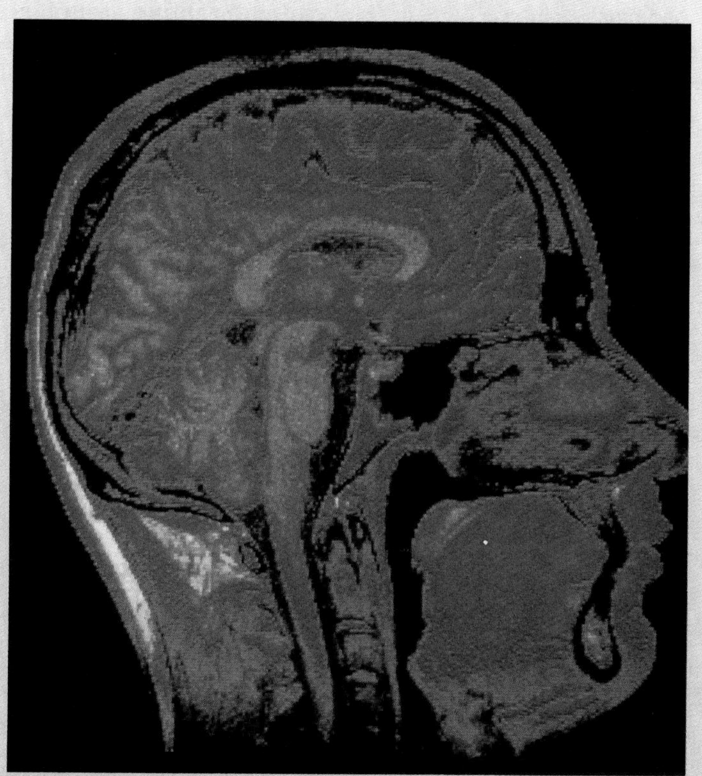

29.1 CONCEPT REVIEW

8. How do the following hydrocarbons differ?
   - **a.** aromatic and aliphatic
   - **b.** chain and cyclic
   - **c.** saturated and unsaturated
9. How is the numbering of the parent chain determined in naming a branched hydrocarbon?
10. What are the general formulas of the alkane, alkene, and alkyne homologous groups? Name the compound in each series with the lowest molecular mass.
11. **Apply** 1,2- and 1,3-dimethylbenzenes are xylenes used in producing industrial chemicals, plastics, and in enriching gasoline. Write their structural formulas.

OBJECTIVES

- identify and explain the inductive effect.
- differentiate among the functional groups of substituted hydrocarbons, organic oxygen compounds, and organic nitrogen compounds.
- recognize, name, and write structural formulas of representative halogen substituted hydrocarbons, organic oxygen compounds, and organic nitrogen compounds.

29.2 Other Organic Compounds

Atoms other than carbon and hydrogen can be substituted for part of a hydrocarbon molecule. When this substitution occurs, the chemical reactivity of the hydrocarbon is generally increased. The nonhydrocarbon part of the molecule is called a functional group. Most of the chemical reactivity of the substituted hydrocarbon is due to the functional group.

Halogen Derivatives

One family of substituted hydrocarbon molecules has a halogen atom substituted for a hydrogen atom. For example, if we substitute a bromine atom for a hydrogen atom on methane, we obtain CH_3Br. The name of this compound is bromomethane. It is also possible to replace more than one hydrogen atom by halogen atoms. In the compound $CHCl_3$, three chlorine atoms have been substituted for three of the hydrogen atoms in a methane molecule. The name of this compound is trichloromethane. You may know this compound by its common name, *chloroform*. Chloroform has been widely used as a solvent. It was once used as an anesthetic. In the compound CCl_4, four chlorine atoms have been substituted for the four hydrogen atoms in a methane molecule. The common name of this tetrachloromethane compound is carbon tetrachloride. These compounds are named as derivatives of the hydrocarbons.

```
      H                H                Cl
      |                |                |
Cl — C — Cl       H — C — Br      Cl — C — Cl
      |                |                |
      Cl               H                Cl
trichloromethane  bromomethane  tetrachloromethane
```

In large chains, we number the carbon atoms to avoid any confusion in naming the compounds. Suppose we have a chain that contains both a double bond and a halogen. In this case, begin the numbering at the end closer to the double bond. Thus, the name of the following compound is 1,4-dichloro-1-butene.

```
     H    H
     |    |
H — C⁴ — C³ — C² = C¹ — H
     |    |    |    |
     Cl   H    H    Cl
```

In aromatic compounds, it is necessary to indicate the relative positions of the various substituent groups on the ring. If two or more substituents are attached to the benzene ring, it is necessary to assign position numbers to the carbon atoms of the ring. The atoms in the benzene ring are numbered to give the smallest position numbers to the substituents. For example, the name of the following compound is 1,3-dibromobenzene, not 1,5-dibromobenzene.

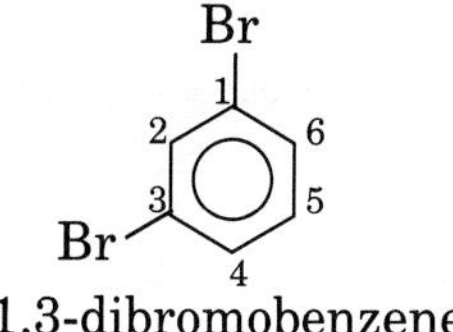

1,3-dibromobenzene

There are four 1-positions possible in each molecule of naphthalene. The 1-position that gives the lowest numbers to substituents is always used. The 1-position is next to the atom without a hydrogen atom attached. The numbering system for naphthalene is as follows.

naphthalene   2-chloronaphthalene   1,3-dibromonaphthalene

Several more examples that illustrate the naming of substituted hydrocarbons are

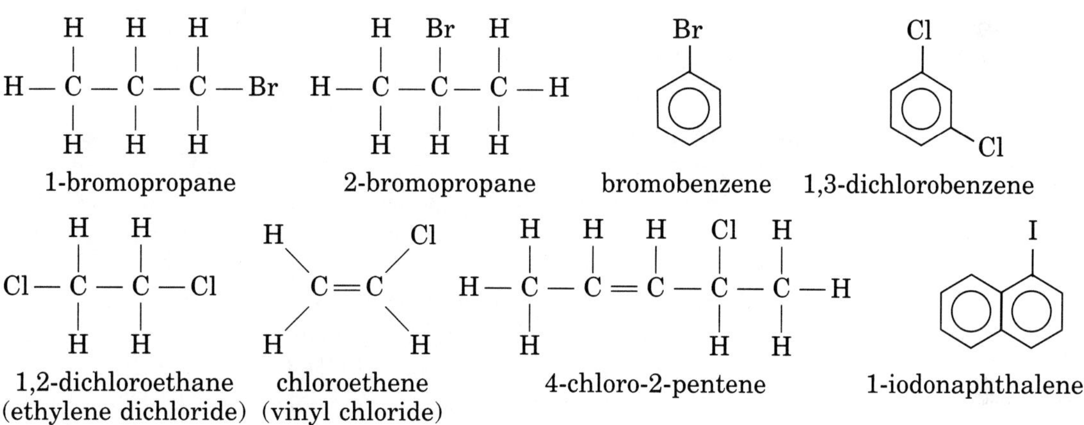

1,2-dichloroethane (common name *ethylene dichloride*) is manufactured in large quantities from ethene. In turn, it is converted to chloroethene (common name *vinyl chloride*). The vinyl chloride is used to make a plastic, polyvinyl chloride (PVC). PVC has numerous applications, such as water and waste piping.

Figure 29.6 Plumbing pipes are manufactured from polyvinyl chloride (PVC), a compound that is produced from halogen derivatives of ethene.

Table 29.3

Organic Compounds Containing Oxygen*			
Class	**General Formula**	**Ending**	**Example**
Alcohol	$R—O—H$	*-ol*	CH_3OH methanol
Ether	$R—O—R'$	*-oxy-*	$CH_3OCH_2CH_3$ methoxyethane
Aldehyde	$R—\overset{\overset{O}{\|}}{C}—H$	*-al*	CH_3CH_2CHO propanal
Ketone	$R—\overset{\overset{O}{\|}}{C}—R'$	*-one*	$CH_3COCH_2CH_3$ butanone
Acid	$R—\overset{\overset{O}{\|}}{C}—O—H$	*-oic acid*	$CH_3CH_2CH_2CH_2COOH$ pentanoic acid
Ester	$R—\overset{\overset{O}{\|}}{C}—O—R'$	*-yl -oate*	$CH_3CH_2CH_2OOCCH_3$ propyl ethanoate

*R— and R'— each represent a hydrocarbon radical. The two radicals may be the same, or they may be different.

Organic Oxygen Compounds

Hundreds of thousands of organic compounds contain oxygen as well as hydrogen and carbon. Many of these compounds are familiar household items. Many more are important solvents and reactants in industry. Table 29.3 lists the names of the principal classes of oxygen-containing compounds. It also gives the general formula of the family.

The symbol *R*— represents any hydrocarbon radical. Thus, *R*—OH is the general formula for alcohols. If *R*— represents CH_3—, the alcohol has the formula CH_3OH. The endings in the third column indicate the method of naming compounds of each class. For example, alcohols have the ending *-ol*. The alcohol CH_3OH is methanol.

Figure 29.7 Rubbing alcohol is an aqueous solution of 2-propanol, commonly called isopropyl alcohol, and acts as a disinfectant.

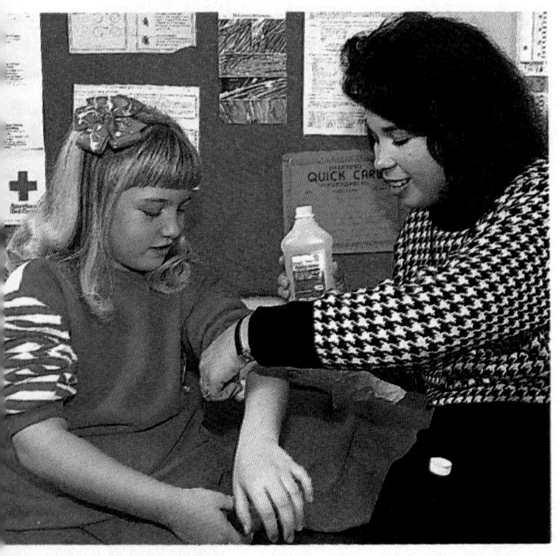

Alcohols and Ethers

Alcohols contain the hydroxyl group, —OH. Yet alcohols are neither acidic nor basic. The hydrogen atom is displaced only by active metals. Methanol, CH_3OH, is manufactured from CO and H_2 and is used extensively in producing plastics and fibers. Ethanol, CH_3CH_2OH, is produced from ethene and is one of the most important solvents as well as a raw material for other chemical products. $CH_3CHOHCH_3$, 2-propanol, is produced from propene and is converted to acetone, an important solvent. A solution of 2-propanol and water is known as "rubbing alcohol."

The lower molecular mass alcohols are soluble in water through hydrogen bonding with the water. Alcohols with four or more carbon atoms, however, are not soluble in water. Their molecules are largely nonpolar because they consist mostly of nonpolar hydrocarbon radicals.

It is possible to have more than one hydroxyl group in a single molecule.

$$\begin{array}{cc} CH_2 - & CH_2 \\ | & | \\ OH & OH \end{array}$$

1,2-ethanediol
(ethylene glycol)

This is the structure of a compound called 1,2-ethanediol. Its common name is *ethylene glycol* and, as such, it is sold in large amounts as automobile cooling system antifreeze.

When hydroxyl groups are attached to the benzene ring, the resulting compounds tend to be slightly acidic. The simplest of these compounds is phenol.

$$C_6H_5-OH$$

phenol

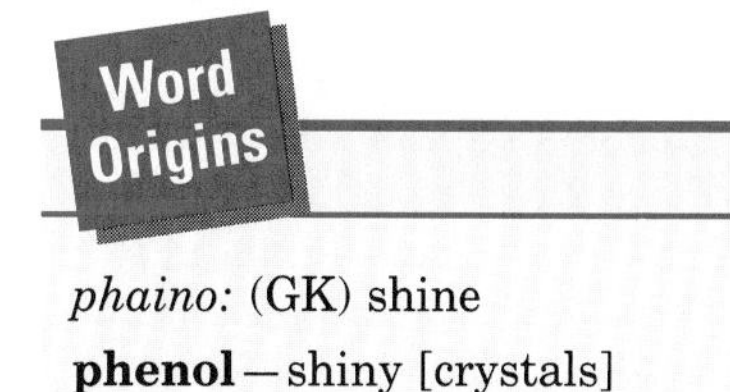

phaino: (GK) shine
phenol—shiny [crystals]

Phenol is produced from cumene and is used to make plastics, drugs, and fibers. Because aromatic hydroxyl compounds differ somewhat in properties from most alcohols, they are classed as **phenols.**

Ethoxyethane (diethyl ether) was used for many years as an anesthetic. The cyclic ether, ethylene oxide,

$$H_3C-CH_2-O-CH_2-CH_3$$

diethyl ether

$$\begin{array}{c} CH_2-CH_2 \\ \diagdown\ \diagup \\ O \end{array}$$

ethylene oxide

is synthesized from ethene. In addition to being converted to ethylene glycol, it is also used in making fibers, films, and detergents. Propylene oxide,

$$\begin{array}{c} CH_3-CH-CH_2 \\ \diagdown\ \diagup \\ O \end{array}$$

propylene oxide

produced from propene, is used in making plastics, cellophane, and hydraulic fluids.

Figure 29.8 1,2-ethanediol (ethylene glycol) is commonly used in automobile antifreeze.

Aldehydes and Ketones

Methanal, better known as *formaldehyde,* is manufactured from methanol and is used to make plastics and adhesives. A 37% solution of methanal in water is known as formalin. There are no other aldehydes of great industrial significance. Both aldehydes and ketones are characterized by the carbonyl group, $>C=O$. The most important ketone is propanone, or acetone. Acetone is made from 2-propanol and is used in plastics, solvents, and making other chemicals.

$$H-\overset{\overset{O}{\|}}{C}-H \qquad CH_3-\overset{\overset{O}{\|}}{C}-CH_3$$

methanal — propanone (acetone)

MINUTE LAB

Not all Mirrors are Silvered

Put on an apron and goggles. Place 2 cm^3 of 0.05*M* acidified tin(II) chloride solution in a clean test tube. Stopper it and shake well. Pour the contents into a recovery container provided by your teacher. Repeat the procedure using 2 cm^3 of 0.06*M* silver nitrate solution. **CAUTION:** *Silver nitrate is corrosive and will stain.* Pour the contents into a recovery container provided by your teacher. Mix 2 cm^3 of basic sodium potassium tartrate solution with 2 cm^3 of copper (II) sulfate in the test tube. **CAUTION:** *Copper sulfate is poisonous and a skin irritant.* Add 2 cm^3 of 40% glyoxal. **CAUTION:** *Glyoxal is corrosive and an eye irritant.* Stopper the tube and swirl to mix. Remove the stopper and place the test tube in a hot water bath. When the mixture begins to bubble, remove the test tube from the bath using a test tube clamp. Record your observations. Pour the contents of the test tube into the beaker provided by your teacher. Rinse the test tube thoroughly with water. How might copper-mirrored glass be used in the construction industry?

Acids and Esters

Most organic acids are characterized by the carboxylic acid group, $-COOH$, and are weak acids. However, the strength of these acids is strongly influenced by other atoms in the molecule. Because the strength of the acid depends upon the breaking of the $O-H$ bond by having the hydrogen ion removed by a water molecule, anything that would weaken that bond results in a stronger acid. The effect of one functional group on another functional group is called an **inductive effect.** If a chlorine atom bonds to a carboxylic acid molecule, the acid becomes stronger. Chlorine, with its high electronegativity, tends to attract electrons to itself from other parts of the molecule. The chlorine draws electrons away from the acid group. As an example, consider the K_a values for the acids in Table 29.4. Note that the value of the ionization constant increases as the number of chlorine atoms in the acid increases.

Table 29.4

Ionization Constants of Some Organic Acids

Name	Formula	Ionization Constant
Acetic acid	CH_3COOH	1.75×10^{-5}
Chloroacetic acid	$ClCH_2COOH$	1.36×10^{-3}
Dichloroacetic acid	$Cl_2CHCOOH$	5.50×10^{-2}
Trichloroacetic acid	Cl_3CCOOH	3.02×10^{-1}

You have used ethanoic acid *(acetic acid)* in the laboratory. This acid is produced in large quantities from methanol and carbon monoxide. Vinegar is a 5% aqueous solution of acetic acid made by fermenting fruit or grain. Acetic acid is also used in the manufacture of fibers and plastics. Hexanedioic acid *(adipic acid)* is synthesized from cyclohexane and is used in making nylon. The structure of hexanedioic acid is $HOOC(CH_2)_4COOH$.

The compound, 1,4-xylene is used to make the dicarboxylic acid, terephthalic acid.

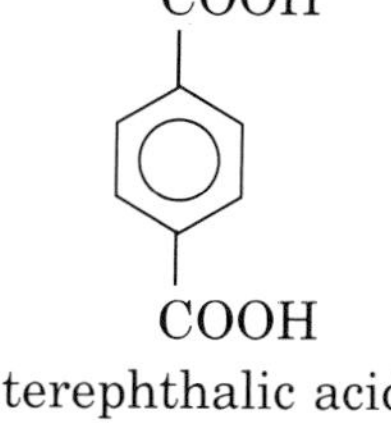

terephthalic acid

Fibers, films, and bottles made of polyester all contain terephthalic acid.

It is possible to have anhydrides of organic acids just as for inorganic acids. Thus, acetic anhydride can be made from acetic acid, as shown below. It is used to produce rayon, a synthetic fiber, and plastics.

$$\underset{\text{acetic acid}}{CH_3-\overset{\overset{\displaystyle O}{\|}}{C}-O-H + H-O-\overset{\overset{\displaystyle O}{\|}}{C}-CH_3} \rightarrow \underset{\text{acetic anhydride}}{CH_3-\overset{\overset{\displaystyle O}{\|}}{C}-O-\overset{\overset{\displaystyle O}{\|}}{C}-CH_3} + \underset{\text{water}}{H_2O}$$

Esters can be considered as having been made from an acid and an alcohol. For example, consider the reaction between ethanol and acetic acid.

$$\underset{\text{ethanol}}{CH_3CH_2-O-H} + \underset{\text{acetic acid}}{H-O-C(=O)CH_3} \rightarrow \underset{\text{ethyl acetate}}{CH_3CH_2-O-C(=O)CH_3} + \underset{\text{water}}{H_2O}$$

When they react, they produce water and ethyl acetate. The $^{18}_{8}O$ nuclide can be used to determine the path of oxygen in the reaction. For example, assume we have some ethanol that contains $^{18}_{8}O$ atoms. This ethanol is then reacted with acetic acid containing the common $^{16}_{8}O$. The products are separated after the reaction and separately analyzed by mass spectrometer. We find the $^{18}_{8}O$ in the ethyl acetate. We conclude that the oxygen in the ester came from the alcohol, not from the acid.

Ethenyl ethanoate *(vinyl acetate)*, $CH_3COOCH{=}CH_2$, is an ester produced from ethyne and ethanoic acid. It is used to make adhesives and paints.

Figure 29.9 Beeswax, $C_{15}H_{31}COOC_{30}H_{61}$, is an ester of palmitic acid, $C_{15}H_{31}COOH$, and myricyl alcohol, $C_{30}H_{61}OH$.

Organic Nitrogen Compounds

Many organic compounds of biological importance contain nitrogen. Because nitrogen has more than one oxidation state, it can combine with organic radicals in a number of ways.

Amines are organic compounds in which a nitrogen atom is bound to alkyl groups and hydrogen atoms. Amines are derivatives of ammonia. If only one hydrogen atom of ammonia is replaced by an alkyl group, the compound is a primary amine. If two hydrogen atoms are replaced, the compound is a secondary amine. If all three are replaced, the compound is a tertiary amine.

EVERYDAY CHEMISTRY

Dyes

Think how boring life would be without color. We all gaze in awe at beautiful sunrises, at leaves in autumn, and at the blue of the sea. In addition to the colors we perceive in nature, we try to brighten our lives by coloring many things we use. The coloring matter for these products is called a dye.

There are hundreds of dyes, but their molecules share many common structural features. Most dye molecules consist of two parts, one to give color and one to control solubility and dyeing properties. The coloring part of the molecule is usually one or more benzene rings containing functional groups, such as

$>C{=}C$

$>C{=}S$

$>C{=}O$

$—NO_2$

$>C{=}NH$

$—N{=}N—$

Functional groups causing the dye molecule to adhere to the fibers of a material are

$$—NH_2 \qquad —OH \qquad —COOH$$

Because many dyes are structurally similar, they can be manufactured from a relatively few types of chemical reactions. Standard procedure is to react functional groups with small molecules to form intermediates. Two or more different intermediates are then reacted to form the dye molecules.

Exploring Further

1. Are most organic dyes aliphatic or aromatic hydrocarbons?
2. Use reference materials to identify natural substances that were once used as commercial dyes.

In general, chemists use the symbol $R—$ to represent an alkyl group. General formulas for amines are

H—N—H with H on N	H—N—R with H on N	R—N—R′ with H on N	R—N—R′ with R″ on N
ammonia	primary amine	secondary amine	tertiary amine

In addition to using *-amine* as a suffix (Chapter 23), amines are named in two other ways. Amines can be named as derivatives of ammonia. For example, the name of the following compound is diethylamine.

$$CH_3—CH_2—\underset{\displaystyle H}{\underset{|}{N}}—CH_2—CH_3$$

diethylamine

Amines may also be named as amino-substituted hydrocarbons. The carbon chain is numbered to give the amine group the lowest number. For example, the name of the following compound is 1,4-diaminobutane.

$$H_2N - \overset{1}{C}H_2 - \overset{2}{C}H_2 - \overset{3}{C}H_2 - \overset{4}{C}H_2 - NH_2$$

1,4-diaminobutane

Amines are Lewis bases because the nitrogen atom possesses an unshared pair of electrons.

Amides are characterized by a carbonyl group and an amine group. An amine-like compound of importance in both industry and living systems is urea, $H_2N-CO-NH_2$. Some organisms excrete their nitrogenous wastes in the form of urea. Commercially, urea is formed from ammonia and carbon dioxide. It is used as a fertilizer, a raw material for plastics production, and a livestock feed supplement.

Other important classes of nitrogen-containing compounds are listed in Table 29.5. *G*— is used in place of *R*— with amino acids because some amino acid radicals have other elements in addition to C and H. We will study amino acids in more detail in Chapter 30.

Nitriles are characterized by a carbon-nitrogen triple bond. One important nitrile is propenenitrile (acrylonitrile). It is manufactured from propene, ammonia, and oxygen and is used in fibers, plastics, and synthetic rubber.

Table 29.5

Organic Compounds Containing Nitrogen

Class	General Formula	Example
Amines	$R-NH_2$	$CH_3CH_2NH_2$ ethanamine
Amides	$R-\overset{\overset{O}{\parallel}}{C}-NH_2$	CH_3CONH_2 ethanamide
Amino acids	$G-\underset{}{\overset{\overset{NH_2}{\vert}}{C}}H-COOH$	$CH_3CH(NH_2)COOH$ alanine (2-aminopropanoic acid)
Nitriles	$R-C\equiv N$	$CH_3CH_2CH_2CN$ butanenitrile
Nitro compounds	$R-NO_2$	$C_6H_5NO_2$ nitrobenzene

29.2 CONCEPT REVIEW

12. Does a halogen atom form a single or double bond with a carbon atom?

13. How does the functional group of organic alcohols differ from that of aldehydes and ketones?

14. How do the data in Table 29.4 illustrate the inductive effect?

15. Apply Two compounds which have particularly bad odors are putrescine, $C_4H_{12}N_2$ and cadaverine, $C_5H_{14}N_2$. The latter contributes to bad breath. How are these compounds classified?

Summary

29.1 Hydrocarbons

1. Organic compounds may have carbon atoms linked in chains or in rings. All chain or ring hydrocarbons that do not contain a benzene ring are aliphatic compounds.
2. Saturated chain hydrocarbons, or alkanes, contain only single bonds between carbon atoms. Unsaturated chain hydrocarbons, alkenes or alkynes, contain one or more double or triple bonds.
3. Alkanes form a homologous series, each successive member differing from the last by the structural unit $—CH_2—$. Alkanes have the general formula C_nH_{2n+2}.
4. A hydrocarbon radical is a hydrocarbon that has lost one hydrogen atom with its associated electron.
5. Names of branched-chain hydrocarbons are based on the name of the longest (parent) chain.
6. Most organic compounds have isomers, compounds with the same number and kinds of atoms but linked in a different arrangement.
7. Cyclic saturated hydrocarbons have the general formula C_nH_{2n}.
8. Alkenes constitute a homologous series and contain double bonds between carbon atoms. Alkenes have the general formula C_nH_{2n}.
9. In naming alkenes, parent chains are numbered so that the first double-bonded carbon has the lowest number.
10. Alkynes are a homologous series and have triple bonds between carbon atoms. Alkynes have the general formula C_nH_{2n-2}.
11. In naming alkynes, parent chains are numbered so that the first triple-bonded carbon has the lowest number.
12. Aromatic compounds derive stability from a system of delocalized electrons.

29.2 Other Organic Compounds

13. A functional group is a nonhydrocarbon part of an organic molecule.
14. Principal classes of oxygen-containing compounds are alcohols, ethers, aldehydes, ketones, acids, and esters.
15. Inductive effect is the effect of one functional group on another. Electron-withdrawing functional groups tend to make organic acids stronger.
16. Principal classes of nitrogen-containing compounds are amines, amides, amino acids, nitriles, and nitro compounds.

Key Terms

aromatic hydrocarbon
aliphatic hydrocarbon
alkane
saturated hydrocarbon
homologous series
radical
branch
substituent
unsaturated hydrocarbon
alkene
alkyne
alcohol
phenol
inductive effect
amine
amide

Review and Practice

16. Distinguish between chain and cyclic hydrocarbon compounds.
17. What is the main structural characteristic of alkanes?
18. Explain why alkanes are called saturated hydrocarbons.
19. Write the formulas and structures for unbranched alkanes having 3, 5, and 7 carbon atoms. Write formulas for those having 10, 16, and 19 carbon atoms.
20. What is a homologous series? How do alkanes fit the definition of a homologous series?

Chapter 29 REVIEW

21. What is a hydrocarbon radical, and how is it formed?

22. What is the chemical formula for the heptyl radical? The decyl radical?

23. What rules are followed in labeling the positions of the substituent groups when naming hydrocarbons?

24. What is a substituent group? How is a substituent group named?

25. Write condensed structural formulas for the following compounds.
    a. 3-ethylpentane
    b. 2,2-dimethylbutane
    c. 3-ethyl-2-methylhexane
    d. 2,4,5-trimethylheptane
    e. 3-ethyl-2,2-dimethylhexane

26. Write the general formula for the cycloalkane series.

27. Write condensed structural formulas for the following compounds.
    a. cyclopentane
    b. methylcyclobutane
    c. ethylcyclohexane
    d. 1,3-dimethylcyclobutane
    e. 1-ethyl-2-propylcyclopentane
    f. 1,1-dimethylcyclopropane

28. What are unsaturated hydrocarbons?

29. Write the general formula for the alkenes.

30. Using the general formula for the alkenes, write the formula for heptene, octene, and decene.

31. Write condensed structural formulas for the following compounds.
    a. 2-methylpropene
    b. 3-hexene
    c. 2,5-heptadiene
    d. 4-propyl-4-octene
    e. 1,4-cyclohexadiene

32. Name and give commercial uses for three alkenes.

33. What are the main structural characteristics of aromatic compounds?

34. Write condensed structural formulas for the following compounds.
    a. 1,2-dimethylbenzene
    b. 2-phenylbutane
    c. 1-phenylpropene
    d. 4-ethyl-1-methylbenzene
    e. 2-phenyl-2-butene

35. What is a functional group? How does a functional group generally affect the chemical activity of a hydrocarbon?

36. Write condensed structural formulas for the following compounds.
    a. 1,4-dichlorohexane
    b. 2-bromo-3-chlorobutane
    c. 1,2-dichloro-2-pentene
    d. 1,4-diiodobenzene
    e. 1,3,7-trichloronaphthalene

37. What functional group characterizes alcohols?

38. Classify each of the following as an alcohol, an ether, an aldehyde, an acid, or a ketone.

    a. $-CH_2CH_2OH$

    b. $-OH$

    c. $-C(H)=O$

    d. $-C(=O)-CH_3$

    e.
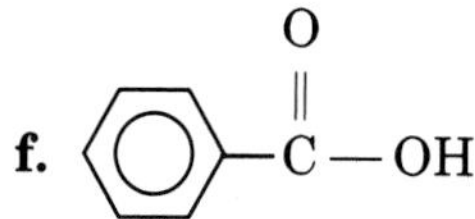

    f. $-C(=O)-OH$

39. What functional group characterizes amines? How do amines and amides differ?

Chapter 29
REVIEW

40. Write condensed structural formulas for the following compounds.
- **a.** 2-butanol
- **b.** 1-ethoxypropane
- **c.** 3-pentanone
- **d.** butanoic acid
- **e.** dipropylamine
- **f.** nitroethane

41. Write the structural formula for the explosive TNT, 2,4,6-trinitrotoluene.

42. Classify each of the following as an amine, an amide, a nitrile, or an amino acid.

a. 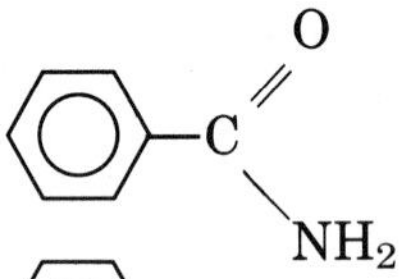 $C_6H_5{-}C(=O){-}NH_2$

b. $C_6H_5{-}NH_2$

c. $C_6H_5{-}C{\equiv}N$

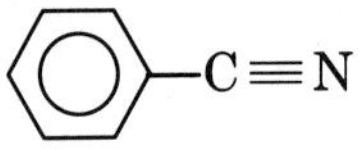

d. 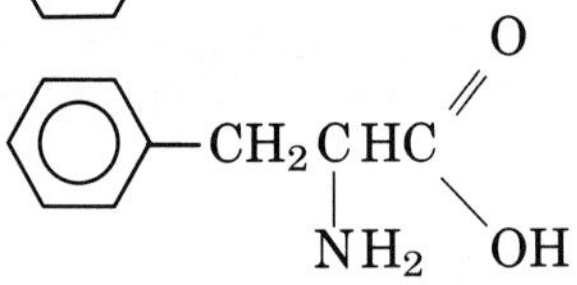 $C_6H_5{-}CH_2CH(NH_2)C(=O){-}OH$

43. Write the general formula for an alkyl radical.

44. Write a chemical formula for each of the following.
- **a.** a 6-carbon alkene
- **b.** an 8-carbon alkane
- **c.** a 10-carbon alkyne

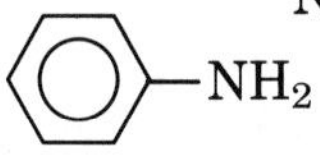

45. Name the following compound.

$$CH_2{=}CH{-}CH_2{-}C{\equiv}CH$$

46. Why is the name 2-ethylpentane an incorrect name? What would be the correct name for the hydrocarbon this name represents?

47. Name the following compounds.

a. 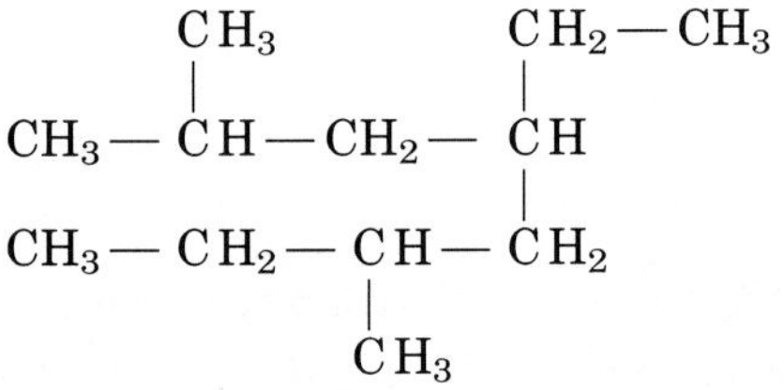
```
       CH3            CH2—CH3
        |              |
CH3— CH—CH2—  CH
                       |
CH3— CH2— CH— CH2
              |
             CH3
```

b. $CH_3{-}CH{=}CH{-}CH_2{-}CH_3$

c. 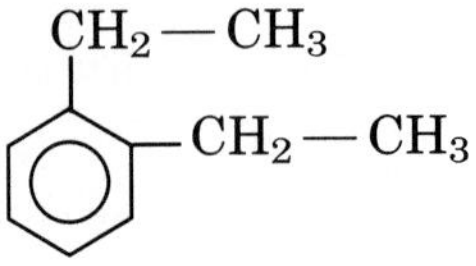 benzene ring bearing $CH_2{-}CH_3$ and $CH_2{-}CH_3$ on adjacent carbons

d. 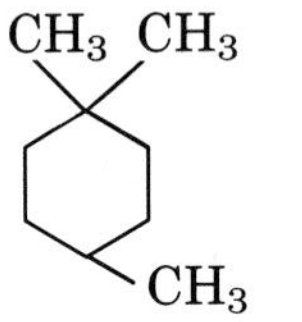 cyclohexane ring bearing CH_3 CH_3 on one carbon and CH_3 on the opposite carbon

e.

$$CH{\equiv}C{-}C(CH_3)_2{-}CH_3$$

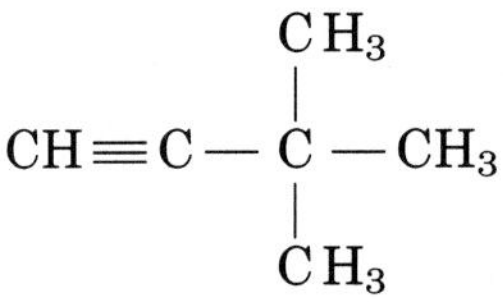

48. Write condensed structural formulas for the following compounds.
- **a.** 3-ethylhexane
- **b.** 1,2,3,4,5,6-hexamethylcyclohexane
- **c.** 2-methyl-2-butene
- **d.** 1-hexyne
- **e.** 2-methylnaphthalene

49. Name the following compounds.

a. $CH_3{-}CH_2{-}CH_2{-}OH$

b. 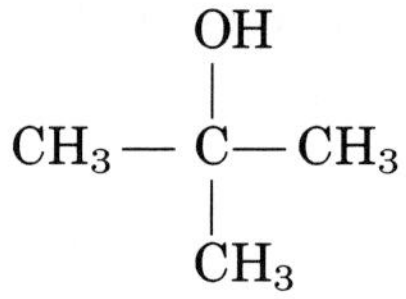 $CH_3{-}C(OH)(CH_3){-}CH_3$

c. $CH_3{-}CH_2{-}CH_2{-}CH_2{-}CH_2{-}O{-}CH_2{-}CH_2{-}CH_2{-}CH_2{-}CH_3$

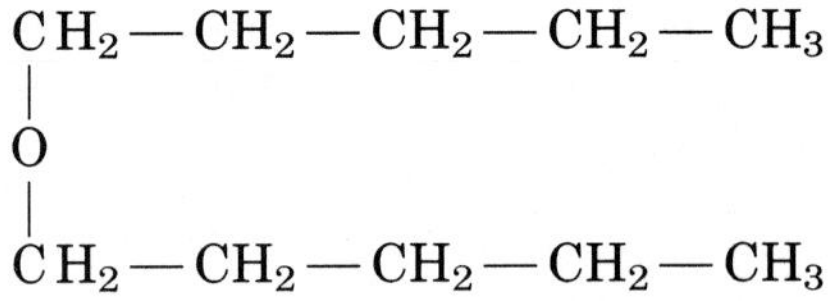

d. 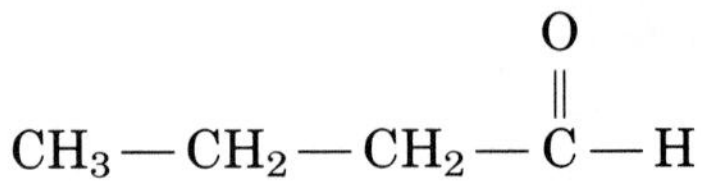 $CH_3{-}CH_2{-}CH_2{-}C(=O){-}H$

e. 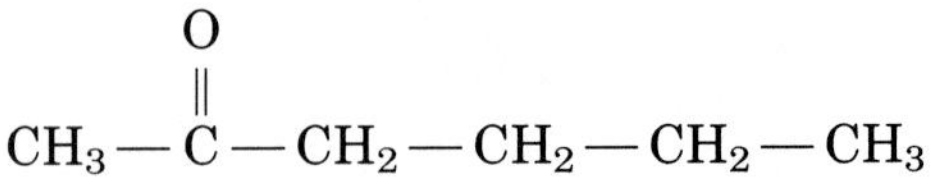 $CH_3{-}C(=O){-}CH_2{-}CH_2{-}CH_2{-}CH_3$

f. 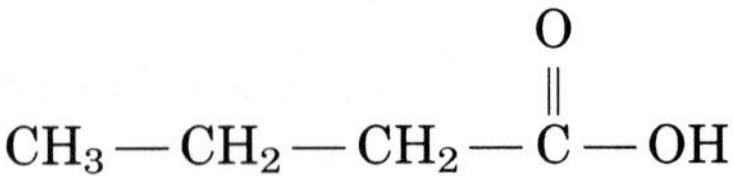 $CH_3{-}CH_2{-}CH_2{-}C(=O){-}OH$

g. 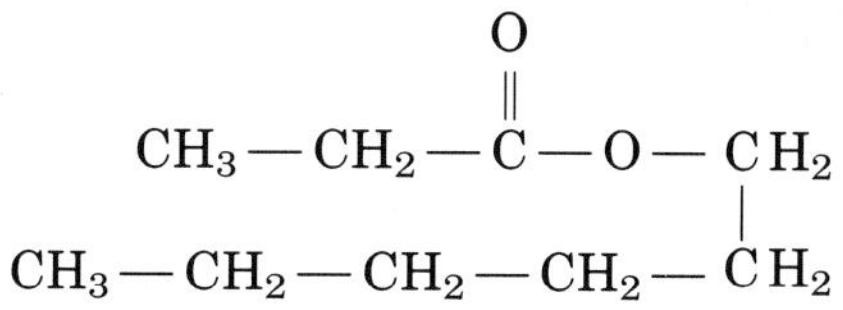
$$\begin{array}{l} \qquad\qquad\qquad\quad O \\ \qquad\qquad\qquad\quad \| \\ CH_3-CH_2-C-O-CH_2 \\ \qquad\qquad\qquad\qquad\qquad\quad | \\ CH_3-CH_2-CH_2-CH_2-CH_2 \end{array}$$

h. 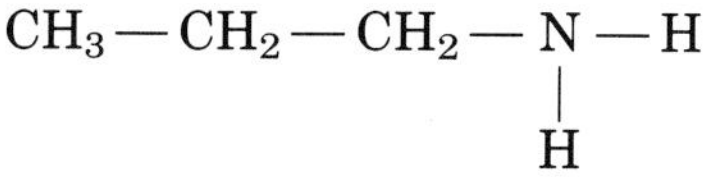
$$\begin{array}{l} CH_3-CH_2-CH_2-N-H \\ \qquad\qquad\qquad\qquad | \\ \qquad\qquad\qquad\qquad H \end{array}$$

50. Draw condensed structural formulas for the following compounds.

- **a.** 1-butanol
- **b.** 2-methyl-1-propanol
- **c.** 1-propoxybutane
- **d.** ethanal
- **e.** 3-hexanone
- **f.** propyl methanoate
- **g.** methyl propanoate
- **h.** trimethylamine
- **i.** propanoic acid
- **j.** propanamide
- **k.** nitromethane

51. Name the following compounds.

a. 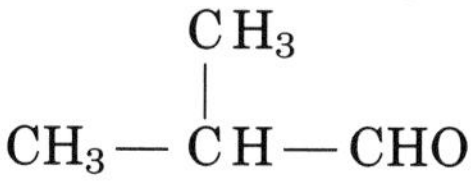
$$\begin{array}{l} \qquad\quad CH_3 \\ \qquad\quad | \\ CH_3-CH-CHO \end{array}$$

b. 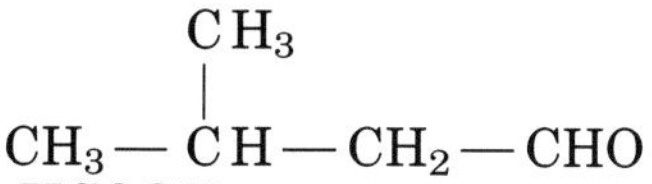
$$\begin{array}{l} \qquad\quad CH_3 \\ \qquad\quad | \\ CH_3-CH-CH_2-CHO \end{array}$$

c. $HCOOH$

d. $HOOC-COOH$

e. CH_3NH_2

f. CH_3-Cl

52. Write condensed structural formulas for the following compounds.

- **a.** 1-phenylcyclobutene
- **b.** 1,2,4-trimethylbenzene
- **c.** iodobenzene
- **d.** 1-bromopentane
- **e.** 3-methyl-3-pentanol
- **f.** 4-methylphenol

53. Write condensed structural formulas for the following compounds.

- **a.** cyclopentanone
- **b.** 3-pentene-2-one
- **c.** butyl butanoate
- **d.** ethyl chloroethanoate
- **e.** ethanenitrile
- **f.** 2-amino-3-phenylpropanal
- **g.** 1,2-dibromo-1-phenylethane
- **h.** hexadecane

54. Name the following compounds.

a. 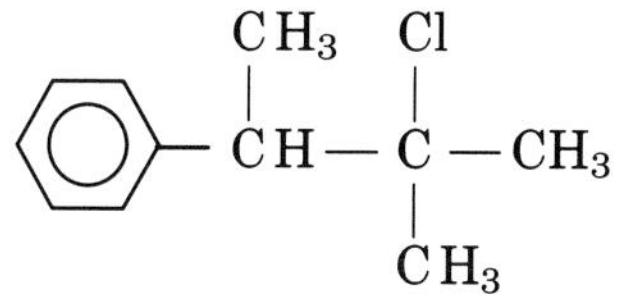
$$\begin{array}{l} \qquad CH_3 \quad Cl \\ \qquad | \qquad\ \ | \\ -CH-C-CH_3 \\ \qquad\qquad\ \ | \\ \qquad\qquad\ \ CH_3 \end{array}$$

b. CH_3, Cl, Cl

c.
$$\begin{array}{l} \qquad\qquad\qquad\qquad O \\ \qquad\qquad\qquad\qquad \| \\ CH_3-CH_2-O-C- \end{array}$$

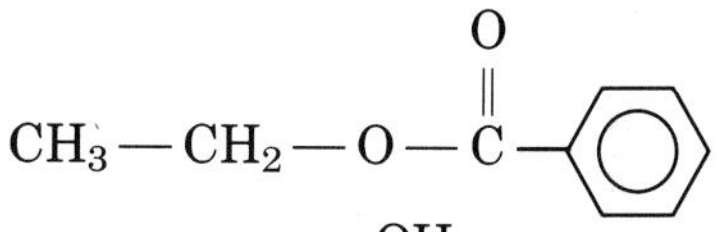

d. 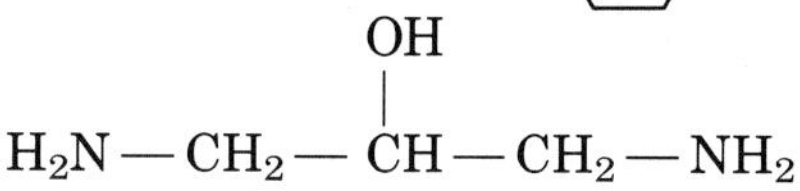
$$\begin{array}{l} \qquad\qquad\qquad\quad OH \\ \qquad\qquad\qquad\quad | \\ H_2N-CH_2-CH-CH_2-NH_2 \end{array}$$

e. Cl—

f. $-C\equiv C-CH_3$

g.
$$\begin{array}{l} \qquad\quad CH_3 \quad O \\ \qquad\quad | \qquad \| \\ CH_3-CH-C-NH_2 \end{array}$$

h.
$$\begin{array}{l} \qquad\qquad\qquad\qquad O \\ \qquad\qquad\qquad\qquad \| \\ CHCl_2-CH_2-C-H \end{array}$$

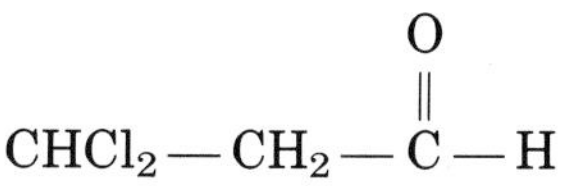

i. 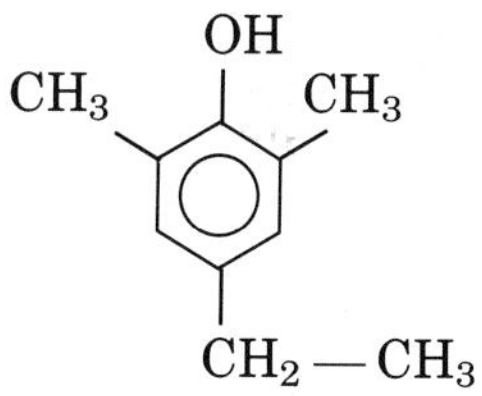 OH, CH_3, CH_3, CH_2-CH_3

j. 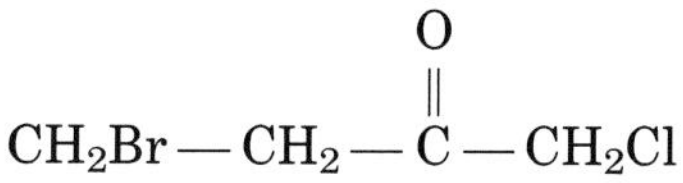
$$\begin{array}{l} \qquad\qquad\qquad\qquad O \\ \qquad\qquad\qquad\qquad \| \\ CH_2Br-CH_2-C-CH_2Cl \end{array}$$

k. 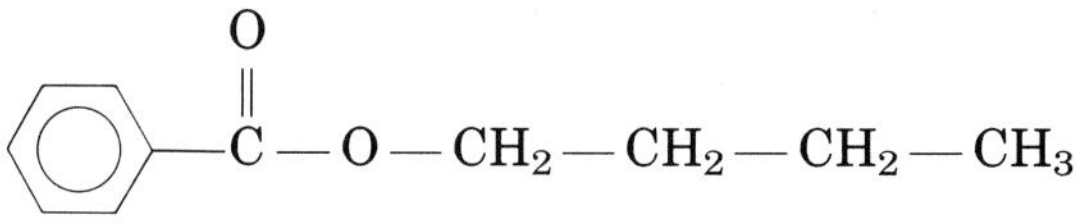
$$\begin{array}{l} \quad O \\ \quad \| \\ -C-O-CH_2-CH_2-CH_2-CH_3 \end{array}$$

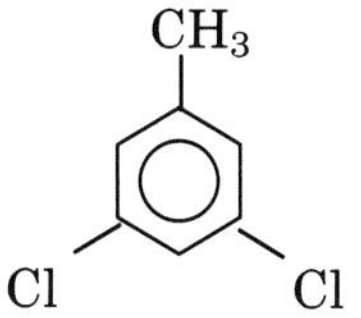

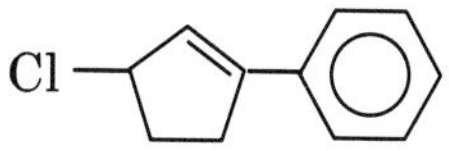

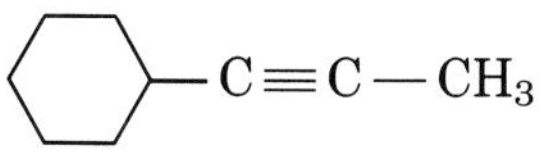

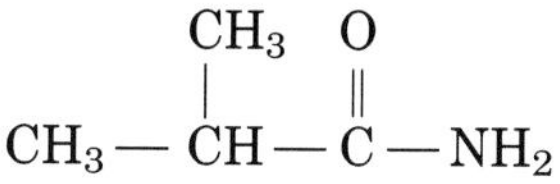

REVIEW

Concept Mastery

55. What structural combination is common to aldehydes, ketones, carboxylic acids, and esters?

56. Explain why the addition of chlorine atoms to a carboxylic acid makes the acid stronger.

57. If the structural unit of alkanes is $-CH_2-$, containing one carbon and two hydrogen atoms, why is the general formula for alkanes C_nH_{2n+2}?

58. Why is it impossible for the correct name of an alkane to begin "1-methyl"?

59. There is a compound whose name is 2,4-dimethylhexane. Can there be a compound named 2,4-dimethylcyclohexane? Explain.

60. In alkenes and alkynes, the carbon atoms that are doubly or triply bonded are generally more reactive than other carbon atoms in the molecule. How can you explain this observation?

61. Write equations for the following using structural formulas.
  - **a.** the formation of propanoic anhydride from propanoic acid
  - **b.** the formation of butyl propanoate from an alcohol and an acid

62. Which of the following compounds, octane, pentene, or heptyne, would you predict to be most reactive? Why?

63. How can you account for the difference in K_a values for the following acids?

$CH_3CH_2CHClCOOH$
2-chlorobutanoic acid
$K_a = 1.39 \times 10^{-3}$

$ClCH_2CH_2CH_2COOH$
4-chlorobutanoic acid
$K_a = 2.96 \times 10^{-5}$

Application

64. Name and write structural formulas for the nine isomers of heptane.

65. Write names and structural formulas for all alkynes having the formula C_5H_8.

66. Give a practical use for each of the following compounds.
  - **a.** 1,2-ethanediol
  - **b.** 2-propanol
  - **c.** ethylene oxide

67. In climates where seasonal temperatures vary to a great extent, refiners produce gasoline to be sold in winter that contains hydrocarbons with lower average molecular mass than those used in gasoline sold in the summer. Explain.

Critical Thinking

68. Give reasons to explain why the boiling points of alkanes increase as their molecular masses increase.

69. Which of the following would be the most stable? Why?
  - **a.** cyclohexane
  - **b.** cyclopropane
  - **c.** cyclobutane
  - **d.** cyclopentane

70. The reaction of graphite with a strong oxidizing agent such as potassium permanganate produces mellitic acid.

COOH
HOOC COOH
HOOC COOH
COOH

mellitic acid

Treatment of mellitic acid with a dehydrating agent causes adjacent carboxyl groups to form anhydrides with each other, resulting in mellitic anhydride, a triple anhydride.
  - **a.** Draw the structure for mellitic anhydride.
  - **b.** In the nineteenth century, chemists inferred the nature of graphite from the structure of mellitic acid. What evidence in the structure of mellitic acid points to that of graphite?

Readings

Arrigo, Joseph T. "The Mystique of Musk." *ChemMatters* 9, no. 2 (April 1991): 12-15.

Berlfein, Judy. "Alcohol in Your Tank." *ChemMatters* 6, no. 4 (December 1988): 10-12.

Gough, Michael. "Dioxin: Part II." *ChemMatters* 6, no. 2 (April 1988): 15-19.

Philp, R. Paul. "Geochemistry in the Search for Oil." *Chemical and Engineering News* 64, no. 6 (February 10, 1986): 28-43.

Stone, Judith. "Bovine Madness." *Discover* 10, no. 2 (February 1989): 38-41.

Weisburd, Stefi. "Radical Dangers Up in Smoke." *Science News* 132, no. 11 (September 12, 1987): 169.

Cumulative Review

1. Compare the particles emitted by naturally radioactive materials in terms of charge and mass.

2. Why is binding energy per particle an indication of nuclear stability?

3. Describe the mechanism of a fission chain reaction.

4. Suggest a method that might be used to produce element 114.

5. Predict the voltage to be expected at 25°C from a cell constructed as follows:

$Co \mid Co^{2+}(0.100M) \parallel Cu^{2+}(2.00M) \mid Cu.$

6. If a voltaic cell produces a potential difference of 1.00 volt under standard conditions, what is the equilibrium constant for the reaction taking place in the cell? What is the free energy change for the reaction? Assume one electron is transferred in the balanced equation.

7. A piece of metal alloy with mass 0.8128 gram is to be analyzed for its nickel content. The metal is dissolved in acid and any ions likely to interfere are removed by appropriate chemical treatment. The resulting solution is then subjected to electrolysis using a platinum cathode. The cathode has mass 12.3247 g before the electrolysis and 12.6731 g after the electrolysis. What is the percentage of nickel in the alloy?

8. The half-life of an Ir isotope is 74.2 days. How long will it take 6.30×10^{11} atoms of that nuclide to disintegrate to 2.46×10^{9} atoms?

9. Complete the following equation.

$$^{240}_{92}U \rightarrow {}^{0}_{-1}e + ?$$

10. What is the molecular mass of a 40-carbon saturated hydrocarbon?

11. Methanol and ethanol react quite differently with oxygen in the body. Methanol and oxygen form formic (methanoic) acid, which paralyzes the optic nerve and causes blindness. Ethanol and oxygen form carbon dioxide and water. Using structural formulas, write equations for these two reactions.

CHAPTER Preview

CONTENTS

Organic Reactions and Biochemistry

CHAPTER 30

The human hand is an intricate mechanism. Engineers designing robots such as the hand shown here have found the action of the human hand virtually impossible to duplicate with even the most advanced technology. The action of the human hand depends upon the behavior of the individual tissues and cells making up the hand, the cells of the nervous system, and the substances carried to and from the hand by the blood.

We usually think of body tissues as solid matter, but blood is also considered a tissue. It contains both red and white blood cells, which are living, as well as thousands of organic and inorganic substances in solution. In this chapter, you will learn about how organic substances react, and about the products made from those substances.

EXTENSIONS AND CONNECTIONS

OBJECTIVES

- describe substitution, addition, elimination, esterification, saponification, addition polymerization, and condensation polymerization reactions.
- describe the processing of petroleum and the octane rating of gasolines.
- explain how synthetic fibers, plastics, and elastomers are produced industrially.

30.1 Organic Reactions and Products

In addition to being organic, what do garbage bags, antifreeze, laundry detergent, garden hose, and aspirin have in common? They are all manufactured, at least in part, from ethene, or ethylene ($CH_2{=}CH_2$). How can so great a variety of substances be produced from a simple raw material? Organic compounds undergo an enormous variety of reactions. However, there are a few common types of reactions that are frequently found in industrial processes. We will investigate some of these reactions in this section. In addition, we will look at some of the more common consumer products composed of organic substances.

Substitution Reactions

A reaction in which either a hydrogen atom of a hydrocarbon is replaced by a functional group or a functional group is replaced by another functional group is called a **substitution reaction.** Alkanes react with chlorine in sunlight to produce chloro-substituted compounds. The product is a mixture of different isomers with very similar properties. A number of aromatic substitution reactions can be controlled to produce specific products. For example, benzene reacts with nitric acid in the presence of concentrated sulfuric acid to form nitrobenzene.

$$C_6H_6(l) + HNO_3(l) \xrightarrow[H_2SO_4]{conc.} C_6H_5NO_2(l) + H_2O(l)$$

benzene, nitric acid, nitrobenzene, water

Alkyl groups and halogen atoms can also be substituted easily onto a benzene ring.

In the second type of substitution reaction, one functional group replaces another. Alcohols undergo substitution reactions with hydrogen halides to form alkyl halides. For example, 2-propanol reacts with hydrogen iodide to form 2-iodopropane. The reaction is reversible and reaches equilibrium. By introducing a substance to absorb the water, the reaction can be forced to the right, producing more product.

$$CH_3{-}CH(OH){-}CH_3(l) + HI(aq) \rightarrow CH_3{-}CHI{-}CH_3(l) + H_2O(l)$$

2-propanol, hydrogen iodide, 2-iodopropane, water

Alkyl halides, in turn, react with ammonia to produce amines. For example, bromoethane reacts with ammonia to produce ethylamine.

$$CH_3{-}CH_2{-}Br(l) + NH_3(aq) \rightarrow CH_3{-}CH_2{-}NH_2(aq) + HBr(aq)$$

bromoethane, ammonia, ethylamine, hydrogen bromide

PRACTICE PROBLEMS

Complete and balance each substitution reaction.

1. $HOH + (CH_3)_3CI \rightarrow$

2. $CH_3CH_2CH_3 + Cl_2 \xrightarrow{\text{sunlight}}$

3. (benzene ring) $+ Br_2 \rightarrow$

4. $CH_3CH_2OH + HF \rightarrow$

5. $CH_3Cl + NH_3 \rightarrow$

CHEMISTS AND THEIR WORK

Bertram O. Fraser-Reid (1934-)

Bertram Fraser-Reid is a James Duke Distinguished Professor of Chemistry at Duke University. His major research efforts over the years have been in organic chemistry, notably sugars. He has worked on biological regulators for immune systems and has developed methods for producing pheromones, insect sex attractants, from glucose. These pheromones were used in place of DDT to control timber-destroying insects by disrupting their mating patterns. Fraser-Reid's work in sugars has led him to believe that many products now made from petroleum can also be made from sugar—a renewable resource.

Addition Reactions

In a double bond between two carbon atoms, each carbon atom contributes two electrons to the bond. Suppose one bond is broken and the other remains intact. Each carbon atom then has one electron available to bond with some other atom. A number of substances will cause one bond of a double bond to break by adding on at the double bond. This type of reaction is called an **addition reaction.** An example is the addition of bromine to the double bond of ethene. The product of this reaction is 1,2-dibromoethane.

$$\underset{\text{ethene}}{H_2C{=}CH_2(g)} + \underset{\text{bromine}}{Br_2(l)} \rightarrow \underset{\text{1,2-dibromoethane}}{BrH_2C{-}CH_2Br(l)}$$

Atoms of many substances can be added at the double bond of an alkene. Some common addition agents are the halogens (except fluorine), the hydrogen halides, and sulfuric acid. The double bonds in the benzene ring of aromatic compounds are so stabilized by delocalization that addition reactions do not occur readily in these compounds.

PRACTICE PROBLEMS

Complete and balance each addition reaction.

6. $CH_3CH_2CH{=}CH_2 + Br_2 \rightarrow$

7. $CH_2{=}CHCH_3 + HCl \rightarrow$

8. $CH_2{=}CH_2 + H_2SO_4 \rightarrow$

Elimination Reactions

We have seen that under certain circumstances, atoms can be "added on to" a double bond. It is also possible to remove certain atoms from a molecule to create a double bond. Such a reaction is known as an **elimination reaction.** In the most common elimination reactions, a water molecule is removed from an alcohol. A hydrogen atom is removed from one carbon atom, and a hydroxyl group is removed from the adjacent carbon atom. For example, propene can be made from 1-propanol by the removal of water. Sulfuric acid is used as the dehydrating agent.

$$\underset{\text{1-propanol}}{H-\overset{\overset{H}{|}}{\underset{\underset{H}{|}}{C}}-\overset{\overset{H}{|}}{\underset{\underset{H}{|}}{C}}-\overset{\overset{H}{|}}{\underset{\underset{OH}{|}}{C}}-H(l)} \xrightarrow{H_2SO_4} \underset{\text{propene}}{H-\overset{\overset{H}{|}}{\underset{\underset{H}{|}}{C}}-\overset{\overset{H}{|}}{C}=\overset{\overset{H}{|}}{C}-H(g)} + \underset{\text{water}}{H_2O(l)}$$

Esterification and Saponification Reactions

When an alcohol reacts with either an organic acid or an organic acid anhydride, an ester is formed. This type of reaction is called **esterification.** For example, in the reaction between acetic acid and methanol, methyl acetate and water are formed. Methyl acetate is an ester.

$$\underset{\text{acetic acid}}{CH_3-\overset{\overset{O}{\|}}{C}-OH(l)} + \underset{\text{methanol}}{CH_3-OH(l)} \rightarrow \underset{\text{methyl acetate}}{CH_3-\overset{\overset{O}{\|}}{C}-OCH_3(l)} + \underset{\text{water}}{H_2O(l)}$$

An ester can be split into an alcohol and a carboxylic acid by hydrolysis. The hydrolysis reaction is the reverse of the reaction shown above. However, if a solution of a metallic base is used for the hydrolysis instead of water, the metallic salt of the carboxylic acid is obtained, not the acid. This process is called **saponification.** Saponification is the process used in making soaps. Since ancient times, soaps have been made from vegetable and animal oils and fats cooked in bases (KOH, NaOH). Soap is a metallic salt of a fatty acid. The natural fat or oil is an ester. In the following saponification reaction, the natural fat is a glyceride ester. The products of the reaction are soap and glycerol (the alcohol of the glyceride).

HOME EC CONNECTION

Soaps and Detergents

A soap molecule is a nonpolar hydrocarbon on one end with an ionic charge at the other end. This dual nature is the key to the cleansing action of soap. Most "dirt" contains oils, greases, and similar nonpolar compounds. Water, our principal cleaning agent, is highly polar. Consequently, water will not dissolve most dirt. However, upon the application of soap, the dirt is easily removed. The charged end of the soap particle dissolves readily in the water, while the long nonpolar hydrocarbon chain dissolves readily in the dirt. The whole complex can be washed away by water.

Unfortunately, water supplies in many locations contain ions, such as Ca^{2+} and Mg^{2+}, which react with soap to form a precipitate. Water containing these ions is called "hard water". Clothes washed with soap in hard water don't look clean because the precipitate deposits on the fabric. To overcome this problem, chemists developed substances called detergents. Detergents have a polar end and a nonpolar end. Thus, synthetic detergents clean by the same mechanism as soap. However, detergents do not form precipitates with calcium or magnesium ions and can be used freely in hard water.

Figure 30.1 Most of the characteristic flavors and odors of fruits and flowers come from esters.

$$\begin{array}{l} \quad\quad\quad\quad O \\ \quad\quad\quad\quad \| \\ CH_2-O-C-(CH_2)_{16}-CH_3 \\ | \quad\quad\quad\; O \\ | \quad\quad\quad\; \| \\ CH\;-O-C-(CH_2)_{16}-CH_3 \\ | \quad\quad\quad\; O \\ | \quad\quad\quad\; \| \\ CH_2-O-C-(CH_2)_{16}-CH_3 \end{array} + 3NaOH \rightarrow \begin{array}{ccc} CH_2- & CH- & CH_2 \\ | & | & | \\ OH & OH & OH \end{array} + 3[CH_3-(CH_2)_{16}-\overset{\overset{\displaystyle O}{\|}}{C}O^-Na^+]$$

glyceryl tristearate (a fat) base glycerol sodium stearate (soap)

PRACTICE PROBLEMS

9. Complete and balance this elimination reaction.
$CH_3CH_2CH_2CH_2OH \xrightarrow{H_2SO_4}$

10. Complete and balance this esterification reaction.
$CH_3CH_2OH + CH_3CH_2CH_2COOH \rightarrow$

11. Write the balanced equation for the esterification reaction that takes place between methanol and methanoic acid.

MINUTE LAB

Time to be Inventive

You have been asked to invent a new brand of soap. Put on an apron and goggles. Place 25 g of solid vegetable shortening in a 250 cm^3 beaker. Add 10 cm^3 of ethanol and 5 cm^3 of 6*M* NaOH solution (1.2 g NaOH in 5 cm^3 water). Using a hot plate, heat the mixture and stir constantly for 15 minutes. **CAUTION:** *Ethanol is flammable.* While heating, add something to color the soap. Think of a way to add a pleasant scent and a way to make your soap unique (such as insect-repellant soap). Add small amounts of both ingredients at the end of the heating time. Cool the mixture in a cold water bath, then add 25 cm^3 of a saturated NaCl solution. The soap should now appear as curds. Collect it by filtering through several layers of cheese cloth and press it into an evaporating dish. Rinse your hands until they no longer feel slippery. Dry the soap overnight. If it isn't solid, bottle it and call it shampoo.

Polymerization

All of us in today's society are constantly dealing with very large molecules called polymers. **Polymers** are huge molecules made by combining hundreds, thousands, even tens of thousands of small molecules. In most cases, the small molecules, called monomers, are alike. In some other cases, two different monomers will be combined, and in a few cases as many as twenty different monomers may be combined.

There are two ways to produce a polymer from monomers. The reaction that takes place depends on the class of compound that is reacting. We have already seen that compounds containing double bonds can undergo addition reactions. In **addition polymerization,** compounds with double bonds add on to each

Figure 30.2 Here, molten polyethylene is being formed into a film. This low-density polyethylene is most widely used in making the tough transparent film used for packaging many consumer goods and for wrapping food products.

other, end-to-end, to form a polymer. Chains (or cross-linked chains) of very large molecular size and mass are formed. Consider the reaction between two molecules of ethene.

$$CH_2{=}CH_2(l) + CH_2{=}CH_2(l) \rightarrow -CH_2-CH_2-CH_2-CH_2-(amor)$$

This process can continue almost indefinitely. The bonds at the end of a chain can cross-link to other chains or react with a substance deliberately introduced to end a chain. In this case, the polymer is called polyethene or polyethylene.

Another method of making polymers is to combine two compounds, one of which will lose a hydrogen atom and the other a hydroxyl group. The lost groups then combine to form water. The reaction is therefore called **condensation polymerization.** Note that in order to continue the chain, each monomer must contain two functional groups. There are very few condensation polymerization reactions in which the small molecule produced is not water. We will study only those reactions in which water is produced. Look at the following reaction.

$$HO-CH_2-COOH + HO-CH_2-COOH \rightarrow HO-CH_2-CO-O-CH_2-COOH + H_2O$$

As you can see, this same process could also continue indefinitely, with the hydroxyl group of the product molecule reacting with another acid group, while the acid group of the product could react with another hydroxyl group.

Petroleum

The chief source of organic compounds is the naturally occurring mixture called petroleum. Other important sources are coal tar, natural gas, wood, and the fermentation of natural materials. Petroleum is a mixture of hydrocarbons along with small amounts of compounds of nitrogen, oxygen, and sulfur. The hydrocarbons are mainly alkanes and cycloalkanes. The initial treatment of petroleum in a refinery is a fractional distillation. This treatment takes place in a distillation tower and separates the mixture into portions having different boiling ranges. Figure 30.3 shows this process.

At the top of the distillation tower, gases that were dissolved in the petroleum are removed. These gases are principally the one-carbon to six-carbon alkanes. The lighter gases are compressed and sold in cylinders as liquefied petroleum gas (LPG). The heavier gases are often added to gasoline in the winter because they are very volatile and, therefore, give the gasoline better starting characteristics. The next group of fractions consists of gasoline, naphtha (a solvent), jet fuel, kerosene, and light heating oil. The third group of fractions removed from the tower contains heavy fuel oils, diesel oil, and gas oil (raw material for further processing). The final group of fractions is made of heavy mineral oil, lubricating oil, and oil for froth-flotation processing of minerals. The residue remaining in the tower consists of lubricating oil, fuel oil, petrolatum, road oil, asphalt, and coke.

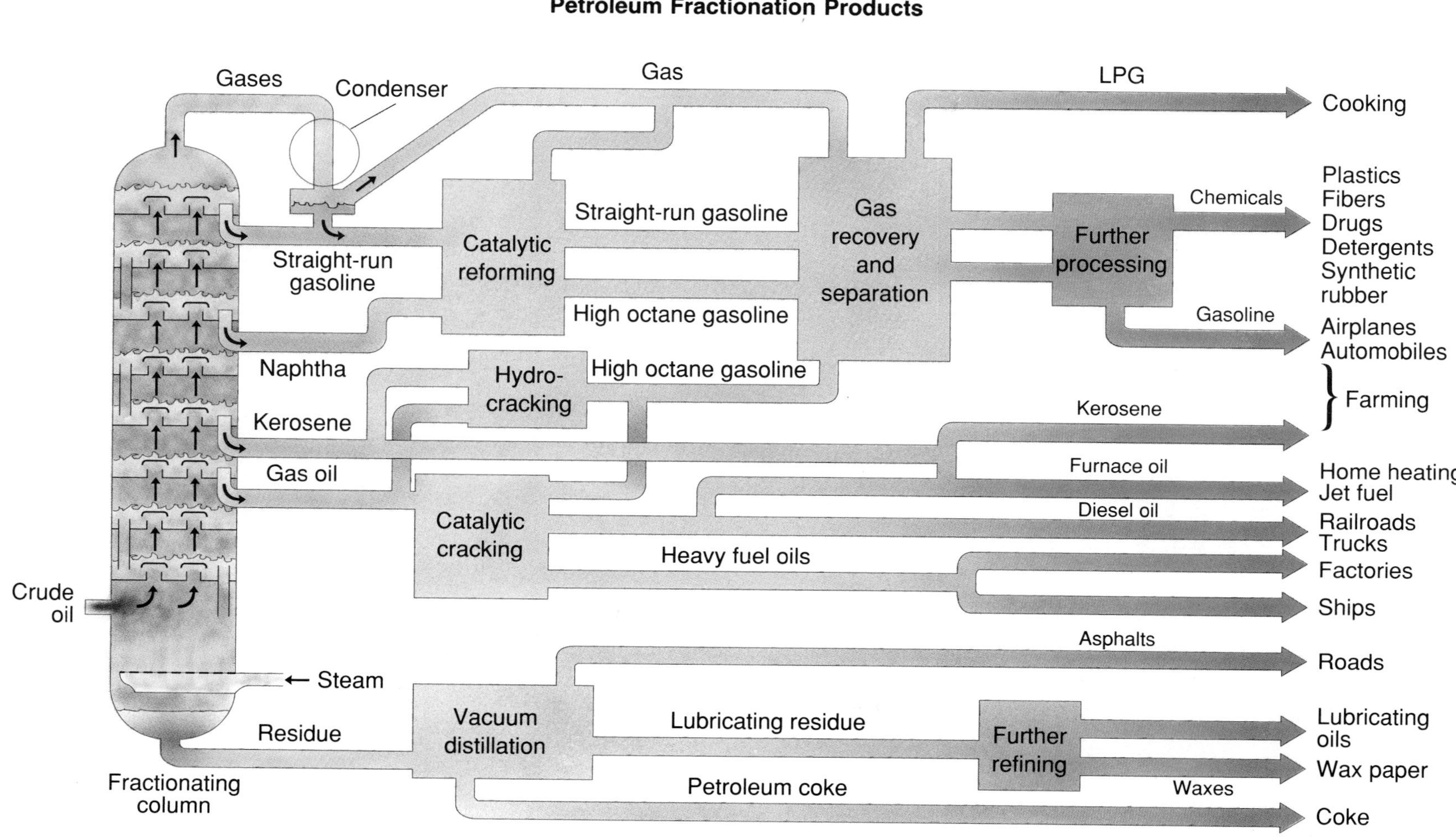

Figure 30.3 The products of petroleum fractionation and some of their uses are shown in this generalized diagram. The supply of petroleum is critical to many industries.

Figure 30.4 The higher the octane rating of a gasoline, the better the gasoline's performance in an automobile engine. An isomer of octane, 2,2,4-trimethylpentane, has a rating of 100 while heptane has a rating of zero. Benzene has an octane number of 106.

The products from the distillation cannot be marketed in the form in which they come from the distillation tower. They must be refined to remove undesirable substances, particularly sulfur compounds. Additives are blended with gasoline and other products to improve their performance. Many by-products of gasoline production are used as raw materials for the production of plastics, synthetic fibers, and elastomers.

The yield of gasoline can be improved in a number of ways. In one process, smaller molecules from the lower boiling fractions are joined to form larger molecules. In another process, the larger molecules of the higher boiling fractions are cracked, or broken, into smaller molecules.

Gasolines are rated on a scale known as **octane rating.** The basis for this scale is the property of some fuels to cause "knocking" in engines. The knocking occurs when some of the fuel explodes suddenly instead of burning evenly. High-octane gasolines contain more ring compounds and highly-branched hydrocarbons than do low-octane gasolines. Heptane is assigned an octane rating of 0, while 2,2,4-trimethylpentane (an isomer of octane) is assigned a rating of 100. Gasolines are rated in accordance with the blend of these two substances giving the same performance characteristics as the gasoline under test. Tetraethyl lead was at one time the major additive used to improve the octane rating of gasoline. The pollution of the atmosphere by the lead compounds from gasoline engine exhaust has led to the gradual phasing out of leaded gasoline use. The principal octane-enhancing additive used today is 2-methoxy-2-methylpropane (methyl *tert*-butyl ether).

Word Origins

elastikos: (GK) get smaller

elastomer—parts get smaller after stretching

Elastomers

Natural rubber and many kinds of synthetic rubber are classified together as elastomers. **Elastomers** are substances that can be deformed under the influence of an outside force, but will return to their original shape once the force is removed. An example is found in an ordinary rubber ball. When the ball strikes a surface, the force of the collision causes the molecules to become more tightly folded, a higher energy state. As the molecules return to their equilibrium arrangement, they stretch out again and the ball rebounds.

The producers of synthetic substitutes for rubber have built a thriving industry based on organic chemistry. No one synthetic elastomer can as yet replace natural rubber. However, there are many synthetics that can perform a particular job as well as or better than natural rubber. The production of synthetic elastomers depends primarily upon addition polymerization.

Natural rubber is a polymer of 2-methylbutadiene. The 2-methylbutadiene molecule is the monomer of natural rubber. Many types of synthetic elastomers can be made from various monomers. Often they are made by polymerizing two or more substances containing double bonds. A polymer that is made from

more than one substance is called a copolymer. The largest selling synthetic elastomer is a copolymer of 1,3-butadiene and styrene. It is used chiefly in the production of automobile tires.

Figure 30.5 The materials that make up rubber balls (right) are elastomers. The balls deform when they are bounced but then return to their original shape. The major use of synthetic rubber is the manufacture of tires (left).

$$CH_2{=}\overset{\overset{\displaystyle CH_3}{|}}{C}{-}CH{=}CH_2 \qquad CH_2{=}CH{-}CH{=}CH_2$$

2-methylbutadiene 1,3-butadiene

Plastics

The production of plastics has created another important organic chemical industry. Some common plastics made by addition polymerization are polyethylene, polypropylene, polyvinyl chloride (PVC), polyvinyl acetate, polystyrene, Teflon, and acrylics.

Plastics can also be made by condensation polymerization. Polyesters and polyurethane are made this way. Other plastics, such as cellophane and celluloid, are made by the chemical treatment of cellulose, a natural polymer.

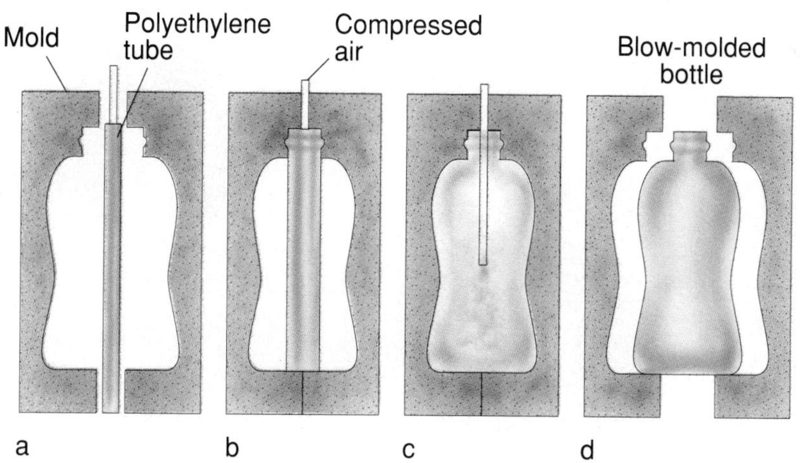

Figure 30.6 The major use of high-density polyethylene (HDPE) is to make blow-molded products, such as bottles for milk and other consumer products. In the process, a tube of HDPE is placed in a mold (a), and the mold is closed, sealing the bottom of the tube (b). Compressed air is then blown into the polyethylene tube and the warm tube expands to take the shape of the mold (c). The bottle is then removed from the mold (d).

Figure 30.7 The models that come in hobby kits are made of polystyrene, which is produced by addition polymerization.

Most polymers made by addition polymerization can be softened by heating. They are thermoplastic materials. When softened, they can be molded into useful shapes. In contrast, cross-linked polymers formed by condensation reactions get harder when heated. The heating causes more cross-links to form. To force these materials into a mold would break covalent bonds. Thus, they are called thermosetting polymers. In the shaping process, they are first ground to a powder. Then, fillers are added, along with colorants and a plasticizer, to improve flow. The whole mixture is injected into a mold and formed by heat and high pressure as cross-links are reformed.

Structural diagrams for the monomer and polymer units of two typical plastics are shown below.

Addition polymerization plastic:

$$\underset{\text{C}_6\text{H}_5}{\text{CH}}\!=\!\text{CH}_2(\text{l}) \rightarrow \cdots-\underset{\text{C}_6\text{H}_5}{\text{CH}}-\text{CH}_2-\underset{\text{C}_6\text{H}_5}{\text{CH}}-\text{CH}_2-\underset{\text{C}_6\text{H}_5}{\text{CH}}-\text{CH}_2\cdots$$

phenylethene or styrene — polystyrene (amor)

Condensation polymerization plastic:

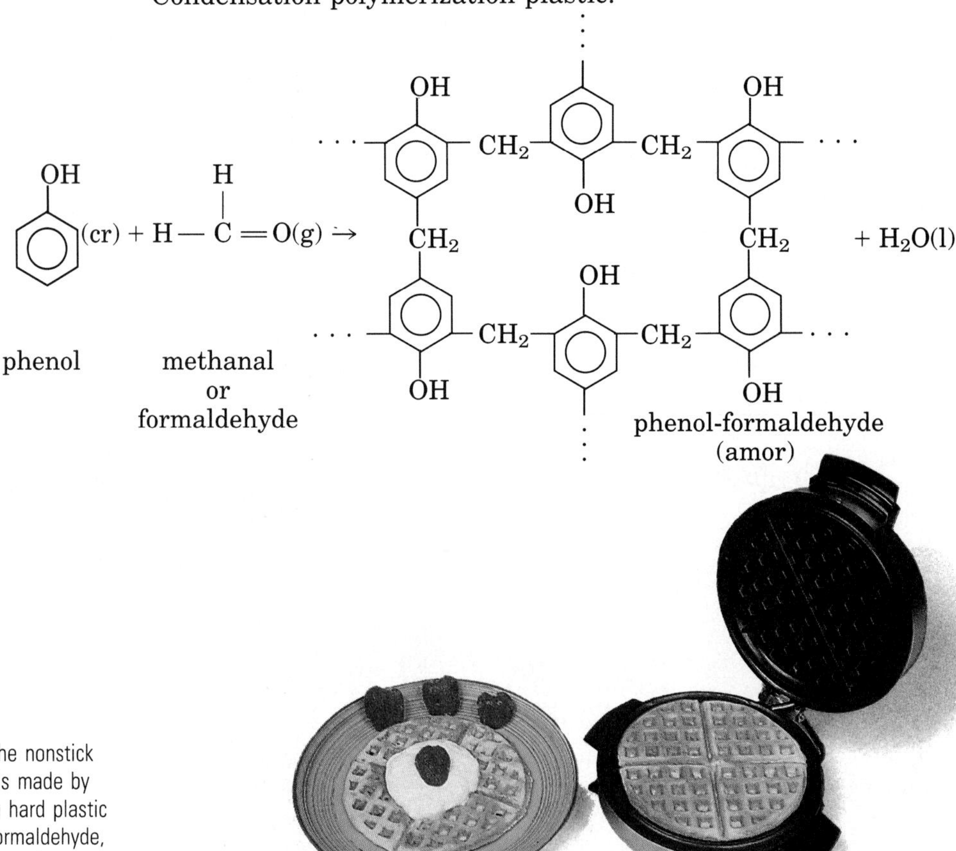

Figure 30.8 The Teflon in the nonstick coating on this waffle iron is made by addition polymerization. The hard plastic handle is made of phenol-formaldehyde, which is produced by condensation polymerization.

Figure 30.9 Rayon fibers (left) are produced from cellulose. The cell walls of trees, as shown in this photomicrograph (right), are the source of this cellulose.

Synthetic Fibers

One of the earliest synthetic fibers produced was rayon. Rayon is simply reconstituted cellulose, the principal structural component of plants. Cellulose is a natural polymer of glucose, $C_6H_{12}O_6$.

Glucose

Cellulose

Cellulose in the form of purified wood pulp is soaked in a sodium hydroxide solution. The cellulose is converted to sodium cellulose whose structure is unknown. The sodium cellulose is then treated with carbon disulfide, CS_2, forming sodium cellulose xanthate, which is soluble in sodium hydroxide solution. This solution, called "viscose," is squeezed through small holes into a solution of sulfuric acid. The xanthate reacts with the sulfuric acid to regenerate the cellulose fiber.

Nylon is a name for a whole group of polyamide fibers. Nylon 66, made from adipic acid (hexanedioic acid) and hexamethylenediamine (1,6-diaminohexane), is the most common form. The amino group of the amine and the carboxyl group of the acid form an amide by condensation.

$$\underset{\text{1,6-diaminohexane}}{\mathrm{H-\overset{\overset{H}{|}}{N}-(CH_2)_6-\overset{\overset{H}{|}}{N}-H}} + \underset{\text{hexanedioic acid}}{\mathrm{H-O-\overset{\overset{O}{\|}}{C}-(CH_2)_4-\overset{\overset{O}{\|}}{C}-O-H}} \rightarrow \cdots \underset{\text{amide link}}{\mathrm{-\overset{\overset{H}{|}}{N}-\overset{\overset{O}{\|}}{C}-}} \cdots + H_2O$$

Since each reactant has a functional group on each end, a polymer can be formed by condensation polymerization. The polymer is melted and forced through small holes into a stream of cool air where the polymer solidifies to form fibers.

A similar polymer, Nylon 6, is produced from the cyclic compound, caprolactam. Nylon 6 has a lower melting point than Nylon 66 but is more resistant to light, heat, and fatigue. Kevlar, another type of nylon, is a copolymer of 1,4-diaminobenzene and 1,4-benzenedicarbonyldichloride. Kevlar has extremely high strength and is used, among other things, for tire cords.

Figure 30.10 Nylon can be made easily in the laboratory using solutions of hexanedioic acid and 1,6-diaminohexane. Nylon-66 forms at the interface of the two immiscible solutions.

30.1 *CONCEPT REVIEW*

12. What type of inorganic reaction is similar to substitution reactions?

13. What relationship do addition and elimination reactions have to each other?

14. How do addition polymerization and condensation polymerization differ from each other? How do plastics formed by these two reactions differ from each other?

15. What is an elastomer?

16. Apply Why would a car engine run better with high-octane gasoline than with low-octane gasoline?

30.2 Biochemistry

OBJECTIVES

- state the structure and function of the four main biomolecules: proteins, carbohydrates, lipids, and nucleic acids.
- describe the progress in and problems associated with the biomaterials industry.

The study of the substances and the chemical reactions involved in life processes is called **biochemistry.** Many of the most important molecules found in living systems are polymers. Even in an organism as complex as a human, most of the compounds can be grouped into four main types. We will look at them in greater detail. Three of these types—proteins, carbohydrates, and nucleic acids—are polymers. Lipids, although not usually considered polymers, are often composed of several simpler molecules that have joined together through condensation reactions.

Proteins

Approximately one-half of your non-water mass consists of proteins. **Proteins** are biological polymers that are linked by amide groups. These substances make up some structural parts of your body, such as cartilage and tendons. Cartilage and tendons are composed principally of the protein collagen. Three molecules of collagen are always found twisted together to make a sort of cable. The entwined molecules are tightly locked together and are therefore rigid and unelastic.

About one-half of the protein in your body is used within the cells as catalysts. Biological catalysts are called **enzymes.** Organisms, which are composed of cells, require energy. When we think of the word *fuel*, we think of burning a substance to obtain energy. "Burning" fuels would destroy the cells. Instead, enzymes present in the cells enable reactions to occur at temperatures that are not injurious to the cells.

Proteins are made of **amino acids.** There are 20 common amino acids. These acids are listed in Table 30.1. Earlier, we found that carboxylic acids react with amines to form amides.

$$-\overset{\overset{\displaystyle O}{\|}}{C}-\overset{\overset{\displaystyle H}{|}}{N}-$$

Amino acids are linked through the same condensation process. However, biochemists call the amide link a **peptide bond.** The new molecule formed from two amino acids is called a dipeptide. Three amino acids form a tripeptide. Many amino acids condense to a polypeptide. If a polypeptide has a biological function, it is called a protein.

Proteins differ from each other in several ways. The first difference, and the most important, is the sequence of the amino acids composing the protein. Another way they differ is the way the polymer chain is coiled, folded, and twisted. Finally, the type of bonding holding the polymer in a particular shape can vary. Proteins contain from about 30 to several thousand amino acids. The 20 amino acids listed in Table 30.1 can form an enormous number of different proteins.

Word Origins

proteios: (GK) first

protein—the primary (first) structural material of animals

en: (GK) in

zymos: (GK) leaven

enzyme—something inside that causes a change, as does yeast in the leavening of bread

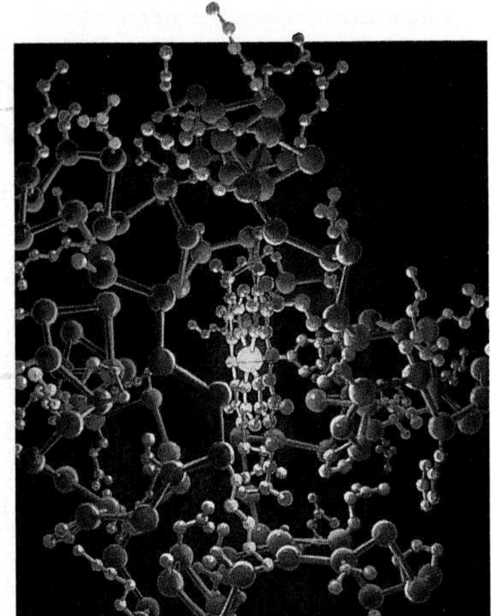

Figure 30.11 This model of the protein cytochrome c (top) shows that protein chains are folded and coiled into a three-dimensional shape. Cytochrome c is present in muscle tissue such as this steak (bottom).

Table 30.1

Amino Acids

Amino acids have the form

$$H-\overset{\displaystyle COOH}{\underset{\displaystyle G}{\overset{|}{\underset{|}{C}}}}-NH_2$$

In the table, only the composition of *G* is represented.

Name	G	Symbol
Glycine	$H-$	Gly
Alanine	CH_3-	Ala
Valine	$CH_3-\underset{\displaystyle CH_3}{\underset{\vert}{CH}}-$	Val
Leucine	$CH_3-\underset{\displaystyle CH_3}{\underset{\vert}{CH}}-CH_2-$	Leu
Isoleucine	$CH_3-CH_2-\underset{\displaystyle CH_3}{\underset{\vert}{CH}}-$	Ile
Tryptophan	(indole ring, N–H) $-CH_2-$	Trp
Lysine	$H_2N-CH_2-CH_2-CH_2-$	Lys
Arginine	$H_2N-\underset{\displaystyle NH}{\underset{\Vert}{C}}-NH-CH_2-CH_2-CH_2-$	Arg
Phenylalanine	$C_6H_5-CH_2-$ (benzene ring)	Phe
Histidine	(imidazole ring, N, NH) $-CH_2-$	His
Asparagine	$O=\underset{\displaystyle NH_2}{\underset{\vert}{C}}-CH_2-$	Asn
Glutamine	$O=\underset{\displaystyle NH_2}{\underset{\vert}{C}}-CH_2-CH_2-$	Gln
Serine	$HO-CH_2-$	Ser
Threonine	$CH_3-\underset{\displaystyle OH}{\underset{\vert}{CH}}-$	Thr
Aspartic acid	$HOOC-CH_2-$	Asp
Glutamic acid	$HOOC-CH_2-CH_2-$	Glu
Tyrosine	$HO-C_6H_4-CH_2-$ (benzene ring)	Tyr
Methionine	$CH_3-S-CH_2-CH_2-$	Met
Cysteine	$HS-CH_2-$	Cys
Proline (an exception to the general formula)	ring: H_2C-CH_2, H_2C, $CH-COOH$, N–H	Pro

Two different dipeptides can be made from two amino acids, depending on which ends of the molecules react. For example,

$$H_2N-CH_2-COOH + H_2N-CH(CH_3)-COOH \rightarrow H_2N-CH_2-CO-NH-CH(CH_3)-COOH + H_2O$$

glycine, alanine, glycylalanine dipeptide

$$H_2N-CH(CH_3)-COOH + H_2N-CH_2-COOH \rightarrow H_2N-CH(CH_3)-CO-NH-CH_2-COOH + H_2O$$

alanine, glycine, alanylglycine dipeptide

Reactions utilizing catalysts consume or produce energy. Some of the energy produced appears as heat in the cell. However, most of the energy must be converted to forms other than heat or the cell will die. In cells, this excess energy is used to produce a product whose synthesis requires the input of energy. This compound is usually adenosine-5′-triphosphate, or ATP for short. All organisms, from single-celled bacteria to humans, use ATP as their energy transfer molecule. When an enzymatic reaction requiring energy occurs, some ATP can be decomposed to provide that energy. For example, glucose is an important energy source for organisms. The first step in the use of glucose is the formation of glucose-6-phosphate. The ΔG for glucose reacting with phosphate ion is +12 600 kJ/mol. However, the ΔG for glucose reacting with ATP to form glucose-6-phosphate and ADP (adenosine diphosphate) is −21 400 kJ/mol. The structure of ATP is

$$HO-P(=O)(OH)-O-P(=O)(OH)-O-P(=O)(OH)-O-CH_2-\text{(ribose)}-\text{(adenine)}$$

triphosphate group

adenine

ribose

Carbohydrates

Carbohydrates are also important to living systems. These compounds contain the elements carbon, hydrogen, and oxygen. The word *carbohydrate* literally means "hydrate of carbon." Almost all **carbohydrates** are either simple sugars or condensation polymers of sugars. The most common simple sugar is glucose. Another common simple sugar is fructose.

MINUTE LAB

Sugar Fermentation – It's a Gas

In a beaker, dissolve 5 teaspoons of sucrose in 250 cm^3 of warm water. Place 1 teaspoon of dry yeast and 3 teaspoons of warm water in a 500-cm^3 soda bottle. Add the sucrose solution. Fit a large, flat balloon over the mouth of the bottle. Place 3 cm^3 of distilled water in a small test tube, 3 cm^3 of vinegar in a second tube, and 3 cm^3 of household ammonia in a third tube. Add 1 drop of 0.1% bromothymol blue to each tube. Observe the color in neutral, acidic, and basic solutions. Fill a large test tube with distilled water. Add bromothymol blue until the solution is pale green. Pinch the balloon closed and remove it from the bottle. Place a plastic straw in the large tube. Hold the open end of the balloon tightly around the straw, allowing the CO_2 in the balloon to bubble through the solution. Is the solution acidic, basic, or neutral? Write an equation for the reaction.

Figure 30.12 When concentrated sulfuric acid is poured over sucrose, carbon is produced. Sulfuric acid acts as a dehydrating agent to pull water out of sucrose, a carbohydrate with the formula $C_{12}(H_2O)_{11}$ or $C_{12}H_{22}O_{11}$.

glucose

fructose

Combining two simple sugars produces a disaccharide. In this case, combining glucose and fructose produces common table sugar, sucrose.

sucrose

The combination of many simple sugars in one molecule is a polysaccharide. There are a number of biologically important polysaccharides formed from the monomer, glucose. The chief storage form for carbohydrates in plants is starch, while in animals it is glycogen. Cellulose is yet another polymer of glucose.

Glycogen, starch, and cellulose differ simply in the way the monomers are bonded to each other. The links in the chain and the manner in which the chains are cross-linked determine which substance is produced by the condensation polymerization.

The process of breaking down carbohydrates to carbon dioxide and water is the chief energy source of an organism. Carbohydrate storage (starch/glycogen) is the energy reserve of the organism. The oxidation of glucose is the "burning" of the cell's "fuel." However, the enzymes in the cell permit this oxidation to take place in carefully controlled steps.

Lipids

Proteins and carbohydrates tend to be more soluble in water than in nonpolar solvents. **Lipids** are those biological compounds more soluble in nonpolar solvents than in water. Lipids can be divided into several different groups. Among these groups are fats and oils.

Fats are esters formed from glycerol and fatty acids. Fatty acids are carboxylic acids with 12 to 20 carbon atoms in the chain. The number of carbons in the chain is an even number. The most abundant fatty acids are composed of 16 and 18 carbon atom chains. Some fatty acids are saturated, while others have as many as four double bonds. In general, animal fats are more saturated and plant oils are more unsaturated. There are a number of other lipids derived from glycerol in addition to fats. Many of these other lipids are used to build cell membranes.

Another class of lipids consists of compounds called steroids. All steroids contain the tetracyclic ring system, sometimes called the steroid nucleus.

One such steroid is cholesterol.

cholesterol

Cholesterol is found in bile and is an important constituent of cell membranes.

Some vitamins are lipids. **Vitamins** are substances used by cells to aid enzymatic reactions. For example, vitamin A aids in the synthesis of eye pigments needed for vision. The compounds chlorophyll and heme are also sometimes considered lipids. Chlorophyll (page 242) is the green pigment used by plants in the process of converting carbon dioxide and water to carbohydrates. Heme is a component of hemoglobin, the red pigment utilized by humans and some other animals as an oxygen carrier in the blood.

heme portion of hemoglobin

retinol
(vitamin A)

Figure 30.13 Oil, a lipid, is not soluble in water. When oil is spilled into water, an oil slick results.

vita: (L) life

vitamin—an amine vital to life

Figure 30.14 The pink skin of these pigs is due to hemoglobin molecules to which oxygen molecules are attached.

HISTORY CONNECTION

The First Synthetic Fiber

In 1884, Hilaire Chardonnet, a French chemist, produced a synthetic nitrocellulose fiber made from cotton. The nitrocellulose was then hyrolyzed to cellulose. Chardonnet had been a student of Louis Pasteur when Pasteur had studied silk and silkworms. Chardonnet called his fiber "artificial silk." It was not named rayon until 1924, when the term was derived from *ray*-, from the sheen of the fiber and *-on*, showing it was derived from cotton.

Nucleic Acids

Even though they are present in living things in small quantities, nucleic acids are an important group of biological polymers. **Nucleic acids** are polymers that form a code determining our genetic inheritance. They are able to replicate, or duplicate, themselves. Nucleic acids control the cell by controlling the synthesis of all proteins.

Nucleic acids are polymers of units called nucleotides. Each **nucleotide** has three parts: a nitrogen base, a sugar, and a phosphate group. ATP, shown on page 783, is a type of nucleotide. Only two sugars are found in nucleic acids, ribose and deoxyribose.

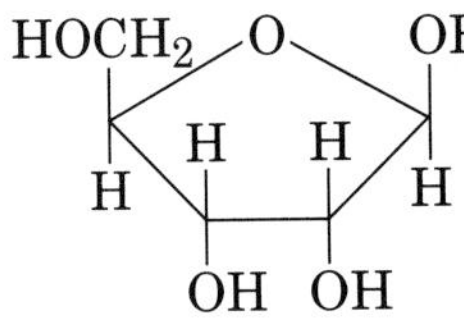

$HOCH_2$ O OH H H H H OH OH
ribose

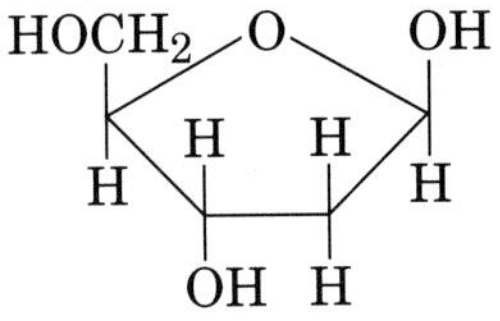

$HOCH_2$ O OH H H H H OH H
deoxyribose

Ribose is found in ribonucleic acid (RNA) and deoxyribose is found in deoxyribonucleic acid (DNA). Of the five nitrogen bases found in nucleic acids, DNA contains adenine, cytosine, guanine, and thymine. RNA also contains adenine, cytosine, and guanine, but contains uracil in place of thymine. Our genetic makeup is determined by the sequences of these bases in the nucleic acid polymers.

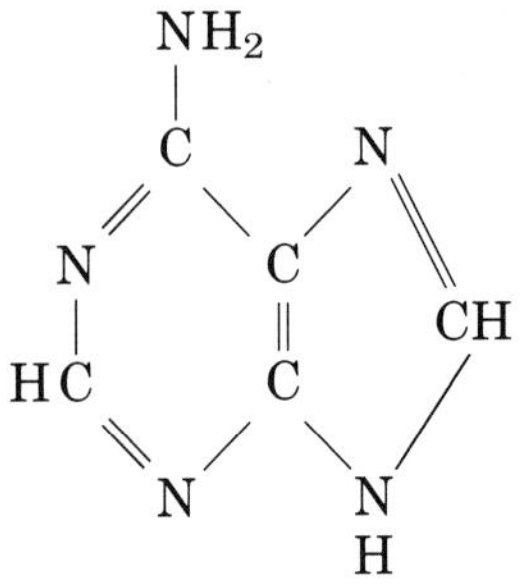

NH_2 C N N C CH HC C N N H
adenine

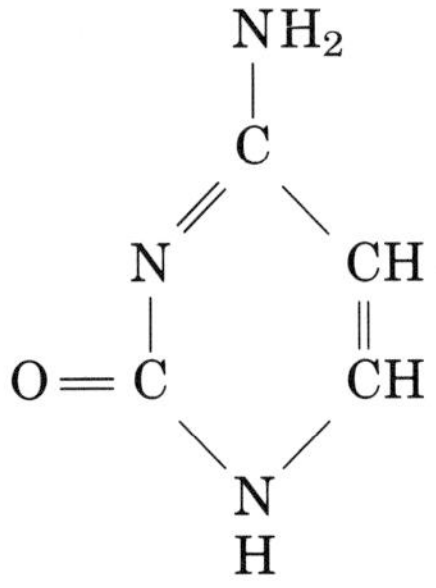

NH_2 C N CH O=C CH N H
cytosine

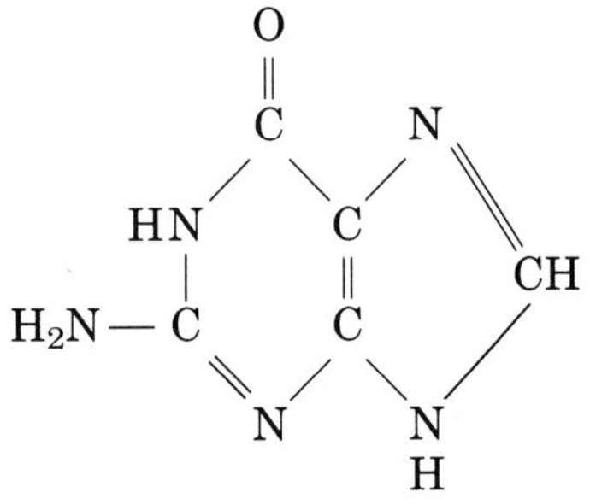

O C N HN C CH H_2N—C C N N H
guanine

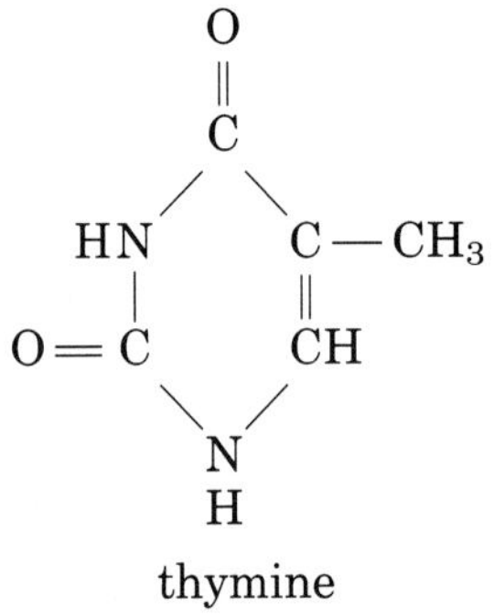

O C HN C—CH_3 O=C CH N H
thymine

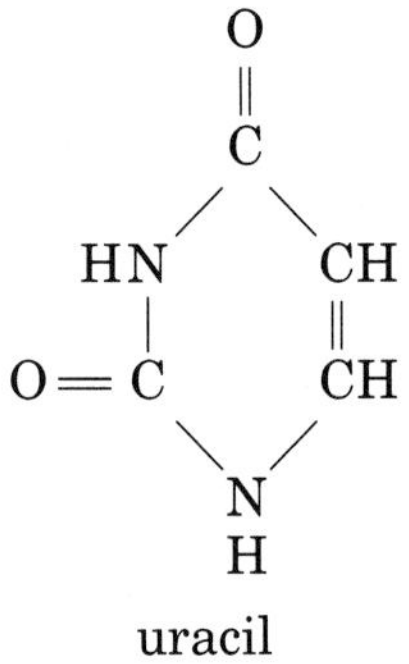

O C HN CH O=C CH N H
uracil

DNA transfers genetic information from one generation to the next because it can duplicate itself. Within a cell, the DNA is used to make several kinds of RNA. The genetic code in the DNA is copied onto the RNA, which in turn is used in the manufacture of

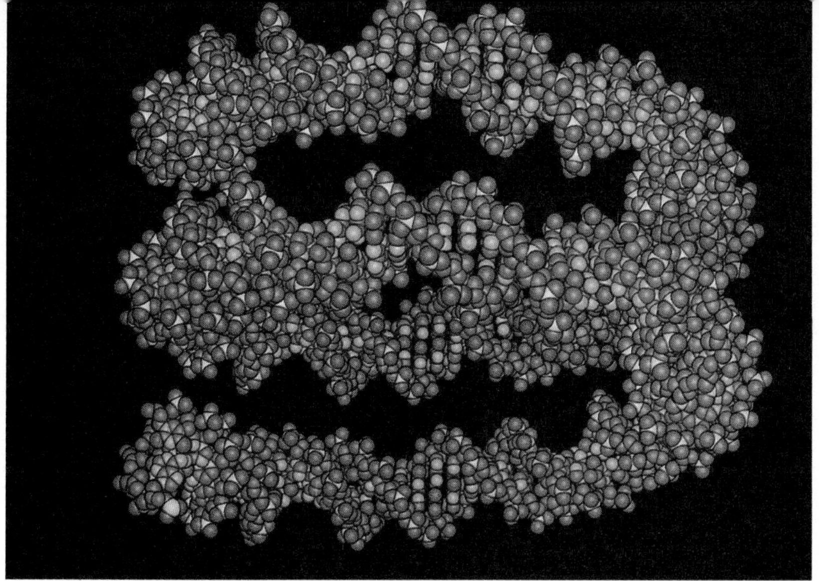

Figure 30.15 Shown here is a model of a segment of DNA. A single DNA molecule may contain many thousands of nucleotides.

proteins. Thus, DNA serves as the master code that controls the synthesis of all proteins in the body. The process is more complicated than presented here, and not all of it is fully understood. Biochemical investigation is perhaps the most exciting area of emphasis in current research.

FRONTIERS

DNA Fingerprinting

DNA fingerprinting is a method of identifying a person's DNA by determining the length of the DNA segments in regions of human genes called minisatellites. Minisatellites consist of sequences of repeated nucleotides in the DNA molecule. The number of sequences and hence the length of the minisatellites is unique to each individual. To analyze the minisatellites, DNA is obtained from a blood or hair sample of a crime suspect and from a sample found at the scene of the crime. The DNA molecules from each sample are broken into specific segments by enzymes. The segments are then separated by gel electrophoresis. The sequence of nucleotides is determined by treating the segments with radioactively-labeled DNA that binds to the segments. The positions of the separated segments are located with X-ray film. The film displays a DNA fingerprint as a series of vertical lines that look like the bar codes on consumer products. The chance of two people having identical DNA fingerprints ranges from 1 in 20 billion to 1 in 30 billion.

Advocates of DNA fingerprinting predict that this technique will soon be a standard procedure in crime investigations. They see a time when large data banks of DNA fingerprints similar to those of fingerprints will be used to identify individuals. In the future, the swirls of your fingerprint may be replaced by the line sequence of your DNA.

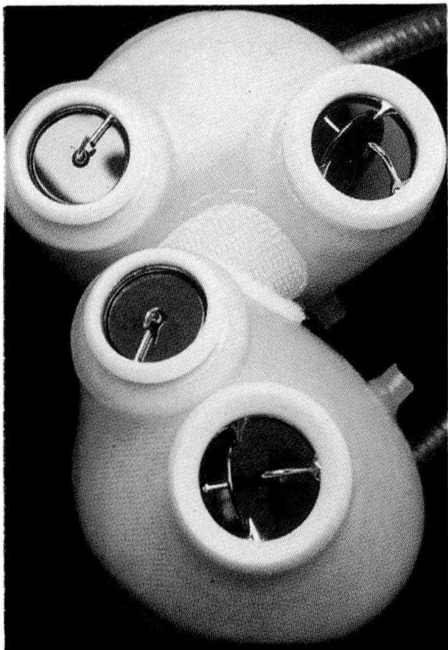

Figure 30.16 This artificial heart, the Jarvik-7, is made of rubber and the plastic polyurethane. It was the first artificial heart to be implanted in a human. More sophisticated hearts have replaced this type in modern transplant surgery.

Biomaterials

Fifty years ago, a person with a leaky heart valve had little chance of living a normal life. Today, surgeons around the world replace malfunctioning natural valves with artificial valves more than 75 000 times a year. Biomedical engineers have made and continue to build replacements for various parts of the human body. Artificial heart valves have been under development for a long time. However, other artificial body parts, such as ears, limbs, and organs, have been attempted only recently. One problem to be overcome by the design engineer is the tendency of the body to reject foreign substances. The body treats an artificial organ as an invader and reacts as if the organ were a bacterium or virus. Materials appropriate for artificial organs must not trigger any of the body's defense systems.

The biggest problem with implanted materials is their tendency to cause blood clots. The rate of blood flow is critical in the design of implanted devices. If the flow is either too fast or too slow, clots tend to form. Sudden changes in flow patterns, caused by seams or connections, also tend to cause clots to form.

Another problem in using biomaterials concerns the formation of bonds between natural tissue and the replacement part. Many of the substances used in biomaterials cause the growth of fibrous tissue that separates the natural tissue from the implant. This fibrous tissue may harden, causing inflammation, pain, and the destruction of natural tissue. Most biomaterials research concerns controlling any chemical reactions that may occur at the interface between the biomaterial and natural tissue, including blood. Mechanical failure of moving parts is also a concern of the biomedical engineer. The moving parts of an artificial heart flex about 50 000 000 times per year.

One material that has found wide usage in medicine is silicone rubber (Silastic). It is a polymer with a structure similar to the simple silicone shown here.

$$\left[-O-\overset{\displaystyle CH_3}{\underset{\displaystyle CH_3}{\overset{|}{\underset{|}{Si}}}}- \right]_n$$

polydimethylsiloxane

Membranes made of silicone or polyurethane are being used as artificial skin in the treatment of burn victims.

Figure 30.17 The use of artificial skin has saved the lives of many burn victims. Artificial skin is a membrane made of silicone.

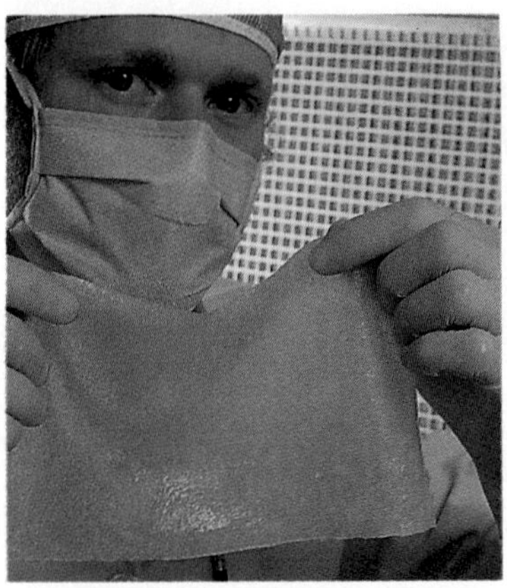

Many other tissue-compatible materials are also polymers. Their molecular shapes are such that they imitate safe biological materials. Some polymers are almost shapeless; that is, their surfaces are relatively smooth. Thus, they are ignored by the body's defenses. Other polymers have deliberately rough surfaces over which the body will build a protective layer that is then ignored by passing molecules.

Artificial blood vessels are usually made of Dacron, the most common polyester, or Teflon.

$$\left[-O-\overset{\overset{O}{\|}}{C}-C_6H_4-\overset{\overset{O}{\|}}{C}-O-CH_2-CH_2- \right]_n$$

Dacron
(polyethylene terephthalate)

$$\left[-\underset{F}{\overset{F}{\underset{|}{\overset{|}{C}}}}-\underset{F}{\overset{F}{\underset{|}{\overset{|}{C}}}}- \right]_n$$

Teflon
(polytetrafluoroethylene)

The flexible parts, or diaphragms, of artificial hearts are often made of either Biomer or Hexyn. Biomer is a copolymer of methylenebis(4-phenylisocyanate) with poly(tetramethylene glycol). Hexyn is a polymer made from 95% 1-hexene, 3% 4-methyl-1,4-hexadiene, and 2% 5-methyl-1,4-hexadiene.

One substance now in wide use as a biomaterial is a noncrystalline form of carbon called pyrolytic carbon. It is presently the most common substance used in artificial heart valves. Titanium is another nonpolymeric material frequently used in medicine where a rigid, strong material is needed, as in bone implants.

There are also applications for biomaterials outside the human body. In open heart surgery, the body's blood is passed through an oxygenator, the so-called heart-lung machine. All the surfaces that the blood touches in the oxygenator must be biocompatible so that blood clots do not form. One of many substances used as membranes for the exchange of gases in oxygenators is polypropylene.

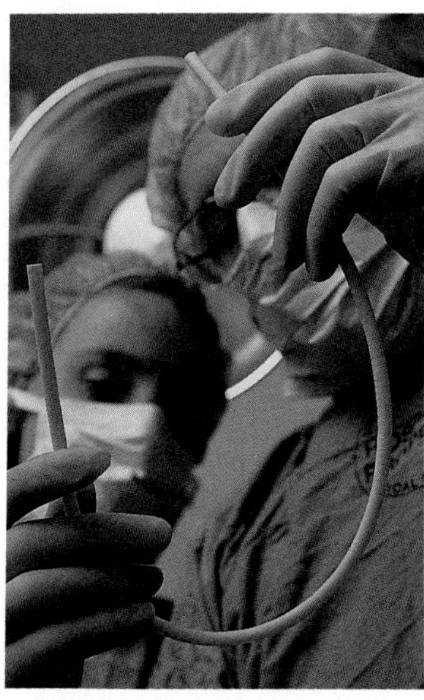

Figure 30.18 Artificial blood vessels, such as the one shown here, are made of Dacron or Teflon.

$$\left[-CH_2-\overset{\overset{CH_3}{|}}{CH}- \right]_n$$

polypropylene

Dialysis and plasmapheresis are two other processes in which blood is taken from the body, processed to purify it, and then returned to the body. Dialysis is used in the case of kidney failure and plasmapheresis in the case of the body's inability to process certain natural but unwanted substances such as nitrogenous wastes. In both cases, membranes are necessary for the removal of unwanted substances. In dialysis, polymethyl methacrylate and polyacrylonitrile are two of several membrane substances. Many of the same substances are used as plasmapheresis membranes. Polyvinyl alcohol and polyethylene are also used.

30.2 CONCEPT REVIEW

17. What characteristic makes a substance suitable for use as a biomaterial?

18. What are the roles of proteins, carbohydrates, and nucleic acids in the body?

19. How are lipids distinguished from proteins and carbohydrates?

20. Apply What kind of bonding would you expect to be best on the surface of a biomaterial?

CHEMISTRY AND
SOCIETY

Genetic Engineering

There are many substances produced commercially by the action of microorganisms. Yeast produce ethanol from grain, and bacteria produce cheese from milk. Bacteria are also used in the production of propanone (acetone). Citric acid can be obtained from molds. Biochemists have now developed methods for "custom tailoring" organisms to produce pharmaceuticals and substances of industrial importance. The substances that can be produced by a microorganism are determined by the organism's nucleic acids. By modifying the DNA of a microorganism, the substances the organism produces can be changed.

Most bacterial DNA is found in the cell chromosome. Some DNA, however, is found outside the chromosome in the form of circles called plasmids. Using appropriate enzymes, a plasmid may be split open, leaving two "sticky ends." Part of a DNA molecule from another organism is then inserted and the loop closed. This recombinant-DNA plasmid is taken in by a bacterial cell. The spliced nucleic acid containing the new section directs the production of the desired substance in the bacterium and its future generations. The model shows gene-splicing using *E. coli* bacteria. This microorganism is often used in recombinant DNA studies because of its comparatively simple genetic makeup.

Biochemical production of insulin and interferon by recombinant DNA microorganisms is a reality. Many new products in the health science field are being developed. Scientists are hoping recombinant DNA research will provide answers to such problems as genetic disorders and cancer.

In order to accomplish such advanced treatment techniques, scientists must first learn where each gene is located in the chromosomes of human cells. A human cell contains 23 pairs of chromosomes. These chromosomes are made of billions of atoms. A major research project called the Human Genome Initiative is being funded by the United States government. Its purpose is to locate and describe the structure of every gene in the human cell. There are between 50 000 and 100 000 genes in human chromosomes. Only a small number of these genes have been located and few of them have had their structure fully determined. It is expected that the Initiative will cost $3 billion over a period of 15 years. The knowledge to be gained about hereditary diseases and defects is the major driving force behind this massive undertaking. There are critics of the project, however. They believe that the money could be spent more wisely in other areas of science. There is the fear that enormous projects such as this one will deprive individual researchers of federal funding for their work in other areas.

Analyzing the Issue

1. How can results from Human Genome Initiative research be used in genetic engineering projects?
2. Do you think the federal government should fund the Human Genome Initiative to such a large extent that other projects are not funded?
3. Find out more about the Human Genome Initiative. Do you think its goals are worth its price?

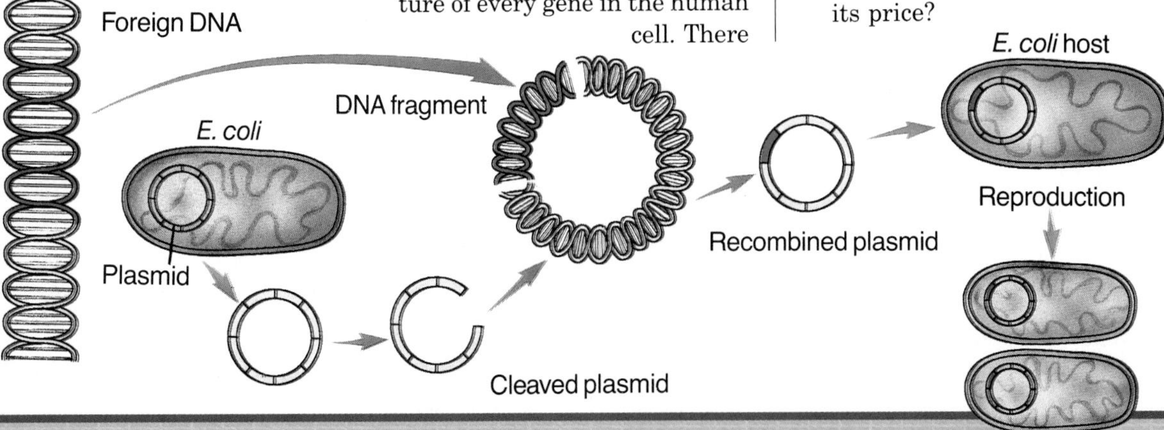

Chapter 30
REVIEW

Summary

30.1 Organic Reactions and Products

1. Organic compounds may undergo many types of reactions.
2. In a substitution reaction, a functional group replaces a hydrogen atom or another functional group of a hydrocarbon.
3. In addition reactions, a substance breaks one bond of a double bond and adds new atoms to the carbons on either side.
4. In elimination reactions, atoms or groups are removed from two adjacent carbon atoms to form a double bond.
5. Esterification is the formation of an ester from an alcohol and an acid or acidic anhydride.
6. The splitting of an ester by a metallic base into the alcohol and the salt of the acid is called saponification.
7. Polymers are produced by combining small molecules of substances called monomers into very large molecules.
8. Addition polymerization is a reaction in which many molecules containing double bonds add to each other to form a large chain or a cross-linked chain.
9. Condensation polymerization is a reaction in which monomers combine to produce a large molecule and many small molecules, usually water.
10. Petroleum is the principal source of organic chemicals. It is composed mostly of hydrocarbons.
11. During refining, fractions of petroleum are separated according to boiling point.
12. The yield of gasoline from petroleum may be improved by cracking larger hydrocarbons into smaller molecules.
13. The octane rating of fuel is enhanced by the addition of substances that help prevent knocking in engines.
14. Synthetic fibers, plastics, and elastomers are all made by addition or condensation polymerization reactions.
15. Elastomers can absorb energy by deforming, and release energy by resuming the original shape.

30.2 Biochemistry

16. Proteins, carbohydrates, lipids, and nucleic acids are important compounds in living organisms.
17. Proteins are polymers of amino acids. They act as structural parts of the organism and as catalysts called enzymes.
18. Carbohydrates are the materials utilized by cells to produce energy.
19. Polysaccharides are polymers of simple sugar molecules. Starch and glycogen act as energy reserves.
20. Lipids are more soluble in nonpolar solvents than in water.
21. The roles of nucleic acids are protein synthesis and the transfer of genetic information from one generation to the next.
22. Nucleic acids are polymers of nucleotides, each of which consists of a nitrogen base, a sugar, and a phosphate group.
23. Polymers and other materials that imitate biological materials are currently being used in medicine.

Key Terms

substitution reaction
addition reaction
elimination reaction
esterification
saponification
polymer
addition polymerization
condensation polymerization
octane rating
elastomer
biochemistry
enzyme
protein
amino acid
peptide bond
carbohydrate
lipid
fat
vitamin
nucleic acid
nucleotide

Review and Practice

21. Describe the two kinds of substitution reactions.
22. What takes place in addition reactions?
23. Describe the process of esterification. What kinds of substances are involved? What combination of atoms is characteristic of an ester?
24. What is the chief source of organic compounds? What are three other sources?
25. What are the two most abundant types of hydrocarbons in petroleum?
26. Why must the products produced by fractional distillation of petroleum be further refined?
27. Complete and balance the following equations. To the left of each balanced equation, place an A or an S to indicate whether the reaction is an addition or a substitution reaction.
    a. $CH_3CH_2CH{=}CH_2 + H_2SO_4 \rightarrow$
    b. $CH_3CH_2CH_2Br + NH_3 \rightarrow$
    c. (benzene ring) $+ Br_2 \rightarrow$
    d. $CH_3CH{=}CH_2 + Cl_2 \rightarrow$
    e. $CH_3OH + HCl \rightarrow$
    f. $CH_3CH_3 + Cl_2 \rightarrow$
    g. $CH_3CH{=}CHCH_3 + HI \rightarrow$
28. What occurs in an elimination reaction?
29. What occurs in the process of saponification? How does saponification differ from hydrolysis using only water?
30. Write balanced equations for esterification reactions involving each of the following. Use condensed structural formulas.
    a. propanoic acid and ethanol
    b. butanoic acid and 2-propanol
31. An ester is named first for the alcohol group and then for the acid group that formed the ester. Name the esters formed in Problem 30.
32. What are polymers? How are polymers formed?
33. What are monomers?
34. What two alkanes serve as reference numbers for the octane scale?
35. Give examples of plastics formed by each of the following.
    a. addition polymerization
    b. condensation polymerization
36. Draw the complete structure of the molecule formed by the first condensation reaction between adipic acid and 1,6-diaminohexane leading to the production of nylon.
37. What is a peptide bond?
38. List the four main types of organic compounds found in living systems, and identify which of these are polymers.
39. List the functions of proteins in the body.
40. Describe the function of enzymes.
41. Describe the structure and chemical makeup of proteins.
42. How can only 20 amino acids form so many different proteins?
43. List three ways in which proteins may differ from each other.
44. What compound is used by the body to transfer energy?
45. What three elements are common to all carbohydrates?
46. Describe the general chemical structure of carbohydrates.
47. Name two common simple sugars.
48. What is a disaccharide? Name a common disaccharide.
49. What is a polysaccharide?
50. List the three classes of compounds found in the human body that are lipids or contain lipids.
51. What role do vitamins play in the cell?
52. Describe the chemical structure of fats.

53. What are nucleic acids? List the components of nucleotides.

54. What characteristic of nucleic acids makes them important in determining heredity?

55. How do DNA and RNA differ in chemical composition?

56. Predict the product of the reaction between ethene and hydrogen iodide.

57. Write an equation for the reaction of cyclopentene with chlorine. Name the product.

58. Write an equation for the saponification of ethyl butanoate in aqueous KOH. Name the products.

59. Using structural formulas, write equations for two successive condensation reactions in the polymerization of 3-hydroxypropanoic acid.

60. Using condensed structural formulas, write balanced equations for each of the following.
   a. the reaction of chlorine and propene
   b. preparation of pentyl ethanoate
   c. elimination of water from 1-butanol
   d. saponification of glyceryl trioctadecanoate with NaOH
   e. the formation of an amide with propanoic acid and 2-aminopropane

61. Write balanced equations for esterification reactions involving each of the following.
   a. propanoic anhydride and ethanol
   b. benzoic acid and 1-propanol
   c. 2,3-dichlorobutanoic acid and 2,2-dimethyl-3-pentanol

62. Predict the structure of the copolymer of ethene and phenylethene.

63. Draw the structural formula for the tripeptide formed from alanine, valine, and tryptophan, in that order. Assume the acid group of alanine reacts.

64. Draw the structural diagram for the polypeptide formed from glycine, cysteine, tyrosine, and lysine, in that order. Assume the acid group of glycine reacts.

65. Write a balanced equation for the formation of the ester phenyl butanoate.

Concept Mastery

66. Why do unsaturated aliphatic compounds undergo addition reactions more readily than do aromatic compounds?

67. Concentrated sulfuric acid is a good dehydrating agent. What is the function of a dehydrating agent?

68. Compare the process of addition polymerization with the process of condensation polymerization. What kinds of molecules are involved in each process?

69. What type of polymerization process would most likely be used to form polymers from each of the following?
   a. $CH_3CH{=}CH_2$
   b. $HOCH_2CH_2COOH$

70. What does octane rating indicate in terms of engine performance? Compare the composition of high octane gasoline to that of low octane gasoline.

71. What is an elastomer? Describe the physical process that occurs at the molecular level when an elastomer deforms and rebounds.

72. What functional group characterizes the peptide bond in all proteins? What type of reaction forms the peptide linkage?

73. Are all polypeptides classed as proteins? Explain your answer.

74. What characteristics do starch, glycogen, and cellulose have in common?

75. Would it be desirable to have no cholesterol in your body? Explain.

76. Nucleic acids control most cell activities. How do they exert this control?

77. What is the principal criterion for materials to be used in artificial organs?

78. What does the term *tissue compatible* mean when it is used in describing biomaterials?

79. In what way is the formation of an amide from the reaction of an amine with an organic acid similar to esterification? In what way is it different?

80. What similarities exist between nylon and protein?

81. Why is it possible to form four different dipeptides using only two amino acids? Illustrate using valine and cysteine.

Application

82. "Polyunsaturated" fats are generally regarded as more acceptable in the diet than saturated fats. What is meant by polyunsaturated fats? In general, what is the typical source of each type of fat?

83. Rayon has been used for many years as a fiber in clothing. What is the composition of rayon? What is the raw material used in the manufacture of rayon? How is the processed material formed into fibers?

84. Cellulose is used in plants for structural support. What are the uses of starch in plants and glycogen in animals?

85. The administration of certain steroids to athletes has been strictly forbidden by athletic associations.
    a. What is a steroid?
    b. Name one steroid that occurs naturally in the body.

86. What is the actual "octane" referred to in octane rating? Draw its structure. How much of this substance must actually be present in gasoline? Explain.

87. The compounds adenine, guanine, cytosine, thymine, and uracil are members of a broad class of organic compounds called heterocyclics. Compare their structures with cyclic hydrocarbons such as benzene and cyclohexane. Look up the meaning of the prefix *hetero-* in a dictionary. From this information, propose a definition for heterocyclic. Explain why the nitrogen bases of DNA fall into this category.

Critical Thinking

88. Why do many important intermediates used in manufacturing organic chemicals have double or triple bonds?

89. In polymerization, why is it necessary to introduce a substance to end a chain?

90. Explain why the petroleum fractions with the lowest boiling points are obtained at the top of a fractional distillation column.

91. Examine the structure on page 778 of the polymer formed by the condensation reaction between formaldehyde and phenol. Would you say that the polymer is more likely to be hard and rigid or soft and flexible? Give reasons.

92. Suggest a chemical method that could break up a condensation polymer.

93. Vitamins are divided into two groups—fat soluble and water soluble. The fat-soluble vitamins are those generally classified as lipids. Look at the structure of vitamin A on page 785. From its structure, explain why it would be fat soluble rather than water soluble.

94. Many organic compounds have two names. One is a formal name that follows the rules of organic naming. The other is an informal common name. An example is acetic acid and ethanoic acid. Although great effort has been put into establishing a uniform system of formal naming, usage of common names continues, especially in industry. Suggest reasons for this continuation. Why would common names continue more in organic chemistry than in inorganic chemistry?

READINGS

Allcock, Harry R. "Chemical Synthesis at the Boundary between Polymer Chemistry and Inorganic Materials." *The Chemist* 67, no.1 (January 1990): 10-16.

Alper, Joseph. "Diesel Under Pressure." *ChemMatters* 9, no. 1 (February 1991): 12-14.

Baden, Daniel G. "Toxic Fish: Why They Make Us Sick." *Sea Frontiers* 36, no. 3 (May-June 1990): 8-15.

Baugh, Mark. "Oil Changes." *ChemMatters* 7, no. 4 (December 1989): 7-9.

Benson, Kim D. "Fast Fat." *ChemMatters* 8, no. 1 (February 1990): 13-15.

Borman, Stu. "New High-Strength Fiber Finding Innovative Uses in Protective Clothing." *Chemical and Engineering News* 67 (October 8, 1989): 23-24.

Kanatzidis, Mercouri G. "Conductive Polymers." *Chemical and Engineering News* 68, no. 49 (December 3, 1990): 36-54.

Kyle, Linda Davis. "Contact Lenses." *ChemMatters* 9, no. 2 (April 1991): 7-11.

Skinner, Karen J. "The Chemistry of Learning and Memory." *Chemical and Engineering News* 69, no. 40 (October 7, 1991): 24-41.

Smith, Trevor. "Distance Running." *ChemMatters* 7, no. 1 (February 1989): 12-15.

Stinson, Stephen C. "Total Synthesis of Huge Polytoxin Molecule Achieved at Harvard." *Chemical and Engineering News* 67 (September 18, 1989): 23-24.

Thayer, Ann M. "Solid Waste Concerns Spur Plastic Recycling Efforts." *Chemical and Engineering News* 67, no. 5 (January 30, 1989): 7-15.

Thornton, John I. "DNA Profiling." *Chemical and Engineering News* 67 (November 20, 1989): 18-30.

Weiss, Rick. "Organic Origami." *Science News* 132, no. 22 (November 28, 1987): 344-346.

West, Robert. "Two Myths of Organosilicon Chemistry." *The Chemist* 66, no. 1 (January 1989): 10-11, 21, 23.

Witzel, Eric. "Poison Ivy." *ChemMatters* 8, no. 3 (October 1990): 4-5.

Zubay, Geoffrey. *Biochemistry*. New York: Macmillan Publishing Co., 1988.

Cumulative Review

1. Balance the following equation for a redox reaction in acidic solution.
$Bi + NO_3^- + H^+ \rightarrow Bi^{3+} + NO + H_2O$

2. What voltage should be generated by the following cell under standard conditions?

$$Co|Co^{2+}||O_2|H_2O_2$$

3. What products would you expect from the electrolysis of an aqueous solution of iron(II) sulfate?

4. What voltage would you expect from the following cell at standard temperature and pressure? Assume $[H^+] = 1.00M$.
$Cu|Cu^+||MnO_4^-|Mn^{2+}(0.0100M)$

5. What is the free energy change for a reaction if it shows a potential difference in a cell of 1.94 volts and $1e^-$ is transferred?

6. What mass of bromine will be produced by a current of 2.60 amperes flowing for 2.17 hours through molten potassium bromide?

7. What is the amount of $^{100}_{46}Pd$ left at the end of 16 days if 270 g existed originally and the half-life is 4.0 days?

8. Complete the following nuclear equation.

$$^{129}_{55}Cs + {}^{0}_{-1}e \rightarrow ?$$

What kind of nuclear decay does this equation represent?

9. Write structural formulas for the following compounds.

a. 1-bromo-1-chloropropene
b. 1,3,5-octatriene
c. 3-chlorocyclopropene
d. 1-bromo-1,2-diphenylbutane

A Science Experiment

The keys to science are observation and measurement, which are often used together in experiments. Experiments are carried out to test hypotheses that attempt to explain the world around us. Also, experiments can lead to new hypotheses.

Chemistry experiments are often designed to gather information about what cannot be directly observed. The purpose of this activity is to demonstrate how an experiment can provide information about something that cannot be seen.

Objectives

- **Record data** for repeated trials of an experiment.
- **Determine** the unknown size of an object without direct measurement.

Materials

spheres, 7 per lab group	meter stick
	masking tape

Procedure

1. Use masking tape to make a line approximately 60 cm long across your lab table as shown in the diagram. Measure and record the exact length of the line in a data table like the one shown. The line will be called your field of action and should not include any faucets, power boxes, or gas jets.
2. Place 6 spheres along your marked line across the width of the table.
3. Place the remaining sphere about one meter away from the lined up spheres.
4. Without looking, one partner will roll the single sphere toward the line of spheres. The other partner will note on a separate sheet of paper if there is a hit or miss. This partner will replace a hit sphere back into position and return the bombarding sphere to the other partner for another trial. Continue for 100 trials.
5. Switch partners and repeat step 4.
6. Total the number of trials and the hits for you and your partner. Record these totals in your data table.
7. Collect the total numbers of hits and trials from each lab group, sum them up with your own totals, and record them.
8. Measure the diameter of the spheres by placing a sphere against a meter stick.

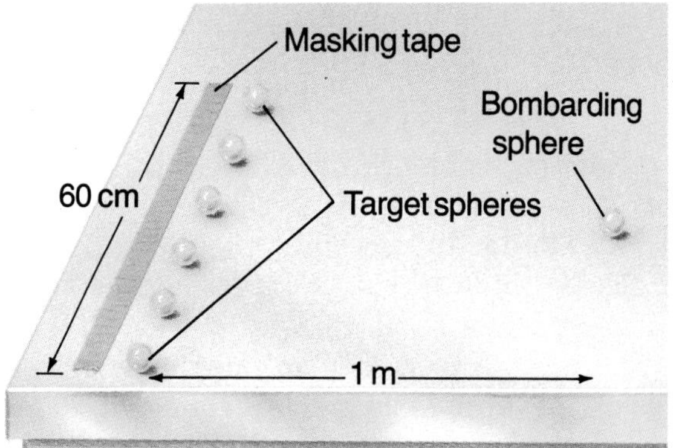

Data and Observations

Observations	Your Data	Class Data
A. Width of field (cm)		—
B. Number of target spheres		—
C. Total number of hits		
D. Total number of trials		
E. Calculated sphere diameter (cm)		—
F. Actual diameter of sphere (measured) (cm)		—

Analysis and Conclusions

1. Were there more hits or misses in your own trials?

2. On what does the hit/miss ratio depend?
3. The calculated diameter of the spheres is determined by using the formula:

$$\text{Diameter} = \frac{\text{Field width (A)} \times \text{Hits (C)}}{2 \times \text{\# of target spheres (B)} \times \text{Trials (D)}}$$

   Calculate the sphere diameter from your own numbers of hits and trials. Record the result.
4. Calculate the sphere diameter from the total hits and trials for the whole class. Record the result.
5. Were the individual or the class average results closest to the actual value for the diameter of the sphere?
6. What would have happened to your data if the size of the spheres were twice as large? Half as large?
7. How does the number of trials affect the results?

Extension and Application

Investigate the gold foil experiment carried out by Hans Geiger and Ernest Marsden. How was that experiment similar to this one? What conclusion was drawn from the Geiger and Marsden experiment?

Chapter 2 ChemActivity 1

Micro

A Density Balance

Have you ever played on a teeter-totter or seesaw? If two children want to use one together, how can they do it if one child is much heavier than the other? Both children take up the same space, but one child's mass is much greater than that of the other. The heavier child must sit closer to the center of the board in order for the children to play. The lighter one child is, the closer to the center the heavier child must sit. The same relation between mass and distance can be used to determine density.

The density of water is 1.0 g/mL. Most liquids have densities less than that of water, and no other common liquid has a density of 1.0 g/mL. Therefore, even if samples of water and another liquid have the same volume, they have different masses. The purpose of this activity is to use a balance to compare the densities of some liquids.

Objectives

- **Compare** the densities of several liquids with that of water.
- **Calculate** the densities of the liquids.

Materials

- ruler with center channel
- thin stem pipets (2)
- ethyl alcohol
- n-butyl alcohol
- isopropyl alcohol
- pencil
- water
- cooking oil
- ethyl acetate
- oleic acid

Procedure

1. Set up a density balance by placing the ruler on the pencil at the 6 inch (15.25 cm) mark of the ruler as shown in Figure A. Move the ruler back and forth slightly until it is balanced. Samples of water and another liquid will be placed on opposite ends of the ruler. **Hypothesize** which of the two samples will have to be moved to restore balance and which way it will have to be moved.
2. Place a pipet filled with water on one end of the ruler. Be sure the pipet is filled with liquid all the way into the stem.
3. On the other end of the ruler, place a pipet filled all the way to the stem with a different liquid. The volumes of the two liquids should be the same.

4. Move the pipet containing the water toward the pencil until the ruler is balanced.
5. Measure the distances of the water and liquid pipets, as indicated by A and B in Figure B, from the center of the ruler. Enter these data in a table like the one shown.

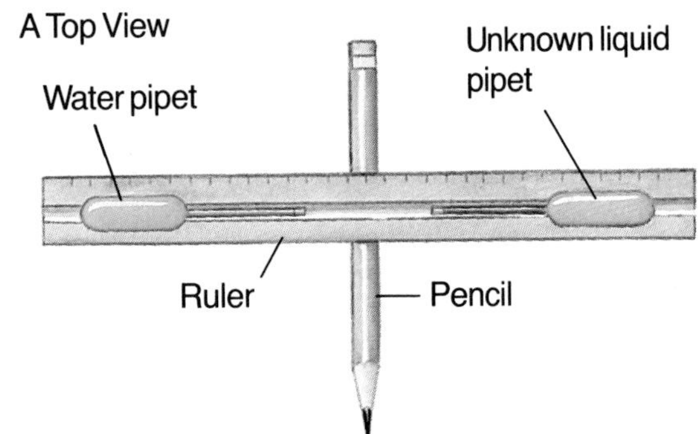

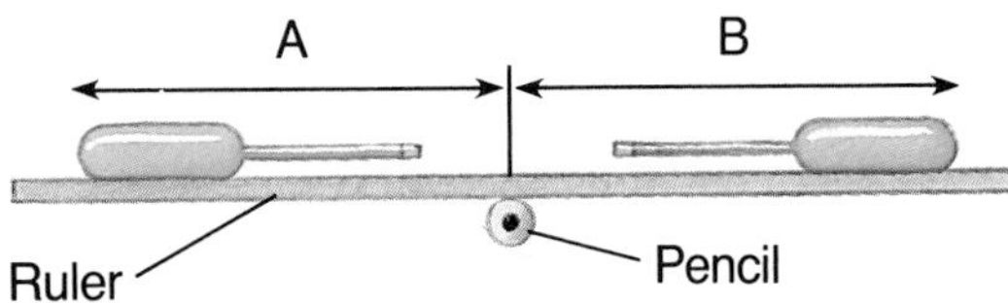

Data and Observations

Liquid	Distance		Density
	A	B	
Ethyl alcohol			
n-butyl alcohol			
Isopropyl alcohol			
Cooking oil			
Ethyl acetate			
Oleic acid			

Analysis and Conclusions

1. The ratio of distance A to distance B equals the ratio of the density of the unknown liquid to the density of water.

$$\frac{A}{B} = \frac{\text{density of unknown liquid}}{\text{density of water}}$$

This relation is used to find the density of the liquid. For example, if the unknown liquid in pipet B is 15.5 cm from the center of the pencil, the water pipet is 12.5 cm from the center of the pencil, and because the density of water is 1.0 g/mL, the density of unknown B can be calculated as

$$\frac{12.5 \text{ cm (A)}}{15.5 \text{ cm (B)}} \times 1.0 \text{ g/mL} = 0.80 \text{ g/mL}$$

Calculate the density of each liquid in this way and enter your results in your data table.

2. Arrange the liquids in a list in the order of increasing density. Compare your results with a list of actual densities.
3. Write the equation for density and describe what it means in your own words.
4. Which factor in the density equation is held constant in this experiment? How can you verify that this is constant? What did you do in the experiment to be sure this factor was constant?

Extension and Application

1. How is the density balance similar to the playground see-saw or teeter-totter?
2. What was the percent error of your results? Percent error is calculated by using the following formula:

$$\% \text{ error} = \frac{\text{your density} - \text{correct density}}{\text{correct density}}$$

3. What mistakes could increase your percent error?

A Graphical Determination of Density

Density is mass per unit volume. The mass of any size sample of a material divided by the volume of that sample will give the density of that material. A good scientist would not be satisfied, however, with a value based on one set of measurements. A more accurate value is found by averaging the results of several experiments.

Another approach is the use of a graphical method to find an average value. The purpose of this experiment is to determine the density of metal shot by a graphical method.

Objectives

- **Use** significant digits in making measurements and in calculations.
- **Graph** a series of laboratory measurements.
- **Determine** density from a mass vs. volume graph.

Materials

different-sized samples of the same metal shot (BBs), 4 samples
100-mL graduated cylinder
30-mL beaker
balance
water

Procedure

Work in teams of four. Each student should carry out the following procedure for a different-sized sample of the metal shot.

1. Mass the beaker to the nearest 0.01 g. Record the mass in a data table like the one shown.
2. Place your shot in the beaker and mass them together. Record this mass.
3. Subtract the mass of the beaker from the mass of the beaker + shot to obtain the mass of the shot sample. Record this mass.
4. Half fill the 100-mL graduated cylinder with tap water. Determine and record this volume of water to the nearest 0.5 mL.
5. Slowly pour the metal shot into the water in the graduated cylinder. Record the new volume to the nearest 0.5 mL.
6. Slowly pour off the water into the sink. Place the wet metal shot between two layers of paper towels and dry the shot before returning it to your teacher.
7. Share your data with your lab team and record their data in the other columns of your data table.

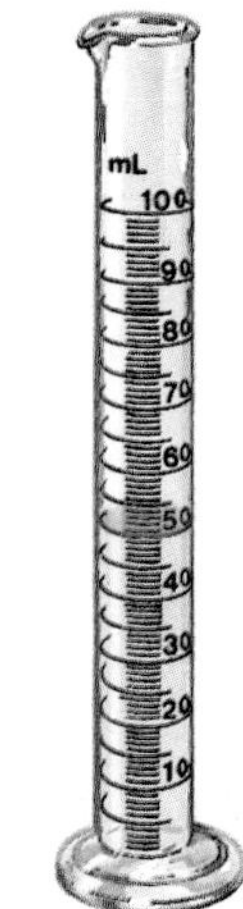

Data and Observations

	Trial			
	1	2	3	4
Mass of beaker + shot				
Mass of empty beaker				
Mass of metal shot (y)				
Total volume of water + metal shot				
Initial volume of water in cylinder				
Volume of metal shot (x)				

Analysis and Conclusions

1. For each trial, subtract the initial volume of water from the total volume of water + shot. Record the resulting values for the volume of the shot in cm^3.
2. Using graph paper, plot the mass of the metal shot on the y-axis and the volume of the metal shot on the x-axis. Locate and mark four points. Draw a straight line that passes through as many points as possible. Points may fall on either side of the line due to experimental error.
3. Determine the x and y values of two points on the line. Calculate the slope, m, of the line by using the following equation:

$$m = \frac{(y_2 - y_1)}{(x_2 - x_1)}$$

   The slope is equal to the average density for all four trials. Record this average density.
4. Determine and record the density from the mass and volume found in your own trial.

$$\text{Density} = \frac{\text{mass of metal shot}}{\text{volume of metal shot}}$$

5. How does the density from your own trial compare with the average density found from the slope of the plot?

Extension and Application

1. How many significant digits did you record in the density calculated for your trial? Explain how you decided on this number of digits.
2. The percent error for each trial is the absolute value of the difference between the density from the trial and the average density, divided by the average density. Calculate your percent error.

$$\%\text{ error} = \frac{\text{density from trial} - \text{average density}}{\text{average density}} \times 100$$

3. How could the experiment be changed to decrease the experimental error?
4. Suppose you wanted to use a graphical method to find the number of miles your car can travel on a gallon of gasoline. Briefly describe a procedure you could use.

Chapter 3 ChemActivity

Properties of Chalk and Baking Soda

Every kind of matter in our world is a mixture, an element, or a chemical compound. Each of these kinds of matter has its own set of physical and chemical properties by which it can be identified.

A chemical change is a change in the nature of a substance. When a chemical compound undergoes a reaction, it changes in identity. The physical properties of the reacting compound are replaced by the physical properties of the product. For this reason, observing a physical change often indicates that a chemical change has occurred.

Some physical changes that signal a chemical change are a color change, formation of bubbles, a change in crystal form, disappearance of a solid, or formation of a solid where there was none. The purpose of this activity is to observe physical changes of two everyday substances and relate them to chemical changes.

Objectives

- **Observe** physical changes when chalk and baking soda are mixed with water or hydrochloric acid.
- **Examine** changes in the form of crystals produced from these mixtures.
- **Infer** if chemical reactions have occurred.

Materials

- glass slides (3 per group)
- hand lens
- grease pencil or crayon
- hot plate
- test-tube holder
- toothpick
- piece of chalk (1 per group)
- baking soda
- distilled water
- 3*M* hydrochloric acid (HCl)

Procedure

1. From your knowledge of substances, **hypothesize** whether or not a physical change in a substance *always* signals that a chemical change has occurred.
2. Obtain a glass slide and with a grease pencil or crayon draw a line dividing the slide into two halves. Mark one side H_2O (water) and the other HCl (hydrochloric acid), as in Figure A.
3. Using another glass slide, scrape equal quantities of chalk (calcium carbonate) onto both sides of the slide, Figure B.

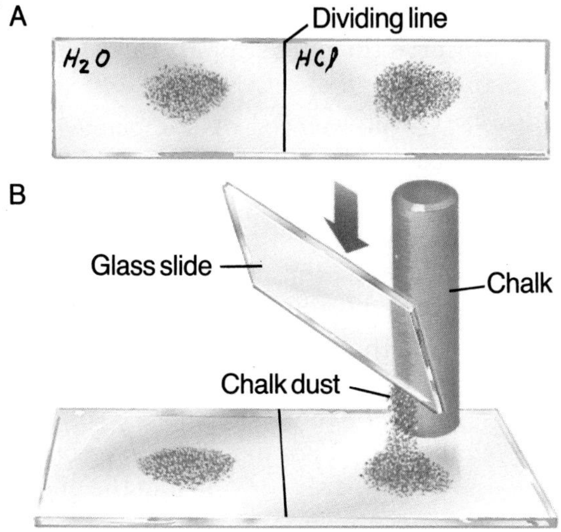

4. Examine the crystal structure of chalk with the hand lens and record its appearance.
5. Add 10 drops of water to the powdered chalk on the side marked H_2O. Stir the water and chalk together with a toothpick. Record any physical changes in a data table like the one shown.
6. Add 10 drops of hydrochloric acid to the powdered chalk on the other side of the slide. Stir the hydrochloric acid and chalk together with the same toothpick. Record any physical changes that you observe.
7. Evaporate the liquid on the slide by placing the slide on a hot plate set on low heat.
8. Use a test-tube holder to remove the slide from the heat when the liquid has evaporated. Allow the slide to cool by placing it on the base of the ring stand or on a separate piece of wire gauze.
9. Use the hand lens to observe the crystal structure of the two solids on the slide and record their appearances.
10. Using another slide, repeat steps 2-9 with baking soda (sodium hydrogen carbonate) instead of chalk. Use a chemical scoop to place the solid baking soda on the slide.

Data and Observations

	With Water	With HCl
Chalk on mixing		
after evaporation		
Baking soda on mixing		
after evaporation		

Analysis and Conclusions

1. How did the samples of chalk change?
2. How did the samples of baking soda change?
3. List the four mixtures and classify each as resulting in a physical change or a chemical change. Was your hypothesis correct? Explain.
4. Which observations proved when a chemical change occurred and when it did not?

5. What can you infer from your observations of the substances remaining after hydrochloric acid was mixed with chalk and baking soda?

Extension and Application

1. What is the difference between a chemical and a physical change?
2. Does fizzing always indicate the occurrence of a chemical change? Explain why or why not.
3. Consider the water cycle that you studied in biology or Earth science. The water cycle is the process in which water moves from the surface of Earth into the atmosphere and back again. Identify any physical and chemical changes of water in this cycle.
4. List the physical and chemical changes that occur when gasoline enters an automobile engine and burns.

Chapter 4 ChemActivity

Atomic Spectra

Each element in the Periodic Table has a specific number of electrons in specific energy levels. When atoms of any element absorb energy, electrons move from lower to higher energy levels. As these electrons fall back into their original energy levels, the excess energy is released as electromagnetic radiation.

The characteristic patterns of radiation for different elements are called atomic spectra. For some elements, this radiation is in the visible region of the spectrum. These spectra can be seen when compounds containing the elements are heated in a flame. The purpose of this activity is to compare the atomic spectra of different elements.

Objectives

- **Observe** the characteristic atomic emission spectra of some elements.
- **Diagram** the observed spectra.
- **Analyze** a mixture from its spectrum.

Materials

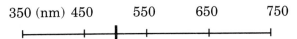

laboratory burner
hand spectroscope
plastic bottles containing anhydrous carbonates of copper, lithium, and sodium
plastic bottle containing an unknown mixture of two of the above salts

Procedure

1. Form a **hypothesis** about the appearance of spectra of three different compounds.
2. Work in pairs. Place a laboratory burner in a fume hood. Light and adjust the flame of the burner until the flame is all blue.
3. Observe the burner flame through the slit in the hand spectroscope.
4. Shake the covered plastic bottle containing copper carbonate. **CAUTION:** *Copper carbonate is toxic if ingested.*
5. Loosen the top of the bottle, hold the bottle close to the collar (air inlet) of the burner, and remove the lid.
6. Record the color of the flame. Then, observe the flame with the hand spectroscope and sketch the specific color bands that you see. For example, to record a line at 500 nm, draw a scale like the one below and place a vertical line at 500 nm:

   350 (nm) 450 550 650 750

   If your spectroscope does not have a built-in scale, compare your observed bands with those on a standard spectral chart like the one below to estimate wavelengths.
7. Repeat steps 4-6 with the plastic bottles

containing lithium carbonate and sodium carbonate. **CAUTION:** *Sodium carbonate can irritate the skin. Lithium carbonate dissolves in water to form a strong base.*

8. Repeat steps 4-6 with your unknown mixture of salts.

Data and Observations

Element:	Copper	Sodium	Lithium
Flame Color:			

350 (nm) 450 550 650 750

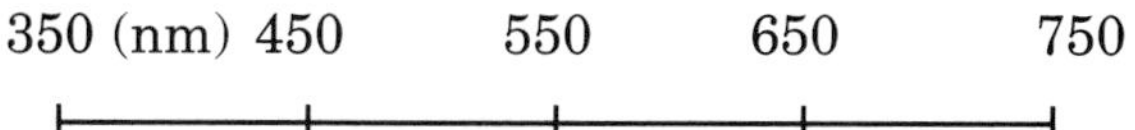

Analysis and Conclusions

1. Compare the bright lines you observed for copper, sodium, and lithium with the lines you recorded for your mixture. Identify which elements are in your mixture.
2. Were any of your spectra identical? Similar? Was your hypothesis correct?
3. Why did each element give a different observed flame color?
4. How can you identify the composition of a mixture of any two salts?

Extension and Application

1. Distance between the observer and the glowing element does not play a major factor in the examination of spectra. How could atomic spectra be useful to an astronomer observing stars?
2. As the concentration of an element in a mixture increases, the intensity of the color of that element's lines in the spectrum increases. Investigate the principle of an Atomic Absorption Spectrometer. How does this laboratory instrument make it possible to determine the concentrations in mixtures?

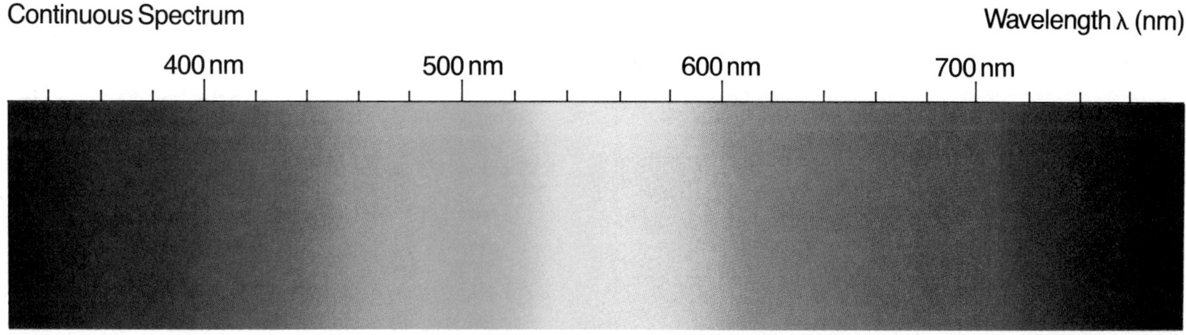

Chapter 5
ChemActivity

Felt-tip Electron Distribution

The position of an electron in an atom at a given moment cannot be predicted. The region of space in which the electron can probably be found is, however, predictable. This region is often called an "electron cloud" and is represented by a fuzzy shape with the nucleus at the center. The shape is determined by mathematical calculations using the wave-mechanical model of the atom. The electron cloud is not absolute. Infrequently, the electron may be found outside the cloud.

The purpose of this activity is to demonstrate a probable distribution of locations around a central point.

Objectives

- **Graph** the distribution of marks on a target.
- **Observe** the region of highest probability of marks.

Materials

fine-point felt-tip marker
paper target

Procedure

1. Place your target on a notebook on the floor.
2. Drop your marker from a height of about 1 m onto the target so that it makes a mark. Try to hit the center. Repeat 100 times.
3. Count the number of marks in each numbered region of the target and record the numbers in a data table like the one shown.

Data and Observations

Area	Number of marks
1	
2	
3	
4	
5	

Analysis and Conclusions

1. Plot your results on graph paper, with the number of marks on the *y* axis and the numbered area on the *x* axis.
2. Determine from your graph which target area had the highest probability of a hit.
3. How does the shape of your graph compare with the shape of the graph for the probability of finding an electron in a hydrogen atom as in Figure 5.2, page 113.

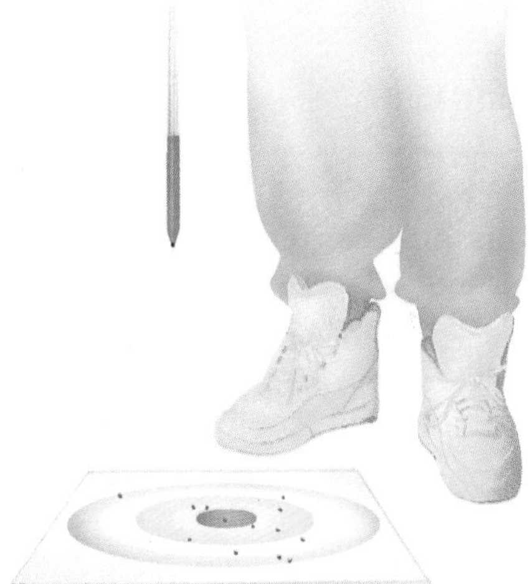

Extension and Application

1. How are orbital pictures similar to your target? How are they different?
2. Explain why a photo of the blur of a fast-moving fan blade is a better analogue to an electron cloud picture than your target with its marks.

Chapter 6
ChemActivity

An Alien Periodic Table

Your ship has arrived at a planet at the edge of the galaxy. You have discovered what appears to be an ancient laboratory containing many artifacts of scientific investigations. Among the artifacts is information on what was obviously a periodic table of the elements.

Your ship's computer has been able to decipher the numerical values of each alien element's boiling point and melting point and a description of one of the elements that is labelled Σ. Fortunately, the compounds formed by this one element with the other elements have also been indicated in the alien table. Your mission is to arrange the elements in a table similar to your own planet's periodic table, keeping elements with similar properties in the same column.

Objectives

- **Use information** about a group of elements to arrange them into correct periods and groups.
- **Relate** a model periodic table to the actual periodic table.

Materials

copy of alien table of information for a group of elements
scissors

Alien Elements

œ (g) –157°C* –153°C**	Σ-Σ Σ green-yellow gas –101°C –34°C	$-\Sigma_2$ ‡ (cr) 650°C 1105°C	¥ (g) –112°C –108°C	π (g) –248°C –246°C	$-\Sigma_2$ å (cr) 776°C 1412°C
$-\Sigma_2$ ß (cr) 841°C 1500°C	$-\Sigma_4$ ∂ (cr) 3600°C 4200°C	*f* (g) –189°C –186°C	• (g) –270°C –268°C	$-\Sigma_3$, $-\Sigma_5$ Δ (cr) 44°C 280°C	$-\Sigma$ æ (cr) 40°C 697°C
$-\Sigma$ Ω (cr) 63°C 766°C	$-\Sigma$ ≈ (cr) 113°C 184°C	$-\Sigma$ ç (g) –219°C –188°C	$-\Sigma_3$, $-\Sigma_5$ √ (cr) 630°C 1635°C	$-\Sigma$ ∫ (cr) 98°C 897°C	$-\Sigma_3$, $-\Sigma_5$ µ (g) -210°C -195°C
$-\Sigma_3$ ≤ (cr) 2080°C 3865°C	$-\Sigma_2$, $-\Sigma_4$ ≥ (cr) 232°C 2623°C	$-\Sigma$ 8 (cr) 180°C 1347°C	$-\Sigma_2$ * (g) –218°C –183°C	$-\Sigma_2$, $-\Sigma_4$ ● (cr) 221°C 685°C	$-\Sigma$ ((l) –7°C 59°C
$-\Sigma_2$, $-\Sigma_4$ ¡ (cr) 945°C 2850°C	$-\Sigma_3$, $-\Sigma_5$ ¢ (cr) 816°C 615°C	$-\Sigma$ ∞ (g) –259°C –252°C	$-\Sigma_3$ § (cr) 660°C 2517°C	$-\Sigma_2$, $-\Sigma_4$ ¶ (cr) 450°C 990°C	$-\Sigma$, $-\Sigma_3$ ª (l) 30°C 2203°C
$-\Sigma$, $-\Sigma_3$ Å (cr) 156°C 2080°C	$-\Sigma_2$ £ (cr) 1287°C 2468°C	$-\Sigma_2$, $-\Sigma_4$ ± (cr) 115°C 444°C	$-\Sigma_4$ ¿ (cr) 1411°C 3231°C		

* The top number gives the melting point.
** The bottom number gives the boiling point.

Procedure

1. Study the information on your table of alien elements. Notice that the element Σ is a green-yellow gas. The number of atoms of element Σ that combine with each of the unknown elements is shown at the top left corner of each square. For example, $-\Sigma_2$ means that the unknown element combines with two atoms of Σ.
2. Cut out the box for each element and arrange the boxes into a periodic pattern like the one shown. Take into account that in families of metals, melting points and boiling points generally decrease as elements get heavier, but the opposite is true for families of nonmetals. Also note that the way elements combine with other elements tends to vary with regularity across the periodic table. Another important factor to notice is the state of each element.

Data and Observations

Periodic pattern:

Periodic Pattern

Analysis and Conclusions

1. Explain how you decided the order of the families and elements.
2. Which is the lightest element on the alien periodic table? Which is the heaviest?
3. How are the methods you used similar to or different from the methods employed by Dmitri Mendeleev?

Extension and Application

Mendeleev was able to predict the existence and properties of ekasilicon from his knowledge of the periodic table.

Chemists today use periodic relationships to predict the properties of elements heavier than element 103. What elements give clues to the properties of elements 104 and 118?

Formulas and Oxidation Numbers

Oxidation numbers and the charges of ions give the information needed to write the formulas of many chemical compounds. Only a few guidelines are needed:

1. In a neutral compound, the charges on ions, or the oxidation numbers, balance out to zero.
2. One positive charge balances one negative charge.
3. Atoms with positive charges or positive oxidation numbers are written first.
4. Subscripts show the relative numbers of atoms or ions in a compound.
5. To show more than one of a polyatomic ion, the symbol is enclosed in parentheses and the subscript follows, for example, $Al_2(SO_4)_3$.

The purpose of this activity is to use paper models to show how chemical formulas are derived from oxidation numbers.

Ion Models

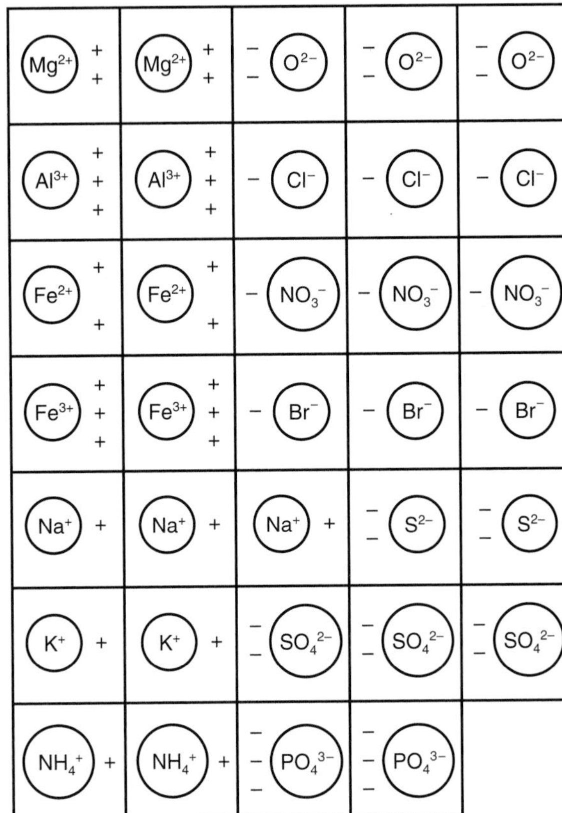

Objectives

- **Write** formulas of chemical compounds.
- **Name** chemical compounds.

Materials

scissors, paper, pencil, sheet of ion models

Procedure

1. Cut out each of the ion squares on your sheet of ion models.
2. Assemble the ions for a compound containing magnesium and chlorine. To do this, place the Mg^{2+} ion on a piece of paper. Place enough Cl^- ions alongside the Mg^{2+} ion to balance the charges.
3. Record the formula and name of the compound of magnesium and chlorine in a data table like the one shown.
4. Assemble the ions for five compounds from the list below and record their formulas and names in your data table. Use the rules listed in the introduction and in your textbook for writing formulas and naming compounds. Aluminum and bromine; sodium and oxygen; iron(II) and sulfur; aluminum and nitrate ion; potassium and sulfate ion; iron(III) and chlorine; ammonium ion and sulfur; aluminum and oxygen; iron(III) and sulfate ion; sodium and phosphate ion.

Data and Observations

Combining Substances	Chemical Formula	Name of Compound
magnesium and chlorine	$MgCl_2$	magnesium chloride

Analysis and Conclusions

1. Some compounds are described as "binary compounds." What does this mean? List the formulas and names of any binary compounds you have constructed.
2. Which element on your list forms ions in two different oxidation states?
3. Parentheses must be used to show more than one of a polyatomic ion. List the formulas of any compounds on your list where this was necessary.
4. Should the formulas you have written be described as empirical formulas or molecular formulas?

Extension and Application

1. Some elements have more than one oxidation number. To show the oxidation number of such elements in a compound, either a prefix or a Roman numeral is given in the name of the compound. Give both names for the following compounds:
   a. UF_5 **c.** $SeCl_2$
   b. UF_6 **d.** Se_2Cl_2
2. Manganese has oxidation number 4+ in a number of compounds. Write the formulas and names of compounds of 4+ manganese with oxygen and bromine.
3. Which of the following are molecular formulas?
   H_2O $C_4H_8O_2$ NaBr $MnSO_4$ CH_4
4. Hydrogen peroxide and water both contain the same two elements. Using reference materials, write the chemical formulas for these two substances and describe their properties and uses.

Chapter 8 ChemActivity 1

Nuts and Bolts Chemistry

Atoms of different elements combine in simple whole number ratios to form chemical compounds. The same two elements may form several different compounds by combining in different ratios.

Chemical formulas such as H_2O show by subscripts the number of atoms of each element in a compound. The simplest whole number ratios by which elements combine are given in what is called the empirical formula of a compound. The actual number of atoms of each element in the compound is given in molecular formulas. The purpose of this activity is to use a physical model to illustrate the meaning of chemical formulas.

Objectives

- Use nuts and bolts to **model** combinations of atoms to form compounds.
- **Determine** the percent composition of "compounds" made of nuts and bolts.
- **Demonstrate** how empirical formulas can be determined from the percent composition.

Materials

nuts and bolts (10 per group)
balance

Procedure

1. Mass an individual nut and bolt separately and enter the two masses in a data table like the one shown.
2. Combine a single nut and bolt to form the compound BtNt. Determine the mass of the nut and bolt combination. Enter the mass in your data table.
3. Combine two nuts and one bolt to form the compound $BtNt_2$. Mass this combination and enter the mass in the data table.
4. Combine two bolts with one nut to form the compound Bt_2Nt. Mass this combination and enter the mass in the data table.

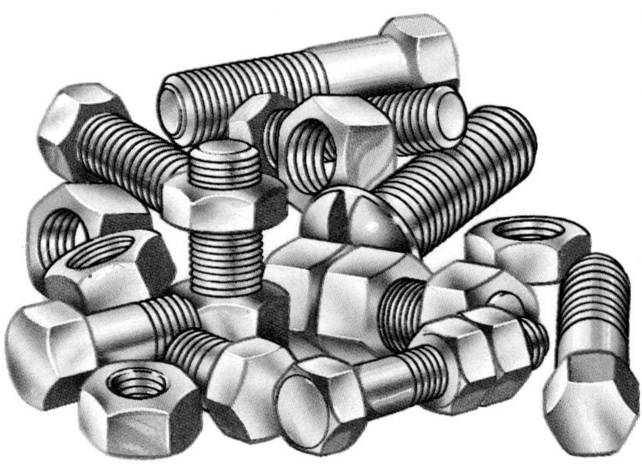

5. Make any combination of nuts and bolts you wish. Mass this combination and enter the mass in your data table.

Data and Observations

A. Mass of bolt	___ g
B. Mass of nut	___ g
C. Mass of compound BtNt	___ g
D. Mass of compound $BtNt_2$	___ g
E. Mass of compound Bt_2Nt	___ g
F. Mass of your compound Bt_xNt_y	___ g

Analysis and Conclusions

1. Find the percent Nt and Bt in each of your nut and bolt compounds by using the formulas below. Enter the results in a table like the one shown.

$$\% \text{ bolts} = \frac{\text{mass of bolts}}{\text{total mass of compound}}$$

$$\% \text{ nuts} = \frac{\text{mass of nuts}}{\text{total mass of compound}}$$

	BtNt	$BtNt_2$	Bt_2Nt	Bt_xNt_y
% Bt				
% Nt				

2. For BtNt, carry out the calculations below:

$$\frac{\% \text{ bolts}}{\text{mass of one bolt}} \qquad \frac{\% \text{ nuts}}{\text{mass of one nut}}$$

Divide the larger of these two calculated numbers by the smaller one. The result is the ratio of bolts to nuts in the empirical formula. Write the empirical formula.

3. Repeat Analysis step 2 for each of your other compounds. Remember that the empirical formula must be the simplest whole number ratio. If your ratios are not whole numbers, they must be converted to whole numbers (e.g., $Bt_{0.5}Nt_{1.0}$ becomes $BtNt_2$) or rounded to whole numbers (e.g., $Bt_{0.9}Nt_{1.1}$ becomes BtNt). Write the empirical formulas found from your data.

Extension and Application

1. Using your measured masses, calculate the percent composition of Bt_2Nt_2. Compare it with the percent composition of BtNt that you calculated. Assuming you can't see the combined nuts and bolts, what information could you use to decide whether your invisible compound is BtNt or Bt_2Nt_2?
2. Obtain the percent composition of the Bt_xNt_y compound of another lab group, calculate the empirical formula of their Bt_xNt_y, and check to see if your answer agrees with their compound.
3. The atomic masses of the known elements can be found in the Periodic Table of the Elements. Determine the empirical formulas for the following compounds:

   a. H = 6.0%, S = 94%
   b. Na = 21.7%, Cl = 33%, O = 45%

4. What is the empirical formula for a compound that has the analysis below?

   2.3 g = Na
   1.2 g = C
   1.6 g = O
   ___
   5.1 g total sample

Aluminum Foil Thickness

How thick is aluminum foil in centimeters? How many atoms thick is this? The small size of any one atom gives a clue to the relatively large number of atoms in a sample of matter that we can pick up and measure. The purpose of this activity is to relate the size of an aluminum atom to the thickness of a piece of aluminum foil.

Objectives

- **Determine** the density of aluminum.
- **Compute** the thickness of aluminum foil.
- **Use** the factor-label method correctly.

Materials

string	250-mL graduated cylinder
scissors	aluminum block
water	aluminum foil
metric ruler	

Procedure

1. **Form a hypothesis** as to what directly observable data are necessary to compute the thickness of a piece of aluminum foil in centimeters.
2. Mass the block or cylinder of aluminum metal. Record the mass in grams in a data table like the one shown.
3. Fill a 250-mL graduated cylinder with water to the 100-mL mark.
4. Tie a string to the aluminum block or cylinder and lower it into the water in the graduated cylinder until it is completely immersed. Read and record the new volume of water. Compute and record the volume of the aluminum.
5. Remove the aluminum block from the water, dry it, and return it to your teacher.
6. Cut a piece of aluminum foil approximately 12 cm × 12 cm.
7. Measure and record the length and width of the piece of aluminum foil to the nearest 0.1 cm.
8. Mass the piece of aluminum foil and record your result.

Data and Observations

A. Mass of piece of aluminum (g)	
B. Volume of water in cylinder	100 mL
C. Volume of water plus aluminum (mL)	
D. Volume of piece of aluminum (mL) (C – B)	
E. Length (l) of aluminum foil (cm)	
F. Width (w) of aluminum foil (cm)	
G. Mass of aluminum foil (g)	
H. Density of aluminum (A/C) (g/cm^3)	
I. Volume of foil (cm^3) (G/H)	
J. Height (thickness) of foil (cm)	
K. Moles of aluminum in foil	
L. Atoms of aluminum in foil	

Analysis and Conclusions

1. What is the purpose of the aluminum block or cylinder?
2. Determine the density of aluminum and record it in your data table.

   Density = mass (A)/volume (D)
3. Find the volume of the foil and record it in cubic centimeters ($1\ mL = 1\ cm^3$).

$$\text{Volume} = \frac{\text{mass in g (G)}}{\text{density in g/cm}^3\text{ (H)}}$$

4. Find the height (thickness) of the foil (h) in cm from the relation

$$\text{Volume} = l \times w \times h$$

and record the result. Was your hypothesis about the data needed to compute the height of the foil correct?

5. One aluminum atom is 2.5×10^{-8} cm thick. Find the thickness of the foil in atoms, which is given by

$$\text{Number of atoms thick} = \text{height (I)} \times \frac{1 \text{ atom}}{2.5 \times 10^{-8} \text{ cm}}$$

6. Compute the number of moles of aluminum and the total number of atoms of aluminum in your piece of aluminum foil.

Extension and Application

If the population of the world is 5.6×10^9 individuals, how many atoms of aluminum could you distribute to each person from your sample of aluminum foil?

Chapter 9
ChemActivity

Reaction of Metal Salts with Sodium Hydroxide

Chemicals react in the simple whole number ratios shown by the coefficients in balanced equations. The exact amounts of reactants and products shown by a balanced chemical equation are called stoichiometric equivalents. For example, barium nitrate reacts with sulfuric acid as shown below:

$$Ba(NO_3)_2(aq) + H_2SO_4(aq) \rightarrow BaSO_4(cr) + 2HNO_3(aq)$$

Exactly one mole of barium nitrate reacts with one mole of sulfuric acid to form one mole of barium sulfate and two moles of nitric acid. These are the stoichiometric equivalents. Anything different from a one-to-one ratio of the two reactants will leave some excess chemicals unreacted in solution. The purpose of this activity is to compare the stoichiometry of the reactions of three different metal salts when combined with sodium hydroxide.

Objectives

- **Determine** the stoichiometric relationships in chemical reactions.
- **Compare** the stoichiometry of reactions of an element in two different oxidation states.

Materials

microplate, 96-well
plastic micropipets
0.1M sodium hydroxide, NaOH
0.1M solutions of the following (with 0.5% phenolphthalein added):
  nickel(II) nitrate, $Ni(NO_3)_2$
  iron(II) sulfate, $FeSO_4$
  iron(III) nitrate, $Fe(NO_3)_3$
Microplate Data Form

Procedure

Part A

Reaction of nickel(II) nitrate with sodium hydroxide

1. The reaction of nickel(II) nitrate with sodium hydroxide is a double displacement reaction that yields a precipitate. **Form a hypothesis** about what you will see to indicate the reaction of stoichiometric equivalents.
2. Arrange your microplate so that letters for rows are on the left and numbers for columns are on the top.
3. Using a pipet, place one drop of $Ni(NO_3)_2$ solution in well A1. **CAUTION:** *Several solutions used in this activity are toxic, corrosive, or skin irritants. Any spills on skin or clothes should be immediately rinsed with water.*
4. Place two drops of $Ni(NO_3)_2$ solution in well A2. Continue to add $Ni(NO_3)_2$ solution to the wells in row A, increasing the amount added by one drop in each subsequent well until you have added nine drops to well A9.
5. Using a clean pipet, add one drop of NaOH to the $Ni(NO_3)_2$ in well A9.
6. Add two drops of NaOH to well A8. Continue to add NaOH to the wells in row A in reverse order from A7 to A1, increasing the amount added by one drop in each subsequent well until you have added nine drops to A1.
7. Record your observations in your Microplate Data Form. Note any color changes across the rows. Observe the levels of precipitate by viewing the row from the side of the plate and record which well contains the most precipitate.
8. Discard the chemicals and rinse the microplate and micropipets with water.

Part B

Reactions of iron(II) sulfate and iron(III) nitrate with sodium hydroxide

1. Using a clean pipet, repeat steps 2-4 of Part 1 by adding iron (II) sulfate, $FeSO_4$, to wells A1 to A9.
2. Repeat steps 5 and 6 of Part A by adding NaOH to wells A9 to A1.
3. Record in the Microplate Data Form your observations of color and the well with the most precipitate.
4. Using clean pipets and row H in your microplate, repeat steps 2-7 of Part A by adding iron(III) nitrate, $Fe(NO_3)_3$, and NaOH to wells H1 to H9.
5. Discard the chemicals and rinse the microplate and micropipets with water.

No. Drops NaOH	9	8	7	6	5	4	3	2	1
No. Drops Metal Salt	1	2	3	4	5	6	7	8	9
	1	2	3	4	5	6	7	8	9

$Ni(NO_3)_2$ A

$FeSO_4$ B

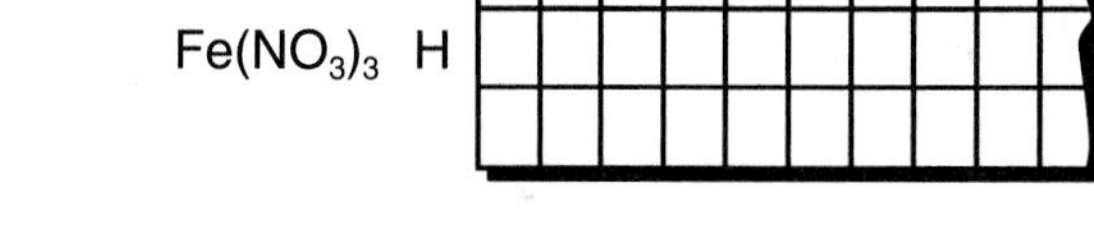

$Fe(NO_3)_3$ H

Data and Observations

Record your observations in a Microplate Data Form.

Analysis and Conclusions

1. Phenolphthalein turns pink in a solution that contains excess hydroxide ion. How does this property help you determine the stoichiometry of the reactions you performed? With each salt, where was the hydroxide ion used up?
2. What property of the hydroxides of nickel and iron helped in this experiment? What did it show about when stoichiometric equivalents had reacted? Was your hypothesis from Procedure step 1 correct?

3. Nickel(II) ion has a green tint in solution. What did you observe about the color of nickel ions in Part A? What did this observation indicate?
4. Based on your observations, compare the stoichiometry of the reactions of the two iron compounds. Are they the same or different?
5. Based on your observations, which two of the three reactions have the same stoichiometry?
6. Show the stoichiometry of the reactions you carried out by balancing the following equations:

$$Ni(NO_3)_2 + NaOH \rightarrow Ni(OH)_2 + NaNO_3$$
$$FeSO_4 + NaOH \rightarrow Fe(OH)_2 + Na_2SO_4$$
$$Fe(NO_3)_3 + NaOH \rightarrow Fe(OH)_3 + NaNO_3$$

7. Make a table to show the following. For each salt, which well contained the most precipitate? What was the ratio of NaOH to metal salt for this well? Compare this ratio with the balanced equations and decide if this was the ratio closest to that of the stoichiometric equivalents.

Extension and Application

1. Write the balanced equation for the reaction of aluminum nitrate with sodium hydroxide. What would you observe when the experiment you have done is carried out for this reaction?
2. Why is the stoichiometry of a chemical reaction important in the chemical manufacturing industry?

Chapter 10
ChemActivity

Periodicity of Halogen Properties

The elements in Group 17 (VIIA) of the periodic table are known as the halogens. The word *halogen* means "salt-former." The chemical property of forming salts, along with other common chemical and physical characteristics, reflects the placement of the halogens in the same periodic table group. The halogens have common chemical reactions, but their chemical activity varies. A more active halogen will displace the ion of a less active halogen, for example,

$$F_2(g) + 2NaCl(aq) \rightarrow 2NaF(aq) + Cl_2(g)$$

The purpose of this activity is to determine the difference in activity of the halogens by observing single displacement reactions.

Objectives

- **Observe** some properties of chlorine, bromine, iodine, and their ions in solution.
- **Determine** the order of activity of the halogen elements.

Materials

microplate, 24-well
white paper, one sheet
plastic micropipets
TTE (trichlorotrifluorethane)
Row solutions
  6*M* HCl
  5% sodium hypochlorite (NaOCl; bleach)
  bromine (Br_2) water
  iodine/potassium iodide (I_2/KI)
Column solutions
  2*M* solutions of
  sodium fluoride (NaF)
  sodium chloride (NaCl)
  sodium bromide (NaBr)
  sodium iodide (NaI)
Microplate Data Form

Procedure

1. Consult the periodic table and **hypothesize** in what order the activity of the halogen elements will increase.

2. Use a sheet of white paper to make a microplate template by copying the diagram below. Place a 24-well microplate on your template with the numbered columns at the top and the lettered rows to the left.
3. Add approximately 1/2 pipet of NaF solution to each of the wells in column 1, rows A through D.
4. Repeat step 3 by adding approximately 1/2 pipet of NaCl solution to the wells in column 2, NaBr to the wells in column 3, NaI to the wells in column 4.
5. Add 10 drops of 6*M* HCl to each well in row B, columns 1 through 4.
6. Add approximately 1/2 pipet of NaOCl solution to each well in row B, columns 1 through 4. The wells in row B now contain chlorine (Cl_2) in solution.
7. Add approximately 1/2 pipet of bromine water to the wells in row C, columns 1 through 4. **CAUTION:** *Do not inhale the bromine fumes from the solution.*
8. Finally, add approximately 1/2 pipet of I_2/KI solution to the wells in row D, columns 1 through 4.

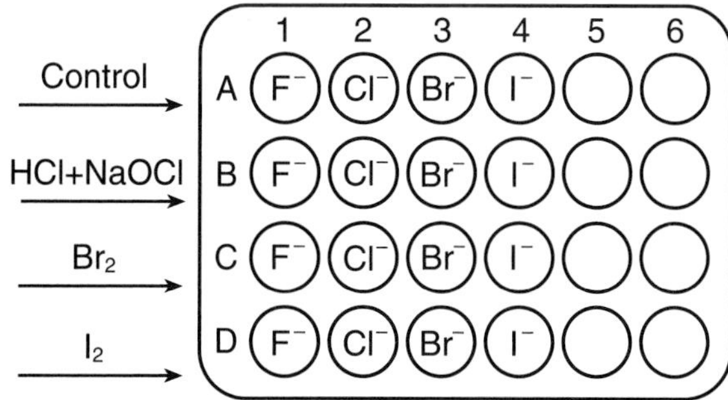

9. Compare all column 1 solutions to well A1; compare all column 2 solutions to well A2; and so on. Record your observations in a Microplate Data Form.
10. Add 1/4 pipet of TTE to any well that showed a change in appearance. **CAUTION:** *TTE is an irritant. Avoid contact with skin, nose, and eyes.*
11. One by one, draw up into separate plastic pipets the contents of each well to which you added TTE. Invert the pipet and be sure no drop of liquid remains in the stem. Shake the pipet to ensure complete mixing. Record your observations in your Microplate Data Form.

Data and Observations

Label the rows and columns of your Microplate Data Form the same way as on your template. Use it to record your observations.

1. What were the physical properties of the column solutions before the row solutions were added?
2. What were the physical properties of the row solutions before adding them to the column solutions?
3. Which combinations of solutions showed a change?
4. What happened when TTE was added to wells in which a change was noted?

Analysis and Conclusions

1. Which row of solutions showed the greatest number of changes?
2. Which row of solutions showed the least number of changes?
3. If a physical change is a sign of a chemical reaction, which halogen is the most active, that is, which displaces other halogen ions most easily? Which is least active?
4. How is the observed order of reactivity of the halogens related to position in the periodic table? Was your hypothesis from Procedure step 1 correct?

Extension and Application

1. Write the chemical equations for each of the single displacement reactions that you observed.
2. Do you think a single displacement reaction will occur between fluorine and sodium iodide? Why?
3. All of the halogens are harmful to living things. Describe how this property has been put to use for each of the halogens.

Preparation and Properties of Hydrogen

Hydrogen is the most common element in the universe, and water is the most common hydrogen compound on the surface of Earth. Hydrogen is a colorless, odorless, tasteless gas. It is less dense than air and is the least dense of all the elements.

The reaction of hydrogen with oxygen releases a large amount of energy. For this reason, hydrogen can be used as a fuel. Liquid hydrogen is the fuel of choice for rockets like the space shuttle. The purpose of this activity is to demonstrate how hydrogen gas can be prepared and tested for.

Objectives

- **Set up** apparatus for generating and collecting hydrogen gas.
- **Prepare** hydrogen gas by a displacement reaction.
- **Demonstrate** a chemical test for hydrogen gas.

Materials

thin stem pipet	toothpicks
micropipets (3)	magnesium ribbon (1 cm)
clear tape	1*M* hydrochloric acid (HCl)
matches	distilled water
24-well microplate	250-mL beaker
scissors	
matches	

Procedure

1. Cut a small slit in the middle of the bulb of the thin-stem pipet as illustrated in Figure A. This will be the generator pipet, pipet G.
2. Insert a piece of magnesium into pipet G through the slit.
3. Cover the slit in the pipet with clear tape to seal the magnesium inside the pipet bulb.
4. Trim the stem of pipet G to 2.5 cm.

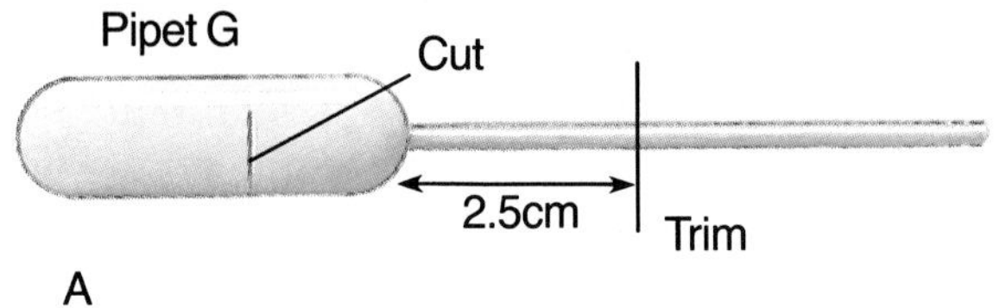

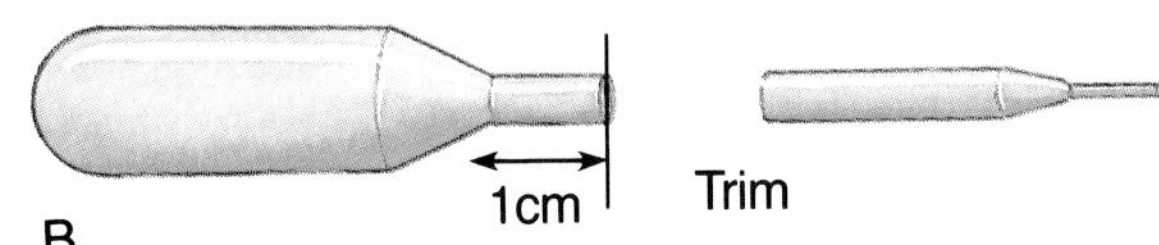

5. Cut the stems of the two micropipets to a length of 1 cm as in Figure B. These will be the collector pipets, C1 and C2.
6. Submerge pipets C1 and C2 under water with the stems pointing up in a 250-mL beaker. Squeeze the bulbs repeatedly until no more air is expelled. The pipets should be completely filled with water.
7. Using a new micropipet, place about 1/2 pipet of 1*M* HCl in well C3 in the 24-well microplate.
8. Turn pipet G so the tip is pointing down. Squeeze the air out of the pipet.
9. Place the stem of pipet G in well C3 and draw up the acid in the well.
10. Immediately, turn the pipet over and insert the stem of pipet G into the stem of water-filled pipet C1. Be sure to place the stem of pipet G into pipet C1 as far as it will go. Stand pipets G and C1 in the microplate as shown in Figure C.
11. Replace pipet C1 with C2 when C1 is almost full of hydrogen gas. Allow a "plug" of water to remain in the neck of the pipet before removing it from pipet G.
12. Stand the gas-filled pipets in wells of the microplate as shown in Figure D.
13. Light a toothpick with a match.
14. Quickly invert pipet C1 to dislodge the plug of water from the neck of the pipet.

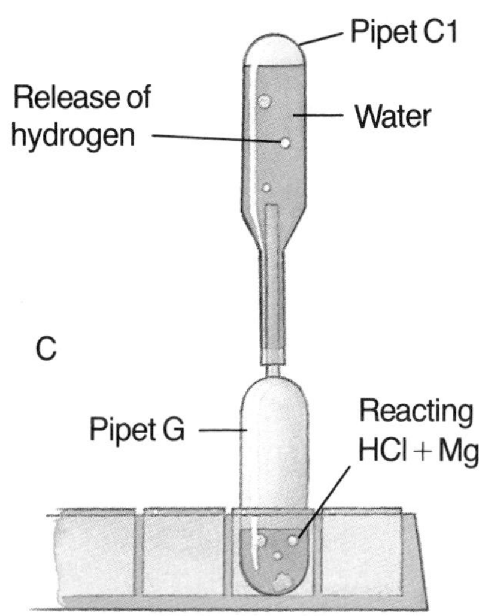

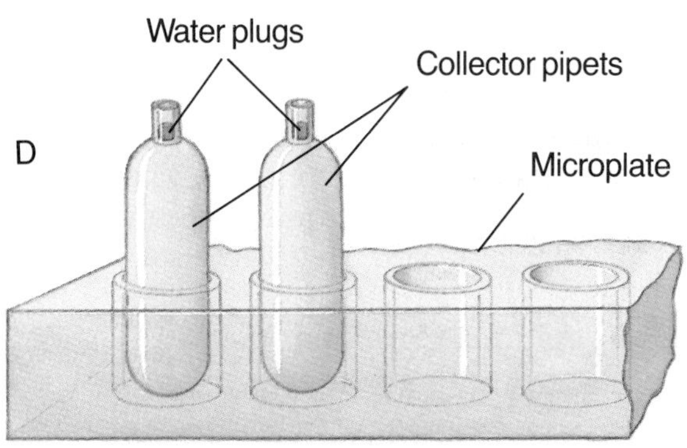

15. Place the flaming end of the toothpick close to the mouth of pipet C1.
16. A small pop indicates the presence of hydrogen gas. Repeat the test for hydrogen with pipet C2.

Data and Observations

1. What are the physical properties of hydrogen gas that you observed?
2. What happened to hydrogen gas when it was exposed to the flame?

Analysis and Conclusions

1. What compound do you think will be left in pipet G after the reaction of magnesium metal and hydrochloric acid? How could you isolate this product?
2. What are the products when hydrogen gas is burned in air?

Extension and Application

Commercial drain cleaners sometimes use aluminum metal and sodium hydroxide as active agents. What is the purpose of both the metal and the sodium hydroxide? What makes commercial drain cleaners flammable?

Chapter 11 ChemActivity 2

Transition Metals

The elements with atomic numbers 21 through 30 exhibit many properties different from those of other elements in the same period. The elements directly below in the next two periods, however, have similar properties. Together, these elements in Groups 3 to 12 are known as the transition metals. The purpose of this activity is to compare the chemical reactions of some transition metal ions with those of non-transition metals from the same period.

Objectives

- **Observe** physical and chemical properties of transition metal ions in aqueous solution.
- **Observe** results of mixing three different chemicals (NH_3, KSCN, HCl) with metal ions.
- **Compare** chemical reactions of transition metal ions with those of other metal ions.

Materials

96-well microplate
Microplate Data Form
plastic micropipets
toothpicks
0.1*M* solutions of compounds containing period 4 metals: KNO_3, $Ca(NO_3)_2$, NH_4VO_3, $Cr(NO_3)_2$, $Mn(NO_3)_2$, $Co(NO_3)_2$, $Fe(NO_3)_3$, $Ni(NO_3)_2$, $Cu(NO_3)_2$, $Zn(NO_3)_2$
6.0*M* ammonia, NH_3
1*M* potassium thiocyanate, KSCN
6*M* hydrochloric acid, HCl
white paper, one sheet

Procedure

1. Review the list of metals that you will be testing. **Hypothesize** which three are likely to have properties different from the others.
2. Use a sheet of white paper to make a microplate template by copying the diagram below. Place a 96-well microplate on your template with the numbered columns at the top and the lettered rows to the left.
3. Place 5 drops of KNO_3 in each of wells A1, B1, C1, and D1.
4. Place 5 drops of $Ca(NO_3)_2$ in each of wells A2, B2, C2, and D2.
5. Continue to repeat this process of placing 5 drops of solutions of metal compounds in subsequent columns of the microplate. Use chemicals in the order in which they are given in the materials list. The last drops will be in column 10.
6. Add 5 drops of 6*M* NH_3 to each well in row A. Mix well with a toothpick. **CAUTION:** *Ammonia solution is caustic.*
7. Add 5 drops of KSCN solution to each well in row B. Mix well with a toothpick.
8. Finally, add 5 drops of HCl to each well in row C. **CAUTION:** *Do not mix HCl and KSCN. Clean up all spills immediately.* Mix well. Use row D as a control for comparing with the other rows.

Data and Observations

Label the rows and columns of your Microplate Data Form the same way as your template.

1. Observe row D of your microplate. Record your observations in the Data Form.
2. Compare the solutions in the wells in rows A, B, and C in each column to the solution in row D in that column. Where there was a change, record what you observe.

Analysis and Conclusions

1. Compare the order of the metal ions in the wells from left to right in the rows of your microplate. Note the relationship of your metal ions to the positions of the metals in the periodic table.

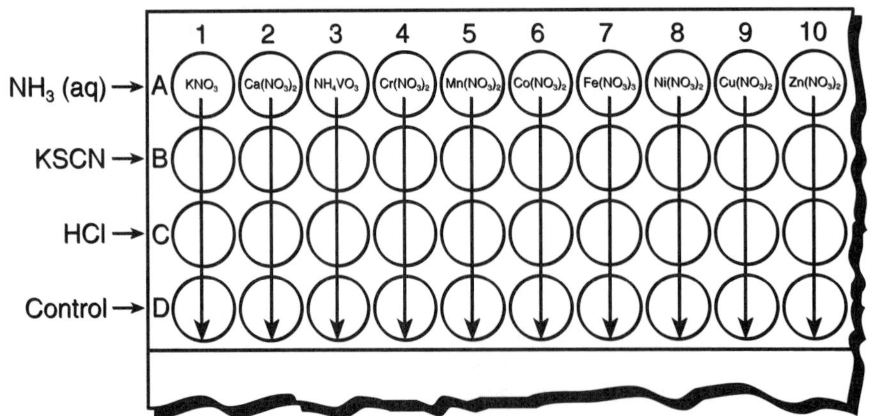

2. What did you observe in columns 1, 2, and 10 in your microplate? What do the metal ions in these columns have in common?
3. Review what you observed in columns 3 through 9 of your microplate. What do the results in these columns have in common? What do the metal ions in these columns have in common? Is there any evidence for the occurrence of more than one type of chemical reaction?
4. Was your hypothesis correct? Why or why not?

Extension and Application

1. What physical or chemical properties would help to identify a salt as containing a transition metal ion?
2. Modern atomic theory provides an explanation for the similarities of the transition elements. Describe the basis for this explanation. How are the properties of zinc accounted for?

Chapter 12
ChemActivity

A 3-D Periodic Table

The periodic table is the most valuable tool in chemistry. The organization of the elements according to atomic number and electron configuration gives insights into the chemical and physical properties of the elements and the compounds that they form.

Transition element properties vary less than those of other elements. Therefore, this activity is limited to properties of the elements in Groups 1, 2, and 13-18 (Groups IA to VIIIA), which are known as the main-group elements. The purpose of this activity is to show periodic relationships among properties of the main-group elements.

Objectives

- **Assemble** a three-dimensional scale for a property of the elements.
- **Examine** properties of the elements according to their periodic relationships.
- **Identify** properties with similar periodic relationships.

Materials

96-well microplate
soda straws (25 each group)
metric ruler
large-square graph paper
scissors
grease pencil

Procedure

1. Your teacher will select a property for you to study from the list below:

   density | boiling point
   atomic radius | melting point
   ionization energy | electronegativity

2. Locate the data for your property of elements in Groups IA through VIIIA in tables in Chapter 10, Chapter 12, or Appendix Table A-3 in your textbook.
3. Measure the length of a soda straw; determine one-half the length of the straw, and record the results in a data table like the one shown.
4. Select an appropriate scale so that the largest value of your property can be represented by one-half the length of the soda straw. For example, suppose the highest value of your property is 35 units. If the straw is 24 cm long, half of which is 12 cm, the scale for your property would be

$$\frac{12 \text{ cm}}{35 \text{ units}} = 0.34 \text{ cm/unit}$$

The length of the straw representing the element with the value of 35 units would be 0.34 cm/unit × 35 units = 12 cm. Enter the scale for your property in your data table.

5. Cut soda straws to the proper lengths to represent your assigned property for each of the main-group elements. Mark each straw near the top with the symbol of the element.

6. By comparing your 96-well microplate with a periodic table, locate the well that represents each main-group element. Place the cut straw for that element in the well. Be sure to preserve the location of the elements relative to each other as shown in the diagram.

7. Using a grease pencil, label the outside of your plate with the name of the physical or chemical property represented by your 3-D table.

8. Record in your data table how your property varies across the rows and down the families.

9. Observe the microplates of other students in your class. In one sentence, summarize in your data table your observations about the properties studied.

Data and Observations

Length of soda straw (cm)	
Half the straw length (cm)	
Largest value of property to be investigated	
Scale for straw length vs. property (cm): ½ straw length ÷ largest value of property	
Observation notes on trend for property:	

Analysis and Conclusions

1. Choose two pairs of properties. For each pair, draw conclusions about the relationship between the trends. For example, what seems to be the relationship between electronegativity and ionization energy?
2. Can several of the properties be grouped according to similar periodic trends?

Extension and Application

1. While you are not looking, have your lab partner remove one of the straws from the set you have constructed. Cut another soda straw so that it will fit into the general trend of your 3-D table. Predict and describe to your partner the property value for the unknown element. Compare your description to the actual value.
2. Reverse roles and repeat the exercise.
3. Switch plates with another group and repeat Extension 1.
4. How is the periodic table useful in identifying chemical elements and their properties?

Electron Clouds

Electrons occupy space about the nucleus. The exact path or location of electrons in an atom cannot be known. Instead, we speak about the probability of locating an electron in an electron cloud—a volume determined by the energy of the electron. Balloons allow us to model the electron clouds of bonding electron pairs whose arrangement determines the shapes of molecules.

Round balloons will be used to represent the spherical electron cloud occupied by one or two *s* electrons. Pear-shaped balloons will be used to represent the electron clouds occupied by pairs of bonding electrons. The purpose of this activity is to provide you with a model of how atoms join together to form molecules.

Objectives

- **Use** balloons to model the arrangement of electron pairs around the nucleus of an atom.
- **Observe** a model of bonding in three dimensions.

Materials

round and pear-shaped balloons
clear tape
string

Procedure

1. Blow up five round balloons to the same size. Tie their ends closed.
2. Make a loop of tape on each of the balloons to represent an electron. It takes a pair of loops to make a chemical bond.
3. Stick two of the balloons together at the two tapes to form a model of a covalent bond as shown in Figure A. This is a model of two hydrogen atoms and an example of linear bonding.
4. Blow up four pear-shaped balloons. Tie three together at their knotted ends as shown in Figure B.

A

Linear bonding

B

Planar bonding

5. Place a loop of tape at the end of each balloon opposite where they are knotted together. Attach a round balloon by its loop of tape on the end of each pear-shaped balloon. This is an example of planar bonding, as shown in Figure B.
6. Add the fourth pear-shaped balloon to the other three pear-shaped balloons by tying it where the others are knotted together.
7. Place a loop of tape at the end of the fourth pear-shaped balloon.
8. Place a round balloon at the end of the fourth pear-shaped balloon, as shown in Figure C. This is a model of CH_4, and an example of tetrahedral bonding.

Data and Observations

1. Compare your three models and picture the location of the bonded atoms. Using atomic symbols and straight lines for the bonds, diagram each type of molecule.

2. What are the angles between the bonds in your models of BCl_3 and CH_4?

C

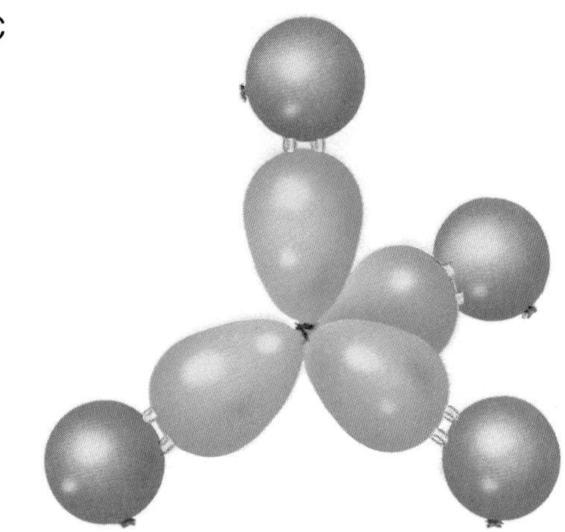

Tetrahedral bonding

Analysis and Conclusions

1. Explain in your own words how the balloons represent the forces that shape molecules.
2. Which of the molecules listed below will be linear? planar? tetrahedral?
   (a) CH_3Cl **(b)** Cl_2 **(c)** CCl_4
   (d) HCl **(e)** BF_3

Extension and Application

1. Use your round and pear-shaped balloons to make a model of $BeCl_2$. What are the bond angle and the molecular shape of this molecule?
2. Blow up two round balloons four times larger than before. Replace two round balloons in the tetrahedral molecule. What happens? When does something similar happen in a real molecule?
3. In determining molecular shape by the repulsion of electron pairs, double or triple bonds have the same effect as single bonds. Are ethylene, $H_2C{=}CH_2$, and acetylene, $H{-}C{\equiv}C{-}H$, planar or three-dimensional? What will be the bond angles of the carbon-carbon-hydrogen bonds in each molecule?

Chapter 14
ChemActivity 1

Paper Chromatography

Chromatography is a technique of separation. There are many forms of chromatography, but in this activity you will be using paper chromatography. The separation of compounds in paper chromatography depends on how strongly the components of a mixture are attracted to a solvent and to the paper.

In paper chromatography, the *stationary phase* is a piece of paper. The *mobile phase* is the mixture to be separated, often a solvent in which a sample of unknown composition is dissolved. Your unknown sample will be the black ink from a marking pen. The purpose of this activity is to observe the effect of solvent polarity on the separation of a mixture by paper chromatography.

Objectives

- **Set up** a paper chromatography experiment.
- **Observe** the separation of the dyes in an ink sample.
- **Examine** the relationship between separation and solvent polarity.

Materials

1×9 cm strips of filter paper (3)
plastic microtip pipet
24-well microplate
black, water-soluble marker
heavy plastic bag
distilled water
ethyl alcohol

Procedure

1. Place ¼ pipet of water and ¼ pipet of ethyl alcohol in well A1 of a 24-well microplate.
2. Place ½ pipet of water in well B1.
3. Place ½ pipet of ethyl alcohol in well C1.
4. One by one place each strip of filter paper in an empty well and draw a pencil line on it even with the top of the well.
5. Remove each strip and make a small circle on the pencil line with the black marker as shown in Figure A.

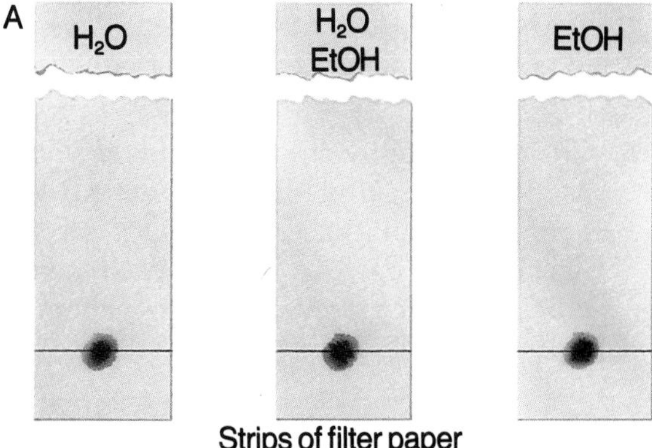

Strips of filter paper

6. Allow the mark to dry and mark the spot again.
7. In pencil, label the tops of the paper strips with the solvents to be used: A1, the water-ethyl alcohol mixture (H_2O-EtOH); B1, water (H_2O); C1, ethyl alcohol (EtOH). See Figure A.
8. Place the marked strips in the appropriate wells, A1, B1, and C1.
9. Fold over the marked strips and place the entire assembly in a plastic bag as shown in Figure B.

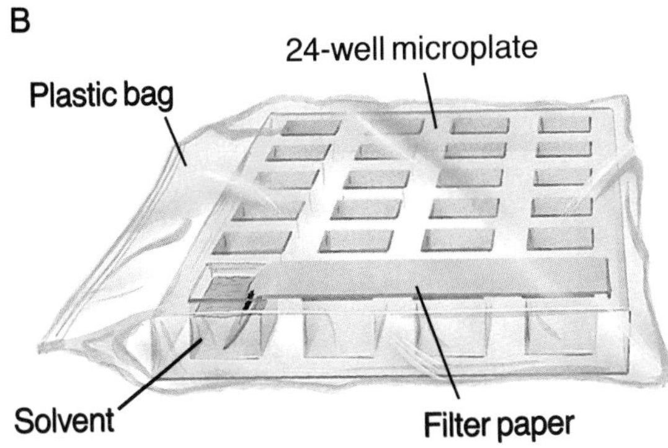

10. Allow the chromatogram to "develop" until the solvent comes close to the top of the paper.
11. Remove the paper chromatograms.
12. Choose spots of the same color on each chromatogram for your analysis. Measure the distance that the spot of this color moved from the original spot on each chromatogram, as in Figure C. Enter this distance in a data table like the one shown.

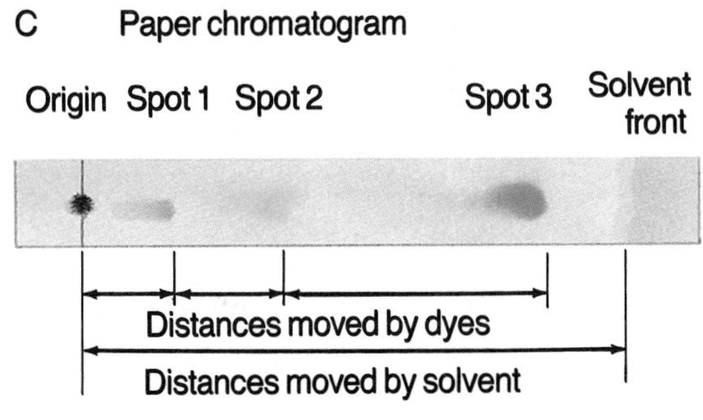

Data and Observations

Well	Solvent	Distance to solvent front	to spot 1	to spot 2	R_f Value
A1	H_2O + EtOH				
B1	H_2O				
C1	EtOH				

13. Measure the distance that the solvent moved from the origin to the solvent front on each chromatogram. Enter this distance in your data table.

Data and Observations

1. How many colored spots do you have? What are the components of the mixture?
2. How are the three chromatograms similar? How are they different?
3. The degree of separation is shown by the distance between the dye spots on the paper. Which solvent was best at separating the different dyes?

Analysis and Conclusions

1. Determine the R_f values for each of your dye colors by using the formula

$$R_f = \frac{\text{Distance dye moved}}{\text{Distance to solvent front}}$$

Enter your calculated R_f values in your data table.

2. Compare the R_f values for your spots. Do any two of these spots have the same R_f value? Explain your answer.
3. Water is more polar than ethyl alcohol and the mixture is intermediate between the two in polarity. The most polar molecules will move farthest in the most polar solvent. What can you conclude about the polarity of the dye in the spots you analyzed?

Extension and Application

1. Explain how a polar solvent works in the separation of polar dyes.
2. How could you prove that two dye spots of the same color but with different R_f values in different solvents are the same compound?
3. Chromatography of many kinds is widely used to isolate and identify chemical compounds and mixtures. How might chromatography be used in a chemical research lab? In a police forensic lab? In a chemical manufacturing plant?

Chapter 14
ChemActivity 2

Miscibility of Liquids

The solubility of a substance depends on several factors. One is the polarity of both the substance and the solvent. A general guide is that "like dissolves like." A solvent that is polar will dissolve polar substances. A solvent that is nonpolar will dissolve nonpolar substances.

Water is a polar liquid and TTE (trichlorotrifluoroethane) is a nonpolar liquid. The purpose of this activity is to classify liquids as more polar or less polar based on their solubilities in water and TTE.

Objectives

- **Observe** the solubility of liquids in water and TTE.
- **Classify** liquids as more or less polar.

Materials

96-well microplate (3)
plastic microtip pipets (8 per group)
small beaker or plastic cup
toothpicks
construction paper, white
distilled water
TTE (trichlorotrifluoroethane)

liquids to be tested:
- 1-butanol
- 2-butanol
- *tert*-butanol
- 1-propanol
- 2-propanol
- hexane

food coloring, blue
Microplate Data Form

Procedure

1. Place a 96-well microplate with the numbered columns away from you and the lettered rows to the left.
2. Place 5 drops of 1-butanol in each of wells A1 to A10.
3. With a clean microtip pipet, place 5 drops of 2-butanol in each of wells H1 to H10.
4. Using a clean microtip pipet for each liquid, repeat steps 2 and 3 with pairs of remaining test liquids and two more microplates. Use only rows A and H in each microplate so that you can observe the test mixture from the sides of the microplates.
5. Place 20 mL of distilled water in a small plastic cup or beaker. Add 3 drops of blue food coloring.
6. In the first microplate, using a clean microtip pipet, add 1 drop of the colored water to wells A1 and H1. Add 2 drops of colored water to wells A2 and H2. Continue adding more colored water until you reach wells A5 and H5. Mix the contents of each row with a separate toothpick.
7. Next, with another clean microtip pipet, add 1 drop of TTE to wells A10 and H10 of the first microplate. Add 2 drops of TTE to wells A9 and H9. Continue adding more water to each successive well until you reach wells A6 and H6. Using clean toothpicks, mix the contents of each row of wells with a separate toothpick.
8. Hold the plate over a piece of white paper. Observe the wells from the side of the plate. Note your observations in a Microplate Data Form like the one shown.
9. Repeat steps 6–8 for the four liquids in the other two microplates.

Data and Observations

Relabel your Microplate Data Form so that you can record the results from all six liquids on the same form:

Test Liquids	H_2O					TTE				
	1	2	3	4	5	6	7	8	9	10
1-butanol A										
2-butanol H										
tert-butanol A										
1-propanol H										
2-propanol A										
Hexane H										

1. Two liquids that mix together in all proportions are said to be *miscible,* meaning they are soluble in each other. The formation of two distinct layers means that two liquids are not miscible. Which combinations of liquids formed two layers?
2. The formation of a white cloudy mixture that does not separate means that the two liquids are partially miscible. Which combinations of liquids formed cloudy mixtures?
3. Which combinations of liquids formed clear solutions, indicating that they are miscible?
4. Were any of the liquids tested miscible with both water and TTE?

Analysis and Conclusions

1. Which were the most polar liquids tested? Which were the least polar?
2. What do you think would be the result of mixing colored water and TTE? Try it.

Extension and Application

1. The formulas of three of the liquids in your experiment are shown below:

$CH_3CH_2CH_2CH_2OH$ 1-butanol

$CH_3CH_2CH_2OH$ 1-propanol

$CH_3CH_2CH_2CH_2CH_2CH_3$ hexane

What do you see in the structures that might explain the results of your experiment with these substances?

2. Examine the three structures below and predict which of these substances would be miscible or immiscible with water.

CH_3OH methanol

CH_3CH_2OH ethanol

$CH_3CH_2CH_2CH_2CH_2CH_2CH_2CH_3$ octane

Chapter 15
ChemActivity

Gas Pressure

The volume of a gas depends on three factors. The first is the number of moles of the gas, the second is the temperature at which the volume is measured, and the third is the pressure. For measurements at constant temperature, there is a simple mathematical relationship between the volume of a gas and the pressure. This relationship is known as Boyle's law in honor of Sir Robert Boyle, the British chemist who first recognized it about 300 years ago. The purpose of this activity is to discover the relationship between the pressure and the volume of a gas.

Objectives

- **Observe** the effect of increased pressure on the volume of a confined gas.
- **Graph** the pressure to volume relationship.
- **Describe** the pressure to volume relationship in mathematical terms.

Materials

thin stem pipet	matches
plastic cup	glass rod
graph paper	metric ruler
textbooks (8 of equal size and mass)	water
	methylene blue

Procedure

1. **Form a hypothesis** to explain what would happen to the volume of a gas as the pressure is increased or decreased.
2. Fill a plastic cup with 20 mL of water.
3. Add a few drops of methylene blue solution to the water.
4. Fill only the bulb of a thin-stem plastic pipet with the tinted water.
5. Soften the tip of the filled pipet by slightly heating it in a match flame. **CAUTION:** *Be careful not to heat the tip of the pipet more than necessary to soften it; it could catch fire.*
6. Seal the tip by pressing down on it with a glass rod.
7. Allow the pipet tip to cool.
8. Place two equal-sized books on the bulb of the pipet. The stem of the pipet should be visible as shown in the diagram.
9. Measure the column length of the trapped air in the stem of the pipet in millimeters and enter this measurement in a data table like the one shown.
10. Place another book on top of the first two, and measure and record the length of the air column again.

11. Continue to add the remaining books to the stack one-by-one. Measure and record each new air-column length.

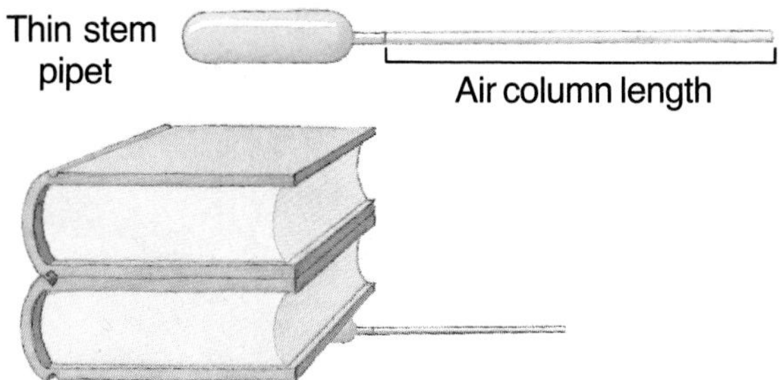

Data and Observations

Pressure (books)	Length of air column (mm)	1/Pressure
2		
3		
4		

Analysis and Conclusions

1. Plot a graph of your data. Use the pressure in books as the x variable and the length of the air column as the y variable. Is the relationship between gas pressure and volume linear?
2. Was your hypothesis correct? Explain why or why not.
3. Calculate the inverse of the pressure, 1/pressure, in books. Enter your calculations in your data table.
4. Plot a second graph. Use the inverse, 1/pressure, as the x variable and the length of the air column as the y variable. Is the relationship between 1/pressure and the volume of the gas linear?
5. What generalization can you make about the effect of increased book pressure on the volume of the confined gas?

Extension and Application

1. Write a mathematical equation for the relationship between the pressure and volume of a gas at constant temperature. Use P for pressure, V for volume, and k for any necessary constant.
2. What can happen to a gas under extremely high pressures or extremely low temperatures?
3. If a pressurized gas is released from a vessel, ice can be seen forming on the outside of the vessel. Why does this happen?

Chapter 16
ChemActivity

Crystals and their Structure

Crystals are classified according to their unit cells, the simplest repeating units of atoms, ions, or molecules. A crystal that is large enough to see is a collection of microscopic unit cells.

Crystals can be many sizes. The shape of a large crystal gives a hint to the characteristic shape of the unit cell. The purpose of this activity is to study the possible shapes of crystals by constructing unit-cell models and comparing the models with real crystals.

Objectives

- **Construct** some unit-cell models.
- **Observe** the process of crystallization.
- **Observe** and **classify** the unit cells of some common crystals.

Materials

gumdrops (3 colors)
toothpicks (16)
protractor
plastic microtip pipet
24-well microplate
chemical scoop
glass microscope slides (4)

Materials (continued)

- hand lens or low-power microscope
- sodium chloride $NaCl(cr)$
- sodium nitrate, $NaNO_3(cr)$
- benzoic acid ($C_7H_6CO_2$)
- aspirin (acetylsalicyclic acid, $C_9H_8O_4$) (1 tablet)
- ethanol (C_2H_5OH)

Procedure

Unit-cell models

1. Obtain 14 gumdrops from your teacher. Take 4 of one color, 4 of another color, and 6 of a third color.
2. Place one gumdrop of the first color on your lab table. Using a protractor to measure angles, gently push three toothpicks into the gumdrop at right angles to each other as shown in Figure A.

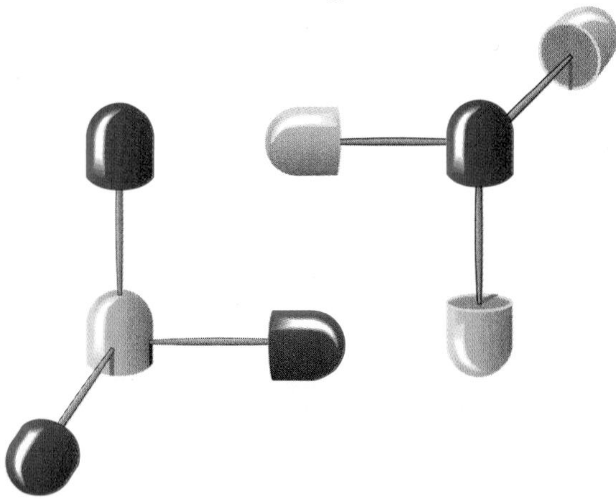

A

3. Place a gumdrop of the second color at the end of each of the three toothpicks.
4. Repeat steps 2 and 3 using the opposite color scheme.
5. Place the two models around a common center so that the central gumdrops of each model are situated at opposite corners of a cube. Join opposite colored gumdrops to each other with toothpicks to form a cube as shown in Figure B.

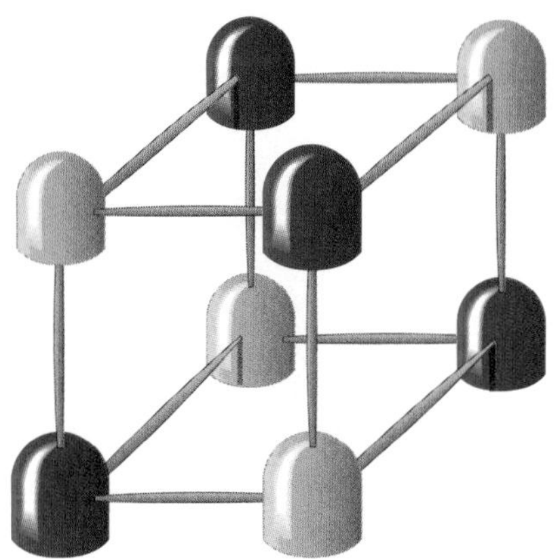

B Simple cubic unit cell

6. In a data table like the one below, enter the value of the three angles of this simple cubic unit cell indicated in the table.
7. Place a gumdrop of the third color in the center of your cubic unit cell and secure it with the toothpicks of the four corners of the cube. This is a model of a body-centered cubic unit, Figure C.

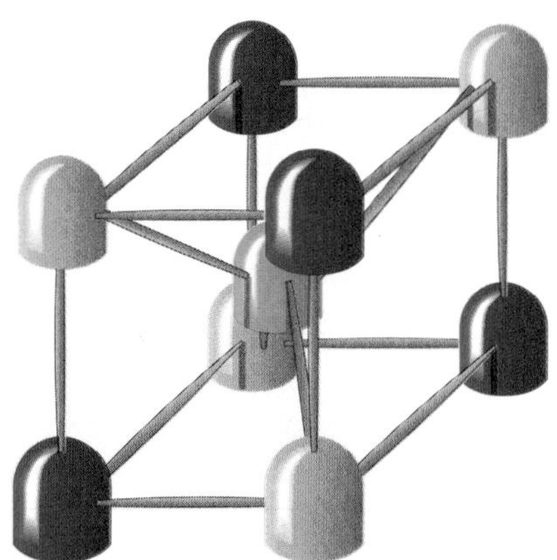

C Body-centered cubic unit cell

8. In your data table, enter the values of the three angles for a body-centered cubic unit.
9. Remove the center gumdrop.
10. Relocate the center gumdrop in the center of one of the faces of the cube. Place other gumdrops of the same color at the center of each of the other faces to make a model of a face-centered cubic unit cell like that shown in Figure D.

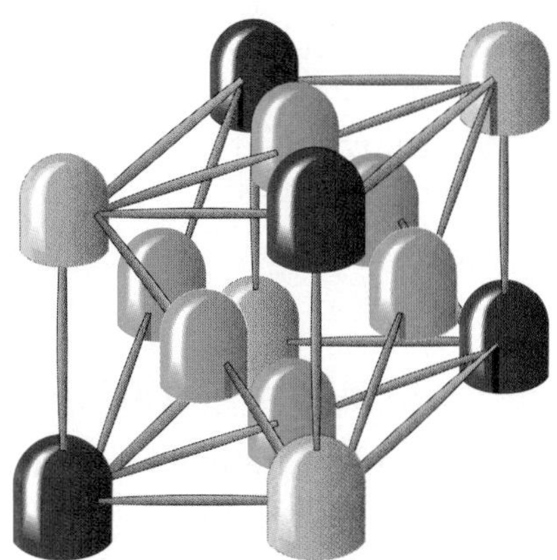

D Face-centered cubic unit cell

11. Record the indicated three angles of this face-centered cubic unit cell in your data table.
12. Obtain eight more gumdrops of two colors and assemble a simple cubic unit cell as you did in steps 2–5.
13. Make a rhombohedral unit cell from your simple cubic unit cell. Hold the bottom of the new cubic unit cell and push the top four gumdrops to one side about 3 cm, Figure E. Measure the three angles and record them in your data table.

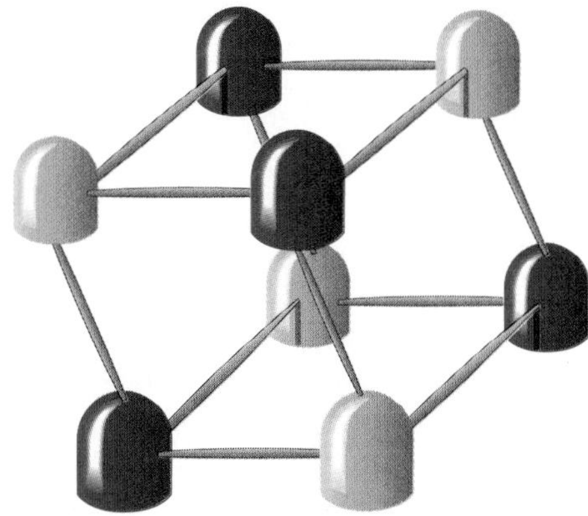

E Rhombohedral unit cell

Crystals

14. Place ½ pipet of ethanol in each of wells A1 and B1 of a 24-well microplate.
15. Add 0.2 g of benzoic acid to well A1. Add 1 aspirin to well B1.
16. Stir the contents of each well with a separate toothpick. The aspirin will seem not to dissolve. The insoluble material is starch, not the aspirin itself.
17. Allow any solids to settle to the bottom of each well.
18. Place a clean glass slide on a microscope or white piece of paper.
19. Using a plastic pipet, transfer three drops of liquid from well A1 to the slide.
20. Observe the slide under a microscope or with a hand lens. Describe the shapes of crystals that you observe in a data table like the one shown.
21. Repeat steps 18–20 using the liquid in well B1.
22. Use a chemical scoop to place a few crystals of sodium chloride on another glass slide. Observe the slide under a microscope or with a hand lens. Describe the crystal shapes you observe.
23. In the same way, observe crystals of sodium nitrate. Describe the crystal shapes in your data table.

Data and Observations

Unit-cell models

3-D Unit Cell	Angle 1	Angle 2	Angle 3
Simple Cubic			
Body-centered cubic			
Face-centered cubic			
Rhombo-hedral			

Crystals

Compound	Description of Crystals
Benzoic acid	
Aspirin	
Sodium chloride	
Sodium nitrate	

Analysis and Conclusions

1. How did the benzoic acid and aspirin crystals compare with the sodium chloride and sodium nitrate crystals?
2. What might be the type of unit cell for sodium chloride?
3. What might be the type of unit cell for sodium nitrate?
4. What might be the shape and angles of a benzoic acid unit cell?

Extension and Application

1. How could a unit-cell model be expanded into a macroscopic crystal model?
2. What would be the physical effects on a crystal if impurities were added?
3. The study of crystals has been aided by the development of X-ray crystallography. How do scientists study crystals using X rays? What information do they obtain by these studies?

Chapter 17
ChemActivity 1

Enthalpy of Fusion of Ice

When energy is added to ice at its melting point of 0°C, the temperature of the ice does not increase. Instead, the added energy allows the water molecules to pull away from each other and enter the liquid state. The energy required to melt one gram of a substance at its melting point is known as the enthalpy of fusion of that substance. The purpose of this activity is to determine the enthalpy of fusion of ice at atmospheric pressure.

Objectives

- **Determine** the enthalpy of fusion of ice.
- **Compare** an experimental value for enthalpy of fusion of ice with the actual value.

Materials

- 50-mL graduated cylinder
- plastic ketchup cups with two tops (4)
- rubber band
- balance
- labels
- thermometer
- crushed ice
- distilled water

Procedure

1. A calorimeter, an instrument used to measure energy changes in chemical systems, must exchange as little heat with the air around it as possible. **Hypothesize** how accurate your measurements are likely to be using a simple plastic calorimeter.
2. Construct a plastic calorimeter by placing a rubber band around one ketchup cup, and then placing this cup inside the second cup as shown here. With a paper punch, make one hole in a ketchup-cup top.

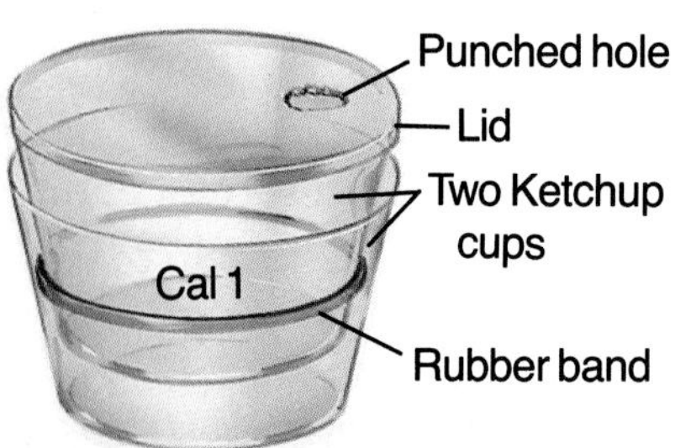

3. Repeat step 2 to make a second plastic calorimeter.
4. Label the two plastic calorimeters *Cal 1* and *Cal 2*.
5. Measure the mass of Cal 1 with its lid and record the mass in a data table like the one shown.
6. Place 20 mL of distilled water in Cal 1. Cover it and measure the total mass of Cal 1 with the water. Record your result.
7. Insert a thermometer into Cal 1 and record this initial water temperature.
8. Fill Cal 2 with crushed ice and thoroughly drain off any water.
9. Remove the thermometer from Cal 1, and add the water from Cal 1 to the ice in Cal 2.
10. Swirl the water/ice mixture for 1 minute. Pour all the water from Cal 2 into Cal 1.
11. Insert the thermometer and measure the temperature of the water in Cal 1. Record this final temperature.
12. Replace the lid on Cal 1, and measure and record the total mass of Cal 1, which now contains the original amount of water plus the melted ice.
13. Discard the water and ice from both calorimeters. Dry the two calorimeters, and repeat the activity starting from Step 4 in order to get another set of data so that you can average the results of two runs.

Analysis and Conclusions

1. Complete the data table by finding the values of C, F, and H for each run. Disregard the minus sign in data F because only the amount of temperature change is of concern here.
2. The heat that melted the ice came from the original water sample, causing a drop in water temperature. This quantity of heat is found from the equation

$$q\text{, in joules} = m(\Delta T)C_p$$

where m is the initial mass of water (data C), ΔT is the change in temperature of this water (data F), and C_p is the specific heat of water, which is 4.18 J/g·C°. Calculate q for each of your runs and record the values for data I in your data table.
3. The enthalpy of fusion is the heat needed, in joules, to melt one gram of ice. To find the experimental values for your two runs, divide the results of analysis step 2 (data I) by the mass of ice melted (data H). Record the data in your data table.
4. Find the average enthalpy of fusion from your two runs (data K). Compare the result of your experiment with the actual value for the enthalpy of fusion of water, which is 334 J/g. Suggest some reasons for the differences. Was your hypothesis supported by your data?

Data and Observations

Observations	Run 1	Run 2
A. Mass of empty Cal 1 (g)		
B. Mass of Cal 1 + water (g)		
C. Mass of water (B – A) (g)		
D. Initial temperature of water in Cal 1 (°C)		
E. Final temperature of water in Cal 1 (°C)		
F. Change in temperature of water (D – E) (°C)		
G. Mass of Cal 1, water, and melted ice (g)		
H. Mass of ice melted (G – B) (g)		
I. Heat to melt ice (J)		
J. Enthalpy of fusion (I/H) (J/g)		
K. Average enthalpy of fusion (J/g)		

Extension and Application

1. When is ice always 0°C?
2. What is dry ice? How is dry ice different from water ice?
3. Ice is used as a coolant. What are the advantages and disadvantages in the use of water ice as a coolant instead of dry ice?
4. In the equilibrium between ice and water as shown by the equation

$$H_2O(cr) \rightleftharpoons H_2O(l)$$

the change in the forward direction is melting and the reverse change is freezing or crystallization. How much heat do you think is released when 1 gram of liquid water freezes?

Chapter 17
ChemActivity 2

Enthalpy of Vaporization of Water

When energy is added to water at its boiling point of 100°C, the temperature of the water does not increase. Instead, the added energy allows the water molecules to pull away from each other completely and enter the gaseous state. The energy required to vaporize one gram of a substance at its boiling point is known as the enthalpy of vaporization of that substance. An equal amount of energy is released when 1 g of the vapor condenses to a liquid at its boiling point. The purpose of this activity is to determine the enthalpy of vaporization of water at atmospheric pressure by measuring the heat released in the condensation of steam.

Objectives

- **Determine** the enthalpy of vaporization of water.
- **Compare** the enthalpy of vaporization of water with the enthalpies of vaporization of other liquids.

Materials

plastic ketchup cups with one lid (2)
rubber band
125-mL Erlenmeyer flask
one-hole rubber stopper
glass bend
hot plate
50-mL graduated cylinder
thermometer
distilled water

Procedure

1. Construct a plastic calorimeter from two ketchup cups as described in ChemActivity 17-1, procedure Step 2.
2. Measure the mass of the calorimeter and its lid. Enter this mass (A) in a data table like the one shown.
3. Half-fill the 125-mL Erlenmeyer flask with water.
4. Fit the flask with a glass bend and rubber stopper as shown here.
5. Heat the flask on a hot plate as shown.
6. Place 20 mL of distilled water in the calorimeter. Measure and record the total mass of the calorimeter and water (B).

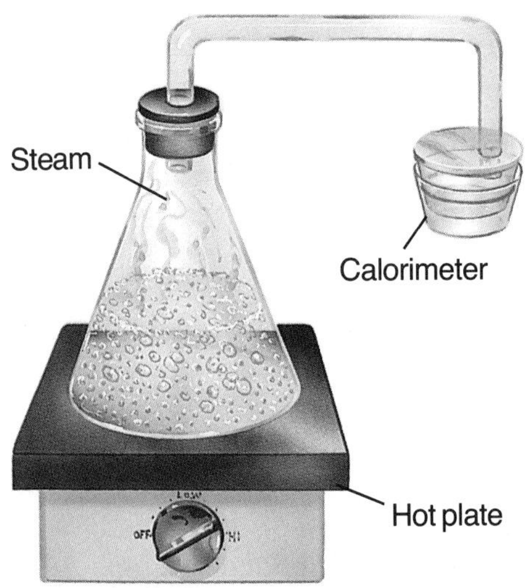

7. Place a thermometer in the calorimeter and record this initial temperature of the water (D).
8. When vapor is seen coming out of the glass bend, insert the tip of the glass bend through the hole in the lid and down into the water in the calorimeter. Hold it in this position for 30-45 seconds. **CAUTION:** *Wear a thermal glove to avoid scalds from the water vapor.*
9. Remove the calorimeter from under the glass bend. Place the thermometer in the water in the calorimeter and record this final temperature (E).
10. Measure the mass of the calorimeter with its lid, water, and condensed water vapor and record the mass (G).

Data and Observations

A. Mass of the empty calorimeter (g)	
B. Mass of the calorimeter and water (g)	
C. Mass of water (B – A) (g)	
D. Initial temperature of water in the calorimeter (°C)	
E. Final temperature of water in the calorimeter (°C)	
F. Change in temperature of water (E – D) (C°)	
G. Mass of the calorimeter, water, and condensed water vapor (g)	
H. Mass of condensed water vapor (G – B) (g)	
I. Heat (q) released in condensation (J)	
J. Enthalpy of vaporization (J/g)	

Analysis and Conclusions

1. Complete your data table by finding the values of C, F, and H.
2. The heat released when the steam condensed was added to the water in the calorimeter, causing its temperature to increase. This quantity of heat is found from the equation

$$q\text{, in joules} = m(\Delta T)C_p$$

where m is the initial mass of water (data C), ΔT is the change in temperature of this water (data F), and C_p is the specific heat of water, which is 4.18 J/g·C°. Calculate q for your experiment.

3. The enthalpy of vaporization is the heat needed, in joules, to vaporize one gram of water. It is equal in value to the heat released per gram of water in condensation. Divide the results of analysis step 2 (data I) by the mass of steam condensed (data H). Record the results.
4. The enthalpy of vaporization of water is 2260 J/g. How does your experimental enthalpy of vaporization of water compare with the actual value? How does it compare with the enthalpy of fusion of water (334 J/g)? Explain the difference.
5. Use a chemistry handbook to look up the value for the enthalpies of vaporization of isopropyl alcohol and benzene. How do these values compare with the enthalpy of vaporization of water? How would you explain the difference?

Extension and Application

1. What is superheated steam? How is this substance used for practical purposes?
2. How is the enthalpy of vaporization related to the boiling point of a liquid and the strength of intermolecular forces in the liquid?
3. A burn caused by exposure to steam is more harmful than a burn caused by exposure to boiling water. Why do you think this is so?

Graham's Law of Diffusion

Molecules are constantly in motion, and their kinetic energy depends on their mass and velocity. At the same temperature and pressure, the average kinetic energy of the molecules in any gas is the same. For this to be true, more massive molecules must move more slowly. One way to observe this difference is in the rate of diffusion of gases. A gas of lower molar mass diffuses faster than a gas of higher molar mass. The purpose of this activity is to discover the relationship between the mass and velocity of two different gases.

Objectives

- **Observe** the color change of universal acid-base indicator exposed to hydrochloric acid and ammonia (a base).
- **Determine** the relative velocities of ammonia and hydrogen chloride molecules.
- **Identify** the mathematical relationship between the mass and velocity of two different gases.

Materials

96-well microplate
plastic microtip pipets (3)
soda straws
plastic sandwich bag with zipper seal
concentrated ammonia solution, $NH_3(aq)$
concentrated hydrochloric acid, HCl (aq)
universal indicator

Procedure

1. **Hypothesize** which will move faster, NH_3 molecules or HCl molecules.
2. Fill all the wells in a 96-well microplate with water by placing the plate under a tap of slowly running water.
3. Place the filled microplate on a flat surface with the numbered columns away from you and the lettered rows to the left.
4. Remove the water from wells D1 and D12 with a plastic pipet.
5. Add a single drop of universal indicator to each of the wells in the microplate except wells D1 and D12. Note the color of the indicator in a data table like the one shown.
6. Cut 6 pieces of soda straw, each approximately 3 cm long.
7. Place the soda straws in wells D1, D12, A1, A12, H1 and H12, as shown in the diagram. The straws will prevent a plastic bag from coming in contact with the surface of the liquid in the plate.
8. Hold the plate horizontally, slip it into a sandwich bag, and seal the bag.
9. Puncture the plastic bag with a pencil or pen point just above wells D1 and D12.
10. One partner should obtain a few drops of ammonia solution in a plastic pipet.
11. The other partner will obtain a few drops of hydrochloric acid in a separate plastic pipet. **CAUTION:** *Both ammonia and hydrochloric acid solutions are toxic and corrosive. Avoid inhalation of fumes or contact with skin or eyes.*
12. While one partner places the ammonia pipet into the hole above well D1, the other partner should place the hydrochloric acid pipet into the hole above well D12.
13. Simultaneously add the acid and ammonia to wells D1 and D12. Wait for 30 seconds.
14. Remove the plate from the plastic bag.

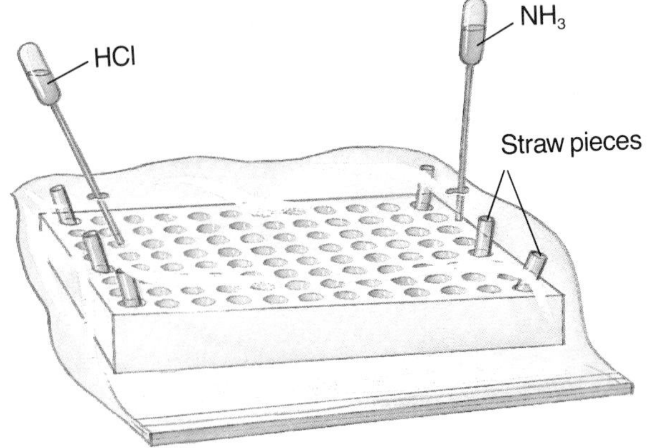

15. Record the number of wells that changed to red or yellow due to the reaction of HCl with the indicator (B in your data table).
16. Record the number of wells that changed to blue or violet due to the reaction of NH_3 with the indicator (C).
17. Dispose of the solutions in your microplate according to the instructions given by your teacher.

Data and Observations

A. Initial color of universal indicator in water	
B. Wells with red or yellow color from exposure to HCl	
C. Wells with blue or violet color from exposure to NH_3	
D. NH_3/HCl velocity ratio (C/B)	
E. Molar mass of NH_3 (g/mol)	
F. Molar mass of HCl (g/mol)	

Analysis and Conclusions

1. Find the ratio of blue or violet wells to red or yellow wells. This is the velocity ratio:

$$\frac{\text{Blue or violet wells}}{\text{Red or yellow wells}} = \frac{NH_3 \text{ velocity}}{\text{HCl velocity}}$$

   Which gas moved faster? Did your results support your hypothesis?

2. Calculate the molar masses of ammonia (E) and hydrogen chloride (F) and record them in your data table.
3. Estimate the NH_3/HCl velocity ratio (D) and enter this in your data table.
4. Calculate the four molar mass ratios for NH_3 and HCl listed below:

$$\text{Mass ratio 1} = \frac{\text{molar mass } NH_3}{\text{molar mass HCl}}$$

$$\text{Mass ratio 2} = \frac{\text{molar mass HCl}}{\text{molar mass } NH_3}$$

$$\text{Mass ratio 3} = \sqrt{\frac{\text{molar mass } NH_3}{\text{molar mass HCl}}}$$

$$\text{Mass ratio 4} = \sqrt{\frac{\text{molar mass HCl}}{\text{molar mass } NH_3}}$$

5. Compare each of the molar mass ratios with the velocity ratio. Which mass ratio is most like the estimated velocity ratio? Write the equation for the relationship of the molar masses and molecular velocities of two different gases.

Extension and Application

1. Suggest a method for determining the molar mass of a gas based on Graham's law of diffusion. Use your data to apply this method, assuming that the molar mass of HCl is unknown.
2. Uranium has several isotopes. Only one of these isotopes is useful in nuclear fission. How does Graham's law of diffusion aid in the separation of these isotopes?
3. Oxygen and nitrogen gases are common in Earth's atmosphere. Hydrogen is not. Explain this difference using the kinetic molecular theory and Graham's law of diffusion.

Molar Mass of Butane

The ideal gas equation gives the relationships among the pressure, volume, temperature, and number of moles of a gas sample.

$$PV = nRT$$

Because n, the moles of gas, is equal to the mass of the gas divided by its molar mass, the ideal gas equation provides a method for finding molar mass. It is necessary to measure the mass and volume of a gas sample and the temperature and pressure at which the volume was measured. The only unknown remaining in the equation is then the molar mass of the gas. The purpose of this activity is to determine the molar mass of butane using the ideal gas equation.

Objectives

- **Measure** the mass and volume of a sample of butane gas.
- **Measure** the temperature and pressure at which the volume was measured.
- **Determine** the molar mass of butane gas using the ideal gas equation.

Materials

thermometer barometer
100-mL graduated cylinder
plastic sandwich bag with zipper seal
empty coffee can (2- or 3-lb size)
disposable butane lighter

Procedure

1. Measure the mass of the butane lighter. Record the mass (A) in a data table like the one shown.
2. Put the lighter into a plastic bag. Place the lighter and bag on a flat surface, squeeze as much air out of the bag as possible, and seal the bag.
3. Set a large coffee can in the bottom of a sink and fill it to the top with water.
4. Find the release valve on the lighter and allow butane to fill the bag.
5. Place the filled bag into the can of water. Allow water to spill over the top of the can until the entire bag is below the surface.
6. Remove the bag from the can.
7. Determine the volume of water that spilled out of the coffee can. To do this, use a graduated cylinder to measure the necessary volume of water, and refill the can to the top. Enter this volume (D) in your data table.
8. Open and empty the bag of gas in the fume hood or near an open window.
9. Remove the lighter from the bag. Measure the mass of the lighter again and record the value (B).
10. Measure out 50 mL of water in the graduated cylinder. Lower the lighter into the graduated cylinder and record the new volume (F). Remove the lighter.

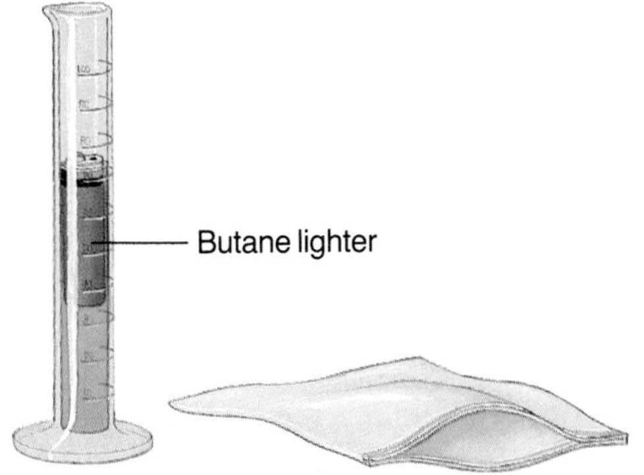

11. Refill the graduated cylinder with water up to the 80-mL mark.
12. Flatten your plastic bag. Fold the bag into a convenient shape to fit into the graduated cylinder.
13. Immerse the bag completely and record the new volume (I).
14. Use your thermometer to measure and record the temperature of the room (L).
15. Read the atmospheric pressure in the room from a barometer and record the pressure in millimeters of mercury (M).

Data and Observations

A. Mass of butane lighter (g)	
B. Mass of lighter after releasing gas (g)	
C. Mass of butane used (A − B) (g)	
D. Volume of water displaced (volume of butane, uncorrected) (mL)	
E. Volume of water in graduated cylinder	50 mL
F. Volume of water and lighter	
G. Volume of lighter (Volume − 50 mL)	
H. Volume of water in graduated cylinder	80 mL
I. Volume of water and plastic bag (mL)	
J. Volume of bag (Volume − 80 mL)	
K. Volume of gas in bag (D − (G + J)) (mL)	(____ L)
L. Temperature of the room (°C)	
M. Barometric pressure (mm Hg)	(____ kPa)
N. No. of moles of gas (mol)	
O. Molar mass of gas (g/mol)	

Analysis and Conclusions

1. Find the values of data table entries C, G, J, and K.
2. The data must be converted to appropriate units for use in the ideal gas equation with $R = 8.31$ L·kPa/mol·K. Convert the pressure (M) in mm Hg to pressure in kPa (1 kPa = 7.50 mm Hg).
3. Convert the volume of the gas sample (K) to liters (1 L = 1000 mL).
4. Convert the temperature (L) to Kelvin (C° + 273 = Kelvin).
5. Using the values of *P*, *V*, and *T* from analysis steps 2, 3, and 4 in the ideal gas equation, calculate the number of moles of gas in your sample and record the value (N).
6. Use the mass of gas in your sample (C) and the number of moles of gas (N) to calculate the molar mass of butane (O).

Extension and Application

1. Write the form of the ideal gas equation used to find the density of a gas. At the same temperature and pressure, which of the gases—hydrogen, hydrogen chloride, helium, or oxygen—will be most dense? Which will be least dense?
2. Gases are usually stored and transported under pressure as liquids. Butane is sold in small, light containers (cigarette lighters) but helium for balloons is sold in heavy metal tanks. Why?
3. At the same temperature and pressure, which weighs more, 1 L of ammonia or 1 L of hydrogen chloride gas?
4. At the same temperature and pressure, which will occupy a larger volume, 100 g of helium or 100 g of oxygen?

Colloids

Colloids are intermediate between solutions and suspensions. Solutions are homogeneous mixtures of solvent and solute. Suspensions are mixtures of a solvent and insoluble particles so large that they do not remain in suspension but eventually settle out. Colloids are mixtures of a continuous phase (like the solvent) and a dispersed phase composed of insoluble particles so small that they do not settle out. The purpose of this activity is to compare suspensions with colloids.

Objectives

- **Use** chemical reactions to produce suspensions or colloids.
- **Compare** the properties of suspensions and colloids.

Materials

plastic microtip pipets (5)
24-well microplate
toothpicks
hand lens or low-power microscope
0.1*M* iron(III) nitrate, $Fe(NO_3)_3$
0.1*M* aluminum nitrate, $Al(NO_3)_3$
0.3*M* sodium hydroxide, NaOH
0.15*M* sodium thiosulfate, $Na_2S_2O_3$
6*M* hydrochloric acid, HCl(aq)
Microplate Data Form

Procedure

1. Place a 24-well microplate on a flat surface with the numbered columns away from you and the lettered rows to the left.
2. Use a microtip pipet to place 30 drops of 0.1*M* iron(III) nitrate solution in well A1.
3. Use a clean pipet to place 30 drops of 0.1*M* aluminum nitrate solution in well A2.
4. Place 30 drops of 0.15*M* sodium thiosulfate solution in well A3.
5. Add 30 drops of 0.3*M* sodium hydroxide solution to well A1 and 30 drops to well A2.
6. Stir the contents of wells A1 and A2 with different toothpicks.
7. Observe the contents of wells A1 and A2 with a hand lens or low-power microscope. Record your observations in a Microplate Data Form like the one shown here.

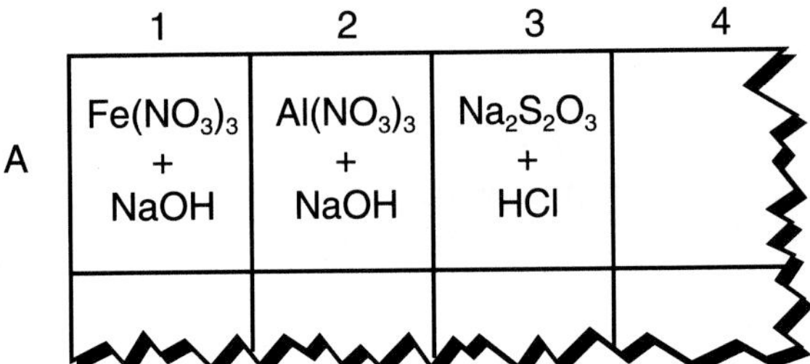

8. Add 30 drops of 6*M* hydrochloric acid to well A3. Stir with a different toothpick.
9. Observe the contents of well A3 with a hand lens or low-power microscope. Record your observations.
10. After 10 minutes, reexamine the contents of all three wells and record your observations. Make certain that you also observe the mixtures from the side of the plate.

Data and Observations

1. Complete and balance the following equations:
   Reaction in well A1
   $Fe(NO_3)_3(aq) + NaOH\ (aq) \longrightarrow ?$
   Reaction in well A2
   $Al(NO_3)_3(aq) + NaOH(aq) \longrightarrow ?$
   Reaction in well A3
   $Na_2S_2O_3(aq) + HCl(aq) \longrightarrow S(cr) + SO_2(g) + ?$

Analysis and Conclusions

1. How could you tell that a chemical reaction occurred in each well?
2. In which well(s) did the reaction produce a suspension? Explain why you reached this conclusion.
3. In which well(s) did the reaction produce a colloid? Explain why you reached this conclusion.

4. Identify the continuous and dispersed phases in your colloid(s).

Extension and Application

1. Describe two more tests with larger samples that you could use to distinguish between your suspensions and colloids. What would be the results of each with your three mixtures?
2. Mayonnaise is a type of colloid known as an emulsion. Look up the definition of an emulsion. Of the three main ingredients in mayonnaise—oil, water, and egg yolk—identify the emulsifying agent, the continuous phase, and the dispersed phase.

Chapter 21
ChemActivity

Super Steam

Pure water boils at 100°C at one atmosphere of pressure. The addition of soluble inorganic or organic compounds changes the boiling point of water. This phenomenon is an example of a colligative property of a solvent. The purpose of this activity is to observe the effects of a solute on the boiling point of water.

Objectives

- **Observe** boiling point elevation.
- **Determine** the effect of solute concentration on the boiling point of a solvent.
- **Determine** experimentally the value of the molal boiling point constant for water.

Materials

notebook paper
100-mL beaker
hot plate
thermometer
stirring rod
graph paper
sodium chloride

Procedure

1. **Hypothesize** what will happen to the boiling point of water as the concentration of added sodium chloride is increased.
2. Prepare five 0.58-g samples of sodium chloride on individual pieces of paper.
3. Measure the mass of a 100-mL beaker and record the mass in a data table like the one shown.
4. Place 50 mL of water in the 100-mL beaker. Measure the mass of the beaker and water. Record this new mass.
5. Heat the water in the beaker on a hot plate until the water boils.
6. Measure and record the temperature of the boiling water to the nearest 0.1°C.
7. Add one of the prepared samples of NaCl. Stir until the NaCl is completely dissolved.
8. Measure the temperature again and record the results.
9. Repeat steps 7 and 8 until all the NaCl samples have been used.

Data and Observations

	Trial 1	Trial 2	Trial 3
NaCl mass (grams)	0	0.58	1.16
Temp. boiling water/NaCl mixture (°C)			
Moles of NaCl			
Moles of ions			
Moles of ions/kilogram of water			

A. Mass of beaker (g)	
B. Mass of beaker and water (g)	
C. Mass of water (B – A) (g)	(_____ kg)

1. What happened to the temperature of the water/NaCl mixture as more NaCl was added?
2. What happened to the boiling water when more NaCl was added?

Analysis and Conclusions

1. For each trial, convert grams of NaCl to the number of moles of NaCl and record the results.
2. For each trial, convert the number of moles of NaCl to the number of moles of ions (moles NaCl × 2 ions/mole). Record the results.
3. How many kilograms of water did you use (Data C)(1000 g/kg)?
4. Complete the data table entries for moles of ions per kilogram of water.
5. Construct a graph of your results. Plot total moles of ions/kg of water on the x-axis and the temperature of the boiling water on the y-axis.
6. Draw the best-fitting straight line through your points.
7. Find the slope of this line. What are the units of the slope of the line? The slope is the molal boiling point constant. Compare your value with the actual value of 0.515 C°/m.

Extension and Application

1. How would the results differ if you repeated this activity with acetone instead of sodium chloride?
2. Look up the meaning of the term *azeotrope,* a special kind of solution of two liquids. How is an azeotrope different from other solutions?

Chapter 22
ChemActivity

Preparation and Properties of Oxygen

Oxygen gas, O_2, can be prepared in the laboratory by releasing the element from a compound that contains oxygen. Decomposition reactions usually require the addition of energy as heat or electrical current in order to take place. In some cases, addition of a catalyst will cause decomposition to occur. Sodium hypochlorite decomposes to give oxygen and sodium chloride in a reaction catalyzed by the cobalt ion, Co^{2+}:

$$2NaOCl\ (aq) \xrightarrow{Co^{2+}} 2NaCl\ (aq) + O_2\ (g)$$

Sodium hypochlorite is found as a 5 percent solution in household bleach. The purpose of this activity is to prepare oxygen gas from sodium hypochlorite.

Objectives

- **Prepare** oxygen from a compound containing oxygen.
- **Observe** the action of a catalyst.
- **Investigate** the properties of oxygen.

Materials

thin stem micropipet
microtip pipets (2)
24-well microplate
household bleach (5% NaOCl)
0.1M Cobalt(II) nitrate, $Co(NO_3)_2$ (aq)
toothpicks
matches

Procedure

1. Use a microtip pipet to place 30 drops of household bleach in well A1 of a 24-well microplate. Wash the pipet with water.

2. Trim the stem of a thin-stem pipet to a length of about 2.5 cm. Label this pipet G; it is the generator pipet.
3. Draw up the 30 drops of household bleach into the bulb of pipet G. Stand the pipet up in well A6 until you are ready to use it, as shown in Figure A.
4. Using another microtip pipet, place 10 drops of cobalt nitrate solution in well A2 of the microplate. Wash the pipet with water.
5. Cut the stems of the two microtip pipets to a length of 1 cm. These will be the collector pipets, C1 and C2.
6. Place pipets C1 and C2 under water with the stems pointing up in a large beaker. Squeeze the bulbs repeatedly until no more air is expelled. The pipets should be completely filled with water. Stand the two pipets up in wells C1 and C2 as shown in Figure A.

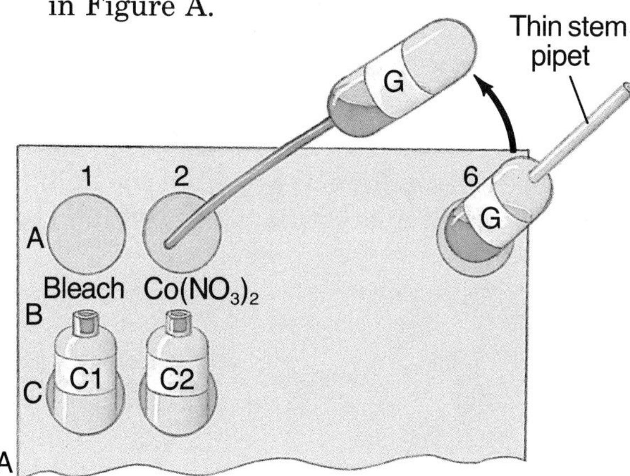

7. Hold pipet G, containing the NaOCl solution, with the stem pointing up. Carefully squeeze the air out of pipet G and insert the stem into well A2. Draw up the cobalt nitrate solution from the well.
8. Immediately, remove pipet G, insert the short tube of pipet G into the stem of pipet C1, and stand pipet G with C1 in well C3 as shown in Figure B. As the reaction continues and oxygen gas collects in pipet C1, the displaced water will collect in microplate well C3.

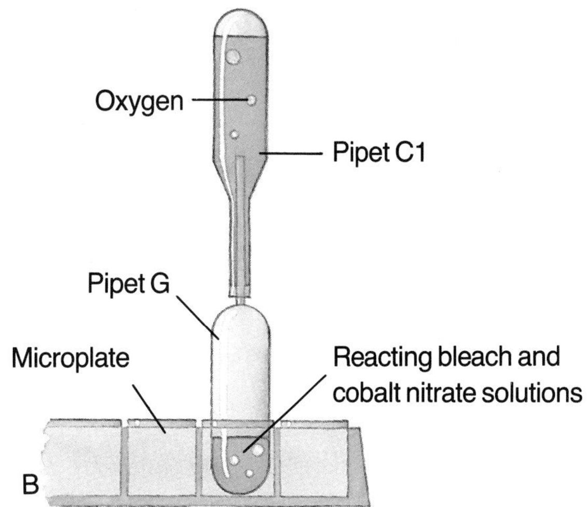

9. Replace pipet C1 with C2 when C1 is almost filled with oxygen gas.
10. Allow a small "plug" of water to be trapped in the neck of the pipet as shown in Figure C. Stand the filled pipet in an empty well of the microplate.

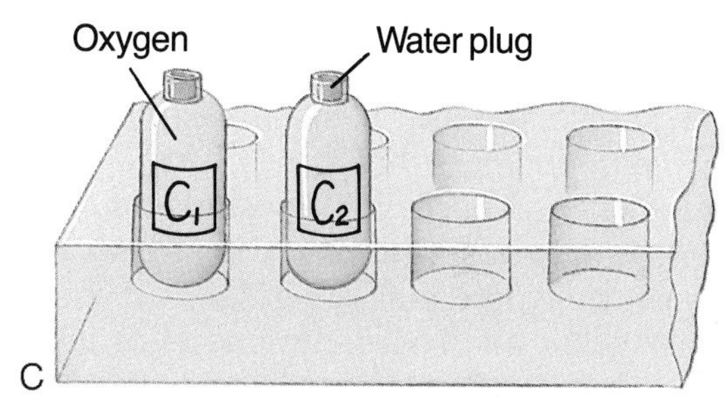

11. Light a toothpick with a match. Gently blow out the flame so that the tip of the toothpick remains glowing.
12. Hold the stem of the gas-filled pipet close to the toothpick and squeeze the bulb of the pipet to force the oxygen gas over the glowing end.

Data and Observations

1. What physical properties of oxygen did you observe?
2. What happened when the glowing toothpick was placed in pure oxygen?

Analysis and Conclusions

1. Is oxygen gas soluble in water? What evidence can you offer to support your answer?
2. Does oxygen gas burn?
3. How would it be possible to demonstrate that the cobalt ion was a catalyst in this reaction?
4. Explain what happened to the reaction rate when the glowing toothpick was exposed to oxygen and explain why it happened.

Extension and Application

1. What other compounds could be used to isolate oxygen? Suggest another method of preparing oxygen, using a reference source if necessary.
2. Oxygen gas is often administered to athletes who have been playing at a rapid pace. Why is this done?
3. Great quantities of oxygen gas are required for missiles such as the Saturn rocket, which propelled astronauts to the moon. How is this large amount of oxygen stored in a rocket?

Chapter 23
ChemActivity

Acidic and Basic Anhydrides

When dissolved in water, the oxide of a metal forms a basic solution. The oxide of a nonmetal, when dissolved in water, forms an acidic solution. Unless water is added to them, the oxides of both metals and nonmetals are anhydrides, that is, compounds without water. Since a metallic oxide in water forms a basic solution, a metallic oxide is called a basic anhydride. The purpose of this activity is to form acidic and basic solutions using an acidic anhydride and a basic anhydride.

Objectives

- **Use** the oxide of a metal to form a basic solution.
- **Use** the oxide of a nonmetal to form an acidic solution.
- **React** an acid and a base to form a salt.

Materials

thin stem micropipet	tap water
24-well microplate	2*M* HCl
universal indicator	scissors
calcium oxide (CaO)	toothpick
marble chips	clear tape
Microplate Data Form	

Procedure

1. **Form a hypothesis** about whether the solutions formed by dissolving a calcium oxide in tap water and a carbon oxide in tap water will be acidic or basic.
2. Use scissors to make a slit in the bulb of a thin stem micropipet, as shown in Figure A.

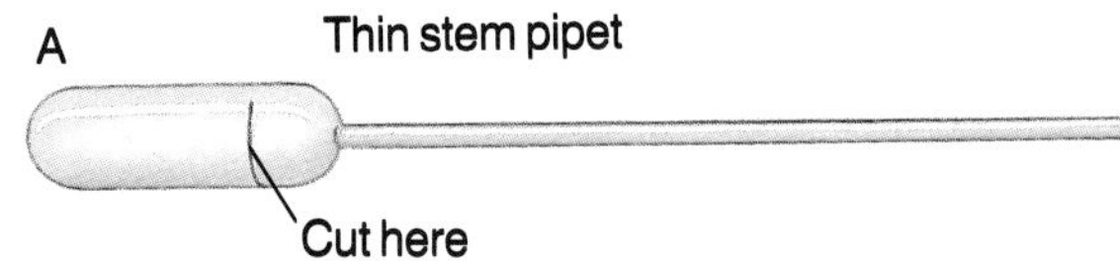

3. Push a small piece of marble through the slit into the bulb.
4. Seal the slit with a piece of clear tape.
5. Use Figure B as a guide to the following procedures.
6. Using a clean micropipet, place 1/2-pipet of tap water in wells C2, C3, and C4 of the microplate.
7. Add 2 drops of universal indicator to wells C2, C3, and C4. Note the color of the universal indicator and the approximate pH in your Microplate Data Form.

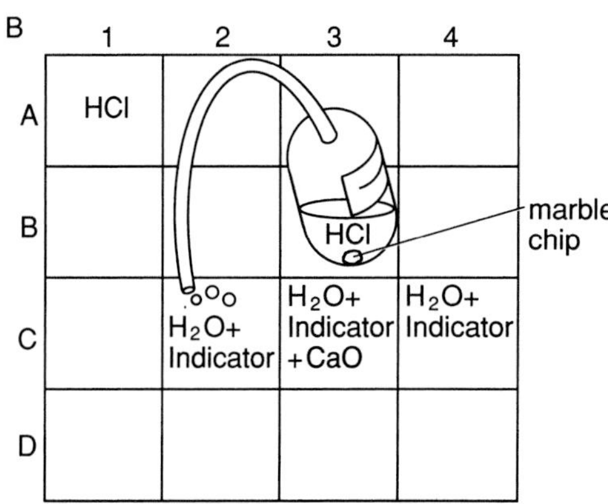

8. Using a clean pipet, place 1/2-pipet of HCl in well A1 of the microplate.
9. Squeeze the bulb of the micropipet containing the piece of marble to eject the air from the pipet. Continue to hold the bulb so that air cannot enter the pipet.
10. Invert the micropipet and place the stem into well A1. Release the bulb of the micropipet and draw up approximately 1/4-pipet of HCl into the bulb.
11. Place the bulb of the micropipet in well B3 and direct the stem into well C2.
12. Allow the gas produced to bubble through well C2. Note the color of the universal indicator-water solution in your Microplate Data Form.
13. Using a chemical scoop, place a few crystals of calcium oxide into well C3. Note the color of the universal indicator-water solution in your Microplate Data Form.
14. Allow the solid to settle out in well C3. Remove approximately 1/8-pipet of the liquid from well C3.
15. Place 10 drops of liquid from well C3 into well D1.
16. Using a clean micropipet, remove approximately 1/8-pipet of liquid from well C2. Place 10 drops of the liquid from well C2 into well D1.
17. Mix the liquid in D1 thoroughly with a toothpick. Note the color of the universal indicator-water solution in your Microplate Data Form.

Data and Observations

1. Label a Microplate Data Form as indicated by Figure B and use it to record your observations.
2. What color was the universal indicator in tap water alone?
3. What was the color of the universal indicator when the gas was bubbled through the water in well C2?
4. What was the color of the universal indicator when calcium oxide was added to the water in well C3?
5. What happened to the universal indicator when the two solutions were mixed in well D1?

Analysis and Conclusions

1. Balance this equation, which describes the reaction in the micropipet containing the marble chip.

$$CaCO_3 + HCl \rightarrow CO_2 + H_2O + CaCl_2$$

2. Balance this equation, which describes what happened to the water in well C2.

$$CO_2 + H_2O \rightarrow H_2CO_3$$

Look at your Microplate Data Form and determine whether CO_2 is an acid anhydride or a basic anhydride.

3. Balance this equation, which describes what happened when calcium oxide was added to the water in well C3.

$$CaO + H_2O \rightarrow Ca(OH)_2$$

Determine whether CaO is an acid anhydride or a basic anhydride.

4. Balance this equation, which describes the reaction in well D1.

$$Ca(OH)_2 + H_2CO_3 \rightarrow CaCO_3 + H_2O$$

Extension and Application

When a salt is dissolved in water, the salt solution almost never has a neutral pH. Why? What are the products when a salt is dissolved in water? How is the pH of a solution changed by a salt?

Buffers

When an acid reacts with a base, a salt and water are formed. A solution that contains either a weak acid and the salt of a weak acid or a weak base and the salt of a weak base is called a buffer system. Buffer systems have special properties that are of great importance in certain biological and chemical processes. The purpose of this activity is to compare the reactions of strong acid and base combinations with the reactions of buffer systems.

Objectives

- **Develop** a buffer system.
- **Investigate** the properties of a buffer system.

Materials

24-well microplate	0.1*M* solutions of the following:
96-well microplate	HCl (strong acid)
microtip pipets (4)	NaOH (strong base)
drinking straws	$HC_2H_3O_2$ (weak acid)
scissors	NH_3 (aq) (weak base)
distilled water	
universal indicator (with color scale)	
toothpick	
labels	

Procedure

Part A

1. With drinking straws, construct a ringstand, following the instructions in the diagram. Set up the ringstand in well A1 in the 96-well microplate.
2. Label four microtip pipets HCl, NaOH, $HC_2H_3O_2$, and NH_3 (aq).
3. Fill the labeled pipets with HCl, NaOH, $HC_2H_3O_2$, or NH_3 (aq) solution. Then store the pipets, tip up, in row D of the 24-well microplate.
4. Place 20 drops of distilled water in well A1 of the 24-well plate.
5. Add 10 drops of HCl to the distilled water in well A1.
6. Place the NaOH pipet in the straw ringstand, as shown in the diagram.
7. Add 2 drops of universal indicator to well A1 of the 24-well plate.
8. Record in the data table the pH of the solution in well A1.
9. Add 2 drops of NaOH solution to well A1. Stir with a toothpick.
10. Record in the data table the pH of the solution in A1.
11. Repeat steps 9 and 10 until a pH of 12 or greater is reached.

Part B

Repeat the procedure in steps 4–11 of Part A, using HCl in well A2 and NH_3 (aq) in the pipet.

Part C

Repeat the procedure in steps 4–11 of Part A, using $HC_2H_3O_2$ in well A3 and NH_3 (aq) in the pipet.

Part D

Repeat the procedure in steps 4–11 of Part A, using $HC_2H_3O_2$ in well A4 and NaOH in the pipet.

Data and Observations

Drops of Base	pH of Solution PART			
	A	B	C	D
0				
2				
4				
6				

1. What is the general effect of a base on an acid?
2. Was the number of drops of base required to neutralize the acid the same, regardless of which acid and which base were involved?

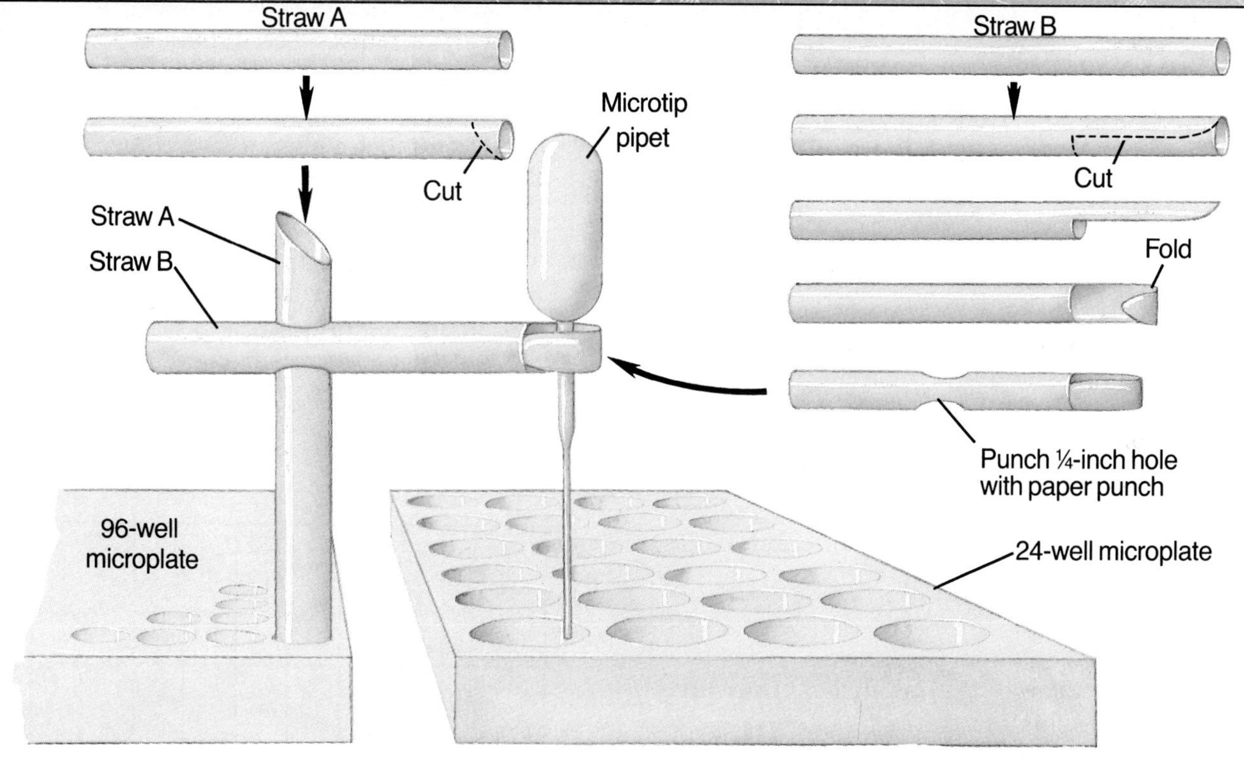

Analysis and Conclusions

1. Make a graph of your data. Make the x-axis the number of drops of base used and the y-axis the pH of the solution. Graph each set of data on the same piece of graph paper.
2. How does neutralization of strong and weak acids and strong and weak bases compare?
3. What data supports your conclusions?
4. Look at your graphs. Which combinations of acids and bases produced a buffer system?
5. What is the special property of a buffer system?

Extension and Application

1. Why might buffer systems be important in living things?
2. What other combinations of acids and bases would produce a buffer system?

Oxidation/Reduction of Vanadium

Vanadium, element 23 on the periodic table, is one of the transition elements of period 4. One characteristic of transition elements is that they have variable oxidation states. This means that elements 21 through 29 and the elements directly below them on the periodic table have several stable oxidation states. For example, the vanadium atom in an ionic form can have oxidation states of 5+, 4+, 3+ and 2+. The purpose of this activity is to prepare ions of an element that are in different oxidation states.

Objectives

- **Prepare** ions of the same element that have different oxidation states.
- **Compare** some properties of the element in different oxidation states.

Materials

24-well microplate	0.1*M* NH_4VO_3
microtip pipets (2)	household bleach (diluted 1:1)
scissors	white paper
zinc granules	
micropipet	

Procedure

Part A Reduction of Vanadium

1. **Form a hypothesis** about whether vanadium ions in different oxidation states will have different physical and chemical properties.
2. Place a 24-well microplate on a piece of white paper with the numbered columns away from you and the lettered rows to the left.

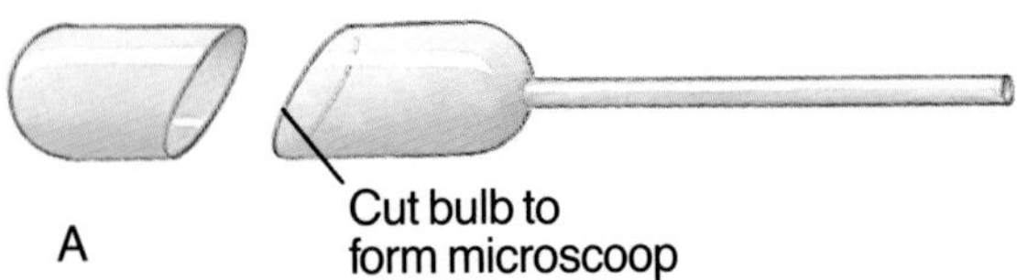

3. Make a microscoop by cutting the bulb of of a micropipet, as shown in Figure A.
4. Place a half microscoop of zinc metal into well A1.
5. Using a microtip pipet, place 1/2-pipet of NH_4VO_3 in well A2. Ammonium vanadate contains vanadium in the 5+ state. Well A2 will serve as a control well. Use the diagram of your microplate in Figure B as a guide to the steps that follow.

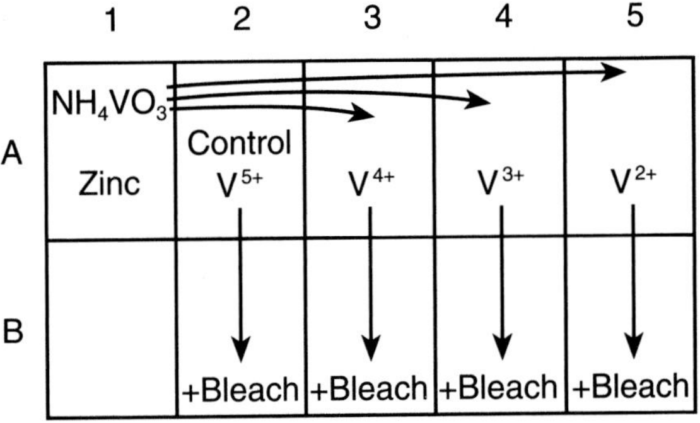

B

6. Three quarters fill a microtip pipet with NH_4VO_3 solution. Place the NH_4VO_3 in well A1 with the zinc metal.
7. Draw the liquid in well A1 back up into the microtip pipet. Then return the liquid to well A1. This is one pass. Repeat this process for as many passes as necessary until the solution turns color. The solution of vanadium is now in a V^{4+} state. Record the number of times you must return the V^{5+} solution to the zinc for a color change in a data table like the one shown.
8. Place 20 drops of the V^{4+} solution into well A3.
9. Place the rest of the V^{4+} solution remaining in the pipet back into well A1 with the zinc metal.
10. Draw the liquid in well A1 back up into the pipet. Then return the liquid to well A1. Repeat passes until the solution turns color. The solution of vanadium in A1 is now in the V^{3+} state. Record the number

of times you must return the solution to the zinc for a color change in your data table.

11. Place 20 drops of the V^{3+} solution into well A4.
12. Place the rest of the V^{3+} solution remaining in the pipet back into well A1 with the zinc metal.
13. Draw the liquid in well A1 back up into the pipet. Then return the liquid to well A1. Repeat passes until the solution turns color. The solution of vanadium in A1 is now in the V^{2+} state. Record the number of times you must return the solution to the zinc for a color change in your data table.
14. Place 20 drops of the V^{2+} solution in well A5. Return any remaining solution to well A1.

Part B Oxidation of Vanadium

1. Rinse your micropipet and transfer 10 drops of the contents of well A5 to well B5. Rinse your micropipet again.
2. Transfer 10 drops of the contents of well A4 to well B4. Rinse your micropipet.
3. Transfer 10 drops of the contents of well A3 to well B3.
4. Fill a clean micropipet with diluted household bleach.
5. Drop by drop, add the diluted bleach to well B5 until a color change occurs. Record your observations and the number of drops required in the column labeled "Color with Bleach" in your data table.
6. Following the procedure in step 5, add diluted bleach to wells B4, B3, and B2. Record your observations.

Data and Observations

Well	Number of Passes	Color	Oxidation State	Color with Bleach
A2				
A3				
A4				
A5				

Analysis and Conclusions

1. How many changes in oxidation state did vanadium go through in Parts A and B?
2. Which change in oxidation state required the greatest number of passes through the zinc metal?
3. What happened to the solutions in the A row when the bleach was added?
4. If V^{2+} is allowed to stand for any length of time, it reverts to V^{5+}. Explain why you think this happens.

Extension and Application

Iron as Fe^{2+} is an essential nutrient, but iron as Fe^{3+} has no value as a nutrient. Some iron-enriched cereals contain iron filings. What is the advantage of using iron in this form?

Lemon Battery

A battery is a device in which chemical energy is converted to electrical energy. The energy obtained from a battery is produced by a difference in activity of two different metals. When two metals are placed in an electrolyte and are connected by a conductor, electrons flow from the more active metal through the conductor to the less active metal. The flow of electrons is an electrical current and can be made to do work. The purpose of this activity is to investigate the activity of different metals in various combinations in a simple battery. Note that carbon is a conductor and will be considered a metal in this activity.

Objectives

- **Make** a simple battery, using a lemon.
- **Compare** the activity of metals used in different combinations in the battery.

Materials

lemon (1/4 per group)
carbon rod (pencil lead will do)
voltmeter or multimeter
short lengths of connecting wire
alligator clips — small knife
chemical scoop
small strips of the following metals:
magnesium — lead
zinc — copper

Procedure

1. Using a chemical scoop, pierce the flesh of a piece of lemon in two places approximately 1 cm apart.
2. Select a strip of two different metals.
3. Insert each strip of metal into a different slit in the flesh of the lemon, as shown in the Figure.
4. With connecting wire and alligator clips, connect each metal strip to the voltmeter. Observe the needle and scale of the voltmeter as you complete the connection. If the needle does not move or you get a negative reading, reverse the connections to the metals. Note the polarity of the metals, that is, whether the metal is positive or negative. Record the voltmeter reading and the polarity (+ or −) of each metal in a data table like the one shown.
5. Remove the metal strips from the lemon and rinse them in tap water.
6. Repeat steps 2–5 until you have tested each of the ten possible combinations.

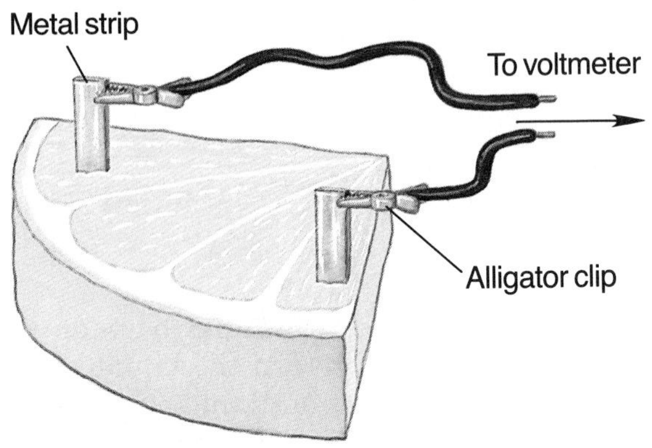

Data and Observations

Polarity of Metals				
Pb	C	Zn	Mg	
				Cu
			Mg	
		Zn		
	C			

Analysis and Conclusions

1. Which pair of metals gave the highest reading on the voltmeter?
2. Which pair of metals gave the lowest reading on the voltmeter?

Extension and Application

Compare your battery with a commercially produced battery. In what ways are the two batteries similar? How do they differ?

Physical and Chemical Thermodynamics

Chemical reactions are accompanied by two driving forces—the tendency to reach minimum energy (enthalpy) and the tendency to reach maximum disorder (entropy). Although some reactions tend toward more order or higher energy, these reactions are the exceptions. All chemical reactions eventually cease when maximum entropy and minimum enthalpy are reached. The purpose of this activity is to investigate the flow of energy and the effects of energy on physical and chemical systems.

Objectives

- **Explore** enthalpy and entropy, using a physical system as a model for chemical reactions.
- **Classify** reactions as exothermic or endothermic and as entropic or more orderly.

Materials

- thick rubber band
- hot plate
- 250-mL beakers (2)
- 1- or 2-kg mass
- ringstand and iron ring
- 24-well microplate
- scissors
- ruler
- ammonium chloride (solid)
- sodium hydrogen carbonate (solid)
- hydrochloric acid (6*M*)
- thermometer
- thin stem micropipets (2)
- toothpicks
- Microplate Data Form

Procedure

Part A

1. Half-fill a 250-mL beaker with water and place the beaker on a hot plate to heat.
2. While the water is heating, put a thick rubber band around both of your index fingers and hold the fingers apart so that the rubber band is at its full length, *but is not stretched.* Briefly touch the rubber band to your upper lip and sense the "temperature" of the band.
3. Move your index fingers apart to stretch the rubber band as much as you can. Briefly touch the stretched rubber band to your lip. Notice whether you feel a change in the "temperature" from your earlier observation. Record your observations in a data table like the one shown.
4. Slip the rubber band onto the ring attached to the ringstand and let it dangle. Hang a mass weighing 1 to 2 kg from the end of the rubber band, as shown in the diagram.
5. With a ruler, measure the length of the rubber band, and enter the length in your data table.
6. Place an empty 250-mL beaker below the mass and the rubber band.
7. Wearing thermal mitts, carefully pour hot water from the other beaker over the rubber band, collecting the water in the beaker beneath. Measure the length of the rubber band and record this new length.

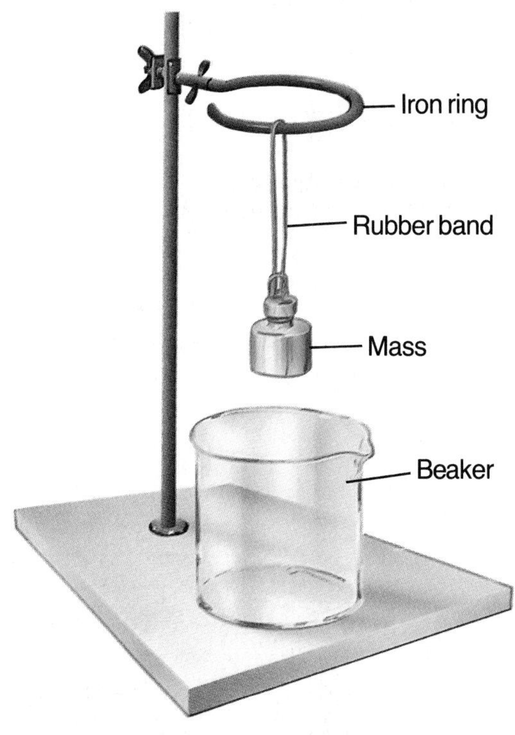

Part B

1. Make a chemical microscoop by cutting the end off the bulb of a micropipet. (See ChemActivity 25, Procedure Step 3.)
2. Place 1/2-microscoop of solid ammonium chloride in well A1 and well B1 of a 24-well microplate. Add 1/2-microscoop of solid sodium hydrogen carbonate to well A2 and well B2.
3. With the thermometer, measure the temperature of the solid chemicals in wells A1, A2, B1, and B2. Record these data in your Microplate Data Form.
4. Using the thin stem pipet, place 1/2-pipet of water in wells A1 and A2. Stir with a toothpick. Enter any physical changes you observe in your Microplate Data Form.
5. With the thin stem pipet, add HCl to wells B1 and B2. Stir with a toothpick. Enter any physical changes you observe in your Microplate Data Form.
6. Use the thermometer to measure the temperature of the chemicals as they dissolve in wells A1, A2, B1, and B2. Be sure to rinse your thermometer in cold water between each reading. Record these data in your Microplate Data Form.

	1	2
A	NH_4Cl + H_2O	$NaHCO_3$ + H_2O
B	NH_4Cl + HCl	$NaHCO_3$ + HCl

Data and Observations

Part A

A	The stretched rubber band felt warmer/cooler than the unstretched rubber band.	
B	Length of stretched rubber band at room temperature (cm)	
C	Length of stretched and heated rubber band (cm)	

Part B

Record your observations in the appropriate boxes of your Microplate Data Form as indicated here.

Analysis and Conclusions

Use your data and the symbols in the table below to analyze the reactions in Part A and Part B. Write the reaction and replace each question mark by the appropriate sign for ΔH, ΔS, and ΔG.

$\Delta H+$	=	endothermic (cool)
$\Delta H-$	=	exothermic (warm)
$\Delta S+$	=	more disorder
$\Delta S-$	=	less disorder
$\Delta G+$	=	not spontaneous
$\Delta G-$	=	spontaneous

1. unstretched band → stretched band
   ΔH? ΔS? ΔG?
2. $NH_4Cl(cr) \rightarrow NH_4^+(aq) + Cl^-(aq)$
   ΔH? ΔS? ΔG?
3. $NaHCO_3(cr) \rightarrow Na^+(aq) + HCO_3^-(aq)$
   ΔH? ΔS? ΔG?
4. $NaHCO_3(cr) + HCl \rightarrow Na^+(aq) + Cl^-(aq) + CO_2(g) + H_2O$
   ΔH? ΔS? ΔG?
5. Which reaction took place spontaneously?
6. What would be the reverse reaction with the rubber band?

Extension and Application

1. Find out about the familiar reactions named below. What are the signs (+ or −) for ΔG, ΔH, and ΔS for each system?
   a. photosynthesis
   b. rusting of a car
   c. formation of a diamond

A Half-Life Model

Radioactive isotopes are unstable atoms that decompose spontaneously to the atoms of a different element. The breakdown of atoms takes place at a set rate, called half-life, which differs for each radioactive isotope. Half-life is the amount of time it takes for one-half the atoms in a sample of a radioactive isotope to decay to the atoms of a different element. Because it is not practical for you to study the half-life of a real radioactive isotope, you will use a model of such an element in this activity. The purpose of this activity is to explore the phenomenon of half-life.

Objectives

- **Use a model** to study half-life.
- **Construct** a decay curve of the model atoms.
- **Use a model** to evaluate the effect of a catalyst on half-life.

Materials

- split peas (100 per group)
- lima beans (10 per group)
- 250-mL beakers (2 per group)
- clock or watch
- graph paper

Procedure

Part A

1. Label the beakers *Decayed Atoms* and *Undecayed Atoms*. Make a table like the one shown.
2. Count out 100 split peas and place them in the beaker labeled Undecayed Atoms. Each pea will represent one atom of the imaginary element peanium.
3. Note the time in your data table.
4. Shake the Undecayed Atoms beaker and dump the split peas out onto the table.
5. Note that some peas landed curved side down and the others landed curved side up. Separate the peas into two groups, according to the position, up or down, of the curved side. The group with the curved side down will represent the peanium atoms that have decayed to the atoms of a different element. The group with the curved side up will represent the undecayed peanium atoms.
6. Count the undecayed peanium atoms and record the number in your data table.
7. Place the decayed peanium atoms in the Decayed Atoms beaker. Return the undecayed atoms to the Undecayed Atoms beaker.
8. Repeat steps 4–7 until there are no more atoms in the Undecayed Atoms beaker. Note the time in your data table.

Part B

Repeat Part A, but this time beans will represent beanium atoms. Add 10 lima beans to the peanium atoms in step 2. Since the sides of the lima beans do not differ significantly, always sort the beanium atoms into the group of undecayed peanium atoms.

Data and Observations

	Part A	Part B
Time: start end total		
Number of runs		
Number of Undecayed Atoms		
Run #	Part A	Part B
1 (Start)		
2		
3		
4		
5		

Analysis and Conclusions

1. Make a graph of your results. Place time (the run number), on the x-axis and the number of remaining undecayed peanium atoms on the y-axis.
2. Regraph your data using time on the x-axis and the natural logarithm of the number of remaining peanium atoms on the y-axis. Remember, the natural logarithm of a number is 2.3 times the logarithm of the number ($\ln x = 2.3 \log x$).
3. Which of the graphs is linear?
4. What is the relationship between time and the number of remaining peanium atoms?
5. In your experiment, how did beanium affect the rate of decomposition of peanium?
6. Assuming that lima beans are a valid model for the atoms of any catalyst, what statement can you make about the role of catalysts in the decay of radioisotopes?

Extension and Application

1. The rate of decay of every radioactive element is a constant logarithmically. How do scientists use this constant to determine the age of natural objects, such as rocks and bones, and materials made by humans, such as cloth and baskets?
2. Find out if any physical processes, such as heat or pressure, can change the half-life of an element.
3. Nuclear power plants use radioactive materials to produce power. Some of the waste products from these plants are radioactive. Research the topic of radioactive waste management and report to the class on the problems associated with nuclear waste.

Chapter 29
ChemActivity

Formation of Ethyl and Butyl Acetates

When an organic acid reacts with an alcohol, an organic compound called an ester is formed. Esters are noted for their aromas and for their often distinctive flavors. Esters are responsible for the odors and flavors of many fruits. The artificial flavors and fragrances that are added to many food and toiletry products are also usually esters. The general reaction for the formation of an ester is:

$$R\text{—}CH_2OH + HOOC\text{—}R' \rightarrow R\text{—}CH_2OOC\text{—}R' + H_2O$$

The symbols R and R′ represent chains of carbon atoms. Different combinations of acids and alcohols produce different esters. Some are given in the following table

Acid	Alcohol	Ester	Aroma
butyric	ethyl	ethyl butyrate	pineapple
acetic	amyl	amyl acetate	banana
acetic	ethyl	ethyl acetate	nail polish remover

The purpose of this activity is to prepare an ester.

Objectives

- **Prepare** an organic compound, an ester.
- **Compare** the properties of the ester with those of the compounds from which the ester was prepared.

Materials

- 100-mL glass beaker
- thermometer
- 10-mL graduated cylinder
- ethyl alcohol
- acetic acid
- sulfuric acid (concentrated)
- hot plate
- anhydrous sodium sulfate
- plastic micropipet
- plastic cup
- water
- small test tube
- paper towel

CAUTION: *Do not use a laboratory burner for this experiment. Do not allow vapors from this experiment to come into contact with an open flame from any source.*

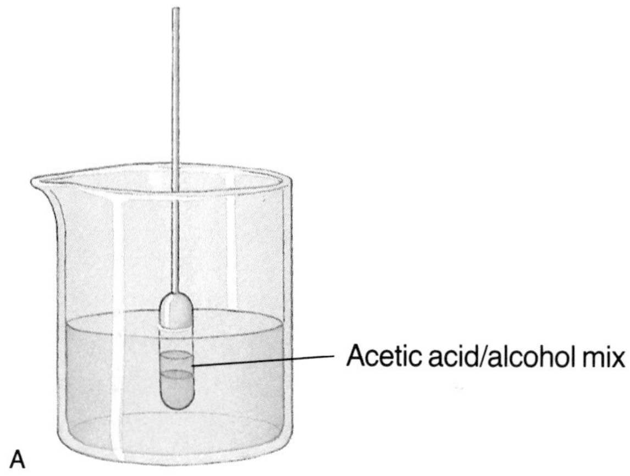

Procedure

1. Half-fill the 100-mL beaker with water.
2. Begin heating the water in the beaker to 50°C while you proceed with the ChemActivity.
3. Place 1.0 mL each of ethyl alcohol and acetic acid in a plastic cup. Note the odor of both the alcohol and the acid as you work with these compounds. Record your observations in a data table like the one shown.
4. Add 5 drops of sulfuric acid to the mixture in the plastic cup. **CAUTION:** *Handle sulfuric acid with great care. It can cause severe burns if it touches the skin and will damage clothing if it comes into contact with it.*
5. Draw up the mixture of chemicals in the plastic cup into the micropipet.
6. Rinse the plastic cup in tap water and dry with a paper towel.
7. Place the pipet with the stem pointing upward into the heated water, as shown in Figure A.
8. Maintain the temperature of the water bath at 45–50°C for 10 minutes.
9. At the end of the 10 minutes, remove the pipet from the warm water bath and place it stem upward in the plastic cup.
10. Discard the warm tap water.
11. Refill the beaker up to half-full with cold tap water.
12. Holding the pipet with the stem pointing upward, squeeze the bulb to force air out of the bulb and the stem. Continue to hold the bulb so that it cannot expand or allow air to enter.
13. Invert the pipet and place its stem in the beaker of cold water. Release the bulb so that cold water will be drawn into the pipet.
14. Holding the pipet by the bulb, swirl the pipet to mix its contents. The mixture in the pipet now has an aqueous layer at the bottom and a crude ester layer above the aqueous layer, as shown in Figure B.

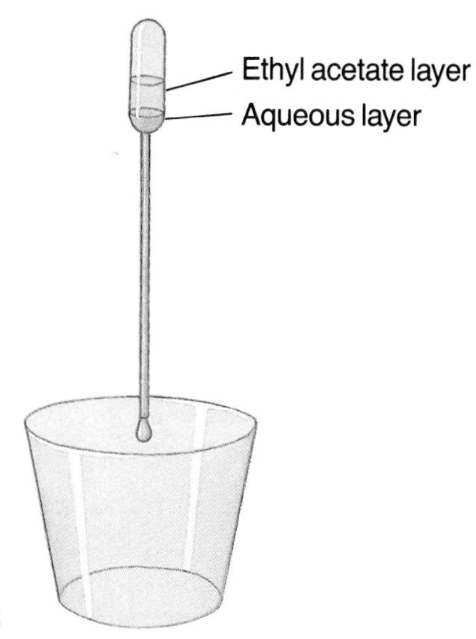

15. Separate the aqueous layer from the ester layer by holding the pipet stem downward over the beaker of water. Squeeze the bulb gently to expel only the layer of water.
16. Place 2 grams of anhydrous sodium sulfate in the plastic cup.
17. Holding the pipet stem downward over the plastic cup, squeeze the bulb to expel the ester layer and add it to the sodium sulfate. Swirl the plastic cup until the ester is clear.
18. Carefully sniff the odor of the ester and record it in your data table. Also record any other physical properties of the ester, the acid, and the alcohol.

Analysis and Conclusions

1. Did the ester formed have an odor similar to either the alcohol or the acid?
2. Of what common substance did the odor of acetic acid remind you? Ethyl alcohol? The ester, ethyl acetate?
3. What physical property showed that a new chemical compound had been formed when ethyl alcohol reacted with acetic acid?

Data and Observations

Odor and Physical Properties		
	Odor	Physical Properties
Ethyl alcohol		
Acetic acid		
Ethyl acetate		

Extension and Application

In a supermarket, look at products containing artificial flavors. If the names of the chemicals that produced the flavors are given, make a list of them. Also check reference books and textbooks to find the chemical names for artificial flavors. Choose one name that includes *ester* and find out from which acid and alcohol the ester is made.

Chapter 30
ChemActivity

Production of Aspirin

Aspirin is one of the most commonly used medicines in the world. It is effective in relieving pain such as headache and can reduce fever and inflammation. Its chemical name is acetylsalicylic acid. The purpose of this activity is to produce aspirin and to isolate it.

Objectives

- **Prepare** and **isolate** an organic compound.
- **Observe** the formation of crystals from a chemical reaction.

Materials

- 100-mL glass beakers (2)
- hot plate
- ice
- thermometer
- 10-mL graduated cylinder
- acetic anhydride
- salicylic acid
- sulfuric acid (concentrated)
- ethyl alcohol
- plastic micropipets (2)
- plastic cup
- toothpick
- scissors
- microscale
- paper towel

Procedure

1. Half-fill a 100-mL beaker with water.
2. Begin heating the water to 60°C while you proceed with the activity.
3. Place 2.0 mL of acetic anhydride in a plastic cup.
4. Add 5 drops of sulfuric acid to the acetic acid in the plastic cup. **CAUTION:** *Handle sulfuric acid with great care. It can cause severe burns if it touches the skin and will also damage clothing.*
5. Place 1.0 g of salicylic acid in the plastic cup. Mix all three chemicals together with a toothpick.
6. Draw up the mixture of chemicals in the plastic cup into a micropipet.
7. Rinse the plastic cup in tap water and dry with a paper towel.
8. Place the pipet with its stem pointing upward into the heated water.
9. Maintain the temperature of the water bath at 55°–60°C for 25 minutes.
10. At the end of the 25 minutes, remove the pipet from the warm water bath and place it stem upward in the plastic cup.
11. Do not discard the warm tap water.
12. Fill another 100-mL beaker with cold tap water and ice.
13. Holding the pipet with the stem pointing upward, place the bulb in the ice-cold water bath.
14. Swirl the bulb to cool the contents of the pipet. You may begin to notice white crystals forming. These crystals are aspirin.
15. Separate the liquid, or aqueous, layer (the reaction mixture) from the aspirin by holding the pipet downward over the beaker of water and squeezing the bulb gently.
16. Draw approximately ½-pipet of ethyl alcohol into a clean pipet. Place the pipet stem upward in the beaker containing warm water.
17. Heat the ethyl alcohol for 5 minutes.
18. When the ethyl alcohol is warm, eject it from the pipet into the plastic cup.
19. Draw up the warm ethyl alcohol from the plastic cup into the pipet containing the aspirin crystals. Mix well.
20. Holding the pipet with the stem pointing upward, squeeze the air out of the pipet.
21. Invert the pipet and place the stem in the ice-cold water. Draw up approximately ½-pipet of water. Mix well.
22. Allow the pipet to cool.
23. Cut open the pipet to reveal the crystals of aspirin.
24. Mass the crystals of aspirin and enter the mass in a data table like the one shown.

Data and Observations

Compound	Mass (g)	Moles
Acetic anhydride		$\frac{\text{mass}}{102\text{g/mole}} =$
Salicylic acid		$\frac{\text{mass}}{138\text{g/mole}} =$
Aspirin		$\frac{\text{mass}}{180\text{g/mole}} =$

Analysis and Conclusions

1. Complete the data table. First, enter the mass of each compound. Then calculate the moles of the compounds you used to produce the aspirin and the moles of aspirin you obtained.
2. What compounds reacted to form aspirin?
3. Why do you think sulfuric acid was added to the reacting compounds?
4. What was the purpose of the ice-water bath?

Extension and Application

Find out when aspirin first came into use as a medicine. What ancient folk remedy is aspirin related to?

APPENDIX A

Table A-1

Definitions of Standards

1 ampere is the constant current which, if maintained in two straight parallel conductors of infinite length, of negligible circular cross-section, and placed 1 meter apart in a vacuum, would produce a force of 2×10^{-7} newton per meter of length between these conductors.

1 candela is the luminous intensity, in the perpendicular direction, of a surface of 1/600 000 m^2 of a blackbody at the temperature of freezing platinum at a pressure of 101 325 pascals.

1 cubic decimeter is equal to 1 liter.

1 kelvin is 1/273.16 of the thermodynamic temperature of the triple point of water.

1 kilogram is the mass of the international prototype kilogram.

1 meter is the distance light travels in 1/299 792 458 of a second.

1 mole is the amount of substance containing as many elementary entities as there are atoms in 0.012 kilogram of carbon-12.

1 second is equal to 9 192 631 770 periods of the natural electromagnetic oscillation during that transition of ground state $^2S_{1/2}$ of cesium-133 which is designated $(F = 4, M = 0) \leftrightarrow (F = 3, M = 0)$.

Avogadro constant = $6.022\,136\,7 \times 10^{23}$

1 electronvolt = $1.602\,177\,33 \times 10^{-19}$ J

Faraday constant = 96 485.309 C/mole e^-

Ideal gas constant = 8.314 471 J/mol · K = 8.314 471 dm^3 · kPa/mol · K

Molar gas volume at STP = 22.414 10 dm^3

Planck's constant = $6.626\,075 \times 10^{-34}$ J · s

Speed of light = $2.997\,924\,58 \times 10^8$ m/s

Table A-2

SI Prefixes

Prefix	Symbol	Meaning	Multiplier (Numerical)	Multiplier (Exponential)
		Greater than 1		
exa	E	quintillion	*1 000 000 000 000 000 000	1×10^{18}
peta	P	quadrillion	1 000 000 000 000 000	1×10^{15}
tera	T	trillion	1 000 000 000 000	1×10^{12}
giga	G	billion	1 000 000 000	1×10^{9}
mega	M	million	1 000 000	1×10^{6}
kilo	k	thousand	1 000	1×10^{3}
hecto	h	hundred	100	1×10^{2}
deka	da	ten	10	1×10^{1}
		Less than 1		
deci	d	tenth	0.1	1×10^{-1}
centi	c	hundredth	0.01	1×10^{-2}
milli	m	thousandth	0.001	1×10^{-3}
micro	μ	millionth	0.000 001	1×10^{-6}
nano	n	billionth	0.000 000 001	1×10^{-9}
pico	p	trillionth	0.000 000 000 001	1×10^{-12}
femto	f	quadrillionth	0.000 000 000 000 001	1×10^{-15}
atto	a	quintillionth	0.000 000 000 000 000 001	1×10^{-18}

*Spaces are used to group digits in long numbers. In some countries, a comma indicates a decimal point. Therefore, commas will not be used.

Table A-3

Some Properties of the Elements

Element	Symbol	Atomic Number (Z)	Atomic Mass* (u)	Melting Point (°C)	Boiling Point (°C)	Density (g/cm³) (gases measured at STP)	Atomic Radius (pm)	First Ionization Energy (kJ/mol)	Standard Reduction Potential (V) (for elements from or to oxidation state indicated)	Enthalpy of Fusion (kJ/mol)	Specific Heat (J/g · C°)	Enthalpy of Vaporization (kJ/mol)	Abundance in Earth's Crust (%)	Major Oxidation States
Actinium	Ac	89	[227.0278]	1050	3300	10.07	203	666	(3+)−2.13	14.3	0.120	293	trace	3+
Aluminum	Al	13	26.981539	660.37	2517.6	2.699	143	577.5	(3+)−1.67	10.71	0.9025	290.8	8.1	3+
Americium	Am	95	[243.0614]	994	2600	13.67	183	579	(3+)−2.07	10	—	238.5	—	2+, 3+, 4+
Antimony	Sb	51	121.757	630.7	1635	6.697	161	834	(3+)+0.15	19.5	0.2072	193	2×10^{-5}	3+, 5+
Argon	Ar	18	39.948	−189.37	−185.86	0.001784	191	1521	—	1.18	0.52033	6.52	4×10^{-6}	
Arsenic	As	33	74.92159	816 (2840 kPa)	615 (sublimes)	5.778	121	947	(3+)+0.24	27.7	0.3289	— (sublimes)	1.9×10^{-4}	3+, 5+
Astatine	At	85	[209.98037]	300	350	—	—	916	(1−)+0.2	23.8	—	90.3	trace	1−, 5+
Barium	Ba	56	137.327	726.9	1845	3.62	222	502.9	(2+)−2.92	8.012	0.2044	140	0.039	2+
Berkelium	Bk	97	[247.0703]	986	—	14.78	170	601	(3+)−2.01	—	—	—	—	3+, 4+
Beryllium	Be	4	9.012182	1287	2468	1.848	112	899.5	(2+)−1.97	7.895	1.824	297.6	2×10^{-4}	2+
Bismuth	Bi	83	208.98037	271.4	1564	9.808	151	703	(3+)+0.317	10.9	0.1221	179	8×10^{-7}	3+, 5+
Boron	B	5	10.811	2080	3865	2.46	85	800.6	(3+)−0.89	50.2	1.026	504.5	9×10^{-4}	3+
Bromine	Br	35	79.904	−7.25	59.35	3.1028	119	1139.9	(1−)+1.065	10.571	0.47362	29.56	2.5×10^{-4}	1−, 1+, 3+, 5+
Cadmium	Cd	48	112.411	320.8	770	8.65	151	867.7	(2+)−0.4025	6.19	0.2311	100	1.6×10^{-5}	2+
Calcium	Ca	20	40.078	841.5	1500.5	1.55	197	589.8	(2+)−2.84	8.54	0.6315	155	4.66	2+
Californium	Cf	98	[251.0796]	900	—	—	186	608	(3+)−2	—	—	—	—	3+, 4+
Carbon	C	6	12.011	3620	4200	2.266	77	1086.5	(4−)+0.132	104.6	0.7099	711	0.018	4−, 2+, 4+
Cerium	Ce	58	140.115	804	3470	6.773	181.8	541	(3+)−2.34	5.2	0.1923	313	0.007	3+, 4+
Cesium	Cs	55	132.90543	28.4	674.8	1.9	262	375.7	(1+)−2.923	2.087	0.2421	67	2.6×10^{-4}	1+
Chlorine	Cl	17	35.4527	−101	−34	0.003214	91	1255.5	(1−)+1.3583	6.41	0.47820	20.41	0.013	1−, 1+, 3+, 5+
Chromium	Cr	24	51.9961	1860	2679	7.2	128	652.8	(3+)−0.74	20.5	0.4491	339	0.01	2+, 3+, 6+
Cobalt	Co	27	58.9332	1495	2912	8.9	125	758.8	(2+)−0.277	16.192	0.4210	382	0.0028	2+, 3+
Copper	Cu	29	63.546	1085	2570	8.92	128	745.5	(2+)+0.34	13.38	0.38452	304	0.0058	1+, 2+
Curium	Cm	96	[247.0703]	1340	3540	13.51	174	581	(3+)−2.06	—	—	—	—	3+, 4+
Dysprosium	Dy	66	162.5	1407	2600	8.536	178.1	572	(3+)−2.29	10.4	0.1733	250	6×10^{-4}	2+, 3+
Einsteinium	Es	99	[252.0828]	860	—	—	186	619	(3+)−2	—	—	—	—	3+
Erbium	Er	68	167.26	1497	2900	9.045	176.1	589	(3+)−2.32	17.2	0.1681	293	3.5×10^{-4}	3+
Europium	Eu	63	151.965	826	1439	5.245	208.4	547	(3+)−1.99	10.5	0.1820	176	2.1×10^{-3}	2+, 3+
Fermium	Fm	100	[257.0951]	—	—	—	—	627	(3+)−1.96	—	—	—	—	2+, 3+
Fluorine	F	9	18.9984032	−219.7	−188.2	0.001696	69	1681	(1−)+2.87	0.51	0.8238	6.54	0.0544	1−
Francium	Fr	87	[223.0197]	24	650	—	280	375	—	2	—	63.6	trace	1+
Gadolinium	Gd	64	157.25	1312	3000	7.886	180.4	592	(3+)−2.29	15.5	0.2355	311.7	6.3×10^{-4}	3+
Gallium	Ga	31	69.723	29.77	2203	5.904	134	578.8	(3+)−0.529	5.59	0.3709	256	0.0018	1+, 3+
Germanium	Ge	32	72.61	945	2850	5.323	123	761.2	(4+)+0.124	31.8	0.3215	334.3	1.5×10^{-4}	2+, 4+

*[ ] indicates mass of longest-lived isotope

Element	Symbol	Atomic Number (Z)	Atomic Mass* (u)	Melting Point (°C)	Boiling Point (°C)	Density (g/cm^3) (gases measured at STP)	Atomic Radius (pm)	First Ionization Energy (kJ/mol)	Standard Reduction Potential (V) (for elements from or to oxidation state indicated)	Enthalpy of Fusion (kJ/mol)	Specific Heat (J/g · C°)	Enthalpy of Vaporization (kJ/mol)	Abundance in Earth's Crust (%)	Major Oxidation States
Gold	Au	79	196.96654	1064	2808	19.32	144	889.9	(3+)+1.52	12.4	0.12905	324.4	3×10^{-7}	1+, 3+
Hafnium	Hf	72	178.49	2227	4691	13.28	159	654.4	(4+)−1.56	29.288	0.1442	661	3×10^{-4}	4+
Helium	He	2	4.002602	−269.7 (2536 kPa)	−268.93	0.00017847	122	2372	—	0.02	5.1931	0.084	—	
Holmium	Ho	67	164.9032	1461	2600	8.78	176.2	581	(3+)−2.33	17.1	0.1646	251	1.5×10^{-4}	3+
Hydrogen	H	1	1.00794	−259.19	−252.76	0.0000899	78	1312	(1+) 0.0000	0.117	14.298	0.904	—	1−, 1+
Indium	In	49	114.82	156.61	2080	7.29	167	558.2	(3+)−0.3382	3.26	0.2407	231.8	2×10^{-5}	1+, 3+
Iodine	I	53	126.90447	113.6	184.5	4.93	138	1008.4	(1−)+0.5355	15.517	0.21448	41.95	4.6×10^{-5}	1−, 1+, 5+, 7+
Iridium	Ir	77	192.22	2447	4550	22.65	135.5	880	(4+)+0.926	26.4	0.1306	563.6	1×10^{-7}	3+, 4+, 5+
Iron	Fe	26	55.847	1536	2860	7.874	126	759.4	(3+)−0.4	13.807	0.4494	350	5.8	2+, 3+
Krypton	Kr	36	83.8	−157.2	−153.35	0.0037493	201	1351	—	1.64	0.2480	9.03	—	
Lanthanum	La	57	138.9055	920	3420	6.17	187	538	(3+)−2.37	8.5	0.1952	402	0.0035	3+
Lawrencium	Lr	103	[260.1054]	—	—	—	—	—	(3+)−2.06	—	—	—	—	3+
Lead	Pb	82	207.2	327	1746	11.342	175	715.6	(2+)−0.1251	4.77	0.1276	178	0.0013	2+, 4+
Lithium	Li	3	6.941	180.5	1347	0.534	156	520.2	(1+)−3.045	3	3.569	148	0.002	1+
Lutetium	Lu	71	174.967	1652	3327	9.84	173.8	524	(3+)−2.3	11.9	0.1535	414	8×10^{-5}	3+
Magnesium	Mg	12	24.305	650	1105	1.738	160	737.8	(2+)−2.356	8.477	1.024	127.4	2.76	2+
Manganese	Mn	25	54.93805	1246	2061	7.43	127	717.5	(2+)−1.18	12.058	0.4791	219.7	0.1	2+, 3+, 4+, 6+, 7+
Mendelevium	Md	101	[258.0986]	—	—	—	—	635	—	—	—	—	—	2+, 3+
Mercury	Hg	80	200.59	−38.9	357	13.534	151	1007	(2+)+0.8535	2.2953	0.13950	59.1	2×10^{-6}	1+, 2+
Molybdenum	Mo	42	95.94	2623	4679	10.28	139	685	(6+) 0.114	36	0.2508	590	1.2×10^{-4}	4+, 5+, 6+
Neodymium	Nd	60	144.24	1024	3111	7.003	181.4	530	(3+)−2.32	7.13	0.1903	283.7	0.004	2+, 3+
Neon	Ne	10	20.1797	−248.61	−246.05	0.0008999	131	2081	—	0.34	1.0301	1.77	—	
Neptunium	Np	93	237.0482	640	3900	20.45	155	597	(5+)−0.91	9.46	—	336	—	2+, 3+, 4+, 5+, 6+
Nickel	Ni	28	58.6934	1455	2883	8.908	124	736.7	(2+)−0.257	17.15	0.4442	375	0.0075	2+, 3+, 4+
Niobium	Nb	41	92.90638	2477	4858	8.57	146	664.1	(5+)−0.65	26.9	0.2648	690	0.002	4+, 5+
Nitrogen	N	7	14.00674	−210	−195.8	0.0012409	71	1402	(3−)−0.092	0.72	1.0397	5.58	0.002	3−, 2−, 1−, 1+, 2+, 3+, 4+, 5+
Nobelium	No	102	[259.1009]	—	—	—	—	642	(2+)−2.5	—	—	—	—	2+, 3+
Osmium	Os	76	190.2	3045	5025	22.57	135	840	(4+)+0.687	31.7	0.130	627.6	2×10^{-7}	4+, 6+, 8+
Oxygen	O	8	15.9994	−218.8	−183	0.001429	60	1313.9	(2−) 0.815	0.44	0.91738	6.82	45.5	2−, 1−
Palladium	Pd	46	106.42	1552	2940	11.99	137	805	(2+) 0.915	17.6	0.2441	362	3×10^{-7}	2+, 4+
Phosphorus	P	15	30.973762	44.2	280.5	1.823	109	1012	(3−)−0.063	0.659	0.76968	49.8	0.11	3−, 3+, 5+
Platinum	Pt	78	195.08	1769	3824	21.41	138.5	868	(4+)+1.15	19.7	0.1326	510.4	1×10^{-6}	2+, 4+
Plutonium	Pu	94	[244.0642]	640	3230	19.86	162	585	(4+)−1.25	2.8	0.138	343.5	—	3+, 4+, 5+, 6+
Polonium	Po	84	[208.9824]	254	962	9.4	164	813	(4+)+0.73	3.81	0.125	103	—	2−, 2+, 4+, 6+
Potassium	K	19	39.0983	63.2	766.4	0.862	231	418.8	(1+)−2.925	2.334	0.7566	76.9	1.84	1+
Praseodymium	Pr	59	140.90765	935	3343	6.782	182.4	522	(3+)−2.35	11.3	0.1930	332.6	9.1×10^{-4}	3+, 4+
Promethium	Pm	61	[144.9128]	1168	2460	7.2	183.4	536	(3+)−2.29	8.17	—	293	trace	3+

*[ ] indicates mass of longest-lived isotope

Element	Symbol	Atomic Number (Z)	Atomic Mass* (u)	Melting Point (°C)	Boiling Point (°C)	Density (g/cm³) (gases measured at STP)	Atomic Radius (pm)	First Ionization Energy (kJ/mol)	Standard Reduction Potential (V) (for elements from or to oxidation state indicated)	Enthalpy of Fusion (kJ/mol)	Specific Heat (J/g · C°)	Enthalpy of Vaporization (kJ/mol)	Abundance in Earth's Crust (%)	Major Oxidation States
Protactinium	Pa	91	231.03588	1552	4227	15.37	163	568	(5+)−1.19	14.6	—	481	trace	3+, 4+, 5+
Radium	Ra	88	226.0254	700	1630	5	228	509.1	(2+)−2.916	8.36	—	136.8	—	2+
Radon	Rn	86	[222.0176]	−71	−62	0.00973	232	1037	—	16.4	—	16.4	—	
Rhenium	Re	75	186.207	3180	5650	21.232	137	760	(7+)+0.34	33.4	0.1368	707	1×10^{-7}	3+, 4+, 6+, 7+
Rhodium	Rh	45	102.9055	1960	3727	12.39	134	720	(3+)+0.76	21.6	0.2427	494	1×10^{-7}	3+, 4+, 5+
Rubidium	Rb	37	85.4678	39.5	697	1.532	248	403	(1+)−2.925	2.19	0.36344	69.2	0.0078	1+
Ruthenium	Ru	44	101.07	2310	4119	12.41	134	711	(4+)+0.68	25.5	0.2381	567.8	—	2+, 3+, 4+, 5+
Samarium	Sm	62	150.36	1072	1800	7.536	180.4	542	(3+)−2.3	8.9	0.1965	191	7×10^{-4}	2+, 3+
Scandium	Sc	21	44.95591	1539	2831	3	162	631	(3+)−2.03	15.77	0.5677	304.8	0.0022	3+
Selenium	Se	34	78.96	221	685	4.79	117	940.7	(2−)−0.67	5.43	0.3212	26.3	5×10^{-6}	2−, 2+, 4+, 6+
Silicon	Si	14	28.0855	1411	3231	2.336	118	786.5	(4−)−0.143	50.2	0.7121	359	27.2	2+, 4+
Silver	Ag	47	107.8682	961	2195	10.49	144	730.8	(1+)+0.7991	11.65	0.23502	255	8×10^{-6}	1+
Sodium	Na	11	22.989768	97.83	897.4	0.968	186	495.9	(1+)−2.714	2.602	1.228	97.4	2.27	1+
Strontium	Sr	38	87.62	776.9	1412	2.6	215	549.5	(2+)−2.89	7.4308	0.301	137	0.0384	2+
Sulfur	S	16	32.066	115.2	444.7	2.08	103	999.6	(2−)−0.45	1.7272	0.7060	9.62	0.03	2−, 4+, 6+
Tantalum	Ta	73	180.9479	2980	5505	16.65	146	760.8	(5+)−0.81	36.57	0.1402	737	2×10^{-4}	4+, 5+
Technetium	Tc	43	97.9072	2200	4567	11.5	136	702	(6+)+0.83	23.0	—	577	—	2+, 4+, 6+, 7+
Tellurium	Te	52	127.6	450	990	6.25	138	869	(2−)−1.14	17.4	0.2016	50.6	2×10^{-7}	2−, 2+, 4+, 6+
Terbium	Tb	65	158.92534	1356	2800	8.272	177.3	564	(3+)−2.31	10.3	0.1819	293	1×10^{-4}	3+, 4+
Thallium	Tl	81	204.3833	303.5	1457	11.85	170	589.1	(1+)−0.3363	4.27	0.1288	162	7×10^{-5}	1+, 3+
Thorium	Th	90	232.0381	1750	4787	11.78	179	587	(4+)−1.83	16.11	0.1177	543.9	8.1×10^{-4}	4+
Thulium	Tm	69	168.93421	1545	1727	9.318	175.9	596	(3+)−2.32	18.4	0.1600	213	5×10^{-5}	—
Tin	Sn	50	118.71	232	2623	7.265	141	708.4	(4+)+0.064	7.07	0.2274	296	2.1×10^{-4}	2+, 4+
Titanium	Ti	22	47.88	1666	3358	4.5	147	658.1	(4+)−0.86	14.146	0.5226	425	0.63	2+, 3+, 4+
Tungsten	W	74	183.85	3680	6000	19.3	139	770.4	(6+)−0.09	35.4	0.1320	806	1.2×10^{-4}	4+, 5+, 6+
Unnilennium	Une	109	[266]	—	—	—	—	—	—	—	—	—	—	—
Unnilhexium	Unh	106	[263]	—	—	—	—	—	—	—	—	—	—	—
Unniloctium	Uno	108	[265]	—	—	—	—	—	—	—	—	—	—	—
Unnilpentium	Unp	105	[262]	—	—	—	—	—	—	—	—	—	—	—
Unnilquadium	Unq	104	[261]	—	—	—	—	—	—	—	—	—	—	—
Unnilseptium	Uns	107	[262]	—	—	—	—	—	—	—	—	—	—	—
Uranium	U	92	238.0289	1130	3930	19.05	156	584	(6+)−0.83	12.6	0.11618	423	2.3×10^{-4}	3+, 4+, 5+, 6+
Vanadium	V	23	50.9415	1917	3417	6.11	134	650.3	(4+)−0.54	22.84	0.4886	459.7	0.0136	2+, 3+, 4+, 5+
Xenon	Xe	54	131.29	−111.8	−108.09	0.0058971	218	1170	—	2.29	0.15832	12.64	—	
Ytterbium	Yb	70	173.04	824	1427	6.973	193.3	603	(3+)−2.22	7.66	0.1545	155	3.4×10^{-4}	2+, 3+
Yttrium	Y	39	88.90585	1530	3264	4.5	180	616	(3+)−2.37	17.15	0.2984	393	0.0035	3+
Zinc	Zn	30	65.39	419.6	907	7.14	134	906.4	(2+)−0.7626	7.322	0.3884	115	0.0076	2+
Zirconium	Zr	40	91.224	1852	4400	6.51	160	659.7	(4+)−1.7	20.92	0.2780	590.5	0.0162	4+

*[ ] indicates mass of longest-lived isotope

Table A-4

Names and Charges of Polyatomic Ions			
1−	**2−**	**3−**	**4−**
Amide, NH_2^-	Hexachloroplatinate, $PtCl_6^{2-}$	Arsenate, AsO_4^{3-}	Hexacyanoferrate(II), $Fe(CN)_6^{4-}$
Astatate, AtO_3^-	Hexafluorosilicate, SiF_6^{2-}	Arsenite, AsO_3^{3-}	Orthosilicate, SiO_4^{4-}
Azide, N_3^-	Molybdate, MoO_4^{2-}	Borate, BO_3^{3-}	Diphosphate, $P_2O_7^{4-}$
Benzoate, $C_6H_5COO^-$	Peroxydisulfate, $S_2O_8^{2-}$	Citrate, $C_6H_5O_7^{3-}$	
Bismuthate, BiO_3^-	Phosphite, HPO_3^{2-}	Hexacyanoferrate(III), $Fe(CN)_6^{3-}$	
Bromate, BrO_3^-	Ruthenate, RuO_4^{2-}		
Formate, $HCOO^-$	Selenate, SeO_4^{2-}		
Hypobromite, BrO^-	Selenite, SeO_3^{2-}		
Hypophosphite, $H_2PO_2^-$	Tellurate, TeO_4^{2-}		
Perbromate, BrO_4^-	Tellurite, TeO_3^{2-}		
Periodate, IO_4^-	Tungstate, WO_4^{2-}		
Perrhenate, ReO_4^-			
Thiocyanate, SCN^-			
Vanadate, VO_3^-			
1+	**2+**		
Neptunyl(V), NpO_2^+	Neptunyl(VI), NpO_2^{2+}		
Plutonyl(V), PuO_2^+	Plutonyl(VI), PuO_2^{2+}		
Uranyl(V), UO_2^+	Uranyl(VI), UO_2^{2+}		
Vanadyl(V), VO_2^+	Vanadyl(IV), VO^{2+}		

Table A-5

Specific Heat Values (J/g · K)					
Substance	C_p	**Substance**	C_p	**Substance**	C_p
AlF_3	0.8948	Fe_3C	0.5898	$NaVO_3$	1.540
$BaTiO_3$	0.79418	$FeWO_4$	0.37735	$Ni(CO)_4$	1.198
BeO	1.020	HI	0.22795	PbI_2	0.1678
CaC_2	0.9785	K_2CO_3	0.82797	SF_6	0.6660
$CaSO_4$	0.7320	$MgCO_3$	0.8957	SiC	0.6699
CCl_4	0.85651	$Mg(OH)_2$	1.321	SiO_2	0.7395
CH_3OH	2.55	$MgSO_4$	0.8015	$SrCl_2$	0.4769
CH_2OHCH_2OH	2.413	MnS	0.5742	Tb_2O_3	0.3168
CH_3CH_2OH	2.4194	Na_2CO_3	1.0595	$TiCl_4$	0.76535
CdO	0.3382	NaF	1.116	Y_2O_3	0.45397
$CuSO_4 \cdot 5H_2O$	1.12				

Table A-6

Thermodynamic Properties (at 25°C and 100.000 kPa)

$\Delta H_f°$ (kJ/mol) $\Delta G_f°$ (kJ/mol) S° (J/mol · K)
(concentration of aqueous solutions is 1*M*)

Substance	$\Delta H_f°$	$\Delta G_f°$	S°	Substance	$\Delta H_f°$	$\Delta G_f°$	S°
Ag(cr)	0	0	42.55	H_3PO_3(aq)	−964.4	—	—
AgCl(cr)	−127.068	−109.789	96.2	H_3PO_4(aq)	−1279.0	−1119.1	110.50
AgCN(cr)	146.0	156.9	107.19	H_2S(g)	−20.63	−33.56	205.79
Al(cr)	0	0	28.33	H_2SO_3(aq)	−608.81	−537.81	232.2
Al_2O_3(cr)	−1675.7	−1582.3	50.92	H_2SO_4(aq)	−909.27	−744.53	20.1
$BaCl_2$(aq)	−871.95	−823.21	122.6	$HgCl_2$(cr)	−224.3	−178.6	—
$BaSO_4$(cr)	−1473.2	−1362.2	132.2	Hg_2Cl_2(cr)	−265.22	−210.745	192.5
Be(cr)	0	0	9.50	Hg_2SO_4(cr)	−743.12	−625.815	200.66
BeO(cr)	−609.6	−580.3	—	I_2(cr)	0	0	116.135
Bi(cr)	0	0	56.74	K(cr)	0	0	64.18
$BiCl_3$(cr)	−379.1	−315.0	177.0	KBr(cr)	−393.798	−380.66	95.90
Bi_2S_3(cr)	−143.1	−140.6	200.4	$KMnO_4$(cr)	−837.2	−737.6	171.71
Br_2(l)	0	0	152.231	KOH(cr)	−424.764	—	—
CH_4(g)	−74.81	−50.72	186.264	LiBr(cr)	−351.213	—	—
C_2H_2(g)	+226.73	+209.20	200.94	LiOH(cr)	−484.93	−438.95	42.80
C_2H_4(g)	+52.26	+68.15	219.56	Mn(cr)	0	0	32.01
C_2H_6(g)	−84.68	−32.82	229.60	$MnCl_2$(aq)	−555.05	−490.8	38.9
CO(g)	−110.525	−137.168	197.674	$Mn(NO_3)_2$(aq)	−635.5	−450.9	218
CO_2(g)	−393.509	−394.359	213.74	MnO_2(cr)	−520.03	−465.14	53.05
CS_2(l)	+89.70	+65.27	151.34	MnS(cr)	−214.2	—	—
Ca(cr)	0	0	41.42	N_2(g)	0	0	191.61
$Ca(OH)_2$(cr)	−986.09	−898.49	—	NH_3(g)	−46.11	−16.45	192.45
Cl_2(g)	0	0	223.066	NH_4Br(cr)	−270.83	−175.2	113
Co_3O_4(cr)	−891	−774	—	NO(g)	+90.25	86.55	210.761
CoO(cr)	−237.94	−214.20	52.97	NO_2(g)	+33.18	+51.31	240.06
Cr_2O_3(cr)	−1139.7	−1058.1	81.2	N_2O(g)	+82.05	+104.20	219.85
CsCl(cr)	−443.04	−414.53	101.17	Na(cr)	0	0	51.21
Cs_2SO_4(cr)	−1443.02	−1323.58	211.92	NaBr(cr)	−361.062	—	—
CuI(cr)	−67.8	−69.5	96.7	NaCl(cr)	−411.153	−384.138	72.13
CuS(cr)	−53.1	−53.6	66.5	$NaNO_3$(aq)	−447.48	—	—
Cu_2S(cr)	−79.5	−86.2	120.9	NaOH(cr)	−425.609	—	—
$CuSO_4$(cr)	−771.36	−661.8	109	Na_2S(aq)	−447.3	—	—
F_2(g)	0	0	202.78	Na_2SO_4(cr)	−1387.08	−1270.16	149.58
$FeCl_3$(cr)	−399.49	—	—	O_2(g)	0	0	205.138
FeO(cr)	−272.0	—	—	P_4O_6(cr)	−1640.1	—	—
Fe_2O_3(cr)	−824.2	−742.2	87.40	P_4O_{10}(cr)	−2984.0	−2697.7	228.86
Fe_3O_4(cr)	−1118.4	−1015.4	146.4	$PbBr_2$(cr)	−278.7	−261.92	161.5
H(g)	+217.965	—	114.713	$PbCl_2$(cr)	−359.41	−314.10	136.0
H_2(g)	0	0	130.684	S(cr)	0	0	31.80
HBr(g)	−36.40	−53.45	198.695	SO_2(g)	−296.830	−300.194	248.22
HCl(g)	−92.307	−95.299	186.908	SO_3(g)	−454.51	−374.21	70.7
HCl(aq)	−167.159	−131.228	56.5	SrO(cr)	−592.0	−561.9	54.4
HCN(aq)	+150.6	+172.4	94.1	Ti(cr)	0	0	30.63
HCHO(g)	−108.57	−102.53	218.77	TiO_2(cr)	−939.7	−884.5	49.92
HCOOH(l)	−424.72	−361.35	128.95	TlI(cr)	−123.8	−125.39	127.6
HF(g)	−271.1	−273.2	173.779	UCl_4(cr)	−1019.2	−930.0	197.1
HI(g)	+26.48	+1.70	206.594	UCl_5(cr)	−1059	−950	242.7
H_2O(l)	−285.830	−237.129	69.91	Zn(cr)	0	0	41.63
H_2O(g)	−241.818	−228.572	188.825	$ZnCl_2$(aq)	−488.19	−409.50	0.8
H_2O_2(l)	—	−120.35	109.6	ZnO(cr)	−348.28	−318.30	43.64
H_3PO_2(l)	−595.4	—	—	$ZnSO_4$(aq)	−1063.15	−891.59	−92.0

Table A-7

Solubility Rules

You will be working with water solutions, and it is helpful to have a few rules concerning what substances are soluble* in water. The more common rules are listed below.

1. All common salts of the Group 1(IA) elements and ammonium ion are soluble.
2. All common acetates and nitrates are soluble.
3. All binary compounds of Group 17(VIIA) elements (other than F) with metals are soluble except those of silver, mercury(I), and lead.
4. All sulfates are soluble except those of barium, strontium, lead, calcium, silver, and mercury(I).
5. Except for those in Rule 1, carbonates, hydroxides, oxides, sulfides, and phosphates are insoluble.

*A substance is considered soluble if more than 3 grams of the substance dissolve in 100 mL of water.

Table A-8

Molal Freezing–Point Depression and Boiling–Point Elevation Constants

Substance	K_{fp} (C° kg/mol)	Freezing Point (°C)	K_{bp} (C° kg/mol)	Boiling Point (°C)
Acetic Acid	3.90	16.66	2.530	117.90
Benzene	5.12	5.533	2.53	80.100
Camphor	37.7	178.75	5.611	207.42
Cyclohexane	20.0	6.54	2.75	80.725
Cyclohexanol	39.3	25.15	—	—
Nitrobenzene	6.852	5.76	5.24	210.8
Phenol	7.40	40.90	3.60	181.839
Water	1.853	0.000	0.515	100.000

Table A-9

Ionization Constants

Substance	Ionization Constant	Substance	Ionization Constant	Substance	Ionization Constant
$HCOOH$	1.77×10^{-4}	HBO_3^{-2}	1.58×10^{-14}	HS^-	1.26×10^{-13}
CH_3COOH	1.75×10^{-5}	H_2CO_3	4.37×10^{-7}	HSO_4^-	1.02×10^{-2}
$CH_2ClCOOH$	1.36×10^{-3}	HCO_3^-	4.68×10^{-11}	H_2SO_3	1.29×10^{-2}
$CHCl_2COOH$	5.50×10^{-2}	HCN	6.17×10^{-10}	HSO_3^-	6.17×10^{-8}
CCl_3COOH	3.02×10^{-1}	HF	6.61×10^{-4}	$HSeO_4^-$	2.19×10^{-2}
$HOOCCOOH$	5.36×10^{-2}	HNO_2	7.24×10^{-4}	H_2SeO_3	2.29×10^{-3}
$HOOCCOO^-$	5.35×10^{-5}	H_3PO_4	7.08×10^{-3}	$HSeO_3^-$	5.37×10^{-9}
CH_3CH_2COOH	1.34×10^{-5}	$H_2PO_4^-$	6.31×10^{-8}	$HBrO$	2.51×10^{-9}
C_6H_5COOH	6.25×10^{-5}	HPO_4^{2-}	4.17×10^{-13}	$HClO$	2.88×10^{-8}
H_3AsO_4	6.03×10^{-3}	H_3PO_3	6.31×10^{-2}	HIO	2.29×10^{-11}
$H_2AsO_4^-$	1.05×10^{-7}	$H_2PO_3^-$	2.00×10^{-7}	NH_3	1.74×10^{-5}
H_3BO_3	5.75×10^{-10}	H_3PO_2	5.89×10^{-2}	H_2NNH_2	8.71×10^{-7}
$H_2BO_3^-$	1.82×10^{-13}	H_2S	1.07×10^{-7}	H_2NOH	8.91×10^{-9}

Table A-10

Solubility Product Constants (at 25°C)					
Substance	K_{sp}	**Substance**	K_{sp}	**Substance**	K_{sp}
$AgBr$	5.01×10^{-13}	$BaSO_4$	1.10×10^{-10}	Li_2CO_3	2.51×10^{-2}
$AgBrO_3$	5.25×10^{-5}	$CaCO_3$	2.88×10^{-9}	$MgCO_3$	3.47×10^{-8}
Ag_2CO_3	8.13×10^{-12}	$CaSO_4$	9.12×10^{-6}	$MnCO_3$	1.82×10^{-11}
$AgCl$	1.78×10^{-10}	CdS	7.94×10^{-27}	$NiCO_3$	6.61×10^{-9}
Ag_2CrO_4	1.12×10^{-12}	$Cu(IO_3)_2$	7.41×10^{-8}	$PbCl_2$	1.62×10^{-5}
$Ag_2Cr_2O_7$	2.00×10^{-7}	CuC_2O_4	2.29×10^{-8}	PbI_2	7.08×10^{-9}
AgI	8.32×10^{-17}	$Cu(OH)_2$	2.19×10^{-20}	$Pb(IO_3)_2$	3.24×10^{-13}
$AgSCN$	1.00×10^{-12}	CuS	6.31×10^{-36}	$SrCO_3$	1.10×10^{-10}
$Al(OH)_3$	1.26×10^{-33}	FeC_2O_4	3.16×10^{-7}	$SrSO_4$	3.24×10^{-7}
Al_2S_3	2.00×10^{-7}	$Fe(OH)_3$	3.98×10^{-38}	$TlBr$	3.39×10^{-6}
$BaCO_3$	5.13×10^{-9}	FeS	6.31×10^{-18}	$ZnCO_3$	1.45×10^{-11}
$BaCrO_4$	1.17×10^{-10}	Hg_2SO_4	7.41×10^{-7}	ZnS	1.58×10^{-24}

Table A-11

Acid-Base Indicators			
Indicator	**Lower Color**	**Range**	**Upper Color**
Methyl violet	yellow-green	0.0–2.5	violet
Malachite green HCl	yellow	0.5–2.0	blue
Thymol blue	red	1.0–2.8	yellow
Naphthol yellow S	colorless	1.5–2.6	yellow
p-Phenylazoaniline	orange	2.1–2.8	yellow
Methyl orange	red	2.5–4.4	yellow
Bromophenol blue	orange-yellow	3.0–4.7	violet
Gallein	orange	3.5–6.3	red
2,5-Dinitrophenol	colorless	4.0–5.8	yellow
Ethyl orange	salmon	4.2–4.6	orange
Propyl red	pink	5.1–6.5	yellow
Bromocresol purple	green-yellow	5.4–6.8	violet
Bromoxylenol blue	orange-yellow	6.0–7.6	blue
Phenol red	yellow	6.4–8.2	red-violet
Cresol red	yellow	7.1–8.8	violet
m-Cresol purple	yellow	7.5–9.0	violet
Thymol blue	yellow	8.1–9.5	blue
Phenolphthalein	colorless	8.3–10.0	dark pink
o-Cresolphthalein	colorless	8.6–9.8	pink
Thymolphthalein	colorless	9.5–10.4	blue
Alizarin yellow R	yellow	9.9–11.8	dark orange
Methyl blue	blue	10.6–13.4	pale violet
Acid fuchsin	red	11.1–12.8	colorless
2,4,6-Trinitrotoluene	colorless	11.7–12.8	orange

Table A-12

Symbols and Abbreviations	
α = rays from radioactive materials, helium nuclei	K_{eq} = equilibrium constant
β = rays from radioactive materials, electrons	K_{sp} = solubility product constant
γ = rays from radioactive materials, high-energy quanta	kg = kilogram
Δ = change in	M = molarity
λ = wavelength	m = mass, molality
ν = frequency	m = meter (*length*)
Π = osmotic pressure	mol = mole (*amount*)
A = ampere (*electric current*)	min = minute (*time*)
Bq = becquerel (*nuclear disintegration*)	N = newton (*force*)
°C = Celsius degree (*temperature*)	N_A = Avogadro's number
C = coulomb (*quantity of electricity*)	n = number of moles
c = speed of light	P = pressure, power
cd = candela (*luminous intensity*)	Pa = pascal (*pressure*)
C_p = specific heat	$\boldsymbol{p}$ = momentum
D = density	q = heat
E = energy, electromotive force	R = gas constant
F = force, Faraday	S = entropy
G = free energy	s = second (*time*)
g = gram (*mass*)	Sv = sievert (*absorbed radiation*)
Gy = gray (*radiation*)	T = temperature
H = enthalpy	U = internal energy
Hz = hertz (*frequency*)	u = atomic mass unit
h = Planck's constant	V = volume
h = hour (*time*)	V = volt (*electromotive force*)
J = joule (*energy*)	v = velocity
K = kelvin (*temperature*)	W = watt (*power*)
K_a = ionization constant (acid)	w = work
K_b = ionization constant (base)	x = mole fraction

APPENDIX B

LOGARITHMS

A logarithm or log is an exponent. We will work with exponents given in terms of base 10.

$$N = b^x$$

$$\text{number} = \text{base}^{\text{exponent or logarithm}}$$

$$100 = 10^{2.0000}$$

For the log 2.000, the part of the numeral to the left of the decimal point is the characteristic. The part to the right of the decimal point is the mantissa.

$$\text{Log } 100 = \underset{\text{characteristic}}{2}.\underset{\text{mantissa}}{000}$$

SAMPLE PROBLEM

How to Find a Logarithm

Find the log of 657.

(a) Write the number in scientific notation, 6.57×10^2.

(b) Look in the table under the column (*N*). Find the first two digits, (65).

(c) Look to the right and find the mantissa that is in the vertical column under the third digit of the number (7). It is .8176.

(d) From the scientific notation, write the power of ten as the characteristic, to the left of the decimal point.

(e) Write the four digits from the table as the mantissa to the right of the characteristic and the decimal point, 2.8176.

$$\text{thus } 657 = 10^{2.8176} \text{ or } \log 657 = 2.8176$$

When given a logarithm and asked to find the number it represents, we use the table to find the first three digits for the number. We use the characteristic to determine where to locate the decimal point with respect to these digits.

SAMPLE PROBLEM

How to Find the Antilogarithm

Given the logarithm 2.8176, find the number it represents (antilog).

(a) In the log table, find the mantissa that is closest to .8176.

(b) We find by looking under the column (*N*) that this mantissa corresponds to 65. The third digit is found at the top of the column in which the mantissa appears, 7. (657).

(c) Write the three digits (657) in scientific notation, 6.57×10^x.

(d) The characteristic will be the power of ten.

$$\text{antilog } 2.8176 = 6.57 \times 10^2 \text{ or } 657.$$

SAMPLE PROBLEMS

1. Logarithms of Numbers Less Than 1

Find the log 0.00657.

(a) Write the number in scientific notation, 6.57×10^{-3}

(b) Look in the table under the column N for the first two digits, 6.5, and to the right in the column under the third digit, 7, for the mantissa. Note that the mantissa is always a positive number, .8176

(c) From the scientific notation, we get the negative characteristic, -3.

(d) Add the negative characteristic and the positive mantissa $(-3.0000) + (+.8176) = -2.1824$. This value is more commonly represented as 7.8176-10. However, the negative logarithm -2.1824 is more useful in pH calculations.

2. Antilog of a Negative Logarithm

Find the antilog of -2.1824.

(a) We ask ourselves what number would we add to the next, lesser integer, $-3.$, to get the log -2.1824. It would be 0.8176.

$$\begin{array}{lr} & -3.0000 \\ \text{subtract} & -2.1824 \\ \hline & 0.8176 \end{array}$$

We know that logarithm tables do not give mantissas for negative numbers. So, we have changed the -2.1824 into the sum of a negative characteristic and a positive mantissa. The characteristic is always the next negative number. The positive mantissa was determined by asking ourselves what positive number would we add to the negative characteristic to get -2.1824.

$$-2.1824 = -3. + 0.8176$$

(b) Antilog -2.1824 = antilog $-3.$ $\times$ antilog 0.8176
We know the antilog of $-3.$ is 10^{-3}. From the table, we find that the antilog 0.8176 = 6.57. Therefore the antilog of $-2.1824 = 6.57 \times 10^{-3}$.

SIGNIFICANT DIGITS AND LOGARITHMS

The characteristic of a logarithm simply tells us how many decimal places the number has. The mantissa, on the other hand, represents the actual value of the number. Thus, there may be only as many significant digits in the mantissa as there are significant digits in the number whose logarithm is being obtained. Thus,

$$\log 2.34 \times 10^7 = 7.369$$

but

$$\log 2.34 = 0.369$$

When taking antilogarithms of numbers, the same principle applies. The antilog contains as many significant digits as the mantissa. Thus,

$$\text{antilog } 4.357 = 22\ 800$$

and

$$\text{antilog } 0.357 = 2.28$$

Table B-1

Logarithms of Numbers										
N	**0**	**1**	**2**	**3**	**4**	**5**	**6**	**7**	**8**	**9**
10	0000	0043	0086	0128	0170	0212	0253	0294	0334	0374
11	0414	0453	0492	0531	0569	0607	0645	0682	0719	0775
12	0792	0828	0864	0899	0934	0969	1004	1038	1072	1106
13	1139	1173	1206	1239	1271	1303	1335	1367	1399	1430
14	1461	1492	1523	1553	1584	1614	1644	1673	1703	1732
15	1761	1790	1818	1847	1875	1903	1931	1959	1987	2014
16	2041	2068	2095	2122	2148	2175	2201	2227	2253	2279
17	2304	2330	2355	2380	2405	2430	2455	2480	2504	2529
18	2553	2577	2601	2625	2648	2672	2695	2718	2742	2765
19	2788	2810	2833	2856	2878	2900	2923	2945	2967	2989
20	3010	3032	3054	3075	3096	3118	3139	3160	3181	3201
21	3222	3243	3263	3284	3304	3324	3345	3365	3385	3404
22	3424	3444	3464	3483	3502	3522	3541	3560	3579	3598
23	3617	3636	3655	3674	3692	3711	3729	3747	3766	3784
24	3802	3820	3838	3856	3874	3892	3909	3927	3945	3962
25	3979	3997	4014	4031	4048	4065	4082	4099	4116	4133
26	4150	4166	4183	4200	4216	4232	4249	4265	4281	4298
27	4314	4330	4346	4362	4378	4393	4409	4425	4440	4456
28	4472	4487	4502	4518	4533	4548	4564	4579	4594	4606
29	4624	4639	4654	4669	4683	4698	4713	4728	4742	4757
30	4771	4786	4800	4814	4829	4843	4857	4871	4886	4900
31	4914	4928	4942	4955	4969	4983	4997	5011	5024	5038
32	5051	5065	5079	5092	5105	5119	5132	5145	5159	5172
33	5185	5198	5211	5224	5237	5250	5263	5276	5289	5302
34	5315	5328	5340	5353	5366	5378	5391	5403	5416	5428
35	5441	5453	5465	5478	5490	5502	5514	5527	5539	5551
36	5563	5575	5587	5599	5611	5623	5635	5647	5658	5670
37	5682	5694	5705	5717	5729	5740	5752	5763	5775	5786
38	5798	5809	5821	5832	5843	5855	5866	5877	5888	5899
39	5911	5922	5933	5944	5955	5966	5977	5988	5999	6010
40	6021	6031	6042	6053	6064	6075	6085	6096	6107	6117
41	6128	6138	6149	6160	6170	6180	6191	6201	6212	6222
42	6232	6243	6253	6263	6274	6284	6294	6304	6314	6325
43	6335	6345	6355	6365	6375	6385	6395	6405	6415	6425
44	6435	6444	6454	6464	6474	6484	6493	6503	6513	6522
45	6532	6542	6551	6561	6571	6580	6590	6599	6609	6618
46	6628	6637	6646	6656	6665	6675	6684	6693	6702	6712
47	6721	6730	6739	6749	6758	6767	6776	6785	6794	6803
48	6812	6821	6830	6839	6848	6857	6866	6875	6884	6893
49	6902	6911	6920	6928	6937	6946	6955	6964	6972	6981
50	6990	6998	7007	7016	7024	7033	7042	7050	7059	7067
51	7076	7084	7093	7101	7110	7118	7126	7135	7143	7152
52	7160	7168	7177	7185	7193	7202	7210	7218	7226	7235
53	7243	7251	7259	7267	7275	7284	7292	7300	7308	7316
54	7324	7332	7340	7348	7356	7364	7372	7380	7388	7396

N	0	1	2	3	4	5	6	7	8	9
55	7404	7412	7419	7427	7435	7443	7451	7459	7466	7474
56	7482	7490	7497	7505	7513	7520	7528	7536	7543	7551
57	7559	7566	7574	7582	7589	7597	7604	7612	7619	7627
58	7634	7642	7649	7657	7664	7672	7679	7686	7694	7701
59	7709	7716	7723	7731	7738	7745	7752	7760	7767	7774
60	7782	7789	7796	7803	7810	7818	7825	7832	7839	7846
61	7853	7860	7868	7875	7882	7889	7896	7903	7910	7917
62	7924	7931	7938	7945	7952	7959	7966	7973	7980	7987
63	7993	8000	8007	8014	8021	8028	8035	8041	8048	8055
64	8062	8069	8075	8082	8089	8096	8102	8109	8116	8122
65	8129	8136	8142	8149	8156	8162	8169	8176	8182	8189
66	8195	8202	8209	8215	8222	8228	8235	8241	8248	8254
67	8261	8267	8274	8280	8287	8293	8299	8306	8312	8319
68	8325	8331	8338	8344	8351	8357	8363	8370	8376	8382
69	8388	8395	8401	8407	8414	8420	8426	8432	8439	8445
70	8451	8457	8463	8470	8476	8482	8488	8494	8500	8506
71	8513	8519	8525	8531	8537	8543	8549	8555	8561	8567
72	8573	8579	8585	8591	8597	8603	8609	8615	8621	8627
73	8633	8639	8645	8651	8657	8663	8669	8675	8681	8686
74	8692	8698	8704	8710	8716	8722	8727	8733	8739	8745
75	8751	8756	8762	8768	8774	8779	8785	8791	8797	8802
76	8808	8814	8820	8825	8831	8837	8842	8848	8854	8859
77	8865	8871	8876	8882	8887	8893	8899	8904	8910	8915
78	8921	8927	8932	8938	8943	8949	8954	8960	8965	8971
79	8976	8982	8987	8993	8998	9004	9009	9015	9020	9025
80	9031	9036	9042	9047	9053	9058	9063	9069	9074	9079
81	9085	9090	9096	9101	9106	9112	9117	9122	9128	9133
82	9138	9143	9149	9154	9159	9165	9170	9175	9180	9186
83	9191	9196	9201	9206	9212	9217	9222	9227	9232	9238
84	9243	9248	9253	9258	9263	9269	9274	9279	9284	9289
85	9294	9299	9304	9309	9315	9320	9325	9330	9335	9340
86	9345	9350	9355	9360	9365	9370	9375	9380	9385	9390
87	9395	9400	9405	9410	9415	9420	9425	9430	9435	9440
88	9445	9450	9455	9460	9465	9469	9474	9479	9484	9489
89	9494	9499	9504	9509	9513	9518	9523	9528	9533	9538
90	9542	9547	9552	9557	9562	9566	9571	9576	9581	9586
91	9590	9595	9600	9605	9609	9614	9619	9624	9628	9633
92	9638	9643	9647	9652	9657	9661	9666	9671	9675	9690
93	9685	9689	9694	9699	9703	9708	9713	9717	9722	9727
94	9731	9736	9741	9745	9750	9754	9759	9763	9768	9773
95	9777	9782	9786	9791	9795	9800	9805	9809	9814	9818
96	9823	9827	9832	9836	9841	9845	9850	9854	9859	9863
97	9868	9872	9877	9881	9886	9890	9894	9899	9903	9908
98	9912	9917	9921	9926	9930	9934	9939	9943	9948	9952
99	9956	9961	9965	9969	9974	9978	9983	9987	9991	9996

APPENDIX C

SOLUTIONS TO IN-CHAPTER PRACTICE PROBLEMS

Chapter 2

4. **a.** centimeter **b.** micrometer **c.** kilogram **d.** deciliter

5. **a.** millimeter **b.** microsecond **c.** centigram **d.** picosecond

6. Student 2 has the precise data because the largest difference in values is 0.03 g while Student 1 has a 0.59-g difference.

7. No, the balance error would subtract out.

8. $\% \text{ error} = \dfrac{11.342 \text{ g} - 10.95 \text{ g}}{11.342 \text{ g}}(100) = 3.5\%$

9. $\% \text{ error} = \dfrac{59.35°\text{C} - 40.6°\text{C}}{59.35°\text{C}}(100) = 31.6\%$

10. **a.** 1 **b.** 2 **c.** 2 **d.** 2 **e.** 3 **f.** 1 **g.** 2 **h.** 3 **i.** 4 **j.** infinite

11. $D = \dfrac{m}{V} = \dfrac{8.76 \text{ g}}{3.07 \text{ cm}^3} = 2.85 \text{ g/cm}^3$

12. $D = \dfrac{m}{V} = \dfrac{26.8 \text{ g}}{14.5 \text{ cm}^3} = 1.85 \text{ g/cm}^3$

13. $D = \dfrac{m}{V} = \dfrac{0.61 \text{ g}}{0.26 \text{ cm}^3} = 2.3 \text{ g/cm}^3$

14. $m = DV = \left(\dfrac{2.72 \text{ g}}{\text{cm}^3}\right) \times (24.9 \text{ cm}^3) = 67.7 \text{ g}$

15. $V = \dfrac{m}{D} = (7.91 \text{ g}) \times \left(\dfrac{\text{cm}^3}{2.50 \text{ g}}\right) = 3.16 \text{ cm}^3$

16. $m = DV = \left(\dfrac{1.84 \text{ g}}{\text{cm}^3}\right) \times (7.62 \text{ cm}^3) = 14.0 \text{ g}$

17. $(795 \text{ kg})\left(\dfrac{1000 \text{ g}}{1 \text{ kg}}\right)\left(\dfrac{\text{cm}^3}{0.788 \text{ g}}\right)\left(\dfrac{\text{m}^3}{10^6 \text{ cm}^3}\right) = 1.01 \text{ m}^3$

18. $D = \dfrac{m}{V} = \dfrac{2580 \text{ g}}{(4.05 \text{ cm})(8.85 \text{ cm})(164 \text{ cm})}$
$= 0.439 \text{ g/cm}^3$

19. $D = \dfrac{m}{V} = \dfrac{m}{(\pi d^2/4)(L)} = \dfrac{51.6 \text{ g}}{\pi(0.622^2 \text{ cm}^2/4)(22.1 \text{ cm})}$
$= 7.68 \text{ g/cm}^3$

20. The first balance permits measurement within 0.1 g; the second balance permits measurement within 1 g.

21. **a.** 5000 mg
b. 0.15 m
c. 0.2 s
d. 5 cm
e. 0.06 kg
f. 2000 μm

22. 21°C, 3 m of tape, 1.4 kg, 90 s

Chapter 3

5. **a.** 38 g/100 g H_2O
b. 230 g/100 g H_2O
c. 46 g/100 g H_2O
d. 95 g/100 g H_2O

10. **a.** $(1980 \text{ J})\left(\dfrac{1}{4.184 \text{ J/cal}}\right) = 473 \text{ cal}$

b. $(1.11 \text{ Cal})(1000 \text{ cal/Cal})(4.184 \text{ J/cal}) = 4640 \text{ J}$

c. $(800 \text{ cal})\left(\dfrac{1}{1000 \text{ cal/Cal}}\right) = 0.8 \text{ Cal}$

d. $(3.40 \text{ J})\dfrac{1}{(4.184 \text{ J/cal})(1000 \text{ cal/Cal})}$
$= 0.000\,813 \text{ Cal or } 8.13 \times 10^{-4} \text{ Cal}$

e. $(47.0 \text{ cal})(4.184 \text{ J/cal}) = 197 \text{ J}$

11. $q = m(\Delta T)C_p$

$q = (854 \text{ g})(61.5 \text{ C°})\left(\dfrac{4.18 \text{ J}}{\text{g·C°}}\right)$

$= 220\,000 \text{ J or } 2.20 \times 10^5 \text{ J}$

12. $q = m(\Delta T)C_p$

$q = (96.7 \text{ g})(37.5 \text{ C°})\left(\dfrac{0.874 \text{ J}}{\text{g·C°}}\right) = 3170 \text{ J or } 3.17 \times 10^3 \text{ J}$

13. $q = m(\Delta T)C_p$

$q = (10.35 \text{ g})(24.3 \text{ C°})\left(\dfrac{0.856 \text{ J}}{\text{g·C°}}\right) = 215 \text{ J}$

14.

$$q_{\text{lost}} = q_{\text{gained}}$$

$$m(\Delta T)C_p = m(\Delta T)C_p$$

$$(3.90 \text{ g})(99.3°\text{C} - T_f)\left(\frac{0.903 \text{ J}}{\text{g·C°}}\right) = (10 \text{ g})(T_f - 22.6°\text{C})\left(\frac{4.18 \text{ J}}{\text{g·C°}}\right)$$

$$T_f = 28.6°\text{C}$$

15.

$$q_{\text{lost}} = q_{\text{gained}}$$

$$m(\Delta T)C_p = m(\Delta T)C_p$$

$$(65.6 \text{ g})(100.0 \text{ °C} - T_f)\left(\frac{0.231 \text{ J}}{\text{g·C°}}\right) = (25.0 \text{ g})(T_f - 23.0 \text{ °C})\left(\frac{4.18 \text{ J}}{\text{g·C°}}\right)$$

$$T_f = 32.7 \text{ °C}$$

16.

$$q_{\text{lost}} = q_{\text{gained}}$$

$$m(\Delta T)C_p = m(\Delta T)C_p$$

$$(23.8 \text{ g})(67.50 \text{ C°})(C_p) = (50.0 \text{ g})(8.5 \text{ C°})\left(\frac{4.18 \text{ J}}{\text{g·C°}}\right)$$

$$C_p = 1.1 \text{ J/g·C°}$$

Chapter 4

1. 23 protons

2. 47

3. 11

4. rubidium

5. barium

6. **a.** 89 protons, 89 electrons, 132 neutrons
b. 12 protons, 12 electrons, 13 neutrons
c. 45 protons, 45 electrons, 60 neutrons
d. 57 protons, 57 electrons, 76 neutrons

12. $\dfrac{5(176) + 19(177) + 27(178) + 14(179) + 35(180.0)}{100}$

$= 179$ u

13. $\dfrac{(92.21 \times 27.977) + (4.70 \times 28.976) + (3.09 \times 29.974)}{100}$

$= 28.09$ u

Chapter 5

6. The greatest number of electrons in a given energy level can be determined by finding the value of $2n^2$ where n is the energy level involved; for
$n = 2,\ 2n^2 = 2(2)^2 = 8$
$n = 3,\ 2n^2 = 2(3)^2 = 18$
$n = 5,\ 2n^2 = 2(5)^2 = 50$
$n = 7,\ 2n^2 = 2(7)^2 = 98$
7. **a.** 1 **b.** 3 **c.** 4
8. **a.** An s sublevel contains 1 orbital.
   b. A p sublevel contains 3 orbitals.
   c. A d sublevel contains 5 orbitals.
   d. An f sublevel contains 7 orbitals.
14.

1	$1s^1$	11	$1s^22s^22p^63s^1$
2	$1s^2$	12	$1s^22s^22p^63s^2$
3	$1s^22s^1$	13	$1s^22s^22p^63s^23p^1$
4	$1s^22s^2$	14	$1s^22s^22p^63s^23p^2$
5	$1s^22s^22p^1$	15	$1s^22s^22p^63s^23p^3$
6	$1s^22s^22p^2$	16	$1s^22s^22p^63s^23p^4$
7	$1s^22s^22p^3$	17	$1s^22s^22p^63s^23p^5$
8	$1s^22s^22p^4$	18	$1s^22s^22p^63s^23p^6$
9	$1s^22s^22p^5$	19	$1s^22s^22p^63s^23p^64s^1$
10	$1s^22s^22p^6$	20	$1s^22s^22p^63s^23p^64s^2$

15. **a.** $1s^22s^22p^63s^23p^64s^23d^8$ Ni:
   b. $1s^22s^22p^63s^23p^6$:Ar:
   c. $1s^22s^22p^63s^23p^4$ ·S:
   d. $1s^22s^22p^63s^23p^64s^23d^{10}4p^65s^24d^9$ Ag:
   (actually $4d^{10}5s^1$ Ag·)
   e. $1s^22s^22p^63s^23p^64s^1$ K·
   f. $1s^22s^22p^63s^23p^64s^23d^{10}4p^2$ ·Ge:

Chapter 6

6. **a.** nonmetal **b.** metal **c.** metalloid **d.** metal **e.** metal **f.** metal **g.** metal

Chapter 7

1. **a.** $CaCl_2$ **b.** NaCN **c.** MgO **d.** BaO **e.** NaF **f.** $Al(NO_3)_3$ **g.** ZnI_2 **h.** $CoCO_3$
2. **a.** AgF **b.** NiS **c.** $CrBr_3$ **d.** $Pb_3(PO_4)_2$ **e.** $(NH_4)_2C_2O_4$ **f.** SrI_2 **g.** Li_2O
3. **a.** KH **b.** $Hg(CN)_2$ **c.** $ZnC_4H_4O_6$ **d.** $CdSiO_3$ **e.** $(NH_4)_2Cr_2O_7$ **f.** $Pb(NO_3)_2$ **g.** $Cu(ClO_4)_2$ **h.** $Na_2B_4O_7$
8. **a.** barium sulfide **b.** bismuth(III) iodide **c.** magnesium nitride **d.** lead(II) bromide **e.** zinc fluoride
9. **a.** calcium hydride **b.** sodium phosphide **c.** calcium sulfide **d.** thallium(I) iodide **e.** cobalt(II) bromide
10. **a.** barium chloride **b.** sodium bromide **c.** aluminum fluoride **d.** lithium carbonate **e.** potassium chloride **f.** mercury(II) iodide **g.** zinc nitrate **h.** barium hydroxide
11. **a.** diphosphorus pentoxide **b.** phosphorus pentachloride **c.** sulfur hexafluoride **d.** phosphorus trichloride
12. **a.** ammonium nitrate **b.** acetic acid or hydrogen acetate **c.** sodium phosphate **d.** hydrochloric acid or hydrogen chloride **e.** nitric acid or hydrogen nitrate **f.** copper(II) acetate **g.** potassium oxide **h.** sulfuric acid or hydrogen sulfate
13. **a.** 4 **b.** 7 **c.** 3 **d.** 9 **e.** 8 **f.** 5
14. **a.** hexane **b.** cyclobutane
15. **a.** NO_2 **b.** $C_3H_4O_3$ **c.** CH_3 **d.** CH_4 **e.** HgI **f.** C_4H_9
16. **a.** 1 **b.** 3 **c.** 2 **d.** 4 **e.** 6 **f.** 5 **g.** 3 **h.** 3

Chapter 8

1. $(0.143\ \cancel{h})\left(\dfrac{60\ \cancel{min}}{1\ \cancel{h}}\right)\left(\dfrac{60\ s}{1\ \cancel{min}}\right) = 515\ s$

2. $(0.84\ \cancel{m})\left(\dfrac{100\ cm}{1\ \cancel{m}}\right) = 84\ cm$

3. $(31.5\ \cancel{cg})\left(\dfrac{1\ \cancel{g}}{100\ \cancel{cg}}\right)\left(\dfrac{1000\ mg}{1\ \cancel{g}}\right) = 315\ mg$

4. $(65.22\ \cancel{mg})\left(\dfrac{1\ g}{1000\ \cancel{mg}}\right) = 0.065\ 22\ g$

5. $(531\ \cancel{cm^3})\left(\dfrac{1\ \cancel{m^3}}{100^3\ \cancel{cm^3}}\right)\left(\dfrac{10^3\ dm^3}{1\ \cancel{m^3}}\right) = 0.531\ dm^3$

6. $(718\ \cancel{nm})\left(\dfrac{1\ \cancel{m}}{10^9\ \cancel{nm}}\right)\left(\dfrac{100\ cm}{1\ \cancel{m}}\right) = 7.18 \times 10^{-5}\ cm$

7. $\dfrac{0.032\ \cancel{g}}{} \left|\ \dfrac{1000\ mg}{1\ \cancel{g}}\right. = 32\ mg$

8. $\dfrac{0.436\ \cancel{m^3}}{} \left|\ \dfrac{100^3\ cm^3}{1\ \cancel{m^3}}\right. = 436\ 000\ cm^3$ or $4.36 \times 10^5\ cm^3$

9. $\dfrac{302.1\ \cancel{mL}}{} \left|\ \dfrac{1\ cm^3}{1\ \cancel{mL}}\right. = 302.1\ cm^3$

10. $\dfrac{0.693\ \cancel{dm^3}}{} \left|\ \dfrac{1\ \cancel{m^3}}{10^3\ \cancel{dm^3}}\right| \dfrac{100^3\ cm^3}{1\ \cancel{m^3}} = 693\ cm^3$

11. $\dfrac{9.06\ \cancel{km}}{\cancel{h}} \left|\ \dfrac{1000\ m}{1\ \cancel{km}}\right| \dfrac{1\ \cancel{h}}{60\ min} = 151\ m/min$

12. $\dfrac{0.307\ \cancel{mg}}{cm^3} \left|\ \dfrac{1\ g}{1000\ \cancel{mg}}\right. = 3.07 \times 10^{-4}\ g/cm^3$

13. $\dfrac{822\ \cancel{dm^3}}{\cancel{s}} \left|\ \dfrac{1\ L}{1\ \cancel{dm^3}}\right| \dfrac{60\ \cancel{s}}{1\ min} = 49\ 300\ L/min$
or 4.93×10^4 L/min

14. $\dfrac{0.78\ \cancel{L}}{\cancel{min}} \left|\ \dfrac{1000\ \cancel{mL}}{1\ \cancel{L}}\right| \dfrac{1\ cm^3}{1\ \cancel{mL}} \left|\ \dfrac{1\ \cancel{min}}{60\ s}\right. = 13\ cm^3/s$

15. $\dfrac{0.848\ \cancel{kg}}{\cancel{L}} \left|\ \dfrac{1000\ \cancel{g}}{1\ \cancel{kg}}\right| \dfrac{1000\ mg}{1\ \cancel{g}} \left|\ \dfrac{1\ \cancel{L}}{1000\ \cancel{mL}}\right| \dfrac{1\ \cancel{mL}}{1\ cm^3}$
$= 848\ mg/cm^3$

16. $\dfrac{81.42\ \cancel{nm}}{\cancel{s}} \left|\ \dfrac{1\ \cancel{m}}{10^9\ \cancel{nm}}\right| \dfrac{100\ cm}{1\ \cancel{m}} \left|\ \dfrac{60\ \cancel{s}}{1\ min}\right.$
$= 4.885 \times 10^{-4}\ cm/min$

17. $\dfrac{7.56\ \cancel{mm^3}}{\cancel{s}} \left|\ \dfrac{1\ \cancel{m^3}}{1000^3\ \cancel{mm^3}}\right| \dfrac{10^3\ dm^3}{1\ \cancel{m^3}} \left|\ \dfrac{60\ \cancel{s}}{1\ min}\right.$
$= 4.54 \times 10^{-4}\ dm^3/min$

18. $\dfrac{0.03\ \cancel{cm}}{\cancel{s}} \left|\ \dfrac{1\ \cancel{m}}{100\ \cancel{cm}}\right| \dfrac{1\ km}{1000\ \cancel{m}} \left|\ \dfrac{60\ \cancel{s}}{1\ \cancel{min}}\right| \dfrac{60\ \cancel{min}}{1\ h}$
$= 0.001\ km/h$

19. $\dfrac{0.0775\ \cancel{\mathrm{cg}}}{\cancel{\mathrm{cm}^3}} \Bigg| \dfrac{1\ \mathrm{g}}{100\ \cancel{\mathrm{cg}}} \Bigg| \dfrac{100^3\ \cancel{\mathrm{cm}^3}}{1\ \mathrm{m}^3} = 775\ \mathrm{g/m}^3$

20. $\dfrac{0.95\ \cancel{\mathrm{kg}}}{\cancel{\mathrm{cm}^3}} \Bigg| \dfrac{1000\ \cancel{\mathrm{g}}}{1\ \cancel{\mathrm{kg}}} \Bigg| \dfrac{1000\ \mathrm{mg}}{1\ \cancel{\mathrm{g}}} \Bigg| \dfrac{100^3\ \cancel{\mathrm{cm}^3}}{1\ \cancel{\mathrm{m}^3}} \Bigg| \dfrac{1\ \cancel{\mathrm{m}^3}}{1000^3\ \mathrm{mm}^3}$
$= 950\ \mathrm{mg/mm}^3$ or $9.5 \times 10^2\ \mathrm{mg/mm}^3$

21. $0.0753\ \mathrm{m}^2$
22. $1.1 \times 10^6\ \mathrm{mm}^2$
23. $7.4 \times 10^2\ \mathrm{g/dm}^3$
24. 763 g
25. 9.40 kg
26. $1.33 \times 10^{-5}\ \mathrm{cm}^2$
27. 3.59×10^{12} cm

32.

a. 1 C atom	1×12.0	=	12.0 u
4 H atoms	4×1.01	=	4.04 u
1 O atom	1×16.0	=	16.0 u
			32.0 u
b. 2 C atoms	2×12.0	=	24.0 u
6 H atoms	6×1.01	=	6.06 u
			30.1 u
c. 12 C atoms	12×12.0	=	144 u
22 H atoms	22×1.01	=	22.2 u
11 O atoms	11×16.0	=	176 u
			342 u
d. 3 C atoms	3×12.0	=	36.0 u
8 H atoms	8×1.01	=	8.08 u
1 O atom	1×16.0	=	16.0 u
			60.1 u

33.

a. 1 Ta atom	1×181	=	181 u
1 C atom	1×12.0	=	12.0 u
			193 u
b. 1 Al atom	1×27.0	=	27.0 u
1 N atom	1×14.0	=	14.0 u
			41.0 u
c. 4 P atoms	4×31.0	=	124 u
3 S atoms	3×32.1	=	96.3 u
			2.20×10^2 u
d. 3 Ca atoms	3×40.0	=	120.0 u
2 P atoms	2×31.0	=	62.0 u
8 O atoms	8×16.0	=	128 u
			3.10×10^2 u
e. 1 Ba atom	1×137	=	137 u
2 Cl atoms	2×35.5	=	71.0 u
2 O atoms	2×16.0	=	32.0 u
			2.40×10^2 u

34. $\dfrac{0.638\ \cancel{\mathrm{mol\ Ba(CN)_2}}}{} \Bigg| \dfrac{189\ \mathrm{g\ Ba(CN)_2}}{1\ \cancel{\mathrm{mol\ Ba(CN)_2}}} = 121\ \mathrm{g\ Ba(CN)_2}$

35. $\dfrac{50.4\ \cancel{\mathrm{g\ CaBr_2}}}{} \Bigg| \dfrac{1\ \mathrm{mol\ CaBr_2}}{2.00 \times 10^2\ \cancel{\mathrm{g\ CaBr_2}}} = 0.252\ \mathrm{mol\ CaBr_2}$

36. $\dfrac{1.26\ \cancel{\mathrm{mol\ NbI_5}}}{} \Bigg| \dfrac{727.4\ \mathrm{g\ NbI_5}}{1\ \cancel{\mathrm{mol\ NbI_5}}} = 917\ \mathrm{g\ NbI_5}$

37. $\dfrac{86.2\ \cancel{\mathrm{g\ C_2H_4}}}{} \Bigg| \dfrac{1\ \mathrm{mol\ C_2H_4}}{28.0\ \cancel{\mathrm{g\ C_2H_4}}} = 3.08\ \mathrm{mol\ C_2H_4}$

38. $\dfrac{0.943\ \cancel{\mathrm{mol}}\ \mathrm{H_2O}}{} \Bigg| \dfrac{6.02 \times 10^{23}\ \mathrm{molecules}}{1\ \cancel{\mathrm{mol}}}$
$= 5.68 \times 10^{23}$ molecules $\mathrm{H_2O}$

39. $\dfrac{7.74 \times 10^{26}\ \cancel{\mathrm{formula\ units}}\ \mathrm{Al_2O_3}}{} \Bigg| \dfrac{1\ \mathrm{mol}}{6.02 \times 10^{23}\ \cancel{\mathrm{formula\ units}}} = 1.29 \times 10^3\ \mathrm{mol\ Al_2O_3}$

40. $\dfrac{91.9\ \cancel{\mathrm{g\ NH_4IO_3}}}{} \Bigg| \dfrac{1\ \cancel{\mathrm{mol}}\ \mathrm{NH_4IO_3}}{193\ \cancel{\mathrm{g\ NH_4IO_3}}} \Bigg| \dfrac{6.02 \times 10^{23}\ \mathrm{formula\ units}}{1\ \cancel{\mathrm{mol}}}$
$= 2.87 \times 10^{23}$ formula units $\mathrm{NH_4IO_3}$

41. $\dfrac{6.63 \times 10^{23}\ \cancel{\mathrm{molecules}}\ \mathrm{C_6H_{12}O_6}}{} \Bigg| \dfrac{1\ \mathrm{mol}}{6.02 \times 10^{23}\ \cancel{\mathrm{molecules}}}$
$= 1.10\ \mathrm{mol\ C_6H_{12}O_6}$

42. $\dfrac{5.23\ \cancel{\mathrm{g\ Fe(NO_3)_2}}}{100.00\ \cancel{\mathrm{cm^3}}\ \mathrm{soln}} \Bigg| \dfrac{1\ \mathrm{mol\ Fe(NO_3)_2}}{1.80 \times 10^2\ \cancel{\mathrm{g\ Fe(NO_3)_2}}} \Bigg| \dfrac{1000\ \cancel{\mathrm{cm^3}}}{1\ \mathrm{dm^3}}$
$= 0.291M\ \mathrm{Fe(NO_3)_2}$

43. $\dfrac{8.55\ \cancel{\mathrm{g\ NH_4I}}}{50.0\ \cancel{\mathrm{cm^3}}\ \mathrm{soln}} \Bigg| \dfrac{1\ \mathrm{mol\ NH_4I}}{145\ \cancel{\mathrm{g\ NH_4I}}} \Bigg| \dfrac{1000\ \cancel{\mathrm{cm^3}}}{1\ \mathrm{dm^3}}$
$= 1.18M\ \mathrm{NH_4I}$

44. $\dfrac{9.94\ \cancel{\mathrm{g\ CoSO_4}}}{2.50 \times 10^2\ \cancel{\mathrm{cm^3}}\ \mathrm{soln}} \Bigg| \dfrac{1\ \mathrm{mol\ CoSO_4}}{155\ \cancel{\mathrm{g\ CoSO_4}}} \Bigg| \dfrac{1000\ \cancel{\mathrm{cm^3}}}{1\ \mathrm{dm^3}}$
$= 0.257M\ \mathrm{CoSO_4}$

45. $\dfrac{44.3\ \cancel{\mathrm{g\ Pb(ClO_4)_2}}}{250.0\ \cancel{\mathrm{cm^3}}\ \mathrm{soln}} \Bigg| \dfrac{1\ \mathrm{mol\ Pb(ClO_4)_2}}{406\ \cancel{\mathrm{g\ Pb(ClO_4)_2}}} \Bigg| \dfrac{1000\ \cancel{\mathrm{cm^3}}}{1\ \mathrm{dm^3}}$
$= 0.436M\ \mathrm{Pb(ClO_4)_2}$

46. $\dfrac{1.00\ \cancel{\mathrm{dm^3\ soln}}}{} \Bigg| \dfrac{3.00\ \cancel{\mathrm{mol\ NiCl_2}}}{1\ \cancel{\mathrm{dm^3\ soln}}} \Bigg| \dfrac{1.30 \times 10^2\ \mathrm{g\ NiCl_2}}{1\ \cancel{\mathrm{mol\ NiCl_2}}}$
$= 3.90 \times 10^2\ \mathrm{g\ NiCl_2}$

Dissolve 3.90×10^2 g $\mathrm{NiCl_2}$ in enough water to make 1.00 $\mathrm{dm^3}$ of solution.

47. $\dfrac{2.50 \times 10^2\ \cancel{\mathrm{cm^3\ soln}}}{} \Bigg| \dfrac{4.00\ \cancel{\mathrm{mol\ CoCl_2}}}{1\ \cancel{\mathrm{dm^3\ soln}}} \Bigg| \dfrac{1.30 \times 10^2\ \mathrm{g\ CoCl_2}}{1\ \cancel{\mathrm{mol\ CoCl_2}}} \Bigg| \dfrac{1\ \cancel{\mathrm{dm^3}}}{1000\ \cancel{\mathrm{cm^3}}} = 1.30 \times 10^2\ \mathrm{g\ CoCl_2}$

Dissolve 1.30×10^2 g $\mathrm{CoCl_2}$ in enough water to make $2.50 \times 10^2\ \mathrm{cm^3}$ of solution.

48. $\dfrac{0.500\ \cancel{\mathrm{dm^3\ soln}}}{} \Bigg| \dfrac{1.50\ \cancel{\mathrm{mol\ AgF}}}{1\ \cancel{\mathrm{dm^3\ soln}}} \Bigg| \dfrac{127\ \mathrm{g\ AgF}}{1\ \cancel{\mathrm{mol\ AgF}}} = 95.3\ \mathrm{g\ AgF}$

Dissolve 95.3 g AgF in enough water to make 0.500 $\mathrm{dm^3}$ of solution.

49. $\dfrac{2.50 \times 10^2\ \cancel{\mathrm{cm^3\ soln}}}{} \Bigg| \dfrac{0.002\,00\ \cancel{\mathrm{mol}}\ \cancel{\mathrm{Cd(IO_3)_2}}}{1\ \cancel{\mathrm{dm^3\ soln}}} \Bigg| \dfrac{462\ \mathrm{g\ Cd(IO_3)_2}}{1\ \cancel{\mathrm{mol\ Cd(IO_3)_2}}} \Bigg| \dfrac{1\ \cancel{\mathrm{dm^3}}}{1000\ \cancel{\mathrm{cm^3}}} = 0.231\ \mathrm{g\ Cd(IO_3)_2}$

Dissolve 0.231 g $\mathrm{Cd(IO_3)_2}$ in enough water to make $2.50 \times 10^2\ \mathrm{cm^3}$ of solution.

50. $\mathrm{NH_3}$:

1 N atom	1×14.0 u =	14.0 u
3 H atoms	3×1.01 u =	3.03 u
		17.0 u

Percentage of N in $\mathrm{NH_3} = \dfrac{\text{mass N}}{\text{mass NH}_3} \times 100$

$= \dfrac{14.0\ \cancel{\mathrm{u}}}{17.0\ \cancel{\mathrm{u}}} \times 100 = 82.4\%$

$CO(NH_2)_2$:

1 C atom	$1 \times 12.0\ \text{u} =$	12.0 u
1 O atom	$1 \times 16.0\ \text{u} =$	16.0 u
2 N atoms	$2 \times 14.0\ \text{u} =$	28.0 u
4 H atoms	$4 \times 1.01\ \text{u} =$	4.04 u
		60.0 u

$$\text{Percentage of N in } CO(NH_2)_2 = \frac{\text{mass 2N}}{\text{mass } CO(NH_2)_2} \times 100$$

$$= \frac{28.0\ \cancel{\text{u}}}{60.0\ \cancel{\text{u}}} \times 100 = 46.7\%$$

51.

2 Al atoms	$2 \times 27.0\ \text{u} =$	54.0 u
3 S atoms	$3 \times 32.1\ \text{u} =$	96.3 u
		1.50×10^2 u

$$\text{Percentage of Al} = \frac{\text{mass 2Al}}{\text{mass } Al_2S_3} \times 100$$

$$= \frac{54.0\ \cancel{\text{u}}}{1.50 \times 10^2\ \cancel{\text{u}}} \times 100 = 36.0\%$$

$$\text{Percentage of S} = \frac{\text{mass 3S}}{\text{mass } Al_2S_3} \times 100$$

$$= \frac{96.3\ \cancel{\text{u}}}{1.50 \times 10^2\ \cancel{\text{u}}} \times 100 = 64.2\%$$

52.

1 Ni atom	$1 \times 58.7\ \text{u} =$	58.7 u
2 I atoms	$2 \times 127\ \text{u} =$	254 u
		313 u

$$\text{Percentage of Ni} = \frac{\text{mass Ni}}{\text{mass } NiI_2} \times 100$$

$$= \frac{58.7\ \cancel{\text{u}}}{313\ \cancel{\text{u}}} \times 100 = 18.8\%$$

$$\text{Percentage of I} = \frac{\text{mass 2I}}{\text{mass } NiI_2} \times 100$$

$$= \frac{254\ \cancel{\text{u}}}{313\ \cancel{\text{u}}} \times 100 = 81.2\%$$

53.

1 Ca atom	$1 \times 40.1\ \text{u} =$	40.1 u
2 C atoms	$2 \times 12.0\ \text{u} =$	24.0 u
2 N atoms	$2 \times 14.0\ \text{u} =$	28.0 u
		92.1 u

$$\text{Percentage of Ca} = \frac{\text{mass Ca}}{\text{mass } Ca(CN)_2} \times 100$$

$$= \frac{40.1\ \cancel{\text{u}}}{92.1\ \cancel{\text{u}}} \times 100 = 43.5\%$$

$$\text{Percentage of C} = \frac{\text{mass 2C}}{\text{mass } Ca(CN)_2} \times 100$$

$$= \frac{24.0\ \cancel{\text{u}}}{92.1\ \cancel{\text{u}}} \times 100 = 26.1\%$$

$$\text{Percentage of N} = \frac{\text{mass 2N}}{\text{mass } Ca(CN)_2} \times 100$$

$$= \frac{28.0\ \cancel{\text{u}}}{92.1\ \cancel{\text{u}}} \times 100 = 30.4\%$$

54.

1 Cu atom	$1 \times 63.5\ \text{u} =$	63.5 u
2 Cl atoms	$2 \times 35.5\ \text{u} =$	71.0 u
8 O atoms	$8 \times 16.0\ \text{u} =$	128 u
		263 u

$$\text{Percentage of Cu} = \frac{\text{mass Cu}}{\text{mass } Cu(ClO_4)_2} \times 100$$

$$= \frac{63.5\ \cancel{\text{u}}}{263\ \cancel{\text{u}}} \times 100 = 24.1\%$$

$$\text{Percentage of Cl} = \frac{\text{mass 2Cl}}{\text{mass } Cu(ClO_4)_2} \times 100$$

$$= \frac{71.0\ \cancel{\text{u}}}{263\ \cancel{\text{u}}} \times 100 = 27.0\%$$

$$\text{Percentage of O} = \frac{\text{mass 8O}}{\text{mass } Cu(ClO_4)_2} \times 100$$

$$= \frac{128\ \cancel{\text{u}}}{263\ \cancel{\text{u}}} \times 100 = 48.7\%$$

55.

2 N atoms	$2 \times 14.0\ \text{u} =$	28.0 u
8 H atoms	$8 \times 1.01\ \text{u} =$	8.08 u
2 C atoms	$2 \times 12.0\ \text{u} =$	24.0 u
4 O atoms	$4 \times 16.0\ \text{u} =$	64.0 u
		124 u

$$\text{Percentage of N} = \frac{\text{mass 2N}}{\text{mass } (NH_4)_2C_2O_4} \times 100$$

$$= \frac{28.0\ \cancel{\text{u}}}{124\ \cancel{\text{u}}} \times 100 = 22.6\%$$

$$\text{Percentage of H} = \frac{\text{mass 8H}}{\text{mass } (NH_4)_2C_2O_4} \times 100$$

$$= \frac{8.08\ \cancel{\text{u}}}{124\ \cancel{\text{u}}} \times 100 = 6.52\%$$

$$\text{Percentage of C} = \frac{\text{mass 2C}}{\text{mass } (NH_4)_2C_2O_4} \times 100$$

$$= \frac{24.0\ \cancel{\text{u}}}{124\ \cancel{\text{u}}} \times 100 = 19.4\%$$

$$\text{Percentage of O} = \frac{\text{mass 4O}}{\text{mass } (NH_4)_2C_2O_4} \times 100$$

$$= \frac{64.0\ \cancel{\text{u}}}{124\ \cancel{\text{u}}} \times 100 = 51.6\%$$

56.

$$\frac{1.67\ \cancel{\text{g Ce}}}{} \left| \frac{1\ \text{mol Ce}}{1.40 \times 10^2\ \cancel{\text{g Ce}}} \right. = 0.0119\ \text{mol Ce}$$

$$\frac{4.54\ \cancel{\text{g I}}}{} \left| \frac{1\ \text{mol I}}{127\ \cancel{\text{g I}}} \right. = 0.0357\ \text{mol I}$$

$0.0357/0.0119 \approx 3$; 1:3 ratio; $\therefore CeI_3$

57.

$$\frac{0.556\ \cancel{\text{g C}}}{} \left| \frac{1\ \text{mol C}}{12.0\ \cancel{\text{g C}}} \right. = 0.0463\ \text{mol C}$$

$$\frac{0.0933\ \cancel{\text{g H}}}{} \left| \frac{1\ \text{mol H}}{1.01\ \cancel{\text{g H}}} \right. = 0.0924\ \text{mol H}$$

$0.0924/0.0463 \approx 2$; 1:2 ratio; $\therefore CH_2$

58. Assume 100.0 g sample.

$$\frac{68.8\ \cancel{\text{g C}}}{} \left| \frac{1\ \text{mol C}}{12.0\ \cancel{\text{g C}}} \right. = 5.73\ \text{mol C}$$

$$\frac{4.95\ \cancel{\text{g H}}}{} \left| \frac{1\ \text{mol H}}{1.01\ \cancel{\text{g H}}} \right. = 4.90\ \text{mol H}$$

$$\frac{26.2\ \cancel{\text{g O}}}{} \left| \frac{1\ \text{mol O}}{16.0\ \cancel{\text{g O}}} \right. = 1.64\ \text{mol O}$$

$5.73/1.64 \approx 3.5$; $4.90/1.64 \approx 3$; 7:6:2 ratio; $\therefore C_7H_6O_2$

59. Assume 100.0 g sample

$$9.93\ \cancel{\text{g C}} \times \frac{1\ \text{mol C}}{12.0\ \cancel{\text{g C}}} = 0.828\ \text{mol C}$$

$$58.6\ \cancel{\text{g Cl}} \times \frac{1\ \text{mol Cl}}{35.5\ \cancel{\text{g Cl}}} = 1.65\ \text{mol Cl}$$

$$31.4\ \cancel{\text{g F}} \times \frac{1\ \text{mol F}}{19.0\ \cancel{\text{g F}}} = 1.65\ \text{mol F}$$

$1.65/0.828 \approx 2$; 1:2:2 ratio; $\therefore$ CCl_2F_2

60. $CH = 12.0 + 1.01 = 13.0$ u

$$\frac{78\ \cancel{\text{u}}}{13.0\ \cancel{\text{u}}} = 6.0,\ 6(CH) = C_6H_6$$

61. $CHOCl = 12.0 + 1.01 + 16.0 + 35.5 = 64.5$ u

$$\frac{129\ \cancel{\text{u}}}{64.5\ \cancel{\text{u}}} = 2.00$$

$2(CHOCl) = C_2H_2O_2Cl_2$ or $HCCl_2COOH$

62. $CClN = 12.0 + 35.5 + 14.0 = 61.5$ u

$$\frac{184\ \cancel{\text{u}}}{61.5\ \cancel{\text{u}}} = 3.00,\ 3(CClN) = C_3Cl_3N_3$$

63. Assume 100.0 g sample.

$$40.9\ \cancel{\text{g C}} \times \frac{1\ \text{mol C}}{12\ \cancel{\text{g C}}} = 3.41\ \text{mol C}$$

$$4.58\ \cancel{\text{g H}} \times \frac{1\ \text{mol H}}{1.01\ \cancel{\text{g H}}} = 4.53\ \text{mol H}$$

$$54.5\ \cancel{\text{g O}} \times \frac{1\ \text{mol O}}{16.0\ \cancel{\text{g O}}} = 3.41\ \text{mol O}$$

$4.53/3.41 = 1.33$, 3:4:3 ratio
$\therefore$ $C_3H_4O_3$ is the empirical formula.
$C_3H_4O_3 = 36.0 + 4.04 + 48.0 = 88.0$ u
176 u/88.0 u = 2
$\therefore$ $C_6H_8O_6$ is the molecular formula.

64. Assume 100.0 g sample.

$$60.0\ \text{g} \times \frac{1\ \text{mol}}{12.0\ \text{g}} = 5.00\ \text{mol C}$$

$$4.48\ \text{g} \times \frac{1\ \text{mol}}{1.01\ \text{g}} = 4.44\ \text{mol H}$$

$$35.5\ \text{g} \times \frac{1\ \text{mol}}{16.0\ \text{g}} = 2.22\ \text{mol O}$$

$$\frac{5.00}{2.22} = 2.25,\ \frac{4.44}{2.22} = 2.00,\ \frac{2.22}{2.22} = 1.00$$

multiply by 4 to get $C_9H_8O_4$; formula mass $= 1.80 \times 10^2$ u
$\therefore$ $C_9H_8O_4$ is also the molecular formula.

65. **a.** sodium thiosulfate pentahydrate
b. calcium sulfate dihydrate

66. $Na_2B_4O_7 \cdot 10H_2O$ (borax)

67. Li_2SiF_6: molecular mass $= 2(6.94) + 28.1 + 6(19.0) = 156$ u

$$0.391\ \cancel{\text{g } Li_2SiF_6} \times \frac{1\ \text{mol}}{156\ \cancel{\text{g } Li_2SiF_6}} = 0.002\ 51\ \text{mol}$$

H_2O: molecular mass $= 2(1.01) + 16.0 = 18.0$ u

$$0.0903\ \cancel{\text{g } H_2O} \times \frac{1\ \text{mol}}{18.0\ \cancel{\text{g } H_2O}} = 0.005\ 02\ \text{mol}$$

$0.005\ 02/0.002\ 51 = 2.00$; 1:2 ratio
$\therefore$ $Li_2SiF_6 \cdot 2H_2O$

68.

$$0.737\ \cancel{\text{g } MgSO_3} \times \frac{1\ \text{mol}}{104\ \cancel{\text{g } MgSO_3}} = 0.007\ 09\ \text{mol}$$

$$0.763\ \cancel{\text{g } H_2O} \times \frac{1\ \text{mol } H_2O}{18.0\ \cancel{\text{g } H_2O}} = 0.0424\ \text{mol } H_2O$$

$0.0424/0.007\ 09 \approx 6.00$; 1:6 ratio
$\therefore$ $MgSO_3 \cdot 6H_2O$

69. Assume 100.0 g of sample.
76.9 g $CaSO_3$, 23.1 g H_2O

$$76.9\ \cancel{\text{g } CaSO_3} \times \frac{1\ \text{mol}}{120\ \cancel{\text{g } CaSO_3}} = 0.641\ \text{mol}$$

$$23.1\ \text{g} \times \frac{1\ \text{mol}\ \cancel{H_2O}}{18.0\ \cancel{\text{g } H_2O}} = 1.28\ \text{mol}$$

$1.28/0.641 \approx 2.00$; 1:2 ratio
$\therefore$ $CaSO_3 \cdot 2H_2O$

70. Assume 100.0 g of material.
$\therefore$ 89.2 g $BaBr_2$, 10.8 g H_2O

$$89.2\ \cancel{\text{g } BaBr_2} \times \frac{1\ \text{mol}}{297\ \cancel{\text{g } BaBr_2}} = 0.300\ \text{mol}$$

$$10.8\ \cancel{\text{g } H_2O} \times \frac{1\ \text{mol}}{18.0\ \cancel{\text{g } H_2O}} = 0.600\ \text{mol}$$

$0.600/0.300 = 2.00$; 1:2 ratio
$\therefore$ $BaBr_2 \cdot 2H_2O$

Chapter 9

1. balanced
2. $Al(NO_3)_3 + 3NaOH \rightarrow Al(OH)_3 + 3NaNO_3$
3. $2KNO_3 \rightarrow 2KNO_2 + O_2$
4. $2Fe + 3H_2SO_4 \rightarrow Fe_2(SO_4)_3 + 3H_2$
5. $3O_2 + CS_2 \rightarrow CO_2 + 2SO_2$
6. balanced
7. $3Mg + N_2 \rightarrow Mg_3N_2$
8. $CuCO_3 \rightarrow CuO + CO_2$
9. $2Na + 2H_2O \rightarrow 2NaOH + H_2$
10. $2Cu + S \rightarrow Cu_2S$
11. $2AgNO_3 + H_2SO_4 \rightarrow Ag_2SO_4 + 2HNO_3$
12. $2C_2H_6 + 7O_2 \rightarrow 4CO_2 + 6H_2O$
13. single displacement
14. decomposition
15. synthesis
16. combustion
17. single displacement
18. double displacement
19. $2Al + 3Cu(NO_3)_2 \rightarrow \underline{3Cu} + \underline{2Al(NO_3)_3}$
20. $2Hg + O_2 \rightarrow \underline{2HgO}$
21. $H_2SO_4 + 2KOH \rightarrow \underline{K_2SO_4} + \underline{2H_2O}$
22. $2C_5H_{10} + 15O_2 \rightarrow \underline{10CO_2} + \underline{10H_2O}$

28.

$$12.5\ \cancel{\text{g } C_6H_{12}O_6} \times \frac{1\ \cancel{\text{mol } C_6H_{12}O_6}}{1.80 \times 10^2\ \cancel{\text{g } C_6H_{12}O_6}} \times \frac{6\ \cancel{\text{mol } O_2}}{1\ \cancel{\text{mol } C_6H_{12}O_6}} \times \frac{32.0\ \text{g } O_2}{1\ \cancel{\text{mol } O_2}} = 13.3\ \text{g } O_2$$

29. $\frac{3.41\ \text{g H}_2}{} \times \frac{1\ \text{mol H}_2}{2.02\ \text{g H}_2} \times \frac{2\ \text{mol NH}_3}{3\ \text{mol H}_2} \times \frac{17.0\ \text{g NH}_3}{1\ \text{mol NH}_3}$
$= 19.1\ \text{g NH}_3$

30. $3KClO_3 \rightarrow 2KCl + 3O_2$

$\frac{80.5\ \text{g O}_2}{} \times \frac{1\ \text{mol O}_2}{32.0\ \text{g O}_2} \times \frac{2\ \text{mol KCl}}{3\ \text{mol O}_2} \times \frac{74.6\ \text{g KCl}}{1\ \text{mol KCl}}$
$= 125\ \text{g KCl}$

31. $2Al + 6HCl \rightarrow 2AlCl_3 + 3H_2$

$\frac{9.23\ \text{g Al}}{} \times \frac{1\ \text{mol Al}}{27.0\ \text{g Al}} \times \frac{3\ \text{mol H}_2}{2\ \text{mol Al}} \times \frac{2.02\ \text{g H}_2}{1\ \text{mol H}_2} = 1.04\ \text{g H}_2$

32. a. $\frac{2.50\ \text{g K}_2\text{PtCl}_4}{} \times \frac{1\ \text{mol K}_2\text{PtCl}_4}{415\ \text{g K}_2\text{PtCl}_4} \times \frac{1\ \text{mol PtCl}_2(\text{NH}_3)_2}{1\ \text{mol K}_2\text{PtCl}_4} \times \frac{3.00 \times 10^2\ \text{g PtCl}_2(\text{NH}_3)_2}{1\ \text{mol PtCl}_2(\text{NH}_3)_2} = 1.81\ \text{g Pt(NH}_3)_2\text{Cl}_2$

b. $\frac{2.50\ \text{g K}_2\text{PtCl}_4}{} \times \frac{1\ \text{mol K}_2\text{PtCl}_4}{415\ \text{g K}_2\text{PtCl}_4} \times \frac{2\ \text{mol NH}_3}{1\ \text{mol K}_2\text{PtCl}_4} \times \frac{17.0\ \text{g NH}_3}{1\ \text{mol NH}_3} = 0.205\ \text{g NH}_3$

33. percentage yield $= \frac{39.7\ \text{g}}{65.6\ \text{g}} \times 100 = 60.5\%$

34. $N_2(g) + 3H_2(g) \rightarrow 2NH_3(g)$

$\frac{5.50\ \text{g H}_2}{} \times \frac{1\ \text{mol H}_2}{2.02\ \text{g H}_2} \times \frac{2\ \text{mol NH}_3}{3\ \text{mol H}_2} \times \frac{17.0\ \text{g NH}_3}{1\ \text{mol NH}_3}$
$= 30.9\ \text{g NH}_3$, theoretical yield

percentage yield $= \frac{20.4\ \text{g NH}_3}{30.9\ \text{g NH}_3} \times 100 = 66.0\%$

35. $NH_3 + HBr \rightarrow NH_4Br + 188.32\ \text{kJ}$

$\frac{193\ \text{g NH}_4\text{Br}}{} \times \frac{1\ \text{mol NH}_4\text{Br}}{97.9\ \text{g NH}_4\text{Br}} \times \frac{188.32\ \text{kJ}}{1\ \text{mol NH}_4\text{Br}} = 371\ \text{kJ}$

$q_r = -371\ \text{kJ}$

36. $\frac{0.772\ \text{g CoCO}_3}{} \times \frac{1\ \text{mol CoCO}_3}{119\ \text{g CoCO}_3} \times \frac{81.6\ \text{kJ}}{1\ \text{mol CoCO}_3} = 0.529\ \text{kJ}$

$q_r = 0.529\ \text{kJ}$, or 529 J

37. $\frac{0.0663\ \text{g Br}_2}{} \times \frac{1\ \text{mol Br}_2}{160\ \text{g Br}_2} \times \frac{100.18\ \text{kJ}}{1\ \text{mol Br}_2} = 0.0415\ \text{kJ}$

$q_r = -0.0415\ \text{kJ}$, or $-41.5\ \text{J}$

38. $\frac{6.18\ \text{kJ}}{} \times \frac{1\ \text{mol CrO}_3}{5.4\ \text{kJ}} \times \frac{99.99\ \text{g CrO}_3}{1\ \text{mol CrO}_3} = 110\ \text{g CrO}_3$

Chapter 10

1. **a.** Ar **b.** B **c.** P **d.** Cl **e.** Ca
2. **a.** O **b.** Mg^{2+} **c.** Te **d.** Ti^{4+}

Chapter 12

1. Electronegativity values:
Sb: 1.82, F: 4.10, In: 1.49, Se: 2.48
∴ In, Sb, Se, F
indium, antimony, selenium, fluorine

2. Electronegativity values:
Fr: 0.86, Ga: 1.82, Ge: 2.02, P: 2.06, Zn: 1.66
∴ Fr, Zn, Ga, Ge, P
francium, zinc, gallium, germanium, phosphorus

3. **a.** $|1.47 - 1.74| = 0.27$ covalent
b. $|0.97 - 3.50| = 2.53$ ionic
c. $|2.50 - 2.20| = 0.30$ covalent
d. $|0.97 - 2.44| = 1.47$ covalent
e. $|1.04 - 2.06| = 1.02$ covalent
f. $|2.01 - 1.01| = 1.00$ covalent
g. $|1.04 - 2.83| = 1.79$ ionic
h. $|4.10 - 2.44| = 1.66$ covalent
i. $|2.74 - 0.89| = 1.85$ ionic

4. **a.** $|2.20 - 2.21| = 0.01$ covalent
b. $|1.96 - 1.47| = 0.49$ covalent
c. $|1.70 - 4.10| = 2.40$ ionic
d. $|2.83 - 2.01| = 0.82$ covalent
e. $|2.74 - 1.08| = 1.66$ covalent
f. $|1.04 - 4.10| = 3.06$ ionic

9. **a.** sum of covalent radii = 118 pm + 99.5 pm = 218 pm
b. 37.1 pm + 133 pm = 170 pm
c. 70 pm + 110 pm = 180 pm
d. 119 pm + 103 pm = 222 pm
e. 82 pm + 71.5 pm = 154 pm

Chapter 13

1. **a.** H:Te: (H) **b.** :F:P:F: (F) **c.** :I:N:I: (I) **d.** :Br:C:Br: (Br, Br)

2. 1c—one unshared; 1d—none
3. **a.** one unshared pair; < 109.5°
b. two unshared pairs; < 109.5°
c. no unshared pairs; > 109.5°
d. Form is trigonal bipyramidal: As in center, one F atom at each end of polar axis, and three F atoms on equator.
Five shared pairs; no unshared pairs; equatorial-equatorial angles are > 109.5°, axial-equatorial angle is < 109.5°.
e. There are four shared pairs of electrons. The outer shell of Te is expanded to 10 electrons. The form is nearly tetrahedral, with some angles > 109.5°, and some angles < 109.5°.
f. The Xe atom shares all eight of its outer electrons with the four O atoms. The angle is 109.5°.

9. **a.** propyne
b. cyclobutene

10. **a.** **b.**

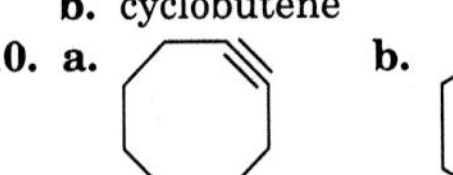

Chapter 14

1.

	Bond	*Electronegativity difference*
a.	Al—P	$|1.47 - 2.06| = 0.59$
b.	Cl—C	$|2.83 - 2.50| = 0.33$
c.	Mo—Te	$|1.30 - 2.01| = 0.71$
d.	H—S	$|2.20 - 2.44| = 0.24$
e.	P—S	$|2.06 - 2.44| = 0.38$
f.	Cl—Si	$|2.83 - 1.74| = 1.09$

In order of decreasing bond polarity: f, c, a, e, b, d

2. a. HBr is polar because of the concentration of negative charge near the Br atom and of positive charge near the H atom.
   b. SO_2 is polar because oxygen has a greater electronegativity than sulfur and the molecule is asymmetric.

7. a. $[Cd(NH_3)_2]^{2+}$ diamminecadmium ion
   b. $[Cr(H_2O)_4]^{2+}$ tetraaquachromium(II) ion
   c. $[Co(NH_3)_6]^{2+}$ hexaamminecobalt(II) ion
   d. $[Cu(NH_3)_4]^{2+}$ tetraamminecopper(II) ion
   e. $[PtBr(NH_3)_3]^{+}$ triamminebromoplatinum(II) ion

8. a. hexaiodoplatinate(IV) ion
   b. hexafluorogermanate(IV) ion
   c. hexahydroxoantimonate(V) ion
   d. tetrachloropalladate(II) ion
   e. hexacyanoferrate(II) ion

9. a. hexafluoroantimonate(V) ion
   b. tetrafluoroborate ion
   c. hexafluoroaluminate ion
   d. tetraammineaquachlorocobalt(III) ion
   e. tetraamminecadmium ion

10. a. $[Pt(NH_3)_4]^{2+}$
    b. $[Co(NH_3)_6]^{3+}$
    c. $[Ir(H_2O)_6]^{3+}$
    d. $[Pd(NH_3)_4]^{2+}$

11. a. $[PdCl_6]^{2-}$
    b. $[PtCl_3(NH_3)]^{-}$
    c. $[AuI_4]^{-}$
    d. $[Au(CN)_4]^{-}$

12. a. $[W(CN)_8]^{3-}$
    b. $[Sb(OH)_6]^{-}$
    c. $[Fe(CO)_5(NO)]^{2+}$
    d. $[Co(CO)_4]^{+}$

13. a. tetraamminediaquanickel(II) nitrate
    b. pentaamminechlorocobalt(III) chloride
    c. potassium hexacyanoferrate(III)
    d. potassium hexahydroxostannate(IV)
    e. diamminesilverperrhenate

14. a. $[Co(H_2O)_6]I_2$
    b. $[Co(NH_3)_6]Cl_2$
    c. $K_2[Ni(CN)_4]$
    d. $PdCl_2(NH_3)_2$
    e. $Ru(CO)_5$

Chapter 15

1. $\frac{62\ \text{mm}}{} \times \frac{1\ \text{k Pa}}{7.50\ \text{mm}} = 8.3\ \text{kPa}$

   97.7 kPa − 8.3 kPa = 89.4 kPa

2. $\frac{691\ \text{mm}}{} \times \frac{1\ \text{kPa}}{7.50\ \text{mm}} = 92.1\ \text{kPa}$

3. $\frac{38\ \text{mm}}{} \times \frac{1\ \text{kPa}}{7.50\ \text{mm}} = 5.1\ \text{kPa}$

   96.3 kPa + 5.1 kPa = 101.4 kPa

4. $\frac{86.0\ \text{mm}}{} \times \frac{1\ \text{kPa}}{7.50\ \text{mm}} = 11.5\ \text{kPa}$

9. a. 86 K − 273 = −187°C
   b. 191 K − 273 = −82°C
   c. 533 K − 273 = 2.60×10^{2}°C
   d. 321 K − 273 = 48°C
   e. 894 K − 273 = 621°C

10. a. 23°C + 273 = 296 K
    b. 58°C + 273 = 331 K
    c. −90°C + 273 = 183 K
    d. 18°C + 273 = 291 K
    e. 25°C + 273 = 298 K

11. a. 872 K − 273 = 599°C
    b. 690 K − 273 = 417°C
    c. 384 K − 273 = 111°C
    d. 20 K − 273 = −253°C
    e. 60 K − 273 = −213°C

12. a. N_2, because it has the lowest molecular mass.

Chapter 17

1. a. 50°C b. 99°C c. 70°C d. 48°C
2. $CHCl_3$
3. a. $T_c = 33$ K
   b. $P_c = 1290$ kPa
   c. $T_t = 14$ K
   d. $P_t = 5$ kPa
   e. $T_m = 14$ K
   f. $T_b = 18$ K
4. a. gas b. solid c. liquid d. liquid
5. a. $46.0\ \text{g} \times 58.0\ \text{C}^\circ \times \frac{2.06\ \text{J}}{\text{g}\cdot\text{C}^\circ} = 5.50 \times 10^{3}\ \text{J}$

   b. $46.0\ \text{g} \times \frac{334\ \text{J}}{\text{g}} = 15\ 400\ \text{J}$

   c. $46.0\ \text{g} \times 100\ \text{C}^\circ \times \frac{4.18\ \text{J}}{\text{g}\cdot\text{C}^\circ} = 19\ 200\ \text{J}$

   d. $46.0\ \text{g} \times \frac{2260\ \text{J}}{\text{g}} = 104\ 000\ \text{J}$

   e. $46.0\ \text{g} \times 14\ \text{C}^\circ \times \frac{2.02\ \text{J}}{\text{g}\cdot\text{C}^\circ} = 1300\ \text{J}$

6. $25.4\ \text{g} \times \frac{61.7\ \text{J}}{\text{g}} = 1570\ \text{J}$

7. $4.24\ \text{g} \times \frac{162\ \text{J}}{\text{g}} = 687\ \text{J}$

8. $5.58\ \text{kg} \times \frac{1000\ \text{g}}{1\ \text{kg}} \times \frac{0.4494\ \text{J}}{\text{g}\cdot\text{C}^\circ} \times 980.0\ \text{C}^\circ \times \frac{1\ \text{kJ}}{1000\ \text{J}} = 2460\ \text{kJ}$

9. $70.0\ \text{g} \times 64\ \text{C}^\circ \times \frac{2.06\ \text{J}}{\text{g}\cdot\text{C}^\circ} = 9200\ \text{J}$

   $70.0\ \text{g} \times \frac{334\ \text{J}}{\text{g}} = 23\ 400\ \text{J}$

   $70.0\ \text{g} \times 100.0\ \text{C}^\circ \times \frac{4.18\ \text{J}}{\text{g}\cdot\text{C}^\circ} = 29\ 300\ \text{J}$

   $70.0\ \text{g} \times \frac{2260\ \text{J}}{\text{g}} = 158\ 000\ \text{J}$

   $70.0\ \text{g} \times 422\ \text{C}^\circ \times \frac{2.02\ \text{J}}{\text{g}\cdot\text{C}^\circ} = 59\ 700\ \text{J}$

   9200 J + 23 400 J + 29 300 J + 158 000 J + 59 700 J = 2.80×10^{5} J

10. $28.9\ \text{g} \times (1085 - 25)\ \text{C}^\circ \times \frac{0.384\ 52\ \text{J}}{\text{g}\cdot\text{C}^\circ} = 11\ 800\ \text{J}$

    = 1.18×10^{4} J

Chapter 18

1. $$\frac{4.41\text{ dm}^3 \mid 94.2\ \cancel{\text{kPa}}}{\mid 101.3\ \cancel{\text{kPa}}} = 4.10\text{ dm}^3$$

2. $$\frac{101\text{ kPa} \mid 5.0\ \cancel{\text{dm}^3}}{\mid 10.0\ \cancel{\text{dm}^3}} = 51\text{ kPa}$$

3. a. $$\frac{844\text{ cm}^3 \mid 98.5\ \cancel{\text{kPa}}}{\mid 101.3\ \cancel{\text{kPa}}} = 821\text{ cm}^3$$

b. $$\frac{273\text{ cm}^3 \mid 59.4\ \cancel{\text{kPa}}}{\mid 101.3\ \cancel{\text{kPa}}} = 1.60 \times 10^2\text{ cm}^3$$

c. $$\frac{116\text{ m}^3 \mid 90.0\ \cancel{\text{kPa}}}{\mid 101.3\ \cancel{\text{kPa}}} = 103\text{ m}^3$$

d. $$\frac{77.0\text{ m}^3 \mid 105.9\ \cancel{\text{kPa}}}{\mid 101.3\ \cancel{\text{kPa}}} = 80.5\text{ m}^3$$

4. a. $$\frac{338\text{ cm}^3 \mid 86.1\ \cancel{\text{kPa}}}{\mid 104.0\ \cancel{\text{kPa}}} = 2.80 \times 10^2\text{ cm}^3$$

b. $$\frac{0.873\text{ m}^3 \mid 94.3\ \cancel{\text{kPa}}}{\mid 102.3\ \cancel{\text{kPa}}} = 0.805\text{ m}^3$$

c. $$\frac{31.5\text{ cm}^3 \mid 97.8\ \cancel{\text{kPa}}}{\mid 82.3\ \cancel{\text{kPa}}} = 37.4\text{ cm}^3$$

d. $$\frac{524\text{ cm}^3 \mid 110.0\ \cancel{\text{kPa}}}{\mid 104.5\ \cancel{\text{kPa}}} = 552\text{ cm}^3$$

5. $P_{gas} = 101.1\text{ kPa} - 8.6\text{ kPa} = 92.5\text{ kPa}$

$$\frac{596\text{ cm}^3 \mid 92.5\ \cancel{\text{kPa}}}{\mid 101.3\ \cancel{\text{kPa}}} = 544\text{ cm}^3$$

6. a. $P_{gas} = 93.3\text{ kPa} - 1.6\text{ kPa} = 91.7\text{ kPa}$

$$\frac{888\text{ cm}^3 \mid 91.7\ \cancel{\text{kPa}}}{\mid 101.3\ \cancel{\text{kPa}}} = 804\text{ cm}^3$$

b. $77.5\text{ kPa} - 1.8\text{ kPa} = 75.7\text{ kPa}$

$$\frac{30.0\text{ cm}^3 \mid 75.7\ \cancel{\text{kPa}}}{\mid 101.3\ \cancel{\text{kPa}}} = 22.4\text{ cm}^3$$

c. $82.4\text{ kPa} - 2.1\text{ kPa} = 80.3\text{ kPa}$

$$\frac{34.0\text{ m}^3 \mid 80.3\ \cancel{\text{kPa}}}{\mid 101.3\ \cancel{\text{kPa}}} = 27.0\text{ m}^3$$

d. $78.3\text{ kPa} - 1.4\text{ kPa} = 76.9\text{ kPa}$

$$\frac{384\text{ cm}^3 \mid 76.9\ \cancel{\text{kPa}}}{\mid 101.3\ \cancel{\text{kPa}}} = 292\text{ cm}^3$$

e. $87.3\text{ kPa} - 3.6\text{ kPa} = 83.7\text{ kPa}$

$$\frac{8.23\text{ m}^3 \mid 83.7\ \cancel{\text{kPa}}}{\mid 101.3\ \cancel{\text{kPa}}} = 6.80\text{ m}^3$$

7. $$\frac{60.0\text{ cm}^3 \mid 273\ \cancel{\text{K}}}{\mid 309\ \cancel{\text{K}}} = 53.0\text{ cm}^3$$

8. a. $$\frac{617\text{ cm}^3 \mid 273\ \cancel{\text{K}}}{\mid 282\ \cancel{\text{K}}} = 597\text{ cm}^3$$

b. $$\frac{609\text{ cm}^3 \mid 273\ \cancel{\text{K}}}{\mid 356\ \cancel{\text{K}}} = 467\text{ cm}^3$$

c. $$\frac{942\text{ cm}^3 \mid 273\ \cancel{\text{K}}}{\mid 295\ \cancel{\text{K}}} = 872\text{ cm}^3$$

d. $$\frac{7.12\text{ m}^3 \mid 273\ \cancel{\text{K}}}{\mid 988\ \cancel{\text{K}}} = 1.97\text{ m}^3$$

9. a. $$\frac{2.90\text{ m}^3 \mid 296\ \cancel{\text{K}}}{\mid 226\ \cancel{\text{K}}} = 3.80\text{ m}^3$$

b. $$\frac{7.91\text{ m}^3 \mid 538\ \cancel{\text{K}}}{\mid 325\ \cancel{\text{K}}} = 13.1\text{ m}^3$$

c. $$\frac{667\text{ cm}^3 \mid 314\ \cancel{\text{K}}}{\mid 431\ \cancel{\text{K}}} = 486\text{ cm}^3$$

d. $$\frac{4.82\text{ m}^3 \mid 304\ \cancel{\text{K}}}{\mid 295\ \cancel{\text{K}}} = 4.97\text{ m}^3$$

14. $$\frac{2.23\text{ dm}^3 \mid 4.85\ \cancel{\text{kPa}} \mid 278.5\ \cancel{\text{K}}}{\mid 1.38\ \cancel{\text{kPa}} \mid 271.6\ \cancel{\text{K}}} = 8.04\text{ dm}^3$$

15. a. $$\frac{7.51\text{ m}^3 \mid 273\ \cancel{\text{K}} \mid 59.9\ \cancel{\text{kPa}}}{\mid 278\ \cancel{\text{K}} \mid 101.3\ \cancel{\text{kPa}}} = 4.36\text{ m}^3$$

b. $$\frac{351\text{ cm}^3 \mid 309\ \cancel{\text{K}} \mid 82.5\ \cancel{\text{kPa}}}{\mid 292\ \cancel{\text{K}} \mid 94.5\ \cancel{\text{kPa}}} = 324\text{ cm}^3$$

c. $$\frac{7.03\text{ m}^3 \mid 273\ \cancel{\text{K}} \mid 111\ \cancel{\text{kPa}}}{\mid 304\ \cancel{\text{K}} \mid 101.3\ \cancel{\text{kPa}}} = 6.92\text{ m}^3$$

d. $$\frac{955\text{ cm}^3 \mid 349\ \cancel{\text{K}} \mid 108.0\ \cancel{\text{kPa}}}{\mid 331\ \cancel{\text{K}} \mid 123.0\ \cancel{\text{kPa}}} = 884\text{ cm}^3$$

e. $$\frac{960.0\text{ cm}^3 \mid 286\ \cancel{\text{K}} \mid 107.2\ \cancel{\text{kPa}}}{\mid 344\ \cancel{\text{K}} \mid 59.3\ \cancel{\text{kPa}}} = 1440\text{ cm}^3$$

16. $$\frac{v_{He}}{v_{Ar}} = \sqrt{\frac{m_{Ar}}{m_{He}}} = \sqrt{\frac{39.9}{4.00}} = 3.16$$

17. $$\frac{v_{Ar}}{v_{Rn}} = \sqrt{\frac{m_{Rn}}{m_{Ar}}} = \sqrt{\frac{222}{39.9}} = 2.36$$

18. $$\frac{v_{hydrogen}}{v_{oxygen}} = \frac{\sqrt{m_{O_2}}}{\sqrt{m_{H_2}}} = \sqrt{\frac{32.0}{2.02}} = 3.98$$

19. $$\frac{v_{He}}{v_{Rn}} = \frac{\sqrt{m_{Rn}}}{\sqrt{m_{He}}} = \frac{\sqrt{222\text{ g}}}{\sqrt{4.00\text{ g}}} = 7.45$$

20.

$$\frac{v_{He}}{v_{O_2}} = \sqrt{\frac{m_{O_2}}{m_{He}}}$$

$$\sqrt{\frac{32.0\ \cancel{\text{g}}}{4.00\ \cancel{\text{g}}}} = 2.83$$

$v_{He} = 2.83 \times 0.0760\text{ m/s} = 0.215\text{ m/s}$

Chapter 19

1. $$P = \frac{nRT}{V} = \frac{0.622\ \cancel{\text{mol}} \mid 8.31\ \cancel{\text{dm}^3}\cdot\text{kPa} \mid 289\ \cancel{\text{K}}}{9.22\ \cancel{\text{dm}^3} \mid \cancel{\text{mol}}\cdot\cancel{\text{K}} \mid} = 162\text{ kPa}$$

2. $$n = \frac{PV}{RT} = \frac{66.7\ \cancel{\text{kPa}} \mid 0.486\ \cancel{\text{dm}^3} \mid \text{mol}\cdot\cancel{\text{K}}}{\mid 284\ \cancel{\text{K}} \mid 8.31\ \cancel{\text{dm}^3}\cdot\cancel{\text{kPa}}}$$

$= 0.0137\text{ mol}$

3. $$V = \frac{nRT}{P} = \frac{0.684\ \cancel{\text{mol}} \mid 8.31\text{ dm}^3\cdot\cancel{\text{kPa}} \mid 282\ \cancel{\text{K}}}{99.1\ \cancel{\text{kPa}} \mid \cancel{\text{mol}}\cdot\cancel{\text{K}} \mid} = 16.2\text{ dm}^3$$

4. $$T = \frac{PV}{nR} = \frac{100.4\ \cancel{\text{kPa}} \mid 604\ \cancel{\text{cm}^3} \mid \cancel{\text{mol}}\cdot\text{K}}{0.0851\ \cancel{\text{mol}} \mid \mid 8.31\ \cancel{\text{dm}^3}\cdot\cancel{\text{kPa}}}$$

$$\frac{1\ \cancel{\text{dm}^3}}{1000\ \cancel{\text{cm}^3}} = 85.8\text{ K} = -187°\text{C}$$

5. $P = \frac{nRT}{V} = \frac{0.003\ 06\ \text{mol}}{25.9\ \text{cm}^3} \times \frac{8.31\ \text{dm}^3\cdot\text{kPa}}{\text{mol}\cdot\text{K}} \times 282\ \text{K} \times \frac{1000\ \text{cm}^3}{1\ \text{dm}^3} = 277\ \text{kPa}$

6. $M = \frac{mRT}{PV} = \frac{0.922\ \text{g}}{107.0\ \text{kPa}} \times \frac{8.31\ \text{dm}^3\cdot\text{kPa}}{\text{mol}\cdot\text{K}} \times \frac{372\ \text{K}}{0.1500\ \text{dm}^3}$
$= 178$ g/mol

7. $M = \frac{mRT}{PV} = \frac{3.59\ \text{g}}{99.2\ \text{kPa}} \times \frac{8.31\ \text{dm}^3\cdot\text{kPa}}{\text{mol}\cdot\text{K}} \times \frac{304\ \text{K}}{4.34\ \text{dm}^3}$
$= 21.1$ g/mol

8. $M = \frac{mRT}{PV} = \frac{0.858\ \text{g}}{106.3\ \text{kPa}} \times \frac{8.31\ \text{dm}^3\cdot\text{kPa}}{\text{mol}\cdot\text{K}} \times \frac{275\ \text{K}}{150.0\ \text{cm}^3} \times \frac{1000\ \text{cm}^3}{1\ \text{dm}^3} = 123\ \text{g/mol}$

9. $M = \frac{mRT}{PV} = \frac{8.11\ \text{g}}{109.1\ \text{kPa}} \times \frac{8.31\ \text{dm}^3\cdot\text{kPa}}{\text{mol}\cdot\text{K}} \times \frac{283\ \text{K}}{2.38\ \text{dm}^3}$
$= 73.5$ g/mol

15. $N_2(g) + 3H_2(g) \rightarrow 2NH_3(g)$

$\frac{14.0\ \text{g N}_2}{} \times \frac{1\ \text{mol N}_2}{28.0\ \text{g N}_2} \times \frac{2\ \text{mol NH}_3}{1\ \text{mol N}_2} \times \frac{22.4\ \text{dm}^3}{1\ \text{mol}} \times \frac{1000\ \text{cm}^3}{1\ \text{dm}^3} = 22\ 400\ \text{cm}^3\ \text{NH}_3$

$= 22\ 400\ \text{cm}^3\ \text{NH}_3$

16. $Zn(cr) + H_2SO_4(aq) \rightarrow ZnSO_4(aq) + H_2(g)$

$\frac{28.0\ \text{g Zn}}{} \times \frac{1\ \text{mol Zn}}{65.4\ \text{g Zn}} \times \frac{1\ \text{mol H}_2}{1\ \text{mol Zn}} \times \frac{22.4\ \text{dm}^3}{1\ \text{mol}} \times \frac{1000\ \text{cm}^3}{1\ \text{dm}^3}$

$= 9590\ \text{cm}^3\ \text{H}_2$

17. $H_2(g) + Br_2(g) \rightarrow 2HBr(g)$

$\frac{5600\ \text{cm}^3\ \text{H}_2}{} \times \frac{1\ \text{dm}^3}{1000\ \text{cm}^3} \times \frac{1\ \text{mol}}{22.4\ \text{dm}^3} \times \frac{2\ \text{mol HBr}}{1\ \text{mol H}_2} \times \frac{80.9\ \text{g HBr}}{1\ \text{mol HBr}} = 40.5\ \text{g HBr}$

18. $2Sb(cr) + 3Cl_2(g) \rightarrow 2SbCl_3(cr)$

$\frac{3570\ \text{cm}^3\ \text{Cl}_2}{} \times \frac{1\ \text{dm}^3}{1000\ \text{cm}^3} \times \frac{1\ \text{mol}}{22.4\ \text{dm}^3} \times \frac{2\ \text{mol SbCl}_3}{3\ \text{mol Cl}_2} \times \frac{228\ \text{g SbCl}_3}{1\ \text{mol SbCl}_3} = 25.4\ \text{g SbCl}_3$

19. $2C_4H_{10}(g) + 13O_2(g) \rightarrow 8CO_2(g) + 10H_2O(g)$

$\frac{401\ \text{cm}^3\ \text{C}_4\text{H}_{10}}{} \times \frac{13\ \text{cm}^3\ \text{O}_2}{2\ \text{cm}^3\ \text{C}_4\text{H}_{10}} = 2610\ \text{cm}^3\ \text{O}_2$

20. $\frac{75.2\ \text{dm}^3\ \text{Cl}_2}{} \times \frac{1\ \text{dm}^3\ \text{Br}_2}{1\ \text{dm}^3\ \text{Cl}_2} = 75.2\ \text{dm}^3\ \text{Br}_2$

21. $C_7H_{16}(g) + 11O_2(g) \rightarrow 7CO_2(g) + 8H_2O(g)$

$\frac{917\ \text{cm}^3\ \text{C}_7\text{H}_{16}}{} \times \frac{11\ \text{cm}^3\ \text{O}_2}{1\ \text{cm}^3\ \text{C}_7\text{H}_{16}} = 10\ 100\ \text{cm}^3\ \text{O}_2$

volume of air = 10 100/0.209 = 48 300 cm^3 air

22. $2NO(g) + O_2(g) \rightarrow 2NO_2(g)$

$\frac{499\ \text{cm}^3\ \text{NO}}{} \times \frac{1\ \text{cm}^3\ \text{O}_2}{2\ \text{cm}^3\ \text{NO}} = 2.50 \times 10^2\ \text{cm}^3\ \text{O}_2$

23. $\frac{941\ \text{m}^3\ \text{C}_6\text{H}_6}{} \times \frac{4\ \text{m}^3\ \text{H}_2}{1\ \text{m}^3\ \text{C}_6\text{H}_6} = 3770\ \text{m}^3\ \text{H}_2$

24. $2NaBr(aq) + 2H_2SO_4(aq) + MnO_2(cr) \rightarrow Br_2(l) + MnSO_4(aq) + 2H_2O(l) + Na_2SO_4(aq)$

$\frac{2.10\ \text{g NaBr}}{} \times \frac{1\ \text{mol NaBr}}{103\ \text{g NaBr}} = 0.0204\ \text{mol NaBr}$

$\frac{9.42\ \text{g H}_2\text{SO}_4}{} \times \frac{1\ \text{mol H}_2\text{SO}_4}{98.1\ \text{g H}_2\text{SO}_4} = 0.0960\ \text{mol H}_2\text{SO}_4$

∴ NaBr is the limiting reactant.

$\frac{0.0204\ \text{mol NaBr}}{} \times \frac{1\ \text{mol Br}_2}{2\ \text{mol NaBr}} \times \frac{159.8\ \text{g Br}_2}{1\ \text{mol Br}_2}$

$= 1.63\ \text{g Br}_2$

25. $2Ca_3(PO_4)_2(cr) + 6SiO_2(cr) + 10C(amor) \rightarrow P_4(cr) + 6CaSiO_3(cr) + 10CO(g)$

$\frac{4.14\ \text{g Ca}_3(\text{PO}_4)_2}{} \times \frac{1\ \text{mol Ca}_3(\text{PO}_4)_2}{310\ \text{g Ca}_3(\text{PO}_4)_2}$

$= 0.0134\ \text{mol Ca}_3(\text{PO}_4)_2$

$\frac{1.20\ \text{g SiO}_2}{} \times \frac{1\ \text{mol SiO}_2}{60.1\ \text{g SiO}_2} = 0.0200\ \text{mol SiO}_2$

SiO_2 is the limiting reactant, since the ratio is 3:1 in the equation.

$\frac{0.0200\ \text{mol SiO}_2}{} \times \frac{10\ \text{mol CO}}{6\ \text{mol SiO}_2} \times \frac{22.4\ \text{dm}^3}{1\ \text{mol}} = 0.747\ \text{dm}^3\ \text{CO}$

26. $H_2S(g) + I_2(aq) \rightarrow 2HI(aq) + S(cr)$

$\frac{4.11\ \text{g I}_2}{} \times \frac{1\ \text{mol I}_2}{254\ \text{g I}_2} = 0.0162\ \text{mol I}_2$

$\frac{317\ \text{cm}^3\ \text{H}_2\text{S}}{} \times \frac{1\ \text{dm}^3}{1000\ \text{cm}^3} \times \frac{1\ \text{mol}}{22.4\ \text{dm}^3} = 0.0142\ \text{mol H}_2\text{S}$

H_2S is the limiting reactant.

$\frac{0.0142\ \text{mol H}_2\text{S}}{} \times \frac{1\ \text{mol S}}{1\ \text{mol H}_2\text{S}} \times \frac{32.1\ \text{g S}}{1\ \text{mol S}} = 0.456\ \text{g S}$

27. $2Cl_2(g) + HgO(cr) \rightarrow HgCl_2(cr) + Cl_2O(g)$

$\frac{116\ \text{cm}^3\ \text{Cl}_2}{} \times \frac{1\ \text{dm}^3}{1000\ \text{cm}^3} \times \frac{1\ \text{mol}}{22.4\ \text{dm}^3} = 0.005\ 18\ \text{mol Cl}_2$

$\frac{7.62\ \text{g HgO}}{} \times \frac{1\ \text{mol HgO}}{217\ \text{g HgO}} = 0.0351\ \text{mol HgO}$

Cl_2 is the limiting reactant.

$\frac{116\ \text{cm}^3\ \text{Cl}_2}{} \times \frac{1\ \text{cm}^3\ \text{Cl}_2\text{O}}{2\ \text{cm}^3\ \text{Cl}_2} = 58.0\ \text{cm}^3\ \text{Cl}_2\text{O}$

Chapter 20

1. Suggested reagents include:
   a. H_3PO_4 **c.** H_2CO_3
   b. H_2SO_4 **d.** NH_4OH

2. Suggested reagents include:
   a. $Pb(CH_3COO)_2$ **c.** $SrCl_2$
   b. $Hg_2(NO_3)_2$ **d.** $AgNO_3$

3. **a.** $2HF(aq) + Hg_2(NO_3)_2(aq) \rightarrow Hg_2F_2(cr) + 2HNO_3(aq)$
   b. $LiI(aq) + AgNO_3(aq) \rightarrow AgI(cr) + LiNO_3(aq)$
   c. $K_2S(aq) + Co(CH_3COO)_2(aq) \rightarrow CoS(cr) + 2KCH_3COO(aq)$
   d. $6NaOH(aq) + Cr_2(SO_4)_3(aq) \rightarrow 3Na_2SO_4(aq) + 2Cr(OH)_3(cr)$

4. $MgBr_2(s) \rightarrow Mg^{2+} + 2Br^-$

$$\frac{193\ \cancel{\text{g MgBr}_2}}{5.00 \times 10^2\ \cancel{\text{cm}^3}\text{ soln}} \left| \frac{1\ \cancel{\text{mol MgBr}_2}}{184\ \cancel{\text{g MgBr}_2}} \right| \frac{1000\ \cancel{\text{cm}^3}}{1\ \text{dm}^3} \left| \frac{2\text{ mol Br}^-}{1\ \cancel{\text{mol MgBr}_2}} = 4.20M\ \text{Br}^- \right.$$

5. $Ca(C_5H_9O_2)_2(s) \rightarrow Ca^{2+} + 2C_5H_9O_2^-$

$$\frac{8.28\ \cancel{\text{g Ca(C}_5\text{H}_9\text{O}_2)_2}}{2.50 \times 10^2\ \cancel{\text{cm}^3}\text{ soln}} \left| \frac{1\ \cancel{\text{mol Ca(C}_5\text{H}_9\text{O}_2)_2}}{242\ \cancel{\text{g Ca(C}_5\text{H}_9\text{O}_2)_2}} \right| \frac{1000\ \cancel{\text{cm}^3}}{1\ \text{dm}^3} \left| \frac{1\text{ mol Ca}^{2+}}{1\ \cancel{\text{mol Ca(C}_5\text{H}_9\text{O}_2)_2}} = 0.137M\ \text{Ca}^{2+} \right.$$

6. $CaCl_2(s) \rightarrow Ca^{2+} + 2Cl^-$

$$\frac{0.523\ \cancel{\text{mol CaCl}_2}}{1\ \cancel{\text{dm}^3}} \left| \frac{2.00\ \cancel{\text{dm}^3}}{} \right| \frac{1\text{ mol Ca}^{2+}}{1\ \cancel{\text{mol CaCl}_2}}$$

$= 1.05\text{ mol Ca}^{2+}$

7. $$\frac{199\ \cancel{\text{g NiBr}_2}}{5.00 \times 10^2\ \cancel{\text{g H}_2\text{O}}} \left| \frac{1\text{ mol NiBr}_2}{219\ \cancel{\text{g NiBr}_2}} \right| \frac{1000\ \cancel{\text{g H}_2\text{O}}}{1\text{ kg H}_2\text{O}}$$

$= 1.82m\ \text{NiBr}_2$ (mol/kg = m)

8. $$\frac{92.3\ \cancel{\text{g KF}}}{1.00 \times 10^3\ \cancel{\text{g H}_2\text{O}}} \left| \frac{1\text{ mol KF}}{58.1\ \cancel{\text{g KF}}} \right| \frac{1000\ \cancel{\text{g H}_2\text{O}}}{1\text{ kg H}_2\text{O}}$$

$= 1.59m$ KF

9. $$\frac{0.059\ \cancel{\text{g KF}}}{0.272\ \cancel{\text{g H}_2\text{O}}} \left| \frac{1\text{ mol KF}}{58.1\ \cancel{\text{g KF}}} \right| \frac{1000\ \cancel{\text{g H}_2\text{O}}}{1\text{ kg H}_2\text{O}} = 3.7m\ \text{KF}$$

10. $$\frac{12.3\ \cancel{\text{g C}_4\text{H}_4\text{O}}}{} \left| \frac{1\text{ mol C}_4\text{H}_4\text{O}}{68.0\ \cancel{\text{g C}_4\text{H}_4\text{O}}} = 0.181\text{ mol C}_4\text{H}_4\text{O} \right.$$

$$\frac{1.00 \times 10^2\ \cancel{\text{g C}_2\text{H}_6\text{O}}}{} \left| \frac{1\ \cancel{\text{mol C}_2\text{H}_6\text{O}}}{46.1\ \cancel{\text{g C}_2\text{H}_6\text{O}}} = \frac{2.17\text{ mol C}_2\text{H}_6\text{O}}{2.35\text{ mol soln}} \right.$$

$$\frac{0.181}{2.35} = 0.0760\text{ mole fraction C}_4\text{H}_4\text{O}$$

$$\frac{2.17}{2.35} = 0.923\text{ mole fraction C}_2\text{H}_6\text{O}$$

11. $$\frac{156\ \cancel{\text{g C}_{12}\text{H}_{22}\text{O}_{11}}}{} \left| \frac{1\text{ mol C}_{12}\text{H}_{22}\text{O}_{11}}{342\ \cancel{\text{g C}_{12}\text{H}_{22}\text{O}_{11}}} \right.$$

$= 0.456\text{ mol C}_{12}\text{H}_{22}\text{O}_{11}$

$$\frac{3.00 \times 10^2\ \cancel{\text{g H}_2\text{O}}}{} \left| \frac{1\text{ mol H}_2\text{O}}{18.0\ \cancel{\text{g H}_2\text{O}}} = \frac{16.7\text{ mol H}_2\text{O}}{17.2\text{ mol soln}} \right.$$

$$\frac{0.456}{17.2} = 0.0265\text{ mole fraction C}_{12}\text{H}_{22}\text{O}_{11}$$

$$\frac{16.7}{17.2} = 0.971\text{ mole fraction H}_2\text{O}$$

12. $$\frac{75.6\ \cancel{\text{g C}_{10}\text{H}_8}}{} \left| \frac{1\text{ mol C}_{10}\text{H}_8}{128\ \cancel{\text{g C}_{10}\text{H}_8}} = 0.591\text{ mol C}_{10}\text{H}_8 \right.$$

$$\frac{6.00 \times 10^2\ \cancel{\text{g C}_4\text{H}_{10}\text{O}}}{} \left| \frac{1\text{ mol C}_4\text{H}_{10}\text{O}}{74.0\ \cancel{\text{g C}_4\text{H}_{10}\text{O}}} \right.$$

$$= \frac{8.11\text{ mol C}_4\text{H}_{10}\text{O}}{8.70\text{ mol soln}}$$

$$\frac{0.591}{8.70} = 0.0679\text{ mole fraction C}_{10}\text{H}_8$$

$$\frac{8.11}{8.70} = 0.932\text{ mole fraction C}_4\text{H}_{10}\text{O}$$

13. $$\frac{199\text{ g}}{199\text{ g} + (5.00 \times 10^2)\text{ g}} \times 100 = 28.5\%\ \text{NiBr}_2$$

$$\frac{92.3\text{ g}}{92.3\text{ g} + (1.00 \times 10^3)\text{ g}} \times 100 = 8.45\%\ \text{KF}$$

14. $$\frac{0.059\text{ g}}{0.059\text{ g} + 0.272\text{ g}} \times 100 = 18\%\ \text{KF}$$

$$\frac{12.3\text{ g}}{12.3\text{ g} + (1.00 \times 10^2)\text{ g}} \times 100 = 10.9\%\ \text{C}_4\text{H}_4\text{O}$$

15. $$\frac{156\text{ g}}{156\text{ g} + (3.00 \times 10^2)\text{ g}} \times 100 = 34.2\%\ \text{C}_{12}\text{H}_{22}\text{O}_{11}$$

$$\frac{75.6\text{ g}}{75.6\text{ g} + (6.00 \times 10^2)\text{ g}} \times 100 = 11.2\%\ \text{C}_{10}\text{H}_8$$

Chapter 21

1. v.p. H_2O at 25°C = 3.2 kPa
   v.p. soln = (3.2 kPa)(1.000 − 0.163) = 2.7 kPa
2. v.p. H_2O = (101.325 kPa)(0.900) = 91.2 kPa
3. v.p. ethanal = (86.3 kPa)(0.300) = 25.9 kPa
   v.p. methanol = (11.6 kPa)(0.700) = 8.12 kPa
   v.p. soln = 25.9 kPa + 8.12 kPa = 34.0 kPa

8. $$\frac{25.5\ \cancel{\text{g C}_7\text{H}_{11}\text{NO}_7\text{S}}}{1.00 \times 10^2\ \cancel{\text{g}}\text{ H}_2\text{O}} \left| \frac{1\text{ mol C}_7\text{H}_{11}\text{NO}_7\text{S}}{253\ \cancel{\text{g C}_7\text{H}_{11}\text{NO}_7\text{S}}} \right| \frac{1000\ \cancel{\text{g}}}{1\text{ kg}}$$

$= 1.01m\ \text{C}_7\text{H}_{11}\text{NO}_7\text{S}$

$$\Delta T_{fp} = mK_{fp} = \frac{1.01\cancel{m}}{} \left| \frac{1.853\text{ C}°}{\cancel{m}} = 1.87\text{ C}° \right.$$

$$\Delta T_{bp} = \frac{1.01\cancel{m}}{} \left| \frac{0.515\text{ C}°}{\cancel{m}} = 0.520\text{ C}° \right.$$

Freezing point = 0°C − 1.87 C° = −1.87°C
Boiling point = 100°C + 0.520 C° = 100.520°C

9. $$\frac{1.00 \times 10^2\ \cancel{\text{g C}_{10}\text{H}_8\text{O}_6\text{S}_2}}{1.00 \times 10^2\ \cancel{\text{g}}\text{ H}_2\text{O}} \left| \frac{1\text{ mol C}_{10}\text{H}_8\text{O}_6\text{S}_2}{288\ \cancel{\text{g C}_{10}\text{H}_8\text{O}_6\text{S}_2}} \right| \frac{1000\ \cancel{\text{g}}}{1\text{ kg}}$$

$= 3.47m\ \text{C}_{10}\text{H}_8\text{O}_6\text{S}_2$

$$\Delta T_{fp} = \frac{3.47\cancel{m}}{} \left| \frac{1.853\text{ C}°}{\cancel{m}} = 6.43\text{ C}° \right.$$

$$\Delta T_{bp} = \frac{3.47\cancel{m}}{} \left| \frac{0.515\text{ C}°}{\cancel{m}} = 1.79\text{ C}° \right.$$

Freezing point = 0°C − 6.43 C° = −6.43°C
Boiling point = 100°C + 1.79 C° = 101.79°C

10. $NiSO_4 \rightarrow Ni^{2+} + SO_4^{2-}$

$$\frac{21.6\ \cancel{\text{g NiSO}_4}}{1.00 \times 10^2\ \cancel{\text{g}}\text{ H}_2\text{O}} \left| \frac{1\ \cancel{\text{mol NiSO}_4}}{155\ \cancel{\text{g NiSO}_4}} \right| \frac{1000\ \cancel{\text{g}}}{1\text{ kg}} \left| \frac{2\text{ mol ions}}{1\ \cancel{\text{mol NiSO}_4}} \right.$$

$= 2.79m$ ions

$$\Delta T_{fp} = \frac{2.79\cancel{m}}{} \left| \frac{1.853\text{ C}°}{\cancel{m}} = 5.17\text{ C}° \right.$$

$$\Delta T_{bp} = \frac{2.79\cancel{m}}{} \left| \frac{0.515\text{ C}°}{\cancel{m}} = 1.44\text{ C}° \right.$$

Freezing point = 0°C − 5.17 C° = −5.17°C
Boiling point = 100°C + 1.44 C° = 101.44°C

11. $Mg(ClO_4)_2 \rightarrow Mg^{2+} + 2ClO_4^-$

$$\frac{77.0\ \cancel{\text{g Mg(ClO}_4)_2}}{2.00 \times 10^2\ \cancel{\text{g}}\text{ H}_2\text{O}} \left| \frac{1\ \cancel{\text{mol Mg(ClO}_4)_2}}{223\ \cancel{\text{g Mg(ClO}_4)_2}} \right| \frac{1000\ \cancel{\text{g}}}{1\text{ kg}} \left| \frac{3\text{ mol ions}}{1\ \cancel{\text{mol Mg(ClO}_4)_2}} = 5.18m\text{ ions} \right.$$

$$\Delta T_{fp} = \frac{5.18\cancel{m}}{} \left| \frac{1.853\text{ C}°}{\cancel{m}} = 9.60\text{ C}° \right.$$

$$\Delta T_{bp} = \frac{5.18\cancel{m}}{1} \cdot \frac{0.515\ \text{C}°}{\cancel{m}} = 2.67\ \text{C}°$$

Freezing point = 0°C − 9.60 C° = −9.60°C
Boiling point = 100°C + 2.67 C° = 102.67°C

12. $$\frac{41.3\ \text{g}\ \cancel{C_{15}H_9NO_4}}{1.00 \times 10^2\ \text{g}\ C_6H_5NO_2} \cdot \frac{1\ \text{mol}\ C_{15}H_9NO_4}{267\ \cancel{\text{g}\ C_{15}H_9NO_4}} \cdot \frac{1000\ \text{g}}{1\ \text{kg}}$$
$= 1.55m\ C_{15}H_9NO_4$

$$\Delta T_{fp} = \frac{1.55\cancel{m}}{1} \cdot \frac{6.852\ \text{C}°}{\cancel{m}} = 10.6\ \text{C}°$$

$$\Delta T_{bp} = \frac{1.55\cancel{m}}{1} \cdot \frac{5.24\ \text{C}°}{\cancel{m}} = 8.12\ \text{C}°$$

Freezing point = 5.76°C − 10.6 C° = −4.8°C
Boiling point = 210.8°C + 8.12 C° = 218.9°C

13. $$\frac{8.02\ \text{g solute}}{861\ \text{g}\ H_2O} \cdot \frac{1000\ \text{g}}{1\ \text{kg}} = 9.31\ \text{g solute/kg}\ H_2O$$

$$m = \frac{\Delta T_{fp}}{K_{fp}} = \frac{0.430\ \cancel{\text{C}°}}{1} \cdot \frac{m}{1.853\ \cancel{\text{C}°}}$$
$= 0.232$ mol solute/kg H_2O

$$\frac{9.31\ \text{g}\ \cancel{\text{solute}}}{\cancel{\text{kg}\ H_2O}} \cdot \frac{\cancel{\text{kg}\ H_2O}}{0.232\ \text{mol}\ \cancel{\text{solute}}} = 40.1\ \text{g/mol}$$

14. $$\frac{64.3\ \text{g solute}}{3.90 \times 10^2\ \text{g}\ H_2O} \cdot \frac{1000\ \text{g}}{1\ \text{kg}} = 165\ \text{g solute/kg}\ H_2O$$

$$m = \frac{0.680\ \cancel{\text{C}°}}{1} \cdot \frac{m}{0.515\ \cancel{\text{C}°}} = 1.32\ \text{mol solute/kg}\ H_2O$$

$$\frac{165\ \text{g}\ \cancel{\text{solute}}}{\cancel{\text{kg}\ H_2O}} \cdot \frac{\cancel{\text{kg}\ H_2O}}{1.32\ \text{mol}\ \cancel{\text{solute}}} = 125\ \text{g/mol}$$

15. $$\frac{20.8\ \text{g solute}}{128\ \text{g}\ CH_3COOH} \cdot \frac{1000\ \text{g}}{1\ \text{kg}} = 163\ \text{g solute/kg}\ CH_3COOH$$

$\Delta T_{fp} = 16.66°\text{C} - 13.5°\text{C} = 3.2\ \text{C}°$

$$m = \frac{3.2\ \cancel{\text{C}°}}{1} \cdot \frac{m}{3.90\ \cancel{\text{C}°}} = 0.82\ \text{mol solute/kg}\ CH_3COOH$$

$$\frac{163\ \text{g}\ \cancel{\text{solute}}}{\cancel{\text{kg}\ CH_3COOH}} \cdot \frac{\cancel{\text{kg}\ CH_3COOH}}{0.82\ \text{mol}\ \cancel{\text{solute}}} = 2.0 \times 10^2\ \text{g/mol}$$

16. $$\frac{10.4\ \text{g solute}}{164\ \text{g phenol}} \cdot \frac{1000\ \text{g}}{1\ \text{kg}} = 63.4\ \text{g solute/kg phenol}$$

$\Delta T_{fp} = 40.90°\text{C} - 36.3°\text{C} = 4.6\ \text{C}°$

$$m = \frac{4.6\ \cancel{\text{C}°}}{1} \cdot \frac{m}{7.40\ \cancel{\text{C}°}} = 0.62\ \text{mol solute/kg phenol}$$

$$\frac{63.4\ \text{g}\ \cancel{\text{solute}}}{\cancel{\text{kg phenol}}} \cdot \frac{\cancel{\text{kg phenol}}}{0.62\ \text{mol}\ \cancel{\text{solute}}} = 1.0 \times 10^2\ \text{g/mol}$$

17. $$\frac{2.53\ \text{g solute}}{63.5\ \text{g nitrobenzene}} \cdot \frac{1000\ \text{g}}{1\ \text{kg}}$$
= 39.8 g solute/kg nitrobenzene

$\Delta T_{fp} = 5.76°\text{C} - 3.40°\text{C} = 2.36\ \text{C}°$

$$m = \frac{2.36\ \cancel{\text{C}°}}{1} \cdot \frac{m}{6.852\ \cancel{\text{C}°}}$$
= 0.344 mol solute/kg nitrobenzene

$$\frac{39.8\ \text{g solute}}{\cancel{\text{kg nitrobenzene}}} \cdot \frac{\cancel{\text{kg nitrobenzene}}}{0.344\ \text{mol}\ \cancel{\text{solute}}} = 116\ \text{g/mol}$$

18. $$M = \frac{\Pi}{RT} = \frac{780\ \cancel{\text{kPa}}}{298\ \cancel{\text{K}}} \cdot \frac{\text{mol}\cdot\cancel{\text{K}}}{8.31\ \text{dm}^3\cdot\cancel{\text{kPa}}}$$
$= 0.31M\ C_6H_{12}O_6$

19. $$\Pi = MRT = \frac{mRT}{MV};\qquad M = \frac{mRT}{\Pi V}$$

$$M = \frac{0.300\ \text{g hemoglobin}}{30.0\ \cancel{\text{cm}^3}} \cdot \frac{1000\ \cancel{\text{cm}^3}}{1\ \cancel{\text{dm}^3}} \cdot \frac{8.31\ \cancel{\text{dm}^3\cdot\text{kPa}}}{\text{mol}\cdot\cancel{\text{K}}} \cdot \frac{298\ \cancel{\text{K}}}{0.40\ \cancel{\text{kPa}}} = 62\ 000\ \text{g/mol}$$

20. $$\Pi = \frac{0.490\ \cancel{\text{mol}}}{1.00\ \cancel{\text{dm}^3}} \cdot \frac{8.31\ \cancel{\text{dm}^3}\cdot\text{kPa}}{\cancel{\text{mol}\cdot\text{K}}} \cdot \frac{298\ \cancel{\text{K}}}{1}$$
$= 1.21 \times 10^3$ kPa

Chapter 22

1. a. doubled $[2H_2] = 2[H_2]$
b. eight times faster $[2NO]^2[2H_2] = 8[NO]^2[H_2]$
c. slows down

2. a. Rate = $k[H_2][I_2]$
b. $0.2 = k(1)(1)$
$k = 0.2\ \text{dm}^3/\text{mol}\cdot\text{s}$
$0.8 = k(2)(2)$
$k = 0.2\ \text{dm}^3/\text{mol}\cdot\text{s}$
Rate = 0.2(0.5)(0.5)
Rate = 0.05(mol/dm³)/s

3. rate = $k[CH_3COCH_3]$ (Iodine has no effect)

4. $H^+ + CH_3{-}C({=}O){-}CH_3 \rightarrow CH_3{-}C^{\oplus}(OH){-}CH_3$ slow

Mechanisms for other steps:

$CH_3{-}C^{\oplus}(OH){-}CH_3 \rightarrow CH_3{-}C(OH){=}CH_2 + H^+$ fast

$CH_3{-}C(OH){=}CH_2 + I_2 \rightarrow CH_3{-}C^{\oplus}(OH){-}CH_2I + I^-$ fast

$CH_3{-}C^{\oplus}(OH){-}CH_2I + I^- \rightarrow CH_3{-}C({=}O){-}CH_2I + HI$ fast

10. a. $$K_{eq} = \frac{[CO_2][NH_3]^2}{[NH_2COONH_4]}$$

b. $$K_{eq} = \frac{[Cl_2]^2[H_2O]^2}{[HCl]^4[O_2]}$$

c. $$K_{eq} = \frac{[NH_3][H_2S]}{[NH_4HS]}$$

d. $$K_{eq} = \frac{[CuSO_4][H_2O]^5}{[CuSO_4{\cdot}5H_2O]}$$

11. $$K_{eq} = \frac{[H_2][CO_2]}{[CO][H_2O]} = \frac{(0.32)(0.42)}{(0.200)(0.500)} = 1.3$$

12. $$K_{eq} = \frac{[H_2]^2[S_2]}{[H_2S]^2}$$

$$= \frac{(2.22 \times 10^{-3})^2(1.11 \times 10^{-3})}{(7.06 \times 10^{-3})^2} = 1.10 \times 10^{-4}$$

13. $[HI]^2 = \frac{[H_2][I_2]}{K_{eq}}$

$$[HI]^2 = \frac{(2.00 \times 10^{-4})(2.00 \times 10^{-4})}{1.40 \times 10^{-2}} = 2.86 \times 10^{-6}$$

$[HI] = 1.69 \times 10^{-3}M$

14. $[CO] = \frac{[H_2][CO_2]}{K_{eq}[H_2O]}$

$$[CO] = \frac{(0.32)(0.42)}{(2.40)(0.500)} = 0.11M$$

15. $K_{eq} = \frac{[NO_2]^2}{[N_2O_4]}$

$$[NO_2]^2 = K_{eq}[N_2O_4] = (8.75 \times 10^{-2})(1.72 \times 10^{-2}) = 1.51 \times 10^{-3}$$

$[NO_2] = 3.88 \times 10^{-2}M$

Chapter 23

1. **a.** acid — HNO_3
   base — NaOH
   conjugate acid — H_2O
   conjugate base — NO_3^- in $NaNO_3$
   b. acid — HCl
   base — $NaHCO_3$
   conjugate acid — H_2CO_3
   conjugate base — Cl^- in NaCl
2. **a.** hydrogen sulfite ion
   b. hydrogen carbonate ion
   c. amide ion
   d. fluoride ion
3. **a.** Lewis base
   b. Lewis base
   c. Lewis acid
   d. Lewis base
4. $Ag^+ + 2NH_3 \rightarrow Ag(NH_3)_2^+$
5. $Cu^{2+} + 4NH_3 \rightarrow Cu(NH_3)_4^{2+}$
6. $Co^{3+} + 6NH_3 \rightarrow Co(NH_3)_6^{3+}$
7. $Fe^{3+} + 6CN^- \rightarrow Fe(CN)_6^{3-}$
8. $Zn^{2+} + 4OH^- \rightarrow Zn(OH)_4^{2-}$
9. **a.** hydrobromic acid
   b. hydrofluoric acid
10. **a.** selenous acid
    b. nitrous acid
    c. phosphoric acid
    d. arsenous acid
    e. iodic acid
11. **a.** H_2CO_3
    b. HNO_3
    c. H_3AsO_4
    d. H_2SeO_4
    e. HIO
12. **a.** calcium hydroxide
    b. potassium hydroxide
    c. aluminum hydroxide
    d. methanamine
    e. rubidium hydroxide
13. **a.** CsOH
    b. $CH_3CH_2CH_2NH_2$
    c. $CH_3CH_2CH_2CH_2NH_2$
    d. LiOH
14. **a.** basic
    b. basic
    c. acidic
    d. basic
    e. acidic
15. **a.** BaO
    b. I_2O_7
    c. TeO_3
    d. Al_2O_3
    e. ZnO
21. **a.** sodium hydrogen sulfate
    b. potassium hydrogen tartrate
    c. sodium dihydrogen phosphate
    d. sodium hydrogen sulfide
    e. aluminum hydroxide silicate
22. **a.** $NaHCO_3$
    b. Na_2HPO_4
    c. NH_4HS
    d. $KHSO_4$
    e. $Sn(OH)NO_3$
23. $2Ag^+(aq) + HSO_4^-(aq) \rightarrow Ag_2SO_4(cr) + H^+(aq)$
24. $H_4SiO_4(aq) + 4OH^-(aq) \rightarrow SiO_4^{4-}(aq) + 4H_2O(l)$
25. $2Cl^-(aq) + 2Cr^{2+}(aq) + 2Hg^{2+}(aq) \rightarrow 2Cr^{3+}(aq) + Hg_2Cl_2(cr)$
26. $2Mn^{2+}(aq) + 5NaBiO_3(cr) + 14H^+(aq) \rightarrow 2MnO_4^-(aq) + 5Bi^{3+}(aq) + 7H_2O(l) + 5Na^+(aq)$
27. $2Cu^{2+}(aq) + SO_4^{2-}(aq) + 2CNS^-(aq) + H_2SO_3(aq) + H_2O(l) \rightarrow 2CuCNS(cr) + 2HSO_4^-(aq) + 2H^+(aq)$
28. $HClO + H_2O \rightleftarrows H_3O^+ + ClO^-$

$$K_a = \frac{[H_3O^+][ClO^-]}{[HClO]} = 2.88 \times 10^{-8} = \frac{x^2}{0.0200 - \otimes} \leftarrow \text{neglect}$$

$x = 2.40 \times 10^{-5}M$

29. $N_2H_4 + H_2O \rightleftarrows N_2H_5^+ + OH^-$

$$K_b = \frac{[N_2H_5^+][OH^-]}{[N_2H_4]} = \frac{(1.23 \times 10^{-3})^2}{0.499}$$

$K_b = 3.03 \times 10^{-6}$

30. $K_a = \frac{[CH_3COO^-][H_3O^+]}{[CH_3COOH]}$

$$1.75 \times 10^{-5} = \frac{x^2}{0.200 - \otimes} \leftarrow \text{neglect}$$

$x^2 = 3.50 \times 10^{-6}$
$x = 1.87 \times 10^{-3}$

$$\% = \frac{1.87 \times 10^{-3}}{0.200} \Bigg| \frac{100}{1} = 0.935\%$$

31. $K_a = \frac{[A^-][H_3O^+]}{[HA]} = \frac{(0.0200)^2}{0.980} = 4.08 \times 10^{-4}$

32. $HNO_2 + H_2O \rightleftarrows H_3O^+ + NO_2^-$

$$K_a = \frac{[H_3O^+][NO_2^-]}{[HNO_2]}$$

$$7.24 \times 10^{-4} = \frac{x(\otimes + 0.03)}{(0.150 - \otimes)} \leftarrow \text{neglect}$$

$x = 3.26 \times 10^{-3}M$

33. $HBrO + H_2O \rightleftharpoons H_3O^+ + BrO^-$

$$K_a = \frac{[H_3O^+][BrO^-]}{[HBrO]}$$

$$2.51 \times 10^{-9} = \frac{x(\otimes + 0.000\ 195)}{0.403 - \otimes \leftarrow \text{neglect}}$$

$x = 5.19 \times 10^{-6}M$

34. Add either HCO_3^- or CO_3^{2-} in the form of soluble salts; add H_3O^+.

35. b, d, e, f

36. oxalic acid $\dfrac{5.36 \times 10^{-2}}{5.35 \times 10^{-5}} = 1.00 \times 10^3$

arsenic acid $\dfrac{6.03 \times 10^{-3}}{1.05 \times 10^{-7}} = 5.74 \times 10^4$

boric acid $\dfrac{5.75 \times 10^{-10}}{1.82 \times 10^{-13}} = 3.16 \times 10^3$

carbonic acid $\dfrac{4.37 \times 10^{-7}}{4.68 \times 10^{-11}} = 9.34 \times 10^3$

phosphoric acid $\dfrac{7.08 \times 10^{-3}}{6.31 \times 10^{-8}} = 1.12 \times 10^5$

phosphorus acid $\dfrac{6.31 \times 10^{-2}}{2.00 \times 10^{-7}} = 3.16 \times 10^5$

hydrogen sulfide $\dfrac{1.07 \times 10^{-7}}{1.26 \times 10^{-13}} = 8.49 \times 10^5$

sulfurous acid $\dfrac{1.29 \times 10^{-2}}{6.17 \times 10^{-8}} = 2.09 \times 10^5$

selenous acid $\dfrac{2.29 \times 10^{-3}}{5.37 \times 10^{-9}} = 4.26 \times 10^5$

The first ionization constant is considerably larger than the second ionization constant.

Chapter 24

1. a. $K_{sp} = [Pb^{2+}][I^-]^2$
b. $K_{sp} = [Cu^{2+}]^3[PO_4^{3-}]^2$

2. $AgI \rightleftharpoons Ag^+ + I^-$
$K_{sp} = [Ag^+][I^-]$
$[Ag^+] = [I^-] = x$
$x^2 = 8.32 \times 10^{-17}$
$x = 9.12 \times 10^{-9}M = [Ag^+]$

3. $D_2A \rightleftharpoons 2D^+ + A^{2-}$
$K_{sp} = [D^+]^2[A^{2-}]$
$2[A^{2-}] = [D^+] = 1.00 \times 10^{-5}$
$[A^{2-}] = 2.00 \times 10^{-5}$
$K_{sp} = (2.00 \times 10^{-5})^2(1.00 \times 10^{-5})$
$K_{sp} = 4.00 \times 10^{-15}$

4. $Be(OH)_2 \rightleftharpoons Be^{2+} + 2OH^-$
$K_{sp} = [Be^{2+}][OH^-]^2$
$1.58 \times 10^{-22} = x(2x)^2$
$x = 3.41 \times 10^{-8}M$

5. $PbI_2 \rightleftharpoons Pb^2 + 2I^-$
$K_{sp} = [Pb^{2+}][I^-]^2$
$K_{sp} = (1.21 \times 10^{-3})(2.42 \times 10^{-3})^2$
$K_{sp} = 7.09 \times 10^{-9}$

6. $TlBr \rightleftharpoons Tl^+ + Br^-$
$K_{sp} = [Tl^+][Br^-]$

$$[Br^-] = \frac{0.050 \text{ mol}}{500.0 \text{ cm}^3} \left| \frac{100^3 \text{ cm}^3}{1 \text{ m}^3} \right| \frac{1 \text{ m}^3}{10^3 \text{ dm}^3} = 0.100M$$

$3.39 \times 10^{-6} = [Tl^+](0.100)$
$[Tl^+] = 3.39 \times 10^{-5}M$

7. $$[Fe^{3+}] = \frac{0.100 \text{ mol}}{1 \text{ dm}^3} \left| \frac{0.250 \text{ dm}^3}{0.500 \text{ dm}^3} \right. = 5.00 \times 10^{-2}M$$

$$[OH^-] = \frac{0.010 \text{ mol}}{1 \text{ dm}^3} \left| \frac{0.250 \text{ dm}^3}{0.500 \text{ dm}^3} \right. = 5.0 \times 10^{-3}M$$

$[Fe^{3+}][OH^-]^3 = (5.00 \times 10^{-2})(5.0 \times 10^{-3})^3 = 6.3 \times 10^{-9}$
Since the K_{sp} of $Fe(OH)_3$ is much smaller than this ($K_{sp} = 3.98 \times 10^{-38}$), a precipitate will form.

8. $$[Li^+] = \frac{0.100 \text{ mol}}{1 \text{ dm}^3} \left| \frac{0.500 \text{ dm}^3}{1.000 \text{ dm}^3} \right. = 5.00 \times 10^{-2}M$$

$$[CO_3^{2-}] = \frac{0.100 \text{ mol}}{1 \text{ dm}^3} \left| \frac{0.500 \text{ dm}^3}{1.000 \text{ dm}^3} \right. = 5.00 \times 10^{-2}M$$

$[Li^+]^2[CO_3^{2-}] = (5.00 \times 10^{-2})^2(5.00 \times 10^{-2})$
$= 1.25 \times 10^{-4}$
Since the K_{sp} of Li_2CO_3 is larger than this ($K_{sp} = 2.51 \times 10^{-2}$), no precipitate will form.

9. $H_2O + H_2O \rightleftharpoons H_3O^+ + OH^-$
$K_w = [H_3O^+][OH^-]$
$1.00 \times 10^{-14} = x(6.80 \times 10^{-10})$
$x = 1.47 \times 10^{-5}M$

10. $H_2O + H_2O \rightleftharpoons H_3O^+ + OH^-$
$K_w = [H_3O^+][OH^-]$
$1.00 \times 10^{-14} = x(5.67 \times 10^{-3})$
$x = 1.76 \times 10^{-12}M$

11. a. $-\log(1.00 \times 10^{-3}) = 3.000$
b. $-\log(1.00 \times 10^{-6}) = 6.000$
c. $-\log(6.59 \times 10^{-10}) = 9.181$
d. $-\log(7.01 \times 10^{-6}) = 5.154$
e. $-\log(9.47 \times 10^{-8}) = 7.024$
f. $-\log(6.89 \times 10^{-14}) = 13.162$

12. a. $\text{antilog}(-3.000) = 1.00 \times 10^{-3}M$
b. $\text{antilog}(-10.000) = 1.00 \times 10^{-10}M$
c. $\text{antilog}(-6.607) = 2.47 \times 10^{-7}M$
d. $\text{antilog}(-2.523) = 3.00 \times 10^{-3}M$
e. $\text{antilog}(-6.149) = 7.10 \times 10^{-7}M$
f. $\text{antilog}(-7.622) = 2.18 \times 10^{-8}M$

13. a. $pH = 14.00 - pOH = 14.00 - 2.00 = 12.00$
b. $pH = 14.00 - pOH = 14.00 - 7.00 = 7.00$
c. $pH = 14.00 - pOH = 14.00 - 1.263 = 12.737$
d. $pH = 14.00 - pOH = 14.00 - 4.976 = 9.024$
e. $pH = 14.00 - pOH = 14.00 - 9.714 = 4.286$
f. $pH = 14.00 - pOH = 14.00 - 3.004 = 10.996$

14. a. $[H_3O^+] = \dfrac{1.00 \times 10^{-14}}{1.00 \times 10^{-4}} = 1.00 \times 10^{-10}M;$
$pH = -\log(1.00 \times 10^{-10}) = 10.000$
b. $[H_3O^+] = \dfrac{1.00 \times 10^{-14}}{1.00 \times 10^{-6}} = 1.00 \times 10^{-8}M;$
$pH = -\log(1.00 \times 10^{-8}) = 8.000$
c. $[H_3O^+] = \dfrac{1.00 \times 10^{-14}}{2.64 \times 10^{-13}} = 3.79 \times 10^{-2}M;$
$pH = -\log(3.79 \times 10^{-2}) = 1.422$
d. $[H_3O^+] = \dfrac{1.00 \times 10^{-14}}{3.45 \times 10^{-8}} = 2.90 \times 10^{-7}M;$
$pH = -\log(2.90 \times 10^{-7}) = 6.538$
e. $[H_3O^+] = \dfrac{1.00 \times 10^{-14}}{4.97 \times 10^{-10}} = 2.01 \times 10^{-5}M;$
$pH = -\log(2.01 \times 10^{-5}) = 4.696$

f. $[H_3O^+] = \frac{1.00 \times 10^{-14}}{2.93 \times 10^{-2}} = 3.41 \times 10^{-13}M;$

$pH = -\log(3.41 \times 10^{-13}) = 12.467$

15. $[H_3O^+] = \text{antilog}\ (-pH)$

a. antilog $(-12.00) = 1.00 \times 10^{-12}M$
b. antilog $(-7.00) = 1.00 \times 10^{-7}M$
c. antilog $(-12.737) = 1.83 \times 10^{-13}M$
d. antilog $(-9.024) = 9.46 \times 10^{-10}M$
e. antilog $(-4.286) = 5.18 \times 10^{-5}M$
f. antilog $(-10.966) = 1.01 \times 10^{-11}M$

16. a. neutral **b.** basic **c.** acidic

17. a. acidic **d.** acidic
b. acidic **e.** basic
c. acidic **f.** basic

24.
$$\frac{21.4\ \cancel{\text{cm}^3\ \text{HCl}}}{} \left|\ \frac{0.106\ \cancel{\text{mol HCl}}}{1000\ \cancel{\text{cm}^3\ \text{HCl}}}\ \right|\ \frac{1\ \cancel{\text{mol NaOH}}}{1\ \cancel{\text{mol HCl}}}\ \left|\ \frac{1000\ \text{cm}^3\ \text{NaOH}}{0.0947\ \cancel{\text{mol NaOH}}} = 24.0\ \text{cm}^3\ \text{NaOH}\right.$$

25.
$$\frac{21.7\ \cancel{\text{cm}^3\ \text{soln}}}{} \left|\ \frac{0.500\ \cancel{\text{mol HBr}}}{1000\ \cancel{\text{cm}^3\ \text{soln}}}\ \right|\ \frac{1\ \text{mol LiOH}}{1\ \cancel{\text{mol HBr}}}\ \left|\ \frac{}{26.4\ \cancel{\text{cm}^3}\ \text{soln}}\ \right|\ \frac{1000\ \cancel{\text{cm}^3}}{1\ \text{dm}^3} = 0.411M\ \text{LiOH}$$

26.
$$\frac{23.4\ \cancel{\text{cm}^3\ \text{soln}}}{} \left|\ \frac{0.551\ \cancel{\text{mol NaOH}}}{1000\ \cancel{\text{cm}^3\ \text{soln}}}\ \right|\ \frac{1\ \text{mol HCl}}{1\ \cancel{\text{mol NaOH}}}\ \left|\ \frac{}{50\ \cancel{\text{cm}^3}\ \text{soln}}\ \right|\ \frac{1000\ \cancel{\text{cm}^3}}{1\ \text{dm}^3} = 0.258M\ \text{HCl}$$

Chapter 25

1. a. $2(1+) + x + 3(2-) = 0;\ x = 4+$
b. $(1+) + x + 4(2-) = 0;\ x = 7+$
c. $(2+) + 2x + 6(-2) = 0;\ x = 5+$
d. $2(1+) + x + 3(2-) = 0;\ x = 4+$
e. $x + 2(2-) = 0;\ x = 4+$
f. $(1+) + x + 4(2-) = 1-;\ x = 6+$
g. $2(1+) + 2x + 7(2-) = 0;\ x = 6+$
h. $2(3+) + 3x = 0;\ x = 2-$
i. $x + 2(1-) = 0;\ x = 2+$
j. $12x + 22(1+) + 11(2-) = 0;\ x = 0$

2.

	Oxidizing agent	Reducing Agent
a.	Cu^{2+}	Mg
b.	Sn^{4+}	Fe
c.	S	Na

3. a. Mg; 2 electrons
b. Fe; 2 electrons
c. Na; 1 electron per atom

4. yes
H is oxidized and H_2 is the reducing agent.
N is reduced and N_2 is the oxidizing agent.

5. yes
C is oxidized and the reducing agent
H is reduced
H_2O is oxidizing agent

6. not a redox reaction

7. not a redox reaction

8. not a redox reaction

9. yes
O is reduced
S is oxidized
H_2O_2 is oxidizing agent
PbS is reducing agent

10. not a redox reaction

11. yes
N is reduced
P is oxidized
HNO_3 is oxidizing agent
H_3PO_3 is reducing agent

12. yes
N is reduced
I is oxidized
HNO_3 is oxidizing agent
I_2 is the reducing agent

13. not a redox reaction

14. yes
N is reduced
Fe^{2+} is oxidized and the reducing agent
NO_3^- is oxidizing agent

15. yes
Br_2 is reduced and the oxidizing agent
Fe^{2+} is oxidized
$FeBr_2$ is reducing agent

16. yes
I is reduced
S is oxidized
$S_2O_3^{2-}$ is the reducing agent
I_2 is the oxidizing agent

17. yes
Mn is reduced
O is oxidized
MnO_4^- is oxidizing agent
H_2O_2 is reducing agent

23. a. $K_2Cr_2O_7$ is the oxidizing agent
Na_2SO_3 is the reducing agent
b. MnO_2 is the oxidizing agent
KI is the reducing agent
c. H_2O_2 is the reducing agent
MnO_2 is the oxidizing agent
d. K_2MnO_4 is the oxidizing agent
HCl is the reducing agent
e. Al is the reducing agent
HCl is the oxidizing agent

24. a. Reduction half-reaction:

Skeleton	$Ag^+ \rightarrow Ag$
Electrons	$Ag^+ + e^- \rightarrow Ag$
Check Charge	$0 = 0$

Oxidation half-reaction:

Skeleton	$Cu \rightarrow Cu^{2+}$
Electrons	$Cu \rightarrow Cu^{2+} + 2e^-$
Check Charge	$0 = 0$

Multiply reduction half-reaction by 2, add, and simplify:

$$2Ag^+ + 2e^- \rightarrow 2Ag$$
$$Cu \rightarrow Cu^{2+} + 2e^-$$
$$\overline{2Ag^+ + Cu \rightarrow 2Ag + Cu^{2+}}$$

b. Reduction half-reaction:

Skeleton	$NO_3^- \rightarrow NO$
Electrons	$NO_3^- + 3e^- \rightarrow NO$
Oxygen	$NO_3^- + 3e^- \rightarrow NO + 2H_2O$
Hydrogen	$NO_3^- + 3e^- + 4H^+ \rightarrow NO + 2H_2O$
Check Charge	$0 = 0$

Oxidation half-reaction:

Skeleton	$H_2S \rightarrow S$
Electrons	$H_2S \rightarrow S + 2e^-$
Hydrogen	$H_2S \rightarrow S + 2e^- + 2H^+$
Check Charge	$0 = 0$

Multiply to balance electrons, add, and simplify:

$$2NO_3^- + 6e^- + 8H^+ \rightarrow 2NO + 4H_2O$$
$$3H_2S \rightarrow 3S + 6e^- + 6H^+$$
$$\overline{2H^+ + 2NO_3^- + 3H_2S \rightarrow 3S + 2NO + 4H_2O}$$

c. Reduction half-reaction:

Skeleton	$PbO_2 + HSO_4^- \rightarrow PbSO_4$
Electrons	$PbO_2 + HSO_4^- + 2e^- \rightarrow PbSO_4$
Oxygen	$PbO_2 + HSO_4^- + 2e^- \rightarrow PbSO_4 + 2H_2O$
Hydrogen	$PbO_2 + HSO_4^- + 2e^- + 3H^+ \rightarrow PbSO_4 + 2H_2O$
Check Charge	$0 = 0$

Oxidation half-reaction:

Skeleton	$Pb + HSO_4^- \rightarrow PbSO_4$
Electrons	$Pb + HSO_4^- \rightarrow PbSO_4 + 2e^-$
Hydrogen	$Pb + HSO_4^- \rightarrow PbSO_4 + 2e^- + H^+$
Check Charge	$1- = 1-$

Add and simplify:

$$PbO_2 + HSO_4^- + 2e^- + 3H^+ \rightarrow PbSO_4 + 2H_2O$$
$$Pb + HSO_4^- \rightarrow PbSO_4 + 2e^- + H^+$$
$$\overline{Pb + PbO_2 + 2HSO_4^- + 2H^+ \rightarrow 2PbSO_4 + 2H_2O}$$

25. $(8H^+ + 5e^- + MnO_4^- \rightarrow Mn^{2+} + 4H_2O) \times 2$
$(H_2O + H_2SO_3 \rightarrow HSO_4^- + 2e^- + 3H^+) \times 5$
$16H^+ + 10e^- + 2MnO_4^- \rightarrow 2Mn^{2+} + 8H_2O$
$5H_2O + 5H_2SO_3 \rightarrow 5HSO_4^- + 10e^- + 15H^+$
$\overline{H^+ + 2MnO_4^- + 5H_2SO_3 \rightarrow 2Mn^{2+} + 5HSO_4^- + 3H_2O}$

26. $(14H^+ + 6e^- + Cr_2O_7^{2-} \rightarrow 2Cr^{3+} + 7H_2O)$
$(2I^- \rightarrow I_2 + 2e^-) \times 3$
$14H^+ + 6e^- + Cr_2O_7^{2-} \rightarrow 2Cr^{3+} + 7H_2O$
$6I^- \rightarrow 3I_2 + 6e^-$
$\overline{Cr_2O_7^{2-} + 14H^+ + 6I^- \rightarrow 2Cr^{3+} + 3I_2 + 7H_2O}$

27. $(5OH^- + NH_3 \rightarrow NO + 5e^- + 4H_2O) \times 4$
$4H_2O + 4e^- + O_2 \rightarrow 2H_2O + 4OH^-$
Simplifying the 2nd equation gives:
$2H_2O + 4e^- + O_2 \rightarrow 4OH^-$
$(2H_2O + 4e^- + O_2 \rightarrow 4OH^-) \times 5$
$20OH^- + 4NH_3 \rightarrow 4NO + 20e^- + 16H_2O$
$10H_2O + 20e^- + 5O_2 \rightarrow 20OH^-$
$\overline{4NH_3 + 5O_2 \rightarrow 4NO + 6H_2O \text{ (basic)}}$

28. $(5H_2O + As_2O_3 \rightarrow 2H_3AsO_4 + 4e^- + 4H^+) \times 3$
$(4H^+ + 3e^- + NO_3^- \rightarrow NO + 2H_2O) \times 4$
$15H_2O + 3As_2O_3 \rightarrow 6H_3AsO_4 + 12e^- + 12H^+$
$16H^+ + 12e^- + 4NO_3^- \rightarrow 4NO + 8H_2O$
$\overline{3As_2O_3 + 4H^+ + 4NO_3^- + 7H_2O \rightarrow 6H_3AsO_4 + 4NO}$

29. $(2e^- + I_2 \rightarrow 2I^-) \times 1$
$(H_2O + H_2SO_3 \rightarrow HSO_4^- + 2e^- + 3H^+) \times 1$
$\overline{I_2 + H_2SO_3 + H_2O \rightarrow 2I^- + HSO_4^- + 3H^+}$

30. $(8H^+ + 8e^- + H_3AsO_4 \rightarrow AsH_3 + 4H_2O) \times 1$
$(Zn \rightarrow Zn^{2+} + 2e^-) \times 4$
$8H^+ + 8e^- + H_3AsO_4 \rightarrow AsH_3 + 4H_2O$
$4Zn \rightarrow 4Zn^{2+} + 8e^-$
$\overline{H_3AsO_4 + 8H^+ + 4Zn \rightarrow AsH_3 + 4H_2O + 4Zn^{2+}}$

31. $(MnO_4^{2-} \rightarrow MnO_4^- + e^-) \times 2$
$(4H^+ + 2e^- + MnO_4^{2-} \rightarrow MnO_2 + 2H_2O) \times 1$
$2MnO_4^{2-} \rightarrow 2MnO_4^- + 2e^-$
$\overline{3MnO_4^{2-} + 4H^+ \rightarrow 2MnO_4^- + MnO_2 + 2H_2O}$

32. $(8H^+ + 5e^- + MnO_4^- \rightarrow Mn^{2+} + 4H_2O) \times 2$
$(2H_2O + SO_2 \rightarrow SO_4^{2-} + 2e^- + 4H^+) \times 5$
$16H^+ + 10e^- + 2MnO_4^- \rightarrow 2Mn^{2+} + 8H_2O$
$10H_2O + 5SO_2 \rightarrow 5SO_4^{2-} + 10e^- + 20H^+$
$\overline{2MnO_4^- + 5SO_2 + 2H_2O \rightarrow 2Mn^{2+} + 5SO_4^{2-} + 4H^+}$

33. $e^- + NO_2 \rightarrow NO_2^-$
$2OH^- + NO_2 \rightarrow NO_3^- + e^- + H_2O$
$\overline{2NO_2 + 2OH^- \rightarrow NO_2^- + NO_3^- + H_2O \text{ (basic)}}$

34. $(4Cl^- + HgS \rightarrow S + 2e^- + HgCl_4^{2-}) \times 3$
$(4H^+ + 3e^- + NO_3^- \rightarrow NO + 2H_2O) \times 2$
$12Cl^- + 3HgS \rightarrow 3S + 6e^- + 3HgCl_4^{2-}$
$8H^+ + 6e^- + 2NO_3^- \rightarrow 2NO + 4H_2O$
$\overline{8H^+ + 3HgS + 12Cl^- + 2NO_3^- \rightarrow 3HgCl_4^{2-} + 3S + 2NO + 4H_2O}$

Chapter 26

6. -1.676 V + 0.257 V = -1.419 V; no reaction
7. 0.7991 V + 0.277 V = 1.076 V; reaction occurs
8. -0.154 V + 0.1251 V = -0.029 V; no reaction
9. 1.52 V + 0.44 V = 1.96 V; reaction occurs

10.

a.	$Zn \rightarrow Zn^{2+} + 2e^-$	0.7626 V
	$Fe^{2+} + 2e^- \rightarrow Fe$	−0.44 V
		0.32 V
b.	$Mn \rightarrow Mn^{2+} + 2e^-$	1.18 V
	$Br_2 + 2e^- \rightarrow 2Br^-$	1.0652 V
		2.25 V
c.	$H_2C_2O_4 \rightarrow 2CO_2 + 2H^+ + 2e^-$	0.475 V
	$MnO_4^- + 8H^+ + 5e^- \rightarrow Mn^{2+} + 4H_2O$	1.51 V
		1.99 V
d.	$Ni \rightarrow Ni^{2+} + 2e^-$	0.257 V
	$Hg_2^{2+} + 2e^- \rightarrow 2Hg$	0.7960 V
		1.053 V
e.	$Cu \rightarrow Cu^{2+} + 2e^-$	−0.340 V
	$Ag^+ + e^- \rightarrow Ag$	0.7991 V
		0.459 V
f.	$Pb \rightarrow Pb^{2+} + 2e^-$	0.1251 V
	$Cl_2 + 2e^- \rightarrow 2Cl^-$	1.35828 V
		1.4834 V

11. a. H_2 and O_2—Hydrogen reduction requires less energy than that of sodium; sulfate cannot be oxidized.
b. Cu and O_2—Copper reduction requires less energy than that of hydrogen; oxygen oxidation requires less energy than that of chlorine.
c. Co and O_2—Cobalt reduction requires less energy than that of hydrogen; oxygen oxidation requires less energy than that of fluorine.
d. Pb and O_2—Lead reduction requires less energy than that of hydrogen; nitrate cannot be oxidized.
e. H_2 and O_2—Hydrogen reduction requires less energy than that of lithium; oxygen oxidation requires less energy than that of bromine.
f. H_2 and O_2—Hydrogen reduction requires less energy than that of sodium; oxygen oxidation requires less energy than that of chlorine.

12. $Pb + Sn^{4+} \rightarrow Pb^{2+} + Sn^{2+}$

$$E = E° - \frac{0.059\,16}{n}\log K = E° - \frac{0.059\,16}{n}\log\frac{[Pb^{2+}][Sn^{2+}]}{[Sn^{4+}]}$$

$$E = 0.279 - \frac{0.059\,16}{2}\log\frac{(0.458)(0.346)}{(0.652)} = 0.296 \text{ V}$$

13. $Ni + Cu^{2+} \rightarrow Ni^{2+} + Cu$

$$E = E° - \frac{0.059\,16}{n}\log K = E° - \frac{0.059\,16}{n}\log\frac{[Ni^{2+}]}{[Cu^{2+}]}$$

$$E = 0.597 - \frac{0.059\,16}{2}\log\frac{(1.00)}{(0.0100)} = 0.538 \text{ V}$$

14. $Fe + 2Ag^+ \rightarrow Fe^{2+} + 2Ag$

$$E = E^\circ - \frac{0.059\ 16}{n}\log K = E^\circ - \frac{0.059\ 16}{n}\log\frac{[Fe^{2+}]}{[Ag^+]^2}$$

$$E = 1.24 - \frac{0.059\ 16}{2}\log\frac{(0.720)}{(0.785)^2} = 1.24\text{ V}$$

15. $Ag^+(aq) + e^- \rightarrow Ag(cr)$

$$\frac{1.00\text{ A}}{} \left|\ \frac{9650\text{ s}}{}\right| \frac{1\text{ mol }e^-}{96\ 485\text{ A·s}} \left|\ \frac{1\text{ mol Ag}}{1\text{ mol }e^-}\right| \frac{108\text{ g Ag}}{1\text{ mol Ag}}$$

$= 10.8$ g Ag

16. $Ag^+(aq) + e^- \rightarrow Ag(cr)$

$$\frac{5.00\text{ A}}{} \left|\ \frac{10.0\text{ min}}{}\right| \frac{60\text{ s}}{1\text{ min}} \left|\ \frac{1\text{ mol }e^-}{96\ 485\text{ A·s}}\right| \frac{1\text{ mol Ag}}{1\text{ mol }e^-} \left|\ \frac{108\text{ g Ag}}{1\text{ mol Ag}}\right. = 3.36\text{ g Ag}$$

17. $Cu^{2+}(aq) + 2e^- \rightarrow Cu(cr)$

$$\frac{1.75\text{ mol Cu}}{6.24\text{ min}} \left|\ \frac{2\text{ mol }e^-}{1\text{ mol Cu}}\right| \frac{96\ 485\text{ A·s}}{1\text{ mol }e^-} \left|\ \frac{1\text{ min}}{60\text{ s}}\right. = 902\text{ A}$$

18. $2Cl^-(aq) \rightarrow Cl_2(g) + 2e^-$

$$\frac{1.00\text{ mol }Cl_2}{5.00\text{ A}} \left|\ \frac{2\text{ mol }e^-}{1\text{ mol }Cl_2}\right| \frac{96\ 485\text{ A·s}}{1\text{ mol }e^-} = 3.86 \times 10^4\text{ s}$$

$= 643$ min $= 10.7$ h

19. $Ca^{2+}(aq) + 2e^- \rightarrow Ca(cr)$

$$\frac{0.375\text{ g Ca}}{3.93\text{ A}} \left|\ \frac{1\text{ mol Ca}}{40.1\text{ g Ca}}\right| \frac{2\text{ mol }e^-}{1\text{ mol Ca}} \left|\ \frac{96\ 485\text{ A·s}}{1\text{ mol }e^-}\right| \frac{1\text{ min}}{60\text{ s}}$$

$= 7.65$ min

Chapter 27

1. $\Delta U = q + w = 350 + 230 = 580$ J

2. $w = \Delta U - q = -2.37 - 0.65 = -3.02$ kJ

3. $\Delta U = q + w = 419 + (-389) = 3.0 \times 10^1$ kJ

4. $\Delta U = q + w = -196 + (+4.20 \times 10^2) = 224$ kJ

5. $2NO(g) + O_2(g) \rightarrow 2NO_2(g)$

$\Delta H_r^\circ = [2(33.18)] - [2(90.25) + 0]$

$\Delta H_r^\circ = -144.14$ kJ

6. $4FeO(cr) + O_2(g) \rightarrow 2Fe_2O_3(cr)$

$\Delta H_r^\circ = [2(-824.2)] - [4(-272.0) + 0]$

$\Delta H_r^\circ = -560.4$ kJ

7. $H_2SO_4(l) \rightarrow SO_3(g) + H_2O(l)$

$BaO(cr) + SO_3(g) \rightarrow BaSO_4(cr)$

$BaO(cr) + SO_3(g) + H_2SO_4(l) \rightarrow BaSO_4(cr) + SO_3(g) + H_2O(l)$

$\Delta H_r^\circ = 78.2$ kJ

$\Delta H_r^\circ = -213.4$ kJ

$\Delta H_r^\circ = -135.2$ kJ

12. a. positive

b. negative

c. negative

d. positive

13. a. positive

b. negative

14. $\Delta G_r^\circ = \Sigma\Delta G_{f\,(products)}^\circ - \Sigma\Delta G_{f\,(reactants)}^\circ$

$$= \left[\frac{1\text{ mol }BaSO_4}{} \left|\ \frac{(-1362.2)\text{ kJ}}{1\text{ mol }BaSO_4}\right. + \frac{2\text{ mol HCl(aq)}}{} \left|\ \frac{(-131.2\text{ kJ}}{1\text{ mol HCl(aq)}}\right.\right] - \left[\frac{1\text{ mol }BaCl_2(aq)}{} \left|\ \frac{(-823.2)\text{ kJ}}{1\text{ mol }BaCl_2(aq)}\right. + \frac{1\text{ mol }H_2SO_4(aq)}{} \left|\ \frac{(-744.5)\text{ kJ}}{1\text{ mol }H_2SO_4(aq)}\right.\right]$$

$= -56.9$ kJ; Yes, the reaction is spontaneous.

15. $\Delta H_r^\circ = \Sigma\Delta H_{f\,(products)}^\circ - \Sigma\Delta H_{f\,(reactants)}^\circ$

$$= \left[\frac{1\text{ mol }Ca(OH)_2}{} \left|\ \frac{(-986.09)\text{ kJ}}{1\text{ mol }Ca(OH)_2}\right. + \frac{1\text{ mol }H_2}{} \left|\ \frac{0\text{ kJ}}{1\text{ mol }H_2}\right.\right] - \left[\frac{1\text{ mol Ca}}{} \left|\ \frac{0\text{ kJ}}{1\text{ mol Ca}}\right. + \frac{2\text{ mol }H_2O}{} \left|\ \frac{(-285.830)\text{ kJ}}{1\text{ mol }H_2O}\right.\right]$$

$= -414.43$ kJ

$\Delta G_r^\circ = \Sigma\Delta G_{f\,(products)}^\circ - \Sigma\Delta G_{f\,(reactants)}^\circ$

$$= \left[\frac{1\text{ mol }Ca(OH)_2}{} \left|\ \frac{(-898.49)\text{ kJ}}{1\text{ mol }Ca(OH)_2}\right. + \frac{1\text{ mol }H_2}{} \left|\ \frac{0\text{ kJ}}{1\text{ mol }H_2}\right.\right] - \left[\frac{1\text{ mol Ca}}{} \left|\ \frac{0\text{ kJ}}{1\text{ mol Ca}}\right. + \frac{2\text{ mol }H_2O}{} \left|\ \frac{(-237.129)\text{ kJ}}{1\text{ mol }H_2O}\right.\right]$$

$= -424.23$ kJ

$$\Delta G_r^\circ = \Delta H_r^\circ - T\Delta S_r^\circ;\ \Delta S_r^\circ = \frac{\Delta H_r^\circ - \Delta G_r^\circ}{T}$$

$$\Delta S_r^\circ = \frac{-414.43\text{ kJ} - (-424.23)\text{ kJ}}{298\text{ K}}$$

$= 32.9$ J/K

16. $\Delta G_r^\circ = \Sigma\Delta G_{f\,(products)}^\circ - \Sigma\Delta G_{f\,(reactants)}^\circ$

$$= \left[\frac{1\text{ mol }H_2}{} \left|\ \frac{0\text{ kJ}}{1\text{ mol }H_2}\right. + \frac{1\text{ mol }F_2}{} \left|\ \frac{0\text{ kJ}}{1\text{ mol }F_2}\right.\right] - \left[\frac{2\text{ mol HF}}{} \left|\ \frac{(-273.2)\text{ kJ}}{1\text{ mol HF}}\right.\right]$$

$= +546.4$ kJ, not spontaneous

17. $\Delta S_r^\circ = \Sigma S^\circ_{(products)} - \Sigma S^\circ_{(reactants)}$

$$\Delta S_r^\circ = 463.6\frac{\text{J}}{\text{mol·K}} - \left(462.8\frac{\text{J}}{\text{mol·K}} + 130.7\frac{\text{J}}{\text{mol·K}}\right)$$

$$= -129.9\frac{\text{J}}{\text{mol·K}}$$

18. $\log K_{eq} = -\dfrac{\Delta G}{2.30RT} = -\dfrac{28.95}{(2.30)(8.31 \times 10^{-3})(298)}$

$= -5.0828$

antilog$(-5.0828) = 8.26 \times 10^{-6}$

19. $\Delta G = -2.30RT(\log K_{eq})$

$= (-2.30)(0.008\ 31)(900)(\log 3.16 \times 10^{-4})$

$= 60.2$ kJ/mol

20. $\Delta G = -2.30RT(\log K_{eq})$

$= (-2.30)(0.008\ 31)(473)(\log 0.457)$

$= 3.07$ kJ/mol

21. $E^\circ = (0.52 + 0.8535)\text{V} = 1.37$ V

$\Delta G = -nFE^\circ = (-2)(96\ 485)(1.37) = -264$ kJ

22. $E^\circ = (0.9110 + 0.7626)\text{V} = 1.6736$ V

$\Delta G = -nFE^\circ = (-2)(96\ 485)(1.6736) = -322.95$ kJ

Chapter 28

1. ${}^{27}_{12}\text{Mg} \rightarrow {}^{0}_{-1}e + {}^{27}_{13}\text{Al}$
2. ${}^{49}_{24}\text{Cr} \rightarrow {}^{0}_{+1}e + {}^{49}_{23}\text{V}$
3. ${}^{76}_{36}\text{Kr} + {}^{0}_{-1}e \rightarrow {}^{76}_{35}\text{Br}$
4. ${}^{213}_{88}\text{Ra} \rightarrow {}^{4}_{2}\text{He} + {}^{209}_{86}\text{Rn}$
5. ${}^{231}_{90}\text{Th} \rightarrow {}^{4}_{2}\text{He} + {}^{227}_{88}\text{Ra}$
6. From Table 28.1, the half-life of ${}^{69}_{32}\text{Ge}$ is 36 hours.

   $$\text{Number of half-lives} = \frac{15\ \text{days}}{36\ \text{h}} \cdot \frac{24\ \text{h}}{1\ \text{day}} = 10$$

   $\left(\frac{1}{2}\right)^{10} = \frac{1}{1024}$ of the original sample remains after 15 days.

7. $\frac{5.32 \times 10^{9}}{5.20 \times 10^{6}} = 1020 = 2^{10} = 10$ half-lives

   $10(32.1\ \text{h}) = 321\ \text{h} = 13.4$ days

8. $\frac{3.03 \times 10^{6}}{3.79 \times 10^{5}} = 7.99 = 8$ half-lives

   $2^{8} = 256; \frac{5.80 \times 10^{28}}{256} = 2.27 \times 10^{26}$ atoms

Chapter 29

1. c. 2,3-dimethylpentane
2. a. 2,2-dimethylbutane
   b. 3-methylpentane
   c. hexane
   d. 3-ethyl-2-methylhexane
3.
```
            CH3
            |
a.  CH3 — CH — CH3

        CH3   CH3   CH3
        |     |     |
b.  CH3—CH — CH — CH — CH2 — CH2 — CH2 — CH3

              CH3
              |
        CH3   CH2
        |     |
c.  CH3—CH — CH — CH2 — CH2 — CH3

                     CH2CH3
                     |
                     CH2
                     |
d.  CH3—CH2—CH2—CH—CH2—CH2—CH2—CH3
```
4. a. 2,3-dimethylpentane
   b. octane
   c. 2,2,3-trimethylheptane
   d. 2,3,4-trimethylhexane
   e. 2,2-dimethylbutane
5.
```
            CH3
            |
a.  CH3 — CH — CH2 — CH2 — CH2 — CH2 — CH3

          CH3 CH3
          |   |
b.  CH3 — C — C — CH3
          |   |
          CH3 CH3

          CH3        CH3
          |          |
c.  CH3 — C — CH2 — CH — CH3
          |
          CH3

                CH3
                |
          CH3   CH2
          |     |
d.  CH3 — CH — CH — CH2 — CH3

e.  CH3 — CH2 — CH — CH2 — CH2 — CH3
                |
                CH2
                |
                CH3
```
6.
```
C—C—C—C—C—C                  C
                             |
    C                    C — C — C — C
    |                        |
C — C — C — C — C            C
        C                    C   C
        |                    |   |
C — C — C — C — C        C — C — C — C
```
   (Note: carbon forms four bonds. Hydrogen atoms are not shown because of space limitations.)
7. hexane
   2-methylpentane
   3-methylpentane
   2,2-dimethylbutane
   2,3-dimethylbutane

Chapter 30

1. $H_2O + (CH_3)_3CI \rightarrow HI + (CH_3)_3COH$
2. $CH_3CH_2CH_3 + Cl_2 \rightarrow CH_3CH_2CH_2Cl + HCl$ or any of fifteen other products through perchloropropane, $CCl_3—CCl_2—CCl_3$
3. (benzene) $+ Br_2 \rightarrow$ (bromobenzene, ring with Br) $+ HBr$
4. $CH_3CH_2OH + HF \rightarrow CH_3CH_2F + H_2O$
5. $CH_3Cl + NH_3 \rightarrow CH_3NH_2 + HCl$
6. $CH_3CH_2CH{=}CH_2 + Br_2 \rightarrow CH_3CH_2CHBrCH_2Br$
7. $CH_2{=}CHCH_3 + HCl \rightarrow CH_3CHClCH_3$
8. $CH_2{=}CH_2 + H_2SO_4 \rightarrow CH_3CH_2OSO_3H$
9. $CH_3CH_2CH_2CH_2OH \xrightarrow{H_2SO_4} CH_3CH_2CH{=}CH_2 + H_2O$
10. $CH_3CH_2OH + CH_3CH_2CH_2COOH \rightarrow$
    $CH_3CH_2OCCH_2CH_2CH_3 + H_2O$ (with C=O on the carbonyl carbon)
11. $CH_3OH + HCOOH \rightarrow CH_3OCH + H_2O$ (with C=O on the carbonyl carbon)

GLOSSARY

A

absolute zero: the temperature at which all molecular motion should cease
accelerator: a device that is used to accelerate charged particles to high speeds
accuracy: the relationship between the graduations on a measuring device and the actual standard for the quantity being measured
acid: a substance that produces hydrogen ions in water solution; a proton donor
acidic anhydride: a nonmetallic oxide that can react with water to form an acid
actinoid series: fourteen elements beginning with actinium in which the arrow diagram predicts the highest energy electrons to be in the 5*f* sublevel
activated complex: an assembly of atoms in an excited state between reactants and products in a chemical reaction
activation energy: the energy required to form the activated complex
activity (ion): the effective concentration of a species
addition polymerization: the formation of a polymer through addition reactions
addition reaction: the combining of two or more molecules through adding on at the double or triple bond of an unsaturated organic compound
adiabatic system: system in which heat neither leaves nor enters
adsorption: the process of one substance being attracted and held to the surface of another substance
alcohol: one of a class of organic compounds characterized by the presence of the hydroxyl group, $-OH$
aldehyde: one of the class of organic compounds characterized by the presence of the carbonyl group ($>C=O$) on the end of the carbon chain (RCHO)
aliphatic: hydrocarbons consisting of chains or nonaromatic rings
alkali metal: an element in Group 1(IA)
alkaline earth metal: an element in Group 2(IIA)
alkane: an aliphatic hydrocarbon having only single carbon-carbon bonds
alkene: an aliphatic hydrocarbon having one or more double bonds
alkyne: an aliphatic hydrocarbon having one or more triple bonds
allotrope: form of an element differing in crystal or molecular structure
alloy: a mixture of a metal and one or more other elements, usually metals
alpha particle: a helium nucleus
amide: an organic compound containing the $-CO-NH_2$ group; an inorganic compound containing NH_2^-
amine: an organic compound derived from ammonia by replacement of one or more hydrogen atoms by hydrocarbon radicals
amino acid: an organic compound characterized by the presence of an amino group and a carboxyl group on the same carbon atom
amorphous: a noncrystalline material that appears solid but without long-range order; supercooled liquid
ampere: the unit of electric current equal to one coulomb per second
amphoteric: having the ability to act as either an acid or a base
amplitude: the maximum value attained by a wave
anhydrous: without water
anion: a negative ion
anode: the positive electrode (general); the electrode at which oxidation occurs (electrochemical)
antiparticle: a particle identical to a second particle in all respects except for opposite charge and magnetic moment
aromatic compound: an organic ring compound containing one or more benzene rings
arrow diagram: a system for predicting the order of filling energy sublevels with electrons
atom: the smallest particle of an element
atomic mass: the mass of an atom in atomic mass units; the average mass of the atoms of an element
atomic mass unit: one-twelfth the mass of the $^{12}_{6}C$ atom
atomic number: the number of protons in the nucleus of an atom
atomic radius: the distance from the center of an atom to the 90% probability surface of the electron cloud
atomic theory: the body of knowledge concerning the existence of atoms and their characteristic structure
Avogadro constant: the number of objects in a mole; $6.022\ 136\ 7 \times 10^{23}$
Avogadro's principle: Equal volumes of gases at the same temperature and pressure contain the same number of molecules.

B

balance: an instrument used to measure mass
barometer: a manometer used to measure atmospheric pressure
baryon: a subatomic particle classified as a heavy hadron
base: a substance that produces hydroxide ions in water solution; a proton acceptor
basic anhydride: a metallic oxide that will react with water to form a base
beta particle: an electron (−) or positron (+)
binary acid: an acid containing only hydrogen and one other element
binary compound: a compound composed of only two elements

binding energy: the energy required to split the nucleus into separate nucleons

biochemistry: the study of the substances and reactions involved in life processes

body-centered cubic: having a unit cell with a particle at each vertex of a cube and a particle in the center of the cube

Bohr atom: planetary atom model

boiling point: the temperature at which the vapor pressure of a liquid equals the atmospheric pressure

bond: the force holding atoms together in a compound or molecule

bond angle: the angle between two bond axes extending from the same atom

bond axis: the imaginary line connecting the nuclei of two bonded atoms

bond character: the relative ionic or covalent nature of a chemical bond

bond length: the distance between the nuclei of bonded atoms

bond strength: the energy required to break a bond

Boyle's law: The volume of a specific amount of gas varies inversely as the pressure if the temperature remains constant.

branch: a carbon group, named as a radical, that is attached to the main carbon chain in an organic compound

Brownian motion: the random motion of colloidal particles due to their bombardment by the molecules of the dispersing medium

buffer system: a solution that can receive moderate amounts of either acid or base without significant change in its pH

C

calorimeter: a device for measuring the transfer of heat during a chemical or physical change

capillary rise: the tendency of a liquid to rise in a tube of small diameter due to the surface tension of the liquid

carbide ion: a carbon atom that has gained four electrons; C^{4-}

carbohydrate: compounds of carbon, hydrogen, and oxygen that are mostly simple sugars or condensation polymers of sugars

carboxylic acid: the class of organic compounds characterized by the carboxyl group ($-COOH$)

catalysis: the speeding up of chemical reactions by the presence of a substance that is unchanged after the reaction

catalyst: a substance that speeds a chemical reaction without being permanently changed itself

catenation: the joining of like atoms in chains

cathode: the negative electrode (general); the electrode at which reduction occurs (electrochemical)

cathode rays: the beam of electrons in a gas discharge tube

cation: a positive ion

cell potential: the voltage obtained from a voltaic cell

cellulose: a polymer of glucose

Celsius scale: the temperature scale based on the boiling point of water as 100 degrees and the freezing point of water as 0 degrees

chain reaction: a reaction in which the product from each step acts as a reactant for the next step

chalcogen: an element in Group 16(VIA)

Charles's law: The volume of a specific amount of gas varies directly as the absolute temperature if the pressure remains constant.

chemical change: a rearrangement of atoms and/or molecules to produce one or more new substances with new properties

chemical formula: the notation using symbols and numerals to represent the composition of substances

chemical property: a property characteristic of a substance when it is involved in a chemical change

chemical reaction: a change in which one or more substances are changed into one or more new substances

chemical symbol: a notation using one to three letters to represent an element

chemistry: the study of the structure and properties of matter

chromatography: the separation of a mixture using a technique based upon differential adsorption between a stationary phase and a mobile phase

closest packing: the crystal structure in which space between particles is minimized

coefficient: a numeral, representing the number of formula units of the substance, placed before a formula

colligative properties: the properties of solutions that depend only on the number of particles present without regard to type

colloid: a dispersion of particles from 1 nm to 100 nm in at least one dimension in a continuous medium

colloid chemistry: the study of colloids, especially of surfaces

column chromatography: chromatography in which the stationary phase is held in a column

combustion: burning, or reaction with oxygen producing heat and usually light

common ion effect: an equilibrium phenomenon in which an ion common to two or more substances in a solution shifts the point of equilibrium away from itself

complex ion: a central positive ion surrounded by bonded ligands

compound: a substance composed of atoms of two or more elements linked by chemical bonds

concentrated solution: a solution with a high ratio of solute to solvent

concentration: the ratio of the amount of solute to the amount of solvent or solution

condensation polymer: the formation of a polymer and a small molecule, usually water, from monomers

condensed state: solid or liquid

conductivity: the relative ability to conduct heat and electricity

conjugate acid: the particle obtained after a base has gained a proton

conjugate base: the particle remaining after an acid has donated a proton

conjugated system: a group of four or more adjacent atoms in a molecule with an extended π-bonding system

consumer product: an item for sale to the general public

contact catalyst: a catalyst that functions by adsorbing one of the reactants on its surface; a heterogeneous catalyst
containment vessel: a reinforced concrete and steel structure designed to contain any leakage from a nuclear reactor
continuous phase: the dispersing medium in a colloid
control rod: the neutron-absorbing substance used to control the rate of reaction in a nuclear reactor
coordinate covalent bond: a covalent bond in which both electrons of the shared pair come from the same atom
coordination number: the number of points at which ligands are attached to the central atom or ion in a complex ion or coordination compound
corrosion: the chemical interaction of a metal with its environment
coulomb: the quantity of electricity equal to the flow of one ampere for one second
counting number: natural number; any cardinal number except zero
covalent bond: a bond characterized by the sharing of one or more pairs of electrons between two atoms
covalent radius: the radius of an atom along the bond axis
critical pressure: the pressure needed to liquefy a gas at its critical temperature
critical temperature: the highest temperature at which the vapor and liquid states of a substance can exist in equilibrium
crystal: a solid in which the particles are arranged in a regular, repeating pattern
crystal defect: an imperfection in a crystal lattice
crystallization: the forming of crystals by evaporation or cooling
crystalloid: a substance that can penetrate a semipermeable membrane
cubic closest packing: face-centered cubic
cyclic: consisting of atoms bonded in a closed ring
cycloalkanes: hydrocarbons in which the carbon atoms are bonded in a ring and all bonds are single bonds

D

Dalton's law: In a mixture of gases, the total pressure of the mixture is the sum of the partial pressures of each component gas.
de Broglie's hypothesis: Particles may have the properties of waves.
decomposition: a reaction in which a compound breaks into two or more simpler substances
degenerate: having the same energy
dehydrating agent: a substance that can absorb water from other substances
deliquescent: the property of a solid to absorb sufficient water from the air to form a liquid solution
delocalization: the concept in which bonding electrons are not confined to the region between two atoms, but may be spread over several atoms or a whole piece of metal
density: mass per unit volume
desiccant: a drying agent
didentate: a ligand that attaches to the central ion in a complex in two places
diffusion: the spontaneous spreading of particles throughout a given volume until they are uniformly distributed
dilute solution: a solution with a low ratio of solute to solvent
dipeptide: two amino acids linked by an amide bond
dipole: a polar molecule
dipole-dipole force: an attraction between dipoles; component of van der Waals forces
dipole-induced dipole force: an attraction between a dipole and a nonpolar molecule that has been induced to become a dipole; component of van der Waals forces
dipole moment: the strength of a dipole expressed as charge multiplied by distance
dislocation: a crystal defect
dispersed phase: colloidal particles distributed throughout the continuous phase
dispersion forces: the forces between particles that are not permanent dipoles; component of van der Waals forces
dissociation: the separation of ions in solution
distillation: a separation method based on the evaporation of a liquid and the condensation of its vapor
doping: the addition of impurities to a semiconductor to increase electrical conductivity
double bond: a covalent bond in which two atoms share two pairs of electrons
double displacement: a chemical reaction in which the positive part of one compound combines with the negative part of another compound, and vice versa
drift tube: an uncharged tube through which particles being accelerated travel while the decelerating part of an electromagnetic wave passes
ductility: the ability of a substance to be drawn out into a thin wire
dynamic equilibrium: the state in which two opposite changes take place simultaneously and at the same rate so that there is no overall change in the system

E

edge dislocation: crystal defect in which an extra layer of atoms is found between unit cells
effusion: the movement of gas through a small opening
elastic: describing collisions in which kinetic energy is conserved
elastomer: a substance that can be deformed under the influence of an outside force but will return to its original shape once the force is removed
electric current: the flow of charged particles
electrochemistry: the study of the interaction of electric current and chemical reactions
electrode potential: the potential of a reduction half-cell compared to that of the standard hydrogen half-cell
electrolysis: a chemical change caused by an electric current
electrolyte: a substance whose aqueous solution conducts electricity

electrolytic cell: an electric cell in which passage of an electric current causes a chemical reaction
electrolytic conduction: the migration of ions in solution
electromagnetic energy: radiant energy; energy transferred by electromagnetic waves
electron: an elementary particle with unit negative charge
electron affinity: the attraction of an atom for an electron
electron cloud: the space effectively occupied by an electron in an atom
electron configuration: a description of the arrangement of the electrons in an atom
electronegativity: the relative attraction of an atom for a shared pair of electrons
electronic conduction: the flow of electrons in a metal
electrophoresis: the migration of colloidal particles in an electric field
element: a substance whose atoms all have the same number of protons in the nucleus
elimination reaction: an organic reaction in which a small molecule is removed from a larger molecule leaving a double bond in the larger molecule
empirical formula: the formula giving the simplest ratio between the atoms of the elements present in a compound
endergonic: a process having an increase in Gibbs free energy
endothermic: a change that takes place with the absorption of heat
endpoint: the point in a titration where equivalent amounts of reactants are present
energy: a property of matter that can be converted to work under the proper circumstances
energy level: a specific energy or group of energies that may be possessed by electrons in an atom
energy sublevel: a specific energy that may be possessed by electrons within an energy level in an atom
enthalpy: that part of the energy of a substance that is due to the motion of its particles added to the product of its volume and pressure
enthalpy of formation: the net amount of energy produced or consumed when a mole of a compound is formed from its elements
enthalpy of fusion: the energy required to change 1 gram of a substance from solid to liquid
enthalpy of reaction: the change in enthalpy accompanying a chemical reaction
enthalpy of solution: the change in enthalpy when one substance is dissolved in another
enthalpy of vaporization: the energy needed to change 1 gram of a substance from liquid to gas
entropy: the degree of disorder in a system
enzyme: a biological catalyst
equation: a symbolic expression representing a chemical change
equilibrium: a state in which no net change takes place in a system
equilibrium constant: a mathematical expression giving the ratio of the product of the concentrations of the products to the product of the concentrations of reactants in a chemical reaction
ester: an organic compound characterized by the functional group $R-CO-O-R'$
esterification: the production of an ester by the reaction of an alcohol with a carboxylic acid
ether: an organic compound characterized by the functional group $R-O-R'$
evaporation: the process by which surface particles of liquids escape into the vapor state
excess reactant: reactant remaining when all of some other reactant has been consumed
exergonic: a process having a decrease in Gibbs free energy
exothermic: a change that produces heat
experiment: a test of a hypothesis under controlled conditions
extensive property: a property dependent on the amount of matter present

face-centered cubic: having a cubic unit cell with the addition of a particle in the center of each face
factor-label method: a problem-solving method in which units (labels) are treated as factors
family: the elements composing a vertical column of the periodic table
fat: a biological ester of glycerol and a fatty acid
first ionization energy: the energy required to remove the most loosely held electron from an atom
fission: the splitting of an atomic nucleus into two approximately equal parts
fluid: a material that flows (liquid or gas)
formula: the symbolic representation of a chemical compound
formula mass: the sum of the atomic masses of the atoms in a formula
formula unit: the amount of a substance represented by its formula
fractional crystallization: a separation method based on the difference in the solubility of substances
fractional distillation: a separation method based on the difference in the boiling points of substances
fractionation: separating a whole into its parts, a mixture into its components
free electrons: the delocalized electrons that are in a metal
freezing point: the temperature equal to the melting point of a pure substance
frequency: the number of complete wave cycles per unit of time
functional group: an atom other than hydrogen or carbon introduced into an organic molecule
functional isomers: organic compounds with the same formula, but with the nonhydrocarbon part of the molecule bonded in different ways
fusion reaction: nuclear reaction in which small nuclei are combined to make a larger nucleus

galvanizing: coating iron with a protective layer of zinc
galvanometer: an instrument used to detect an electric current

gamma ray: a quantum of energy of very high frequency and very short wavelength
gas: the state of matter in which particles are far apart and moving randomly
gas chromatography: a chromatographic method in which a carrier gas (inert) distributes the vapor being analyzed in a packed column
geometric isomers: compounds with the same formula but different arrangement of substituents around a double bond
Gibbs free energy: the chemical reaction potential of a substance or system
gluon: a theoretical massless particle exchanged by quarks
glycogen: a biological polymer of glucose
Graham's law: The ratio of the relative rates of diffusion of gases is equal to the square root of the inverse ratio of their molecular masses.
gray: the unit of absorbed dose of radiation equivalent to 1 J/kg of living tissue
ground state: the state of lowest energy of a system
group: the elements of a vertical column in the periodic table

H

hadrons: a class of heavy subatomic particles
half-cell: the part of an electrochemical cell in which either the oxidation or reduction reaction is taking place; single electrode in contact with the solution of an electrolyte
half-life: the length of time necessary for one-half an amount of a radioactive nuclide to disintegrate
half-reaction: either the oxidation or the reduction part of a redox reaction
halogen: an element in Group 17(VIIA)
heat: energy transferred due to differences in temperature
Heisenberg uncertainty principle: It is impossible to know exactly both the position and momentum of an electron at the same instant.
Henry's law: The mass of gas that will dissolve in a specific amount of a liquid varies directly with the pressure.
hertz: the unit of frequency equal to one cycle per second
Hess's law: The enthalpy change for an overall reaction is equal to the sum of the enthalpy changes for all steps of the reaction.
heterogeneous: composed of more than one phase
heterogeneous catalyst: a catalyst in a phase different from that of the reactants
heterogeneous mixture: a combination of two or more substances that are not uniformly dispersed
heterogeneous reaction: a reaction in which not all reactants are in the same phase
hexagonal closest packing: having a crystal structure in which space between particles is minimized; found in most metals
high performance liquid chromatography (HPLC): a type of column chromatography in which the surface area of the particles in the stationary phase is increased
homogeneous: uniform throughout
homogeneous catalyst: a catalyst in the same phase as the reactants
homogeneous reaction: a reaction in which the reactants are in the same phase
homologous series: compounds that differ from each other by a specific structural unit
hybrid orbitals: equivalent orbitals formed from orbitals of different energies
hybridization: the merging of two or more unlike orbitals to form an equal number of identical orbitals in an atom
hydrate: a compound (crystalline) in which the ions are attached to one or more water molecules
hydrated ion: complex ion in which the ligands are water molecules
hydration: the adhering of water molecules to dissolved ions
hydride ion: a hydrogen atom that has gained an electron; H^-
hydrocarbon: compound containing only the elements hydrogen and carbon
hydrogen bonding: a very strong dipole-dipole interaction involving molecules in which hydrogen is bonded to a highly electronegative element (N, O, F)
hydrogen ion: a hydrogen atom that has lost its electron; H^+; a proton
hydrolysis: the reaction of a salt with water to form a weak acid or weak base, or both
hydronium ion: H_3O^+
hygroscopic: absorbing water from the air
hypothesize: to propose an explanation based on observations

I

ideal gas: a model in which gas particles are mass points and exert no attraction for each other
ideal gas equation: $PV = nRT$
ideal solution: a solution in which all intermolecular forces are roughly equal
immiscible: a property of two liquids that will not dissolve in each other at all
indicator: a weak organic acid whose color differs from that of its conjugate; used to indicate the pH of a solution
induced dipole: a nonpolar molecule that is transformed into a dipole by an electric field
inductive effect: the influence of one functional group on another
inertia: the tendency of an object to resist any change in its velocity
infrared spectroscopy: the study of the behavior of matter when it is exposed to infrared radiation
inhibitor: a substance that stops or retards a chemical reaction by forming a complex with a reactant
inner transition elements: those elements that fall between numbers 57 and 70 (the lanthanoids) and between numbers 89 and 102 (the actinoids) of the periodic table
inorganic compound: a molecular compound that does not contain carbon

inorganic substance: a substance that is not a hydrocarbon or a derivative of a hydrocarbon
insulator: a material that does not conduct heat or electricity
intensive property: a property of a substance that is independent of the amount of matter present
interface: the area of contact between two phases
intermediate: the material that is produced from raw materials and processed further to produce some consumer products
intermolecular force: the force holding molecules to each other
internal energy: that energy of a system that is altered by the absorption or release of heat and by doing work or having work done on it; energy of a system due to the energy of its constituent particles, but excluding the kinetic and potential energy of the system as a whole
internuclear distance: the distance between the nuclei of two atoms or ions
intramolecular force: the force holding atoms together in a molecule
ion: an atom or molecule that has gained or lost one or more electrons
ion chromatography: a type of column chromatography in which the column is packed with an ion exchange resin
ion product constant of water: the product of the hydronium and hydroxide ion concentrations in water solutions, equal to 1.00×10^{-14} at 25°C
ionic bond: the electrostatic attraction between ions of opposite charge
ionic compound: a compound that is formed by ionic bonds
ionic radius: the radius of an ion
ionization constant: the equilibrium constant for the ionization of a weak electrolyte
ionization energy: the energy required to remove an electron from an atom
irreversible thermodynamic change: change in volume or pressure in which some energy is lost to an entropy change
isobaric process: a process taking place at constant pressure
isomer: a substance that has the same molecular formula as another substance, but differs in structure; a substance that exhibits isomerism with another substance
isomerism: the property of having more than one structure for the same formula
isomorphism: condition of two or more compounds having the same crystalline structure
isothermal process: a process taking place at constant temperature
isotope: one of two or more atoms having the same number of protons but different numbers of neutrons

J

joule: the SI unit of energy; 1 kg·m^2/s^2
Joule-Thomson effect: the cooling effect observed when a compressed gas is allowed to expand rapidly through a small opening

K

kelvin: the SI unit of temperature; 1/273.16 of the interval between absolute zero and the triple point of water
Kelvin scale: the temperature scale with 0 equal to absolute zero and 273.16 equal to the triple point of water
ketone: an organic compound characterized by the functional group R — CO — R′
kilogram: the SI unit of mass
kinetic energy: the energy of an object due to its motion
kinetic theory: the group of ideas explaining the interaction of matter and energy due to particle motion
kinetically stable: property of a compound for which the activation energy for decomposition is so high that reaction proceeds too slowly for a change to be detected

L

lanthanoid series: fourteen elements beginning with lanthanum in which the arrow diagram predicts the highest energy electrons to be in the 4*f* sublevel
law of conservation of energy: Energy is conserved in all nonnuclear changes; it cannot be created or destroyed.
law of conservation of mass: Mass is conserved in all nonnuclear changes; it cannot be created or destroyed.
law of conservation of mass-energy: Although they can be interconverted, the total amount of mass and energy in the universe is constant.
law of definite proportions: The elements composing a compound are always found in the same ratio by mass.
law of multiple proportions: The masses of one element that combine with a fixed amount of another element to form more than one compound are in the ratio of small whole numbers.
law of octaves: The same properties appear every eighth element when the elements are listed in order of their atomic masses.
Le Chatelier's principle: If a system at equilibrium is subjected to a stress, the system will adjust so as to relieve the stress.
length: the distance between two points
leptons: light subatomic particles
Lewis electron dot diagram: the representation of an atom, ion, or molecule in which an element symbol stands for the nucleus and all inner level electrons while dots stand for outer level electrons
ligand: a negative ion or polar molecule attached to a central ion in a complex
limiting reactant: the reactant that is consumed completely in a chemical reaction
linear accelerator: a device for accelerating particles in a straight line
lipid: a biological molecule that is soluble in nonpolar solvents
liquefaction: condensing a gas to a liquid

liquid: the state of matter characterized by its constituent particles appearing to vibrate about moving points

liquid crystal: a substance that has order in the arrangement of its particles in only one or two dimensions

liter: one cubic decimeter

London forces: dispersion forces

M

macromolecule: a crystal composed of a single molecule with all atoms covalently bonded in a network fashion

magnetohydrodynamics: the study of the behavior of plasmas in magnetic fields

malleability: the property of a substance that allows it to be beaten into thin sheets

manometer: a device for measuring gas pressure

mass: measure of the amount of matter

mass defect: the difference between the mass of an atom and the sum of the masses of the particles composing it

mass-energy problem: a problem in which the amount of energy absorbed or released during a reaction can be calculated from the mass of materials

mass-mass problem: a problem in which the mass of one substance is provided and the mass of another substance must be calculated

mass number: the total number of protons and neutrons in an atom

mass spectrometry: an analysis of substances on the basis of the behavior of their ionized forms in magnetic and electric fields

material: a specific kind of matter

matter: anything that exhibits the property of inertia

mean free path: the average distance a particle travels between collisions

melting point: the temperature at which the vapor pressures of the solid and liquid phases of a substance are equal

meson: a subatomic particle classified as a hadron

metal: an element that tends to lose electrons in chemical reactions

metallic bond: a force holding metal atoms together and characterized by free or delocalized electrons

metallic conduction: electronic conduction; flow of electrons

metalloid: an element that has properties characteristic of a metal and a nonmetal

metastable: the state in which no change will occur unless acted upon by an outside force, but not the most stable state

meter: the SI unit of length

miscibility: the ability of two liquids to dissolve in each other in all proportions

mixture: a material consisting of two or more substances

mobile phase: the fluid containing the mixture to be fractionated in chromatography

model: an arrangement analogous to, and useful for, understanding a system in nature, but existing only in one's mind

moderator: a substance used to slow neutrons in a nuclear reactor

molal boiling point constant: the change of the boiling point of a solvent in a one-molal solution

molal freezing point constant: the change of the freezing point of a solvent in a one-molal solution

molality: a unit of concentration equal to the number of moles of solute per kilogram of solvent

molar heat capacity: the energy necessary to raise the temperature of one mole of a substance by one Celsius degree

molar mass: the mass in grams of one mole of a substance

molar volume: the volume occupied by one mole of a substance; equal to 22.414 10 dm^3 for a gas at standard temperature and standard atmospheric pressure

molarity: a unit of concentration equal to the number of moles of solute in a cubic decimeter of solution

mole: the Avogadro constant number of objects

mole fraction: a unit of concentration equal to the number of moles of component per mole of solution

molecular formula: a formula indicating the actual number of atoms of each element making up a molecule

molecular mass: the mass found by adding the atomic masses of the atoms comprising the molecule

molecule: a neutral group of atoms held together by covalent bonds

momentum: the product of mass and velocity

N

nematic substance: a liquid crystal with one dimension of order

net ionic equation: a chemical equation with spectator ions eliminated

network crystal: a crystal in which each atom is covalently bonded to all its nearest neighbors, so that the entire crystal is one molecule

neutral: neither acidic nor basic (electrolytes); neither positive nor negative (electricity)

neutralization: combining equivalent amounts of acid and base

neutralization reaction: the double displacement reaction between an acid and a base to produce salt and water

neutrino: a neutral particle associated with leptons

neutron: a neutral subatomic particle; a hadron

Newtonian mechanics: the laws of mechanics applicable in the macroscopic world

nitrile: an organic compound characterized by the functional group $—CN$

nitro: the functional group $—NO_2$

noble gas: an element in Group 18(VIIIA)

noble gas configuration: an arrangement of eight electrons in the outer energy level, except for helium with two electrons in the outer level

nonmetal: an element that tends to gain electrons in chemical reactions

nonvolatile: does not evaporate easily

normal boiling point: the temperature at which the vapor pressure of a liquid is equal to standard atmospheric pressure

nuclear force: the force holding nucleons together in a nucleus

nuclear magnetic resonance spectroscopy (NMR): the analysis of the structure of a substance by the behavior of its nuclei in a magnetic field

nuclear reactor: a device engineered to run a controlled nuclear reaction

nucleic acid: an organic compound containing nitrogenous bases, sugars, and phosphate groups; a compound that either transfers genetic information (DNA) or synthesizes biomolecules (RNA)

nucleon: a particle found in the nucleus of an atom; a proton or a neutron

nucleotide: a substance containing a nitrogenous base, a sugar, and a phosphate group

nuclide: an atom of a specific energy with a specified number of protons and a specified number of neutrons in its nucleus

O

observe: to note with the senses, aided or unaided

octahedral: the shape in which six objects are equally spaced about a central object

octane rating: a system of rating gasoline based upon the proportions of heptane and 2,2,4-trimethylpentane in the mixture

octet: an especially stable arrangement of four pairs of electrons in the outer energy level of an atom

octet rule: the tendency of atoms to gain or lose electrons so that they acquire eight electrons in their outer level

ohm: the SI unit of electrical resistance; one volt per ampere

olefin: an alkene

optimum conditions: the conditions maximizing the product of an equilibrium reaction

orbital: the space that can be occupied by 0, 1, or 2 electrons with the same energy level, energy sublevel, and spacial orientation

organic: pertaining to carbon compounds

organic chemistry: the chemistry of the compounds of carbon

organic compound: a compound containing carbon, with a very few exceptions

organic oxidation reaction: the conversion of an organic compound to carbon dioxide, water, and other appropriate oxides

organic substance: a compound that contains the element carbon; a few carbon compounds are considered inorganic

osmotic pressure: the pressure developed across a semipermeable membrane by differential diffusion through the membrane

oxidation: the loss of electrons

oxidation number: the apparent charge on an atom if the electrons in a compound are assigned according to established rules

oxidation-reduction reaction: a chemical reaction in which electrons are transferred

oxidizing agent: a substance that tends to gain electrons

P

packing: the adsorbent in a chromatographic column

pair repulsion: a model used to predict molecular shape based on the mutual repulsion of electron clouds

paper chromatography: a chromatographic method that uses paper as the stationary phase; the mobile phase moves by capillary action

parent chain: the longest continuous chain of carbon atoms in an organic compound

partially miscible: property of two liquids that dissolve in each other to some extent, but not completely

pascal: the SI unit of pressure; 1 N/m^2

Pauli exclusion principle: No two electrons in an atom can have the same set of quantum numbers.

peptide bond: an amide link; $-CO-NH-$

percent of ionization: the amount ionized divided by the original amount, multiplied by 100

percentage composition: the mass of an element in a compound divided by the mass of the compound, multiplied by 100

percentage yield: the mass of product actually obtained from a chemical reaction divided by the amount of product expected from a mass-mass calculation, multiplied by 100

period: a horizontal row of the periodic table

periodic law: The properties of the elements are a periodic function of their atomic numbers.

periodic property: a property of elements that appears periodically when the elements are arranged in order of their atomic numbers

periodic table: a pictorial arrangement of the elements based upon their atomic numbers and electron configurations

petroleum: a raw material consisting chiefly of a complex mixture of hydrocarbons

pH: $-\log[H_3O^+]$

pH meter: an electronic device for the determination of pH values in solutions

pH scale: a logarithmic scale expressing degree of acidity or basicity

phase: a physically distinct section of matter with uniform properties set off from the surrounding matter by physical boundaries

phase diagram: a graphical representation of the equilibrium relationships of the phases of a substance

phenol: C_6H_5OH; any compound having a hydroxyl group attached to a benzene ring

photoelectric effect: ejection of electrons from a surface exposed to light

photon: quantum of radiant energy

physical change: a change in which the same substance is present before and after the change

physical property: a property that can be observed without a change of substance

pi bond: a bond formed by the sideways overlap of *p* orbitals

planetary model: the model of the atom in which the sun represents the nucleus and the planets represent the electrons

plasma: a state composed of electrons and positive ions that have been knocked apart by collisions at very high temperatures

pOH: $-\log[OH^-]$
point mass: an ideal gas particle with mass but no dimensions
polar covalent: a bond formed by a shared pair of electrons that are more strongly attracted to one atom than to the other
polarity: property of a molecule caused by an unsymmetrical charge distribution
polyatomic ion: a group of atoms covalently bonded but possessing an overall charge
polymer: a very large molecule made from the same simple units repeated many times
polymerization: the formation of a polymer from monomers
polymorphism: the property of a substance whereby it exists in more than one crystalline form
polyprotic acid: an acid with more than one ionizable hydrogen atom
positional isomers: two or more molecules having the same formula but having a functional group in different positions on the parent chain
positron: the antiparticle of the electron
potential difference: the difference in electric potential
potential energy: the energy of an object due to its position
precipitate: a solid, produced by a reaction, that separates from a solution
precision: the measure of the reproducibility of measurements within a set
pressure: force per unit area
principal quantum number: the quantum number designating energy level and electron cloud size
probability: mathematical expression of "chance" or "odds"
product: a substance produced as the result of a chemical change
protein: a biological polymer of amino acids linked by amide groups
proton: positive nucleon

Q

qualitative: concerning the kinds of matter present
quantitative: concerning the amounts of matter present
quantum: a discrete "packet" of energy (*plural:* quanta)
quantum mechanics: the laws of mechanics concerning the interaction of matter and radiation at the atomic and subatomic level
quantum number: a number describing a property of an electron in an atom
quantum theory: the concept that energy is transferred in discrete units
quark: a theoretical particle believed to be elementary and a constituent of a hadron

R

rad: 0.01 gray
radiant energy: energy being transferred between objects by electromagnetic waves
radical: a fragment of a molecule; neutral, yet at least one atom lacking its octet of electrons
radioactivity: spontaneous nuclear decay
Raoult's law: The vapor pressure of a solution of a nonvolatile solute is the product of the vapor pressure of the pure solvent and the mole fraction of the solvent.
rate determining step: the slowest step in a reaction mechanism
raw material: a crude, unprocessed material found in nature and used to make intermediates or consumer products
reactant: a starting substance in a chemical reaction
reaction mechanism: the series of steps through which the reactants pass in being converted to the products in a chemical reaction
reaction rate: the rate of disappearance of a reactant or the rate of appearance of a product
real gas: a gas with particles of finite volume and van der Waals forces between particles
redox reaction: an oxidation-reduction reaction
reducing agent: a substance that tends to give up electrons
reduction: the gain of electrons
rem: 0.01 sievert
reversible change: a change that can also go in the opposite direction
reversible reaction: a reaction in which the products may react to produce the original reactants
reversible thermodynamic change: an ideal change in which the difference in pressure is infinitesimal
Rutherford-Bohr atom: the planetary atom model

S

salt: a compound formed from a positive ion other than hydrogen and a negative ion other than hydroxide
salt bridge: an ionic solution used to complete an electric circuit in a voltaic cell
saponification: the reaction of an ester with a strong aqueous base to form a soap and glycerol
saturated: (vapor) the gaseous phase of a system with equilibrium between a substance and its vapor
saturated compound: a compound having only single bonds between carbon atoms
saturated hydrocarbon: a hydrocarbon in which all carbon-carbon bonds are single bonds
saturated solution: a solution in which undissolved solute is in equilibrium with dissolved solute
science: the systematic investigation of nature
scientific notation: the expression of numbers in the form $M \times 10^n$ where $1.00 \leq M < 10$ and n is an integer
screw dislocation: a crystal defect due to improperly aligned unit cells
second: the SI unit of time
semiconductor: a substance that conducts electricity, but poorly
semipermeable membrane: a barrier allowing the passage of small ions and molecules but blocking passage of large particles

shared pair: a pair of electrons bonding two atoms together by being shared by the two atoms

shielding effect: the decrease in the attraction between outer electrons and the nucleus due to the presence of other electrons between them

SI units: the internationally accepted set of standards for measurements

side chain: a branch on the parent chain of an organic molecule

sievert: the SI unit used to measure the absorbed dose of radiation; ionizing radiation equal to 100 rem

sigma bond: a bond formed by the direct or end-to-end overlap of atomic orbitals

significant digits: the reliable digits in a measurement based on the accuracy of the measuring instrument

silicates: compounds containing silicon and oxygen

simple cubic: having a unit cell with one particle centered on each vertex of a cube

single displacement: a reaction in which one element replaces another in a compound

smectic substance: a liquid crystal having two dimensions of order

solid: the state of matter characterized by particles that appear to vibrate about fixed points

solubility: the quantity of solute that will dissolve in a specified amount of solvent at a specific temperature

solubility product constant: the equilibrium constant for the dissolving of a slightly soluble salt

solute: the substance present in lesser quantity in a solution

solution: a homogeneous mixture composed of solute and solvent

solution equilibrium: the state in which solute is dissolving at the same rate that solute is coming out of solution

solvation: the attaching of solvent particles to solute particles

solvent: the substance present in the greater amount in a solution

space lattice: the arrangement pattern of the unit cells in a crystal

specific heat: the amount of energy required to raise the temperature of one gram of a substance by one Celsius degree

specific rate constant: a constant relating the rate of a reaction to reactant concentrations

spectator ion: an ion present in a solution but not taking part in a chemical reaction

spectroscopy: the study of the interaction of matter and radiant energy

spectrum: a unique set of wavelengths absorbed or emitted by a substance

spin: a property of subatomic particles corresponding to rotation on an axis

spontaneous: occurring without outside influence

square planar: the arrangement in which four objects are at the corners of a square around a fifth object in the center

stability: the ability of a substance to remain undecomposed

standard atmospheric pressure: 101.325 kPa

standard solution: a solution whose concentration is known with a high degree of accuracy

standard state: thermodynamic reference conditions, 25°C, 100 kPa, 1*M*

standard temperature: 0°C for gases; 25°C for thermodynamics

starch: a biological polymer of glucose

state: the particle arrangement in a phase as solid, liquid, gas, or plasma

state function: a thermodynamic quantity that is determined solely by the conditions, not the method of arriving at those conditions

stationary phase: the adsorbent in chromatography

stoichiometry: mass and volume relationships in chemical changes

STP: standard temperature and atmospheric pressure (273 K and 101.325 kPa)

strong (acid or base): a completely ionized electrolyte

structural isomers: two or more compounds with the same formula but differing arrangements of the parent carbon chain

subatomic particle: a particle smaller than an atom

sublevel: energy subdivision of an energy level

sublimation: the change directly from solid to gas

substance: a material with a constant composition

substituent: a hydrocarbon branch or functional group attached to the parent chain of an organic compound

substitution reaction: a reaction of organic compounds in which a hydrogen atom or functional group is replaced by another functional group

supercooled liquid: a liquid cooled below its normal freezing point without having changed state to the solid form

supersaturated solution: a solution containing more solute than would a saturated solution at the same temperature

surface tension: the apparent "skin" effect on the surface of a liquid or solid due to unbalanced forces on the surface particles

suspension: a dispersion of particles > 100 nm in a continuous medium

synchrotron: a device for accelerating particles in a circular path

synthesis: the formation of a compound from two or more substances

synthetic element: an element not occurring in nature

system: that part of the universe under consideration

T

technology: the practical applications of scientific discoveries

temperature: a measure of the average kinetic energy of the particles composing a material

temporary dipole: a dipole formed from a nonpolar molecule for a brief period due to the presence of an electric field

ternary: composed of three elements

ternary acid: an acid containing hydrogen, usually oxygen, and one other element

tetradentate: describing a ligand that attaches to the central ion in four locations

tetrahedral: four objects equally spaced in three dimensions around a fifth object
theory: an explanation of a phenomenon
thermodynamic stability: the stability of a substance due to a positive change in the Gibbs free energy for the decomposition of the substance
thermodynamics: the study of the flow of energy in systems
thermometer: a device for measuring temperature
thin layer chromatography: a method of chromatography utilizing an adsorbent spread over a flat surface in a thin layer
time: the interval between two occurrences
titration: a laboratory technique for measuring the relative concentrations of solutions
tracer: a radioactive nuclide used to follow the progress of a reaction or a process
transistor: an electronic device made from a doped semiconductor
transition element: an element whose highest energy electron is in a *d* sublevel
transmutation: the conversion of one element into another
transuranium element: an element with an atomic number greater than that of uranium
triad: a group of three elements with similar properties
tridentate: a ligand that attaches to the central ion in three locations
tripeptide: three amino acids joined by amide links
triple bond: a bond in which two atoms share three pairs of electrons
triple point: the temperature and pressure at which all three states of a substance are in equilibrium
Tyndall effect: the scattering of light by colloidal or suspended particles

U

ultraviolet spectroscopy: the study of the interaction of matter and ultraviolet radiation
unit cell: the simplest unit of repetition in a crystal
unsaturated compound: an organic compound containing one or more multiple bonds
unsaturated hydrocarbon: hydrocarbon containing one or more multiple bonds
unsaturated solution: a solution containing less than the saturated amount of solute
unshared pair: a pair of electrons in an orbital belonging to one atom

V

van der Waals forces: weak forces of attraction between molecules
van der Waals radius: radius of closest approach of a nonbonded atom
vapor: the gaseous state of a substance that is liquid or solid at room temperature and pressure
vapor equilibrium: the equilibrium state between a liquid and its vapor
vapor pressure: the pressure exerted by a vapor in equilibrium with its liquid
velocity: the speed and direction of motion
viscosity: the resistance of a liquid to flow
visible spectroscopy: the study of the interaction of matter with visible radiation
vitamin: a group of biochemicals that are necessary for some enzymatic reactions to take place
volatile: an easily evaporated liquid
volt: the SI unit of electric potential difference
voltaic cell: a cell in which a chemical reaction generates an electric current

W

wave: a periodic disturbance in a medium
wave equation: the equation describing the behavior of the electron as a wave
wavelength: the distance between two successive crests of a wave
wave-particle duality of nature: All particles have wave properties and all waves have particle properties.
weak (acids and bases): a slightly ionized electrolyte
weak forces: an attraction of molecules for each other through the action of dipoles
weight: the gravitational attraction of Earth or a celestial body for matter
work: a force moving through a distance

INDEX

A

D

E

F

G

H

I

J

K

L

M

N

O

P

R

S

T

U

V

W

X

Y

Z

Cover, Filser/The Image Bank; iii, Aaron Haupt; iv, (t) First Image, (bl) Les Losego/Crown Studios, (bc) MAK-1 Photodesign, (br) E.R. Degginger; v, First Image; vi, (l) Doug Martin, (r) image by Tom Palmer, Visualization Group, NC Supercomputing Center/based on work by Jerzy Bernholc, Qiming Zhang, Jae-Yel Yi, Charles Brabec, NC State University; vii, (t) Spencer Swanger/Tom Stack & Associates, (bl) simulations and image: Jerzy Bernholc, Qiming Zhang, Jae-Yel Yi, Charles Brabec, NC State University/Tom Palmer, Cray Research, NC Supercomputing Center, (br) Tom Van Sant/Eyes On Earth; viii, (t) Robert Mathena/Fundamental Photographs, (b) F. Stuart Westmorland/Tom Stack & Associates; ix, (l, c) Matt Meadows, (r) T.J. Florian from Rainbow; x, (l) R. Feldman from Rainbow, (r) Julie Houck/Stock Boston; xi, (l) GE Research & Technology Center, (r) Science Source/Photo Researchers; xii, courtesy Tecktronics, Inc.; xiii, William McCoy from Rainbow; 1, Aaron Haupt; 2, Matt Meadows; 3, Norbert Wu/Tony Stone Worldwide; 4, Doug Martin; 5, 6, Bob Daemmrich; 7, Los Alamos National Laboratory; 8, (t) Les Lesego/Crown Studios, courtesy Bath & Brass Emporium, (b) Bob Thomason/Tony Stone Worldwide; 9, Matt Meadows; 10, Paul Silverman/Fundamental Photographs; 11, John Durham/Science Photo Library/Photo Researchers; 12, (l) courtesy Dr. Bruce N. Ames, (c) Andy Sacks/Tony Stone Worldwide, (r) D. Wilder/Tom Stack & Associates; 13, Fred Ward/Black Star; 14, (t) Dr. Jeremy Burgess/Science Photo Library/Photo Researchers, (b) Dr. Gerardo B. Staal, Sandoz Crop Protection Corp., Zoecon Research Institute, Palo Alto, CA; 15, 16, First Image; 17, Brent Turner/BLT Productions; 19, Julie Houck/Stock Boston; 20, Stacey Pick/Stock Boston; 21, First Image; 22, Hank Morgan/Rainbow; 23, Matt Meadows; 25, Bettmann Archive; 26, Bob Daemmrich/Stock Boston; 27, Bill Everitt/Tom Stack & Associates; 30, (l) National Bureau of Weights and Measures, (r) First Image; 31, Chip Clark; 32, Skip Comer Photography; 33, 34, Matt Meadows; 36, Doug Martin; 37, Matt Meadows; 39, Chip Clark; 40, Skip Comer Photography; 41, (l) Chip Clark, (r) Les Losego/Crown Studios; 46, Doug Martin, courtesy Ohio Historical Society; 47, MAK-1 Photodesign; 48, (tl) Aaron Haupt, (bl) Doug Martin, (r) MAK-1 Photodesign; 49, Eastcott-Momatiuk/Woodfin Camp & Associates; 51, Chip Clark; 52, GE Research & Technology Center; 53, MAK-1 Photodesign; 54-55, from "The Periodic Systems of the Elements", author P. Menzel. © by Ernst Klett Schulbuch Verlag GmbH, Stuttgart, Germany. Charts available from Science Imports, P.O. Box 465, Sillery, Quebec Canada G1T 2R8; 56, Rod Joslin, courtesy Columbus Cultural Arts Center; 57, © Zev Radovan; 58, Manfred Kage/Peter Arnold Inc.; 59, Chip Clark; 60, (l) Chip Clark, (r) Klaus D. Franke/Peter Arnold, Inc.; 61, Chip Clark; 66, NASA; 67, Wesley Bocxe/Photo Researchers; 70, William McCoy from Rainbow; 76, Tony Stone Worldwide; 77, Doug Martin; 79, Chip Clark; 81, Skip Comer Photography; 86, Stuart B. Anderson/Caltech; 88, Chip Clark; 89, Dan McCoy from Rainbow; 93, Bud Fowle; 98, Dan McCoy from Rainbow; 99, Les Losego/Crown Studios; 100, Sven-Olof Lindblad/Photo Researchers; 102, Chris Jones/The Stock Market; 108, 109, Pictures Unlimited; 111, (l) Philippe Plailly/Science Photo Library/Photo Researchers, (r) Doug Martin; 115, William E. Ferguson; 116, Skip Comer Photography; 118, Sepp Dietrich/Tony Stone Worldwide;125, © 1980 Anglo-Australian Telescope Board, photo by David F. Malin; 126, courtesy C. Li and Arthur Freeman, Northwestern University; 130, E.R. Degginger; 136, MAK-1 Photodesign, courtesy Camelot Music; 137, First Image; 140, courtesy National Foundation for History of Chemistry; 147, Chip Clark; 149, Stanford University; 150, Mark Burnett; 151, Doug Martin, courtesy Allen's Coin Shop, Westerville, OH; 153, Argonne National Lab; 155, courtesy Burroughs Wellcome Co.; 156, Matt Meadows; 157, StudiOhio; 161, Doug Martin; 162, Andy Sacks/Tony Stone Worldwide; 164, Matt Meadows; 167, Larry Koons; 170, Matt Meadows; 171, Chip Clark; 172, MAK-1 Photodesign; 173, Chip Clark; 174, (t) Matt Meadows, (b) Ted Rice; 176, MAK-1 Photodesign; 177, Bud Fowle; 178, courtesy Burroughs Wellcome Co.; 179, (l) Bud Fowle, (r) Matt Meadows; 181, (l) Dan McCoy from Rainbow, (r) courtesy John P. Collier; 184, Panographics/L.S. Stepanowicz 186, Doug Martin; 188, MAK-1 Photodesign; 189, Matt Meadows; 191, Les Losego/Crown Studios; 192, (l) Donald Johnston/Tony Stone Worldwide, (r) Doug Martin; 194, (l) courtesy IBM Corporation, (r) file photo; 195, Les Losego/Crown Studios; 197, Aaron Haupt/Glencoe Photo; 198, MAK-1 Photodesign; 200, Matt Meadows; 201, 204, Chip Clark; 207, Riley Caton/FPG; 208, Grant Heilman from Grant Heilman Photography; 209, E.R. Degginger; 211, Gerard Photography; 214, Chip Clark; 222, NASA; 224, Liane Enkelis/Stock Boston; 229, (tl) Tom Pantages, (tr) Matt Meadows, (b) Chip Clark; 230, (t) Ken O'Donoghue, (b) Doug Martin; 232, First Image; 233, Chip Clark; 234, 235, Bob Daemmrich; 238, file photo; 239, courtesy General Motors Corp.; 240, John Elk III/Stock Boston; 248, Focus On Sports; 249, Les Losego/Crown Studios; 256, Chip Clark; 257, Matt Meadows; 260, courtesy Tecktronics, Inc.; 262, Skip Comer Photography; 263, Chip Clark; 264, NASA; 268, Doug Martin; 270, David Stoecklein/The Stock Market; 271, Science Photo Library/Photo Researchers; 272, Chip Clark; 274, H. Armstrong Roberts; 275, courtesy of Medtronic, Inc.; 276, (t) Tim Courlas, (b) Runk/Schoenberger from Grant Heilman; 277, Dr. D. Eastburn, University of Illinois; 278, Tony Stone Worldwide; 279, GE Research & Development Center; 280, Chris Kovach; 281, Sobel-Klonsky/The Image Bamk; 282, (l) courtesy IRECO Inc., (r) Terry Farmer/Tony Stone Worldwide; 284, (t) Matt Meadows, (b) Tony Stone Worldwide; 285, (t) Chip Clark, (b) Doug Martin; 288, (l) Tom Pantages, (r) Brian Parker/Tom Stack & Associates; 289, David M. Dennis/Tom Stack & Associates; 291, First Image; 292, Les Losego/Crown Studios; 293, Philippe Plailly/Science Photo Library/Photo Researchers; 300, Aaron Haupt; 302, Doug Martin; 304, Ken O'Donoghue; 305, First Image; 309, (l) David Hiser/The Image Bank, (r) Chip Clark; 310, Matt Meadows; 312, courtesy Airbus Industries of North America; 320, F. Stuart Westmorland/Tom Stack & Associates; 321, image by Tom Palmer, Visualization Group, NC Supercomputing Center, based on work by Jerzy Bernholc, Qiming Zhang, Jae-Yel Yi & Charles Brabec, NC State University; 324, First Image; 328, First Image; 333, Robert Feldman from Rainbow; 338, First Image; 340, simulations and image: Jerzy Bernholc, Qiming Zhang, Jae-Yel Yi & Charles Brabec, NC State University/T. Palmer, Cray Research, NC Supercomputing Center.; 342, NASA; 343, Les Losego/Crown Studios, courtesy Sawmill Athletic Club; 348, Jon Feingersh/The Stock Market; 349, Doug Martin; 351, Tom Pantages; 352, John Shaw/Tom Stack & Associates; 359, Chris Rogers/The Stock Market; 360, Ken O'Donoghue; 361, 362, Doug Martin;

363, Ken O'Donoghue; 367, Les Losego/Crown Studios; 368, Karen Miller, courtesy Prof. Gary Means, Dept. of Biochemistry, Ohio State University; 369, courtesy Culligan International Co.; 370, Ralf-Finn Hestoff/Third Coast Stock Source; 376, Mark Reinstein/Tony Stone Worldwide; 377, Matt Meadows; 378, Hickson-Bender Photography; 381, Nigel Snowden/Tony Stone Worldwide; 382, (l, c) Matt Meadows, (r) Doug Martin; 387, courtesy Leybold AG; 388, Tom Tracy/Allstock; 389, B.S.I.P./Custom Medical Stock Photo; 394, Stephen Frisch/Stock Boston; 395, Doug Martin; 396, (l) Doug Martin, (r) First Image; 397, (tl) E.R. Degginger, (tr, bl, br) Chip Clark, (bc) Lee Boltin; 402, (l) Chip Clark, (r) Doug Martin; 404, P. Cabrol Jacana/Photo Researchers; 405, An Pang Tsai, Akihisa Inoue and Tsuyoshi Masumoto, Tokoku University; 406, Pictures Unlimited; 409, Aldo Mastrocola/Lightwave; 410, Ken Frick; 411, Hickson-Bender Photography; 412, MAK-1 Photodesign; 413, Chip Clark; 414, MAK-1 Photodesign; 420, Skip Comer Photography; 421, Tom Van Sant/Eyes on Earth; 422, Matt Meadows; 424, Gerard Vandystadt/Photo Researchers; 427, Matt Meadows; 430, (t) Bob Daemmrich, (b) Spencer Swanger/Tom Stack & Associates; 431, Bud Fowle; 433, 440, Dan McCoy from Rainbow; 441, Matt Meadows; 442, courtesy Mead Fine Paper; 443, (tl) David M. Dennis/Tom Stack & Associates, (tr) Martin Dohrn/Science Photo Library/Photo Researchers, (b) Matt Meadows; 450, Chris Harvey/Tony Stone Worldwide; 451, Les Losego/Crown Studios, courtesy ASK Water Sports; 452, D.E. Cox/Tony Stone Worldwide; 455, First Image; 457, Eric Reynolds/Adventure Photo; 458, Spencer Swanger/Tom Stack & Associates; 460, Matt Meadows; 462, courtesy Carrier Corp.; 464, courtesy Kodansha Ltd.; 465, Matt Meadows; 466, U.S. Department of Energy; 468, Skip Comer Photography; 470, Les Losego/Crown Studios; 476, Matt Meadows; 477, courtesy Fuji Film Co.; 478, Matt Meadows; 480, MAK-1 Photodesign; 484, Tom Pantages; 485, Chip Clark; 486, (l) Michael Ponzini/Focus On Sports, (r) Barry L. Runk from Grant Heilman; 487, Chip Clark; 489, Hal McKusick/DPI; 490, MAK-1 Photodesign; 491, (l) David Frazier Photolibrary, (r) Les Losego/Crown Studios; 493, 496, Doug Martin; 498, Matt Meadows; 499, Les Losego/Crown Studios; 502, Chip Clark; 503, (t) MAK-1 Photodesign, (b) Chip Clark; 504, Ken O'Donoghue; 506, (tl) Matt Meadows, (tc) Tom Pantages, (tr) Chip Clark, (b) Tim Courlas; 507, (t) Chip Clark, (b) E.R. Degginger; 509, Doug Martin; 510, Wolfgang Kaehler; 511, StudiOhio; 513, (l) Matt Meadows, (r) Ken O'Donoghue; 514, Matt Meadows; 515, Aaron Haupt; 520, Thomas Kitchin/Tom Stack & Associates; 524, courtesy Du Pont Merck Pharmaceutical Co.; 526, (l) Doug Martin, (r) Greg Vaughn/Tom Stack & Associates; 527, Doug Martin; 528, Science Source/Photo Researchers; 529, Les Losego/Crown Studios; 531, National Science Foundation; 534, Matt Meadows; 536, (l) H. Richard Johnston/Tony Stone Worldwide, (r) Bud Fowle; 542, Chip Clark; 543, Robert Mathena/Fundamental Photographs; 545, MAK-1 Photodesign; 546, Steve Lissau; 547, MAK-1 Photodesign; 549, Hank Morgan/VHSID Lab/ECE Dept. at U. MA. Amherst/Science Source/Photo Researchers; 554, AC Rochester/Division of General Motors; 555, Matt Meadows; 563, courtesy American Lime Association; 565, Chip Clark; 566, 567, StudiOhio; 572, Dan McCoy from Rainbow; 573, MAK-1 Photodesign; 574, E.R. Degginger; 575, 578, Chip Clark; 582, E.R. Degginger; 583, Gary Milburn/Tom Stack & Associates; 585, MAK-1 Photodesign; 586, (l) Chip Clark, (r) Charles Kennard/Stock Boston; 588, Chip Clark; 589, First Image; 592, J.H. Robinson/Photo Researchers; 595, John Zoiner; 596, Doug Martin; 600, Mike Severns/Tom Stack & Associates; 601, F. Stuart Westmorland/Tom Stack & Associates; 604, StudiOhio; 606, Bob Daemmrich; 608, (l) Stanley Schoenberger from Grant Heilman, (r) Grant Heilman Photography; 610, (t) Bob Daemmrich, (b) Matt Meadows; 611, Walter Frerck/UPI/Bettmann Newsphotos; 613, MacDonald Photography/Third Coast Stock Source; 614, Runk/Schoenberger from Grant Heilman; 615, Chip Clark; 616, Matt Meadows; 618, Tom Pantages; 622, Doug Martin; 624, Andy Levin/Photo Researchers; 625, First Image; 626, (l) Hickson-Bender Photography, (r) Dr. Paul A. Zahl/Photo Researchers; 627, J.T. Miller/The Stock Market; 628, John Running/Stock Boston; 629, Dawson Jones/Tony Stone Worldwide; 630, Matt Meadows; 633, Doug Martin; 634, NASA; 636, courtesy Solar Energy Research Institute; 638, E.R. Degginger; 640, 641, Chip Clark; 642, First Image; 644, Ken O'Donoghue; 645, David Frazier Photolibrary; 650, Thomas J. Edwards/Third Coast Stock Source; 651, 654, Matt Meadows; 658, courtesy Aluminum Corp. of America; 659, Liane Enkelis/Stock Boston; 660, Doug Martin; 661, Chip Clark; 664, Doug Martin; 666, Stephen Frisch/Stock Boston; 668, (l) Chip Clark, (r) Hickson-Bender Photography; 669, Doug Martin; 674, (t) Matt Meadows, (b) Tom Tracy/FPG; 684, Larry Lee/Westlight; 685, Les Losego/Crown Studios; 686, Matt Meadows; 691, Aaron Haupt, garments courtesy NeutraTherm Corporation; 692, (l) Matt Meadows, (r) Stephen Homer/First Light; 695, Brian Milne/First Light; 696, Doug Martin; 697, Chip Clark; 698, UPI/Bettmann; 699, Tersch; 700, Peter Menzel/Stock Boston; 702, Doug Martin; 704, courtesy Jennie R. Patrick; 710, Gordon Gahor/Lawrence Livermore Laboratory; 711, NASA; 712, U.S. Department of Energy; 714, Brookhaven National Laboratory; 715, Los Alamos National Laboratory; 718, courtesy Glenn Seaborg; 719, Batelle Pacific Northwest Laboratories; 723, Hank Morgan from Rainbow; 725, Doug Martin; 727, (t) Les Losego/Crown Studios, (b) Dan McCoy from Rainbow; 728, Matt Meadows; 729, SIU/Photo Researchers; 730, (t) Nuclear Division, Union Carbide Corporation, (b) Sandia National Laboratories, photo by Walter Dickenman; 731, California Institute of Technology; 732, Jerry Mason/Science Photo Library/Photo Researchers; 738, Bob Daemmrich; 739, StudiOhio; 744, Eugene Gilliom; 750, W. Eastep/The Stock Market; 753, Prof. J. Bories/CNRI/Science Photo Library/Photo Researchers; 755, 756, Bob Daemmrich; 757, StudiOhio; 759, Stephen Dalton/Photo Researchers 760, First Image; 768, Doug Martin; 769, T.J. Florian from Rainbow; 771, courtesy Dr. Bart Fraser-Reid; 772, Aaron Haupt; 773, (l) Chris Springman/The Stock Market, (r) Ken Frick; 776, Doug Martin; 777, (l) Peter M. Fisher/The Stock Market, (r) Doug Martin; 778, (t) Les Losego/Crown Studios, (b) MAK-1 Photodesign; 779, (l) Ted Horowitz/The Stock Market, (r) James Bell/Photo Researchers; 780, Chip Clark; 782, (t) Irving Geis/Science Source/Photo Researchers, (b) Les Losego/Crown Studios; 784, Chip Clark; 785, (t) Calvin Larson/Photo Researchers, (b) Thomas Hovland from Grant Heilman; 787, (t) LLNL/Science Source/Photo Researchers, (b) Sinclair Stammers/Science Photo Library/Photo Researchers; 788, 789, Dan McCoy from Rainbow.

International Atomic Masses

Element	Symbol	Atomic number	Atomic mass	Element	Symbol	Atomic number	Atomic mass
Actinium	Ac	89	227.027 8*	Neon	Ne	10	20.179 7
Aluminum	Al	13	26.981 539	Neptunium	Np	93	237.048 2
Americium	Am	95	243.061 4*	Nickel	Ni	28	58.6934
Antimony	Sb	51	121.757	Niobium	Nb	41	92.906 38
Argon	Ar	18	39.948	Nitrogen	N	7	14.006 74
Arsenic	As	33	74.921 59	Nobelium	No	102	259.100 9*
Astatine	At	85	209.987 1*	Osmium	Os	76	190.2
Barium	Ba	56	137.327	Oxygen	O	8	15.999 4
Berkelium	Bk	97	247.070 3*	Palladium	Pd	46	106.42
Beryllium	Be	4	9.012 182	Phosphorus	P	15	30.973 762
Bismuth	Bi	83	208.980 37	Platinum	Pt	78	195.08
Boron	B	5	10.811	Plutonium	Pu	94	244.064 2*
Bromine	Br	35	79.904	Polonium	Po	84	208.982 4*
Cadmium	Cd	48	112.411	Potassium	K	19	39.098 3
Calcium	Ca	20	40.078	Praseodymium	Pr	59	140.907 65
Californium	Cf	98	251.079 6*	Promethium	Pm	61	144.912 8*
Carbon	C	6	12.011	Protactinium	Pa	91	231.035 88
Cerium	Ce	58	140.115	Radium	Ra	88	226.025 4
Cesium	Cs	55	132.905 43	Radon	Rn	86	222.017 6*
Chlorine	Cl	17	35.452 7	Rhenium	Re	75	186.207
Chromium	Cr	24	51.996 1	Rhodium	Rh	45	102.905 50
Cobalt	Co	27	58.933 20	Rubidium	Rb	37	85.467 8
Copper	Cu	29	63.546	Ruthenium	Ru	44	101.07
Curium	Cm	96	247.070 3*	Samarium	Sm	62	150.36
Dysprosium	Dy	66	162.50	Scandium	Sc	21	44.955 910
Einsteinium	Es	99	252.082 8*	Selenium	Se	34	78.96
Erbium	Er	68	167.26	Silicon	Si	14	28.085 5
Europium	Eu	63	151.965	Silver	Ag	47	107.868 2
Fermium	Fm	100	257.095 1*	Sodium	Na	11	22.989 768
Fluorine	F	9	18.998 403 2	Strontium	Sr	38	87.62
Francium	Fr	87	223.019 7*	Sulfur	S	16	32.066
Gadolinium	Gd	64	157.25	Tantalum	Ta	73	180.947 9
Gallium	Ga	31	69.723	Technetium	Tc	43	97.907 2*
Germanium	Ge	32	72.61	Tellurium	Te	52	127.60
Gold	Au	79	196.966 54	Terbium	Tb	65	158.925 34
Hafnium	Hf	72	178.49	Thallium	Tl	81	204.383 3
Helium	He	2	4.002 602	Thorium	Th	90	232.038 1
Holmium	Ho	67	164.930 32	Thulium	Tm	69	168.934 21
Hydrogen	H	1	1.007 94	Tin	Sn	50	118.710
Indium	In	49	114.82	Titanium	Ti	22	47.88
Iodine	I	53	126.904 47	Tungsten	W	74	183.85
Iridium	Ir	77	192.22	Unnilennium†	Une	109	266*
Iron	Fe	26	55.847	Unnilhexium†	Unh	106	263*
Krypton	Kr	36	83.80	Unniloctium†	Uno	108	265*
Lanthanum	La	57	138.905 5	Unnilpentium†	Unp	105	262*
Lawrencium	Lr	103	260.105 4*	Unnilquadium†	Unq	104	261*
Lead	Pb	82	207.2	Unnilseptium†	Uns	107	262*
Lithium	Li	3	6.941	Uranium	U	92	238.028 9
Lutetium	Lu	71	174.967	Vanadium	V	23	50.941 5
Magnesium	Mg	12	24.305 0	Xenon	Xe	54	131.29
Manganese	Mn	25	54.938 05	Ytterbium	Yb	70	173.04
Mendelevium	Md	101	258.098 6*	Yttrium	Y	39	88.905 85
Mercury	Hg	80	200.59	Zinc	Zn	30	65.39
Molybdenum	Mo	42	95.94	Zirconium	Zr	40	91.224
Neodymium	Nd	60	144.24				

*The mass of the isotope with the longest known half-life.
†Names for elements 104-109 have been approved for temporary use by the IUPAC. The USSR has proposed Kurchatovium (Ku) for element 104, and Bohrium (Bh) for element 105. The United States has proposed Rutherfordium (Rf) for element 104, and Hahnium (Ha) for element 105.